PROCESS DYNAMICS IN ENVIRONMENTAL SYSTEMS

WALTER J. WEBER, JR.
The University of Michigan
Ann Arbor, Michigan

FRANCIS A. DIGIANO
The University of North Carolina
Chapel Hill, North Carolina

A WILEY-INTERSCIENCE PUBLICATION
JOHN WILEY & SONS, INC.

To order books or for customer service please, call 1(800)-CALL-WILEY (225-5945).

Library of Congress Cataloging in Publication Data:

Weber, Walter J., Jr., 1934–
Process dynamics in environmental systems / by Walter J. Weber, Jr. and Francis A. DiGiano.
p. cm.—(Environmental science and technology)
"Updates and expands on the book Physicochemical processes for water quality control"—Pref.
Includes bibliographical references and index.
ISBN 0-471-01711-6
1. Environmental engineering. 2. Environmental chemistry. 3. Biochemistry. I. DiGiano, Francis A. II. Weber, Walter, J., Jr., 1934– Physicochemical processes for water quality control. III. Title.
TD177.W43 1995
628—dc20 94-49669

. . . To Those with Whom We Share the Excitement and Rewards of Learning . . .

आपोसमं, राहुल, भावनं भावेहि।
आपो मम, सत्था।

SERIES PREFACE

Environmental Science and Technology

We are in the third decade of the Wiley Interscience Series of texts and monographs in Environmental Science and Technology. It has a distinguished record of publishing outstanding reference texts on topics in the environmental sciences and engineering technology. Classic books have been published here, graduate students have benefited from the textbooks in this series, and the series has also provided for monographs on new developments in various environmental areas.

As new editors of this Series, we wish to continue the tradition of excellence and to emphasize the interdisciplinary nature of the field of environmental science. We publish texts and monographs in environmental science and technology as it is broadly defined from basic science (biology, chemistry, physics, toxicology) of the environment (air, water, soil) to engineering technology (water and wastewater treatment, air pollution control, solid, soil, and hazardous wastes). The series is dedicated to a scientific description of environmental processes, the prevention of environmental problems, and to preservation and remediation technology.

There is a new clarion for the environment. No longer are our pollution problems only local. Rather, the scale has grown to the global level. There is no such place as "upwind" any longer; we are all "downwind" from somebody else in the global environment. We must take care to preserve our resources as never before and to learn how to internalize the cost to prevent environmental degradation into the product that we make. A new "industrial ecology" is emerging that will lessen the impact our way of life has on our surroundings.

In the next 50 years, our population will come close to doubling, and if the developing countries are to improve their standard of living as is needed, we will require a gross world product several times what we currently have. This will create new pressures on the environment, both locally and globally. But there are new opportunities also. The world's people are recognizing the need for sustainable development and leaving a legacy of resources for future generations at least equal to what we had. The goal of this series is to help understand the environment, its functioning, and how problems can be overcome;

the series will also provide new insights and new sustainable technologies that will allow us to preserve and hand down an intact environment to future generations.

Jerald L. Schnoor
Alexander J. B. Zehnder

PREFACE

This text updates and expands on the book *Physicochemical Processes for Water Quality Control* (W. J. Weber, Jr., Wiley-Interscience, 1972). Our knowledge of the sciences and engineering of environmental control processes, particularly of concepts underlying the dynamics of these processes, has become increasingly sophisticated since the publication of that book more than 20 years ago. This text, *Process Dynamics in Environmental Systems*, acknowledges and reflects that enhanced sophistication. The topic was covered in the first chapter of the earlier book, but only by way of an introduction to the subject; it is developed here to the point of comprising a stand-alone text. We believe that the results more fully reflect the manner in which the principles of process dynamics should be *taught and employed* in the description and design of environmental systems. These principles find wide application in both natural and engineered systems. Their importance is nowhere more evident than in addressing contemporary problems dealing with the complex continua of component interactions and relationships associated with multiphase systems.

The book consists of twelve chapters focusing on various aspects of process dynamics. It begins with an overview of the subject and a formulation of the approach to process engineering that we use throughout the book. It concludes with illustrations of how process models developed in the text can be employed for interpretation and design of widely varied process operations. Each chapter includes detailed example problems, supplementary reading recommendations, and assigned problems, many with answers provided. Several chapters include appendices addressing issues specific to those chapters, and several general appendices are also provided. We feel that all of the material covered in *Process Dynamics* is essential for students receiving their *first professional degree* in environmental engineering or science. How the material is best covered will vary from one program to another; suggestions in this regard are made later in this *Preface*.

In the *overview* provided in *Chapter 1* of the book we introduce factors and features of systems and processes that must be considered in the analysis and interpretation of environmental transformations. It describes the general nature of fundamental processes, the significant characteristics of different types of systems in which they occur, and the way in which these factors come together to influence the dynamics of processes and their effects on environmental systems.

In *Chapter 2* we develop a *rationale* and *protocol* for process characterization and analysis. The development of this protocol leads to identification of certain key features of processes and systems that must be understood thoroughly to facilitate its application. The core elements of process analysis—*energetics, kinetics, and reactor dynamics*—are characterized in Chapter 2, and the role of *process modeling* as a means for incorporating these elements in descriptions of natural systems and in designs of engineered systems is explained. The most fundamental and requisite tool for rigorous process analysis, the *material balance equation*, is introduced in Chapter 2, and its role as a foundation for the quantitative modeling of environmental systems is described. The chapter closes with a discussion of applications of modeling approaches for *characterization* and *design* of environmental systems.

Because macroscopic and microscopic transport phenomena are fundamental considerations for analysis and characterization of processes, and because these topics build directly on the continuity relationships developed in Chapter 2, discussions of reactions more detailed than those presented in Chapter 2 are preceded in the book by considerations of transport processes. To this end, *Chapter 3* focuses on the mechanics of macroscopic transport processes and *Chapter 4* on microscopic transport processes. In these two chapters we develop concepts drawn upon subsequently for detailed considerations of reaction phenomena in Chapters 5 through 8 and of reactor engineering in Chapters 9 through 11.

The "middle" chapters of *Process Dynamics* expand the reaction and mass transfer concepts introduced in Chapters 1 and 2 and integrate the issues of macroscopic and microscopic mass transport and transfer introduced in Chapters 3 and 4. In *Chapter 5* we present a basis for interpreting environmental reaction phenomena in the context of chemical species and their transformations. This establishes a foundation for a comprehensive understanding of the role of *chemical energetics*, or *thermodynamics*, in reaction processes. In Chapter 5 we focus for the most part on *homogeneous* (single-phase) systems, for which the notions of energy balances and flows associated with species and species transformations are relatively straightforward. Now that this groundwork has been established, *Chapter 6* goes on to address the more complicated issues associated with energy balances and flows in *heterogeneous* (multiphase) systems.

The same logic is applied in our approach to *reaction rates*, or *kinetics.* A molecular basis for reaction kinetics in homogeneous systems is established in *Chapter 7* and the rationale associated with applications of kinetic theory to single-phase systems of environmental interest is extended. The more complex issues of reaction rates in heterogeneous systems are presented in *Chapter 8.* The term *kinetics* is generally less appropriate as a descriptor for the types of rate phenomena considered in Chapter 8 than it is for reaction rates in homogeneous systems. While rates of transformation in homogeneous systems are largely *chemically* or *kinetically* determined, overall rates in heterogeneous systems are more commonly controlled by *microscopic mass transport phenomena.*

The approaches to process characterization, analysis, and modeling laid out in Chapter 2 are brought together in Chapters 9 through 11 with the mass transport processes detailed in Chapters 3 and 4 and the reaction processes considered in Chapters 5 through 8. All of the material in Chapters 9 through 11 is focused on *analysis* and *design* applications *in environmental systems*; i.e., on *environmental reactor engineering.* These chapters detail the role of *transport phenomena in reactor dynamics.* At the *macroscopic scale*, transport phenomena influence mass distributions within systems, and mass distributions influence reaction energetics and rates. Beyond that, however, the interactions among molecules required for reactions to occur may depend upon transport considerations at a much smaller scale than that of the system itself, that is, at the *molecular* or *microscopic scale.*

Engineering design considerations associated with reactor systems in which a single phase is involved; (e.g., the oxidation of *dissolved* organic contaminants by *dissolved* ozone in drinking water treatment operations) are discussed in *Chapter 9*. In *Chapter 10* we examine engineering and design considerations associated with reactor systems involving transformations among and between the components of two or more phases: for example, liquid-gas mass transfer and reaction in systems employed for air-stripping of contaminants from groundwaters, and in natural systems involving the evolution of volatile organic and inorganic contaminants from surface waters into the atmosphere.

In Chapters 9 and 10 we consider only one particular *state* of reactor operation, the *steady state.* Developments of analysis and design models for this state involve relatively straightforward applications of the material balance protocol developed in Chapter 2. Many real system operations, particularly for engineered systems, approximate this state, and thus allow assumptions of negligible net mass accumulation or depletion over time to simplify mathematical descriptions. There are, however, important circumstances in which such assumptions are not valid. Aspects of reactor engineering that must be considered and quantified for circumstances that involve *time-variable*, or *nonsteady*, conditions are treated in *Chapter 11.*

A synopsis of how various elements of the text come together to prepare the reader to apply concepts of process dynamics to the interpretation and design of environmental systems is given in *Chapter 12.* The salient principles of process dynamics are reviewed and a synopsis of process models presented. This chapter further builds upon those principles to illustrate how they can be extended to an even wider array of environmental systems and circumstances than specifically covered in the first eleven chapters.

Several approaches to study of the material presented in *Process Dynamics* are possible, as are several ways of teaching it in various academic programs. For example, one can first cover homogeneous systems using Chapters 1 through 3, 5, 7, and 9, and then return to Chapters 4, 6, 8, 10, and 11 for a logical transition to the treatment of heterogeneous systems. In some programs it may be desirable to apply this approach over a two-course presentation of the subject material, interspersing it liberally with coverage of various specific units of operation. In an aggressive program with well-prepared seniors or first-year graduate students, much of the material in the book might be covered in an "upfront" one-semester course. The approach used at the Universities of Michigan and North Carolina is to provide a course *dedicated* to the principles and applications of material balance relationships, mass transfer, and process and reactor dynamics in environmental systems early in a student's program. This enables the student to subsequently draw upon and apply these fundamental concepts in a variety of intermediate and advanced level courses.

The intended and, we feel, best use of *Process Dynamics in Environmental Systems* as a teaching text is in a *series* of courses, beginning at the junior level and continuing throughout advanced studies at various graduate degree levels. Again, depending upon the particulars of a given program, it may be used as a single text for one or more courses *and* as a companion text for several others. Examples of chapter combinations comprising sole and companion text requirements for particular courses commonly included in undergraduate and graduate level programs in environmental sciences and engineering are presented in the table provided at the end of this *Preface.* As the table suggests, a variety of approaches to ensuring adequate coverage of the content of this text are both possible and reasonable.

To make the book useful as a multicourse text and reference, we have included a significant amount of working data and information in appendices to individual chapters as well as in *Appendices I–III*, a set of general appendices to the book. The chapter and general appendices bring together material that is generally scattered throughout the literature, and thus not often readily available in a single source, particularly not a teaching text. Notation and associated dimensions employed in the book are detailed in Appendix I. The parameters used throughout the text and summarized in Appendix I are drawn from a number of disciplines and sources that often employ different notations for the same parameters and, just as often, the same notations for different parameters. To provide accuracy, consistency, and continuity of notation, we have attempted a *homogenization* across disciplinary subject coverage, resulting in a fairly extensive system of alphabetic and font styles, and parameter scripting. The information presented in Appendices II and III is extensive and thorough, but it is not advanced as either exhaustive or absolute. These appendices are intended simply as convenient sources of material for use in a variety of environmental science and engineering courses, particularly in conjunction with open-ended problems for student assignments and examinations.

We are indebted to many for their contributions to the development and production of this book. Our students have challenged us, asking the questions we attempt here to answer. Our efforts have been nourished in the rich intellectual environments provided by our programs and colleagues at the Universities of Michigan and North Carolina, and our families have given levels of support and understanding without which our efforts on the book would not have moved from their beginning to this conclusion.

Particular and specific contributions were made by Dr. Kevin P. Olmstead, who provided support in developing appendix information, assigned problems, and graphics, by Dr. James E. Kilduff in developing detailed solutions to a number of the examples and many of the assigned problems, and by Patricia Lum in helping to process portions of the text and related graphics. Stefan Grimberg, Phillip Heck, and Christopher Hull also helped develop examples and assigned problems. Finally, the successful completion of this book owes much to Rebecca Rangel Mullin and Susan DeZeeuw. Their *beyond-the-call* levels of industry, care, diligence, and dedication over the past several years in the synthesizing of hieroglyphic scribbles, sketches, and notes, and polyglot formats and fonts into readable text contributed immensely to making *Process Dynamics* a reality.

WALTER J. WEBER, JR.
FRANCIS A. DIGIANO

Ann Arbor, Michigan
Chapel Hill, North Carolina

Suggestions for Incorporating *Process Dynamics* into Academic Programs in Environmental Sciences and Engineering

Type of Course	Year[a]	Chapters	As the Only Text	As a Companion Text
Introduction to Environmental Sciences and Engineering	3	1 through 4		×
Contaminant Hydrology, Fate, and Transport	3–5	2 through 4 and 7 and 8		×
Environmental Chemistry	4–5	5 through 8		×
Dynamics of Environmental Systems	3–5	2 through 9, 12	×	
Mass Transport and Reactor Principles	3–5	1 through 4, and 7 through 9	×	
Advanced Topics in Mass Transport and Reactor Principles	5–6	6, 8 through 12	×	
Water and Wastewater Treatment	4–5	2, 3, 9, 10		×
Industrial Waste Treatment	4–5	2, 3, 9, 10		×
Biological Processes	4–6	7 through 12		×
Physicochemical Processes	4–6	7 through 12		×

Note: The material presented in this book presumes a certain prior knowledge of the elementary concepts of chemistry, physics, and mathematics. We suggest a level of preparation in these topics typical of that normally attained in the first three years of an undergraduate program in engineering or physical sciences, i.e., two semesters each of college level chemistry and physics, and calculus through differential equations.

[a]3, Junior; 4, senior; 5–6, graduate student.

CONTENTS

PROCESS DYNAMICS IN ENVIRONMENTAL SYSTEMS

1

OVERVIEW

1.1 INTRODUCTION

The term *process dynamics* is both descriptive and definitive. A *process* involves one or more phenomena that produce some end result. The *dynamics* of any process relate to its driving forces and associated patterns of change. Engineers and scientists who deal with *environmental systems* are concerned ultimately with changes that result from processes that occur within them.

Examples of changes, and their causes in different environmental systems, include (1) *acidification* of rainfall by power plant emissions to the atmosphere, (2) *removal* of gaseous sulfur oxides from power plant emissions by wet limestone scrubbing, (3) *contamination* of groundwater and subsurface soils by seepage from landfills, (4) *removal* of contaminants from water supplies by treatment with activated carbon, (5) *consumption* of dissolved oxygen by microbial degradation of organic matter in rivers and lakes, and (6) *reduction* of the biochemical oxygen demand (BOD) of a wastewater by biological treatment prior to discharge to a receiving water. Each change in these examples is underlain by a specific process. Some of these processes take place in *natural* environmental systems while others occur in *engineered* systems. Six different types of environmental systems have been used as examples, but only three fundamentally different processes are represented. These processes are phase transfer (gas–liquid) and acid-base reactions in the first two examples, phase transfer (liquid–solid) reactions in the second two, and biological oxidation-reduction reactions in the last two examples.

The important point to recognize is that although environmental systems are virtually limitless in number, change is controlled by a relatively small number of fundamental processes. Each specific process is modified in its extent and effect by the nature of the system in which it occurs. For example, microbial exertion of BOD takes place more rapidly and efficiently in a biological treatment plant than it does in a river because the plant is specifically designed to facilitate the process.

A number of processes common to a variety of environmental systems are identified and described in Table 1-1. For each process examples are given of a common engineering application and a related phenomenon broadly observed in natural environmental systems. The nature of the reaction(s) underlying each process and the potential involvement of *microscopic* or *molecular-level mass transfer* phenomena in controlling the reaction(s) are also indicated in Table 1-1. These topics are dealt with in detail in subsequent chapters. For now it suffices to recognize that most processes of interest in environmental systems involve both reaction and mass transfer considerations, *irrespective* of the systems in which they take place. The system itself, however, imposes important additional mass transport considerations at a larger scale which we will refer to as *macroscopic mass transport* phenomena.

The examples given in Table 1-1 focus on systems of interest in water quality transformations. These transformations, however, generally involve components representing all three environmental phases: gases, liquids, and solids. As such, they are among the most complex of environmental systems. Notwithstanding their complexity, need and practice have led in many instances to a reasonable understanding of their associated process dynamics. Most important, the basic principles of process dynamics are common to all types of environmental systems and should therefore be broadly applicable to all environmental systems.

Each of the processes identified in Table 1-1 underlies environmental quality changes and transformations over a wide range of systems. To illustrate this, Table 1-2 describes changes caused by different classes of sorption processes alone in different types of engineered and natural systems, along with associated reactions and microscopic mass transfer phenomena.

Many processes of common environmental interest involve transformations

Table 1-1 Processes Common to Natural and Engineered Systems

Process	Engineered Systems	Natural Systems	Nature of Reaction(s)	Common Microscopic Mass Transfer Phenomena
Absorption	Aeration of biological treatment units (e.g., activated sludge systems)	Uptake of atmospheric oxygen by lakes, streams, and estuaries	Gas–liquid mass transfer and dissolution of molecular oxygen	Molecular diffusion of oxygen at air–water interfaces
Adsorption	Removal of organic contaminants from waters and wastes by activated carbon treatment	Uptake of organic contaminants from groundwaters by soil mineral surfaces (e.g., sand)	Interactions and accumulation of solutes at solid surfaces	Interfacial and intraparticle diffusion of dissolved solutes
Biochemical oxidation	Degradation of organic contaminants in biological treatment units (e.g., activated sludge, trickling filter systems)	Exertion of biochemical oxygen demand (BOD) in surface waters	Enzyme-mediated transformations of chemical species by electron transfer	Diffusion of substrates, nutrients, and metabolic products across microbial cell walls
Chemical oxidation	Transformation of cyanide to cyanate by chlorine in industrial waste treatment systems; oxidation of organic compounds by ozone in contaminated surface or subsurface water supplies	Photooxidation of dissolved organic contaminants in surface waters	Homogeneous phase transformations of chemical species by electron transfer reactions	None
Chemical oxidation with catalysis	Transformation of cyanide to CO_2 by oxygen in the presence of copper and activated carbon	Photooxidation of organic compounds sorbed at mineral surfaces	Heterogeneous phase transformation of chemical species by electron transfer reactions (e.g., oxidations at reactive surfaces)	Molecular diffusion at the interfaces of and within solid catalysts

Table 1-1 (*Continued*)

Process	Engineered Systems	Natural Systems	Nature of Reaction(s)	Common Microscopic Mass Transfer Phenomena
Chemical reduction	Transformation of hexavalent chromium to trivalent chromium for subsequent precipitation	Decomposition of ozone by CFCs in the upper atmosphere	Homogeneous phase transformations of chemical species by electron transfer	None
Coagulation	Destabilization of turbidity and suspended solids by iron, aluminum, lime, or organic polyelectrolytes, coagulants in water, and wastewater treatment	Destabilization of colloids by natural salts in marine estuaries, or by natural biopolymers in fresh waters	Modifications of particle surface and near-surface chemistry to reduce particle–particle repulsions	Molecular diffusion of dissolved species into particle double layers and to particle surfaces
Disinfection (pathogenic organisms) and sterilization (all organisms)	Destruction or inactivation of organisms using chemicals, heat, or shortwave irradiation	Destruction or inactivation of organisms by naturally occurring chemical conditions, heat, or irradiation (e.g., sunlight)	Enzyme inactivation, protein denaturing, or cell lysis	Mass or heat transfer across cell membranes
Electrochemical transformation	Galvanic and anodic protection of metal components employed in heat transfer and reactor systems	Dissolution of metallic contaminants from ores, pipes, and other metal structures	Heterogeneous phase transformations of chemical species by electron transfer at electrode surface	Diffusion of ions and corrosion inhibitors through inorganic and organic films at metal–water interfaces
Equalization	Moderation of flow or concentration transients to facilitate downstream process control	Natural chemical and flow moderations effected by lakes, embayments, and estuaries receiving river or stream discharges	Flow and mass buffering provided by expanded spatial and temporal scales	None

Extraction: liquid–liquid	Removal of organic contaminants from aqueous streams using immiscible solvents	Dissolution of organic contaminants from nonaqueous phase liquids (NAPLs) in surface water or groundwater systems	Phase transfer of species by dissolution	Mass transfer at liquid–liquid interfaces
Extraction: liquid–solid	Soil washing with surfactant solutions; extraction of contaminants from solids using supercritical fluids (e.g., CO_2)	Leaching of minerals and organic species from natural solids (e.g., soil, rocks) by water	Phase transfer of species by dissolution	Mass transfer at liquid–solid interfaces
Filtration	Removal of suspended solids from waters and wastes by deep-bed or septum filtration	Clogging of aquifers by individual and aggregated bacteria and other colloids in subsurface systems	Particles or aggregate interception and accumulation at solid surfaces	Microscopic particle transport and interfacial deposition
Flocculation	Aggregation of chemically destabilized colloids induced by fluid mixing processes	Aggregation of bacteria and other colloids in surface water and groundwater systems	Particle–particle collisions	Particle–particle transport and interactions in macroscopic flow fields
Flotation	Removal of greases, oils, and low-density particles in water and waste treatment	Flotation of particulate solids attached to gas bubbles in natural aquatic systems	Physical separation of immiscible phases	Particle–particle interactions in near-quiescent flow fields
Hydrolysis	Decomposition of cyanate to ammonia and CO_2 under acidic conditions	Hydrolysis of metal ions discharged or dissolved into natural water bodies	Transformations of chemical species involving proton and/or hydroxide interactions	None
Incineration	High-temperature oxidation of organic wastes in gas phase	Volcanic eruptions and forest fires	Transformation of chemical species by electron transfer reactions at elevated temperatures in heterogeneous phase systems	Gas–solid mass and heat transfer

Table 1-1 (*Continued*)

Process	Engineered Systems	Natural Systems	Nature of Reaction(s)	Common Microscopic Mass Transfer Phenomena
Ion exchange	Removal of metals from water and wastes by ion exchange (e.g., softening, demineralization, recovery of precious metals)	Multivalent cation uptake and retardation by clayey soils (e.g., Ca^{2+}, Cd^{2+}, Pb^{2+})	Exchange of ions of like but unequal charge at charged surfaces	Interfacial and intraparticle ion diffusion for solid ion-exchange materials; none for liquid ion-exchange materials, which are subsequently removed by ultrafiltration, adsorption, or precipitation
Membrane separation	Desalination of marine and brackish waters by reverse osmosis and electrodialysis	Separation of dissolved oxygen from water by the gill membranes of fish	Selective separations of molecular species by microporous barriers	Molecular diffusion at solid–water interfaces and within microporous membranes
Neutralization	Adjustment of pH from excessively low or high values for downstream process control or product quality	The acid-base reactions of carbonic species (e.g., CO_2, H_2CO_3, HCO_3^-, and CO_3^{2-}), which function to poise the natural waters at pH levels ~ 8.3	Proton and related coordination-partner transfer reactions	None for homogeneous phase (liquid) and/or base additions; interfacial mass transfer when gaseous or solid phase acids or bases used
Polymerization	Generation of metal–ion and activated silica polyelectrolytes in water and waste coagulation	Formation of polynuclear humic and fulvic substances in concentrated sediment pore waters and at solid surfaces	Homogeneous interactions of dissolved monomeric species to form polymers or macromolecules	Diffusion of monomers across polymer–water interfaces and within amorphous polymer matrices

Precipitation	Removal of heavy metals and phosphates from wastewaters; removal of hardness (e.g., Ca^{2+} and Mg^{2+}) in water treatment operations	Precipitation of iron oxides at surfaces and wetted interfaces near groundwater outcrops; deposition of calcium carbonates and magnesium silicates on submerged surfaces in natural aquatic systems	Phase transformations of chemical species by coordination-partner exchange reactions	Interfacial and intraparticle diffusion associated with crystallization and particle growth phenomena
Supercritical water oxidation	Oxidation of organic wastes at high temperatures and pressures in supercritical water phase	Geothermal activity in deep waters (e.g., beneath ocean floors)	Chemical transformations of chemical species by electron transfer at elevated temperatures in homogeneous phase systems	None
Thermal desorption	Volatilization of organic matter adsorbed to soils and other solid phases	Sun-baking of surface soils	Phase transformation by volatilization at elevated temperatures	Mass and heat transfer at phase interfaces
Volatilization	Stripping of taste and odor compounds from drinking waters, ammonia from wastes, and volatile organic contaminants from groundwaters by vigorous mixing	Escape of hydrogen sulfide from benthic deposits and overlying waters into the atmosphere	Phase transformations by volatilization at moderate (ambient) temperature	Mass transfer at phase interfaces

Table 1-2 Absorption and Adsorption Processes in Natural and Engineered Systems

Process	Engineered Systems	Natural Systems	Nature of Reaction(s)	Associated Microscopic Mass Transfer Phenomena
Absorption	Air stripping of volatile organic chemicals from groundwater, or ammonia from wastewater	Hydrogen sulfide evolution from stagnant ponds	Gas–liquid mass transfer of hydrogen sulfide	Molecular diffusion at air–water interfaces
	Solvent extraction of organic contaminants from industrial wastes	Dissolution of nonaqueous-phase liquid (NAPL) contaminants into groundwaters	Liquid–liquid mass transfer of organic species	Molecular diffusion at solvent–water interfaces
	Soil washing with solubility enhanced (e.g., detergent-added) aqueous phases	Uptake of dissolved contaminants into organic matrices associated with sediments and soils	Solid–liquid mass transfer of organic species	Molecular diffusion of water–organic interfaces and within organic phases
	Heat exchange systems in distillation and freezing operations	Absorption of geothermal or atmospheric heat by surface water and groundwater	Heterogeneous phase transfer of thermal energy	Diffusion of heat across phase interfaces and boundaries
Adsorption	Froth flotation of industrial wastewaters for removal of colloids and emulsions	Surface tension reduction and foaming by natural macromolecules at water surfaces	Accumulation of surfactants at phase interfaces	Molecular diffusion in interfacial boundary layers
	Attached-film biological reactors (e.g., trickling filters, rotating disk biological contactors)	Buildup of biofilms at solid surfaces in aquatic ecosystems	Attachment of microbial species at water–solid interfaces	Molecular diffusion of substrates and nutrients through biofilms
	Adhesion of dissolved species at heat exchange surfaces	Coating of geothermal interfaces by attached biofilms and algae	Heterogeneous accumulation at phase boundaries	Retardation of heat transfer across phase boundaries
	Concentration of organic substrates at biomass interfaces for facilitated biodegradation (e.g., biological activated-carbon processes)	Facilitated transport of hydrophobic contaminants sorbed to colloids	Trace contaminant concentration on natural colloids and biological solids	Molecular diffusion at particle interfaces

and mass transfer phenomena that occur at molecular or particle scales, as enumerated in Tables 1-1 and 1-2. These are essentially processes within processes, fundamentally unaltered in character by the size and configuration of the system in which they occur. Conversely, the degree to which such processes operate to change the character of a system in which they occur is markedly dependent on their spatial and temporal distributions (e.g., how they are distributed along the length of a river or within a series of carbon adsorption beds). This dependence on scale relates to the fact that the extent or influence of every one of these fundamental processes is dependent on the masses of materials reacted or transferred. The distribution of these masses is, in turn, dependent on transport phenomena taking place at the system scale.

The quantification of environmental changes must consider first the nature of fundamental processes involved and then the interpretation of the way in which system properties influence the nature of those processes. Fundamental processes involve reactions among system components, and systems involve movement and transport of those components in space and time. Example 1-1 illustrates in a very elementary way how a given process, in this case the oxidation of an organic compound by ozone, can be influenced by the nature of the system in which it occurs.

Example 1-1

- **Situation:** Ozone and a moderately volatile organic compound are added in identical amounts to three sealed 1-L jars. The jars, or "systems," differ only in the following regards:

  System 1: contains ozone, the organic compound, and air.

  System 2: contains ozone, the organic compound, and water.

  System 3: same as system 2, but there is a magnetic stirrer in the jar to mix its contents continuously.

- **Question(s):** In which system will the organic compound be most rapidly oxidized by the ozone, and in which system will it be most slowly oxidized if the systems are all at the same temperature of the room in which the jars are stored?

- **Assumption(s):** Assume that the organic compound and the ozone are both completely soluble in water. Assume also that the mechanism by which the molecules of these substances react with each other is the same whether they are in a gaseous state or a dissolved state, but in both cases that mechanism requires physical contact between the molecules.

- **Logic and Answer(s):** Physical contact between molecules is required for the oxidation reaction to occur. Molecules contained in a gas phase are free to move about rapidly under the influence of Brownian motion and thermal-gradient mixing, with little resistance to this motion being caused by other gas molecules. The reaction will therefore be most rapid in system 1 because there will be more molecular contacts per unit time in this system. Water, on the other hand, presents a substantial resistance to Brownian motion and requires large thermal gradients to cause mixing of the water molecules and thus more frequent contacts between dissolved molecules. System 3 is physically mixed, enhancing molecular contacts and causing a greater number of contacts per unit time than will occur in system 2, which is otherwise identical. The rate of oxidation of the organic compound will thus be lowest in system 2.

We distinguish from the outset between those elements of change that are rooted in reactions and mass transfer processes at the molecular level, and those that relate to the more macroscopic character and properties of the systems in which the processes occur. The drawing of this distinction is an overriding theme in our analyses of process dynamics in environmental systems. We pursue each aspect of process dynamics in detail, and persistently underscore the way in which they interact to motivate and control environmental change.

We turn first to a consideration of the similarities and differences between two broad categories of environmental systems, natural and engineered, and then to a consideration of the temporal and spatial scales associated with those two categories of systems.

1.2 ENVIRONMENTAL SYSTEMS

1.2.1 Natural and Engineered Systems

We are commonly concerned in *natural systems* with *understanding and describing changes* in quality parameters, while in *engineered systems* we are more generally concerned with *the selection of conditions required to accomplish specific changes* in quality parameters. As illustrated extensively in Tables 1-1 and 1-2 and depicted schematically in Figure 1-1, however, there are common aspects of these concerns and of how they are addressed.

For example, dissolved oxygen is a critical parameter for maintenance of desirable ecological conditions in natural aquatic systems as well as for engineered biological treatment systems. A typical process analysis for a natural aquatic system might involve measurement and characterization of the physical, chemical, and biological features of a lake, or a section or reach of a river, to assess how a waste discharge containing a biologically degradable organic substance will change dissolved oxygen levels. This analysis would involve several of the individual processes listed in Tables 1-1 and 1-2. In this case the analysis leads to a quantitative description, or *model*, of how dissolved oxygen varies temporally and spatially in that lake or section of a river.

An engineered-system counterpart to the natural system described above might, for example, involve the design of a biological or chemical treatment

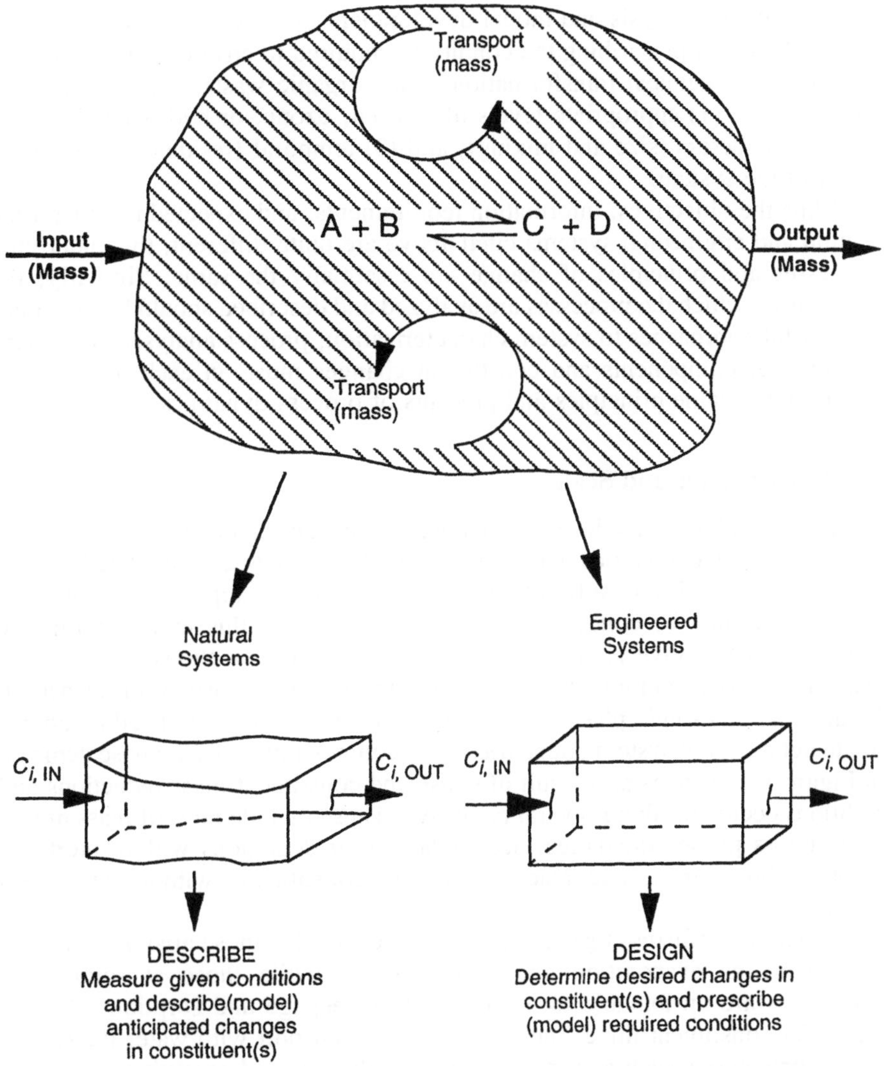

Figure 1-1 Transport and transformation processes in natural and engineered systems.

operation to reduce the oxygen-consuming properties of a waste prior to its discharge to a surface water. The process by which the waste is treated may well be similar to that by which it consumes oxygen in the natural aquatic system, that is, by aerobic (oxygen using) microbial transformation of organic matter. In this situation the same set of individual processes selected from Tables 1-1 and 1-2 would be involved. Conversely, the waste may be chemically transformed using an activated form of oxygen, such as ozone or hydrogen peroxide.

The process analysis and model development in the case of the engineered system focus on prescribing the conditions required to ensure desired levels of microbial or chemical transformation of the organic waste. These considerations include (1) appropriate levels of oxygen, ozone, or hydrogen peroxide; (2) spatial factors (size and mixing); and (3) temporal (residence time) properties of the system.

While the objectives, information requirements, and expected results for the natural and engineered systems cited above are quite different, the underlying processes and principles of change are essentially the same. Similarly, the methods by which the processes are analyzed and described should be the same. Successful approaches to system characterization, process analyses, and quantification of components and constituent changes must, in each instance, be based on the same principles and precepts of process dynamics.

1.2.2 Character and Scale

At the most elementary level we distinguish system character and properties on the basis of boundaries and size. Figure 1-2 depicts a generalized system of gaseous, liquid, and solid phases. In any particular application of process analysis we may be concerned with a system having this entire multiphase character. Alternatively, we may be concerned with, or choose to limit our analysis to, considerations of changes within only one phase. With respect to Figure 1-1, this single-phase characterization can be accommodated by appropriate definition of system boundaries. The composition of each phase depicted in Figure 1-2 changes as a result of phase and constituent mass reductions and additions occurring along several pathways. These include rates of mass movement across phase boundaries, accumulations or depletions within interfaces at phase boundaries, and reactions among constituents within the system boundaries.

Typical water fluxes at gas-phase boundaries involve those arising from such phenomena as evaporation and condensation, while those at solid-phase boundaries are attributable to events such as seepage and sorption. Driving forces for constituent fluxes include cocurrent transport with water fluxes and phase conversions such as vaporization, dissolution, and precipitation. Accumulations at phase boundaries, or interfaces, generally relate to the chemical incompatibility of a constituent with both phases and thus its tendency to form an "intermediate" phase. Depletions from these interfaces usually result from excess accumulation, or from the creation of conditions within one phase or another which enhance the compatibility of that phase for the constituent. Transformation of constituents within individual phases can involve a variety of physical, chemical, and biological reactions.

The system shown in Figure 1-2 is schematic in terms of its components and configuration and unspecified in terms of its spatial scale, or size. As suggested in Figure 1-3, the environment is a continuum of systems involving similar processes over a remarkable range of different temporal and spatial

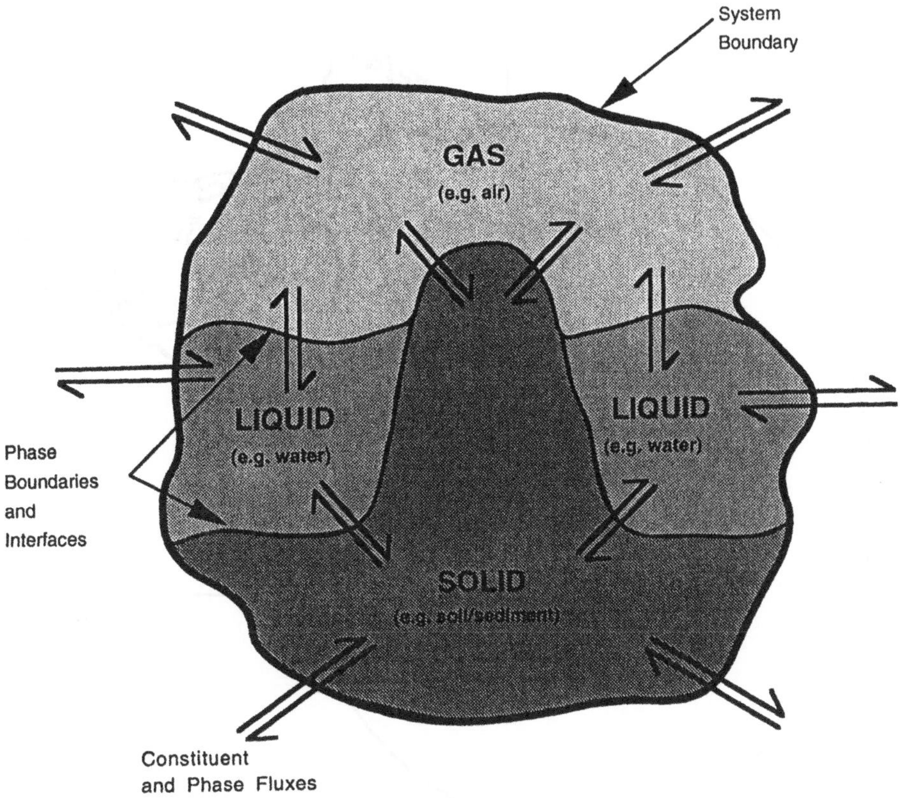

Figure 1-2 Generic multiphase environmental system.

scales. The schematic given in Figure 1-2 might in the context of Figure 1-3, for example, be a representation of an extremely large *megasystem*, such as Lakes Superior, Michigan, Huron, Erie, and Ontario: the Great Lakes. Here the gas phase would be the atmosphere, the aqueous phase the hydrosphere, and the solid phase the lithosphere. We may be concerned in such a system with the long-term exchange of dissolved substances, such as semivolatile organic pesticides, between the water column, the overlying atmosphere, and the underlying and suspended sediments. The boundaries and volumes of such systems are inherently difficult to describe. Significant changes in composition occur only over extended periods of time: typically, decades and centuries. Because megasystems typically lack precise spatial and temporal detail, their descriptions generally warrant no more than correspondingly coarse quantification of the transport and transformation processes that take place at or within their boundaries.

The *microsystem* counterpart of the megasystem example might be, for example, a sediment particle suspended in the nearshore waters of Lake Michi-

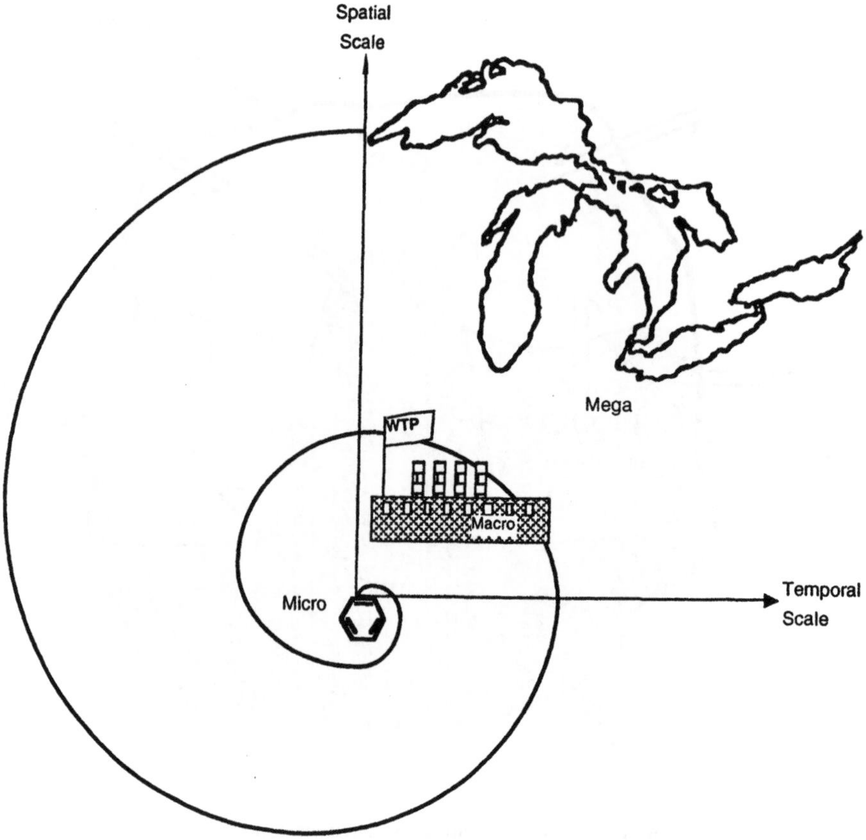

Figure 1-3 Environmental continuum.

gan. The local environment of that particle might well be comprised by the same components as those in Figure 1-2, with the particle itself being a complex "solid" phase suspended in the aqueous medium. The gas phase might be attached air bubbles generated by abiotic or biotic gas production, within or on the surfaces of the particle, or by cavitation induced by wave action. This particle-scale system is subject to precise boundary quantification, and within it a variety of transformations might take place over short periods.

If we are concerned in this microsystem with the same reaction phenomena as we were in the case of the megasystem, we should consider organic pesticides dissolved in the water or volatilized in the air bubbles in the immediate vicinity of the particle. Specifically, we should consider them in the context of their fluxes across bubble–particle, bubble–water, and water–particle interfaces measured in distances of microns, and sorption reactions at the surface of the particle measured in seconds and minutes. The temporal and spatial refinement of the microsystem affords precise definition and quantification of related flux and reaction processes for description of process dynamics.

Our focus in this book will relate primarily to systems that have time and space scales smaller than those of megasystems. The boundaries of the systems in which we have most interest are usually better defined than those of global systems and commonly are measured in fractions or multiples of meters. Transformations of interest occur within time periods ranging from tens of minutes to days in the case of most engineered operations to weeks and months for most natural systems. The behavior of these more local *macrosystems* is often markedly dependent on processes that occur within them at the microsystem scale.

A macrosystem in the context of the organic pesticides of our earlier example might be comprised of an activated carbon system designed and engineered for removing traces of pesticides from Lake Michigan water in a water treatment plant (WTP) prior to its use as a drinking water supply. In this case the scale of the carbon adsorbers is on the order of meters, and the residence time of water within them is on the order of minutes or hours. At the same time, this macrosystem is comprised by a multitude of particles of granular activated carbon of millimeter or micron scale. Like the particles suspended in the nearshore waters of Lake Michigan, each of these particles represents a microsystem wherein the pesticide interacts at the molecular level. We can thus further define, within the carbon particle, transport and reaction scales of microscopic and molecular levels, as suggested in Figure 1-4.

All systems are thus comprised by subsystems: megasystems by macrosystems, and macrosystems by microsystems. This is why many of the processes described in Table 1-1 can be influenced at the macroscopic scale by microscopic mass transfer phenomena. The most fundamental analysis of any system has its origins ultimately at the molecular level and must provide a continuity of principle from this scale to its full scale. The description, analysis, or design of a megasystem or a macrosystem frequently involves characterization and quantification of processes at several levels of scale. The design of the activated carbon adsorber for removal of organic pesticides from a water supply or wastewater entails description of transport processes at both macroscopic and microscopic scales, as suggested schematically in Figure 1-5a. At the macroscopic scale of the adsorber, flow into and out of the adsorber carries organic matter to the external surfaces of the carbon particles. Once the organic matter reaches the external surfaces of the porous particles of activated carbons, it begins to diffuse molecularly into the pore structure of these particles, ultimately to be adsorbed on internal surfaces. In practice, this second level of molecular or microscopic mass transport often controls the overall rate and residence time requirements for the process. These same considerations are appropriate for the sorption of organic contaminants by soil and other aquifer particles and aggregates, and for the ultimate design of remediations such as soil vapor extraction of volatile organic contaminants, as depicted in Figure

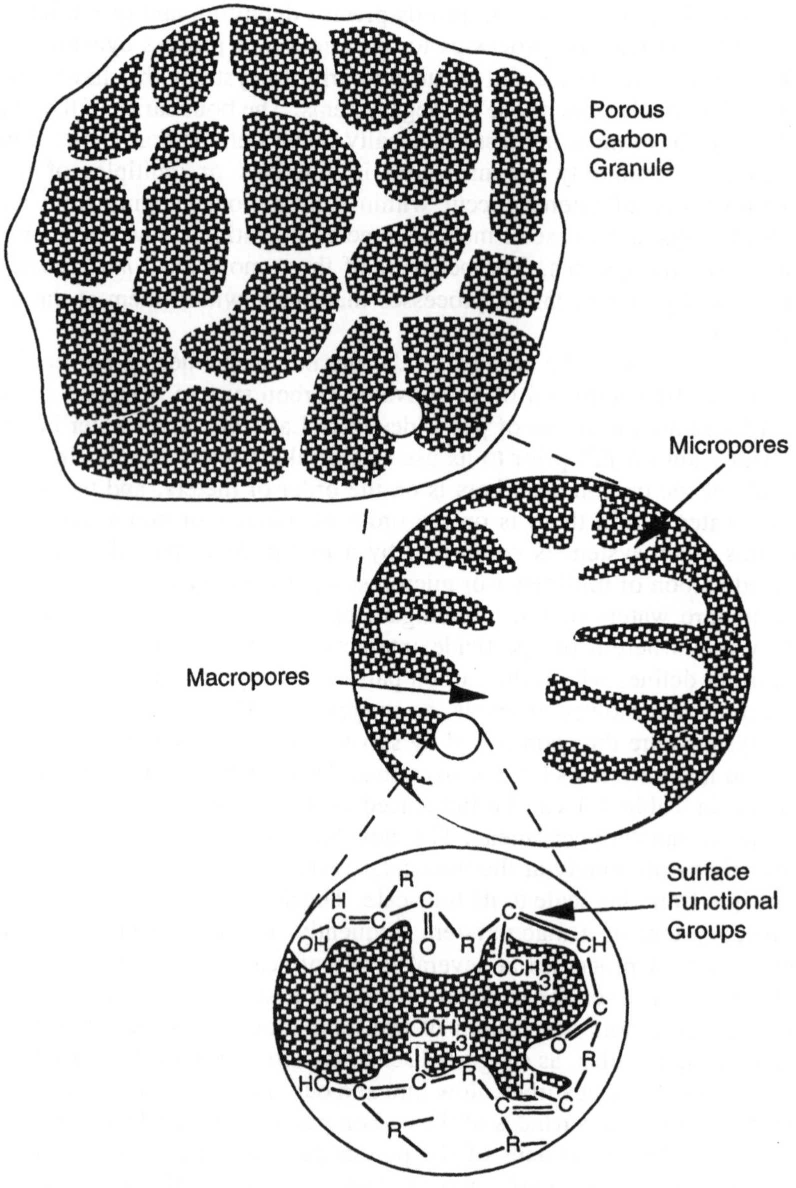

Figure 1-4 All systems are comprised of subsystems.

1-5b. Any analysis of the process for purposes of description or design must couple descriptions of phenomena at the appropriate microscale with those of phenomena at the appropriate system scale in a continuous and consistent manner.

Many natural and engineered environmental systems involve multilevel scale

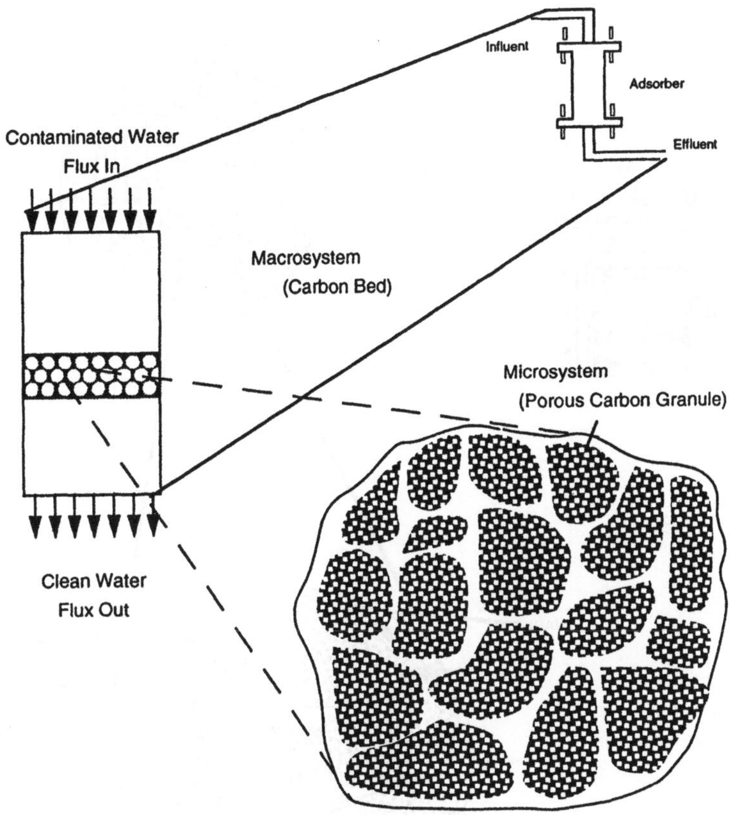

(a) Activated Carbon Adsorber

Figure 1-5 Macroscopic and microscopic aspects of two familiar environmental systems.

and transport considerations similar to those described for the activated carbon adsorber. The necessity for these considerations often arises because more than one phase is involved in most environmental systems. Therefore, interphase mass transfer, which implies microscopic scales and molecular phenomena, is involved. We must appreciate the role of microsystem process dynamics and understand how to incorporate information on processes at this scale in the characterization, analysis, interpretation, and design of environmental macrosystems.

Process characterization and analysis must be rigorous at all scales involved in gathering data and information for process description, and at all scales to which those data and information are subsequently applied. For example, we can obtain accurate and detailed information about process reactions from

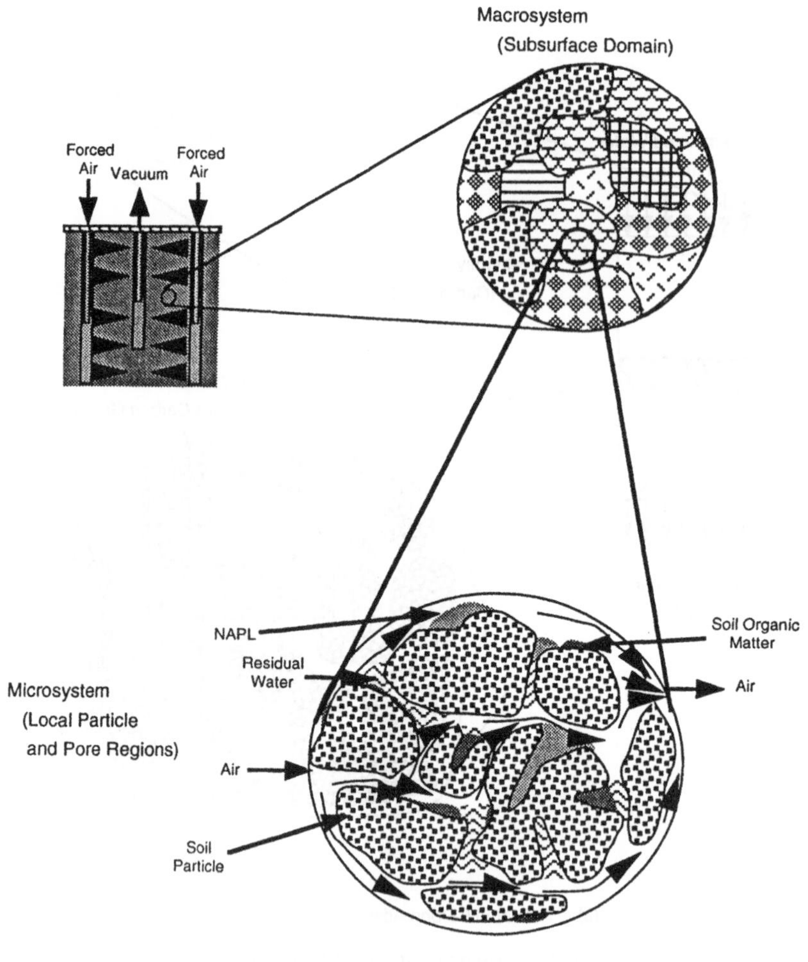

(b) Soil Vapor Extraction System

Figure 1-5 (*Continued*)

small-scale laboratory observations, but only if the constraints of the systems in which such observations are made are accounted for and applied properly in the analysis of resulting data. The effects of system size and configuration on mass transport and distribution must be properly recognized and separated from those aspects of a process pertaining to reactions only. The subsequent use of reaction rate data so obtained may then be applied to other systems of any scale or configuration as long as proper account is again made of the size, configuration, and properties of the system to which the information is applied.

Example 1-2 provides an illustration of how the scale of a system in which a process occurs can affect its rate.

Example 1-2

- **Situation:** In Example 1-1 we considered the dependence of the rate at which a given process occurred on the conditions of the system in which it took place. To illustrate some of the points we have since discussed relative to scale, consider a fourth type of system, a series of three sealed vessels, each filled half with air and half with water. The first is a 1-L jar, the second a 100-L carboy, and the third a 10,000-L tank. All three vessels are cylindrically shaped and dimensioned such that the ratio of cross-sectional area to volume is the same in each vessel. In each system there is a single two-bladed rotating mixer located exactly in the middle of that volume of tank containing the water. The blades are the same size in each reactor, and all operate at the same number of revolutions per minute.
- **Question(s):** Will the overall rate of oxidation of the organic compound per unit volume of reactor be the same in each system? If not, in which system will it be greatest and in which system lowest? Assume that the gas/liquid mass ratios for both the ozone and the organic compound are 2 : 1 in each system at all times and that simultaneous reactions take place in both phases.
- **Logic and Answer(s):** A thoughtful analysis of the information presented suggests that the rates of oxidation per unit volume of reactor will not be the same in each system, for several reasons. First, it is clear that no purposeful and quantitative mixing of the gas phase is provided in any of the reactors. Thus, as the volume of the gas phase in the series of three systems increases, there is an increase in the distances over which incidental mixing of that phase caused by shear at the water interface, thermal gradients, and Brownian motion must operate. As these distances increase, the probability of molecule–molecule interactions will decrease, particularly as the total number of molecules in each system is decreased by reaction. This is even more the case for the ozone and organic molecules dissolved in the water. A second readily evident difference among the reactors is that the effectiveness of mixers of equal size and rotation rate in establishing completely mixed conditions will decrease with increasing size of the reactor. This will, of course, affect rates of molecule–molecule interactions and thus rates of reaction per unit volume of reactor. Finally, it is apparent that the gas-phase reaction will be more rapid than the liquid-phase reaction in each tank. Because both the ozone and organic compound are volatile, each will tend to move from the liquid phase to the gas phase as the latter is depleted to maintain the stipulated condition of a 2 : 1 molecular distribution of each reactant be-

tween the two phases. Movement of the molecules through the greater depths of liquid in the larger systems will mean a slower replenishment of molecules in the gas phase. For these several reasons, the 1-L system will most closely approximate complete mixing, providing the greatest opportunity for molecular interaction and highest rate of reaction per unit volume. All of the observations above should be qualitatively intuitive. As we move through our developments of process dynamics in this book, we will learn ways in which to quantify them.

1.2.3 Components and Change

Like the properties of processes and changes that occur within them, the components of environmental systems are frequently categorized as physical, chemical, and biological. Examples of components common to the systems discussed above include particulate solids; biochemically and chemically degradable organic substances, either dissolved or contained in particulate solids; microbial organisms that use dissolved oxygen as the electron acceptor in oxidation of these chemicals; and other, more active forms of oxygen, such as ozone or hydrogen peroxide, which do not require biological mediation. In the most elementary sense, all these components are either discrete chemical species or more complex arrangements of chemical matter. Inorganic particulates are largely mineral structures such as clays or sands, and organic solids and microbial organisms are highly structured arrangements of organic compounds.

A constituent of a system that neither changes in composition nor causes change in any other component of the system is *environmentally stable*. We distinguish here subtle but important differences between chemical stability and environmental stability. Chloride, for example, a constituent of most natural waters and wastewaters, is a chemically and environmentally stable form of the element chlorine. It remains virtually constant in concentration and has little or no effect on changes in the environmental quality of waters or wastes in most treatment processes, save those specifically designed for removal of dissolved salts. Changes in its concentration may signal other changes in composition but do not result from actions of the chloride itself to modify water quality. Chlorine, on the other hand, is both chemically and environmentally unstable, reacting readily with other constituents of waters and wastewaters to change their form and to undergo change itself. Like ozone and hydrogen peroxide, chlorine is a strong oxidant because of its high degree of instability and reactivity and is commonly employed for oxidation of inorganic and organic contaminants.

Certain chlorine-containing substances, such as polychlorinated biphenyls (PCBs), are nearly as chemically stable as chloride but not nearly as stable environmentally. PCBs are chemically or thermodynamically stable in the sense that they do not readily undergo change in their chemical form under normal environmental conditions, but they can cause major changes in environmental quality by inducing toxic and carcinogenic reactions in living cells.

When present in water or wastewaters, environmentally reactive compounds are generally specifically targeted for *change*. Such change may be accomplished by *removal* in some phase separation process, such as precipitation, adsorption, or volatilization, or *transformation* in a chemical or biochemical oxidation process. The choice of the process depends on certain properties of the targeted compounds, their reactive tendencies, and their resultant susceptibility to specific separation or transformation phenomena. The type of system chosen should be one that maximizes the effectiveness of the process selected, tempered by any constraints that may be dictated by particular circumstances in a given application. Ways to maximize effectiveness are illustrated in Example 1-3.

Example 1-3

- **Situation:** A drinking water supply contains a volatile organic contaminant that can be destroyed effectively by ozone. The conditions surrounding this reaction are similar to those described in Examples 1-1 and 1-2.
- **Question(s):** Discuss several ways in which one might consider designing a system to maximize the effectiveness of the oxidation process.
- **Logic and Answer(s):** We have shown that the reaction can be carried out more rapidly in the gas phase than in the liquid phase in the systems examined. This is true because the mixing is more complete, and thus the opportunity for reaction is greater in the gas phase. This circumstance may be used to advantage in the design of a treatment system by carrying out the bulk of the transformation process in the gas phase. Specific considerations might include:

    1. Ways in which to configure the system to maximize surface area and minimize the depth of water.
    2. Mixing the water to facilitate movement of molecules to the liquid surface (gas–liquid interface).
    3. Dispersing gas bubbles through the liquid to increase the gas–liquid interfacial area and reduce the liquid-phase distances over which molecules have to move to reach the gas phase.
    4. Collecting the gas phase continuously to keep the contaminant concentration in that phase low and thus increase the rate of its movement into the gas to maintain a 2 : 1 distribution.
    5. Compressing the collected gas to bring contaminant molecules into more immediate contact with one another to speed up the reaction rate per unit volume of gas phase.

The problem posed in this example is a fairly common one for decontamination of groundwaters containing volatile compounds such as gasoline hydrocarbons and chlorinated solvents. The process may be restricted in some cases simply to "stripping" the organic contaminants from the water and dispersing them to the atmosphere. Systems to accomplish this may involve bubbling air through the water, as indicated above, or accomplishing the same effect by allowing the water to drain over the surfaces of a loosely packed and highly porous tower of solid media while air is forced through the tower in the opposite direction. In other cases the gas must be treated before release to destroy the organic contaminant, as in the example above. Because gases are compressible, concentrations of contaminants and reactants can be increased by increased pressure, and this is frequently done to provide higher reaction rates and smaller treatment systems.

1.2.4 Measures of Quantity and Concentration

The properties of a system that characterize its inherent mass, energy, or momentum relationships can be divided into two general categories: *extensive* and *intensive*. We assume for the purpose of this discussion that these properties do not vary spatially within the system (i.e., the system is completely mixed). An extensive property of a system is one whose magnitude depends on the size of the system or of a sample taken from the system; examples include mass, volume, heat capacity, and calories. Conversely, an intensive property of a system does not depend on the size of the system or on that of any sample taken from it (e.g., temperature and pressure).

The *quantity* of a *substance* (defined here as any matter of definite and recognizable composition and specific properties) may be expressed in terms of mass or weight, volume, or number of moles. The term *concentration* expresses the quantity of a substance, i, present in a certain *quantity of a particular phase, j*. This *ratio of quantities* remains constant regardless of the system or sample size. Concentration is thus an intensive property. In general, substance i and phase j may be any solid, liquid, or gas present in the system under study. Phase j often may represent a mixture of solids, liquids, or gases. For example, air pollution scientists are commonly concerned with the amounts of CO_2 (substance i, gas) or particulate matter (substance i, solids) present in *air* (phase j, a *mixture* of gases). In the field of water quality, we are often concerned with the amount of dissolved solids (substance i) present in water or wastewater (phase j). Many water or wastewater treatment operations involve the production and management of sludges, for which we are concerned with the amount of solid (substance i) present in the *sludge* (substance j, a *mixture* of water and solids). The selection of substance i and phase j may depend on the particular system of interest. For instance, if the sludge from a treatment operation is dewatered in preparation for landfilling, it may be more convenient to characterize the amount of water (substance i) in the sludge (phase j).

Quantities can be expressed in terms of mass or weight, volume, or moles. Concentrations may be given as mass fractions, volume fractions, mole fractions, mass per unit volume, moles per unit volume, moles per unit mass, and equivalent charge per unit volume. It is expected that the reader is familiar with various units for expression of quantities and concentrations. A brief review of this subject, including examples and exercises, is presented in Appendix A.1.1.

1.3 ENVIRONMENTAL PROCESSES

1.3.1 Transformation

If all the components of a closed system are comprised of chemical substances, all changes in the composition of that system must, in the most elementary sense, involve chemical reactions. These reactions result in the *transformation* of one chemical form, species, or state to another. Microbial degradations of organic substances, for example, involve chemical transformations of substances that serve as fuel or food sources for microbes. These chemical materials are converted biochemically to other chemicals, which are then released as waste products. The biochemical transformations are facilitated by other chemicals, termed *enzymes*.

To say that all environmental transformations are no more than chemical reactions would be an oversimplification, but it is true that virtually all environmental transformations are chemically mediated. Chemical concepts and relationships provide perspectives that aid the understanding of environmental processes and facilitate their description; these may be categorized primarily as chemical, physical, physicochemical, biological, or biochemical processes. Chemical concepts and relationships permit development of models by which events and processes in natural systems can be interpreted and predicted. These same models allow us to select, diagnose, and facilitate processes and transformations in engineered reactors to accomplish specific environmental quality objectives.

The chemically mediated reactions by which transformations occur can be broadly categorized as either *homogeneous* or *heterogeneous*, depending on the number of phases they involve. As stressed earlier, environmental systems are typically heterogeneous in the sense that they involve more than one phase. Nevertheless, the transformation reactions that occur within them are not necessarily heterogeneous in nature.

A macroscale elaboration of the three-phase system depicted in Figure 1-2 is given in Figure 1-6. This is a system from which water is drawn continuously for use and into which used water is discharged continuously. The boundaries of such a system are defined by particular circumstances, such as the need to consider a specific reach of a river, or by specification of the size and configuration of a holding tank or treatment unit. Generalized cocurrent

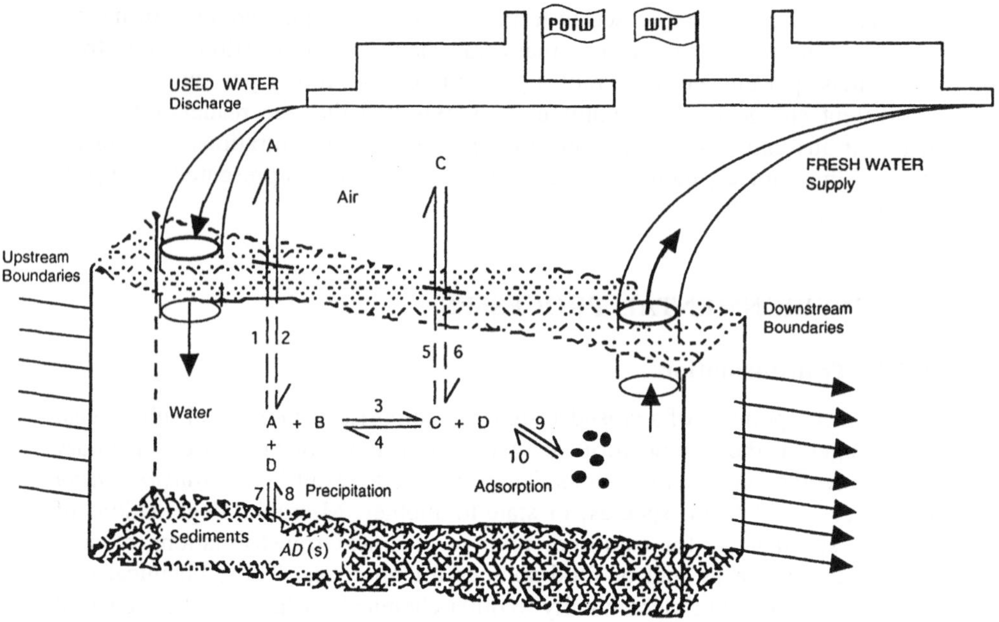

Figure 1-6 Natural environmental macrosystem.

mass fluxes of water and other constituents into and out of the system are indicated by the bold arrows, and the specific fluxes and reactions of several generic constituents, A, B, C, and D, by the single-headed arrows. In this example the particular chemical species of interest undergo transformations that change their state within the aqueous phase. These transformations involve reaction of species A with species B to form species C and D. The species involved can also change their distributions among the three phases of the system. Species A might well represent dissolved oxygen in a river system or activated sludge basin, or ozone in a pretreatment unit of a water treatment plant, and species B, an organic compound. Certain of the transformations of species A and B that occur along pathways 1 through 10 in this system are homogeneous, while others are heterogeneous. Species A and reaction product C, which may, for example, represent oxygen and carbon dioxide respectively, are transformed from dissolved states in the water phase to gaseous states in the air phase along reaction pathways 1 and 5; A and C may be retransformed to dissolved states, respectively, along reaction pathways 2 and 6. Reactions 1, 2, 5, and 6 are heterogeneous, involving two different phases and two correspondingly different states of species A and C. Reactions 7 and 8 are also heterogeneous because they involve the reaction of product D with A to form a precipitate, a separate solid phase, and the redissolution of this solid phase. Similarly, reactions 9 and 10 are heterogeneous because they involve interfacial sorption and desorption reactions of reaction product D with suspended solids, such as clays or other sediments. In contrast, reactions 3 and 4 are homogeneous reactions because they involve changes only in the aqueous-phase states of the species involved.

Homogeneous reactions are governed by relationships which require that certain entropy, enthalpy, and energy balances be obeyed among the components or species participating in transformation processes that occur within a single phase. These relationships yield characteristic thermodynamic properties and corresponding *equilibrium constants*. Although having fundamentally the same origin for such different classes of transformations as acid-base reactions, oxidative-reduction reactions, and complex formation reactions, these constants are commonly given somewhat different names, such as *acidity constants*, *electrochemical potentials*, and *stability constants*, respectively.

The chemical relationships governing the heterogeneous reactions of species are different from those for homogeneous reactions, but they also have their fundamental origins in chemical thermodynamics. Reactions 1, 2, 5, and 6, for example, adhere to the following general proportional relationships between concentrations in each phase:

$$P_A = \mathcal{K}_{H,A} C_A \tag{1-1}$$

$$P_C = \mathcal{K}_{H,C} C_C \tag{1-2}$$

where P_A and P_C are the concentrations of A and C in the gas phase, expressed in terms of their respective *pressures* (e.g., atmospheres), and C_A and C_C are the *molar* concentrations (e.g., moles per cubic meter) of the same species in the liquid (aqueous) phase.

The constants of proportionality $\mathcal{K}_{H,A}$ and $\mathcal{K}_{H,C}$ in Equations 1-1 and 1-2 are constants defining the relative preferences of species A and C, respectively, for the gas and liquid phases at a point of thermodynamic stability (i.e., the condition of equilibrium).

Equations 1-1 and 1-2 characterize the *equilibrium* state of phase *partitioning* processes for species A and B between aqueous and gas phases. These equations are commonly termed Henry's law relationships, and $\mathcal{K}_{H,A}$ and $\mathcal{K}_{H,C}$ are termed Henry's constants. Further details are given in Chapters 5 and 6. It suffices for now to state that every chemical species has a known Henry's constant to characterize its tendency to escape from dilute aqueous solutions into air. Oxygen, for example, has a relatively high Henry's constant. Thus it is abundant in the atmosphere, but found in only relatively low concentrations in water. Carbon dioxide, on the other hand, has a low Henry's constant, preferring more to partition from gas phase into water. Table 1-3 provides a listing of Henry's constant values for a number of inorganic gases of common interest for environmental systems. Example 1-4 gives an illustration of how these constants can be determined and applied.

Table 1-3 Values of Henry's Constant for Inorganic Gases in Aqueous Solutions

	$\mathcal{K}_H \left(\frac{\text{atm-m}^3}{\text{mol}} \times 10^3\right)$ at:			
Gas	10°C	15°C	20°C	25°C
Carbon dioxide (CO_2)	18.7	22.0	25.6	29.5
Carbon monoxide (CO)	796	880	965	1044
Chlorine (Cl_2)	8.8	10.7	13.3	16
Hydrogen (H_2)	1145	1190	1229	1273
Hydrogen sulfide (H_2S)	6.6	7.6	8.7	9.8
Nitric oxide (NO)	392	436	475	517
Nitrogen (N_2)	1202	1328	1447	1557
Nitrous oxide (N_2O)	25.4	29.9	35.6	40.5
Oxygen (O_2)	589	655	722	788
Ozone (O_3)	44.6	51.8	67.7	82.3
Sulfur dioxide (SO_2)	0.43	0.49	0.62	0.72

Source: Numerical values calculated from constants and solubility data in Perry et al. (1984).

Similar relationships exist for reactions involving the partitioning of chemical species from water into organic liquids, solids, or biological life forms; their precipitation from solution to form solid phases; and their adsorption onto solid surfaces. The relative tendencies of species to undergo phase transformations are characterized by a set of corresponding thermodynamic constants. Prime examples of such constants are the *solubility limit*, C_S, and the *solubility constant*, $\mathcal{K}_s$; *vapor pressure*, P^v; the *octanol–water partitioning coefficient* $\mathcal{K}_{o,w}$; the *bioconcentration factor*, $\mathcal{K}_B$; the *Freundlich adsorption coefficients*, $\mathcal{K}_F$ and n; and so on.

Table 1-4 presents a list of different types of phase transformation constants for a series of organic compounds of environmental concern. Although it is not expected that the exact physical significance of each of these constants is obvious to the reader at this point, there are certain intuitive relationships that become evident by even cursory examination of their ranked values. Examination of these values reveals, for example, distinct trends in the expected environmental behavior of the compounds listed. To illustrate these trends, compare the two aromatic hydrocarbons, benzene and ethylbenzene, and note from the respective values of C_S, P^v, $\mathcal{K}_{o,w}$, $\mathcal{K}_F$, and $\mathcal{K}_B$, the consistently greater tendency for benzene to accumulate in aqueous phases. The consistency of the trend is not surprising because most of these constants are interrelated in one way or another. In subsequent considerations of the thermodynamic properties of transformation reactions in Chapters 5 and 6 we elaborate more fully on various thermodynamic constants and their interpretation and significance as measures of the environmental tendencies of different chemical and biochemical species.

Table 1-4 Properties and Constants of Organic Contaminants at 25°C

Compound	Aqueous Solubility, C_S (mg/L)	Vapor Pressure, P^v (atm)	Henry's Constant, $\mathcal{K}_H$ $\left(\frac{\text{atm-m}^3}{\text{mol}} \times 10^3\right)$	Octanol–Water Partition Coefficient, $\mathcal{K}_{o,w}$(log)	Freundlich Adsorption Coefficients[a]		Bioconcentration Factor, $\mathcal{K}_B$ (log)	
					$\mathcal{K}_F$	n	Calculated[b]	Experimental[c]
Benzene	1750	1.25×10^{-1}	5.49	1.95	0.5	0.46	0.96	—
Toluene	534	3.74×10^{-2}	6.66	2.69	1.2	0.47	1.59	—
Ethylbenzene	152	1.25×10^{-2}	8.73	3.15	1.6	0.39	1.98	—
o-Xylene	175	8.68×10^{-3}	5.27	3.12	1.1	0.47	1.95	—
Chlorobenzene	488	1.382×10^{-2}	3.71	2.84	1.8	0.40	1.71	—
1,2-Dichlorobenzene	145	1.32×10^{-3}	2.96	3.38	3.2	0.41	2.17	2.40
1,2,4-Trichlorobenzene	30	5.53×10^{-4}	2.32	4.26	6.2	0.44	2.92	3.01
Nitrobenzene	1900	1.97×10^{-4}	0.023	1.85	4.0	0.39	0.87	—
Lindane	7.3	4.34×10^{-7}	0.00043	3.72	3.2	0.35	2.46	2.75

Source: Adapted from Weber et al. (1987).

[a]Freundlich coefficients for Equation 6-67 ($q_e = \mathcal{K}_F C_e^n$). Determined for Hydrodarco C activated carbon for q_e = μg/mg and C_e = μg/L.

[b]From empirical correlation $\log \mathcal{K}_B = 0.85 \log \mathcal{K}_{o,w} - 0.70$ (see Equation 6-52 and Section 6.7.2.5).

[c]Determined experimentally by Weber et al. (1987) for activated sludge biomass.

Example 1-4

- **Situation:** Experimental determination and application of Henry's constant.
- **Question(s):** (a) Devise an experiment to measure Henry's constant for oxygen in water in contact with air at 20°C and 1 atm pressure. (b) Determine how much oxygen will be dissolved in water at 20°C if pure oxygen gas is bubbled through the water until it is saturated.
- **Logic and Answer(s):**

(a) Henry's constant is a constant of proportionality relating to the distribution of a gas between a dilute aqueous phase and a gaseous phase at equilibrium. The concentration of oxygen in air, expressed as a partial pressure, p_{O_2}, is 0.21; that is, air is comprised by 21% oxygen on a molar basis. If we bubble a large amount of air through a liter of water until it is saturated with the air, the amount of oxygen dissolved in the water at this condition of equilibrium is dictated by Henry's law (Equation 1-1). If this amount of oxygen is measured, Henry's constant can be calculated.

1. For example, if the aqueous phase oxygen concentration is measured and found to be 9.3 mg/L, the mass ratio of oxygen in the water is

$$X_{O_2,H_2O} = 9.3 \times 10^{-6} \text{ g/g}$$

2. Given that 1 mol of gas occupies 24.05 L of volume at 1 atm pressure according to the ideal gas law ($V = n\mathcal{R}T/P$ and $\mathcal{R} = 8.21 \times 10^{-5}$ atm-m^3/mol-K), and that the density of air at 20°C (293 K) is 1.2 g/L, the mass ratio of oxygen in the air is

$$X_{O_2,\text{Air}} = \frac{p_{O_2} \text{ (32 g } O_2\text{/mol)}}{24.05 \text{ (L/mol) (1.2 g/L)}}$$

$$= \frac{0.21\text{(32 g } O_2\text{/mol)}}{24.05 \text{ (L/mol) (1.2 g/L)}} = 0.233 \text{ g/g}$$

3. Henry's constant in nondimensional form is then

$$\mathcal{K}_H^\circ = \frac{0.233}{9.3 \times 10^{-6}} \left(\frac{\text{g } O_2\text{/g air}}{\text{g } O_2\text{/g water}} \right) \times \frac{1.2 \times 10^3 \text{ g/m}^3 \text{ air}}{0.998 \times 10^6 \text{ g/m}^3 \text{ water}}$$

$$= 30.12 \frac{\text{mol } O_2\text{/m}^3 \text{ air}}{\text{mol } O_2\text{/m}^3 \text{ water}}$$

or, in units of atm-m^3/mol,

$$\mathcal{K}_H = \mathcal{K}_H^\circ \mathcal{R}T$$

$$= 30.12(8.21 \times 10^{-5})(293) = 0.725 \text{ atm-m}^3\text{/mol}$$

4. It will be noted that this value for $\mathcal{K}_H$ is close to that given for oxygen in Table 1-3.

(b) If pure oxygen is used to saturate the water, the dissolved oxygen concentration can then be determined from the Henry's constant calculated above as follows:

1. From Henry's law (e.g., Equation 1-1):

$$P_{O_2} = \mathcal{K}_H C_{O_2}$$

2. For pure oxygen, $P_{O_2} = 1$ atm; thus

$$C_{O_2} = \frac{P_{O_2}}{\mathcal{K}_H} = \frac{1.0 \text{ atm}}{0.725 \text{ atm-m}^3/\text{mol}}$$

$$= 1.38 \text{ mol/m}^3 = \frac{1.38 \text{ mol}(32 \text{ g/mol})}{\text{m}^3}$$

$$= 44.1 \text{ g/m}^3 = 44.1 \text{ mg/L}$$

1.3.2 Transport

A complete description and quantification of compositional changes in a system, such as those depicted schematically in Figures 1-1 and 1-6, must account for (1) changes related to the movement of components into or out of the system, (2) distributions of components within its boundaries, and (3) distribution of components across and within the boundaries of subsystems comprising the overall system. Movements of components across and within system boundaries affect and are affected by a number of different processes, which earlier in this chapter we broadly categorized as *macroscopic* and *microscopic* transport processes. Macroscopic transport processes occur at the scale of the system or reactor, whereas those that are microscopic occur at the scale of subcomponents (e.g., suspended particles, phase interfaces, etc.).

Transport processes occur as the result of specific driving forces. In essence, these driving forces are *spatial energy gradients*. In the case of the macroscopic transport of a constant-density fluid between two hydraulically connected tanks, for example, the driving force may be as simple as the difference in the free-surface elevations of water in the tanks. This *elevation head* causes the fluid to flow from the tank having the higher free-surface elevation to the one of lower free-surface elevation. Alternatively, a gradient in *density* induced by a difference in temperature or in the concentration of a dissolved salt in a single quiesient tank at a fixed elevation will cause fluid elements to move from one point in the tank to another. *Pressure* differences in systems having no free surfaces (e.g., pipe flowing full) exert the same influence on the movement of a fluid and thus on the macroscopic transport of substances dissolved or suspended in that fluid.

Physical or mechanical macroscopic transport phenomena may in many instances dominate overall process performance. It is essential to include them in the analysis of any chemical, biochemical, or physicochemical process. Reaction models alone, no matter how relevant and well understood they may be, are not adequate *in and of themselves* for description of anticipated process behavior in a system or for an explanation of observed behavior. The same reaction occurring in two different systems (or reactors) of different configuration will probably produce different extents of conversion. This is true because reactor configuration influences fluid mixing patterns and thus influences local concentrations of reactants. This point was illustrated in Example 1-2. We define movement of dissolved and suspended material that occurs as the result of bulk fluid movement as *advective transport* in our discussions of macroscopic transport processes in Chapter 3.

Even in the absence of bulk fluid motion, dissolved and suspended matter may migrate under the influence of spatial energy gradients that relate to other system properties. Most common among such transport processes are those that occur as a result of spatial gradients in *concentration*. These processes generally involve much lower rates (velocities) of migration than do advective processes and thus involve much smaller spatial scales and distances. They are for this reason termed *microscopic processes*. The purest form (smallest scale) of microscopic mass transport is molecular diffusion, or *diffusive transport*. We learn in Chapter 3 that there is another important form of transport, referred to as *dispersive transport*, which derives from spatial gradients in *momentum*, and which thus lies intermediate between advective and diffusive transport in both scale and mechanism.

There are many instances in environmental systems in which both microscopic and macroscopic transport phenomena must be considered; a large number of examples were given in Tables 1-1 and 1-2. Such systems generally involve separations of components among phases, or reaction phenomena that require molecular-level transport of species to reactive components contained within, or at the surfaces of, another phase. The activated carbon system depicted in Figure 1-5a and discussed in our consideration of system scales is, for example, a system in which both macroscopic and microscopic transport must be characterized and quantified, the first in the context and scale of the *reactor*, and the second in the context and scale of the *reaction*. The system process model must then interface appropriate characterizations of macroscopic and microscopic transport and reaction phenomena to ensure process continuity.

Briefly stated, then, spatial gradients in energy are associated with all transport phenomena, and different types of transport are caused by spatial gradients in different forms of energy. Elevation, pressure, and density differences represent *gradients in mechanical energy*, while concentration differences represent *chemical energy gradients*. Similarly, spatial differences in electrical potential can cause fluid or solute transport as a result of *electrical energy gradients*. Spatial energy gradients are rarely *singular* in environmental sys-

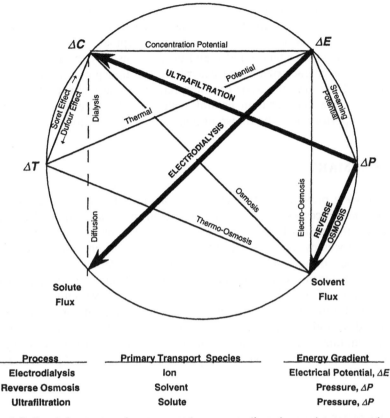

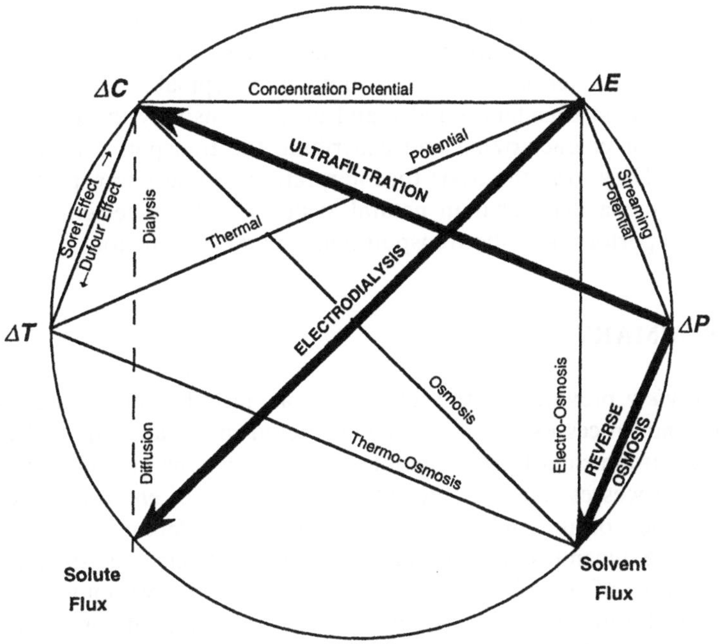

Process	Primary Transport Species	Energy Gradient
Electrodialysis	Ion	Electrical Potential, ΔE
Reverse Osmosis	Solvent	Pressure, ΔP
Ultrafiltration	Solute	Pressure, ΔP

Figure 1-7 Coupled transport phenomena and energy gradients in membrane separation processes.

tems. Net transport is most commonly determined by some combination of different types of energy gradients. In fact, transport induced by one type of energy gradient may produce other types of energy gradients and thus different (additional) types of transport.

Coupled gradients and transport processes are most readily evident in separation processes involving membranes (e.g., electrodialysis, reverse osmosis, ultrafiltration) but occur for a wide range of other heterogeneous environmental processes as well. Figure 1-7 provides a schematic illustration of coupled energy gradients and transport processes for the three different membrane separation processes referenced above, each of which has one energy gradient as a primary driving force but also involves a number of different secondary gradients and processes. Note that other energy gradients, such as ΔC and ΔT, can result similarly in either solute flux or solvent flux. Note also that when they occur under the influence of one primary energy gradient, such fluxes create other energy gradients.

Regardless of the specific energy gradient(s) and primary and secondary process(es) involved, all transport phenomena can be grouped into one of the three major categories defined above (advective, dispersive, diffusive), and then quantified in terms of relatively simple conceptual models. These models are developed in Chapters 3 and 4 and used extensively throughout the book as a basis for characterizing and describing the transport aspects of process dynamics. They form the basis for interpreting complex transport and separation phenomena that occur in natural systems and for designing engineered systems to perform specific transport and separation functions.

1.4 SUMMARY

The *overview* provided in this chapter is an introduction to the features of systems and processes that must be considered in the analysis and interpretation of environmental transformations. It describes the general nature of fundamental processes, the character of different types of environmental systems in which they occur, and the way in which these factors come together to influence the dynamics of the processes and their effects on systems. Adequate considerations of such factors are prerequisite to the development of rigorous approaches for the analysis, characterization, and design of environmental processes and operations such as those described in Tables 1-1 and 1-2. Having gained from this overview an appreciation of the important features of systems and processes, we turn now to more rigorous considerations of process dynamics.

1.5 REFERENCES AND SUGGESTED READINGS

1.5.1 Citations and Sources

Olsen, R. L., and M. C. Kavanaugh, 1993, "Can Groundwater Restoration Be Achieved?" *Water Environment and Technology*, March, 42–47. (Problem 1-7)

Perry, R. H., C. H. Chilton, and S. D. Kirkpatrick, 1984, *Chemical Engineers' Handbook*, 4th Edition, Section 14, McGraw-Hill Book Company, New York. (Table 1-3)

Weber, W. J., Jr., B. E. Jones, and L. E. Katz, 1987, "Fate of Toxic Organic Compounds in Activated Sludge and Integrated PAC Systems," *Water Science and Technology*, *19*, 471–482. (Table 1-4)

1.5.2 General Introductions to Environmental Science and Engineering

The following books provide an introductory treatment to environmental engineering. Although they do not use a process approach, they are useful for their treatment of contemporary environmental issues, including risk assessment, global atmospheric change, hazardous and radioactive wastes, and noise pollution.

Masters, G. M., 1991, *Introduction to Environmental Engineering and Science*, Prentice Hall, Englewood Cliffs, NJ.

Vesilind, P. A., J. J. Peirce, and R. Weiner, 1988, *Environmental Engineering*, Butterworth Publishers, Stoneham, MA.

1.5.3 Background Reading

Davis, M. L., and D. A. Cornwell, 1991, *Introduction to Environmental Engineering*, 2nd Edition, McGraw-Hill Book Company, New York. An overview of environmental and resource management systems and a discussion of the major elements of environmental engineering, including air and water quality, solid and hazardous wastes, and environmental legislation, are provided in Chapter 1.

Henry, J. G., and G. W. Heinke, 1989, *Environmental Science and Engineering*, Prentice Hall, Englewood Cliffs, NJ. Environmental problems are discussed in Chapter 1 from the perspective of the interaction of systems (water–air–land) and the processes involved in the global cycling of wastes.

Jorgensen, S. E., and W. J. Mitsch, 1989, "Ecological Engineering Principles," in *Ecological Engineering* (W. J. Mitsch and S. E. Jorgensen, Eds.), John Wiley & Sons, Inc., New York. Ecological engineering principles, including ecosystem structure, "forcing functions," chemical cycling in ecosystems, and the spatial and temporal scales of ecosystem processes, are discussed in Chapter 3.

James J. Montgomery Consulting Engineers, 1985, *Water Treatment Principles and Design*, John Wiley & Sons, Inc., New York. The history of water treatment is discussed in Chapter 1 and the principles of process engineering from a water treatment perspective in Chapter 5.

Stumm, W., and J. J. Morgan, 1995, *Aquatic Chemistry*, 3rd Edition, John Wiley & Sons, Inc., New York. Environmental processes and systems are discussed from a chemical perspective in Chapter 1.

Wanielista, M. P., Y. A. Yousef, J. S. Taylor, and C. D. Cooper, 1984, *Engineering and the Environment*, Brooks/Cole Publishing Co., Monterey, CA. Chapter 1 presents several case studies of environmental pollution, and the cycling of water, energy, and nutrients from an ecosystem point of view is described in Chapter 2.

1.6 PROBLEMS

1-1 The solubility of a gas in solution can be expressed in terms of its gas-phase partial pressure by Henry's law. From Table 1-3 we note that Henry's constant for oxygen is 722×10^{-3} atm-m^3/mol at 20°C. The pressure of a gas in a mixture of gases, is given by $P_i = X_g P_T$, where X_g is the mole fraction of the gas in the atmosphere. For an atmospheric pressure (P_T) of 1.0 atm, calculate the solubility of oxygen in water at 20°C expressed on a mass per unit volume basis, mg/L.

Answer: *9.33 mg/L (or for $\rho_w = 1.0$ g/cm^3, 9.33 ppm).*

1-2 The solubility of nitrogen gas at 20°C is $C_{S,N_2} = 19$ mg/L.

(a) Calculate the maximum molar and mass concentrations of N_2 in water equilibrated with air.

(b) Discuss whether the Pacific Ocean is in equilibrium with the overlying atmosphere with respect to N_2. Describe how you arrived at your answer and why your answer makes sense in terms of the systems involved.

Answer: *(a) 0.54 × 10^{-3} mol/L, or 15.2 mg/L.*

1-3 A water treatment plant that uses soda ash (Na_2CO_3) and unslaked lime (CaO) for softening wishes to produce caustic soda for magnesium removal [as $Mg(OH)_2$] according to the reactions

$$CaO(s) + H_2O \rightleftharpoons Ca(OH)_2(s)$$

$$Na_2CO_3 + Ca(OH)_2 \rightleftharpoons 2NaOH + CaCO_3(s)$$

(a) Identify the process(es) listed in Table 1-1 that are involved in this operation. Discuss the bases for your choices.

(b) Calculate how much NaOH can be obtained per ton of Na_2CO_3.

(c) Discuss circumstances that might compel the plant engineer to decide to use the product of the second reaction, NaOH, rather than that of the first, $Ca(OH)_2$, to precipitate $Mg(OH)_2$.

(d) These circumstances might also make the $CaCO_3$ precipitated in the second reaction a useful by-product rather than a waste sludge. Discuss why and how this might be so.

(*Note:* The answers to parts (c) and (d) are not immediately obvious from the material presented in this chapter, but intuition and thoughtful consideration of the nature of raw water sources and of the processes described in Table 1-1 should yield reasonable responses.)

Answer: *(b) 1,510 lb per short ton (2000 lb), or 755 kg per metric ton.*

1-4 The water treatment plant described in Problem 1-3 uses chlorine gas for disinfection. A municipal swimming pool requests that the plant provide 500 kg of sodium hypochlorite (NaClO) for its use, the hypochlorite ion (ClO^-) being a more stable form of active chlorine than Cl_2 gas. NaClO can be prepared as follows:

$$Cl_2 + 2NaOH \rightleftharpoons Na^+ + Cl^- + NaClO + H_2O$$

Other forms of stable hypochlorite, such as $Ca(ClO)_2$, can be prepared in a similar manner.

(a) Determine how much Cl_2 will be required to produce 500 kg of NaClO, assuming 85% conversion efficiency.

(b) Reflect on Problem 1-3 and suggest an alternative disinfectant that can meet the requirement of the municipal pool operator for a stable disinfectant, yet be somewhat more compatible with the operating scheme at the water treatment plant.

(c) Specify what quantity of this alternative disinfectant would be required to have the same disinfection potential as the 500 kg of NaClO.

Answers: (a) 561 kg of $Cl_2(g)$*; (c) 481 kg of* $Ca(OCl)_2$.

1-5 A packed-tower stripping column (the packing is used to promote air–water mass transfer) is to be used as a pretreatment to remove ammonia from an industrial wastewater. The feed stream flow rate is 500,000 gallons per day (gpd), and has an NH_3 concentration of 500 mg NH_3-N/L. The discharge permit allows a concentration of 100 mg NH_3-N/L in the effluent stream.

(a) Calculate how many moles of NH_3 must be stripped per day.

(b) Describe and briefly discuss three different types of systems in which the stripping process might be configured.

Answer: (a) 54,000 mol NH_3*/day.*

1-6 The ammonia stripping column in Problem 1-5 must be analyzed to determine if an air standard of 10 ppm (*volume* per *volume*) can be met. The least amount of air that can be used in the stripping process is that which yields equilibrium between the air and liquid phase as the air leaves the air–water contacting device. If the packed tower is operated in countercurrent fashion, the ammonia in the exiting air is equilibrated with that in the *incoming* water; if operated in cocurrent fashion, ammonia in the exiting air is equilibrated with that in the *exiting* water. The Henry's constant for ammonia at 20°C is 1.37×10^{-5} atm-m^3/mol.

(a) Compare the minimum airflow rates for countercurrent and cocurrent operation to determine if either mode meets the air standard for ammonia.

(b) Discuss what can be done if neither of the stripper operations described meets the air standard.

Answers: (a) Neither mode of operation meets the standard for the operating conditions described. (b) Increase the airflow rate to increase the liquid–air mass transfer driving force throughout column.

1-7 Any meaningful analysis of an environmental system or problem should include the fundamental concepts discussed in this chapter. Review, for example, the paper by Olsen and Kavanaugh cited in Section 1.5.1 and prepare a brief annotated list of fundamental concept applications given in that paper.

APPENDIX A.1.1 METHODS FOR EXPRESSING QUANTITIES AND CONCENTRATIONS

A.1.1.1 Terms and Units

Mass fraction, a dimensionless quantity, may be expressed as a *fraction (0–1)*, a *percentage (0–100)*, or as some other mass or weight ratio [e.g., *parts per million (ppm)* or *parts per billion (ppb)*]. This concentration unit is the mass of the substance of interest per total system mass. This measure is particularly useful for expressing concentrations of solids. For example, if 2 kg of fine silt is mixed into 8 kg of water, the silt represents a solids concentration of 20% or 200,000 ppm. In many practical situations, the amount of solid (dissolved or suspended) present in an aqueous phase does not appreciably change the density of that phase. The total system mass is for all intents and purposes then equal to the mass of water (i.e., it remains essentially constant). For the case in which the density of the aqueous phase is equal to 1000 kilograms per cubic meter (kg/m^3, or 1 g/cm^3), 1 ppm is the same as 0.001 kg/m^3, or 1 mg/L. In practice, therefore, we often see the ppm and mg/L units of concentration used interchangeably. Keep in mind that they are not equivalent units *a priori*; they may be used interchangeably only when the conditions outlined above apply.

Volume fraction is a dimensionless quantity that expresses the volume of substance per total system volume. For example, if 1 L of methanol is mixed with 3 L of water, the resulting volume fraction of methanol is 25%. Conversely, the volume fraction of water in the solution is 75%. In hydrology, soil moisture content is often expressed as a volume fraction $V_i/(V_i + V_j)$, where V_i and V_j are the water and solid volumes of a total wet soil sample, $V_T = V_i + V_j$.

Mole fraction is the thermodynamically most rigorous concentration unit, expressing the number of moles of the substance of interest per total system moles. Recall that, in general, 1 mol is simply Avogadro's number, 6.022×10^{23} of atoms, molecules, or ions [i.e., a specific number of the most basic atomic or molecular units of a substance, or 1 gram molecular weight (GMW)]. Molecular weight is represented here by the symbol $W_{g,mo}$. An object's weight is not the same as its mass. It is a measure of force given by mass times the gravitational acceleration constant, g (i.e., $W_g = M \times g$).

Because it is nearly impossible (or at least very expensive) to identify the individual species that comprise most environmental systems, the *total* number of moles in a system cannot in general be calculated exactly. For dilute aqueous solutions, however, the number of moles of water may be assumed to approximate closely the total number of moles in the system. Using this assumption, an approximate mole fraction may be computed.

Mass per unit volume is a common concentration unit expressing the mass of substance i per total system volume. For example, if 1 g of sucrose is added to a volumetric flask, which is then filled to the 1-L mark with water, the resulting concentration is 1 g per liter of sucrose solution, or 1000 mg/L. In general, the mass-per-volume concentration unit can be related to the mass fraction unit if the density of phase j (the sucrose solution) is known. The presence of the sucrose means that the volume of water added to the volumetric flask to reach the 1-L mark will be slightly less than 1 L, even though the sucrose dissolves. Thus the concentration of water in the sucrose solution is less than 1000 mg/L and the density of the resulting solution is greater than the density of pure water.

Concentration as *moles per unit volume* is an expression for the number of moles of substance i present in a volume of phase j. The term *molarity* (M) refers to the number of moles of substance i per liter of phase j. Thus we could find the molarity of the sucrose solution in the example above by dividing the mass per volume concentration (g/L) by the gram molecular weight (g/mol) of sucrose (substance i). This computation would yield the number of moles of sucrose per liter of sucrose solution. A 1-molar (1 M) sucrose solution would be prepared by adding 1 mol of sucrose to a volumetric flask and then filling the flask to the 1-L mark with solvent.

The molar concentration of dissolved species in dilute aqueous systems is often small. To facilitate working with these small numbers, it is convenient to define *p-notation*; a familiar example is pH. The notation pC_i is defined as the negative logarithm (base 10) of the *molarity* of substance i; thus pH refers to the negative logarithm of the *molar* hydronium ion (hydrated proton) concentration.

The *moles per unit mass* concentration parameter expresses the number of moles of substance i per unit mass of phase j. When the mass of phase j is taken as 1 kg, the term *molality* is used (i.e., molality is the number of moles of solute per kilogram of solvent). In contrast to the molarity concentration unit, phase j is defined as the solvent used to make up the solution; thus phase j is not a mixture. To make up a 1-molal (1 m) sucrose solution, the volumetric flask would first be filled with 1 kg of water, then a mass of sucrose corresponding to 1 mol would be added to the flask. The final volume will be somewhat greater than 1 L. (Note that 1 L of water weighs 1 kg at 4°C.)

Molarity and molality are not equivalent: a 1 M sucrose solution refers to 1 mol of sucrose per liter of sucrose *solution*, while a 1 m sucrose solution refers to 1 mol of sucrose per kilogram of *solvent*. The numerical difference between molar and molal measures differs by less than 1% in dilute aqueous solutions at 20°C. Molarity is more commonly used in practice because molality is not a convenient analytical concentration unit.

We have to this point discussed quantities in terms of mass, volume, and moles and concentration units as ratios of these quantities. An additional quantity used in practice is the *equivalent* (eq). The use of this quantity can simplify calculations relating to chemical reactions. If a reaction is expressed in terms of equivalents, the number of equivalents of each reacting species equals the number of equivalents of product.

One equivalent weight, $W_{g,e}$, of a substance per liter of solution defines the concentration unit of *normality*. Because equivalents are used in the context of specific chemical reactions, the definition of an equivalent weight depends on the reaction involved. In general, the equivalent weight of a substance is equal to its *gram molecular weight* divided by an integer factor which depends on the reaction involved. This may equal the charge of an ion in an ionic reaction, the number of protons or hydroxyl ions exchanged in an acid-base reaction, or the number of electrons transferred in an oxidation-reduction reaction. The same substance may therefore have up to three different equivalent weights. Example A.1.1.2 illustrates the equivalent weight concept.

A.1.1.2 Example

- **Situation:** Phosphate is a limiting nutrient in many natural waters and must therefore be controlled in waste discharges to sensitive receiving waters. Phosphate can be precipitated with calcium, added as unslaked lime $CaO(s)$, according to the following reaction:

$$3CaO(s) + H_2O + 2HPO_4^{2-} \leftrightarrow Ca_3(PO_4)_2(s) + 4OH^-$$

Adjustment of pH to ensure optimal precipitation can be accomplished by addition of caustic (NaOH) or a mineral acid (e.g., H_2SO_4).

- **Question(s):** What are the respective molarities and normalities of the reagents involved in this process?

- **Logic and Answer(s):** The equivalent charge concentration unit expresses the number of moles of charge units per liter of solution. The equivalent weight ($W_{g,e}$) is defined as the molecular weight ($W_{g,mo}$) divided by the ionic charge. GMW values can be calculated from the information given in Appendix A.5.1.

1. For the lime–phosphate reaction, the equivalent weight of calcium in lime $[\{Ca^{2+}\}\{O^{2-}\}](s)$ is

$$W_{g,e} = \frac{W_{g,mo}}{2} = \frac{40.08 \text{ g/mol}}{2 \text{ Eq/mol}} = 20.04 \frac{\text{g}}{\text{Eq}}$$

$$1\ M\ Ca^{2+} = \frac{1 \text{ mol}}{\text{L}} = \frac{1\ W_{g,mo}}{\text{L}} = \frac{40.08 \text{ g}}{\text{L}}$$

$$1\ M\ CaO(s) = 40.08 + 15.99 = 56.07 \text{ g/L}$$

$$1\ W_{g,mo}\ Ca^{2+} = 2 \text{ mol of charge units}$$

$$1\ N\ Ca^{2+} = \frac{1 \text{ Eq } Ca^{2+}}{\text{L}} = \frac{40.08 \text{ g}}{2(\text{L})} = \frac{20.04 \text{ g}}{\text{L}}$$

$$1\ N\ CaO(s) = \frac{56.07 \text{ g}}{2 \text{ L}} = \frac{28.04 \text{ g}}{\text{L}}$$

2. For the caustic reaction:

$$NaOH + H_3O^+ \rightleftharpoons Na^+ + 2H_2O$$

$$1 \text{ mol NaOH} = 1 \text{ mol } Na^+ = 1 \text{ mol of charge units}$$

$$1\ M \text{ NaOH} = 1\ N \text{ NaOH} = 22.99 + 15.99 + 1.01 = 39.99 \text{ g/L}$$

3. For the sulfuric acid reaction:

$$H_2SO_4 + 2H_2O \rightleftharpoons 2H_3O^+ + SO_4^{2-}$$

$$1 \text{ mol } H_2SO_4 = 2 \text{ mol } H^+ = 2 \text{ mol of charge units}$$

$$1\ M\ H_2SO_4 = 2(1.01) + 32.06 + 4(15.99) = 98.04 \text{ g/L}$$

$$1\ N\ H_2SO_4 = \frac{98.04 \text{ g/L}}{2 \text{ Eq/mol}} = 49.02 \text{ g/L}$$

A.1.1.3 Review Exercises

1. Calculate the quantities (in grams) of the following reagents per liter of solution required to make up the following solutions:

(a) 0.4 M NaBr.

(b) 0.3 M $K_2Cr_2O_7$.

(c) 10^{-2} M $Fe(NH_4)_2(SO_4)_2$.

Answers: *(a) 41.16 g; (b) 88.26 g; (c) 2.84 g.*

2. Given the following composition of dry air at sea level:

Molecule	Mole Fraction (%)
O_2	20.95
N_2	78.09
Ar	0.93
CO_2	0.03

Assume that air is an ideal gas and a system in which the atmospheric pressure is exactly 1.0 atm and the temperature is 20°C. The ideal gas law is given by $PV = n\mathcal{R}T$, where $\mathcal{R} = 8.21 \times 10^{-5}$ atm-m^3/mol-K.

(a) Compute the density of the dry air at these conditions. Express your answer in g/m^3.

(b) Determine the concentration of argon in air, expressed on a mass basis, in ppm.

Answers: *(a) 1204 g/m^3; (b) 12,791 ppm.*

3. The pH of a popular soft drink is about 3.8, and the sucrose ($C_{12}H_{22}O_{11}$) concentration, C_{Su}, of this drink is about 40,000 mg/L.

(a) Calculate the hydronium ion concentration $[H_3O^+]$ in mol/L.

(b) Determine p[Su].

Answers: *(a) 1.59 × 10^{-4} mol/L; (b) 0.93.*

4. (a) Calculate the molarity of pure water at 20°C.

   (b) Calculate the molarity of water in a 5.0-g/L NaCl solution at 20°C given that the mass of 1.0 L of this solution is 1000.2 g/L.

   (c) Determine the percentage by which the molarity of pure water is decreased in the salt solution.

   Answers: *(a) 55.44 mol/L; (b) 55.29 mol/L; (c) 0.3%.*

5. It is desired to produce lime (CaO) for use in a water-softening process from the combustion of a sludge containing calcium carbonate ($CaCO_3$). The sludge has the dry-basis weight composition of 80% $CaCO_3$ and 20% inert material. The reaction proceeds according to the equation

$$CaCO_3 \rightleftharpoons CO_2 + CaO$$

Determine how many tons of CaO plus inert material will be produced from 100 tons of a wet sludge containing 40% water on a weight basis.

Answer: *38.9 tons.*

2

PROCESS CHARACTERIZATION AND ANALYSIS

2.1 PROCESS CHARACTERIZATION

Processes of interest in environmental systems are conveniently divided into those affecting the transformation and those affecting the transport of particular constituents. *Transformation processes* include chemical, biochemical, and physicochemical reactions. *Transport processes* include those affecting movement of constituents in the bulk of a system as well as at its boundaries. Transformation and transport processes act together to produce compositional change in a system. Once these processes have been characterized, the objective of process analysis is to determine the response of a particular system to the presence of some constituent under different circumstances. The success of any process analysis is measured by the accuracy with which this response can be predicted.

Process characterization must begin at a fundamental level and should be as complete as possible. To illustrate the consequence of less than rigorous characterization, consider the design of a chemical oxidation process to destroy aqueous-phase organic contaminants. It is generally known that strong chemical oxidants such as chlorine and ozone will eventually oxidize most organic compounds given sufficient dosage and residence time. In the face of constraints on time and/or design resources, it may, in any instance, seem reasonable to defer to this knowledge and do less than a fully detailed characterization of the chemical nature of these contaminants and their potential reactions with different oxidants. If proper characterization is compromised, however, it may result in selection of a less effective oxidant than others available, suboptimal conditions for process operations, inappropriate size and configuration of treatment units, failure to achieve adequate reduction of organic contaminants, and, quite possibly, environmentally harmful products (e.g., halogenated hydrocarbons).

The fundamental approaches that we take to process characterization (and subsequently, to analysis) are in most regards similar for natural and engineered systems. The underlying goal is to understand the cause of change in any system. As discussed in Chapter 1, naturally or anthropologically induced change may occur in an uncontrolled manner in natural systems such as a river of subsurface aquifer. Alternatively, change is usually controlled to accomplish a specific result in engineered treatment systems.

2.1.1 Reactions

Transformation processes have their bases in the reactions and interactions of chemical species with one another. The constituents of environmental systems are all fundamentally chemical in structure and behavior; thus changes or transformations in them can be described in terms of *chemical reactions*. We are careful here to differentiate between *transformation processes* and *reactions*. Although it is true that transformation processes are comprised by reactions, they may additionally involve considerations that depend more on the system in which a reaction takes place than on the reaction itself. In particular, mass transfer considerations may be important.

Chemical reaction phenomena adhere to a rigorous set of governing conditions and relationships, which in turn establishes a basis for quantification of environmental transformations. Certain of these conditions and relationships embody the principles of thermodynamics and control the *energetics* of reactions. Others are additionally involved in control of the *intrinsic rates*, or *kinetics*, of reactions.

The most fundamental governing conditions or relationships for reactions are the rules of stoichiometry, which derive from the fact that every chemical transformation involves a discrete amount of energy and, because energy and mass are inextricably related, a discrete amount of mass of each reacting substance.

Consider, for instance, the example given in Chapter 1 (Figure 1-6) involving the reaction of species A with species B to yield species C and D:

$$\mathrm{A} + \mathrm{B} \rightleftharpoons \mathrm{C} + \mathrm{D} \tag{2-1}$$

This reaction can be written more generally as

$$\gamma_A \mathrm{A} + \gamma_B \mathrm{B} \rightleftharpoons \gamma_C \mathrm{C} + \gamma_D \mathrm{D} \tag{2-2}$$

where the coefficients γ_A and γ_B define the respective quantities of the reacting substances, A and B, required to form the quantities γ_C and γ_D of products, C and D, respectively. These coefficients each have a value of 1 in Equation 2-1 and always have fixed integer values in any specific reaction, as generalized in Equation 2-2. Fixed integer values are necessary because each chemical unit of each reactant possesses a well-defined potential, or amount of available energy for reaction, and each reaction requires a fixed amount of energy to occur. These coefficients, termed *stoichiometric coefficients*, prescribe the respective *number of units* (e.g., molecules or moles) of each chemical substance required to accomplish a specific change in specific reactants to form specific products.

With each unit of change there is an associated transfer of a *discrete* amount of energy, either produced by a reaction that is spontaneous, or required as input to induce a nonspontaneous reaction. The amount of associated energy, whether produced spontaneously or required to make the reaction occur, is termed the *free-energy change*, ΔG_r, of the reaction. This free-energy change has a characteristic and prescribed value which relates to the number of chemical units involved in the reaction, which in turn, because each unit of a chemical substance has a well-defined mass, relates to the masses of the reacting substances. The incremental change in the concentration, ΔC_i, of a substance to be expected in a transformation process is thus proportional to ΔG_r, or, in the limit,

$$dC_i = \phi(dG_r) = \phi(\text{masses of reacting constituents}) \tag{2-3}$$

The relationships given in Equations 2-2 and 2-3 form the building blocks for structured approaches to process characterization and analysis. Every process involves a change in chemical constituents, and a specific amount of energy is involved for each set of reactants and products. For example, in organic oxidation reactions, a given amount of a particular organic chemical requires a specific amount of dissolved oxygen or other chemical oxidant (e.g., ozone, hydrogen peroxide, or chlorine) to be transformed into a specific amount of non-oxygen-consuming product, that is, to be "mineralized" to a stable inorganic form of carbon (e.g., CO_2). Example 2-1 provides some simple illustrations of the practical significance of the stoichiometric coefficients defined in Equation 2-1 and of the reaction energy concepts discussed above.

Example 2-1

- **Situation:** The operating board of a water-softening plant that uses soda ash (Na_2CO_3) and unslaked lime (CaO) to precipitate calcium hardness (Ca^{2+}) as calcium carbonate $CaCO_3$ is considering recovery of the lime for reuse by calcining (heating) the limestone sludge, which is 90% pure $CaCO_3$.
- **Question(s):** (a) Is the conversion of $CaCO_3$ to CaO a process that will occur naturally? If not, why not? (b) How much lime can be recovered per unit weight of sludge produced? (c) The soda ash purchased by the plant is of commercial grade. You are requested to determine whether a particular shipment meets the minimum purity specifications of 90% Na_2CO_3.
- **Logic and Answer(s):**

(a) This is an exercise in stoichiometry as well as a simple illustration of the relationship of reaction feasibility to available energy. The reaction involved in transformation of $CaCO_3$ to unslaked lime is

$$CaCO_3 \rightleftharpoons CaO + CO_2$$

The free-energy change for this reaction is unfavorable; that is, there is more energy associated with the compounds on the right-hand side of the reaction than there is with $CaCO_3$ (see Equation 5-62 and related discussion). Because the reaction would thus not lead to a release of energy, it will not occur naturally. It can be made to occur, however, by an input of energy, such as heat; that is, the reaction as written below will occur when the heat input becomes larger than that required to overcome the free-energy difference between the products and the reactant:

$$CaCO_3 + \text{heat} \rightleftharpoons CaO + CO_2$$

(b) The molecular weights of $CaCO_3$ and CaO can be determined to be 100 and 56.1 g/mol, respectively (see Appendix A.5.1). By a molar balance, 1 mol of $CaCO_3$ (100 g) yields 1 mol of CaO (56.1 g) plus 1 mol of CO_2 (43.9 g). The maximum amount (100% efficiency) of CaO to be recovered per 100 kg of limestone sludge is then

$$(100\text{ kg limestone})\left(\frac{0.9\text{ kg CaCO}}{\text{kg limestone}}\right)\left(\frac{56.1\text{ kg CaO}}{100\text{ kg CaCO}_3}\right) = 50.5\text{ kg/kg}$$

(c) The purity of the Na_2CO_3 shipment can be determined by carrying out a reaction of known stoichiometry. For example, the soda ash can be reacted with $CaCl_2$ and the amount of the resulting precipitate determined (i.e., a gravimetric analysis). The reaction involved is

$$Na_2CO_3 + CaCl_2 \leftrightarrow CaCO_3 + 2NaCl$$

Suppose that 6.0 g of the soda ash is reacted with a solution of $CaCl_2$ and the amount of $CaCO_3$ precipitated, after filtration and drying, is determined to be 5.1 g. Assume that this precipitate does not contain any significant amount of impurities. The number of moles (remember, reaction stoichiometry is based on molar quantities) of $CaCO_3$ formed is determined to be

$$[CaCO_3](s) = \frac{5.1\text{ g CaCO}_3}{100\text{ g CaCO}_3/\text{mol}} = 0.051\text{ mol}$$

$$= 51 \times 10^{-3}\text{ mol} = 51\text{ mmol}$$

Then, since the stoichiometric coefficients for $NaCO_3$ and $CaCO_3$ are the same (i.e., 1),

$$[Na_2CO_3](s) = [CaCO_3](s) = 51 \times 10^{-3}\text{ mol}$$

The molecular weight of Na_2CO_3 is 106.0 g Na_2CO_3/mol. The mass of pure Na_2CO_3 in the sample assayed can then be determined:

$$[Na_2CO_3](s) = 51 \times 10^{-3}\text{ mol }(106.0) = 5.4\text{ g}$$

and the percentage Na_2CO_3 purity of the commercial-grade soda ash is calculated:

$$\%\text{ purity} = \frac{5.4\text{ g}}{6.0\text{ g}}(100) = 90.1\%$$

The shipment is acceptable.

A second fundamental element of processes upon which we structure our approach to their characterization lies in the fact that reaction rate, r_i, as well as reaction energy, is functionally dependent on the numbers of reacting units and the masses of reacting substances; that is,

$$r_i = \frac{dC_i}{dt} = \phi(\text{masses of reacting constituents}) = \phi(dG_r) \qquad (2\text{-}4)$$

This relationship has its origins in the *mass law*, which states that rates of reactions are *proportional* to the masses of reacting substances. For a *specific* reaction, therefore, rate will increase with available free energy. Because the coefficients of proportionality may vary from reaction to reaction, it does not imply that the relative rates of distinctly different reactions can be related to their respective free energies.

In the case of simple *elementary* reactions, the coefficients of proportionality are given directly by the stoichiometric coefficients (i.e., to γ_A through γ_D in the reaction illustrated in Equation 2-2). This is seldom the case for *nonelementary* reactions. In complex microbial reactions that involve oxidation of organic matter for example, the rate of consumption of oxygen or other chemical oxidant is often proportional to the concentrations (or masses) of the organic matter and of the oxidant, but the proportion is not necessarily defined by the stoichiometry. Regardless of whether such reactions occur in a river or engineered treatment system, their energy requirements are exactly the same. Rates of organic matter conversion in treatment systems, can, however be manipulated more readily by varying the masses of the reactants to reduce required reaction times and sizes of reactor systems.

Knowledge of the rate dependence of reactions on process parameters is essential for describing process dynamics. This information also provides insight into how processes may be optimized and reactor sizes and costs minimized. In complex systems the dependencies of reaction rates on the numbers of reacting units are generally different than the dependencies of chemical energy, although they often increase and decrease in similar directions among reacting constituents. These rate dependencies must be quantified in process characterization. Reaction rates frequently depend on more than simple molecular or elementary reaction kinetics. They often involve complex phenomena, such as microtransport processes, related to the characteristics of the systems in which the reactions occur. These characteristics must also be identified and quantified in process analysis. In this sense we broadly define systems of different characteristics as different types of *reactors*.

2.1.2 Reactors

Transport and reactions of constituents take place within the three-dimensional space of environmental systems. We refer to these spaces as *reactors*, regardless of whether they are described by highly irregular geometry (typical of natural systems) or by regular geometry (typical of engineered systems). Characterizations of transport phenomena depend on reactor scale, configuration, and boundary conditions, all of which are embodied by the term *reactor dynamics*.

Assume, for example, that the reactor in Figure 1-6 represents a 20-km river stretch in which we are interested in changes in the concentration of dissolved oxygen in the water. The reactor system then includes the overlying atmosphere, underlying sediments, and all other inputs and outputs involved over that stretch of the river which contain dissolved oxygen or materials that can effect changes in dissolved oxygen. If, on the other hand, we are interested in treatment of an oxygen-demanding waste discharge to the river, the reactor system reduces to a smaller, less complex, better defined, and more controllable scale. The reactor reduces further in structure if we are interested in conducting a laboratory or pilot-scale test to assess the feasibility of using a particular process to change the oxygen-demanding character of the waste. Differences in temporal and spatial scales, configurations and boundaries, and conditions of mass distribution within and across reactor system boundaries can cause significant differences in the extent and rates at which processes will occur among those systems. This is true even if the basic processes and constituents of the systems are similar.

Primary factors determining how mass is distributed in a system can be subdivided into two categories: (1) the ways in which mass enters and leaves the system across its boundaries, and (2) the ways in which mass is mixed within the system boundaries. These are both primarily transport considerations, although they eventually dictate the "efficiency" of a transformation reaction in a given process operation. It is possible, for example, in engineered systems to cause flows of constituents across the systems, and mixing of constituents within them, which will effect much higher efficiencies (greater extent, higher rate) of contaminant conversion than might occur by the same fundamental reaction(s) in natural river systems.

The major difference between natural systems and engineered reactors is that in the latter we select and control temporal and spatial scale, boundaries, and conditions of mass distribution within and across system boundaries. Engineered reactors lend themselves well to description, design, and control to achieve specific process behavior and accomplish specific treatment results. Different types of processes require different reactor configurations and conditions of operation to achieve optimal behavior and cost-effective results. Reactor behavior and design considerations comprise a major aspect of process analysis, and reactor engineering thus comprises a major focus of this book.

2.1.3 Process Complexity

The levels of detail that are possible or practical in the characterization of processes vary greatly with the complexity of the reactions and the systems in which they take place. Reaction complexity is determined by reaction molecularities and by the number of components (species and phases) involved. System complexity is determined by temporal and spatial scales, boundary conditions, and initial conditions.

Figure 2-1 links the concept of complexity (both of a reaction and the system in which it occurs) with the concept of scale. The concept of micro-, macro-, and megasystems depicted in Figure 2-1 was introduced in Chapter 1 to define scales at which we are interested in processes. Although such definitions are admittedly arbitrary, they offer a useful perspective on the relative sensitivities of different systems to particular sets of conditions. The complexity of reactions and/or of systems may influence decisions on where to com-

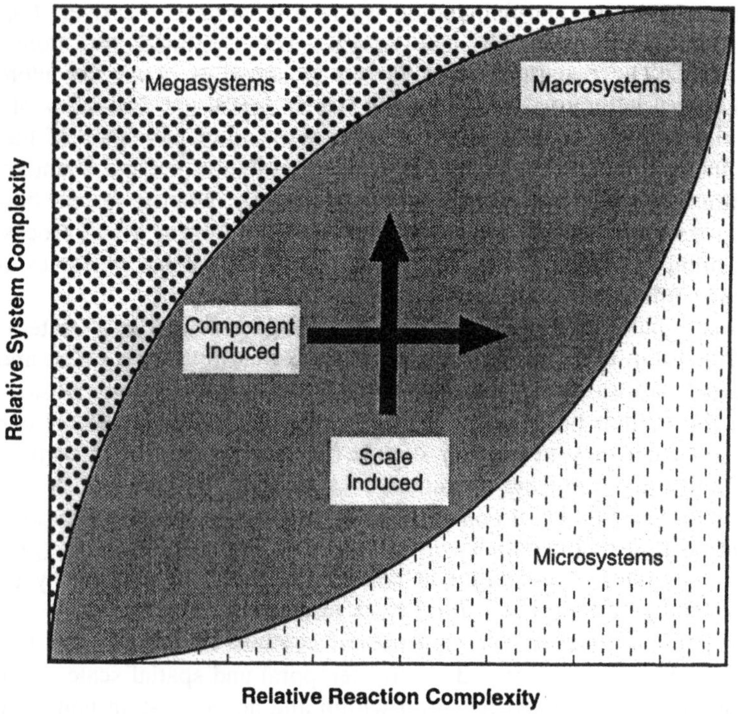

Figure 2-1 Process complexity as a function of system and reaction complexities.

promise accurate characterization of processes in order to move more directly toward an engineering solution to the problem at hand. As illustrated in Figure 2-1, megasystem complexity is scale induced rather than reaction induced. That is, change that occurs due to a simple reaction in a large spatial scale may be difficult to analyze because transport processes are not easily quantified; changes induced by more complex reactions in a megasystem only exacerbate the problem. At the other extreme of system scale, transport in microsystems is easily described, so that complexity is reaction induced rather than scale induced.

Complexity relates to the levels of detail that must be specified rigorously to define a given process and system. Levels of detail relate either to the required number of parameters or their levels of accuracy, or both. Reaction complexity ranges on the left from elementary monomolecular reactions that occur along single pathways (e.g., the autodecomposition of ozone to oxygen, or the decay of a radionuclide to a daughter product) to nonelementary, multimolecular, multipath reactions on the right (e.g., multiple-step catalytic oxidations of complex organic compounds by microbially generated enzymes). Molecularly induced complexities do not by themselves singularly determine process complexity. Processes and their relative complexity are inherently dependent on the scale of the systems in which they take place, which in turn is determined by the specific problems or issues being addressed. Quantification of oxygen transfer rates between gas and liquid phases is obviously a much more complex problem when viewed from the perspective of an ocean environment than it is from the perspective of an aerated waste treatment lagoon. Many more processes and conditions must be characterized and quantified in the case of the ocean than in the case of the lagoon.

The arbitrary scale definitions micro-, macro-, and mega- as used in Figure 2-1 may be interpreted on the one hand as characterizing the relative scales of *different processes* and *different systems*. They may be interpreted as well, however, to describe scale variations among *similar systems* involving essentially *identical overall processes*. For example, the chemical oxidation of a specific organic compound by ozone (e.g., Examples 1-1 through 1-3) at various stages of development and design is of interest to us at several alternative system scales: in a liter-sized laboratory reactor used for a feasibility investigation, in a pilot-scale field test unit having a total volume of several cubic meters, or in a basin of several million cubic meters capacity in a full-scale water or waste treatment plant.

Transport of the reacting materials, ozone and the organic compound, is different in each of the three cases cited above, as very likely will be the observed results or performance of the process itself. The rules and laws that govern the motion, velocity, and distribution of the respectives masses of the primary medium of the system (water) and of the organic compound and the ozone will dictate the influence of transport phenomena on the process. These phenomena are governed by the rules and laws of physics.

2.1.4 Process Models

A rational engineering approach to process analysis generally includes some system analog, or model, to facilitate process characterization and description. Physical analogs that reproduce the essential features of a system on a smaller and more manageable scale are useful in some circumstances, but the most flexible and economical approach more commonly involves characterization and description of a system by a mathematical analog. The development and use of *mathematically based models* to describe the components, composition, and related transport and transformation processes of any physical system involves a logical sequence of procedures:

- Characterization of system boundaries.
- Selection, definition, and description of system components.
- Description of major transport and transformation processes by equations that define the way in which each affects the variables of interest.
- Development of simplifying assumptions.
- Formulation of equations comprising the overall system model.

- Development of algorithms for solution of these system equations.
- Calibration of the model equations with data specific to the system.
- Verification of the model with independent data to determine whether it adequately describes the behavior of the system for conditions other than those to which it was calibrated.
- Application of the model for analysis, prediction, or design.

The equations used to structure a mathematical model for an environmental system must meet certain criteria. First and foremost, the transport and transformation processes involved must combine and interact in a manner that satisfies the fundamental principles of thermodynamics and continuity, and this must be reflected in the equations comprising the model. These principles dictate that any change in a system must be such that the energy, mass, and momentum of that system are conserved, although each may change in form.

When the changes that occur within a system involve transformation of one or more of its constituents, the overriding continuity relationships are driven by changes in the number of units (e.g., molecules or moles) of specific chemical species. Each unit has a well-defined weight (e.g., molecular or molar weight) or mass. Although the numbers of units of individual species comprising a system are changed by a transformation process, the summed mass of those species must at any time remain constant if the system is closed, that is, if no mass enters or leaves the system in the course of the transformation. This is inherent in the stoichiometric relationship given in Equation 2-2. Equation 2-2 relates only to transformation reactions in *closed systems*. As indicated in Figures 1-1, 1-2, and 1-6, most environmental systems are *open systems*; that is, they involve transport of mass across their boundaries. We are able to characterize the net effects of these transport and transformation processes only if we can fully account for the changes in the units or masses of individual species and can balance those factors that add units or masses of components to those that decrease them. Because such balances may be stated in terms of either molecular units or mass, they are commonly referred to as *material balances*. The term *material balance* is more general because transport and reaction phenomena occur in units of chemical entities; for example, molecules, atoms, ions, or moles (1 mol = 6.022×10^{23} molecules, atoms, or ions, i.e., *Avogadro's number*, $\mathfrak{N}_{Av}$).

2.2 MATERIAL BALANCE RELATIONSHIPS

Material balance relationships comprise the basic framework for the mathematical modeling of physical systems. They are the foundation and logical starting point for integrating the three core elements of process analysis (energetics, kinetics, and reactor dynamics) into a quantitative description of process dynamics.

2.2.1 Control Volumes

The mass or material balance relationship for a constituent within a reactor accounts for changes in the composition and concentration of that constituent that occur by both transport of materials across the reactor boundaries and reactions or processes that take place within the reactor. The material relationships associated with these phenomena translate, for specific systems, to mass relationships by virtue of stoichiometry and molecular (atomic or ionic) structure. Fluid balances can best be described in terms of mass (i.e., mass of water, mass of air, etc.). In this book we use primarily the more general term *material balance*. On appropriate occasions, cross-reference may be made to the more specific term *mass balance*. Interchangeable uses of the terms should present no confusion as long as the points of difference noted above are kept in mind.

The control volume approach employed in elementary fluid mechanics (i.e., the *Reynolds transport theorem*) to derive the continuity equations for fluid mass, momentum, and energy is also applicable for any component, i, within a fluid. The general "word equation" for balance of a material, i, within any control volume, expressed in terms of its chemical units or moles, is given by

net rate of transport of i through the control volume	±	net rate of transformation of i within the control volume	=	net rate of change of i within the control volume

(2-5)

[molar rate of input of i across the control volume boundaries]	−	[molar rate of output of i across the control volume boundaries]	±	[molar rate of chemical reaction of i within the control volume]	=	[net molar rate of accumulation (or depletion) of i within the control volume]

The most fundamental mathematical elaboration of Equation 2-5 is developed by application of the principles of continuum mechanics to the general control volume, V_C, presented in Figure 2-2; more specifically, to a differential element of that control volume, dV. The differential element is fundamentally a *point form* representation of the control volume, that is, the control volume reduced to an infinitely small scale. The point form is considered to be the largest truly *homogeneous* form of the control volume—a necessary, although not completely sufficient condition for application of continuum theory. As indicated in Figure 2-2, a similar approach is employed in a more macroscopic sense for characterizing complex *heterogeneous* systems using *volume averaging*. In this less rigorously correct but nonetheless effective application of

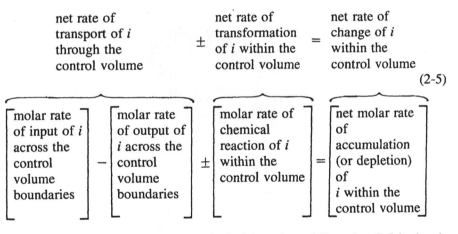

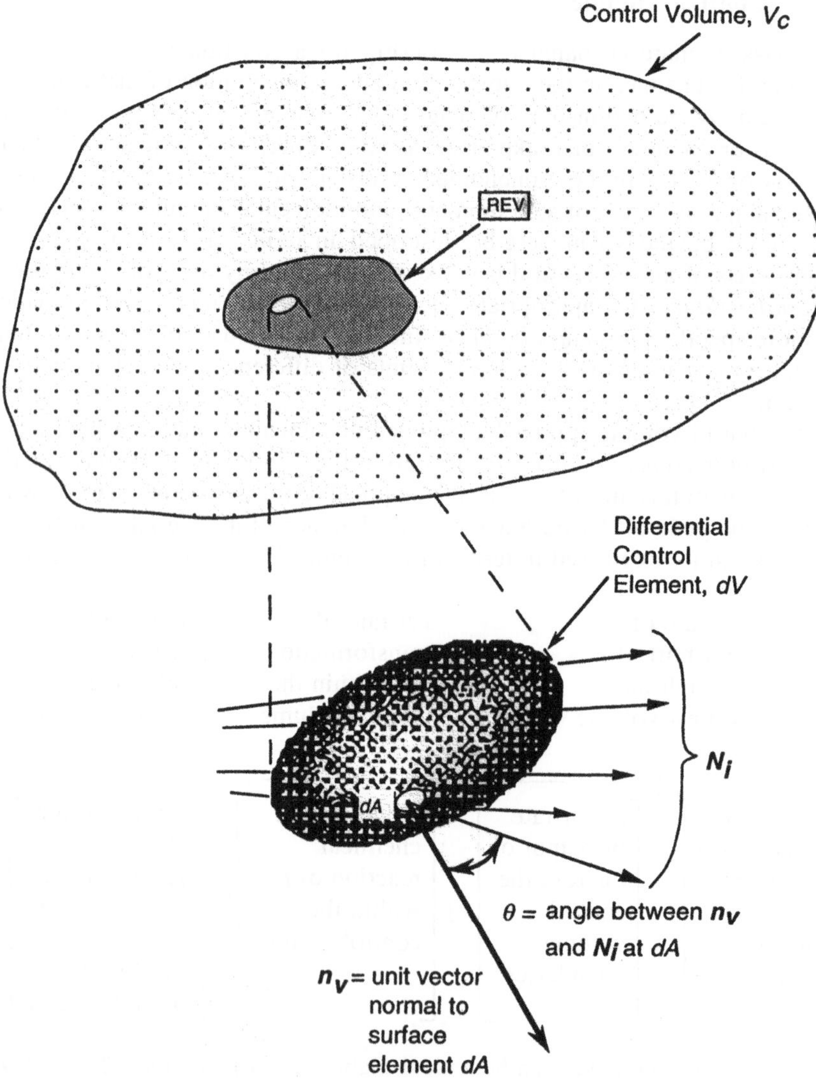

Figure 2-2 Fluid flow through an arbitrary control volume.

continuum modeling, a *representative elementary volume* (REV) is defined for a particular system.

The term N_i in Figure 2-2 refers to the molar flux of component i, that is, *the number of moles of i transported per unit cross-sectional area per unit time*. This flux term is a vector quantity that can be represented as $N_i = N_{ix}\boldsymbol{i}_x + N_{iy}\boldsymbol{i}_y + N_{iz}\boldsymbol{i}_z$, where N_{ix} is the flux component in the x direction and $\boldsymbol{i}_x$ is the unit vector in the x direction. The rate at which component i leaves the control volume (moles/time) is then given by projecting the flux in a direction normal to the differential surface area, dA:

$$\begin{pmatrix} \text{magnitude of the molar flux normal to} \\ \text{a differential element of surface area, } dA \end{pmatrix} = |N_i| \cos \theta = N_i \cdot n_v \tag{2-6}$$

where $|N_i|$ is the *magnitude* of the flux vector N_i, and the angle, θ, is measured between N_i and the outward-directed vector, n_v, normal to dA. A brief review of the use of vector notation in the analysis of transport through control volumes is given in Appendix A.2.1.

Detailed descriptions of transport processes responsible for flux are provided in subsequent chapters. As we have already noted, considerations of transport involve both macroscopic transport at the system or reactor scale and microscopic transport at the reaction scale. It suffices for the present discussion to state these processes as (1) *advection*, sometimes referred to as *convection*; (2) *dispersion*; (3) *diffusion*; and (4) *interphase mass transfer*. The last of these describes transport across phase boundaries. If the control volume under consideration is a single homogeneous phase contiguous with other different phases at its boundaries, interphase mass transfer relationships are incorporated in the *macroscopic flux terms* of Equation 2-5. Alternatively, if the control volume is heterogeneous and represented in its smallest form by the REV, interphase mass transfer phenomena take place at the boundaries of the REV subsystems as well and as such are generally incorporated in the *reaction terms* of Equation 2-5. In each instance, flux is a vector quantity because it is produced by transport of component i either with directional velocity streamlines or under the influence of directional gradients in concentration. Consider, for example, solute i at concentration C_i in a stream of water flowing at a velocity v. For this advective transport, the flux of solute, N_i, is given by the product vC_i. The term N_i is used in this book to denote flux across a spatial boundary within a given phase.

2.2.2 Integral Form of the Mass Continuity Relationship

Integration of the molar rate of output per element of surface area depicted in Figure 2-2 and quantified in Equation 2-6 over the entire control surface area gives

$$(\text{molar rate of output for entire control volume}) = \int_{SC} (N_i \cdot n_v)\, dA \tag{2-7}$$

where, from vector algebra, the quantity $(N_i \cdot n_v)\, dA$ is a scalar product. If this integral is positive, there is a net molar rate of output from the control volume; that is, $\cos \theta$ is positive in Equation 2-6 because flux is directed outward across the control surface, as shown in Figure 2-2. The derivation of net flow rate out of a control volume in fluid mechanics is developed analogously; in this instance, N_i is the volumetric flux of the fluid.

For a net molar rate of input to the control volume, the integral is negative in Equation 2-7; that is, $\cos\theta$ is negative because flux is directed inward rather than outward. In the material balance written in Equation 2-5 we are interested in the net molar rate of input, or

$$\text{(net molar rate of input)} = -\int_{SC} (N_i \cdot n_v)\, dA \tag{2-8}$$

The molar rate of chemical reaction of i needed in the material balance is defined as the time rate of change in the concentration of i in the control volume, V_C, due to reaction alone. However, the reaction rate, r_i, usually depends on concentration, and concentration can vary within the control volume. The most general material balance relationship therefore accounts for a reaction rate which varies with space in the control volume such that

$$\text{(rate of chemical reaction within } V_C) = \int_{VC} r_i\, dV \tag{2-9}$$

Finally, the net rate of accumulation of mass (expressed in moles) in the control volume is obtained from the change in mass of component i over an increment of time, $\Delta t = t_2 - t_1$, such that

$$\text{(rate of accumulation within } V_C) = \frac{\int_{VC} C_i\, dV\big|_{t=t_2} - \int_{VC} C_i\, dV\big|_{t=t_1}}{\Delta t} \tag{2-10}$$

At *steady state*, the above term is *zero*. Replacing the word description of the material balance given in Equation 2-5 with Equations 2-8, 2-9, and 2-10, and taking the limit as Δt approaches zero gives the *integral form* of the mass continuity equation:

$$-\int_{SC} (N_i \cdot n_v)\, dA + \int_{VC} r_i\, dV = \frac{\partial}{\partial t}\left(\int_{VC} C_i\, dV\right) \tag{2-11}$$

As we illustrate in our discussions of mass transport in Chapter 3, the integral form of the continuity equation for mass of component i in the fluid is useful conceptually because it is analogous to those for the mass, energy, and momentum of the fluid itself. It is valid for any arbitrary control volume and is not dependent on the operative form of either the mass transport or reaction rate expressions. It further accommodates variations in flux with control surface area and therefore conveniently describes the dynamics of such natural systems as lakes and groundwaters, where flux may vary with surface area selected for the control volume. Total system or reactor descriptions and quantifications require superposition of appropriate reaction processes on transport processes. In other words, we must develop equations that simultaneously describe not only bulk transport by fluid flow but also any pertinent chemical reactions. Both types of processes involve rather complicated dynamics.

A simple application of Equation 2-11 to a reactor of constant volume and homogeneous content (e.g., a completely mixed tank) yields

$$QC_{i,\mathrm{IN}} - QC_{i,\mathrm{OUT}} + r_i V_R = V_R \frac{dC}{dt} \tag{2-12}$$

In this material balance, the "net molar rate in" is equal to $QC_{i,\mathrm{IN}} - QC_{i,\mathrm{OUT}}$, where Q is the steady uniform flow rate through the volume, V_R. Example 2-2 provides an illustration of the differences and interrelationships between the reaction and transport elements of Equation 2-5.

Example 2-2

- **Situation:** A simple reactor system (system 1) in which air is bubbled through an organic waste contained in an open concrete tank is used to strip a volatile organic compound to the atmosphere. To eliminate air pollution problems associated with this treatment, two alternative types of systems are considered. The first (system 2) is one in which the tank is sealed without any headspace and pure ozone gas is substituted for air and bubbled at a rate that is just sufficient for it to completely dissolve and to oxidize the organic contaminant in the water phase (i.e., there are no reactor off-gases). The second alternative (system 3) is identical to the sealed tank except that water is introduced at one end of the tank and withdrawn at the other at a flow rate of Q. In both cases the contents of the tank are kept completely mixed by a mechanical mixer.
- **Question(s):** Develop simple material balance relationships for each of these two systems.
- **Logic and Answer(s):** Equation 2-5 can be written for each of the three systems as detailed below:

  System 1

  - Rate of input of volatile organic contaminant (VOC) = 0
  - Rate of output of VOC = rate of volatilization
  - Rate of chemical oxidation of VOC = 0

  Although no chemical oxidation occurs, there is a rate of volatilization or mass transport to the gas phase, r_v. Thus the material balance for a reactor volume V_R is

$$r_v V_R = V_R \left(\frac{dC_{\text{voc}}}{dt}\right)_v$$

The net rate of volatilization is proportional to the interfacial surface area per unit volume (A/V_R) of reactor and to the concentration of VOC in the reactor:

$$r_v = -k_f \frac{A}{V_R} C_{\text{voc}}$$

where k_f is an effective mass transfer coefficient (Lt^{-1}; see Chapter 4 for details). The material balance expression for this *unsteady state* is

$$\left(\frac{dC_{\text{voc}}}{dt}\right)_v = -k_f \frac{A}{V_R} C_{\text{voc}}$$

Integration is straightforward by separation of variables so that the decrease in C_{voc} with time can be calculated readily.

System 2

- Rate of input of VOC = 0
- Rate of output of VOC = 0
- Rate of chemical oxidation of VOC = r_{ox}

If the reaction rate is second order, that is, proportional to the concentration of the ozone and the VOC, the material balance is

$$r_{ox} V_R = V_R \frac{dC_{\text{voc}}}{dt} = (-k_{ox} C_{\text{voc}} C_{\text{ozone}}) V_R$$

where k_{ox} is the reaction rate coefficient. If the ozone is added and dissolved in a large excess of the amount required to oxidize the VOC, its residual concentration can be assumed nearly constant in time, so that C_{ozone} is incorporated in a pseudo-first-order reaction rate coefficient, $(k_{ox})_{ps}$:

$$\frac{dC_{\text{voc}}}{dt} = -(k_{ox})_{ps} C_{\text{voc}}$$

As with system 1, an *unsteady state* condition exists for which the concentration decrease with time is easily calculated.

System 3

- Rate of input of VOC = QC_{IN}
- Rate of output of VOC = QC_{OUT}
- Rate of chemical oxidation of VOC = $r_{ox} = -(k_{ox})_{ps} C_{\text{OUT}}$

Note that the control volume in this particular example is the entire reactor volume, and thus the expressions for rates of input and output are given by the product of the volumetric flow rate and concentration of VOC at the entrance and exit. The resulting material balance is

$$QC_{\text{IN}} - QC_{\text{OUT}} + r_{ox}V_R = V_R \frac{dC_{\text{OUT}}}{dt}$$

or, for *steady-state* operation,

$$QC_{\text{IN}} - QC_{\text{OUT}} - (k_{ox})_{ps}C_{\text{OUT}}V_R = 0$$

The word description of the material balance relationship given in Equation 2-5 thus translates into a variety of mathematical forms that vary with the type of system in which the process is embodied.

2.2.3 Point Form of the Mass Continuity Relationship

The point form of the continuity equation can be obtained by localizing the integral form given in Equation 2-11 with respect to the control volume, V_C. The *Gauss divergence theorem* is used to first convert the surface area integral representing net molar rate of input (Equation 2-8) to a volume integral:

$$\int_{SC} (\boldsymbol{N}_i \cdot \boldsymbol{n}_v)\, dA = \int_{VC} \nabla \cdot \boldsymbol{N}_i\, dV \tag{2-13}$$

where the *del operator*,

$$\nabla = \boldsymbol{i}_x \frac{\partial}{\partial x} + \boldsymbol{i}_y \frac{\partial}{\partial y} + \boldsymbol{i}_z \frac{\partial}{\partial z}$$

Because V_C is independent of time, Leibnitz's formula may be used to transform the order of integration and differentiation for the net rate of accumulation term (Equation 2-9):

$$\frac{\partial}{\partial t} \int_{VC} C_i\, dV = \int_{VC} \frac{\partial C_i}{\partial t}\, dV \tag{2-14}$$

Collecting all three integrals over the control volume yields

$$\int_{VC} \left(-\nabla \cdot \boldsymbol{N}_i + r_i - \frac{\partial C_i}{\partial t} \right) dV = 0 \tag{2-15}$$

Equation 2-15 is integrated over the entire volume of a system and is thus valid for a system of any size or configuration. If the volume over which it is integrated is the infinitesimally small differential control volume, Equation 2-15 can be localized to the point form

$$-\nabla \bullet N_i + r_i = \frac{\partial C_i}{\partial t} \tag{2-16}$$

Because this equation describes reactor conditions at one point in space and time, it is expressed in moles per unit volume per unit time. On the other hand, the integral form of the continuity equation (Equation 2-11) expresses reactor conditions within a control volume and is therefore expressed in moles per unit time.

An alternative approach to development of the point form of the mass continuity equation for component i in rectangular (orthogonal) coordinates can be obtained by making the control volume a small but discrete element ($\Delta x\ \Delta y\ \Delta z$), as shown in Figure 2-3. This approach is sometimes referred to as a *shell balance*. If flux is assumed to occur normal to each of the three planes of this discrete element, we can write the corresponding material balance as

$$[N_{ix}|_{(x)} - N_{ix}|_{(x+\Delta x)}]\ \Delta y\ \Delta z + [N_{iy}|_{(y)} - N_{iy}|_{(y+\Delta y)}]\ \Delta x\ \Delta z + [N_{iz}|_{(z)} - N_{iz}|_{(z+\Delta z)}]\ \Delta x\ \Delta y + r_i\ \Delta x\ \Delta y\ \Delta z = \frac{\Delta C_i}{\Delta t}(\Delta x\ \Delta y\ \Delta z) \tag{2-17}$$

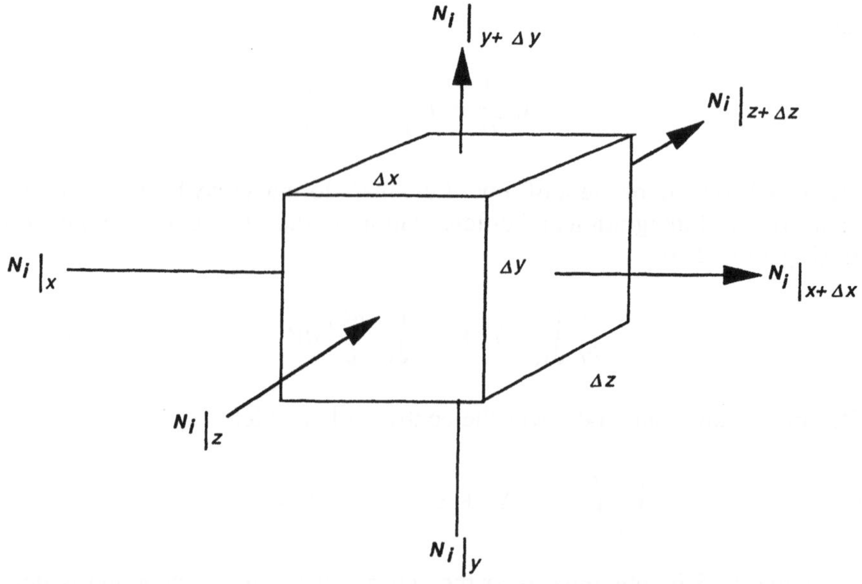

Figure 2-3 A discrete control volume element in rectangular coordinates.

The point form of the continuity equation derives immediately by dividing both sides by the volume ($\Delta x\ \Delta y\ \Delta z$) of the discrete element, and taking the limit as this volume approaches zero; that is, as $\Delta x \rightarrow 0$, $\Delta y \rightarrow 0$, and $\Delta z \rightarrow 0$,

$$-\left(\frac{\partial N_{ix}}{\partial x} + \frac{\partial N_{iy}}{\partial y} + \frac{\partial N_{iz}}{\partial z}\right) + r_i = \frac{\partial C_i}{\partial t} \tag{2-18}$$

which is equivalent to the point form arrived at in Equation 2-16 by differentiating Equation 2-11.

It may be evident from the two different approaches given above for developing the point form of the continuity equation that they converge as the volume of the discrete element approaches that of the differential element (i.e., as $\Delta V \rightarrow dV$). The differential form of the continuity equation develops directly from a balance on the differential element itself, whereas the discrete control volume approach (or shell balance approach) involves development of the corresponding differential continuity equation by taking the limits of a finite difference relationship. This is elaborated for a one-dimensional case in Appendix A.2.2. For clarity and simplicity of equation development, we use the *differential element* (e.g., dx, dV, dC, dt, etc.) approach in this book. The *finite difference* notation (e.g., $\Delta x = x_2 - x_1$, $\Delta V = V_2 - V_1$, $\Delta C = C_2 - C_1$, $\Delta t = t_2 - t_1$, etc.) is reserved for situations in which discrete differences in quantities are represented.

Figure 2-4 provides an illustration of a one-dimensional homogeneous phase system involving plug-flow-like fluid movement through a reactor with simultaneous reaction of the components of the fluid. This particular condition is for a steady state with respect to the balance of flow and reaction (i.e., the concentration profile along the reactor is constant in time). For this case the right-hand side of Equation 2-18 is zero, as are the second and third terms on the left-hand side of the equation if there is only one-dimensional transport.

2.2.4 Alternative Forms of the Mass Continuity Relationship

The point-form statement of the continuity relationship given in Equation 2-16 is a rigorous representation of the most fundamental basis for description of natural systems and design of engineered reactors. This form of the continuity equation requires integration to determine concentration patterns as functions of distance and time. Its application as written is often problematic in environmental systems because of their complexities and associated deficiencies in system data and details. Approximations are thus frequently employed to simplify use of the relationship expressed by Equation 2-16. For situations in which flux occurs dominantly in only one direction (e.g., in the x dimension, normal to the y–z plane), the terms $\partial N_{iy}/\partial y$ and $\partial N_{iz}/\partial z$ may be dropped in Equation 2-18 to obtain a much more readily solved approximation. This approach is illustrated in Example 2-3, in which the behavior of a complex re-

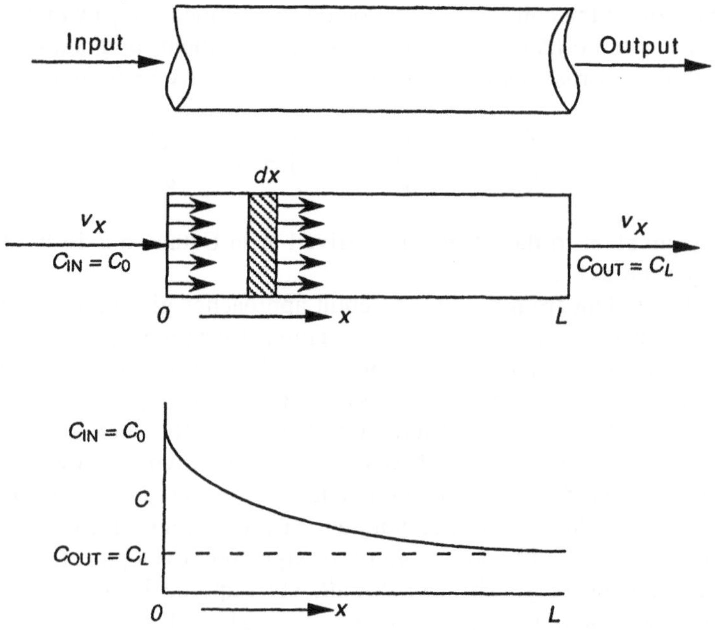

Figure 2-4 One-dimensional, homogeneous-phase, steady-state, ideal plug flow reactor behavior.

actor system is approximated by a relatively simplified form of its material balance relationship.

Example 2-3

- **Situation:** An anaerobic biological reactor (a high-density biofilm device) is selected for treatment of a high-strength industrial waste. Because the wastewater does not contact air in anaerobic treatment, there is no volatilization of organic chemicals in the biological treatment step. The volatile organic chemicals (VOCs) are not completely biodegraded during treatment, and some residual VOCs thus remain.
- **Question(s):** Develop the steady-state material balance relationship and process model to describe the loss of a residual VOC as the treated wastewater passes from the biological treatment unit through a long, narrow open channel to a polishing lagoon before discharge to a river.
- **Logic and Answer(s):**

1. We will assume that the flow in the channel is like that in a pipe (see Figure 2-4) such that the flux of any VOC (subscript i) is one-dimensional and defined by

$$N_i = \frac{QC_i}{A} = v_x C_i$$

where Q is the flow rate, A is the cross-sectional area of the channel filled with water, and v_x is the velocity in the direction of channel length. The concentration of the volatile component is a function of position, x, along the channel, and changes due to losses to the air as the water travels as a "slug" down the channel; thus a point form material balance is appropriate.

2. Volatilization is a transport process that occurs at the interface between the air and water (see also Example 2-2). It represents a rate of loss of VOC from the water to air as given by

$$r_{v,i} = -k_{f,i}(a_s^\circ)_R\,(C_i - C_{S,i})$$

where $r_{v,i}$ is the rate of volatilization (M/L^3-t), $k_{f,i}$ is the mass transfer coefficient (Lt^{-1}), $(a_s^\circ)_R$ is the surface area available for mass transfer per unit volume of the "reactor" (L^{-1}), and $C_{s,i}$ is the concentration of the VOC that is predicted by Henry's law to exist in the water when equilibrium is reached with the atmosphere (see Table 1-4 for illustrative values). Because the VOC transferred to the air above the channel is being greatly diluted, the partial pressure of the VOC is very close to zero and thus, by Henry's law $C_{s,i}$ must also be close to zero (see Chapter 1).

3. Steady state is a reasonable assumption if the VOC concentration entering the channel and the wastewater flow rate remain the same for all time. The point form of the material balance is obtained either by writing a shell balance over a distance x of the channel (see Example 2-4 for an illustration) or by writing directly an expression for the change in flux over a differential element, dx, the latter approach being illustrated below:

$$-dN_i A + r_{v,i} A\,dx = 0$$

Dividing through by $A\,dx$, the volume of the differential element, gives

$$-\frac{dN_i}{dx} + r_{v,i} = 0$$

4. Substituting for the flux (step 1) in the x direction and the volatilization rate (step 2) provides the final form of the material balance:

$$-v_x \frac{dC_i}{dx} + k_{f,i}(a_s^\circ)_R(C_{S,i} - C_i) = 0$$

or given that $C_{S,i} = 0$,

$$-v_x \frac{dC_i}{dx} - k_{f,i}(a_s^\circ)_R C_i = 0$$

5. A process model is obtained by separation of variables and integration:

$$\frac{C_i}{C_{i,0}} = \exp\left(\frac{-k_{f,i}(a_s^\circ)_R x}{v_x}\right)$$

where $C_{i,0}$ is the concentration of VOC entering the channel ($x = 0$).

In some instances we are interested in writing material balances for transport and reaction in spheres (e.g., adsorbent or ion-exchange particles) and cylinders (e.g., reactive surfaces on tubular membranes or pipe walls). The three-dimensional point forms of the continuity equation for spherical and cylindrical geometries are derived by spatial transformation of the gradient of flux, $\nabla \cdot N_i$, from rectangular (Equation 2-16) to either spherical or cylindrical coordinates by use of vector mathematics. Details of these transformations are beyond the scope of the present discussion but are readily available in several of the mathematically and transport-oriented background reading references given in Section 2.5.2.

The resulting three-dimensional point forms of the continuity equation will be presented first. Simpler, one-dimensional forms result when flux is normal to the surface (i.e., radially directed). These forms are also derived in Appendix A.2.2.

Spatial transformation of the gradient of flux in *spherical* coordinates results in

$$\nabla \cdot N_i = \frac{1}{r^2}\frac{\partial}{\partial r}(r^2 N_{ir}) + \frac{1}{r \sin\theta}\frac{\partial}{\partial \theta}(N_{i\theta}\sin\theta) + \frac{1}{r\sin\theta}\frac{\partial N_{i\phi}}{\partial \phi} \qquad (2\text{-}19)$$

where $r = r_s$, θ, and ϕ are the spherical coordinates illustrated in Figure 2-5 and N_{ir}, $N_{i\theta}$, and $N_{i\phi}$ are the flux components in those coordinate directions. For a system such as a spherical adsorbent particle, where fluxes are solely in the radial direction, $N_{i\theta}$ and $N_{i\phi}$ will be zero and $N_i = N_{ir}$. Thus the second and third terms in Equation 2-19 vanish. Substitution in Equation 2-16 yields a continuity relationship for one-dimensional transport with spherical geometry:

$$-\frac{1}{r^2}\frac{\partial}{\partial r}(r^2 N_i) + r_i = \frac{\partial C_i}{\partial t} \qquad (2\text{-}20)$$

The same relationship is derived in Appendix A.2.2 starting with a material balance that includes only the radial component of flux. Alternatively, the first term may be expanded to produce

$$-\left(\frac{\partial N_i}{\partial r} + \frac{2}{r}N_i\right) + r_i = \frac{\partial C_i}{\partial t} \qquad (2\text{-}21)$$

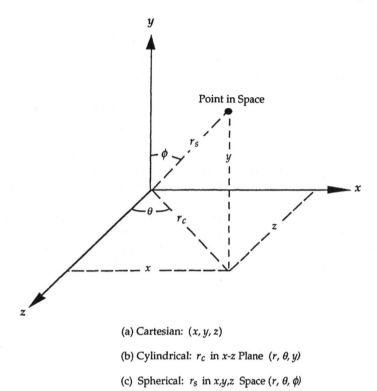

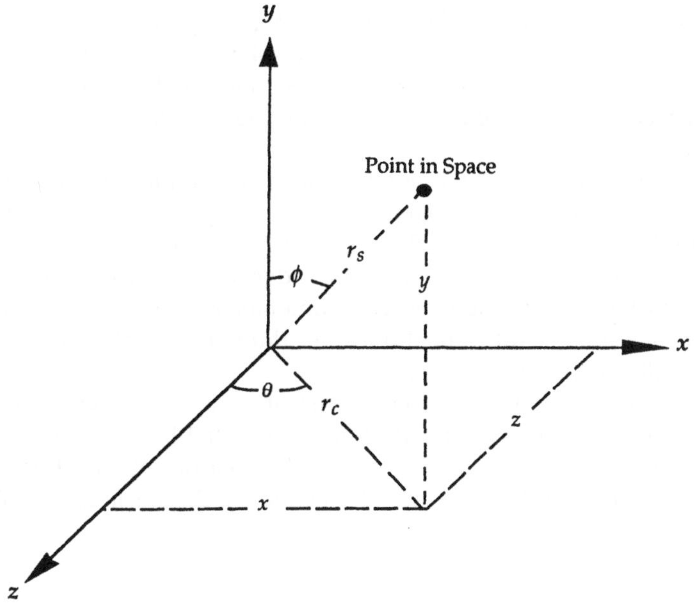

(a) Cartesian: (x, y, z)

(b) Cylindrical: r_c in x-z Plane (r, θ, y)

(c) Spherical: r_s in x,y,z Space (r, θ, ϕ)

Figure 2-5 Point location in alternative coordinate systems.

The spatial transformation of the gradient of flux in *cylindrical* coordinates can be derived similarly. The result is

$$\nabla \cdot N_i = \frac{1}{r}\frac{\partial}{\partial r}(rN_{ir}) + \frac{1}{r}\frac{\partial N_{i\theta}}{\partial \theta} + \frac{\partial N_{ix}}{\partial x} \tag{2-22}$$

where $r = r_c$, θ, and x are the cylindrical coordinates illustrated in Figure 2-5 and N_{ir}, $N_{i\theta}$, and N_{ix} are the flux components in those directions. As for the case of spherical coordinates, $N_{i\theta}$ and N_{ix} are zero and $N_i = N_{ir}$ when the only flux is radial. Substitution of the first term of Equation 2-22 into Equation 2-16 yields

$$-\frac{1}{r}\frac{\partial}{\partial r}(rN_i) + r_i = \frac{\partial C_i}{\partial t} \tag{2-23}$$

or, alternatively,

$$-\left(\frac{\partial N_i}{\partial r} + \frac{1}{r}N_i\right) + r_i = \frac{\partial C_i}{\partial t} \tag{2-24}$$

The same result is obtained in Appendix A.2.2 by starting with a material balance that considers only radial flux. The spherical and cylindrical forms of the continuity equation are somewhat more complex than for rectangular geometries because additional differential terms are introduced when N_i is a function of radial distance.

Another modification of the point form of the continuity equation obtains for circumstances in which the gain or loss of component i occurs by transport normal to the direction of axial flow. For example, consider the material balance needed to describe the macroscopic advective transport and reaction of component i within a plastic pipe with simultaneous loss or gain of that component by microscopic diffusional transport through the polymeric structure of the pipe wall, as shown in Figure 2-6.

The control volume is a section of pipe of radius R for which we need to include both axial flux, N_{ix}, and axisymmetric radial flux, N_{ir}, along with the rates of reaction and accumulation within the control volume. For this situation, the first and third flux terms of Equation 2-22 are nonzero. Thus, the material balance is written

$$-\frac{\partial N_{ix}}{\partial x} - \frac{\partial N_{ir}}{\partial r} - \frac{1}{r} N_{ir} + r_i = \frac{\partial C_i}{\partial t} \tag{2-25}$$

Now suppose that we wish to average Equation 2-25 across the cross section of the pipe:

$$\frac{1}{\pi R^2} \int_0^R \left[-\frac{\partial N_{ix}}{\partial x} - \frac{1}{r}\frac{\partial}{\partial r}(rN_{ir}) + r_i - \frac{\partial C_i}{\partial t} \right] 2\pi r\, dr = 0 \tag{2-26}$$

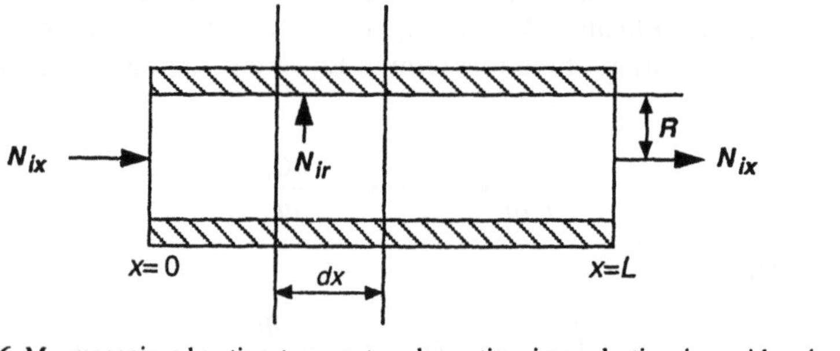

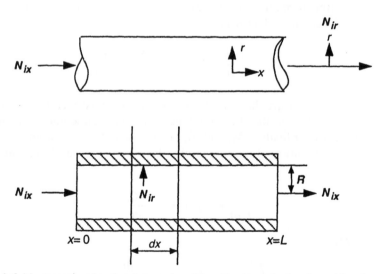

Figure 2-6 Macroscopic advective transport and reaction in a plastic pipe with microscopic transport through the pipe walls.

Here we have integrated over an annular element of area $2\pi r\, dr$ and divided through by the cross-sectional area of the pipe (πR^2). If it is further assumed that N_{ix}, r_i, and C_i are independent of r, we can easily integrate Equation 2-26 to obtain

$$-\frac{\partial N_{ix}}{\partial x} - \frac{2}{R} N_{ir}|_R + r_i = \frac{\partial C_i}{\partial t} \qquad (2\text{-}27)$$

Typically, the flux N_{ix} is determined by macrotransport processes through the reactor, whereas the flux N_{ir} is determined by microtransport processes across the interface (e.g., diffusion across the thickness of a pipe wall). As written, Equation 2-27 expresses the loss of material from the pipe due to radial transport *outward*. An alternative approach to Equation 2-27 for radial transport of material *into* the pipe is presented in Example 2-4.

Example 2-4

- **Situation:** Leakage of gasoline from underground storage tanks has contaminated the groundwater in the vicinity of a submerged plastic pipe that carries irrigation water to the fields. Various components of gasoline are known to "soften" plastic pipe material such that it becomes permeable. Among the major components of gasoline is toluene.
- **Question(s):** Develop a material balance–based model to predict the toluene concentration in the irrigation line after it passes through the region of contaminated soil.
- **Logic and Answer(s):**

1. The first step in analysis is to develop a definition sketch in order to visualize the components needed in the material balance.

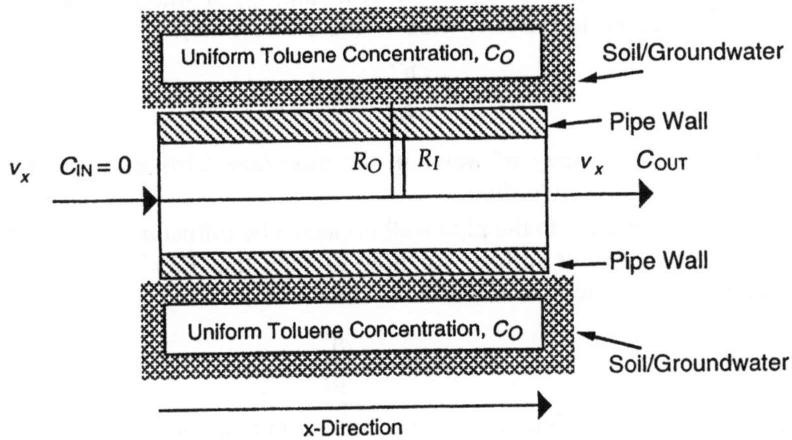

2. The material balance must include diffusive transport of toluene through the pipewall along the length of pipe as well as advective transport once inside the pipe. The integral form of the material balance equation is not appropriate because concentration is changing all along the length of pipe. The point form can be obtained in two ways: (a) by representing the pipe volume in cylindrical coordinates and averaging the diffusive flux that occurs radially through the pipe wall (from outside to inside) as given in the differential element approach (Equation 2-27); or (b) by representing the pipe volume in rectangular coordinates and accounting for diffusive flux that occurs into the pipe at the wall, for which the shell balance is

$$(N|_x - N|_{x+\Delta x})\pi R_I^2 + N_{R_I}(2\pi R_I \,\Delta x) = \frac{\partial C_x}{\partial t}\,\pi R_I^2\,\Delta x$$

where the flux of toluene in the direction of pipe flow is designated by the subscript x and the diffusive flux normal to the pipe wall (i.e., in the radial direction) is designated by the subscript R_I. Dividing through by $\pi R_I^2\,\Delta x$ and taking the limit as $x \rightarrow 0$ yields

$$-\frac{\partial N_x}{\partial x} + N_{R_I}\frac{2}{R_I} = \frac{\partial C_x}{\partial t}$$

The equation above is the same as Equation 2-27, with the exception of the sign for radial flux, which is positive to account for gain inside the pipe. Steady state is a reasonable assumption if the toluene concentration surrounding the pipe wall is large and is not depleted significantly by diffusion.

$$-\frac{\partial N_x}{\partial x} + N_{R_I}\frac{2}{R_I} = 0$$

3. The flux of toluene in the direction of pipe flow is defined as the mass of toluene carried along by the flow rate, Q, of water per unit area of the pipe normal to the flow (flux = mass/area-time):

$$N_x = \frac{QC_x}{\pi R_I^2} = v_x C_x$$

where v_x is the velocity of water in the pipe (see Chapter 3 for more information on advective flux).

4. The flux normal to the pipe wall is caused by diffusion (see Chapter 4 for more details). It suffices for now to state that Fick's first law of diffusion governs the flux in the radial direction:

$$N_r = -\mathfrak{D}_m \frac{dC_r}{dr}$$

where $\mathfrak{D}_m$ is the diffusivity of toluene through the pipe material (i.e., the characteristic constant of proportionality between the concentration gradient and the flux). The material balance for diffusive flux through a differential element of the pipe wall of unit length is

$$(N|_r - N|_{r+\Delta r})2\pi r = \frac{\partial C_r}{\partial t} 2\pi r\ \Delta r$$

Dividing through by $2\pi r\ \Delta r$ and letting $\Delta r \to 0$ gives

$$-\frac{1}{r}\frac{\partial}{\partial r}(rN_r) = \frac{\partial C_r}{\partial t}$$

The expression above is the same as Equation 2-23 for the situation where no reaction occurs within the pipe wall. Because the concentration of toluene outside the pipe wall, C_0, is assumed constant, the diffusive flux at any location along the pipe wall, x, must be at steady state; that is,

$$-\frac{1}{r}\frac{\partial}{\partial r}(rN_r) = 0$$

The product, rN_r, must accordingly be a constant (see Appendix A.2.2). This simplification together with Fick's law allows the diffusive flux at $r = R_I$ to be expressed in terms of diffusive flux at $r = R_O$:

$$R_O N_{RO} = R_I N_{R_I}$$

$$R_I N_{R_I} \int_{Ro}^{R_I} \frac{dr}{r} = -\mathfrak{D}_m \int_{C_0}^{C_I} dC_r$$

$$N_{R_I} = -\frac{\mathfrak{D}_m(C_I - C_0)}{R_I \ln (R_I/R_O)}$$

5. For this problem, $C_0 > C_I$, so that the direction of positive radial flux from step 4 is outward. Substituting the advective flux expression (step 3) and the diffusive flux expression (step 4) with the proper sign to account for flux *into* the pipe, the material balance for toluene in the pipe (step 2) leads to

$$-v_x \frac{dC_x}{dx} - \frac{2\mathfrak{D}_m}{R_I \ln (R_O/R_I)}(C_x - C_0) = 0$$

An implicit assumption is that concentration in the radial direction at the inside edge of the pipe wall (C_I) is equal to the concentration everywhere in the cross section of the pipe at a particular location, x. The solution to this differential equation is easily obtained by separation of variables:

$$\frac{C_x}{C_0} = 1 - \exp\left(-\frac{\kappa x}{v_x}\right)$$

where

$$\kappa = \frac{2\mathfrak{D}_m}{R_I \ln (R_O/R_I)}$$

6. The predicted pattern of fractional concentration (C/C_0) change with distance along the pipeline is shown below.

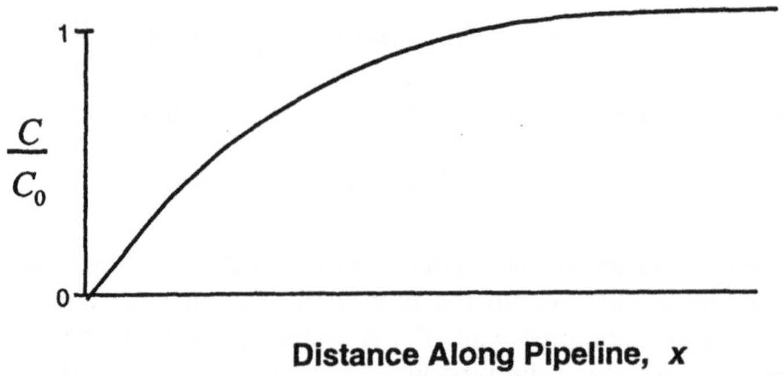

As a practical example, suppose that toluene has contaminated the soil and groundwater over a length of 300 m. The following additional information is available:

$$\mathfrak{D}_m = 1 \times 10^{-9} \text{ cm}^2/\text{s}$$

$$R_O = 10 \text{ cm}$$

$$R_I = 9.9 \text{ cm}$$

$$v_x = 5 \text{ cm/s}$$

The predicted fractional concentration pattern in the pipeline is shown below.

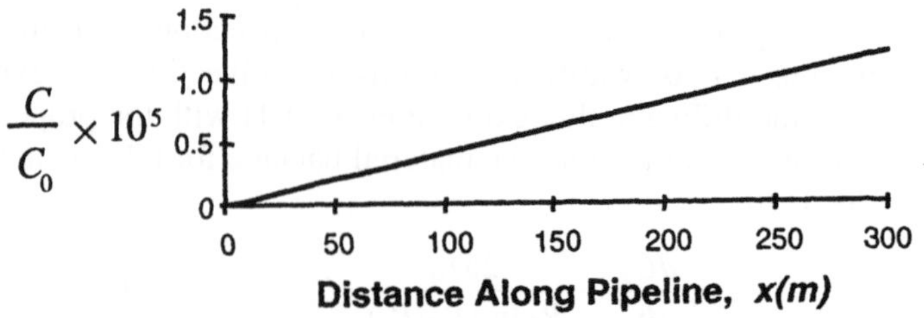

The fractional concentration is not large but it still may be significant, depending on the concentration in the surrounding groundwater.

7. Closer examination of the predictive equation shows that for thin pipe walls,

$$R_I^2 \ln \frac{R_O}{R_I} \approx R_I^2$$

such that the fractional concentration will increase as the diffusivity in the pipe material increases and as the pipe radius squared decreases.

2.2.5 Initial and Boundary Conditions

A key consideration in applying the foregoing material balance relationships is the specification of appropriate conditions existing at system boundaries and at initial points in time from which system behavior is evaluated. The equations we have developed to describe material balances are differential equations in certain specified dependent variables. Applications of any differential equation in dependent variables require that the variables satisfy the differential equation throughout some domain of their corresponding independent variables and at the boundaries of the domain. Initial and boundary conditions are selected to reflect physical phenomena and assumptions within a system and along its interfaces with surrounding environments. *Initial conditions* (ICs) involve specification of state variables, such as concentration and temperature, at time $t = 0$ for all points in the system. This specification implies a knowledge of the initial value of spatial derivatives (e.g., dC/dx) as well.

A *boundary condition* (BC) specifies conditions at a particular boundary of the domain which must be satisfied by the dependent variable, say u, while an initial condition (IC) specifies the value of u throughout the domain such that the initial value of that variable is defined by

$$u_0 = \phi(x, y, z) \tag{2-28}$$

Boundary conditions involve specification of state variables along the system boundaries for all times considered. Boundary conditions include the absolute value of the variable as well as spatial derivatives when necessary. The number of spatial derivatives required corresponds to one less than the order of the material balance equation. For example, if the model equation is second order (i.e., including d^2C/dx^2), boundary values for C and dC/dx are necessary.

Equilibrium requirements for concentrations at an interface, such as Henry's law at the air–water interface or a Freundlich adsorption isotherm at a particle surface, generally lead to specifications of boundary value concentrations. Conservation requirements for material fluxes at interfaces generally lead to specification of spacial derivatives. The appropriate conservation equations are similar to the general material balance relationship given in Equation 2-5, having the form

$$\begin{bmatrix} \text{rate of flux} \\ \text{to the} \\ \text{boundary} \end{bmatrix} - \begin{bmatrix} \text{rate of flux} \\ \text{away from the} \\ \text{boundary} \end{bmatrix} = \begin{bmatrix} \text{rate of} \\ \text{accumulation at} \\ \text{the boundary} \end{bmatrix} \tag{2-29}$$

When mass transport is by diffusion, the driving force for transport between two points in space is a function of the difference in concentration of the diffusing species between those two points divided by the distance between them (i.e., the concentration *gradient*). This leads to a boundary condition involving the first derivative of concentration. System geometry and the location of *inputs* and *sinks* may also define boundary conditions.

One or more of three different boundary conditions apply to most environmental systems:

- The type 1, or *Dirichlet* condition
- The type 2, or *Neumann* condition
- The type 3, or *Robin* condition

For the type 1 condition the value of the unknown function, u, is specified at the boundary, either as a constant or as a function of time. The specification of a constant value for u at the boundary generally implies a steady-state condition for the dependent variable at that spatial location. Let us reconsider, for example, the reactor system shown schematically in Figure 2-4. Suppose that an oxidation reaction occurs between a strong oxidant and organic matter. The reaction rate may be considered as first order with respect to the organic matter concentration if the oxidant has been added in excess and is not significantly depleted. Flow through the reactor follows steady "plug flow" behavior at constant velocity v_x along one primary dimension, x. The initial concentration of organic matter, C_0, decreases along the length of the reactor as shown in Figure 2-4, at a rate proportional to the residual concentration C [i.e., $r = \phi(C)$]. If there is no change in the inlet concentration, flow rate, or temperature, the reaction rate should not change. Thus the concentration pattern along the length of the reactor should be constant with time, so the system is said to be at steady state [i.e., $(\partial C/\partial t)_x = 0$]. Beginning with Equation 2-18 and substituting $v_x C$, the material balance for this reactor reduces to

$$-v_x \frac{dC}{dx} + r = 0 \qquad (2\text{-}30)$$

The type 1 or Dirichlet condition is the appropriate boundary condition for this case. Thus, specifying the value of C at the boundary, we have

$$C = C_0 \qquad \text{at } x = 0 \qquad (2\text{-}31)$$

If r is given by a first-order reaction such as kC, the solution of the differential equation describing the concentration profile along the reactor is

$$C = C_0 e^{-kx/v_x} \qquad (2\text{-}32)$$

The type 2 condition involves specification of the value of the derivative of the function u normal to the domain boundary. A typical application of this type of BC is the "no flux" boundary (i.e., $\partial u/\partial \boldsymbol{n}_v = 0$ at all points along the boundary, where $\boldsymbol{n}_v$ is the normal vector shown in Figure 2-2). For the one-dimensional reactor shown in Figure 2-6,

$$\frac{dC}{dx} = 0 \qquad \text{at } x = L \qquad (2\text{-}33)$$

Finally, the type 3 condition characterizes values of $\alpha u + du/dn_v$ at the boundaries, where α may be a constant or a function of corresponding independent variables. In transport problems, a type 3 BC is typically invoked to specify values of boundary flux attributable to macroscopic transport phenomena. The type 3 BC is appropriate to use, for example, to describe the inlet condition at $x = 0$ for the reactor pictured in Figure 2-7, where it is assumed that transport immediately at $x = 0$ out of the completely mixed sump occurs

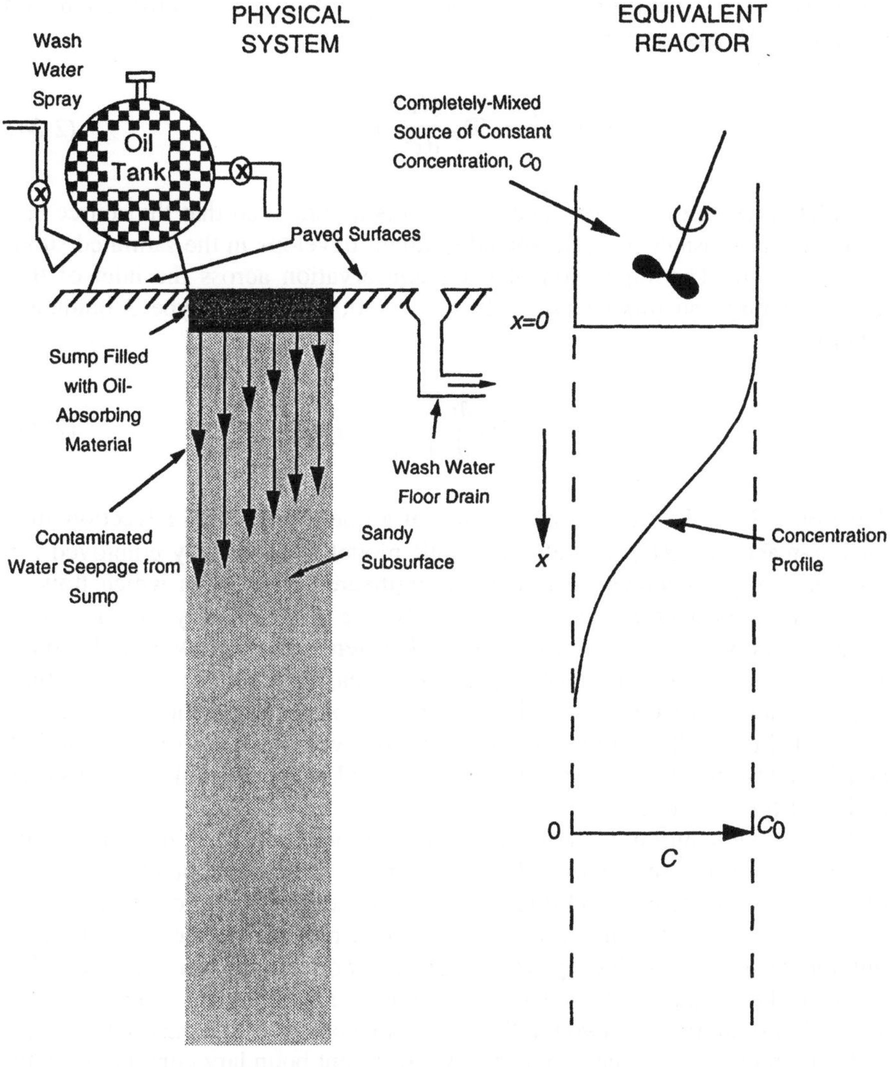

Figure 2-7 Modeling constant-source contamination of a semi-infinite domain with fluid dispersion along the axis of flow.

by advection only, whereas transport downgradient of $x = 0$ occurs by a combination of advection and internal mixing (see Equation 2-35 below).

Internal mixing, which can be induced by eddy current development, causes dispersion of fluid elements along the x axis so that some move faster than the average velocity, v_x, and some move more slowly, the distribution of differential velocities being approximately *Gaussian*, or *normal*. Such deviations from plug flow can be characterized in terms of a "dispersion" coefficient, $\mathfrak{D}_d$ (L^2t^{-1}), and as we shall demonstrate in ensuing chapters discussing reactor design relationships, the flux, N_i, is then given by the sum of $v_xC - \mathfrak{D}_d(dC/dx)$. The material balance relationship for operation of this system at steady state (i.e., assuming continuous leakage of oil) from substitution into Equation 2-18 has the form

$$-v_x \frac{\partial C}{\partial x} + \mathfrak{D}_d \frac{\partial^2 C}{\partial x^2} + r = 0 \qquad (2\text{-}34)$$

Complete mixing of the zone above $x = 0$ is assumed so that the source can be treated as constant (i.e., no spatial gradient develops in the saturated sump above $x = 0$). The stipulation of mass conservation across the inlet of this reactor requires no mixing precisely at $x = 0$, yielding a type 3 boundary condition:

$$\left[v_xC - \mathfrak{D}_d \frac{\partial C}{\partial x}\right]_{x=0} = v_xC_0 \qquad (2\text{-}35)$$

Equation 2-34 is the steady-state version of a one-dimensional advection–dispersion–reaction (ADR) model. The ADR model is commonly employed for description of contaminant behavior in engineered reactors in which flow is specifically controlled to occur in essentially one direction only. It is also frequently employed for description of natural systems, such as rivers and certain subsurface flow domains, where it provides a reasonably close approximation to a condition that would otherwise require complex and potentially unnecessary modeling detail. Criteria and assumptions invoked in the use of simplified models to describe complex systems are discussed in more detail in the ensuing section of this chapter.

Because the solution of the material balance equation for a reactor system is a function of the boundary conditions specified, we must in practice be careful to choose boundary conditions that match the physical problem(s) being modeled. If boundary conditions are not correct physically, the mathematical solutions generated by descriptive and design models may very well also be incorrect. The impact of boundary conditions in mathematical solutions is illustrated in Example 2-5; two totally different solutions of the material balance model are presented for the system for two different boundary conditions, only one of which is correct.

Example 2-5

- **Situation:** An environmental contamination event and its equivalent one-dimensional semi-infinite reactor are pictured in Figure 2-7. In this simplified situation the absorbent material filling the sump beneath the tank spigot completely captures and retains the small amount of oil spilled during filling and withdrawal operations of the tank. The area is washed down frequently, and some of this wash water is collected and held in the sump, draining from it into the underlying soil column at a slow but essentially constant velocity (v_x) and contaminant saturation level (i.e., $C_0 = C_S$).
- **Question(s):** Specify the appropriate model equations and boundary conditions for describing this situation.
- **Logic and Answer(s):** For this example we cannot assume steady state because the concentration profile of the contaminant is changing in space and time. Thus the right-hand term in Equation 2-34 is not zero:

$$-v_x \frac{\partial C}{\partial x} + \mathfrak{D}_d \frac{\partial^2 C}{\partial x^2} + r = \left(\frac{\partial C}{\partial t}\right)_x$$

The reaction rate term, r, is assumed to be zero. If the reactor is free of solute at the beginning of the simulation, the initial $C(x, t)$ condition is given by

$$C(x, 0) = 0$$

For the inlet boundary condition a type 3 condition identical to that given by Equation 2-35 would apply, while for the outlet boundary condition a type 2 condition is appropriate, so that

$$\frac{\partial C}{\partial x} = 0 \qquad \text{as } x \rightarrow \infty, \text{ all } t$$

Solution of the unsteady-state form of Equation 2-34 for these initial and boundary conditions yields (Gershon and Nir, 1969)

$$\frac{C_x}{C_S} = 0.5 \operatorname{erfc}\left[\frac{x - v_x t}{2(\mathfrak{D}_d t)^{0.5}}\right] - 0.5 \exp\left(\frac{v_x x}{\mathfrak{D}_d}\right) \operatorname{erfc}\left\{\left[\frac{x + v_x t}{2(\mathfrak{D}_d t)^{0.5}}\right]\right.$$
$$\cdot \left.\left[1 + \frac{v_x(x + v_x t)}{\mathfrak{D}_d}\right]\right\} + \left[v_x\left(\frac{t}{\pi \mathfrak{D}_d}\right)^{0.5}\right]$$
$$\cdot \exp\left\{\frac{v_x x}{\mathfrak{D}_d} - 0.5\left[\frac{x + v_x t}{(\mathfrak{D}_d t)^{0.5}}\right]^2\right\}$$

Discussion of the error function, erf(z), and error function complement, erfc(z), of a variable, z, is provided in Chapters 4 and 8, and values are tabulated in Appendix A.8.1. It should be noted that an entirely different solution to Equation 2-34 is obtained if a different inlet boundary condition is chosen; in other words, the solution to the material balance equation for a particular system is a function of the boundary conditions chosen. If, for example, a type 1 boundary condition is chosen, so that $C(0, t) = C_0 = C_S$, the solution for Equation 2-34 becomes

$$\frac{C_x}{C_S} = 0.5 \text{ erfc}\left[\frac{x - v_x t}{2(\mathfrak{D}_d t)^{0.5}}\right] + 0.5 \exp\left(\frac{v_x x}{\mathfrak{D}_d}\right) \text{erfc}\left[\frac{x + v_x t}{2(\mathfrak{D}_d t)^{0.5}}\right]$$

For this problem, the type 1 boundary condition is not correct because it does not conserve mass across the reactor inlet. Problem 2-13 provides an exercise in comparing concentration profiles predicted by these two different sets of boundary conditions.

2.3 CONSIDERATIONS IN APPLYING PROCESS MODELS

2.3.1 Types and Levels of Process Analysis

Equation 2-5 is a general description of the material balance relationship, the writing of which is the first and most important step in process analysis. The mathematical elaboration of this relationship varies from one model to another depending on the nature and level of information known about the system (model structure and input) and the nature and level of information desired from the modeling effort (output requirements). In this book we consider many different applications of process analysis based on the general material balance principle and the development of specific types of process models. In certain situations the nature of reactor operation, the choice of scale, or the specification of temporal or spatial limits on a model drives one or more of the terms in Equation 2-5 to zero, as we have seen, for example, for assumptions of steady-state conditions. We also encounter a variety of reaction rate expressions and reactor geometries.

2.3.2 Material Balance–Based Models

The most sophisticated environmental models are usually comprised of one or more material balance relationships, written in the form of differential equations describing the change of one system variable as a function of another, the latter most commonly being either time or distance. This level of modeling requires a thorough conceptual knowledge of processes operating within a system, as well as a sufficient data base on which to evaluate all the equations involved in the model structure. A data base may be very useful to determine model parameters for use in material balance–based models; well-developed empirical correlations that are often appropriate for practice have evolved. If this level of knowledge and data is available and incorporated in a mass balance model, the resulting model should be versatile with respect to the examination of a variety of conditions, and its output or predictions can be expected to be accurate and precise.

A general approach for development and use of mathematical models for description of physical systems was given in Section 2.1.4. That approach can be expanded upon based on our intervening discussions to develop a protocol for application of the material balance relationships to the modeling of any environmental system. At a minimum, such a protocol would require that we:

- Identify the purpose and goals of the modeling effort.
- Visualize the system and its components:
  - ~ Develop system representation.
  - ~ Assign notation.
  - ~ Categorize variables as input, operating parameters, or output.
  - ~ List underlying and simplifying assumptions.
  - ~ Refine system representation.
- Develop differential equations:
  - ~ Identify unit or control volume.
  - ~ Write material balances in differential forms.
  - ~ Rearrange and take limits to convert to differential forms.
- Evaluate system parameters:
  - ~ Develop and list correlations.
  - ~ Substitute where appropriate.
- Establish boundary and/or initial conditions:
  - ~ Develop and list.
- Express equations in dimensionless forms.
- Solve equations:
  - ~ Analytical solutions.
  - ~ Numerical solutions.
- Calibrate with data set.
- Conduct parameter sensitivity analyses.
- Verify with second data set for different conditions.
- Apply for prediction and design.

Once the overall purpose and goals of the modeling effort have been established, the system and its components can be visualized and pictured. Notation is assigned and underlying and simplifying assumptions are noted. The material balance equation is written for the appropriate control volume and converted to differential form. Equations defining model parameters are identified and substituted where appropriate. The essential boundary and initial conditions are then determined and listed.

The working equations are often most conveniently represented in dimensionless form, particularly if the ultimate solution algorithm is numerical. If analytical techniques are employed, this conversion may be deferred until after the solution equation is derived. We demonstrate in numerous examples in this book that the conversion of systems of differential equations describing complex processes to dimensionless form generalizes those equations and extends their utility to broader ranges of application.

Finally, the equations must be solved in the appropriate manner. Throughout the entire modeling process, the available data base influences model development, form, and detail. Ultimately, the model results feed back into the data that are available, indicating where efforts regarding future experiments and data collection should be directed.

2.3.3 Alternative Modeling Approaches

Some environmental systems are not sufficiently well defined to warrant such *deterministic* levels of model sophistication. In other instances the level of information or prediction required may be no more than approximations or estimates, for which mathematical elegance is neither necessary nor cost-effective. The material balance equation(s) written in terms of descriptive relationships predicated on empirical observations or statistical correlations are often sufficient in such instances.

The general preference for deterministic models based on the material balance equation is logical because this equation is firmly rooted in physical concepts and readily derived from a consideration of basic mass transport and reaction phenomena. Moreover, appropriate solution techniques, both analytical and numerical, are fully developed and readily available. Not all mathematical models that are based on material balance relationships, however, lead inexorably to a set of differential equations. In particular, *probablistic* models can be developed in which the material balance relationship is satisfied by the requirement that probabilities associated with different system states must sum to 1. The method of solution generally involves integration over a probability domain to determine *expected* values, such as *first moments* (means) and other *statistically related moments* (e.g., second moments, or variances) of the output desired. These models may require additional assumptions related to system processes, but often lead to simplification of solution procedures. Simplification leads to a less general model output, providing, for example, only means and variances rather than entire time histories of the parameters of interest. In certain situations, however, the analyst may only require such output.

The discussion above suggests that when water quality processes are too complex for adequate representation of exact material balance relationships, it may be both necessary and sufficient to fall back upon empirical or statistical correlations as a basis for modeling. Empirical equations and nomagraphs have a long history of use in water and waste treatment practice. Relationships between process performance (e.g., pollutant removal) and the flow rates at which unit processes are operated, for example, are commonly developed from *pilot-plant* experiments. This practice is to be distinguished from the use of empirical correlations to determine model parameters for a material balance–based model.

Statistical regressions that relate water quality at a water supply inlet to upland rainfall, industrial activity, and land use provide another example. These models are often simpler to develop than rigorous material balance models and are easier to apply. In some circumstances, their predictive capabilities may equal or exceed those of a material balance model. This is particularly the case when data required to address the full complexity of the more rigorous model(s) are lacking. Because such models are not rooted in physical principles and processes, however, they have definite limitations and drawbacks. They are more likely to yield inaccurate results outside their range of calibration because of conceptual misrepresentation. In addition, many of the insights gained by developing and applying material balance models—for example, exposition of transport paths, delineation of source effects, and sensitivity to model parameters—are not realized when an empirically derived equation is used for modeling purposes. As such, the statistical-empirical model may be viewed as an approach of last resort. There are, nonetheless, a number of applications for which it is necessary and appropriate.

Each of the modeling steps itemized in Section 2.3.2 should be performed from the perspective of maximizing their respective relationships to the goals and constraints of the modeling effort. It is always possible to build great complexity into a mathematical model. Such complexity, however, may be counterproductive if the level of detail in the model exceeds that required in the decision-making process that depends on interpretation of the model results. The modeler must be certain, even in cases where additional complexity is useful, that the additional data requirements of a more complex model can be justified. Inevitably, a balanced approach that considers the model complexity, the data requirements, the accuracy of the results, the intended use of the results, and, of course, the cost of its development and application must be established.

2.3.4 Evaluation Criteria

Any set of system boundaries, processes, assumptions, and equations used to structure a model for an environmental system should be evaluated according to three criteria:

- The level of detail established should be consistent with the intended use of the model.
- Acceptably accurate field measurements should be obtainable at the level of detail established by the model.
- System process must be understood and quantifiable at the level of data established.

To illustrate the first criterion, consider the engineering of a system to upgrade the performance of a municipal water treatment plant to meet standards for taste and odor. Suppose that taste and odor are generated intermittently by organic substances during periods of algal activity in a lake that serves as the raw water source. Let us further suppose that the taste- and odor-causing compounds can be treated readily by addition of a slurry of powdered activated carbon to the rapid-mix chamber of the coagulation unit of the plant, on an as-needed basis. The adsorption process to be engineered is thus one of relatively straightforward design and inherent flexibility. In this case the model required to engineer and design the adsorption system to produce acceptable estimates of cost and predictions of performance can be simple in both structure and function. The dosages of powdered carbon needed under worst conditions can readily be measured and projected with reasonable accuracy given the seasonal occurrence of taste and odor; moreover, dosages can be changed periodically as needed. In addition, the reactor systems are already in place to affect treatment: the rapid-mix basin, the coagulation and flocculation units, the sedimentation basins, and the final filters. Their performance is well known. Finally, the unlikely worst-case situation is failure of the planned adsorption system, the outcome of which is temporarily irate consumers who object to the taste or odor of the water.

Consider the contrasting situation in which a lake is found to contain organic pesticides during certain times of the year, and as a result of runoff from surrounding streets and paved lots, is subject periodically to transient contamination by aromatic gasoline components and combustion products (e.g., benzene, toluene, xylene, and selected polynuclear aromatic hydrocarbons). The carbon adsorption system required to protect the public from these potentially toxic and carcinogenic chemical contaminants must be much more sophisticated and more rigorously designed and operated than for control of occasional taste and odor problems. To achieve more effective treatment, carbon adsorption contactors may be installed following the filtration step in the plant. Continuous protection is required and thus these contactors will be a regular on-line unit operation. Both frequency of appearance and concentration of pesticides and other contaminants may not easily be predicted. The presence of many contaminants will result in competition for adsorption sites and, potentially, the displacement of weakly adsorbed components. Because the adsorption system will be continuously in operation and have stringent performance requirements, carbon replacement and regeneration will be frequent, and these needs and costs must be predicted accurately.

A failure of the adsorption system in treating organic contaminants such as pesticides and gasoline components for any reason in the situation described above could lead to serious degradation not only of the palatability of the product water but of its potability as well. Some redundancy must therefore be built into the system as a contingency against failure. This system would require more careful consideration of an appropriate model for process engineering and design than would a system to handle occasional taste and odor problems. In addition, the data required to calibrate and verify that model properly must correspondingly be more detailed and more accurate.

The second criterion for model structure is that acceptably accurate field measurements should be obtainable at the level of detail established by the model. We are often forced to use a simpler analysis than that demanded by the intended use of a model because of the lack of sufficient and appropriate data. For instance, there would be little value in developing a multidimensional model to predict speciation, concentration, and behavior of each organic substance contributing to taste and odor removed by carbon adsorption if we neither have, nor can make, detailed measurements of such parameters on a regular basis.

Typically, an engineer or scientist in charge of a project to design a treatment process must make decisions on the allocation of project resources between modeling and sampling. The ability to gather data may also be limited by fundamental considerations. We saw in Figure 1-4, for example, that the transport of a toxic substance within the pore structure of an activated carbon granule is of concern in the design of adsorption systems. There are, however, inherent limitations associated with making direct measurements at the microscopic scale. We are thus forced to rely on data collected from experimental systems containing fairly large populations of carbon particles and to attempt extrapolation of these data to behavior at the single-particle scale. Other measurement limitations at a fundamental level are associated with investigations of highly complex but not readily characterized systems, such as heterogeneous subsurface environments, or unstable systems in which individual chemical species change in a rapid and nonmeasurable manner.

Finally, the third criterion for model development requires that system processes be understood and quantifiable at the level of data established. It is not uncommon in the modeling of environmental systems to encounter "frontiers" of available knowledge regarding a complex transport and transformation process. We may be forced to rely on obvious simplifications of complex phenomena simply because we do not understand the processes involved sufficiently to incorporate more detail in a model. A model developed to predict the frequency and concentrations of gasoline components present in the influent to the water treatment plant of our second example might have to utilize empirical relationships to describe runoff and lake mixing conditions. Runoff and mixing may be very important in determining the time of appearance and the concentrations of various components at the water intake crib. However, empirical relationships would ignore known differences in the uptake of the individual components by suspended solids in the runoff and in the lake. Unfortunately, the present state of the art is lacking in information both to quantify these differences for all of the expected components and to specify how the distributions of these components change in time.

2.3.5 Philosophy of Approach

The modeling of an environmental system and its associated processes involves reduction of related mass transport and transformation or separation processes to a logical system of mathematical analogs. The system of equations comprising the model is then resolved, calibrated with appropriate data pertaining to the physical system, verified on a prototype basis with data other than those employed for calibration, and used for the descriptions, predictions, or other design applications for which it was devised.

The respective roles and interactions of physical systems and mathematical analogs in the modeling process are presented schematically in Figure 2-8. A physical system is generally first visualized or conceptualized in terms of its constituent components, compartments, and processes. Each of these components is then described by an appropriate equation or set of equations. The modeling process brings these descriptive equations together into a system of equations comprising the mathematical analog of the physical system. Adjustment and tuning of this system model is done by *calibrating* it to one or more sets of data from laboratory or controlled field measurements. The calibrated model is used to design a prototype system or experiment. The latter is employed to collect an independent set of data from a field application, which if it conforms to model predictions, serves to *validate* the model.

For large natural environmental systems such as lakes or rivers, and for complex existing engineered systems such as water or wastewater treatment plants, the system and the prototype depicted in Figure 2-8 are usually one and the same. It is seldom possible, or necessary, to construct a representative physical prototype for such systems. Instead, the calibration data and verification data are generally collected from field measurements of the physical system at different times and under different conditions.

A much different circumstance exists if the physical system is a component treatment unit of a pollution control facility that is to be designed and constructed. In such situations there is usually an opportunity to build a pilot-scale prototype based on laboratory bench-scale studies. The pilot unit is then operated under particular and closely controlled sets of conditions. The design model can be calibrated with data from the independent bench-scale studies and verified rigorously with data developed in the pilot-scale study. In this regard, pilot-scale investigations can provide important information for full-scale design. In fact, the most effective use of pilot-plant investigations is as a means for intermediate verification of design models rather than for collection of fundamental information and data.

Both the mathematical model and the pilot-scale prototype are valuable conceptual and operational tools for environmental scientists and engineers. Too frequently, however, these tools are misused in specific applications. The value of a mathematical model is constrained by the degree to which it accounts reliably for the physical characteristics, transport processes, and transformation processes comprising a particular system—in other words, by how accurately it describes, or *captures*, the physical system. If a model is oversimplified or incorrectly structured, or if calibrated to one system and indiscriminately applied to another without appropriate modification and revalidation, it may be more detrimental than beneficial to the ultimate objectives of the modeling process.

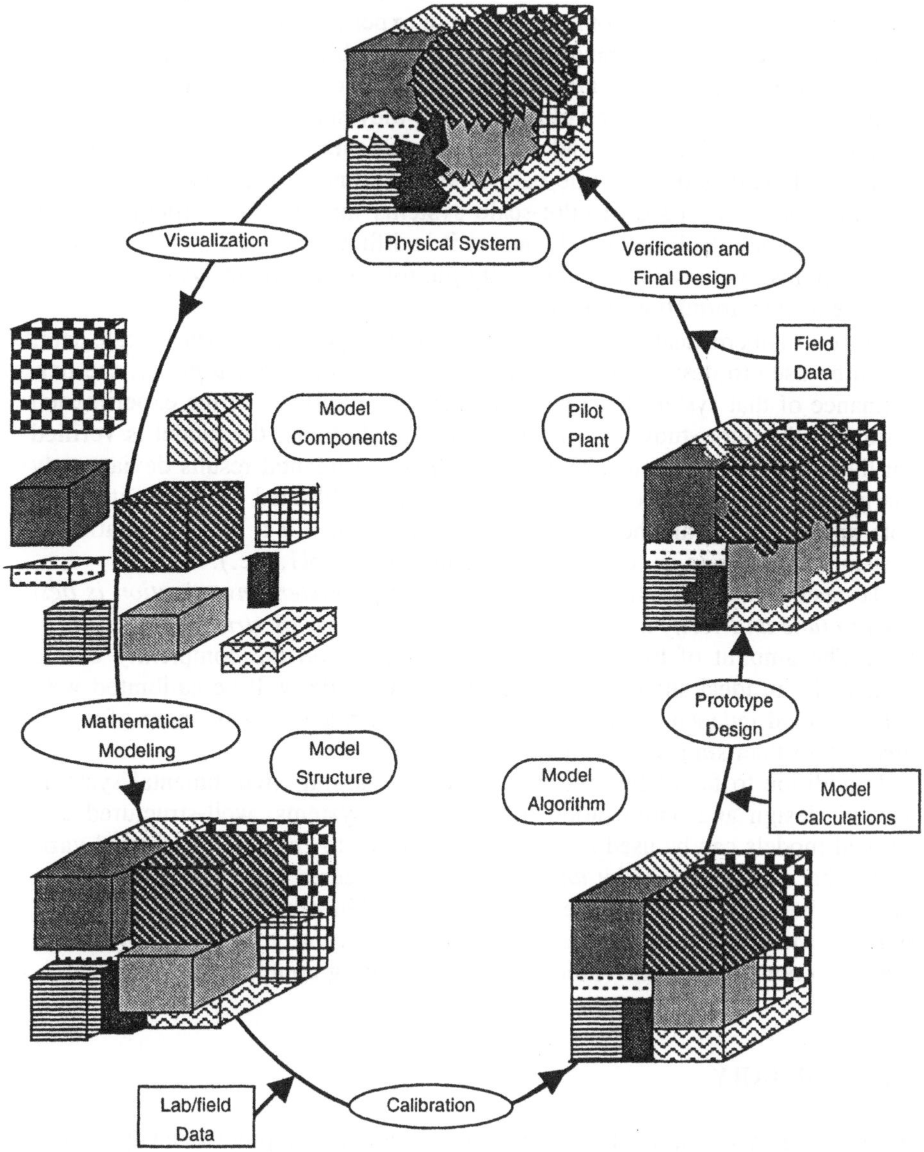

Figure 2-8 Process modeling: physical systems and mathematical analogs.

The best approach to environmental system description and process design is to integrate the use of mathematical models and prototype studies. Proper integration can be accomplished by first structuring mathematical models on the basis of sound process concepts and principles. Appropriate data for initial model calibration can then be obtained from bench-scale experiments. Because of their ease of operation, relatively low cost, small sample requirements, and easily characterized conditions, bench-scale experiments afford examination of a large number and broad range of pertinent variables. Prototype pilot or field studies, on the other hand, are usually lengthy, costly, and require elaborate equipment and large sample volumes. Thus they seldom permit extensive investigation of the full scope and range of significant variables. Moreover, the transport dynamics of full-scale natural systems and of the types and sizes of reactors commonly used in pilot-scale systems may not be particularly well defined. Without well-defined dynamics it is difficult, if not impossible, to use such systems to characterize sufficiently the roles of pertinent system variables relative to transformation processes.

Once a mathematical model is calibrated with experimental bench-scale data, it can be used to design a pilot-scale system and to predict, *a priori*, the performance of that system for a given matrix of conditions. If the experimental results of the pilot study confirm the model predictions, the model is verified for the range of conditions examined. If the experimental results deviate only marginally from predictions, the model can be recalibrated or *tuned* with these data and tested on another set of pilot-system data for a different matrix of system conditions (e.g., concentrations, flow rates, pH, etc.).

To restate for emphasis, *the pilot or prototype system investigation is best used to tune and verify a mathematical model, not for its structure or calibration*. The amount of tuning required will depend on the complexity of the system. In an ideal situation, a verified design model will be calibrated with data obtained independent of the prototype system and thus be more of a *predictive* tool than simply a *simulator* of observed data.

In addition to facilitating the description of natural environmental systems and the design and optimization of engineered systems, well-structured and verified models can be used to evaluate responses to changes in selected variables, that is, for *sensitivity analyses*. For engineered systems, such models can also be used as interpretive elements in *feedfoward* and *feedback control systems*. Modeling for purposes of control is particularly useful for processes subject to non-steady-state (time-variable) operating conditions.

2.4 SUMMARY

We have in this chapter defined the core elements of process analysis and identified how these elements are integrated and interpreted to characterize systems and processes. The most rigorous means for accomplishing this integration and interpretation is provided by the principles of continuity: specifically, by mass continuity or material balance relationships. These relationships have been shown to incorporate both the transport and transformation aspects of process dynamics. We have learned, in a qualitative way, how to express transport and transformation relationships and how to incorporate them into material balance equations. To fully exploit the potential of the material balance approach to process analysis and design, however, we must delve in greater detail into the character and forms of both transport processes and reaction processes. We begin closer examination of the former class of process in the next two chapters, first in the context of *macrotransport* phenomena in Chapter 3 and then in the context of *microtransport* and *mass transfer phenomena* in Chapter 4, with particular emphasis on multiphase systems.

2.5 REFERENCES AND SUGGESTED READINGS

2.5.1 Citations and Sources

Gershon, N. D., and A. Nir, 1969, "Effects of Boundary Conditions of Models on Tracer Distribution in Flow Through Porous Mediums," *Water Resources Research*, *5*(4), 830–839. (Example 2-5)

2.5.2 Background Reading

Bennett, C. O., and J. E. Myers, 1982, *Momentum, Heat, and Mass Transfer*, McGraw-Hill Book Company, New York. The mass balance equations for fluid flow are developed in detail using both the integral approach (Chapter 3) and the differential approach (Chapter 7).

Bird, R. B., W. E. Stewart, and E. N. Lightfoot, 1960, *Transport Phenomena*, John Wiley & Sons, Inc., New York. A classic text in transport phenomena. The use of vectors and tensors to make transformations of the continuity equation from rectangular to cylindrical and spherical coordinates is discussed in Chapter 3 and summarized in Appendix A.6. The shell balance (differential) approach to developing the material balance equations with application to several diffusion problems is described in Chapter 17. Boundary conditions are also discussed.

Cussler, E. L., 1984, *Diffusion: Mass Transfer in Fluid Systems*, Cambridge University Press, New York. General approaches to modeling diffusion problems are discussed in Chapter 1, and the mass balance equations using the differential approach are developed in Chapter 2 together with examples in diffusive transport.

Fogler, H. S., 1986, *Elements of Chemical Reaction Engineering*, Prentice Hall, Englewood Cliffs, NJ. Reaction rates are described in Chapter 1, together with a simplified development of the mole balance equation using the integral approach. The general mole balance equation is applied to ideal reactor configurations, and design equations are developed. A section illustrating some industrial reactor configurations is included.

Gershon, N. D., and A. Nir, 1969, "Effects of Boundary Conditions of Models on Tracer Distribution in Flow Through Porous Mediums," *Water Resources Research*, *5*(4), 830–839. Analytical solutions are provided for a wide range of unsteady-state situations involving transformations and transport.

Greenberg, M. D., 1988, *Advanced Engineering Mathematics*, Prentice Hall, Englewood Cliffs, NJ. Detailed derivations of the continuity equation in cylindrical and spherical coordinates using scalar and vector field theory are provided in Chapter 9.

Horvath, I., and N. W. Schmidtke, 1983, "Scale-Up and Scale-Down Concepts and Problems," in *Scale-Up of Water and Wastewater Treatment Processes* (N. W. Schmidtke and D. W. Smith, Eds.), Butterworth Publishers, Woburn, MA. Concepts of process scaling, with applications to water and wastewater treatment, are discussed.

Levenspiel, O., 1972, *Chemical Reaction Engineering*, Second Edition, John Wiley & Sons, Inc., New York. The material balance equation is discussed in Chapter 5 and applied to ideal reactor configurations to develop reactor design equations.

Masters, G. M., 1991, *Introduction to Environmental Engineering and Science*, Prentice Hall, Englewood Cliffs, NJ. A basic introduction to material balance relationships with applications to steady-state well-mixed systems is provided in Chapter 1.

James J. Montgomery Consulting Engineers, 1985, *Water Treatment Principles and Design*, John Wiley & Sons, Inc., New York. The principles of separation process engineering from a water treatment perspective are presented in Chapter 5; material balance equations for several different geometries are developed using the differential method; applications to ideal reactor configurations, including airstripping reactors, are provided. The material balance approach is used to develop mathematical process models for adsorption processes in Chapter 8 and gas-transfer processes in Chapter 11.

Schwarzenbach, R. P., P. M. Gschwend, and D. M. Imboden, 1993, *Environmental Organic Chemistry*, John Wiley & Sons, Inc., New York. Modeling concepts in the context of organic pollutants in lake systems are presented in Chapter 15 with an emphasis on the "box-model" (completely mixed reactor) approach.

2.6 PROBLEMS

2-1 A reactor is to be designed in which the oxidation of cyanide (CN^-) to cyanate (CNO^-) is to occur by the following reaction:

$$\tfrac{1}{2}O_2 + CN^- \rightleftharpoons CNO^-$$

The reactor is to be a tank which is vigorously stirred so that its contents are completely mixed, and into and out of which there is a constant flow of waste and treated effluent, respectively. The feed stream flow rate is 378 m^3/day, and the feed stream contains 15,000 mg/L CN^-. The desired reactor effluent concentration is 10 mg/L CN^-. Assume that oxygen is in excess and that the reaction is directly proportional to the cyanide concentration, with a rate constant of $k = 1\ s^{-1}$. Determine the volume of reactor required to achieve the desired treatment objective. (*Hint:* See Example 2-2.)

Answer: *6.56 m^3.*

2-2 Repeat Problem 2-1 for a reactor comprised by a long tubular chamber (e.g., a large-diameter pipe) through which flow passes laminarly and there is no fluid mixing. (*Hint:* See Example 2-3.)
Answer: $0.032\ m^3$.

2-3 Explain why the volume obtained in Problem 2-2 is much smaller than that obtained for the same reaction in Problem 2-1.
Answer: Because the concentration driving force for the reaction is maintained at a higher level over the course of the reaction rather than being dropped immediately to the desired effluent concentration. This is discussed in much greater detail in Chapter 3.

2-4 Example 2-3 presented the derivation of the material balance relationship for a long, narrow channel (reactor) exhibiting plug-flow characteristics. A modification that is often made to this relationship is the inclusion of dispersion, which allows the reactor to be nonideal. Dispersion is included in the differential equation in an additional term involving the product of $\mathfrak{D}_d$, the dispersion coefficient, and the local spatial concentration gradient, $\partial C(x)/\partial x$. Derive the steady-state material balance equation for the reactor presented in Example 2-3, including dispersion. Write the resulting material balance equation in dimensionless form by letting $c^\circ = C/C_S$ and $\zeta^\circ = x/L$, where C_S is the saturation concentration of the solute and L is the channel length.
Answers: $-v_x\dfrac{\partial C}{\partial x} + \mathfrak{D}_d\dfrac{\partial^2 C}{\partial x^2} = -k_f(a_s^\circ)_R(C_S - C)$; $-\dfrac{\partial c^\circ}{\partial \zeta^\circ} + \dfrac{\mathfrak{D}_d}{v_x L}\dfrac{\partial^2 c^\circ}{\partial \zeta^{\circ 2}} = -\dfrac{Lk_f(a_s^\circ)_R}{v_x}(1 - c^\circ)$. *Dispersion is discussed in general in Chapter 3 and in detail with respect to reactors in Chapter 9.*

2-5 Equalization tanks are commonly used in industrial waste treatment to smooth variations in flow and concentration that are experienced in typical batch productions. Suppose that the concentration of a contaminant in the influent to a tank is 10 mg/L and the concentration within the tank is also uniformly 10 mg/L. This contaminant does not undergo reaction in the tank. Suddenly, the influence concentration increases to 20 mg/L. Determine the concentration change in the effluent of the tank over time if the flow rate is 5 m^3/min and the volume of the tank is 500 m^3. State all assumptions.

2-6 A well-mixed reactor is being operated to treat a hazardous waste chemically. Catalytic reagent is delivered to the inlet of the reactor by a chemical feed pump. This pump is supposed to maintain a concentration of 25 mg/L in the wastewater stream as it enters the reactor. Being a catalyst, this chemical is not depleted within the reactor but merely facilitates the reaction. The treatment process can operate effectively as long as the concentration of this catalyst does not drop below 15 mg/L. Operation is at steady state until suddenly the feed pump malfunctions and begins to deliver a concentration of 10 mg/L at the inlet to the reactor instead of 25 mg/L.

(a) Determine how much time the operators have before the treatment process becomes ineffective.

$$\text{Volume, } V_R\text{, of the reactor} = 400 \text{ m}^3$$

$$\text{Flow rate, } Q\text{, of process water} = 200 \text{ m}^3/\text{h}$$

Note: The flow rate of the feed pump delivering the chemical is very small compared to the wastewater flow rate and can thus be ignored.

(b) Write a process model to describe this situation if the catalyst is also undergoing deactivation at a rate r_c (mol/L-h) within the reactor.

2-7 An experiment is performed in the laboratory to investigate the effectiveness of oxidizing a hazardous chemical in a newly developed process. The test solution is added to a vessel with no inflow nor outflow, and the concentration of the hazardous chemical is measured with time. The results obtained at two different starting concentrations are shown below.

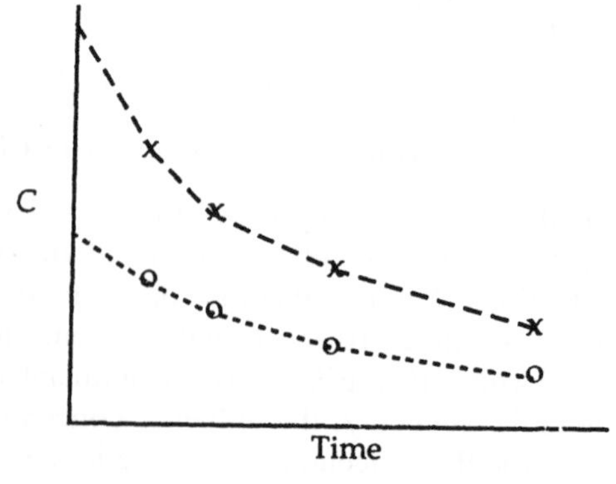

Write the material balance to describe such a system and discuss what *general* inferences you could make from inspection of these data with regard to the variation in reaction rate (this requires no inferences about specific kinetic formulations) over time in each case.

2-8 A dye is discharged uniformly across the width of a channel at a depth y. Because of dispersion, the dye will spread in the vertical (y) and horizontal (x) direction as it travels downstream with the flow (advective transport). The purpose of the study is to measure dye concentration downstream and at different depths and from these data to determine the coefficients of dispersion in each direction. Set up a differential equation describing the concentration of dye as a function of time and space if horizontal transport is by advection and dispersion and vertical transport is by dispersion only. Discuss your answer.

2-9 A schematic of a hollow fiber membrane used for water purification is shown below.

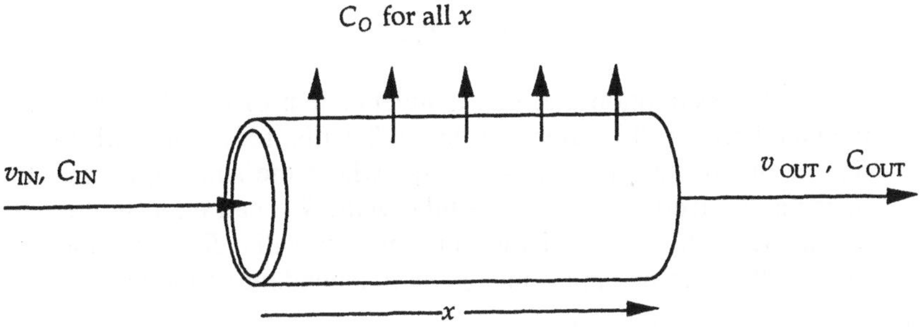

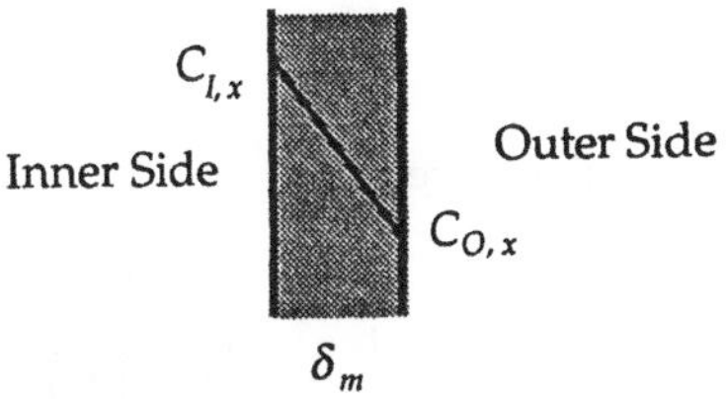

The flow of water is laminar along the length of the tube (no dispersion). In response to the applied pressure, product water flows radially through the walls of the tube, thus producing purified water. Most of the contaminant is rejected by the membrane, but the membrane is not completely impervious to contaminant molecules. For membranes containing very small pores (nanometers or less in diameter), contaminant transport through the membrane can be modeled simplistically as a diffusion-driven process (see the cross section of the wall). In this model, the pressure-driven, advective transport of pure water is an independent process and does not account for transport of contaminant. The steady-state diffusion flux (see Chapter 4 for more detail) is given by

$$N_m = \mathfrak{D}_m \frac{C_{I,x} - C_{O,x}}{\delta_m}$$

in which N_m is the flux ($ML^{-2}t^{-1}$) through the membrane, $\mathfrak{D}_m$ is the diffusion constant within the membrane material, $C_{I,x}$ is the contaminant concentration on the inner side of the membrane at distance x, $C_{O,x}$ is the contaminant concentration on the outer side of the membrane, and

δ_m is the thickness of the membrane (see the note below). Assume that $C_{O,x}$ is the same down the entire length of the membrane tube. Develop a mathematical model and show the general shape of the function describing the contaminant concentration on the inner side of the membrane as a function of membrane length, x. Do this first by assuming that v_x is a constant. Repeat by assuming that the loss of water through the membrane is linear with length such that

$$v_x = v_{\text{IN}} - \kappa x$$

Note: The assumption that flux is independent of radial position within the membrane wall is often justifiable. That is, for a thin wall, the radial flux at steady state is $R_I N_I = R_O N_O$, where the subscripts refer to the inside and outside radius of the tube wall. We can express the outside flux as $N_O = N_I(R_I/R_O)$. However, for thin walls, $R_I \approx R_O$ and we see that the flux can be approximated as independent of radius.

2-10 Another aspect of membrane treatment is the development of a concentration polarization layer on the inner edge of the membrane wall as depicted below.

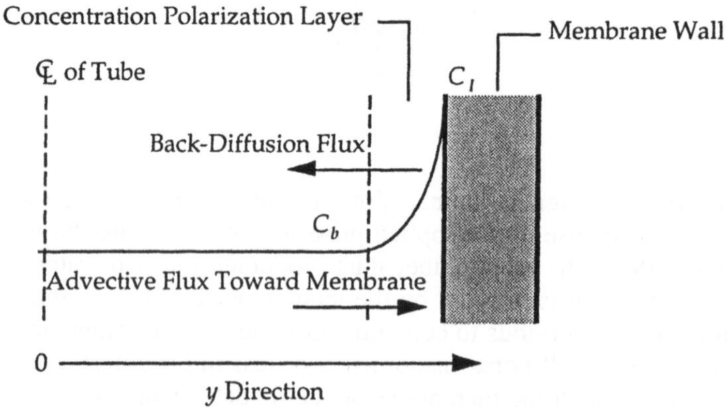

The concentration is higher at the wall due to the rejection of contaminant by the membrane. The concentration is thus higher at the inner edge of the wall than in the bulk water inside the membrane. This concept is used to estimate the velocity, v_y, of the water moving toward the membrane. There is an advective flux of contaminant within the concentration polarization layer in the direction of the wall given by $v_y C_y$. In addition, we have a contaminant backflux into the tube which is predicted by Fick's law of diffusion:

$$N = -\mathfrak{D}_l \frac{dC_y}{dy}$$

where $\mathfrak{D}_l$ is the diffusivity of the contaminant in the water. Set up a differential equation to describe the net flux toward the wall and find the contaminant concentration as a function of distance, y, in the concentration polarization layer. If we know the value of C_l (based on an estimated molecule packing near the wall) and C_b, show how the model can be used to calculate the water flux. State all assumptions and discuss what you think the profile of C_y may look like.

2-11 Spherical-shaped "blobs" of a non-aqueous-phase liquid (NAPL) have contaminated a saturated subsurface system. This NAPL can dissolve into the groundwater. At the microscale of the interstitial pore volume, dissolution is a diffusion process, as shown below.

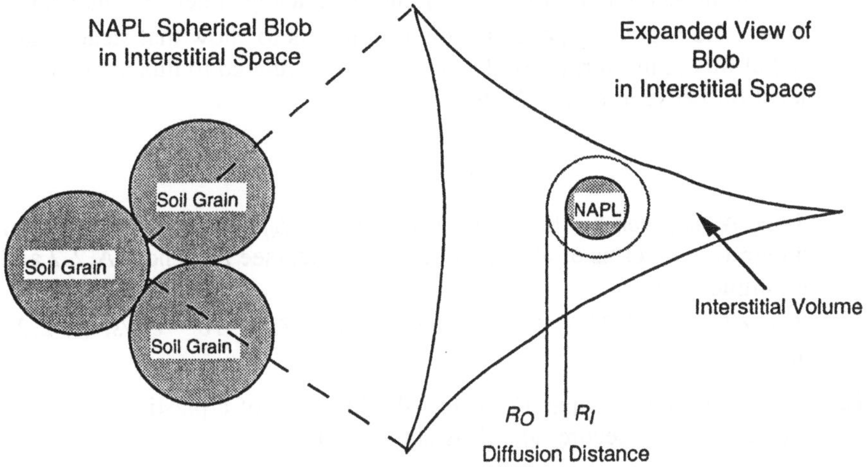

The diffusion flux in the radial direction is given by

$$N_r = -\mathfrak{D}_l \frac{\partial C}{\partial r}$$

The following assumptions are made: (1) the volume of the NAPL sphere is very small relative to the interstitial volume, V_I; (2) the solubility limit, C_S, of the NAPL in water is reached in the water just at the outer-edge surface of NAPL sphere ($r = R_I$); (3) the NAPL diffuses through the distance between $r = R_I$ and $r = R_O$ and then becomes well mixed into the interstitial water; and (4) the initial diffusion flux can be approximated as steady state (this is known as the *quasi-steady-state* assumption) because C_S is constant, the concentration in the bulk interstitial water is close to zero, and R_I does not change rapidly (so the diffusion distance does not change significantly over the period of concern).

(a) Find the initial rate of change in concentration (mg/cm^3-s) in the interstitial volume, V_I, caused by diffusion of the NAPL if C_S = 10 mg/L, $\mathfrak{D}_l = 1 \times 10^{-5}$ cm^2/s (typical value), R_I = 0.01 cm, R_O = 0.015 cm, and V_I = 0.01 cm^3.

(b) Estimate the initial rate of change of NAPL radius, dR_I, if the density of the NAPL is 0.8 g/cm^3.

(c) Assuming that the flux remains constant (i.e., that the NAPL radius is changing very slowly and that the change in concentration of NAPL at R_O is small enough to ignore), find the concentration in the interstitial volume after 10 min.

(d) Discuss how you may still use the quasi-steady-state model to estimate flux and account for the effect of an increasing concentration in the interstitial volume (C_I) and decreasing sphere radius (R_I).

Hint: The beginning point for this problem is to show that steady-state radial flux at the surface of the sphere can be related to flux at any other location between R_I and R_O by

$$N_{R_I} R_I^2 = N_r (r)^2$$

Since N_r is given by the diffusion equation, N_{R_I} can be expressed after integration in terms of R_I, R_O, C_s, C_I, and $\mathfrak{D}_l$ (see Appendix A.2.2 and Example 2-4).

Answers: *(a)* 3.77×10^{-6} *mg/cm*^{3}*-s; (b)* 3.75×10^{-8} *cm/s; (c) 2.27 mg/L.*

2-12 The experimental apparatus to test the integrity of a plastic liner material for use in a secure landfill is shown below.

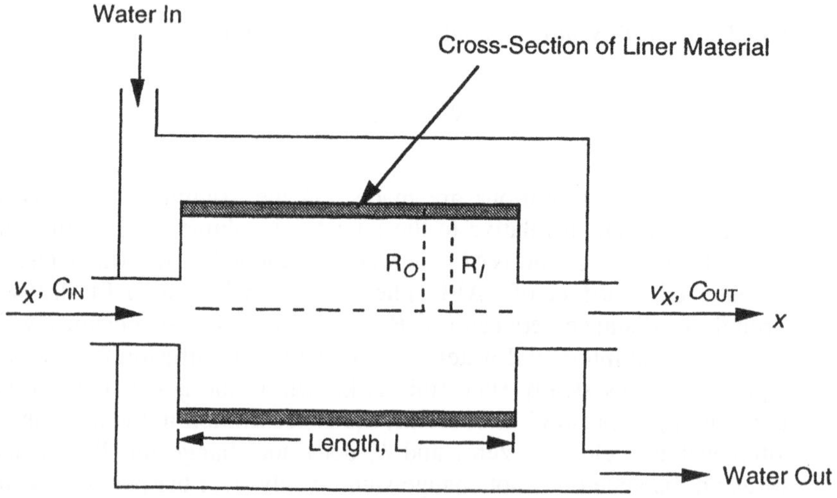

A piece of liner material is first preexposed to simulate environmental conditions of concern. The material is then rolled up to form a hollow tube by sealing it to plastic disks at either end. A hole in each disk allows the solution containing the test chemical to pass through the tube. The tube is jacketed by a vessel through which water is passed continuously. The flow of water around the outside of the tube prevents the concentration of the test chemical, which may diffuse through the liner, from increasing significantly above zero during the test run. A test is conducted by passing a solution of the test chemical through the tube and measuring the concentrations, C_{IN} and C_{OUT}. Assuming advective flow within the tube and steady-state diffusion through the liner, develop a mathematical model that will allow calculation of the effective diffusion coefficient through the plastic liner.

Answer: $$\mathfrak{D}_l = \frac{-vR_I^2\left(\ln \dfrac{R_O}{R_I}\right)\left(\ln \dfrac{C_{OUT}}{C_{IN}}\right)}{2L}$$

2-13 The material balance required to describe contaminant transport in the subsurface from the onset of leakage from a storage tank was presented in Example 2-5. The concepts of unsteady-state transport and boundary conditions were introduced. Suppose that the water is moving downward with a velocity of 10 m/day and we wish to find the concentration at a distance of 10 m from the ground surface as a function of time. Compare the results for assumption of a type 3 boundary condition with that for a type 1 boundary condition at the surface for the following values of the dispersion coefficient: (1) $\mathfrak{D}_d = 0.5$ m^2/day and (2) $\mathfrak{D}_d = 10$ m^2/day. The reader is encouraged to try other combinations of parameter values to gain more insight into the nature of such unsteady-state solutions. [*Note:* The erf (z) and erfc (z) can be easily calculated using a spreadsheet. The important properties of error functions are erf $(-z) = -\text{erf}(z)$, erfc $(-z) = 1 - \text{erf}(-z)$, and erfc $(-z) = 1 + \text{erf}(z)$.]

APPENDIX A.2.1 USE OF VECTOR NOTATION IN ANALYSIS OF TRANSPORT THROUGH CONTROL VOLUMES

The flux, N_i, of component i as it passes through a finite element of a control volume is shown in Figure 2-3. For notational convenience and simplicity in this illustration, we drop the subscript i and for single-component systems consider a flux that is uniform over the area of the control volume. Then, in the simplified representation given below, flux enters through surface area A_1 and leaves through surface area A_2.

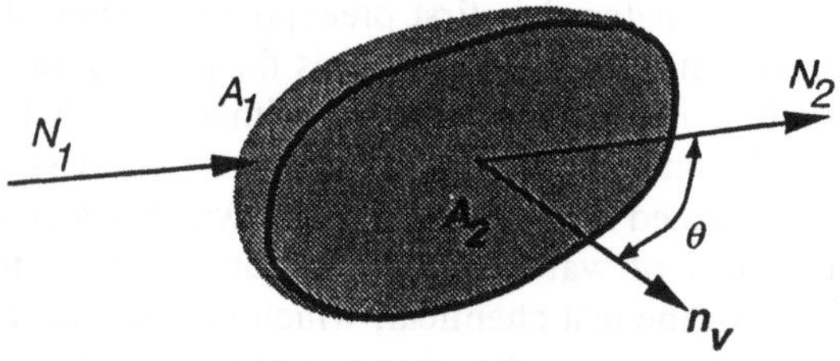

In developing the integral form of the continuity equation we need to evaluate the net molar rate of component flow into the control volume. By definition, flux is defined normal to a surface, and we must therefore consider either the area normal to the direction of flux or the component flux normal to the actual surface area. Both approaches produce the same mathematical result; that is, the flux is $N \cos \theta$, or, in vector notation, $\boldsymbol{N} \cdot \boldsymbol{n}_v$. For convenience of notation, we use N to denote this vector. Each of the surfaces of the control volume has two sides, one facing in and the other facing out. The surface area is thus a *vector quantity* also because it has *direction*. The convention generally adopted in considerations of fluid flow through control volumes is that an area vector points outward from the control volume. The vectors in our example are thus:

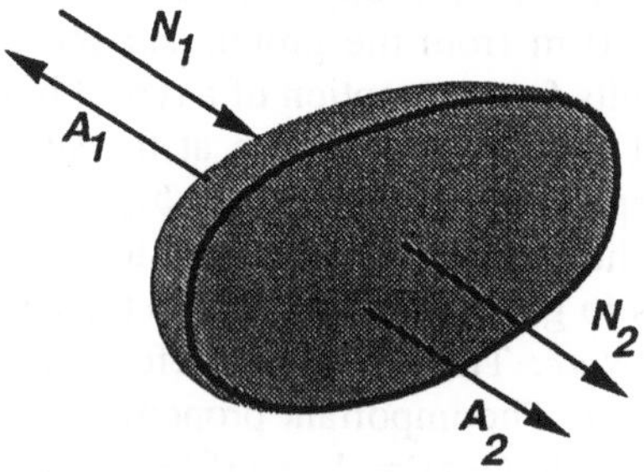

To determine the net molar rate into the control volume we have, in scalar terms,

$$(\text{molar rate in} - \text{molar rate out}) = N_1 \cos \theta\, A_1 - N_2 \cos \theta\, A_2 \quad \text{(A.2.1-1)}$$

or, in vector terms,

$$(\text{molar rate in} - \text{molar rate out}) = -\boldsymbol{N}_1 \cdot \boldsymbol{A}_1 - \boldsymbol{N}_2 \cdot \boldsymbol{A}_2 \quad \text{(A.2.1-2)}$$

Note that the signs in Equation A.2.1-2 originate from the cosine of the angle between N_1 and A_1 being -1, and that of the angle N_2 and A_2 being $+1$.

Finally, we write

$$\begin{aligned}&(\text{molar rate in} - \text{molar rate out})\\ &\quad = -(N_1 \cdot A_1 + N_2 \cdot A_2) = -\sum_{Sc} N \cdot A\end{aligned} \tag{A.2.1-3}$$

Equation A.2.1-3 states that we sum the dot products, $N \cdot A$, to obtain the component's net molar flow rate. If the flux in and out are equal (i.e., no reaction and no accumulation), the dot product is zero. If the dot product is positive, there is a net flux out (because of the negative sign in front of the summation), and if the dot product is negative, there is a net flux in. We have for this illustration assumed that flux is uniform over the area of the control surface. Equation A.2.1-3 can be written in more general form, however, to account for the possibility that flux is not constant over the entire control surface. Thus we have

$$(\text{net molar rate in}) = -\sum_{Sc} N \cdot dA \tag{A.2.1-4}$$

which is the equivalent of Equation 2-8.

APPENDIX A.2.2 SHELL BALANCE APPROACH TO DEVELOPMENT OF THE CONTINUITY EQUATION IN ONE DIMENSION

A more intuitive approach to the development of the continuity relationship (Equation 2-16) may be made by considering a material balance on a finite volume. This is referred to as the *shell balance* approach. Consider again, for example, Figure 2-3. The *discrete* elemental control volume, ΔV, has the finite dimensions $\Delta x = x_2 - x_1$, $\Delta y = y_2 - y_1$, and $\Delta z = z_2 - z_1$. Now suppose that the volume represented in Figure 2-3 is sealed in the x–y and x–z planes, so that flux of component i occurs only across the finite element area $\Delta A_{yz} = \Delta y\, \Delta z$ in a direction perpendicular to the y–z plane, as in Figure A.2.2-1.

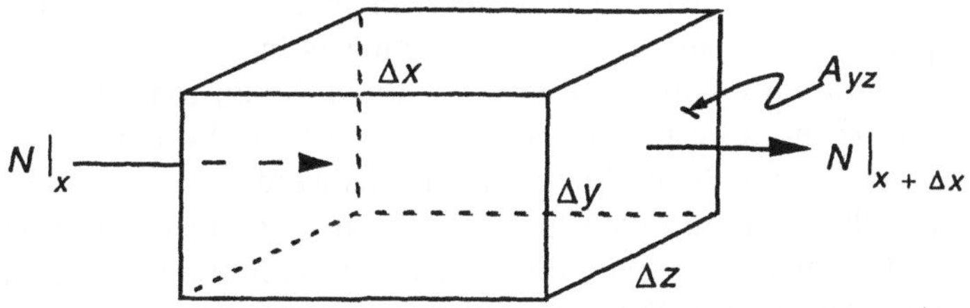

Figure A.2.2-1 Discrete element control volume.

Assuming that A_{yz} and ΔV are constant, we can write a mass balance for the finite volume, ΔV, from Equation 2-5, again dropping the subscript i to simplify notation for a single-component illustration:

$$[\text{mass in}] - [\text{mass out}] + \begin{bmatrix}\text{mass change}\\ \text{due to reaction}\end{bmatrix} = \begin{bmatrix}\text{mass change}\\ \text{within } \Delta V\end{bmatrix} \qquad \text{(A.2.2-1)}$$

or, in mathematical terms,

$$(N|_x - N|_{x+\Delta x})A_{yz}\,\Delta t + r\,\Delta V\,\Delta t = (C|_{t+\Delta t} - C|_t)\,\Delta V \qquad \text{(A.2.2-2)}$$

where the respective terms have the following units:

$$(ML^{-2}t^{-1})(L^2)(t) + (ML^{-3}t^{-1})(L^3)(t) = (ML^{-3})(L^3)$$

Equation A.2.2.-2 can be divided through by A_{yz}, Δt, and Δx, to yield

$$\left[\frac{N|_x - N|_{x+\Delta x}}{\Delta x}\right] + r = \left[\frac{C|_{t+\Delta t} - C|_t}{\Delta t}\right] \qquad \text{(A.2.2-3)}$$

where $\Delta V = A_{yz}\,\Delta x$.

We now let the finite volume, ΔV, and the time increment, Δt, shrink to zero to develop a continuity equation. Note that in Equation A.2.2-3 the terms in brackets resemble that in the definition of the derivative of any function, $f(u)$: namely,

$$\frac{d\phi\,(u)}{d(u)} = \lim_{\Delta u \to 0}\left[\frac{\phi\,(u + \Delta u) - \phi\,(u)}{\Delta u}\right] \qquad \text{(A.2.2-4)}$$

Taking the limits of Equation A.2.2-4 as Δx and $\Delta t \to 0$ thus yields

$$-\frac{\partial N}{\partial x} + r = \frac{\partial C}{\partial t} \qquad \text{(A.2.2-5)}$$

The shell balance approach can be used similarly to develop continuity equations for one-dimensional radial transport in spherical and cylindrical coordinates (Equations 2-20 and 2-23). The starting points for the earlier derivation of these continuity equations were the three-dimensional forms of the continuity equation in the respective coordinate systems (Equations 2-19 and 2-22, respectively). These three-dimensional forms were presented without derivation because they require rather tedious, vector mathematics to transform the gradients of flux in the x, y, and z directions to those in the r, θ, ϕ directions (for spherical coordinates) or the r, θ, and z directions (for cylindrical coordinates).

The shell balance approach to development of the continuity equations for spherical and cylindrical coordinates (see Figure 2-5) with *radial transport only* is presented below:

Spherical coordinates:

$$(N|_r - N|_{r+\Delta r})4\pi r^2 + \kappa(4\pi r^2\,\Delta r) = \frac{\partial(C \cdot 4\pi r^2\,\Delta r)}{\partial t} \quad \text{(A.2.2-6)}$$

Dividing through by $4\pi r^2\,\Delta r$ and taking the limit as $\Delta r \to 0$ gives

$$-\frac{1}{r^2}\frac{\partial}{\partial r}(r^2 N) + \kappa = \frac{\partial C}{\partial t} \quad \text{(A.2.2-7)}$$

which is equivalent to Equation 2-20.

Cylindrical coordinates (assume unit length along the axis of the cylinder):

$$(N|_r - N|_{r+\Delta r})2\pi r + \kappa(2\pi r\,\Delta r) = \frac{\partial(C \cdot 2\pi r\,\Delta r)}{\partial t} \quad \text{(A.2.2-8)}$$

Dividing through by $2\pi r\,\Delta r$ and taking the limit as $\Delta r \to 0$ gives

$$-\frac{1}{r}\frac{\partial}{\partial r}(rN) + \kappa = \frac{\partial C}{\partial t} \quad \text{(A.2.2-9)}$$

which is equivalent to Equation 2-23.

The shell balance is a useful intuitive approach for describing axial flux along the length (x) of a hollow cylinder (e.g., a pipe or a fiber membrane) of radius, R, coupled with radial transport through the wall (see Figure 2-6):

$$(N|_z - N|_{z+\Delta z})\pi R^2 + \kappa(\pi R^2\,\Delta z) + N|_R \cdot 2\pi R\,\Delta z = \frac{\partial(C\pi R^2\,\Delta z)}{\partial t}$$

(A.2.2-10)

Dividing through by $\pi R^2\,\Delta z$ and letting $\Delta z \to 0$ gives

$$-\frac{\partial N_x}{\partial x} + \kappa + \frac{2}{R}N_r|_R = \frac{\partial C}{\partial t} \quad \text{(A.2.2-11)}$$

which, noting that radial flux is directed inward here rather than outward, is equivalent in form to Equation 2-27. It is possible to use a steady-state approximation for radial diffusion *without* reaction in spherical and cylindrical coordinates in certain situations (see Problems 2-10 and 2-11). Radial diffusive flux through a sphere is described by Fick's law. According to Equation A.2.2-7, the steady-state diffusive flux at any specific radial position of interest, r_1, is related to that at any general value of the position variable, r, by

$$(N)_{r_1} = \left(\frac{r}{r_1}\right)^2 (N)_r = -\left(\frac{r}{r_1}\right)^2 \mathfrak{D}_l \frac{dC}{dr} \qquad \text{(A.2.2-12)}$$

If concentrations are specified at the two specific radial positions r_1 and r_2, the flux at r_1 can be obtained by integration of Equation A.2.2-12.

Steady-state diffusion *without* reaction in cylindrical coordinates (see Example 2-4 and Problem 2-11) can be analyzed in the same way as for spherical coordinates. Beginning with Equation A.2.2-9, the following expression is obtained:

$$(N)_{r_1} = \frac{r}{r_1} (N)_r = -\frac{r}{r_1} \mathfrak{D}_l \frac{dC}{dr} \qquad \text{(A.2.2-13)}$$

Equation A.2.2-13 can be applied, for example, in Equation A.2.2-11, where an expression is needed for radial flux at the inner edge of a pipe or membrane wall ($r_1 = R_I$). The concentration at the inner edge is the same as that throughout the internal cross section at location x. If a steady-state concentration is specific at the outer edge (e.g., concentration may equal zero at all x), it is possible to integrate Equation A.2.2-13 and obtain a simple expression for $N_r|_{R_I}$.

3

MACROTRANSPORT PROCESSES

3.1 TRANSPORT PROCESSES

Changes in the concentrations of substances in environmental systems can be caused by two principal types of processes: transport and reaction processes. Transport processes are in turn divided into two broad categories: *macroscopic or macrotransport processes*, the subject of this chapter, *and microscopic or microtransport processes*, the subject of Chapter 4. Figure 3-1 illustrates these transport processes schematically in two dimensions. The fluid element in this illustration contains a dissolved substance that is moved by macroscopic and microscopic processes in both the x and y directions in the bulk fluid, and by microscopic processes in the y direction at the boundary between the element and a solid surface. The relative magnitudes of the several bulk-phase transport processes for typical flow conditions are suggested by the widths of the corresponding arrows.

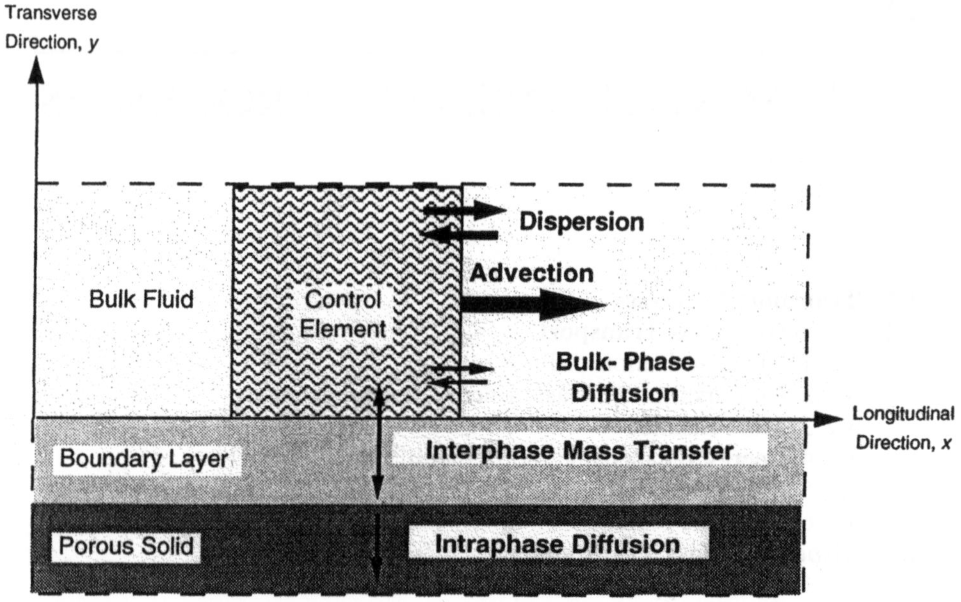

Figure 3-1 Mechanisms of mass transport and transfer.

Macrotransport is comprised principally of two processes: *advection* and *dispersion*. *Advective transport* is the bulk movement of substance mass as a direct result of cocurrent movement of the fluid elements in which it is contained; that is, substance mass is carried along with fluid mass. This type of transport, referred to in Chapter 2 as *plug flow*, is characterized directly by the bulk flow conditions of a system.

Dispersive transport can result from several different physical phenomena operating at several different scales, and can thus be described by several different names. The dispersion represented in Figure 3-1, termed *turbulent* or *eddy dispersion*, involves the movement of substance mass as a result of the macroscopic exchange of momentum between fluid elements in a turbulent flow field. Other types of dispersion include *Taylor dispersion* and *hydrodynamic* (or *mechanical*) *dispersion*. These three different types of dispersion can all be described by the same mathematical relationship, which, fortuitously, has the same form as that for *bulk-phase diffusion*, a microscopic transport process. As pictured in Figure 3-1, bulk-phase diffusion can function in concert with advection and dispersion. The diffusion process is *molecularly* driven rather than fluid-element driven, however, and thus differs mechanistically from turbulent dispersion.

At steady state, bulk-phase diffusion is described by Fick's first law, which states that solute flux is proportional to the spatial gradient of solute concentration. This functional relationship derives from the assumption that the movement of molecules under the influence of Brownian forces in a quiescent space is essentially random and that the distribution of forces acting on the molecules is *normal*, or *Gaussian*. Net movement of the molecules of substance i between two points separated by a distance dx will thus occur only if there is a net difference in the concentration, C_i, of molecules at those two points, and thus a gradient in C_i along dx. In the introduction to Equation 2-34 it was noted that dispersion caused by fluid element movement is also proportional to a spatial gradient in solute concentration. The functional form of the relationship for solute movement caused by diffusive and dispersive transport is therefore the same (i.e., the flux of solute i in one dimension along a path of length dx is in both cases given by $N_i \propto dC_i/dx$). Although two distinctly different transport processes may have identical mathematical forms and thus be treated as additive, they have different roles and exert remarkably different effects in environmental systems. Most important, the coefficients of proportionality between flux and spatial concentration gradient (i.e., the diffusion and dispersion coefficients) generally differ by orders of magnitude. The terms *diffusion* and *dispersion* should thus not be used casually, and certainly not interchangeably.

We noted that dispersion is generally classified as *Taylor*, *turbulent* (*eddy*), or *hydrodynamic* (*mechanical*). *Taylor* dispersion occurs in laminar flow in pipes and narrow channels, flow characterized by parabolic distributions of velocities resulting from the action of fluid drag forces at boundary surfaces (i.e., different rates of fluid element movement along parallel flow paths). These parabolic velocity distributions in turn result in parabolic-like distributions of solute concentration fronts in the *axial* (*longitudinal*) direction, creating concentration gradients in the *radial* (*transverse*) direction (i.e., normal to the principal flow direction). *Diffusion* of solute in the transverse direction thus occurs as the bulk laminar flow proceeds in the longitudinal direction. *Taylor dispersion coefficients* are unique in two regards: (1) they are inverse functions of the *free-liquid molecular diffusion coefficient* of a solute, and (2) they can be calculated explicitly from known properties of the flow field and the diffusion coefficient.

Turbulent or *eddy* dispersion, the type illustrated in Figure 3-1, is an important transport process in large channels, rivers, streams, and lakes. Dispersion of this type results from velocity fluctuations created by fluid turbulence acting across large advection-dominated flow fields. These velocity fluctuations create momentum and solute mass exchanges among fluid elements. The velocity fluctuations are distributed in an essentially Gaussian manner in terms of their ensemble averages, and the mass exchanges of solute are thus similarly distributed. *Eddy dispersion coefficients* are related to fluid viscosity and are generally orders of magnitude larger than molecular diffusion coefficients. They are also highly system specific and must therefore be evaluated experimentally.

Finally, *hydrodynamic* and *mechanical* are two terms commonly used to define types of dispersion phenomena that are important in flow through porous media (e.g., in sand or carbon filters and in groundwater systems). In some circumstances, hydrodynamic or mechanical dispersion is similar to Taylor dispersion. Such dispersion may occur, for example, in small pores and fissures in fractured rock. At a somewhat larger scale, hydrodynamic dispersion may result from more irregular disruptions of flow around obstacles such as sand, carbon grains, or soil aggregates. The irregularly varied flow paths that result in such cases lead to varied flow velocities, which like velocity fluctuations in large-scale turbulent flow fields are normally distributed (Gaussian) over large travel distances. The distributed flow paths and velocities resulting from such microscale "eddy" developments lead in turn to distributed solute concentration fronts, and thus to solute flux in directions normal to the irregular flow paths. In yet other circumstances, hydrodynamic dispersion refers to various combinations of these different laminar flow types of dispersion. Hydrodynamic dispersion coefficients are generally much larger than molecular diffusion and Taylor dispersion coefficients, in many cases being the same order of magnitude as eddy dispersion coefficients.

As a general rule, we can say that solute spreading effects in *quiescent* fluid systems result from *molecular diffusion*, while in any system that involves *advective flow, either laminar or turbulent*, such spreading is the result of *dispersion*. If advective flow is *turbulent*, the dispersion can be referred to as *eddy dispersion*, whereas spreading effects in macroscopically *laminar* advective flow are attributable to *hydrodynamic dispersion*.

Turbulent dispersive transport and bulk-phase diffusion are both pictured in Figure 3-1 as acting only in the x direction, an intentionally simplistic representation. Dispersion and bulk diffusion can and generally do occur in three dimensions. Moreover, the extent of dispersion taking place in each direction is often unequal. Temperature stratification of fluid in the y direction, for example, would cause turbulence, and thus dispersion, in this direction to be different from that in the x direction.

Transport processes that limit local reactions within a system by limiting mass transfer are termed microtransport processes. In Figure 3-1 these are depicted as *interphase mass transfer* and *intraphase diffusion*. Interphase mass transfer involves microtransport across a boundary between phases or, in the example above, across the hydrodynamic boundary between a fluid and a solid. Diffusion is the only means of transport through this boundary. Interphase mass transfer must be treated differently from simple molecular diffusion, however, because the length of the diffusional path generally is dependent on the characteristics of the macrotransport processes and is not measurable. As illustrated in Figure 3-1, molecular diffusion may be important also after a substance is transported across the interface and enters a second, nonhomogeneous phase (e.g., a matrix of solid and fluid volumes). This transport process is referred to as *intraphase diffusion*. It may occur along solid surfaces (*surface diffusion*) or within the fluid contained in the pore spaces of the matrix (*pore diffusion*).

Each of the four mass transport mechanisms discussed above has a unique mathematical representation as a flux term, N_i, in the point form of the continuity equation developed in Chapter 2:

$$-\nabla \cdot N_i + r_i = \frac{\partial C_i}{\partial t} \tag{2-16}$$

One critical step in process dynamics is to express Equation 2-16 in a form that can be integrated to provide information on the spatial and temporal variations of the concentration of a substance. This requires specification of appropriate mathematical expressions for both transport and transformation processes, the latter being represented in Equation 2-16 by the reaction rate term, r_i.

In the remainder of this chapter we develop a characterization of fluid flow and macrotransport processes employing the principles of conservation of mass, energy, and momentum. While we will focus on macrotranspórt processes, the mass, energy, and momentum balances used for describing fluid flow and the material balances used for describing constituent transport within flowing fluids are derived at a point in space (i.e., they are the point forms of the balance equations). To appreciate this blending of macro- and microscale concepts, consider briefly the flux term in the *point form* of the continuity equation given by Equation 2-16. This equation, although it represents transport and reaction processes occurring in a *microscopic* fluid control volume (i.e., a point in space), is a required starting point as well for description of *macrotransport* movements of constituents in fluids. Because macrotransport flux is produced by advection and dispersion, velocity and dispersivity are usually averaged over some length scale. This perspective is essential for our subsequent developments of mass and momentum balances on fluids in which constituents are transported. We write the *point forms* of the mass and momentum balances so we can describe fluid velocity and pressure in the *space continuum*. For practical computations, however, it ultimately becomes necessary to *average* these velocities and pressures over some *finite scale* that is consistent with the particular macrotransport processes being described.

Detailed discussions of microtransport processes building on the descriptions of fluid flow presented here will be left to Chapter 4. Following Chapter 4, the energetics and equilibrium principles underlying transformations of substances dissolved within (homogeneous systems) or in contact with (heterogeneous systems) fluids are discussed in Chapters 5 and 6. Similarly, in Chapters 7 and 8 we consider rate relationships for homogeneous and heterogeneous systems. Finally, in Chapters 9, 10, 11, and 12 we bring all these elements together in a comprehensive characterization and discussion of reactor analysis.

3.2 ADVECTIVE MASS TRANSPORT

Advective transport involves cocurrent movement within a fluid element that leads to flux given by

$$N_{x,i} = v_x C_i \tag{3-1}$$

where v_x is the velocity of fluid flow in the x direction and C_i is the concentration of substance i. For the ensuing discussion, it will be helpful to consider fluid velocity as a volumetric flux of fluid, the dimensions of which are volume of fluid per unit area per unit time ($L^3L^{-2}t^{-1} = Lt^{-1}$). Thus the molar flux, $N_{x,i}$ (moles of i · area^{-1} · time^{-1}), of component i in a given element of fluid, is the product of fluid flux and the concentration of substance i within that element. To describe a three-dimensional situation, we use the generalized vector form

$$\mathbf{N}_i = \mathbf{v}C_i \tag{3-2}$$

where $\mathbf{N}_i$ and $\mathbf{v}$ are the resultant vectors of flux and velocity, respectively, in three dimensions.

3.2.1 Characterization of Flow

The character of fluid flow, and thus of cocurrent advective transport of a dissolved or suspended substance, varies widely among different types of systems. It may, for example, be *laminar* or *turbulent*, *steady* or *unsteady*, and *uniform* or *nonuniform*. Laminar flow, as the name implies, involves the transport of fluid elements within discrete parallel laminae along smooth flow paths. The greater the viscosity of the fluid and the lower the velocity at which it moves, the greater the tendency for flow to be laminar. Such flow eventually becomes unstable as (1) viscosity decreases, (2) velocity increases, and/or (3) the dimensions and scale of the flow field increase. When this occurs, individual fluid elements comprising the flow move in irregular pathways, with a resulting irregular transfer of mass and momentum between fluid elements and between different coordinate points in the system. From the fluid-flow perspective, this causes a larger and more rapid proportional conversion of the inherent mechanical energy, or *head*, of a system to thermal energy, or frictional *head loss*, than occurs under conditions of laminar flow, a significant consideration for process design and operation. Rate of head loss varies linearly with fluid velocity for laminar flow and approximately as the square of velocity for turbulent flow.

Most natural water systems and treatment operations involve turbulent flow conditions and, therefore, irregular patterns of advective mass transport. Irrespective of whether flow is laminar or turbulent, however, Newton's laws of motion holds for every fluid element at every time, and the integrity of appropriate continuity relationships must be maintained. As discussed in more detail in ensuing sections, certain assumptions and approximations can be made in the engineering of real systems to facilitate characterization and modeling of mass transport phenomena in the presence of turbulent flow fields on both macroscopic and microscopic scales.

The terms *steady* and *unsteady* flow relate to the *temporal* variability of advective velocity and other flow conditions at any coordinate point in a system, while *uniform* and *nonuniform* flow address the *spatial* variability of flow conditions. Steady flow exists when all pertinent flow conditions at any point in the flow field are constant over time, and flow is unsteady when such conditions vary with time. These qualifications relate to both the magnitude and direction of the advective velocity vector, $\boldsymbol{v}$. Thus, in terms of time, t, and an arbitrary space variable, ζ, the specifications for different flow conditions are given as:

$$\text{Steady flow:} \quad \frac{\partial \boldsymbol{v}}{\partial t} = 0 \tag{3-3}$$

$$\text{Unsteady flow:} \quad \frac{\partial \boldsymbol{v}}{\partial t} \neq 0 \tag{3-4}$$

$$\text{Uniform flow:} \quad \frac{\partial \boldsymbol{v}}{\partial \zeta} = 0 \tag{3-5}$$

$$\text{Nonuniform flow:} \quad \frac{\partial \boldsymbol{v}}{\partial \zeta} \neq 0 \tag{3-6}$$

The definition of steady flow given by Equation 3-3 must be generalized somewhat to accommodate the irregular movement of fluid elements in turbulent advective flow fields. In this case it is necessary to consider the time variability of the temporal mean velocity in each dimension; for example,

$$v_{x,t} = \frac{\int_{t_1}^{t_2} v_x \, dt}{\int_{t_1}^{t_2} dt} \tag{3-7}$$

For turbulent conditions, then, steady flow implies that $\partial v_{x,t}/\partial t = 0$.

3.2.2 Conservation of Mass, Energy, and Momentum

Fluid motion or *fluid kinematics* is governed by the laws of conservation of mass, energy, and momentum. This motion determines the advective mass transport of a material by cocurrent flow both within the bulk of the fluid and in the boundary layer between the fluid and a surface. With regard to the latter, the law of conservation of momentum is particularly important for understanding the nature of boundary layers and interphase mass transport. For these reasons we undertake a brief review of fluid kinematics.

The control volume concept is the cornerstone of fluid kinematic analysis. This concept was introduced in Chapter 2 to develop a *mass* or *material* balance for a component within a fluid and, subsequently, the differential and integral forms of the *mass continuity equation*. Continuity equations for mass, energy, and momentum of the fluid itself are derived similarly. Rather than describe the motion of a fluid by tracking individual fluid particles as they move through time and space, the *Lagrangian perspective for fluid movement*, the selection of a control volume fixes a position in space at which to account for the fluid passing through it; the latter is referred to as the *Eulerian perspective*. The differences between these two perspectives are evident in the situation described in Example 3-1.

Example 3-1

- **Situation:** Treated wastewater is discharged to the ocean through a submerged outfall located several kilometers offshore. The outfall consists of a series of diffusers that direct the discharge horizontally. New effluent standards are to include consideration of aquatic toxicity. The responsible regulatory agency in this case has experience with setting standards for discharges to rivers but not to oceans. For river discharges, a "mixing zone" is allowed in determining the dilution of the wastewater with the river water. Standards are set based on a specified percent survival of the test organisms after selecting a worst-case dilution factor of the wastewater with the stream based on historical stream flow data (e.g., the consecutive 7 days of low flow occurring once in 10 years, or "7Q10"). However, the discharge of wastewater into the ocean presents an entirely different hydrodynamic situation.
- **Question(s):** What factors should be considered in analysis of mixing and dilution to set an aquatic toxicity standard?
- **Logic and Answer(s):**

1. Both the Eulerian and Lagrangian perspectives of a control volume are important when analyzing the potential aquatic toxicity effects of a discharge plume rising to the ocean surface, as illustrated below.

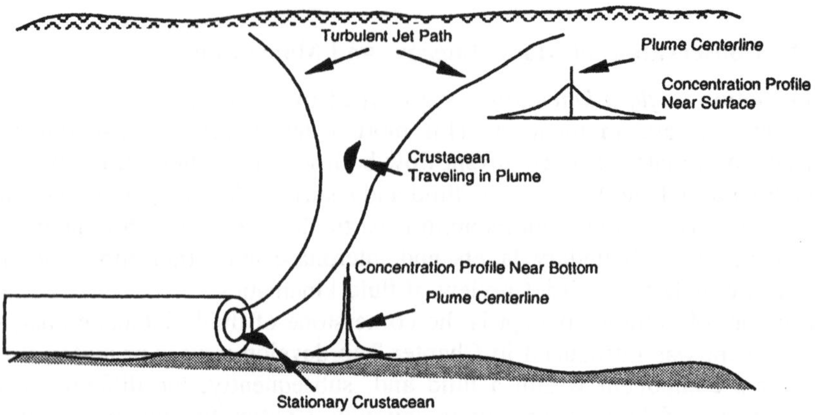

2. The size and shape of the plume depicted above depends on the initial velocity and density of the jet (advective flow), flow induced by ocean currents, and macroscale turbulence (dispersion) that causes mixing of the wastewater into the ocean water. At any vertical distance, the concentration of potential toxicants (C) in the plume decreases from the centerline due to dispersion, in the general fashion shown by the inset graph.

3. The Eulerian perspective requires the selection of a control volume fixed in space. This viewpoint is important for analysis of the exposure of a stationary crustacean to the wastewater discharge. If the concentration of potential toxicants in the discharge is constant with time, the crustacean that remains near the discharge point is constantly exposed to a fixed concentration.

4. In contrast, the Lagrangian perspective is important for assessing the exposure of the traveling crustacean to toxicants because this organism travels with the plume and thus experiences a decreasing concentration of toxicants with time of travel to the surface. Accordingly, a toxicity standard for the fixed crustacean requires determination of the initial dilution of wastewater with ocean water and an emphasis on long-term chronic exposure, whereas a toxicity standard for the traveling crustacean requires devising a short-term toxicity test (i.e., an acute toxicity test) that simulates the change in dilution of the wastewater over the time of rise of the plume. Further illustration of use of the Lagrangian perspective is provided in Problem 3-8.

5. Mathematical models to describe rising plumes clearly must incorporate mass and momentum balances on the fluid to represent these two diverse circumstances properly. These fundamentals are discussed next.

A macroscopic control volume is shown in Figure 3-2. In the most general of terms, we are interested in developing mass balance equations for three properties of a fluid; i.e., its *mass*, *energy* and *momentum*. These are *extensive* properties because they are each additive; i.e., the addition of two volume elements of fluid which each contains the same amount of mass, energy or

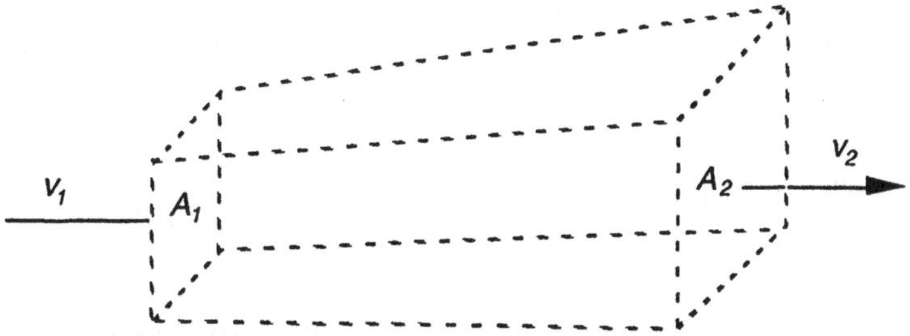

Figure 3-2 Flow through a macroscopic control volume.

momentum doubles these quantities. By contrast *intensive* properties are not additive and do not vary with the size or volume of a system. For example, fluid density expresses the *mass of fluid per unit volume of fluid*. If two volume elements of fluid at the same temperature are added together the density does not change. Temperature and pressure are similarly intensive properties, while heat and mass are extensive. Extensive and intensive properties apply equally well to any substance within a fluid. The number of moles of substance i dissolved in a fluid system is an extensive property of that system, whereas its concentration (*moles per volume of fluid*) is an intensive property of the system.

A generalized conservation or balance equation can be written as

$$\begin{bmatrix}\text{rate of}\\ \text{change of any}\\ \text{extensive property}\\ \text{of a system}\end{bmatrix} = \begin{bmatrix}\text{net rate of}\\ \text{accumulation of mass,}\\ \text{energy, or momentum}\\ \text{within the system}\end{bmatrix}$$

$$+ \begin{bmatrix}\text{rate of mass,}\\ \text{energy, or momentum}\\ \text{output across the boun-}\\ \text{daries of the system}\end{bmatrix} - \begin{bmatrix}\text{rate of mass,}\\ \text{energy, or momentum}\\ \text{input across the boun-}\\ \text{daries of the system}\end{bmatrix} \qquad (3\text{-}8)$$

Equation 3-8 is a statement of the *Reynolds transport theorem*. The left-hand side represents the change in a particular extensive property for an entire *system*, which includes the control volume and its surroundings. It is zero if the extensive property is mass because by definition the mass of fluid cannot change within the system that contains the control volume. However, if the extensive property is energy, the first law of thermodynamics requires that we account for a change in energy, which is the difference between heat addition from the surroundings of the control volume and work done by the control volume on its surroundings. Similarly, if the extensive property is momentum, Newton's second law of motion requires that we account for the time rate of change of momentum, which is equal to the external forces acting on the control volume.

The rate of accumulation and net rate of output of mass, energy, or momentum in Equation 3-8 can be expressed as the product of corresponding extensive and intensive properties:

$$\frac{\partial(\text{extensive property})}{\partial t} = \frac{\partial}{\partial t}\int_{VC} (\text{intensive property})\, dV$$

$$+ \int_{SC} (\text{intensive property } (\boldsymbol{v} \cdot \boldsymbol{n}_v)\, dA \qquad (3\text{-}9)$$

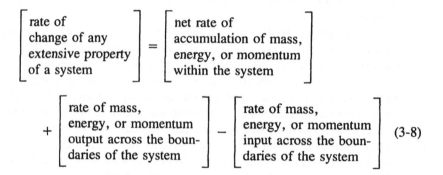

where V_C is the control volume and S_C is the control surface normal to the velocity of the fluid. The extensive property in the first term on the right side of Equation 3-9 is the differential volume, dV. The extensive property in the second term on the right side of the equation is the volumetric rate of fluid transport, which is expressed as the product of the volumetric flux of fluid (i.e., the velocity vector, $\boldsymbol{v} \cdot \boldsymbol{n}_v$, and the differential area normal to flow, dA). The intensive property in the mass balance (or fluid continuity equation) is ρ, the fluid density (mass/volume); that in the energy balance is the product $e_m \rho$, where e_m is the energy per unit mass; and that in the momentum balance is $v\rho$, where v can be thought of as *momentum per unit mass of fluid*.

We can discuss the material balance presented in Chapter 2 for any substance within a fluid in similar terms of extensive and intensive properties. The integral form of the continuity equation for any substance within a fluid, as given by Equation 2-11, expresses the change in the *number of moles of component i per unit time* (i.e., a time rate of change of an extensive property). The corresponding intensive property is the *concentration of component i in the control volume*. For example, if advection accounts for transport of component i through the control volume, the flux of i is given by the product of concentration of i and velocity of the fluid (see Equation 3-1). After rearrangement, the integral form of Equation 2-11 is

$$\int_{V_C} r_i \, dV = \frac{\partial}{\partial t} \int_{V_C} C_i \, dV + \int_{S_C} C_i(\boldsymbol{v} \cdot \boldsymbol{n}_v) \, dA \tag{3-10}$$

The term on the left represents the change in the number of moles of substance i within the system due to some type of reaction; r_i is an intensive property, being some function of concentration. This equation is of the same form as the generic statement of the Reynolds transport theorem for mass, energy, and momentum of the fluid in which the substance is transported. Accounting for transformation of a substance by reaction is, therefore, analogous to accounting for a change of energy in a system due to external work, or a change in momentum caused by external forces.

3.2.3 Mass Balance

If we describe the transport of fluid along velocity streamlines that are normal to the control surface area in Figure 3-2, Equation 3-9 takes the form

$$0 = \frac{\partial}{\partial t} \int_{V_C} \rho \, dV + \int_{S_C} \rho(\boldsymbol{v} \cdot \boldsymbol{n}_v) \, dA \tag{3-11}$$

For steady advective flow, the first term on the right-hand side must be zero, giving

$$0 = \int_{S_C} \rho(\boldsymbol{v} \cdot \boldsymbol{n}_v) \, dA \tag{3-12}$$

This is a formal statement of the mass continuity equation for steady flow. If we analyze mass flow of an incompressible fluid such as water through a control volume under isothermal conditions, mass density is constant. Further, if uniform flow exists over the cross section, Equation 3-12 states that

$$v_\zeta \, dA_1 = v_\zeta \, dA_2 = \text{constant flow rate} \tag{3-13}$$

where ζ is an arbitrary coordinate along a streamline, and the subscripts 1 and 2 refer to two positions normal to the streamlines defining the increments of control volume, and thus the increments of cross section, $dA|_1$ and $dA|_2$. This is a formal way of stating the well-known equation of flow continuity.

The point form of the mass continuity equation in *Cartesian* (rectangular) coordinates is derived by writing a mass balance using Equation 3-8 for a differential control element having sides dx, dy, and dz, as shown in Figure 3-3. This yields

$$0 = \frac{\partial \rho}{\partial t} \, dx\, dy\, dz + \frac{\partial(\rho v_x)}{\partial x} \, dx\, dy\, dz + \frac{\partial(\rho v_y)}{\partial y} \, dy\, dx\, dz + \frac{\partial(\rho v_z)}{\partial z} \, dz\, dx\, dy \tag{3-14}$$

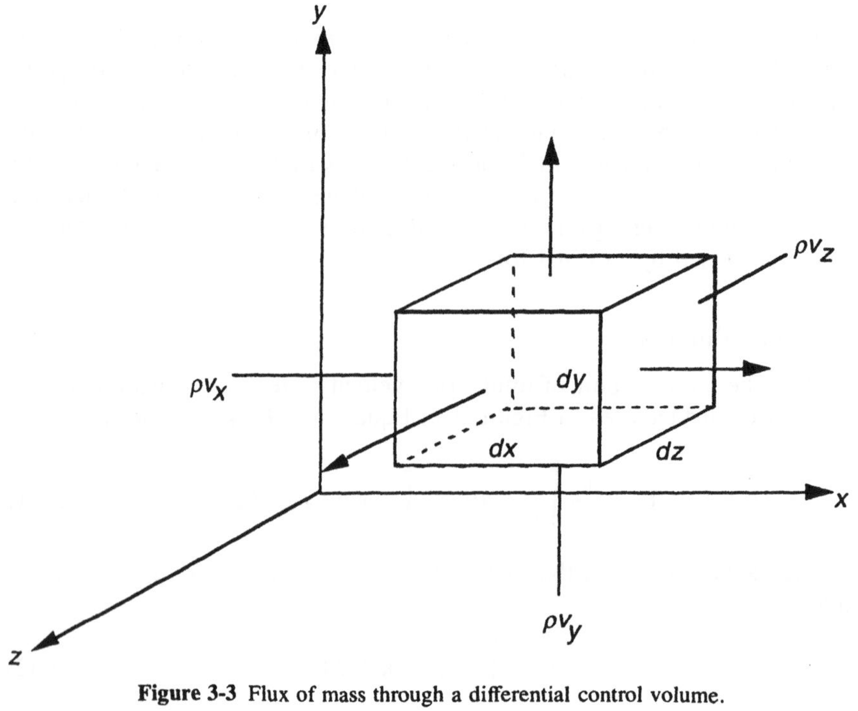

Figure 3-3 Flux of mass through a differential control volume.

Dividing by the differential volume $dx\,dy\,dz$ gives

$$\frac{\partial \rho}{\partial t} + \frac{\partial \rho v_x}{\partial x} + \frac{\partial \rho v_y}{\partial y} + \frac{\partial \rho v_z}{\partial z} = 0 \qquad (3\text{-}15)$$

which for incompressible flow reduces to

$$\frac{\partial v_x}{\partial x} + \frac{\partial v_y}{\partial y} + \frac{\partial v_z}{\partial z} = 0 \qquad (3\text{-}16)$$

or, when written in vector form,

$$\nabla \cdot \boldsymbol{v} = 0 \qquad (3\text{-}17)$$

It is important to note that Equation 3-17 is valid regardless of whether the flow is steady or unsteady.

3.2.4 Energy Balance and the Bernoulli Equation

The first law of thermodynamics describes the change in total energy, dE, in a system in terms of the sum of its change in heat, dQ_H, and the change of work, dW, so that the left-hand side of Equation 3-8 is

$$\begin{bmatrix}\text{rate of change}\\ \text{of total energy } (E)\\ \text{in a system}\end{bmatrix} = \begin{bmatrix}\text{rate of addition of heat}\\ (Q_H) \text{ to the control volume}\\ \text{from its surroundings}\end{bmatrix} - \begin{bmatrix}\text{rate of work } (W) \text{ done}\\ \text{by the control volume on}\\ \text{its surroundings}\end{bmatrix}$$

or

$$\frac{dE}{dt} = \frac{dQ_H}{dt} - \frac{dW}{dt} \qquad (3\text{-}18)$$

The work done can be subdivided into flow work and shaft work. Our focus is on *flow work*, the energy needed to move a unit volume of fluid across the system boundary. If the fluid is viscous, it also includes the shear work done to overcome shear stress (this represents a system loss in mechanical energy). *Shaft work*, a term derived from hydromachinery applications, is typically that done by the control volume on its surroundings to cause a shaft to rotate (e.g., as water passes through a turbine).

Substituting Equation 3-18 and the product of the intensive properties, e_m (energy per unit mass) and ρ (mass per unit volume), into Equation 3-9 we obtain the energy equation:

$$\frac{dQ_H}{dt} - \frac{dW}{dt} = \frac{\partial}{\partial t}\int_{VC} e_m\rho \, dV + \int_{SC} e_m\rho(\boldsymbol{v} \cdot \boldsymbol{n}_v) \, dA \tag{3-19}$$

where $e_m\rho = E$. We observe in Equation 3-19 that

$$\text{energy flux} \left(\frac{\text{energy}}{\text{area} \cdot \text{time}}\right) = e_m\rho(\dot{\boldsymbol{v}} \cdot \boldsymbol{n}_v) \tag{3-20}$$

This contrasts with the mass balance given by Equation 3-11, wherein

$$\text{mass flux} = \rho(\boldsymbol{v} \cdot \boldsymbol{n}_v) \tag{3-21}$$

The *Bernoulli equation* is a restricted form of the energy equation. To obtain this form we assume (1) an ideal fluid, which means a nonviscous (inviscid) fluid such that there is no friction and no shear work; (2) no shaft work; (3) steady, incompressible, and isothermal flow; and (4) no heat transfer to the fluid. Accordingly, Equation 3-19 reduces to

$$-\frac{dW}{dt} = \int_{SC} e_m\rho(\boldsymbol{v} \cdot \boldsymbol{n}_v) \, dA \tag{3-22}$$

The work or energy term is due to the pressure forces acting on the system to move the system through space. Because work is the product of force × distance, the work done on a unit of fluid volume to move it across the system boundary is the product of the pressure force, PA, and the distance, V/A, moved by the volume, or

$$\text{flow work} = PA\frac{V}{A} = PV \tag{3-23}$$

This is, in fact, the thermodynamic definition of work. Because we wish to express flow work *per unit time* in Equation 3-22, we substitute the time derivative of flow work from Equation 3-23 to give

$$\frac{dW}{dt} = -\int_{SC} P(\boldsymbol{v} \cdot \boldsymbol{n}_v) \, dA \tag{3-24}$$

That is, $(\boldsymbol{v} \cdot \boldsymbol{n}_v)\, dA$ represents the rate of movement of a differential volume of fluid. The Bernoulli equation is then obtained by substituting Equation 3-24 into Equation 3-22:

$$\int_{SC} \rho\left(e_m + \frac{P}{\rho}\right)(\boldsymbol{v} \cdot \boldsymbol{n}_v) \, dA = 0 \tag{3-25}$$

The energy per unit mass, e_m, is the sum of the kinetic, potential, and internal energy:

$$e_m = \frac{v^2}{2} + gy + U \qquad (3\text{-}26)$$

where y is the distance above the datum, g is the acceleration of gravity, and U is the internal energy. In reference to the control volume depicted in Figure 3-2, mass enters only at section 1 and leaves only at section 2. Thus the surface integral is

$$\left(\frac{v_2^2}{2} + gy_2 + U_2 + \frac{P_2}{\rho}\right)(\rho v_2 A_2) - \left(\frac{v_1^2}{2} + gy_1 + U_1 + \frac{P_1}{\rho}\right)(\rho v_1 A_1) = 0 \qquad (3\text{-}27)$$

The internal energy, U, in the Bernoulli equation is constant if there is no heat transfer, no shear work, and isothermal flow. After assuming steady flow ($\rho v_2 A_2 = \rho v_1 A_1$), the final general form of the Bernoulli equation is

$$\frac{v^2}{2g} + y + \frac{P}{\rho g} = \kappa \qquad (3\text{-}28)$$

where ρg is the specific weight, W_g, of the fluid. The first term describes the kinetic energy head attributable to any velocity differential (*velocity head*); the second represents the energy head relating to differences in elevation of the fluid mass across the element (*static or elevation head*); and the last term is the energy head contribution of any pressure differential across the element (*pressure head*). Bernoulli's equation thus states that *total* mechanical energy, E, remains constant along any flow streamline, although the forms in which this total available energy exists may change.

The assumptions associated with development of the Bernoulli equation are not particularly restrictive with respect to its engineering application to aqueous systems. Little error is introduced, for example, in its application to unsteady flow situations in which changes in conditions of flow occur gradually. Moreover, total system energy content in such systems is usually distributed fairly evenly among parallel streamlines of flow; the constant, κ, in Equation 3-28 is thus essentially the same for parallel streamlines.

Water is a viscous fluid, and its flow causes shear stresses that convert mechanical energy to thermal energy. This thermal energy is rarely reconverted to mechanical energy and thus represents unavailable or lost energy. Equation 3-28 can be modified to reflect the magnitude of this loss between any two points along the flow path. Consider, for example, coordinate points 1 and 2 in a one-dimensional advective flow field. The mechanical energy balance for a given element of water moving from point 1 to point 2 is then

$$E_1 = E_2 + h_l \tag{3-29}$$

where E is available energy and h_l is the loss of energy, or head loss, between points 1 and 2. Combining Equations 3-29 and 3-28 gives

$$y_1 + \frac{P_1}{\rho g} + \frac{v_1^2}{2g} = y_2 + \frac{P_2}{\rho g} + \frac{v_2^2}{2g} + h_l \tag{3-30}$$

3.2.5 Momentum Balance

Newton's second law of motion constitutes the basis for development of momentum relationships for advective flow fields. It is fundamental to our understanding of the transport of components both in bulk flow and at boundaries. The momentum equation states that the force component, $\Sigma \boldsymbol{F}$, acting on a fluid element is the result of a time rate of change of momentum. The total force on the element includes forces resulting from spatial location in a force field (e.g., forces resulting from gravitational and/or velocity fields) and forces resulting from direct contact and interaction of the fluid with its surroundings (e.g., surface forces resulting from friction at a boundary).

The time rate of change of momentum in a system and the definition of the intensive property (momentum per unit mass) given earlier are substituted into the general form of the Reynolds transport theorem equation (Equation 3-9) to obtain

$$\Sigma \boldsymbol{F} = \frac{\partial}{\partial t}\int_{Vc} \boldsymbol{v}\rho \, dV + \int_{Sc} \boldsymbol{v}\rho(\boldsymbol{v} \cdot \boldsymbol{n}_v) \, dA \tag{3-31}$$

Analogous to the energy flux in the energy balance, the momentum flux is given by

$$\text{momentum flux} = \boldsymbol{v}\rho(\boldsymbol{v} \cdot \boldsymbol{n}_v) \tag{3-32}$$

The similarities among (1) the continuity equation for component i within a fluid (Equation 2-11), (2) the continuity equation for the mass of fluid (Equation 3-11), (3) the energy equation (Equation 3-19), and (4) the momentum equation (Equation 3-31) are noteworthy. This once again indicates the generality of the continuity and conservation concepts.

The differential or point form of the momentum equation may be developed analogously to the differential form of the continuity equation (Equation 3-15). The complete derivation for viscous and inviscid fluids can be found in texts on fluid mechanics and mass, momentum, and heat transfer. Some highlights will be presented here to reinforce use of the control volume concept.

We begin by applying the generic word form of the Reynolds transport theorem (Equation 3-8) for the analysis of momentum change in a differential control element, as depicted in Figure 3-4. The external forces acting in each

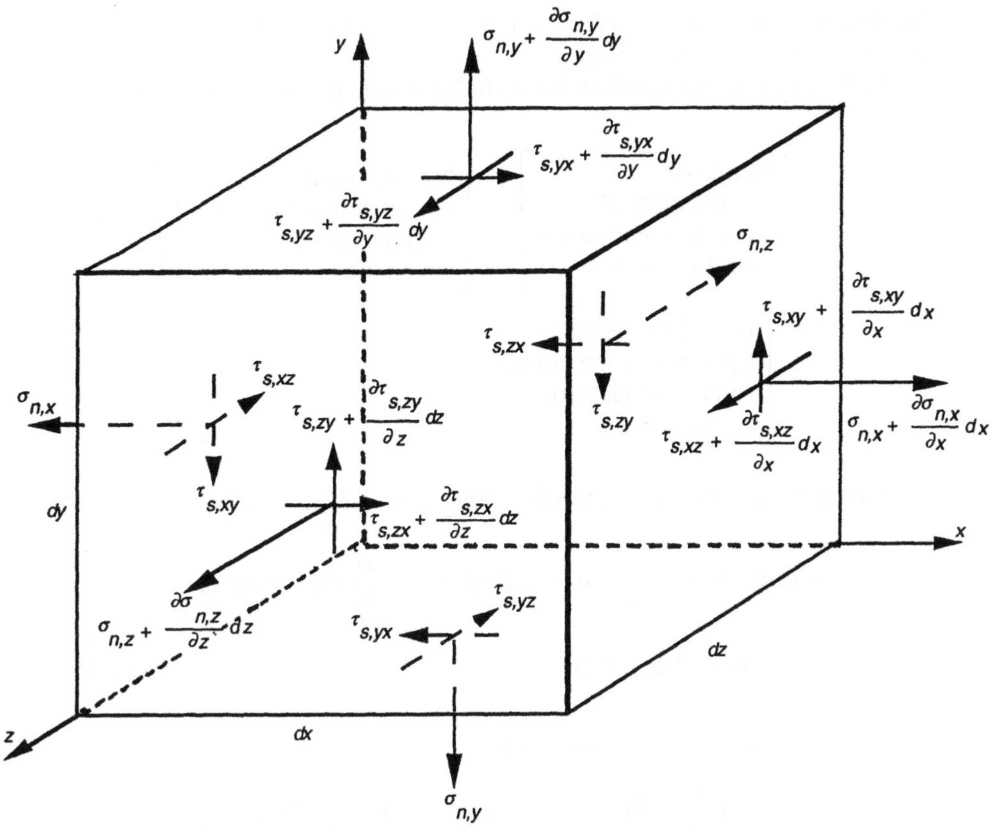

Figure 3-4 Normal and tangential surface forces acting on a differential control volume of a viscous fluid. [Adapted from Olson and Wright (1990).]

direction are due to normal stress (σ_n) and shear stress (τ_S); in addition, there is the force of gravity (g) acting on the volume. Shear stress, $\tau_{s,ij}$, refers to stress acting in the plane normal to the i axis and acting in the j direction. Thus the summation of forces in the x direction, for example, is given by

$$\Sigma F_x = \rho g_x \, dx \, dy \, dz + \frac{\partial \sigma_{n,x}}{\partial x} dx \, dy \, dz + \frac{\partial \tau_{s,yx}}{\partial y} dy \, dx \, dz + \frac{\partial \tau_{s,zx}}{\partial z} dz \, dx \, dy \tag{3-33}$$

where g_x is the x component of gravitational acceleration. After dividing through by the differential volume $dV = dx \, dy \, dz$, Equation 3-33 becomes

$$\frac{\Sigma F_x}{dV} = \rho g_x + \frac{\partial \sigma_{n,x}}{\partial x} + \frac{\partial \tau_{s,yx}}{\partial y} + \frac{\partial \tau_{s,zx}}{\partial z} \tag{3-34}$$

The forces acting in the y and z directions can be summed in like manner to give equations of the same form.

The terms to be evaluated on the right-hand side of Equation 3-8 are

$$= \begin{bmatrix} \text{net rate of} \\ \text{accumulation of} \\ \text{momentum within the} \\ \text{control volume} \end{bmatrix} + \begin{bmatrix} \text{rate of momentum} \\ \text{output across the control} \\ \text{volume surfaces or} \\ \text{boundaries} \end{bmatrix}$$

$$- \begin{bmatrix} \text{rate of momentum} \\ \text{input across the control} \\ \text{volume surfaces or} \\ \text{boundaries} \end{bmatrix}$$

Using the definition of momentum flux (Equation 3-32), we obtain

$$\frac{\partial}{\partial t}(\boldsymbol{v}\rho)\,dx\,dy\,dz + \frac{\partial}{\partial x}(\boldsymbol{v}\rho v_x)\,dx\,dy\,dz + \frac{\partial}{\partial y}(\boldsymbol{v}\rho v_y)\,dy\,dx\,dz$$
$$+ \frac{\partial}{\partial z}(\boldsymbol{v}\rho v_z)\,dz\,dx\,dy \tag{3-35}$$

Upon differentiation Equation 3-35 yields

$$\rho\frac{\partial \boldsymbol{v}}{\partial t} + \boldsymbol{v}\frac{\partial \rho}{\partial t} + \rho\left(v_x\frac{\partial \boldsymbol{v}}{\partial x} + v_y\frac{\partial \boldsymbol{v}}{\partial y} + v_z\frac{\partial \boldsymbol{v}}{\partial z}\right)$$
$$+ \boldsymbol{v}\left(\frac{\partial \rho v_x}{\partial x} + \frac{\partial \rho v_y}{\partial y} + \frac{\partial \rho v_z}{\partial z}\right) \tag{3-36}$$

Equation 3-36 can be simplified by substitution of the continuity equation (Equation 3-15), giving

$$\rho\frac{\partial \boldsymbol{v}}{\partial t} + \rho\left(v_x\frac{\partial \boldsymbol{v}}{\partial x} + v_y\frac{\partial \boldsymbol{v}}{\partial y} + v_z\frac{\partial \boldsymbol{v}}{\partial z}\right) \tag{3-37}$$

We can now equate the summation of force expressions in each direction with the appropriate directional components of the vector equation for the rate of change of momentum to complete the statement of Newton's second law:

$$\rho g_x + \frac{\partial \sigma_{n,x}}{\partial x} + \frac{\partial \tau_{s,yx}}{\partial y} + \frac{\partial \tau_{s,zx}}{\partial z} = \rho\left(\frac{\partial v_x}{\partial t} + v_x\frac{\partial v_x}{\partial x} + v_y\frac{\partial v_x}{\partial y} + v_z\frac{\partial v_x}{\partial z}\right) \tag{3-38}$$

$$\rho g_y + \frac{\partial \tau_{s,xy}}{\partial x} + \frac{\partial \sigma_{n,y}}{\partial y} + \frac{\partial \tau_{s,zy}}{\partial z} = \rho\left(\frac{\partial v_y}{\partial t} + v_x \frac{\partial v_y}{\partial x} + v_y \frac{\partial v_y}{\partial y} + v_z \frac{\partial v_y}{\partial z}\right) \tag{3-39}$$

$$\rho g_z + \frac{\partial \tau_{s,xz}}{\partial x} + \frac{\partial \tau_{s,yz}}{\partial y} + \frac{\partial \sigma_{n,z}}{\partial z} = \rho\left(\frac{\partial v_z}{\partial t} + v_x \frac{\partial v_z}{\partial x} + v_y \frac{\partial v_z}{\partial y} + v_z \frac{\partial v_z}{\partial z}\right) \tag{3-40}$$

The right-hand side of each of the equations above defines what is referred to as the *substantial* or *total derivative* of velocity (i.e., the time rate of change of velocity at a point as it moves through space) in a given dimension (x, y, or z). When taken together these equations thus yield

$$\frac{D'\boldsymbol{v}}{D't} = \left(\frac{\partial}{\partial t} + v_x \frac{\partial}{\partial x} + v_y \frac{\partial}{\partial y} + v_z \frac{\partial}{\partial z}\right)\boldsymbol{v} \tag{3-41}$$

That is, the total acceleration of the fluid in a given direction is the sum of the local acceleration at a point and the convective acceleration (i.e., the change in velocity from point to point).

Equations 3-38, 3-39, and 3-40 can be written more conveniently in vector form. Normal stress, σ_n, is fluid pressure normal to each direction such that the vector term is $\boldsymbol{P}$. Additional normal stress terms which involve dynamic viscosity (μ_v) and velocity gradients result from Stokes' hypothesis, but these are combined with the shear stress terms and some are eliminated for incompressible fluid flow. Multidimensional shear stress, $\tau_{s,ij}$, is described by Stokes' viscosity relations (an extension of Newton's one-dimensional viscosity relation, $\tau_s = \mu_v dv/dx$). The derivatives of shear stress with respect to distance appear in Equations 3-38, 3-39, and 3-40; thus the result is a vector term, $\mu_v \nabla \boldsymbol{v} \cdot \nabla \boldsymbol{v}$ or $\mu_v \nabla^2 \boldsymbol{v}$. The final form of the momentum equation is usually expressed as

$$\rho \frac{D'\boldsymbol{v}}{D't} = \rho \boldsymbol{g} - \nabla \boldsymbol{P} + \mu_v \nabla^2 \boldsymbol{v} \tag{3-42}$$

which is known as the *Navier–Stokes equation*.

The Navier–Stokes equation is the point form of the momentum equation for incompressible, constant-viscosity fluids under laminar flow conditions. Its fundamental nature is evident from dimensional homogeneity; that is, each term in Equation 3-42 (after dividing by ρ) has the same dimensions (Lt^{-2}). The physical meaning of the left side is the total acceleration (both spatial and temporal), or in other words, the *inertial forces* of the fluid. This acceleration is determined by the sum of the forces on the right-hand side of Equation 3-42 (i.e., the gravity, pressure, and viscous forces).

We can further observe from the definition of the substantial derivative in Equation 3-41 that the Navier–Stokes equation can be written as

$$\rho\left(\frac{\partial \boldsymbol{v}}{\partial t} + \boldsymbol{v} \cdot \nabla \boldsymbol{v}\right) = \rho \boldsymbol{g} - \nabla \boldsymbol{P} + \mu_v \nabla^2 \boldsymbol{v} \qquad (3\text{-}43)$$

For incompressible fluids such as water, the continuity equation reduces to $\nabla \cdot \boldsymbol{v} = 0$ (see Equation 3-17), and Equation 3-43 thus reduces to

$$\rho \frac{\partial \boldsymbol{v}}{\partial t} = \rho \boldsymbol{g} - \nabla \boldsymbol{P} + \mu_v \nabla^2 \boldsymbol{v} \qquad (3\text{-}44)$$

This equation is solved together with the continuity equation and boundary conditions to find velocity and pressure as a function of time and space in natural and engineered water systems. The resulting system of equations is nonlinear, and exact solutions are therefore limited to a few special situations (e.g., steady and uniform flow in one or two dimensions so that the acceleration term on the left-hand side can be set equal to zero). There are, however, many practical situations in which such simplifying assumptions are reasonable. For example, the *Hagen–Poiseuille equation* for pressure drop in a pipe when laminar flow exists is obtained from the momentum balance. Example 3-2 carries development of the Hagen–Poisseuille equation further to show that it also explains the *Darcy equation* for flow through porous media.

The development of the momentum balance here has been for an elemental volume for which Cartesian coordinates are appropriate. Momentum balances similarly can be written for cylindrical or spherical çoordinates to facilitate applications to fluid flow where these geometries are appropriate. For example, the analysis of groundwater flow in the vicinity of a well generally lends itself best to a description based on a cylindrical coordinate system in which flow is radially directed toward the wellhead.

The Navier–Stokes equation can be extended to include the influence of an electrical field on fluid transport. This extension is useful in gaining a fundamental understanding of the permeation of water (and electrolytes) through charged, capillary tubes, which is often how reverse-osmosis membranes are envisioned. For steady and uniform flow, the Navier–Stokes equation is written as

$$0 = -\nabla \boldsymbol{P} + \mu_v \nabla^2 \boldsymbol{v} - \sum_{i=1}^{N} z_i^\circ \mathfrak{F} C_i \nabla \Phi \qquad (3\text{-}45)$$

The last term has been added to account for the influence of electrical body forces on transport of the fluid through the pore. The parameters associated with this term are: z_i°, the valence of the ions; $\mathfrak{F}$, Faraday's constant (coulombs per equivalent); C_i, the concentrations of *i*th ions in the pore; and $\nabla\Phi$, the gradient of electrical potential in the pore. The description of fluid velocity in the pore that is provided by this equation can be coupled with the Nernst–Planck equation to describe diffusive transport of ions caused by concentration and electrical potential gradients and advective transport caused by movement with the bulk flow (see Section 3.3.3). The description of the potential gradient requires more detailed analysis of the electrostatic potential due to the charge on the pore wall and the electrical potential that arises due to applied electric fields and ion diffusivity and concentration differences.

The Navier–Stokes equation applies also for turbulent flow fields because turbulent stress forces are inherently included in the momentum flux term. An application for analysis of steady, three-dimensional incompressible flow in two differently baffled tanks is shown in Figure 3-5a. If the space within these tanks is discretized, or divided into meshes, to facilitate use of a finite-element numerical solution, the velocity fields obtained for the two differently baffled systems by solution of the Navier–Stokes equations will reveal distinct differences. Baffle system A will exhibit more uniform velocity fields than baffle system B, and, therefore, far less mixing of the flow. Less mixing implies a greater tendency for plug-like flow behavior and less fluid dispersion. Figure 3-5b shows how this difference would be reflected in different transport behavior for dissolved components of the fluid. The curves shown here would typically result if a slug of dye were injected instantaneously at the inlet of each system and its concentration measured at the outlet with time. We note that for baffle system A the dye is much less spread out in time and arrives at the outlet very close to the tank hydraulic detention time, $\bar{t}$ (i.e., the time for advective transport without dispersion).

In Chapter 4 we apply the Navier–Stokes equation to determine the thickness of boundary layers. Boundary layers are important in interphase mass transport; this includes processes involving flow over reactive or adsorbent surfaces, through porous media, and through membranes.

For inviscid fluids, the last term in the Navier–Stokes equation is zero, and Equation 3-42 becomes the *Euler equation*:

$$\rho \frac{\partial \boldsymbol{v}}{\partial t} = \rho g - \nabla \boldsymbol{P} \tag{3-46}$$

It can be shown that the Bernoulli equation (Equation 3-28) represents an integration of the Euler equation along a streamline in a steady incompressible flow field.

Both the energy balance and the Euler equation (a special case of the momentum balance) describe advective transport of fluid elements and associated components under the assumed conditions of an ideal fluid and frictionless flow. In certain instances, as in the Bernoulli equation, friction effects and losses are considered (see Equation 3-30), but not related directly to the flow characteristics of real fluids. The principal difference between ideal and real fluids is that the latter exhibit the property of viscosity.

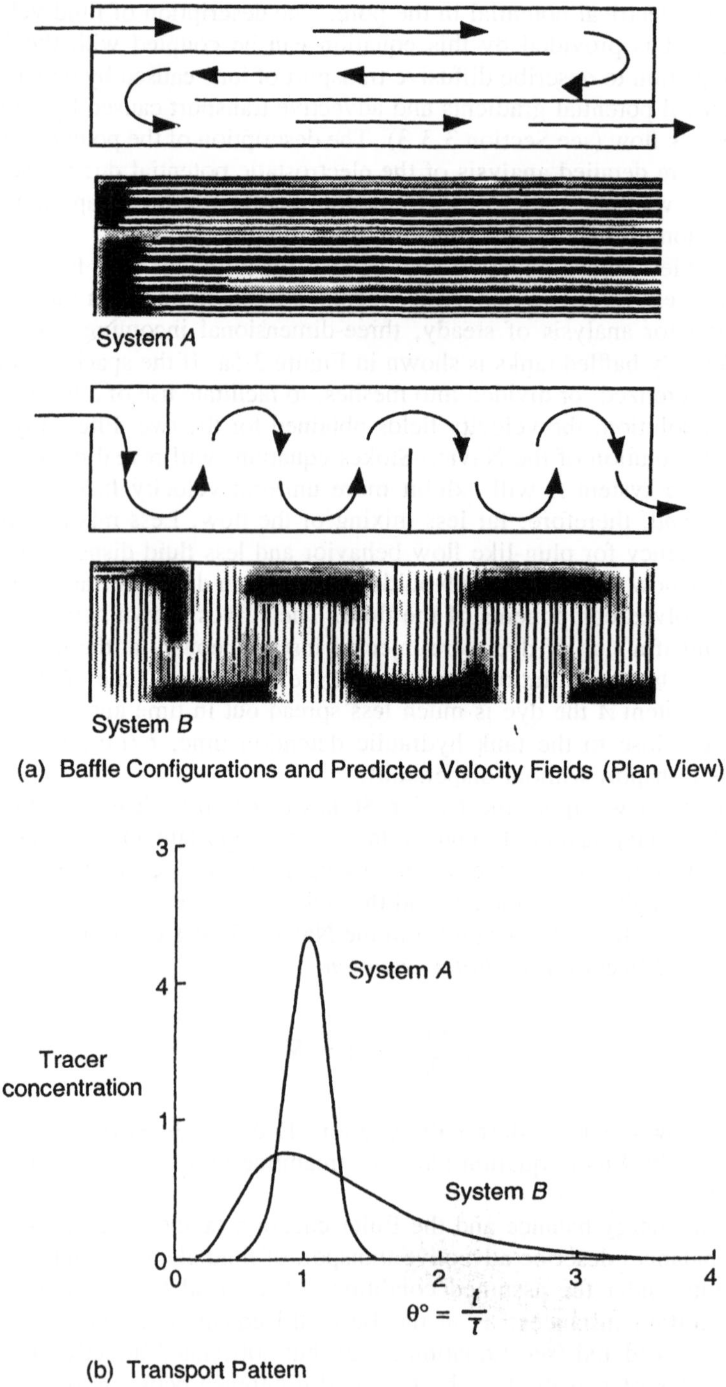

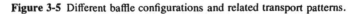

Figure 3-5 Different baffle configurations and related transport patterns.

The Navier–Stokes equation and the Bernoulli equation should make clear the important relationships among viscosity, stress, and energy loss. The Navier–Stokes equation derives from a momentum balance that includes shear stress caused by viscosity. On the other hand, the Bernoulli equation derives from an energy balance on an inviscid fluid; consistent with this assumption, the internal energy (U) remains constant. The extension of Bernoulli's equation to a viscid fluid implies that internal energy is no longer constant (i.e., shear stress converts available mechanical energy to unavailable thermal energy). Example 3-2 shows that Darcy's law can be derived from the Navier–Stokes equation.

Example 3-2

- **Situation:** Consider the flow of water through horizontally oriented channels of thickness, δ, in the subsurface fractured rock domain illustrated schematically below.

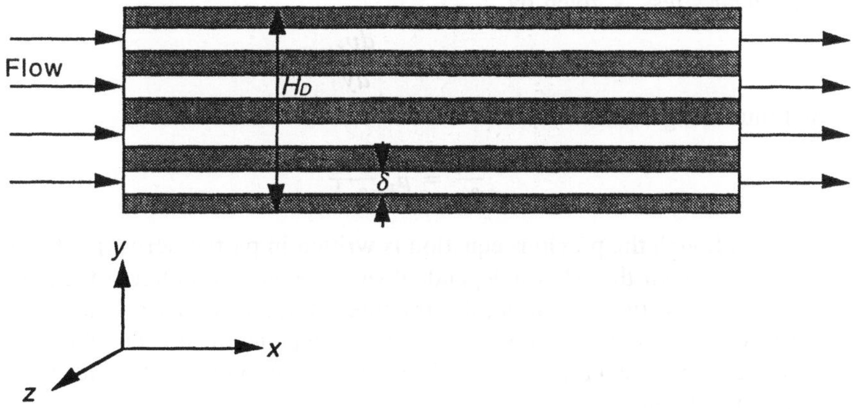

- **Question(s):** Develop an expression for flow in the x direction.
- **Assumption(s):** Flow is viscous and laminar. The flow lines are parallel in the x direction. The control volume of fluid is defined by a differential volume within the fracture space having width, Z_W in the z direction.

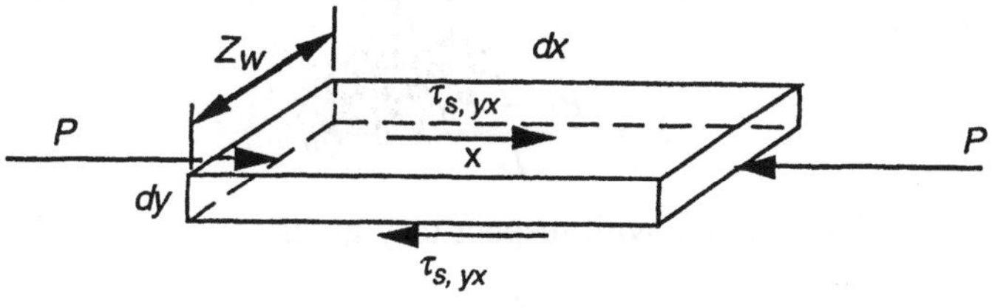

- **Logic and Answer(s):**

1. The Navier–Stokes equation (Equation 3-42) applies.

2. The left-hand side of Equation 3-42 is zero because flow is steady and uniform. The forces on the right-hand side can be analyzed over the differential control volume to give

$$0 = -\frac{\partial}{\partial x}(P)Z_W\, dx\, dy + \frac{\partial}{\partial y}(\tau_{s,yx})\, Z_W\, dy\, dx$$

3. The component of gravitational force in the x direction (see Equation 3-42) is zero because the fractures are not inclined. Note the consistency in signs based on direction of shear stresses shown in the sketch above (positive shear stresses act in the positive coordinate direction on *positive surfaces*, i.e., those with normal vectors in the positive coordinate direction). Dividing the equation above by $Z_W\, dx\, dy$ gives

$$-\frac{\partial P}{\partial x} + \frac{\partial \tau_{s,yz}}{\partial y} = 0$$

The shear stress is given by

$$\tau_{s,yx} = \mu_v \frac{dv_x}{dy}$$

and thus we get

$$\frac{\partial P}{\partial x} = \mu_v \frac{\partial^2 v_x}{\partial y^2}$$

4. Although the previous equation is written in partial derivative form, it can be shown that P is independent of y because the effect of gravity is the same everywhere along the fracture. Also, v_x is not a function of x if flow is constant. Because y and x are independent of each other, the only way for the equation above always to be true is for both sides to be constant, so that

$$\frac{dP}{dx} = \kappa \qquad \text{and} \qquad \mu_v \frac{d^2 v_x}{dy^2} = \kappa$$

Integrating pressure over distance gives

$$P = \kappa x + \kappa_1$$

The specific solution is obtained by letting $P = P_0$ at $x = 0$ and $P = P_L$ at $x = L$; thus

$$P = \frac{P_L - P_0}{L} x + P_0$$

Integrating velocity twice with respect to y gives

$$\mu_v v_x = \kappa \frac{y^2}{2} + \kappa_1 y + \kappa_2$$

The velocity at both $y = 0$ and $y = \delta$ must be zero; thus

$$v_x = \frac{\kappa}{2\mu_v}(y^2 - \delta y)$$

Note that the velocity profile within the fracture is parabolic just as it is for the analogous flow problem in circular pipes. The constant κ is known from integrating P over the distance x, so that

$$v_x = \frac{P_L - P_O}{2\mu_v L}(y^2 - \delta y)$$

5. If there are n parallel fractures over a depth of media, H_D, the porosity of the domain is defined as

$$\epsilon_d = \frac{n\delta}{H_D}$$

and total flow rate, Q_T, through the fracture is

$$Q_T = nQ_F$$

Flow rate within a given fracture, Q_F, is obtained by integrating over the fracture height, δ, so that the total flow rate is

$$Q_T = n\int_0^{\delta} Z_W\, v_x\, dy = \frac{nZ_W\,\delta^3(P_O - P_L)}{12\mu_v L}$$

After substitution for n from the relationship for ϵ_d given above, we obtain

$$Q_T = \frac{H_D Z_W \epsilon_d \delta^2(P_O - P_L)}{12\mu_v L}$$

6. This result reinforces the notion that shear stress results in energy loss and thus a pressure drop. The equivalent equation for pipe flow (see Problem 3-7), derived for radial (y, r, θ) coordinates, is

$$Q_F = \int_0^{2\pi}\int_0^{R_P} v_y r\, dr\, d\theta = \frac{\pi R_P^4}{8\mu_v}\frac{\Delta P}{L}$$

where R_P is the pipe radius. This is the Hagen–Poiseuille equation. The expression for Q_T arrived at in step 5 can also be written as a Darcy's law formulation:

$$\frac{Q_T}{A} = -\frac{\kappa_p}{\mu_v}\frac{\Delta P}{\Delta x}$$

where A is the cross-sectional area normal to the flow and κ_p is a proportionality constant between the pressure drop and velocity defined by

$$\kappa_p = \frac{\epsilon_d \delta^2}{12}$$

The negative sign in the Darcy equation reflects flow moving in the direction of high to low pressure. Note that κ_p accounts for media characteristics (ϵ_d and δ).

7. If the fractures are vertical instead of horizontal, the Navier–Stokes equation requires the addition of a gravity force term, so that

$$0 = -\rho g\, Z_W\, dx\, dy - \frac{\partial}{\partial y}(P)\, Z_W\, dy\, dx + \frac{\partial}{\partial x}(\tau_{s,xy})\, Z_W\, dx\, dy$$

which, after simplification, is

$$\frac{\partial P}{\partial y} = \frac{\partial \tau_{s,xy}}{\partial x} - \rho g$$

The solution procedure to obtain an expression for v_x is the same as described in step 4, the result being

$$Q_T = n \int_0^\delta Z_W\, v_x\, dy = \frac{n Z_W \delta^3}{12\mu_v}\left(\frac{P_0 - P_L}{L} - \rho g\right)$$

or, making use of the appropriate definition of porosity for parallel vertical fractures in step 5,

$$\frac{Q_T}{A} = \frac{\epsilon_d \delta^2}{12\mu_v}\left(\frac{P_0 - P_L}{L} - \rho g\right)$$

where A is the cross-sectional area normal to the downward direction of flow. The three-dimensional representation of this equation is

$$\frac{Q_T}{A} = -\frac{\epsilon_d \delta^2}{12\mu_v}(\nabla P + \rho g\, \nabla y) = -\frac{\epsilon_d \delta^2 \rho g}{12\mu_v}(\nabla h)$$

where h, the hydraulic head, is given by

$$h = \frac{P}{\rho g} + y$$

8. The derivation above illustrates that applying the Navier–Stokes equation to flow through porous media leads to the same formulation obtained experimentally by Darcy. The constant of proportionality between the hydraulic head gradient and the velocity (Q/A), referred to as the hydraulic conductivity, has the dimensions of Lt^{-1}.

3.2.6 Reynolds Number

The Reynolds number is a simple, dimensionless grouping of variables, yet a powerful tool for characterization of flow. An appreciation of the origin of the Reynolds number from an analysis of momentum exchange in the Navier–Stokes equation makes it clear why this is so. Momentum exchange occurs along the continuum from laminar to turbulent flow. Laminar flow involves only molecular interchange of momentum. Tendencies for instability and turbulence are damped by viscous shear forces, which resist relative motion of fluid elements in parallel flow laminae. Turbulent flow conversely involves erratic movement of fluid elements, with significant transverse exchange of momentum among these elements. We have seen that the Navier–Stokes equation provides a complete mathematical description of flow dynamics. It was Reynolds, however, who showed experimentally that geometrically similar flow systems are dynamically similar if they also exhibit the same relative significance, or same ratio, of shear stresses resulting from turbulence and from viscosity. Reynolds modified the Navier–Stokes equation to include turbulent flow. He did this by representing both velocity and pressure as the sum of two terms, an average value term plus a fluctuating-value term, to produce a time-averaged velocity and pressure in the Navier–Stokes equations. For many practical applications, characteristic flow dynamics lie somewhere between those of discretely laminar and those of discretely turbulent flow. These conditions produce irregular patterns of advective transport of component or constituent masses.

The Reynolds number appears in the nondimensional form of the Navier-Stokes equation. The substantial derivative given in Equation 3-42 is first rewritten as

$$\frac{D'\boldsymbol{v}}{D't} = \boldsymbol{g} - \frac{\nabla P}{\rho} + \nu_v \nabla^2 \boldsymbol{v} \tag{3-47}$$

where ν_v is kinematic viscosity (μ_v/ρ) and $\boldsymbol{g}$ is a gravity vector. De-dimensionalization of the Navier–Stokes equation leads to the isolation of dimensionless parameters having distinct physical meanings. Substitution of the definition of the substantial derivative (Equation 3-41) into Equation 3-47 yields

$$\frac{\partial \boldsymbol{v}}{\partial t} + \boldsymbol{v} \cdot \nabla \boldsymbol{v} = -\nabla\left[\frac{P}{\rho} - (g_x x + g_y y + g_z z)\right] + \nu_v \nabla^2 \boldsymbol{v} \tag{3-48}$$

Because the overall dimensions of both sides of Equation 3-48 are Lt^{-2}, the equation can be de-dimensionalized by selecting a characteristic velocity, v_c, and a characteristic length, L_c, such that

$$\boldsymbol{v}^\circ = \frac{\boldsymbol{v}}{v_c}$$

$$\zeta_x^\circ = \frac{x}{L_c} \qquad \zeta_y^\circ = \frac{y}{L_c} \qquad \zeta_z^\circ = \frac{z}{L_c}$$

$$\theta^\circ = \frac{t v_c}{L_c}$$

$$\nabla^\circ = L_c \nabla \quad \text{and} \quad \nabla^{\circ 2} = L_c^2 \nabla^2 \tag{3-49}$$

Characteristic velocity and length are system specific. For example, although v_c is often defined as the bulk velocity, it may also be the shear velocity at a boundary. Similarly, L_c may be defined depending on system geometry (e.g., as the diameter of a cylinder, diameter of a sphere, or length along a flat surface).

Substitution of the dimensionless parameters above into Equation 3-48 gives

$$\frac{\partial \boldsymbol{v}^\circ}{\partial t^\circ} + \boldsymbol{v}^\circ \cdot \nabla^\circ \boldsymbol{v}^\circ = -\nabla^\circ P^\circ + \frac{\nu_v}{v_c L_c} \nabla^{\circ 2}(\boldsymbol{v}^\circ) \tag{3-50}$$

where

$$P^\circ = \frac{P - \rho L_c(g_x \zeta_x^\circ + g_y \zeta_y^\circ + g_z \zeta_z^\circ)}{\rho v_c^2} \tag{3-51}$$

and where the Reynolds number appears as

$$\text{Reynolds number, } \mathfrak{N}_{\text{Re}} = \frac{\text{inertial force}}{\text{viscous force}} = \frac{L_c v_c}{\nu_v} \tag{3-52}$$

The final form of the dimensionless Navier–Stokes equation includes the Reynolds number:

$$\frac{\partial \boldsymbol{v}^\circ}{\partial t^\circ} + \boldsymbol{v}^\circ \cdot \nabla^\circ \boldsymbol{v}^\circ = -\nabla^\circ P^\circ + \frac{1}{\mathfrak{N}_{\text{Re}}} \nabla^{\circ 2}(\boldsymbol{v}^\circ) \tag{3-53}$$

Two other dimensionless parameters are included in dimensionless pressure, P°:

$$\mathfrak{N}_{\text{Fr}}, \text{ Froude number} = \frac{\text{inertial force}}{\text{gravity force}} = \frac{v_c^2}{g L_c} \tag{3-54}$$

$$\mathfrak{N}_{\text{Eu}}, \text{ Euler number} = \frac{\text{pressure force}}{\text{inertial force}} = \frac{P}{\rho v_c^2} \tag{3-55}$$

These are defined by statement of the boundary conditions needed to solve the Navier–Stokes equation; they are important only in situations where free surface effects on flow must be included. The explicit appearance of the Reynolds number in the Navier–Stokes equation emphasizes its importance in determining the temporal and spatial distribution of velocity and pressure.

The Reynolds number is widely regarded as the most important dimensionless group in fluid mechanics. The larger the Reynolds number, the more the general character of flow will be determined by turbulence. The absolute value at which viscous effects become less significant than turbulent effects depends entirely on the characteristic properties of the fluid and the characteristic geometry of the system in which flow takes place. For flow in pipes, a value of $\mathfrak{N}_{Re} \approx 2000$ usually defines the point above which flow is no longer discretely laminar. Flow in such systems is usually considered discretely turbulent above $\mathfrak{N}_{Re} \approx 4000$.

The importance of the Froude number is not to be overlooked. A useful illustration is provided by the notion of specific energy in open channel flow. The *specific energy head*, h_{se}, is defined by letting the datum in the Bernoulli equation (Equation 3-28) be the bottom of the channel so that the water depth is $y = H_D$ and $h_{se} = v^2/2g + H_D$. The Bernoulli and continuity of flow ($Q = Av = H_D Z_W v$) equations for a channel of width Z_W are solved together for a constant discharge per unit width, Q/Z_W. The resulting relationship between h_{se} and H_D shows a minimum in specific energy at a depth referred to as the *critical depth*, $H_{D,c}$; two alternative depths (one higher and the other lower than critical depth) are possible at all other specific energies. After setting the derivative, dh_{se}/dH_D, to *zero* at critical depth, the Froude number (with characteristic length being the depth) is obtained and its value is 1. A Froude number value of $\mathfrak{N}_{Fr} > 1$ defines *supercritical* flow (i.e., high inertial flow that occurs when $H_D < H_{D,c}$). Flow at the same specific energy but at the alternative depth, $H_D > H_{D,c}$, is characterized as *subcritical* because $\mathfrak{N}_{Fr} < 1$. Channel transitions that produce hydraulic jumps give vivid practical meaning to the Froude number. Flow is supercritical (high inertia) and changes to subcritical as a result of a sudden and large rise in a channel bottom that causes a large decrease in specific energy. It has been shown experimentally that the extent of turbulence downstream of a hydraulic jump increases as the Froude number upstream increases.

Much use is made of the Reynolds and Froude numbers and other dimensionless parameters in the development of physical hydraulic models of large systems (e.g., rivers and harbors) to facilitate application of experimental data from the models to prototype systems. The first requirement for dynamic similarity or *similitude* is that the systems have geometric similarity (i.e., equal ratios of corresponding dimensions). Second, the ratio of forces in the model and prototype must be equal. For example, any combination of length, velocity, and kinematic viscosity that gives the same Reynolds number will have the same flow characteristics if viscous forces dominate over gravity and pressure forces.

Dynamic similitude is a critical concept in aerodynamic and hydrodynamic design problems (e.g., the hull shape of a ship having the least fluid drag). It is also vitally important in the physical design of bench- and pilot-scale models for environmental treatment processes and small-scale simulations of natural surface and subsurface systems. In these applications it is essential to account not only for dynamic similitude but for process similitude as well. In most systems of interest in process engineering, viscosity forces dominate over gravity and pressure forces such that similarities in the Froude and Euler numbers are not as important when developing a physical model as are similarities in the Reynolds number. The Froude number, however, can be important if forces at free surfaces are significant (e.g., wave action). It may not always be easy to achieve similitude in both the Reynolds and Froude numbers, as illustrated in Example 3-3.

Example 3-3

- **Situation:** A lagoon treatment system is used to process wastewater from a particleboard fabrication facility. It is important to remove fine particles in the lagoon to comply with effluent limitations on suspended solids. The lagoon is subject to significant wind action. The physical and chemical processes governing particle settling are difficult to predict with a mathematical model, especially in such a complex hydrodynamic setting. In view of this, a decision is made to construct a small-scale physical hydraulic model of the essentially rectangular lagoon to obtain experimental information on the settling process.
- **Question(s):** What considerations must be involved in the design of the hydraulic scale model to ensure reasonable simulation of the settling characteristics of the full-scale lagoon?
- **Logic and Answer(s):**

1. Dynamic similitude is important because fluid flow directly affects the trajectory of particles. If wave action is important, dynamic similitude requires that we consider both free surface forces ($\mathfrak{N}_{\text{Fr}}$) and viscous forces ($\mathfrak{N}_{\text{Re}}$).

2. The characteristic length of the lagoon is L_{actual}. The characteristic length of the scale model is L_{model} and $L_{\text{model}} = R_L L_{\text{actual}}$, where R_L is the length scaling factor ($R_L << 1$). The widths of the lagoon and scale model are $Z_{W,\text{actual}}$ and $Z_{W,\text{model}}$, respectively. Equating the Froude numbers (Equation 3-54) for the model and the actual lagoon, we obtain

$$(v_c)_{\text{model}} = R_L^{0.5}(v_c)_{\text{actual}}$$

and, using the relationship above to equate the respective Reynolds numbers (Equation 3-52) while equating the Froude numbers,

$$(\nu_v)_{\text{model}} = R_L^{1.5}(\nu_v)_{\text{actual}}$$

3. Because $R_L << 1$, fluid velocity and viscosity must be decreased for the small-scale tests, the latter from one to two orders of magnitude. Although it is not difficult to decrease velocity, it is impossible to decrease viscosity by the amount needed, even by using another fluid. Even were it possible to decrease viscosity, it would be further necessary, according to Stokes' law, to employ smaller particles and/or particles of much lower density in the model test than are present in the actual lagoon system in order to simulate the proper settling velocity. The effects of

the free surface forces on the settling of particles thus cannot be determined easily.

4. While dynamic hydraulic similarity is important, the resulting test model must also be appropriate in terms of reasonably simulating the settling process operative in the full-scale lagoon. As a beginning point, we examine factors governing settling under ideal hydraulic conditions (advective transport along parallel laminar streamlines in a fluid of uniform density). For discrete particles it is known that particle removal increases as the ratio of flow rate to surface area of the tank (the so-called overflow rate) decreases; for flocculent particles, removal increases with increasing depth and thus with detention time [see, for example, Weber (1972)]. The scale model should, therefore, reproduce the overflow rates and detention times associated with the actual lagoon.

5. Suppose that $R_L = 0.1$ and we wish to maintain similitude only with respect to the Reynolds number. According to the first equation in step 2, the velocity in the scale model is $0.32 v_{\text{actual}}$. If we also wish to maintain geometric similarity, the model flow rate is

$$Q_{\text{model}} = 0.32 v_{\text{actual}} (0.1 Z_{W,\text{actual}})(0.1 H_{D,\text{actual}}) = 0.0032 Q_{\text{actual}}$$

where $H_{D,\text{actual}}$ is the depth of wastewater in the lagoon. The hydraulic residence time (HRT) in the scale model is thus

$$(\text{HRT})_{\text{model}} = \frac{0.001}{0.0032} (\text{HRT})_{\text{actual}} = 0.31 (\text{HRT})_{\text{actual}}$$

The model overflow rate would be

$$\frac{0.0032 Q_{\text{actual}}}{(0.1\ Z_{W,\text{actual}})(0.1\ L_{\text{actual}})} = 0.32\ (\text{overflow rate})_{\text{actual}}$$

The result is a shorter HRT which may or may not compensate for the smaller value of overflow rate in determining the settling efficiency. Thus, the geometric and Reynolds number similitude conditions required to achieve a reasonable hydraulic model lead to a scale model which may not simulate the removal of particles properly if both detention time and overflow rate are in fact important.

6. An alternative is to keep the respective overflow rates and Reynolds numbers of the two systems equal. If

$$Z_{W,\text{model}} = 0.1 Z_{W,\text{actual}} \quad \text{and} \quad L_{\text{model}} = 0.1 L_{\text{actual}}$$

then equal overflow rate means $Q_{\text{model}} = 0.01(Q_{\text{actual}})$. To obtain equal values for the Reynolds number requires that:

$$H_{D,\text{model}} = \frac{0.01 Q_{\text{actual}}}{(0.1 Z_{W,\text{actual}})(0.1 v_{\text{actual}})} = H_{D,\text{actual}}$$

7. If Froude number similitude were the goal, and the viscosity of the test scale system could not be changed, Reynolds number similarity could not be achieved; that is,

$$(\mathfrak{N}_{\text{Re}})_{\text{model}} = \frac{R_L^{0.5}(v_c)_{\text{actual}}L_{\text{model}}}{(\nu_v)_{\text{actual}}} = 0.32(\mathfrak{N}_{\text{Re}})_{\text{actual}}$$

In addition, the HRT would be $0.1L_{\text{actual}}/0.32v_{\text{actual}}$, or 0.31 times the actual value. A lower Reynolds number in the model would mean less turbulence and more effective settling, whereas the shorter detention time would mean less effective settling. It is more likely that the effect of shorter detention time would more than offset that of smaller Reynolds number, leading to an underestimate of the extent of settling.

- **Conclusion(s):** The hydrodynamics of settling tanks are complex, owing to the effects of geometry, inlet structures, outlet structures, density currents, wind action, and so on. These factors cause recirculation zones, dead zones and scouring, all of which affect the settling of particles. None of the mathematical models of settling tanks put forth to account for these complexities in real tanks has gained wide acceptance and, as seen here, correct physical models are also difficult to construct, owing to difficulties in achieving dynamic similarity while satisfying the geometry needed to simulate accurately settling processes occurring in the full-scale system.

3.3 DISPERSIVE MASS TRANSPORT

Bulk flow in large tanks and open water systems is more likely to be turbulent than laminar. In such cases, complex motions involving irregular velocity fluctuations, or *eddies*, transport elements of fluid in an irregular manner, somewhat analogous to the random behavior involved in molecular diffusion. Earlier in this chapter we referred to this process as *dispersive mass transport*. We also noted that dispersive mass transport may occur when flow is laminar at the macroscopic scale, which is generally the case for flow through packed beds of filter media or adsorbents and around soil grains. The cause of dispersion in such cases is the disruption of flow lines by obstacles or surfaces. The resulting local variations in fluid velocities can lead to dispersion at the smallest scale, even that of an individual pore channel (e.g., Taylor dispersion). Losses of energy in a fluid as it encounters obstructions cause flow velocities to decrease near surfaces. In cases of flow path disruptions by regular surfaces, this causes parabolic distributions of velocities. In cases of sharp and irregular flow-path disruptions by irregular obstacles, microscale eddies form. The former effect was illustrated in Example 3-2, where laminar flow through fractured channels in a subsurface system was analyzed by the Navier–Stokes equation and velocity distributions within channels were found to be parabolic rather than linear. Such flow conditions will result in an instantaneous injection of some component into the flow at one end of the channel being spread out as it travels down the channel. In other words, although the flow is laminar in a macroscopic sense, there is dispersion due to the distribution of velocities within the flow profile.

The term *hydrodynamic (mechanical) dispersion* is used to describe mass transport caused by all fluid mixing processes that occur at the microscale. The microscale may be the size of an individual grain of soil. For example, at slightly larger (but still microscopic) scale, fluid mixing in subsurface environments and in water filters occurs at intersections of adjacent pore channels of different size having different attendant flow velocities. At an even larger microscopic scale, dispersion takes place as a result of tortuous paths taken by flow through porous media, since not all interparticle pore channels are continuous. It also includes molecular diffusion resulting from Brownian motion, a thermally induced kinetic energy effect rather than a mechanical energy effect. As noted earlier in this discussion, molecular diffusion is an important contributor to overall dispersion/diffusion effects only if fluid flow velocities are very low.

3.3.1 Prandtl Hypothesis

Prandtl (1925) proposed the mixing-length hypothesis in an effort to characterize turbulent flow regimes. In the context of the two-dimensional fixed rectangular coordinate system given in Figure 3-1, the Prandtl hypothesis considers an instantaneous x-component velocity fluctuation, $\tilde{v}_x$, internal to the fluid element, attributable to a y-directional movement of a turbulent eddy through a distance l_m, the mixing length. Figure 3-6 presents a schematic characterization of the mixing length and its associated velocity and spatial relationships.

If the fluid eddy depicted in Figure 3-6 has a mean velocity, $v_{x,y}$, and the adjacent streamline it intercepts a mean velocity $v_{x,y+l_m}$, the magnitude of the velocity fluctuation is

$$\tilde{v}_x = (v_{x,y+l_m} - v_{x,y}) \tag{3-56}$$

For small values of l_m, the velocity fluctuation is given by

$$\tilde{v}_x = l_m \frac{dv_x}{dy} \tag{3-57}$$

Similar instantaneous velocity fluctuations, $\tilde{v}_y$, will occur in the y dimension in turbulent flow fields. Continuity relationships dictate that the x and y components of such instantaneous fluctuations be proportional and that their ab-

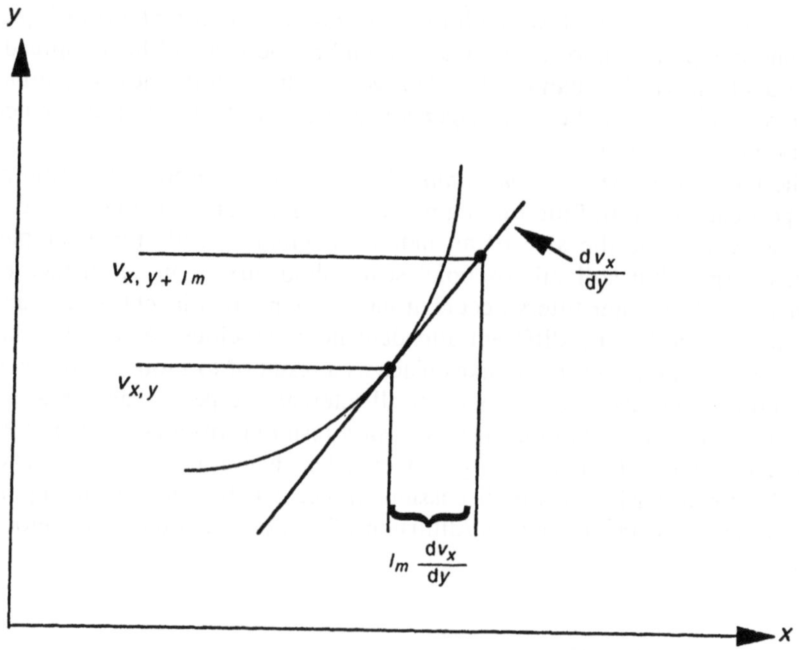

Figure 3-6 Prandtl mixing-length concept.

solute values be similar; thus

$$|\tilde{v}_x| \sim |\tilde{v}_y| \sim l_m \frac{dv_x}{dy} \tag{3-58}$$

Velocity fluctuations act to introduce turbulent shear stresses within fluid elements. If we envision layers of fluid adjacent in the x direction, we can appreciate that some of the momentum flux in any given layer will be lost to an adjacent layer. Although velocity fluctuations are typically time dependent, we can use a steady-flow velocity, v_x, as the mean condition. The translation of x-directional momentum to the y direction with velocity, v_y, as a result of momentum flux lost to an adjacent fluid layer is depicted in Figure 3-7. The mean, steady-state form of the momentum balance (Equation 3-31) for a control volume with the surface area of interest normal to the x direction is

$$\Sigma F_x\big|_{\text{turbulence}} = \int_{\text{top}} \tilde{v}_y \rho (v_x + \tilde{v}_x)\, dA \tag{3-59}$$

Because $\rho\tilde{v}_y v_x$ is much less than $\rho\tilde{v}_y\tilde{v}_x$, this momentum flux can be ignored. The analogous natures of microscopic exchange of momentum in laminar flow and macroscopic exchange of momentum in turbulent flow suggest that $\rho\tilde{v}_y\tilde{v}_x$ can be considered as a *shear-stress* term. Thus, in a momentum balance that

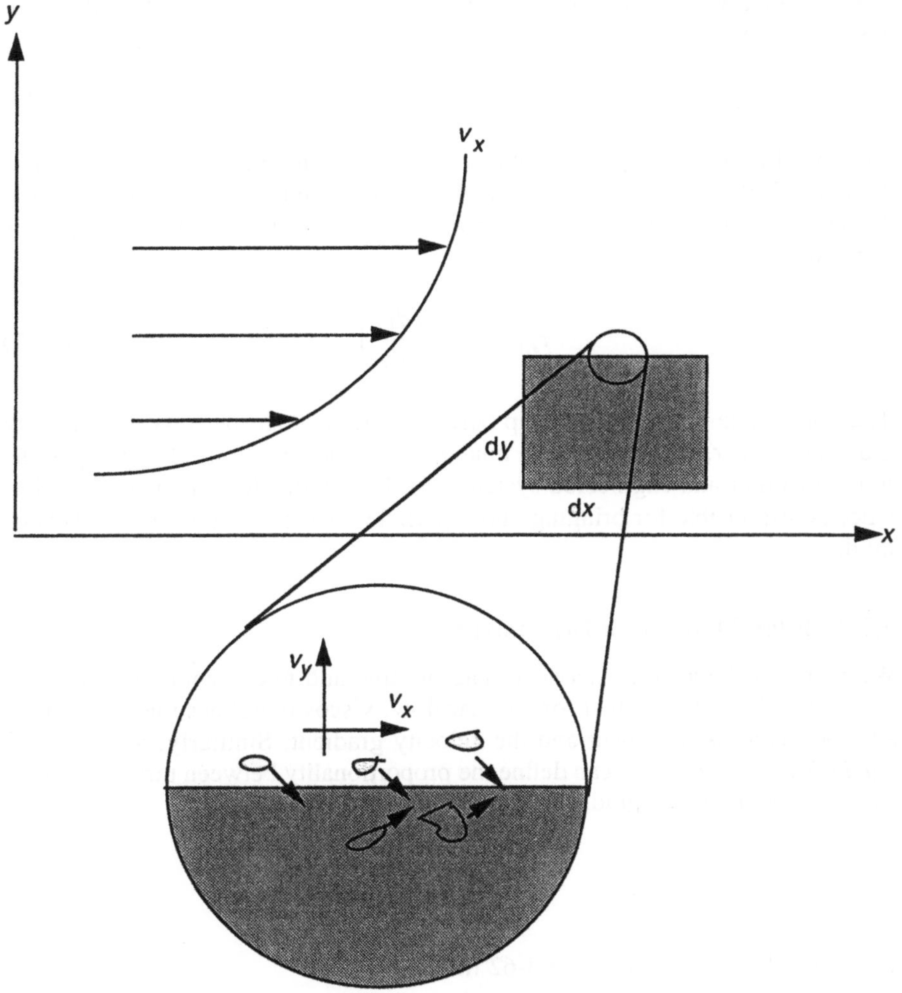

Figure 3-7 Momentum exchange at the surface of a differential control volume.

includes both microscopic and macroscopic forces acting on a control volume, $\rho\tilde{v}_y\tilde{v}_x$ can be transposed from the right side of Equation 3-59 and incorporated as a force-like shear stress in the left-hand side of the equation, yielding

$$\tau_{s,yx} = \mu_v \frac{dv_x}{dy} - \rho\tilde{v}_y\tilde{v}_x \tag{3-60}$$

Total shear stress can therefore be viewed as the summation of momentum flux due to (1) molecular exchange of momentum in laminar flow, and (2) macroscopic exchange in turbulent flow. The latter term is referred to as *turbulent shear stress*, which when expressed in terms of the absolute values of the instantaneous velocities is

$$\tau_{s,t} = \rho|\tilde{v}_x\tilde{v}_y| \tag{3-61}$$

Finally, the unknown proportionality of velocity to the velocity gradient in Equation 3-58 can be incorporated with the unknown mixing-length parameter, l_m, and then substituted into the turbulent shear stress relationship (Equation 3-61) to give

$$\tau_{s,t} = \rho l_m^2 \left(\frac{dv_x}{dy}\right)^2 \tag{3-62}$$

The importance of the relationship between turbulent shear stress and velocity gradient is clear, for example, in fundamental studies of particle aggregations in both natural and engineered systems. In this application, microscale turbulence is responsible for bringing about particle contacts in adjacent packets of fluid.

3.3.2 Eddy Momentum Dispersivity

We return to the analogy between macroscopic and microscopic exchange of momentum by recalling that for laminar flow, viscosity defines the proportionality between viscous shear and the velocity gradient. Similarly, a *time-averaged eddy viscosity*, $\mu_{v,e}$, can define the proportionality between turbulent shear and the time-averaged gradient of turbulent velocity; that is,

$$\tau_{s,t} = \mu_{v,e} \frac{dv_x}{dy} \tag{3-63}$$

Thus, it follows from Equation 3-62 that

$$\mu_{v,e} = \rho l_m^2 \frac{dv_x}{dy} \tag{3-64}$$

Eddy viscosity is not a property of the fluid alone, but rather, combines a fluid property (ρ) and flow properties l_m^2 (dv_x/dy). As such, it is a function of location, or a spatially dependent variable. In the immediate vicinity of phase interfaces or fluid boundaries, for example, the value of the eddy viscosity term must approach zero as l_m approaches zero. Indeed, the Prandtl hypothesis does not uniquely define the relationship of l_m to y boundary dimensions. This point will be addressed later when discussing the turbulent boundary layer and its relationship to interphase mass transport in Chapter 4.

Fluid properties and flow properties are separated by defining a *time-aver-*

aged, kinematic eddy viscosity constant, $\nu_{v,e} = \mu_{v,e}/\rho$, which is a function only of flow properties, such that, from Equation 3-64,

$$\nu_{v,e} = \frac{\mu_{v,e}}{\rho} = l_m^2 \frac{dv_x}{dy} \tag{3-65}$$

Transposing Equation 3-65 and substituting into Equation 3-63 yields the following expression for turbulent shear stress:

$$\tau_{s,t} = \rho \nu_{v,e} \frac{dv_x}{dy} \tag{3-66}$$

Kinematic eddy viscosity is frequently referred to as *eddy momentum diffusivity*. It has the same dimensions as molecular diffusivity (L^2t^{-1}). We prefer to refer to it as *eddy momentum dispersivity*, or simply *eddy dispersivity*. As we will see in Chapter 4, molecular diffusion accounts for the microscale transport of component i within a fluid, and it is functionally dependent on spatial differences in its concentration, C_i, at the microscale. The following empirical expression was first proposed by Fick in 1855 to describe this transport process:

$$N_{i,x} = -\mathfrak{D}_{l,i} \frac{dC_i}{dx} \tag{3-67}$$

where $N_{i,x}$ is diffusional flux in moles per unit area per unit time and $\mathfrak{D}_{l,i}$ is the free-liquid diffusion coefficient or *free-liquid diffusivity*. The similarities between the mathematical expressions for molecular diffusion and eddy dispersion are once again apparent. Both free-liquid diffusivity and eddy dispersion have the same dimensions. Free-liquid diffusivity is the proportionality between mass flux and the concentration gradient, whereas eddy momentum dispersion is the proportionality between momentum flux and the turbulent velocity gradient.

The preceding discussion pertains to transport of *fluids* by eddies, but we can similarly conceptualize concurrent transport of any *component* contained within a fluid. The Prandtl mixing-length concept is again invoked, and simply by replacing velocity with concentration, Figure 3-6 serves to describe turbulence-induced fluctuations in the concentration of component i:

$$\tilde{C}_i = C_i|_{y+l_m} - C_i|_y = l_m \frac{dC_i}{dy} \tag{3-68}$$

where $\tilde{C}_i$ is the *instantaneous* concentration fluctuation, C_i is the *time-averaged* concentration at each location in adjacent fluid layers, and dC_i/dy is the mean concentration gradient. The flux of C_i in the y direction due to molecular diffusion plus turbulent transport is

$$N_{y,i} = -\mathfrak{D}_{l,i}\frac{dC_i}{dy} - \tilde{v}_y l_m \frac{dC_i}{dy} \tag{3-69}$$

The second term on the right side of Equation 3-69 represents the instantaneous flux in the y direction, that is, the product of the velocity fluctuation induced by turbulence and the concentration fluctuation (Equation 3-68). Neither $\tilde{v}_y$ nor l_m are measurable, and the two terms are therefore grouped together to define the eddy dispersion coefficient, or *eddy dispersivity*, $\mathfrak{D}_\epsilon$. Turbulence involves random motion fluid elements just as molecular diffusion involves random motion of molecules. The total flux of component i due to the combined influence of molecular and eddy diffusion in any direction may thus be written in terms of its concentration gradient in that direction as

$$N_{y,i} = -(\mathfrak{D}_{l,i} + \mathfrak{D}_\epsilon)\frac{dC_i}{dy} \tag{3-70}$$

The relationship between *eddy mass dispersivity* and *eddy momentum dispersivity* merits further discussion. We recall that velocity fluctuations are given by

$$\tilde{v}_y \approx \tilde{v}_x \approx l_m \frac{dv_x}{dy} \tag{3-58}$$

and further that the eddy mass dispersivity can be defined from Equation 3-69 as

$$\mathfrak{D}_\epsilon = \tilde{v}_y l_m \tag{3-71}$$

Substitution of Equation 3-58 into Equation 3-71 yields another way to express eddy mass dispersivity:

$$\mathfrak{D}_\epsilon = l_m^2 \frac{dv_x}{dy} \tag{3-72}$$

This expression is identical to the definition of eddy momentum dispersivity given by Equation 3-65 and it serves to introduce an important analogy between momentum and mass transfer (the *Reynolds analogy*) which will be further explored at the end of Chapter 4.

3.3.3 Dispersion Coefficients and Numbers

As noted in Equation 3-70, the contributions of molecular diffusion and eddy dispersion to mass transport in turbulent flow can be lumped together because both mechanisms are random processes and thus depend on the concentration gradient. The resulting *dispersion coefficient*, $\mathfrak{D}_d$, is then

$$\mathfrak{D}_d = \mathfrak{D}_{l,i} + \mathfrak{D}_\epsilon \tag{3-73}$$

The free-liquid diffusivity will usually be orders of magnitude smaller than the eddy dispersivity in turbulent flow situations, and under such conditions, mass transport contributions due to molecular diffusion can be neglected.

Unlike free-liquid diffusivity, which can be estimated from the molecular characteristics of the component and the viscosity and temperature of the fluid, eddy dispersivity depends on flow conditions and on the scale at which fluid mixing is measured. The dispersion coefficient is therefore necessarily a *system-specific constant* and usually impossible to determine *a priori* for real systems.

In most systems of practical concern, macrotransport of components occurs by a combination of advection and dispersion. That is, components are carried along by the bulk flow while being dispersed by momentum exchange of fluid elements due to turbulence. In the two-dimensional fixed-coordinate system depicted in Figure 3-1, the total molar flux in the x direction is the sum of the advective flux and dispersive flux, or

$$N_i|_T = vC_i - \mathfrak{D}_d \frac{dC_i}{dx} \tag{3-74}$$

In natural systems, dispersion occurs in two and three dimensions. Dispersion coefficients generally do not have the same value in each direction in such cases. Even if the medium is *isotropic* (i.e., has the same dispersive properties in each direction in space), the fact that there is a principal direction to bulk flow means that more dispersion is created in that direction (*longitudinal dispersion*) than in directions normal to the flow (*transverse dispersion*). A simple example of directional differences in dispersion for an *isotropic* medium is evident in the spreading of a wastewater discharge plume in a river.

If the medium is *anisotropic*, it has different diffusion properties in different directions. For example, lakes stratify vertically due to development of thermoclines, and this limits vertical dispersion. Dispersion in the subsurface is even more complex because of three-dimensional heterogeneity (e.g., different orientations of soil grains and layering of different soil types). Thus the dispersive flux in each of three directions is additionally influenced by the concentration gradients of solutes and dispersion coefficients in the other two directions; that is,

$$-N_x = \mathfrak{D}_{d11} \frac{\partial C}{\partial x} + \mathfrak{D}_{d12} \frac{\partial C}{\partial y} + \mathfrak{D}_{d13} \frac{\partial C}{\partial z} \tag{3-75}$$

$$-N_y = \mathfrak{D}_{d21}\frac{\partial C}{\partial x} + \mathfrak{D}_{d22}\frac{\partial C}{\partial y} + \mathfrak{D}_{d23}\frac{\partial C}{\partial z} \tag{3-76}$$

$$-N_z = \mathfrak{D}_{d31}\frac{\partial C}{\partial x} + \mathfrak{D}_{d32}\frac{\partial C}{\partial y} + \mathfrak{D}_{d33}\frac{\partial C}{\partial z} \tag{3-77}$$

Accounting for both advective and dispersive flux, we may write the advection–dispersion equation in compact vector form as

$$N_i|_T = \boldsymbol{v}C_i - (\mathfrak{D}_d \cdot \nabla)C_i \tag{3-78}$$

where it is understood that $\mathfrak{D}_d$ is a *second-order tensor*, and has both magnitude and three first-order tensor (ordinary vector) components. If the directions of flow correspond exactly to the directions of the coordinate system, $\mathfrak{D}_d$ values need be specified only for each of three directions. However, in subsurface systems, groundwater may flow along tilted beds. In this instance, the *local* direction of transport is along the bedding planes while the *global* direction is along the horizontal. It is now necessary to account for nine components of dispersion (three for each of three directions) to obtain their contributions in the local direction of transport. Accordingly, $\mathfrak{D}_d$ is referred to as a *hydrodynamic dispersion tensor.*

Dispersive transport is not restricted to movement under the influence of *concentration* gradients. It can be motivated, for example, by any of the various types of energy gradients illustrated earlier in Figure 1-7. As a case in point, an *electrical potential* gradient ($\Delta\Phi$) produced by the charged surfaces of a porous medium (e.g., the surfaces of a membrane comprised of parallel capillary tubes) can induce dispersive transport of a charge solute species. The equation to describe transport under the influence of a gradient in a potential field is referred to as the *Nernst–Planck equation*:

$$N_i|_T = \boldsymbol{v}C_i - (\mathfrak{D}_d \cdot \nabla)C_i - \frac{z_i^\circ \mathfrak{F}}{\mathfrak{R}T}(\mathfrak{D}_d \cdot \nabla)C_i \cdot \Delta\Phi \tag{3-79}$$

This equation is solved together with the Navier–Stokes equation (Equation 3-45), as discussed earlier, to determine the velocity and concentration profiles in porous media.

An additional complication arises in describing hydrodynamic dispersion in porous media. That is, longitudinal dispersion has been shown by experiments to be scale related over long length scales. Thus for the one-dimensional case of dispersion in the x direction,

$$\mathfrak{D}_d = \mathfrak{D}_{l,i} + \phi(x)\mathfrak{D}_\epsilon \tag{3-80}$$

The larger the distance, x, over which dispersion is measured, the larger the dispersion. This scale effect is caused by the heterogeneity of the soil media.

The *advection–dispersion–reaction* (ADR) *equation* is written most generally by substitution of the total flux equation (Equation 3-78) into the point form of the material balance (Equation 2-16), yielding

$$\nabla \cdot (\mathfrak{D}_d \cdot \nabla C_i) - (v \cdot \nabla)C_i + r_i = \frac{\partial C_i}{\partial t} \tag{3-81}$$

This equation allows spatial and temporal determination of the concentration of component *i* in systems involving advective and dispersive transport coupled with homogeneous reactions. More terms may be added for heterogeneous reactions (i.e., reactions involving components of more than one phase).

The one-dimensional form of Equation 3-81 is

$$\mathfrak{D}_d \frac{\partial^2 C_i}{\partial x^2} - v_x \frac{\partial C_i}{\partial x} + r_i = \frac{\partial C_i}{\partial t} \tag{3-82}$$

By making this equation nondimensional (see Chapter 9), we obtain the dimensionless *dispersion number:*

$$\text{dispersion number} = \mathfrak{N}_d = \frac{\mathfrak{D}_d}{v_x L} \tag{3-83}$$

where L is the total length in the x direction. The dispersion number is important in reactor engineering and for the description of mass transport in natural systems. The larger $\mathfrak{N}_d$, the more dominant is dispersive transport over advective transport. Example 3-4 provides a solution to the ADR equation at steady state.

Example 3-4

- **Situation:** The outfall from a municipal wastewater treatment plant is located on a river 10 km above a bathing beach. A plant malfunction causes the coliform bacteria count in the river to rise almost instantaneously to 400 cells per 100 mL. The velocity of the river is 10 km per day and the dispersion coefficient is 10 km^2 per day. The die-off rate of bacteria is first order with respect to bacterial count, with a rate constant $k = 0.5$ per day. The bathing standard is $<$200 cells per 100 mL.
- **Question(s):** Will the bathing standard be violated at the beach?
- **Logic and Answer(s):**
  1. The ADR equation (Equation 3-82) can be applied to this analysis.
  2. The reaction rate term in the ADR equation is in this case

$$r = -kC = -0.5C$$

3. Steady-state conditions can be assumed to exist if the river flow and bacteria count at the discharge point are reasonably constant in time. The equation to be solved is then

$$\mathfrak{D}_d \frac{d^2C}{dx^2} - v_x \frac{dC}{dx} - kC = 0$$

4. Integration of this expression with appropriate initial and boundary conditions (see Chapter 9) yields

$$\frac{C}{C_{\text{IN}}} = \frac{4\beta_D \exp\left(0.5 \dfrac{v_x L}{\mathfrak{D}_d}\right)}{(1 + \beta_D)^2 \exp\left(0.5\beta_D \dfrac{v_x L}{\mathfrak{D}_d}\right) - (1 - \beta_D)^2 \exp\left(-0.5\beta_D \dfrac{v_x L}{\mathfrak{D}_d}\right)}$$

$$\beta_D = \left(1 + 4 \frac{kL}{v_x} \frac{\mathfrak{D}_d}{v_x L}\right)^{0.5}$$

where C_{IN} is the concentration of bacteria in the river at the outfall. The value of the dispersion number (Equation 3-83) which appears in this equation is equal to 0.1. The resulting fractional concentration is 0.619. Thus the value of C is (0.619)(400) = 248 cells per 100 mL, which violates the standard.

5. To examine the relative effect of dispersion, we can calculate a value of C for which dispersion is ignored entirely. The ADR equation given in step 3 then reduces to

$$-v_x \frac{dC}{dx} - kC = 0$$

which can readily be solved by separation of variables to give

$$\frac{C}{C_{\text{IN}}} = \exp\left(\frac{-kx}{v_x}\right)$$

The fractional concentration at $x = 10$ km is 0.606, which is slightly lower than that obtained by including dispersion. The effect of dispersion is thus minimal in this instance.

6. The effect of higher levels of dispersion can be examined by increasing $\mathfrak{N}_d$ to a value of 1.0. Substitution into the concentration ratio relationship developed in step 4 gives

$$\frac{C}{C_{\text{IN}}} = 0.645$$

An increase in dispersion to this level means that more elements of the fluid reach the bathing area in a shorter time than would be true if transport were by advection only. The fractional concentration therefore increases because less time is available for bacteria die-off.

3.4 SUMMARY

The transport of component mass through a fluid system may involve a variety of processes, including (1) advection, (2) dispersion, (3) bulk-phase molecular diffusion, and (4) interphase mass transfer. In this chapter we have focused on the *macrotransport* processes of advection and dispersion. The nature of fluid flow is in these cases determined by mass, energy, and momentum balances, which lead to the well-known equations developed by Bernoulli and Navier–Stokes and equally well-known dimensionless parameters, such as the Reynolds and Froude numbers. These equations and dimensionless numbers have great utility for analyzing flows in engineered reactors, rivers, lakes, oceans, and subsurface systems.

Under purely laminar flow conditions, components are transported advectively at the velocities of fluid elements comprising the flow streamlines. Environmental systems more commonly involve some degree of turbulence at various scales, however, such that the macrotransport of components becomes a complex process that cannot be quantified in terms of fluid velocity alone. Whether in the form of large-scale eddies in bulk fluids or in the form of small-scale disturbances to flow patterns caused by fluid shear near solid surfaces, turbulence produces dispersive transport. Although existing theories cannot quantify turbulence exactly, the Prandtl mixing-length concept has proven useful for relating shear and velocity gradients to describe fluid eddies. The description of dispersive transport of components within fluid eddies then follows from the Prandtl mixing-length concept and from the definition of eddy dispersivity. Eddy dispersivity is, however, still very much an empirical notion.

Further empiricism is introduced when eddy dispersivity and molecular diffusivity are summed to define the dispersion coefficient. Although these proportionality constants are summed, the scale at which the associated concentration gradients are operative may be very different. In fact, it is well known that the scale at which experimental measurements of concentration gradients are made greatly influences the magnitude of the resulting dispersion coefficient. Despite these fundamental shortcomings, the dispersion coefficient is an essential tool for estimating macrotransport in practical situations. In particular, the *dispersion number*, an important dimensionless group that incorporates the dispersion coefficient, provides a means for quantifying the relative importance of dispersive and advective macrotransport in engineered and natural systems. It is intuitive that the dispersion number and the Reynolds number should be closely related. The Reynolds number quantifies turbulence, and we have seen that turbulence is a factor that can significantly influence dispersive transport. The connection between these two dimensionless parameter groups will be explored quantitatively in Chapter 9.

3.5 REFERENCES AND SUGGESTED READINGS

3.5.1 Citations and Sources

Bishop, M. M., J. M. Morgan, B. Cornwell, and D. K. Jamison, 1993, "Improving the Disinfection Detention Time of a Water Treatment Plant Clearwell," *Journal of the American Water Works Association*, *85*(3), 68–75. (Problem 3-5)

Cwirko, E., and R. G. Carbonnell, 1992, "Interpretation of Transport Coefficients in Nafion Using a Parallel Pore Model," *Journal of Membrane Science*, *67*, 227–247. (Equations 3-45 and 3-79 and related discussion)

de Marsily, G., 1986, *Quantitative Hydrogeology*, Academic Press, Inc., New York (source material for Example 3-2)

Olson, R. M., and S. J. Wright, 1990, *Essentials of Engineering Fluid Mechanics*, 5th Edition, Harper & Row, Publishers, New York (Figure 3-4)

Prandtl, L. Z., 1925, "Bericht uber Untersuchungen zur ausgebildeten turbulent," *Angewandte Mathematik und Mechanik*, *5*, 136. (Section 3.3.1)

Weber, W. J., Jr., 1972, *Physicochemical Processes for Water Quality Control*, Wiley-Interscience, New York. (Example 3-3)

3.5.2 Background Reading

Freeze, R. A., and J. A. Cherry, 1979, *Groundwater*, Prentice Hall, Englewood Cliffs, NJ. Chapter 2 provides a good introduction to the isotropy (homogeneous and heterogeneous) and anistropy (homogeneous and heterogeneous) in the context of subsurface flow conditions. Various scales of hydrodynamic dispersion are discussed.

Olson, R. M., and S. J. Wright, 1990, *Essentials of Engineering Fluid Mechanics*, 5th Edition, Harper & Row, Publishers, New York. Open-channel flow analysis is developed in Chapter 12. The specific energy concept and critical depth in relation to the Froude number are detailed. Numerous examples and photographs illustrate hydraulic jump situations.

Probstein, R. F., 1994, *Physicochemical Hydrodynamics*, 2nd Edition, Butterworth Publishers, Stoneham, MA. A broad range of fluid flow effects on physical, chemical, and biochemical processes and, conversely, the effects of these processes on fluid flow are discussed. The behavioral similarities of different physical and chemical phenomena are demonstrated by use of transport relations for heat, mass, and charge. Dimensionless forms of the continuity, momentum, convective diffusion, and energy equations are derived, from which many useful dimensionless parameters, including the Reynolds, Peclet, Prandtl, and Schmidt numbers, are interrelated.

Roberson, J. A., and C. T. Crowe, 1985, *Engineering Fluid Mechanics*, 3rd Edition, Houghton Mifflin Company, Boston. The Lagrangian and Eulerian viewpoints of fluid motion are discussed in Chapter 4, as is a useful perspective on area and velocity vectors in development of the control volume for analysis of flow.

Welty, J. R., C. E. Wicks, and R. E. Wilson, 1984, *Fundamentals of Momentum, Heat, and Mass Transfer*, 3rd Edition, John Wiley & Sons, Inc., New York. Chapters 5 and 6 cover derivation of momentum and energy balances, and the Navier–Stokes equation is derived in detail in Chapter 9; practical applications are given. Turbulent shear stress and Prandtl's mixing-length hypothesis are discussed in Chapter 13.

White, F. M., 1986, *Fluid Mechanics*, 2nd Edition, McGraw-Hill Book Company, New York. Chapters 3 and 4 provide detailed development of the control volume concept and the use of integral and differential approaches to mass, energy, and momentum balances from which derive the Bernoulli and Navier–Stokes equations. Chapter 5 is a good treatment of dimensional analysis, including the Buckingham pi theory, the concepts of geometric and kinetic similarity, and the derivation of Reynolds, Froude, and Euler numbers from the dimensionless form of the Navier–Stokes equation.

3.6 PROBLEMS

3-1 The tank shown below is filled initially with an industrial solvent.

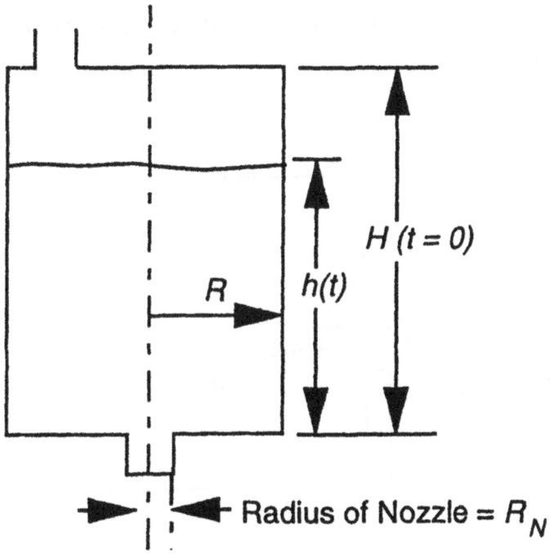

In ordinary operation, the nozzle at the bottom of the tank is opened and closed electronically. A malfunction occurs and the nozzle remains open. Derive an expression to predict the time for the tank to empty. [*Hint:* The beginning point is an energy balance assuming that atmospheric pressure exists at the top of the fluid and at the exit point of the fluid.

The relationship sought for exit nozzle velocity, v_N, is known as Torricelli's law, $v_N = (2gh)^{0.5}$, which needs to be combined with a mass balance on the fluid. Chapter 6 in the book by Welty, Wicks, and Wilson (Background Reading) is helpful.]

Answer: $t = \left(\frac{R}{R_N}\right)^2 \left(\frac{2H}{g}\right)^{0.5}$

3-2 The Bernoulli equation was derived from the energy balance by assuming constant internal energy, U, so that between two points, $U_1 - U_2 = 0$. *However*, we then added a head-loss term, h_l, in the Bernoulli equation to account for the *change* in internal energy. Water flows through a sudden expansion as shown below. Considering only normal stress (neglect wall stress), write the momentum balance between the points 1 and 2 and write the energy balance *without* making the assumption of constant internal energy. Use these two equations to show that

$$U_2 - U_1 = \frac{v_1^2}{2}\left(1 - \frac{A_1}{A_2}\right)^2$$

Is this *change* in internal energy consistent with the head-loss term that we write in the Bernoulli equation and with your knowledge of what happens at pipe expansions? Discuss how this equation compares to well-known empirical relationships for minor head losses incurred at pipe expansions. [*Hint:* Chapter 6 in Welty, Wicks, and Wilson (Background Reading) is helpful.]

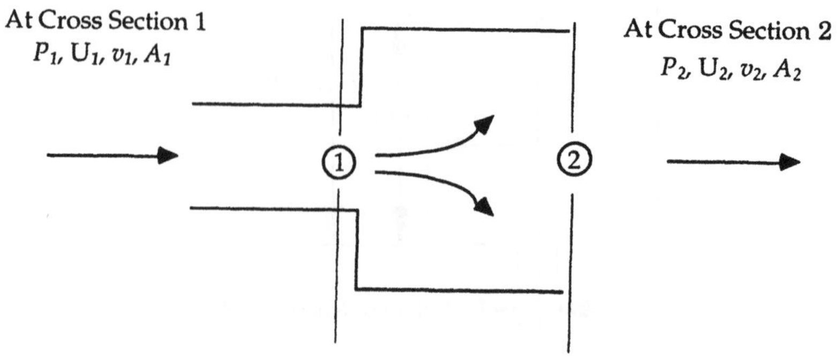

3-3 Consider the open-channel flow problem in which we have a rise in channel bed, the so-called *rapidly varying flow problem*, due to channel transition. Show that (a) the critical depth corresponds to a Froude number, $\mathfrak{N}_{Fr}$, of 1; (b) $\mathfrak{N}_{Fr}$, > 1 for supercritical flow; and (c) $\mathfrak{N}_{Fr}$ < 1 for subcritical flow. What is the physical significance of this observation?

3-4 Use the data in Example 3-4 to make a plot of the dependence of the residual bacterial cell count at the bathing beach on the dispersion number (ranging from 0.1 to 1.0). Repeat this plot for a die-off rate of 1.0 per day instead of 0.5 per day. Discuss these results relative to those for an assumption of complete mixing.

3-5 Bishop et al. (1993) used physical models to simulate flow patterns in full-scale disinfection clearwells. In light of Example 3-3, explain their approach to scaling and the evidence they present to show that similitude was achieved. Discuss whether the bacteria kill would necessarily have been the same as that expected for the full-scale system if both a disinfectant and bacteria had been introduced in the influent to the model clearwell and the bacteria remaining in the effluent measured.

3-6 A deep-injection well disposal has been proposed for a brine waste. A study is planned to investigate dispersion in which a salt slug will be injected instantaneously and deep bore holes will be used to measure salt concentration in space. Assume that injection well is a *line source* of salt, as shown in the accompanying figure. That is, the salt concentration is uniform in the vertical direction (y direction) and there is advective and dispersive transport in the direction of groundwater flow (x direction) and dispersive transport in the transverse direction (z direction).

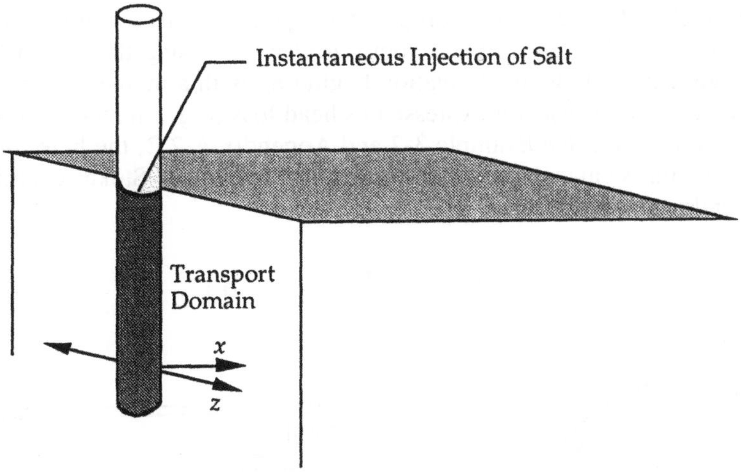

(a) Write a material balance to describe the transport of salt in the direction of groundwater flow and in the transverse direction. State all of your assumptions.

(b) The mathematical solution of the resulting partial differential equation to describe two-dimensional transport for an instantaneous source is

$$C(x, z, t) = \frac{W_g/L}{4\pi t(\mathfrak{D}_{\epsilon,x}\mathfrak{D}_{\epsilon,z})^{0.5}} \exp\left[-\frac{(x - v_x t)^2}{4\mathfrak{D}_{\epsilon,x}t} - \frac{y^2}{4\mathfrak{D}_{\epsilon,z}t}\right]$$

in which $\mathfrak{D}_{\epsilon,x}$ and $\mathfrak{D}_{\epsilon,z}$ are the dispersion coefficients in the x and z directions, v_x the velocity in the x direction, and W_g/L the amount (weight) of salt released per unit length of bore hole. You wish to estimate the location of sample bore holes to capture the salt concentration profile after injection and from which measurements of dispersion coefficients may be obtained. Sampling will be done once a month for 6 months. For estimating purposes, the typical dispersion coefficients expected are: $\mathfrak{D}_{\epsilon,x} = 1.0\ \text{m}^2/\text{day}$ and $\mathfrak{D}_{\epsilon,z} = 0.5\ \text{m}^2/\text{day}$. Assume further that $v_x = 0.5$ m/day and $W_g/L = 9$ kg/m. Select three or four sampling locations and compute the concentration profiles that might be expected for monthly sampling.

3-7 The head loss in laminar, steady, uniform pipe flow is given by the Hagen–Poiseuille equation:

$$\frac{\Delta P}{L} = \frac{32\mu_v v}{d_P^2}$$

in which ΔP is the pressure drop, L the pipe length, d_P the pipe diameter, μ_v the viscosity, and v the velocity ($4Q/\pi d_P^2$). Using the sketch below for guidance, show the derivation beginning with a momentum balance. Pressure drop is usually expressed as head loss (e.g., in meters of head). Explain. [*Hint:* See Example 3-2 and Appendix A.2.2, the latter providing the mass balance approach needed for the Navier–Stokes equation in cylindrical coordinates:

$$-\frac{\partial P}{\partial z} + \frac{1}{r}\frac{\partial}{\partial r}(r\tau_{s,rz}) = 0.]$$

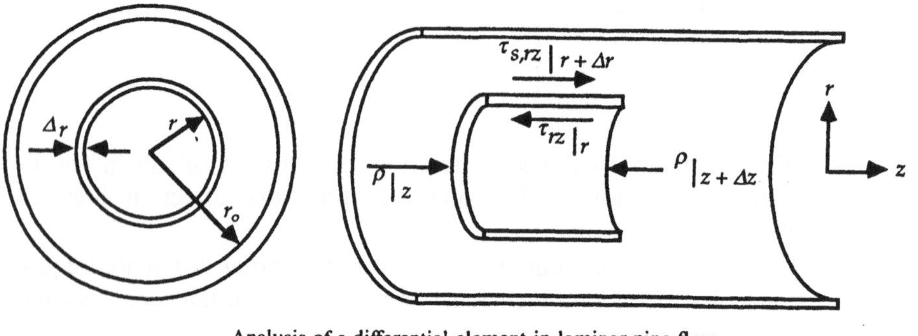

Analysis of a differential element in laminar pipe flow.

3-8 An explosion at a fuel storage tank farm located just upstream of a water intake produces a spill that increases the concentration of benzene in the river to 50 μg/L for several days. Water is transported from the intake to the water treatment plant through a 3000-m pipeline at a velocity v_x of 1 m/s. The dispersion coefficient, $\mathfrak{D}_d$, in the pipeline is 2 m^2/s, and this causes spreading of the originally sharp benzene "front." To facilitate appropriate preventive action, estimate when to expect the benzene concentration to reach 0.5 μg/L at the water treatment plant. Comment on the influence of pipeline dispersion on the arrival time of the benzene at the treatment plant. [*Hint:* The transport problem is similar to that in Example 2-5. However, the Lagrangian frame of reference provides a more convenient solution than in Example 2-5. An "observer" travels with the velocity of water through the pipeline and observes the spread in benzene concentration due to dispersion. The relative velocity is now zero in the advection–dispersion equation found in step 1, which then simplifies to

$$\mathfrak{D}_d \frac{\partial^2 C}{\partial x^2} = \frac{\partial C}{\partial t}$$

The analytical solution (see Chapters 8 and 11 for more detail for this situation) is

$$\frac{C}{C_0} = 0.5 \operatorname{erfc}\left[\frac{x_R}{(4\mathfrak{D}_d t)^{0.5}}\right]$$

where x_R is the position relative to the position of the observer traveling at the velocity of the water in the pipeline at which the benzene concentration is calculated from the equation above. The initial and boundary conditions are

$$\text{IC:} \quad C = C_0(x_R \leq 0) \qquad C = 0(x_R \geq 0)$$

$$\text{BC:} \quad C = C_0(x_R = -\infty) \qquad C = 0(x_R = \infty)$$

The following specific relationship exists between x_R and t for calculation of the benzene concentration at the water treatment plant:

$$x_R = x - v_x t = 3000 - v_x t.]$$

Answer: Benzene at a concentration of 0.5 μg/L arrives 0.765 h after the spill and it leads the observer traveling at the water velocity by a distance of 245 m. Ignoring the effects of dispersion (i.e., assume plug flow conditions), the concentration of benzene at the water intake would reach a level of 50 μg/L and would arrive 0.833 h after the spill.

3-9 A highly treated industrial wastewater from a small manufacturing process is injected into a 30-cm-diameter well which fully penetrates a 3-m-thick confined aquifer. The recharge pumping rate is 200 m^3/day. If 50 μg/L of a solvent is present in the injected wastewater and the solvent is degraded by microbial action in the subsurface according to a first-order rate process (k = 0.2 per day), find the solvent concentration at a radial distance of 10 m from the injection well. Assume steady-state conditions and that transport is dominated by one-dimensional radial advection. The aquifer porosity, ϵ_d, is 0.2. [*Hint:* See Chapter 2 and Appendix A.2.2 for a description of the material balance in the radial direction. The radial seepage velocity is the pumping rate divided by the cross-sectional area of the confined aquifer (accounting for porosity) normal to the radial direction of flow and is therefore a function of radius.]

Answer: *7.1 μg/L.*

4

MICROTRANSPORT PROCESSES

4.1 DIFFUSIVE MASS TRANSPORT

4.1.1 Concentration-Dependent Transport Processes

If a system contains a component that is not homogeneously distributed—that is, if one or more spatial concentration gradients exist—natural forces act to transfer the mass of that component in such manner as to minimize its concentration differences. The diffusion of oxygen from bubbles in an aeration basin, the removal of water from chemical and biological sludges, and the air stripping of dissolved gases from wastewaters are typical examples of transport processes effected by concentration gradients. These examples involve gradients and mass transport across phase boundaries; that is, *interphase* mass transport processes. Spatially variable distributions of mass within a given phase can similarly lead to *intraphase* mass transport processes, which are important considerations in many water quality systems. Examples of intraphase mass transfer processes include diffusion in the interstitial waters of soils and other porous media, diffusion of adsorbed substances within the pores of activated carbon, the diffusive mass transport of salt in estuarine systems, and diffusion of oxygen through the water column of a stream or lake after it has crossed the air–water interface during reaeration.

The time required for a substance to distribute within a system can vary widely. In a totally quiescent system, the mechanism of intraphase mass transport leading to reduction of concentration gradients is the random molecular motion of individual molecules. If the system is not quiescent, the dynamic characteristics of fluid movement affect groups or clusters of molecules and thus tend to diminish concentration gradients more rapidly. These two distinctly different modes of transport of dissolved substances within and among the fluid elements of a water system are distinguished by the terms *diffusive mass transport* and *dispersive mass transport*, respectively. On a fundamental level, both involve the exchange of momentum but at different scales. Diffusion occurs at the microscale (the molecular level), whereas dispersion occurs at the macroscale and involves the mixing of fluid elements, as described in Chapter 3.

4.1.2 Molecular Diffusion

The behavior of the individual molecules of a substance dissolved in a dilute solution contained in a sealed and totally quiescent vessel provides a convenient model to illustrate natural diffusive processes. Because of their respective kinetic energies, the solute and solvent molecules comprising this system continuously undergo *Brownian motion*, and frequent collisions occur among them. As a result of these collisions, each solute molecule travels a course that changes direction continuously and randomly. The resulting path may carry an individual molecule away from a region of high concentration or away from a region of low concentration. The movement of a particular molecule in such a system is random, and the process is frequently referred to as a *random-walk process*. The fact that there are greater numbers of molecules moving about in regions of higher concentration, however, means that there must be a net random movement of molecules out of these regions into regions of lower concentration. The magnitude of such net diffusive transport of a substance i is thus functionally dependent on spatial differences in its concentration, C_i. While this description of diffusive transport is probabilistic in character, the early development of diffusion relationships was largely empirical, and ostensibly deterministic. We shall explore these early developments first to develop terminology and a clearer perception of the process and then reexamine the probabilistic character of diffusion.

4.1.2.1 Fick's First Law In 1855, Adolf Fick proposed the following empirical relationship to describe one-dimensional (x) steady-state diffusion of a component i in a dilute solution in response to an observed spatial concentration gradient dC_i/dx:

$$J_{i,x} = -\mathfrak{D}_{l,i} \frac{dC_i}{dx} \tag{4-1}$$

Fick's law states that the *diffusive* flux, $J_{i,x}$, is directly proportional to the concentration gradient. The coefficient of proportionality, $\mathfrak{D}_{l,i}$, termed the *free-liquid diffusion coefficient*, has dimensions of L^2t^{-1}. Equation 4-1 includes a negative sign to reflect the fact that the flux is positive when the concentration is decreasing in the positive x direction.

Strictly speaking, the term $J_{i,x}$ (Equation 4-1) defines the flux of i *relative* to that of the molar average flux of a mixture of two components, i and j. $J_{i,x}$ is therefore a flux measured relative to a *moving* coordinate system. For most applications of interest, the second component is water, the solvent in which i is dissolved. As we discuss in more detail in Section 4.1.2.4, the notation N_i is used to describe the flux of i relative to a *fixed* coordinate system. N_i can be thought of as the sum of the diffusional flux, J_i, and that caused by bulk motion. In a quiescent system the latter flux is significant only if the concentration of i relative to j is very large. Thus, J_i and N_i are used interchangeably when diffusion of i occurs in dilute solutions, the most common situation for contaminants dissolved in water.

The similarity of Fick's first law of diffusion to the one-dimensional form of Fourier's law for heat flux should be noted:

$$N_{H,x} = -k_c \frac{dT}{dx} \tag{4-2}$$

where $N_{H,x}$ is the heat flux in calories per unit area per unit time, k_c the thermal conductivity, and dT/dx the spatial temperature gradient. This similarity is discussed further in Section 4.2.8.

In generalized vector notation, Equation 4-1 becomes

$$J_i = -\mathfrak{D}_{l,i} \nabla C_i \tag{4-3}$$

Any flux term is inherently a vector quantity, as is the spatial gradient in the potential that induces the flux. Fick's first law, given in its most common form in Equation 4-1, usually defines flux with respect to a specific directional component (i.e., x, y, or z) and thus is a scalar quantity. When so written, the fact that the flux and the concentration gradient are directional components of their respective vectors is implied, if not specifically stated. A practical application of the steady-state diffusion equation is given in Example 4-1.

Example 4-1

- **Situation:** Passive dosimetry is being considered as a potentially inexpensive method to detect the presence of volatile organic chemicals in the vapor phase above the groundwater table in the vicinity of underground storage tanks. Passive dosimetry has been used for years in the field of industrial hygiene to measure the exposure of workers to airborne contaminants in the workplace. A schematic diagram of a dosimeter is shown below.

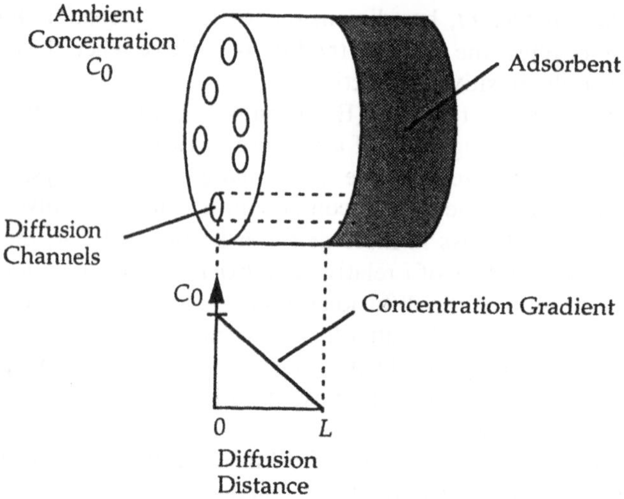

Passive dosimeters take the form of "badges" worn by workers on their outer garments. If we assume that the ambient contaminant concentration, C_0, remains constant over the dosimeter exposure period and that the concentration remains at zero at the far end of the diffusion channel in the badge due to the presence of the large mass of adsorbent, this diffusion process can be assumed to occur under steady-state conditions. At the end of the exposure period, the adsorbent is removed from the badge and extracted to measure the total mass of contaminant collected during the exposure period.

- **Question(s):** Analyze the performance characteristics of a passive dosimeter of the type described above and discuss its applicability to this problem.

- **Logic and Answer(s):**

1. If the contaminant is adsorbable and can be desorbed for measurement of the mass collected, a passive dosimeter could be lowered into a sampling well, the walls of which are perforated above the groundwater table to allow equilibration with the surrounding vapor phase.

2. In such a case, Fick's first law would apply if the ambient concentration remains relatively constant over the sampling time and the adsorbent mass is sufficiently large to ensure that the concentration at the end of the diffusion channel nearest the adsorbent is approximately zero.

3. The total molar flux, N_i, of contaminant with respect to the fixed-coordinate system represented by the normal plane of the dosimeter face is, according to Fick's first law, given by

$$N_i = -\mathfrak{D}_{g,i} \frac{C_0 - 0}{0 - L}$$

where $\mathfrak{D}_{g,i}$ is the gas-phase free diffusion coefficient for component i.

4. The total mass of contaminant collected by the dosimeter in time t is

$$M_i(t) = N_i A t = \mathfrak{D}_{g,i} \frac{C_0 A t}{L}$$

where A is the total cross-sectional area of the diffusion channels exposed to the ambient vapor phase. The mass of contaminant collected by the dosimeter is measured by extracting the adsorbent material. The critical dosimeter dimensions, A and L, and the sampling time (t) can be selected to achieve the desired sensitivity once the minimum detectable mass (M_i) is known.

5. The concentration, C_0, is, therefore, given by

$$C_0 = \frac{M_i(t) L}{\mathfrak{D}_{g,i} A t}$$

The same concept may be applicable to long-term monitoring of the ambient concentrations of contaminants in rivers and lakes; in this instance, a dosimeter of selected critical dimensions is submerged in the water for a prescribed length of time, the important diffusion process being liquid-phase diffusion and the appropriate diffusion coefficient being $\mathfrak{D}_{l,i}$ [see DiGiano et al. (1988)].

4.1.2.2 Fick's Second Law Fick's first law does not relate changes of concentration at a particular point in a system to changes in time and thus defines only the steady-state condition for diffusion. An expression to describe time-varying (transient) conditions is obtained by writing a mass balance equation for a differential control volume of unit cross section, A, and of length dx along the diffusion path, as shown in Figure 4-1. This balance yields

$$\left[J_{i,x} - \left(J_{i,x} + \frac{\partial J_{i,x}}{\partial x} dx\right)\right] A = \frac{\partial C_i}{\partial t} A\, dx \tag{4-4}$$

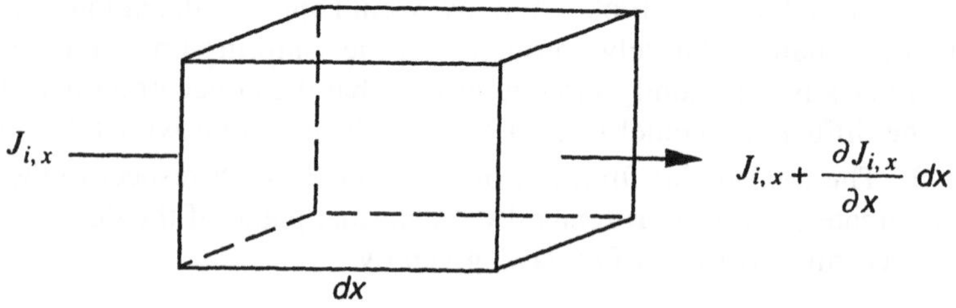

Figure 4-1 Differential control volume for analysis of diffusional flux.

Dividing Equation 4-4 by the volume, $A\,dx$, and dropping the directional subscript, x, for convenience then gives

$$-\frac{\partial J_i}{\partial x} = \frac{\partial C_i}{\partial t} \tag{4-5}$$

The flux, J_i, is given by Fick's first law (Equation 4-1). Assuming that $\mathfrak{D}_{l,i}$ is not a function of x, Equation 4-5 can then be rewritten as

$$\frac{\partial}{\partial x}\left(\mathfrak{D}_{l,i}\frac{\partial C_i}{\partial x}\right) = \frac{\partial C_i}{\partial t} \tag{4-6}$$

which is usually referred to as *Fick's second law* of diffusion. For diffusion in isotropic multidimensional space, this transient process is described by the relationship

$$\mathfrak{D}_{l,i}\,\nabla^2 C_i = \frac{\partial C_i}{\partial t} \tag{4-7}$$

4.1.2.3 Probabilistic Character of Diffusion The probabilistic nature of the diffusion process was investigated by Einstein in 1905, about 50 years after Fick had proposed his empirical relationship. Einstein explained the Brownian motion of particles in fluids by use of a probability density function. It is significant that he arrived at the equivalent mathematical statement of Fick's second law. In his approach, Einstein assumed that the motion of each particle is independent of all other particles and, further, that the motion in succeeding time intervals does not depend on motion in the previous time interval (i.e., the process is truly random). The distribution in the size of incremental movements, δ_p, for each particle, was described by a frequency distribution, $\phi_D(\delta_p)$ (i.e., some particles move farther than others in a given time interval). The number of particles moving an absolute distance δ_p (both positive and negative in direction) in some time interval is

$$dn = n\phi_D(\delta_p)\,d\delta_p \tag{4-8}$$

where n is the total number of particles and the distribution of movement increments is

$$\int_{-\infty}^{\infty} \phi_D(\delta_p)\, d\delta_p = 1 \tag{4-9}$$

The number of particles present in an incremental diffusion length, dx, is a measure of particle concentration, C_p, at that distance, x. Further, the new concentration within dx after the next small time interval, Δt, can be found by starting with the distribution of particles at the end of the previous time as obtained by the frequency distribution function. The resulting equation to describe particle concentration, C_p, is

$$C_p(x, t + \Delta t)\, dx = dx \int_{-\infty}^{\infty} C_p(x + \delta_p, t)\phi_D(\delta_p)\, d\delta_p \tag{4-10}$$

Einstein, assuming Δt to be sufficiently small to allow approximation of the time interval by dt, used certain criteria of classical calculus [i.e.,

$$C_p(x, t + \Delta t) = C_p(x, t) + \Delta t \frac{\partial C_p}{\partial t}$$

$$C_p(x + \delta_p, t) = C_p(x, t) + \delta_p \frac{\partial C_p(x, t)}{\partial x} + \frac{(\delta_p)^2}{2!} \frac{\partial^2 C_p(x, t)}{\partial x^2} + \cdots$$

and $\phi_D(x) = \phi_D(-x)$] to show that the relationship given in Equation 4-10 can be written

$$\frac{\partial C_p}{\partial t} = \mathfrak{D}_l \frac{\partial^2 C_p}{\partial x^2} \tag{4-11}$$

where, for dilute aqueous solutions, the diffusion coefficient, $\mathfrak{D}_l$, is given by

$$\mathfrak{D}_l = \frac{1}{\Delta t} \int_{-\infty}^{\infty} \frac{\delta_p^2}{2} \phi_D(\delta_p)\, d\delta_p \tag{4-12}$$

In other words, the diffusion coefficient increases as the frequency function specifies larger incremental movements of particles in a given increment of time.

The random nature of the diffusion process is discussed further in Chapters 8, 9, and 11, wherein solutions to the partial differential equation specified by Fick's second law (or Einstein's statistical derivation) are presented. In many physical situations, the boundary conditions needed to solve Fick's second law result in the following solution:

$$\frac{C(x, t)}{C_0} = \text{erfc}\left[\frac{x}{2(\mathfrak{D}_l t)^{0.5}}\right] \tag{4-13}$$

where C_0 is the concentration at $x = 0$ and erfc(z) is a standard mathematical function known as the *complementary error function* [erfc(z) = 1 − erf(z)]. Recall that this function appeared in the solution to the non-steady-state transport equation discussed in Example 2-5. The error function, which is described in detail in Table A.8.1, defines the cumulative probability distribution, once again supporting the description of diffusion as a random process. More applications of the complementary *error function* to diffusion processes will be presented in Sections 8.6.3 and 11.4. A variety of environmental process applications involving steady-state or unsteady-state diffusion as an overall rate-limiting condition are discussed throughout this book, as are cases of diffusion accompanied by chemical reaction, partitioning, and adsorption phenomena in various types of diffusion domains.

4.1.2.4 Diffusional Flux in the Context of Total Flux We have defined diffusional flux in terms of the movement of a component, i, of a mixture with respect to the overall movement of that mixture. To examine this moving-coordinate system definition in more detail and to place it in the context of the *total* flux of i relative to a fixed coordinate system, let's now consider the mean molecular or molar velocity of component i.

The mean molar velocity, $\overline{v}_{mo,i}$, of a substance in a system containing n molecules or moles of that substance is given by

$$\overline{v}_{mo,i} = \frac{\sum_{k=1}^{n} \overline{v}_{mo,k}}{n} \tag{4-14}$$

where $\overline{v}_{mo,k}$ is the velocity of the kth molecule (or mole) of i. Extending this notion to a two-component mixture of solute i dissolved in solvent j, the molar average velocity, $\overline{v}_{mo,m}$, of the mixture is

$$\overline{v}_{mo,m} = \frac{\overline{v}_{mo,i}C_i + \overline{v}_{mo,j}C_j}{C_i + C_j} \tag{4-15}$$

The average diffusional flux, $\bar{J}_i$, is measured relative to the molar average flux of the mixture and given by

$$\bar{J}_i = C_i(\bar{v}_{mo,i} - \bar{v}_{mo,m}) \tag{4-16}$$

This diffusional flux is also defined by Fick's law (Equation 4-1). Combining Equations 4-1 and 4-16 yields

$$C_i\bar{v}_{mo,i} = C_i\bar{v}_{mo,m} - \mathfrak{D}_{l,i}\frac{dC_i}{dx} \tag{4-17}$$

where C_i is expressed in molar units and $\mathfrak{D}_{l,i}$ is the free diffusion coefficient of the solute *i dissolved* in the *liquid* solvent *j*. If both *i* and *j* are gases, the diffusion coefficient would be expressed as $\mathfrak{D}_{g,i}$. We note that $\bar{v}_{mo,i}C_i$ and $\bar{v}_{mo,j}C_j$ in Equation 4-15 are fluxes relative to a fixed coordinate system and are designated by N_i and N_j, respectively, thus:

$$\bar{v}_{mo,m} = \frac{N_i + N_j}{C_T} \tag{4-18}$$

An expression for the *total flux* of *i* relative to a *fixed* coordinate system can then be obtained by combining Equations 4-17 and 4-18:

$$N_i = -\mathfrak{D}_{l,i}\frac{dC_i}{dx} + \frac{C_i}{C_T}(N_i + N_j) \tag{4-19}$$

where C_T is the sum of C_i and C_j. This equation is often written as

$$N_i = -C_T\mathfrak{D}_{l,i}\,\nabla X_i + X_i(N_i + N_j) \tag{4-20}$$

in which $X_i = C_i/C_T$ is the *mole fraction of i* in the mixture. Thus total flux includes transport resulting from gradients in concentration (or mole fraction) and from bulk solvent flow.

In dilute solutions, X_i is very small and the bulk transport term can be ignored, so that N_i can be closely approximated in terms of Fick's law. When written using the concentration gradient of component *i* instead of the mole fraction gradient, this expression is given by Equation 4-3 or, more simply for one-dimensional diffusion, by Equation 4-1. Although many situations in process engineering for water quality control involve solutions that are sufficiently dilute to permit these approximations, there are notable exceptions. For instance, some subsurface contaminants have limited solubility in water and frequently exist as *non-aqueous-phase liquids* (NAPLs). When the NAPLs are less dense than water (e.g., benzene, toluene, and xylene), they will form floating layers. Volatilization from the floating layer frequently can result in subsequent migration of these contaminants through the vapor phase in overlying unsaturated zones. Because the mole fraction in the vapor phase just above the NAPL layer may well be high in such instances, the contribution of bulk flow transport to diffusion may not be ignored. Example 4-2 provides more detail on calculation of diffusive flux when bulk transport cannot be ignored.

Example 4-2

- **Situation:** An organic solvent has drained through the vadose zone from a surface spill site and has reached the surface of the groundwater table. Because it is less dense than water and very insoluble, the solvent forms a "light" non-aqueous-phase liquid (LNAPL), which floats on the groundwater. In addition to slow dissolution into the groundwater, this LNAPL can be transported by volatilization and diffusion through that fraction of the interstitial pore volume which is comprised of air in the partially saturated vadose zone above it. The latter process is depicted below.

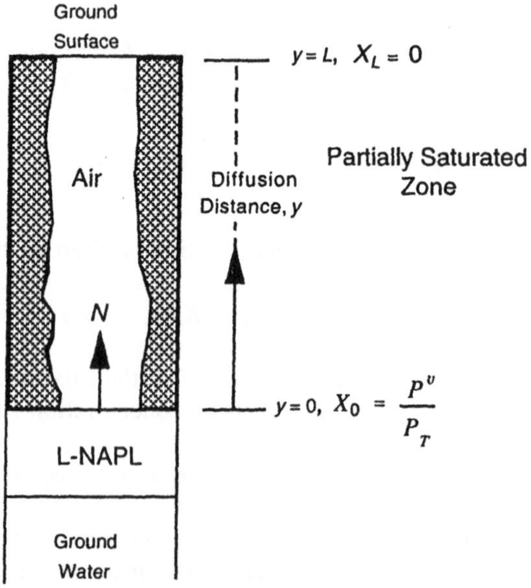

- **Question(s):** Develop an expression for the rate of volatilization of the solvent.
- **Logic and Answer(s):**

1. The flux of volatile liquid, N, through the pore volume filled with air occurs by both diffusion and diffusion-induced bulk transport. The

latter transport mechanism is significant because the liquid has a higher vapor pressure (i.e., a high mole fraction, X_0, exists at the liquid–air interface). Because of dilution with the open atmosphere at ground surface, the mole fraction of the LNAPL is zero and this produces a steady-state rate of diffusion. In this simple model formulation, the pore volume filled with air is assumed constant throughout the length of the partially saturated zone. In reality, this volume would become greater nearer the surface. In addition, we are assuming that the solvent in the vapor phase has equilibrated with the water phase throughout the partially saturated zone.

2. N is given by rearranging Equation 4-20 with $N_j = N_{\text{air}} = 0$ (it is assumed that the air remains stationary in the subsurface):

$$N(1 - X) = -C_T \mathfrak{D}_g \frac{dX}{dy}$$

This can be readily integrated as follows:

$$N \int_0^L dy = -C_T \mathfrak{D}_g \int_{X_0}^{X_L} \frac{dX}{1 - X}$$

from which we obtain the required result:

$$N = \frac{-C_T \mathfrak{D}_g}{L} \ln (1 - X_0)$$

The mole fraction in the vapor phase at the liquid–air interface (at $y = 0$) is given by P^v/P_T, where P^v is the vapor pressure of the NAPL and P_T is the total pressure (usually 1 atm). The value of C_T is obtained from the ideal gas law ($C_T = P_T/\mathcal{R}T$). We should also note that the flux per unit area of ground surface is $\epsilon_d N$, where ϵ_d is the average void volume (filled with air) fraction of the total volume of the partially saturated soil matrix.

3. The natural logarithm term of the flux expression given in step 2 can be written as the following infinite series: $X_0 + (X_0)^2/2 + (X_0)^3/3 + \cdots + (X_0)^m/m$. If the vapor pressure of i is low, $X_0 \lll 1$, so only the first term of the expansion is important. The flux is therefore given by

$$N = \frac{C_T \mathfrak{D}_g X_0}{L},$$

which is equivalent to the analysis of dilute diffusion wherein the second term on the right-hand side of Equation 4-20 can be ignored.

4.1.2.5 Stokes–Einstein Equation A more rigorous deterministic interpretation of diffusive transport than that provided by Fick's law considers gradients in chemical potential or partial molar free energy, rather than in concentration, as the motivating forces behind the diffusive migration of substances. In this treatment the net diffusive force is equal directly to the negative spatial gradient in chemical potential. As discussed in more detail in Chapter 5, chemical potential, μ_i, is related to chemical activity, a_i, which in turn, for sufficiently dilute solutions, is closely approximated by chemical concentration. Molar chemical potential is then related to concentration by

$$\mu_i = \mu_i^\circ + \mathcal{R}T \ln C_i \tag{4-21}$$

where: μ_i° is the standard-state partial molar free energy, a constant reference value for μ_i defined for a specific set of standard conditions; $\mathcal{R}$ is the universal gas constant; and T is absolute temperature.

Einstein proposed that the one-dimensional *diffusive transport force*, $F_{mo,i,x}$, acting on one molecule of a substance i in the x direction is given by dividing the negative gradient of the molar chemical potential by Avogadro's number, $\mathcal{N}_{\text{Av}}$, the number of molecules per mole:

$$F_{mo,i,x} = -\frac{1}{\mathcal{N}_{\text{Av}}} \frac{\partial \mu_i}{\partial x} \tag{4-22}$$

$F_{mo,i,x}$ is expressed as a *force per molecule* because μ_i is an expression of energy per mole, and in turn, energy is a product of a *force* and a *distance*. If the velocity at which the molecules are moving in the absence of any bulk fluid motion is given by $v_{mo,i,x}$, the number of moles transported per unit time across a unit area perpendicular to x is the product of this velocity and the molar concentration:

$$N_{i,x} = v_{mo,i,x} C_i \tag{4-23}$$

where $N_{i,x}$ is the molar flux in the x direction.

Molecular mobility, $m_{v,i}$, is defined as a molecular velocity per unit field force. Therefore, the molecular velocity, $v_{mo,i,x}$, is $m_{v,i,x}F_x$. We can now substitute this definition of molecular velocity into Equation 4-23 and replace $F_{i,x}$ in Equation 4-22 to give

$$N_{i,x} = -\frac{m_{v,i,x} C_i}{\mathcal{N}_{\text{Av}}} \frac{d\mu_i}{dx} \tag{4-24}$$

The partial derivative of molar chemical potential with respect to x is obtained from Equation 4-21:

$$\frac{\partial \mu_i}{\partial x} = \frac{\mathcal{R}T}{C_i} \frac{\partial C_i}{\partial x} \tag{4-25}$$

When Equation 4-25 is substituted into Equation 4-24, we obtain an equivalent expression for diffusive flux as given by Fick's first law:

$$N_{i,x} = -\frac{m_{v,i,x}\mathcal{R}T}{\mathfrak{N}_{\mathrm{Av}}}\frac{\partial C_i}{\partial x} = -m_{v,i,x}\kappa_B T\frac{\partial C_i}{\partial x} \tag{4-26}$$

where κ_B is the *Boltzmann constant*; i.e., the universal gas constant *per molecule*.

Stokes further modified the expression for $N_{i,x}$ given in Equation 4-26 by drawing an analogy between the motion of a solute molecule and that of a rigid solid sphere. The movement of the latter, or more specifically the resistance to its movement, through a fluid is related to the ratio of the force acting on it and its net velocity. For a sphere of diameter d and a fluid of viscosity μ_v, this resistance, R_F°, commonly defined as the molecular friction factor, is given by

$$R_F^\circ = 3\pi\mu_v d_{mo} \tag{4-27}$$

The analogy proposed by Stokes suggests that

$$R_F^\circ = \frac{F_{mo,i,x}}{v_{mo,i,x}} = \frac{1}{m_{v,i,x}} \tag{4-28}$$

which, when substituted into the relationship given in Equation 4-26, produces the *Stokes–Einstein diffusion equation* for a spherical molecule of diameter d_i:

$$N_{i,x} = -\frac{\kappa_B T}{3\pi\mu_v d_{mo,i}}\frac{\partial C_i}{\partial x} \tag{4-29}$$

Accordingly, the free-liquid diffusivity or diffusion coefficient of solute i is defined as

$$\mathfrak{D}_{l,i} = \frac{\kappa_B T}{3\pi\mu_v d_{mo,i}} \tag{4-30}$$

Based on its derivation, the Stokes–Einstein diffusion coefficient describes the random movement of noncharged spherical molecules in a fluid; this movement is referred to as *Brownian* motion. Although the Stokes–Einstein equation provides a good conceptual framework for understanding diffusion, its application is quite limited. For example, values of $\mathfrak{D}_{l,i}$ calculated for macromolecules (e.g., humic substances) and inorganic colloids (e.g., clay particles) can be considered only approximations. Such entities, which are of common interest in water quality control processes, are not rigorously defined as spheres nor as chemicals for which the chemical potential is known or even clearly defined. Moreover, they often possess surface charges, which influence their mobility, and the ionic strength of a given solution will similarly influence their sizes and configurations.

A slight modification of Equation 4-30 provides a more convenient means than diameter for expressing the size of molecules. If a molecule is assumed to be spherical and unhydrated, then

$$d_{mo,i} = \left(\frac{6}{\pi} \frac{W_{g,mo,i}}{\rho_{mo,i} \mathfrak{N}_{Av}}\right)^{1/3} \tag{4-31}$$

where $\rho_{mo,i}$ is the density of the molecule and $W_{g,mo,i}$ is its molecular weight. Substituting Equation 4-31 into Equation 4-30 and simplifying ($\kappa_B = 1.38 \times 10^{-16}$ erg/K and $\mathfrak{N}_{Av} = 6.022 \times 10^{23}$ molecules per mole) yields

$$\mathfrak{D}_{l,i} = \frac{10^{-9} T}{\mu_v} \left(\frac{\rho_{mo,i}}{W_{g,mo,i}}\right)^{1/3} \tag{4-32}$$

where μ_v is in units of g/cm-s (poise) and $\rho_{mo,i}$ in g/cm^3.

The relationships given in Equations 4-30 and 4-32 do not as such afford very useful means for determining values for the diffusion coefficients of solutes of interest in complex environmental systems. This is due in part to the violation of certain assumptions involved in their development, as noted above, and in part to the lack of information about certain of the parameters involved in the equations themselves. We shall therefore shortly consider alternative means for estimation of diffusion coefficients. Before doing so, however, we can note some important properties of diffusivity from inspection of Equation 4-32. Diffusivity increases directly with temperature and decreases with the viscosity of the fluid. These effects are intuitively reasonable. Diffusivity should increase with temperature because the kinetic energy of a molecule increases with temperature. Similarly, diffusivity should decrease with viscosity because viscosity expresses the resistance offered by the fluid to movement of molecules. We note further that the term $\rho_{mo,i}/W_{g,mo,i}$ is equal to the inverse of *molar volume*. Diffusivity therefore decreases with the one-third power of molar volume, which is not unexpected because, as shown in Equation 4-31, the diameter of a molecule is proportional to the one-third power of its molar volume. Thus molecules with larger molar volumes have smaller diffusivities because they are larger. Finally, as indicated by the numerical coefficient (10^{-9}) in Equation 4-32 and illustrated by Example 4-3, diffusion is a very slow process.

Example 4-3

- **Situation:** Humic substances can be described as long-chain, coiled macromolecules composed of hydrophobic cores of aromatic and aliphatic species to which hydrophilic functional groups such as carbohydrates and phenols are attached. Their removal in water treatment is of interest because of their reactions with chlorine to form by-products of health concern. Phase separation by adsorption or membrane processes are two possible treatment strategies. Both depend on mass transport by diffusion.

- **Question(s):** Estimate the diffusivity of humic substances at 25°C and calculate the flux assuming a linear concentration gradient over the domain shown below.

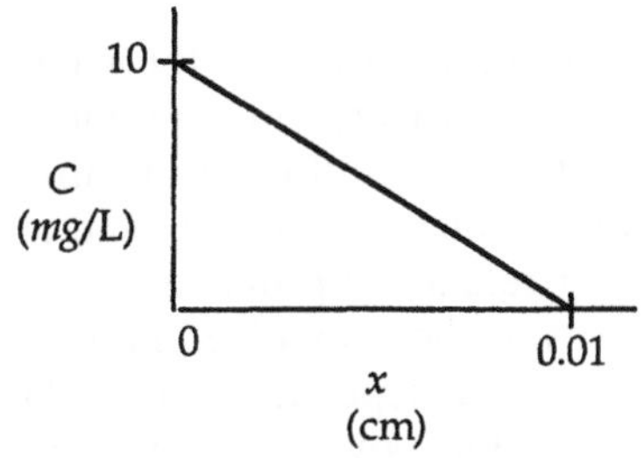

- **Logic and Answer(s):**

1. Molecular weights of humic substances range from less than 1000 to 100,000. Let us assume a molecular weight of 50,000 and a density of 0.8 g/cm³.

2. For a temperature of 25°C, $\mu_v = 8.9 \times 10^{-3}$ g/cm-s and Equation 4-32 gives

$$\mathfrak{D}_{l,i} = \frac{10^{-9}T}{\mu_v}\left(\frac{\rho_{mo,i}}{W_{g,mo,i}}\right)^{1/3} = \frac{298 \times 10^{-9}}{8.9 \times 10^{-3}}\left(\frac{0.8}{5 \times 10^4}\right)^{1/3}$$

$$= 8.2 \times 10^{-7} \text{ cm}^2/\text{s}$$

3. The diffusional flux is then

$$N_i = -\mathfrak{D}_{l,i}\frac{dC_i}{dx} \approx -\mathfrak{D}_{l,i}\frac{\Delta C_i}{\Delta x}$$

$$= 8.5 \times 10^{-7}\,\frac{\text{cm}^2}{\text{s}}\left(\frac{0.01 \text{ mg/cm}^3}{0.01 \text{ cm}}\right)$$

$$= 8.5 \times 10^{-7} \text{ mg/cm}^2\text{-s}$$

4.1.2.6 Methods for Estimating Diffusivity The diffusivities of gases can be predicted with reasonable accuracy using the Chapman–Enskog equation, which derives from the kinetic theory of binary collisions of rigid spheres. The physical and chemical interactions between solute and solvent molecules that affect molecular motion in solutions are not as well understood as are those for gases, however. The Stokes–Einstein equation introduced above is fundamentally sound but, as we noted, limited in its application to very large uncharged molecules (i.e., to molecules with molecular weights greater than about 1000). As a result of these and other limitations of theoretically based relationships, several empirical equations for predicting free-liquid diffusivities have evolved.

The Wilke–Chang equation is perhaps the most widely used relationship for estimating free-liquid diffusivities. This equation has the form

$$\mathfrak{D}_{l,i} = 7.4 \times 10^{-8}(\phi_S^{\circ} W_{g,mo,s})^{0.5} \frac{T}{\mu_v V_{mo,i}^{0.6}} \qquad (4\text{-}33)$$

where ϕ_S° is a solvent association term (2.6 for water), $W_{g,mo,s}$ the molecular weight of the solvent, μ_v its viscosity (in centipoise), and T its temperature (K). According to this relationship, diffusivity is directly proportional to temperature and inversely proportional to viscosity and to the 0.6 power of molar volume. The Wilke–Chang relationship thus assumes the same dependencies on temperature and viscosity as those involved in the Stokes–Einstein equation (Equation 4-32), with the inverse dependence on molar volume adjusted to the 0.6 power rather than the 0.333 power, and incorporates an added term to account for solute–solvent interactions. Based on the data to which the equation was calibrated, the coefficient in the Wilke–Chang equation is 7.4×10^{-8}, which is close to that in the Stokes–Einstein equation (note that μ_v is expressed in centipoise in the Wilke–Chang equation rather than in poise as in the Stokes–Einstein equation).

Table 4-1 Molecular Volume Increments

Atom in Molecule	Volume Increment (cm^3/mol)
Carbon	14.8
Chlorine, terminal as R—Cl	21.6
Chlorine, medial as —CHCl—	24.6
Fluorine	8.7
Helium	1.0
Hydrogen	3.7
Mercury	15.7
Nitrogen in primary amines	10.5
Nitrogen in secondary amines	12.0
Oxygen in ketones and aldehydes	7.4
Oxygen in methyl esters and ethers	9.1
Oxygen in higher esters and ethers	9.9
Oxygen in acids	12.0
Oxygen bonded to S, P, or N	8.3
Phosphorus	27.0
Sulfur	25.6
For organic cyclic compounds	
Three-membered ring	−6.0
Four-membered ring	−8.5
Five-membered ring	−11.5
Six-membered ring	−15.0
Naphthalene	−30.0
Anthracene	−47.5

Source: Numerical values from Perry and Chilton (1973).

The estimation of the molar volume of a solute as its molecular weight divided by its density is not as accurate as that obtained by an analysis of its structural features. In this alternative, the molar volume of a molecule is calculated by adding together the atomic volumes of the elements comprising it. This volume is then adjusted for specific molecular features of the molecule, such as its being a member of a ring structure. A list of atomic volumes and adjustment factors for structural features is provided in Table 4-1 and a numerical illustration of the Wilke–Chang equation is given in Example 4-4.

Example 4-4

- **Situation:** A phase-separation process involving adsorption by activated carbon is frequently used to achieve removal of phenolic compounds from water supplies. The process depends on diffusive mass transport of the phenolic compounds from the bulk water phase to the internal pore structure of the adsorbent.
- **Question(s):** Determine the free-liquid diffusivity of phenol at 25°C. Because humic substances may also be targeted for removal by adsorption, it is of interest to compare the diffusivities of phenol and humic substances. Diffusion will often limit the effectiveness of the adsorption process.
- **Logic and Answer(s):**

1. From Table 4-1, the molar volume for phenol is obtained as follows:

$$\begin{array}{lrr} 6\ \mathrm{C} = 6 \times 14.8 & = & 88.8 \\ 6\ \mathrm{H} = 6 \times 3.7 & = & 22.2 \\ 1\ \mathrm{O}\ (\text{in OH}) & = & \underline{12.0} \\ & & 123.0 \\ \text{less 15 for ring} & & \underline{-15.0} \\ V_{mo,i} & = & 108.0 \end{array}$$

2. Using Equation 4-33, we obtain

$$\mathfrak{D}_{l,i} = 7.4 \times 10^{-8} \, (\phi_S^\circ W_{g,mo,s})^{0.5} \frac{T}{\mu V_{mo,i}^{0.6}}$$

$$= 7.4 \times 10^{-8} \, (2.6 \times 18)^{0.5} \left[\frac{298}{8.9 \times 10^{-1}(108)^{0.6}}\right]$$

$$= 1.02 \times 10^{-5} \ \mathrm{cm^2/s}$$

3. The diffusivity of humic substances was estimated from the Stokes-Einstein equation in Example 4-1 as 8.2×10^{-7} cm^2/s, or about one order of magnitude smaller. Thus humic substances will diffuse more slowly than phenol into the adsorbent.

4.2 INTERPHASE MASS TRANSFER

Rarely are water quality transformations in natural and engineered systems restricted to single homogeneous phases. Interphase mass transfer is therefore a frequently encountered microtransport process in such systems. As we will see, characterizations of this mode of transport are also structured upon concepts of molecular diffusion.

Mass transport considerations relating to the movement of material between boundary surfaces and moving fluids or between two immiscible fluids are extremely important in many treatment processes. The resistance to mass transfer at an interface may limit the supply of reactants reaching the phase in which reaction occurs, therefore limiting the rate of reaction and thus controlling reactor design. Typical examples of mass transfer–controlled processes include transport of gases and volatile organic compounds across interfaces between air and water; organic solutes across interfaces between water and solid adsorbents such as activated carbon; ionic species across interfaces between water and solid-ion exchange materials, such as clays and synthetic resins; and particles and molecules across interfaces between water and filter materials such as sands, screens, and porous membranes.

4.2.1 Transport Across Boundaries and Interfaces

Bulk flow in most natural water systems is turbulent, and in treatment operations a high degree of mixing is usually provided to ensure that high rates of mass transfer obtain. In heterogeneous phase systems, however, such bulk turbulence is commonly damped in the regions of boundaries and phase interfaces, as it is at the boundaries of homogeneous systems. The movement of a fluid past a surface or past another fluid brings about the development of a boundary layer wherein mass transport occurs through a combination of molecular and turbulent diffusion. Molecular diffusion occurs close to the surface, where a viscous sublayer exists, while farther away from the surface, eddies move randomly and transfer reactant by turbulent diffusion.

Turbulence can be dampened at interfaces between the water phase and a second phase by cocurrent local movement of the two phases if the second phase travels advectively at essentially the same velocity as the fluid element in which it is suspended. This is characteristic of highly dispersed two-phase systems such as small oxygen bubbles and/or clay particles dispersed in advection-driven water columns, or in rapidly stirred but not well-mixed reactors. Thus, although bulk mixing and macroturbulence may be great in such systems, the microturbulence at phase interfaces is low. Systems in which the second phase is fixed in space with respect to the fluid phase are at the opposite end of the physical spectrum but subject to the same net result. In a fixed bed of adsorbent or a sand filter, for example, boundary conditions dictate zero relative velocity at the immediate interface between the liquid and solid phases and thus no interfacial microturbulence. This is a limiting deficiency of the

Prandtl hypothesis discussed in Chapter 3 with respect to its application to interfaces.

For multiphase systems such as those described above, the transport of mass at relatively quiescent interfaces between phases reduces to a process similar to that of molecular diffusion. Transport of material from one phase to the other is controlled by the gradient in concentration (rigorously, chemical potential) of the material across the interface, in a manner analogous to that described by either Equation 4-1 or 4-3. It differs, however, from homogeneous-phase molecular diffusion in the sense that the impedances of the domains through which diffusion occurs, and thus resistances to diffusion are not those of a well-defined single fluid but rather those of system-specific interfacial conditions. The coefficient of the concentration gradient thus becomes a lumped parameter which must account for more than molecular resistance. For example, it must also account for relative fluid velocities in the interface and, potentially, for differences in the standard states of the diffusing substance(s) in different phases.

The description of a concentration gradient, and hence of a driving force for a material being transferred from one phase to another, must relate to differences between its respective values of chemical potential, μ_i, in the two phases. Unlike molecular diffusion in homogeneous phases, this difference in chemical potential is not simply given by the concentration difference because the standard-state potential, μ_i°, of the substance may differ from one phase to another. As discussed in depth in Chapter 6, one phase is likely to have a higher or lower thermodynamic affinity for the substance being transferred. Indeed, such differences in thermodynamic potential and affinity comprise the basis for phase separation processes such as gas transfer, adsorption, ion exchange, membrane separations, and solvent extractions. By way of simple analogy, differences in the chemical potential of a substance between two different phases can be compared to differences between the heat capacity of a homogeneous metal rod suspended in a fluid and that of the fluid itself. Expanding on this analogy, diffusive mass transfer from one phase to another can be likened to the conductance of heat between the rod and the fluid in which it is suspended.

The nature of microturbulence at interfaces cannot be characterized precisely. Interfacial turbulence relates to relative phase velocities, roughness of surfaces at the interface, frictional and adhesive forces, surface tensions, and several other parameters. Description of each of these parameters is difficult and tedious at best; characterization of their interactions in terms of microturbulence is impossible. From an engineering perspective, it is therefore necessary to conceptualize the interface in one of two ways: (1) a finite domain with properties that can be correlated to the hydrodynamic properties of the bulk phases it partitions or (2) a molecular layer to which mass is transported advectively by infinitesimally small fluid elements circulating in the bulk phase. The first of these conceptual micromodels, termed the *stagnant film model* or more simply the *film model*, involves a steady-state description of interphase mass transport. The second, the *penetration* or *surface-renewal model*, describes convective mass transport in non-steady-state terms. Both models facilitate quantification of mass flux by correlating microturbulence at interfaces to overall bulk-phase turbulence.

4.2.2 Boundary Layers

Reactor designs to accomplish phase transformations, particle separations, or solute separations implicitly involve the analysis of mass transport across boundary layers. Examples already cited include gas absorption and stripping, filtration, membrane separations, adsorption and ion exchange, and solids drying operations. Thus boundary layers of interest involve those that form between gases and liquids, solids and liquids, and gases and solids.

Before discussing engineering approaches to quantification of interphase mass transport, an understanding of the fundamental concept of a boundary layer should be developed. A flow boundary layer occurs whenever a flow pattern is disrupted by a surface. This surface produces shear stress, which causes velocity to change in the vicinity of the boundary. Flow boundary layers affect interphase mass transport processes, and thus both cause and affect concentration boundary layers. Mathematical relationships to describe flow and concentration boundary layers can be combined to provide a fundamental appreciation for the dependence of the mass transfer coefficient on fluid and solute properties.

Equations are available to predict the thicknesses and velocity distributions of laminar and turbulent boundary layers. These were initially developed to describe flow over smooth boundaries such as that of an idealized interface between water and a channel bottom or between air and the surface of a wing. A detailed discussion of boundary layer theory is beyond the scope of this book, but some highlights will be presented to reinforce the importance of the continuity and momentum equations for fluid flow and their applications to boundary layer processes.

A general sketch of the flow regime near a boundary is given in Figure 4-2. For illustrative purposes, this boundary could be at the macroscale of a river channel bottom. The objective of our present analysis is to determine the thickness of the boundary layer and the velocity distribution within it. A boundary layer is initiated at $x = 0$ and grows in the x direction because shear stress at the surface retards adjacent *fluid elements*, which in turn retard the motion of other fluid elements farther from the surface. The boundary layer becomes thicker as x increases, and flow eventually becomes unstable, thus moving through a transition region and eventually giving way to turbulence. The physical description of the turbulent boundary domain is more complex than that of a strictly laminar boundary region. In the former domain, a laminar sublayer adjacent to the surface gives way to a turbulent layer as distance from the surface (y) increases. The velocity distribution for the small laminar subregion is easy to describe using Newton's viscosity law ($dv_x/dy = \tau_{s,0}/\mu_v$), where $\tau_{s,0}$

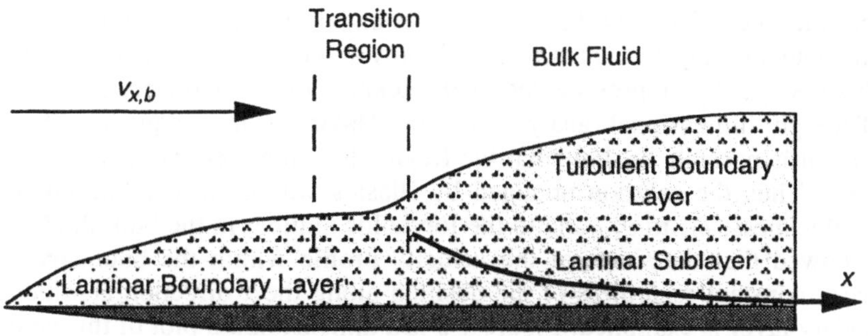

Figure 4-2 Boundary layer development on a sharp-edged flat plate.

is the shear stress at the surface. A different relationship exists farther away from the surface, however. Mathematical description of the velocity distribution for the entire turbulent boundary layer must draw upon a combination of relationships.

Boundary layer theory originates with the classic work of Prandtl early in the twentieth century. Blasius, a student of Prandtl, solved the Navier–Stokes equation (Equation 3-44) and the continuity equation (Equation 3-16) for laminar flow over a flat plate having a sharp leading edge, such as that depicted in Figure 4-2. The two-dimensional forms of the continuity and Navier–Stokes equation to be solved for incompressible laminar steady-state flow are

$$\frac{\partial v_x}{\partial x} + \frac{\partial v_y}{\partial y} = 0 \tag{4-34}$$

$$v_x\frac{\partial v_x}{\partial x} + v_y\frac{\partial v_y}{\partial y} = -\frac{1}{\rho}\frac{\partial P}{\partial x} + \nu_v\left(\frac{\partial^2 v_x}{\partial x^2} + \frac{\partial^2 v_x}{\partial y^2}\right) \tag{4-35a}$$

$$v_x\frac{\partial v_x}{\partial x} + v_y\frac{\partial v_y}{\partial y} = -\frac{1}{\rho}\frac{\partial P}{\partial x} + \nu_v\left(\frac{\partial^2 v_y}{\partial x^2} + \frac{\partial^2 v_y}{\partial y^2}\right) \tag{4-35b}$$

where ν_v is kinematic viscosity (μ_v/ρ). At the surface of the boundary ($y = 0$), velocity is zero in both the x and y directions. For y values exceeding the full thickness of the boundary layer, the velocity in the x direction is constant and the pressure gradient in the x direction is zero ($dP/dx = 0$); that is, flow is a potential flow and viscous effects can be ignored. With these boundary conditions and simplifications, it is possible, although tedious, to solve the Navier–Stokes equation in two dimensions (x and y). Profiles of velocity in the flow direction, $v_x = \phi(y)$, and transverse velocity, $v_y = \phi(y)$, are obtained at each location, x. The latter shows an increase in velocity outward from the boundary, which is logical because as the boundary layer increases in thickness with x, flow along the edge of the plate must displace fluid in the y direction. However, the former is more important for purposes here because $v_x = \phi(y)$

determines the velocity profile within the boundary layer and thus the thickness of the boundary layer. That is, the position y at which v_x becomes equal to the bulk velocity, $v_{x,b}$, represents the total thickness of the boundary layer.

Through dimensional analysis of the Navier–Stokes equation (Section 3.2.6), it is possible to show that the Reynolds number characterizes the flow regime. Thus the solution obtained by Blasius was used to relate boundary layer thickness, δ_L, to $v_{x,b}$ and to the Reynolds number in the bulk fluid. Blasius showed that the shape of the velocity profile within the boundary layer remains constant regardless of the position along the x direction at which the boundary layer is examined. Figure 4-3 is a dimensionless plot of this velocity profile. The abscissa is the velocity within the boundary layer (v_x) relative to the velocity in the bulk flow ($v_{x,b}$) and the ordinate is $(y_x \mathfrak{N}_{Re})^{0.5}/x$, where y_x is the distance above the surface (y) at any x value and the Reynolds number, $\mathfrak{N}_{Re}$, is calculated from the bulk fluid velocity. This observation is the basis for the similarity solution of the boundary layer problem. It is apparent from Figure 4-3 that for any x value, the velocity increases linearly with y near the surface, becomes curvilinear farther away, and then asymptotically approaches

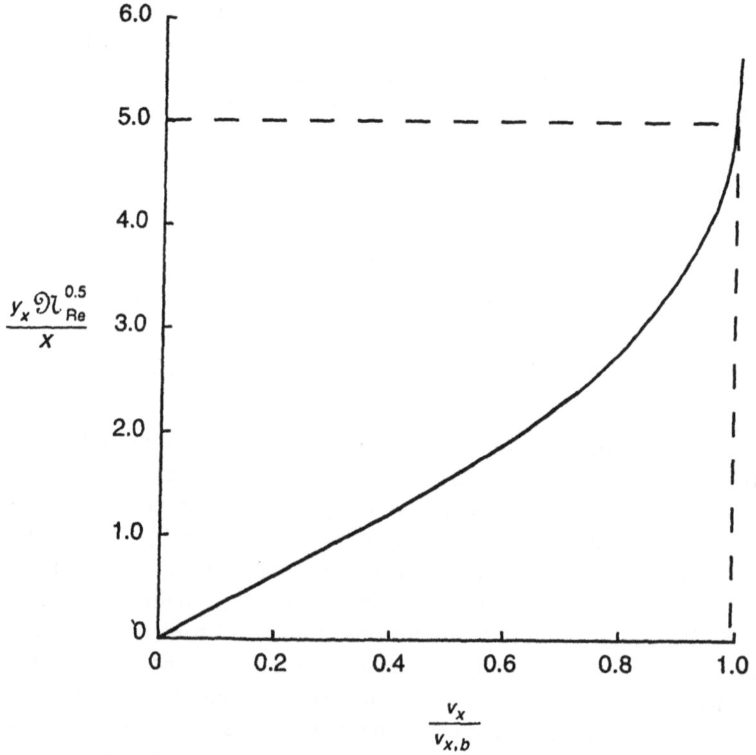

Figure 4-3 Blasius solution to the Navier–Stokes equation for a laminar boundary layer. [Adapted from Welty et al. (1984).]

a terminal velocity. The thickness of the laminar boundary layer, δ_L, can be estimated by evaluating the ordinate when $v_x/v_{x,b}$ is 0.99, for which the value of the ordinate is 5.0, or

$$\delta_L = \frac{5x}{\mathfrak{N}_{\text{Re}}^{0.5}} \tag{4-36}$$

This equation states that the thickness of the laminar boundary layer increases with distance along the boundary and decreases with the square root of the Reynolds number. An inverse relationship between boundary layer thickness and Reynolds number is intuitive. A direct relationship between boundary layer thickness and x follows from our earlier explanation of Figure 4-2; that is, as a result of viscous effects, the boundary layer is expected to increase with x because of increased retardation of fluid elements farther away from the surface.

The mathematical analysis of a turbulent boundary layer is more difficult than that for a laminar boundary layer because, as explained in Chapter 3, velocity in such cases becomes a function of time as well as space. The development of a more appropriate relationship is aided by a closer examination of the nature of this turbulent boundary layer.

The laminar sublayer of the turbulent boundary layer is shown in Figure 4-2. If we assume that $\tau_{s,0}$ is constant throughout this sublayer, the velocity gradient is constant, and velocity thus increases linearly with distance from the surface [i.e., $v_x = (\tau_{s,0}/\mu_v)y$]. The velocity distribution in the remaining portion of the boundary layer is more complex, however. Mathematical descriptions are based in part on Prandtl's mixing-length concept to describe turbulent shear stress, $\tau_{s,t}$, where from Equation 3-62,

$$\tau_{s,t} = \rho l_m^2 \left(\frac{dv_x}{dy}\right)^2 \tag{3-62}$$

Prandtl showed that mixing length, l_m, is proportional to distance from the surface ($l_m = \mathfrak{Z}_u y$) for regions close to the surface. The coefficient of proportionality, $\mathfrak{Z}_u$, referred to as the *universal turbulence constant*, or *von Kármán's constant*, must be determined experimentally (typically, this value is about 0.4). After taking the square root of both sides of Equation 3-62 and then integrating, it follows that

$$\frac{v_x}{v_s} = \frac{1}{\mathfrak{Z}_u} \ln y + \kappa \tag{4-37}$$

where v_s, or $(\tau_{s,0}/\rho)^{0.5}$, is referred to as the *shear velocity* and κ is a constant of integration. The equation is not valid in the laminar sublayer, where the velocity profile is linear, and thus is restricted to the domain $y > \delta_{L,s}$, where

$\delta_{L,s}$ is the thickness of the laminar sublayer. For smooth surfaces the constant of integration can be obtained by matching the velocities determined from Equation 4-37 and $v_x = (\tau_{s,0}/\mu_v)y$, where $y = \delta_{L,s}$. The selection of $\delta_{L,s}$ is based on experimental measurements. Rough surfaces add further empiricism. For the purpose of this discussion, however, only the general form of Equation 4-37 is important. It shows that the thickness of the boundary layer increases exponentially with increased velocity; the term *logarithmic zone* is thus used commonly as a descriptor for this domain.

An empirical power-law equation was suggested by Prandtl to capture the general shape of the velocity distribution given by Equation 4-37, namely

$$\frac{v_x}{v_{x,b}} = \left(\frac{y}{\delta_T}\right)^{0.143} \tag{4-38}$$

This relationship cannot be used for the laminar sublayer because the velocity gradient (dv_x/dy, at $y = 0$) would be infinite. It is therefore again necessary to match the velocities given by Equation 4-37 to that for the outer edge of the laminar sublayer, $v_x = (\tau_{s,0}/\mu_v)\delta_{L,s}$. The general shape of the entire velocity profile is shown in Figure 4-4.

Having approximated the velocity distribution with Equation 4-37, the *von Kármán integral* approach is used to obtain an equation to describe the thickness of the turbulent boundary layer. Although less rigorous than the Navier–Stokes equation, the von Kármán integral still makes use of the continuity and momentum equations to describe fluid motion. We restrict the control volume

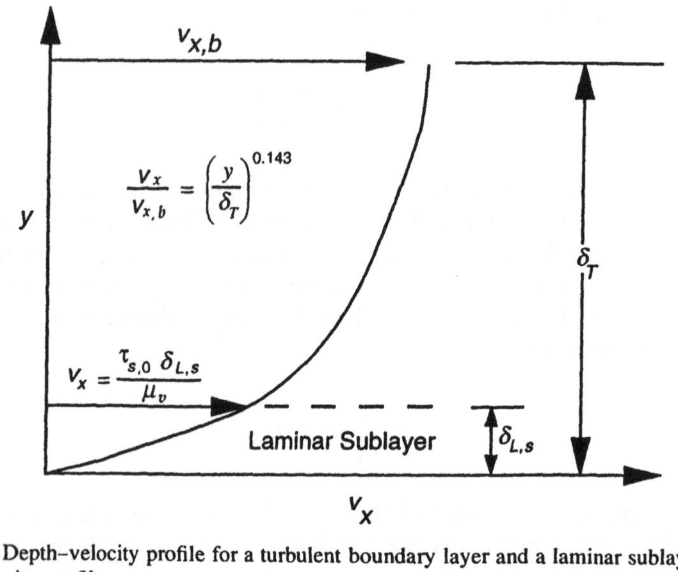

Figure 4-4 Depth–velocity profile for a turbulent boundary layer and a laminar sublayer having a linear velocity profile.

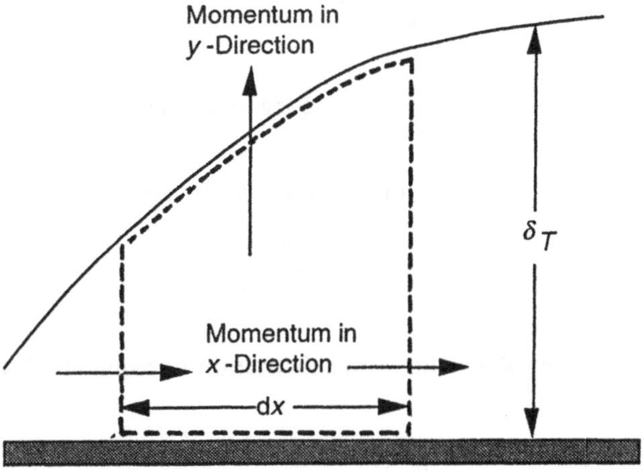

Figure 4-5 Control volume for analysis of a fluid boundary layer [Adapted from Welty et al. (1984).]

to the thickness of the boundary layer, δ_T, as presented in Figure 4-5. Beyond this thickness, the velocity is that of the bulk flow, $v_{x,b}$. The force acting in the x direction in the momentum balance is the shear force, $\tau_{s,0}$, generated at the boundary between the fluid and the solid surface. Because we are restricting the control volume, we must account for the loss of both momentum and mass through the top of the boundary layer. Assuming steady-state conditions and two-dimensional (x, y) flow, the mass and momentum balance statements and their corresponding equations are:
Mass balance:

$$\begin{bmatrix}\text{rate of mass flow}\\ \text{out of control}\\ \text{volume}\end{bmatrix} - \begin{bmatrix}\text{rate of mass flow}\\ \text{into control}\\ \text{volume}\end{bmatrix} = 0$$

$$\int_0^{\delta_T} \frac{\partial \rho v_x}{\partial x}\, dx\, Z_W\, dy + \rho v_y Z_W\, dx = 0 \tag{4-39}$$

Momentum balance:

$$\begin{bmatrix}\text{rate of momentum}\\ \text{flow out of control}\\ \text{volume}\end{bmatrix} - \begin{bmatrix}\text{rate of momentum}\\ \text{flow into control}\\ \text{volume}\end{bmatrix} = \begin{bmatrix}\text{sum of forces}\\ \text{acting on}\\ \text{control volume}\end{bmatrix}$$

$$\int_0^{\delta_T} v_x \frac{\partial \rho v_x}{\partial x}\, dx\, Z_W\, dy + \rho v_{x,b} v_y Z_W\, dx - 0 = \tau_{s,0} Z_W\, dx \tag{4-40}$$

where Z_W is the width of the boundary layer in the z direction. Dividing Equation 4-40 through by $Z_W\,dx$ gives

$$\tau_{s,0} = \frac{d}{dx}\int_0^{\delta_T} \rho v_x^2\,dy + \rho v_{x,b} v_y \tag{4-41}$$

and, performing the same steps on Equation 4-39, yields

$$v_y = -\frac{d}{dx}\int_0^{\delta_T} v_x\,dy \tag{4-42}$$

The von Kármán integral results from substitution of Equation 4-42 for v_y into Equation 4-41:

$$\tau_{s,0} = \frac{d}{dx}\int_0^{\delta_T} (v_{x,b} - v_x)\rho v_x\,dy \tag{4-43}$$

An empirical expression developed by Blasius for shear stress at the boundary, $\tau_{s,0}$, in turbulent flow has the form

$$\tau_{s,0} = 0.0233\rho v_{x,b}^2 \left(\frac{\nu_v}{v_{x,b}\delta_T}\right)^{0.25} \tag{4-44}$$

where ν_v is the kinematic viscosity, μ_v/ρ. Substitution of Equation 4-44 into Equation 4-43 and replacement of v_x on the right-hand side of Equation 4-43 with the empirical expression for the velocity distribution as a function of y proposed by Prandtl (Equation 4-38) yields

$$0.0233\rho v_{x,b}^2 \left(\frac{\nu_v}{v_{x,b}\delta_T}\right)^{0.25} = \rho v_{x,b}^2 \frac{d}{dx}\int_0^{\delta_T} \left[1 - \left(\frac{y}{\delta_T}\right)^{0.143}\right]\left(\frac{y}{\delta_T}\right)^{0.143} dy \tag{4-45}$$

After integrating the right-hand side of Equation 4-45, we have

$$0.0233\left(\frac{\nu_v}{v_{x,b}\delta_T}\right)^{0.25} = \frac{d\delta_T}{dx} \tag{4-46}$$

Finally, after separation of variables and integration (assuming $x = 0$, $\delta_T = 0$), Equation 4-46 leads to the following simple expression for the turbulent boundary layer thickness:

$$\delta_T = \frac{0.37x}{\mathfrak{N}_{\text{Re}}^{0.2}} \quad \text{for } 5 \times 10^5 < \mathfrak{N}_{\text{Re}} < 10^7 \tag{4-47}$$

Equation 4-47 was derived by assuming that the boundary layer is turbulent from the leading edge of the surface, whereas, in fact, a laminar boundary layer should exist for some small distance. The distance at which the transition from laminar to turbulent flow occurs can be estimated using the Reynolds number, $\mathfrak{N}_{\mathrm{Re}} = xv_{x,b}\rho/\mu_v$, from which the value of x can be calculated given the bulk fluid velocity, density, and viscosity. An application of Equation 4-47 to the evaluation of the development of turbulent boundary layer is given in Example 4-5.

Example 4-5

- **Situation:** Polychlorinated biphenyls (PCBs) are strongly hydrophobic, and their fate (and for that matter, the fate of other hydrophobic chemicals) once they have been discharged into rivers is determined largely by sorption onto suspended particles. Larger suspended particles may settle fairly readily, forming the bulk of river sediments. However, colloidal size particles (on the order of a few microns), which also possess a surface charge, remain in suspension much longer, sometimes indefinitely. Natural water chemistry can induce a lessening of surface charge and thus colloids are able to aggregate. In this way, some colloids eventually become large enough to settle to river sediment. However, association with the sediment may be tenuous given both the hydrodynamic and chemical (i.e., charge restabilization) forces at work. A large-scale open-channel experiment is devised to measure the transport rate of colloids from sediment to overlying water:

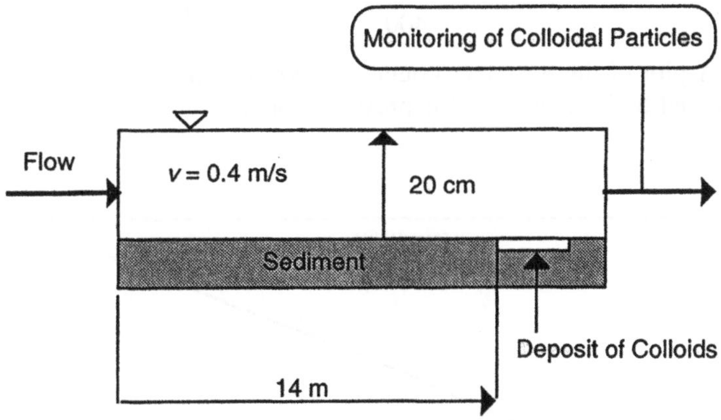

- **Question(s):** Determine whether this experimental system can provide measurements of the interfacial mass transfer of colloidal particles that relate to the hydrodynamics of established flow in a river segment.

- **Logic and Answer(s):**

1. Regardless of the specific model selected for interfacial mass transfer (several approaches will be discussed subsequently in this chapter), it is essential to describe the hydrodynamic regime near the sediment–water boundary. The hydrodynamic regime is in turn determined by development of the boundary layer in the water above the sediment along the channel length. Unless the boundary layer has become fully developed throughout the channel depth, the hydrodynamic regime does not represent that of established flow in a river segment.

2. To estimate boundary layer development in turbulent flow, we can use Equation 4-47:

$$\delta_T = \frac{0.37x}{\mathfrak{N}_{\text{Re}}^{0.2}} \qquad \text{for } 5 \times 10^5 < \mathfrak{N}_{\text{Re}} < 10^7$$

This equation is an estimate because the experimental system does not match the ideal physical system it is intended to describe (i.e., the surface of the sediment is not smooth and the entry point to the channel is not sharp edged).

3. The Reynolds number is calculated from

$$\mathfrak{N}_{\text{Re}} = \frac{v_{x,b}\rho x}{\mu_v} = \frac{(400 \text{ cm/s})\ (1 \text{ g/cm}^3)x}{0.01 \text{ g/cm-s}}$$

and the distance, x, at which the transition occurs from laminar to turbulent flow is estimated by setting the $\mathfrak{N}_{\text{Re}}$ to 5×10^5:

$$x = \frac{5 \times 10^5 \times 0.01}{400} = 125 \text{ cm}$$

4. A plot of the turbulent boundary layer thickness for $x > 125$ cm determined by Equation 4-47 is presented below.

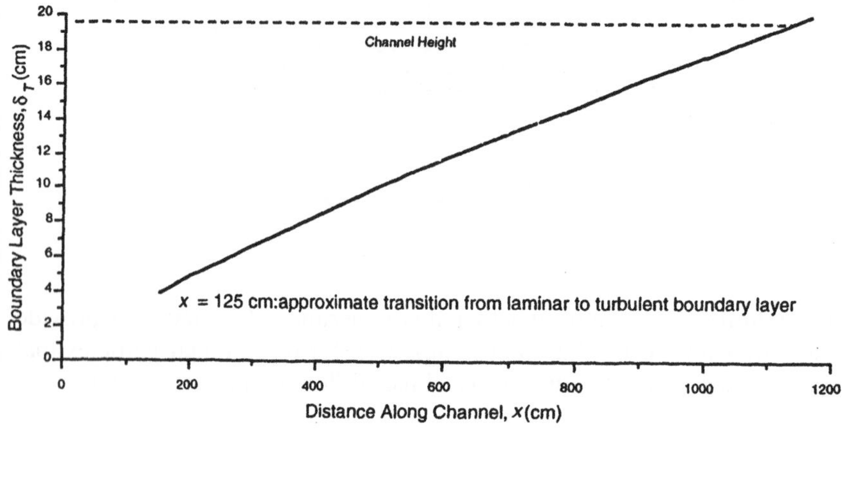

This figure indicates that the boundary layer is fully developed in a distance of approximately 12 m from the inlet to the channel. The colloidal particles have been deposited at a distance greater than 12 m to conduct the mass transfer experiment and thus the resulting data should represent the transport process properly. The Reynolds number corresponding to this distance is 5.2×10^6.

The von Kármán integral approach can also be used to obtain an expression for the thickness of a laminar boundary layer. The Blasius solution to the Navier–Stokes equation reveals that the shape of the velocity profile in the boundary layer is independent of x. This shape can thus be approximated by a simple polynomial expression in y (e.g., $v_x = \alpha_0 + \alpha_1 y + \alpha_2 y^2 + \alpha_3 y^3 + \cdots$), which, when forced to conform to four boundary conditions ($y = 0$, $v_x = 0$; $y = 0$, $\partial^2 v_x/\partial y^2 = 0$; $y = \delta_L$, $v_x = v_{x,b}$; and $y = \delta_L$, $\partial v_x/\partial y = 0$) gives

$$\frac{v_x}{v_{x,b}} = \frac{3}{2}\left(\frac{y}{\delta_L}\right) - \frac{1}{2}\left(\frac{y}{\delta_L}\right)^3 \tag{4-48}$$

This expression can be substituted into the right side of Equation 4-43. The shear stress term on the left side of the von Kármán integral given in Equation 4-43 is simply $\tau_{s,0} = \mu_v(dv_x/dy)|_{y=0}$, where dv_x/dy can now be determined from Equation 4-48. Performing the same mathematical steps as outlined above for analysis of the turbulent layer then leads to an expression for the thickness of the laminar layer:

$$\delta_L = \frac{4.64x}{\mathfrak{N}_{\text{Re}}^{0.5}} \qquad \text{for } \mathfrak{N}_{\text{Re}} < 10^5 \tag{4-49}$$

Comparison of Equations 4-49 and 4-36 reveals that the laminar boundary layer equation derived from the approximation provided by the von Kármán integral compares very well to the more exacting one derived by Blasius. It is important to remember that the turbulent and laminar expressions for boundary layer thickness derived herein are applicable strictly only to *smooth* surfaces.

A comparison of the velocity profiles for turbulent and laminar boundary layers at a Reynolds number of 5×10^5 is given in Figure 4-6. The relative thickness of the turbulent to laminar boundary layer (δ_T/δ_L) is calculated from the ratio of Equation 4-47 to Equation 4-49. The velocity profiles for turbulent and laminar boundary layers are calculated from Equations 4-38 (with δ_T expressed relative to δ_L) and 4-48, respectively. We note that the turbulent boundary layer is thicker than the laminar boundary layer and also has a higher mean velocity. Consequently, a turbulent boundary layer involves significantly greater momentum and energy than does a laminar layer.

4.2.3 Film Models for Mass Transfer

4.2.3.1 One-Film Model Mass transfer from one phase to another is driven by concentration (chemical potential) gradients existing across the hy-

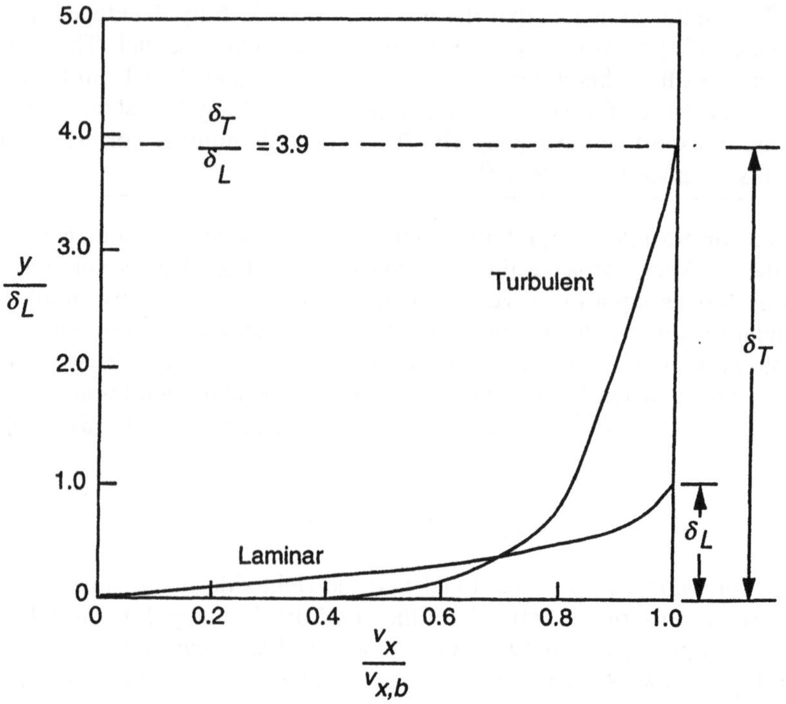

Figure 4-6 Depth–velocity profiles for laminar and turbulent boundary layers [Adapted from Welty et al. (1984.)]

drodynamic boundary layer that separates the two phases. The one-film theory is used to describe this hydrodynamic boundary layer in a simplistic way as illustrated in Figure 4-7. The concentration of component i in solution immediately adjacent to the surface is designated by C_i°. The thickness of the boundary layer or film is δ and the concentration, $C_{i,\delta}$, is the same as that in bulk solution. A variety of surfaces are possible. For instance, the surface may be a precipitated form of solute such that C_i° is given by its solubility limit. Alternatively, it may be the liquid form of an organic solvent, which then evaporates into a flowing airstream; C_i° would then be given by the vapor pressure of the solvent.

Mass transfer through the film shown in Figure 4-7 occurs by diffusion. The diffusional flux is in the y direction and is described by Equation 4-19:

$$N_i = -\mathfrak{D}_{l,i} \frac{dC_i}{dy} + \frac{C_i^\circ}{C_i} (N_i + N_j) \tag{4-50}$$

The second term on the right side is the diffusion-induced flux and is significant only if C_i° is large; we characterize situations where this term is ignored as those in which *dilute mass transfer* takes place. Although dilute mass transfer is certainly of most interest in environmental applications, there are some not-

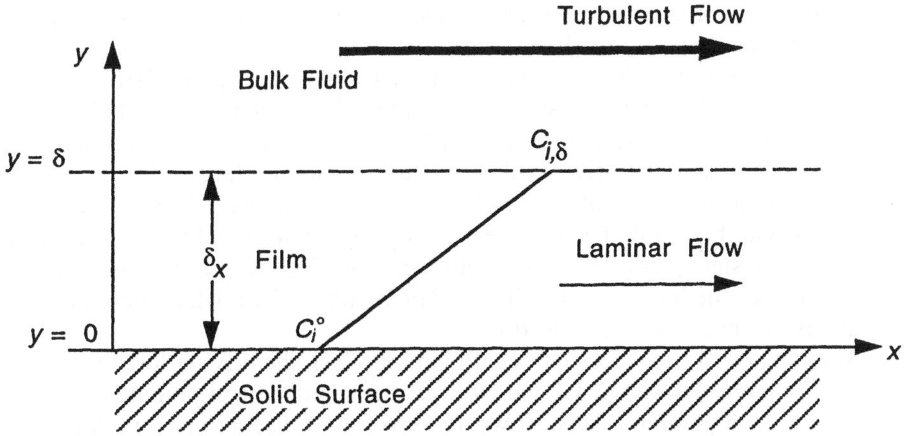

Figure 4-7 Film model for interphase mass transfer.

able exceptions that make it instructive to consider the analysis of Equation 4-50 in further detail. Substitution of the definition of molar average velocity (Equation 4-18) into Equation 4-50 gives

$$N_i = -\mathfrak{D}_{l,i}\frac{dC_i}{dy} + C_i^\circ \overline{v}_{mo} \tag{4-51}$$

where $\overline{v}_{mo}$ can be either away from (the positive direction) or toward (the negative direction) the surface.

The concentration gradient, dC_i/dy, in Equation 4-51 is not known because the thickness of the concentration boundary layer, δ, is not measurable. Thus Equation 4-51 must be simplified to

$$N_i^\circ = k_{f,i}\,\Delta C_i + C_i^\circ \overline{v}_{mo} \tag{4-52}$$

where N_i° is the flux measured at the interface, ΔC_i is the difference in concentration between the surface and the bulk fluid ($C_i^\circ - C_{i,\delta}$), and $k_{f,i}$ is a mass transfer coefficient which quantifies the relationship between the flux of i and the driving force. This coefficient, which has the dimensions of *length per unit time*, is defined by

$$k_{f,i} = \frac{\mathfrak{D}_{l,i}}{\delta} \tag{4-53}$$

For dilute solutions, which correspond to low mass transfer rates, $C_i^\circ \overline{v}_{mo}$ is small. Equation 4-52 can be simplified further for such conditions to

$$N_i^\circ = k_{f,i}\,\Delta C_i \tag{4-54}$$

Several distinctions between the diffusion coefficient and the mass transfer coefficient are noteworthy. First, the dimensions of $\mathfrak{D}_{l,i}$ are L^2t^{-1}, whereas those for $k_{f,i}$ are Lt^{-1}. Phenomenologically, $k_{f,i}$ depends upon relative velocities in the vicinity of the interface and upon the thickness, δ, of the film. Although it is not possible to measure film thickness, we can reason that it is an inverse function of microturbulence at the phase interface, which, in turn, is directly dependent on bulk fluid-phase turbulence. Thus δ decreases and $k_{f,i}$ increases with bulk turbulence, at least to certain limiting values. If bulk flow is reasonably steady, then δ is essentially constant. Further, if the bulk solution is well mixed, the bulk concentration of i in the phase from which net transport occurs is essentially constant in time.

Additionally, as noted earlier (see Equation 4-24), the driving force for mass transfer is more rigorously defined by a gradient in chemical potential than by a concentration gradient. This, however, involves a discontinuity in chemical potential immediately at the surface of the second phase.

Common treatments of interphase mass transfer describe C_i in Equation 4-54 as the difference between the concentration in one phase and the *limiting* concentration in that phase were it in equilibrium with the second phase. For instance, suppose that mass transfer controls the rate of dissolution of precipitated solids. The dissolution rate will be determined by the difference between the *saturation* concentration and the actual concentration of the soluble ions in the bulk liquid. Similarly, consider subsurface contamination by a relatively immiscible organic solvent that becomes entrapped in some fraction of the pore space and exists there as ganglia (or "blobs") of non-aqueous-phase liquid (NAPL). This NAPL subsequently undergoes dissolution into the groundwater (see Example 4-9) until, according to equilibrium principles, a maximum dissolved concentration in the water is determined by the aqueous-phase *solubility* of the organic solvent. For this case the mass transfer expression given by Equation 4-54 is $k_f(C_S - C_b)$, where C_S is the solubility limit of the solvent and C_b is the bulk concentration of that material dissolved in the water.

Another important feature of the film model is the interfacial area of the phase into which mass transfer is occurring. When this parameter is specified, mass transfer can be expressed in a manner analogous to a rate of reaction, namely:

$$\mathcal{r}_{i,mt} = N_i(a_s^\circ)_R \tag{4-55}$$

where $\mathcal{r}_{i,mt}$ is the overall rate of mass transfer with dimensions of $ML^{-3}t^{-1}$, and $(a_s^\circ)_R$ is the specific interfacial area per unit volume of reactor, with dimensions of L^{-1}. The overall mass transfer rate depends upon two system constants, k_f and $(a_s^\circ)_R$. It may be possible to determine k_f experimentally if $(a_s^\circ)_R$ can be specified (see Example 4-6). The difficulty, however, is that in many instances $(a_s^\circ)_R$ cannot be determined easily. For example, it is difficult to quantify the total interfacial surface area formed by bubbles of air rising through a column of water. This will depend on, among other factors, the distribution of bubble sizes produced by the sparger apparatus and on the depth of the overlying water column. It is even more difficult to determine the interfacial area of blobs of NAPL entrapped between soil grains and in contact with groundwater. Moreover, to varying degrees in both cases, the interfacial areas will change as constituents of the gas and NAPL phases dissolve into the water.

The difficulty in specification of $(a_s^\circ)_R$ is usually overcome by assessing a lumped system-specific constant, $\hat{k}_v = k_f(a_s^\circ)_R$, referred to as the *overall volumetric mass transfer coefficient* (t^{-1}). Because $\hat{k}_v$ is system specific, its value must be evaluated separately for each system of interest (see, e.g., Examples 4-6 through 4-9).

The dimensionless *Sherwood number* provides a useful means for quantifying mass transfer relationships in a way that allows extrapolation of information regarding these relationships from one system to another. The Sherwood number for a given system comprised by the elements of mass transfer depicted in Figure 4-7 can be obtained by equating the mass transport flux across δ to the diffusive flux at the surface of that system:

$$k_{f,i}(C_i^\circ - C_{i,\delta}) = -\mathfrak{D}_{l,i} \left. \frac{dC_i}{dy} \right|_{y=0} \tag{4-56}$$

For a constant surface concentration, C_i°, the diffusive flux can also be written as

$$-\mathfrak{D}_{l,i} \left. \frac{dC_i}{dy} \right|_{y=0} = \mathfrak{D}_{l,i} \left. \frac{d(C_i^\circ - C_i)}{dy} \right|_{y=0} \tag{4-57}$$

Combining Equations 4-56 and 4-57 and multiplying both sides of the resulting equation by the thickness of the film then yields

$$\frac{k_{f,i}\delta}{\mathfrak{D}_{l,i}} = \frac{\left. \dfrac{d(C_i^\circ - C_i)}{dy} \right|_{y=0}}{\dfrac{C_i^\circ - C_{i,\delta}}{\delta}} \tag{4-58}$$

The dimensionless term on the left is the Sherwood number, $\mathfrak{N}_{\mathrm{Sh}}$, with δ being the characteristic length (L_c). The numerator of the term on the right represents the driving force for molecular diffusion and the denominator of that term is the driving force for interfacial transport. The inverse of the numerator is the *impedance* or *resistance* to diffusion and the inverse of the denominator is the *impedance* to interfacial mass transfer. Hence the Sherwood number is defined as

$$\mathfrak{N}_{\mathrm{Sh}} = \frac{\text{interfacial mass transfer impedance}}{\text{molecular diffusion impedance}} \tag{4-59}$$

The impedance to interfacial mass transfer depends upon system hydrodynamics and upon the diffusion properties of the component undergoing transfer. As we will show later, development of appropriate relationships between the Sherwood number for a system and the dimensionless parameters that describe the hydrodynamics (the *Reynolds number*) and diffusion properties (the *Schmidt number*) of that system is possible from boundary layer theory. However, only simple boundary layer conditions are easily analyzed with mass and momentum balances (e.g., laminar flow over a flat plate). Other boundary layers, such as those created by fluid flow around irregularly shaped grains of sand or adsorbent media in packed beds, are more difficult to describe from first principles. Consequently, experimental correlations between the Sherwood number and other dimensionless parameters must be developed (see Section 4.2.7) for each type of boundary layer. The utility of such correlations is that they allow estimation of the mass transfer coefficient for a system once its hydrodynamic and interfacial conditions have been specified. The determination of mass transfer coefficient for an aqueous-solid phase system is described in Example 4-6.

Example 4-6

- **Situation:** A fixed bed of activated carbon is to be used to adsorb a contaminant from water. A small adsorption column is set up in the laboratory for a preliminary test. The column has a diameter of 2.5 cm (cross-sectional area of 4.9 cm^2) and length of 5 cm. It is packed with activated carbon particles having an equivalent diameter of 0.8 mm. The resulting bed porosity is 0.4. The feed concentration is 100 μg/L and the superficial velocity of flow is 0.5 cm/s. It is noted that the exit concentration remains constant and equal to 50 μg/L for the first several hours of operation and then begins to increase.
- **Question(s):** How can the data obtained from the preliminary test be used to estimate a mass transfer coefficient?
- **Logic and Answer(s):**

1. The macroscopic and microscopic aspects of fixed-bed adsorbers were depicted in Figure 1-5. Analysis of the adsorption rate process at the microscopic level leads to a mathematical model at the macroscopic level that is used to predict the concentration of the contaminant as a function of position in the adsorption bed and time. It is generally assumed that the adsorption process itself is almost instantaneous. The rate of solute uptake is, instead, *controlled* by the rate of transport across the external film surrounding the particle (see Figure 1-5) and/or the rate of transport to an adsorption site within the porous structure. Further details of adsorption rate analysis are presented in Chapters 8 and 11. For now it suffices to state that the mass transfer coefficient determines the rate of

mass transfer across the external film and that intraparticle diffusion coefficients determine internal transport (both within the pore and on the surface). Each of these parameters is experimentally determined.

2. In this experiment, the detention time of water in the bed is very short:

$$\text{detention time} = \frac{0.4(5\text{ cm})}{0.5\text{ cm/s}} = 4\text{ s}$$

This limits the amount of mass transfer that can occur. In addition, the hourly rate at which contaminant is being introduced is very low: (0.5 cm/s) (3600 s/h) (4.9 cm^2) (100 μg/L) (10^{-3} L/cm^3) or 882 μg/h. If a typical adsorptive capacity is 0.1 g/g and adsorbent density is 1.4 g/cm^3, the adsorbent will equilibrate with (1 − 0.4) (4.9 cm^2) (5 cm) (1.4 g/cm^3) (0.1 g/g), or with about 2 g of contaminant. Thus the amount of contaminant introduced during the first few hours is an extremely small fraction of the adsorptive capacity. Given the excess of adsorptive capacity during this time, the local concentration in the fluid adjacent to the adsorbent surface, C_i°, should be nearly equal to zero.

3. If external mass transfer is a very fast process, the exit concentration of the column at startup would also have been zero because the bulk fluid concentration, C_i, would be equal to the concentration adjacent to the surface, C_i°. However, the data show that the exit concentration is large at startup. This can be explained only if external diffusion is a relatively slow process that limits the amount of contaminant being transferred to the adsorptive surface. We can also reason that the exit concentration will remain constant temporarily because C_i° will remain very close to zero due to rapid adsorption of each molecule that passes through the external film.

4. The mass transfer rate is therefore at steady state. As these adsorption sites near the surface become occupied, C_i° must rise and the concentration gradient across the external film must decrease, which steadily lowers the mass transfer rate. This produces an increase in the exit concentration and an unsteady-state condition.

5. The steady-state material balance can be written over an elemental control volume of the adsorber unit as

$$-v_s \frac{dC_i}{dx} A\,dx + N_i(a_s^\circ)_R A\,dx = 0$$

in which ϵ_B is the bed porosity or void ratio, v_s the superficial or Darcy velocity (flow rate/cross-sectional area), A the cross-sectional area of the bed, N_i is

$$N_i = k_{f,i}(C_i - C_i^\circ) = k_{f,i}(C_i - 0)$$

and $(a_s^\circ)_R$ is the specific surface area available for mass transfer per unit volume of reactor, or

$$(a_s^\circ)_R = \left(\frac{\text{surface area of sphere}}{\text{volume of sphere}}\right)\left(\frac{\text{volume of spheres}}{\text{volume of bed}}\right)$$

$$= \frac{6}{d_p}(1 - \epsilon_B) = \frac{6}{0.08 \text{ cm}}(0.6) = 45 \text{ cm}^{-1}$$

6. Dividing the material balance relationship given in step 5 by $\epsilon_B A\, dx$ and substituting the flux relationship above yields

$$-v_s \frac{dC_i}{dx} - k_{f,i}(a_s^\circ)_R C_i = 0$$

7. To solve for the mass transfer coefficient, $k_{f,i}$, we must integrate the equation above, which in turn requires that we specify the boundary condition (i.e., at $x = 0$, $C_i = 100\ \mu$g/L). The result is

$$\frac{C_i}{100} = \exp\left[-\frac{k_{f,i}(a_s^\circ)_R}{v_p}x\right]$$

from which we find that

$$k_{f,i} = -\frac{v_s}{(a_s^\circ)_R x}\ln\frac{C_i}{100}$$

$$= -\frac{0.5 \text{ cm/s}}{(45 \text{ cm}^{-1})(5 \text{ cm})}\ln\frac{50}{100}$$

$$= 1.54 \times 10^{-3} \text{ cm/s}$$

8. Mass transfer coefficients are typically in the range 10^{-2} to 10^{-3} cm/s. Other mathematical models are available to determine the mass transfer coefficient and the internal diffusion coefficients simultaneously when mass transport is not limited solely by transfer across the external film and when the effect of limitations in sorptive capacity become important. These models rely on a more complicated, unsteady-state analysis.

Mass transfer in concentrated solutions such as brines, sludges, and slurries requires consideration of the diffusion-induced bulk transport term, $C_i^\circ \overline{v}_{mo}$, incorporated in Equation 4-52. For example, the concentration may be rather large on the reject side of a membrane used for solute separations. The effect of diffusion-induced bulk transport on mass transfer can be examined by integrating Equation 4-51 over the thickness of the concentration boundary layer, δ. If steady-state diffusion is assumed, N_i and $\overline{v}_{mo}$ are constant, and the resulting integration of this nonhomogeneous, ordinary differential equation is

$$N_i = \frac{\bar{v}_{mo} C_i^\circ e^{\bar{v}_{mo}\delta/\mathfrak{D}_{l,i}} - C_{i,\delta}\bar{v}_{mo}}{e^{\bar{v}_{mo}\delta/\mathfrak{D}_{l,i}} - 1} \tag{4-60}$$

The separate flux contributions of diffusion and diffusion-induced convection can be recognized by rearrangement of Equation 4-60 in the following way:

$$N_i^\circ = \frac{\bar{v}_{mo}}{e^{\bar{v}_{mo}\delta/\mathfrak{D}_{l,i}} - 1}(C_i^\circ - C_{i,\delta}) + C_i^\circ \bar{v}_{mo} \tag{4-61}$$

The flux is written specifically for the location immediately adjacent to the surface, but with the assumption of steady state, it is the same everywhere in the film.

Comparison of Equations 4-61 and 4-52 shows that the mass transfer coefficient for concentrated solutions is then defined as

$$(k_{f,i})_{\text{conc}} = \frac{\bar{v}_{mo}}{e^{\bar{v}_{mo}\delta/\mathfrak{D}_{l,i}} - 1} \tag{4-62}$$

Replacing δ in Equation 4-62 with $\mathfrak{D}_{l,i}/k_{f,i}$ provides a means for comparing mass transfer coefficients for concentrated and dilute solutions:

$$\frac{(k_{f,i})_{\text{conc}}}{(k_{f,i})_{\text{dilute}}} = \frac{\bar{v}_{mo}/k_{f,i}}{e^{\bar{v}_{mo}/k_{f,i}} - 1} \tag{4-63}$$

Equation 4-63 allows for adjustment of the mass transfer coefficient determined for a dilute mass transfer for use when the concentration immediately adjacent to the surface is significant. A power series expansion of the denominator of the right-hand side of Equation 4-63 would show that the mass transfer coefficient for very concentrated solutions will always be smaller than for dilute systems (i.e., for diffusive mass transfer in the direction of the bulk fluid flow). Example 4-7 illustrates an application of this concept for mass transfer of volatile subsurface contaminant into the soil vapor phase.

Example 4-7

- **Situation:** Example 4-2 presented a situation in which toluene exists as a light non-aqueous-phase liquid (LNAPL) floating on a groundwater surface. Suppose that a pump-and-treat remediation scheme has been used to remove this toluene but that pockets of pure LNAPL remain trapped in the unsaturated zone (these pockets are referred to as blobs or ganglia). Soil venting is proposed to strip this remaining toluene from the subsurface. This process requires pulling of an airstream through the subsurface by installation of a vacuum pump at the surface.

- **Question(s):** Develop a conceptual model to explain mass transfer of toluene from the LNAPL into the flowing air stream.
- **Logic and Answer(s):** A schematic representation of the situation is shown below.

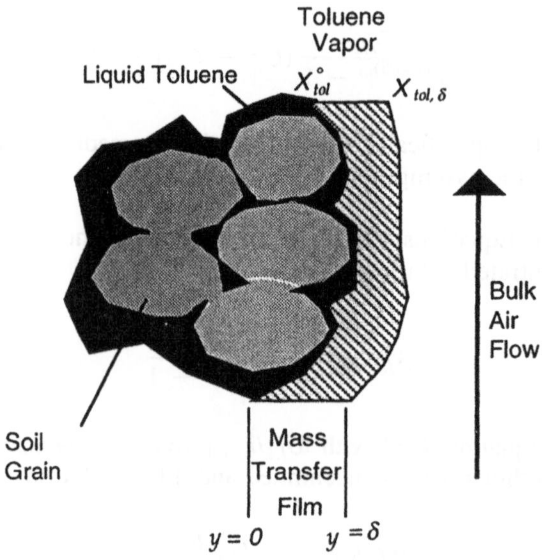

1. The toluene exists as a pure liquid in the interstitial region near soil grains. Toluene exists in the vapor phase adjacent to the liquid phase. Airflow establishes a hydrodynamic boundary layer or "film" and mass transfer of vapor-phase toluene occurs through this film. The mole fraction at the interface ($y = 0$) is X°_{tol}. The mole fraction in the moving bulk airstream is assumed equal to zero.

2. The expression to describe flux due to mass transfer is analogous to that used to investigate diffusion of toluene through a stagnant column of air in the unsaturated zone (see Example 4-2). In this situation, however, the diffusional distance is defined as the thickness of the boundary layer or film that separates the toluene vapor from the bulk airstream. Dilute mass transfer is not an appropriate description of this system because the mole fraction of toluene in the film may be very high. Assuming that the flux of air in the y direction is zero, the total flux of toluene (tol) is given by Equation 4-51:

$$N_{\text{tol}} = -\mathfrak{D}_{g,\text{tol}} \frac{dC_{\text{tol}}}{dy} + C^{\circ}_{\text{tol}} \overline{v}_{mo}$$

where the average molar velocity (Equation 4-18) is

$$\overline{v}_{mo} = \frac{N_{\text{tol}}}{C_T}$$

3. The flux in the "concentrated" toluene system from Example 4-2 is

$$N_{tol} \int_0^{\delta} dy = -C_T \mathfrak{D}_{g,tol} \int_{X_{tol}^{\circ}}^{0} \frac{dX_{tol}}{1 - X_{tol}}$$

$$N_{tol|conc} = \frac{-\mathfrak{D}_{g,tol} C_T}{\delta} \ln(1 - X_{tol}^{\circ})$$

where $X_{tol}^{\circ} = C_{tol}^{\circ}/C_T$.

4. This flux can be compared to a "dilute" toluene system wherein

$$N_{tol|dilute} = (k_f)_{dilute} C_{tol}^{\circ}$$

Recalling the definition of this mass transfer coefficient, $(k_f)_{dilute} = \mathfrak{D}_g/\delta$, the concentrated-to-dilute system ratio of flux is

$$\frac{N_{tol|conc}}{N_{tol|dilute}} = -\frac{1}{X_{tol}^{\circ}} \ln(1 - X_{tol}^{\circ})$$

5. The graph below shows that the flux ratio increases with increasing mole fraction. This additional flux is caused by diffusion-induced convective transport. The flux can be over 2.5 times larger than that in a dilute system.

6. The ratio of concentrated to dilute system mass transfer coefficients is given by Equation 4-63.

$$\frac{(k_f)_{conc}}{(k_f)_{dilute}} = \frac{e^{\bar{v}_{mo}/k_f}}{e^{\bar{v}_{mo}/k_f} - 1}$$

where $\bar{v}_{mo}/(k_f)_{dilute}$ is obtained from steps 2 and 3:

$$\frac{\bar{v}_{mo}}{k_f} = \frac{N_{tol}}{k_f C_T} = -\ln(1 - X_{tol}^{\circ})$$

The graph below shows that Equation 4-63 predicts a mass transfer coefficient that decreases relative to that in the dilute system as mole fraction of toluene at the interface increases.

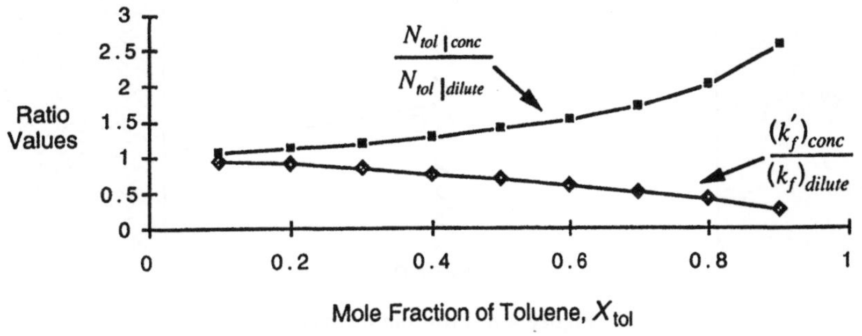

4.2.3.2 Two-Film Model Mass transfer across the interfaces between gases and liquids is often described using a *two-film model*. As shown in Figure 4-8, mass transfer of a component from the bulk gas phase to the bulk liquid phase is envisioned in this model to occur across two films. The component is first transferred from the bulk gas phase, where its concentration is given by the pressure, P_b, to the interface, where its concentration is P_I. The resistance to this transfer is referred to as the *gas-side film* impedance. The interface represents the exact position at which the gas and liquid phases coexist, and at which there is instantaneous equilibrium between gas- and liquid-phase concentrations. The component is then transferred from this interface to the bulk liquid, where its concentration is C_b. The resistance to this second mass transfer step is referred to as the *liquid-side* impedance.

At steady state, the fluxes through the gas- and liquid-side films are equal. We can use the film model with the same units for concentration and mass transfer coefficients in both phases to write

$$\frac{k_{f,g}}{\mathcal{R}T}(P_b - P_I) = k_{f,l}(C_I - C_b) \tag{4-64}$$

It is convenient to express the bulk gas-phase concentration in terms of the liquid phase concentration, C_S, with which it will equilibrate according to Henry's law (i.e., the solubility limit corresponding to a given partial pressure). Similarly, P_I can be replaced by C_I with the aid of Henry's law to give

$$N_l = \frac{k_{f,g}\mathcal{K}_H}{\mathcal{R}T}(C_S - C_I) = k_{f,l}(C_I - C_b) \tag{4-65}$$

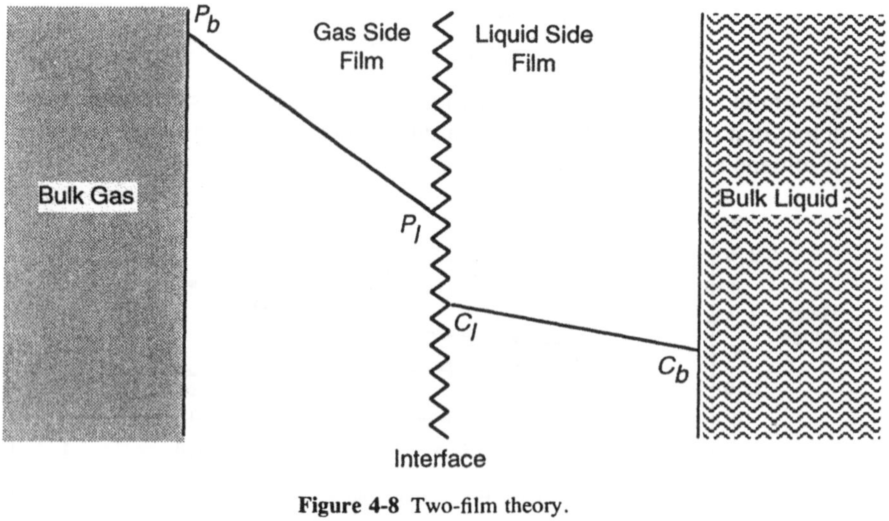

Figure 4-8 Two-film theory.

Because the interfacial concentration, C_I, cannot be measured, it is eliminated by rearranging Equation 4-65 to solve for C_I:

$$C_I = \frac{k_{f,l} C_{b,l} + \frac{k_{f,g}\mathcal{K}_H}{\mathcal{R}T} C_S}{k_{f,l} + \frac{k_{f,g}\mathcal{K}_H}{\mathcal{R}T}} \tag{4-66}$$

Replacing C_I in Equation 4-65 with Equation 4-66 gives

$$N_l = \hat{k}_{f,l}(C_{S,l} - C_{b,l}) \tag{4-67}$$

The overall mass transfer coefficient, $\hat{k}_{f,l}$ (Lt^{-1}), relative to the liquid form of component i is given by

$$\hat{k}_{f,l} = \frac{1}{\frac{\mathcal{R}T}{k_{f,g}\mathcal{K}_H} + \frac{1}{k_{f,l}}} \tag{4-68}$$

The exchange of oxygen at an air–water interface during aeration or reaeration provides a convenient illustration of this two-film mass transfer process. In this instance the transfer of oxygen from air to water is assumed to involve diffusive flux across an interfacial zone between the two phases. The driving force or potential gradient for this flux at any time is the difference between the average bulk solution concentration of dissolved oxygen, C_b, at any time, t, and the saturation concentration, C_S, were the air and water concentrations of oxygen in thermodynamic equilibrium.

The overall mass transfer coefficient depends on specific conditions of the air–water interface, including the interface molecular mobility of the component being transferred, the thickness of the hypothetical quiescent interfacial domain, and the degree of microturbulence within that domain. Inspection of Equation 4-68 shows that for gases that are highly soluble in water (e.g., ammonia), $\mathcal{K}_H$ is sufficiently small that $\hat{k}_{f,l}$ can be approximated by $k_{f,g}\mathcal{K}_H/\mathcal{R}T$. Mass transfer in this case is controlled essentially by the gas-side resistance alone. The only way to increase mass transfer is to decrease the impedance on the gas side of the interface (e.g., increase the velocity of the gas to decrease the thickness of the gas-side film). On the other hand, for sparingly soluble gases such as oxygen, $\mathcal{K}_H$ is sufficiently large that $\hat{k}_{f,l}$ can be approximated by $k_{f,l}$. Mass transfer in this case is controlled essentially by the liquid-side impedance, and it is necessary to increase the velocity of the liquid to increase mass transfer. For gases of intermediate solubility, both gas- and liquid-side resistances are important.

The overall impedance to mass transfer can also be understood as the inverse of the overall mass transfer coefficient, which from rearrangement of Equation

4-68 is

$$\frac{1}{\hat{k}_{f,l}} = \frac{\mathcal{R}T}{\mathcal{K}_H}\frac{1}{k_{f,g}} + \frac{1}{k_{f,l}} \qquad \text{(4-69)}$$

A mass transfer relationship in terms of gas-phase concentrations rather than liquid-phase concentrations can also be obtained. The liquid concentrations at the interface and the bulk liquid concentration in Equation 4-64 are expressed in terms of gas-phase concentrations, $P_b/\mathcal{K}_H$ and $P_S/\mathcal{K}_H$, respectively. The latter represents the gas-phase concentration which would exist in equilibrium with the bulk liquid concentration. The resulting flux is

$$N_g = \frac{k_{f,g}}{\mathcal{R}T}(P_b - P_I) = k_{f,l}\left(\frac{P_I}{\mathcal{K}_H} - \frac{P_S}{\mathcal{K}_H}\right) \qquad \text{(4-70)}$$

Equation 4-70 can be used to eliminate P_I. Further, we can use the ideal gas law ($PV = n\mathcal{R}T$) to express gas-phase concentrations in the same units as liquid-phase concentration and obtain

$$N_g = \hat{k}_{f,g}(C_{b,g} - C_{S,g}) \qquad \text{(4-71)}$$

The overall mass transfer coefficient, $\hat{k}_{f,g}$ (Lt^{-1}), relative to the gaseous form of component i is given by

$$\hat{k}_{f,g} = \frac{1}{\dfrac{1}{k_{f,g}} + \dfrac{\mathcal{K}_H}{k_{f,l}\mathcal{R}T}} \qquad \text{(4-72)}$$

Equations 4-68 and 4-72 can be combined to show that

$$\hat{k}_{f,g} = \frac{\hat{k}_{f,l}\mathcal{R}T}{\mathcal{K}_H} \qquad \text{(4-73)}$$

In a qualitative sense, the two-film model provides a vehicle for assessing how best to "engineer" a mass transfer system. If appropriate calculations indicate that the gas side of the film is controlling, it makes good sense to minimize the thickness of the gas side of the film and, conversely, to minimize the liquid side of the film if it is controlling. For example, in reactors in which a countercurrent flow of air is used to strip ammonia or some other highly soluble gas from water, the thickness of the gas-side film can be reduced by increasing the air velocity. Similarly, in surface impeller systems used to transfer oxygen, the thickness of the liquid-side film can be reduced by increasing the rotational speed of the impeller. A practical application of the two-film model to an engineered system is provided in Example 4-8.

Example 4-8

- **Situation:** The last step of treatment at a particular wastewater treatment is nitrate removal by an anoxic (i.e., without oxygen) biological process commonly termed denitrification. Following denitrification, it is necessary to reaerate the effluent before final discharge to the receiving water. A floating mechanical surface aerator consisting of rapidly rotating impellers is used for this purpose. The detention time in the aeration unit under present operating conditions is 20 min and the dissolved oxygen (DO) concentration leaving the unit is 2 mg/L. The DO saturation concentration, C_S, according to Henry's law is ~9 mg/L. The present performance is judged unsatisfactory by regulatory officials, who request that the DO level be increased to 4 mg/L prior to discharge.
- **Question(s):** What alternatives are available to meet the new aeration requirements?
- **Logic and Answer(s):**

1. We begin by writing a material balance on oxygen assuming that the aeration device is well mixed:

$$QC_{\mathrm{IN}} - QC_{\mathrm{OUT}} + N(a_s^\circ)_R V_R = 0$$

2. The concentration of DO in the feed to the aerator is zero and the flux, N, is given by the two-film theory of mass transfer (Equation 4-67). The specific interfacial area, $(a_s^\circ)_R$, per unit volume of reactor is that produced by mechanically mixing of the air–water interface. Given these conditions the above equation can be rewritten as

$$-QC_{\mathrm{OUT}} + \hat{k}_{f,l}(a_s^\circ)_R V_R(C_S - C_{\mathrm{OUT}}) = 0$$

Note that the bulk concentration in the mass transfer relationship is equal to the outlet concentration of the aeration device because complete mixing has been assumed. Note further that we assume that no other reaction is occurring to consume DO.

3. Solving for the outlet concentration then gives

$$C_{\mathrm{OUT}} = \frac{\hat{k}_{v,l} C_S}{\hat{k}_{v,l} + \dfrac{Q}{V_R}}$$

where $\hat{k}_{v,l} = \hat{k}_{f,l}(a_s^\circ)_R$, the overall volumetric mass transfer coefficient.

4. We see that the outlet concentration depends on $\hat{k}_{v,l}$, the saturation concentration, C_S, and the hydraulic retention time, V_R/Q. Although $\hat{k}_{v,l}$ was not given, we can find it from the present operating conditions for which C_S, V_R/Q, and C_{OUT} are specified:

$$\hat{k}_{v,l} = \frac{C_{\mathrm{OUT}}}{(C_S - C_{\mathrm{OUT}})\dfrac{V_R}{Q}}$$

$$= \frac{2}{(9 - 2) \times 20 \text{ min}} = 0.014 \text{ min}^{-1}$$

5. Having determined $\hat{k}_{v,l}$, we can find the additional aeration volume needed if the same $\hat{k}_{v,l}$ is maintained by adding proportionally more mechanical aerators of the same design. Rearranging the relationship developed for C_{OUT} in step 3 to solve for detention time needed to reach 4 mg/L, we obtain

$$\frac{V_R}{Q} = \frac{C_{\mathrm{OUT}}}{\hat{k}_{v,l}(C_S - C_{\mathrm{OUT}})}$$

$$= \frac{4 \text{ mg/L}}{0.014 \text{ min}^{-1} \text{ } (9 - 4) \text{ mg/L}}$$

$$= 57 \text{ min}$$

The volume and number of surface aerators thus need to be approximately tripled to reach the new performance objective.

6. Other alternatives involve ways to increase mass transfer without increasing tank volume. This may be accomplished by (a) increasing the $\hat{k}_{f,l}(a_s^\circ)_R$ of the existing aeration units by rotating the surface impellers at a higher speed which would decrease the boundary layer thickness and increase $\hat{k}_{f,l}$, and possibly $(a_s^\circ)_R$, (b) adding more aeration units to increase $(a_s^\circ)_R$; and (c) replacing existing aerators with units that produce a higher $\hat{k}_{f,l}(a_s^\circ)_R$.

4.2.4 Penetration and Surface Renewal Models

4.2.4.1 Penetration Model The penetration model for interphase mass transfer was proposed by Higbie in 1935. In this model, liquid elements or "packets" are envisioned as being transported to the gas–liquid interface, where they remain for some time before returning to the bulk liquid phase. The mass transfer of gas into the liquid element while it resides at the interface is pictured in Figure 4-9 as diffusion into a liquid film falling along a vertical wall of length, L, and width, Z_W. Note here that the orientation of x and y axes are shown to coincide with the vertical direction of advection and horizontal direction of diffusion, respectively. This physical model is less restrictive than the film model because we no longer must assume a linear concentration gradient across a film of unknown thickness. Instead, the concentration gradient changes as the liquid element moves down the vertical wall and the gas diffuses into it. On a microscale, the falling liquid film represents a continuous flow of

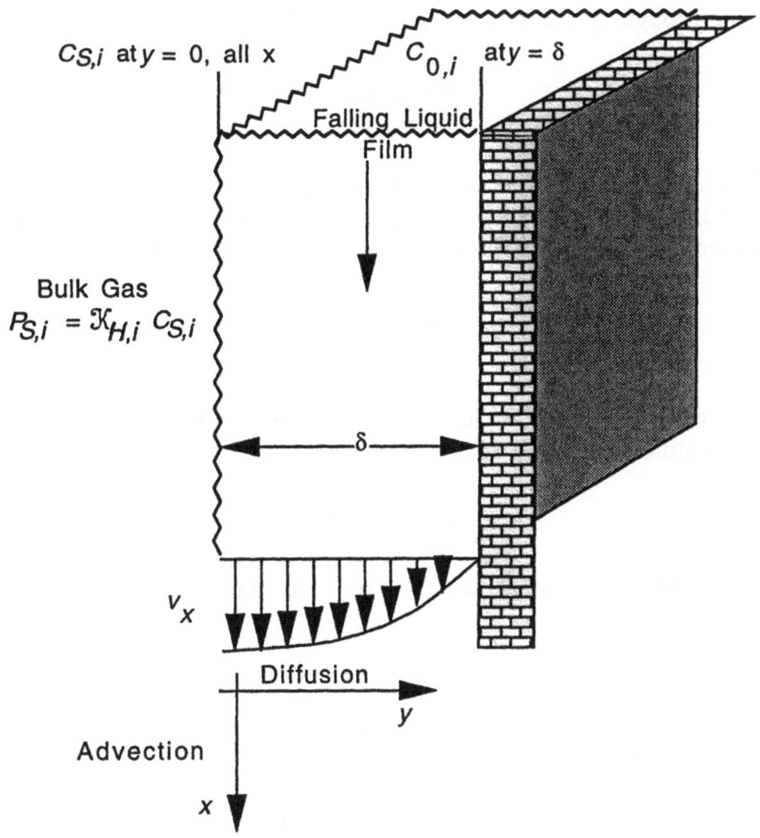

Figure 4-9 Falling-film analysis of penetration theory. (The falling film is continuously generated by flow.)

liquid elements past an interface. On a macroscale, this view of gas penetration into a liquid also describes, for example, the diffusion of oxygen into a film of water flowing over a solid surface, a diffusive mass transfer process that is significant for any attached film biological treatment operation such as a trickling filter, a rotating biological contactor, or a biologically active carbon system.

The penetration model can be analyzed by aid of a material balance over a differential volume (*dx dy dz*) of the system pictured in Figure 4-9. The concentration of the gas is the same everywhere along the interface. As each element passes in the x direction, gas diffuses into the liquid in the y direction. The material balance is then written as

$$\frac{\partial N_i}{\partial x}\, dx\, dy\, Z_W + \frac{\partial N_i}{\partial y}\, dy\, Z_W\, dx = 0 \qquad (4\text{-}74)$$

Dividing through by the volume element $dx\, dy\, Z_W$ gives

$$\frac{\partial N_i}{\partial x} + \frac{\partial N_i}{\partial y} = 0 \qquad (4\text{-}75)$$

The flux in the x direction is caused by advective flow (i.e., dispersion in this direction is assumed to be unimportant). For laminar flow, velocity increases with distance away from the wall in the following manner:

$$v_x = v_{\text{max}}\left[1 - \left(\frac{y}{\delta}\right)^2\right] \qquad (4\text{-}76)$$

For penetration distances of less than about 0.32δ, $v_x > 0.9v_{\text{max}}$, and this allows simplification of the advective flux term to

$$N_{i,x} = v_{\text{max}} C_i \qquad (4\text{-}77)$$

The flux in the y direction is described by Fick's law:

$$N_{i,y} = -\mathfrak{D}_{l,i}\frac{\partial C_i}{\partial y} \qquad (4\text{-}78)$$

Equation 4-75 thus becomes

$$v_{\text{max}}\frac{\partial C_i}{\partial x} = \mathfrak{D}_{l,i}\frac{\partial^2 C_i}{\partial y^2} \qquad (4\text{-}79)$$

This is an alternative expression of Fick's second law (Equation 4-6) wherein the time for diffusion is given by x/v_{max} (i.e., the time of travel of the liquid down the wall or the length of time the liquid is in contact with the interface). To solve Equation 4-79, we need to apply the appropriate initial condition, $C_i(y > 0, x = 0) = C_{0,i}$, and boundary conditions, $C_i(y = 0, x > 0) = C_{S,i}$ and $\partial C_i/\partial y = 0$ at $y = \delta$, where $C_{S,i}$ is the saturation concentration of the gas in the liquid given by Henry's law and $C_{0,i}$ is the original gas concentration in the liquid, which is often assumed to be zero. A relatively simple analytical solution obtains by assuming that the gas penetrates only a short distance, so that we can replace the second boundary condition by $C_i(y \to \infty, x > 0) = C_{0,i}$. The solution is given by

$$\frac{C_i - C_{0,i}}{C_{S,i} - C_{0,i}} = \text{erfc}\left[y\left(\frac{4\mathfrak{D}_{l,i}x}{v_{\text{max}}}\right)^{-0.5}\right] \qquad (4\text{-}80)$$

Figure 4-10 illustrates the general shape of the liquid-phase concentration profile given by Equation 4-80 as shown at two different positions along the wall.

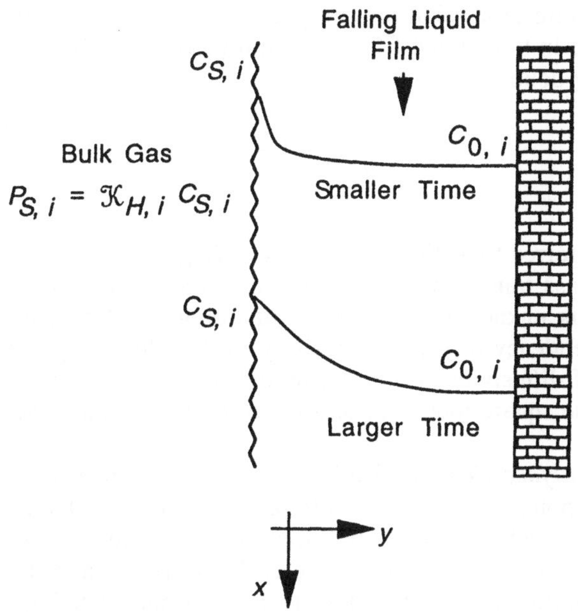

Figure 4-10 Concentration profiles in a falling liquid film.

Fick's law determines the flux of gas into the liquid film at the gas–liquid interface ($y = 0$):

$$N_i = N_i^\circ = -\mathfrak{D}_{l,i} \left.\frac{\partial C_i}{\partial y}\right|_{y=0} \tag{4-81}$$

The concentration gradient, $\partial C_i/\partial y$, is obtained by taking the derivative of Equation 4-80. After substitution into Equation 4-81, the flux at the interface is given by

$$N_i^\circ = -\left(\frac{\mathfrak{D}_{l,i} v_{\max}}{\pi x}\right)^{0.5} (C_{S,i} - C_{0,i}) \tag{4-82}$$

We can now calculate the average interfacial flux, $\overline{N}_i^\circ$, over the width of the wall, Z_W, and the distance, L, through which the liquid film has fallen:

$$\overline{N}_i^\circ = \frac{1}{Z_W L} \int_0^L \int_0^{Z_W} N_i^\circ \, dz \, dx \tag{4-83}$$

which, after substitution for N_i° from Equation 4-82 and integration, yields

$$\overline{N}_i^\circ = 2\left(\frac{\mathfrak{D}_{l,i} v_{\max}}{\pi L}\right)^{0.5} (C_{S,i} - C_{0,i}) \tag{4-84}$$

The foregoing expression for interphase mass transfer has the same form as that presented in Equation 4-54 for the film model, but the mass transfer coefficient is now defined as

$$k_{f,i} = 2\left(\frac{\mathfrak{D}_{l,i}}{\pi t_c}\right)^{0.5} \quad (4\text{-}85)$$

where t_c is referred to as the *contact time* ($L/v_{\max}$) between the gas and liquid during the mass transfer process. Like film thickness in the film model, the contact time is unknown. Comparison of Equations 4-53 and 4-85 reveals a different dependency of $k_{f,i}$ on diffusivity in the film and penetration models; a doubling of diffusivity in the film model doubles $k_{f,i}$, whereas it increases $k_{f,i}$ by only the square root of 2, or a factor of 1.4, in the penetration model.

4.2.4.2 Surface Renewal Model The surface renewal model of Danckwerts builds upon penetration theory to provide a more realistic physical model of the interface and of contact time. Instead of assuming a fixed contact time between liquid packets and the interface, the surface renewal model incorporates a distribution of contact times. The initial and boundary conditions are the same as those used in the penetration model; however, the initial condition now corresponds to $t = 0$ rather than $z = 0$. Thus the flux at the interface is analogous to that given by Equation 4-82; the contact time is now defined as the *average residence time*, $\bar{t}$:

$$N_i^\circ = 2\left(\frac{\mathfrak{D}_{l,i}}{\pi t}\right)^{0.5} (C_{S,i} - C_{0,i}) \quad (4\text{-}86)$$

The interface between the two phases is viewed as being formed by random movement of eddies from the bulk fluid to the interface. Therefore, contact time between these eddies and the surface is given by a probability distribution:

$$\phi_D(t) = \frac{e^{-(t/\bar{t})}}{\bar{t}} \quad (4\text{-}87)$$

where t is the residence time and $\phi_D(t)$ is the distribution of residence times at the surface. The probability that an eddy spends some time, t, at the surface is $\phi_D(t)\,dt$. From a standard table of integrals we can determine that $\int_0^\infty e^{-x}\,dx = \pi^{0.5}/2$. Thus the average flux is given by

$$\overline{N}_i^\circ = \int_0^\infty \phi_D(t) N_i^\circ dt = \left(\frac{\mathfrak{D}_{l,i}}{t}\right)^{0.5} (C_{S,i} - C_{0,i}) \quad (4\text{-}88)$$

and the mass transfer coefficient is

$$k_{f,i} = \left(\frac{\mathfrak{D}_{l,i}}{\bar{t}}\right)^{0.5} \tag{4-89}$$

We observe that the mass transfer coefficient increases with decreasing average residence time of eddies at the surface. In other words, the faster the surface renewal, the more rapid is mass transfer. The penetration and surface renewal theories both suggest the same dependency of the mass transfer coefficient on diffusivity.

4.2.5 Comparison of Mass Transfer Coefficients Developed from Film, Penetration, and Surface Renewal Theories

Relationships between the free-liquid diffusion coefficient and mass transfer coefficients derived for gas–liquid systems from film, penetration, and surface renewal theories are compared in Table 4-2. Direct use of any of these relationships to calculate the mass transfer coefficient is difficult for two reasons. First, each of these three models includes a "system" constant that is not measurable directly; these are δ, t_c, and ϕ_D, respectively. Accordingly, mass transfer coefficients must be obtained from experimental measurements or from experimental correlations with certain system variables (e.g., intensity of mixing). The development of *mass transfer correlations* will be explored in much greater detail in ensuing sections of this chapter.

Given that experimental data or correlations are needed, the second difficulty that arises is in assessing the exact dependence of the mass transfer coefficient on the free-liquid diffusivity. As illustrated in Table 4-2, diffusivity is raised to the first power in the film theory and to the one-half power in the penetration and surface renewal models. In fact, experimental measurements of mass transfer coefficients suggest that the dependence on diffusivity is often somewhere between the one-half and first powers. Further, the dependence appears to be affected by the degree of turbulence in the mass transfer system, being closer to the first power predicted by the film model in laminar flow and

Table 4-2 Mass Transfer Coefficients Derived from Different Theoretical Concepts

Concept	Coefficient
One-film theory	$k_{f,i} = \dfrac{\mathfrak{D}_{l,i}}{\delta}$
Penetration theory	$k_{f,i} = 2\left[\dfrac{\mathfrak{D}_{l,i}}{\pi t_c}\right]$ where $t_c = \dfrac{L}{v_{\max}}$
Surface renewal theory	$k_{f,i}\left[\dfrac{\mathfrak{D}_{l,i}}{\bar{t}}\right]^{0.5}$ where $\delta_D(t) = \dfrac{e^{-t/\bar{t}}}{\bar{t}}$

closer to the one-half power predicted by the penetration (or surface renewal) theory for turbulent flow.

If the proper dependence on free-liquid diffusivity is known, the k_f or, more commonly, the $k_f(a_s^\circ)_R$, of an easily measured reference component (designated by the subscript r) can be used to predict the mass transfer coefficient of any other component, i:

$$(k_f(a_s^\circ)_R)_i = (k_f(a_s^\circ)_R)_r \left(\frac{\mathfrak{D}_{l,i}}{\mathfrak{D}_{l,r}}\right)^{\psi} \quad (4\text{-}90)$$

where the exponent ψ is a system-specific empirical factor, the magnitude of which depends on reactor configuration and mixing conditions.

The notion of using a reference component is particularly appealing when the component of interest is difficult or expensive to analyze. However, it is essential that the reference component have a Henry's constant that is similar to the component of interest in order to assure that the overall mass transfer coefficient of both components is determined by similar contributions of gas- and liquid-phase impedance. An illustration of a practical application of the relationship given by Equation 4-90 is given in Example 4-9.

Example 4-9

- **Situation:** New air pollution regulations require an evaluation of the release of trichloroethylene (TCE) into the atmosphere during aerobic biological treatment (mechanical surface aerators) of an industrial wastewater. The following data are available on the operation of the biological reactor:

  - Flow rate, $Q = 160\ \text{m}^3/\text{day}$
  - Hydraulic retention time, $\bar{t} = V/Q = 20\ \text{h}$
  - O_2 mass transfer requirement per unit volume of wastewater is $M/V = 400\ \text{g/m}^3$
  - Oxygen transfer rating (field conditions) of the mechanical aerators is 3 kg O_2/kW–h

Unfortunately, the analytical capabilities are very limited at the facility, and measurements of TCE volatilization would require considerable investment of time and equipment. However, prior laboratory tests have been conducted to measure the volumetric mass transfer coefficient, $\hat{k}_{f,l}(a_s^\circ)_R$, for oxygen. Note that the two-film theory applies here. These tests were conducted in a vessel that was fitted with vertical baffles attached to the sides. A shaft was inserted with a ring-guarded turbine and three upward-curved blades at the water surface and a flat-bladed impeller positioned near the bottom of the vessel.

The water was initially stripped of oxygen (by bubbling nitrogen gas) and then measurements were made of dissolved oxygen (DO) with time at a fixed mixer speed. The results of one such laboratory test for a volume of water, $V = 8$ L, a power input of $P_w = 2$ W, and an oxygen saturation concentration of $C_S = 9.0$ mg/L are given below.

Time (s)	0	50	100	200	400	600	800	1200
DO (mg/L)	0	1.6	3.1	5.0	7.3	8.3	8.8	8.9

- **Question(s):** Estimate the percentage removal of TCE in the actual aerobic biological reactor.
- **Logic and Answer(s):**

1. The overall volumetric mass transfer coefficient is given by $\hat{k}_{v,l} = \hat{k}_{f,l}(a_s^\circ)_R$. The first step is to determine $\hat{k}_{v,l}$ for oxygen from the experimental data provided under the conditions noted. Because of the intense mixing, it is reasonable to assume that the experimental device is well mixed such that

$$\frac{dC}{dt} = (a_s^\circ)_R N = \hat{k}_{f,l}(a_s^\circ)_R(C_S - C)$$

Integration of this mass transfer expression gives

$$\int_0^C \frac{dC}{C_S - C} = \hat{k}_{v,l} \int_0^t dt$$

or

$$\ln \frac{C_S - C}{C_S} = -\hat{k}_{v,l}\, t$$

2. The experimental data are plotted below according to this linearized form of the mass transfer relationship.

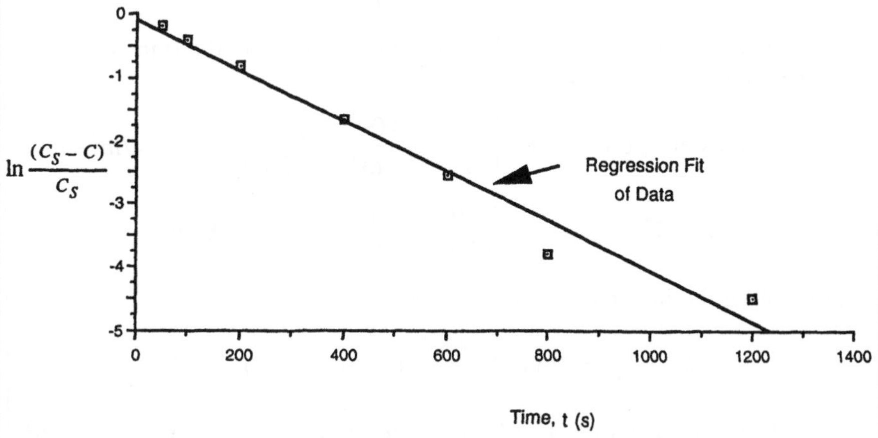

The slope of this line gives $\hat{k}_{v,l} = 0.004\ \text{s}^{-1}$.

3. Equation 4-90 can then be used to estimate $\hat{k}_{v,l}$ for TCE as follows:

$$(\hat{k}_{v,l})_{TCE} = (\hat{k}_{v,l})_{O_2}\left(\frac{\mathfrak{D}_{l,TCE}}{\mathfrak{D}_{l,O_2}}\right)^{\psi}$$

The diffusivities of TCE and O_2 are obtained by the Wilke–Chang correlation (Equation 4-33). An exponent value of $\psi = 0.6$ has been determined experimentally for a similar experimental system (Roberts and Dandliker, 1983) by measuring $\hat{k}_{f,l}(a_s^\circ)_R$ for a series of volatile organic chemicals and oxygen as the reference compound under turbulent mixing conditions. This value is much closer to that predicted by the penetration theory (0.5) than to that predicted by film theory (1.0), as should be expected given that mass transfer was measured under turbulent conditions. Substituting for ψ, the diffusivities (in cm^2/s) of TCE and oxygen and the $\hat{k}_{f,l}(a_s^\circ)_R$ experimentally determined for oxygen:

$$(\hat{k}_{v,l})_{TCE} = 0.004\left(\frac{0.86 \times 10^{-6}}{1.88 \times 10^{-6}}\right)^{0.6} = 0.00156\ s^{-1}$$

4. The value of $\hat{k}_{f,l}(a_s^\circ)_R$ given above is valid only for the mixing conditions used in the laboratory batch experiment. Mixing is quantified by the power input per unit volume (P_w/V) and $\hat{k}_{f,l}(a_s^\circ)_R$ has been shown to be approximately proportional to $P_w V$. Thus, as a first-cut estimate for the actual biological reactor:

$$(\hat{k}_{v,l})_{actual} = (\hat{k}_{v,l})_{lab}\frac{(P_w/V)_{actual}}{(P_w/V)_{lab}}$$

From the information provided with laboratory data:

$$(P_w/V)_{lab} = \frac{2\ W}{8 \times 10^{-3}\ m^3} = 250\ W/m^3$$

$(P_w/V)_{actual}$ is estimated from the O_2 requirements and the transfer rating of the mechanical aerators:

$$\text{mass transfer rate of } O_2 = \frac{M}{V}\frac{1}{t} = \left(\frac{400\ g}{m^3}\right)\left(\frac{1}{20\ h}\right) = 20\ g/m^3\text{–}h$$

$$(P_w/V)_{actual} = \frac{20 \times 10^{-3}\ kg/m^3\text{–}h}{3\ kg/kW\text{–}h} = 6.7\ W/m^3$$

Combining the information above gives

$$(\hat{k}_{v,l})_{actual} = 0.00156\left(\frac{6.7}{250}\right) = 4.2 \times 10^{-5}\ s^{-1}$$

5. The percent removal of TCE in the biological reactor is obtained by writing a material balance assuming completely mixed conditions:

$$Q[\text{TCE}]_{\text{IN}} - Q[\text{TCE}]_{\text{OUT}} + \hat{k}_{v,l}([\text{TCE}]_S - [\text{TCE}]_{\text{OUT}})V = 0$$

Due to dilution, the partial pressure of TCE in the atmosphere above the reactor is approximately zero and thus $[\text{TCE}]_S = 0$. After simplification and rearrangement of the material balance, the desired result is

$$\left\{1 - \frac{[\text{TCE}]_{\text{OUT}}}{[\text{TCE}]_{\text{IN}}}\right\} \times 100 = \frac{\hat{k}_{v,l}(V/Q)}{1 + \hat{k}_{v,l}(V/Q)} \times 100 = 75\%$$

These results show that significant loss of TCE occurs due to volatilization. Even more volatilization would occur if the power input from the mechanical aerators was larger, because the mass transfer coefficient is proportional to the power input. Greater power input may be needed, for example, to assure not only that the oxygen transfer requirement is met, but also that mixing is sufficient to prevent biomass settling.

6. The procedure above for estimating $\hat{k}_{v,l}$ depends on the assumption that the overall mass transfer coefficient for both O_2 and TCE is controlled by liquid-side resistance, or from Equation 4-69,

$$\frac{1}{\hat{k}_{f,l}} = \frac{\mathcal{R}T}{\mathcal{K}_H}\frac{1}{k_{f,g}} + \frac{1}{k_{f,l}} \approx \frac{1}{k_{f,l}}$$

This assumption can be checked by calculation of the ratio of liquid- to gas-side impedances, I_l°/I_g°:

$$\frac{I_l^\circ}{I_g^\circ} = \frac{\mathcal{K}_H}{\mathcal{R}T}\frac{k_{f,g}}{k_{f,l}}$$

Hsieh et al. (1993b) found that the $k_{f,g}/k_{f,l}$ for oxygen ranged from 38 to 110. Using the lower ratio, and given that $\mathcal{K}_H$ for O_2 and TCE are 0.79 (see Tables 1-3 and 6-1, respectively) and 1.2×10^{-2} atm-m^3/mol, respectively, we can estimate for 25°C that

$$\left(\frac{I_l^\circ}{I_g^\circ}\right)_{O_2} = 1250 \quad \text{and} \quad \left(\frac{I_l^\circ}{I_g^\circ}\right)_{\text{TCE}} = 18$$

These results indicate that $\hat{k}_{f,l}(a_s^\circ)_R$ is determined by the liquid-side resistance for both components, thus confirming the validity of the approach used.

4.2.6 Development of the Mass Transfer Coefficient from Boundary Layer Theory

A mass transfer relationship can be derived by coupling the concept of a *flow boundary layer* with that of a *concentration boundary layer*. This derivation is important because, unlike the previous film, penetration, and surface renewal modeling approaches, boundary layer theory provides a realistic physical model of interfaces. Thus a relationship can be obtained directly between rate of mass transfer and hydrodynamic conditions at the boundary, rather than indirectly by way of experimental correlations. Unfortunately, practical applications of boundary layer theory are limited because it is difficult to describe interfacial regions around complex geometric shapes and for turbulent flow regimes.

We will consider the simple example of mass transport away from a "soluble" dissolving plate during laminar flow. The velocity and concentration distributions in the boundary layer are depicted in Figure 4-11. If these distributions can be expressed mathematically, the mass transfer flux at any location, y, within the boundary layer can be viewed as being produced by advective transport of component i (i.e., $N_{i,y} = v_{x,y}C_{i,y}$).

The approximate solution method for boundary layer problems given by the von Kármán integral earlier in this chapter is used to develop a mass transfer model. For a laminar boundary layer, the thickness of the flow boundary layer is given by Equation 4-49 and the velocity profile in the boundary layer by Equation 4-48. The concentration boundary layer is determined by diffusive transport normal to the dissolving plate. Like the velocity profile, the concentration profile can be described by a simple polynomial in y ($C_i = \alpha_0 + \alpha_1 y + \alpha_2 y^2 + \alpha_3 y^3 + \cdots$). After applying the appropriate boundary conditions ($d^2C_i/dy^2 = 0$ at $y = 0$ and $C_i = 0$ and $dC_i/dy = 0$ and $y = \delta$), we have

$$\frac{C_i}{C_{S,i}} = 1 - \frac{3}{2}\left(\frac{y}{\delta}\right) + \frac{1}{2}\left(\frac{y}{\delta}\right)^3 \tag{4-91}$$

The material balance for steady state in an elemental volume ($Z_W\,dx\,dy$) of the fluid in the concentration boundary layer, as shown in Figure 4-11, is

$$\int_0^L \frac{\partial C_i v_x}{\partial x}\,dx\,Z_W\,dy - N_i Z_W\,dx = 0 \tag{4-92}$$

where Z_W is the width of the control volume and L is measured in the y direction. We can assume that $C_i = 0$ for $y \geq \delta$, and therefore the limit of integration is actually from 0 to δ. The first term in Equation 4-92 is the net advective transport in the x direction, while the second term accounts for diffusive transport out of the boundary layer in the y direction. Replacing the flux term, N_i, by Fick's law, substituting δ for L and dividing by $Z_W\,dx$, and then rearranging transforms Equation 4-92 to

$$\frac{\partial}{\partial x}\int_0^\delta C_i v_x\,dy = -\mathfrak{D}_{l,i}\left.\frac{\partial C_i}{\partial y}\right|_{y=0} \tag{4-93}$$

If it is assumed that $\delta < \delta_L$, the left side of Equation 4-93 can be integrated and the right side can be differentiated because both $v_x/v_{x,0}$ and $C_i/C_{\delta,i}$ are

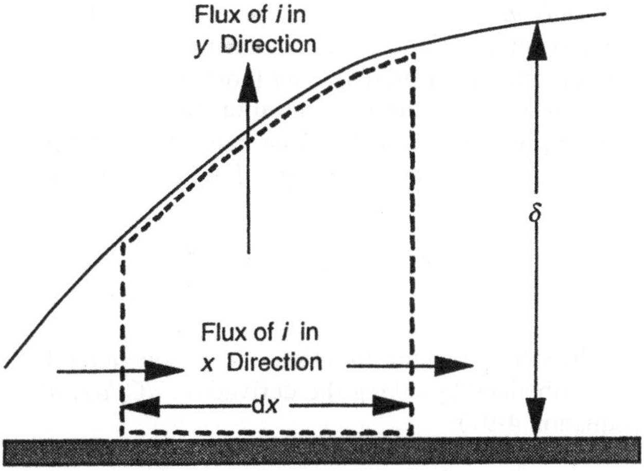

Figure 4-11 Control volume for analysis of a concentration boundary layer.

known functions of y/δ_L and y/δ (Equations 4-48 and 4-91), respectively. The next step is to differentiate the left-hand side with respect to x. However, differentiation cannot proceed before including the dependence of δ_L on x through Equation 4-49. After simplification it can be shown that

$$\left(\frac{\delta_L}{\delta}\right)^3 = \frac{\mathcal{D}_{l,i}\rho}{\mu_v} \tag{4-94}$$

Finally, substitution for δ_L from Equation 4-49 gives an expression to calculate the thickness of the concentration boundary layer:

$$\delta = 4.64x\left(\frac{\mu_v}{xv_{x,b}\rho}\right)^{1/2}\left(\frac{\mathcal{D}_{l,i}\rho}{\mu_v}\right)^{1/3} \tag{4-95}$$

or

$$\delta = \frac{4.64x}{(\mathfrak{N}_{\mathrm{Re}})^{1/2}(\mathfrak{N}_{\mathrm{Sc}})^{1/3}} \tag{4-96}$$

where $\mathfrak{N}_{\mathrm{Re}}$ is the familiar Reynolds number and $\mathfrak{N}_{\mathrm{Sc}}$ is the Schmidt number, defined as the ratio of the kinematic viscosity to molecular diffusivity. Equation 4-96 makes intuitive sense. The thickness of the concentration boundary layer should decrease with increasing values of Reynolds and Schmidt numbers. A higher Reynolds number implies that inertial forces are more important than viscous forces, a fact that is caused by greater momentum transfer in the fluid. We recall from Chapter 3 that kinematic viscosity, ν_v, represents the proportionality between shear stress and the momentum gradient; in other words, it is the *eddy momentum dispersivity*. A high Schmidt number, therefore, implies dominance of macroscale momentum transfer over microscale momentum transfer, the latter produced by molecular diffusion.

Knowing the thickness of the concentration boundary layer allows us to obtain a relationship for mass transfer. The mass transfer flux is equal to the diffusive flux given by Fick's law and applied at $y = 0$, namely,

$$N_i = k_f(C_{S,i} - 0) = -\mathfrak{D}_{l,i} \left.\frac{dC_i}{dy}\right|_{y=0} \tag{4-97}$$

in which $C_{S,i}$ is the saturation concentration of component i at the surface. The diffusional flux is obtained by taking the derivative, dC_i/dy, of the concentration profile (Equation 4-91):

$$N_i = \frac{3}{2}\left(\frac{\mathfrak{D}_{l,i}C_{S,i}}{\delta}\right) \tag{4-98}$$

We can eliminate δ by substituting Equation 4-95 into Equation 4-98. Finally, through incorporation of Equation 4-97, we obtain

$$\frac{k_f x}{\mathfrak{D}_{l,i}} = 0.323\left(\frac{x v_{x,b}\rho}{\mu_v}\right)^{1/2}\left(\frac{\mu_v}{\rho\mathfrak{D}_{l,i}}\right)^{1/3} \tag{4-99}$$

The foregoing equation, derived from boundary layer theory, has important practical implications. The dimensionless term on the left is termed the *local Sherwood number*, local because it applies to a specific location, x. The dimensionless groups on the right have already been identified as the Reynolds and Schmidt numbers. This expression is often written as the *average* value, $\bar{k}_f$, of the mass transfer coefficient over the characteristic length L_c:

$$\frac{\bar{k}_f L_c}{\mathfrak{D}_{l,i}} = 0.626\left(\frac{L_c v_b \rho}{\mu_v}\right)^{1/2}\left(\frac{\mu_v}{\rho\mathfrak{D}_{l,i}}\right)^{1/3} \tag{4-100}$$

The exponents that appear above are given as *fractions* rather than as *decimal values*, to emphasize their *origin* from *theoretical* analysis of a boundary layer instead of from *experiment*. We see from the above that the mass transfer coefficient for dissolution of a solute from a sharp-edged plate into a laminar flow should be predicted from (1) a length, L_c, which characterizes the mass transfer system; (2) the systems fluid flow properties through the Reynolds number; and (3) the molecular diffusivity of the component being transferred.

It is important to note that the results given in Equations 4-96, 4-99, and 4-100 were obtained through characterization of a realistic physical model (the boundary layer), by application of fundamental principles of continuity, conservation of momentum and mass transport (i.e., it does not include empirical constants). While considerable experimental data to show the validity of these relationships are available, the application of boundary layer theory to more complex mass transfer situations is limited by our lack of understanding of the physical details of such situations. Nonetheless, we have shown through application of a rigorous model development that mass transfer relationships for different systems can be interrelated by relatively simple dimensionless groupings of related parameters. This leads us logically then to a fuller exposition of empirical approaches using mass transfer correlation procedures.

4.2.7 Development of Mass Transfer Correlations

The film, penetration, and surface renewal models each provide a way to determine mass transfer coefficients from measurements of flux and concentration in any system, regardless of its hydrodynamic and geometric complexity. In contrast to the boundary layer theory from which Equation 4-99 was developed, however, these models do not relate the mass transfer process to a characteristic dimension of the system for mass transfer, nor to the hydrodynamics of the system or the diffusivity of the component being transferred. Operational correlations among these parameters and processes are therefore needed, and dimensional analysis plays an important role in the development of such relationships. In this approach, the general form of a mass transfer relationship is formulated through dimensional analysis. Experiments are then designed to obtain empirical information that is specific for a given type of system.

Dimensional analysis depends on starting with a reasonably good understanding of the variables that influence a given process. The objective is then to organize these variables into the smallest number of dimensionless groups. The Buckingham Pi theorem, developed in 1915, states that the smallest number of dimensionless groups, each designated as Π_i° (hence the name *Pi theorem*), is given by

$$\Pi_1^\circ = \phi(\Pi_2^\circ, \Pi_3^\circ, \ldots, \Pi_{n-m}^\circ) \tag{4-101}$$

where n is the number of variables and m is the number of basic dimensions included in the variables. We can use the results from boundary layer theory to gain an appreciation for the fundamental variables that affect mass transport. We found that k_f is a function of fluid velocity, fluid density, fluid viscosity, molecular diffusivity, and a characteristic mass transfer length. Given that the basic dimensions are length (L), mass (M), and time (t), we conclude that the minimum number of dimensionless groups is $5 - 3 = 2$.

The so-called *exponent method* of dimensional analysis will be used here. We begin by proposing a general relationship between k_f and all other significant system variables:

$$k_f = \phi(v^{\kappa_1}, \rho^{\kappa_2}, \mu_v^{\kappa_3}, \mathfrak{D}_{l,i}^{\kappa_4}, L^{\kappa_5}) \tag{4-102}$$

Because k_f has the dimensions Lt^{-1}, each of the terms making up the final dimensionless groups must also have dimensions of Lt^{-1}. Thus the exponent to which each variable is raised must yield the final dimensions of length per time, or

$$\frac{L}{T} = \left(\frac{L}{T}\right)^{\kappa_1} \left(\frac{M}{L^3}\right)^{\kappa_2} \left(\frac{M}{LT}\right)^{\kappa_3} \left(\frac{L^2}{T}\right)^{\kappa_4} (L)^{\kappa_5} \tag{4-103}$$

Thus,

$$\sum M = 0 = \kappa_1 + \kappa_3$$

$$\sum L = 1 = \kappa_1 - 3\kappa_2 - \kappa_3 + 2\kappa_4 + \kappa_5$$

$$\sum T = -1 = -\kappa_1 - \kappa_3 - \kappa_4$$

With three equations and five unknowns, we need to select two exponents and thus two dimensionless groups with which to express k_f. Choosing a and c gives

$$k_f = \phi[v^{\kappa_1} \rho^{-\kappa_3} \mu_v^{\kappa_1} \mathfrak{D}_{l,i}^{(1-\kappa_1-\kappa_3)} L^{(-1+\kappa_1)}] \tag{4-104}$$

Dimensionless groups are then formed by grouping variables having the same exponents. This leads to

$$\frac{k_f L}{\mathfrak{D}_{l,i}} = \phi\left[\left(\frac{vL}{\mathfrak{D}_{l,i}}\right)^{\kappa_1} \left(\frac{\mu_v}{\rho \mathfrak{D}_{l,i}}\right)^{\kappa_3}\right] \tag{4-105}$$

The term on the left side of Equation 4-105 is recognized as the Sherwood number, while the second term on the right side is the Schmidt number. The first term on the right side is the product of the Reynolds and Schmidt numbers. This product constitutes another dimensional group, known as the *Peclet number*:

$$(\mathfrak{N}_{Re})\,(\mathfrak{N}_{Sc}) = \frac{vL\rho}{\mu_v}\,\frac{\mu_v}{\rho \mathfrak{D}_{l,i}} = \frac{vL}{\mathfrak{D}_{l,i}} = \mathfrak{N}_{Pe} \tag{4-106}$$

Because the Schmidt number appears in both dimensionless groupings in Equation 4-105, we can combine them to give

$$\mathfrak{N}_{Sh} = \phi[(\mathfrak{N}_{Re})^{\kappa_1}\,(\mathfrak{N}_{Sc})^{\kappa_3-\kappa_1}] \tag{4-107}$$

The form of this functional relationship is referred to as a *mass transfer correlation*:

$$\mathfrak{N}_{Sh} = \psi_1(\mathfrak{N}_{Re})^{\psi_2}(\mathfrak{N}_{Sc})^{\psi_3} \tag{4-108}$$

where ψ_1, ψ_2, and ψ_3 are empirical constants for description of a particular experimental data set. Experimental measurements of k_f for a wide range of values of the important system variables allow empirical determination of the exponents of the Reynolds and Schmidt numbers and the value of the constant, ψ_1. Fluid velocity, type of solute, temperature, and characteristic length of the mass transfer surface (e.g., the diameter of spheres in a packed bed) are typical experimental variables. Use of a mass transfer correlation for evaluating dissolution of NAPL in the subsurface is given in Example 4-10.

Example 4-10

- **Situation:** The release of immiscible liquids to the subsurface is a common cause of groundwater contamination. These non-aqueous-phase liquids (NAPLs) are retained in the interstitial spaces between soil particles. Interphase mass transfer from such "sources" of contaminant to the aqueous phase, the solid phase, and the vapor phase are possible.
- **Question(s):** Using dimensional analysis, explore the system constants that affect interphase mass transfer of solute between an immobile NAPL and the aqueous phase and design an experiment to measure the mass transfer coefficient.
- **Logic and Answer(s):**

1. The inherently dimensionless parameters of interest are porosity, the percent NAPL saturation of the soil, and the contact angle between the solid phase and the NAPL, $\cos\theta_c$. In addition,

$$\hat{k}_{v,l} = \hat{k}_{f,l}(a_s^\circ)_N = \phi(\mu_{v,a}, \mu_{v,n}, \rho_a, \rho_n, d_p, v_p, \mathfrak{D}_l, \mathfrak{D}_l, \sigma_{n,a}^\circ, g)$$

where $\hat{k}_{f,l}(a_s^\circ)_N$ is the product of the mass transfer coefficient and the specific interfacial surface area per unit volume of NAPL, considered together owing to the difficulty of experimentally measuring $(a_s^\circ)_N$. The other terms in the expression above are: μ_v = viscosity; ρ = density; d_p = diameter of a particle of the porous media; v_p = mean aqueous-phase pore velocity; $\mathfrak{D}_l$ = diffusivity of the solute in water; $\sigma_{n,a}^\circ$ = interfacial tension; g = acceleration of gravity; and the subscripts a and n refer to the aqueous and nonaqueous phases, respectively.

2. Dimensional analysis using the Buckingham Pi theorem yields seven dimensionless groups. In addition to the $\mathfrak{N}_{\mathrm{Re}}$, $\mathfrak{N}_{\mathrm{Sc}}$, and $\mathfrak{N}_{\mathrm{Sh}}$ numbers, these are $\mathfrak{N}_{\mathrm{Ca}}$, the capillary number; $\mathfrak{N}_{\mathrm{Bo}}$, the Bond number; ϕ_{vm}°, the viscosity mobility ratio; and $\mathfrak{N}_{\mathrm{Go}}$, the Goucher number:

$$\mathfrak{N}_{\mathrm{Ca}} = \frac{v_p \mu_a}{\sigma_{n,a}^\circ} \qquad \mathfrak{N}_{\mathrm{Bo}} = \frac{(\rho_a - \rho_n) g d_p^2}{\sigma_{n,a}^\circ} \qquad \phi_{vm}^\circ = \frac{\mu_n}{\mu_a} \qquad \mathfrak{N}_{\mathrm{Go}} = \frac{d_p^2 \rho_a g}{\sigma_{n,a}^\circ}$$

Although 10 dimensionless groups are possible, not all need to be included in initial testing of a correlation. NAPL saturation and porosity can be combined to form the volumetric fraction, $\phi^{\circ}_{V,N}$.

3. The form of correlation to be tested is that suggested by Miller et al. (1990) and presented in Appendix A.4.1:

$$\mathfrak{N}'_{\mathrm{Sh}} = \psi_1 \mathfrak{N}_{\mathrm{Re}}^{\psi_2} (\phi^{\circ}_{V,N})^{\psi_3} \mathfrak{N}_{\mathrm{Sc}}^{0.5}$$

where the modified Sherwood number is

$$\mathfrak{N}'_{\mathrm{Sh}} = \frac{(\hat{k}_{f,l}(a_s^{\circ})_N d_p^2}{\mathfrak{D}_l}$$

The exponent of $\mathfrak{N}_{\mathrm{Sc}}$ is assumed from the literature.

4. The mass transfer experiment consists of a laboratory-scale column to which glass beads and a specific volume of NAPL are added. The beads are mixed to give a uniform initial distribution of the NAPL and the NAPL saturation is determined by the amount of NAPL added. Distilled, deionized water is pumped through the column until a steady-state concentration of the solute in the aqueous phase is reached in the effluent. The point form of the material balance (see Chapter 2) on solute in the water phase leads to

$$\mathfrak{D}_d \frac{\partial^2 C}{\partial x^2} - v_a \frac{dC}{dx} + \hat{k}_{f,l}(a_s^{\circ})_N (C_S - C) = 0$$

where $\mathfrak{D}_d$ is the dispersion coefficient (discussed in Chapter 3) in the x direction (along the column length), v_a is the flow velocity of the aqueous phase and C_S is the aqueous solubility of the solute. The analytical solution is

$$\frac{C(x)}{C_S} = 1 - \exp\left[\frac{x}{2\mathfrak{D}_d}(v_a - (v_a^2 + 4\mathfrak{D}_d \hat{k}_{f,l}(a_s^{\circ})_N)^{0.5})\right]$$

5. The dispersion coefficient, $\mathfrak{D}_d$, is measured independently (see Chapter 9 for test procedures using tracers). For any experiment, v_a is known and $C(x)$ is measured at the end of the column; thus $\hat{k}_{f,l}(a_s^{\circ})_N$ can be calculated directly by running a series of experiments in which v_a and d_p are varied and with different initial amounts of NAPL (this determines $\phi^{\circ}_{V,N}$). The coefficients ψ_0, ψ_1, and ψ_2 of the correlation given in step 3 are then determined.

6. Given a typical $\mathfrak{N}_{\mathrm{Re}}$ number (based on d_p), $\phi^{\circ}_{V,N}$ and $\mathfrak{N}_{\mathrm{Sc}}$, $\hat{k}_{f,l}(a_s^{\circ})_N$ is on the order of 2000 day^{-1}. With typical velocity and dispersivity values, the equation given in step 4 predicts that the solute reaches saturation in the groundwater in a time of travel of less than 0.1 day.

Mass transfer correlations are often expressed in terms of a *Colburn factor*, $\mathcal{J}_D$, where

$$\mathcal{J}_D = \frac{\mathfrak{N}_{\text{Sh}}}{\mathfrak{N}_{\text{Sc}}^{0.333}\,\mathfrak{N}_{\text{Re}}} \tag{4-109}$$

The Colburn factor brings together all three dimensionless groups. It includes an assumption about the dependency of mass transfer on the Schmidt number that is familiar from the derivation of a mass transfer relationship for flat plates using boundary layer theory (see Equation 4-100). $\mathcal{J}_D$ is usually correlated with the Reynolds number; for example,

$$\mathcal{J}_D = \psi_1 \mathfrak{N}_{\text{Re}}^{\psi_2} \tag{4-110}$$

whereupon a plot of ln $\mathcal{J}_D$ against ln $\mathfrak{N}_{\text{Re}}$ would yield a straight line. A composite of various tests of a correlation of this type is illustrated in Figure 4-12 for liquid flow through packed beds. If we substitute the definition of the Colburn factor, $\mathcal{J}_D$, and rearrange, we find that

$$\mathfrak{N}_{\text{Sh}} = 0.76\ \mathfrak{N}_{\text{Re}}^{0.34}\ \mathfrak{N}_{\text{Sc}}^{0.333} \tag{4-111}$$

Equation 4-111, which is based on a collection of correlations of experimental data for various mass transfer situations, compares fairly closely to the equation derived theoretically from boundary layer theory for a very different and much simpler mass transfer situation, that of solute dissolving from a flat

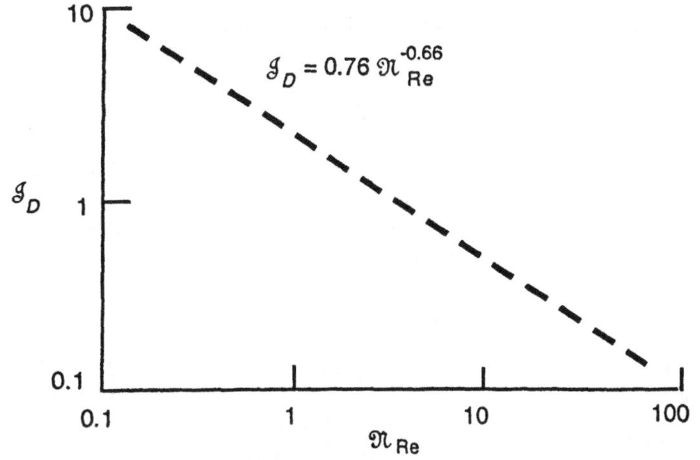

Figure 4-12 Composite of mass transfer correlations for liquids in packed beds. [Adapted from Sherwood et al. (1975).]

plate. This is evident by substituting the dimensionless number terms into Equation 4-100 to obtain

$$\mathfrak{N}_{Sh} = 0.626(\mathfrak{N}_{Re})^{0.5}(\mathfrak{N}_{Sc})^{0.333} \tag{4-112}$$

Another insight into the general form of the mass transfer correlation obtains from rearrangement of Equation 4-108 to solve for the mass transfer coefficient:

$$k_f = \mathfrak{D}_{l,i}^{(1-\psi_3)} \frac{\psi_1}{L} \left(\frac{vL\rho}{\mu_v}\right)^{\psi_2} \left(\frac{\mu_v}{\rho}\right)^{\psi_3} \tag{4-113}$$

According to the one-film model, the exponent of the diffusion coefficient, $\mathfrak{D}_{l,i}$, should be unity (i.e., $\psi_3 = 0$), whereas the penetration and surface-renewal models imply square root relationships such that $\psi_3 = 0.5$. For most mass transfer systems ψ_3 is observed experimentally to fall in the range 0.3 to 0.7, which suggests that neither conceptual model is universally applicable. Generally speaking, the one-film model gives a somewhat better description of convective mass transfer at solid–liquid interfaces, while the penetration or surface-renewal model is more descriptive of convective mass transfer at gas–liquid interfaces.

A few common mass transfer correlations for flow of liquids past flat plates and spheres as well as gases and liquids in packed and fluidized beds are given in Table 4-3. Additional examples of alternative correlations for similar systems and for correlations relating to different types of environmental circumstances are listed in Appendix A.4.1.

Table 4-3 Examples of Mass Transfer Correlations

Application	Equation	Limits
Liquid flow past parallel flat plates	$\mathfrak{N}_{Sh} = 0.99\mathfrak{N}_{Re}^{0.5}\mathfrak{N}_{Sc}^{0.333}$	$600 < \mathfrak{N}_{Re} < 50{,}000$
Liquid flow past a single sphere	$\mathfrak{N}_{Sh} = 2 + 0.95\mathfrak{N}_{Re}^{0.5}\mathfrak{N}_{Sc}^{0.333}$	$2 < \mathfrak{N}_{Re} < 2000$
	$\mathfrak{N}_{Sh} = 0.347\mathfrak{N}_{Re}^{0.62}\mathfrak{N}_{Sc}^{0.333}$	$2000 < \mathfrak{N}_{Re} < 17{,}000$
	$\mathfrak{N}_{Re} = \frac{d_p v \rho_l}{\mu_{v,l}}$; $\mathfrak{N}_{Sh} = \frac{k_f d_p}{\mathfrak{D}_l}$	
	d_p = diameter of particle	
Liquids in a packed bed of spheres	$\mathfrak{N}_{Sh} = \frac{0.250}{\epsilon_B}\mathfrak{N}_{Re}^{0.69}\mathfrak{N}_{Sc}^{0.333}$	$55 < \mathfrak{N}_{Re} < 1500$ $165 < \mathfrak{N}_{Sc} < 10{,}690$
Gases and liquids in fluidized beds of spheres	$\mathfrak{J}_D = 0.010 + \frac{0.863}{\mathfrak{N}_{Re}^{0.58} - 0.483}$	

Source: Adapted from Geankoplis (1978). See Appendix A.4.1 for additional examples.

Numerical values of coefficients associated with mass transfer correlations are determined from experimental data and are therefore subject to experimental error. The mass transfer coefficients calculated from these correlations should thus be considered as *average values* with *associated confidence intervals*. Further, the range of conditions (e.g., Reynolds number, Schmidt number, type of fluids and type of solutes), as well as the amount of data collected to develop the coefficients, are *correlation specific*. These are all reasons to proceed cautiously when selecting a correlation for use. At the very least, a mass transfer correlation must be restricted to the range of Reynolds and Schmidt numbers for which it has been calibrated by rigorous and extensive experimental data.

Correlations such as those given in Table 4-3 and Appendix A.4.1 have a variety of applications in engineered and natural systems. Mass transfer coefficients are important, for example, in the design of packed beds of activated carbon and ion-exchange materials and in the design of air stripping towers for removal of volatile organic chemicals. Similar correlations are applicable to determination of the fate of volatile organic chemicals in surface waters and the fate of immiscible organic liquids (e.g., originating from leaking underground storage tanks) in subsurface systems. The application of a given mass transfer correlation must be restricted to the range of Reynolds and Schmidt numbers for which it has been calibrated by rigorous and extensive experimental data. Various correlations have been developed by different researchers for the same physical systems; an illustrative comparison is provided in Example 4-11.

Example 4-11

- **Situation:** A packed bed of activated carbon is employed to remove trichloroethylene (TCE) from a contaminated groundwater. While the physical characteristics of the packed bed and its operation are known, there is presently no quantitative means to estimate its performance under different operation conditions. Mass transfer to the adsorption sites on the activated carbon particles is limited in part due to transport of TCE across the laminar liquid layer surrounding the particles.

- **Question(s):** You are asked to examine available mass transfer correlations in order to calculate the mass transfer coefficient. You are given the following system properties:

  - Flow velocity, $v = 0.6$ cm/s
  - Characteristic diameter of activated carbon particles, $d_p = 0.5$ cm
  - Bed porosity $\epsilon_B = 0.5$

- **Logic and Answer(s):**

  1. Compare k_f using the Williamson et al. (1963) and Wilson and Geankoplis (1966) mass transfer correlations (Appendix A.4.1) and ver-

ify that the system meets the constraint for Reynolds and Schmidt numbers.

2. From Appendix A.4.1 we determine that the Williamson correlation is

$$\mathfrak{N}_{\mathrm{Sh}} = 2.4\ \epsilon_B\ \mathfrak{N}_{\mathrm{Re}}^{0.34}\ \mathfrak{N}_{\mathrm{Sc}}^{0.42}$$

which is valid for $0.08 < \mathfrak{N}_{\mathrm{Re}} < 125$ and $150 < \mathfrak{N}_{\mathrm{Sc}} < 1300$; and that the Wilson and Geankoplis correlation is

$$\mathfrak{N}_{\mathrm{Sh}} = 1.09\ \epsilon_B^{-0.667}\ \mathfrak{N}_{\mathrm{Re}}^{0.333}\ \mathfrak{N}_{\mathrm{Sc}}^{0.333}$$

which is valid for $0.0016 < \epsilon_B N_{\mathrm{Re}} < 55$ and $950 < \mathfrak{N}_{\mathrm{Sc}} < 70{,}000$.

3. Other necessary data to use these correlations are:

- Fluid viscosity, $\mu_v = 1.5 \times 10^{-2}$ g/cm-s
- Fluid density, $\rho = 1.0$ g/cm^3
- Free-liquid diffusivity of TCE, $\mathfrak{D}_l = 1.02 \times 10^{-5}$ cm^2/s

4. From the definitions:

$$\mathfrak{N}_{\mathrm{Re}} = \frac{v\, d_p \rho}{\mu_v} = 20$$

$$\mathfrak{N}_{\mathrm{Sc}} = \frac{\mu_v}{\rho \mathfrak{D}_l} = 1471$$

The applicability range criteria are thus met for both correlations.

5. The Williamson correlation gives $\mathfrak{N}_{\mathrm{Sh}} = 71.1$ and so

$$k_f = 71.1\ \frac{\mathfrak{D}_l}{d_p} = 1.45 \times 10^{-3} \text{ cm/s}$$

The Wilson and Geankoplis correlation gives $\mathfrak{N}_{\mathrm{Sh}} = 53.2$ and so

$$k_f = 53.2\ \frac{\mathfrak{D}_l}{d_p} = 1.09 \times 10^{-3} \text{ cm/s}$$

6. We thus see that the correlations vary by approximately 25% (lowest to highest values). This is not atypical of the range of mass transfer coefficient values estimated by different correlations meeting the same basic applications criteria. In fact, this would usually be considered a reasonable agreement.

4.2.8 Analogies Among Mass, Momentum, and Energy Transfer

Much has been written in the scientific and engineering literature about analogies among mass, momentum, and energy transport phenomena. The practical significance of such analogies relates to the way in which they can be used to mathematical prediction of the transfer rate, for example, of mass transfer from that of energy (heat) transfer if the physical descriptions of their respective transport situations are analogous to each other. A few important examples of such analogies are presented here to provide an appreciation of the interrelationship among the many concepts presented in this chapter.

The analogies among diffusivities of mass, momentum, and heat are readily apparent by reexamining their respective governing transport relationships.

Mass flux (mass/area/time, $ML^{-2}t^{-1}$):

$$N_{i,x} = -\mathfrak{D}_{l,i} \frac{dC_i}{dx} \tag{3-67}$$

Momentum flux (mass × velocity/area/time, $ML^{-1}t^{-2}$, or force/area, FL^{-2}), which, as a generalization of Equation 3-63, is:

$$\tau_{s,yx} = \mu_v \frac{dv_x}{dy} \tag{4-114}$$

Heat flux (heat/area/time, Mt^{-3}):

$$N_{H,x} = -k_c \frac{dT}{dx} \tag{4-2}$$

Heat flux can be generalized in terms of any energy flux of any type by substituting the appropriate intensive variable for the temperature variable, T.

As we have seen earlier in this chapter, Fick's and Newton's laws determine the flux for mass and momentum, respectively. The corresponding relationship for heat flux, N_H, is known as *Fourier's law*, where the proportionality between heat flux by conduction and the temperature gradient is given by k_c, the thermal conductivity of the medium. The similarity of these relationships is best seen by expressing the intensive parameter included in the gradients governing momentum and heat transport in the same form as that for mass transport (i.e., as momentum per volume and energy per volume, respectively).

Momentum flux:

$$\tau_{s,yx} = \frac{\mu_v}{\rho} \frac{d(\rho v_x)}{dy} \tag{4-115}$$

Heat flux:

$$N_{H,x} = -\frac{k_c}{\rho Q_H^\circ} \frac{d(\rho Q_H^\circ T)}{dx} \tag{4-116}$$

where Q_H° is the specific heat capacity of the medium (heat/mass/unit of temperature, $(ML^2t^{-2}T^{-1})$. Therefore, the respective diffusion or dispersion coefficients for mass, momentum, and heat must all have dimensions of L^2t^{-1}, as shown below:

$$\text{Mass diffusion} = \mathfrak{D}_l \tag{4-117}$$

$$\text{Momentum dispersion} = \frac{\mu_v}{\rho} = \nu_v \tag{4-118}$$

$$\text{Heat diffusion} = \frac{k_c}{\rho Q_H^\circ} \tag{4-119}$$

Using the foregoing consistent definitions of mass and heat diffusivity and momentum dispersivity, it is possible to give some physical meaning to several important dimensionless groups. The Schmidt number was defined earlier (see Equation 4-95) as $\mathfrak{N}_{Sc} = \mu_v/\rho\mathfrak{D}_l$. From the definitions presented in Equations 4-117 and 4-118, the Schmidt number can now be expressed as

$$\mathfrak{N}_{Sc} = \frac{\mu_v}{\rho}\frac{1}{\mathfrak{D}_l} = \frac{\text{momentum dispersion}}{\text{mass diffusion}} \tag{4-120}$$

Using the definitions given in Equations 4-118 and 4-119, respectively, the Prandtl number, an important dimensionless group for heat transfer problems, can similarly be characterized as

$$\mathfrak{N}_{Pr} = \nu_v\left(\frac{\rho Q_H^\circ}{k_c}\right) = \frac{\text{momentum dispersion}}{\text{heat diffusion}} \tag{4-121}$$

It is evident that the Prandtl number is the heat transfer analog of the Schmidt number.

A further appreciation for the analogy between heat and mass convection from a surface is gained by defining heat convection is

$$N_H = k_h(T_S - T_F) \tag{4-122}$$

where k_H is the heat transfer coefficient and $(T_S - T_F)$ is the difference in temperature between the surface and the fluid. The same type of analysis that was done to relate mass transfer and diffusion by way of the film model (Equation 4-56) can be done to relate heat transfer and conductivity. This analysis results in the definition of the Nusselt number for heat transfer, $\mathfrak{N}_{Nh}$, as

$$\mathfrak{N}_{Nh} = \frac{k_h L_c}{k_c} = \frac{\text{total heat transfer}}{\text{conductive heat transfer}} \tag{4-123}$$

We recall from Equation 4-59 that

$$\mathfrak{N}_{Sh} = \frac{\text{interfacial mass transfer impedance}}{\text{molecular diffusion impedance}} \tag{4-59}$$

Clearly, the Nusselt number for heat transfer is the analog of the Sherwood number for mass transfer. In fact, as noted in Appendix III, which provides a list and definitions for a variety of dimensionless numbers of use in analysis and quantification of environmental processes, the Sherwood number is also known as the mass transfer Nusselt number.

4.3 SUMMARY

Descriptions of microtransport processes draw upon a broad array of principles from the fields of fluid mechanics, chemical kinetics and thermodynamics, and even probability theory. We have seen that diffusion of molecules can be explained using empirical, statistical, or chemical potential approaches. Similarly, mass transport across interfaces can be explained in terms of molecular diffusion, with several somewhat different approaches (film, penetration, and surface renewal) available to quantify the process. Taking an even broader view, the analogies among mass, momentum, and heat transport provide reassurance that the theoretical foundation upon which these relationships are structured is sound. The outcome of such analysis is not a set of immutable physical constants governing molecular diffusion and interphase mass transport, however. Our knowledge of hydrodynamics and molecular movement at the microscale is imperfect, so we must depend ultimately upon empirical observations and experimental data. The Wilke–Chang equation for free-liquid diffusivity, and the Sherwood, Reynolds, and Schmidt number correlations for mass transfer coefficients, illustrate this dependence on experimental data very clearly. These engineering-oriented questions and microtransport constants are very useful in process analysis. However, the limitations of the data base upon which these engineering tools have been derived must be recognized. To ensure proper characterization and description of microtransport in any particular system to which they are to be applied, then, the chemical and hydrodynamic properties of that system must be examined carefully.

4.4 REFERENCES AND SUGGESTED READINGS

4.4.1 Citations and Sources

Ahmed, T., and M. Semmens, 1992, "The Use of Independently Sealed Microporous Hollow Fiber Membranes for Oxygenation of Water: Model Development," *Journal of Membrane Science*, *69*, 11–20 (Appendix A.4.1)

Ahn, B. S., and W. K. Lee, 1990, "Simulation and Experimental Analysis of Mass

Transfer in a Liquid–Liquid Stirred Tank Extractor,'' *Industrial Engineering Chemistry Research*, *29*, 1927–1935. (Appendix A.4.1)

Calderbrook, P. H., and M. B. Moo-Young, 1961, ''The Continuous Phase Heat and Mass Transfer Properties of Dispersions,'' *Chemical Engineering Science*, *16*, 39–54. (Appendix A.4.1)

Christakos, G., 1992, *Random Field Models in Earth Sciences*, Academic Press, Inc., San Diego, CA. (Equations 4-8 to 4-12) Chapter 1 of this citation gives a translation of Einstein's analysis of Brownian motion of particles in fluids using a probability density function to arrive at an equivalent statement of Fick's second law.

Churchill, M. A., H. C. Elmore, and R. A. Buckingham, 1962, ''Prediction of Stream Reaeration Rates,'' *Journal of the Sanitary Engineering Division, American Society of Civil Engineers*, *88*, SH4, 1 (Appendix A.4.1)

Cussler, E. L., 1984, *Diffusion: Mass Transfer in Fluid Systems*, Cambridge University Press, London, pp. 288–304. (Equations 4-60 to 4-63 and 4-91 to 4-100)

DiGiano, F. A., D. Elliot, and D. Leith, 1988, ''Application of Passive Dosimetry for Detection of Trace Organic Contaminants,'' *Environmental Science and Technology*, 22, 1365–67. (Example 4-1 and Problem 4-1)

Fogler, H. S., 1992, *Elements of Chemical Reaction Engineering*, Prentice Hall, Englewood Cliffs, NJ. (Appendix A.4.1)

Geankoplis, C. J., 1978, *Transport Processes and Unit Operations*, Allyn and Bacon, Inc., Boston. (Table 4-3, Appendix A.4.1)

Gnielinski, V. T., 1978, ''Gleichung zue Berechnung des Wärme-and Stoffaustauches in durchströmten ruhenden Kugelschüttungen bei mittleren und grossen Pecletzahlen,'' VT *Verfahrenstechnik*, *12*, 363–66. (Appendix A.4.1)

Grasso, D., L. Fujikawa, and W. J. Weber, Jr., 1990, ''Ozone Mass Transfer in a Gas-Sparged Turbine Reactor,'' *Journal of the Water Pollution Control Federation*, *62*, 246–252. (Appendix A.4.1)

Hallappa Gowda, T. P., and J. D. Lock, 1985, ''Volatilization Rates of Organic Chemicals of Public Health Concern,'' *Journal of the Environmental Engineering Division, American Society of Civil Engineers*, *111*(6), 755–776. (Appendix A.4.1)

Hsieh, C., R. W. Babcock, and M. K. Stenstrom, 1993a, ''Estimating Emissions of 20 VOCs. II. Diffused Aeration,'' *Journal of the Environmental Engineering Division, American Society of Civil Engineers*, *111*(6), 1099–1118. (Appendix A.4.1)

Hsieh, C., K. S. Ro, and M. Stenstrom, 1993b, ''Estimating Emissions of 20 VOCs. I. Surface Aeration,'' *Journal of the Environmental Engineering Division, American Society of Civil Engineers*, *119*(6), 1077–1098. (Example 4-9 and Appendix A.4.1)

Levins, B. E., and J. R. Glastonbury, 1972, ''Particle–Liquid Hydrodynamics and Mass Transfer in a Stirred Vessel. II. Mass Transfer,'' *Transactions of the Institution of Chemical Engineers*, *50*, 132–146. (Appendix A.4.1)

Miller, C. T., M. M. Poirier-McNeill, and A. S. Mayer, 1990, ''Dissolution of Trapped Nonaqueous Phase Liquids: Mass Transfer Characteristics,'' *Water Resources Research*, *26*, 11, 2783–2796. (Example 4-10, Problem 4-6, and Appendix A.4.1)

O'Connor, D. J., and W. Dobbins, 1956, ''The Mechanism of Reaeration in Natural Streams,'' *Journal of the Sanitary Engineering Division, American Society of Civil Engineers*, *82*, SA6, 1115. (Appendix A.4.1)

Onda, K., H. Takeuchi, and Y. Okumoto, 1968, ''Mass Transfer Coefficients Between Gas and Liquid Phases in Packed Columns,'' *Journal of Chemical Engineering, Japan*, *1*, 56–62. (Appendix A.4.1)

Perry, R. H., and C. H. Chilton, 1973, *Handbook of Chemical Engineering*, 5th Edition, McGraw-Hill Book Company, New York, pp. 3–234. (Table 4-1)

Powers, S. E., L. M. Abriola, and W. J. Weber, Jr., 1992, ''An Experimental Investigation of Nonaqueous Phase Liquid Dissolution in Saturated Subsurface Systems: Steady-State Mass Transfer Rates,'' *Water Resources Research*, *28*(10), 2691–2705. (Appendix A.4.1)

Roberts, P. V., and P. G. Dandliker, 1983, ''Mass Transfer of Volatile Organic Contaminants from Aqueous Solution to the Atmosphere During Surface Aeration,'' *Environmental Science and Technology*, *17*, 484–489. (Example 4-9 and Problem 4-7)

Rushton, J. H., E. W. Costich, and H. J. Everett, 1950, ''Power Characteristics of Mixing Impellers,'' *Chemical Engineering Progress Part II*, *46*, 467. (Appendix A.4.1)

Schroeder, E. D., 1977, *Water and Wastewater Treatment*, McGraw-Hill Book Company, New York (Appendix A.4.1)

Sherwood, T. K., and F. A. L. Holloway, 1940, ''Performance of Packed Towers: Liquid Film Data for Several Packings,'' *Transactions of the American Institute of Chemical Engineers*, *36*, 39–71. (Appendix A.4.1)

Sherwood, T. K., R. L. Pigford, and C. R. Wilke, 1975, *Mass Transfer*, McGraw-Hill Book Company, New York, p. 244. (Figure 4-12)

Sundstrom, D. W., and H. E. Klei, 1979, *Wastewater Treatment*, Prentice Hall, Englewood Cliffs, NJ. (Appendix A.4.1)

Taffinder, G. G., and B. Batchelor, 1993, ''Measurement of Effective Diffusivities in Solidified Wastes,'' *Journal of the Environmental Engineering Division, American Society of Civil Engineers*, *119*, 17–33. (Problem 4-2)

Thoma, G. J., D. D. Reible, K. T. Vaisaraj, and L. J. Thibodeaux, 1993, ''Efficiency of Capping Contaminated Sediments in Sites 2. Mathematics of Diffusion: Adsorption in the Capping Layer,'' *Environmental Science and Technology*, *27*, 2412–2419. (Appendix A.4.1)

Welty, J. R., C. E. Wicks, and R. E. Wilson, 1984, *Fundamentals and Momentum, Heat, and Mass Transfer*, John Wiley & Sons, Inc., New York. (Figures 4-3, 4-5, and 4-6)

Wilke, C. R., and P. C. Chang, 1955, ''Correlation of Diffusion Coefficients in Dilute Solutions, *AIChE Journal*, *1*, 264–270. (Equation 4-33).

Williamson, J. E., K. E. Bazaire, and C. J. Geankoplis, 1963, ''Liquid-Phase Mass Transfer at Low Reynolds Numbers,'' *Industrial and Engineering Chemistry, Fundamentals*, *2*(2), 126–129. (Example 4-11, Appendix A.4.1)

Wilson, E. J., and C. J. Geankoplis, 1966, ''Liquid Mass Transfer at Very Low Reynolds Numbers, *Industrial and Engineering Chemistry, Fundamentals*, *5*(1) 9–12. (Example 4-11, Appendix A.4.1)

4.4.2 Background Reading

Bird, R. B., W. E. Stewart, and E. N. Lightfoot, 1960, *Transport Phenomena*, John Wiley & Sons, Inc., New York. A very complete theoretical and mathematical development is provided for engineering description of diffusive transport in dilute vs. concentrated solutions and for interfacial mass transfer by film, penetration, surface renewal theory, and boundary layer theory. Examples are included.

Cussler, E. L., 1984, *Diffusion: Mass Transfer in Fluid Systems*, Cambridge University Press, London. An insightful and entertaining approach to engineering analysis of diffusion and mass transfer problems of practical interest, including dilute and concentrated solutions and application to membrane systems. The Stokes–Einstein diffusion equation is derived and compared to other diffusivity relationships. Material and momentum balances are used throughout to derive as well as illustrate the application of many useful equations (e.g., for diffusion coefficients in diffusion cell experiments and for mass transfer coefficients from boundary layer theory with the von Kármán integral approach as a simple approximation to the Navier–Stokes equation).

Schlichting, H., 1979, *Boundary Layer Theory*, 7th Edition, translated by J. Kestin, McGraw-Hill Book Company, New York. An excellent treatise on all aspects of boundary layer theory. Order-of-magnitude analysis for simplification of the solution to the Navier–Stokes equation for a boundary layer on a flat plate is described in Chapter 7; an approximate relationship between boundary layer thickness and Reynolds number is easily obtained. A detailed description of the exact solution derived by H. Blasius is also provided.

Welty, J. R., C. E. Wicks, and R. E. Wilson, 1984, *Fundamentals and Momentum, Heat, and Mass Transfer*, 3rd Edition, John Wiley & Sons, Inc., New York. Chapter 17 covers the analogies among heat, mass, and momentum. A fundamental description of mass transfer coefficients, a comparison of the film, penetration, and surface renewal theories, and a derivation of the momentum and mass balances for boundary layers, which leads to comparison of equations to predict boundary layer thickness for laminar and turbulent conditions, are provided in Chapter 28. Chapter 11 is devoted to various methods of dimensional analysis.

4.5 PROBLEMS

4-1 A passive dosimeter is to be designed for use in obtaining monthly average concentrations of atrazine in a stretch of river. Typical concentrations range from 10 to 50 μg/L. Atrazine, a herbicide commonly found in runoff from agricultural lands, has the following structure:

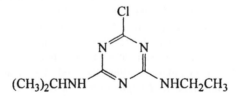

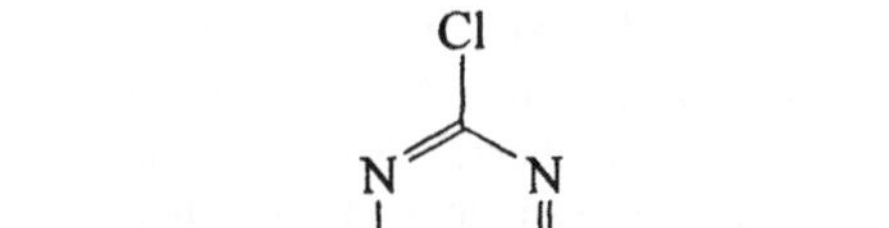

The dosimeter (see the sketch in Example 4-1) will use activated carbon as the adsorbent (DiGiano et al., 1988). To maximize diffusion rate, the diffusion barrier adjacent to the activated carbon is to be a polycarbonate membrane filter that has pores with a nominal diameter of 1 μm and a total surface area of pores that is estimated as 0.635 cm^2. The length of the diffusion channels is the thickness of the membrane (10 μm). Tests have shown that atrazine adsorbs very strongly on activated carbon such that 2 g of activated carbon should be sufficient to ensure that the concentration remains close to zero at the surface of the activated carbon. Given the percent recovery of adsorbed atrazine (using a solvent extraction), the volume of solvent and the sensitivity of the analytical method (a gas chromatographic procedure), the minimum mass of atrazine accumulated by the activated carbon that is capable of detection is 10 μg.

(a) Determine whether this dosimeter design will be adequate to detect atrazine given the sampling time and the typical concentration range mentioned above.

(b) Discuss the limitations of this monitoring approach.

4-2 An attractive treatment technique for inorganic hazardous chemicals (e.g., heavy metals) is solidification/stabilization. However, consideration must be given to the possibility of diffusion of ions through the porous structure of cement under the worst-case scenario of the cement slab being buried below the groundwater table. Taffinder and Batchelor (1993) describe experimental methods they used to determine the effectiveness of solidification/stabilization as a function of the water-to-cement ratio, the curing time, and the thickness of cement. They calculated a MacMullin number, $\mathfrak{N}_M$, defined as

$$\mathfrak{N}_M = \frac{\mathfrak{D}_l}{\mathfrak{D}_e}$$

where $\mathfrak{D}_l$ is the free-liquid diffusivity that would be observed in the pore water of the cement and $\mathfrak{D}_e$ is the effective diffusivity that accounts for a tortuous path (i.e., pores are not lined up horizontally through the slab) and adsorption of ions to the cement itself. A classical diffusion cell was used, as shown schematically below.

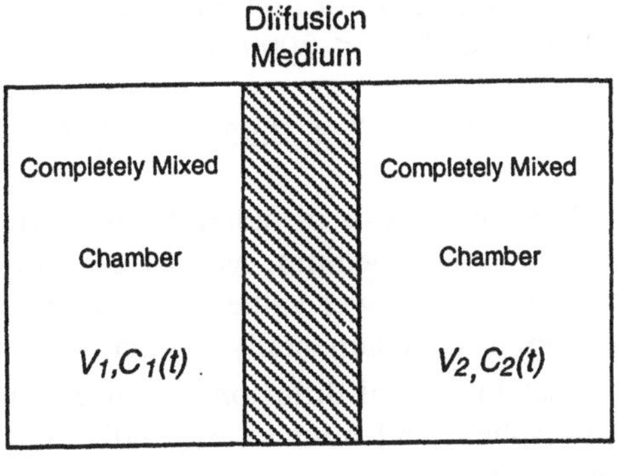

Molar solutions of potassium bromide and lithium chloride were placed in chambers 1 and 2, respectively, on opposite sides of the cement test slab and the concentration of Br and Li were measured with time. The results for bromide in chamber 2 are shown below.

t (days)	C_2 (mg/L)
0	0
6	11,155
8	13,321
9	14,225
11	17,996
31	30,479
34	32,061

The diffusion cell has $V_1 = 1000\ \text{cm}^3$, $V_2 = 1000\ \text{cm}^3$, area (A) of cement slab normal to diffusion direction $= 127.5\ \text{cm}^2$, and diffusion length (L) of cement slab $= 0.44$ cm. The molar volume of Br can be estimated as 50 cm^3/mol.

(a) Find an expression for C_2 for bromide as a function of t, V_1, V_2, $\mathfrak{D}_e$, A, its initial concentration, $C_{1,0}$ in chamber 1 and L, assuming quasi-steady-state diffusion (i.e., concentration changes slowly in each chamber such that Fick's first law is appropriate to describe flux through the distance, L).

(b) Find the MacMullin number and discuss its implications.

[*Hint:* Material balances can be written to express (1) the loss of Br from V_1 due to *steady-state diffusion* through the cement ($V_1 dC_1/dt =$ NA); (2) the gain of Br in V_2 due to steady-state diffusion through the cement; and (3) $C_1V_1 + C_2V_2 = C_{1,0}V_1$.]

4-3 The importance of diffusion-induced convective transport for situations where the mole fraction, X, of the diffusing substance is significant was illustrated in Example 4-2. Make a plot of X against the dimensionless diffusion distance, x/L, and compare this concentration profile with that expected for dilute diffusion.

4-4 The mass transfer correlation for liquid flow past a single sphere at low Reynolds number ($2 < \mathfrak{N}_{\text{Re}} < 2000$) is given in Table 4-3 as

$$\mathfrak{N}_{\text{Sh}} = 2 + 0.95\mathfrak{N}_{\text{Re}}^{0.5}\mathfrak{N}_{\text{Sc}}^{0.333}$$

(a) Apply steady-state analysis of radial diffusion from a sphere into a surrounding quiescent medium, as was used in Problem 2-11, to show that the limiting lower value of the Sherwood number in the correlation above is fundamentally correct.

[*Hint:* Assume that the radial diffusion distance is much greater than the radius of the sphere and refer to Appendix A2.2 for guidance in writing the material balance.]

(b) Reconsider the "blob" of NAPL in the interstitial volume of subsurface shown in Problem 2-11 and determine its dissolution rate (g/h) if the groundwater velocity is 1 m/day, the diffusivity of the NAPL is 1×10^{-5} cm^2/s, the NAPL density is 0.8 g/cm^3, the diameter of the NAPL blob is 100 μm, and the temperature of groundwater is 10°C. Assume that the volume of water is very large compared to the volume of NAPL and the solubility limit is low such that the concentration in the bulk liquid is close to zero.

4-5 An aeration system is to be designed for a municipal wastewater treatment process. A laboratory test system is used to obtain oxygen transfer data. It consists of a 30-cm-diameter cylinder with a water depth of 200 cm. Air is introduced in fine bubbles through a porous bottom plate, and average bubble sizes are determined by photographs. This is referred to as a *semibatch reactor* because the volume of water is fixed but the gas flow is continuous. The oxygen transfer rate is given by $\hat{k}_{f,l}(a_s^\circ)_R(C_S - C)$, where $(a_s^\circ)_R$ is the interfacial area per unit volume of aeration tank and C_S is the saturation concentration of oxygen in water (7.8 mg/L for these conditions). The oxygen concentrations (mg/L) with time for the two experiments at two different air flow rates, Q_g, and two different bubble diameters, d_b, are given below.

	Oxygen (mg/L)	
Time (s)	$Q_g = 7.5$ L/min $d_b = 0.2$ cm	$Q_g = 4.5$ L/min $d_b = 0.16$ cm
0	0	0
100	1.59	1.21
200	2.83	2.21
300	3.80	3.05
600	5.60	4.80
900	6.44	5.82
1200	6.84	6.40

(a) Determine the $\hat{k}_{f,l}(a_s^\circ)_R$ value (make sure to give units) for these two data sets.

(b) The following mass transfer correlation is assumed to hold for bubble aeration (see Appendix A.4.1):

$$\hat{k}_{f,l}(a_s^\circ)_R = \frac{6\psi H_D^{0.667} Q_g}{d_b V_R (\mathfrak{N}_{Sc})^{0.5}}$$

where ψ is an experimentally determined constant for the aeration equipment (porous plate in this instance), H_D is the depth of water, V_R the volume of water in the reactor, and $\mathfrak{N}_{Sc}$ the Schmidt number (assume that viscosity = 0.01 g/cm-s, density = 1 g/cm^3, and diffusivity of oxygen = 2.1×10^{-5} cm^2/s). Calculate ψ to determine if the data collected fit this correlation.

(c) The depth of the full-scale aeration tank is 3.66 m, the volume is 10^6 L, and the bubble diameter is 0.1 cm. Find the relationship between hydraulic residence time (volume of tank/flow rate) and gas flow rate that corresponds to the objective of reaching 50% saturation of oxygen. Assume that the tank is well mixed, that bubble diameter is not a function of flow rate in this particular case and that no reaction occurs to remove oxygen simultaneously. In actual treatment systems, oxygen is also removed at some rate, r, in the process of biodegradation. Indicate how you could account for this in the calculation.

4-6 The solubilization of NAPLs into groundwater is an excellent example of the importance of interphase mass transfer (see Example 4-10). Suppose that toluene is the NAPL. The volume fraction of NAPL held within the porous media, $\theta^{\circ}_{V,N}$, is 0.008, the average diameter of the porous media particles or grains is 0.400 mm, and the aqueous-phase pore velocity is 1 m/day. Assume that viscosity is 0.01 g/cm-s and free-liquid diffusivity of toluene is 1.1×10^{-5} cm^2/s. Use the correlation provided by Miller et al. (1990) given in Appendix A.4.1 to determine the mass transfer coefficient.

(a) Show the fraction of the toluene saturation in the groundwater, C/C_S, as a function of distance if the dispersion coefficient is 0.1 m^2/day.

(b) Suppose that dispersion is absent. Derive the material balance to describe advection and interphase mass transfer and use it to obtain C/C_S as a function of distance. Compare the results from parts (a) and (b).

4-7 Refer to the Roberts and Dandliker (1983) article cited in Example 4-9 and explain the following:

(a) Equation 5 in that article in relation to Equation 4-72.

(b) The observed effect of power input on mass transfer.

(c) The use of oxygen as a reference compound and the dependence of "K_La" on diffusivity, in light of what you would expect from the film, penetration, and surface renewal theories.

(d) Conditions for which assumptions of liquid-phase control of mass transfer are reasonable.

4-8 A batch stripping experiment is performed to obtain process design data for removal of perchloroethylene (PCE), a volatile organic chemical of regulatory concern. The interfacial area for mass transfer, $(a_s^{\circ})_R$, is unknown and thus must be combined with the mass transfer coefficient, $\hat{k}_{f,l}$. The data are as follows:

t (s)	C (μg/L)
0	100
60	80
125	60
230	40
400	20

These data are to be applied to PCE stripping in a full-scale, completely mixed basin (open to the atmosphere) for which the power input by mechanical aeration is only 10% of that used in the batch experiment.

(a) Justify your selection of a volumetric mass transfer coefficient for the full-scale system.

(b) Find the percent removal of PCE in the full-scale system if

$$\text{PCE entering the basin} = 50\ \mu\text{g/L}$$

$$\text{volume of the basin, } V_R = 40\ \text{m}^3$$

$$\text{flow rate, } Q = 160\ \text{m}^3/\text{day}$$

(c) Find the mass of PCE discharged to the air each day.

4-9 The dissolution of lead from Pb-soldered joints in household water piping is of concern. New drinking water regulations require that Pb not exceed 15 μg/L. The dissolution of Pb can be considered a mass-transfer-limited process. If the water is corrosive, Pb dissolves on the surface of the joint. Pb is then transported through a hydrodynamic boundary layer into the bulk water as shown below.

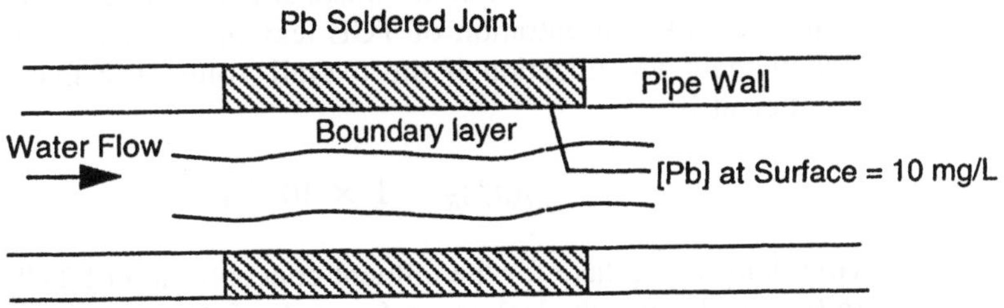

The mass transfer correlation for flow in pipes (Appendix A.4.1) is given by

$$\mathfrak{N}_{\text{Sh}} = 0.023\mathfrak{N}_{\text{Re}}^{0.83}\mathfrak{N}_{\text{Sc}}^{0.33}$$

where the characteristic length in the Sherwood and Reynolds numbers is pipe diameter, d_P.

(a) Find the steady-state concentration of Pb in μg/L in the bulk water downstream of the Pb-soldered joint if:

$$\text{Pb concentration at the surface of the joint} = 10 \text{ mg/L}$$

$$\text{length of the Pb-soldered joint} = 5 \text{ cm}$$

$$\text{velocity of water, } v = 20 \text{ cm/s}$$

$$\text{diameter of pipe, } d_P = 2.5 \text{ cm}$$

$$\text{free-liquid diffusivity of Pb, } \mathfrak{D}_l = 1 \times 10^{-5} \text{ cm}^2\text{/s}$$

$$\mu_v = 0.01 \text{ g/cm-s}$$

$$\rho = 1 \text{ g/cm}^3$$

(b) Suppose that an identical Pb-soldered joint exists some distance farther along the pipe, say for example, 4 m, and Pb is not removed by any mechanism between the two joints. Assuming that the same corrosive conditions exist, determine the concentration of Pb downstream of this second joint.

4-10 Trichloroethylene begins leaking from a 55-gallon drum at a chemical production facility and enters the floor drain, where it mixes with other process wastes to yield a constant concentration of 2 mg/L in the wastewater stream for a period of 6 h. This wastewater stream with a flow rate 80 m^3/day then enters a well-mixed aerated tank (volume of 20 m^3) in the industry's biological wastewater treatment plant. The presence of TCE raises concerns about toxicity to the microorganisms in the treatment plant and about violation of the discharge permit for TCE.

(a) Assuming that TCE is not in the waste stream before the leak, compare the peak concentration of TCE leaving the aerated tank with and without volatilization if the overall volumetric mass transfer coefficient is

$$\hat{k}_{f,l}(a_s^\circ)_R = 1 \times 10^{-4} \text{ s}^{-1}$$

(b) How long will it take for the effluent concentration of TCE to drop to 0.1 mg/L after the leak stops for each of the two assumptions?

Answers: (a) without volatilization, 1.26 mg/L, and with volatilization, 0.23 mg/L; (b) without volatilization, 15.2 h, and with volatilization, 1.6 h.

4-11 Napthalene spheres (mothballs) have been buried below the water table at a disposal site, where they are being dissolved by the groundwater flowing past them. The conditions are: groundwater temperature, 10°C; groundwater velocity, 100 cm/day; aquifer porosity, 0.3; mothball diameter, 1.0 cm; free-liquid diffusivity of napthalene, 0.5×10^{-5} cm^2/s; and one-dimensional advective transport (ignore dispersion, although this is unrealistic). The following mass transfer correlations is available (see Appendix A.4.1):

$$\mathfrak{N}_{\mathrm{Sh}} = \frac{1.09}{\epsilon_B} \mathfrak{N}_{\mathrm{Re}}^{0.333} \mathfrak{N}_{\mathrm{Sc}}^{0.333} \qquad \text{for } 0.0016 < \mathfrak{N}_{\mathrm{Re}} < 55$$

[*Note:* The appropriate velocity to use is the "superficial" velocity (i.e., the velocity as though entire cross-sectional area were available for flow when in actuality, flow can only pass through pore space) and the appropriate length scale is the particle diameter, d_P.] Given 160 mothballs per cubic meter of aquifer, determine the length of aquifer through which initially clean water must travel before the concentration of naphthalene reaches 1% of napthalene solubility, C_S.

Answer: 3.1 m.

APPENDIX A.4.1 Mass Transfer Correlations for Liquids and Gases

Application	Equation	Notes and Limits	Reference
Gases in packed beds of spheres	$\mathcal{N}_{Sh} = \frac{2.06}{\epsilon_B} \mathcal{N}_{Re}^{0.425}\mathcal{N}_{Sc}^{0.333}$ where ϵ_B = bed porosity	$90 < \mathcal{N}_{Re} < 4000$	Geankoplis, 1978
Liquids in packed beds of spheres	$\mathcal{N}_{Sh} = \frac{1.09}{\epsilon_B}\mathcal{N}_{Re}^{0.333}\mathcal{N}_{Sc}^{0.333}$	$0.0016 < \mathcal{N}_{Re} < 55$ $950 < \mathcal{N}_{Sc} < 70{,}000$	Wilson and Geankoplis, 1966
	$\mathcal{N}_{Sh} = 2.4\epsilon_B\mathcal{N}_{Re}^{0.34}\mathcal{N}_{Sc}^{0.42}$	$0.08 < \mathcal{N}_{Re} < 125$ $150 < \mathcal{N}_{Sc} < 1300$	Willamson et al., 1963
	$\mathcal{N}_{Sh} = (2 + 0.644\mathcal{N}_{Re}^{0.5}\mathcal{N}_{Sc}^{0.333})\,[1 + 1.5(1 - \epsilon_B)]$	$\mathcal{N}_{Re} < 100$	Gnielinski, 1978
Gases or liquids in fixed or fluidized beds	$\epsilon_B\mathcal{J}_D = \frac{0.765}{\mathcal{N}_{Re}^{0.82}} + \frac{0.365}{\mathcal{N}_{Re}^{0.386}}$	For gases, $\mathcal{N}_{Re} > 10$; for liquids, $\mathcal{N}_{Re} > 0.01$ $\mathcal{J}_D = \frac{\mathcal{N}_{Sh}}{\mathcal{N}_{Sc}^{0.33}\mathcal{N}_{Re}}$ For nonspherical particles, of diameter d_p: $d_p = \left(\frac{A_P}{\pi}\right)^{0.5}$	Fogler, 1992
Gases or liquids in pipes	$\mathcal{N}_{Sh} = 0.023\,\mathcal{N}_{Re}^{0.83}\mathcal{N}_{Sc}^{0.33}$ $\mathcal{N}_{Re} = \frac{d_p v \rho}{\mu_v}$ where d_P = pipe diameter	For turbulent flow, $\mathcal{N}_{Re} > 2100$, $0.6 < \mathcal{N}_{Sc} < 3000$	Geankoplis, 1978

Bubble aeration	$$\hat{k}_{v,l} = \frac{6\psi H_D^{0.667} Q_g}{d_b V_R \mathfrak{N}_{\text{Sc}}^{0.5}}$$ where $\hat{k}_{v,l} = \hat{k}_{f,l}(a_s^\circ)_R$ = overall volumetric mass transfer coefficient (t^{-1}); $(a_s^\circ)_R$ = interfacial area per unit volume of reactor (L^{-1}); ψ = determined experimentally constant ($L^{0.333}$); H_D = depth to sparger; Q_g = airflow rate; d_b = bubble diameter; V_R = volume of reactor; all units should be consistent when expressing ψ	Value of ψ experimentally determined in small-scale system; relationship is then used for scale-up	Schroeder, 1977
	$$\mathfrak{N}_{\text{Sh}} = 0.65\mathfrak{N}_{\text{Pe}}^{0.5}\left(1 + \frac{\mathfrak{N}_{\text{Re}}}{2}\right)^{0.5}$$	$10 < \mathfrak{N}_{\text{Re}} < 100$	Sundstrom and Klei, 1979
	$$\mathfrak{N}_{\text{Sh}} = 1.13(1 - 2.9\mathfrak{N}_{\text{Re}}^{-0.5})\mathfrak{N}_{\text{Pe}}^{0.5}$$	$100 < \mathfrak{N}_{\text{Re}} < 1000$	
	$$\mathfrak{N}_{\text{Sh}} = 1.13\mathfrak{N}_{\text{Pe}}^{0.5};\ \hat{k}_{f,l}(a_s^\circ)_R = 1.13\left(\frac{\mathfrak{D}_l v_t}{d_b}\right)^{0.5}$$	$\mathfrak{N}_{\text{Re}} > 1000$	
	where v_t = terminal rise velocity of bubbles (cm/s) $$v_t = 27.3\,\frac{\rho_l d_b^2}{\mu_v} \quad \text{for } \mathfrak{N}_{\text{Re}} < 300$$ $$= \left(\frac{2\sigma^\circ}{d_b \rho_l} + \frac{g d_b}{2}\right)^{0.5} \quad \text{for } 300 < \mathfrak{N}_{\text{Re}} < 4000$$ σ° = surface tension (dyn/cm)	Correlations account for rising bubbles not being rigid spheres	
Mass transfer of volatile organic chemicals in diffused aeration	$$\frac{\hat{k}_{f,l}(a_s^\circ)_R}{\dfrac{Q_G}{V_R}} = \psi_1(\mathcal{K}_H^\circ)^{\psi_2}$$ where Q_g = gas flow rate (units consistent with V_R), V_R = volume of aeration tank, $\mathcal{K}_H^\circ$ = dimensionless Henry's constant (see Table 6-2)	$\psi_1 = 2.9$ and $\psi_2 = 1.04$ based on 20 volatile organic chemicals with $\mathcal{K}_H^\circ < 0.3$	Hsieh et al., 1993a
Countercurrent gas–liquid contact in packed beds for both gas- and liquid-phase resistances	$$k_{f,l}\left(\frac{\rho_l}{\mu_{v,l} g}\right)^{0.333} = 0.0051\left[\frac{N_l}{(a_s^\circ)_W \mu_{v,l}}\right]^{0.667}\left(\frac{\mu_{v,l}}{\rho_l \mathfrak{D}_l}\right)^{-0.5}[(a_s^\circ)_T d_p]^{0.4}$$ $$\frac{k_{f,g}}{(a_s^\circ)_T \mathfrak{D}_G} = 5.23\left[\frac{N_g}{(a_s^\circ)_T \mu_{v,g}}\right]^{0.7}\left(\frac{\mu_{v,g}}{\rho_g \mathfrak{D}_g}\right)^{0.333}[(a_s^\circ)_T d_p]^{-2}$$		Onda et al., 1968

APPENDIX A.4.1 *(Continued)*

Application	Equation	Notes and Limits	Reference
	$\frac{(a_s^\circ)_W}{(a_s^\circ)_T} = 1 - \exp\left\{-1.45\left(\frac{\sigma_c^\circ}{\sigma^\circ}\right)^{0.75}\left[\frac{N_l}{(a_s^\circ)_T\mu_{v,l}}\right]^{0.1}\left[\frac{N_l^2(a_s^\circ)_T}{\rho_l^2 g}\right]^{-0.05}\left[\frac{N_l^2}{\rho_l\sigma^\circ(a_s^\circ)_T}\right]^{0.2}\right\}$ where N_l, N_g = liquid and gas mass fluxes (kg-m^2-s); ρ = density (kg-m^3); μ_v = absolute viscosity (Pa-s); $\mathfrak{D}_l$ and $\mathfrak{D}_g$ = liquid- and gas-phase diffusion coefficients (m^2/s); d_p = particle diameter of packing material (m); $(a_s^\circ)_W$ and $(a_s^\circ)_T$ = wetted and total specific surface area (L^{-1}); σ° = liquid surface tension (N/m); σ_c° = critical surface tension of packing material (N/m); g = gravitational acceleration constant (9.81 m/s^2)		
Countercurrent gas–liquid contact in packed beds including liquid-phase resistance only	$\frac{\hat{k}_{f,l}(a_s^\circ)_R}{\mathfrak{D}_l} = 10.76\psi_1\left(\frac{0.3048N_l}{\mu_{v,l}}\right)^{1-\psi_2}\left(\frac{\mu_{v,l}}{\rho_l\mathfrak{D}_l}\right)^{0.5}$ (same units as above)	Typical values of ψ_1 and ψ_2 are 150 and 0.28, respectively, for saddle-type packing	Sherwood and Holloway, 1940
Mass transfer of oxygen from hollow fiber membrane (gas inside) to water (outside)	$\hat{k}_{f,l} = 0.018\,\frac{\mathfrak{D}_g}{d_e}\left(\frac{v_l d_e}{\mu_v}\right)^{0.81}\left(\frac{\mu_v}{\mathfrak{D}_g}\right)^{0.33}$ where $d_e = \frac{4(\text{cross-sectional area of flow})}{\text{wetted perimeter}}$, $\mathfrak{D}_g$ = diffusivity of oxygen in air, v_l = liquid velocity	Correlation obtained using a large number of hollow fibers fitted inside a 50-mm-diameter pipe through which water flowed	Ahmed and Semmens, 1992
Mass transfer to/from small particles (e.g., oxygen from gas bubbles into water, oxygen to surface of microorganisms)	$\frac{\hat{k}_{f,l}}{\mathfrak{D}_l} = \frac{2.0}{d_b} + 0.31\left(\frac{\Delta\rho g}{\mathfrak{D}_l\mu_{v,l}}\right)^{0.333}$ $\hat{k}_{f,l}$ (m/s) $\mathfrak{D}_l$ = liquid phase diffusivity (m^2/s) $\Delta\rho = (\rho_l - \rho_g)$ (kg/m^3)	For gas bubbles: Impeller speed must be sufficient to shear gas flow into fine bubbles; otherwise, orifice size must be included in	Calderbrook and Moo-Young, 1961

$\mu_{v,l}$ = liquid dynamic viscosity (kg/m-s)	correlation; bubble diameter < 0.7 mm to be considered rigid sphere; specific bubble interfacial area, $(a_s^\circ)_P$, determined for the following condition; $\mathfrak{N}_{Re}^{0.7}\left(\frac{\omega_r d_{imp}}{v_g}\right)^{0.3} \leq 30,000$ $\mathfrak{N}_{Re} = \frac{d_{imp}^2 \omega_r \rho_l}{\mu_{v,l}}$	Grasso et al., 1990

Prediction of gas bubble diameter:

$$d_b = \text{bubble diameter (m)}$$

$$= 10^{-1.815}\left[\frac{(\sigma^\circ)^{0.6} V_R^{0.4}}{P_g^{0.4}(g_{s,l}^\circ)^{0.2}}\right](\phi_g^\circ)^{0.4}\left(\frac{\mu_{v,g}}{\mu_{v,l}}\right)^{0.25}$$

$$P_g = \text{gas pressure (N-m/s)}$$

$$= P_w\left[1 - 12.2\frac{Q_g}{\omega_r d_{imp}^3}\right] \text{ for } \frac{Q_g}{\omega_r d_{imp}^3} < 0.037$$

$$= P_w\left[0.062 - 1.85\frac{Q_g}{\omega_r d_{imp}^3}\right] \text{ for } \frac{Q_g}{\omega_r d_{imp}^3} > 0.037$$

P_w = power input (N-m/s); Q_g = gas flow (m^3/s); ω_r = impeller rotational speed (rev/s); d_{imp} = impeller diameter (m); $g_{s,l}^\circ$ = liquid specific gravity

$$\phi_g^\circ = \text{fractional gas holdup} = \left[10^{-0.268}\left(\frac{\mu_{v,g}}{\mu_{v,l}}\right)\left(\frac{v_g}{v_t}\right)^{0.5}\right]^{1.67}$$

v_t = terminal rise velocity of bubbles (m/s) $= \frac{g d_b^2 \Delta\rho}{18\mu_{v,l}}$; v_g = superficial gas velocity (m/s); V_R = liquid volume in reactor (m^3)

$$(a_s^\circ)_R = 1.14\left[\left(\frac{P_g}{V_R}\right)^{0.4}\left(\frac{g_{s,l}^\circ}{(\sigma^\circ)^3}\right)^{0.2}\right]\left(\frac{v_g}{v_t}\right)^{0.5}$$

APPENDIX A.4.1 (*Continued*)

Application	Equation	Notes and Limits	Reference
Mass transfer between particles and solution in stirred reactors	$k_f \mathfrak{N}^{0.667}\mathrm{Sc} = 0.13\left[\frac{(P_w/V_R)\mu_v}{\rho_l^2}\right]^{0.25}$	Various heat and mass transfer systems	Calderbrook and Moo-Young, 1961
	k_f (cm/s); P_w/V_R = power/volume (dyn/cm^2-s); μ_v (g/cm-s); ρ_l (g/cm^3) $P_w/V_R = 2.05 \times 10^{-6}(\omega_r)^{3.11}$ ω_r = rotational speed (rpm)	For power input between 8 and 500 W/m^3; axial flow, marine-type impellers; baffled reactors	Hsieh et al., 1993b
	Alternatively, $P_w = 0.95 d_{imp}^5\, \omega_r^3$ (dyn-cm/s) d_{imp} = impeller diameter (cm)	Propeller pitch twice the impeller diameter; 4 baffles, each 0.1 tank diameter	Rushton et al., 1950
Mass transfer between dispersed NAPL droplets and water in stirred reactors	Continuous phase (water) for NAPL of similar-density: $\frac{k_{f,l} d_d}{\mathfrak{D}_l} = 2 + 0.47\left[d_d^{1.333}\left(\frac{P_w}{V_R \rho_l}\right)^{0.333} \frac{\rho_l}{\mu_v}\right]^{0.62}\left(\frac{d_{imp}}{d_R}\right)^{0.17}\left(\frac{\mu_v}{\rho_l \mathfrak{D}_l}\right)^{0.36}$	Spherical droplet assumed	Ahn and Lee, 1990
	Continuous phase (water) for NAPL of significantly different density: $\frac{k_{f,l} d_d}{\mathfrak{D}_l} = 2 + 0.44\left(\frac{d_d \rho_l v_{d,l}}{\mu_v}\right)^{0.5}\left(\frac{\mu_v}{\rho_l \mathfrak{D}_l}\right)^{0.38}$		Levins and Glastonbury, 1972
	Dispersed phase (NAPL): $k_{f,m} \approx \frac{2}{3}\pi^2 \kappa^\circ \frac{\mathfrak{D}_m}{d_d}$	Experimental results show that $\kappa^\circ \rightarrow 1$	
	Ratio of mass transfer coefficients: $\frac{k_{f,m}}{k_{f,l}} \approx 14\,\kappa^\circ\left(\frac{d_d}{d_{imp}}\right)^{0.17}\left(\frac{\mu_v}{\rho_l}\right)^{0.26}\left(\frac{\mathfrak{D}_l}{\mathfrak{D}_m}\right)^{1.36} d_d^{-0.83}\left(\frac{P_w}{V_R \rho_l}\right)^{-0.21}$		

	d_d = droplet diameter (cm) d_{imp} = stirrer diameter (cm) d_R = reactor diameter (cm) $\mathfrak{D}_l$ = free-liquid diffusivity of solute in water (cm^2/s) $\mathfrak{D}_m$ = diffusivity solute in NAPL matrix (cm^2/s) $k_{f,m}$ = coefficient of mass transfer for NAPL (cm/s) $k_{f,l}$ = coefficient of mass transfer for water (cm/s) P_w = power input ($g\text{-}cm^2/s^3$) $v_{d,l}$ = relative droplet-water velocity V_R = water volume (cm^3) ρ_l = density of water (g/cm^3) μ_v = viscosity (poise) κ° = correlation parameter for internal circulation in droplet (dimensionless)		
Mass transfer from NAPLs to aqueous phase in porous media	$$\mathfrak{N}'_{Sh} = \psi_1 \mathfrak{N}_{Re}^{\psi_2} (\phi^\circ_{V,N})^{\psi_3} \mathfrak{N}_{Sc}^{0.5}$$ $\phi^\circ_{V,N}$ = NAPL volume fraction $\mathfrak{N}'_{Sh} = \dfrac{\hat{k}_{f,l}(a^\circ_s)_N d_p^2}{\mathfrak{D}_l}$ $\mathfrak{N}_{Re}$ based on diameter of media particles or grains	For toluene: $\psi_1 = 12 \pm 2$; $\psi_2 = 0.75 \pm 0.08$; $\psi_3 = 0.60 \pm 0.21$, and $\mathfrak{N}_{Re} < 0.12$	Miller et al., 1990
	$$\mathfrak{N}_{Sh} = 8.41 \mathfrak{N}_{Re}^{0.61} (\phi^\circ_d)^{0.64} (I_u)_i^{0.41}$$ where $\mathfrak{N}_{Sh} = \dfrac{[\hat{k}_{f,l}(a^\circ_s)_N] d_p^2}{\mathfrak{D}_l}$ is based on a lumped-mass transfer coefficient, $\mathfrak{N}_{Re}$ is based on interstitial velocity, $\hat{k}_{f,l}$ is the overall mass transfer coefficient with respect to the water side of the water-NAPL interface, and (I_u) = uniformity index of particle size ϕ°_d = particle diameter ratio; and $= \dfrac{d_{p,50}}{d_{p,M}}$ where $d_{p,50}$ = median grain size $d_{p,M}$ = USDA "medium" sand diameter = 0.05 cm	Tested for styrene and TCE for uniform and graded sand	Powers et al., 1992

APPENDIX A.4.1 (*Continued*)

Application	Equation	Notes and Limits	Reference
Mass transfer of volatile chemicals between river water surfaces and the atmosphere	$\hat{k}_{f,l} = 2.70 g^{0.35} \nu_v^{-1.85} \mathfrak{D}_l^{1.89} Z_W^{0.4} H_D^{-0.25} v^{0.26} (L \cdot s)^{0.16}$ g (7.3206 × 10^{10} m/day^2); ν_v = kinematic viscosity (m^2/day); $\mathfrak{D}_l$(m^2/day); Z_W = channel width (m); H_D = mean depth (m); v = mean velocity (m/day); L = length of reach (m); s = bed slope (m/m)	Tested for benzene, toluene, chloroform, methyl chloride, and ethylene	Hallappa Gowda and Lock, 1985
Mass transfer of solutes from sediment surface to overlying water	$k_{f,l} = 1.165 \left(\frac{Q_l \mathfrak{D}_{l,i}^2}{H_D^2 L Z_W} \right)^{0.333}$ Q_l = volumetric flow rate (cm^3/s) $\mathfrak{D}_{l,i}$ = free-liquid diffusivity (cm^2/s) H_D = depth of water (cm) Z_W = width of sediment surface (cm) L = length of mass transfer zone (cm)	For fully developed laminar flow	Thoma et al., 1993
Reaeration in streams	$\hat{k}_{v,l} = \left(\frac{\mathfrak{D}_l v}{H_D^3} \right)^{0.5}$ $\hat{k}_{v,l} = \hat{k}_{f,l}(a_s^\circ)_R$ = overall volumetric mass transfer coefficient $\mathfrak{D}_l$ = free-liquid diffusivity (cm^2/s) v = stream velocity (cm/s) H_D = depth of water (cm)	Based directly on surface renewal theory	O'Connor and Dobbins, 1956
	$\hat{k}_{v,l} = 11.6 v^{0.97} H_D^{-1.67}$ v (ft/s) H_D (ft) $\hat{k}_{v,l}$ (day^{-1})	Analysis of data in natural streams	Churchill and Buckingham, 1962

5

ENERGY RELATIONSHIPS: CONCEPTS AND APPLICATIONS TO HOMOGENEOUS SYSTEMS

5.1 REACTION CONCEPTS

5.1.1 Aqueous-Phase Chemical Reactions

At the outset of this book we defined process dynamics as dealing with changes in constituents that occur because of transport and transformation processes. Having completed our characterization of processes responsible for transport (Chapters 3 and 4), we now turn our attention to characterization of transformation processes. More specifically, this chapter deals with transformations by aqueous-phase chemical reactions. Although these reactions may be biologically mediated and their rates altered by physical conditions (e.g., mass transfer between the aqueous and other phases), the intrinsic *thermodynamic relationships* among chemical species ultimately govern the extent of change that is possible. The energy relationships of interest in this chapter are those for *homogeneous systems* and, in particular, those pertinent to the aqueous phase. In Chapter 6 we extend these energy relationships to *heterogeneous systems* wherein the focus is on the distribution of chemical species between the aqueous phase and other phases (i.e., vapor and solids of various types). Reaction phenomena in heterogeneous systems cannot be considered alone but rather in conjunction with transport processes. Reactions at interfaces cause depletion or accumulation of species at phase boundaries and species are transferred from one phase to another.

Regardless of whether we are dealing with change in homogeneous or heterogeneous systems, energy relationships determine the extent but not necessarily the rate of that change; that is, they are rooted primarily in equilibrium concepts. The corresponding reaction rate relationships (i.e., *kinetics*) for homogeneous and heterogeneous systems are treated in Chapters 7 and 8, respectively.

A thermodynamic rationale for describing any aqueous system begins with examination of the chemical properties of its principal components. The characteristics of these components lead inherently to transformation, accumulation, and separation reactions that must be identified. Finally, the reactions must be quantified with respect to the appropriate energy balances and thermodynamic properties of each particular system.

5.1.2 Water and the Aqueous Phase

Homogeneous phase reactions of interest to water quality specialists generally take place in dilute aqueous systems. Maximum levels of contaminants in natural waters and water supplies seldom exceed several milligrams per liter (mg/L), and in wastewaters normally not more than a few hundred mg/L. Even the concentrations of residual precipitates and sludges are rarely greater than 10,000 to 30,000 mg/L or about 1 to 3% by weight. Water is clearly the principal component of aqueous systems because it is commonly present in concentrations several orders of magnitude larger than the concentrations of other reacting species.

The very high concentration of water in aqueous environmental systems has a number of implications for water quality processes. In homogeneous systems, water is likely to control the behavior of dissolved substances and to participate in or mediate their reactions with each other. In heterogeneous systems, the properties and characteristics of water will determine the extent to which other chemical species will remain dissolved in the aqueous phase, volatilize to a gas phase, or precipitate to form solid phases. Moreover, most solid phases with which water comes into contact are comprised by chemical structures exhibiting reactivity at their phase boundaries, or surfaces. As a result, water is likely to affect surface phenomena as significantly as it does dissolved-phase processes. Water may be actively involved in the reactions of other components in homogeneous aqueous phases. Its mass concentration will not generally change significantly, however, because it is present in such large excess with respect to any other species. Its solution-phase reactivity cannot, therefore, be assessed adequately in terms of changes in its mass or concentration.

The properties of aqueous environmental systems can be attributed largely to the properties of pure water, the constituent atoms of which are *hydrogen* and *oxygen*. The reactivity and energy relationships of chemical species relate to the structure and form of the atoms which comprise them, a circumstance that is not unique to water. What is different about water is the uniqueness of its constituents, hydrogen and oxygen. In a rudimentary sense, the atoms of chemical elements owe their reactivity to particular numbers and arrangements of *neutrons*, *protons*, and most important, *electrons*. Atoms of all elements other than hydrogen, the smallest atom, have at least one neutron and at least one stable *shell* of electrons surrounding its *nucleus*. The hydrogen atom has neither. Helium, an *inert* substance and the second smallest atom, has the capacity of its innermost, or K, shell filled by *two* electrons. This would be the stable or ''inert'' electron configuration for hydrogen as well. The fact that the hydrogen atom does not have an *enclosing* shell of electrons like helium makes it particularly reactive in combining with other atoms to ''surround'' its single proton with two electrons to satisfy its electron capacity. The proton can no more exist as a stable entity than can the electron, or the neutron. To achieve the stability imparted by a complete enclosing electron shell, the hydrogen atom shares its single electron, often with a number of other atoms simultaneously. The bond that it forms with other elements to ''complete'' its K shell is termed the *hydrogen bond.*

All elements undergo reactions that cause them to move to more stable electron configurations and lower energy states. The two-electron configuration of helium is the most stable for the five elements having the lowest atomic numbers. The most stable state for carbon and elements of higher atomic number is characterized by an outer shell containing *eight* electrons. This gives the same electron configuration as that of the inert gas which is closest to that element in total electron number and structure, and therefore closest to it in the *periodic table.* The periodic table (see Appendix A.5.1) is an arrangement of elements according to the number of protons contained in their nucleus and the number of electrons contained in their surrounding shells. Chemical elements have a tendency to undergo reaction continually until they achieve a stable electron configuration. The existence of eight electrons in any shell of an atom does not mean that the electron *capacity* of that shell has been reached, simply that a *stable configuration* has been achieved. Shells farther from the nucleus of an atom may begin to fill when the next innermost shell contains eight electrons. The order of filling relates to the relative stability of electrons in different *orbitals* within different shells. Because inner shells may still have capacities for additional electrons, reactions other than those involving outer-shell, or *valence,* electrons are possible; these reactions, referred to as *complexation reactions,* will be discussed later.

The other primary component of water, oxygen, is also highly reactive. Its reactions comprise the primary mechanisms by which energy is derived from other substances: for example, the respiration of living cells and the combustion of fuels. As an atom, oxygen is deficient with respect to the number (eight) of electrons required in its outermost shell to impart thermodynamic stability. The oxygen atom thus enters reactions with other elements to attain a more stable electron configuration.

In combination to form water, hydrogen and oxygen atoms share electrons through hydrogen bonding. Each oxygen atom has six electrons in its outer shell and requires two additional electrons to configure its electron structure like that of neon. Each hydrogen atom has one electron and requires one more to assume the electron structure of helium. Thus, the smallest conceivable molecular structure for water would be comprised by two atoms of hydrogen and one atom of oxygen, with intermixed shells of orbiting electrons. The geometric configuration of this molecule is asymmetric, resulting in an asymmetric distribution of electrons, or negative charges. Thus, although water has no net electrical charge, its molecules are *polar,* which makes water compatible with other polar and charged (*ionic*) chemical species and incompatible with nonpolar and uncharged (*nonionic*) species. Water is thus an excellent solvent for salts or ions (e.g., Na^+, K^+, Cl^-, SO_4^{2-}), a fair solvent for polar molecules (e.g., CO_2, NH_3, CH_3Cl, C_6H_5OH), and a poor solvent for nonpolar molecules (e.g., N_2, CH_4, C_6H_6, C_6H_{14}).

Although the smallest molecular structure for a water molecule is two hydrogen atoms and one oxygen atom, the strong tendency of the hydrogen atom to share electrons generally results in its bonding with several other oxygen atoms simultaneously. The structure of water is thus more that of a loosely knit polymer than a discrete molecule, that is, $(H_2O)_n$ rather than H_2O. This imparts some interesting properties to water, for example its *density*. It is one of the few compounds for which the solid state (ice) is less dense than the liquid state. Ice is less dense because its molecular structure is more organized than water (i.e., there is a regular orientation between the hydrogen and oxygen atoms). Water is also one of the few liquids for which flow does not involve the sliding of molecules past each other. The flow of water requires a breakage and reformation of hydrogen bonds; this has relevance with respect to the *viscosity* of water.

Water is a good solvent for ionic species because it has a high *dielectric constant*, $\mathcal{K}_D$. The dielectric constant is a measure of the capacity of a fluid to neutralize attractive or repulsive forces between charged units dissolved or dispersed therein. *Coulomb's law* states that

$$F_C = \frac{1}{\mathcal{K}_D} \frac{z_1^\circ z_2^\circ}{L^2} \tag{5-1}$$

where F_C ($\mathrm{MLt^{-2}}$) is the net Coulombic force between two units of electronic charge z_1° and z_2° separated by a distance L. The negative poles of multimolecular associations of water are able to orient around positively charged ions and the positive poles of these associations around negatively charged ions, effectively forming cages that keep positive ions away from negative ions and maintaining these substances in a dissolved state. Although water itself is a poor conductor of electricity, *aqueous solutions* usually have high *conductivity*. The reason is that water acts as a universal solvent for charged species and is thus seldom free of dissolved ionic impurities.

Heat capacity, Q_H° ($\mathrm{ML^2t^{-2}T^{-1}}$ per mole or unit mass), is another property of water that is significantly influenced by its unique structure. The total energy required to break all the hydrogen bonds involved in a multimolecular association of water is large; this accounts for its high boiling and melting points relative to other liquids. The heat of melting or fusion, $Q_{H,m}$ ($\mathrm{ML^2t^{-2}}$ per mole or unit mass), of ice is 334.7 joules (J) per gram at 0°C. The *heat of vaporization*, $Q_{H,v}$, of water is 2.26 kJ per gram at 100°C, which translates to 40.6 kJ (~9.7 kcal or ~38.5 Btu) per mole. Its *average specific heat capacity*, Q_H°, is 1.00 cal/g-K (4.184 J/g-K) and deviations from this between the freezing and boiling points average less than 1%. This translates to approximately 75.4 J (18 cal or ~0.07 Btu) per mole per kelvin. Thus, simply to raise the temperature of 1 liter of water from 20°C to 25°C requires an energy input of nearly 20 btu. For a flow of 3785 m^3 of water per day (1 million gallons per day, or 1 mgd), for example, raising the temperature from 20°C to 25°C would require approximately 76 × 10^6 Btu or 22.3 MW-h/day. Distillation at this same flow would require a daily energy input approaching 9.3×10^9 Btu or 2725 MW-h. Increasing or decreasing the temperature of water to facilitate or accelerate a particular treatment process is clearly an energy-intensive and expensive proposition. Consequently, two of the most straightforward and seemingly obvious means for purifying water, distillation and freezing, are usually prohibitive in cost, as illustrated in Example 5-1.

Example 5-1

- **Situation:** A small water distillation plant is to be built for a remote coastal outpost located on a shallow marine bay and having no access to fresh water. It is planned that ethylene will be used as fuel for the plant.
- **Question(s):** Assuming 90% efficiency, how much seawater at 20°C can be converted to steam at 100°C by each cubic meter of C_2H_4 gas burned?
- **Logic and Answer(s):** The combustion of ethylene involves the following reaction:

$$C_2H_4(g) + 3O_2(g) \rightleftharpoons 2CO_2(g) + 2H_2O(l)$$

1. We determine first from information that is readily available in standard handbooks that the combusion of ethylene yields a heat output of 1410 kJ/mol (Table 5-6, which will be introduced later in this chapter).

2. We know from elementary physics that 1 mol of gas occupies a volume of 22.4 L at standard temperature and pressure (STP) (i.e., 1 atm and 0°C). Thus,

$$\text{number of moles of } C_2H_4 \text{ in } 1 \text{ m}^3 = \frac{(1 \text{ m}^3)(1000 \text{ L/m}^3)}{22.4 \text{ L/mol}} = 44.6 \text{ mol}$$

3. The heat available in the combustion of 1.0 m^3 of C_2H_4 is then

$$(44.6 \text{ mol})(1410 \text{ kJ/mol}) = 6.3 \times 10^4 \text{ kJ}$$

The useful heat is 90% of this, or $(0.90)(6.3 \times 10^4 \text{ kJ}) = 5.7 \times 10^4$ kJ.

4. The overall process involves raising the water from its ambient temperature to 100°C and then vaporizing it. Assuming that the average water temperature at the shallow bay intake is 20°C, we have

$$H_2O(l, 20°C) \rightleftharpoons H_2O(l, 100°C)$$

$$\text{heat required} = (4.184 \text{ kJ/kg-K})(80 \text{ K}) = 334.7 \text{ kJ/kg}$$

$$H_2O(l, 100°C) \rightleftharpoons H_2O(g, 100°C)$$

$$\text{heat required} = 2259 \text{ kJ/kg}$$

The total heat requirement is therefore 2594 kJ/kg.

5. The mass of water converted is equal to the amount of heat available divided by the heat requirement per kilogram, or the amount of H_2O

produced per cubic meter of C_2H_4:

$$\frac{5.7 \times 10^4 \text{ kJ}}{2594 \text{ kJ/kg}} = 21.97 \text{ kg}$$

6. Distillation thus produces only 22 L, or 0.022 m^3, of water per cubic meter of ethylene burned, an obviously energy-intensive process.

5.1.3 Chemical Structure and Reactivity

We have referred to the periodicity of elements and to the role of electron structure in determining the reactivity of chemical species. In the simplest sense, the reactivity of a substance defines the tendency for its atoms to attain the electron configuration and thermodynamically stable state of the nearest inert gas. Stability can be reached by giving up, taking on, or sharing electrons. The fact that atoms and molecules are *electron-active* forms the basis for transformation processes in aquatic systems.

Some elements behave in a very direct and predictable fashion to form a limited number of discrete species and/or combinations. Other elements must undergo more extensive and more complex changes to approach thermodynamic stability. Elemental carbon, C, which exists in its simplest form as graphite, has four electrons in its outer shell. This element commonly attempts to achieve the stable electron configuration of neon by sharing its electrons with other atoms. It is easier to share four electrons than to give them all away in order to approach the electron configuration of helium, or to strip other species of four electrons. There are many atoms with which carbon can share one or more of its four electrons to achieve at least a quasi- or metastable configuration, as evidenced by the large number and wide diversity of natural and synthetic *organic* (carbon-based) compounds.

Atoms that either completely give up or take on electrons are electronically imbalanced with respect to the number of protons in the nucleus and the total number of electrons surrounding the nucleus. They therefore exhibit a charge, and are termed *ions*. Atoms that share electrons with other atoms to form molecules or compounds may or may not impart a charge to the resulting species. For example, the chlorine atom, Cl, can take on an electron to form the chloride ion, Cl^-, or share electrons with another chlorine atom to form the uncharged chlorine molecule, Cl_2. Alternatively, it may share its electron with a hydrogen atom to form an uncharged compound, HCl. Electrons are actually shared in pairs, either with each atom contributing one electron per pair or with one atom contributing both electrons to each shared pair.

An *electronically balanced atom*, an atom in its *elemental state*, has an *oxidation state of zero*. Its oxidation state increases by one for every electron it gives up, and decreases by one for every electron it takes on. The sharing of electrons can lead to intermediate oxidation states. For convenience, however, the more electronegative of the two species is usually considered to have its oxidation state decreased, while that of the more electropositive is increased. For example, the hydrogen atom in HCl is generally considered to be in a unit positive (I) oxidation state, and the chlorine atom in a unit negative (−I) state. Oxidation state is an important property of elements and their reactivity, particularly with respect to complex species and substitutive oxidation reactions.

5.1.4 Reactivity and Reactions

Atoms that have either given up or taken on an electron to gain greater stability as ions will in further reactions tend to combine with other ions of opposite charge. That is, they will be attracted by the electrostatic or Coulombic forces defined by Equation 5-1. The sodium ion, Na^+, which is in an oxidation state of (I), is strongly attracted, for example, to the chloride ion, Cl^-, which is in an oxidation state of (−I); the result is formation of sodium chloride.

$$Na^+ + Cl^- \rightleftharpoons \{Na^+Cl^-\} \tag{5-2}$$

The right-hand side of Equation 5-2 is written in a form that emphasizes the attractive force and resulting bond between these ions as being strictly *electrostatic*. That is, each substance retains its original electron configuration in the resulting combination or compound. This type of bond is termed *electrovalent*. Conversely, when the hydrogen atom reacts with the chlorine atom to form hydrochloric acid, HCl, the two atoms are bonded by *sharing* a pair of electrons. This type of bond is termed either *covalent* or *coordinate covalent*, depending upon the origin(s) of the shared electrons. The true oxidation states of the hydrogen and chlorine atoms in HCl are intermediate between (I) and (−I), whereas those for Na^+ and Cl^- in NaCl are clearly (I) and (−I), respectively.

Remarkable differences in chemical reactivity are imparted to combinations of atoms by virtue of the manner of their bonding. The electrovalently bonded salt, NaCl, exhibits totally different reactivity than does the hydrogen-bonded mineral acid, HCl. A similar difference can be noted between the electrovalently bonded sodium salt, NaOH, of the hydroxide ion, and hydrogen-bonded water, HOH. As a general rule, electrovalently bonded compounds have high melting and boiling points and high solubility in water. Conversely, *covalently* bonded compounds, in which each partner sharing a pair of electrons provides one electron, generally exhibit low melting and boiling points and are less soluble in water.

Electrovalent reactions between oppositely charged ions can also involve ions of different oxidation states, as illustrated below:

$$Mg^{2+} + 2Cl^- \rightleftharpoons \{Mg^{2+}\,Cl_2{}^-\} \tag{5-3}$$

Examples of predominantly covalent reactions and products are

$$C + O_2 \rightleftharpoons \{O{=}C{=}O\} = CO_2(g) \tag{5-4}$$

and

$$C + 2H_2 \rightleftharpoons \left\{\begin{array}{c} H \\ | \\ H-C-H \\ | \\ H \end{array}\right\} = CH_4(g) \tag{5-5}$$

where each dash (bond) between atoms represents one pair of shared electrons. Equations 5-4 and 5-5 illustrate that the atoms constituting the oxygen (O=O) and hydrogen (H—H) molecules, respectively, are themselves sharing electrons. This is typical of the way in which the atoms of a substance combine to form molecules. Other examples include Br_2, Cl_2, and N_2.

The third type of bonding arrangement, *coordinate* or *coordinate-covalent* bonding, is similar in its eventual form to covalent bonding, in that two atoms share one or more pairs of electrons. In contrast to covalent bonding, however, *one* of the two atoms involved in the sharing of an electron pair contributes *both* of the electrons. That is, one atom is a *donor of an electron pair* to the sharing arrangement, and the other is an *electron-pair acceptor*. The number of pairs of electrons brought to or accepted by a species in such bonding is termed the *coordination number* of that species. The properties of compounds that result from such bonding arrangements are usually intermediate to those of electrovalent and covalent compounds. Examples of coordinate covalent reactions, including the unique case of *hydrogen bonding*, are

$$(O-H)^- + H^+ \rightleftharpoons \{H-O-H\} = H_2O \tag{5-6}$$

$$\left\{\begin{array}{c} H \\ | \\ H-N \\ | \\ H \end{array}\right\} + H^+ \rightleftharpoons \left\{\begin{array}{c} H \\ | \\ H-N-H \\ | \\ N \end{array}\right\}^+ = NH_4^+ \tag{5-7}$$

$$(Cd)^{2+} + 4(OH)^- \rightleftharpoons \left\{\begin{array}{ccc} HO & & OH \\ & (Cd) & \\ HO & & OH \end{array}\right\}^{2-} = \{Cd(OH)_4\}^{2-} \tag{5-8}$$

$$(Fe)^{2+} + 4(HOH) \rightleftharpoons \left\{\begin{array}{ccc} HOH & & HOH \\ & (Fe) & \\ HOH & & HOH \end{array}\right\}^{2+} = \{Fe(H_2O)_4\}^{2+} \tag{5-9}$$

The resulting species are termed *complexes* or, as in Equations 5-8 and 5-9, *complex ions*. The electron-pair acceptors are termed the *central ions* of the resulting complexes, and the electron-pair donating species (e.g., OH^- and H_2O) are *ligands*. The hydrogen ion, or proton, is unique because it accepts only one pair of electrons and, as noted in Equations 5-6 and 5-7, cannot be considered the central ion of the water and ammonium ion complexes. Although the reactions given in Equations 5-6 and 5-7 both qualify as complexation reactions, they are often more specifically referred to as *protolysis* reactions because of the involvement of the proton.

Equations 5-6 and 5-7 represent only *half-reactions* in that, as mentioned earlier, free protons do not exist in nature. Thus, these half-reactions must be coupled with other half-reactions in which protons are released. For example, rearranging and combining the reactions in these two equations yields

$$NH_3 + H_2O \rightleftharpoons NH_4^+ + OH^- \tag{5-10}$$

The cadmium and iron ions in Equations 5-8 and 5-9 have eight electrons in their outer shells and thus satisfy the stable electron configuration of the closest inert gases. Moreover, the product of the reaction given by Equation 5-9 has the same charge characteristics as the central ion because the ligand, H_2O, has no net charge. The product of the reaction given by Equation 5-8 also bears a net charge equal to the sum of the electronic charges of the reactants. These observations point to the fact that the *outer* electron sphere configurations of the two metal ions are not changed in these reactions as they are in simple covalent bonding. In these instances, as for most reactions between heavy metal ions and other species of environmental interest, the electron pairs are shared within *inner* electron spheres. Although the inner spheres may already contain eight electrons, they can accommodate still more. The metal ion retains its outer-shell configuration, and thus its oxidation state, in such reactions.

Equations 5-8 and 5-9 are not quite precise as written. Like the proton, metal ions rarely exist in aqueous phase as simple ions (e.g., as Cd^{2+} and Fe^{2+}). Instead, metal ions have a strong tendency to take on partners in inner-shell coordinate covalent bonding. The product of the reaction shown in Equation 5-9 is actually the closest form to Fe^{2+} in which Fe(II) exists in water, that is, the hydrated metal ion, $[Fe(H_2O)_4]^{2+}$.

The process of ligand exchange by a central ion can be illustrated by considering the four reaction steps that actually constitute Equation 5-8. The cadmium ion, once dissolved in pure water, first takes on water molecules as ligands, as does the iron ion in Equation 5-9. It exists at low-to-moderate pH levels as the complex ion $[Cd(H_2O)_4]^{2+}$. Upon addition of the salt of a strong base, such as NaOH, the salt *dissociates* to release Na^+ and OH^-. The OH^- ion, which has three pairs of free electrons, then competes favorably with water molecules (which have only two pairs of free electrons) to serve as a coordination partner or ligand for the central metal ion, Cd^{2+}:

$$\{Cd(H_2O)_4\}^{2+} + OH^- \rightleftharpoons \{Cd(H_2O)_3(OH)\}^+ + H_2O \tag{5-11}$$

Although by virtue of its three pairs of free electrons, the OH^- ion is a much stronger ligand than H_2O, it is present at much lower concentration. At pH 5, for example, the concentration of OH^- is approximately 10^{-9} mol/L, while that of water is approximately 55.6 mol/L. As pH is increased by the addition of more hydroxide, further ligand transfer reactions occur, as follows:

$$\begin{array}{ccc} \{Cd(H_2O)_3(OH)\}^+ + OH^- & \rightleftharpoons & \{Cd(H_2O)_2(OH)_2\} + H_2O \\ & & \updownarrow \\ & & Cd(OH)_2(s) \end{array} \tag{5-12}$$

$$\begin{array}{ccc} Cd(OH)_2(s) & & \\ \updownarrow & & \\ \{Cd(H_2O)_2(OH)_2\} + OH^- & \rightleftharpoons & \{Cd(H_2O)(OH)_3\}^- + H_2O \end{array} \tag{5-13}$$

$$\{Cd(H_2O)(OH)_3\}^- + OH^- \rightleftharpoons \{Cd(OH)_4\}^{2-} + H_2O \tag{5-14}$$

Because $\{Cd(H_2O)_2(OH)_2\}$ is uncharged, it is less soluble than its precursors and products in the cadmium complexation sequence. This species will therefore precipitate more readily from solution to form a solid, $\{Cd(OH)_2\}(s)$.

Complexation reactions have profound implications with respect to the behavior and bioavailability of heavy metals (e.g., Zn, Cu, Pb, Cd, Fe, Mn) and other common cationic species (e.g., Ca, Mg) in natural waters and in treatment systems. Such reactions alter the solubility relationships of these metals, affect their role as acids, and are important with respect to their corrosion. A large number of inorganic and organic substances found in waters and wastewaters compete with H_2O and OH^- as ligands for complexation of heavy metals, notably compounds or groups that contain basic oxygen and/or nitrogen atoms. Two major examples are:

$$\{Fe(H_2O)_6\}^{3+} + 6CN^- \rightleftharpoons \{Fe(CN)_6\}^{3-} + 6H_2O \tag{5-15}$$

$$\{Fe(H_2O)_6\}^{3+} + (EDTA)^{4-} \rightleftharpoons \{FeEDTA\}^- + 6H_2O \tag{5-16}$$

The product of the reaction between iron(III) and cyanide ions given in Equation 5-15 is the stable ferricyanide ion, an important constituent of certain types of industrial wastewaters. Equation 5-16 depicts a ligand transfer reaction involving the exchange of water by a strong complexing agent, ethylenediaminetetraacetate (EDTA), which has the chemical formula:

$$\begin{array}{ccccc} {}^-OOC-CH_2 & & & & CH_2-COO^- \\ | & & & & | \\ N & -CH_2- & & -CH_2- & N \\ | & & & & | \\ {}^-OOC-CH_2 & & & & CH_2-COO^- \end{array} \tag{5-17}$$

As the structure given in Equation 5-17 indicates, one EDTA molecule contains four basic oxygen atoms and two basic nitrogen atoms, which together can provide a total of six pairs of electrons to satisfy the coordination requirements (coordination number) of Fe(III). Such *multidentate* (multipair) donors, often termed chelates, can do in one reaction step what *monodentate* (single-pair) donors do in several reaction steps. Compare, for example, the four reactions required for complete complexation of Cd(II) by the monodentate hydroxide ligands (Equations 5-11 through 5-14) with the single reaction of the EDTA multidentate ligand in Equation 5-16 to meet all six coordination requirements of Fe(III). Natural waters and wastewaters frequently contain organic substances such as humic and fulvic acids, carboxylic acids, soaps, and a wide range of inorganic electron-pair donors (e.g., PO_4^{3-}, HPO_4^{2-}, $H_2PO_4^-$, NO_3^-, SO_4^{2-}, HSO_4^-, $H_3SiO_4^-$, NH_3, HS^-, S^{2-}, etc.), which can effectively complex heavy metals and thus markedly affect their speciation and behavior in water quality control processes.

Equations 5-6, 5-7, and 5-10 through 5-14 clearly constitute acid-base reactions in that either H^+ or OH^- are released or consumed. Pursuant to the *Lewis theory*, acids are defined as *electron-pair acceptors* and bases as *electron-pair donors*. This definition substantially extends the scope of what we typically think of as acid-base substances and phenomena. Accordingly, Equations 5-15 and 5-16 can be considered acid-base reactions as well. In fact, this definition of acid-base reactivity makes such substances important in virtually all aqueous-phase reactions and, conversely, imparts acid-base behavior to most chemical substances. We can cite numerous examples of acid-base phenomena in the context provided by the Lewis theory to illustrate thermodynamic concepts in both homogeneous and heterogeneous systems.

Example 5-2 illustrates how the oxidation states of elements vary as they interact with one another to form different compounds and complex ions. The example also shows how these oxidation states can be determined based on the composition of a compound or complex ion.

Example 5-2

- **Situation:** It is often helpful when assessing the stability of various chemical compounds to know the oxidation states in which the atoms comprising those compounds exist.

- **Question(s):** Determine the oxidation states of the atoms of the following compounds of sulfur: (a) Na_2SO_4; (b) $FeSO_4$; (c) H_2SO_4; (d) H_2SO_3; (e) H_2S.

- **Logic and Answer(s):** Certain of the atoms involved in these compounds are known to have only one stable oxidation state (see Table A.5.1). These include H(I), Na(I), and O(-II). We can assume that these atoms have achieved their respective stable states in the compounds listed. We know that these compounds are uncharged, so that the net sum of the negative and positive oxidation states of the atoms or ions comprising them must be zero. The oxidation states on the remaining atoms may then be calculated as follows:

(a) The sum of the Na(I) oxidation states $= 2 \times 1 = 2$. The sum of the O(-II) oxidation states $= 4 \times (-2) = -8$. The sum of the sulfur atom oxidation states must be equal and opposite in sign to the sum of the Na(I) and O(-II) oxidation states. Thus, the single atom of sulfur must exist as S(VI) in Na_2SO_4. The sulfate group, SO_4^{2-}, in this compound exists as a complex ion in which the sulfur and oxygen atoms are held together by coordinate covalent bonds. This is similar to the OH^-, NH_4^+, and CN^- complex ions illustrated in Equations 5-10 and 5-15. These are stable species involving one atom of known stable oxidation state, namely O(−II).

(b) The sulfate group in $FeSO_4$, like that in Na_2SO_4, is a complex ion. This SO_4^{2-} complex ion bears a net negative charge of 2, which in the compound $FeSO_4$ must be balanced by an equal positive charge on the iron, indicating that the iron exists as Fe(II) in the compound.

(c) Using the same logic as in parts (a) and (b), the atoms comprising H_2SO_4 exist in the following oxidation states: H(I), S(VI), O(−II).

(d) The sum of the H(I) oxidation states $= 2 \times 1 = 2$. The sum of the O(−II) oxidation states $= 3 \times (-2) = -6$. The sulfur atom in H_2SO_3 must therefore exist as S(IV).

(e) The sum of the H(I) oxidation states $= 2 \times 1 = 2$. The sulfur atom in H_2S must therefore exist as S(−II).

We have thus far considered reactivity and reactions in terms of bonding arrangements and molecular structures. We will return to a more quantitative approach to the estimation of reactivity as a function of structure after we have established a thermodynamic basis for expressing reactivity. Homogeneous phase reactions using proton transfer reactions will serve conveniently to illustrate electron-pair transfer reactions. Some logical extensions of simple proton and ligand transfer reactions to heterogeneous inorganic systems will also be made. In Chapter 6 we expand on the applications of thermodynamics to phase separation reactions and processes, including systems dominated by organic constituents.

5.1.5 Classification of Reactions

Chemical species can interact by directly transferring electrons (*electron transfer reactions*) or by each covalently sharing outer-shell electrons (*electron-sharing reactions*) with each other to achieve more stable states. In yet a third type of reaction (*coordination or complexation reactions*), two reactants combine to share one or more pairs of electrons previously belonging to one of the reactants. The oxidation states of the reacting components change in electron transfer reactions. Although there is generally no discrete change in the oxidation state of either atom in covalent or coordinate-covalent electron sharing, there is generally a change in relative electronegativity or electropositivity of the elements involved.

The reaction between sodium metal and chlorine gas, both having oxidation states of zero, to form sodium ion having an oxidation state of (I) and chloride ion with an oxidation state of (−I), is an example of an electron transfer reaction:

$$2Na + Cl_2 \rightleftharpoons 2Na^+ + 2Cl^- \tag{5-18}$$

Reaction 5-18 is actually the sum of two half-reactions:

$$2Na \rightleftharpoons 2Na^+ + 2e^- \tag{5-19}$$

$$Cl_2 + 2e^- \rightleftharpoons 2Cl^- \tag{5-20}$$

Neither half-reaction can occur alone, again because electrons are not stable as independent entities. In this example sodium increases its oxidation state and is said to be *oxidized*. Chlorine, on the other hand, decreases its oxidation state and is said to be *reduced*. Equation 5-19 thus represents a chemical *oxidation* and Equation 5-20 a chemical *reduction*. The combined electron transfer reaction given by Equation 5-18 is referred to as an *oxidation-reduction*, or *redox*, reaction. All electron transfer reactions are redox reactions, although the changes in oxidation state may not be as obvious as those in the example above.

Equations 5-6 through 5-16 illustrate typical coordination-partner transfer reactions. When the proton, H^+, is involved as the electron-pair acceptor, this type of reaction is referred to as a *proton transfer reaction*. The proton, like the electron, is not independently stable and thus readily reacts with partners having pairs of electrons available for coordinate-covalent bonding. Whereas electron transfer reactions are termed redox reactions, proton transfer reactions are commonly called *acid-base* or *protolysis* reactions. In the most general sense, however, as we have noted previously, all ligand exchange reactions constitute Lewis acid-base reactions.

Acid-base reactions, particularly protolysis reactions, are highly significant for water systems. Not only do they play a major role in governing the balance of natural aquatic ecosystems, but they participate in and mediate most chemical and biochemical transformations in natural systems and in engineered water quality control processes. Table 5-1 presents a listing of several such processes to illustrate their respective acid-base characteristics. The first reaction shown in Table 5-1 involves the first of several ligand exchanges that can occur when aluminum, a multivalent metal ion, dissolves in water. Aluminum has six coordination partners in its dissolved state. The exchange of one type of dominant ligand (H_2O) for another (OH^-) is in essence an acid-base or protolysis reaction. Such ligand exchange reactions are common to many multivalent metal ions, as illustrated in Table 5-2.

The pH variable exerts such a large effect on reactions which occur in water that it can be thought of as a *master variable*, or *control variable*, and the concentrations of most other chemical species as *response variables*. The importance of pH is detailed in entire texts and courses in aquatic chemistry

Table 5-1 Proton Transfer Reactions in Environmental Processes

Process	Reaction	Effect on pH
Coagulation	Protolysis of metal-ion coagulants $\{Al(H_2O)_6\}^{3+} + H_2O \rightleftharpoons \{Al(H_2O)_5(OH^-)\}^{2+} + H_3O^+$	Decrease
Softening	Precipitation of calcium hardness $Ca^{2+} + HCO_3^- + OH^- \rightleftharpoons CaCO_3(s) + H_2O$	Decrease
Recarbonation	Dissolution of carbon dioxide $CO_2(g) + 2H_2O \rightleftharpoons HCO_3^- + H_3O^+$	Decrease
Aerobic biological processes	Biochemical mineralization of organic matter $C_6H_{12}O_6 + 6O_2 \rightarrow 6CO_2 + 6H_2O$	Decrease
	Biochemical nitrification of ammonia $NH_3 + 2O_2 \rightleftharpoons NO_3^- + H_3O^+$	Decrease
Anaerobic biological processes	Biochemical oxidation of organic matter with nitrate reduction $5C_6H_{12}O_6 + 24NO_3^- + 24H_3O^+ \rightarrow 30CO_2 + 12N_2 + 66H_2O$	Increase
	Biochemical oxidation of organic matter with sulfate reduction $C_6H_{12}O_6 + 3SO_4^{2-} + 3H_3O^+ \rightarrow 6CO_2 + 3HS^- + 9H_2O$	Increase
	Biochemical oxidation of organic matter with methane fermentation $C_6H_{12}O_6 + 3CO_2 \rightarrow 3CH_4 + 6CO_2$	Decrease
Odor reduction and corrosion control	Biochemical oxidation of sulfide $HS^- + 2O_2 + H_2O \rightarrow SO_4^{2-} + H_3O^+$	Decrease
	Chemical oxidation of sulfide $HS^- + H_2O_2 + H_3O^+ \rightarrow S + 3H_2O$	Increase
Oxygen production in lakes and oxidation ponds	Photosynthesis by algae and other aquatic plants $6CO_2 + H_2O \rightarrow C_6H_{12}O_6 + 6O_2$	Increase

devoted to its description as the master variable in proton exchange (i.e., acid-base, complexation and precipitation reactions) and electron exchange (i.e., oxidation-reduction reactions). Our intent here is to provide an overview of these reaction concepts, presenting them in well-recognized diagramatic forms that show the equilibrium distributions of chemical species with pH, but leaving detailed explanations to other sources. The role of pH as a master variable is illustrated schematically in Figure 5-1 for the carbonate (H_2CO_3—HCO_3^-—CO_3^{2-}; $C_{Total} = 10^{-3}$ mol/L) system, an acid-base system that profoundly influences natural and engineered aquatic processes. Example 5-3 provides further explanation of how this figure can be used to determine the pH of CO_2-driven, balanced CO_2—CO_3^{-2} and CO_3^{2-}-driven systems.

Figure 5-2 is a similar representation of the distribution of soluble hydroxo complexes of iron(II) as a function of pH, and Figure 5-3, the solubility limits

Table 5-2 First Protolysis Constants for Hydrated Metal Ions[a]

Metal Species		$pK_{a,i}$
Tin	Sn^{2+}	1.7
Iron(III)	Fe^{3+}	2.2
Mercury	Hg^{2+}	2.5
Chromium	Cr^{3+}	2.9
Aluminum	Al^{3+}	5.0
Lead	Pb^{2+}	6.7
Cadmium	Cd^{2+}	7.6
Copper	Cu^{2+}	7.9
Iron(II)	Fe^{2+}	8.3
Zinc	Zn^{2+}	9.6
Manganese	Mn^{2+}	10.6
Nickel	Ni^{2+}	10.6
Magnesium	Mg^{2+}	11.4
Cobalt	Co^{2+}	12.2
Chromium	Cr^{2+}	12.5
Calcium	Ca^{2+}	12.7
Strontium	Sr^{2+}	13.2
Barium	Ba^{2+}	13.4

Source: Numerical values from Freiser and Fernando (1963) and Smith and Martell (1976).

Note: The general reaction scheme presented above includes water molecules as metal ion and proton coordination partners. For ease of notation and convervation of space, these coordination partners are not included in the tabular list of metal ions. For example $Fe^{2+} = \{Fe(H_2O)_4\}^{2+}$ and $H^+ = H_3O^+$, and so on. This notation will be used in all subsequent tables and figures in this book unless otherwise noted.

[a]General reaction for metal M:

$$\{M(H_2O)_m\}^{n+} + H_2O \rightleftharpoons \{M(H_2O)_{m-1}(OH)\}^{(n-1)+} + H_3O^+$$

$$K_{a,1} = \frac{[\{M(H_2O)_{m-1}(OH)\}^{(n-1)+}]\,[H_3O^+]}{[\{M(H_2O)_m\}^{n+}]}$$

$$pK_{a,i} = -\log K_{a,i}$$

of these various iron(II) species as a function of pH. The circled numbers on Figures 5-2 and 5-3 refer to electron and coordination partner transfer reactions listed in Table 5-3. Figures 1–3 apply to equilibrium conditions in *closed* systems (i.e., systems having no input or output of the species represented after the conditions illustrated have been established).

The logic upon which representations of distributions of various species as functions of one major control variable are based will become evident as we develop our understanding of chemical thermodynamics. That pH frequently serves as a useful master or control variable is not surprising given that:

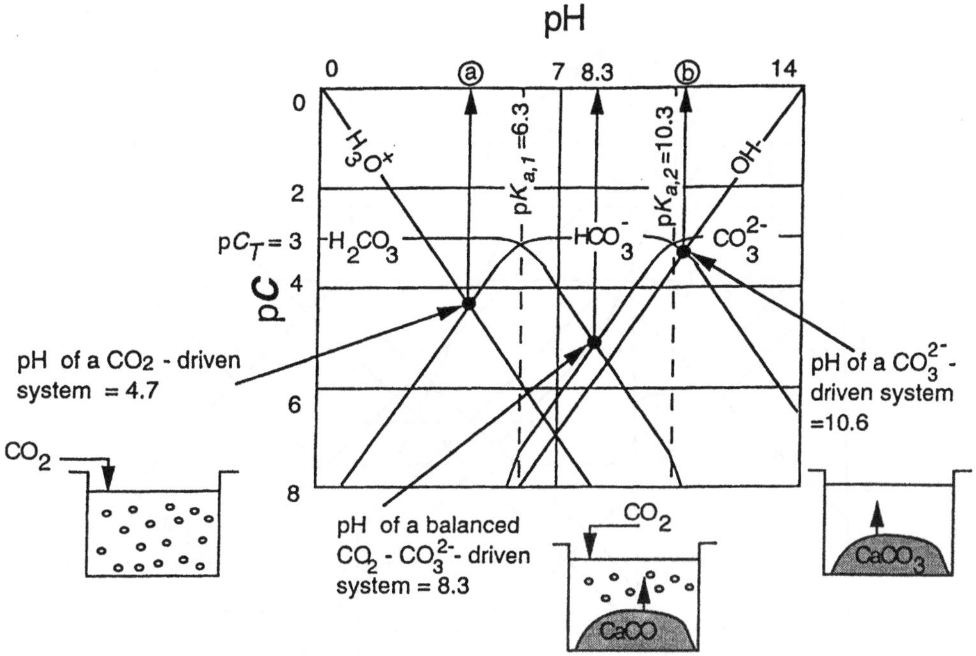

Note: Example 5-3 provides an explanation of positions labeled "a", 8.3, and "b" on the pH scale.

Sample Calculation of Solubility Limit Lines for Equation 1:

A. Mass Balance Relationship:

$$C_T = [H_2CO_3] + [HCO_3^-] + [CO_3^{2-}] = 10^{-3} mol/L$$

B. Mass Law Relationship:

(i) $$H_2CO_3 + H_2O \rightleftharpoons HCO_3^- + H_3O^+; K_{a,1} = 10^{-6.3}$$

$$pK_{a,1} = 6.3 = p[HCO_3^-] + p[H_3O^+] - p[H_2CO_3]$$

(ii) $$HCO_3^- + H_2O \rightleftharpoons CO_3^{2-} + H_3O^+; K_{a,2} = 10^{-10.3}$$

$$pK_{a,2} = 10.3 = p[CO_3^{2-}] + p[H_3O^+] - p[HCO_3^-]$$

Figure 5-1 Variations in species distributions as functions of pH in a closed system (10^{-3} M carbonate).

- pH is an expression of proton concentration (the negative logarithm of proton *activity*).
- The proton is the major reactive component of water and therefore of substances that react with and within aqueous solutions.
- The large majority of reactions that occur in aqueous systems are coor-

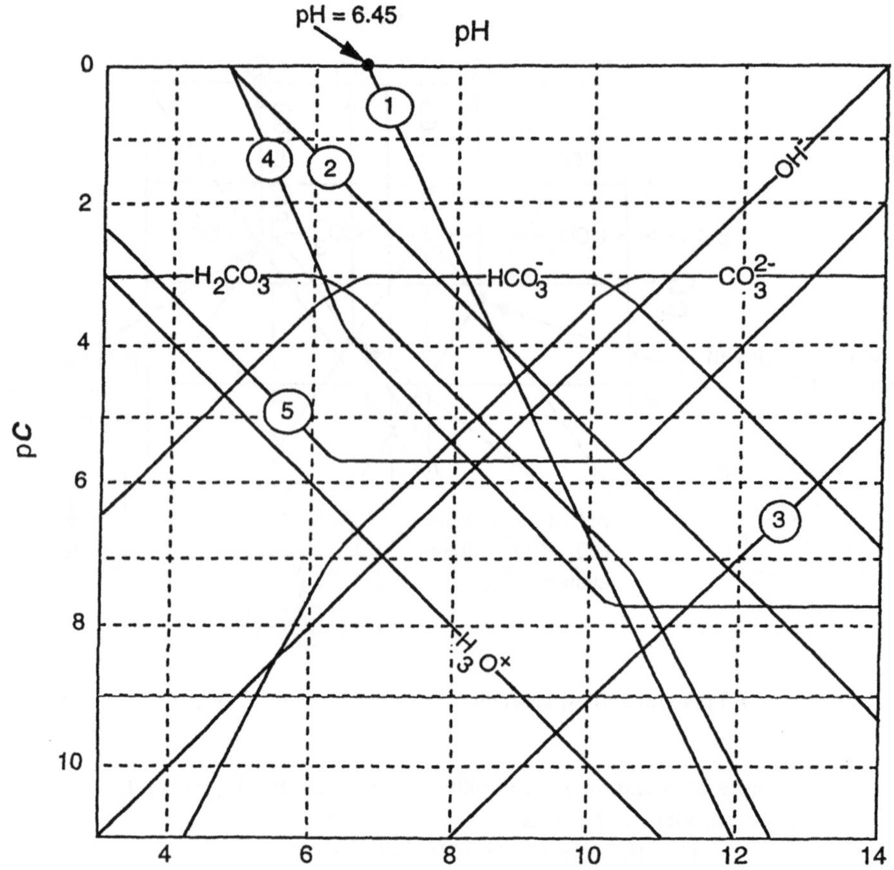

Note: Circled numbers on plot refer to equations 1 through 5 given in Table 5-3.

Sample Calculation of Solubility Limit Lines for Equation 1:

A. Reaction Stoichiometry:

$$\{Fe(OH)_2\}(s) \longleftrightarrow Fe^{2+} + 2OH^-;\ \mathcal{K}_s = 8 \times 10^{-16} = [Fe^{2+}][OH^-]^2$$

B. Mass Law Relationship:

$$\log Fe(II) = \log [Fe^{2+}] = \log \mathcal{K}_s - 2 \log [OH^-];\ \therefore \text{Slope} = 2$$

$$p\,Fe(II) = p\mathcal{K}_s - 2pOH\ ;\ pOH = \frac{p\mathcal{K}_s - pFe(II)}{2}$$

C. Solution:

At intercept for pFe(II) = 0: $pOH = \frac{15.1}{2} = 7.55;\ pH = 14 - 7.55 = 6.45$

$\therefore$ Fe(II) limit line for Equation 1 has an intercept of 6.45 and a slope of 2:1.

Figure 5-2 Solubility limits for individual dissolved Fe(II) species as a function of pH in a closed system (10^{-3} M carbonate).

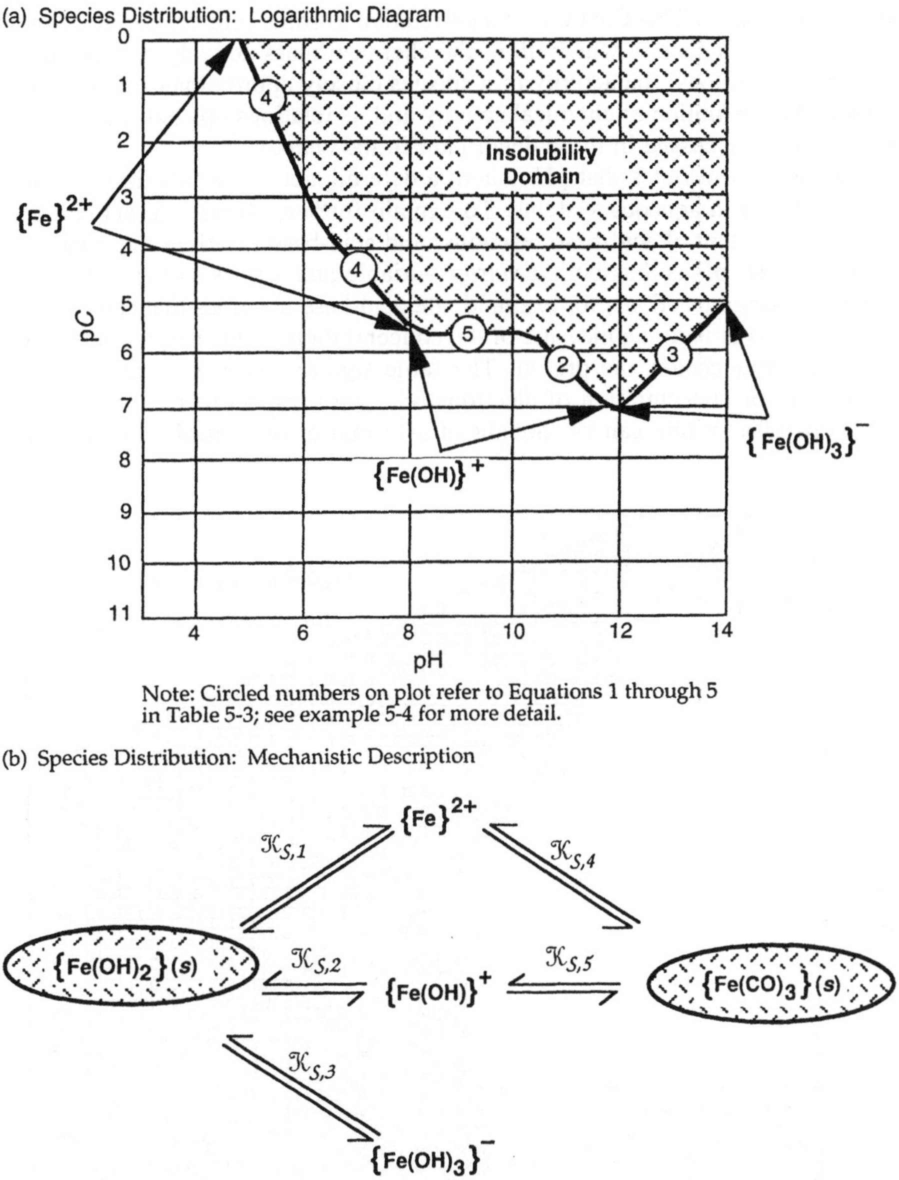

Figure 5-3 Composite solubility of Fe(II) species in a closed system (10^{-3} M carbonate).

dination-partner transfer or exchange reactions, and the proton is highly active in sharing electron pairs provided by other species.

Proton *activity* should be distinguished from proton *concentration*. Chemical activity is a measure of that fraction of the chemical concentration or chemical mass of a substance that *actively participates* in a particular reaction in a particular system. This fraction is taken relative to some reference state of unit activity, (i.e., a state in which *all* of the chemical mass actively participates). For dilute solutions, the activity of a species closely approximates its concentration. The definition of chemical activity and its relationships to other parameters will be discussed in more detail later in the chapter.

Figure 5-1 illustrates that pH is both a determinant and a function of acid-base reactions in aqueous systems. To explain further, changes in pH can determine the extent of acid-base reactions, and acid-base reactions can function to change pH; pH is thus both a *master* variable and a *response* variable. All ligand exchange reactions, for example, are acid-base reactions that can change pH, while pH is in turn a measure of the concentration of one major competing electron-pair acceptor, the proton. The same significance can be assigned to the activity or concentration of electrons, e^-, with respect to reactions. That is, the *pe* of a system can be thought of as a master or control variable, and

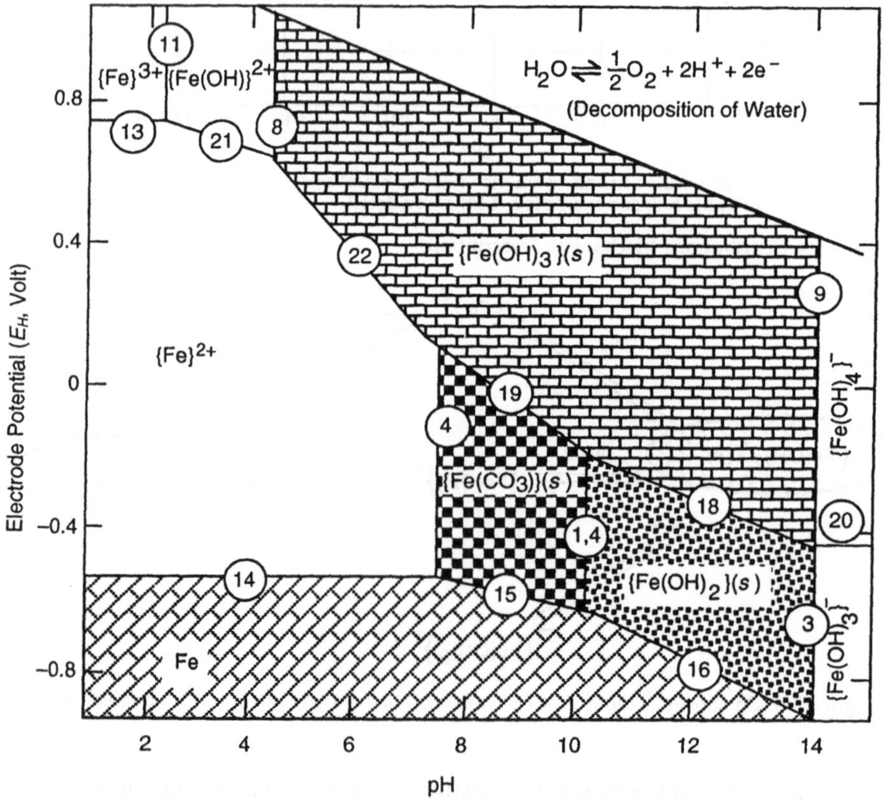

Note: Circled numbers on plot refer to equations given in Table 5-3. The alkalinity of the system is 2×10^{-1} eq/L and the lines denote boundaries along which soluble iron exists at 10^{-5} mol/L.

Figure 5-4 Pourbaix diagram for iron species as functions of E_H and pH in an aqueous system. [Adapted from Stumm and Lee (1960).]

the concentrations of other redox-active chemical species as response variables. By way of example, the *Pourbaix diagram* given in Figure 5-4 is a representation of the distribution of different "stable" oxidation states (O, II, and III) and chemical forms of iron as a function of the electrode potential, E_H, which is related directly to p*e* and is an equivalent but more versatile control variable. The electrode potential, E_H, is subscripted with an *H* to denote that it is referenced to the redox potential of the half reactions of hydro-

Table 5-3 Electron and Coordination-Partner Transfer Reactions and Associated Energy Parameters for Iron in Aqueous Systems

Reaction				Thermodynamic (Equilibrium) Parameter at 25°C	
				Solubility Constant	
Solubility Relationships			Equation	$\mathcal{K}_S$	$p\mathcal{K}_S$
Fe(II)					
$Fe(OH)_2(s)$	⇌	$Fe^{2+} + 2OH^-$	1	7.9×10^{-16}	15.1
$Fe(OH)_2(s)$	⇌	$[Fe(OH)]^+ + OH^-$	2	4.0×10^{-10}	9.4
$Fe(OH)_2(s) + OH^-$	⇌	$[Fe(OH)_3]^-$	3	7.9×10^{-6}	5.1
$FeCO_3(s)$	⇌	$Fe^{2+} + CO_3^{2-}$	4	2.0×10^{-11}	10.7
$FeCO_3(s) + OH^-$	⇌	$[Fe(OH)]^+ + CO_3^{2-}$	5	1.0×10^{-5}	5.0
Fe(III)					
$Fe(OH)_3(s)$	⇌	$Fe^{3+} + 3OH^-$	6	1.6×10^{-39}	38.8
$Fe(OH)_3(s)$	⇌	$[Fe(OH)_2]^+ + OH^-$	7	1.6×10^{-15}	14.8
$Fe(OH)_3(s)$	⇌	$[Fe(OH)]^{2+} + 2OH^-$	8	6.3×10^{-25}	24.2
$Fe(OH)_3(s) + OH^-$	⇌	$[Fe(OH)_4]^-$	9	1.0×10^{-5}	5.0
				Protolysis Constant	
Acid–Base Relationships				K_a	pK_a
$Fe^{2+} + H_2O$	⇌	$[Fe(OH)]^+ + H^+$	10	5.0×10^{-9}	8.3
$Fe^{3+} + H_2O$	⇌	$Fe(OH)^{2+} + H^+$	11	6.3×10^{-3}	2.2
$[Fe(OH)]^{2+} + H_2O$	⇌	$[Fe(OH)_2]^+ + H^+$	12	2.5×10^{-5}	4.6
Redox Relationships				Standard Electrode Potential, E_H° (V)	
$Fe^{3+} + e^-$	⇌	Fe^{2+}	13	+0.77	
$Fe^{2+} + 2e^-$	⇌	Fe	14	−0.44	
$FeCO_3(s) + 2e^-$	⇌	$Fe + CO_3^{2-}$	15	−0.76	
$Fe(OH)_2(s) + 2e^-$	⇌	$Fe + 2OH^-$	16	−0.89	
$[Fe(OH)]^+ + 2e^-$	⇌	$Fe + OH^-$	17	−0.61	
$Fe(OH)_3(s) + e^-$	⇌	$Fe(OH)_2(s) + OH^-$	18	−0.42	
$Fe(OH)_3(s) + CO_3^{2-} + e^-$	⇌	$FeCO_3(s) + 3OH^-$	19	−0.68	
$Fe(OH)_3(s) + e^-$	⇌	$[Fe(OH)_3]^-$	20	−0.72	
$[Fe(OH)]^{2+} + e^-$	⇌	$Fe^{2+} + OH^-$	21	+0.07	
$Fe(OH)_3(s) + e^-$	⇌	$Fe^{2+} + 3OH^-$	22	−1.31	

Source: Adapted from Stumm and Lee (1960).

gen. For standard-state conditions it is superscripted as E_H°. The interrelationships between the proton, electron, E_H, and the energetics of reactions will be demonstrated later in this chapter.

Representations of the type given in Figure 5-4 are particularly useful for describing and understanding such processes as weathering and dissolution, chemical oxidation, corrosion, and corrosion control in environmental systems. The circled numbers on the lines shown in this figure correspond to the reactions listed in Table 5-3. This table illustrates the variety of redox and coordination partner transfer reactions possible for iron, a common constituent of surface and subsurface aquatic systems. Figure 5-4, with its E_H and pH coordinates, thus represents the master and response variables for the two primary classes of reactions occurring in environmental systems, electron and ligand exchange reactions. The underlying concepts and significance of Figure 5-4 are developed more fully in Section 5.4.3.

Each reaction in Table 5-3 has either an equilibrium constant or a standard electrode potential associated with it. These two different thermodynamic parameters have equivalent meanings in the sense that they each characterize the amounts of energy associated with a particular reaction, as we will demonstrate shortly. Example 5-3 provides an illustration of the behavior of one important inorganic ligand for natural waters as well as most water treatment and many wastewater treatment operations (i.e., the carbonate ion).

Example 5-3

- **Situation:** An application of the interrelationships shown in Figure 5-1 for characterization of the pH domains of natural freshwater systems.
- **Question(s):** Demonstrate how the pH values of CO_2-driven, CO_2-CO_3^{2-}-balanced, and CO_3^{2-}-driven systems pictured in Figure 5-1 can be determined by analysis of the information presented. Based on the results of this demonstration, what dominant range of pH would be predicted for most natural aquatic systems?
- **Assumption(s):** (a) The water contains no species other than those shown in the figure; and (b) any component that represents less than 1% of the total composition of the system (excluding H_2O) can be neglected.
- **Logic and Answer(s):** From Figure 5-1:

1. The total concentration of the dissolved carbonate system is

$$C_T = [H_2CO_3](aq) + [HCO_3^-] + [CO_3^{2-}] = 10^{-3} \text{ mol/L}$$

2. The only components that can be present in the system at any time, including Ca^{2+} resulting from dissolution of $CaCO_3(aq)$, are H_3O^+, OH^-, $H_2CO_3(aq)$, HCO_3^-, CO_3^{2-}, and Ca^{2+}.

3. The condition of electroneutrality requires that the system be electrically balanced, that is,

$$\sum \text{positive charges} = \sum \text{negative charges}$$

or

$$[H_3O^+] + 2[Ca^{2+}] = [HCO_3^-] + 2[CO_3^{2-}] + [OH^-]$$

where the brackets denote molar concentrations.

4. If the system was driven entirely by an input of CO_2 to reach $C_T = 10^{-3}$ mol/L, then pH << pK_1 and there is no Ca^{2+} present; thus

$$[H_3O^+] = [HCO_3^-] + 2[CO_3^{2-}] + [OH^-]$$

It is readily apparent for pH << pK_1 that $[OH^-]$ is negligible and both $[H_3O^+]$ and $[HCO_3^-]$ are more than two orders of magnitude larger than $[CO_3^{2-}]$. Therefore, by neglecting $[OH^-]$ and $[CO_3^{2-}]$, we have

$$[H_3O^+] = [HCO_3^-] \quad \text{or} \quad \text{pH} = \text{p}[HCO_3^-]$$

and for pC_T = 3.0,

$$\text{p}[H_3O^+] = \text{p}[HCO_3^-] = 4.7$$

This pH corresponds to position "a" on Figure 5-1

5. For a balanced CO_2–CO_3^{2-} system derived from dissolution of CO_2 and $CaCO_3$, pK_1 < pH < pK_2, $[H_3O^+]$ and $[OH^-]$ are negligible, and the electroneutrality condition is

$$2[Ca^{2+}] = [HCO_3^-] + 2[CO_3^{2-}]$$

To arrive at this balanced condition, equal molar amounts of $CaCO_3$ and CO_2 would have initially dissolved into the system; thus

$$[CO_2]_0 = [CaCO_3]_0 = [Ca^{2+}]_0 = [CO_3^{2-}]_0 = 0.5C_T$$

6. While the concentrations of CO_2 (H_2CO_3) and CO_3^{2-} will change upon subsequent reaction to form HCO_3^-, $[Ca^{2+}]$ will remain constant at $0.5C_T$. The conservation of mass relationship dictates that

$$C_T = [H_2CO_3](aq) + [HCO_3] + [CO_3^{2-}]$$

or, since $[Ca^{2+}] = 0.5C_T$,

$$2[Ca^{2+}] = [H_2CO_3](aq) + [HCO_3^-] + [CO_3^{2-}]$$

7. Substitution into the electroneutrality condition yields

$$[H_2CO_3](aq) + [HCO_3^-] + [CO_3^{2-}] = [HCO_3^-] + 2[CO_3^{2-}]$$

or

$$[H_2CO_3](aq) = [CO_3^{2-}]$$

This condition obtains only at

$$\text{pH} = \frac{\text{p}K_1 + \text{p}K_2}{2} = \frac{6.3 + 10.3}{2} = 8.3$$

regardless of the value of pC_T.
This pH is specifically noted on Figure 5-1

8. For a CO_3^{2-}-driven system, assuming that $CaCO_3$ is the only source of CO_3^{2-}, the electroneutrality condition requires that

$$[H_3O^+] + 2[Ca^{2+}] = [HCO_3^-] + 2[CO_3^{2-}] + [OH^-]$$

In this case, $[H_3O^+]$ can clearly be neglected, and $[Ca^{2+}] = C_T$. Thus

$$2[H_2CO_3](aq) + 2[HCO_3^-] + 2[CO_3^-]$$

$$= [HCO_3^-] + 2[CO_3^{2-}] + [OH^-]$$

Neglecting $[H_2CO_3](aq)$ then gives

$$[HCO_3^-] = [OH^-]$$

and for $pC_T = 3.0$,

$$pOH = p[HCO_3^-] = 3.4$$

$$pH = 14.0 - pOH = 10.6$$

This pH corresponds to position "b" on Figure 5-1

Most natural aquatic systems receive inputs of CO_2 from biological respiration reactions and CO_3^{2-} from the dissolution reactions of calcite ($CaCO_3$), a major constituent of the earth's crust. Thus, the predominant range of pH for many natural aquatic systems is approximately 8.1 to 8.5. These are referred to as carbonate-buffered waters.

As we have noted, this chapter focuses primarily on homogeneous systems, but one of the most straightforward and well-recognized examples of the transition from homogeneous to heterogeneous systems is provided by the carbonate system. The proton-ligand reaction concept can be extended to a simple heterogeneous system in which a solid, $CaCO_3(s)$, is formed. This example is important because Ca^{2+} is the major contributor to the *hardness* of most water supplies. We will briefly consider the heterogeneous aspects of this system, aspects that markedly affect such engineered processes as water softening, neutralization, and the use of lime (e.g., CaO) for metal precipitation and coagulation. The system, which includes $CO_2(g)$ in the gas phase, $H_2CO_3(aq)$, HCO_3^-, and CO_3^{2-} dissolved in the aqueous phase, and $CaCO_3(s)$ as the precipitated solid phase, is frequently the dominant inorganic acid-base system in fresh waters. The primary reactions involved in this three-phase open environmental system are pictured schematically in Figure 5-5. The water column or *aqueous-phase* transformations for the system are

$$H_2CO_3(aq) + H_2O \rightleftharpoons HCO_3^- + H_3O^+ \qquad K_{a,1} \qquad (5\text{-}21)$$

$$HCO_3^- + OH^- \rightleftharpoons CO_3^{2-} + H_2O \qquad K_{a,2} \qquad (5\text{-}22)$$

To quantify these relations properly, appropriate account must be taken of the parallel *interphase* reactions:

$$CO_2(g) + H_2O \rightleftharpoons H_2CO_3(aq) \qquad \mathcal{K}_H \qquad (5\text{-}23)$$

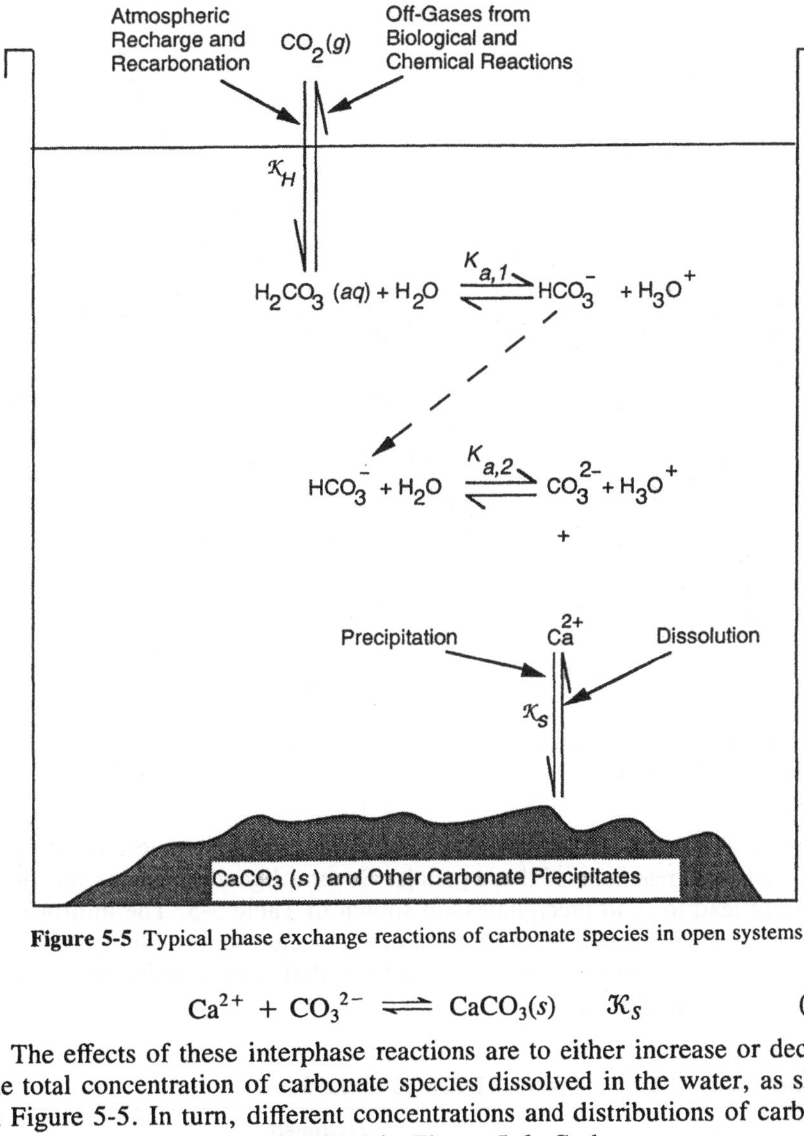

Figure 5-5 Typical phase exchange reactions of carbonate species in open systems.

$$Ca^{2+} + CO_3^{2-} \rightleftharpoons CaCO_3(s) \qquad \mathcal{K}_S \tag{5-24}$$

The effects of these interphase reactions are to either increase or decrease the total concentration of carbonate species dissolved in the water, as shown in Figure 5-5. In turn, different concentrations and distributions of carbonate species are produced, as depicted in Figure 5-6. Carbonate system reactions therefore affect the water quality parameters of *hardness* (e.g., Ca^{2+}) and *alkalinity* (e.g., HCO_3^- and CO_3^{2-}). Figure 5-6 also serves to emphasize that pH may be affected significantly by the total concentration and distribution of carbonate species and by the primary source(s) of their introduction. The computation procedure for pH positions labeled "c" and "d" is the same as that given earlier in Example 5-3 for pH positions "a" and "b."

Similar considerations apply to the interactions of other metal ion species with acid-base-reactive ligands. Table 5-4 identifies solid phases and associated solubility constants and process considerations for a broad range of environmentally significant metals. As discussed above, the insoluble states of

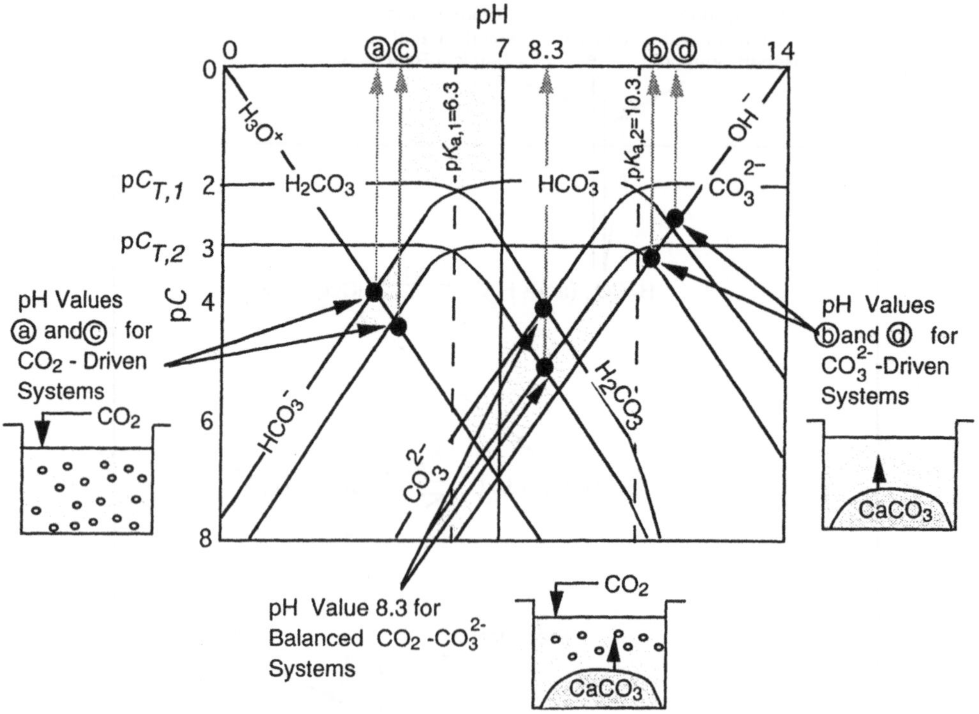

Figure 5-6 Variations in pH and p*C* values as functions of total carbonate species concentration and source in closed systems.

the metals shown in Table 5-4 are actually intermediate states in different ligand exchange reactions of these metals. Several ligand exchange reactions that do not lead to solid precipitates are shown in Table 5-5. The information given in Tables 5-4 and 5-5 can be combined to determine solubility relationships between solid phases and various intermediate coordination species, as illustrated in Example 5-4.

Table 5-4 Solubility Relationships for Hydrated Metal Ions

Precipitate	Application	Hydrated (H_2O-Complexed) Metal Ion	Exchanging Ligand	$p\mathcal{K}_S$ (at 25°C)
$Al(OH)_3(s)$	Coagulation	$\{Al(H_2O)_6\}^{3+}$	OH^-	33.5
$AlPO_4(s)$	Phosphate removal	$\{Al(H_2O)_6\}^{3+}$	PO_4^{3-}	20.6
$CaCO_3(s)$	Softening, corrosion	$\{Ca(H_2O)_4\}^{2+}$	CO_3^{2-}	8.2
$CaF_2(s)$	Fluoridation and defluoridation	$\{Ca(H_2O)_4\}^{2+}$	F^-	10.3
$Ca(OH)_2(s)$	Softening	$\{Ca(H_2O)_4\}^{2+}$	OH^-	5.2
$Ca_3(PO_4)_2(s)$	Softening, phosphate cycle and removal	$\{Ca(H_2O)_4\}^{2+}$	PO_4^{3-}	28.7

Table 5-4 ***(Continued)***

Precipitate	Application	Hydrated (H_2O-Complexed) Metal Ion	Exchanging Ligand	$p\mathcal{K}_S$ (at 25°C)
$Ca_5(PO_4)_3(OH)(s)$	Phosphate cycle and removal	$\{Ca(H_2O)_4\}^{2+}$	$PO_4^{3-} + OH^-$	55.6
$CaHPO_4(s)$	Phosphate cycle and removal	$\{Ca(H_2O)_4\}^{2+}$	HPO_4^{2-}	6.6
$CaSO_4(s)$	Hardness, scale formation	$\{Ca(H_2O)_4\}^{2+}$	SO_4^{2-}	4.6
$Cu(OH)_2(s)$	Algae control	$\{Cu(H_2O)_4\}^{2+}$	OH^-	19.4
$Fe(OH)_2(s)$	Corrosion, iron cycle and removal	$\{Fe(H_2O)_4\}^{2+}$	OH^-	15.1
$FeCO_3(s)$	Iron cycle	$\{Fe(H_2O)_4\}^{2+}$	CO_3^{2-}	10.7
$FeS(s)$	Anaerobic corrosion	$\{Fe(H_2O)_4\}^{2+}$	S^{2-}	18.1
$Fe(OH)_3(s)$	Coagulation, iron cycle and removal	$\{Fe(H_2O)_6\}^{3+}$	OH^-	38.8
$FePO_4(s)$	Phosphate cycle and removal	$\{Fe(H_2O)_6\}^{3+}$	PO_4^{3-}	26.4
$MgCO_3(s)$	Hardness	$\{Mg(H_2O)_4\}^{2+}$	CO_3^{2-}	7.5
$MgCO_3 \cdot 3H_2O(s)$	Hardness	$\{Mg(H_2O)_4\}^{2+}$	CO_3^{2-}	4.7
$MgF_2(s)$	Defluoridation	$\{Mg(H_2O)_4\}^{2+}$	F^-	8.2
$Mg(OH)_2(s)$	Softening	$\{Mg(H_2O)_4\}^{2+}$	OH^-	11.1
$MnCO_3(s)$	Manganese cycle	$\{Mn(H_2O)_4\}^{2+}$	CO_3^{2-}	9.3
$Mn(OH)_2(s)$	Demanganization	$\{Mn(H_2O)_4\}^{2+}$	OH^-	12.8
$Pb(OH)_2(s)$	Lead dissolution, corrosion	$\{Pb(H_2O)_4\}^{2+}$	OH^-	10.9
$SiO_2(s)$	Amorphous silica	$\{Si(IV)\}$	$O(-II)$	2.7
$Zn(OH)_2(s)$	Corrosion	$\{Zn(H_2O)_4\}^{2+}$	OH^-	17.2

Source: Numerical values from Freiser and Fernando (1963) and Smith and Martell (1976).

Table 5-5 Formation Constants for Metal–Hydroxo Complexes

Reaction			pK_f (at 25°C)
$Al^{3+} + OH^-$	$\rightleftharpoons$	$\{Al(OH)\}^{2+}$	−9.0
$2Al(OH)^{2+}$	$\rightleftharpoons$	$\{Al_2(OH)_2\}^{4+}$	−3.0
$Al(OH)_3(s) + OH^-$	$\rightleftharpoons$	$\{Al(OH)_4\}^-$	−5.0
$Fe^{3+} + OH^-$	$\rightleftharpoons$	$\{Fe(OH)\}^{2+}$	−11.8
$Fe(OH)^{2+} + OH^-$	$\rightleftharpoons$	$\{Fe(OH)_2\}^+$	−9.4
$2Fe(OH)^{2+}$	$\rightleftharpoons$	$\{Fe_2(OH)_2\}^{4+}$	−1.5
$Fe(OH)_3(s) + OH^-$	$\rightleftharpoons$	$\{Fe(OH)_4\}^-$	+5.0
$Ca^{2+} + OH^-$	$\rightleftharpoons$	$\{Ca(OH)\}^+$	−1.3
$Mg^{2+} + OH^-$	$\rightleftharpoons$	$\{Mg(OH)\}^+$	−2.6
$Fe^{2+} + OH^-$	$\rightleftharpoons$	$\{Fe(OH)\}^+$	−5.7

Source: Numerical values from Freiser and Fernando (1963) and Smith and Martell (1976).

Example 5-4

- **Situation:** We wish to determine the solubility of Fe(II) as a function of pH in an aqueous carbonate system, in a manner similar to that presented in Figures 5-2 and 5-3 but have information on the solubility constants for only Fe^{2+} with respect to $Fe(OH)_2(s)$ and $FeCO_3(s)$.
- **Question(s):** Calculate solubility constants for the $\{Fe(OH)\}^+$ species of soluble iron with respect to each of the two solid phases involved: (a) ferrous hydroxide and (b) ferrous carbonate.
- **Logic and Answer(s):** We can apply thermodynamic principles to this heterogeneous system by calculating the solubility product for complex species such as $\{Fe(OH)\}^+$ from the primary solubility constants for Fe^{2+} with respect to the two solid phases and the complex formation constants for Fe^{2+} with respect to $\{Fe(OH)\}^+$.

(a) From Table 5-4:

$$Fe(OH)_2(s) = Fe^{2+} + 2OH^- \qquad \mathcal{K}_S = 10^{-15.1}$$

From Table 5-5:

$$Fe^{2+} + OH^- = \{Fe(OH)\}^+ \qquad K_f = 10^{5.7}$$

Summing these equations yields the solubility constant for $\{Fe(OH)\}^+$ with respect to the $Fe(OH)_2(s)$ solid phase:

$$Fe(OH)_2(s) = \{Fe(OH)\}^+ + OH^- \qquad \mathcal{K}_S = 10^{-9.4}$$

(b) From Table 5-4:

$$FeCO_3(s) = Fe^{2+} + CO_3^{2-} \qquad \mathcal{K}_S = 10^{-10.7}$$

From Table 5-5:

$$Fe^{2+} + OH^- = \{Fe(OH)\}^+ \qquad K_f = 10^{5.7}$$

Summing these two equations yields the solubility constant for $\{Fe(OH)\}^+$ with respect to the $FeCO_3(s)$ solid phase:

$$FeCO_3(s) + OH^- = \{Fe(OH)\}^+ + CO_3^{2-} \qquad \mathcal{K}_S = 10^{-5}$$

The values calculated above are observed to be the same as those given in Table 5-3.

5.2 CHEMICAL STATES AND THERMODYNAMIC STABILITY

5.2.1 Stability

According to the concepts of electron structure and chemical reactivity, all atoms exist in a state of *instability* until they have achieved an electron configuration corresponding to that of the closest inert gas. The degree of instability is relative; that is, any particular electron configuration in which an atom can exist may represent a more or less *stable state* than another. Instability relates conceptually to differences in thermodynamic potential, or energy, between states, and operationally to the presence or absence of a facilitating mechanism for moving from one state to another. Figure 5-7 presents an analogy between chemical and gravitational states and stability. Like the block in Figure 5-7, an atom can be temporarily stable, or *metastable*, until such time as it has an opportunity, through facilitating reactions, to move to a lower-energy or more stable state.

The energy levels of the various states of an atom are defined by its nuclear and electron structure. The facility with which it can move from one state to another is dependent on the reactions in which it can participate in a given system, that is, the availability of partners with which it can either exchange or share electrons. In a larger sense, the same is true of each chemical com-

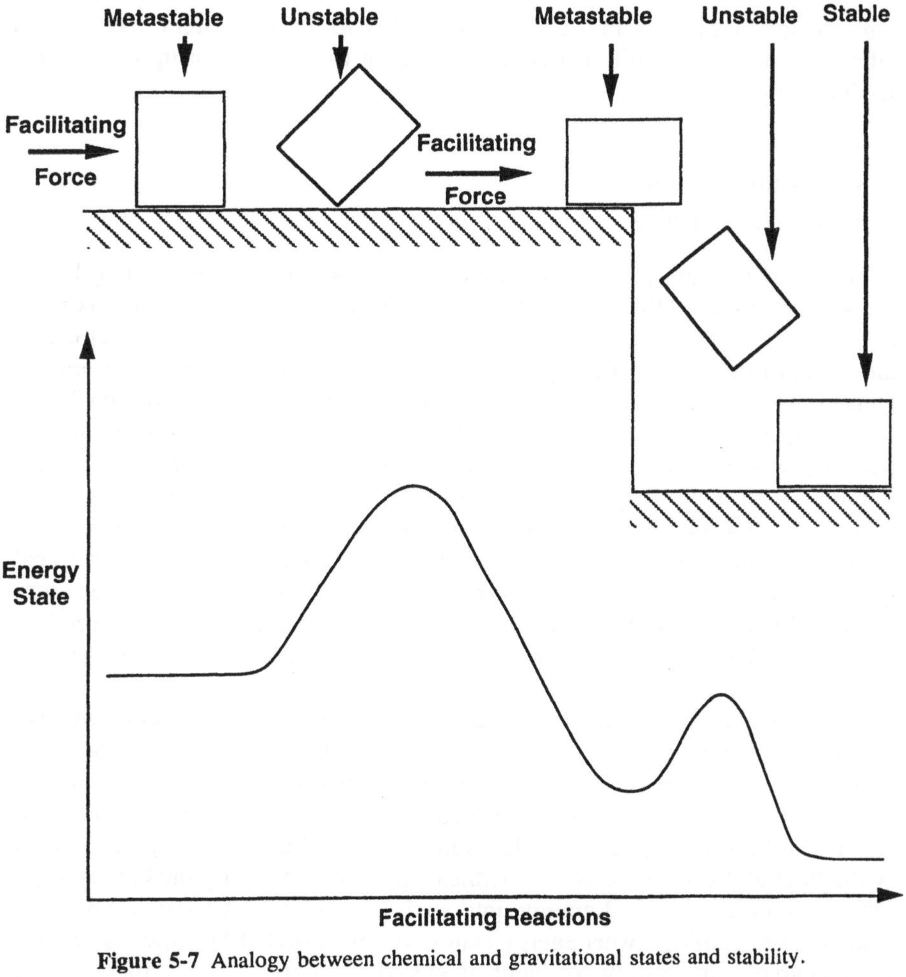

Figure 5-7 Analogy between chemical and gravitational states and stability.

ponent of a given system, and thus of the entire system. Environmental systems can therefore exist in various states of relative chemical stability, and various facilitating reactions can convert them from one state to another.

A reaction takes place when chemical species undergo transformation to different energy states, either by chemical change within a phase or by transfer to another phase. This may occur *exergonically* or *endergonically*, that is, with energy being either released or consumed in the reaction(s). The analogy given in Figure 5-7 depicts changes from higher energy levels to lower levels, thus releasing energy to the system in which those changes occur. Conversely, if energy is input to the system, it is possible to force the reactions in the reverse direction and induce conversion of species from lower-energy states to higher states. To reverse the direction would involve lifting the block in Figure 5-7 from the floor to the table and then standing it upright.

5.2.2 Types of Change

Any change that occurs in the chemical composition of a system may be classified as one of three different types, according to its relationship to the equilibrium state:

- Natural change
- Unnatural change
- Reversible change

A natural change, as the name implies, results from a reaction that brings the system closer to the equilibrium state; that is, moves it toward maximum stability. Conversely, an unnatural change is caused by a reaction that drives the system further from the equilibrium state and toward decreased stability. Unnatural changes are not spontaneous and must therefore be driven by an energy input. The galvanic cell of the Pb–PbO_2 battery pictured in Figure 5-8 tends by spontaneous chemical reaction to an equilibrium state of maximum stability, with a discharge of electrical energy occurring during this natural change. Recharging the battery with an external electrical energy source represents an unnatural change for which the energy input forces the chemical system of the battery away from its equilibrium state. Reversible changes represent a limiting case between natural and unnatural processes. Reversible changes involve the continuous passage of a system in either direction through a series of equilibrium states.

The energy output or input associated with a change in a system may take the form of *chemical energy, mechanical energy, thermal energy,* or *electrical energy*. The inherent energy release from the chemical reactions which take place between the components of a lead (Pb–PbO_2) storage battery like that pictured in Figure 5-8, for example, is manifest as electrical energy generated by the flow of electrons between chemical species comprising the battery. The battery produces electrical energy until such time as its chemical components achieve more stable (lower) energy states. At this point it becomes necessary to "recharge" the battery by input of electrical energy, which functions to

(A) Electrochemical Cells:

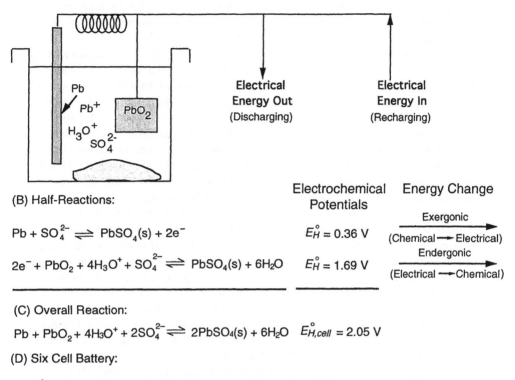

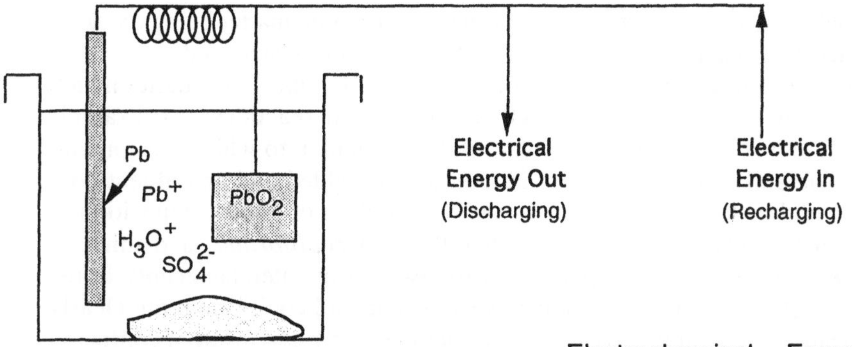

(B) Half-Reactions:

Half-Reactions	Electrochemical Potentials	Energy Change
$Pb + SO_4^{2-} \rightleftharpoons PbSO_4(s) + 2e^-$	$E_H^\circ = 0.36$ V	Exergonic (Chemical → Electrical)
$2e^- + PbO_2 + 4H_3O^+ + SO_4^{2-} \rightleftharpoons PbSO_4(s) + 6H_2O$	$E_H^\circ = 1.69$ V	Endergonic (Electrical → Chemical)

(C) Overall Reaction:

$$Pb + PbO_2 + 4H_3O^+ + 2SO_4^{2-} \rightleftharpoons 2PbSO_4(s) + 6H_2O \quad E_{H,cell}^\circ = 2.05\ \text{V}$$

(D) Six Cell Battery:

$6 \times E_H^\circ = 6 \times 2.05$ V $= 12.3$ V; i.e., a 12-V battery

(E) Nernst Equation:

$$E_{H,r} = E_{H,r}^\circ + \frac{\mathcal{R}T}{ne\mathcal{F}} \ln Q_r = 2.05 + \frac{0.059}{2} \log \frac{1}{[H_3O^+]^4[SO_4^{2-}]^2}$$

Figure 5-8 Electrochemical cell; the lead storage battery.

reconvert the chemical components to higher (less stable) energy states. This recharging process involves reversing the chemical reaction. The reversible character of most strictly chemical reactions is indicated by the double arrows between the reactants and products depicted in Figure 5-8 and in preceding equations in this chapter. Certain reactions, such as the combustion or biological degradation of an organic compound, are not directly reversible.

5.2.3 Equilibrium and Steady States

Any closed chemical system or system for which changes in composition result only from internal reactions will, if it is in a state other than equilibrium, undergo reaction and internal change until it reaches equilibrium. Consider, for example, a sealed tank full of organically contaminated wastewater to which a strong chemical oxidant has been added just before it was sealed. Upon addition of the oxidant, the system is displaced from whatever chemical equilibrium it may previously have attained. So displaced, the contents of that system will continue to react with each other, the organic matter being oxidized and the oxidant reduced, until a new equilibrium state is achieved.

In open systems composition may be changed continuously by addition and/or subtraction of components as well as by internal reactions. An example equivalent to that for the closed system might be a tank into which the contaminated waste and oxidant solutions are continually added and an effluent continually withdrawn. In this type of system a condition of fixed composition in the tank represents a *steady state* rather than an *equilibrium*, a distinction sometimes obscured by improper use of the two terms interchangeably in the description of environmental systems. Steady state and equilibrium are clearly different situations. Failure to distinguish them properly can result in confusion and misinterpretation of system information and behavior.

The reactions within a system are not necessarily at thermodynamic equilibrium for any given steady-state condition; they may, in fact, be very far removed from equilibrium. Because of a balanced input of reactants and output of products, however, the system is poised in a condition for which no net change in its composition is realized from the ongoing internal reactions. In the context of the material balance statement given in Equation 2-5, the net rate of change of the composition of a reactor is zero at steady state; that is, the right-hand term of Equation 2-5 is zero. In a simple sense, a closed system at equilibrium, where forward and reverse internal reaction rates, r_f and r_r, respectively, are equal, is not unlike parallel open systems operated at steady state. These two conditions are illustrated schematically in Figure 5-9 for the reactions

$$\mathrm{A + B \xrightleftharpoons{r_f} C + D} \tag{5-25}$$

$$\mathrm{C + D \xrightleftharpoons{r_r} A + B} \tag{5-26}$$

In terms of energetics, equilibrium is that state in which, for constant temperature and pressure, there is no net energy available for driving a reaction in either its forward or reverse directions. Energy is an *extensive* property because the energy available to a substance for participation in a chemical reaction is a function of its mass or concentration. To be more precise, energy is directly a function of the *reactive fraction* of its mass or concentration, that is, its *chemical activity*. The equilibrium state of minimum available energy or maximum stability for a given chemical system thus relates to a specific composition. The energetics of reactions and processes involve an analysis of component distributions and system conditions for states other than equilibrium; in particular, we must have knowledge of the energies that drive them toward this state. This leads logically to considerations of chemical thermodynamics.

Instability and reactivity must ultimately be referenced to some time scale. A given substance may be highly reactive in terms of energy availability but very slow in terms of the rate at which it reacts. If the rate is so slow that no significant change occurs over a period of practical interest, the substance can-

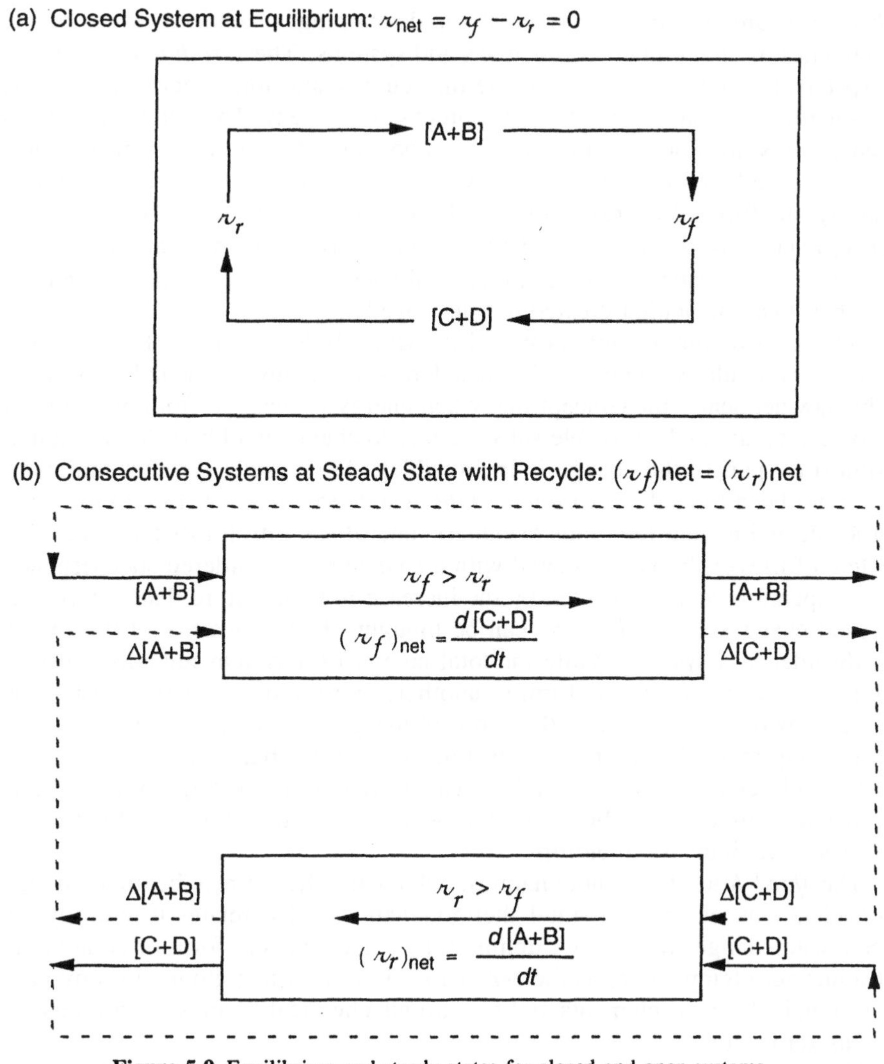

Figure 5-9 Equilibrium and steady states for closed and open systems.

not be thought of as highly reactive despite the energy it has available for reaction. Thus, while the thermodynamic relationships we are about to discuss are essential for assessing reaction and process feasibility, they are not completely sufficient descriptors. Our assessments must ultimately include reaction rates as well.

5.3 THERMODYNAMIC RELATIONSHIPS AND FUNCTIONS

5.3.1 Laws of Thermodynamics

The science underlying classical thermodynamics is structured upon four fundamental laws. Three of these, the *zeroth, first,* and *second laws of thermo-*

dynamics, are of particular interest for developing an understanding of the energetics of environmental processes and systems. The *zeroth and first laws*, respectively, specify criteria for thermal energy and total energy conditions relating to the equilibrium state of a process or system. The zeroth law states that there is no heat flow among the components of systems in thermal equilibrium. The first law extends the concept to total energy, recognizing that heat is only one form of energy. This law further requires the conservation of energy, regardless of the form in which it may exist. That is, heat energy may be converted to chemical energy, electrical energy, mechanical energy, and so on, but it can be neither created nor destroyed.

The *second law* of thermodynamics, which is less intuitive than the first two, is particularly important in considerations of environmental processes. This law delineates the concept that when energy is changed from one form to another, it may not be possible subsequently to change it all back to its original form (i.e., part of the change is irreversible). As a result, systems can attain what has been termed a *disordered state*, a state that is at or close to the most probable and natural (and thus eventual) state of stability for that system. The extent of irreversibility associated with a change to a disordered state depends on temperature and is reflected by the increase in temperature that occurs as a system changes states. *Entropy* can be thought of as a measure of the extent of disorder in a system. While the total energy of a system remains constant as it is converted from one form to another, entropy may remain constant or increase with each change in the form of energy. It remains constant for any completely reversible change in the form of energy, but it increases for any irreversible change. There may be local decreases in entropy within certain components of a system, but these will be offset by other larger local increases to yield a net increase in entropy.

The *third law* of thermodynamics, which has less direct bearing on our considerations of environmental thermodynamics, also relates to entropy. It deals with changes in entropy, defining the perfect crystal form of an atom at absolute zero temperature as the *zero entropy state* of that atom. Entropy values compiled for the elements and for various chemical species are referenced to this zero state.

Entropy is an important concept for energy conversions involving the transformation of chemical species. This concept provides the basis for interrelationships between *chemical mass* or *concentration* and *chemical potential*, *activity*, and *fugacity*. We will demonstrate that chemical potential, activity, and fugacity are themselves closely interrelated and, in many ways, similarly descriptive of the state of particular components within a system. The logical point of beginning for these developments is in the *characteristic state functions* defined in classical thermodynamics.

5.3.2 Characteristic Functions

The five fundamental thermodynamic functions or quantities of systems are given by:

- H, the *enthalpy* (ML^2t^{-2})
- G, the *Gibbs free energy* (ML^2t^{-2})
- A, the *Helmholtz free energy* (ML^2t^{-2})
- U, the *internal energy* (ML^2t^{-2})
- S, the *entropy* ($ML^2t^{-2}T^{-1}$)

These functions are termed *state functions* because changes that occur in them depend only on the initial and final states of that system (i.e., they are independent of the pathways by which the change has occurred). State functions are useful for assessing the overall heats and energies associated with complex multipath environmental processes. They are extensive characteristics of a system, which means that they are dependent on the magnitude or size of the system.

The five characteristic state functions are interrelated in the following ways:

$$H = U + PV \tag{5-27}$$

$$G = H - TS \tag{5-28}$$

$$A = U - TS \tag{5-29}$$

where V is the volume of a system (L^3), P its pressure ($ML^{-1}t^{-2}$), and T its absolute temperature (K). Like the five characteristic functions, volume is extensive, whereas P and T are independent of the magnitude or size of a system and thus *intensive* variables or properties. The terms PV and TS are commonly referred to as the *work* and *entropy gain functions*, respectively.

For single-component systems, or for systems comprised by mixtures of components having constant composition, the first and second laws of thermodynamics are combined into the following single statement: *The net change in the internal energy of the system as it moves from one state to another is given by the difference between the entropy gained and the work done in the change.* If the change is *isothermal* (constant temperature) and *isobaric* (constant pressure), we can combine Equations 5-27 and 5-28 and differentiate to obtain

$$dU = T\,dS - P\,dV \tag{5-30}$$

Changes in the other characteristic functions with respect to changes in the intensive variables of pressure and temperature can be developed similarly from Equations 5-27 through 5-30, yielding

$$dH = T\,dS + V\,dP \tag{5-31}$$

$$dG = -S\,dT + V\,dP \tag{5-32}$$

$$dA = -S\,dT - P\,dV \tag{5-33}$$

The foregoing differential forms of the relationships among characteristic functions are particularly useful for evaluation of heat or energy changes associated with the transformation of a system from one state to another. Because they were obtained from characteristic state functions, values of dH, dG, and dA do not depend on the pathway(s) along which a system is transformed from one state to another. As such, they lend themselves to ready quantification and referencing to some baseline state. Reference values for free energy and enthalpy for a specified *standard state* are given in Table 5-6 and in Appendix A.5.2. The values in these tables are free energies and enthalpies associated with the *formation of chemical species from their elements*. These parameters are denoted by ΔG_f° and ΔH_f°, reflecting that they represent a *finite quantity* (Δ) of energy associated with the *formation* (subscript f) of a *finite number of*

Table 5-6 Thermodynamic Properties of Selected Chemical Species at 25°C

Substance	State[a]	ΔH_f° kJ/mol	ΔG_f° kJ/mol	Substance	State[a]	ΔH_f° kJ/mol	ΔG_f° kJ/mol
Ca^{2+}	aq	−542.7	−553.1	Cs^+	aq	−261.9	−296.0
$CaCO_3$ (calcite)	c	−1207.1	−1128.8	Fe^{2+}	aq	−87.9	−84.9
CaO	c	−635.5	−604.2	Fe^{3+}	aq	−47.7	−10.6
$Ca(OH)_2$	aq	−1002.9	−867.8	$\{Fe(OH)\}^{2+}$	aq		−233.9
CO	g	−110.5	−137.2	$\{Fe(OH)_2\}^+$	aq		−444.3
CO_2	g	−393.7	−394.6	$Fe(OH)_3$	c		−694.5
	aq	−413.0	−386.2	Fe_2O_3 (hematite)	c	−822.2	−741.0
CO_3^{2-}	aq	−676.1	−528.0	H_2	g	0.0	0.0
HCO_3^-	aq	−691.2	−587.0	H_2O	lq	−285.8	−237.2
H_2CO_3	aq	−698.7	−623.4	H_3O^+	aq	−285.8	−237.2
CH_4	g	−74.9	−50.6	H_2O_2	aq	−191.2	−131.8
C_2H_6	g	−84.5	−33.1	I_2	aq		16.3
C_3H_8	g	−103.8	−23.4	HI (I^-)	aq	−56.1	−51.9
C_4H_{10} (n)[6]	g	−124.7	−15.9	I_3^-	aq	−55.9	−51.5
C_5H_{12} (n)[6]	g	−146.4	−8.4	Mg^{2+}	aq	−461.9	−456.1
C_6H_6 (benzene)	g	82.8	129.7	$Mg(OH)_2$	c	−924.7	−833.9
C_6H_{14} (n)[6]	g	−167.4	0.4	Mn^{2+}	aq	−218.8	−223.4
C_8H_{10}				MnO_4^-	aq		−449.4
(ethylbenzene)	g	29.7	130.5	MnO_4^{2-}	aq		−503.8
C_8H_{18} (n)[6]	g	−208.4	17.2	$Mn(OH)_2$	ppt		−614.6
$C_{10}H_{22}$ (n)[6]	g	−249.8	34.3	$MnCO_3$	ppt		−813.0
CH_3OH	g	−201.3	−161.9	NH_3	aq	−80.8	−26.8
	lq	−238.5	−166.1	NH_4^+	aq	−132.6	−79.5
CH_3Cl	g	−82.0	−58.6	NO_2	g	33.9	51.9
CH_2Cl_2	g	−87.9	−58.6	HNO_3 (NO_3^-)	aq	−206.7	−110.5
	lq	−117.2	−63.2	O_2	g	0.0	0.0
$CHCl_3$	g	−100.4	−66.9	O_2	aq	—	16.5
	lq	−131.8	−71.5	O_3	g	142.3	163.6
CN^-	aq	151.0	165.7	OH^-	aq	−230.1	−157.3
CNO^-	aq	−140.2	−98.7	Rb^+	aq	−248.5	−283.0
HCN	aq	105.4	112.1	Rn	g	0.0	0.0
HCNO	aq	−146.9	−120.9	H_2S	aq	−39.3	−27.2
Cl_2	g	0.0	0.0	S^{2-}	aq	32.6	8.6
Cl^-	aq	−167.4	−131.4	HS^-	aq	−17.6	12.6
ClO_2	g	103.3	123.4	Sn^+	aq	−10.0	−24.3
	aq		11.3				
HClO	aq	−118.0	−79.9				

Source: Numerical values calculated from values given in National Bureau of Standards (1952). See Appendix A.5.2 for additional values.

[a]aq, Aqueous; g, gaseous; c, crystalline; lq, liquid; ppt, precipitate. [b](n), n-alkane.

moles (one) of a substance from its elements at a *standard-state condition* (superscript °). The need for a standard-state reference condition for various characteristic thermodynamic functions is inherent in any practical application of the foregoing energy concepts. Before examining the reference and actual states of systems further, however, we consider several of the functions cited earlier which relate the chemical mass or concentration of a species in a system to its thermodynamic properties and behavior.

Example 5-5 reinforces the practical use of thermodynamic calculations to interpret observed environmental processes as well as to indicate how the thermodynamic conditions required to accomplish a desired, or engineered, transformation can be determined.

Example 5-5

- **Situation:** Energy is either released or absorbed even in simple dissolution or acid-base reactions of the type we considered in Example 5-3. This energy frequently takes the form of heat energy, or enthalpy. Such heat releases or absorptions may be important considerations in certain types of natural systems and engineered treatment systems. Consider, for example, the neutralization of a strongly acidic industrial waste containing hydrogen cyanide by a strong base. Cyanide (CN^-) and its conjugate acid (HCN) are common constituents of a number of different types of industrial wastes, such as metal processing and plating wastes.

- **Question(s):** How much heat is either liberated or absorbed in the protolysis of 1 mol of HCN in water? Measurements in the laboratory show that the heat liberated on neutralization of HCN by NaOH under standard-state conditions is 12.1 kJ/mol.

- **Logic and Answer(s):** The most fundamental neutralization reaction for aqueous systems is that involving protolysis of the hydronium ion by the hydroxide ion to yield water:

$$H_3O^+ + OH^- \rightleftharpoons 2H_2O$$

1. Reference-state enthalpy values for the components of this reaction, taken from Table 5-6 and Appendix A.5.2, are as follows:

$$H_2O \text{ and } H_3O^+: \quad \Delta H_f^\circ = -285.8 \text{ kJ/mol}$$

$$(OH^-) \quad \Delta H_f^\circ = -230.1 \text{ kJ/mol}$$

Thus, because enthalpy is an extensive (additive) property of a system or reaction, the hydronium ion protolysis reaction has an enthalpy value of

$$\Delta H_r^\circ = 2(-285.8) - (-285.8) - (-230.1) = -55.7 \text{ kJ}$$

2. The neutralization of HCN by NaOH in aqueous solutions may be thought of as the sum of two processes, protolysis of HCN and neutral-

ization of H_3O^+ with OH^-. NaOH is a strong base, which dissociates completely when added to water. Thus, the amount of OH^- added by addition of NaOH is equivalent on a molar basis. The two reactions and associated heat releases or absorptions may then be summed:

$$HCN + H_2O \rightleftharpoons H_3O^+ + CN^-: \quad \Delta H^\circ_{r,1}$$

$$H_3O^+ + OH^- \rightleftharpoons 2H_2O: \quad \Delta H^\circ_{r,2} = -55.7 \text{ kJ}$$

$$HCN + OH^- \rightleftharpoons H_2O + CN^-: \quad \Delta H^\circ_{r,T} = [\Delta H^\circ_{r,1} + (55.7)] \text{ kJ}$$

3. The enthalpy, $\Delta H^\circ_{r,T}$, for the overall reaction has been measured experimentally as -12.1 kJ (negative because heat was observed to be liberated on neutralization). Thus, applying the principle of additivity to the enthalpy relationships for the individual reactions, the enthalpy for deprotonation of HCN in water under standard-state conditions is

$$\Delta H^\circ_{r,1} = \Delta H^\circ_{r,T} - (-55.7) = (-12.1) - (-55.7) = 43.6 \text{ kJ/mol}$$

4. Because this value is positive, the protolysis reaction is endothermic, and thus enhanced by increased temperature.

5.3.3 Chemical Potential, Activity, Concentration, and Fugacity

The Gibbs free energy, G, is one of the most useful of the characteristic thermodynamic functions. This parameter provides a convenient means for assessing how far, for a given set of conditions, a chemical reaction is from its equilibrium state when its reactants and products are in their standard states. This question commonly confronts environmental engineers and scientists in attempts either to interpret the observed behavior of natural systems or to achieve specific process behavior in engineered systems.

Equation 5-32 for the free-energy change of a system reduces at constant temperature to

$$dG = V\,dP \quad (\text{at } dT = 0) = [V\,dP]_T \tag{5-34}$$

The ideal gas law states that

$$PV = n\mathcal{R}T \tag{5-35}$$

where n is the number of moles of gas, $\mathcal{R}$ the universal gas constant ($\mathcal{R}$ = 8.314 J/mol-K), V the volume of the system, and n/V the concentration of the gas. Substituting Equation 5-35 into 5-34 and rearranging terms yields

$$\left[\frac{dG}{n}\right]_T = d\mu = \frac{\mathcal{R}T}{P}\,dP = \mathcal{R}T\,d\ln P \tag{5-36}$$

where dG/n is the molar free energy, or so-called *chemical potential*, μ, of the pure ideal gas. The pressure, P, is thus an intensive measure of the free energy associated with each mole of ideal gas in a system, or the thermodynamic intensity of the gas. P can be thought of as the tendency of each mole of gas to escape from the volume in which it is contained. The standard-state concept provides a frame of reference for quantifying this intensity as a system changes from one state to another. The standard state for $[dG/n]_T$ in terms of P is defined as

$$\left[\frac{dG^\circ}{n}\right]_T = d\mu^\circ = \mathcal{R}T\, d \ln P^\circ \tag{5-37}$$

The standard-state condition is usually specified for gases as the perfect gas at 1 atm pressure and some specific temperature, T (frequently, 25°C or 298 K).

Equations 5-36 and 5-37 were developed by incorporation of the ideal gas law in the expression for a free-energy change under isothermal conditions, and thus apply only to gases. Different substances have different physical conditions in different states, as illustrated schematically in Figure 5-10 for water. At a pressure of 1 atm, this substance exists variously as a solid ($T \leq 0°C$), a liquid ($0°C < T < 100°C$), and a vapor ($T \geq 100°C$). The relationships given in Equations 5-36 and 5-37 are not particularly useful for such a substance at normal environmental conditions. The form and utility of these expressions may be enhanced by introducing a more general parameter for characterizing the energetic intensity of a substance regardless of the phase in which it may exist. This parameter is termed *fugacity*. The fugacity, f, of an *ideal pure gas* is by definition equal to the pressure, P, of the gas, and has a standard state value of $f^\circ = 1$ (atm).

From Equation 5-32 we note that

$$\left(\frac{dG}{dP}\right)_T = V \tag{5-38}$$

Because $d\mu$ is an expression of dG on a molar basis (Equation 5-36), we may write

$$\left(\frac{d\mu}{dP}\right)_T = V_{mo} \tag{5-39}$$

where $V_{mo} = V/n$ is the *molar volume* of the gas. Combining Equations 5-36 and 5-39 for $f = P$ yields

$$\left(\frac{d \ln f}{dP}\right)_T = \frac{V_{mo}}{\mathcal{R}T} \tag{5-40}$$

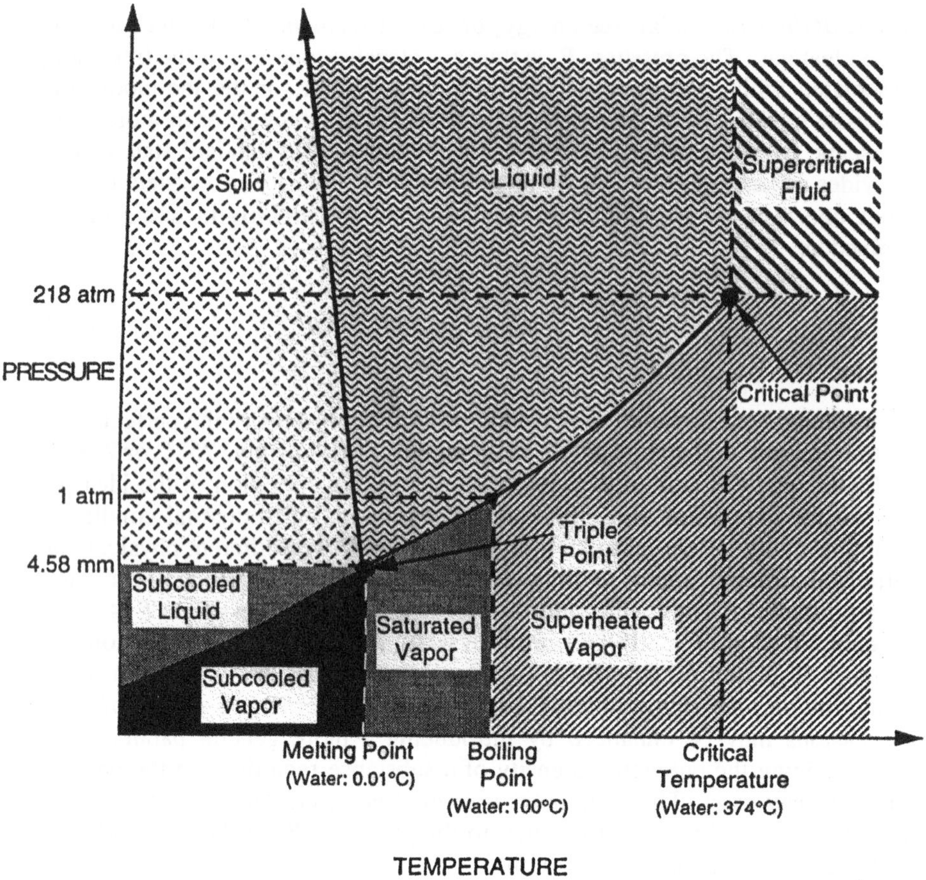

Figure 5-10 Phase diagram for a pure chemical substance with temperatures and pressures illustrated for water.

Integration of Equation 5-40 allows us to determine the fugacity of a gas at any pressure if it is known at some other pressure and if V_{mo} is known as a function of P.

Fugacity is a measure of chemical intensity, which can be related to the chemical potential of a substance regardless of the phase in which it exists. This parameter can be used to characterize not only changes in the state(s) of species within one phase but between phases as well. For example, for a process in which the pressure on a gas is increased isothermally from P_1 to a value, P_2, sufficient to liquefy the gas, there is a corresponding finite (Δ) change in fugacity from f_1 to f_2 such that

$$\Delta\mu = \frac{\Delta G}{n} = \frac{1}{n}[G(T_1, P_2) - G(T_1, P_1)] = \Re T \ln \frac{f_2}{f_1} \qquad (5\text{-}41)$$

This property of fugacity makes it useful for application to thermodynamic analyses of systems comprised by more than one phase and characterized by the movement of species among phases, both of which are common characteristics of environmental systems.

We have defined the relationship between fugacity and its equivalent function, P, for an ideal *pure* gas. If the gas of interest exists at pressure P_i in a *mixture* of other gases having a total pressure P_T, its fugacity is given by its fractional molar contribution to that mixture (i.e., by its partial pressure, $p_i = P_i/P_T$ such that, from Equation 5-36,

$$d\mu_i = \mathcal{R}T\, d \ln p_i = \mathcal{R}T\, d \ln f_i \tag{5-42}$$

We can similarly demonstrate relationships between fugacity and equivalent functions for pure solids or liquids as well as for mixtures of species in different phases. These relationships become more obvious if we expand on the concept of fugacity, a term that derives from the Latin verb *fugere* (*to flee*), as a measure of the tendency of a substance to "escape" or "flee" from the gas state, and thus to be quantified by the pressure, P. Equation 5-41 describes the change in chemical potential associated with each change in the state in which a substance exists. This is true not only for a pure substance changing from one phase to another but also for the movement of a component of a mixture of one phase to a mixture in a different phase.

For two systems to be at equilibrium with each other with respect to a given component, the difference in the chemical potential of that component between the two systems must be zero. Here the distinction between equality in chemical potentials and in concentrations becomes very important. As the concentrations of a chemical substance in two or more contiguous phases approach equilibrium with each other, they do so in response to the thermodynamic requirement that their chemical potentials, not their concentrations, be equal in those two phases. It is awkward, however, to work directly with chemical potentials when a system consists of two or more phases. This is because chemical potentials are often defined differently, and their standard states referenced differently, for different phases. To express chemical potential in other than the gas phase, relationships analogous to that given by Equation 5-36 are required. The analog of the gas-phase relationship for component i in an aqueous solution is given by

$$\left[\frac{dG_i}{n_i}\right]_T = d\mu_i = \mathcal{R}T\, d \ln a_i = \mathcal{R}T\, d \ln (\alpha_i C_i) \tag{5-43}$$

where a_i, the measure of thermodynamic intensity referred to as *chemical activity*, is related to chemical concentration, C_i, by an *activity coefficient*, α_i.

Chemical concentration is an expression of the total mass of component present in a unit volume of solution. The activity coefficient essentially adjusts concentration to account for the fact that the net thermodynamic intensity or reactivity of a substance is somewhat less than it might be if all of its chemical mass was able to participate fully in a particular reaction. Activity coefficients that are less than unity may be encountered in mixtures of different chemical components. Nontransformation interactions (electrostatic, dipole–dipole, etc.) among individual atoms, ions, or molecules of a substance and between those of other substances can attenuate the chemical potential of each component. The extent of this attenuation depends on the specific conditions of the system, including the concentration of the component of interest. Thus the activity coefficient, α_i, of a substance in a mixture is generally less than unity. The fewer the potential nontransformation interactions, the more closely will α_i approach unity. For *dilute* aqueous systems, it is common to assume that $\alpha_i = 1$ and to use the terms *activity* and *concentration* interchangeably in thermodynamic calculations.

In water and wastewater treatment systems we typically deal with chemical mole fraction concentrations on the order of 10^{-9} to 10^{-3}, and in any given situation, with changes of one or two orders of magnitude in these concentrations. The chemical potential of a gas approaches an infinite negative value as its pressure approaches zero (see Equation 5-37). This is even more problematic with respect to chemical potentials in solutions. Additionally, because of the logarithmic relationships of chemical potential to pressure and concentration, even large changes in P_i and C_i are not reflected by significant changes in μ_i. For these reasons chemical potential is not particularly useful for characterizing the chemical energetics of environmental systems.

Vapor pressure, P^v, is an intuitive measure of the escaping tendency of a substance and might very well be usable as a quantitative alternative to chemical potential if every vapor behaved as a perfect gas. Every substance has a finite vapor pressure at a finite temperature. Thus, every substance eventually vaporizes if the pressure on it is decreased sufficiently. If the pressure is decreased beyond this point, the vapor of the substance will eventually approach the condition and behavior of an ideal gas. We might think of fugacity as an "ideal" or corrected vapor pressure for a substance when its vapor does in fact behave like a perfect or ideal gas. This leads logically to a means for extending the concept of fugacity from the gas-phase definition it was assigned in Equations 5-40 through 5-42 to other phases.

5.3.4 Fugacity of Liquids and Solids

The chemical potentials, and thus fugacities, of a substance existing in two or more phases in contact with one another are equal at the condition of equilibrium. A pure solid or liquid phase in equilibrium with its vapor thus has a fugacity that is determined by the vapor pressure of the pure substance. We can assume that the fugacity in both phases is equal to the vapor pressure of the pure solid or liquid, provided that the vapor obeys the ideal gas law within the limits of our ability to make measurements or our willingness to accept a close approximation. Vapor pressure values for common pure substances of environmental interest are readily available in standard reference books and handbooks. For most environmental applications it is acceptable to assume ideal gas behavior and estimate fugacity directly as vapor pressure. For the few cases where this may not be acceptable, corrections for real gas behavior must be made.

Because fugacity is defined with reference to the gaseous state at low pressure, it is the molar weight of a substance in that particular state which should be used in all fugacity calculations and relationships. Gaseous nitrogen dioxide, for example, exists at low pressures as NO_2, but its liquid form is comprised largely by dimer molecules, N_2O_4. The appropriate molecular formula and related molar volume to use in fugacity calculations involving nitrogen dioxide is that of the NO_2 gas. Similarly, chlorine gas exists at ordinary temperatures and pressures as a Cl_2 molecule, and this should be the normal formula employed for fugacity calculations. At very high temperatures and at very low pressures, however, the chlorine molecule dissociates into two chlorine atoms. In certain cases it may therefore be necessary to use the atomic weight and volume as a basis for fugacity determinations. These considerations are particularly important when making fugacity calculations for the components of multiphase systems.

5.3.5 Variations of Fugacity with Temperature

We can compare the chemical potential of a substance in a given state to its chemical potential in the vapor state to gain insight into the effects of temperature on fugacity. Consider, for example, the increase in free energy that occurs when 1 mol of a pure liquid substance is converted into its vapor state:

$$(\mu)_v - (\mu)_l = \mathcal{R}T \ln \frac{f_v}{f_l} \tag{5-44}$$

where the subscripts v and l refer to the vapor and liquid states, respectively. Differentiating Equation 5-44 with respect to temperature at constant pressure yields

$$\left(\frac{d(\mu)_v}{dT}\right)_P - \left(\frac{d(\mu)_l}{dT}\right)_P = \mathcal{R} \ln \frac{f_v}{f_l} + \mathcal{R}T\left(\frac{d \ln f_v}{dT}\right)_P - \mathcal{R}T\left(\frac{d \ln f_l}{dT}\right)_P \tag{5-45}$$

The fugacity of a gas at low pressure is equal to its pressure, so the middle term on the right-hand side of Equation 5-45 has a numerical value of zero. If Equation 5-44 is then substituted in Equation 5-45, we obtain

$$\left(\frac{d(\mu)_v}{dT}\right)_P - \left(\frac{d(\mu)_l}{dT}\right)_P = \frac{(\mu)_v - (\mu)_l}{T} - \mathcal{R}T\left(\frac{d \ln f_l}{dT}\right)_P \tag{5-46}$$

From Equations 5-28 and 5-32 we determine that

$$\left(\frac{\partial G}{\partial T}\right)_P = \frac{G - H}{T} \tag{5-47}$$

Applying this relationship on a molar basis to each of the two terms on the left-hand side of Equation 5-46 and simplifying gives

$$\left(\frac{d \ln f}{dT}\right)_P = \frac{H_v - H_l}{n\mathcal{R}T^2} \tag{5-48}$$

The quantity $(H_v - H_l)/n$ is termed the ideal *molar heat of vaporization* (i.e., the increase in heat content when a mole of a substance escapes into a vacuum).

5.3.6 Raoult's Law, Henry's Law and Fugacity

A number of interrelated parameters predicated on the properties of pure solutes have been developed to describe the behavior of those solutes in different fluids and fluid mixtures of other components. The behavior of many fluids may be closely approximated using the concept of ideal solutions. A gaseous mixture is said to form an ideal solution when the fugacity, f_i, of each component of the mixture is related to its fugacity, f_i°, in the pure state by

$$f_i = f_i^\circ X_{i,v} \tag{5-49}$$

where $X_{i,v}$ is its mole fraction in the vapor-phase mixture. We can obtain an expression for partial molar free energy or chemical potential, μ_i, of species i by integrating Equation 5-42 between any given state and a chosen standard state to give

$$\mu_i = \mu_i^\circ + \mathcal{R}T \ln \frac{f_i}{f_i^\circ} \tag{5-50}$$

Here again the zero superscripts designate the standard state, which by definition is $f_i^\circ = 1$ atm for an ideal gas at $P_i = P_T = 1$ atm (i.e., for a pure gas). The fugacity ratio (f_i/f°) in Equation 5-50 for an ideal gas in a mixture is therefore numerically equal to its fugacity in the given state, which in turn is equal to its activity (a_i) and to its dimensionless partial pressure $(P_i/P_T = p_i)$ in that state. Thus, for an ideal gas,

$$\frac{f_i}{f_i^\circ} = a_i = X_{i,v} = p_i \tag{5-51}$$

For pure solids and liquids, fugacities can be closely approximated by corresponding vapor pressures. For mixtures, we can closely approximate f_i/f_i° by the *partial vapor pressure*, p_i^v:

$$\frac{f_i}{f_i^\circ} = a_i \approx \frac{P_i}{P_i^v} = p_i^v \tag{5-52}$$

For substances dissolved in water, particularly ions and other relatively nonvolatile materials, it is common to state chemical potential and related equilibrium criteria in terms of chemical activity, a_i. This dimensionless variable is in turn given by the product of the activity coefficient, α_i, and the mole fraction concentration of component i in the liquid phase, $X_{i,l}$. The fugacity of substances dissolved in water is thus given by:

$$f_i = f_i^\circ a_i = f_i^\circ \; \alpha_i \; X_{i,l} \approx \alpha_i X_{i,l} P_i^2 \tag{5-53}$$

We have shown that an *ideal gas* mixture or *ideal solution mixture* is one in which the fugacity of each component is directly proportional to its mole fraction (Equation 5-49) or to its activity (Equation 5-53) in that gas or solution phase. This concept is extendable to other fluids as well. Raoult in 1886 articulated the proportionality between fugacity and mole fraction from observations regarding the vapor pressures of pure solutions. *Raoult's law* describes the behavior of pure solutions well and extends reasonably to the principal components of highly concentrated mixtures. Real solutions, however, seldom follow Raoult's law over extended ranges of concentration. Deviations result largely because the nature of interactions among components changes with substantial changes in their relative concentrations.

Dilute solutions provide a case of special interest in water quality. By dilute we mean that the concentration of the solute is so low that each of its molecules is completely surrounded by molecules of the solvent. In this case the minor component is present in a totally uniform environment even though it may form solutions with the major component that are far from ideal at higher concentrations. For such very dilute solutions, the fugacity of the minor component is still proportional to its mole fraction in the solution. However, the proportionality is no longer given by the fugacity, f_i°, in the pure state but by a different proportionality constant we shall refer to as $\mathcal{K}_H$, such that

$$f_i = \mathcal{K}_{H,i} X_{i,l} \tag{5-54}$$

Equations of similar form were referred to in Chapter 1 as expressions of *Henry's law*, and the coefficients of proportionality, $\mathcal{K}_H$, as Henry's constants (see Equations 1-1 and 1-2 as well as Example 1-4). This relationship was established and tested extensively by William Henry in 1803 in a series of experiments designed to study the dependence of gas solubilities in liquids on their partial pressures in overlying gas phases. Although it is most commonly applied to gas-water systems, Henry's law is not restricted to aqueous solutions or indeed to gas–liquid equilibria. It is in fact adhered to by a variety of dilute solutions and by virtually all solutions in the limit of extreme dilution (i.e., solutions approaching "infinite" dilution).

The discussion of dilute and concentrated solutions above leads to two different definitions of standard states that are in common use; these are illustrated in Figure 5-11. With increasing dilution, a solute dissolved in a particular solvent always approaches the ideal behavior specified by Henry's law, and the solvent always approaches the ideal behavior specified by Raoult's law. Henry's law can be thought of as a deviation from Raoult's law. Each molecule of solute in the solvent is surrounded by solvent molecules, and its ease of escape thus becomes dependent upon its affinity for the molecules of solvent rather than the affinity of the solute molecules for each other. The behaviors of modestly concentrated solutions fall somewhere between the behavior predicted by each of the two laws, as illustrated in Figure 5-11. The most con-

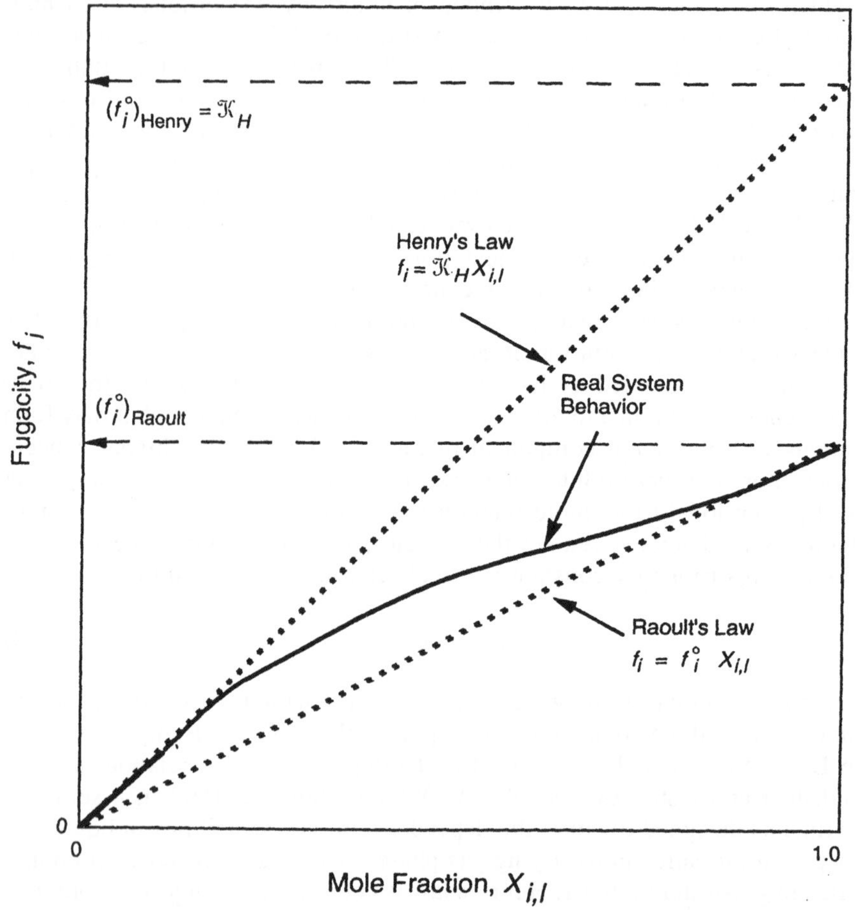

Figure 5-11 Standard states and solution behaviors for Henry's and Raoult's laws.

venient of the two standard-state references can be selected for any component in a particular system as long as that definition is used consistently in its application to that system.

The parallels among pressure, partial pressure, vapor pressure, chemical activity, and fugacity are especially noteworthy. They come about as a result of the similar significance that each of these parameters has for the molar free energy of a substance in different states and phases. This parallelism can be used to great advantage in establishing interrelationships among chemical reactivities in different phases, as we will see in our discussions of heterogeneous systems in Chapter 6.

5.3.7 Standard-State Free Energy of Reaction

Free energy is an extensive, and thus additive, property of a system. When two or more chemical species interact to form a product or products, the free energies of the reactants and products may be respectively summed, and the difference between these values represented as the associated free-energy change of the reaction, dG_r. The *standard-state free-energy change of reaction*, dG_r°, is the free-energy change calculated by subtracting the sum of the free energies of the reactants in their respective standard states at an arbitrary temperature T from the sum of the free energies of the products in their respective standard states at the same temperature. As noted previously, standard states may be selected arbitrarily. The common definition of standard state is the perfect gas state at 1 atm pressure or the perfect solution state at unit activity and 25°C (298 K) for all elements and compounds except carbon, which is referenced as graphite at the same pressure and temperature.

Consider the transformation of species A according to a reaction involving a moles of substance A and b moles of substance B:

$$aA + bB \rightleftharpoons cC + dD \tag{5-55}$$

The stoichiometric coefficients ($\gamma_i = a, b, c,$ and d) illustrate the extensive properties of reaction energies. The standard free energy of reaction for the reaction given in Equation 5-55 can be written in general terms as

$$dG_{r,T}^\circ = \sum \mu_i^\circ \, dn_i = \sum \left(\frac{dG_i^\circ}{n_i}\right) dn_i = \sum \left(\frac{dG_i^\circ}{n_i}\right) \gamma_i \tag{5-56}$$

where dG_i°/n_i is the standard-state free energy of species i per mol and $(dG_i^\circ/n_i)\gamma_i$ is the standard-state free energy of that species *associated with a particular reaction*. Before this equation can be applied, we need to specify the standard free energies of the reactants and products. *Elements* are defined as being in the reference state and arbitrarily assigned standard-state enthalpy and free-energy values of zero. An element exists either as an atom or as a simple molecule (e.g., O_2, N_2, H_2, etc.). Most environmental processes, on the other hand, involve reactions among *nonelemental* species. We must therefore determine free energies associated with *reactions involving the formation* of nonelemental species from their elemental components before we can address the free energies of *environmentally significant reactions*. Moreover, we need to normalize these free energies in a way that will permit their use in *any given* reaction.

5.3.8 Standard-State Free Energy of Formation

The standard-state free energy of formation for a chemical species is defined as that amount of energy involved when the species is formed from its elements under standard-state conditions. Consider, for example, the formation of a compound A_aB_b at *its* standard state by reaction of *its elements* A and B at *their respective* standard states:

$$aA + bB \rightleftharpoons A_aB_b \tag{5-57}$$

The standard free energy of this *formation reaction* is

$$dG^\circ_{r,T} = dG^\circ_{f,T} = \left\{ dG^\circ_{A_aB_b} - a\left(\frac{dG^\circ_A}{n_A}\right) - b\left(\frac{dG^\circ_B}{n_B}\right)\right\}_T \tag{5-58}$$

Because $dG^\circ_A = dG^\circ_B = 0$ by definition (convention), it follows that the energy associated with formation of A_aB_b is given by

$$dG^\circ_{A_aB_b,T} = dG^\circ_{r,T} = dG^\circ_{f,T} \tag{5-59}$$

Further, if $dG^\circ_{f,T}$ is defined in terms of the formation of 1 mol of a substance, it can be expressed as the standard state *molar* free energy of formation, $dG^\circ_{f,T}/n_i$ or $\Delta G^\circ_{f,T}$. This is apparent in the respective definitions of $dG^\circ_{f,T}$ as a general derivative and $\Delta G^\circ_{f,T}$ as a *finite* difference in energy between two specific conditions.

The general approach to determination of the standard free energy of *reaction* for a transformation process involving other than elemental species is then to employ the *standard molar free energies of formation* of the participating species, $\Delta G^\circ_{f,i}$. When used to describe the standard free energy for occurrence of the transformation reaction shown in Equation 5-55 from one specific system condition or state to another, Equation 5-56 can thus be rewritten as

$$\Delta G^\circ_{r,T} = \{c(\Delta G^\circ_{f,C}) + d(\Delta G^\circ_{f,D}) - a(\Delta G^\circ_{f,A}) - b(\Delta G^\circ_{f,B})\}_T \tag{5-60}$$

The aforegoing developments can be used similarly to establish a *standard molar enthalpy of formation*.

5.3.9 Applications of Free-Energy Concepts

To restate for emphasis, the free energy of a reaction is the energy available for that reaction at any particular state of a system, and its minimum value (zero) is reached when the system is in its equilibrium state. We have seen from Equation 5-28 that free energy can be defined in most simple terms for a single chemical substance as the difference between the enthalpic and entropic state functions for that substance. The Gibbs free energy, G, is thus a measure of net work obtainable, or that portion of the heat content or enthalpy, H, which is isothermally available to do work.

To describe a reaction or transformation process the chemical potential term must be added to the basic Gibbs equation. Thus, for a closed system undergoing reaction, the differential change in free energy is given by

$$dG = V\,dP - S\,dT + \Sigma\,\mu_i\,dn_i \tag{5-61}$$

The change in the free energy of a *system* undergoing reaction is thus the sum of a heat function ($S\,dT$), a work function ($V\,dP$), and a chemical function ($\Sigma\,\mu_i\,dn_i$), each of these being the product of a characteristic intensive variable and a related extensive variable.

By convention, the free-energy change associated with a specific *reaction* is the difference between the free-energy changes of its products and its reactants:

$$dG_r = (dG)_{\text{products}} - (dG)_{\text{reactants}} \tag{5-62}$$

For a reaction in a state of equilibrium, dG_r must be zero. In other words, the energy available to drive the reaction to the right $[(dG)_{\text{reactants}}]$ is precisely equal to the energy available to drive it to the left $[(dG)_{\text{products}}]$. If dG_r has a negative value, $(dG)_{\text{reactants}} > (dG)_{\text{products}}$ and the reaction is driven to the right (forward direction) with a net accumulation of product and a net disappearance of reactants. This type of reaction is said to be thermodynamically *spontaneous*. Conversely, if dG_r has a positive value, then $(dG)_{\text{reactants}} < (dG)_{\text{products}}$ and the reaction is driven to the left (reverse direction) with a net disappearance of product and a net accumulation of reactants. Figure 5-12 is a schematic representation of the variation of free energy with extent of reaction, where extent of reaction represents the fraction of a given reactant converted to a corresponding product.

Values for the standard-state molar free energy, ΔG_f°, associated with the formation of a substance from its elements for a selected group of ionic and molecular species of interest, along with corresponding values for the enthalpy of formation, ΔH_f°, were presented in Table 5-6. Additional values are given in Appendix A.5.2. The relative fugacity, f_i/f_i°, of a substance in the aqueous phase is termed its activity, a_i. For the standard state, $a_i^\circ = 1$, and for any other state a_i can be related to chemical potential by modifying Equation 5-50 as follows:

$$\mu_i = \mu_i^\circ + \mathcal{R}T \ln a_i \tag{5-63}$$

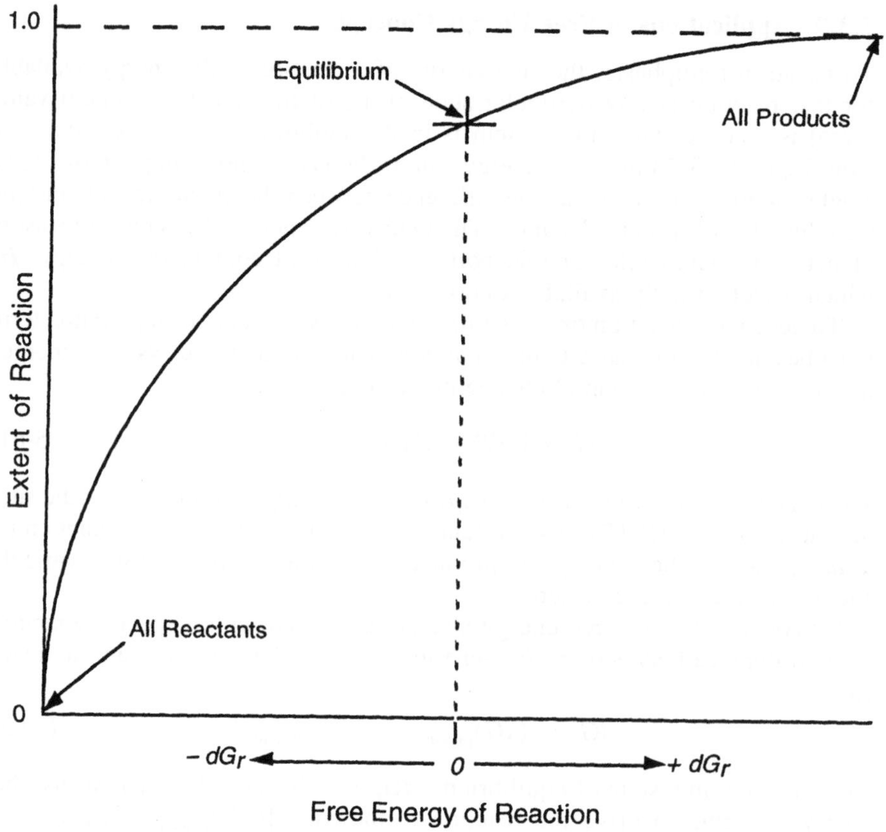

Figure 5-12 Variation of free energy of reaction with extent of reaction.

or

$$a_i = \exp\left(\frac{\mu_i - \mu_i^\circ}{\mathcal{R}T}\right) \tag{5-64}$$

where all terms are as defined previously.

An illustration of how standard-state reference values for the free energies of formation of different species can be used in an assessment of their inter-reaction tendencies is given in Example 5-6. This is a useful approach to a "first-cut" evaluation of process feasibility.

Example 5-6

- **Situation:** Consideration is being given to the potential use of iodine (as I_3^-) to provide continuous disinfection of a public swimming pool. Unlike chlorine, iodine does not react with ammonia to cause eye-irritating trihaloamines. For this reason it is sometimes preferred to chlorine, although it is more expensive. One major problem with this disinfectant is that it cannot be obtained easily as either a pure gas (as can Cl_2) or a solid (as can $Ca(OCl)_2$), and it degenerates with time in aqueous solutions. It therefore cannot be stored for significant periods of time.

- **Question(s):** What is the feasibility of oxidizing iodide (I^-) to iodine by hydrogen peroxide (H_2O_2) in slightly acidic aqueous solution to provide I_3^- on a continuous basis? Both I_3^- and H_2O_2 are effective disinfectants. I_3^- provides residual protection and is thus effective for a longer period than H_2O_2.
- **Logic and Answer(s):** The reaction corresponding to the process proposed is

$$3I^- + H_2O_2 + 2H_3O^+ \rightleftharpoons I_3^- + 4H_2O$$

1. In this analysis of feasibility we will consider the reaction only from a thermodynamic point of view, to illustrate the application of free energy of formation values.

2. From Equation 5-62, and considering for the present only standard-state conditions, $\Delta G_r^\circ = \Sigma\, \Delta G_{f,\text{products}}^\circ - \Sigma\, \Delta G_{f,\text{reactants}}^\circ$

3. From Table 5-6:

Reaction Component	ΔG_f° (kJ/mol)
I^-	−51.9
H_2O_2	−131.8
H_3O^+	−237.2
I_3^-	−51.5
H_2O	−237.2

4. Then

$$\Delta G_r^\circ = [(-51.5 + 4(-237.2)] - [3(-51.9) + (-131.8) + 2(-237.2)]$$

$$= [-1000.4] - [-761.9] = -238.5 \text{ kJ}$$

5. The thermodynamics of the proposed process are thus favorable for it to proceed as written, at least for conditions of standard state. We will shortly see how to extend the analysis to non-standard-state conditions.

The free-energy concepts discussed above apply to the transformation of a chemical species to some other form, regardless of the means by which the transformation occurs. A particular organic compound, for example, may be oxidized chemically at room temperature with oxygen to form carbon dioxide, combusted to CO_2 in an oxygen atmosphere in a furnace, or metabolized by aerobic microorganisms to yield CO_2. The reaction rates are markedly different in these several cases, but the total amount of energy released by the reaction of the compound with oxygen will be exactly the same regardless of the means by which the reaction is facilitated.

Table 5-7 presents a summary of selected biologically mediated chemical reactions and associated free-energy values for dilute aqueous systems at pH 7. These reactions illustrate the point that the fundamental energetics of all reactions are determined by chemical thermodynamics. The reactions are stoichiometrically normalized to a one-electron transfer basis; thus the values

Table 5-7 Biologically Mediated Half-Reactions and Associated Aqueous-Phase Free Energies of Reaction

Half-Reaction[a]	$\Delta G_r(aq)$[b] (kJ/eq)
Organic (Electron Donor) Oxidation	
Carbohydrate (cellulose, starch, sugars) $\frac{1}{4}CH_2O + \frac{1}{4}H_2O \rightleftharpoons \frac{1}{4}CO_2 + H^+ + e^-$	−41.8
Methanol $\frac{1}{6}CH_3OH + \frac{1}{6}H_2O \rightleftharpoons \frac{1}{6}CO_2 + H^+ + e^-$	−37.7
Protein (amino acids, proteins, nitrogenous organics) $\frac{1}{66}C_{16}H_{24}O_5N_4 + \frac{27}{66}H_2O \rightleftharpoons \frac{8}{33}CO_2 + \frac{2}{33}NH_4^+ + \frac{31}{33}H^+ + e^-$	−32.2
Domestic wastewater $\frac{1}{50}C_{10}H_{19}O_3N + \frac{9}{25}H_2O \rightleftharpoons \frac{9}{50}CO_2 + \frac{1}{50}NH_4^+ + \frac{1}{50}HCO_3^- + H^+ + e^-$	−31.8
Ethanol $\frac{1}{12}CH_3CH_2OH + \frac{1}{4}H_2O \rightleftharpoons \frac{1}{6}CO_2 + H^+ + e^-$	−31.8
Benzoate $\frac{1}{30}C_6H_5COO^- + \frac{13}{20}H_2O \rightleftharpoons \frac{1}{5}CO_2 + \frac{1}{30}HCO_3^- + H^+ - e^-$	−28.9
Acetate $\frac{1}{8}CH_3COO^- + \frac{3}{8}H_2O \rightleftharpoons \frac{1}{8}CO_2 + \frac{1}{8}HCO_3^- + H^+ + e^-$	−27.6
Grease (fats and oils) $\frac{1}{46}C_8H_{16}O + \frac{15}{46}H_2O \rightleftharpoons \frac{4}{23}CO_2 + H^+ + e^-$	−27.6
Inorganic (Electron Acceptor) Reduction	
Oxygen $\frac{1}{4}O_2 + H^+ + e^- \rightleftharpoons \frac{1}{2}H_2O$	−78.2
Nitrate $\frac{1}{5}NO_3^- + \frac{6}{5}H^+ + e^- \rightleftharpoons \frac{1}{10}N_2 + \frac{3}{5}H_2O$	−71.6
Carbon dioxide $\frac{1}{8}CO_2 + H^+ + e^- \rightleftharpoons \frac{1}{8}CH_4 + \frac{1}{4}H_2O$	+24.3
Sulfate $\frac{1}{8}SO_4^{2-} + \frac{19}{16}H^+ + e^- \rightleftharpoons \frac{1}{16}H_2S + \frac{1}{16}HS^- + \frac{1}{2}H_2O$	+21.3

[a]The reactions are written for a one-electron exchange. The term $\Delta G_r(aq)$ is thus expressed as the amount of free energy (kJ) exchanged per electron equivalent rather than per mole reacted. The notation (aq) further defines the value as applicable to neutral aqueous solutions (i.e., pH 7).

[b]Although not stated specifically in the source from which these $\Delta G_r(aq)$ values were calculated (Christensen and McCarty, 1975), a reference given in that source suggests that they are based on a temperature 25°C.

for ΔG_r are given in units of kilocalories per electron equivalent. A "generic" oxidation reaction for an organic compound having a general composition defined by the stoichiometric formation coefficients a, b, c, and d is as follows:

$$C_aH_bO_cN_d + (2a - c)H_2O \rightleftharpoons aCO_2 + dNH_4^+ + (4a + b - 2c - 4d)H^+ + (4a + b - 2c - 3d)e^- \quad (5\text{-}65)$$

Microorganisms mediate such reactions to derive energy from them, but they cannot alter inherent energy balances. The organic reactions given in Table 5-7 are all oxidation (electron donor) half-reactions, and the corresponding standard-state free energies are the amounts of energy associated with each mole of electrons released or "donated" by a half-reaction. Conversely, the reactions for the electron acceptors are written as reduction half-reactions. The overall free energy for a given biological transformation can be obtained simply by summing the appropriate half reactions. The energy derived from aerobic and anaerobic oxidation of a substrate will be different because the electron acceptor half-reactions are not the same. The difference can be substantial, as illustrated in Example 5-7.

Example 5-7

- **Situation:** A caustic waste from a food-processing operation contains simple sugars and starches. The waste is to be treated biologically after neutralization with sulfuric acid.
- **Question(s):** Which type of biological treatment process would be most energetically favorable, an aerobic process or an anaerobic process?
- **Logic and Answer(s):** A logical first approach to this question is to compare the relative amounts of energy to be derived from the biological transformation of a simple carbohydrate under aerobic conditions using oxygen as an electron acceptor, and under anaerobic conditions using sulfate as an electron acceptor.

1. Aerobic conditions from Table 5-7:

Reaction	$\Delta G_r(aq)$ (kJ/eq)
Oxidation half-reaction:	
$\frac{1}{4}CH_2O + \frac{1}{4}H_2O \rightleftharpoons \frac{1}{4}CO_2 + H^+ + e^-$	−41.8
Reduction half-reaction:	
$\frac{1}{4}O_2 + H^+ + e^- \rightleftharpoons \frac{1}{2}H_2O$	−78.2
Overall reaction:	
$\frac{1}{4}CH_2O + \frac{1}{4}O_2 \rightleftharpoons \frac{1}{4}CO_2 + \frac{1}{4}H_2O$	−120.0

2. Anaerobic conditions

Reaction	$\Delta G_r(aq)$ (kJ/eq)
Oxidation half-reaction:	
$\frac{1}{4}CH_2O + \frac{1}{4}H_2O \rightleftharpoons \frac{1}{4}CO_2 + H^+ + e^-$	−41.8
Reduction half-reaction:	
$\frac{1}{8}SO_4^{2-} + \frac{19}{16}H^+ + e^- \rightleftharpoons \frac{1}{16}H_2S + \frac{1}{16}H_2S^- + \frac{1}{2}H_2O$	+21.3
Overall reaction:	
$\frac{1}{4}CH_2O + \frac{1}{8}SO_4^{2-} + \frac{1}{4}H^+ \rightleftharpoons \frac{1}{4}CO_2 + \frac{1}{8}H_2S^- + \frac{1}{4}H_2O$	−20.5

3. It is evident that a greater amount of energy can be derived from oxidation of the organic material using oxygen as an electron acceptor. The analysis further tells us that even in the presence of the sulfate resulting from neutralization of the waste with sulfuric acid, thermodynamics dictate that noxious hydrogen sulfide will not be produced by microbial degradation of organic wastes as long as aerobic conditions prevail.

Table 5-7 and Example 5-7 provide an opportunity to reinforce the caution expressed earlier about overinterpretation of thermodynamic spontaneity. A reaction may indeed be thermodynamically spontaneous but its rate very slow. Only in those cases for which equilibrium is readily obtained, as in elementary electron transfer reactions, does a favorable standard free energy (or electrode potential) indicate that the reaction will in fact occur within a *relevant* period of observation. Most oxidations involving organic reactions occur under non-equilibrium conditions, and in such instances thermodynamic favorability does not guarantee that a reaction will occur within a particular time period.

We will learn in Chapter 7 that the most significant factor governing rates of reactions is *activation energy*. Activation energy can be viewed as an energy barrier to potential reactivity. It is an energy level that must be achieved over and above the initial energy level of a substance in order for a reaction to occur. The greater the energy barrier, the smaller will be the fraction of molecules in a system possessing sufficient "excess" energy, and the slower will be the reaction. This point is illustrated vividly by the coexistence of organic matter and atmospheric oxygen in the face of obviously favorable free energies for its oxidation, as typified by the values given in Table 5-7. Activation energy serves in this instance as a fortuitous natural barrier (more accurately, retardant) to the destruction of organic life, sustaining a metastable organic world.

The activation energy barrier is reaction-path dependent, whereas free energy is not. Thus, if an alternative pathway of lower activation energy can be found, an otherwise slow reaction can be speeded up, perhaps so much as to transform a temporally *metastable* system into a *reactive* system. There are a number of chemical and biochemical substances that can provide alternate pathways, particularly for electron exchange reactions. These substances are termed *catalysts*. Enzymes comprise one of the major classes of chemical catalysts. These materials are the substances that allow microbes to accomplish reactions having favorable free energies but otherwise very slow reaction rates, such as those illustrated in Table 5-7.

5.4 ENERGY INTERRELATIONSHIPS AND REACTIVITY

5.4.1 Thermodynamic Basis of the Equilibrium Constant

Thermodynamic relationships are usually incorporated in some operational parameter to facilitate their application. This parameter is termed an *equilibrium constant*. Many types of equilibrium constants are possible, depending on the fundamental nature of the reaction and on whether it is homogeneous or heterogeneous. Relevant examples are acidity constants, complex formation constants, solubility constants, Henry's constants, partitioning coefficients, and Freundlich isotherm constants. To illustrate how such parameters are developed from thermodynamic considerations, consider the general reaction scheme

$$\gamma_A \mathrm{A} + \gamma_B \mathrm{B} + \gamma_C \mathrm{C} + \cdots \rightleftharpoons \cdots + \gamma_X \mathrm{X} + \gamma_Y \mathrm{Y} + \gamma_Z \mathrm{Z} \tag{5-66}$$

where γ_A represents the number of units of substance A involved in the reaction, γ_B the corresponding number of units of substance B, and so on. The units involved may be molecules, moles, or any other molecular multiple. The parameter γ_i is the *stoichiometric coefficient* for component i. The basis of reaction stoichiometry lies in the fact that the ratios of γ_i for the components of a reaction are definite and fixed.

The stoichiometry represented by Equation 5-66 can be expressed as

$$\sum \gamma_i (W_{g,mo})_i = 0 \tag{5-67}$$

where $(W_{g,mo})_i$ is the molecular weight of component i. By convention, values of γ_i are positive for reaction products and negative for reactants. From Equation 5-61 we note that change in the Gibbs free energy of a system at constant temperature and pressure is given by

$$dG|_{T,P} = \sum_i \mu_i \, dn_i \tag{5-68}$$

and that the condition for thermodynamic *equilibrium* in a closed system is

$$dG_r = \sum_i \mu_i \, dn_i = 0 \tag{5-69}$$

The term dn_i in Equation 5-64 can be replaced by $\gamma_i \, d\nu$, where ν, the degree of advance of the reaction, is the same for all components of the reaction ($d\nu$

$= dn_i/\gamma_i$). For each *finite degree* of advance of a reaction (i.e., for $d\nu = 1$), the change in the number of moles, Δn_i, of species i is then equal to the stoichiometric coefficient, γ_i, for that species. Thus, again for constant temperature and pressure, a finite free-energy change for the reaction given in Equation 5-66 can be written:

$$\Delta G_r = \sum \mu_i \gamma_i \tag{5-70}$$

A similar expression can be written for the standard state:

$$\Delta G_r^\circ = \sum \mu_i^\circ \gamma_i \tag{5-71}$$

The finite free-energy change associated with the progression of a specific reaction from a nonequilibrium or non-standard-state condition to an equilibrium or standard state-condition is given by ΔG_r. This energy can be related to the mass-law equilibrium constant, K, for that reaction at constant temperature and pressure utilizing Equations 5-56, 5-63, and 5-70 as follows:

$$\Delta G_r = \sum \mu_i^\circ \gamma_i + \mathcal{R}T \sum \gamma_i \ln a_i \tag{5-72}$$

$$\Delta G_r = \Delta G_r^\circ + \mathcal{R}T \sum \ln (a_i)^{\gamma_i} \tag{5-73}$$

$$\Delta G_r = \Delta G_r^\circ + \mathcal{R}T \ln \prod (a_i)^{\gamma_i} \tag{5-74}$$

For the general reaction scheme described by Equation 5-66 we can now write

$$\Delta G_r = \Delta G_r^\circ + \mathcal{R}T \ln \frac{(a_X)^{\gamma_X} (a_Y)^{\gamma_Y} (a_Z)^{\gamma_Z}}{(a_A)^{\gamma_A} (a_B)^{\gamma_B} (a_C)^{\gamma_C}} \tag{5-75}$$

The quotient in the last term on the right-hand side of Equation 5-75 is termed the *reaction quotient*, Q_r:

$$Q_r = \frac{(a_X)^{\gamma_X} (a_Y)^{\gamma_Y} (a_Z)^{\gamma_Z}}{(a_A)^{\gamma_A} (a_B)^{\gamma_B} (a_C)^{\gamma_C}} \tag{5-76}$$

As the reaction approaches equilibrium, that is, as $\Delta G_r \rightarrow 0$, the value of the reaction quotient, Q_r approaches a constant value, such that

$$\mathcal{R}T (\ln Q_r)_{\text{equilibrium}} = -\Delta G_r^\circ \tag{5-77}$$

The particular value of Q_r attained at equilibrium is constant for any reaction and a stipulated set of standard conditions. This value for Q_r is referred to as the *equilibrium constant*, K, of the reaction. Thus

$$\Delta G_r^\circ = -\mathcal{R}T \ln K \tag{5-78}$$

and

$$\Delta G_r = \Delta G_r^\circ + \mathcal{R}T \ln Q_r = \mathcal{R}T \ln \frac{Q_r}{K} \tag{5-79}$$

The standard-state free-energy change of a reaction can thus be determined by direct computation using a reaction-specific equation such as that given in Equation 5-60 for the reaction in Equation 5-55 and tabulated standard-state molar free-energy values for the reactants and products. Alternatively, it can be determined from experimentally established values for the equilibrium constant, using Equation 5-78. In the absence of experimental data, the equilibrium constant can be calculated from free-energy values such as those given in Tables 5-6 and 5-7 and Appendix A.5.2. Uses of standard-state free-energy values to calculate equilibrium constants and to assess non-standard-state reaction feasibility are illustrated in Example 5-8.

Example 5-8

- **Situation:** An industrial plant that produces a 26,000-mg/L cyanide waste plans to treat this material by oxidizing it to cyanate by alkaline chlorination (pH > 10.5), and then to create appropriate conditions for the cyanate to undergo acid hydrolysis to yield carbonic acid and ammonium ion. The reactions are to be carried out at the ambient temperature of the waste, 25°C.

- **Question(s):** From the perspective of thermodynamics, determine to what pH the cyanate solution resulting from alkaline chlorination must be lowered to reduce the final cyanate concentration to 0.5 μg/L by hydrolysis.

- **Logic and Answer(s):** Alkaline oxidation of cyanide proceeds by the following reaction:

$$CN^- + OCl^- \rightleftharpoons CNO^- + Cl^-$$

The pH is then reduced to cause acid hydrolysis of the cyanate to the hydronium ion and carbonic acid:

$$CNO^- + 2H_3O^+ \rightleftharpoons NH_4^+ + H_2CO_3$$

1. Convert weight concentrations to molar concentrations. The molecular weight of CN^- is 26 g and that of CNO^- is 42 g (see Appendix A.5.1). Thus 26,000 mg/L of CN^- is 1 mol/L, and <0.5 μg/L of CNO^- is essentially $\sim 10^{-8}$ mol/L.

2. The following free energy of formation values can be obtained from Table 5-6:

Species	ΔG_f° (kJ/mol)
CNO^-	−98.7
H_3O^+	−237.2
NH_4^+	−79.5
H_2CO_3	−623.4

3. From Equation 5-79:

$$\Delta G_r = \Delta G_r^\circ + \mathcal{R}T \ln Q_r$$

$$\Delta G_r^\circ = \Sigma\, \Delta G_{f,\text{products}}^\circ - \Sigma\, \Delta G_{f,\text{reactants}}^\circ$$

$$= [-623.4 + (-79.5)] - [-98.7 + 2(-237.2)]$$

$$= -129.8 \text{ kJ/mol}$$

4. At equilibrium, $\Delta G_r = 0$ and $-\Delta G_r^\circ = -(-129.8) = \mathcal{R}T \ln K = 5.7 \log K$:

$$\log K = \frac{129.8}{5.7} = 22.8 \qquad K = 10^{22.8} = \frac{[NH_4^+]\,[H_2CO_3]}{[CNO^-]\,[H_3O^+]^2}$$

$$[H_3O^+]^2 = \frac{[NH_4^+]\,[H_2CO_3]}{[CNO^-]\,K}$$

5. The final pH will determine the distribution between $[H_2CO_3]$ and $[HCO_3^-]$ (see, for example, Figure 5-1). For this example we assume that the contributions of HCO_3^- to the hydrolysis of CNO^- are the same as those of H_2CO_3.

6. We want $[CNO^-] = 10^{-8}$ mol/L and have $[NH_4^+] = [H_2CO_3] = 1$ mol/L. Thus

$$[H_3O^+]^2 = \frac{1}{(10^{-8})\,(10^{22.8})} = \frac{1}{10^{14.8}}$$

$$[H_3O^+]^2 = 10^{-14.8}$$

$$[H_3O^+] = 10^{-7.4}$$

7. To ensure sufficient hydrolysis, the pH of the cyanate solution must therefore be reduced to pH 7.4.

For oxidation-reduction reactions the free energy and the equilibrium constant can be related to a corresponding parameter commonly associated with electron transfer processes, the standard-state *electrode potential*, E_H°:

$$\mathcal{R}T \ln K = n_e \mathcal{F} E_H^\circ = -\Delta G_r^\circ \tag{5-80}$$

where n_e is the number of electrochemical equivalents/mol of reactant, and $\mathfrak{F}$ is the Faraday constant (96,485 C/mol; 96.485 kJ or 23.060 kcal per electron-volt). Substitution of the relationships given in Equation 5-80 into the Gibbs free-energy relationship given in Equation 5-75 yields the *Nernst equation*:

$$E_H = E_H^\circ + \frac{\mathcal{R}T}{n_e \mathfrak{F}} \ln \frac{(a_X)^{\gamma_X} (a_Y)^{\gamma_Y} (a_Z)^{\gamma_Z}}{(a_A)^{\gamma_A} (a_B)^{\gamma_B} (a_C)^{\gamma_C}} \tag{5-81}$$

Values of E_H° for redox reactions are readily available in chemical and chemical engineering reference books and handbooks. Some values for iron were given earlier in Table 5-3. Examples for other reactions of common environmental interest are given in Table 5-8 and in Appendix A.5.3. Standard-state free-energy changes for oxidation-reduction reactions can be computed directly from the standard electrode potential using Equation 5-80. For E_H°, in volts and ΔG_r° in kJ/eq, Equation 5-80 reduces to

$$\Delta G_r^\circ = -96.485 n_e E_H^\circ \tag{5-82}$$

Different sign conventions are used for electrode or redox potentials in different scientific literature. The convention used in this book is one adopted by the International Union for Pure and Applied Chemistry (IUPAC), in which all half-reactions are written as reductions, with an E_H° sign corresponding to the sign of log K for the reduction reaction.

5.4.2 Reaction Feasibility

The thermodynamic relationships and functions developed above are useful for preliminary evaluation of the feasibility of reactions involved in proposed treatment processes and for interpretation of observed transformations in natural and engineered systems. They can be employed also to determine the amount of energy required either to cause a desirable reaction that otherwise would not be feasible or to prevent an undesirable reaction that otherwise might proceed. Such energy inputs can be thermal, mechanical, electrical, or chemical in nature.

All the equations above include temperature as a variable; thus the potential role of thermal energy is clear. Mechanical energy inputs are useful primarily for reactions that respond to pressure or volume changes. Homogeneous reactions of this type include all gas-phase processes and such processes as reverse osmosis in the aqueous phase. Pressure-responsive heterogeneous reactions include virtually all reactions involving gas-liquid, gas-solid, and gas-liquid-solid systems. Equations 5-80 through 5-82 illustrate the relationships between chemical and electrical energy for redox reactions such as chemical oxidation and electrochemical corrosion. Chemical alterations of the free energy of a system and of its electrochemical potential can be made by adding or withdrawing quantities of reactants and/or products, as illustrated earlier in

Table 5-8 Standard Electrode Potentials for Selected Inorganic Species

Reduction Half-Reaction			E_H° (V)[a]
$Ag^+ + e$	$\rightleftharpoons$	Ag	0.80
$As + 3H^+ + 3e^-$	$\rightleftharpoons$	AsH_3	−0.54
$Cd^{2+} + 2e^-$	$\rightleftharpoons$	Cd	−0.40
$Cl_2(g) + 2e$	$\rightleftharpoons$	$2Cl^-$	1.36
$ClO_2 + H^+ + e$	$\rightleftharpoons$	$HClO_2$	1.27
$Co^{2+} + 2e$	$\rightleftharpoons$	Co	−0.28
$Cr^{3+} + 3e^-$	$\rightleftharpoons$	Cr	−0.74
$Cr^{6+} + 3e$	$\rightleftharpoons$	Cr^{3+}	1.10
$Cu^{2+} + 2e^-$	$\rightleftharpoons$	Cu	0.34
$F_2 + 2e^-$	$\rightleftharpoons$	$2F^-$	2.87
$Fe^{2+} + 2e^-$	$\rightleftharpoons$	Fe	−0.41
$Fe^{3+} + e$	$\rightleftharpoons$	Fe^{2+}	0.77
$2H^+ + 2e^-$	$\rightleftharpoons$	H_2	0.00
$H_2O_2 + 2H^+ + 2e^-$	$\rightleftharpoons$	$2H_2O$	1.78
$Hg^{2+} + 2e^-$	$\rightleftharpoons$	Hg	0.85
$I_2 + 2e^-$	$\rightleftharpoons$	$2I^-$	0.54
$Mn^{2+} + 2e^-$	$\rightleftharpoons$	Mn	−1.03
$Mn^{3+} + e^-$	$\rightleftharpoons$	Mn^{2+}	1.51
$N_2O + 2H^+ + 2e^-$	$\rightleftharpoons$	$N_2 + H_2O$	1.77
$Ni^{2+} + 2e^-$	$\rightleftharpoons$	Ni	−0.23
$2NO + 2H^+ + 2e^-$	$\rightleftharpoons$	$N_2O + H_2O$	1.59
$HNO_2 + H^+ + e^-$	$\rightleftharpoons$	$NO + H_2O$	0.99
$NO_3^- + 3H^+ + 2e^-$	$\rightleftharpoons$	$HNO_2 + H_2O$	0.94
$O_2 + 4H^+ + 4e^-$	$\rightleftharpoons$	$2H_2O$	1.23
$O_3 + 2H^+ + 2e^-$	$\rightleftharpoons$	$O_2 + H_2O$	2.07
$Pb^{2+} + 2e^-$	$\rightleftharpoons$	Pb	−0.13
$S + 2e^-$	$\rightleftharpoons$	S^{2-}	−0.51
$Se + 2e^-$	$\rightleftharpoons$	Se^{2-}	−0.78
$Sn^{2+} + 2e^-$	$\rightleftharpoons$	Sn	−0.14
$Zn^{2+} + 2e^-$	$\rightleftharpoons$	Zn	−0.76

Source: Numerical values from *Handbook of Chemistry and Physics* (1982). See Appendix A.5.3 for additional values.

[a] E_H° values are given with respect to the standard hydrogen electrode at 25°C.

Figure 5-8. The importance of reactant and product activity is evident from Equations 5-72 through 5-75 and Equation 5-81.

The potential for manipulating the energy balances of reactions by input or withdrawal of chemical energy is a valuable asset in process design and operation. For example, as suggested by the types of reaction examples given in Table 5-1, the hydronium ion, $a_{H_3O^+}$ appears in the reaction quotient Q_r in Equations 5-75, 5-76, and 5-81 for many reactions in aquatic environmental systems. Changes in pH thus effectively change the energy state of a system; this was illustrated in Example 5-8. Even when the hydronium ion is not involved, it may still be possible to reduce the magnitude of the activity or concentration of one of the products of a particular reaction by involving that product in a secondary reaction, thus preventing a continual decline in energy and reactivity as equilibrium is approached. For example, a metal cation formed as a product in the primary reaction of a process can be removed by precipitation in a secondary reaction with an added anion. Removal of a product will decrease the magnitude of the reaction quotient for the primary reaction (see Equation 5-76) and thus maintain a favorable free-energy or electrochemical potential for that reaction.

The discussion above is designed to encourage use of thermodynamic data and relationships to interpret and modify process dynamics. As always, however, appropriate caution must be taken in reaching conclusions when the information is limited to this singular aspect of process dynamics (i.e., to equilibrium considerations).

5.4.3 Chemical and Electrical Energy; The Proton–Electron Analogy

The reversible redox potential, E_H, for oxidation-reduction reactions like those given in Tables 5-7 and 5-8 and Appendix A.5.3 is related to electron activity (concentration) in a manner analogous to the relationship between pH and proton activity (concentration). Thus E_H is a master variable for redox reactions in the same manner that pH is a master variable for acid-base and complexation reactions. These parallel roles of E_H and pH serve to reinforce the idea that, excluding nuclear reactions, only two fundamental types of reactions occur in environmental systems: electron transfer reactions and coordination partner transfer reactions.

The proton–electron analogy is explored below using the aqueous-phase redox reaction between ferric iron and hydrogen as an example:

$$Fe^{3+} + \frac{1}{2} H_2 \rightleftharpoons Fe^{2+} + H^+ \qquad K_{OR} \tag{5-83}$$

where K_{OR} is the standard-state equilibrium constant. Again, for ease of representation, water molecules associated with the hydrated proton and the iron-hydroxo complexes are not specifically noted in Equation 5-83. Equation 5-83 can be divided into two half-reactions, involving oxidation of the hydrogen atom:

$$\frac{1}{2} H_2 \rightleftharpoons H^+ + e^- \qquad K_O \tag{5-84}$$

and reduction of the ferric iron to ferrous iron (Equation 13 in Table 5-3):

$$Fe^{3+} + e^- \rightleftharpoons Fe^{2+} \qquad K_R \tag{5-85}$$

If the equilibrium constant, K_O, for the hydrogen oxidation half-reaction is taken by thermodynamic convention to be unity (see Table 5-6, $\Delta G_f^\circ = 0$), the equilibrium constant, K_{OR}, for the overall reaction in Equation 5-83 is given by

$$K_{OR} = K_O K_R = K_R = \frac{a_{Fe^{2+}}}{a_{Fe^{3+}}\, a_{e^-}} \tag{5-86}$$

or

$$\log K_{OR} = \log \frac{a_{Fe^{2+}}}{a_{Fe^{3+}}} - \log a_{e^-} \tag{5-87}$$

and

$$pe = \log K_{OR} + \log \frac{a_{Fe^{3+}}}{a_{Fe^{2+}}} \tag{5-88}$$

where a_{e^-} is the electron activity and pe is the negative base 10 logarithm of a_{e^-}. For standard state, $a_{Fe^{3+}} = a_{Fe^{2+}}$, and

$$pe^\circ = \log K_{OR} \tag{5-89}$$

Equation 5-88 can then be written

$$pe = pe^\circ + \log \frac{a_{Fe^{3+}}}{a_{Fe^{2+}}} \tag{5-90}$$

Recalling from Equation 5-78 that $\Delta G_r^\circ = -\mathcal{R}T \ln K$, the terms pe and pe° in Equation 5-90 can be replaced by $-\Delta G_r/2.3\ \mathcal{R}T$ and $-\Delta G_r^\circ/2.3\ \mathcal{R}T$, respectively, to give

$$-\frac{\Delta G_r}{2.3\ \mathcal{R}T} = -\frac{\Delta G_r^\circ}{2.3\ \mathcal{R}T} + \log \frac{a_{Fe^{3+}}}{a_{Fe^{2+}}} \tag{5-91}$$

The relationship between free energy and electrode potential from Equation 5-80 (for $n_e = 1$) then gives

$$\frac{\mathcal{F}E_H}{2.3\ \mathcal{R}T} = \frac{\mathcal{F}E_H^\circ}{2.3\ \mathcal{R}T} + \log \frac{a_{Fe^{3+}}}{a_{Fe^{2+}}} \tag{5-92}$$

or

$$E_H = E_H^\circ + \frac{2.3\ \mathcal{R}T}{\mathcal{F}} \log \frac{a_{Fe^{3+}}}{a_{Fe^{2+}}} \tag{5-93}$$

The development above can now be generalized, again with hydrogen as the standard reference half-reaction:

$$\mathrm{OX} + \frac{n_e}{2}\mathrm{H_2} \rightleftharpoons \mathrm{RED} + n_e\mathrm{H^+} \tag{5-94}$$

where OX is the oxidized state of any particular chemical species and RED is its reduced state. Based on Equation 5-93, the general relationship between electrode potential and chemical activity is

$$E_H = E_H^\circ + \frac{\mathcal{R}T}{n_e\mathcal{F}} \ln \frac{a_{\mathrm{OX}}}{a_{\mathrm{RED}}} = E_H^\circ + \frac{\mathcal{R}T}{n_e\mathcal{F}} \ln \prod \; (a_i)^{\gamma_i} \tag{5-95}$$

The result is essentially the Nernst equation, which was given in Equation 5-81. It is the most common relationship for describing energy relationships for oxidation-reduction reactions. If all half-reactions are defined in a stoichiometry corresponding to an exchange of a single mole of electrons, Equation 5-95 reduces to

$$E_H = E_H^\circ + \frac{\mathcal{R}T}{\mathcal{F}} \ln \prod (a_i)^{\gamma_i} \tag{5-96}$$

The similarity between Equations 5-96 and 5-74 is again to be noted. The relationships above illustrate that p*e* is equivalent for redox or electron transfer reactions to pH for acid-base or proton transfer reactions (more generally, to p*L* for coordination-partner transfer reactions, where *L* is the ligand in the reaction). Thus p*e*, or more commonly its proportional factor, E_H, can function as the master variable for redox reactions in much the same way as does pH for acid-base reactions (Figures 5-1 through 5-3).

The Pourbaix diagram for the aqueous iron–carbonate system introduced in Figure 5-4 provides a practical and important illustration of the p*e*–pH relationship. Here the equilibria of aqueous iron are influenced simultaneously by hydrogen ions and electrons; moreover, electrochemical potential affects the solubility of metal species. For example, if the pH of a 10^{-5} mol/L bivalent iron solution poised at an electrochemical potential of −0.4 V and pH 6 is changed to pH 8, $FeCO_3$ precipitation occurs. If that same solution is, instead, held at pH 6 while its potential is increased to somewhat less than +0.4 V, the ferrous iron oxidizes to ferric iron, and precipitation of $Fe(OH)_3$ occurs. The thermodynamic stability of various solid phases is also shown in Figure 5-4 to vary with electrochemical potential. For example, at pH 8, iron is stable as metallic iron only at very negative electrode potentials (more negative than −0.5 V). From about −0.5 to +0.1 V, $FeCO_3(s)$ is thermodynamically stable, while solid $Fe(OH)_3$ prevails at more positive electrode potentials.

The diagram presented in Figure 5-4 is conceptually correct but not rigorously quantitative. Its principal value is as an aid to interpretation of real system behavior. For example, the diagram shows that elementary iron cannot corrode if the electrode potential is either very negative or very positive. Thus two efficient corrosion-prevention procedures are suggested: *cathodic protection* and *anodic passivation*. In cathodic protection the electrode potential is kept sufficiently negative either by applying an external electromotive force to render the iron cathodic or by providing a *sacrificial anode* in the system. In passivation, the potential is maintained at a value sufficiently positive for the representative point to lie in the passivation domain, where the metal becomes covered with a protection film of Fe_2O_3. For instance, passivation control can be accomplished using oxidizing agents such as chromate. Under favorable conditions, even dissolved oxygen may raise the potential sufficiently to cause formation of protective Fe_2O_3 films.

An illustration of the interrelationships among standard-state free energies of formation, electrochemical potentials, and equilibrium constants is given in Example 5-9. The calculations therein demonstrate how one thermodynamic parameter can be translated into another to meet a particular application need.

Example 5-9

- **Situation:** As indicated at the beginning of this chapter, molecular oxygen is a highly reactive substance that participates in a broad range of environmental oxidation processes. Depending on the nature of a particular process, different thermodynamic parameters may be required to express the reactivity of oxygen.

- **Question(s):** Given the standard-state free energy of reaction of molecular oxygen, O_2, determine (a) the standard-state electrode potential, E_H°; (b) the pe°; and (c) the equilibrium constant for this half-reaction at the same temperature and pressure for which ΔG_r° is specified.

- **Logic and Answer(s):** The half-reaction and associated free-energy change for a one-electron transfer reduction of oxygen are given in Table 5-7 as

$$\frac{1}{4}O_2 + H^+ + e^- \rightleftharpoons \frac{1}{2}H_2O \qquad \Delta G_r^\circ = -78.2 \text{ kJ/eq}$$

(a) From Equation 5-82,

$$\Delta G_r^\circ = -96.485 n_e E_H^\circ$$

$$E_H^\circ = \frac{-\Delta G_r^\circ}{96.485 n_e} = \frac{-78.2}{96.485(1)} = 0.81 \text{ V}$$

This value compares with that for the half-reaction given in Appendix A.5.3 for pH 7:

$$\frac{1}{2}O_2 + 2H^+ + 2e^- \rightleftharpoons H_2O \qquad E_H^\circ = 0.81 \text{ V}$$

Thus we see that the electrode potential is an expression of the free energy per electron equivalent.

(b) From Equations 5-90 and 5-91,

$$pe^\circ = -\frac{\Delta G_r^\circ}{2.3\ \mathcal{R}T} = \frac{-\Delta G_r^\circ}{2.3(8.314 \times 10^{-3})\ (298^\circ)}$$

$$= \frac{-\Delta G_r}{5.7} = \frac{78.2}{5.7} = 13.7$$

(c) From Equation 5-89,

$$pe^\circ = \log K_{OR}$$

$$\log K_{OR} = 13.7$$

$$K_{OR} = 10^{13.7}$$

5.4.4 Enthalpy and the Temperature Dependence of Equilibrium

The free energy of a system and the equilibrium constant defining the distribution of its species are both functions of enthalpy. The connection between free energy and enthalpy is stated explicitly in Equation 5-28, and that between enthalpy and the equilibrium constant, K, is stated explicitly through its relationship to dG_r (Equation 5-78). Enthalpy data for chemical species, like data for free-energy and standard redox potentials, are readily available in a number of reference books. Enthalpy is an extensive variable, and subject to the same type of treatment as that given free energy. Several important formal relationships among enthalpy, free energy, and the equilibrium constant are illustrated below.

Differentiation of the Gibbs free-energy relationship given in Equation 5-32 with respect to temperature at constant pressure yields

$$\left(\frac{\partial G}{\partial T}\right)_P = -S \tag{5-97}$$

Integration of Equation 5-97 requires knowledge of S as a function of temperature T. While this relationship can be established, it is at best a tedious process. An alternative approach was presented in Equation 5-47, which can also be arrived at by combining Equation 5-97 with Equation 5-28. From Equation 5-47 we can write a relationship for the variation of dG_r with temperature in a specific reaction as

$$\left(\frac{\partial(dG_r)}{\partial T}\right)_P = -dS_r = \frac{dG_r - dH_r}{T} \tag{5-98}$$

Equation 5-98, the *Gibbs–Helmholtz equation*, allows us to determine dH_r from knowledge of dG_r. We can show that

$$\frac{\partial}{\partial T}\left(\frac{dG_r}{T}\right) = \frac{1}{T}\frac{\partial(dG_r)}{\partial T} + \frac{dG_r}{T^2} \tag{5-99}$$

The first term on the right-hand side of Equation 5-99 can be replaced with Equation 5-98 to give an alternative form of the Gibbs–Helmholtz equation:

$$\left[\frac{\partial}{\partial T}\left(\frac{dG_r}{T}\right)\right]_P = \frac{-dH_r}{T^2} \tag{5-100}$$

With the aid of differential calculus (i.e., $d\int T^{-2}dT = -dT^{-1}$), Equation 5-100 can be rewritten as

$$\left[\frac{\partial(dG_r/T)}{\partial(1/T)}\right]_P = dH_r \tag{5-101}$$

The van't Hoff equation, which expresses the dependence of the equilibrium constant on temperature, follows directly from Equation 5-101 after replacing dG_r/T from Equation 5-78 with $-\mathcal{R}\ln K$:

$$\left(\frac{d\ln K}{d(1/T)}\right)_P = \frac{-dH_r}{\mathcal{R}} \tag{5-102}$$

or

$$\left(\frac{d\ln K}{dT}\right)_P = \frac{dH_r}{\mathcal{R}T^2} \tag{5-103}$$

According to Equation 5-102, a plot of ln K versus T^{-1} should yield a straight line with a slope of $-dH_r/\mathcal{R}$, as illustrated in Figure 5-13; more correctly, this slope is $-\Delta H_r/\mathcal{R}$ because it involves a *finite* temperature difference. The plot in Figure 5-13 is for a reaction that is endothermic (i.e., the equilibrium constant increases with increasing temperature). The slope is negative, so the enthalpy is positive. Thus we see that endothermic reactions have positive enthalpy and, by extension, exothermic reactions have negative enthalpy.

Another important use of the enthalpy concept is for calculation of an equilibrium constant for a reaction at some temperature given knowledge of the standard-state molar enthalpy, ΔH_r°, and K at another temperature. The value of ΔH_r° can be found by applying the same concepts employed to calculate the standard-state free energy of a reaction, now using tables of standard-state molar enthalpy values, such as Table 5-6 and Appendix A.5.1.

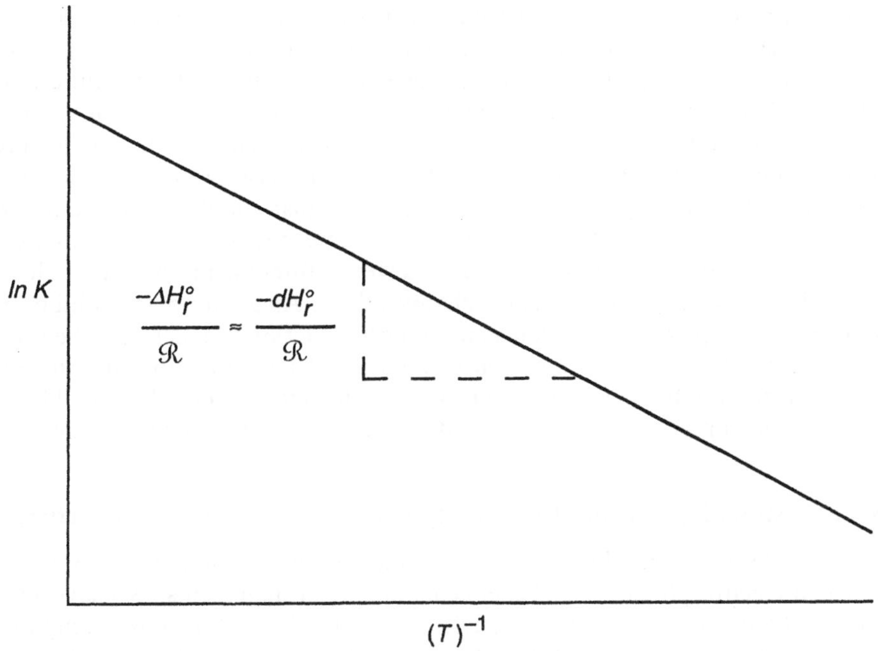

Figure 5-13 Dependence of equilibrium constant on absolute temperature for an endothermic reaction.

If $dH_r^\circ \approx \Delta H_r^\circ$ is assumed constant over a relatively narrow temperature range, Equation 5-103 can be integrated at two temperatures, T_1 and T_2, and the constant of integration taken as the same for both temperatures, to give

$$\ln \frac{K_2}{K_1} = \frac{\Delta H_r^\circ}{\mathcal{R}} \frac{T_2 - T_1}{T_1 T_2} \tag{5-104}$$

Applications of the enthalpy relationships to other characteristic thermodynamic functions of reactions are several fold. The most obvious is for calculation of the temperature dependence of equilibrium constants, thereby eliminating the need to measure K at all temperatures of interest. The temperature dependence of equilibrium constants is, of course, very important in process engineering because it provides information about reaction feasibility, at least from an equilibrium point of view.

An additional use of enthalpy relationships is for interpretation of the mechanistic character of complex reaction processes from experimental measurements of their equilibrium properties. Equation 5-104 relates enthalpy to the ratio of equilibrium constants at different temperatures. This ratio can be replaced by a ratio of values for any measurable equilibrium property (e.g., extent of reaction, or sorption capacity) which is directly proportional to K, as long as the proportionality is temperature independent. Experimental measurements of such a property for a system at several values of temperature thus allows calculation of an enthalpy value for the primary reaction of that system.

Enthalpy values are characteristic of reaction mechanisms and frequently permit more rigorous process analysis than is otherwise possible in complex systems. They can suggest appropriate variables that should be controlled to enhance or retard primary reaction(s). For example, the biological metabolism of an organic substrate may proceed by several different mechanisms, all of which have different characteristic enthalpy values. An appropriate equilibrium property, perhaps simply the concentration of a reaction product, may be used to determine the operative mechanism by way of an estimation of enthalpy from Equation 5-104. Once the enthalpy is calculated, it may be possible to alter system conditions to achieve greater metabolism of the contaminant.

5.4.5 Estimating Chemical Reactivity and Thermodynamic Properties

We have discussed relationships between the reactivities of chemical species and their electron structures and chemical bonding arrangements. We have also identified the manner in which reactivity is affected by interactions among the atoms of molecular species. These relationships can be quantified, at least approximately, by examining the constituent atoms and bonding arrangements of molecules.

Exact measurements of thermodynamic calculations of equilibrium constants should be used to determine reactivities and reactions. In some instances, however, particularly in cases of complex organic species, exact measurements may not be possible. Although methods to estimate thermodynamic properties have limited accuracy, they may provide useful insights into fate and transport predictions. These estimation methods may also bridge the gap between theory and empirical observations.

The most direct approach to estimation of chemical properties from chemical structure and bonding data is by application of linear free-energy relationships (LFERs). These relationships develop from correlations between the standard free energies of species involved in parallel reactions in which identical changes in reactant structure occur under identical conditions. Equation 5-78 specifies that the free energy of a reaction is logarithmically related to its equilibrium constant, K. Thus, if the conditions of LFER theory are met, the logarithms of equilibrium constants associated with changes in reactant structure in parallel reactions should be linearly related to each other. Each reaction may involve the chemical modification of an organic compound by the substitution of the same atom or group of atoms. For some of them the equilibrium constants may be known while the equilibrium constants for others are to be estimated.

As an example, Figure 5-14 illustrates parallel protolysis reactions for two parent (p) and two chlorine-substituted (s) organic species, benzoic and carbolic (phenol) acids. What is known about the effect of chlorine substitution on the reaction behavior of one of the parent precursor acids can be used to estimate the effect of the same substitution on the reaction behavior of the other. This approach is predicated on the concept that it is possible to separate a "core" reaction characteristic from substituent effects. The core aspect of the approach presumes that particular atoms, or combinations of atoms, impart particular and repeatable properties and behavioral characteristics to com-

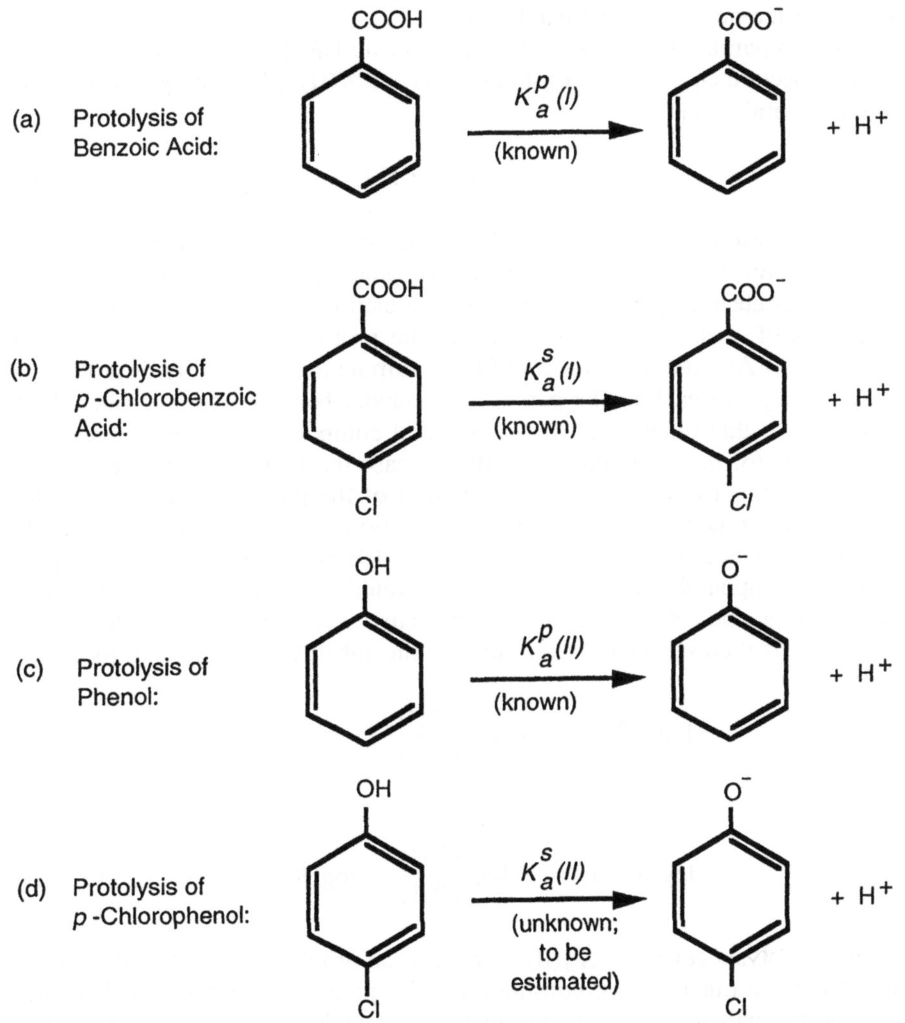

Figure 5-14 Protolysis reactions of parent (p) and chlorine-substituted (s) reference and target organic acids.

pounds. Particular substituent groups are then presumed to contribute additional "fragment-related" properties.

A fragment is an atom, or several atoms in a chemical group, bonding to isolating carbon atoms. Single-atom fragments include isolating carbon atoms themselves (e.g., $-\overset{|}{\underset{|}{C}}-$ in CH_4), hydrogen or heteroatoms (e.g., $-H$ in $H-C{=}C-H$, and $-O-$ in CH_3-O-CH_3). Multiple-atom fragments result from any combination of a nonisolating carbon, hydrogen, or heteroatom. There are hundreds of atom or atom group fragment constants available in various reference books and handbooks.

For two parallel reactions, I and II, the basic LFER representation relating their respective equilibrium constants, $K(\mathrm{I})$ and $K(\mathrm{II})$, is given by a correlation of the general form

$$\log K(\mathrm{II}) = m \log K(\mathrm{I}) + \kappa \tag{5-105}$$

Similar representations can be made for activation energies and rate coefficients, but for now we focus on free-energy and equilibrium constants.

To illustrate the application of LFER to estimation of the thermodynamic properties of chemical species, consider the protolysis reactions depicted in Figure 5-14. Assume for purposes of this example that the acidity or protolysis constant, K_a, for each of the parent compounds, benzoic acid and phenol, is known, as is that for the chlorine-substituted compound *p*-chlorobenzoic acid but not that for the *p*-chlorophenol. In this case the LFER method becomes a substituent-effect approach to the estimation of the protolysis constant of the substituted compound in the second (II) reaction sequence, using the first (I) reaction sequence as a reference. Equation 5-105 is then written for each pair of parent compounds and each pair of substituted compounds. The correlation parameters m and κ are assumed to be the same. The correlation for the parent compound is then subtracted from that for the substituted compound to give

$$\log \frac{K_a^s(\mathrm{II})}{K_a^p(\mathrm{II})} = m \log \frac{K_a^s(\mathrm{I})}{K_a^p(\mathrm{I})} \tag{5-106}$$

or

$$\log K_a^s(\mathrm{II}) = m \log \frac{K_a^s(\mathrm{I})}{K_a^p(\mathrm{I})} + \log K_a^p(\mathrm{II}) \tag{5-107}$$

The protolysis constant, $K_a^s(\mathrm{II})$, for the *p*-chlorophenol can then be calculated from the other known constants for the parent and substituted benzoic acid and the parent phenol. The term $\log K_a^s(\mathrm{I})/K_a^p(\mathrm{I})$ is a measure of the intrinsic effect of the substituent change, while the constant of proportionality, m, is a measure of the relative sensitivity of the I and II reaction sequences to a particular substituent change.

The underlying concept of the LFER approach suggests a broad range of applicability. In reality, however, the accuracy of the approach is affected significantly by the choice of the reference substituent reaction(s). Principal differences among correlations commonly employed for application of LFER concepts relate to differences in the choice of reference reaction sequences used to define substituent parameters.

5.5 SUMMARY

The dynamics of change of any constituent in an environmental system, regardless of the phases involved (water, air, soil, sediment, or other solid phase), depend first on assessment of the thermodynamic feasibility of change, and subsequently on evaluation of the rates at which change will occur. Although in this chapter we have focused on the thermodynamics of homogeneous *aqueous-phase* reactions, the concepts are clearly appropriate for any phase, provided that chemical activity can be estimated in that phase. As we will see in Chapter 7, highly feasible reactions (i.e., *spontaneous* or *highly energetic*) are driven so far toward products as to be treated as irreversible, a desirable simplification from the point of view of developing simple rate relationships. Thermodynamic analysis may provide valuable insights into the relative rates of reaction expected. For example, proton (acid-base), electron (oxidation-reduction), and ligand (complexation) exchange reactions in aqueous phase are so fast that equilibrium is usually a safe assumption for the relevant time scales for macrosystems, and thus rate relationships need not be of concern.

Numerous examples have been presented to emphasize that thermodynamic calculations are practical in assessment of the extent of change. Standard free energies, enthalpies, and entropies are available for a wide array of chemical formations involved in reactions of concern in environmental systems. The outcomes of analyses of these properties are equilibrium constants, associated knowledge of the free energies involved in change, and the dependence of these energies on temperature. Equilibrium constants based on these thermodynamic constants can be determined for many common reactions (e.g., acid-base, precipitation, and complexation reactions).

Another important outcome of the thermodynamic perspective presented in this chapter is the recognition that proton and electron transfer reactions are interrelated (e.g., the Gibbs and Nernst equations). The most vivid example is the dependence of oxidation-reduction reactions and related precipitation reactions on pH (e.g., the Pourbaix diagram). This is particularly important in aquatic systems because pH is the *master variable*, a variable that gives each natural water its unique reaction environment, especially when reactions with anthropogenic agents must be assessed.

Despite the extensiveness of thermodynamic data and the appeal of associated concepts, it would be misleading to suggest that the feasibility of any reaction of environmental interest can be assessed singularly on the basis of thermodynamic information. There remain innumerable reactions of anthropogenic chemicals with each other and with constituents of natural systems for which such calculations have not been made. Linear free-energy relationships, a form of structure–activity relationship, may hold promise for the future as an approach for overcoming limitations in *experimental* thermodynamic data. There is, moreover, an opportunity to extend this thermodynamic perspective to encompass microbially mediated chemical change, processes for which definitive experimental data frequently are lacking.

As will become clearer in Chapter 7 rate relationships are heavily dependent on making accurate experimental measurements, an effort that is both costly and time consuming. It thus makes good sense to invest first in the analysis of the thermodynamic predispositions of reactions and processes whenever possible. Such analysis can usually help determine whether rate relationships are in fact needed, and how the associated data should be collected.

5.6 REFERENCES AND SUGGESTED READINGS

5.6.1 Citations and Sources

Christenson, D. R., and P. L. McCarty, 1975, "Multi-process Biological Treatment Model," *Journal of the Water Pollution Control Federation*, *47*, 11. (Table 5-7)

Freiser, H., and Q. Fernando, 1963, *Ionic Equilibria in Analytical Chemistry*, John Wiley & Sons, Inc., New York. (Tables 5-2, 5-4, and 5-5)

Handbook of Chemistry and Physics, 1982, 62nd Edition, CRC Press, Boca Raton, FL. (Table 5-8 and Appendix A.5.3)

National Bureau of Standards, 1952, *Selected Values of Chemical Thermodynamic Properties*, Circular 500, U.S. Government Printing Office, Washington, DC. (Tables 5-6 and Appendix A.5.2)

Smith, R. M., and A. E. Martell, 1976, *Critical Stability Constants*, Plenum Press, New York. (Tables 5-2, 5-4, and 5-5)

Stumm, W., and G. F. Lee, 1960, "The Chemistry of Aqueous Iron," *Schweizerische Zeitschrift fuer Hydrologie*, Vol. XXII, Fascil Birkhäuser Verlag, Basel. (Figure 5-4 and Table 5-3)

5.6.2 Background Reading

Atkins, P. W., 1990, *Physical Chemistry*, 4th Edition, W.H. Freeman and Company, Publishers, New York. Part 1 treats chemical equilibrium and includes chapters on the properties of gases (Chapter 1), the first and second laws (Chapters 2 through 5), thermodynamics of solutions (Chapter 7), and reaction equilibria (Chapter 9).

Lewis, G. N., and M. Randall, 1961, *Thermodynamics*, 2nd Edition, revised by L. Brewer and K. S. Pitzer, McGraw-Hill Book Company, New York. A classic text in chemical thermodynamics.

Lyman, W. J., W. F. Reehl, and D. H. Rosenblatt, 1982, *Handbook of Chemical Property Estimation Methods: Environmental Behavior of Organic Compounds*, McGraw-Hill Book Company, New York. An informative treatise on linear free-energy relationships and how to apply them for estimating reactivity and thermodynamic properties of chemical species. Also a useful source of substituent group contributions to reactivity.

Morel, F. M. M., and J. G. Hering, 1993, *Principles and Applications of Aquatic Chemistry*, John Wiley & Sons, Inc., New York. An intermediate-level aquatic chemistry text emphasizing geochemical and geobiological cycles of elements in natural waters and featuring the tableau method of solving chemical equilibrium problems. Treats chemical energetics (Chapter 2) with applications to acid-base (Chapter 4), complexation chemistry (Chapter 6), and oxidation-reduction chemistry (Chapter 7).

Pankow, J. F., 1991, *Aquatic Chemistry Concepts*, Lewis Publishers, Chelsea, MI. A didactic approach to aquatic chemistry structured to reinforce the treatment of Stumm and Morgan (2nd Edition, 1981). Chapter 1 provides a discussion of activity and activity coefficients. Thermodynamic principles, including a good discussion of reference states, are introduced in Chapter 2. Acid-base chemistry is treated in Chapters 3 and 10, complexation chemistry in Chapter 18, and redox chemistry in Chapters 19 through 24.

Schwarzenbach, R. P., P. M. Gschwend, and D. M. Imboden, 1993, *Environmental Organic Chemistry*, John Wiley & Sons, Inc., New York. The book provides a comprehensive treatment of *organic* aquatic chemistry. Chapter 3 presents a brief introduction to thermodynamics, including discussions of reference states and fugacity. The thermodynamic aspects of the solubility of organic compounds in water are discussed in Chapter 5. Other topics include organic acids and bases (Chapter 8), oxidation and reduction reactions (Chapter 12), photolysis reactions (Chapter 13), and microbially mediated transformation reactions (Chapter 14).

Smith, J. M., and H. C. Van Ness, 1975, *Introduction to Chemical Engineering Thermodynamics*, McGraw-Hill Book Company, New York. Thermodynamics from a chemical engineering perspective applied to ideal and nonideal systems. In-depth treatment of the first and second laws (Chapters 2 and 5), thermodynamic properties of fluids (Chapter 6), thermodynamic properties of homogeneous mixtures, including a development of fugacity (Chapter 7), chemical reaction equilibria (Chapter 9), and applications to chemical engineering processes.

Snoeyink, V. L., and D. Jenkins, 1980, *Water Chemistry*, John Wiley & Sons, Inc., New York. A first-level text providing a simplified presentation of chemical equilibrium (Chapter 3) with applications to acid-base (Chapter 4), coordination (Chapter 5), and oxidation-reduction chemistry (Chapter 7). A good introduction to constructing p*C*–pH diagrams for the solution of chemical equilibrium problems.

Stumm, W., and J. J. Morgan, 1995, *Aquatic Chemistry*, 3rd Edition, John Wiley & Sons, Inc., New York. A classic treatise on aquatic chemistry. The fundamentals of chemical thermodynamics and kinetics are discussed comprehensively in Chapter 2. Topics pertaining to process energetics include the first, second, and third laws of thermodynamics, Gibbs free energy, chemical potential, chemical reference states, and chemical equilibria. Concepts are developed for homogeneous systems in the context of acid-base reactions (Chapter 3), dissolved carbon dioxide (Chapter 4), coordination chemistry (Chapter 6), precipitation and dissolution reactions (Chapter 7), and redox chemistry (Chapter 8).

5.7 PROBLEMS

5-1 An experimental analysis of the aqueous-phase solubility of $PbSO_4$ is conducted by mixing a 10-g quantity of the salt with 3.78 L of distilled water. After allowing sufficient time for dissolution of $PbSO_4$, 50 mL of the water is decanted and a gravimetric analysis is done, yielding a postevaporation residue of 2.2 mg. A similar analysis on an equal volume of the distilled water yields a residue of 0.3 mg. Determine the solubility product of $PbSO_4$ in the distilled water.

Answer: $p\mathcal{K}_S = 7.8$.

5-2 The pH of an acid mine drainage through which hydrogen sulfide bubbles continuously is measured and found to have a value of pH 3.7. Estimate the concentration of S^{2-} in this water.

Answer: 3×10^{-15} *mol/L.*

5-3 The acid mine drainage of Problem 5-2 is found to contain 56 mg/L of Fe(II). How much more Fe(II) might be able to dissolve in this water before FeS precipitates?

Answer: ~ *18 mg/L.*

5-4 The free copper ion concentration of a $Cu(NO_3)_2$ waste is to be reduced to a maximum level of 0.6×10^{-2} ng/L by complexation with ammonia. The complex formation constant for $\{Cu(NH_3)_4\}^{2+}$ is $pK_f = -11.7$. Neglecting the amount of copper in complexes containing fewer than four ammonia groups per central ion, calculate the amount of ammonia to be added if the original concentration of $Cu(NO_3)_2$ is 188 mg/L.

Answer: 6.47 g/L.

5-5 Determine the amount of NH_4Cl that must be added to a solution containing 170 mg of ammonia and 24.3 mg of Mg(II) per liter to repress the reaction

$$NH_3 + H_2O \rightleftharpoons NH_4^+ + OH^-$$

sufficiently to prevent $Mg(OH)_2$ from precipitating. The pK_a of ammonia is 9.3.

Answer: 66.8 mg/L.

5-6 Calculate $[Cd^{2+}]$ in 1 L of solution prepared by dissolving 10^{-3} mol of $Cd(NO_3)_2$ and 1.5 mol of NH_3. The overall dissociation constant for the $\{Cd(NH_3)_4\}^{2+}$ complex is 1.8×10^{-7}. Neglect complexes containing fewer than four ammonia ligands.

Answer: 3.6×10^{-11} *mol/L.*

5-7 Determine the concentration of free Cd^{2+} in a 5.0-m*M* solution of $CdCl_2$ if the first formation constant $pK_{f,1}$ for chloride complexation of Cd^{2+} is -2.0 and the second complexation step is neglected.

Answer: *315 mg/L.*

5-8 A 1660-mg quantity of dry KI is added to a 1-L solution containing 28 mg of Cd(II). If the first and second formation constants for complexation of Cd^{2+} by I^- are given by $pK_{f,1} = -2.28$ and $pK_{f,2} = -1.64$, determine the percentages of Cd(II) present as Cd^{2+}, CdI^+, and CdI_2 at equilibrium.

Answers: *27%, 51%, 22%, respectively.*

5-9 A 0.2-g quantity of platinum is immersed in 1 g of water in a 50-g glass beaker. Calculate the energies required to heat each of these materials from 15 to 65°C, given that the specific heat capacities for glass and platinum are 0.84 J/g-K and 0.134 J/g-K, respectively.

Answers: *Water, 209 J; platinum, 1.34 J; beaker, 2092 J.*

5-10 In a laboratory combustion test a 5-g sample of coal tar residue is found to raise the temperature of 1 kg of water from 10°C to 47°C. Calculate the heat value of the residue.

Answer: *30.96 kJ/g.*

5-11 Methane is to be used as fuel in a heater in an industrial process for raising the temperature of an aqueous dye bath solution from 15°C to 95°C. The heat of combustion of methane is 891 kJ/mol. Assuming that 50% of the heat is useful, determine how many kilograms of dye bath can be heated to specification by burning 200 L of methane, CH_4, measured at STP.

Answer: *11.9 kg.*

5-12 The dye bath of Problem 5-11 is at a temperature of 50°C after use. To render it more amenable to biological treatment, its temperature is to be lowered by addition of waste ice from a refrigerated storage operation. Estimate the resulting temperature of the waste stream if the ice (0°C) is added on a 1:9 weight basis.

Answer: *37°C.*

5-13 Ten cubic meters of water stored in an uninsulated outdoor water feed tank for a steam generation plant has frozen during a 2-week plant shutdown. A preheater in this tank will be used to convert the ice to water at 50°C, which will then be processed to steam. Determine the total heat required to change the ice at 0°C to steam at 100°C.

Answer: 3.0×10^7 *kJ.*

5-14 The engineer of a water treatment plant that employs *hydrated lime* [$Ca(OH)_2$] for water softening is contemplating recalcination of precipitated $CaCO_3$ to recover *quick lime* (CaO) and CO_2 for recarbonation. The reaction underlying this process is

$$CaCO_3(s) \overset{\text{heat}}{\rightleftharpoons} CaO(s) + CO_2(g)$$

Calculate the enthalpy of decomposition of $CaCO_3$ into CaO and CO_2. Does this value indicate a spontaneous reaction?

Answer: $\Delta H_r^\circ = 177.9$ *kJ/mol of $CaCO_3$. The positive value signifies an endothermic reaction; that is, the reaction is not spontaneous—a heat input of 177.9 kJ/mol is required.*

5-15 When solid zinc is added to an aqueous copper sulfate solution, a sequential redox and solid-phase transformation process occurs, according to the reaction

$$CuSO_4(aq) + Zn(s) \rightleftharpoons ZnSO_4(aq) + Cu(s)$$

Determine the enthalpy change associated with this process and indicate whether the process would be aided or impeded by heating the copper sulfate solution?

Answer: $\Delta H_r^\circ = -216.7$ *kJ/mol. The negative value suggests that the reaction is exothermic and would be impeded by heating.*

5-16 The quick lime produced in the recalcination operation in Problem 5-14 is to be slaked to hydrated lime before use, according to the reaction

$$CaO(s) + H_2O(l) \rightleftharpoons Ca(OH)_2(s)$$

Determine the amount of heat associated with this reaction and indicate whether heat is evolved or required as input.

Answer: $\Delta H_r^\circ = -1180$ *kJ. The negative value indicates that the reaction is exothermic (i.e., heat is evolved).*

5-17 Data obtained from experimental measurements of the amounts of heat evolved or absorbed on the combustion of compounds can be used to determine standard enthalpy values for formation of these compounds. If the heat liberated on complete combustion of 1 mol of CH_4 (*g*) to $CO_2(g)$ and $H_2O(l)$ is 890 kJ and that for combustion of 1 g of starch, $(C_6H_{10}O_5)_x$, into $CO_2(g)$ and $H_2O(l)$ is 17.5 kJ, determine the standard-state enthalpy of formation of 1 mol of CH_4 (*g*) and 1 g of starch.

Answers: $\Delta H^\circ_{r,CH_4} = -75$ *kJ/mol;* $\Delta H^\circ_{r,starch} = -5.88$ *kJ/g.*

5-18 A shallow ponded area at a strip mine in a semiarid region is in equilibrium with underlying gypsum, $CaSO_4(s)$, deposits. An organic pollutant, formaldehyde (CH_2O), is discharged to the pond. Biodegradation of this pollutant may occur by an anaerobic reaction in which sulfate is used as an electron acceptor:

$$SO_4^{2-} + 2CH_2O + 2H_3O^+ \rightleftharpoons H_2S(g) + 2CO_2(g) + 4H_2O$$

or, if sufficient oxygen is present, by the following aerobic reaction:

$$O_2(aq) + CH_2O \rightleftharpoons CO_2(g) + H_2O$$

with $\Delta G_R^\circ = -518$ kJ/mol. A *partial* analysis of the system indicates the following:

$$Ca^{2+} = 200 \text{ mg/L} \qquad O_2(aq) = 5 \text{ mg/L}$$

$$P_{H_2S} = 10^{-5.0} \text{ atm} \qquad [CH_2O] = 10^{-6} \text{ M}$$

$$P_{CO_2} = 10^{-3.5} \text{ atm} \qquad pH = 7$$

Determine whether either or both of these reactions are thermodynamically feasible under the conditions stipulated.

Answers:

$$\Delta G_R \text{ (aerobic)} = -482 \text{ kJ/mol}$$

$$\Delta G_R \text{ (anaerobic)} = -210 \text{ kJ/mol}$$

APPENDIX A.5.1 PERIODIC ARRANGEMENT OF SELECTED CHEMICAL ELEMENTS*

Groups

IA	IIA	IIIA	IVA	VA	VIA	VIIA	← VIIIA	→		IB	IIB	IIIB	IVB	VB	VIB	VIIB	VIII
1	2	3	4	5	6	7	2,3	2,3	2	1	2	3	4	3,5	—	(1)	0
							Principal Oxidation States										
1 H 1.008 $1s^1$																	2 He 4.003 $1s^2$
3 Li 6.939 [He] $2s^1$	4 Be 9.012 [He] $2s^2$											5 B 10.81 [He] $2s^2 2p^1$	6 C 12.01 [He] $2s^2 2p^2$	7 N 14.01 [He] $2s^2 2p^3$	8 O 16.00 [He] $2s^2 2p^4$	9 F 19.00 [He] $2s^2 2p^5$	10 Ne 20.18 [He] $2s^2 2p^6$
11 Na 22.99 [Ne] $3s^1$	12 Mg 24.31 [Ne] $3s^2$											13 Al 26.98 [Ne] $3s^2 3p^1$	14 Si 28.09 [Ne] $3s^2 3p^2$	15 P 30.97 [Ne] $3s^2 3p^3$	16 S 32.06 [Ne] $3s^2 3p^4$	17 Cl 35.45 [Ne] $3s^2 3p^5$	18 Ar 39.95 [Ne] $3s^2 3p^6$
19 K 39.10 [Ar] $4s^1$	20 Ca 40.08 [Ar] $4s^2$	21 Sc 44.96 [Ar] $3d^1 4s^2$	22 Ti 47.90 [Ar] $3d^2 4s^2$	23 V 50.94 [Ar] $3d^3 4s^2$	24 Cr 52.00 [Ar] $3d^5 4s^1$	25 Mn 54.94 [Ar] $3d^5 4s^2$	26 Fe 55.85 [Ar] $3d^6 4s^2$	27 Co 58.93 [Ar] $3d^7 4s^2$	28 Ni 58.71 [Ar] $3d^8 4s^2$	29 Cu 63.54 [Ar] $3d^{10} 4s^1$	30 Zn 65.37 [Ar] $3d^{10} 4s^2$	31 Ga 69.72 [Ar] $3d^{10}$ $4s^2 4p^1$	32 Ge 72.59 [Ar] $3d^{10}$ $4s^2 4p^2$	33 As 74.92 [Ar] $3d^{10}$ $4s^2 4p^3$	34 Se 78.96 [Ar] $3d^{10}$ $4s^2 4p^4$	35 Br 79.91 [Ar] $3d^{10}$ $4s^2 4p^5$	36 Kr 83.80 [Ar] $3d^{10}$ $4s^2 4p^6$
37 Rb 85.47 [Kr] $5s^1$	38 Sr 87.62 [Kr] $5s^2$	39 Y 88.91 [Kr] $4d^1 5s^2$	40 Zr 91.22 [Kr] $4d^2 5s^2$	41 Nb 92.91 [Kr] $4d^4 5s^1$	42 Mo 95.94 [Kr] $4d^5 5s^1$	43 Tc (99) [Kr] $4d^5 5s^2$	44 Ru 101.1 [Kr] $4d^7 5s^1$	45 Rh 102.9 [Kr] $4d^8 5s^1$	46 Pd 106.4 [Kr] $4d^{10}$	47 Ag 107.9 [Kr] $4d^{10} 5s^1$	48 Cd 112.4 [Kr] $4d^{10} 5s^2$	49 In 114.8 [Kr] $4d^{10}$ $5s^2 5p^1$	50 Sn 118.7 [Kr] $4d^{10}$ $5s^2 5p^2$	51 Sb 121.8 [Kr] $4d^{10}$ $5s^2 5p^3$	52 Te 127.6 [Kr] $4d^{10}$ $5s^2 5p^4$	53 I 126.9 [Kr] $4d^{10}$ $5s^2 5p^5$	54 Xe 131.3 [Kr] $4d^{10}$ $5s^2 5p^6$
55 Cs 132.9 [Xe] $6s^1$	56 Ba 137.3 [Xe] $6s^2$	57 La 138.9 [Xe] $5d^1 6s^2$	72 Hf 178.5 [Xe] $4f^{14}$ $5d^2 6s^2$	73 Ta 180.9 [Xe] $4f^{14}$ $5d^3 6s^2$	74 W 183.9 [Xe] $4f^{14}$ $5d^4 6s^2$	75 Re 186.2 [Xe] $4f^{14}$ $5d^5 6s^2$	76 Os 190.2 [Xe] $4f^{14}$ $5d^6 6s^2$	77 Ir 192.2 [Xe] $4f^{14}$ $5d^7 6s^2$	78 Pt 195.1 [Xe] $4f^{14}$ $5d^9 6s^1$	79 Au 197.0 [Xe] $4f^{14}$ $5d^{10} 6s^1$	80 Hg 200.6 [Xe] $4f^{14}$ $5d^{10} 6s^2$	81 Tl 204.4 $[Xe]4f^{14}$ $5d^{10}$ $6s^2 6p^1$	82 Pb 207.2 $[Xe]4f^{14}$ $5d^{10}$ $6s^2 6p^2$	83 Bi 209.0 $[Xe]4f^{14}$ $5d^{10}$ $6s^2 6p^3$	84 Po (209) $[Xe]4f^{14}$ $5d^{10}$ $6s^2 6p^4$	85 At (210) $[Xe]4f^{14}$ $5d^{10}$ $6s^2 6p^5$	86 Rn (222) $[Xe]4f^{14}$ $5d^{10}$ $6s^2 6p^6$
87 Fr (223) [Rn] $7s^1$	88 Ra (226) [Rn] $7s^2$	89 Ac (227) [Rn] $6d^1 7s^2$															

Additional Oxidation States: ( ) = -

As: (3)	Cu: 2	Nb: 3	Po: 2,4	Se: (2),4,6
At: 1,3,5,7	Hg: 1	Ni: 3	Pt: 4	Sn: 2
Au: 3	I: 1,5,7	O: (2)	Re: (1),2,4,6	Te: (2),4,6
Br: 1,5	Ir: 4,6	Os: 4,6,8	Rh: 4	Ti: 3
C: (4),2	Mn: 2,3,4,6	P: (3),4,5	Ru: 4,6,8	Tl: 1
Cl: 1,3,5,7	Mo: 2,3,4,5	Pb: 2	S: (2),2,4,6	V: 2,3,4
Cr: 2,3	N: (3),2,4	Pd: 4	Sb: (3),5	W: 2,3,4,5

Atomic Number	Element Symbol
Atomic Mass	
Electron Configuration	
[Kernel]	
Outer Shell	

*Excludes lanthanide series (atomic numbers 58–71) and elements above atomic number 89.

APPENDIX A.5.2 THERMODYNAMIC PROPERTIES OF SELECTED CHEMICAL SPECIES AT 25°C

Substance	State[a]	ΔH_f° kJ/mol	ΔG_f° kJ/mol
Aluminum			
Al^{3+}	aq	−531.4	−485.3
$AlF_3 \cdot 3H_2O$	aq	−1512.1	—
Arsenic			
As (gray)	c	0.0	0.0
AsO^+	aq	—	−163.6
H_3AsO_3	aq	−741.8	−639.7
H_3AsO_4	aq	−898.7	−769.0
Barium			
Ba	c	0.0	0.0
Ba^{2+}	aq	−538.5	−560.7
$BaCO_3$	aq	−1214.6	−1088.7
$Ba(OH)_2$	aq	−998.3	−875.3
Beryllium			
Be^{2+}	aq	—	−389.1
Bromine			
Br_2	g	30.5	3.3
Br^-	aq	−120.9	−102.9
Cadmium			
Cd (α)	c	0.0	0.0
Cd^{2+}	aq	−72.4	−77.7
$CdCl_2$	aq	−407.5	−340.2
$Cd(NO_2)_2$	aq	−485.3	−298.7
$CdSO_4$	c	−926.3	−820.1
	aq	−979.9	−819.6
Calcium			
Ca	c	0.0	0.0
$Ca(HCO_3)_2$	aq	−1925.1	−1727.2
$CaCl_2$	aq	−877.8	−815.5
$Ca(OH)_2$	c	−986.6	−896.6
$CaHPO_4 \cdot 2H_2O$	c	−2410.0	−2153.1
$CaSO_4$ (anhydrite)	c	−1432.6	−1320.5
$CaSO_4$	c	−1419.2	−1307.5
	aq	−1450.6	−1294.9
Carbon			
C	g	718.4	673.2
C (graphite)	c	0.0	0.0
CO_2	g	−393.7	−394.6
	aq	−413.0	−386.2
C_2H_4 (ethylene)	g	52.3	68.2
C_3H_6 (propylene)	g	20.5	62.8
C_4H_8 (l)[b]	g	1.3	72.0
C_5H_{10} (l)[b]	g	−20.9	78.7

Substance	State[a]	ΔH_f° kJ/mol	ΔG_f° kJ/mol
C_5H_{10} (cyclopentane)	g	−77.4	38.5
C_6H_{12} (cyclohexane)	g	−123.0	31.8
C_7H_8 (toluene)	g	50.2	122.2
C_7H_{12} (l)[b]	g	−62.3	95.8
C_7H_{16} (*n*-heptane)	g	−187.9	8.8
C_8H_8 (styrene)	g	147.7	213.8
C_8H_{16} (l)[b]	g	−82.8	104.2
C_9H_{12} (*n*-propylbenzene)	g	7.9	137.2
C_9H_{12} (isopropylbenzene)	g	3.8	136.8
C_9H_{12} (1,3,5-trimethylbenzene)	g	−15.9	118.0
C_9H_{18} (l)[b]	g	−103.3	112.5
C_9H_{20} (*n*-nonane)	g	−228.9	25.9
$C_{10}H_{10}$ (l)[b]	g	−124.3	121.3
$C_{10}H_{14}$ (*n*-butylbenzene)	g	−13.8	144.8
$C_{10}H_{22}$ (*n*-decane)	g	−249.8	34.3
$HCOO^-$	aq	−410.0	−334.7
CH_2O	g	−115.9	−110.0
CH_3Br	g	−35.6	−25.9
CH_2Br_2	g	−4.2	−5.9
$CHBr_3$	g	25.1	15.9
CH_4ON_2	c	−333.0	−197.1
$CNCl$	g	144.3	137.7
CNS^-	aq	180.7	181.2
CS_2	g	115.5	65.3
	lq	87.9	63.6
$C_2O_4^{2-}$	aq	−824.2	−674.9
$HC_2O_4^-$	aq	−818.0	−699.1
$C_2H_2O_2$	c	−826.8	−697.9
C_2H_4O	g	−235.1	−168.6
	lq	−277.8	−174.9
$C_2H_4Br_2$	lq	−80.8	−20.5
C_2H_5Cl	g	−105.0	−53.1
$C_2H_4Cl_2$	lq	−166.1	−80.3
C_2N_2	g	307.9	296.2
C_2H_3N	g	87.9	105.4
	lq	53.1	100.4
$C_2H_5O_2N$	c	−528.4	−370.7
C_2H_6S	lq	−56.9	14.2
Chlorine			
Cl_2O	g	76.1	93.7
HCl	g	−92.5	−95.4
Chromium			
Cr	c	0.0	0.0
Cr^{3+}	aq	—	−215.5
$Cr(OH)_3$	ppt	—	−859.8
CrO_4^{2-}	aq	—	736.9

Substance	State[a]	ΔH_f° kJ/mol	ΔG_f° kJ/mol
$Cr_2O_7^{2-}$	aq	—	−1319.6
Cr_3C_2	c	−87.9	−88.7
$CrCl_3$	c	−395.8	−356.5
Copper			
Cu	c	0.0	0.0
Cu^+	aq	72.4	50.2
Cu^{2+}	aq	64.4	64.9
$CuSO_4$	c	−769.9	−661.9
	aq	−843.1	−677.0
Fluorine			
F_2	g	0.0	0.0
F^-	aq	−329.7	276.6
Hydrogen			
H_2O*	g	−241.8	−228.4
H_2O_2	lq	−187.4	−118.0
Iodine			
I_2	g	62.3	19.2
HI	g	25.9	1.3
Iron			
Fe	g	404.6	359.0
	c	0.0	0.0
$FeCl_2$	c	−341.0	−302.1
$FeCl_3$	aq	−535.1	−403.8
Fe_2O_3 (hematite)	c	−822.2	−741.0
Fe_3O_4 (magnetite)	c	−1117.1	−1014.2
FeS	c	−95.0	−97.5
FeS_2 (pyrite)	c	−177.8	−166.5
Lead			
Pb	g	193.7	161.1
Pb^{2+}	aq	1.7	−24.3
$PbCO_3$	c	−700.0	−626.3
$PbCl_2$	c	−359.4	−313.8
	aq	−333.5	−286.6
PbO (red)	c	−219.2	−189.5
Lithium			
Li	c	0.0	0.0
Li^+	aq	−278.2	−293.7
LiOH	c	−487.4	−443.9
	aq	−580.4	−451.0
Magnesium			
Mg	g	150.2	115.5
	c	0.0	0.0
$MgCl_2$	c	−641.8	−592.5
	aq	−797.1	−718.4
$MgSO_4$	c	−1278.2	−1173.6
	aq	−1369.4	−1197.9

Substance	State[a]	ΔH_f° kJ/mol	ΔG_f° kJ/mol
Manganese			
Mn	g	285.8	243.5
	c	0.0	0.0
MnO	c	−384.9	−363.2
MnO_2	c	−520.9	−466.1
$MnSO_4$	c	−1063.6	−956.0
Mercury			
Hg	g	60.7	31.8
	lq	0.0	0.0
Hg^{2+}	aq	171.1	164.8
$Hg_2{}^{2+}$	aq	172.4	154.0
Nickel			
Ni	g	425.1	379.9
	c	0.0	0.0
Ni^{2+}	aq	−64.0	−46.4
Nitrogen			
N_2	g	0.0	0.0
NH_3	g	−46.0	−16.7
$(NH_4)_2CO_3$	aq	−941.8	−687.0
NH_4HCO_3	aq	−823.8	−666.5
NOBr	g	82.0	82.4
NOCl	g	52.7	66.5
NO	g	90.4	86.6
N_2O_4	g	9.6	98.3
N_2O	g	81.6	103.8
HNO_3	lq	−173.2	−79.9
Phosphorus			
P	g	314.6	279.1
P(white)	c	0.0	0.0
Potassium			
K	g	90.0	61.1
K^+	aq	−251.0	−282.4
KCl	g	−215.9	−235.1
	c	−436.0	−408.4
	aq	−418.8	−413.4
Silicon			
Si	g	368.2	323.8
	c	0.0	0.0
SiO_2 (quartz)	c	−859.4	−805.0
Silver			
Ag	g	289.1	250.2
	c	0.0	0.0
Ag^+	aq	105.9	77.0
AgBr	c	−99.6	−93.7
Ag_2CO_3	c	−506.3	−437.2
$AgNO_2$	c	−44.4	19.7
	aq	−0.4	41.0

Substance	State[a]	ΔH_f° kJ/mol	ΔG_f° kJ/mol
Sodium			
Na	g	108.8	78.2
	c	0.0	0.0
Na^+	aq	−239.7	−261.9
NaCl	c	−411.0	−384.1
	aq	−407.1	−392.9
$NaNO_3$	c	−424.7	−366.1
	aq	−446.4	−372.4
Na_2SO_4	c	−1384.5	−1266.9
	aq	−1387.0	−1265.7
Strontium			
Sr	g	164.0	110.0
	c	0.0	0.0
Sr^{2+}	aq	−545.6	−557.3
$SrCl_2$	c	−828.4	−781.2
	aq	−880.3	−818.8
$SrSO_4$	c	−1444.7	−1334.3
	aq	−1453.5	−1298.3
Sulfur			
S	g	223.0	182.4
H_2S	g	−20.1	−33.1
H_2SO_4	aq	−907.5	−741.8
SO_3^{2-}	aq	−624.3	−485.8
SO_4^{2-}	aq	−907.5	−741.8
$S_2O_2^{2-}$	aq	−644.3	−532.2
HSO_3^-	aq	−628.0	−527.2
HSO_4^-	aq	−885.8	−752.7
Zinc			
Zn	g	130.5	95.0
	c	0.0	0.0
Zn^{2+}	aq	−152.3	−147.3
$ZnCO_3$	c	−812.5	−731.4
$ZnCl_2$	c	−415.9	−369.4
	aq	−487.4	−409.6

Source: Numerical values calculated from values given in National Bureau of Standards (1952). See Table 5-6 for additional values.

[a]aq, Aqueous; g, gaseous; c, crystalline; lq, liquid; ppt, precipitate.

APPENDIX A.5.3 STANDARD ELECTRODE POTENTIALS FOR SELECTED INORGANIC SPECIES

Reduction Half-Reaction			E_H° (V)[a]
$AgCl + e^-$	$\rightleftharpoons$	$Ag + Cl^-$	0.22
$AgCN + e^-$	$\rightleftharpoons$	$Ag + CN^-$	−0.02
$Al^{3+} + 3e^-$	$\rightleftharpoons$	Al (basic)	−1.71
$H_2AlO_3^- + H_2O + 3e^-$	$\rightleftharpoons$	$Al + 4OH^-$	−2.35
$As_2O_3 + 6H^+ + 6e^-$	$\rightleftharpoons$	$2As + 3H_2O$	0.23
$HAsO_2 + 3H^+ + 3e^-$	$\rightleftharpoons$	$As + 2H_2O$	0.25
$AsO_2^- + 2H_2O + 3e^-$	$\rightleftharpoons$	$As + 4OH^-$	−0.68
$H_3AsO_4 + 2H^+ + 2e^-$	$\rightleftharpoons$	$HAsO_2 + 2H_2O$ (acidic)	0.58
$AsO_4^{3-} + 2H_2O + 2e^-$	$\rightleftharpoons$	$AsO_2^- + 4OH^-$	−0.71
$Au^+ + e^-$	$\rightleftharpoons$	Au	1.68
$Ba^{2+} + 2e^-$	$\rightleftharpoons$	Ba	−2.90
$Ba(OH)_2 + 8H_2O + 2e^-$	$\rightleftharpoons$	$Ba + 2OH^- + 8H_2O$	−2.97
$Be^{2+} + 2e^-$	$\rightleftharpoons$	Be	−1.70
$Be_2O_3^{2-} + 3H_2O + 4e^-$	$\rightleftharpoons$	$2Be + 6OH^-$	−2.28
$Br_2(aq) + 2e^-$	$\rightleftharpoons$	$2Br^-$	1.09
$HBrO + H^+ + 2e^-$	$\rightleftharpoons$	$Br^- + H_2O$	1.33
$Ca^+ + e^-$	$\rightleftharpoons$	Ca	−3.02
$Ca^{2+} + 2e^-$	$\rightleftharpoons$	Ca	−2.76
Calomel electrode, molal KCl			0.28
Calomel electrode, sat'd. KCl or NaCl			0.24
$Ca(OH)_2 + 2e^-$	$\rightleftharpoons$	$Ca + 2OH^-$	−3.02
$Ce^{3+} + 3e^-$	$\rightleftharpoons$	Ce	−2.34
$Ce^{4+} + e^-$	$\rightleftharpoons$	Ce^{3+}	1.44
$HClO + H^+ + e^-$	$\rightleftharpoons$	$\frac{1}{2}Cl_2 + H_2O$	1.63
$HClO + H^+ + 2e^-$	$\rightleftharpoons$	$Cl^- + H_2O$	1.49
$ClO^- + H_2O + 2e^-$	$\rightleftharpoons$	$Cl^- + 2OH^-$	0.90
$ClO_2 + e^-$	$\rightleftharpoons$	ClO_2^-	1.15
$HClO_2 + 2H^+ + 2e^-$	$\rightleftharpoons$	$HClO + H_2O$	1.64
$HClO_2 + 3H^+ + 3e^-$	$\rightleftharpoons$	$\frac{1}{2}Cl_2 + 2H_2O$	1.63
$HClO_2 + 3H^+ + 4e^-$	$\rightleftharpoons$	$Cl^- + 2H_2O$	1.56
$ClO_2^- + H_2O + 2e^-$	$\rightleftharpoons$	$ClO^- + 2OH^-$	0.59
$ClO_2^- + 2H_2O + 4e^-$	$\rightleftharpoons$	$Cl^- + 4OH^-$	0.76
$ClO_2(aq) + e^-$	$\rightleftharpoons$	ClO_2^-	0.95
$ClO_3^- + 2H^+ + e^-$	$\rightleftharpoons$	$ClO_2 + H_2O$	1.15
$ClO_3^- + 3H^+ + 2e^-$	$\rightleftharpoons$	$HClO_2 + H_2O$	1.21
$ClO_3^- + 6H^+ + 5e^-$	$\rightleftharpoons$	$\frac{1}{2}Cl_2 + 3H_2O$	1.47
$ClO_3^- + 6H^+ + 6e^-$	$\rightleftharpoons$	$Cl^- + 3H_2O$	1.45
$ClO_3^- + H_2O + 2e^-$	$\rightleftharpoons$	$ClO_2^- + 2OH^-$	0.35
$ClO_3^- + 3H_2O + 6e^-$	$\rightleftharpoons$	$Cl^- + 6OH^-$	0.62
$ClO_4^- + 2H^+ + 2e^-$	$\rightleftharpoons$	$ClO_3^- + H_2O$	1.19
$ClO_4^- + 8H^+ + 7e^-$	$\rightleftharpoons$	$\frac{1}{2}Cl_2 + 4H_2O$	1.34
$ClO_4^- + 8H^+ + 8e^-$	$\rightleftharpoons$	$Cl^- + 4H_2O$	1.37
$ClO_4^- + H_2O + 2e^-$	$\rightleftharpoons$	$ClO_3^- + 2OH^-$	0.17
$(CN)_2 + 2H^+ + 2e^-$	$\rightleftharpoons$	2HCN	0.37

Reduction Half-Reaction			E_H° (V)[a]
$2HCNO + 2H^+ + 2e^-$	$\rightleftharpoons$	$(CN)_2 + 2H_2O$	0.33
$(CNS)_2 + 2e^-$	$\rightleftharpoons$	$2CNS^-$	0.77
$Cr^{2+} + 2e^-$	$\rightleftharpoons$	Cr	−0.56
$Cr^{3+} + e^-$	$\rightleftharpoons$	Cr^{2+}	−0.41
$Cr^{6+} + 3e^-$	$\rightleftharpoons$	Cr^{3+} (basic)	−0.12
$Cr_2O_7^{2-} + 14H^+ + 6e^-$	$\rightleftharpoons$	$2Cr^{3+} + 7H_2O$	1.33
$CrO_2^- + 2H_2O + 3e^-$	$\rightleftharpoons$	$Cr + 4OH^-$	−1.2
$HCrO_4^- + 7H^+ + 3e^-$	$\rightleftharpoons$	$Cr^{3+} + 4H_2O$	1.20
$CrO_4^{2-} + 4H_2O + 3e^-$	$\rightleftharpoons$	$Cr(OH)_3 + 5OH^-$	−0.12
$Cr(OH)_3 + 3e^-$	$\rightleftharpoons$	$Cr + 3OH^-$	−1.3
$Cs^+ + e^-$	$\rightleftharpoons$	Cs	−2.92
$Cu^+ + e^-$	$\rightleftharpoons$	Cu	0.52
$Cu^{2+} + 2CN^- + e^-$	$\rightleftharpoons$	$Cu(CN)_2^-$	1.12
$Cu^{2+} + e^-$	$\rightleftharpoons$	Cu^+	0.16
$Cu(OH)_2 + 2e^-$	$\rightleftharpoons$	$Cu + 2OH^-$	−0.22
$Fe^{3+} + 3e^-$	$\rightleftharpoons$	Fe	−0.04
$Fe(CN)_6^{3-} + e^-$	$\rightleftharpoons$	$Fe(CN)_6^{4-}$ (basic)	0.46
$Fe(CN)_6^{3-} + e^-$	$\rightleftharpoons$	$Fe(CN)_6^{4-}$ (acidic)	0.69
$FeO_4^{2-} + 8H^+ + 3e^-$	$\rightleftharpoons$	$Fe^{3+} + 4H_2O$	1.90
$Fe(OH)_3 + e^-$	$\rightleftharpoons$	$Fe(OH)_2 + OH^-$	−0.56
$\frac{1}{2}H_2 + e^-$	$\rightleftharpoons$	H^-	−2.23
$HO_2 + H^+ + e^-$	$\rightleftharpoons$	H_2O_2	1.50
$Mg^{2+} + 2e^-$	$\rightleftharpoons$	Mg	−2.38
$Mg(OH)_2 + 2e^-$	$\rightleftharpoons$	$Mg + 2OH^-$	−2.67
$MnO_2 + 4H^+ + 2e^-$	$\rightleftharpoons$	$Mn^{2+} + 2H_2O$	1.21
$MnO_4^- + e^-$	$\rightleftharpoons$	MnO_4^{2-}	0.56
$MnO_4^- + 4H^+ + 3e^-$	$\rightleftharpoons$	$MnO_2 + 2H_2O$	1.68
$MnO_4^- + 8H^+ + 5e^-$	$\rightleftharpoons$	$Mn^{2+} + 4H_2O$	1.49
$MnO_4^- + 2H_2O + 3e^-$	$\rightleftharpoons$	$MnO_2 + 4OH^-$	0.59
$MnO_4^- + 2H_2O + 3e^-$	$\rightleftharpoons$	$MnO_2 + 4OH^-$	0.58
$Mn(OH)_2 + 2e^-$	$\rightleftharpoons$	$Mn + 2OH^-$	−1.47
$N_2 + 2H_2O + 4H^+ + 2e^-$	$\rightleftharpoons$	$2NH_3OH^+$	−1.87
$3N_2 + 2H^+ + 2e^-$	$\rightleftharpoons$	$2HN_3$	−3.1
$N_2H_5^+ + 3H^+ + 2e^-$	$\rightleftharpoons$	$2NH_4^+$	1.27
$H_2N_2O_2 + 2H^+ + 2e^-$	$\rightleftharpoons$	$N_2 + 2H_2O$	2.65
$N_2O_4 + 2e^-$	$\rightleftharpoons$	$2NO_2^-$	0.88
$N_2O_4 + 2H^+ + 2e^-$	$\rightleftharpoons$	$2HNO_2$	1.07
$N_2O_4 + 4H^+ + 4e^-$	$\rightleftharpoons$	$2NO + 2H_2O$	1.03
$Na^+ + e^-$	$\rightleftharpoons$	Na	−2.71
$2NH_3OH^+ + H^+ + 2e^-$	$\rightleftharpoons$	$N_2H_5^+ + 2H_2O$	1.42
$2NO + 2e^-$	$\rightleftharpoons$	$N_2O_2^{2-}$	0.10
$2NO + H_2O + 2e^-$	$\rightleftharpoons$	$N_2O + 2OH^-$	0.76
$2HNO_2 + 4H^+ + 4e^-$	$\rightleftharpoons$	$H_2N_2O_2 + 2H_2O$	0.80
$2HNO_2 + 4H^+ + 4e^-$	$\rightleftharpoons$	$N_2O + 3H_2O$	1.27
$NO_2^- + H_2O + e^-$	$\rightleftharpoons$	$NO + 2OH^-$	−0.46
$2NO_2^- + 2H_2O + 4e^-$	$\rightleftharpoons$	$N_2O_2^{2-} + 4OH^-$	−0.18
$2NO_2^- + 3H_2O + 4e^-$	$\rightleftharpoons$	$N_2O + 6OH^-$	0.15

Reduction Half-Reaction			E°_H (V)[a]
$NO_3^- + 4H^+ + 3e^-$	⇌	$NO + 2H_2O$	0.96
$2NO_3^- + 4H^+ + 2e^-$	⇌	$N_2O_4 + 2H_2O$	0.80
$NO_3^- + H_2O + 2e^-$	⇌	$NO_2^- + 2OH^-$	0.01
$2NO_3^- + 2H_2O + 2e^-$	⇌	$N_2O_4 + 4OH^-$	−0.85
$\frac{1}{2}O_2 + 2H^+(10^{-7}\,M) + 2e^-$	⇌	H_2O	0.81
$O_2 + 2H^+ + 2e^-$	⇌	H_2O_2	0.68
$O_2 + H_2O + 2e^-$	⇌	$HO_2^- + OH^-$	−0.08
$O_2 + 2H_2O + 2e^-$	⇌	$H_2O_2 + 2OH^-$	−0.15
$O_2 + 2H_2O + 4e^-$	⇌	$4OH^-$	0.40
$O_3 + H_2O + 2e^-$	⇌	$O_2 + 2OH^-$	1.24
$O(g) + 2H^+ + 2e^-$	⇌	H_2O	2.42
$OH + e^-$	⇌	OH^-	1.4
$HO_2^- + H_2O + 2e^-$	⇌	$3OH^-$	0.87
$PbO + H_2O + 2e^-$	⇌	$Pb + 2OH^-$	−0.58
$PbO_2 + 4H^+ + 2e^-$	⇌	$Pb^{2+} + 2H_2O$	1.46
$PbO_2 + H_2O + 2e^-$	⇌	$PbO + 2OH^-$	0.28
$PbO_2 + SO_4^{2-} + 4H^+ + 2e^-$	⇌	$PbSO_4 + 2H_2O$	1.69
$PbSO_4 + 2e^-$	⇌	$Pb + SO_4^{2-}$	−0.36
$H_3PO_4 + 2H^+ + 2e^-$	⇌	$H_3PO_3 + H_2O$	−0.28
$PO_4^{3-} + 2H_2O + 2e^-$	⇌	$HPO_3^- + 3OH^-$	−1.05
$Pt^{2+} + 2e^-$	⇌	Pt	1.12
$S + 2H^+ + 2e^-$	⇌	$H_2S(aq)$	0.14
$S + H_2O + 2e^-$	⇌	$HS^- + OH^-$	−0.48
$S_2O_6^{2-} + 4H^+ + 2e^-$	⇌	$2H_2SO_3$	0.6
$S_2O_8^{2-} + 2e^-$	⇌	$2SO_4^{2-}$	2.0
$S_4O_6^{2-} + 2e^-$	⇌	$2S_2O_3^{2-}$	0.09
$Se + 2H^+ + 2e^-$	⇌	$H_2Se(aq)$	−0.36
$H_2SeO_3 + 4H^+ + 4e^-$	⇌	$Se + 3H_2O$	0.74
$SeO_3^{2-} + 3H_2O + 4e^-$	⇌	$Se + 6OH^-$	−0.35
$SeO_4^{2-} + 4H^+ + 2e^-$	⇌	$H_2SeO_3 + H_2O$	1.15
$SeO_4^{2-} + H_2O + 2e^-$	⇌	$SeO_3^{2-} + 2OH^-$	0.03
$Sn^{4+} + 2e^-$	⇌	Sn^{2+}	0.15
$2H_2SO_3 + H^+ + 2e^-$	⇌	$HS_2O_4^- + 2H_2O$	−0.08
$H_2SO_3 + 4H^+ + 4e^-$	⇌	$S + 3H_2O$	0.45
$2SO_3^{2-} + 2H_2O + 2e^-$	⇌	$S_2O_4^{2-} + 4OH^-$	−1.12
$2SO_3^{2-} + 3H_2O + 4e^-$	⇌	$S_2O_3^{2-} + 6OH^-$	−0.58
$SO_4^{2-} + 4H^+ + 2e^-$	⇌	$H_2SO_3 + H_2O$	0.20
$2SO_4^{2-} + 4H^+ + 2e^-$	⇌	$S_2O_6^{2-} + H_2O$	−0.2
$SO_4^{2-} + H_2O + 2e^-$	⇌	$SO_3^{2-} + 2OH^-$	−0.92
$ZnO_2^{2-} + 2H_2O + 2e^-$	⇌	$Zn + 4OH^-$	−1.22

Source: Numerical values from *Handbook of Chemistry and Physics* (1982). See Table 5-8 for additional values.

[a]E°_H values given with respect to the standard hydrogen electrode at 25°C.

6

ENERGY RELATIONSHIPS: APPLICATIONS TO HETEROGENEOUS SYSTEMS

6.1 PHASE TRANSFORMATIONS

While discussions of chemical reactions in Chapter 5 focused primarily on the roles of chemical structure, reactivity, and bonding interactions in *homogeneous* or *single-phase* transformations, several examples of *heterogeneous* or *multiphase* interactions (e.g., Equations 5-12, 5-13, 5-24; Figures 5-2 through 5-6) were also given. As suggested by the examples, general thermodynamic relationships developed for homogeneous systems can be applied to multiphase systems, but somewhat different and additional considerations apply when chemical species change phase or migrate from one phase to another.

Distinctions between different types of phase changes are important in extending our considerations of thermodynamic principles to heterogeneous systems. The most elementary type of phase change involves conversion of pure chemical substances from one physical state to another. Simple examples of these are the freezing and boiling of water to form ice and water vapor, respectively; they are accompanied by significant changes in temperature and pressure (see Figure 5-10). Water and waste treatment processes that rely on such phase changes include (1) the freezing or distillation of water to separate inorganic impurities, (2) the use of supercritical fluids to facilitate the dissolution or oxidative destruction of organic impurities, and (3) the condensation of organic vapors to allow their removal from gas streams. A similar but more complex phase change occurs when two dissolved or aqueous-phase chemical species combine to form a precipitate, a solid phase. Precipitation equilibria were illustrated for metal ions and inorganic ligands in Figures 5-4 through 5-6 and in Example 5-4. The formation of a solid phase is caused by input of chemical energy in the form of increased masses of reactants, or from inputs or extraction of thermal energy that change the solubility relationships between reacting species. This type of phase change occurs in a broad range of environmental processes, particularly those involving reactions of metal ions with organic and inorganic ligands to form solid precipitates.

Another frequent type of phase transformation is one in which a relatively minor component of one phase migrates to another without significant change in the temperature or pressure of either phase. Some common examples are (1) the dissolution of oxygen from air into water; (2) the volatilization of an organic contaminant, such as benzene, from water to air; (3) the precipitation of a heavy metal, such as cadmium, from a dissolved state in water to a solid oxide; or (4) the sorption of a hydrophobic contaminant, such as a polychlorinated biphenyl (PCB), by lake sediments or by activated carbon in an engineered waste treatment system. When minor components of one phase migrate to another, their chemical structure may or may not change. The motivation for migration is instead to achieve a greater degree of compatibility between chemical structures of the component and the phase with which it associates.

Phase transformations cause temporary phase instability. To cite one example, molecular oxygen has a limited solubility in water, as discussed in Chapter 1. If amounts of oxygen that exceed this limited solubility at a given temperature and pressure are forced into water, local pressures increase momentarily and an unstable condition results. This instability eventually causes excess oxygen to migrate to the air phase until a balanced distribution of oxygen between the water and air is achieved. Oxygen is in a zero-oxidation state in both phases, a state that is more thermodynamically compatible with air than with water. As its oxidation state is reduced to −II by reactions in the water, the oxygen becomes more compatible with the polar solvent water than with air.

Because of their form, certain chemical substances are not particularly compatible with any phase to which they have access in a given system. Instead, these substances gather at interfaces between phases, sometimes forming their own interfacial "phase." Interfacial accumulation is typified by surface-active agents, such as detergents, which are only marginally compatible with water and even less so with air. Such substances thus tend to form *foams* at air–water interfaces.

As the foregoing discussion illustrates, chemical form not only affects reactivity within individual phases, but also the distribution of species among and between phases and phase interfaces. Chemical form thus governs the distribution of different types of chemicals among different compartments of natural systems. This explains why on discharge to water large proportions of chemicals such as benzene and trichloroethylene eventually find their way to the atmosphere, while PCBs and polynuclear aromatic hydrocarbons (PAHs) are found primarily in sediments. The tendency of species to distribute differently among phases forms the basis for separation technologies employed in engineered systems. Two examples of frequently used technology are air stripping for removal of benzene and trichloroethylene from contaminated groundwater, and adsorption by activated carbon for removal of PCBs and PAHs from contaminated waste streams.

6.2 EQUILIBRIA IN HETEROGENEOUS SYSTEMS

6.2.1 General Concepts

We are concerned in this chapter with several aspects of the interactions of species among multiple phases in complex environmental systems. Chemical species ranging from molecular level to particle level are discussed. We must consider such a wide span of entities because contaminants and other chemical species exist in various forms in environmental systems. This in turn leads to the existence of various types of phase transfer and separation processes. For instance, molecules of certain soluble species can interact with one another to form polymolecular (polynuclear) associations, or polymers, of sufficient size to transform them from the dissolved state to the macromolecular or colloidal state; proteins behave in this manner. Alternatively, soluble, well-defined molecular species may associate with what are often poorly defined macromolecules or colloids, such as humic and fulvic acids, clays, bacteria, or algae. In such cases the fate and behavior of the molecular species of concern may change markedly, and thermodynamic "stability" in a given environmental system may be affected entirely by parameters and features unique to that system. A good example of species that undergo shifts in thermodynamic stability is high-molecular-weight PCBs. As dissolved species PCBs are unstable in aquatic environments. When attached to a macromolecule or colloid, however, a PCB may remain in a stable or metastable suspended state. Its transport with the flow of water may thus be facilitated.

We begin our considerations of energy relationships in heterogeneous systems with relatively simple systems in which substances are directly transferred from a "dissolved" state in one phase to a "dissolved" state in another phase. We then extend those considerations to the interfaces between phases and to the behavior of macromolecules and colloids. Finally, we use thermodynamic considerations to determine the distributions of different species in different environmental phases, or compartments.

The logical starting point for our discussions lies in criteria that govern phase behavior. These criteria are rooted in thermodynamic principles that translate into familiar constants and properties of different substances with respect to different phases, such as (1) Henry's constant, $\mathcal{K}_H$; (2) the octanol-water partition coefficient, $\mathcal{K}_{o,w}$; (3) aqueous phase solubility, C_S; (4) vapor pressure, P^v; and (5) melting and boiling points, T_m and T_b, respectively. Each of these parameters is linked to particular thermodynamic criteria for phase behavior for which we develop relationships in this chapter. Table 6-1 and Appendix A.6.1 are tabulations of the above referenced constants and properties for selected carbon compounds of particular environmental significance.

6.2.2 Energetic Criteria

Certain thermodynamic requirements exist for establishment of equilibrium conditions between components in different phases of heterogeneous systems. First and most simply, for systems that are closed and for which the phases are each of constant volume and composition, the zeroth law stipulates that these phases must be equal in temperature for thermal equilibrium to exist; that is, there can be no interphase heat flow at equilibrium. Thus, for two phases, j and k,

$$d(S)_{j,k}\Big|_{P,n_i} = d(S)_j - d(S)_k\Big|_{P,n_i} = 0 \qquad T_j = T_k \tag{6-1}$$

Table 6-1 Phase Properties for Selected Organic Compounds[a]

Formula	Compound	$\mathcal{K}_H$ $\left(\frac{\text{atm} - \text{m}^3}{\text{mol}} \times 10^3\right)$	$\log \mathcal{K}_{o,w}$	C_S (mg/L)	P^v (atm × 10^3)	T_m (°C)	T_b (°C)
CH_4	Methane	639 ± 20	1.09	22.7	275.4 × 10^{3b}	−182.5	−161.5
CH_3Cl	Chloromethane (methyl chloride)	9.6	0.91	5,287	5.65 × 10^{3b}	−97.7	−24.2
CH_2Cl_2	Dichloromethane	2.6	1.15	19,442	588.8	−95.1	39.7
$CHCl_3$	Trichloromethane (chloroform)	3.3	1.97	7,709	228.5	−63.5	61.7
CCl_3F	Trichlorofluoromethane	113.3	2.53	1,100	1.05 × 10^3	−111.0	23.8
CCl_4	Tetrachloromethane (carbon tetrachloride)	24.0	2.73	969	151.4	−22.9	76.5
C_2H_6	Ethane	490 ± 20	1.81	61.4	39.8 × 10^{3b}	−183.3	−88.6
$C_2H_4Cl_2$	1,1-Dichloroethane	6.0	1.79	4,960	301.9	−97.0	57.5
$C_2H_4Cl_2$	1,2-Dichloroethane	1.1	1.47	8,425	91.2	−35.4	83.5
C_2H_3Cl	Chloroethane (vinyl chloride)	22.4	0.60	2,790	3.89 × 10^3	−153.8	−13.4
$C_2H_3Cl_3$	1,1,1-Trichloroethane	23.2	2.47	730	126.7	−32.0	113.0
$C_2H_3Cl_3$	1,1,2-Trichloroethane	0.9	1.89	4,500	29.1	−36.5	113.8
$C_2H_2Cl_2$	1,1-Dichloroethylene	154.5	2.13	400	791.2	−122.1	37.0
C_2HCl_3	Trichloroethylene	11.7	2.29	1,100	97.7	−73.0	87.0
C_2Cl_4	Tetrachloroethylene	27.5	2.88	151.2	25.1	−19.0	121.0
C_6H_{14}	*n*-Hexane	768	4.0 ± 0.25	11 ± 1	199.9	−95.4	68.7
C_6H_6	Benzene	5.4 ± 0.3	2.13 ± 0.10	1,770 ± 40	125.7	5.5	80.1
C_6H_6O	Phenol	4.0 × 10^{-4}	1.50 ± 0.05	80,200	0.70	40.6	181.9
C_6H_5Cl	Chlorobenzene	3.7	2.84	472	15.6	−46.5	132.0
$C_6H_5NO_2$	Nitrobenzene	2.4 × 10^{-2}	1.85 ± 0.05	1,900 (20°C)	0.20	5.7	210.6
$C_6H_4Cl_2$	1,4-Dichlorobenzene	1.6	3.40	83.1	0.89	53.1	174.0
$C_6H_3Cl_3$	1,2,4-Trichlorobenzene	3.2	4.00	34.6	0.60	17.0	213.5
C_6Cl_6	Hexachlorobenzene	412.4	5.47 ± 0.02	0.005	2.3 × 10^{-5}	230.0	322.0

Table 6-1 (*Continued*)

Formula	Compound	$\mathcal{K}_H$ $\left(\frac{\text{atm}-\text{m}^3}{\text{mol}} \times 10^3\right)$	log $\mathcal{K}_{o,w}$	C_S (mg/L)	P'' (atm × 10^3)	T_m (°C)	T_b (°C)
C_7H_8	Toluene	6.6 ± 0.4	2.73 ± 0.10	530 ± 20	37.6	−94.9	110.6
C_8H_{18}	*n*-Octane	2,960 ± 490	5.15 ± 0.45	0.71 ± 0.09	18.6	−56.8	125.0
$C_8H_{18}O$	1-Octanol	15.9 × 10^{-3}	3.07 ± 0.10	540 ± 50	6.6 × 10^{-2}	−15.4	195.2
C_8H_{10}	Ethylbenzene	7.9 ± 0.7	3.15 ± 0.20	169 ± 9	12.6	−94.9	136.2
C_8H_{10}	*o*-Xylene	4.9 ± 0.6	3.12 ± 0.20	173 ± 5	8.7	−25.2	144.4
$C_{10}H_{22}$	*n*-Decane	6,900 ± 300	6.25 ± 0.70	0.015	1.74	−29.6	174.1
$C_{10}H_8$	Naphthalene	4.2 × 10^{-1}	3.35 ± 0.10	31 ± 1	0.10	80.2	217.9
$C_{12}H_{26}$	*n*-Dodecane	7,400 ± 2,960	6.80 ± 1.0	0.0037	0.16	−9.6	216.2
$C_{12}H_{10}$	Acenaphthene	2.4 × 10^{-1}	3.92 ± 0.25	3.8 ± 0.2	7.9 × 10^{-3}	95	277.5
$C_{14}H_{10}$	Anthracene	6.0 × 10^{-2}	4.50 ± 0.15	0.062	7.9 × 10^{-6}	217.5	342.0
$C_{14}H_{10}$	Phenanthrene	4.0 × 10^{-2}	4.52 ± 0.15	1.1 ± 0.1	1.6 × 10^{-4}	99.5	340.2
$C_{16}H_{22}O_4$	*n*-Butyl phtalate	1.8 × 10^{-3}	4.72	3.2	1.3 × 10^{-2}	135.0	340.0
$C_{16}H_{10}$	Pyrene	8.9 × 10^{-3}	5.22	0.135	5.9 × 10^{-6}	156.0	360.0
$C_{20}H_{12}$	Benzo[*a*]pyrene	4.6 × 10^{-4}	6.04	4.0 × 10^{-3}	6.9 × 10^{-9}	175.0	—

[a]The numerical values given in this table are presented only by way of general illustration. They are derived from a number of sources (see Section 6.9.1). In some cases they have been transformed in units, and in other cases calculated from tabulated data. Some values for Henry's constant, for example, have been calculated from vapor pressure and solubility data, while some for vapor pressure have similarly been calculated from Henry's constant. It should be noted that significant variations in the numerical values of such constants reported in different literature sources are common. When sufficient numbers of values were available, typical ranges of variation are reported above. Values given for 25°C unless otherwise stated.

[b]Vapor pressure of supercooled liquid (cooled below melting point, but still a liquid).

Second, and more important for our considerations of phase equilibria, if the pressures and temperatures of the two phases are constant, then

$$(\mu_i \, dn_i)_{j,k} \Big|_{P,T} = -(\mu_i \, dn_i)_j + (\mu_i \, dn_i)_k \Big|_{P,T} = 0 \qquad \mu_{i,j} = \mu_{i,k} \tag{6-2}$$

Thus, for constant temperature and pressure, the thermodynamic requirement for a stable equilibrium condition is that the chemical potential of each component must be the same in each phase.

6.2.3 Phase Exchange Relationships

The two ideal-solution laws that govern the equilibrium behavior of chemical species distributed between adjacent gas and liquid phases were discussed in Chapter 5. The first of these, *Raoult's law*, governs the behavior of pure or nearly pure solid or liquid substances (i.e., highly concentrated) with respect to pure vapor phases. The relationships generated by this law, and given earlier in Equation 5-49 for fugacity as a function of mole fraction, were illustrated in Figure 5-11. This figure also shows the relationships predicted for extremely dilute solutions by the second of the ideal solution laws, *Henry's law*.

Figure 6-1a depicts a closed two-phase system comprised by pure water and pure air. Consider the thermodynamic relationships that govern the vapor pressure of water in air and the concentration of oxygen in water. Driven by the vapor pressure of water, a net evolution of water vapor into the air occurs until a phase equilibrium is attained. The primary determinant of the chemical potential of water in the air is the vapor pressure of water. The equilibrium distribution of *water* between the two phases is, therefore, governed by Raoult's law. Oxygen, which has a mole fraction of $X_{O_2} \approx 0.21$ in air, is dissolved in the water until an equilibrium determined largely by the aqueous-phase activity of oxygen is achieved. Because the oxygen is very dilute in the solvent water, with a mole fraction of $X_{O_2} \approx 10^{-5}$, the equilibrium distribution of *oxygen* between the two phases is governed by Henry's law.

Henry's constant, which was introduced in Chapter 1 and discussed in Chapter 5, can be explored further using the fugacity principle. The fugacities of a solute i in the liquid ($f_{i,l}$) and vapor ($f_{i,v}$) phase are equal at equilibrium, so that from Equations 5-52 and 5-53:

$$f_{i,l} = \alpha_{i,l} X_{i,l} P_i^v = f_{i,v} = \alpha_{i,v} P_i \tag{6-3}$$

in which $X_{i,l}$ is the mole fraction of the solute in the liquid, P_i^v is the vapor pressure of the pure solute, P_i is the actual pressure exerted by the solute in the vapor phase, and $\alpha_{i,l}$ and $\alpha_{i,v}$ are the activity coefficients in the liquid and vapor phases, respectively. The activity coefficient in the vapor phase is as-

(a). Two-Phase System: Air and Water

Equilibrium Condition

$\mu_{O_2, air} = \mu_{O_2,H_2O}$

$\mu_{H_2O, air} = \mu_{H_2O,H_2O}$

Governing Law

Henry's Law

Raoult's Law

(b)   Three-Phase System: Air, Xylene, and Water

Equilibrium Condition

$\mu_{xyl, air} = \mu_{xyl, xyl} = \mu_{xyl, H_2O}$

Governing Law

Raoult's Law

(c) A Two-Phase System: Air and Xylene Dissolved in Water

Equilibrium Condition

$\mu_{xyl,air} = \mu_{xyl,H_2O}$

Governing Law

Henry's Law

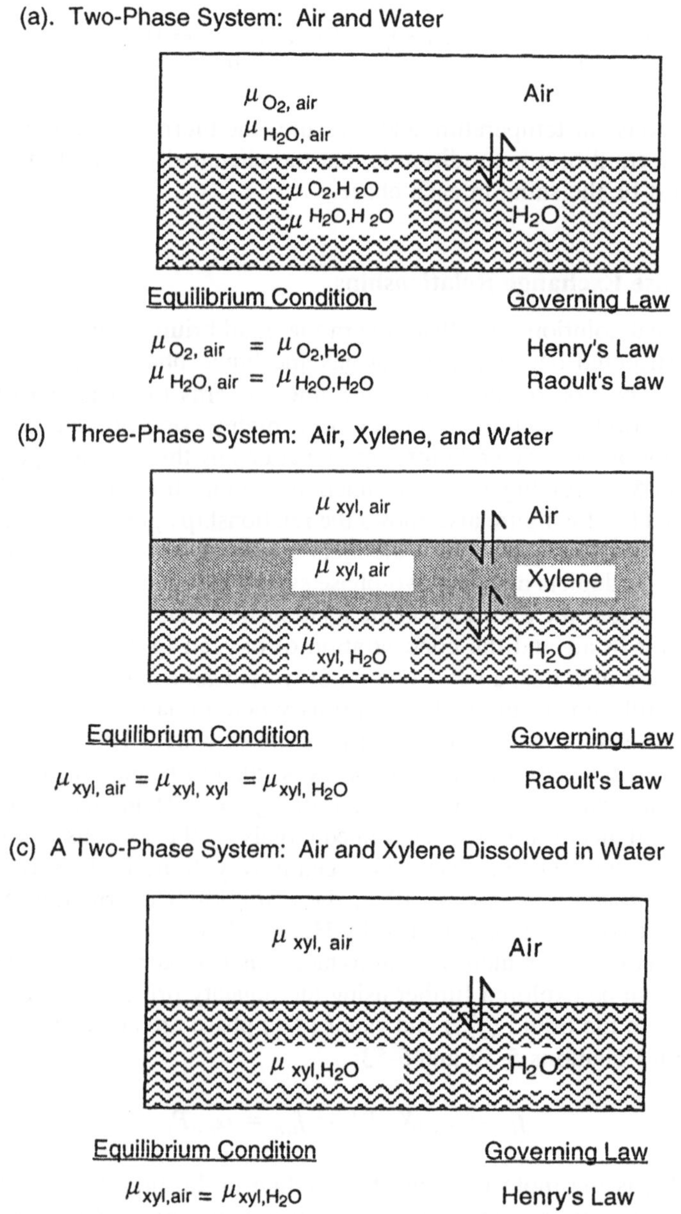

Figure 6-1 Applications of Raoult's and Henry's laws for closed systems.

sumed to be 1, but the activity coefficient in the liquid phase is 1 only for a *pure* solute. The Henry's law relationship (Equations 1-1, 1-2, and 5-54) is thus written

$$P_i = (\alpha_{i,l} P_i^v) X_{i,l} \tag{6-4}$$

from which we conclude that the Henry's constant, $\mathcal{K}_H$, for solute i is

$$\mathcal{K}_{H,i} = \alpha_{i,l} P_i^v \qquad \text{(in atm)}$$

Based on the expression above, Henry's constant can be calculated from data for the activity coefficient of a solute (correlations exist between activity coefficients and molecular properties of solutes) and for vapor pressure.

Another estimation procedure can be used for sparingly soluble solutes. Activity is unity at the solubility limit, and thus it follows that

$$\alpha_{i,l} = \frac{1}{X_{S,i}} \tag{6-5}$$

where $X_{S,i}$ is the mole fraction of the solvent at its aqueous solubility limit. Substituting for the activity coefficient in the Henry's constant defined above gives

$$\mathcal{K}_{H,i} = \frac{P_i^v}{X_{S,i}} \qquad \text{(in atm)} \tag{6-6}$$

This method of calculating Henry's constant has serious limitations, although vapor pressure and solubility data for some environmentally important solutes are available (see, for example, Table 6-1 and Appendix A.6.1). Vapor pressures for many organic solutes of interest are very low at ambient temperatures and thus not easily measured. Errors can occur in extrapolating vapor pressure data from higher temperatures. Moreover, the vapor pressure must be extrapolated through the triple point (see Figure 5-10) if the solute is a solid, and this leads to additional error.

A variety of Henry's law expressions having different units and dimensions are commonly used in the literature, often causing a degree of confusion. The units and associated values vary with the units used to describe component concentrations in the gas and liquid phases. The most common expressions and the relationships among them are presented in Table 6-2. For gas-phase concentrations in atmospheres and solution phase concentrations in moles per cubic meter, the units of $\mathcal{K}_H$ are atm-m^3/mol. For concentrations expressed as dimensionless mass or molar ratios (e.g., partial pressures, p, and mole fractions, X), Henry's constant is dimensionless and designated as $\mathcal{K}_H^\circ$. Values of $\mathcal{K}_H$ in atm-m^3/mol can be converted to dimensionless values of $\mathcal{K}_H^\circ$ (see the

Table 6-2 Common Expressions and Interrelationship for Henry's Law

Expression for Henry's Law	Definition of Terms	Units for Henry's Constant
$P_i = \mathcal{K}_{H,i}X_i$	X_i = mole fraction of component i in liquid phase. P_i = pressure exerted by component i in vapor phase (atm)	atm (per mole fraction)
$P_i = \mathcal{K}_{H,i}C_{i,l}$	$C_{i,l}$ = molar concentration of component i in liquid phase	$\frac{\text{atm} - \text{m}^3}{\text{mol}}$ *Note:* $\mathcal{K}_{H,i}\left(\frac{\text{atm} - \text{m}^3}{\text{mol}}\right) = \frac{\mathcal{K}_{H,i}(\text{atm})\ X_i}{C_{i,l}} = \frac{\mathcal{K}_{H,i}(\text{atm})}{55.6 \times 10^3} = \mathcal{K}_{H,i}(\text{atm}) \times 18 \times 10^{-6}$
$p_i = \mathcal{K}^{\circ}_{H,i}X_i$	p_i = partial pressure of component i in vapor phase, P_i/P_T P_T = total pressure of vapor phase (atm)	Dimensionless (partial pressure) *Note:* $\mathcal{K}^{\circ}_{H,i}(\text{partial pressure}) = \frac{\mathcal{K}_{H,i}(\text{atm})}{P_T}$
$C_{i,v} = \mathcal{K}^{\circ}_{H,i}C_{i,l}$	$C_{i,v}$ = molar concentration of component i in vapor phase, where $C_{i,v} = \frac{P_i}{\mathcal{R}T}$	Dimensionless (molar concentration ratio) *Note:* $\mathcal{K}^{\circ}_{H,i}(\text{conc. ratio}) = \mathcal{K}_{H,i}(\text{atm})\frac{X_i}{\mathcal{R}TC_{i,l}} = \mathcal{K}_{H,i}\left(\frac{\text{atm} - \text{m}^3}{\text{mol}}\right)\frac{1}{\mathcal{R}T}$

last entry in Table 6-2) by dividing by $\mathcal{R}T$, where $\mathcal{R}$ is the universal gas constant and T is absolute temperature. The value of $\mathcal{R}$ is 8.314 J/mol-K = 8.314 Pa-m^3/mol-K, = 1.987 cal/mol-K = 8.21×10^{-5} atm-m^3/mol-K. Thus at 20°C (293 K), $\mathcal{K}^{\circ}_H$ (dimensionless) = $\mathcal{K}_H$ (atm-m^3/mol) × 41.57. One of these expressions may in any given instance be more convenient to use than another, depending on the situation being analyzed.

A closed three-phase system represented by air, water, and xylene is illustrated in Figure 6-1b. Here the concentration of xylene in the air is driven by the vapor pressure of xylene, whereas its concentration in the water is determined by its activity in the aqueous phase. If the pure xylene phase is removed from the system, as depicted in Figure 6-1c, the equilibrium condition for xylene between the two remaining phases is driven by its contribution to the pressure of the air phase and its activity in the aqueous phase; for this situation Henry's law governs the phase distribution. These different governing conditions in phase exchange reactions render the use of fugacity more convenient than activity or chemical potential for multiphase systems.

Various components that are mixed on an intermolecular basis with each other and distributed between gas and liquid phases can be thought of as being *dissolved* or *absorbed* in those phases. Their equilibria are accordingly dictated by the chemical potential and fugacity relationships discussed in Chapter 5. This intraphase solute distribution, illustrated schematically by the sketch on the left-hand side of of Figure 6-2a, is typically referred to as the *absorption*

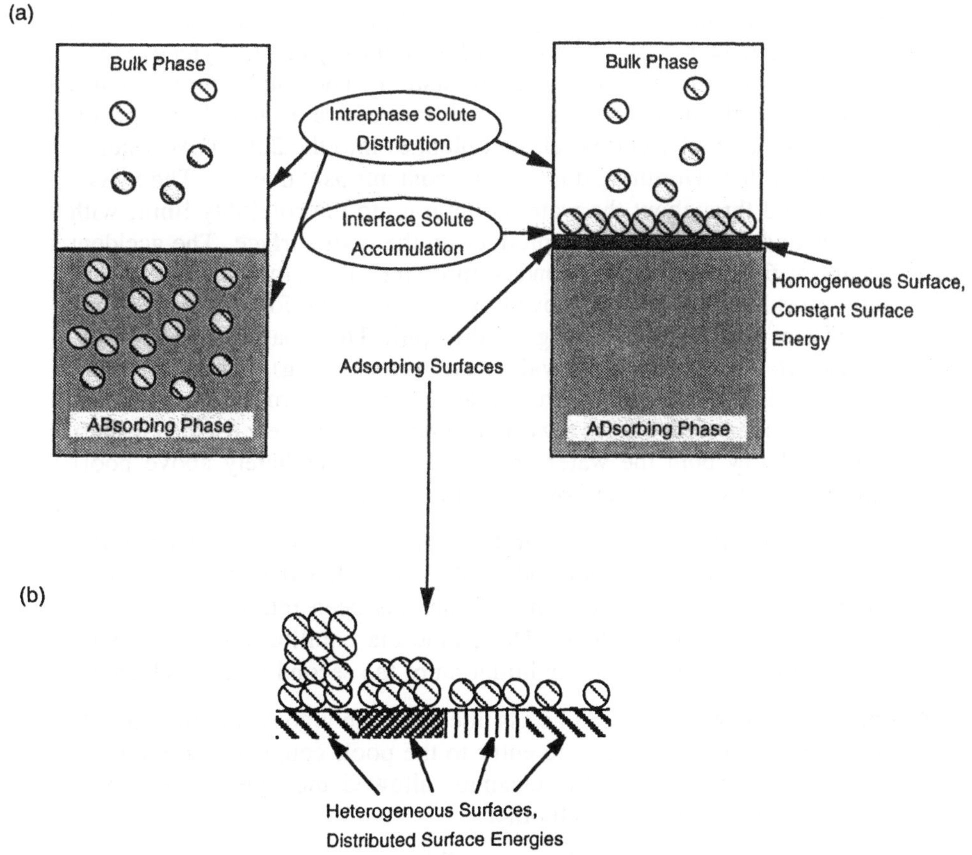

Figure 6-2 Schematic characterizations of absorption and adsorption processes.

of a substance by a phase. Absorption processes can occur between gas and liquid phases, different gas phases, different liquid phases, and the phases of some gas–solid and liquid–solid systems as well. As illustrated by the sketch on the right-hand side of Figure 6-2a, other phase separation processes involving gases, solids, and liquids result in accumulation of a component of one phase at an interface between that phase and another. This type of process is typically referred to as *adsorption*. It is different from absorption, both thermodynamically and phenomenologically. Figure 6-2b illustrates that the energetics of adsorption processes can be much more complex, particularly if the interfaces or surfaces at which adsorption occurs are heterogeneous in character.

Example 6-1 illustrates how phase distributions of substance by absorption phenomena can be quantified by application of linear relationships such as Henry's and Raoult's laws.

Example 6-1

- **Situation:** During a particularly heavy and prolonged rainstorm, the air brakes on a tanker truck containing 10 tons of *o*-xylene (2-xylene) fail as it turns off a highway into a fueling station. The truck jumps a curb, crashes through one wall of a glass enclosure over a heated (25°C) swimming pool at an adjacent motel, and plunges into the 250 m^3 of water in the pool. It then overturns, dumping its contents as it does so. The xylene is dissolved throughout the water column up to the solubility limit, with the remainder floating as free product on the pool surface. The accident occurs at 6:00 A.M. on a Monday morning. The emergency response authorities have the truck removed by 9:00 A.M. the following morning, and at that time begin cleaning up the spill. The cleanup operation involves two stages: first, removal of the free (floating) xylene, and second, treatment of the water with activated carbon to remove the dissolved xylene. It is imperative from a worker safety point of view for the cleanup crews to know both the water and local air (immediately above pool) concentrations of xylene at both stages of operation.

- **Question(s):** (a) Determine the initial concentrations of *o*-xylene in the water of the swimming pool and in the air within the glass enclosure above the pool. (b) After the free xylene has been removed, the water remains saturated with xylene. Determine the equilibrium vapor-phase concentration of *o*-xylene at the initiation of the second stage of cleanup.

- **Logic and Answer(s):** It can be assumed that the vigorous initial mixing caused by the truck's indelicate entry to the pool, coupled with the time elapsed until initiation of the cleanup, allowed the xylene and water phases to equilibrate with each other.

  1. According to Raoult's law at equilibrium:

$$X_{xyl,v} = (P^v_{xyl}/P_T)X_{xyl,l}$$

where $X_{xyl,v}$ is the mole fraction of xylene in the vapor phase, $X_{xyl,l}$ is the mole fraction of xylene in the liquid phase, P_{xyl}^{v} is the vapor pressure of xylene, and P_T is the total pressure of the system.

2. Because the water is saturated with xylene at equilibrium, the aqueous-phase concentration (Table 6-1) is 173 mg/L, which translates into

$$(0.173 \text{ g/L})(1 \text{ mol/106 g}) = 0.00163 \text{ mol/L}$$

3. The xylene floating on the surface (approximately 5 cm of free-floating product) constitutes the liquid phase for Raoult's law; thus $X_{xyl,l} = 1.0$ and

$$X_{xyl,v} = 8.7 \times 10^{-3}$$

4. The vapor pressure of xylene is given in Table 6-1 as 8.7×10^{-3} atm at 25°C. Thus, since the total pressure, P_T, is 1 atm,

$$X_{xyl,v} = 8.7 \times 10^{-3}$$

and the dimensionless vapor-phase concentration, c_v°, for the first stage of cleanup, is

$$c_v^\circ = (8.7 \times 10^{-3})(106 \text{ g xylene/mol/28.8 g air/mol})$$

$$= 3.2 \times 10^{-2} \text{ g xylene/g air}$$

5. After removal of the free xylene, the aqueous-phase concentration remains at the saturation limit of 173 mg/L (0.00163 mol/L). It is now this concentration, and Henry's law, which together determine the new air-phase concentration.

6. One statement or form of Henry's law (Table 6-2) is given by

$$\mathcal{K}_H = P_{xyl}/C_{xyl,l}$$

where $\mathcal{K}_H$ is the Henry's law constant (atm-m^3/mol), P_{xyl} is the pressure of xylene in the vapor phase (atm), and $C_{xyl,l}$ is the solution-phase concentration of xylene (mol/L).

7. From Table 6-1, $\mathcal{K}_H$ for xylene can be taken as 4.9×10^{-3} atm-m^3/mol (4.9 atm-L/mol) at 25°C. It follows that

$$P_{xyl} = \mathcal{K}_H \times C_{xyl,l} = (4.9)(0.00163) = 8.0 \times 10^{-3} \text{ atm}$$

8. Since the total pressure is 1.0 atm,

$$X_{xyl,v} = \frac{P_{xyl}}{P_T} = \frac{8.0 \times 10^{-3}}{1.0} = 8.0 \times 10^{-3}$$

and the dimensionless vapor-phase mass concentration for the second stage of cleanup is

$$c_v^\circ = \frac{(8.0 \times 10^{-3})106 \text{ g xylene/mol}}{28.8 \text{ g air/mol}} = 2.9 \times 10^{-2} \text{ g xylene/g air}$$

6.3 ABSORPTION PROCESSES

6.3.1 General Characterization

Absorption includes a number of processes we have referenced previously in other contexts. Some notable examples are: (1) dissolution of gaseous oxygen from air into water, (2) dissolution of xylene from its pure liquid state into water, (3) vaporization of water and xylene from their respective pure liquid states into air, and (4) volatilization of xylene from a dissolved aqueous state into air.

As depicted in Figure 6-2a, the equilibrium condition for absorption processes mandates a *uniform intraphase distribution* of chemical species among the phases involved. The distribution will probably be different on an *interphase* basis, that is, between the phases; however, it will be uniform within each phase. With specific reference to the phases depicted in Figure 6-1, dissolved oxygen and xylene molecules are each distributed uniformly throughout the water and air phases at equilibrium, and xylene molecules uniformly throughout the pure liquid xylene. Because these substances may be considered as dissolved in their respective phases, their absorption can be related to their solubility in each of these phases. The liquid xylene, for example, dissolves to a greater extent in organic liquids such as hexane than it does in water because organic liquids are more compatible with each other than they are with water. In other words, the hydrocarbon character of xylene is more compatible with the hydrogen–carbon structure of hexane than it is with the hydrogen–oxygen structure of water. Xylene, like many organic compounds, is thus more *organophylic* than it is *hydrophylic*. In fact, many organic compounds are distinctly water-disliking, or *hydrophobic*.

The logic relating to compatibility between a solute and a solution phase can be extended to predict solubility in certain *amorphous solid* and *semisolid* phases. For instance, dodecane is an amorphous solid at ambient temperature and pressure, but it can still actively absorb or ''dissolve'' other organic species more effectively than can water. It is thus an attractive phase into which molecules of substances such as xylene migrate from aqueous phase. The ordering of the structural matrix of a solid can affect the time required to reach equilibrium. For example, dodecane is more highly ordered than liquid hexane; therefore, xylene will take longer to penetrate the molecular structure of dodecane.

Partitioning of a chemical species between different phases depends on their relative thermodynamic compatibility or solubility in those phases. We must be aware, however, that the thermodynamics of systems in which water is one of two or more phases differ significantly from simple homogeneous aqueous solutions. Aqueous phase solubility or saturation implies a fixed and identifiable limiting concentration that does not generally exist as such in multiphase systems. Instead, the solubility of a component in any one phase is dependent on its concentration in the other(s), if there are two or more alternative phases with which it can associate in ''dissolved'' states. The term *phase partitioning* is thus used most properly in the sense of the relative dissolution of a solute in each of two or more different phases.

Unfortunately, phase partitioning is a term often used without regard for the mechanism(s) affecting the phenomenological distribution of a compound between two phases, especially when the phases are as complex as water and a soil or sediment. Adsorption forces and their associated thermodynamic considerations may be at least as important as are those of absorption. In this book we differentiate between adsorption and absorption phenomena and use the term *partitioning* only for description of absorption processes. The term *sorption* is used when both absorption and adsorption may be involved in determining the distribution of a compound between two phases. An illustration of a system that involves both a case of strictly absorption (between air and water) and a case of sorption (between solid particles and water) is given in Example 6-2.

Example 6-2

- **Situation:** A closed vessel having a volume of 12 m^3 is used to collect the intermittent generation of 10 m^3 of industrial waste containing 4.2 mg/L (0.023 mol/m^3) of 1,2,4-trichlorobenzene (MW = 181.45 g/mol) and 100 mg/L (100 g/m^3) of suspended particulate matter. The suspended matter settles and forms a sludge that is withdrawn for landfilling. The supernatant is decanted and diluted with other waste streams for further treatment. Whether sorption of 1,2,4-trichlorobenzene (TCB) occurs onto the suspended matter is unknown. If significant sorption occurs, further treatment of the sludge may be necessary (e.g., by thermal desorption of TCB and return of the liquid stream to the wastewater treatment plant) before the solids can be landfilled. No simple and direct way is available to measure the sorbed-phase concentration. The presence of suspended matter also interferes with determination of the liquid-phase concentration. However, a small air vent in the vessel provides a convenient location for vapor-phase measurements. The vapor-phase concentration is found to be 1.91×10^{-3} mol/m^3 at a temperature of 25°C and pressure of 1 atm.

- **Question(s):** Determine the solid- and liquid-phase concentrations of TCB in the vessel.

- **Logic and Answer(s):**

1. To solve this problem, we assume that equilibrium exists between the suspended solid (ss), liquid (l) and vapor (v) phases. The appropriate material balance relationship is

$$M_T = C_v V_v + C_l V_l + C_{ss} M_{ss}$$

where M_T is the total mass of TCB, M_{ss} is the mass of suspended solids, and the subscripted C values are the concentrations of TCB in the respective phases. At equilibrium, Henry's law applies to relate liquid- and vapor-phase concentrations, so that

$$M_T = C_v V_v + \frac{\mathcal{R}T}{\mathcal{K}_H} C_v V_l + C_{ss} M_{ss}$$

Solving for the solid-phase concentration gives

$$C_{ss} = \left(M_T - C_v V_v - \frac{\mathcal{R}T}{\mathcal{K}_H} C_v V_l \right) \div M_{ss}$$

2. The $\mathcal{K}_H$ value for 1,2,4-TCB is 2.32×10^{-3} atm-m^3/mol (see Table 1-4). The other data needed are

$$M_T = 10 \text{ m}^3 \times 0.023 \text{ mol/m}^3 = 0.23 \text{ mol}$$

$$M_{ss} = 10 \text{ m}^3 \times 100 \text{ gm}^3 = 1000 \text{ g}$$

$$\mathcal{R} = 8.21 \times 10^{-5} \text{ atm-m}^3\text{/mol-K}$$

$$T = 298 \text{ K}$$

$$V_l = 10 \text{ m}^3$$

$$V_v = 2 \text{ m}^3$$

$$C_v = 1.91 \times 10^{-3} \text{ mol/m}^3$$

We then determine from the material balance that

$$C_{ss} = 2.48 \times 10^{-5} \text{ mol/g} = 0.0045 \text{ g/g}$$

The liquid-phase concentration is then calculated from Henry's law:

$$C_l = \frac{\mathcal{R}T}{\mathcal{K}_H} C_v = 20.14 \times 10^{-3} \text{ mol/m}^3 = 3.7 \text{ mg/L}$$

3. Equilibration with the vapor and solid phases has thus reduced the liquid-phase concentration only slightly, from 4.2 mg/L in the feed to 3.7 mg/L in the discharge of the vessel. The small amount of TCB on these particles (0.005 g/g) may suggest absorption into an organic carbon phase surrounding particle surfaces. If a linear sorption were assumed from this single set of measurements, the distribution coefficient would be

$$\mathcal{K}_D = \frac{C_{ss}}{C_l} = \frac{2.48 \times 10^{-5} \text{ mol/g}}{20 \times 10^{-3} \text{ mol/m}^3} = 1.24 \times 10^{-3} \text{ m}^3\text{/g}$$

Note: Example 6-5 explores use of correlations based on the concept described by Equation 6-13 to estimate $\mathcal{K}_D$ from the octanol–water partition coefficient of the sorbate and the organic carbon content of the sorbent. However, more data would first need to be obtained at different solids concentrations and/or TCB concentrations in the vessel to confirm that the sorption relationship is, in fact, linear. We will begin examining such relationships in detail in the next section of this chapter.

6.3.2 Phase Partitioning and Activity Coefficients

The partition coefficient, $\mathcal{K}_P$, quantifies the distribution of a *liquid* or *supercooled liquid* organic solute between two phases at equilibrium; specifically, for phases j and k,

$$\mathcal{K}_P = \mathcal{K}_{j,k} = \frac{C_{i,j}}{C_{i,k}} \tag{6-7}$$

where $C_{i,j}$ and $C_{i,k}$ are the *molar* concentrations of solute i in the two phases. The supercooled liquid state is achieved by lowering a liquid below its melting point without formation of the solid phase. It is a metastable thermodynamic condition because a solid state will form upon prolonged standing or upon addition of some amount of the solid phase.

Where equilibrium exists between two immiscible solvent phases and the same reference or standard state is applicable to both, the requirement of equal chemical potential reduces to equal chemical activity. We can then say that the activities of solute i are equal in each phase:

$$a_{i,j} = a_{i,k} \tag{6-8}$$

As developed in Chapter 5, chemical activity is quantified by the expression

$$a_i = \alpha_i X_i \tag{6-9}$$

where α_i and X_i are the activity coefficient and mole fraction of the solute, respectively. The activity coefficient varies with X_i and a_i. It represents the degree to which the behavior of a substance deviates from its ideal behavior (i.e., behavior in an ideal solution). The two standard states commonly employed for defining the activity coefficient are the *pure liquid* or *supercooled solute* state for Raoult's law, and a hypothetical *infinitely dilute solute solution* state for Henry's law.

The important standard-state conditions are a solute activity equal to 1 and constant in the region near that of the pure component ($X_i = 1$), and conformity with Raoult's law. Under these standard-state conditions Equation 6-9 shows that the activity coefficient is greater than unity in a nonideal aqueous solution in which X_i approaches zero (infinite dilution). That is, the chemical activity of the component per unit of its mass will be greater than that at a reference state defined at $X_i = 1$. The *saturation-limit* activity coefficient is commonly used to define the aqueous solubility of hydrophobic organic compounds. These values differ both from those at infinite dilution and from those near the pure component condition.

For the condition of equilibrium, Equations 6-8 and 6-9 combine to yield

$$(\alpha_i X_i)_j = (\alpha_i X_i)_k \tag{6-10}$$

If the concentration of a solute is very low in each of the two phases, its respective mole fractions are approximated by

$$X_{i,j} = (C_i V_{mo})_j \quad \text{and} \quad X_{i,k} = (C_i V_{mo})_k \tag{6-11}$$

where V_{mo} is the molar volume of the solvent (i.e., its molecular weight divided by its density). The equations above can then be combined to yield an alternative expression for the *k-j* phase partition coefficient (see Equation 6-7) of the solute in terms of its respective activity coefficients in each phase:

$$\mathcal{K}_P = \mathcal{K}_{k,j} = \frac{(\alpha_i V_{mo})_k}{(\alpha_i V_{mo})_j} \tag{6-12}$$

The expression given in Equation 6-12 is rigorous only for totally immiscible phases, that is, when the respective solubility of each phase in the other is insignificant. As we will show in more detail later, the relative hydrophobicity of a compound is commonly characterized by its partitioning between water and a convenient reference solvent. Octanol is the most frequently used reference organic solvent. The hydrophobicity of a compound is then characterized by Equation 6-12 in terms of its *octanol–water* partition coefficient, $\mathcal{K}_{o,w}$. Octanol is not, however, completely immiscible in water. The solubility of octanol in water (4.5×10^{-3} *M*) may significantly alter the activity coefficient of an organic solute, and hence its aqueous solubility. This combined "mutual solubility" effect offers a potential explanation for differences between experimentally observed octanol–water partition coefficients and those predicted by relationships based on activity coefficients, such as that given in Equation 6-12. The influence of mutual solubility is elaborated on in Section 6.3.4.

Relationships exist between the partition coefficients of a given solute for different organic–aqueous phase systems, provided that the solute–organic solvent molecular interactions are limited to weak *London dispersion forces*, that is, to forces arising only from the interactions of induced dipoles over extremely small molecular separation distances (see the discussion accompanying Figure 6-7). Good log-linear correlations have been observed to exist between the $\mathcal{K}_P$ values of hydrophobic compounds for different organic solvent–aqueous phase combinations if all the organic phases contain the same functional groups. The octanol–water partition coefficient has been found to serve as a useful reference correlation, such that

$$\log \mathcal{K}_{o,w} = \alpha_P \log \mathcal{K}_P + \beta_P \tag{6-13}$$

where α_P and β_P are empirical constants relating the partition coefficient of a hydrophobic compound to its octanol–water reference value. This correlation does not necessarily apply to all $\mathcal{K}_P$ values for various organic solvent–water combinations. The noteworthy exceptions are systems involving solutes that interact specifically with the organic phase through polar and/or hydrogen bonding rather than through London dispersion forces.

The relationship given by Equation 6-13 can be considered from a somewhat different perspective as well. In Chapter 5 we introduced the concept of *linear free-energy relationships* (LFERs). This concept was used to approximate equilibrium constants for homogeneous phase reactions. That the environmental partitioning characteristics of an organic compound can also be correlated to some reference solvent is another indication that behavior and chemical structure are interrelated. This is clearly evidenced by the information presented in Table 6-3, which shows how significantly different substituent groups on benzene change the physicochemical properties of the parent compound.

Organic compounds that are similar in chemical type and structure also behave similarly with respect to partitioning among different phases. *Structure–activity* models can then be used to estimate environmental behavior from known and interrelated thermodynamic properties. Figure 6-3 illustrates that the octanol–water partition coefficient of a compound is integrally related to several other important physicochemical properties of that compound, including its aqueous-phase solubility. Moreover, Figure 6-3 reinforces the idea that such descriptive parameters are ultimately related to chemical structure through thermodynamic properties. As discussed in Chapter 5, LFER methods for estimating thermodynamically related constants, such as $\mathcal{K}_{o,w}$ and C_S, from empirically derived atomic or group *fragment constants* and structural features of the arrangement of these units in specific organic chemicals are available.

Structural features which must be considered in methods that employ fragment constants for estimating $\mathcal{K}_{o,w}$ and C_S include degree of saturation, branching, molecular flexibility, and substituent atoms or groups. Values are available for both the fragmentation constants and structural factors such that $\mathcal{K}_{o,w}$ and C_S can be estimated for virtually any chemical compound. Thus, the absence of measured values of these and other phase-partition constants for specific chemicals should not necessarily pose a problem. It is possible, for

Table 6-3 Substituent Group Effects on the Phase Properties of Mono-substituted Benzene Derivatives

Compound	P^v (atm × 10^3)	C_S (mol/L × 10^3)	$\mathcal{K}_H$ $\left(\frac{\text{atm} - \text{m}^3}{\text{mol}} \times 10^3\right)$	log $\mathcal{K}_{o,w}$
Benzene (C_6H_6)	125.7	22.7	5.4	2.1
Toluene (C_6H_5–CH_3)	37.6	5.8	6.6	2.7
Chlorobenzene (C_6H_5–Cl)	15.6	4.5	3.73	2.8
Aniline (C_6H_5–NH_2)	1.3	385.8	0.0033	0.9
Phenol (C_6H_5–OH)	0.7	852.6	0.0004	1.5
Nitrobenzene (C_6H_5–NO_3)	0.2	1900	0.024	1.9

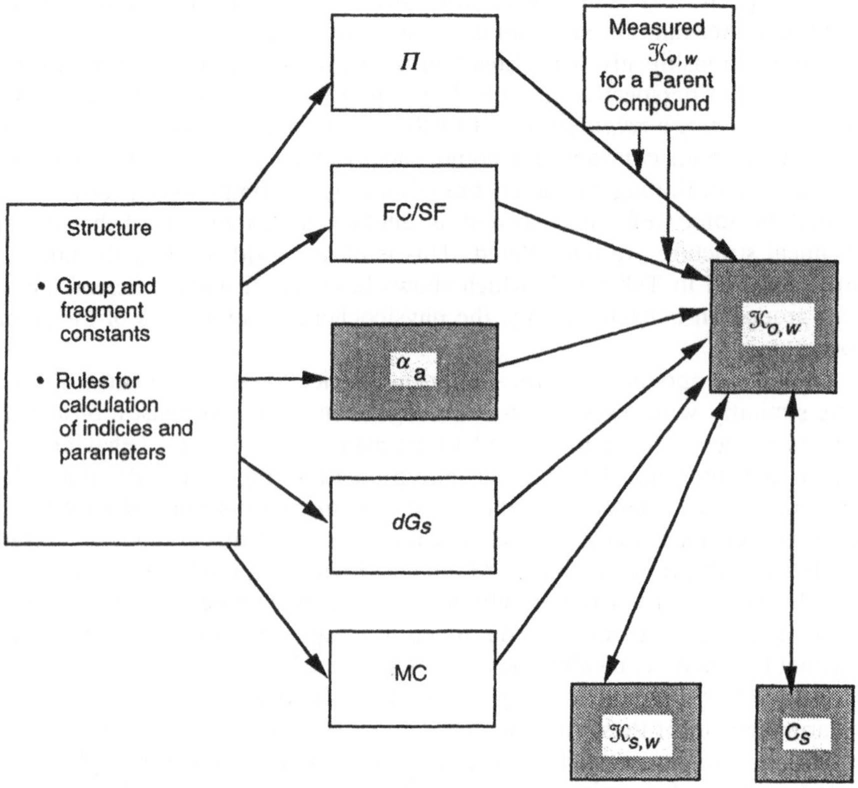

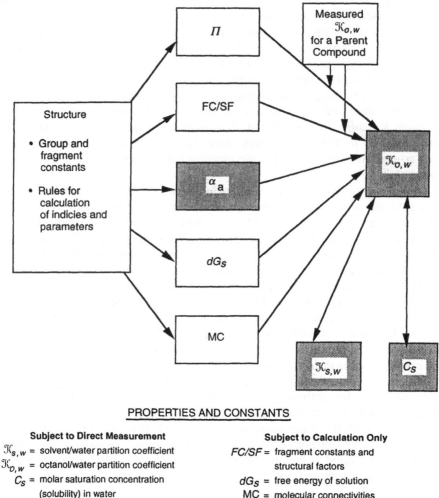

PROPERTIES AND CONSTANTS

Subject to Direct Measurement

$\mathcal{K}_{s,w}$ = solvent/water partition coefficient
$\mathcal{K}_{o,w}$ = octanol/water partition coefficient
C_S = molar saturation concentration (solubility) in water
α_a = aqueous phase activity coefficient

Subject to Calculation Only

FC/SF = fragment constants and structural factors
dG_S = free energy of solution
MC = molecular connectivities
Π = "pi" (substituent) constants

Figure 6-3 Relationships of the octanol–water partition coefficient to other properties and constants of an organic chemical. [Adapted from Lyman et al. (1982).]

example to estimate the environmental behavior of newly synthesized compounds with these procedures.

6.3.3 Aqueous-Phase Solubility

Like its octanol–water partition coefficient, the aqueous-phase solubility of an organic substance is an important measure of its tendency to distribute between phases, specifically between water and its pure liquid or solid state. Substances that are highly soluble also exhibit other properties which allow them to be rapidly dispersed in aquatic systems. Such materials generally do not partition significantly into other phases, and therefore do not sorb or bioconcentrate significantly. Moreover, they are usually more readily degraded by microorganisms than are less soluble compounds.

The interrelationships among solubility and several other major environmental properties of organic substances are reflected in Figure 6-4. This representation for solubility is of the same type as given in Figure 6-3 for the octanol–water partition coefficient. The point is that pathways for estimating the water solubility and octanol–water partitioning of organic substances from other physicochemical properties and structural features are similar.

An aqueous-phase solubility constant characterizes the maximum amount of a substance that can dissolve in a unit volume of pure water at some specified temperature. Concentrations higher than this limit result in saturated aqueous solutions and solid or liquid organic phases. Direct measurement of the aqueous solubility of an organic chemical is more problematic than for an inorganic substance. Although several methods are available, no single method is applicable over the entire range of solubilities of different organic chemicals. Low-solubility (hydrophobic) compounds are particularly difficult because they require long equilibration periods and involve low aqueous-phase concentrations. Different methods for measuring the solubility of such compounds may therefore give different results.

The solubility of an organic compound can in theory be estimated from knowledge of its chemical structure. The procedure should be similar to that discussed in Chapter 5 for estimating equilibrium constants in homogeneous phases. A solubility constant is, after all, another type of equilibrium constant. Indeed, all of the factors cited in the pathways schematized in Figure 6-4 are either compound-specific properties or equilibrium constants for specific reactions of that compound. As evidenced by Figure 6-4, however, only one pathway allows solubility to be estimated directly from structural information. Every other pathway involves an intermediate property or parameter. The method for estimating solubility directly is a correlation procedure, involving the substitution of atomic and structural constants into the equation:

$$-\log c^{\circ}_{S,r} = \kappa^{\circ} + \sum \kappa_i^{\circ} n_i + \sum \kappa_j^{\circ} n_j \tag{6-14}$$

where $c^{\circ}_{S,r}$ is a *dimensionless* expression of solubility, the saturation mass ratio (MM^{-1}) of a compound in water; κ° a correlation parameter related to the type of compound; κ_i° the contribution of atom i; n_i the number of atoms of i contained in the molecule; κ_j° the contribution of various structural elements; and n_j the number of such structural elements contained in the model. Examples of different types of compounds yielding different values for κ° include (1) aromatics, (2) aliphatics, (3) halogen-derivative saturated aliphatics, and (4) cycloaliphatics. Carbon, hydrogen, and chlorine are examples of atoms yielding different κ_i° values. Finally, different structural elements yielding different

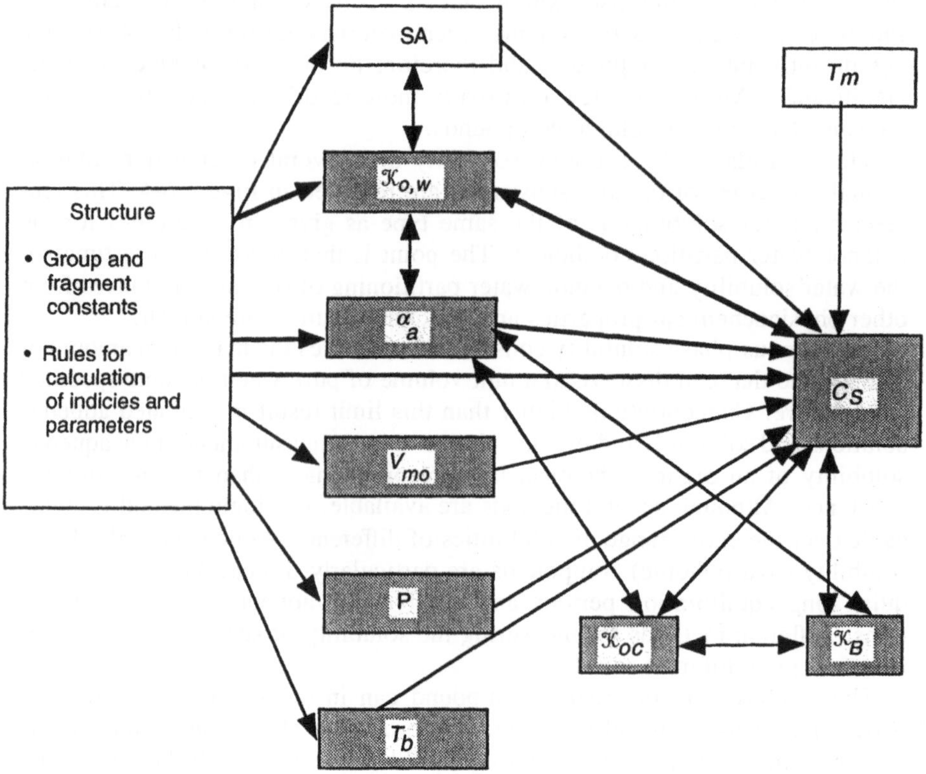

PROPERTIES AND CONSTANTS

Subject to Direct Measurement

$\mathcal{K}_B$ = Bioconcentration factor

$\mathcal{K}_{oc}$ = Organic carbon normalized distribution coefficient

$\mathcal{K}_{o,w}$ = Octanol/water partition coefficient

P = Parachor

C_s = Molar saturation concentration (solubility) in water

T_b = Boiling point

V_{mo} = Molar volume

α_a = Aqueous-phase activity coefficient

Subject to Calculation Only

SA = Molecular surface area

T_m = Melting point

Figure 6-4 Relationships of aqueous solubility to other properties and constants of an organic chemical. [Adapted from Lyman et al. (1982).]

values of κ_j° include (1) double bonds, (2) conjugated bonds, (3) triple bonds, and (4) monohalogenated saturated carbons. Values for κ°, κ_i°, and κ_j° derived from measured solubilities are available in standard references on structure–activity models.

Equation 6-14 yields dimensionless solubility values for different states of different compounds, depending on their state under standard conditions. For example, substances that are gases at a pressure of 1 atm have a $c^{\circ}_{S,r}$ given by the solubility of the liquefied gas at the vapor pressure of both coexisting phases. For substances that are solids at 25°C, $c^{\circ}_{S,r}$ is the mass-ratio solubility of the supercooled liquid state, which can be converted to the mass-ratio solubility of its solid state having a melting point, T_m (°C), using the following approximation:

$$-\log (c^{\circ}_{S,r})_s = -\log (c^{\circ}_{S,r})_l + 0.0095(T_m - 25) \qquad (6\text{-}15)$$

The aqueous solubilities of hydrophobic organic compounds that are normally liquids ($T_m < 25$°C) can be approximated directly from their activity coefficients. Further considerations are required to estimate supercooled solute activities for substances that are normally crystalline solids ($T_m > 25$°C). A good first approximation to the liquid solute activity of a sparingly soluble hydrophobic compound is unity, that is, the same as its pure component activity as specified by the standard state. An activity of unity is reasonable because the solubility of water in the pure liquid solute is negligible. The aqueous solubility limit of a solute in mole fraction units, X_S, is then given in terms of its aqueous-phase activity coefficient by Equation 6-9 as

$$X_S = \frac{1}{\alpha_a} \qquad (6\text{-}16)$$

In molar units, C_S, solubility is the product of the mole fraction solubility and the molarity of water (~55.6 mol/L). The relationship between C_S and X_S is thus log linear and, for an activity coefficient of unity:

$$\log C_S = \log X_S + 1.74 \qquad (6\text{-}17)$$

Estimation of the aqueous solubility of crystalline organic solids having high T_m values requires a more rigorous thermodynamic treatment. If the pure liquid is specified as the standard state, the activity of an organic solute is defined as

$$a = \frac{f_{cs}}{f^{\circ}_l} \qquad (6\text{-}18)$$

where f_{cs} and f°_l are the fugacities of the crystalline solid and the pure liquid standard state, respectively. The solubility of the solid can be determined by estimating its hypothetical supercooled liquid activity. The activity of the supercooled solute is no longer unity but may be estimated from the change in the latent heat of fusion (melting), $dQ_{H,m}$, which for any solute is equal to the difference between the enthalpies of sublimation and vaporization for that solute. The relationship between activity and $dQ_{H,m}$ is given by

$$\log a = \frac{-dQ_{H,m}}{2.3\mathcal{R}} \frac{T_m - T}{T_m T} + \frac{dQ_H^\circ}{2.3\mathcal{R}} \frac{T_m - T}{T} - \frac{dQ_H^\circ}{2.3\mathcal{R}} \log \frac{T_m}{T} \qquad (6\text{-}19)$$

where dQ_H° is the change in the solute heat capacity upon melting and all other variables are as defined previously. The change in the heat capacity of the solute is generally small, at least for many crystalline solid organic compounds, and as a first approximation may be considered negligible. This simplifies Equation 6-19 to

$$\log a = \frac{-dQ_{H,m}}{2.3\mathcal{R}} \frac{T_m - T}{T_m T} \qquad (6\text{-}20)$$

The aqueous solubility of solid organic solutes in mole fraction units can be determined from the relationship

$$\log X_S = \log a - \log \alpha_a \qquad (6\text{-}21)$$

Combining Equations 6-20 and 6-21 yields

$$\log X_S = \frac{-dQ_{H,m}}{2.3\mathcal{R}} \frac{T_m - T}{T_m T} - \log \alpha_a \qquad (6\text{-}22)$$

Molar aqueous solubility can be estimated by combining Equations 6-17 and 6-22:

$$\log C_S = \frac{-dQ_{H,m}}{2.3\mathcal{R}} \frac{T_m - T}{T_m T} - \log \alpha_a + 1.74 \qquad (6\text{-}23)$$

According to Equation 6-23, an accurate determination of the aqueous solubility of an organic compound is dependent upon reliable estimates of the activity coefficient at the saturation limit and the heat of fusion. While the change in heat capacity during solute-phase transition may be insignificant for many compounds, the solute heat capacity term should be incorporated if available. Values of heats of melting (fusion) and heats of vaporization are listed in Table 6-4 for a number of hydrocarbons.

6.3.4 Partition Coefficients and Solubility

The preceding sections have demonstrated that the aqueous-phase molar solubility and the octanol–water partition coefficient for an organic compound are both related to the activity coefficient in a given phase. It follows that C_S and $\mathcal{K}_{o,w}$ are related. The particular relationship for the octanol–water system can

Table 6-4 Heats of Fusion (Melting) and Vaporization for Selected Hydrocarbons

		$dQ_{H,m}$		$dQ_{H,v}$	
Formula	Compound	J/g	kJ/mol	J/g	kJ/mol
CH_4	Methane	58.7	0.96	510.4	8.2
C_2H_6	Ethane	95.1	2.9	489.5	14.7
C_3H_8	Propane	79.9	3.5	426.8	18.8
C_4H_{10}	*n*-Butane	80.2	4.6	385.3	22.4
C_5H_{12}	*n*-Pentane	116.4	8.4	357.3	25.8
C_6H_6	Benzene	125.9	9.8	393.7	30.8
C_6H_{12}	Cyclohexane	31.7	2.7	357.3	30.0
C_6H_{14}	*n*-Hexane	151.2	13.0	334.7	28.8
C_7H_8	Toluene	71.5	6.6	363.2	33.4
C_8H_{10}	Ethylbenzene	286.3	9.2	338.9	35.9
C_8H_{10}	*o*-Xylene	128.1	13.6	346.9	36.8
C_8H_{10}	*m*-Xylene	109.0	11.5	343.1	36.4
C_8H_{10}	*p*-Xylene	161.1	17.1	339.7	36.1
C_8H_{18}	*n*-Octane	181.6	20.8	300.8	34.4
C_9H_{20}	*n*-Nonane	120.6	15.5	286.2	36.7
$C_{10}H_{22}$	*n*-Decane	201.8	28.7	276.1	39.3
$C_{12}H_{26}$	*n*-Dodecane	216.3	36.8	256.5	43.7
$C_{18}H_{38}$	*n*-Octadecane	241.2	61.3	215.5	54.8
$C_{19}H_{40}$	*n*-Nonadecane	170.6	45.8	210.5	56.5

Source: Numerical values from Reid et al. (1987).

be derived by starting with Equation 6-12:

$$\mathcal{K}_{o,w} = \frac{\alpha_a V_{mo,w}}{V_{mo,o}} \tag{6-24}$$

or, in logarithmic form,

$$\log \alpha_a = \log \mathcal{K}_{o,w} + 0.94 \tag{6-25}$$

where $\log (V_{mo,o}/V_{mo,w})$ has a value of 0.94. If the "mutual solubility" effect is taken into consideration, the molar volume of the water-saturated octanol phase, $V_{mo,o-w}$, is 120 cm^3/mol, and the logarithmic form of Equation 6-24 becomes

$$\log \alpha_a = \log \mathcal{K}_{o,w} + 0.82 \tag{6-26}$$

Once the activity coefficient for a compound in a particular solvent has been determined, the ideal solubility can be calculated in the following manner. The enthalpy of melting is approximated from the entropy of melting, dS_m, using the relationship

$$dQ_{H,m} = dS_m T_m \tag{6-27}$$

where T_m is given in Kelvin. Combining Equations 6-20 and 6-27 yields

$$\log a = \frac{-dS_m}{2.3\mathcal{R}T}(T_m - T) \tag{6-28}$$

Further combining of Equations 6-23, 6-25, and 6-28 gives an expression that interrelates the aqueous molar solubility of an organic nonelectrolyte and its octanol–water partition coefficient:

$$\log C_S = \frac{-dS_m}{2.3\mathcal{R}T}(T_m - T) - \log \mathcal{K}_{o,w} + 0.80 \tag{6-29}$$

This model, which assumes that *mutual solubility* is negligible, has been reported to provide good correlations between water solubilities and octanol–water partition coefficients for polycyclic aromatic compounds, halobenzenes, and alkyl *p*-substituted benzoates. If we consider the *mutual saturation* of the two partitioning phases, Equation 6-26 is used in lieu of Equation 6-25 to yield

$$\log C_S = \frac{-dS_m}{2.3\mathcal{R}T}(T_m - T) - \log \mathcal{K}_{o,w} + 0.92 \tag{6-30}$$

An alternative definition of the octanol–water partition coefficient in terms of mole fraction leads to an approximate relationship to solubility. Instead of partitioning between octanol and water being defined by the ratio of molar concentrations in each phase, as in Equation 6-7, we can define it as the ratio of mole fractions, yielding a modified octanol–water coefficient, $\mathcal{K}'_{o,w}$, given by

$$\mathcal{K}'_{o,w} = \frac{X_o}{X_a} \tag{6-31}$$

For low concentrations in each phase, $\mathcal{K}_{o,w}$ and $\mathcal{K}'_{o,w}$ are related by

$$\mathcal{K}_{o,w} = \frac{C_o}{C_a} = \frac{X_o}{X_a}\frac{V_{mo,w}}{V_{mo,o}} = \mathcal{K}_{o,w}\frac{V_{mo,w}}{V_{mo,o}} \tag{6-32}$$

The fugacities must be equal, so that

$$\alpha_o X_o = \alpha_a X_a \tag{6-33}$$

$$\mathcal{K}'_{o,w} = \frac{\alpha_a}{\alpha_o} \tag{6-34}$$

If the assumptions are made that the activity coefficient in the octanol phase and the fugacity in the aqueous phase are each unity (due to having reached the solubility limit), we obtain the following with the aid of Equation 6-5:

$$\mathcal{K}'_{o,w} = \frac{1}{X_S} \tag{6-35}$$

Using the relationship between C_S and X_S from Equation 6-17 and rearranging the expression above, we find that

$$\log C_S = -\log \mathcal{K}'_{o,w} + 1.74 \tag{6-36}$$

or, in terms of $\mathcal{K}_{o,w}$,

$$\log C_S = -\log \mathcal{K}_{o,w} + \log \frac{V_{mo,w}}{V_{mo,o}} + 1.74 \tag{6-37}$$

The molar volumes for water and octanol are 18 and 158.5 cm^3/mol, respectively. Thus the final expression is

$$\log C_S = -\log \mathcal{K}_{o,w} + 0.80 \tag{6-38}$$

Comparing Equation 6-38 to Equation 6-29 or 6-30, we see that this approximation is valid only when the hypothetical supercooled liquid activity contribution to solubility (i.e., the first term in Equations 6-29 and 6-30) is very small.

Thus, we have shown with the foregoing analyses that their mutual dependence on fugacity and the activity coefficient provides relationships among the solubility, octanol–water partition coefficient, and other phase distribution coefficients of organic compounds. An illustration of the estimation of octanol-water partitioning coefficients from known values of water solubility is given in Example 6-3.

Example 6-3

- **Situation:** Traces of three organic hydrocarbons are identified by qualitative analysis in a groundwater contaminated by leakage from an underground gasoline storage tank. The compounds are octane, decane, and dodecane. Accurate quantitative analysis of these materials requires preparatory concentration by extraction into an organic solvent; octanol is being considered for this purpose.

- **Question(s):** You find that aqueous-phase solubility limits are listed in the reference books available in your laboratory but octanol–water partitioning coefficients are not. Use the solubility information to estimate the partitioning coefficients you can expect for extraction of the three contaminants with octanol at an ambient temperature of 25°C.

- **Logic and Answer(s):**

1. Relationships between the aqueous-phase solubilities and octanol-water partition coefficients for organic compounds are given in Equations 6-29 and 6-30.

2. Assume in the cases of these low-solubility paraffinic hydrocarbons that solubility effects between water and octanol must be taken into account. Then, from Equation 6-30,

$$\log C_S = \frac{-dS_m}{2.3\mathcal{R}T}(T_m - T) - \log \mathcal{K}_{o,w} + 0.92$$

3. Solubility data for the compounds of interest can be found in Table 6-1, expressed on a weight basis. The respective molar solubilities predicated on the mean values for mass solubilities can then be calculated, yielding:

Compound	Molecular Weight (g)	C_S (mg/L × 10^3)	C_S (mol/L × 10^5)
Octane (C_8H_{18})	114.2	710	622
Decane ($C_{10}H_{22}$)	142.3	15	10.5
Dodecane ($C_{12}H_{26}$)	170.3	3.7	2.17

4. Melting point (T_m) data are also given in Table 6-1, and molar heat of fusion ($dQ_{H,m}$) data in Table 6-4. From Equation 6-27,

$$dS_m = \frac{dQ_{H,m}}{T_m}$$

the entropies of fusion for the compounds of interest can then be calculated:

Compound	T_m (°C)	T_m (K)	$dQ_{H,m}$ (kJ/mol)	dS_m (J/mol-K)
Octane	−56.8	216.2	20.8	96.2
Decane	−29.6	243.3	28.7	118.0
Dodecane	−9.6	263.4	36.8	139.7

6. The next step is to determine the log $\mathcal{K}_{o,w}$ values, as follows:

(a) Octane:

$$\log \mathcal{K}_{o,w} = -\log(622 \times 10^{-5}\ \text{mol/L}) - \frac{(96.2\ \text{J/mol-K})(216.2\ \text{K} - 298\ \text{K})}{2.3(8.314\ \text{J/mol-K})(298\ \text{K})} + 0.92$$

$$= 2.21 + 1.38 + 0.92$$

$$= 4.51$$

(b) Decane:

$$\log \mathcal{K}_{o,w} = -\log (10.5 \times 10^{-5}) - \frac{(118.0)(243.3\ \text{K} - 298\ \text{K})}{5698} + 0.92$$

$$= 3.98 + 1.13 + 0.92$$

$$= 6.03$$

(c) Dodecane:

$$\log \mathcal{K}_{o,w} = -\log (2.17 \times 10^{-5}) - \frac{(139.7)(263.4\ \text{K} - 298\ \text{K})}{5698} + 0.92$$

$$= 4.66 + 0.85 + 0.92$$

$$= 6.43$$

- **Conclusion:** Comparisons of these estimated values with the experimental values listed in Table 6-1 reveal reasonable agreement, particularly considering the ranges of experimentally determined values for both $\mathcal{K}_{o,w}$ and C_S given in Table 6-1.

6.4 ADSORPTION PROCESSES

6.4.1 General Characterization

Adsorption involves the accumulation of dissolved substances at interfaces of and between phases. Certain adsorption phenomena can be related to the expulsion of solute from one phase and its *indifferent* accumulation at the interface of that phase. Accumulation occurs because the solute has no other compatible phase into which it can partition. The formation and accumulation of foams at air–water interfaces is a manifestation of this type of adsorption. Alternatively, adsorption may occur strictly as the result of the specific attraction of a surface or interface for a chemical species, such as the adsorption of phenol from water by activated carbon. These two types of adsorption are referred to as *solvent-motivated* (entropy-driven) and *sorbent-motivated adsorptions*, respectively. Most adsorption phenomena are comprised by some combination of solvent- and adsorbent-related forces.

Solvent-motivated adsorptions may occur at any interface between two phases. The clearest case of strictly solvent-driven systems is evidenced by the effects of hydrophobic compounds on the *surface tension* of aqueous phases. Surface tension is that particular case of *interfacial tension* for which the second phase is air. Air exerts virtually no effect on the behavior of nonvolatile hydrophobic molecules. The air–water interface is consequently one of the most "indifferent" interfaces found in environmental systems. As such, it deserves special consideration with respect to the causes and manifestations of surface energy relationships.

6.4.2 Solvent-Motivated Adsorption

Adsorption generally occurs at least partly as a result of, and also influences, forces active within phase boundaries. These forces impart characteristic *boundary energies*. A common manifestation of boundary energy can be observed when a drop of water is placed on a flat plate comprised by some material with which the water is incompatible (e.g., a glass plate coated with a layer of wax). As depicted in Figure 6-5, the drop resists spreading and attempts to gain and retain a nearly spherical shape. In so doing, the water minimizes its surface area and thus its free surface energy. The development of this *surface tension*, which keeps the drop from spreading into a thin flat layer results from attractive forces between molecules of water within the drop. The water molecules are more strongly attracted to one another by cohesion forces than they are by adhesion forces to molecules of the solid plate or to molecules of air at the liquid–gas interface. Water does not wet surfaces effectively in such circumstances.

6.4.2.1 Surface Tension Surface tension is equivalent to the amount of work that would be necessary simply to compensate the reduction it causes in

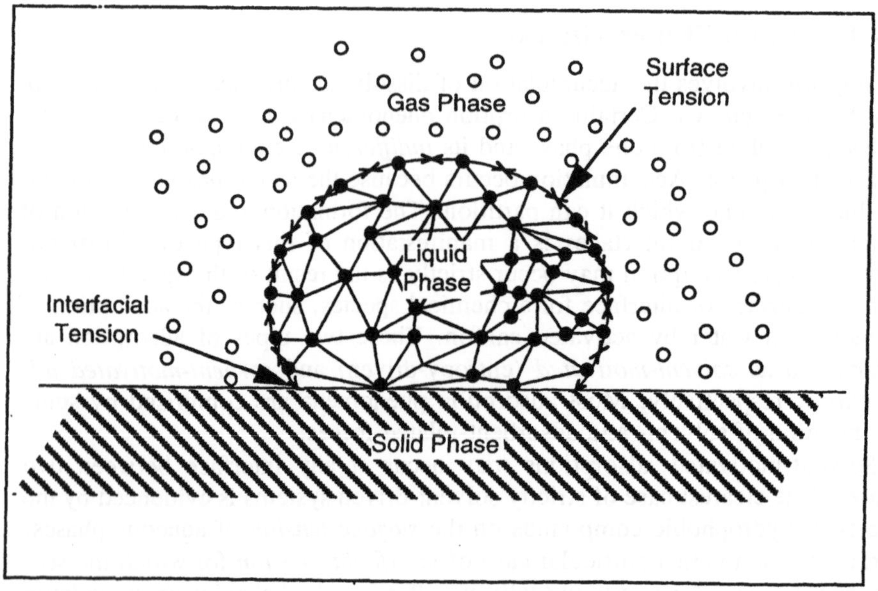

Figure 6-5 Surface and interfacial tensions resulting from intramolecular attraction forces.

free surface energy. From a molecular perspective, however, enlarging a surface requires breaking bonds between molecules comprising the liquid phase, and forming bonds between molecules of different phases. Hence, an input of work in excess of that necessary to compensate the tension at the surface is required to *increase* the surface of a liquid. For example, surface tension must be *overcompensated* to induce frothing at a liquid–gas interface.

Figure 6-6 illustrates the effect of surface tension on interfacial surface area. Each of the three parts of Figure 6-6 represents the contact of a liquid with a

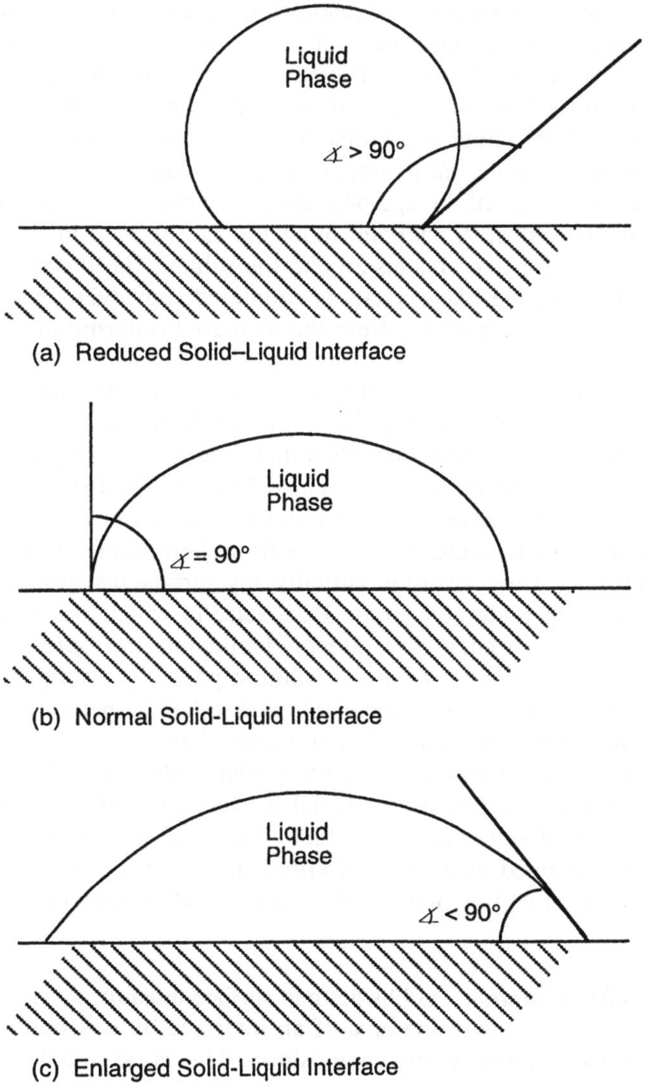

Figure 6-6 Effects of interfacial tension on interfacial areas and contact angles.

solid. In Figure 6-6a, a reduced free surface energy is characterized by a *positive* surface tension, a predominance of cohesion forces, and a reduced area of the solid–liquid interface. In Figure 6-6b, free surface energy is compensated such that cohesion and adhesion forces are balanced, and this results in a normal interfacial contact angle and interfacial area. Finally, in Figure 6-6c, overcompensation of free surface energy is a condition characterized by a *negative* surface tension, predominance of adhesion forces, and enlarged solid–liquid interfaces.

Soluble hydrophobic materials can effectively reduce the surface tension of water by migrating from the bulk phase to its surfaces or interfaces. Detergents and other *surface-active* agents do so readily and are thus termed *wetting* agents. The energy driving such molecules out of solution can be explained in the context of the structure of water. The unit-association concept discussed in Chapter 5 with respect to the molecular structure of water is generally thought to occur as a result of either the tetrahedral coordination of each molecule to four other water molecules *via* hydrogen bonds, yielding a structure similar to that of crystalline ice, or of the agglomeration of more densely packed but less ordered water molecules. A nonpolar molecule can be held in solution by an arrangement of the ice-like crystals around the solute. The favorable enthalpy associated with these ice-like crystals is countered, however, by an unfavorable entropy of solution resulting from the increased ordering of solvent molecules.

Solute molecules may be driven from solution at concentrations below their maximum solubility if the system (including all interfaces) achieves a thermodynamically more favorable state than that in which actual *precipitation* of solute would eventually occur. Chlorinated aromatic hydrocarbons such as PCBs, for example, are adsorbed from water at aqueous-phase concentrations well below those which would normally result in their precipitation if no other phase were present. They adsorb to virtually any surface because the adsorbed state is energetically preferable to the PCB state in which water interactions occur. The reactions in such cases are predominantly *entropic* in character; that is, the solutes are *driven from* the solution phase by energies associated with the ordering of solvent molecules rather than being specifically *attracted* by the phase interface at or in which they accumulate.

Thermodynamically rigorous adsorption relationships can be defined for conditions in which equilibrium interfacial tensions are reduced with increasing concentrations of solute at an interface. These relationships are rooted in an expression developed by J. Willard Gibbs in 1878 for expressing changes in interfacial tension as functions of the extent of adsorption of solutes at interfaces.

6.4.2.2 Gibbs Equation Gibbs defined surface tension, σ°, as the reversible work needed to increase the area of an interface by one unit at constant temperature, pressure, and system composition. This is essentially a change in the free energy of a system in response to a change in its surface area, A°, or

$$\sigma^\circ = \left(\frac{\partial G}{\partial A^\circ}\right)_{T,P,n_i} \tag{6-39}$$

Surface tension is the intensive variable in a free-energy term characterized by a change in the extensive variable, A°. This energy term can be incorporated in the Gibbs relationship given earlier in Equation 5-61 to yield the following expression for free-energy change:

$$dG = \sigma^\circ \, dA^\circ + V\,dP - S\,dT + \sum \mu_i \, dn_i \tag{6-40}$$

If the temperature and pressure of a system are constant, changes in the surface tension of that system can be effected only by changes in the chemical potentials, μ_i, of its constituent components. These changes, in turn, must be reflected in a net accumulation or depletion of components at the system interface; that is,

$$d\sigma^\circ|_{T,P} = -\Sigma\Gamma_i \, d\mu_i \tag{6-41}$$

where the term Γ_i represents the extent of adsorption of component i at the interface, expressed as the concentration of that component in the interface *over and above* its concentration in the bulk phase (i.e., the Gibbs *surface excess*). This surface excess results from an entropy-driven expulsion of solute from solution. It thus actually comprises more an *absorption* of solute *within* an interfacial layer than it does an *adsorption* at an interface.

For dilute solutions of one solute at an *equilibrium molar* concentration of C_e, the Gibbs equation relates change in surface tension to change in solute concentration

$$d\sigma^\circ|_{T,P} = -\Gamma \, d\mu = -\mathcal{R}T\Gamma \, d\ln a = -\mathcal{R}T\Gamma\frac{da}{a} \approx -\mathcal{R}T\Gamma\frac{dC_e}{C_e} \tag{6-42}$$

The Gibbs equation is not readily applicable to environmental sorption reactions, due to the difficulty and uncertainty of required parameter measurements and to limitations imposed by its fundamental assumptions. The equation is presented here because it provides useful insight to energetic considerations associated with adsorption processes. Moreover, the concepts implicit in the Gibbs equation eventually form the basis for one of the more practical models for describing adsorption environmental equilibria, the Freundlich model.

According to Equation 6-42, any solute that reduces the interfacial tension ($d\sigma^\circ/dC_e < 0$) of a liquid results in an increase in Γ; that is, the solute is present at the interface in higher concentration than in bulk solution. For $d\sigma^\circ/dC_e > 0$, negative adsorption is experienced. The surface tension of water at 20°C is ≈ 73 dyn/cm, while that at 20°C for ethanol (ethanol–vapor interface) is ≈ 23 dyn/cm. Solutes that act to *decrease* the surface tension of water may thus very well act to *increase* the interfacial tension of ethanol and other organic solvents. Lowering the surface tension of water causes positive adsorptions, whereas increasing the surface tension of ethanol causes negative adsorptions. It is also evident in Equation 6-42 that a solute which decreases the surface tension of both solvents, but one to a lesser degree than the other, will be more readily and completely adsorbed from the solution in which $d\sigma^\circ/dC_e$ is more negative. Example 6-4 provides an illustration of how such information can be put to practical use, in this case for selecting a polyelectrolyte for destabilizing a suspension of dredge spoil solids.

Example 6-4

- **Situation:** Unpolluted dredge spoil from a navigational channel is to be discharged at a rate of 4 m^3/h from a scow to a deep-water area in a lake, where the ambient temperature is 10°C. To minimize turbidity impacts on biota in the dump area, the spoil will be mixed with an organic polyelectrolyte before discharge to destabilize its colloidal fractions and thus enhance sedimentation. Three cationic polyelectrolytes (*A*, *B*, and *C*) of equivalent cost and molar charge density are considered.

- **Question(s):** Although coagulation data are not available, the surface reduction tendencies of the compounds for water are known, as indicated below.

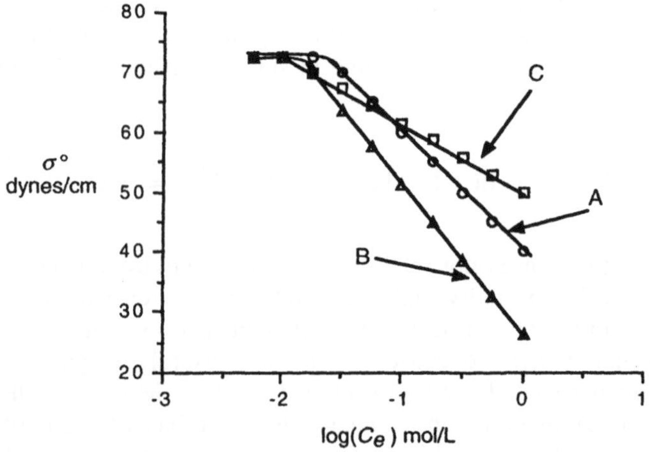

Based on this information, make a first-cut prediction of which of the three polyelectrolytes is likely to function best for the purpose intended.

- **Logic and Answer(s):**

1. Because the molar charge densities (charges/mole) of the three polyelectrolytes are the same, it is likely that the one which adsorbs by

forces additional to electrostatic charges will be most effective in accomplishing total charge neutralization and destabilization of the colloidal particles. This assumes that the extent of adsorption is not great enough in any case to lead to charge reversal and restabilization. The relative adsorption tendencies of the three materials can be roughly estimated from their relative effects on surface tension. The greater the effect on surface tension, the higher will be the surface excess, that is, the higher will be the concentration of polyelectrolyte present at the surface of the colloids relative to the bulk solution. As a first cut, then, the polyelectrolyte with the highest surface excess, Γ, is likely to be the most effective.

2. An expression for calculating Γ from the information presented in the figure above can be developed. From Equation 6-42,

$$d\sigma^\circ = -\Gamma \, d\mu$$

where $\mu = \mu^\circ + \mathcal{R}T \ln a$ and $a \approx C_e$ (molar concentration, at equilibrium) for dilute solutions. Because μ° is constant, $d\mu^\circ = 0$ and

$$d\mu = \mathcal{R}T \, d \ln C_e$$

Therefore,

$$d\sigma^\circ = -\Gamma \mathcal{R}T \, d \ln C_e$$

and

$$\Gamma = \frac{-d\sigma^\circ}{\mathcal{R}T \, d \ln C_e}$$

3. The slope of each line given in the figure above is an expression of $d\sigma^\circ/d$ (log C_e) for that particular polyelectrolyte. The surface tension, σ°, is expressed in dyn/cm, and d (log C_e) $= dC_e/C_e$ is dimensionless. Thus the surface excess in mol/cm^2 can be calculated from the figure as follows:

$$\Gamma = \frac{-d\sigma^\circ}{\mathcal{R}T \, d \ln C_e}$$

$$= \frac{-d\sigma^\circ}{(8.314 \times 10^7 \text{ dyn-cm/mol-K})(283 \text{ K})(2.3) \, d(\log C_e)}$$

$$= \frac{1}{5.4 \times 10^{10}} \times \text{Slope}$$

$$= -1.85 \times 10^{-11} \times \text{Slope (mol/cm}^2\text{)}$$

4. The surface excess for each polyelectrolyte may now be calculated.

(a) For polyelectrolyte A:

$$\frac{d\sigma^\circ}{d \log C_e} = -20 \qquad \Gamma = 1.85 \times 10^{-11}[-(-20)]$$

$$= 3.7 \times 10^{-10} \text{ mol/cm}^2$$

(b) For polyelectrolyte B:

$$\frac{d\sigma^\circ}{d \log C_e} = -25 \qquad \Gamma = 1.85 \times 10^{-11}[-(-25)]$$

$$= 4.63 \times 10^{-10} \text{ mol/cm}^2$$

(c) For polyelectrolyte C:

$$\frac{d\sigma^\circ}{d \log C_e} = -11.4 \qquad \Gamma = 1.85 \times 10^{-11}[-(-11.4)]$$

$$= 2.11 \times 10^{-10} \text{ mol/cm}^2$$

5. Polyelectrolyte B is chosen because its Γ is highest, and its effect on the reduction of surface tension and colloid destabilization is thus most likely to be the greatest of the three compounds per unit of concentration added.

6.4.2.3 Spreading Pressure The force acting per unit length of interface to bring about a decrease in the surface tension of a pure solvent from σ_0° to σ° upon addition of solute is referred to as the *spreading* (or *film*) *pressure*, Π_S:

$$\Pi_S = \sigma_0^\circ - \sigma^\circ \tag{6-43}$$

or

$$d\Pi_S = -d\sigma^\circ \tag{6-44}$$

Spreading pressure, which acts *along a surface*, is a two-dimensional analog of a pressure that acts *over a volume*. A two-dimensional analog to the ideal gas law can thus be developed. The decrease in surface tension of a pure solvent produced by addition of small amounts of solute (i.e., for dilute solutes) is linearly proportional to the equilibrium concentration of the solute, C_e, so that

$$\Pi_S = \kappa C_e \tag{6-45}$$

and

$$d\Pi_S = \kappa \, dC_e \tag{6-46}$$

where κ is a coefficient of proportionality.

The derivative of spreading pressure above can be substituted for the derivative of surface tension in the Gibbs equation (6-42) to give

$$d\Pi_S = \kappa\, dC_e = \mathcal{R}T\Gamma \frac{dC_e}{C_e} \tag{6-47}$$

Combining Equation 6-45 with Equation 6-47 shows that

$$\Pi_S = \Gamma \mathcal{R} T \tag{6-48}$$

which is the two-dimensional analog of the ideal gas law, $P = (n/V)\mathcal{R}T$. The two-dimensional nature of the relationship is seen from surface excess, Γ, which has units of *moles per unit area* in contrast to n/V, which has units of *moles per unit volume.* As will be discussed later, spreading pressure is a key concept in the *ideal adsorbed solution theory* (IAST), which is used to describe competitive adsorption.

Surface excess can also be used to describe the accumulation of solute at the surface of a solid:

$$\Gamma = \frac{q_e}{a^{\circ}_{s,w}} \tag{6-49}$$

where q_e is the amount of solute sorbed per unit weight of the solid at equilibrium (e.g., mol/g) and $a^{\circ}_{s,w}$ is the surface area available per unit weight of the solid (e.g., cm^2/g); that is, solid-phase concentration is expressed per unit area of surface. As we will see, the equilibrium solid-phase concentration is usually some function of the equilibrium solution concentration, C_e. With this definition of Γ, integration of Equation 6-47 defines the spreading pressure as

$$\Pi_S = \frac{\mathcal{R}T}{a^{\circ}_{s,w}} \int_0^{C_e} q_e \frac{dC_e}{C_e} = \frac{\mathcal{R}T}{a^{\circ}_{s,w}} \int_0^{C_e} q_e d\ln C_e \tag{6-50}$$

6.4.3 Sorbent-Motivated Adsorption

Various types of attractive forces exist between solute molecules and the molecules of adsorbing surfaces, and all of these have their origins in electromagnetic interactions of nuclei and electrons. As suggested in Figure 6-7, which identifies some of these forces, three loosely defined categories are traditionally distinguished: (1) *chemical*, (2) *electrostatic*, and (3) *physical*. The energies associated with the attraction of solute molecules by surfaces are largely *enthalpic* in character, in contradistinction to the entropic energies we have discussed relative to the repulsion of solute molecules by a solvent phase.

Chemical adsorption, or chemisorption, involves solute–sorbent interactions having the characteristics of true chemical bonds. Such sorptive phenomena are thus characterized by large heats of adsorption, typically 100 to 400 kJ/mol, although they may be even larger in some cases. Substantial activation energies may be involved in chemisorption reactions, in which cases they can

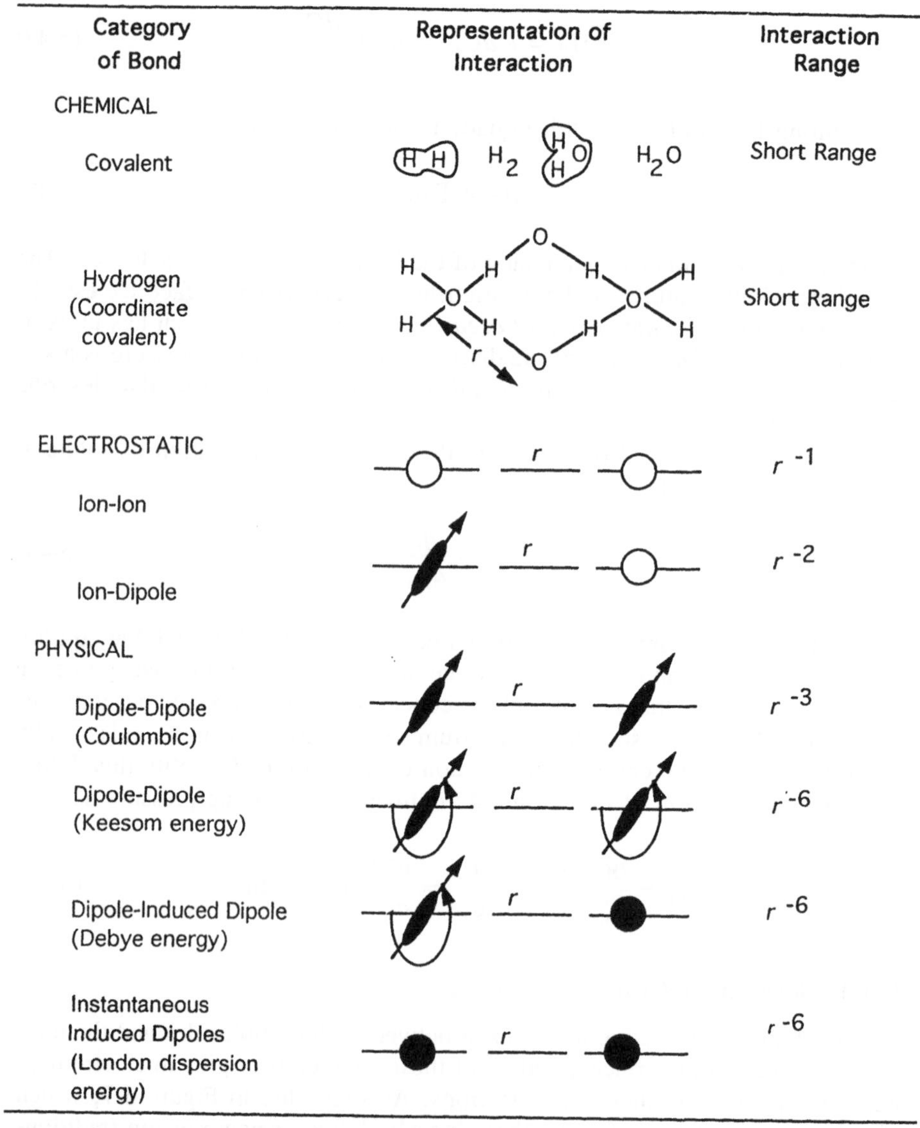
Category of Bond
Representation of Interaction
Interaction Range
CHEMICAL
Covalent
H H
H2
H O H
H2O
Short Range
Hydrogen (Coordinate covalent)
H
O
r
Short Range
ELECTROSTATIC
Ion-Ion
r
r -1
Ion-Dipole
r
r -2
PHYSICAL
Dipole-Dipole (Coulombic)
r
r -3
Dipole-Dipole (Keesom energy)
r
r -6
Dipole-Induced Dipole (Debye energy)
r
r -6
Instantaneous Induced Dipoles (London dispersion energy)
r
r -6

Figure 6-7 Categories and characteristic interactions of adsorption phenomena. [After Weber et al. (1991) as adapted from Israelachvili (1985)]

be induced to occur at greater rates, and thus to greater extents within fixed time intervals, at elevated temperatures.

Electrostatic adsorption also involves high-energy forces, just as electrostatic bonds between charged chemical species involves high-energy bonds. The energies associated with such bonds were discussed in Chapter 5 and defined in Equation 5-1. They relate directly to Coulombic forces of attraction between oppositely charged species, and their differential heats of adsorption may be as large as 200 kJ/mol. Such forces are particularly strong in ion-exchange processes, which occur in both natural (e.g., cation exchange by soils) and engineered processes (e.g., ion-exchange systems for water softening, demineralization, and metal ion removal and recovery). These forces are significant also with respect to the stability of particles suspended in aqueous systems, a topic considered in more detail in a subsequent section of this chapter.

Physical adsorption results from the action of van der Waals forces, comprised by London dispersion forces and classical electrostatic forces. London dispersion forces involve interactions among rapidly fluctuating temporary dipole and quadrupole moments associated with the motion of electrons in their orbitals. As a molecule from bulk solution approaches the surface of an adsorbent, electron distributions interact to induce additional dipole and quadrupole moments and distort to achieve an optimal energy state. The result is a net attraction attributable to dipole–dipole, dipole–quadrupole, and quadrupole–quadrupole interactions. *Dipole–dipole* interactions are generally the most important, varying inversely with the *sixth power* of the distance between molecules, while *dipole–quadrupole* and *quadrupole–quadrupole* interactions are proportional to the *eighth* and *tenth powers* of intermolecular separation distances, respectively. *Repulsive forces* vary with the *twelfth power* of distance and are therefore negligible except at very small intermolecular distances, distances characterized in their limit as the *van der Waals radius*.

Differential heats of adsorption for interactions of the van der Waals type for small molecules are generally on the order of 5 to 10 kJ/mol. This relatively small bonding force is often amplified in the case of hydrophobic molecules by an entropy gradient that functions to drive the molecules out of solution (i.e., a *solvent-motivated* gradient). Although the eventual sorptive bond still involves a van der Waals type of interaction, the combined effect is nonetheless often referred to as *hydrophobic bonding*. Hydrophobic bonds also exhibit differential heats of adsorption of 5 to 10 kJ/mol. Thus combinations of hydrophobic and van der Waals sorptions may have associated heats of adsorption of 10 to 20 kJ/mol.

The second component of physical sorptive forces is comprised by classical electrostatic interactions. For hydrophobic substances, gross electrostatic interactions are usually negligible, but attraction potentials can be developed between polar molecules and ionic or heteropolar solids. The strengths of such attractions are dependent on the strength and characteristics of the surface electric field and the magnitude of the dipole and quadrupole moments of the solute molecules. They are generally not significant for nonpolar hydrophobic compounds. A slight degree of polarity, however, can radically alter the stability of molecules in solution, potentially exerting control over hydrophobic bonding.

Adsorption processes in environmental systems involve various combinations of forces and interactions, and this makes it difficult to assess directly the relative importance of each. Electrostatic interactions are the easiest to identify. Chemisorption can occasionally be distinguished from pure physical adsorption at high temperatures, where the latter becomes less significant. Because of the temperature dependence of solute–solvent interactions, however, this distinction may not be possible when physical adsorption is complemented by hydrophobic bonding. The heterogeneous nature of natural surfaces and engineered adsorbents, such as activated carbon, necessarily precludes any thorough evaluation of surface chemistry, and potential sorbent–sorbate chemical interactions remain largely speculative. Nonetheless, it is possible to draw some inferences by the extent to which observed adsorption data can be described by different types of conceptual models.

6.5 MODELS FOR DESCRIBING SORPTION EQUILIBRIA

Several different models for describing observed sorption phenomena are considered in this section. The general term *sorption* is used to represent either absorption or adsorption, because the distinction is not clear for many environmental systems. Certain phenomenological models are similar to thermodynamically rigorous models developed for pure absorption and adsorption processes. A set of empirical sorption data may be adequately *described* with a model developed from a particular conceptual origin. The model, however, can be considered no more than a fitting tool in the absence of an independent confirmation of sorption *mechanism(s)*.

In practice, sorption studies are conducted by equilibrating known quantities of sorbent with solutions of solute(s). Plots of resulting data relating the variation of solid-phase concentration, or amount of the compound sorbed per unit mass of solid, to the variation of solution-phase concentration are termed *sorption isotherms*. They are referred to as isotherms because the sorption data are collected at constant temperature. Figure 6-8 provides graphical representations of several different forms that such isotherms may take. The term *favorable*, in the context employed in Figure 6-8, refers to sorptions in which capacities for uptake by the solid phase increase sharply from low solution concentrations to high concentrations, generating a *convex isotherm*. Such sorptions are favorable because the amount adsorbed per unit mass of adsorbent decreases only gradually as the concentration of a contaminant decreases; thus efficient use of the sorbent occurs when reducing some process influent value, C_{IN}, to a lower *targeted* process effluent level, C_{OUT}. When such isotherms approach a limiting value at q_e, they are commonly referred to as type I isotherms, and described by the Langmuir model. In the same context, *concave isotherms* can be thought of as reflecting *unfavorable sorption processes* because sorption decreases sharply with lower concentrations; thus the sorbent loses efficiency rapidly as the influent concentration is reduced to the effluent concentration. Isotherms shaped like the unfavorable isotherm depicted in Figure 6-8 are often called type III isotherms. Type II isotherms are shaped like

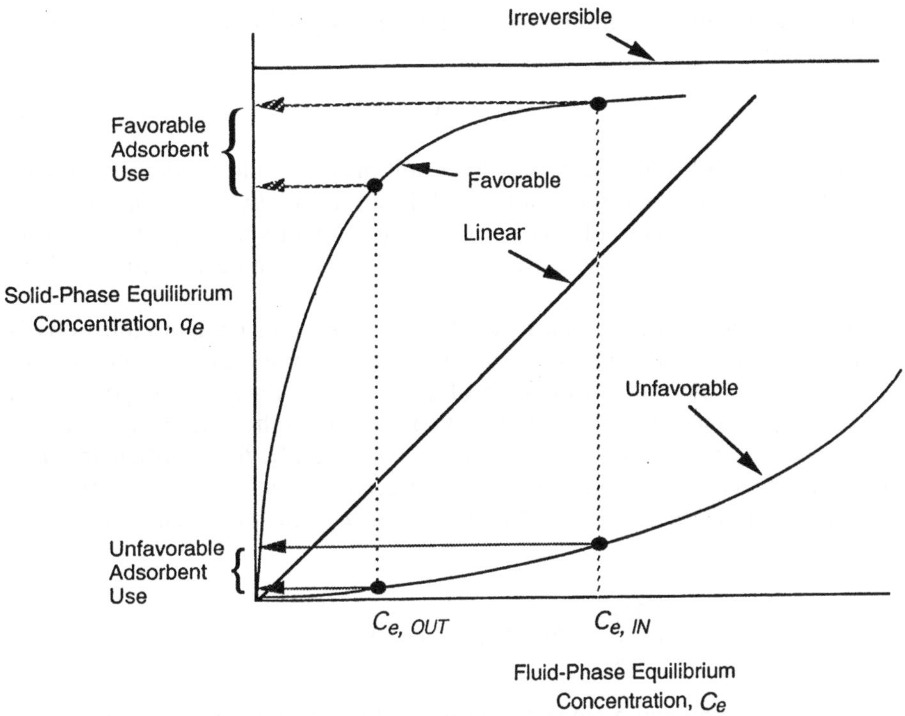

Figure 6-8 Depiction of adsorption isotherm regimes.

Type I isotherms at low solution phase concentrations, but do not approach a q_e plateau at high concentrations.

A variety of models have been developed to characterize the relationships shown in Figure 6-8. Any particular model may be found to describe experimental data accurately under one set of conditions, but not under different conditions. No single model has been found to be generally applicable. This observation is not unexpected given the restrictive assumptions associated with the developments of such models and the complexity and variability of environmental systems and surfaces.

6.5.1 Classical Models

6.5.1.1 Linear Model The evaluation of different models for description of sorption processes in different types of environmental systems has been the focus of significant research and development effort. As a result, a number of generalizations have been advanced. One of the most widely argued is that direct correlations exist between adsorption capacity and solute hydrophobicity. Moreover, in the case of natural sorbents such as soils and sediments, there also exists a correlation between adsorption capacity and the relative organic character of the solid phase. Sorption data are frequently found to be approximated adequately over limited concentration ranges by *linear models* of the general form

$$q_e = \mathcal{K}_D C_e \tag{6-51}$$

where q_e is the mass of solute sorbed per unit mass of solid at equilibrium with a solution of solute concentration C_e, and $\mathcal{K}_D$ is a *distribution coefficient*. This formulation can be deduced from, or related to, any of a number of conceptual models, including the phase partitioning model discussed earlier in this chapter.

An appealing aspect of linear *partitioning analogies* or approximations is that the information required to predict sorption behavior reduces to two readily obtained parameters. In the case of *soils and sediments*, it is possible in many circumstances to normalize the distribution coefficient, $\mathcal{K}_D$, to the *organic content*, or *mass fraction of organic carbon*, ϕ_{oc}°, for a particular solid. This normalized $\mathcal{K}_D$ can then be related directly to some convenient measure of solute hydrophobicity, such as the *octanol–water partition coefficient*, $\mathcal{K}_{o,w}$. Thus

$$\log \frac{\mathcal{K}_D}{\phi_{oc}^{\circ}} = \log \mathcal{K}_{oc} = \alpha_P \log \mathcal{K}_{o,w} + \beta_P \tag{6-52}$$

The coefficients α_P and β_P in correlations of this type have been related to the sorbent properties and to the sorption *efficiency* of the organic carbon fraction of the sorbent for the solute, respectively. The general utility of such correlations is constrained, however, by the system specificity of the calibration coefficients, α_P and β_P.

A further constraint on the conceptual correctness and general utility of correlations based on Equation 6-52 is that they assume and accommodate only linear equilibria. Isotherm nonlinearity is generally evidence of the contributions and potential dominance of sorption processes other than simple partitioning: namely, *adsorption processes*. Nonlinearity can be expected for adsorption of solutes to surfaces over wide solution concentration ranges. It may be exhibited even for absorption into quasi-solid organic matrices if there are interactions between sorbed solute molecules, leading to increases or decreases in their affinity for those organic matrices. Nonetheless, relationships of the type given in Equation 6-52 are often useful in making first-cut approximations to the anticipated environmental behavior of contaminants, as illustrated in Example 6-5. This example also illustrates the system specificity and nongeneral character of such approximation methods.

Example 6-5

- **Situation:** Example 6-1 dealt with prediction of aqueous and gas-phase xylene concentrations after a spill occurred into a motel swimming pool. After cleaning up the spill, the response team notes that in the course of its erstwhile flight to the pool, the truck had hit and ruptured a 55-gallon drum of pesticide which had been sitting on a concrete pad next to a shed alongside the motel. The contents of this drum, now identified as lindane, have been washed by the heavy rainfall into a shallow depression in the ground next to the pad, and have seeped with the runoff water into the temporarily saturated soil.

- **Question(s):** As leader of the response team, you decide to implement an immediate pumping scheme to remove the lindane-contaminated water from the subsurface before it can migrate any farther. What initial determinations must you make to establish reasonable locations for the emergency pumping wells?
- **Logic and Answer(s):**

1. An estimate of the distances to which the lindane has migrated from the point of its entry into the soil column will allow reasonable approximations for locating the emergency pumping wells. The extent of migration relates in part to the water flow which carries the lindane into and within the subsurface, and in part to the extent to which the soil is able to retain the lindane and thus "retard" its migration. The former estimate can be made from knowledge of the surface and subsurface hydrology involved. The latter estimate, of most interest for this particular illustration, involves a projection of the "partitioning" of the lindane from the dissolved aqueous phase to the soil (solid) phase. This can be approached as outlined below.

2. You are compelled in this case to act quickly and thus do not have time to conduct a full sorption isotherm study. You can, however, order a quick determination of the organic carbon content of the soil, and within 2 hours you have results which indicate that the soil ϕ°_{oc} has a value of 0.02.

3. In consulting a handbook of environmental "data facts," you find that a relationship similar to that given in Equation 6-52 for partitioning between water and octanol may be useful for relating a modified soil partitioning coefficient, $\mathcal{K}_{oc}$, to the octanol–water partitioning coefficient, $\mathcal{K}_{o,w}$, of the organic contaminant. In this case, for units of L/kg for $\mathcal{K}_{oc}$, the values of α_P and β_P are given as 0.72 and 0.49, respectively, so that

$$\log \mathcal{K}_{oc} = \log \frac{\mathcal{K}_D}{\phi^{\circ}_{oc}} = 0.72 \log \mathcal{K}_{o,w} + 0.49$$

4. The distribution coefficient of the soil for lindane sorption can thus be estimated as follows:

The octanol–water partitioning coefficient of lindane is shown in Table 1-3 and Appendix A.6.1 as $\log \mathcal{K}_{o,w} = 3.72$; thus $\mathcal{K}_{oc}$ is determined as

$$\log \mathcal{K}_{oc} = 0.72(3.72) + 0.49 = 3.17$$

$$\mathcal{K}_{oc} = 10^{3.17} = 1.48 \times 10^{3}$$

Since $\phi_{oc}^{\circ} = 0.02$ for this soil,

$$\mathcal{K}_D = (1.48 \times 10^3)(0.02) = 29.6 \text{ L/kg}$$

5. Before making estimates of the retention of lindane by the soil you decide to check several other references, in which two alternative correlations between $\mathcal{K}_{oc}$ and $\mathcal{K}_{o,w}$ are found: namely,

$$\log \mathcal{K}_{oc} = \log \mathcal{K}_{o,w} - 0.21$$

$$\log \mathcal{K}_{oc} = 0.38 \log \mathcal{K}_{o,w} + 0.19$$

(*Note:* These two correlations, like the first, are in fact drawn from reports by different authors in the technical literature.)

6. You decide to use these relationships to check your earlier estimates of $\mathcal{K}_D$ values.

(a) For the first alternative correlation,

$$\log \mathcal{K}_{oc} = 3.7 - 0.21 = 3.49 \quad \text{and,} \quad \mathcal{K}_{oc} = 30.9 \times 10^2$$

Thus $\mathcal{K}_D = 61.8$ L/kg.

(b) For the second alternative correlation,

$$\log \mathcal{K}_{oc} = (0.38)(3.7) + 0.19 = 1.60 \quad \text{and,} \quad \mathcal{K}_{oc} = 39.8$$

Thus $\mathcal{K}_D = 0.80$ L/kg.

- **Conclusion:** Empirical correlations between soil "partitioning" and the organic content of soil developed in different investigations may give markedly different estimates for the soil in question. This underscores the need for caution in generalizing relationships or observations which, because of their empirical character, are inherently system specific.

6.5.1.2 Langmuir Model Irving Langmuir, in his work on the oxidation of tungsten elements in electric light bulbs in 1918, developed a model for characterizing the equilibrium adsorption of gases onto solids based on the assumptions that (1) adsorption energy is constant and independent of surface coverage; (2) adsorption occurs only on localized sites, with no interaction between adsorbate molecules; and (3) adsorption is ultimately limited by formation of a monomolecular layer of solute on the surface.

The *Langmuir model* derives from a consideration of the rates of condensation and evaporation of gas molecules at solid surfaces, as depicted schematically in Figure 6-9. If ϕ_m° represents the fraction of complete monolayer coverage of the surface by adsorbed molecules at any time, the rate of evaporation, or desorption, from the surface is proportional to ϕ_m°. Thus

$$\text{rate of desorption} = r_d^{\circ} = k_d \phi_m^{\circ} \tag{6-53}$$

This proportionality derives from mass-law characterization of chemical re-

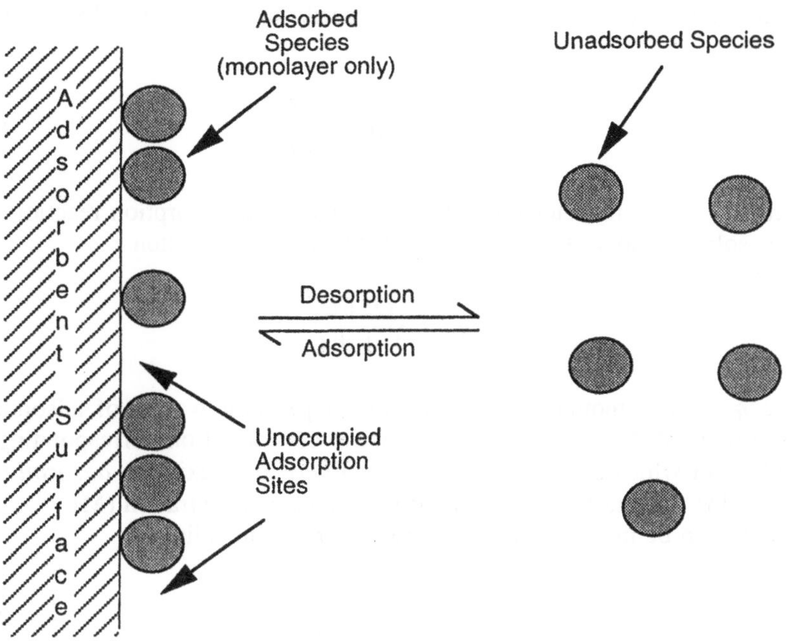

Figure 6-9 Schematic basis for development of the Langmuir adsorption model.

action rates, which are discussed in detail in Chapter 7. Equation 6-53 states that the rate of desorption is a *first-order* (directly dependent) function of the fractional capacity already adsorbed. Similarly, the rate of condensation, or adsorption, of molecules onto the surface is proportional to the fraction of free sites remaining $(1 - \phi_m^\circ)$ and to the rate(s) at which molecules contact the surface. The latter term is proportional to p, the partial pressure of the gas. Thus the overall adsorption rate is described by a *second-order* expression, which is first order in p and first order in $(1 - \phi_m^\circ)$, or

$$\text{rate of adsorption} = \mathscr{r}_a^\circ = k_a p(1 - \phi_m^\circ) \tag{6-54}$$

Equating $\mathscr{r}_a^\circ$ and $\mathscr{r}_d^\circ$ for conditions of equilibrium yields

$$k_d \phi_m^\circ = k_a p(1 - \phi_m^\circ) \quad \text{at} \quad \mathscr{r}_a^\circ = \mathscr{r}_d^\circ \tag{6-55}$$

The terms k_a and k_d in Equations 6-53 through 6-55 are referred to as adsorption and desorption rate constants, respectively. The fraction of surface covered, ϕ_m°, is then

$$\phi_m^\circ = \frac{k_a p}{k_d + k_a p} = \frac{\mathscr{b} p}{1 + \mathscr{b} p} \tag{6-56}$$

The coefficient, $\mathcal{b} = k_a/k_d$, is related primarily to the net enthalpy of adsorption, dH_a:

$$\mathcal{b} = \beta_{a,h} \exp\left(\frac{-dH_a}{\mathcal{R}T}\right) \tag{6-57}$$

where $\beta_{a,h}$ has a characteristic value for a particular adsorption reaction.

For solid–liquid systems, Equation 6-56 is usually written

$$q_e = \frac{Q_a^\circ \mathcal{b} C_e}{1 + \mathcal{b} C_e} \tag{6-58}$$

where q_e is the amount of solute adsorbed per unit weight of adsorbent at equilibrium (MM^{-1}), Q_a° is the solid-phase concentration corresponding to a condition in which all available sites are filled, or maximum *adsorption capacity* (MM^{-1}), and C_e is again the liquid-phase concentration (ML^{-3}). Equation 6-58 can be linearized in a variety of forms, including

$$\frac{C_e}{q_e} = \frac{1}{Q_a^\circ \mathcal{b}} + \frac{C_e}{Q_a^\circ} \tag{6-59}$$

$$\frac{1}{q_e} = \frac{1}{Q_a^\circ} + \frac{1}{\mathcal{b} Q_a^\circ C_e} \tag{6-60}$$

$$q_e = Q_a^\circ - \frac{q_e}{\mathcal{b} C_e} \tag{6-61}$$

While all are equivalent, one particular form may give a more reliable fit to a particular set of data than others, depending on the range and spread of the data to be described. The objective is usually to determine $\mathcal{b}$ and Q_a°. Alternatively, a nonlinear regression technique may be preferred using Equation 6-58.

A specific relationship between spreading pressure, Π_S, and equilibrium solution concentration, C_e, can be obtained for adsorption phenomena described by a Langmuir isotherm. By substitution of Equation 6-58 into Equation 6-50 and integrating, we find that

$$\Pi_S = \frac{\mathcal{R} T Q_a^\circ}{a_{s,w}^\circ} \ln\,(1 + \mathcal{b} C_e) \tag{6-62}$$

Equation 6-62 shows clearly that for a fixed monolayer coverage, Q_a°, greater spreading pressure is exerted at higher values of adsorption energy, $\mathcal{b}$, for any given solution equilibrium concentration, C_e.

Another important thermodynamic implication of $\mathcal{b}$ is in connection with determination of the net enthalpy of adsorption (dH_a). The value of $\mathcal{b}$ obtained in equilibrium tests at different temperatures allows calculation of dH_a by linearization of Equation 6-57:

$$\ln \mathcal{b} = \ln \beta_{a,h} - \frac{dH_a}{\mathcal{R}} \frac{1}{T} \tag{6-63}$$

In such tests, the value of $\mathcal{b}$ is directly determined at 50% surface coverage because Equation 6-58 shows that at this position, $\mathcal{b} = 1/C_e$. The resulting net enthalpy of adsorption is termed the *isoteric heat of adsorption*, given that $\mathcal{b}$ is found by measuring the solution concentration at the same fractional surface coverage (0.5) at different temperatures.

6.5.1.3 BET Model Brunauer and co-workers in 1938 extended the Langmuir model to include the adsorption of multiple layers of molecules. The resulting *Brunauer–Emmett–Teller* (BET) *model* assumes that the first molecules to adhere to a surface do so with an energy comparable to the heat of adsorption for monolayer attachment, while subsequent layers are treated essentially as condensation reactions. If all layers beyond the first have equal energies of adsorption, the BET equation takes the form

$$q_e = \frac{\mathcal{B} C_e Q_a^\circ}{(C_S - C_e)\,[1 + (\mathcal{B} - 1)\,(C_e/C_S)]} \tag{6-64}$$

where C_S is the saturation concentration (solubility limit) of the solute. The parameter $\mathcal{B}$ in the BET model, like $\mathcal{b}$ in the Langmuir model, is expressive of the energy of adsorption, such that

$$\mathcal{B} \approx \exp\left(\frac{-dH_a}{\mathcal{R}T}\right) \tag{6-65}$$

where dH_a is again the net enthalpy of adsorption of the first layer of adsorbate. Equation 6-64 can be linearized to facilitate fitting or calibrating it to a set of empirical adsorption data:

$$\frac{C_e}{(C_S - C_e)q_e} = \frac{1}{\mathcal{B}Q_a^\circ} + \frac{\mathcal{B} - 1}{\mathcal{B}Q_a^\circ} \frac{C_e}{C_S} \tag{6-66}$$

It may be noted that Equation 6-64 reduces to Equation 6-58 if $\mathcal{b} = \mathcal{B}/C_S$, $C_e \ll C_S$, and $\mathcal{B} \gg 1$.

6.5.1.4 Freundlich Model Despite the sound theoretical basis of the Langmuir, BET, and Gibbs models, these isotherms often fail to describe experimental solution sorption data adequately. At approximately the same time as Langmuir was developing the monolayer model for adsorption from gas phase, Heinrich Freundlich and other investigators suggested that data for sorption from solutions are frequently best described by a general exponential concentration-dependent relationship of the form

$$q_e = \mathcal{K}_F C_e^{n} \tag{6-67}$$

where $\mathcal{K}_F$ and n are characteristic constants. Equation 6-67 became associated with Freundlich largely because of his wide use and publication of the model. This equation is linearized by logarithmic transform to give

$$\log q_e = \log \mathcal{K}_F + n \log C_e \tag{6-68}$$

The parameter $\mathcal{K}_F$ can be taken as an indicator of sorption capacity at a specific solution-phase concentration; that is a *specific capacity*. The exponent n can be shown to be a joint measure of the cumulative magnitude and diversity of energies associated with a particular adsorption reaction. Both of the adsorption parameters can be determined by plotting the experimental data according to Equation 6-68 or by performing a non-linear regression fit to Equation 6-67.

Henry developed a theoretical justification for the Freundlich equation as a special case of the Gibbs relationship. If the Gibbs surface excess, Γ_a, is assumed to equal the amount adsorbed, q_e, for dilute solutions, the Gibbs relationship presented in Equation 6-42 becomes

$$\Gamma_a = q_e = -\frac{C_e}{\mathcal{R}T}\frac{d\sigma^\circ}{dC_e} = \frac{C_e}{\mathcal{R}T}\frac{\sigma_0^\circ - \sigma_S}{Q_a^\circ}\frac{dq_e}{dC_e} \tag{6-69}$$

where σ_0° is the initial surface tension of the pure solvent and σ_S° is that of the surface covered with a complete monolayer of solute. The surface free energy, σ°, for fractional surface coverage ϕ_m° is given by

$$\sigma^\circ = \sigma_0^\circ (1 - \phi_m^\circ) + \sigma_S^\circ \phi_m^\circ \tag{6-70}$$

Integration of Equation 6-69 yields the indefinite integral

$$\ln q_e = \frac{\mathcal{R}TQ_a^\circ}{\sigma_0^\circ - \sigma_S^\circ} \ln C_e + \ln \kappa \tag{6-71}$$

which reduces to the Freundlich equation if the constant of integration, κ, is taken as $\mathcal{K}_F$ and

$$\frac{\mathcal{R}TQ_a^\circ}{\sigma_0^\circ - \sigma_S^\circ} = n \tag{6-72}$$

The primary assumption associated with the Henry development of the Freundlich equation, namely that $\Gamma_a = q_e$, is valid only for relatively high surface concentrations and low residual concentrations of solute in solutions.

A Freundlich-type model can also be developed by assuming that the surface is comprised of a continuous series of sites, each of which obeys the Langmuir model but has a different and constant adsorption energy, $\mathcal{b}$. The corresponding model formulation can then be expressed as

$$q_e(C_e) = \int_{\mathcal{b}_{\min}}^{\mathcal{b}_{\max}} q_{e,h}(\mathcal{b}, C_e)\, d\mathcal{b} = \int_{\mathcal{b}_{\min}}^{\mathcal{b}_{\max}} \phi(\mathcal{b}) \frac{Q_a^\circ \mathcal{b} C_e}{1 + \mathcal{b} C_e}\, d\mathcal{b} \tag{6-73}$$

where $q_{e,h}$ represents the isotherm function of a local homogeneous "patch" on a heterogeneous surface. We develop this concept and formulation in a more detailed and general way in Section 6.5.3.1. It is reasonable to assume that the energy function, $\phi(\mathcal{b})$, is related to the variation in net adsorption enthalpy, dH_a, per mole, n, as follows:

$$\phi(\mathcal{b}) \propto e^{-dH_a/nRT} \tag{6-74}$$

Integration of Equation 6-73 then yields

$$q_e = \mathcal{K}_F C_e^{1/n} = \mathcal{K}_F C_e^{\mathcal{n}} \tag{6-75}$$

in which $\mathcal{K}_F \propto \mathcal{R}Tn\beta_{a,h}$ and $\beta_{a,h}$ is defined by Equation 6-57. It is evident from Equation 6-75 that the Freundlich isotherm is convex for $\mathcal{n} < 1$, linear for $\mathcal{n} = 1$, and concave for $\mathcal{n} > 1$. Equation 6-75 reveals the thermodynamic origin of the $1/n$ representation used occasionally for the Freundlich exponent literature. In its broadest use, however, the Freundlich equation is treated as an empirical model, and the exponent can be represented simply as $\mathcal{n}$. As we will see shortly, use of the general parameter $\mathcal{n}$ is particularly appropriate in that the exponential distribution of site energies in Equation 6-74, which led to the original $1/n$ formulation, is only one of many possibilities for heterogeneous surfaces.

The relationship of a Freundlich isotherm to a composite of Langmuir isotherms with different $\mathcal{b}$ values is illustrated graphically in Figure 6-10. Three Langmuir isotherms (dashed lines) are shown together with a summation isotherm (solid points) that was obtained by adding the three q_e values at each C_e. The trace of these solid points is reasonably well approximated by a Freundlich isotherm (solid line) having an exponent $\mathcal{n} = 0.5$. The goodness of fit of a Freundlich equation of a composite system of regularly varied sorption energies tends to increase as the number of contributing sorption components increases.

6.5.1.5 Applications of Classical Models The treatment and analysis of equilibrium sorption data generally begins with selection of an appropriate iso-

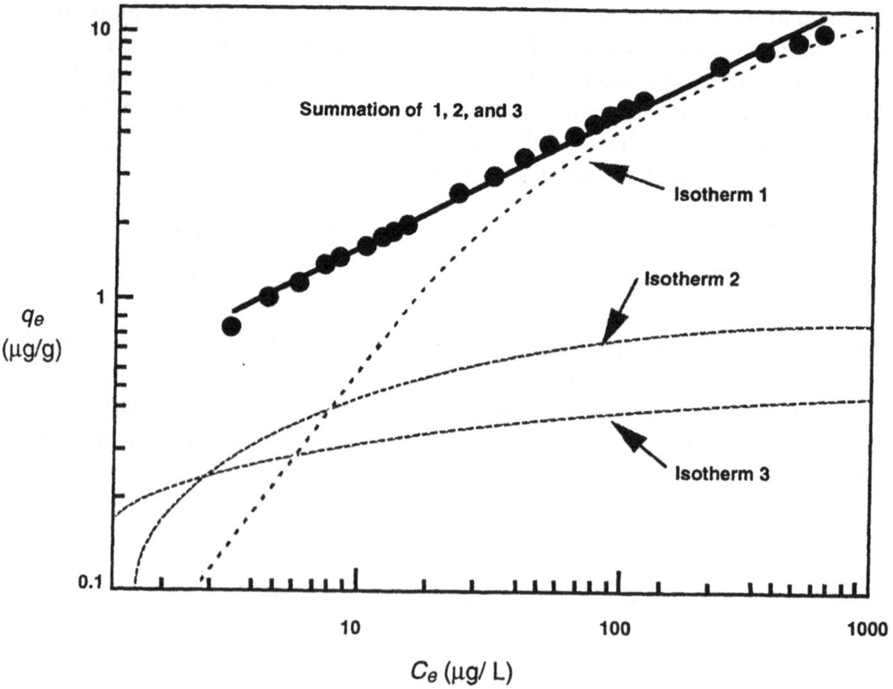

Figure 6-10 Generation of cumulative Freundlich isotherm behavior by multiple Langmuir adsorption phenomena. [Adapted from Weber et al. (1992).]

therm model and an evaluation of the constants for that model. The first step is often influenced by the intended use of the data and model parameters. For example, mathematical simplicity may be the most important consideration if description of the sorption process comprises only a submodel in a larger system model. In this case, any model that adequately describes the data over the concentration range of interest *may* suffice. Plots of the amount sorbed as a function of the equilibrium solution concentration for the several models described above are shown in Figure 6-11. A comparison of similar plots for data from a particular system of interest indicates which of the models is most appropriate for simulation of the sorption process for that system.

The choice of a model may be based in part upon the usefulness of model parameters. For example, the Freundlich $\mathcal{K}_F$ term has been used widely to quantify extent of sorption and provides an easy way to compare different sorbents or conditions for a particular system. Similarly, if an indication of ultimate sorptive capacity is desired, the monolayer saturation term, Q_a°, may be of interest. Figure 6-12 illustrates graphical methods for evaluating the isotherm parameters of the linear, Langmuir, BET, and Freundlich models. Example 6-6 provides an illustration of how such parameters are interpreted in the context of practical applications of adsorption isotherms.

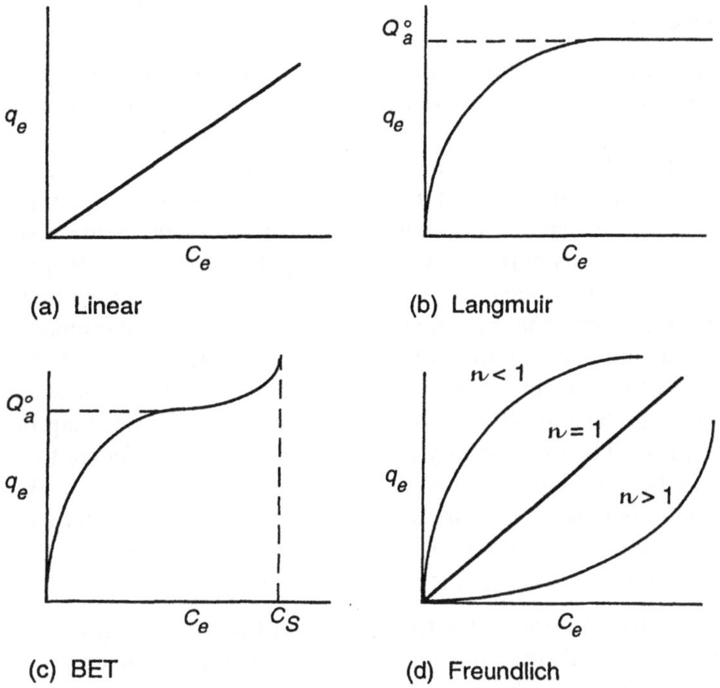

Figure 6-11 Graphical representations of selected isotherm models.

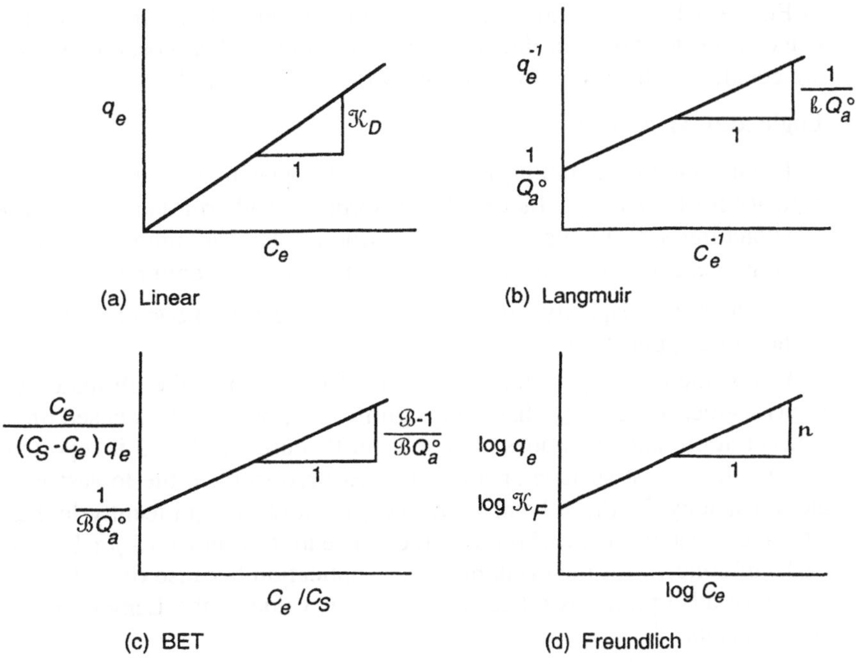

Figure 6-12 Methods for graphical evaluation of isotherm constants.

Example 6-6

- **Situation:** A municipal authority responsible for both water and waste treatment is confronted by a multifaceted organic contamination problem. A halogenated solvent used in the spray cleaning of bus wheels in a municipal terminal has in the past been discharged from a floor drain to the municipal wastewater treatment plant. New pretreatment regulations require that the concentration of this solvent in the discharge be reduced from its current level of 1.0 mg/L to a level of 50 μg/L. Some environmental damage has already been done, however, in that losses of the solvent from the municipal sewer system have contaminated both the shallow and deep aquifers of the region, the latter of which is used as a source of raw water supply for the municipality. The shallow aquifer has an average level of 20 μg/L of contaminant, and the deep aquifer a level of 5 μg/L. The municipality is required by a consent decree to remediate the shallow aquifer to a level of no more than 5 μg/L, and drinking water regulations require no more than 1 μg/L in the treated water supply. The engineering staff of the municipality determine that the best approach to each of these problems is to treat with activated carbon.
- **Question(s):** Samples of activated carbon from three major suppliers are obtained, and adsorption isotherm tests are conducted on each of the three aqueous streams to be treated. Apart from the concentrations of solvent present, the three streams are very similar. For convenience, assume that the measured isotherms for the three carbons are identical to those shown in Figure 6-10 for the three adsorbing components—1, 2, and 3—of the one composite adsorbent illustrated in that figure. If their costs per pound are equal, which of the three carbons would you choose?
- **Logic and Answer(s):**

  1. On first consideration, carbon 1 would appear to have the highest capacity for the solvent. Indeed, the adsorption isotherm data agree with manufacturer's data which shows this carbon to have a high surface area and high capacity as measured by the value of the Langmuir Q_a°.

  2. The high "capacity" of carbon 1 must be put into context for each of the three applications.

  3. For the drinking water application, for example, the ultimate capacity as measured by Q_a° has little significance, because it is based on a concentration range that does not apply for this use. Carbon 3 is reported to have the lowest surface area and is measured to have the lowest isotherm capacity for the solvent. However, it shows a significantly higher effective capacity than carbon over the concentration range (5 μg/L $\rightarrow$ 1 μg/L) of concern for the drinking water application because of its higher energy of adsorption, as reflected by a larger value of the Langmuir energy constant, b.

- **Conclusion(s):** Isotherm constants and manufacturer specifications for adsorbents such as activated carbon must be assessed carefully in the context of a particular application. In the case considered here, the municipality should not select a single carbon for all of its several different applications. Considerations of supply, handling, regeneration, and unit costs being equivalent, carbon 3 would probably be the best choice for the drinking water treatment, carbon 2 for the shallow aquifer remediation, and carbon 1 for pretreatment of the bus-cleaning rinse waters.

Example 6-7 illustrates how Langmuir and Freundlich constants can be obtained from batch equilibrium adsorption data. Example 6-7 also provides a further illustration of the concept presented in Figure 6-10, involving interpretation of a Freundlich isotherm as a composite of two Langmuir isotherms.

Example 6-7

- **Situation:** It is proposed that activated carbon beds be added to a water treatment plant for removal of a targeted synthetic organic contaminant (SOC). Chlorine is currently dosed ahead of the point where the carbon beds would be added. The adsorptive characteristics of surface sites on activated carbon can be affected by contact with oxidants such as chlorine due to formation of surface oxides. Batch adsorption equilibrium tests are required to assess the change in adsorbability of the SOC that might occur after long-term exposure of the activated carbon to chlorine.
- **Question(s):** How can the adsorption data be obtained and interpreted?
- **Logic and Answer(s):**

1. The suggested experiment is as follows. Prior to the equilibrium adsorption tests, a specified mass of activated carbon is exposed to a solution containing a specified mass of chlorine. Different dosages of this carbon are then added to flasks containing a 100-μg/L solution of the SOC. An identical series of dosages of activated carbon particles that have not been exposed to chlorine is then added to another series of flasks having the same initial SOC concentration. Equilibrium is assumed in each flask when there is no further change in SOC concentration with time of contact. The solid-phase concentration, q_e, is calculated from

$$q_e = \frac{C_0 - C_e}{D_o}$$

where $C_0 = 100\ \mu\text{g/L}$ and D_o is dosage (mg/L) of activated carbon added.

2. The results of the experiment above are as follows:

D_o	Test 1: No Exposure to Chlorine		Test 2: After Exposure to Chlorine	
(mg/L)	C_e (μg/L)	q_e (μg/mg)	C_e (μg/L)	q_e (μg/mg)
1	4.8	95.1	13.8	86.3
2	2.3	48.1	6.6	46.7
3	1.5	32.8	4.3	31.9
6	0.7	16.5	2.1	16.3
12	0.4	8.4	1.1	8.2

The increase in C_e values at each dosage in test 2 indicates a significant loss in sorptive capacity due to exposure of activated carbon to chlorine.

3. A Langmuir isotherm can be fit to the data from tests 1 and 2 by linearization according to Equation 6-60.

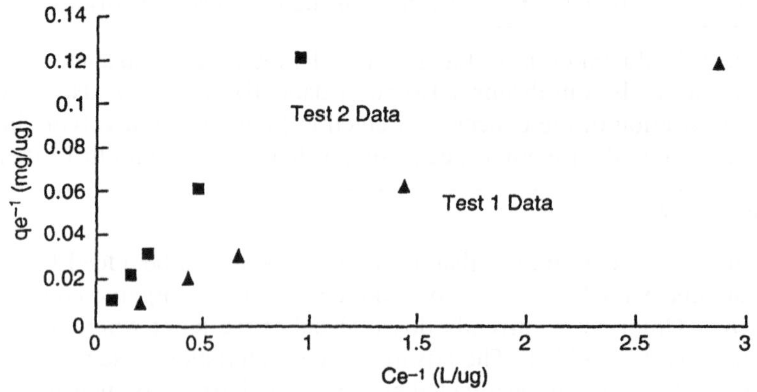

The intercept gives $1/Q_a^\circ$ and the slope gives $1/bQ_a^\circ$. The Langmuir constants are tabulated as follows:

	Test 1	Test 2
Q_a° (μg/mg)	343	400
b (L/μg)	0.07	0.02

These results suggest a significant change in adsorbability as a result of surface oxide formation. This change is reflected by an increase in the sorptive capacity of the carbon, but a lower energy of adsorption.

4. Suppose we assume instead that surface oxide formation causes the complete loss of sorptive sites of some specific sorptive energy. The b and Q_a° values for those lost sites (LS) can be found by a two-parameter search, subject to the constraint that at any given C_e value,

$$q_{e,RS} + q_{e,LS} = q_{e,T}$$

where *RS* refers to the remaining sites, for which Langmuir isotherm constants are given from test 2, and *T* to the total sites, for which Langmuir isotherm constants are given by test 1. The resulting values for Langmuir constants for the lost sites that fit the data from tests 1 and 2 are

$$Q_{a,LS}^\circ = 500 \ \mu\text{g/mg}$$

$$b_{LS} = 0.03 \ \text{L}/\mu\text{g}$$

A calculated total isotherm is obtained by adding together the q_e values for the lost sites and remaining sites at the same C_e, as illustrated below.

C_e (μg/L)	$q_{e,LS}$ (μg/mg), Best Fit	$q_{e,RS}$ (μg/mg), From Test 2	$q_{e,T}$ (μg/mg), Calculated
0.5	7.4	4.0	11.3
1	14.6	7.8	22.4
3	41.3	22.6	63.9
5	65.2	36.4	101.6
10	115.4	66.7	182.1

5. All of the isotherms above are shown below on a ln q_e–ln C_e plot to illustrate that (a) the experimental data for tests 1 and 2 in step 2 can also be fairly well described by a Freundlich isotherm (Equation 6-68), and (b) the hypothesis of the total isotherm consisting of two Langmuir isotherms (for remaining and lost sites) as determined in step 4 is fairly well described by a Freundlich isotherm; this should be expected, given that the Freundlich isotherm represents a patchwork of Langmuir isotherms for each site energy (see Equation 6-73 and also Figure 6-10).

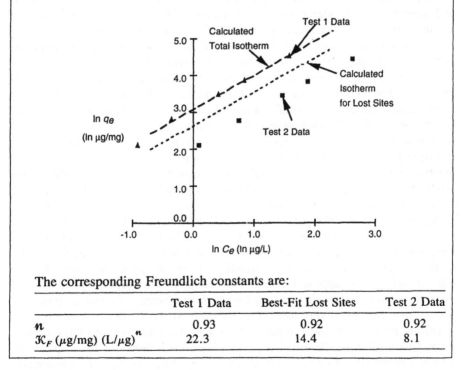

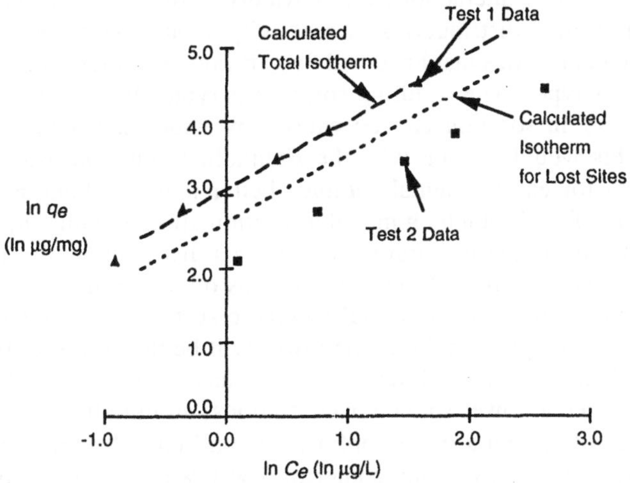

The corresponding Freundlich constants are:

	Test 1 Data	Best-Fit Lost Sites	Test 2 Data
n	0.93	0.92	0.92
$\mathcal{K}_F$ (μg/mg) (L/μg)^n	22.3	14.4	8.1

6.5.2 Nonclassical Nature of Environmental Sorbents and Sorption Processes

The classical models described above are all based on an implicit or explicit assumption about the energetics and distributions of sorption sites on or within a sorbent. More specifically, assumptions are made about the energetics and distribution of *sorption reactions* at sites on or within a sorbent. In the linear model, for example, all the sorption reactions of a specific solute with a specific sorbent are assumed to have the same energy associated with them. Further, the number of sites for such reactions is unlimited. In the Langmuir model, the reaction energy is assumed constant, but the number of sites is limited. Finally, in the Freundlich model, a continuum of varied reaction energies is assumed and sorption sites are hypothetically unlimited.

Most environmental sorption processes involve multiple reaction phenomena and thus are implicitly heterogeneous. Explanations of observed solute uptake and release derived from models predicated on single sorption reactions may therefore be highly system and event specific. This specificity effectively limits their use for predicting sorption behavior in different systems or under different conditions.

Regions of different solids contain different types, amounts, and distributions of surfaces and associated amorphous, condensed, or microcrystalline organic matrices. These regions exist even at the particle scale. The different classes of reactions associated with various types of surfaces were identified in Figure 6-7. Such interactions differ widely with respect to magnitude and mechanism. Varied distributions of reactivity should be expected, therefore, for environmental solids that have widely varying constituent compositions and surface matrix types. The resulting sorption is typically *nonlinear*.

Nonlinearity in sorption can be caused by wide distributions of reaction energies. This would explain why the Freundlich equation usually describes sorption data for environmental samples better than the Langmuir equation. Interpretation of the Freundlich model as a summation of component Langmuir isotherms, as illustrated in Figure 6-10, reveals that nonlinearity may also result from particularly high-affinity interactions of solute molecules with small specific fractions of a sorbent's total available surface. The greater the solute affinity for any particular surface, the lower will be the concentration required to establish high sorbent loadings. This is reflected by sorbent "saturation," being approached at lower solute concentrations, an initially steeper isotherm, a lower value of n, and a more nonlinear overall equilibrium relationship, in other words, a highly favorable isotherm. The lower the affinity of the solute for the sorbent or the lower the variability of site energies, the more closely will the value of n approach unity. Accordingly, the sorption process will come closer to a partitioning or to a low-surface-coverage Langmuir process. Adsorptions that are of very low affinity require an even greater solution *concentration pressure* to force solute onto a surface, thus yielding unfavorable or concave isotherms. Observed degrees of nonlinear behavior for sorption of organic compounds of environmental interest by different types of adsorbents range broadly. Single-isotherm Freundlich n values as low as 0.3 and greater than 1.0 have been reported.

6.5.3 Site Energy Distribution Models

The isotherm models we have discussed thus far involve at least a tacit assumption of an underlying distribution of site energies. If the parameters of any isotherm model can in theory be related to a specific site energy distribution, it should be possible to understand the energetic character of a given sorbent. For example, an arbitrary relationship between q_e and C_e in an adsorption model may have the general form

$$q_e = \phi(C_e, \alpha, \beta, \ldots,) \tag{6-76}$$

where $\alpha, \beta, \ldots$ etc. are arbitrary parameters. Analyses of trends in the changes that occur in such arbitrary isotherm parameters in response to various sorption phenomena might then be used to postulate how a sorbent effectively changes energetically as a result of such sorption phenomena.

6.5.3.1 Theory of Heterogeneous Surfaces The underlying theory of heterogeneous surfaces and distributed reactivity can be characterized in a general way by a mathematical expression of the type:

$$q_e(C_e) = \int_0^{\infty} q_{e,h}(E_n, C_e)\phi_D(E_n)\, dE_n \tag{6-77}$$

This equation defines the total adsorption, q_e, of a solute by a heterogeneous surface as the integral of an energetically homogeneous local isotherm ($q_{e,h}$) multiplied by a site energy frequency distribution function, $\phi_D(E_n)$, over a range of energies, where E_n is, for a given site, the difference in its adsorption energies for a particular solute and the solvent in which the solute is dissolved. The limits on the integral are most appropriately based on the minimum and maximum adsorption energies. These energies, however, are typically not known *a priori*, so they are generally assumed to range from zero to infinity. If the local adsorption isotherm and the site energy distribution function are known in theory, the integral given in Equation 6-77 may be solved to yield a corresponding form for the overall adsorption isotherm. Conversely, if a set of data are found to conform to a particular isotherm model, the integral equation may be solved to derive the corresponding site energy distribution.

As suggested in Equation 6-73, the Langmuir isotherm (Equation 6-58) is commonly used to represent the local isotherm function in Equation 6-77; that is,

$$q_{e,h}(C_e) = \frac{Q_a^\circ b C_e}{1 + b C_e} \tag{6-78}$$

Because it assumes *uniform* site energy, the Langmuir model applies strictly to *homogeneous* surfaces. Its general form, however, is useful for describing *locally homogeneous sites on more generally heterogeneous sorbents*. We employed it in this context in Equation 6-73 and in the analysis of a Freundlich model as a composite of Langmuir isotherms (see Figure 6-10 and Example 6-7).

We can expand on the approach given in Equation 6-73 by introducing a more general local isotherm function which incorporates additional parameters to account for variable site energies; that is,

$$q_{e,h}(C_e) = Q^\circ_{a,g}\left[\frac{(\mathcal{b}_g C_e)^{m}}{1 + (\mathcal{b}_g C_e)^{m}}\right]^{n/m} \tag{6-79}$$

where m and n are termed *heterogeneity parameters*. The term $Q^\circ_{a,g}$ in Equation 6-79 represents maximum sorption capacity and is thus equivalent to Q°_a in Equation 6-58. The term $\mathcal{b}_g$ is analogous to the Langmuir $\mathcal{b}$ defined in Equation 6-57 in terms of adsorption enthalpy and incorporates a *characteristic site energy*, E_o, such that

$$\mathcal{b}_g \propto \exp\left(\frac{E_o}{\mathcal{R}T}\right) \tag{6-80}$$

where E_o is the energy corresponding to the maximum site frequency and determines the *position* of the energy distribution function on the energy axis. For a symmetric, quasi-Gaussian distribution, E_o represents the *mean* site energy, while for an exponential distribution, E_o represents the *minimum* site energy. The *first heterogeneity parameter*, m, characterizes the shape of the site energy distribution in the direction of lower energy values, and the *second heterogeneity parameter*, n, characterizes the shape of the site energy distribution in the direction of higher energy values.

The *generalized Langmuir model* given in Equation 6-79 is a four-parameter isotherm ($Q^\circ_{a,g}$, $\mathcal{b}_g$, m, and n). For various combinations of limiting values of m and n, two useful three-parameter models, commonly referred to as the *Langmuir–Freundlich model* and the *generalized Freundlich model*, can be developed. Like that four-parameter model itself, these specific derivatives of the generalized Langmuir model are convergent at their low-concentration limits on the common two-parameter Freundlich model. The four possible isotherm models that result are given in Table 6-5. In general, the behavior of the heterogeneous isotherm differs significantly from that of the local Langmuir isotherm. The generalized Langmuir model and its two derivatives all embrace the concept of a limiting capacity, but none predicts linear sorption at low concentration. The common Freundlich model neither predicts a linear region at low concentration nor confines sorption to a limiting value at high concentration. It has, nevertheless, proven very useful for describing sorption data over wide ranges of concentration of interest in environmental applications. Heterogeneity parameters equal to unity provide special cases of interest. All of the model equations given in Table 6-5 except the common Freundlich model reduce to the local Langmuir model if both n and m are equal to 1. When its single heterogeneity parameter, n, has a value of 1, the common Freundlich model assumes the low concentration region, or Henry's sorption region (i.e., $\mathcal{b}_g C_e << 1$), approximation of the Langmuir model.

Exact site energy distributions for the relationships given in Table 6-5 have been derived by solving the integral sorption equation for $\phi_D(E_n)$. The Lang-

Table 6-5 Generalized Langmuir Model and Several Derivatives

Model	Isotherm Equation	Heterogeneity Parameter Value Ranges n	m	Approximate Extreme Behaviors Low C_e	High C_e
Generalized Langmuir	$q_e = Q_g^\circ \left[\frac{(b_g C_e)^m}{1 + (b_g C_e)^m} \right]^{n/m}$	0–1	0–1	$Q_g^\circ (b_g C_e)^n$	Q_g°
Langmuir–Freundlich	$q_e = \frac{Q_g^\circ (b_g C_e)^n}{1 + (b_g C_e)^n}$	0–1	$n = m$	$Q_g^\circ (b_g C_e)^n$	Q_g°
Generalized Freundlich	$q_e = Q_g^\circ \left[\frac{b_g C_e}{1 + b_g C_e} \right]^n$	0–1	1	$Q_g^\circ (b_g C_e)^n$	Q_g°
Common Freundlich	$q_e = Q_g^\circ (b_g C_e)^n$	0–1	—	$Q_g^\circ (b_g C_e)^n$	∞

Source: Adapted from Carter et al. (1995).

muir–Freundlich model has been found to give a symmetrical quasi-Gaussian energy distribution centered on E_o. The generalized and common Freundlich models are both characterized by exponential distributions, with E_o serving as the lower energy limit.

6.5.3.2 Approximate Site Energy Distributions Site energy distributions can be approximated using a technique termed the *condensation method*. This method involves generating approximate distribution functions from the isotherm equations. Unlike the exact distributions described in Section 6.5.3.1, the approximate distributions are not normalized. Instead, they are written directly in terms of the isotherm parameters ($Q_{a,g}^\circ$, b_g, m, and n). The area under the distribution (i.e., the sorption capacity) is controlled by $Q_{a,g}^\circ$, the position of the distribution on the energy axis (relative to the reference energy) by b_g, and the spread of the distribution by m and n. This approach facilitates examination of changes in the distributions of site energies on sorbent surfaces resulting from prior sorption or treatment phenomena and how these changes affect the isotherm parameters.

The equilibrium liquid-phase concentration is related to the energy of adsorption by the expression

$$C_e = C_S \exp\left(\frac{-E_n}{\mathcal{R}T}\right) \tag{6-81}$$

where C_S is the maximum solubility of the solute in the solvent and E_n is now explicitly that value of the net sorption energy which corresponds to $C_e = C_S$.

As such, E_n is referenced to the lowest value of physically realizable sorption energy, which depends on the solute but is independent of the sorbent. Because the overall adsorption isotherm is known in terms of C_e, substitution of Equation 6-81 into an isotherm expression results in an equation written in terms of the net sorption energy, E_n. By incorporating Equation 6-81 in Equation 6-77, an approximate site energy distribution, $\phi_D(E_n)$, can be obtained by differentiating the isotherm, $q_e(E_n)$, with respect to E_n:

$$\phi_D(E_n) = \frac{-dq_e(E_n)}{dE_n} \tag{6-82}$$

The resulting site energy distributions are not normalized; hence, the area under the distribution is equal to the maximum adsorption capacity $Q^\circ_{a,g}$:

$$\int_0^\infty \phi_D(E_n)\, dE_n = Q^\circ_{a,g} \tag{6-83}$$

Incorporation of Equation 6-81 into the Langmuir–Freundlich model presented in Table 6-5 gives

$$q_e = \frac{Q^\circ_{a,g}(\mathcal{B}_g C_S)^n \exp\left(\dfrac{-nE_n}{\mathcal{R}T}\right)}{1 + (\mathcal{B}_g C_S)^n \exp\left(\dfrac{-nE_n}{\mathcal{R}T}\right)} \tag{6-84}$$

Differentiating Equation 6-84 with respect to E_n then yields

$$\phi_D(E_n) = \frac{Q^\circ_{a,g} n(\mathcal{B}_g C_S)^n}{\mathcal{R}T} \exp\left(\frac{-nE_n}{\mathcal{R}T}\right)\left[1 + (\mathcal{B}_g C_S)^n \exp\left(\frac{-nE_n}{\mathcal{R}T}\right)\right]^2 \tag{6-85}$$

Like the exact distribution for the Langmuir–Freundlich model, Equation 6-85 corresponds to a quasi-Gaussian site energy distribution.

Application of the condensation approximation method to the generalized Freundlich model yields Equation 6-86:

$$q_e = \frac{(\mathcal{B}_g C_S)^n \exp\left(\dfrac{-nE_n}{\mathcal{R}T}\right)}{\left[1 + (\mathcal{B}_g C_S)^n \exp\left(\dfrac{-E_n}{\mathcal{R}T}\right)\right]^n} \tag{6-86}$$

Differentiation with respect to E_n then gives Equation 6-87, the approximate site energy distribution reflected by this model:

$$\phi_D(E_n) = \frac{Q^\circ_{a,g} n (b_g C_S)^n}{\mathcal{R}T} \exp\left(\frac{-nE_n}{\mathcal{R}T}\right)\left[1 + (b_g C_S) \exp\left(\frac{-E_n}{\mathcal{R}T}\right)\right]^{n+1} \tag{6-87}$$

We recall that the exact site energy distribution for the model is exponential. In contrast, that for the approximate distribution is quasi-Gaussian and widened toward higher energy values.

The Freundlich model, a simplification of the generalized Langmuir model, involves only two independent parameters. Applying the condensation approximation, it can be written in terms of E_n as

$$q_e = Q^\circ_{a,g}(b_g C_S)^n \exp\left(\frac{-nE_n}{\mathcal{R}T}\right) \tag{6-88}$$

The first derivative then yields the approximate site energy distribution in terms of the corresponding adsorption parameters:

$$\phi_D(E_n) = \frac{Q^\circ_{a,g} n (b_g C_S)^n}{\mathcal{R}T} \exp\left(\frac{-nE_n}{\mathcal{R}T}\right) \tag{6-89}$$

Like the exact site energy distribution for the Freundlich model, that for the approximate formulation is exponential.

The formal presentation of the Freundlich model (given in Equation 6-89) shows *three* isotherm parameters. It is, however, effectively a *two-parameter* isotherm because $Q^\circ_{a,g}$ and b_g cannot be determined uniquely from one set of adsorption data. The model is thus more commonly expressed explicitly as the two-parameter isotherm relationship given in Equation 6-67. A comparison of Equation 6-67 with the more rigorous formulation of the Freundlich isotherm given in Table 6-5 provides further insight to the meaning of the isotherm parameters. It shows that the exponent n is consistent in both expressions as an indication of heterogeneity, but that the parameter $\mathcal{K}_F$ is a composite parameter incorporating $Q^\circ_{a,g}$, b_g, and n; that is,

$$\mathcal{K}_F = Q^\circ_{a,g} b_g^n \tag{6-90}$$

*6.5.3.3 **Theoretical Interpretations*** The following analysis shows how to relate site energy distributions to isotherm parameters. The Langmuir–Freundlich model is particularly suitable for this purpose because the symmetrical shape of its site energy distribution permits straightforward illustration of the effects of parameter variations. For this analysis, q_e has the units μg/mg and C_e the units μg/L.

The site energy distribution for the Langmuir–Freundlich model (Equation 6-85) is plotted in Figure 6-13 for variations in $Q^\circ_{a,g}$, b_g, and n, respectively. The areas under these distribution curves can be interpreted as the maximum number of sites available for adsorption. The range of parameter values spec-

(a) As a function of $Q^{\circ}_{a, g}$ for $\ell_g = 10^{-2}\,L/\mu g$ and $\nu = 0.7$

(b) As a function of ℓ_g for $Q^{\circ}_{a, g} = 500\,\mu g/mg$ and $\nu = 0.7$

(c) As a function of ν for $Q^{\circ}_{a, g} = 500\,\mu g/mg$ and $\ell_g = 10^{-2}\,L/\mu g$

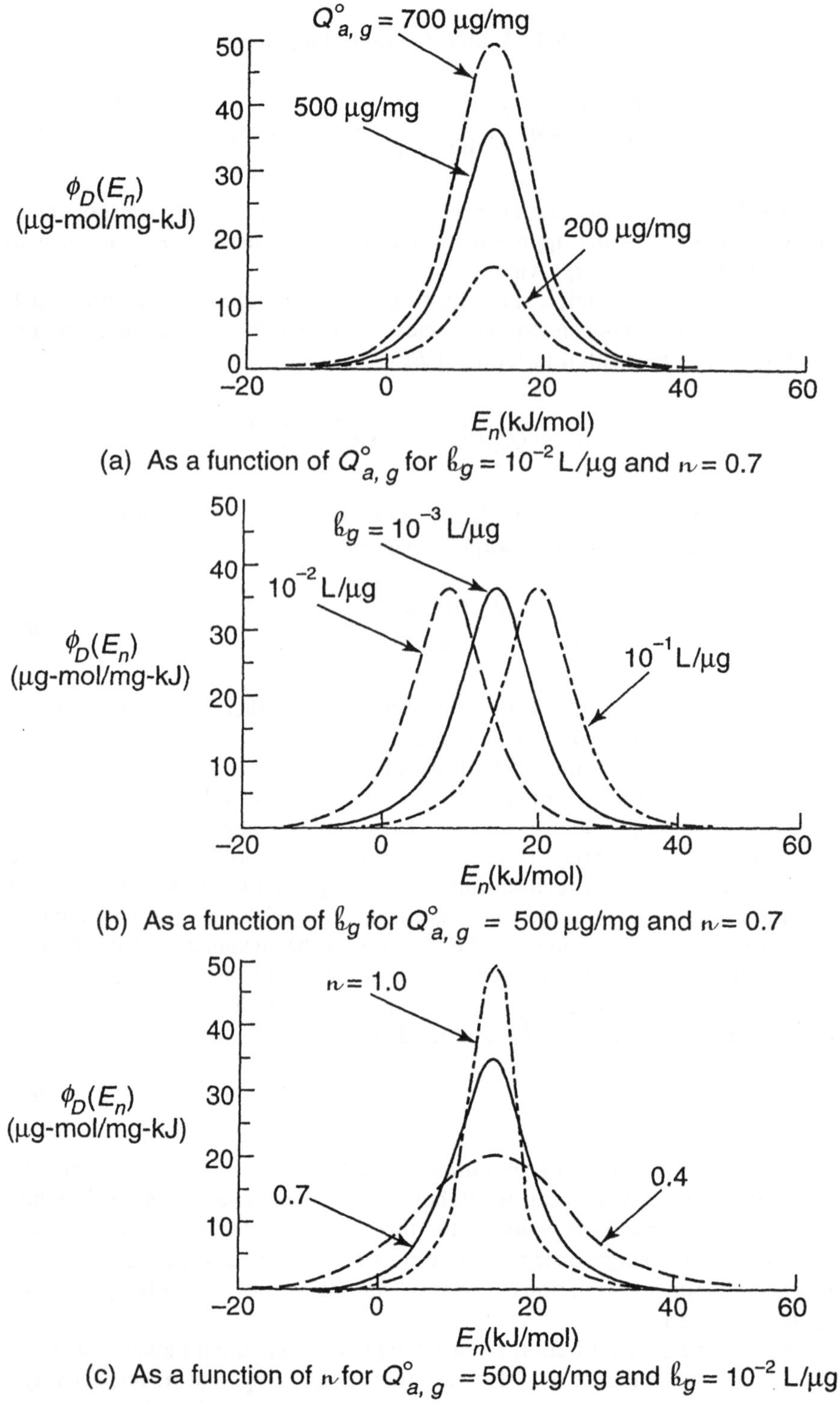

Figure 6-13 Site energy distributions associated with various Langmuir–Freundlich model parameter values. [After Carter et al. (1995).]

ified on this figure are representative of typical, strongly sorbing organic contaminants. The influence of the capacity parameter, $Q^{\circ}_{a,g}$, on the area under the distribution for $Q^{\circ}_{a,g}$ values of 700, 500, and 200 μg/mg is illustrated in Figure 6-13a. The curves in this figure show successive decreases in area with decreasing $Q^{\circ}_{a,g}$ but no changes in either shape or the mean of the distribution. Figure 6-13b depicts changes in the means of the site energy distributions as the affinity parameter, $\mathscr{b}_g$, is varied through values of 10^{-3}, 10^{-2}, and 10^{-1} L/μg. Although the means of the distributions shift to higher values with increases in $\mathscr{b}_g$, neither the shapes nor areas of the curves are altered. The position of the distribution mean on the energy axis can be used as a measure of the affinity of an adsorbate for the adsorbent. Thus, for adsorbents having heterogeneous surfaces, those surfaces having the highest $\mathscr{b}_g$, values will have higher average energies of adsorption and be preferred over lower-affinity surfaces for sorption reactions. Figure 6-13c illustrates the influence of the heterogeneity parameter, $\boldsymbol{n}$, on the width of the site energy distributions. Increases in $\boldsymbol{n}$ cause marked narrowing of the distributions, although the means of the distributions and the area encompassed by the curves do not change. The width of the site energy distribution may be interpreted to relate to the range of energy sites (i.e., to the heterogeneity of surface site energies). Processes that increase the value of $\boldsymbol{n}$ for a given sorbent would therefore decrease the sorptive heterogeneity.

The negative energies shown in Figure 6-13 provide evidence for the effects of parameter variations on site energy distributions, but they do not have true physical significance. Negative values of E_n occur when the absolute energy falls below the energy associated with the maximum solubility of the solute. To illustrate this point, the relationship between solute concentration and energy as given by Equation 6-81 is plotted in Figure 6-14 for a hypothetical solute having an aqueous solubility of $C_S = 10^6$ μg/L. The figure shows that the range of residual solute concentrations over which energy distributions can be compared meaningfully is limited to 1 to 10,000 μg/L; the corresponding E_n values range from about 11 to 34 kJ/mol.

Once E_n is limited to only positive values, the distribution functions implicit to different isotherm models become quite similar over solute concentration ranges of greatest environmental interest. Not surprisingly, a given set of data often can be described reasonably well with any isotherm model that is based on distributed site energies.

6.5.4 Composite Isotherm Models

The site energy distribution models discussed above are each predicated on a particular continuous distribution function. In reality, many environmentally significant sorbents exhibit surface and/or matrix heterogeneities which are discrete and therefore discontinuous. For example, a sand grain of essentially mineral (e.g., SiO_2) structure may have both mineral and organic sorption sites if there are clumps of amorphous or condensed macromolecular organic matter held in crevices or imperfections at some of its surfaces.

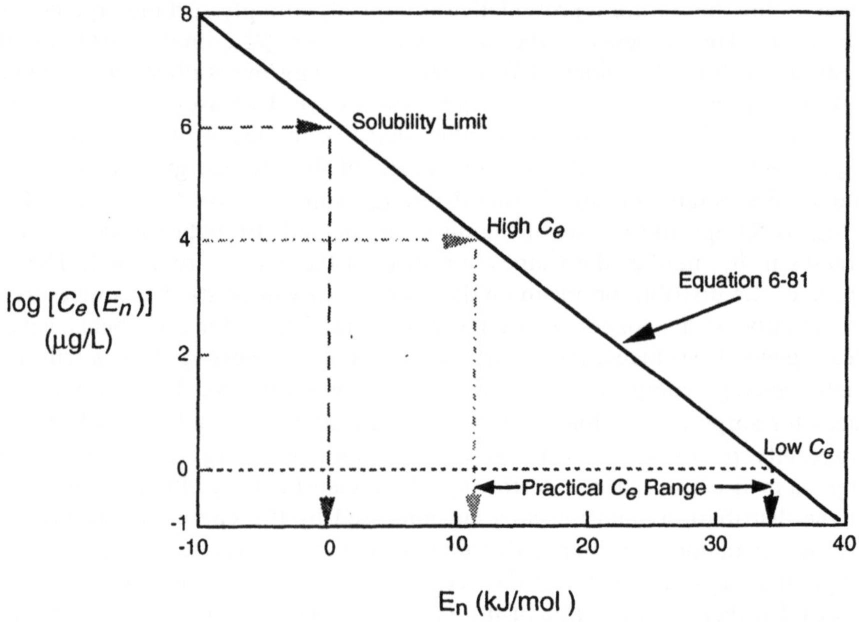

Figure 6-14 Residual solution concentration limits on net adsorption energy ranges. [Adapted from Carter et al. (1995).]

Discrete and discontinuous surface characteristics may well give rise to different combinations of linear and nonlinear "local" sorptions. If overall sorption on sorbent particle regions is nonlinear (i.e., components that exhibit nonlinear behavior are large relative to that of the linearly sorbing regions), the variations of q_e with C_e will probably also be nonlinear. In such cases, a composite isotherm model may be appropriate. In this type of model, a series of different discrete local isotherms are superimposed as an alternative to a single model with one continuous site energy distribution.

A composite isotherm can differ markedly from any particular local isotherm in form and in underlying process as well. If the contributing sorption reactions can be characterized, the composite isotherm can be developed by summing the contributing components. This additivity feature is inherent to a mass balance on the sorbed solute:

$$(q_e)_T = \sum_{i=1}^{k} (X_m)_i (q_e)_i \tag{6-91}$$

where $(q_e)_T$ is the total solute mass sorbed per unit mass of bulk solid, $(X_m)_i$ is the mass fraction of the solid comprising reaction region or component i, and $(q_e)_i$ is the sorbed phase concentration expressed per unit mass of sorbent specifically associated with the ith reaction or sorption mechanism. If each of the individual contributing mechanisms or regions of sorption yields an isotherm that is linear, the composite isotherm will be linear:

$$(q_e)_T = \sum_{i=1}^{k} (X_m)_i(\mathcal{K}_D)_i C_e = \left(\sum_{i=1}^{k} (X_m)_i(\mathcal{K}_D)_i\right) C_e = \overline{\mathcal{K}}_D C_e \quad (6\text{-}92)$$

where $(\mathcal{K}_D)_i$ is the distribution coefficient for reaction i expressed on a per mass of component i basis, and $\overline{\mathcal{K}}_D$ is a mass-averaged distribution coefficient. A strongly sorbing site may therefore dominate overall sorption, even if it comprises a relatively small fraction of the total sorbent mass.

If one or more of the component elements of sorption is governed by a nonlinear relationship between the solution and sorbed phases, the composite isotherm will also be nonlinear. For example, it was illustrated in Figure 6-10 and Example 6-7 that a set of data derived from a composite of Langmuir isotherms may be fit reasonably well by a common Freundlich isotherm. A more probable circumstance for many environmental systems is that observed nonlinearities are caused by a discontinuous array of sorption relationships, some involving linear absorption reactions and others nonlinear adsorption reactions. The following *distributed reactivity model* is an attempt to capture combinations of linear and nonlinear sorptions:

$$(q_e)_T = \overline{X_{m,L}}\overline{\mathcal{K}}_D C_e + \sum_{i=1}^{k} (X_{m,NL})_i(\mathcal{K}_F)_i C_e^{(n)_i} \quad (6\text{-}93)$$

where $\overline{X_{m,L}}$ is the summed mass fraction of solid phase exhibiting *linear* sorptions, $\overline{\mathcal{K}}_D$ is the average distribution coefficient for the summed linear components, and $(X_{m,NL})_i$ is the mass fraction of the ith nonlinearly sorbing component.

Regardless of the particular combination of isotherm elements comprising an overall or composite isotherm, the distributed reactivity model includes linear contributions through use of a mass-averaged distribution coefficient, $\overline{\mathcal{K}}_D$. The complexities of individually quantifying the parameters of each of many individual local isotherms will probably render a fully formulated model impractical for most environmental systems. Multiple linear components will instead have to be grouped, and only two or three major classes of nonlinear components individually included in practical applications of the distributed reactivity model and similar composite models.

6.5.5 Models for Competitive Adsorption

Materials to be adsorbed are commonly mixtures of more than one compound. As will be shown shortly, the compounds comprising such mixtures should not interfere with each other in true *absorption* processes, and solute competition is thus not a factor in such processes. In *adsorption*, however, compounds in mixtures may mutually enhance their respective adsorptions, may act relatively independently, or may interfere with one another. Mutual inhibition of adsorption capacity can be predicted if (1) adsorption is confined to a single or a few molecular layers, (2) the adsorption affinities of the solutes do not differ by several orders of magnitude, and (3) there are no specific interactions between solutes to enhance adsorption. The degree of mutual inhibition of competing sorbates is related to the comparative sizes of the molecules being adsorbed, to their relative adsorptive affinities, and to their respective concentrations.

An isotherm model designed to incorporate the effects of competition should be able to describe the sorption of each of the compounds involved over wide ranges in their respective concentrations. Competitive equilibrium interactions are subject to exact and systematic description, at least theoretically.

6.5.5.1 Langmuir Competitive Model Competition among solutes that each exhibit Langmuir-type isotherms can be predicted from a simple extension of the Langmuir monolayer adsorption model (Equations 6-53 through 6-58). The rate of desorption (Equation 6-53) still depends only on the component. The rate of adsorption (Equation 6-54), however, must account for available sites only [i.e., $(1 - \Sigma_{i=1}^{n} \phi_{m,i}^{\circ})$]. After setting adsorption and desorption rates equal (analogous to the single-solute adsorption model) and performing some algebraic manipulation, the extent of adsorption, $q_{e,i}$ of the ith solute from an n-solute mixture is given by

$$q_{e,i} = Q_{a,i}^{\circ} b_i C_{e,i} \left(1 + \sum_{j=1}^{n} b_j C_{e,j}\right)^{-1} \tag{6-94}$$

For a two-solute mixture of substances A and B, for example, Equation 6-94 written for solute A becomes

$$q_{e,A} = \frac{Q_{a,A}^{\circ} b_A C_{e,A}}{1 + b_A C_{e,A} + b_B C_{e,B}} \tag{6-95}$$

where $Q_{a,A}^{\circ}$, b_A, and b_B are Langmuir constants determined from adsorption measurements in solutions containing each single solute, and $C_{e,A}$ and $C_{e,B}$ are equilibrium concentrations in the mixture of the two solutes (see also Equations 8-22 to 8-25 for a complete derivation). When the concentrations of the two solutes are sufficiently large that surface coverage is substantially complete, the unity term in the denominator of Equation 6-95 may be neglected, allowing the equation to be linearized in the form

$$\frac{C_{e,A}}{C_{e,B} q_{e,A}} = \frac{b_B}{b_A Q_{a,A}^{\circ}} + \frac{C_{e,A}}{Q_{a,A}^{\circ} C_{e,B}} \tag{6-96}$$

Alternatively, the fit of the adsorption data to the Langmuir competitive model may be determined by comparing experimental values of $q_{e,A}$ with those calculated from Equation 6-95.

If the Langmuir competitive model applies to experimental observations, a plot of the quantity $C_{e,A}/C_{e,B}q_{e,A}$ as a function of $C_{e,A}/C_{e,B}$ should be linear, with an intercept of $\mathcal{b}_B/\mathcal{b}_A Q^\circ_{a,A}$ and a slope of $1/Q^\circ_{a,A}$. Similarly, if Equation 6-96 is written in terms of $q_{e,B}$, the linear approximation is

$$\frac{C_{e,B}}{C_{e,A}q_{e,B}} = \frac{\mathcal{b}_A}{\mathcal{b}_B Q^\circ_{a,B}} + \frac{C_{e,B}}{Q^\circ_{a,B}C_{e,A}} \tag{6-97}$$

A thermodynamically more satisfying approach to the description of competitive sorption phenomena is to be found in the *ideal adsorbed solution theory* (IAST) cited earlier in the chapter. The IAST model affords predictions of sorption in multisolute systems based on measurements of single-solute isotherms for each of the components of a mixture.

6.5.5.2 Ideal Adsorbed Solution Theory The ideal adsorbed solution theory provides a thermodynamically based model for competitive adsorption. It is referred to as the ideal *adsorbed* solution theory because the activity coefficients of components in both the solution and adsorbed phases are assumed to be unity. In reality, the model derives from the Gibbs model and thus might more correctly be referred to as the ideal *surface-solution* theory, in that it is rigorously appropriate only for sorption reactions which are primarily entropy driven. For the solution phase, the assumption of unit activity is reasonable, given that most aqueous mixtures of sorbates are dilute. For the sorbed phase, the activity may be less than unity given the high concentrations at the surface; unfortunately, there is no way to measure the activity in this phase. Regardless of the potential weakness of the ideality assumption in the sorbed phase, the IAST is thermodynamically—rather than kinetically—based as is the Langmuir competitive model. Moreover, the IAST does not require that single-solute adsorption be described by any particular model. As a result, it has proven, with appropriate modifications, to be a useful empirical predictor of competitive behavior in a variety of situations.

The underlying assumption of the IAST is that the chemical potential (and the fugacity) of the liquid (l) and the solid (s) phases or states of solute i are equal at equilibrium; this is true for both single- (S) and multisolute (M) systems. As noted above, its thermodynamic basis is the Gibbs free-energy equation, which includes the change in free energy brought about by a change in surface tension and spreading pressure, Π_S. Using the superscripts S and M to represent single and multisolute systems, respectively, the relevant thermodynamic relationships may be written:

$$(\mu_i)_l^S = (\mu_i)_s^S \tag{6-98}$$

$$(\mu_i)_l^M = (\mu_i)_s^M \tag{6-99}$$

$$(\mu_i)_l^S\big|_{T,C^S_{e,i},\Pi_s} = (\mu_i^\circ)_l^S + \mathcal{R}T \ln C^S_{e,i}\big|_{\Pi_s} \tag{6-100}$$

$$(\mu_i)_l^M|_{T,C_{e,i}^M} = (\mu_i^\circ)_l^M|_T + \mathcal{R}T \ln C_{e,i}^M|_{\Pi_s} \tag{6-101}$$

$$(\mu_i)_s^S = \mathcal{R}T \ln (f_i)_s^S \tag{6-102}$$

$$(\mu_i)_s^M|_{T,\Pi_s,(X_i)_s^M} = \mathcal{R}T \ln (X_i)_s^M (f_i)_s^M = (\mu_i)_s^S + \mathcal{R}T \ln (X_i)_s^M \tag{6-103}$$

Equation 6-103, in which $(X_i)_s^M$ is the mole fraction of component i adsorbed from the mixture, is predicated on the assumption that the activity of the sorbed phase is unity.

One of the key relationships for computations using the IAST is derived by first substituting Equations 6-98 and 6-100 into Equation 6-103, yielding

$$(\mu_i)_s^M|_{T,\Pi_s,(X_i)_s^M} = (\mu_i^\circ)_l^S|_T + \mathcal{R}T \ln C_{e,i}^S + \mathcal{R}T \ln (X_i)_s^M \tag{6-104}$$

Because the chemical potentials must be equal in both phases, Equation 6-101 can be substituted for $(\mu_i)_s^M$ to obtain

$$C_{e,i}^M = (X_i)_s^M C_{e,i}^S|_{\Pi_s} \tag{6-105}$$

Equation 6-105 is important because it relates competitive adsorption to single-solute adsorption, the latter of which is of course independent of competition. The particular single-solute concentration of interest is that which produces the *same spreading pressure as the mixture*. The spreading pressure relationship for each single solute is known from Equation 6-50, which, in turn, requires specification of a single-solute adsorption isotherm equation. Methods for application of Equation 6-50 will be discussed in more detail shortly.

Comparison of Equation 6-105 to Raoult's law strengthens the analogy between the two-dimensional form of the ideal gas law, derived from the spreading pressure concept, and the familiar, three-dimensional form of the ideal gas law. In the latter, Raoult's law governs the distribution of each component of a mixture between the gas and liquid phases according to the relationship

$$P_i = X_i P_i^v \tag{6-106}$$

where X_i is the mole fraction of component i in the liquid phase and P_i^v is the vapor pressure that would be produced above a pure solution of that component. The similarity in form and use of Equations 6-105 and 6-106 is readily apparent.

Another essential expression in IAST is that which determines the total number of moles adsorbed in a mixture, $(n_T)^\circ$. This quantity is needed to calculate the amount of each individual component adsorbed $[(X_i)_s^M (n_T)^\circ]$. From the Gibbs free-energy relationship at constant temperature and pressure given in Equation 6-41 we can write

$$A^\circ d\sigma^\circ = -\Sigma (n_i)^\circ d(\mu_i)_s^M \tag{6-107}$$

After substitution of $-d\sigma° = d\Pi_S$ from Equation 6-44, $(n_i)° = (X_i)_s^M[(n_T)°]$, and $d(\mu_i)_s^M = \mathcal{R}Td \ln (X_i)_s^M (f_i)_s^S$, the following equation is obtained:

$$\frac{A°}{(n_T)_s} = \Sigma\, (X_i)_s^M\, \mathcal{R}T \left(\frac{d \ln (f_i)_s^S}{d\Pi_S}\right) = \Sigma\, (X_i)_s^M \left(\frac{d(\mu_i)_s^S}{d\Pi_S}\right) \qquad (6\text{-}108)$$

Further simplification results by first writing Equation 6-108 for adsorption of a single component

$$\frac{d(\mu_i)_s^S}{d\Pi_S} = \frac{A°}{(n_i)°} \qquad (6\text{-}109)$$

Substitution into Equation 6-108 gives

$$\frac{1}{(n_T)°} = \sum_{i=1}^{k} \frac{(X_i)_s^M}{(n_i)°} \qquad (6\text{-}110)$$

Equation 6-110 is usually expressed in terms of solid-phase *concentrations* (moles per unit weight of adsorbent) rather than *number* of moles adsorbed or

$$\frac{1}{q_{e,T}} = \sum_{i=1}^{k} \frac{(X_i)_s^M}{q_{e,i}^S} \qquad (6\text{-}111)$$

where

$$(X_i)_s^M = \frac{q_{e,i}^M}{q_{e,T}} \qquad (6\text{-}112)$$

Equations 6-50, 6-105, and 6-111 must be solved together to determine the equilibrium concentrations of each component in the solution and sorbed phases. The following identities are also useful:

$$\sum_{i=1}^{k} (X_i)_s^M = 1 \qquad (6\text{-}113)$$

$$\sum_{i=1}^{k} q_{e,i}^M = q_{e,T} \qquad (6\text{-}114)$$

The spreading pressure relationship given by Equation 6-50 is usually determined by fitting a Freundlich equation to each of the single-solute adsorption isotherms. Recognizing that for the Freundlich isotherm, $d \ln C_{e,i}^S = d \ln q_{e,i}^S/n$, and further that $d \ln q_{e,i}^S = dq_{e,i}^S/q_{e,i}^S$, the spreading pressure is given by

$$\Pi_{S,i} = \frac{\mathcal{R}T}{A^\circ}\int_0^{q^S_{e,i}} \left(\frac{q^S_{e,i}}{n_i}\right)\frac{dq^S_{e,i}}{q^S_{e,i}} = \frac{\mathcal{R}Tq^S_{e,i}}{n_i A^\circ} \tag{6-115}$$

The procedure suggested above is to fit one Freundlich equation to the entire equilibrium relationship over solid-phase concentrations that range from zero to $q^S_{e,i}$. This is impossible in many instances because of the heterogeneity of adsorption. In addition, the fit of the isotherm for solid-phase concentrations between zero and the lowest experimentally measured value is obviously unknown. For these reasons, Equation 6-115 is only an approximation made to provide a convenient mathematical solution. The magnitude of error will usually not affect the spreading pressure calculation greatly; nonetheless, caution is recommended.

Equation 6-115 is important because it establishes the equilibrium position at which adsorption of each single solute exerts the same spreading pressure as the mixture for use in Equation 6-105. For constant spreading pressure, Equation 6-115 states

$$\frac{q^S_{e,i}}{n_1} = \frac{q^S_{e,2}}{n_2} = \cdots = \frac{q^S_{e,k}}{n_k} \tag{6-116}$$

Further algebraic manipulations of Equations 6-105, 6-111, and 6-116 lead to a useful relationship between the solution concentration of any component in the mixture and that of each in the adsorbed phase:

$$C^M_{e,i} = \frac{q^M_{e,i}}{q_{e,T}}\left(\frac{\sum_{j=1}^{k}\frac{q^M_{e,j}}{n_j}}{\mathcal{K}_{F,i}/n_i}\right)^{1/n_i} \qquad \text{for } i = 1 \text{ to } k \tag{6-117}$$

The equations for IAST can be used in several ways. The computational procedure will depend on which parameters are known and which are to be found. For example, adsorption of a mixture of components from a batch solution can be predicted by combining Equation 6-117 with the material balance equation for each component:

$$q^M_{e,i} = \frac{(C^M_{0,i} - C^M_{e,i})}{D_o} \tag{6-118}$$

in which $C^M_{0,i}$ is the initial concentration and D_o is the dosage of adsorbent added. The result is

$$C^M_{0,i} - D_o\, q_{e,i} - \frac{q_{e,i}}{q_{e,T}}\left(\frac{\sum_{j=1}^{k}\frac{q_{e,j}}{n_j}}{\mathcal{K}_{F,i}/n_i}\right)^{1/n_i} = 0 \qquad \text{for } i = 1 \text{ to } k \tag{6-119}$$

This set of k nonlinear equations must be solved simultaneously to find the values of q_e for each of k components by assuming that all other system constants are known; Newton–Raphson algorithms are available to provide a solution. Once the q_e values are known, the corresponding values of C_e in the mixture are found from Equation 6-118. A simple illustration of the computations involved in the IAST for adsorption of two solutes is given in Example 6-8.

Example 6-8

- **Situation:** A small flow from a chemical production facility contains dichlorobenzene and nitrobenzene. A pretreatment process is used to reduce the concentrations of these materials before biological treatment to prevent potential impacts on the biological processes. Pretreatment consists of a fill-and-draw operation of a slurry contactor that contains powdered activated carbon (PAC). The detention time in the slurry contactor is sufficient to allow equilibrium between the solution and adsorbent phases to be attained. Enough PAC is added to each batch to achieve an equilibrium concentration of 1.5 mg/L (12.2 μM) of dichlorobenzene, the more weakly adsorbed of the two components, prior to discharge. The corresponding equilibrium concentration of nitrobenzene is typically 0.75 mg/L (5.1 μM). Although the primary goal of treatment is to reduce the solution-phase concentration of the two contaminants, the equilibrium solid-phase concentrations on the PAC after contact are also important; if they are too high, the PAC must be sent to a hazardous waste landfill for disposal.

- **Question(s):** The effluent concentrations are known to meet standards. As plant engineer, your job is to determine whether the performance of the treatment system is acceptable (i.e., the solid-phase loading of *o*-xylene and nitrobenzene must both be kept below 0.2 g/g of solids to avoid disposal of the PAC to a hazardous waste landfill).

- **Logic and Answer(s):**

1. The feed concentrations of dichlorobenzene and nitrobenzene during the fill cycle are unknown. No records are kept of the mass of PAC added to each batch.

2. An adsorption isotherm is determined for single-solute solutions of each of the two components. The resulting Freundlich isotherms are found to be:

Component	$\mathcal{K}_F$ (mol/g) (mol/L)^n	$\mathcal{K}_F$ (mg/g) (L/mg)^n	n
Dichlorobenzene	224	68	0.31
Nitrobenzene	526	140	0.43

3. Because the total mass of each contaminant introduced to the treatment system and the dosage of PAC is unknown, it is impossible to cal-

culate the amount adsorbed on the PAC directly from a material balance. However, IAST can be used to find the solid-phase concentrations of *o*-xylene and nitrobenzene given the solution-phase concentrations at equilibrium. A trial-and-error solution can be accomplished with a spreadsheet. The steps are as follows:

4. Let component 1 = dichlorobenzene and component 2 = nitrobenzene.

5. $C^M_{e,1} = 12.2\ \mu M$ and $C^M_{e,2} = 5.1\ \mu M$ (stated mixture equilibrium condition).

6. Estimate $(X_1)^M_s$, and thus $(X_2)^M_s = 1 - (X_1)^M_s$.

7. Calculate the single-solute concentrations of components, $C^S_{e,1}$ and $C^S_{e,2}$, at the same spreading pressure as the mixture using Equation 6-105.

8. Calculate the corresponding values of $q^S_{e,1}$ and $q^S_{e,2}$ from the Freundlich equations provided above.

9. To satisfy the IAST, the values of $q^S_{e,1}$ and $q^S_{e,2}$ calculated in step 8 must meet the criterion specified by Equation 6-116:

$$q^S_{e,2} = \frac{n_2}{n_1} q^S_{e,1} = 1.39 q^S_{e,1}$$

10. Iterate the solution by selecting values of $(X_1)^M_s$ and $(X_2)^M_s$ and repeating steps 6 to 9 until Equation 6-116 is satisfied.

11. Once the correct values of $q^S_{e,1}$ and $q^S_{e,2}$ are determined, calculate $q_{e,T}$ from Equation 6-111 and $q^M_{e,1}$ and $q^M_{e,2}$ from Equation 6-112. The results of iteration are:

Component	$(X_i)^M_s$	$q^M_{e,i}$ (μmol/g)	$q^M_{e,i}$ (g/g)	$q^S_{e,i}$ (μmol/g)
1	0.205	789	0.116	931
2	0.795	203	0.025	1288

This solution meets the criterion within iteration error as given by Equation 6-116:

$$q^S_{e,2} = 1.39 q^S_{e,1} = 1.39(931) = 1291\ \mu\text{mol/g}$$

12. We conclude that neither solid-phase concentration exceeds 0.2 g/g and it is thus not necessary to use a hazardous waste landfill for disposal of the spent PAC.

Note: The solution procedure for more than two components with $C^M_{e,i}$ values ($i = 1$ to k) known cannot begin with estimating $(X_1)^M_s$ because $(X_2)^M_s \cdots (X_n)^M_s$ are not uniquely determined. Instead, the solution begins by estimating the spreading pressure (Equation 6-115), which fixes all values of $q^S_{e,i}$ (Equation 6-116) and $C^S_{e,i}$ (from the Freundlich single-solute isotherms). The $(X_i)^M_s$ values are then calculated from Equation

6-105 and a check is made on the following identity:

$$\sum_{i=1}^{k} (X_i)_s^M = 1$$

The solution is iterated using new estimates of spreading pressure until this identity is satisfied. Once the correct values of $(X_i)_s^M$ are found, q_T is calculated from Equation 6-111 and all values of $q_{e,i}^M$ then follow directly.

Iterative procedures are also straightforward if the values of $q_{e,i}^M$ are known and the values of $C_{e,i}^M$ are sought. Knowing $q_{e,i}^M$ also fixes $q_{e,T}$ and $(X_i)_s^M$. Spreading pressure is iterated until the resulting values of $q_{e,i}^S$ satisfy Equation 6-111:

$$\frac{1}{q_{e,T}} = \sum_{i=1}^{k} \frac{(X_i)_s^M}{q_{e,i}^S}$$

One particularly insightful use of the IAST is to prove that competitive adsorption does not occur if the single-solute isotherms are linear. According to the Langmuir model (Equation 6-58), linear isotherms would be approached at very low, solution-phase concentrations. The spreading pressure relationship (Equation 6-50) for linear isotherms ($q_{e,i}^S = \mathcal{K}_{D,i}C_{e,i}^S$) is

$$\mathcal{K}_{D,1}C_{e,1}^S = \mathcal{K}_{D,2}C_{e,2}^S = \cdots = \mathcal{K}_{D,i}C_{e,i}^S \quad i = 1 \text{ to } k \tag{6-120}$$

The IAST analysis will be illustrated here for a two-component system. Combining Equations 6-105, 6-113, and 6-120 gives

$$\frac{C_{e,1}^M}{C_{e,1}^S} + \frac{C_{e,2}^M}{\dfrac{\mathcal{K}_{D,1}}{\mathcal{K}_{D,2}} C_{e,1}^S} = 1 \tag{6-121}$$

or

$$C_{e,1}^S = C_{e,1}^M + \frac{\mathcal{K}_{D,2}}{\mathcal{K}_{D,1}} C_{e,2}^M \tag{6-122}$$

Equations 6-105 and 6-113 can also be substituted into Equation 6-111 to give

$$q_{e,T} = \frac{(1/\mathcal{K}_{D,2})\,(\mathcal{K}_{D,1}C_{e,1}^S)^2}{\dfrac{\mathcal{K}_{D,1}}{\mathcal{K}_{D,2}} C_{e,1}^M + C_{e,2}^M} \tag{6-123}$$

The denominator can be replaced from Equation 6-122 by $(\mathcal{K}_{D,1}/\mathcal{K}_{D,2})$ $(C_{e,1}^S)$ to give, after simplification,

$$q_{e,T} = \mathcal{K}_{D,1}C_{e,1}^S = \mathcal{K}_{D,2}C_{e,2}^S \tag{6-124}$$

Finally, combining Equations 6-105 and 6-112 with Equation 6-124 gives

$$q_{e,1}^{M} = (X_1)_s^M q_{e,T} = \mathcal{K}_{D,1} C_{e,1}^{M} \tag{6-125}$$

$$q_{e,2}^{M} = (X_2)_s^M q_{e,T} = \mathcal{K}_{D,2} C_{e,2}^{M} \tag{6-126}$$

The results above illustrate that the sorption relationship for both components when sorbed from a mixture is identical to that for each as a single solute if their respective isotherms are each linear. This is rigorously the case, however, only for entropy-driven sorptions. As illustrated by close inspection of the Langmuir equation, this condition is also approximated as solution-phase concentrations become very low:

$$q_{e,1} = \frac{Q_{a,1}^{\circ} \mathcal{b}_1 C_{e,1}}{1 + \mathcal{b}_1 C_{e,1} + \mathcal{b}_2 C_{e,2}} \approx Q_{a,1}^{\circ} \mathcal{b}_1 C_{e,1} \tag{6-127}$$

where $\mathcal{K}_{D,1}$ in Equation 6-125 is equivalent to $Q_{a,1}^{\circ} \mathcal{b}_1$ in Equation 6-127.

6.6 SORPTION PROCESSES AND PARTICLE STABILITY

Our considerations of the energetics of transformation processes in heterogeneous environmental systems have thus far focused on substances which are distributed as molecular entities between two or more distinctly separated phases. We have not yet considered the intermingling of one phase with another to form a stable heterogeneous dispersion. Generally, when one phase is mixed with another, the condition of intermixing is unstable, and once mixing is ceased, the phases separate within a relatively short period of time. Bubbles of air dispersed in water, for example, rise rapidly to the surface and escape into the gas phase. Granular activated carbon particles stirred in water to form a suspension settle from solution when the stirring is stopped. Precipitates such as $CaCO_3$ are dispersed in water when they first form, but relatively soon thereafter settle to form separate solid phases. As discussed in Chapter 1, many impurities and contaminants of aqueous systems are either themselves of macromolecular size or, by sorption processes, are associated with macromolecules or small suspended particles which remain dispersed in water columns for long periods; that is, they comprise *metastable dispersions*. The fate and transport of such contaminants is affected by the stability of the particles with which they are associated. Their transport is facilitated or retarded by the transport of the suspended particles. In turn, the sorption of impurities can affect the stability of suspended particles.

Macromolecules and small particles that do not readily settle from suspension are termed *colloids*. Such small particles—which for example include clays, hydrous metal oxides, bacteria, virus, proteins, and pulp fibers—are typically only a micron or so in size, and thus are affected more by solvation effects and surface forces than they are by gravitational forces. It is, in fact, the intraparticle forces of repulsion and attraction resulting from electrical charges on the surfaces of colloids which generally control the observed stability of suspensions of such particles.

Some suspensions of colloids are stable indefinitely, and some are not. *Thermodynamically stable* colloidal systems are termed *reversible*, in the sense that even if somehow forced to aggregate temporarily, they will subsequently redisperse. In the same sense, *thermodynamically unstable* colloids are *irreversible*. Examples of reversible colloids include soap and detergent micelles, proteins, and starches. Examples of irreversible systems include clays, metal oxides, virus, bacteria, and other microorganisms. Our concern is largely with thermodynamically unstable colloidals, for several reasons. First, they typify systems of environmental interest, and second, because they are fundamentally unstable, we can aggregate and remove them by designing appropriate processes and operations. We distinguish here between true thermodynamic instability and temporal instability. Some irreversible colloidal systems aggregate rapidly and are termed *caducous*. Other thermodynamically unstable colloidal systems aggregate slowly and are termed *diuturnal*. The terms *stable* and *unstable* are often used in a practical sense with reference to diuturnal and caducous colloids, respectively. In this sense, a *practically stable* colloid may *fundamentally* be thermodynamically *unstable* but aggregate at a very slow rate, and thus persist and remain substantially unchanged for very long periods of time.

6.6.1 Surface Charges

The stability of colloids is related in large measure to the effects of electrical surface charges. These charges result from different types of functional groups, including: (1) acidic, basic, and amphoteric OH groups on colloidal oxides, hydroxides, and silicates; (2) localized OH sites and less localized charges relating to oxygen sharing in SiO_4, AlO_6, and AlO_4 groups comprising clay and zeolite colloids; (3) ionizable surface sites on other slightly soluble compounds of weak acid anions, such as sulfides, carbonates, phosphates, and fluorides; and (4) acidic and basic functional groups on organic colloids such as proteins, polysaccharides, and other substituted polymer structures.

Coulombic forces associated with surface charges are similar to those discussed in Chapter 5 and quantified in Equation 5-1 for dissolved ionic species. While they can be either positive or negative, most charges on the surfaces of environmentally significant colloids are negative; this is consistent with their origins. As indicated in Figure 6-15, the similarly negative charges on environmental colloids usually prevent them from aggregating to form particles sufficiently large to settle from suspension.

Surface charges relate in large measure to the ionization or protolysis of functional groups. They are, therefore, usually affected by the pH and ion content of an aqueous phase. The point at which a colloid having acidic func-

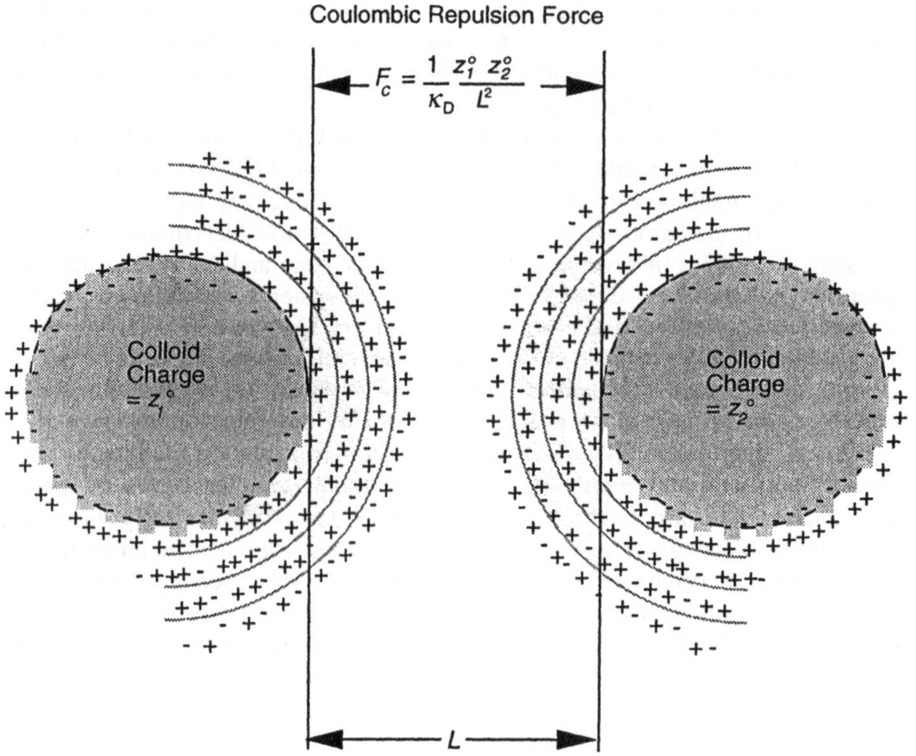

Figure 6-15 Mutual repulsion of like-charged colloids in a medium of dielectric constant κ_D.

tional groups changes from an uncharged state to a negatively charged state in water relates to the pK_a values of those groups. For example, a carboxylic acid functional group on the surface of a protein has a pK_a value in the range 4 to 5. Thus below pH $\approx$ 4 such a colloid is uncharged and relatively unstable, whereas it becomes increasingly negatively charged and more stable as the pH of the solution in which it is suspended is increased to values above pH $\approx$ 5. Carboxylic acid functional groups often dominate the relative stability of colloidal suspensions of microbial species, and such suspensions are commonly found to be unstable in water at values of pH $\approx$ 4 or less.

The pH value at which the net surface charge of a colloid is such that its *zeta potential*, $E_{R,z}$, is zero is termed the *zero point of charge* (ZPC). The zeta potential is the net charge or potential at a particular distance (position) away from the particle, specifically, at the *plane of shear*, as illustrated in Figure 6-16. Thus the ZPC is not a pH at which the surface charge is itself zero. ZPC values for colloids of environmental interest cover a rather broad pH range, as indicated in Table 6-6.

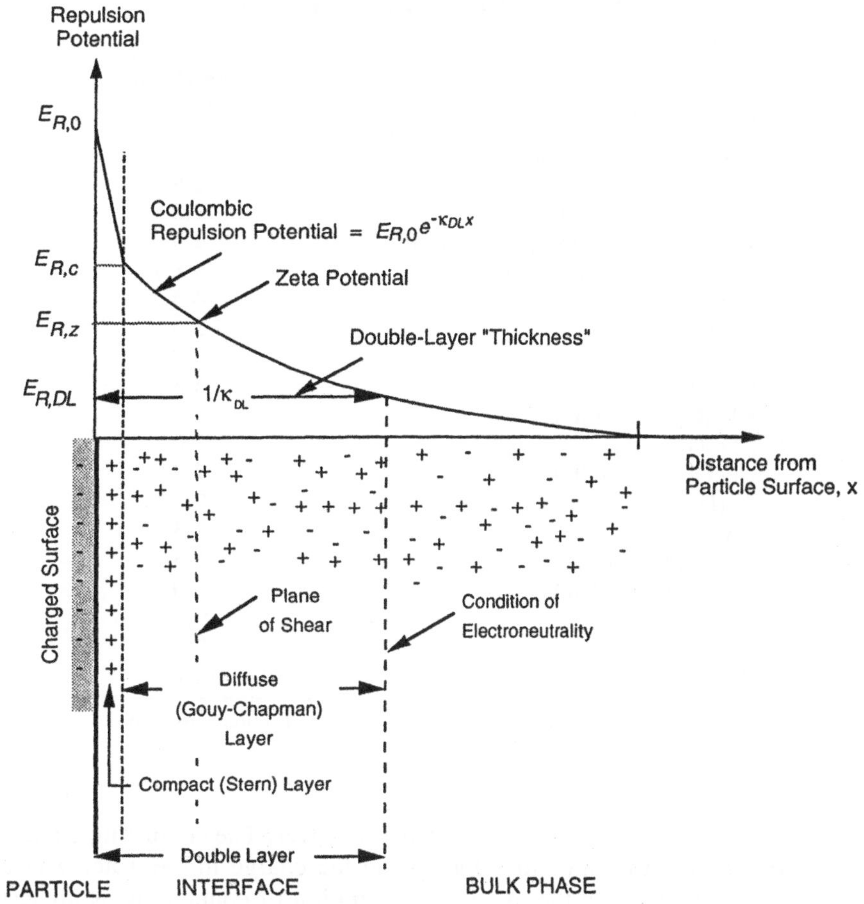

Figure 6-16 Double layer representation of ion distributions and Coulombic repulsion potentials in the vicinity of a charged particle surface.

6.6.2 Repulsive and Attractive Potentials in Colloidal Dispersions

Unlike colloidal *particles*, colloidal *dispersions* do not have net electrical charges, so that the primary charges on particles comprising dispersed phases must be counterbalanced within the dispersing phases. As a result, an *electric double layer* exists at every interface between a colloid and water in aqueous dispersions. This double layer (see Figure 6-16) consists of the charged particle and an equivalent excess of *counterions* of opposite charge that accumulate near the surface of the particle. Counterions are attracted electrostatically to interfacial regions. This attraction gives rise to concentration gradients, in response to which counterions may diffuse away from the surface toward the

Table 6-6 ZPC Values for Selected Environmental Colloids

Composition	Approximate pH_{ZPC}
Albite (NA-fectosilicate)	2.0
α-SiO_2 (quartz)	2.9
δ-MnO_2 (birnessite)	2.2
Montmorillonite (Mg-phyllosilicate)	2.5
Microbes (protein-based)	4.0–5.0
Kaolinite	4.5
$ZrSiO_4$	5.0
α-$Al(OH)_3$	5.0
TiO_2 (rutile)	5.8
α-Fe_3O_4 (magnetite)	6.6
γ-Fe_2O_3	7.0
α-FeOOH (goethite)	7.3
$Ca_3(PO_4)_3OH$ (hydroxyapatite)	7.6
γ-Al_2O_3	8.5
α-Fe_2O_3 (hematite)	8.5
α-Al_2O_3 (corundum)	9.0
$CaCO_3$ (calcite)	9.5
CuO	9.5
MgO	12.4

Source: Numerical values derived from a number of sources listed in Section 6.9.1.

bulk of the solution, where their concentration is lower. The competing process of diffusion and electrostatic attraction spread the charge in the water over a *diffuse layer*, within which the excess concentration of counterions is highest adjacent to the surface of the particle and decreases gradually with increasing distance from the solid–water interface. The manner in which such distributions are affected by the ionic composition of the solution is shown in Figure 6-17a. When the bulk solution contains a high concentration of ionic species (high ionic strength), the diffuse layer is compacted so that it occupies a smaller volume and does not extend as far into the solution.

The primary charge on a colloid surface causes an electrostatic potential (voltage) to develop between the surface of the particle and the bulk of the solution. This electrical potential can be pictured as an *electron pressure* which must be applied to bring a unit charge of the same sign as the primary charge to within a given distance from the particle surface. Schematic representations of such potential distributions are shown in Figure 6-17b. The potential has a maximum value at the particle surface and decreases with distance from the surface. The particular way in which potential decreases with distance is determined by the characteristics of the diffuse layer, and thus by the number

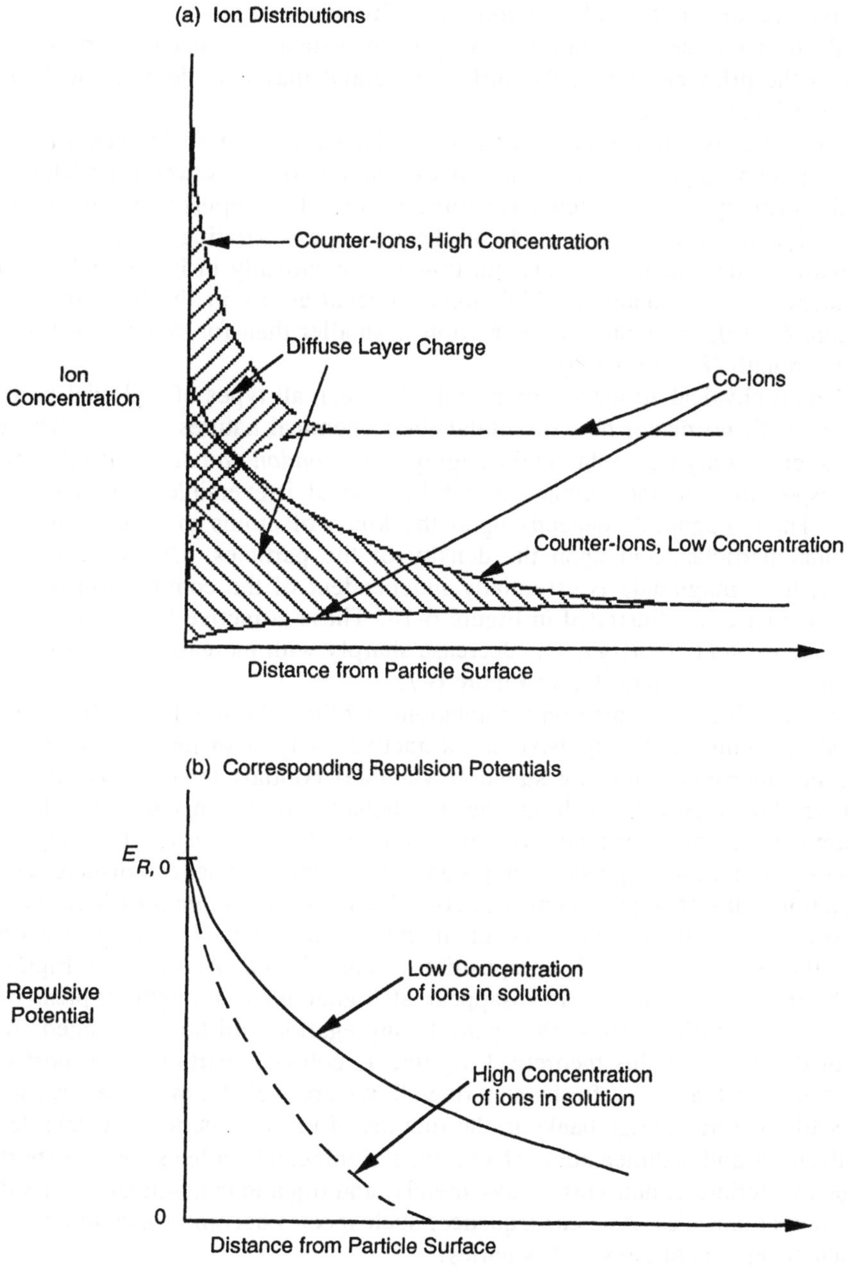

Figure 6-17 Ion distributions and repulsion potentials as functions of indifferent electrolyte concentration.

and type of ions in the bulk solution. At high ionic strength the electric potential decreases to zero within a relatively short distance. Depending on the origin of the primary charge, the surface potential may also decrease with increasing ionic strength.

The respective diffuse layers of two similar colloidal particles begin to interact as they approach each other. If the interaction is between particles of similar primary charge, a repulsive force results. This repulsive potential energy, E_R, increases in magnitude as the distance separating the particles is decreased. Such interactions are illustrated schematically in Figure 6-18. For colloidal systems containing high ionic concentrations in the bulk solution (Figure 6-18b), the repulsive interaction is smaller than for systems of lower ionic strength (Figure 6-18a).

Certain physical attractive forces exist between all types of colloidal particles as well, no matter how dissimilar their chemical natures may be. These forces, commonly referred to in the aggregate as *London-van der Waals forces*, are responsible for the ultimate destabilization of irreversible colloidal systems. Their magnitude depends upon the kinds of atoms that make up the colloidal particles, and upon the density of the particles. Unlike repulsive forces, their magnitude is essentially independent of the composition of the aqueous phase, as illustrated in Figure 6-18. These attractive forces and their associated potential energy, E_A, decrease sharply with increasing distance between particles, as indicated in Figure 6-7.

The net effects of charge on the apparent stability of colloids may be determined by summing the repulsive and attractive energies of interaction ($E_R - E_A$). For thermodynamically stable (reversible) colloids this sum is always positive. For irreversible colloids the sum behaves as shown schematically in Figure 6-18. If ionic concentrations are similar to those of fresh waters, Figure 6-18a shows that net repulsion can predominate over intermediate distances of separation. This net repulsion may be considered an *energy barrier* which must be overcome for aggregation to occur. Its magnitude depends on the charge on the particles and on the ionic composition of the solution. As shown in Figure 6-18b, the energy barrier can disappear at higher ionic strengths typical of brackish and marine waters; thus particle aggregation will be encouraged. In practical support of this theoretical argument, colloidal particles transported into marine estuaries by freshwater river flows are well known to aggregate and settle to form sludge banks in the mouths of those estuaries. Particle destabilization and sedimentation of this type significantly affects the transport of such materials as nutrients, heavy metals, and organic contaminants sorbed on colloidal particles. As a consequence, such species become concentrated in particular regions of bays and estuaries.

The particles of a colloidal dispersion are in constant motion and so possess *kinetic energy*. There is a useful analogy that can be drawn here to the kinetics of chemical reactions, which we will discuss in Chapters 7 and 8. Pursuant to this analogy, a distribution of kinetic energies exists at any instant, with some molecules, or in this case particles, having kinetic energies sufficient to over-

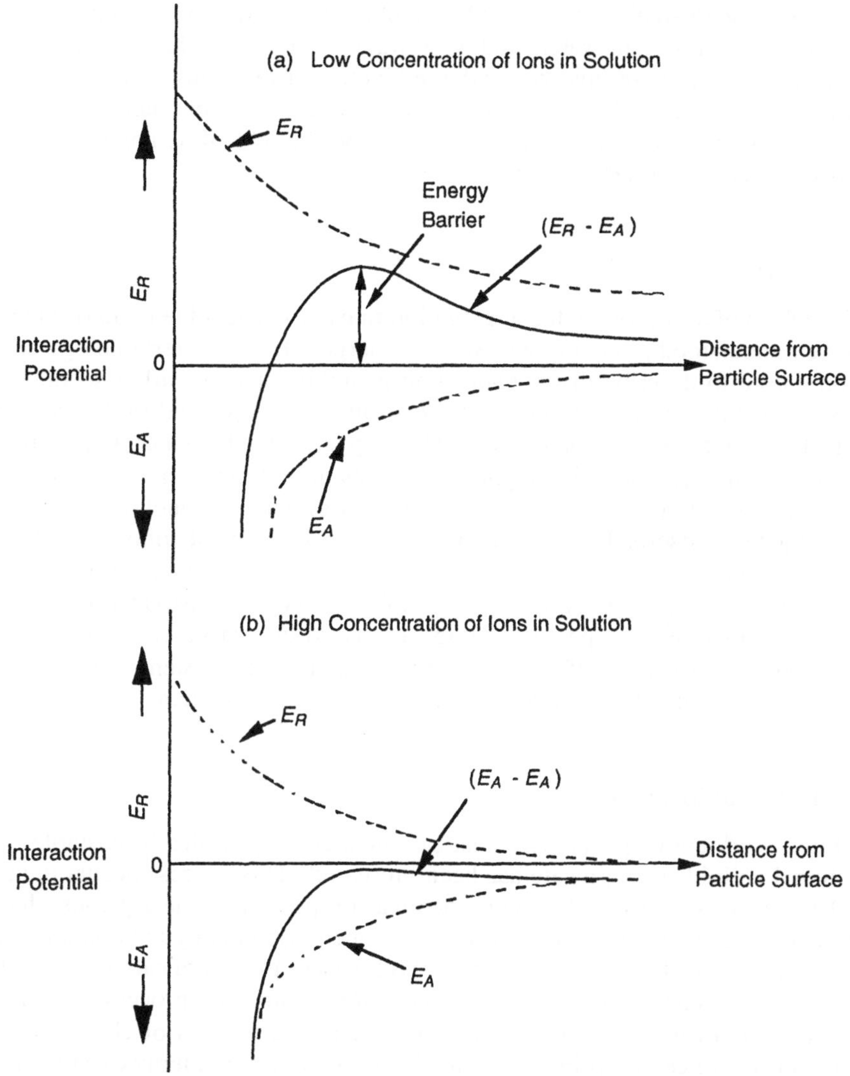

Figure 6-18 Interaction potentials in a colloidal system as functions of indifferent electrolyte concentrations.

come an "activation energy barrier." Rates of aggregation are therefore dependent on the magnitude of the energy barrier and the kinetic energy of the colloidal particles. Irreversible colloidal systems having inherently high energy barriers and low kinetic energies coagulate most slowly, and those having low "activation" energies and high kinetic energies most rapidly. Various combinations lead to intermediate coagulation behavior. These relationships form the basis for designing engineered coagulation–flocculation systems for removal of colloids from waters and wastewaters. As noted, they are analogous in many ways to those involved with the kinetics of reactions in homogeneous solutions. In fact, molecular collision theory and activation energies and catalysis associated with reaction kinetics can be thought of as applicable concepts for particle destabilization rates.

6.6.3 Solvation

The effects of solvation on the aggregation rates of irreversible colloids are not as well understood as are those of surface charges. Even the terminology commonly used to describe this phenomenon is ambiguous. In earlier sections of this chapter the terms *hydrophilic* and *hydrophobic* were used to describe the extent to which a surface or a functional group is wetted by water. On the basis of this definition, strictly hydrophobic colloids do not exist, since all particles are wetted to some extent in aqueous colloidal systems. Monolayers of water are frequently adsorbed chemically on the surfaces of typical colloidal particles such as metal oxides, clays, and quartz, and colloid–water interactions of that type can retard the aggregation of such colloidal systems. Models for the apparent stability of hydrophobic colloids are based on charge effects alone. To the extent that solvation effects may be important in any system, such models cannot completely characterize colloid behavior in that system.

6.6.4 Destabilization

A brief overview of phenomena that permit thermodynamically unstable but diuturnal colloids to aggregate has been presented. This process occurs in natural systems as a result of certain chemical properties of the aqueous phase and is practiced routinely in engineered treatment systems by manipulation of the chemistry of the aqueous phase. In both cases, the chemicals responsible for particle destabilization are termed *coagulants*, and the process is termed *coagulation*. In water and wastewater treatment, the process of chemical coagulation is concerned primarily with the aggregation of thermodynamically unstable but temporally quasi-stable colloids. Coagulation processes are used to increase the rate at which colloidal systems aggregate, transforming quasi-stable systems to rapidly aggregating systems. Different chemical coagulants can accomplish this in different ways. Depending on the conditions under which they are used, some materials can function either as primary coagulants or as adjuncts or aids to another primary coagulant. Certain types of coagulants can achieve destabilization of colloidals by more than one method. In treatment processes, the selection of the proper type and dosage of coagulant(s) for a particular application requires an understanding of how these materials function mechanistically as well as phenomenologically.

6.7 SPECIES DISTRIBUTIONS AMONG PHASES IN ENVIRONMENTAL SYSTEMS

6.7.1 Equilibrium Criteria

Chemical species released to multiphase environmental systems eventually attain different equilibrium concentrations in the phases comprising those systems. Their concentrations in each phase should in theory be quantifiable in terms of the thermodynamic parameters discussed previously. In cases of air-water partitioning, for example, phase concentration ratios can be expressed in terms of the Henry's law constant, $\mathcal{K}_H$ (Equations 6-3 through 6-6 and Table 6-2). For sediment–water partitioning, the parameter $\mathcal{K}_D$ (Equation 6-51) is commonly used, at least as a "first-cut" estimator of the distribution of solute between two phases. Other examples of parameters employed to describe phase distributions include bioconcentration factors, $\mathcal{K}_B$, and, as reference values, octanol–water partition coefficients, $\mathcal{K}_{o,w}$. For each solute, a *partition coefficient* can be defined for each pair of environmental phases. While such partitioning coefficients are useful, it is conceptually more satisfying, and eventually more practical in complex systems, to express equilibrium phase distributions in terms of fundamental quantities that control concentration distributions.

Gibbs demonstrated that an equilibrium in the distribution of a solute between two phases occurs when the chemical potentials of the solute are equal in each of the phases. Partitioning can, however, be generalized in terms of a more convenient criterion for equilibrium between phases than chemical potential: namely, in terms of solute fugacity. We have noted in Chapter 5 and earlier in this chapter that the fugacity of a solute can be viewed as its tendency to escape from (or its *escaping pressure* within) a particular phase. Equilibrium conditions for the distribution of a solute between two phases occurs when the escaping pressure of that solute in one phase is equal to that in the other.

An analogous situation exists for definition of thermal equilibrium between two phases. The criterion for thermal equilibrium between two phases is that they have equal temperatures even though their respective thermal concentrations are different. Like chemical potential, heat is a form of energy (ML^2T^{-2}), an extensive property of a system. Like fugacity, temperature is an intensive parameter that determines the relative state of each of two or more phases, and their respective equilibria among the phases of a system similarly reflect equilibration of their respective thermodynamic potentials. This analogy is pictured schematically in Figure 6-19 for a simple two-phase system. The subscripting of variables in this figure indicates their respective values for vapor (v) and aqueous (a) phases.

Heat concentration and temperature are interrelated by the expression

$$C_H = TQ^{\circ}_{H,V} \tag{6-128}$$

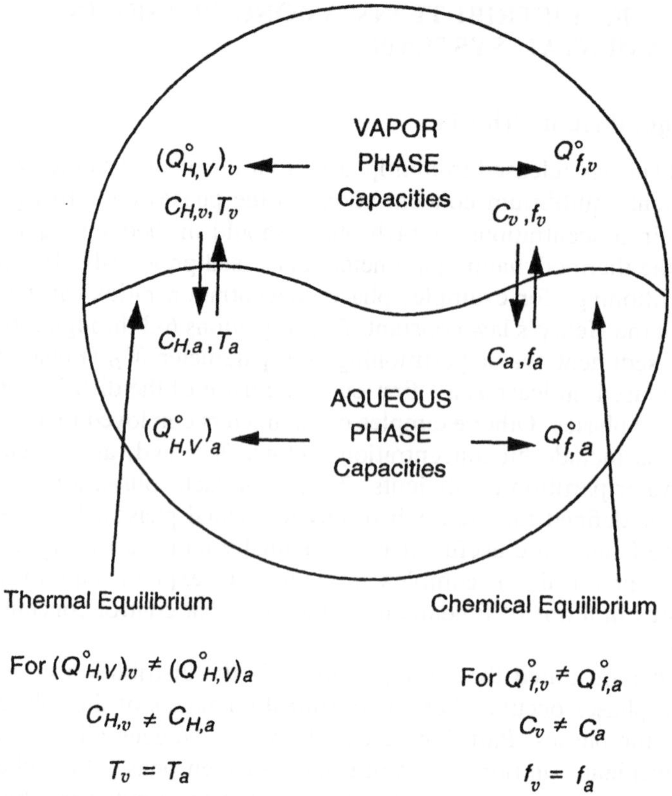

Figure 6-19 Two-phase analogy between heat capacity and fugacity capacity.

where C_H is the "concentration" of heat ($ML^{-1}t^{-2}$), $Q^\circ_{H,V}$ is a *volumetric heat capacity* ($ML^{-1}t^{-2}T^{-1}$) given by the product of the mass heat capacity and the phase density, and T is the absolute or Kelvin temperature (K). The volumetric heat capacity is essentially the thermal energy acquired or released by a unit volume of a phase per unit temperature change. Like temperature and heat capacity, solute concentration and fugacity are usually related linearly at low concentrations: namely,

$$C = fQ^\circ_f \quad \text{or} \quad Q^\circ_f = \frac{C}{f} \tag{6-129}$$

where Q°_f, the *fugacity capacity*, is expressed in terms of concentration per unit escaping pressure (ML^{-3}/pressure $= L^{-2}t^2$). Heat accumulates in phases having high heat capacity, mass accumulates in phases having high fugacity capacity. For example, a copper rod suspended in air in a closed system accumulates much more of the available system heat per unit of its volume than does the air. By analogy, hydrophobic organic compounds partition from aqueous phases into lipid phases because the latter have higher Q_f° values for such compounds.

The equilibrium distribution of a substance between two phases, 1 and 2, is reached when its fugacities in those two phases are equal:

$$f_1 = f_2 \tag{6-130}$$

or, from Equation 6-129,

$$\frac{C_1}{Q_{f,1}^\circ} = \frac{C_2}{Q_{f,2}^\circ} \tag{6-131}$$

and

$$\frac{C_1}{C_2} = \frac{Q_{f,1}^\circ}{Q_{f,2}^\circ} = \mathcal{K}_{1,2} \tag{6-132}$$

where $\mathcal{K}_{1,2}$ is the *partition coefficient*. The partition coefficient, $\mathcal{K}_{1,2}$, controlling the distribution of a solute between two phases is thus the ratio of the fugacity capacities of the two phases for that solute. To characterize and quantify phase distributions, it is therefore necessary to characterize and quantify the respective fugacity capacities of different phases for each solute.

6.7.2 Fugacity Capacities for Different Phases

Environmentally relevant phases generally include (1) vapors, (2) aqueous solutions, (3) sorbed phases, (4) biota, and (5) pure solids and liquids, including the common *organic reference phase, octanol*. The condition of equilibrium for distribution of a solute in a system comprised by these several phases is given by

$$f_v = f_a = f_s = f_b = f_p = f_o = \text{prevailing system fugacity} = f_{\text{system}} \tag{6-133}$$

This condition is pictured schematically in Figure 6-20.

The prevailing system fugacity for a system of k phases is given, from Equation 6-129, as

$$f_{\text{system}} = \frac{C_{\text{system}}}{Q_{f,\text{system}}^\circ} = \left(\sum_{j=1}^{k} \frac{M_j}{V_j}\right)\left(\sum_{j=1}^{k} \frac{1}{Q_{f,j}^\circ}\right) = \frac{M_T}{\sum_{j=1}^{k} Q_{f,j}^\circ V_j} \tag{6-134}$$

where M_j is the number of moles or the mass of solute in phase j, V_j is the volume of that phase, and M_T is the total number of moles or the mass of solute

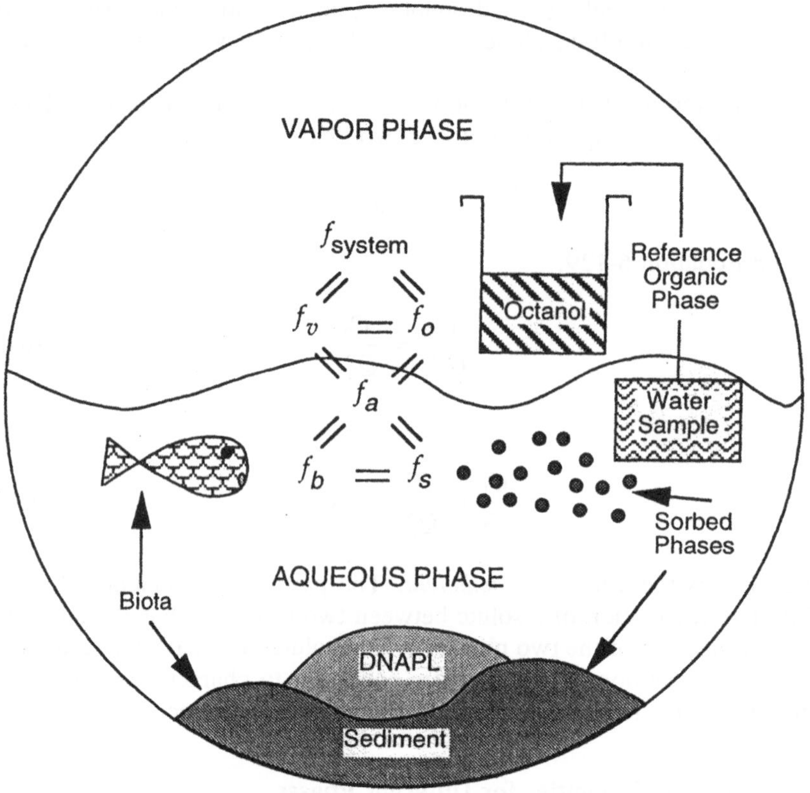

Figure 6-20 Equilibrium in a multiphase environmental system.

in the system. We can draw upon our previous discussions of phase relationships and chemical thermodynamics to define the fugacity of a solute in each of these phases, some more rigorously than others. The most straightforward considerations apply to vapor phases.

6.7.2.1 Vapor Phase The fugacity of a component of a mixed vapor phase is given by

$$f_v = \alpha_f X_v P_T \approx P \tag{6-135}$$

where X_v is the mole fraction of the solute in the vapor phase, P_T and P are the total and component pressures, respectively, and α_f is a dimensionless fugacity coefficient, introduced to account for nonideal behavior. Except for solutes that associate with one another in vapor phases, the coefficient α_f has a value close to 1 at atmospheric pressure.

For the conditions stipulated, and pursuant to the ideal gas law, the concentration of a solute in a vapor phase is given by its pressure contribution, P:

$$C_v = f_v Q^\circ_{f,v} = \frac{n}{V} = \frac{P}{\mathcal{R}T} = \frac{f_v}{\mathcal{R}T} \tag{6-136}$$

The fugacity capacity of the vapor phase for that solute is simply

$$Q^\circ_{f,v} = \frac{1}{\mathcal{R}T} \tag{6-137}$$

Thus, $Q^\circ_{f,v}$ is independent of the nature of the solute or the composition of the vapor for the conditions stated.

6.7.2.2 Aqueous Phase The aqueous phase fugacity of a solute is given by

$$f_a = X_a \alpha_a f_p \tag{6-138}$$

where X_a is the aqueous-phase mole fraction of the solute; α_a is the aqueous phase solute activity coefficient; and f_p is a reference fugacity relating to the *pure liquid* solute at the temperature of the system. The mole fraction X_a is given by the number of solute moles divided by the total number of moles (solute plus water) comprising the aqueous solution. This is closely approximated for dilute solutions as the molar ratio of solute to water. The aqueous-phase solute concentration C_a may thus be approximated closely by $X_a / V_{mo,w}$ where $V_{mo,w}$ is the molar volume of water. Thus

$$Q^\circ_{f,a} = \frac{C_a}{f_a} = \frac{1}{V_{mo,w}\, \alpha_a P^v} \tag{6-139}$$

In the standard state defined by Raoult's law (i.e., when $X_a = 1$) $\alpha_p = 1$ and f_p is given by the pure liquid-state component vapor pressure, P^v. This is problematic for pure solid phases, which are not liquid at the common temperatures of natural systems (i.e., $T < 25°C$). In such cases the reference fugacity is the subcooled liquid vapor pressure, which cannot be measured. Moreover, Equation 6-139 requires specification of the aqueous-phase activity coefficient for the solute.

The difficulties associated with Equation 6-139 can be circumvented by recalling from Equation 6-132 that the vapor–aqueous-phase partition coefficient, $\mathcal{K}_{v,a}$, is defined by the ratio of the vapor- and aqueous-phase fugacity capacities. Further recalling that $\mathcal{K}_{v,a} = Q^\circ_{f,v}/Q^\circ_{f,a} = C_v/C_a = P/\mathcal{R}TC_a = \mathcal{K}_H/\mathcal{R}T$ (see Table 6-2), we can write

$$Q^\circ_{f,a} = \frac{Q^\circ_{f,v}}{\mathcal{K}_{v,a}} = \frac{Q^\circ_{f,v}\mathcal{R}T}{\mathcal{K}_H} = \frac{1}{\mathcal{R}T}\frac{\mathcal{R}T}{\mathcal{K}_H} = \frac{1}{\mathcal{K}_H} \tag{6-140}$$

The aqueous-phase fugacity capacity for a substance can thus be determined directly from Henry's constant for that substance. This agrees with our earlier definition of fugacity in terms of component pressure and with the relationship between Henry's constant and component pressure.

Equation 6-124 was developed for a solute in a truly dissolved aqueous state. Hydrophobic compounds are often sorbed to colloidal or suspended materials, or may even be present as colloidal entities themselves at concentration levels in excess of their solubility limits. In such cases, these solutes are associated with a different phase than water, and thus do not contribute to solution fugacity. Equation 6-140 thus applies only to those solutes, or fractions of solutes, that exist as dissolved species in the aqueous phase.

6.7.2.3 Pure Solids and Liquids According to Raoult's law, the activity coefficient, α_p, for pure solids or liquids is unity. The fugacity of the pure phase is thus given by the vapor pressure, P^v. At the same time, concentrations are defined by the inverse of molar volumes, $V_{mo,p}$ (L^3 mol^{-1}). Fugacity capacity for such pure phases is thus determined by the relationship

$$Q^{\circ}_{f,p} = \frac{C_p}{f_p} = \frac{1}{P^v V_{mo,p}} \tag{6-141}$$

Pure liquid or solid phases occur enviromentally only when the solubility of a solute in another phase is exceeded and "precipitation" occurs. This is a common circumstance in water-saturated subsurface contamination situations, in which nonaqueous phase liquids, NAPLs, or submerged organic solids are often found. Combining Equations 6-132, 6-137, and 6-141 reveals that the partition coefficient of a solute between air and its pure state is dictated essentially by the vapor pressure of that solute; that is,

$$\mathcal{K}_{v,P} = \frac{Q^{\circ}_{f,v}}{Q^{\circ}_{f,p}} = \frac{P^v V_{mo,p}}{\mathcal{R}T} \tag{6-142}$$

where $V_{mo,p}/\mathcal{R}T$ is constant for any given temperature.

Similarly, the water solubility, or aqueous-phase saturation concentration, C_S, of an organic compound, essentially dictates the partition coefficient for water and the pure solute:

$$\mathcal{K}_{a,p} = \frac{Q^{\circ}_{f,a}}{Q^{\circ}_{f,p}} = \frac{P^v V_{mo,p}}{\mathcal{K}_H} = C_S V_{mo,p} \tag{6-143}$$

The aqueous-phase solubility for liquid solutes is

$$(C_S)_{\text{liquid solutes}} = C_{S(l)} = \frac{1}{V_{mo,a}\alpha_a} \tag{6-144}$$

whereas for solid solutes, $C_{S(s)}$,

$$(C_S)_{\text{solid solutes}} = C_{S(s)} = \frac{P^v_{(s)}}{P^v_{(l)} V_{mo,a}\alpha_a} \tag{6-145}$$

where $P^v_{(s)}$ and $P^v_{(l)}$ are the solid and liquid vapor pressures of the solute, respectively.

6.7.2.4 Sorbed Phases The fugacities of a solute sorbed to a solid phase and dissolved in a solution in contact with that solid phase must be equal at equilibrium. Thus, if $Q^\circ_{f,s}$ is the sorbed-phase fugacity capacity, and C_e and q_e are, respectively, the solution phase and solid phase concentrations of the solute at equilibrium,

$$f_a = \frac{C_e}{Q^\circ_{f,a}} = f_s = \frac{q_e}{Q^\circ_{f,s}} \tag{6-146}$$

As discussed earlier in this chapter, sorption equilibria are generally characterized in terms of isotherm equations that relate solution phase concentrations, C_e, to sorbed phase concentrations, q_e. The most simple isotherm model is the linear relationship given in Equation 6-51. To apply the linear isotherm to determination of fugacity capacities for soils and sediments in natural systems, the parameters of Equations 6-51 and 6-130 must be expressed in mass and volume units that are consistent. The sorbed phase concentration, q_e, can be expressed as moles of solute per 10^6 g of sorbent (mol/10^6 g), $\mathcal{K}_D$, as cubic meters of water per 10^6 g of sorbent (m^3/10^6 g), and the sorbent concentration, C_{ss}, as the product of its volume fraction (dimensionless), and its density, ρ_s (g/cm^3). The sorbed-phase concentration expressed in terms of the *volume of sorbent* is then given by $q_e\rho_s$ (mol/m^3).

Putting these units together with Equations 6-146 and 6-140 gives

$$Q^\circ_{f,s} = \frac{q_e\rho_s}{\mathcal{K}_H C_e} = \frac{\mathcal{K}_D\rho_s}{\mathcal{K}_H} \tag{6-147}$$

The product, $\mathcal{K}_D\rho_s$, in Equation 6-147 is actually the dimensionless partition coefficient arrived at by expressing sorbed phase concentration on a mole-per-unit-volume basis. The concentration of dissolved material is $Q^\circ_{f,a}f_s$ or $f_s/\mathcal{K}_H$ moles of solute per m^3 of solution; that is, $f_s/\mathcal{K}_H$ moles are dissolved in the 1 cubic meter. The sorbed phase concentration is $f_sQ^\circ_{f,s}$ or $q_e\rho_s$ moles of solute per cubic meter of sorbent, or a total of $f_sQ^\circ_{f,s}C_{ss}$ or $f_s\mathcal{K}_D\rho_sC_{ss}/\mathcal{K}_H$ moles per m^3 of solution. A mass balance on the solute then yields

$$\text{total solute} = \frac{f_s}{\mathcal{K}_H} + \frac{f_s\mathcal{K}_D\rho_sC_{ss}}{\mathcal{K}_H} = \frac{f_s(1 + \mathcal{K}_D\rho_sC_{ss})}{\mathcal{K}_H} \tag{6-148}$$

where

$$\text{Solute fraction dissolved} = \frac{f_s}{\mathcal{K}_H}\left[\frac{f_s(1 + \mathcal{K}_D\rho_s C_{ss})}{\mathcal{K}_H}\right]^{-1}$$

$$= \frac{1}{1 + \mathcal{K}_D\rho_s C_{ss}} \tag{6-149}$$

$$\text{Solute fraction adsorbed} = \frac{f_s\mathcal{K}_D\rho_s C_{ss}}{\mathcal{K}_H}\left[\frac{f_s(1 + \mathcal{K}_D\rho_s C_{ss})}{\mathcal{K}_H}\right]^{-1}$$

$$= \frac{\mathcal{K}_D\rho_s C_{ss}}{1 + \mathcal{K}_D\rho_s C_{ss}} \tag{6-150}$$

The distribution coefficient, $\mathcal{K}_D$, can be estimated for organic solutes in sediments and soils from correlations with the octanol–water partition coefficient, $\mathcal{K}_{o,w}$, as discussed earlier in this chapter (see Equation 6-52).

6.7.2.5 Biota A bioconcentration factor, $\mathcal{K}_B$, is commonly used as a distribution coefficient for expression of the uptake of solute by biotic phases. When expressed as a ratio of solute concentration in biotic tissue measured on a wet-weight basis to that in water on a volumetric basis, $\mathcal{K}_B$ is identical to $\mathcal{K}_D$. It is also analogous to the group $\mathcal{K}_D\rho_s$, with ρ_s being replaced by the density of the biota, ρ_b; because ρ_b commonly has a numerical value near unity, the difference between $\mathcal{K}_D$ and $\mathcal{K}_D\rho_b$ is generally only dimensional.

6.7.2.6 Octanol We have demonstrated that the octanol–water partition coefficient, $\mathcal{K}_{o,w}$, is a useful indicator of hydrophobicity. It is thus a useful reference for fugacity calculations as well. A value for $Q^\circ_{f,o}$ for a solute in octanol can be specified by analogy to that for a solute in the aqueous phase, as set forth in Equation 6-139:

$$Q^\circ_{f,o} = \frac{1}{V_{mo,o}\alpha_o P^v} \tag{6-151}$$

where α_o is the activity coefficient of the solute in octanol, P^v is the vapor pressure of the liquid solute at the system temperature, and $V_{mo,o}$ is the molar volume of octanol saturated with water. At equilibrium, the fugacities of a solute distributed between octanol and water phases must be equal in the two phases. The vapor pressures cancel to give

$$\mathcal{K}_{o,w} = \frac{C_{e,o}}{C_{e,w}} = \frac{Q^\circ_{f,o}}{Q^\circ_{f,a}} = \frac{V_{mo,a}\alpha_a}{V_{mo,o}\alpha_o} \tag{6-152}$$

where $\mathcal{K}_{o,w}$ is expressed as a concentration ratio. For most hydrophobic organic compounds, the value of $\mathcal{K}_{o,w}$ is dominated by α_a. Because α_a controls aqueous solubility, it follows that solubility and $\mathcal{K}_{o,w}$ are closely related, as we have shown earlier in this chapter (e.g., see Equations 6-29, 6-30, and 6-38).

A detailed illustration of the use of fugacity concepts to determine the distribution of an organic contaminant among the various phases or compartments of an environmental system is given in Example 6-9. Fugacity concepts can be extended to describe contaminant distribution under nonequilibrium conditions. To do so involves incorporating the mass transfer relationship discussed in Chapters 3 and 4 and the reaction rate relationships discussed in Chapters 7 and 8 into the approach illustrated in Example 6-9.

Example 6-9

- **Situation:** The waste from a parts cleaning solvent bath at an airplane repair hanger is to be treated by sedimentation for removal of solids prior to discharge to an airport collection system. The waste contains a low-molecular-weight (100 g/mol) volatile halogenated organic solvent at a concentration of 100 mg/L. The waste is otherwise low in dissolved organic content, but the colloidal and settleable solids are comprised partially by insoluble organic matter. The treatment will be carried out in a modified sump under a steel grating in the hanger floor. The sump is to be operated in a fill-and-draw (sequencing batch reactor) mode to a filled depth of 2 m above sludge level. All of the settleable solids are removed in this operation. The surface dimensions of the sump are 5 × 10 m thus yielding a water volume of 100 m^3. The waste flows into the sump from a submerged discharge pipe. One cubic meter of sludge having a solids concentration of 5% is collected on each cycle of settling, and the residual concentration of colloidal organic matter in the tank effluent is 10 mg/L.
- **Question(s):** You are asked to estimate anticipated concentrations of halogenated solvent in the effluent and in the sludge from the sump in order to prescribe additional treatment requirements.
- **Logic and Answer(s):**

1. The fugacity relationships discussed in this chapter provide a means for making such an estimate, but further information is required to facilitate their application.

2. Further information required:

(a) Volume of air into which the halogenated solvent may volatilize from the sump

(b) Volatility of that solvent

(c) Partitioning relationships of the solvent with respect to the settleable solids and residual colloids in the sump effluent

3. Development of information required:

(a) The hanger building is measured to be 20×50 m^2 in area with a uniform overhead height of 10 m. Air (vapor) volume = 10,000 m^3.

(b) Henry's constant for the solvent is obtained from an appropriate reference (e.g., Table 6-1 and Appendix A.6.1); $\mathcal{K}_H = 10^{-4}$ atm-m^3/mol.

(c) Adsorption isotherms are measured for sorption of the solvent on settleable and colloidal solids. The following results are obtained:

$$\text{Settleable solids, linear isotherm, } \mathcal{K}_D = 10^{-2} \text{ m}^3/\text{g}$$

$$\text{Colloidal solids, linear isotherm } \mathcal{K}_D = 2 \times 10^{-2} \text{ m}^3/\text{g}$$

4. Approach: Equation 6-133 tells us that the fugacities of the compound in the vapor, aqueous, settleable solids, and colloidal phases are equal at equilibrium; that is,

$$f_{\text{system}} = f_v = f_a = (f_s)_{ss} = (f_s)_c$$

If we assume that the fill-and-draw time of the tank allows for establishment of an equilibrium or near-equilibrium condition, the concentration of the compound in each phase is given by the product of the fugacity capacity, Q_f°, for that phase and the prevailing fugacity of the system.

5. Calculations:

(a) Tabulate volumes of phases involved.

$$\text{Air in building} = 10{,}000 \text{ m}^3$$

$$\text{Water in tank} = 100 \text{ m}^3$$

$$\text{Sludge in tank} = 1 \text{ m}^3$$

$$\text{Colloids (in water)} = 100 \text{ m}^3$$

(b) Determine the fugacity capacities of each phase. From Equation 6-137:

$$Q_{f,v}^\circ = \frac{1}{\mathcal{R}T}; \qquad \mathcal{R}T = 0.0244 \text{ m}^3\text{-atm/mol (at 25°C)}$$

$$Q_{f,v}^\circ = \frac{1}{0.0244} = 40.9 \text{ mol/m}^3\text{-atm}$$

From Equation 6-140:

$$Q^{\circ}_{f,a} = \frac{1}{\mathcal{K}_H}; \qquad \mathcal{K}_H = 10^{-4}\ \text{m}^3\text{-atm/mol}$$

$$Q^{\circ}_{f,a} = 10^4\ \text{mol/m}^3\text{-atm}$$

From Equation 6-147:

$$Q^{\circ}_{f,c} = \frac{\mathcal{K}_D \rho_s}{\mathcal{K}_H}; \qquad \mathcal{K}_D = 2 \times 10^{-2}\ \text{m}^3/\text{g}$$

$$= 10\ \text{mg/L} = 10\ \text{g/m}^3$$

$$Q^{\circ}_{f,c} = \frac{(2 \times 10^{-2})(10)}{10^{-4}} = 2 \times 10^3\ \text{mol/m}^3\text{-atm}$$

From Equation 6-147:

$$Q^{\circ}_{f,ss} = \frac{\mathcal{K}_D \rho_s}{\mathcal{K}_H}; \qquad \mathcal{K}_D = 10^{-2}\ \text{m}^3/\text{g}$$

$$\rho_s = 50 \times 10^3\ \text{g/m}^3$$

$$Q^{\circ}_{f,ss} = \frac{(10^{-2})(50 \times 10^3)}{10^{-4}} = 50 \times 10^5\ \text{mol/m}^3\text{-atm}$$

(c) Calculate the prevailing system fugacity. The total amount, M_T, of the compound involved in each fill-and-draw operation is, by specification,

$$M_T = 100\ \text{mg/L} \times 10^5\ \text{L} = 10^4\ \text{g} = 100\ \text{mol}$$

From Equation 6-134 the prevailing system fugacity can then be determined to be

$$f_{\text{system}} = \frac{M_T}{\sum_{j=1}^{k} Q^{\circ}_{f,j} V_j}$$

$$= \frac{100}{(40.9)(10^4) + (10^4)(10^2) + (2 \times 10^3)(10^2) + (50 \times 10^5)(1)}$$

$$= 1.52 \times 10^{-5}\ \text{atm}$$

Compartment	$C_i = fQ^{\circ}_{f,j}$ (mol/m^3)	Total Amount in Each Phase (mol)
Air	6.2×10^{-4}	6.0
Water	1.5×10^{-1}	15.0
Sludge	7.6×10^{1}	76.0
Colloids	3.0×10^{-2}	3.0
	Total amount in all phases =	100.0

6.8 SUMMARY

A fundamentally-based engineering analysis of any environmental system, whether developed to predict the fate and transport of anthropogenic chemicals in natural systems or to design separation processes for such chemicals in engineered systems, must begin with a rigorous analysis of thermodynamic equilibrium relationships. Multiphase compositions and heterogeneities are the rule rather than the exception and these issues must be addressed through pertinent energy relationships. We have emphasized the thermodynamic roots of various phase separation relationships: for example, the thermodynamic basis of solubility constants and of octanol–water partition coefficients. Fundamentally, chemical structure should be related to partitioning behavior. As structure–activity relationships become more highly developed and refined, it may become more feasible to use them in practice as well as in theory. For the present, however, we have to depend quite heavily on experimental data and empirical correlations in our predictions of partitioning behavior.

Fugacity is one thermodynamic parameter that has proven quite useful for quantification of phase-separation predictions. Fugacity describes the tendency of a species to escape or *flee* from one particular phase to another, and thus can be used to estimate equilibrium concentrations of species in each of several contiguous phases. Fugacity relationships for different species, and the corresponding fugacity capacities of various phases for these species, necessarily incorporate different phase separation relationships and their relevant thermodynamics.

Thermodynamic approaches for understanding the accumulation of species at interfaces between phases and the process of adsorption have also been discussed in detail in this chapter. These approaches are rigorous but their applications in practice are confounded by difficulties in quantifying the thermodynamics of environmentally important surfaces such as soils and other natural and synthetic sorbents.

Classical adsorption models, such as those of Langmuir and Freundlich, have sound thermodynamic bases, but the empirical nature of the ways in which they are commonly applied should be made clear. These models, as well as others not discussed specifically, can seldom be used in a truly predictive way because thermodynamically embedded sorbate and sorbent properties generally are not sufficiently well known. Nonetheless, such models are useful for describing experimental data, and fairly accurate empirical correlations between Langmuir or Freundlich constants and sorbate properties for particular classes of chemicals frequently can be developed. Adsorption models have also been extended to incorporate sorbate properties directly by including, for example, solubility, polarity, and polarizability. These have met with limited acceptance because they do not account for sorbent structure.

Interestingly, the lack of a truly predictive, thermodynamically based model for single-solute adsorption has not deterred development of the ideal adsorbed solution theory to predict more complex situations relating to competitive adsorption. In principle, this theory requires only equilibrium data for the single solutes involved, and such data are readily obtained. The complication is that the poorly understood nonideal nature of environmental surfaces can cause significant deviations from predicted behavior. Our research is currently taking us beyond classical adsorption models in attempts to understand and quantify the influences of surface heterogeneity on environmentally important sorbents such as soils, sediments, and common commercial sorbent media.

Colloidal systems introduce another class of interfacial phenomena that involve the effects of accumulated surface charges and resulting particle–particle interactions. Thermodynamics once again provides insights into the energetics, more specifically the stability, of colloids caused by accumulation of surface charges that produce repulsions and attractions between and among colloids. Similarly, thermodynamic relationships help explain processes that can be used to effect the destabilization of colloidal dispersions.

Finally, thermodynamic concepts and engineering goals are connected in at least two important ways. First, reliable predictions of compartmentalization in natural systems are important because they affect regulatory policy concerning discharge of chemicals into the environment. Second, our ability to predict the efficiency of separation technologies affects the development of new approaches for control of environmental contaminants in a host of settings. An understanding of the energetics of heterogeneous systems is essential to achieving these goals; we must caution again, however, that it is not sufficient. As becomes more evident in Chapters 8 and 11, knowledge of accompanying rate processes is equally important.

6.9 REFERENCES AND SUGGESTED READINGS

6.9.1 Citations and Sources

Anon., 1989, *Environmental Health Criteria 93: Chlorophenols Other Than Pentachlorophenol*, World Health Organization, Geneva, Switzerland. (Table 6-1 and Appendix A.6.1)

Ashworth, R. A., G. B. Howe, et al., 1988, "Air–Water Partitioning Coefficients of Organic in Dilute Aqueous Solutions," *Journal of Hazardous Materials*, *18*, 25–36. (Table 6-1 and Appendix. A.6.1)

Boublick, T., V. Fried, and E. Hala, 1973, *The Vapour Pressures of Pure Substances*, Elsevier Science Publishing Co., New York. (Table 6-1 and Appendix A.6.1)

Carter, M. C., J. A. Kilduff, and W. J. Weber, Jr., 1995, "Site Energy Distribution Analysis of Preloaded Adsorbents," *Environmental Science and Technology*, *29*, 7, 1773–1780. (Table 6-5 and Figures 6-13 and 6-14)

Coates, M., D. W. Connell, and D. M. Barron, 1985, "Aqueous Solubility and Octanol to Water Partition Coefficients of Aliphatic Hydrocarbons," *Environmental Science and Technology*, *19*, 628–632. (Table 6-1 and Appendix A.6.1)

Fu, J. K., and R. G. Luthy, 1986, "Aromatic Compound Solubility in Solvent/Water Mixtures," *Journal of Environmental Engineering*, *112*(2), 328–345. (Tables 6-1 and Appendix A.6.1)

Grasselli, J. G., Ed., 1973, *Atlas of Spectral Data and Physical Constants for Organic Compounds*, CRC Press, Inc., Cleveland, OH. (Table 6-1 and Appendix A.6.1)

Israelachvili, J. N., 1985, *Intermolecular and Surface Forces*, Academic Press, Inc. (London) Ltd., London. (Figure 6-7)

James, R. O., and G. A. Parks, 1982, "Characterization of Aqueous Colloids by Their Electrical Double-Layer and Intrinsic Surface Chemical Properties," in *Surface and Colloid Science* (E. Matijevic, Ed.), Vol. 12, pp. 119–216, Plenum Publishing Corporation, London. (Table 6-6)

Jordan, T. E., 1954, *Vapor Pressures of Organic Compounds*, Interscience Publishers, New York. (Table 6-1 and Appendix A.6.1)

Lide, D. R., Ed. in Chief, 1991, *Handbook of Chemistry and Physics*, 72nd Edition, CRC Press, Inc., Boca Raton, FL. (Table 6-1 and Appendix A.6.1)

Lyman, W. J., W. F. Reehl, and D. H. Rosenblatt, 1982, *Handbook of Chemical Property Estimation Methods*, McGraw-Hill Book Company, New York. (Figures 6-3 and 6-4)

Mackay, D., 1991, *Multimedia Environmental Models: The Fugacity Approach*, Lewis Publishers, Inc., Chelsea, MI. (Table 6-1 and Appendix A.6.1)

Mackay, D., and W. Y. Shiu, 1981, "Critical Review of Henry's Constants," *Journal of Physical and Chemical Reference Data*, *10*, 1175–1199. (Table 6-1 and Appendix A.6.1)

Mackay, D., and W. Y. Shiu, 1990, "Physical-Chemical Properties and Fate of Volatile Organic Compounds: An Application of the Fugacity Approach," in *Significance and Treatment of Volatile Organic Compounds in Water Supplies* (N. M. Ram, R. F. Christman, and K. P. Cantor, Eds.), Lewis Publishers, Inc., Chelsea, MI. (Tables 6-1 and Appendix A.6.1)

Mackay, D., W. Y. Shiu, and R. P. Sutherland, 1979, "Determination of the Air-Water Henry's Law Constants for Hydrophobic Pollutants," *Environmental Science and Technology*, *13*, 333–373. (Problem 6-4).

Mailhot, H., and R. H. Peters, 1988, "Empirical Relationships Between the 1-Octanol/Water Partition Coefficient and Nine Physicochemical Properties," *Environmental Science and Technology*, *22*(12), 1479–1488. (Table 6-1 and Appendix A.6.1)

Miller, M. M., S. Ghodbane, et al., 1984, "Aqueous Solubilities, Octanol/Water Partition Coefficients and Entropies of Melting of Chlorinated Benzene and Biphenyls," *Journal of Chemical Engineering Data*, *29*, 184–190. (Table 6-1 and Appendix A.6.1)

Nirmalakhandan, N. N., and R. E. Speece, 1988, "Prediction of Aqueous Solubility of Organic Chemicals Based on Molecular Structure," *Environmental Science and Technology*, *22*, 328–338. (Table 6-1 and Appendix A.6.1)

Reid, R. C., J. M. Prausnitz, and B. E. Poling, 1987, *The Properties of Gases and Liquids*, 4th Edition, McGraw-Hill Book Company, New York. (Table 6-4)

Sangster, J., 1989, "Octanol–Water Partition Coefficients of Simple Organic Compounds," *Journal of Physical and Chemical Reference Data*, *18*(3), 1111–1229. (Table 6-1 and Appendix A.6.1)

Solubility Data Series, Vol. 15, *Alcohols with Water*, 1984, A. F. M. Barton, Ed., A. S. Kertes, Ed. in Chief, Pergamon Press, Inc., Elmsford, MY. (Table 6-1 and Appendix A.6.1)

Solubility Data Series, Vol. 20, *Halogenated Benzene, Toluene and Phenols with Water*, 1985, A. L. Horvath and F. W. Getzen, Eds., Pergamon Press, Inc., Elmsford, NY. (Table 6-1 and Appendix A.6.1)

Solubility Data Series, Vol. 24, 1986, *Propane, Butane, and 2-Methylpropane*, W. Hayduk, Ed., Pergamon Press, Inc., Elmsford, NY. (Table 6-1 and Appendix A.6.1)

Solubility Data Series, Vol. 27/28, 1987, *Methane*, H. L. Clever and C. L. Young, Eds., Pergamon Press, Inc., Elmsford, NY. (Table 6-1 and Appendix A.6.1)

Solubility Data Series, Vol. 37, 1989, *Hydrocarbons with Water and Seawater, Part I: Hydrocarbons C_5–C_7*, D. G. Shaw, Ed., Pergamon Press, Inc., Elmsford, NY. (Table 6-1 and Appendix A.6.1)

Solubility Data Series, Vol. 38, 1989, *Hydrocarbons with Water and Seawater, Part II: Hydrocarbons C_8–C_{36}*, D. G. Shaw, Ed., Pergamon Press, Inc., Elmsford, NY. (Table 6-1 and Appendix A.6.1)

Solubilities of Inorganic and Organic Compounds, Vols. 1 and 2, 1963, H. Stephen and T. Stephen, Eds., Pergamon Press, Inc., Elmsford, NY. (Table 6-1 and Appendix A.6.1)

Verschueren, K., 1983, *Handbook of Environmental Data on Organic Chemicals*, 2nd Edition, Van Nostrand Reinhold, New York. (Table 6-1 and Appendix A.6.1)

Weber, W. J., Jr., Y. P. Chin, and C. P. Rice, 1986, "Determination of Partition Coefficients and Aqueous Solubilities by Reverse Phase Chromatography. I. Theory and Background," *Water Research*, *20*(11), 1433–1442. (Table 6-1 and Appendix A.6.1)

Weber, W. J., Jr., P. M. McGinley, and L. E. Katz, 1991, "Sorption Phenomena in Subsurface Systems: Concepts, Models and Effects on Contaminant Transport," *Water Research*, *25*(5), 499–528. (Figure 6-7)

Weber, W. J., Jr., P. M. McGinley, and L. E. Katz, 1992, "A Distributed Reactivity Model for Sorption by Soils and Sediments. I. Conceptual Basis and Equilibrium Assessments," *Environmental Science and Technology*, *26*(10), 1955–1962. (Figure 6-10)

Yaws, C., H. C. Yang, and X. Pan, 1991, "Henry's Law Constants for 362 Organic Compounds in Water," *Chemical Engineering*, *98*(11), 179–185. (Table 6-1 and Appendix A.6.1)

Yaws, C. L., H. C. Yang, et al., 1990, "Solubility Data for 168 Organic Compounds," *Chemical Engineering*, *97*(7), 115. (Table 6-1 and Appendix A.6.1)

Zakhari, S., et al., 1977, *Isopropanol and Ketones in the Environment*, CRC Press, Inc., Cleveland, OH. (Table 6-1 and Appendix A.6.1)

6.9.2 Background Reading

Adamson, A. W., 1990, *The Physical Chemistry of Surfaces*, 5th Edition, John Wiley & Sons, Inc., New York. Chapter 2 covers Gibbs monolayers and the spreading pressure concept to explain surface tension of different solutes in aqueous solution and Traube's rule. Chapter 9 includes a discussion of adsorption from aqueous solution onto solids and the relation to spreading pressure. The Freundlich equation is explained as a model that accounts for a distribution of site energies, where adsorption on each patch of surface of a given site energy is described by a Langmuir equation.

Atkins, P. W., 1990, *Physical Chemistry*, Fourth Edition, W. H. Freeman and Company, Publishers, New York. Part 1 treats chemical equilibrium and includes chapters on phase transitions and multicomponent systems (Chapter 8).

Lewis, G. N., and M. Randall, 1961, *Thermodynamics*, 2nd Edition, revised by L. Brewer and K. S. Pitzer, McGraw-Hill Company, New York. A classic text in chemical thermodynamics.

Morel, F. M. M., and J. G. Hering, 1993, *Principles and Applications of Aquatic Chemistry*, John Wiley & Sons, Inc., New York. An intermediate-level aquatic chemistry text emphasizing geochemical and geobiological cycles of elements in natural waters and featuring the tableau method of solving chemical equilibrium problems. Treats heterogeneous chemical energetics in the context of solid dissolution and precipitation (Chapter 5), and sorption reactions from a surface complexation perspective (Chapter 8).

Pankow, J. F., 1991, *Aquatic Chemistry Concepts*, Lewis Publishers, Chelsea, MI. A didactic approach to aquatic chemistry. Part III deals with mineral–solution chemistry with chapters covering mineral solubility (Chapter 11), metal carbonates (Chapter 12), existence of multiple solid phases (Chapter 14), and the thermodynamics of solid–solution mixtures (Chapter 16).

Ruthven, D. M., 1984, *Principles of Adsorption and Adsorption Processes*, Wiley-Interscience, New York. A thorough treatment of principles underlying adsorption phenomena and methods for their analysis.

Schwarzenbach, R. P., P. M. Gschwend, and D. M. Imboden, 1993, *Environmental Organic Chemistry*, John Wiley & Sons, Inc., New York. The book provides a comprehensive treatment of organic aquatic chemistry. Air–water partitioning is treated in Chapter 6, organic solvent–water partitioning is treated in Chapter 7, and sorption reactions are introduced in Chapter 11.

Smith, J. M., and H. C. Van Ness, 1975, *Introduction to Chemical Engineering Thermodynamics*, McGraw-Hill Book Company, New York. Thermodynamics from a chemical engineering perspective. The thermodynamics of phase equilibria is treated in Chapter 8.

Snoeyink, V. L., and D. Jenkins, 1980, *Water Chemistry*, John Wiley & Sons, Inc., New York. A first-level text in water chemistry; Chapter 6 treats precipitation and dissolution providing examples of iron, aluminum, calcium, and phosphate chemistry.

Stumm, W., and J. J. Morgan, 1995, *Aquatic Chemistry*, 3rd Edition, John Wiley & Sons, Inc., New York. This classic text and seminal treatise on aquatic chemistry treats heterogeneous chemical energetics in the context of precipitation and dissolution reactions (Chapter 7) and the solid–solution interface (Chapter 9).

6.10 PROBLEMS

6-1 A 0.05-mL volume of carbon tetrachloride (CCl_4) and 0.02 g of *p*-dichlorobenzene (*p*-DCB, $C_6H_4Cl_2$) are added to a 500-mL separatory funnel containing 300 mL of water and 200 mL of octanol. Calculate the concentrations of both compounds in the octanol and water phases after mixing.

6-2 A wash-water waste containing volatile organic solvents is generated at a volumetric rate of Q_l from an automobile paint booth operation. The waste drain from this operation passes through a closed collection chamber *en route* to the plant's central treatment facility. The chamber is vented by air passing over the wastewater at a flow rate of Q_g. The wastewater volume (V_l) and air volume (V_g) can each be considered uniformly mixed. Interphase mass transfer of each solvent is given by

$$N_l = k_{f,l}(C_{S,l} - C_l)$$

where $C_{S,l}$ is the equilibrium liquid-phase concentration of solvent i based on its partial pressure in the air above the vessel and C_l is the liquid-phase concentration within the chamber. Derive a process model to predict the concentrations of the solvent in the air and water leaving the vessel as functions of Q_l, Q_g, $\bar{t}$ ($= V_l/Q_l$), $k_{f,l}a_s^\circ$, the Henry's constant for the solvent, $\mathcal{K}_H$, and the feed concentration of solvent, C_{IN}. *Hint:* Material balances are needed for the water *and* airstreams. The concentration in the airstream is obtained from the ideal gas law ($PV = n\mathcal{R}T$), and the specific interfacial area for mass transfer, a_s°, is the area of interface per volume of wastewater in the vessel; remember, the vessel is closed, so that the solvent partial pressure is not zero.

Answer: $$C_{l,OUT} = \frac{C_{l,IN}}{1 + \dfrac{1}{\dfrac{1}{\bar{t}k_{f,l}} + \dfrac{Q_l\mathcal{R}T}{Q_g\mathcal{K}_H}}}$$

6-3 Consider a modification of Problem 6-2 in which the wastewater stream is generated and discharged to the drain in batches, but the airstream still passes continuously through the collection chamber. Set up the appropriate material balances and derive the model to predict the concentration of solvent remaining in the wastewater and that leaving in the airstream with time.

6-4 Volatilization of chlorine gas from water is of concern in cooling towers because this limits the effectiveness of disinfection. To assess the extent to which such volatilization will occur as a function of cooling water temperature, it is necessary to know Henry's constant for various temperatures. To this end a laboratory stripping column of the type shown below, a design similar to that suggested by Mackay et al. (1979), is devised to measure the Henry's constant.

Associated with the measurements are assumptions that (1) the only loss of solute is by volatilization; (2) the stripping column is well mixed; (3) the gas-phase solute concentration, C_g, in the gas bubbles at the liquid–atmosphere interface (top of the column) is in equilibrium with the liquid-phase solute concentration, C_l; (4) Henry's law describes

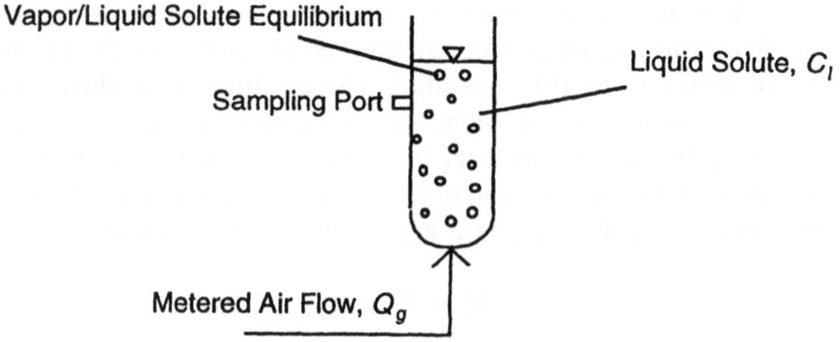

equilibrium; and (5) the liquid volume remains constant. The water volume, V_l, in the stripping column is 5 L and the flow rate of air is 4 L/min. An experiment is performed at 20°C and at pH 5 (to minimize dissociation of HOCl to OCl^-). The data collected for residual chlorine concentration (as Cl_2) in the stripping column as a function of time are given below.

Time (min)	Cl_2 (mg/L)
0	5.0
1200	4.75
2400	4.6
4800	4.2
6000	4.0
7200	3.8
9600	3.5

Determine the Henry's constant at 20°C in atm. *Hint:* Start with the following material balance:

$$-V_l \frac{dC_l}{dt} = Q_g C_g$$

Remember that the gas- and liquid-phase concentrations are in equilibrium as bubbles exit the stripping column. See Table 6-2 for appropriate expression of Henry's law.

Answer: $\mathcal{K}_H = 0.06$ *atm.*

6-5 Ozonation is used in a variety of water and waste treatment operations for oxidation of organic matter. By-products formed in the ozonation process may influence the efficiency of subsequent treatment processes. Research is conducted to evaluate the impact of pre-ozonation on the efficiency of activated carbon for removing organic carbon in the tertiary treatment of the secondary effluent from a wastewater treatment plant. The secondary effluent contains 20 mg/L of total organic carbon (TOC). Adsorption isotherms (expressed in terms of TOC) are measured for the effluent before and after pre-ozonation. Values of q_e and C_e are tabulated below.

	q_e (mg/g)	
C_e (mg TOC/L)	After Ozonation	Before Ozonation
1.0	2.0	5.0
3.0	4.8	7.0
5.0	7.3	8.1
7.0	9.5	9.0
9.0	11.6	9.7
11.0	13.6	10.3
13.0	15.6	10.8
15.0	17.5	11.3
17.0	19.3	11.7
19.0	21.1	12.1

(a) Based on the isotherm data presented, write calibrated Freundlich isotherm models for the two sets of data.

(b) Using the appropriate Freundlich isotherm parameter for each set of data, determine which isotherm is more "favorable." Describe the basis for you response.

6-6 Two major components of a small, intermittently generated, aqueous waste stream are to be adsorbed by activated carbon. Batch treatment with a holding time sufficient for adsorption to reach equilibrium is proposed. The initial concentrations of the two components are $C_{0,1} = 2$ mmol/L and $C_{0,2} = 1$ mmol/L, and the discharge permit limits are 1 mmol/L and 0.1 mmol/L, respectively. Determine the carbon dosage, D_o (mg/L), required to satisfy both of these limits. The Langmuir isotherm constants are $Q^\circ_{a,1} = 100$ mmol/g, $b_1 = 1$ L/mmol, $Q^\circ_{a,2} = 500$ mmol/g, and $b_1 = 2$ L/mmol. [*Hint:* The possible combinations of equilibrium concentrations, $C_{e,1}$ and $C_{e,2}$, can be obtained by solving the Langmuir competitive model together with the material balance for each component; for example, for component 1:

$$q_{e,1} = \frac{C_{o,1} - C_{e,1}}{D_o} = \frac{Q^\circ_{a,1} b_1 C_{e,1}}{1 + b_1 C_{e,1} + b_2 C_{e,2}}$$

where D_o is the dosage of activated carbon. Solving each material balance expression and equating them (D_o is the same) provides a way to relate $C_{e,1}$ to $C_{e,2}$. Meeting the permit limit for one of the components will control the dosage required.]

Answers: $D_o = 21.8$ *mg/L;* $C_{e,1} = 1$ *mmol/L;* $C_{e,2} = 0.09$ *mmol/L.*

6-7 The Freundlich constants that give approximately the same adsorption isotherms in the concentration range 0 to 2 mmol/L as the Langmuir constants in Problem 6-6 are $\mathcal{K}_{F,1} = 48.7$ (mmol/g) (L/mmol)$^{0.68}$ and $n_1 = 0.68$; $\mathcal{K}_{F,2} = 357.8$ (mmol/g) (L/mmol)$^{0.71}$ and $n_2 = 0.71$. Use the values of $C_{e,1}$ and $C_{e,2}$ found in answer to Problem 6-6 to determine the values of $q_{e,1}$ and $q_{e,2}$ from the ideal adsorbed solution theory model and compare these to the corresponding values obtained from the Langmuir competitive model. Expand the comparison by using other combinations of $C_{e,1}$ and $C_{e,2}$ that satisfy the Langmuir competitive model and discuss your results.

Answers: IAST $q_{e,1} = 37.0$, $q_{e,2} = 55.5$; *Langmuir* $q_{e,1} = 45.8$, $q_{e,2} = 65.8$ *(all in mmol/g).*

6-8 2-Methylisoborneol (MIB), a metabolite of algal production, is a common taste- and odor-causing compound in surface waters. Adsorption on activated carbon is one treatment option. A series of batch adsorption equilibrium tests were performed using powdered activated carbon (PAC) as the sorbent. It was of interest to measure the adsorbability over a wide range of MIB concentrations because MIB production varies in surface waters, and the concentration needed after treatment may vary depending on the intensity of the taste and odor problem. The following data were obtained:

C_0 (μg/L)	C_e (μg/L)	D_o (mg/L)
1.0	0.01	20
1.0	0.05	10
1.0	0.10	7
1.0	0.50	2
5.0	1.0	10
5.0	2.0	6
5.0	3.0	3

where C_0 is the initial MIB concentration, C_e is the equilibrium MIB concentration, and D_o is the PAC dosage. Discuss the goodness of fit of the Freundlich and Langmuir adsorption models to these data and determine the corresponding constants for each model.

6-9 A chemical manufacturing company produces a variety of polynuclear aromatic hydrocarbons (PAHs), including fluorene. In a trucking accident involving a shipment of fluorene across the Deep Gorge Bridge, an 800-kg quantity of product is spilled into a seep-fed wetlands at the bottom of the Gorge. The deep and narrow conditions of the gorge inhibit air circulation above the relatively well-mixed waters of the small wetlands. The seep that feeds the wetlands is near its center, and outflow occurs only by seepage return to the subsurface along its perimeters. You are asked to perform an analysis to assess the potential distribution of fluorene in this system if equilibrium conditions are attained. The following issues are of specific interest:

(a) The acute aqueous-phase toxicity limit of the Gorge Lake fiddler toad for fluorene is 100 μg/L. Is the fiddler in immediate danger of chemical poisoning?

(b) Discounting degradation and wash-out, determine what percentage of the total mass spilled into the lake will eventually be taken up by its sediments.

Your staff has compiled the following system data:

Phase volumes (estimated):
Water: 25,000 m^3
Sediment: 1200 m^3; $\rho_{\text{sediment}} = 1200 \text{ kg/m}^3$
Biota: 10 m^3
Air in the gorge above the wetlands: 125,000 m^3
Temperature: $T = 25°\text{C}$

From appropriate references you are further able to determine the following correlations:

$$\log \mathcal{K}_D \text{ (m}^3 \text{ water/kg soil)} = 0.904 \log \mathcal{K}_{o,w} - 4.779$$

$$\log \mathcal{K}_B \text{ (m}^3 \text{ water/kg biota)} = 3.04 - 0.568 \log C_S$$

6-10 A new pesticide, MCTP, is added to a sealed ecosystem containing air, water, and fish. Given the information below, compute the approximate distribution of MCTP in the system for the conditions listed.

$$\text{ecosystem volume} = 200 \text{ m}^3$$

$$\text{volume of water} = 18 \text{ m}^3$$

$$\text{volume of fish} = 0.001 \text{ m}^3$$

Properties of MCTP:

$$\mathcal{K}_{o,w} = 3200$$

$$\mathcal{K}_B = \mathcal{K}_{o,w}/0.7$$

$$\text{molecular weight} = 150 \text{ g/mol}$$

$$\text{aqueous solubility} = 200 \text{ mg/L}$$

$$\text{vapor pressure} = 400 \text{ Pa}$$

$$\text{density} = 1.4 \text{ g/mL}$$

(a) If 0.1 mol of MCTP is added to the system.
(b) If 75.0 mol of MCTP is added to the system.
(c) If [after removing some of the MCTP from the system in part (b)] the equilibrium concentration of MCTP in the ecosystem air is 5×10^{-4} mol/m^3.

APPENDIX A.6.1 PHASE PROPERTIES FOR SELECTED ORGANIC COMPOUNDS[a]

Formula	Compound	$\mathcal{K}_H$ $\left(\frac{\text{atm} - \text{m}^3}{\text{mol}} \times 10^3\right)$	$\log \mathcal{K}_{o,w}$	C_S (mg/L)	P^v (atm) × 10^3	T_m (°C)	T_b (°C)
CH_2O	Formaldehyde	3.3×10^{-4}	0.35 ± 0.10	Miscible	5.12×10^3	−92	−19.5
CCl_3F	Trichlorofluoromethane	113.3	2.53	1100	906.4	−111	23.8
CH_4O	Methanol	7.0×10^{-3}	−0.74 ± 0.1	Miscible	906.8	−97.8	64.7
C_3H_8	Propane	707 ± 24	2.36	66.7	9.33×10^3	−187.7	−42.1
C_3H_6O	Acetone	4.3×10^{-2}	−0.24 ± 0.1	Miscible	281.8	−94.6	56.5
C_3H_9N	Trimethylamine	3.13×10^{-1}	0.16 ± 0.10	4×10^5 (20°C)	2.12×10^3	−117.1	2.9
C_4H_{10}	*n*-Butane	947 ± 40	2.89	72.3	2.45×10^3	−135	−0.5
C_4H_8O	2-Butanone	3.8×10^{-2}	0.29 ± 0.05	249,000	131.2	−85.9	79.6
$C_4H_8O_2$	1,4-Dioxane	6.9×10^{-3}	−0.42	Miscible	50.33	10	101.1
C_5H_{12}	*n*-Pentane	1230 ± 100	3.45 ± 0.22	42 ± 4	691.8	138.4	36.1
C_5H_5N	Pyridine	1.1×10^{-2}	0.65 ± 0.05	72,000	26.3	−4	115.4
C_5H_{10}	1-Pentene	398 ± 20	2.20	148	839.8	—	30.1
C_5H_{10}	Cyclopentane	183 ± 11	3.00 ± 0.30	160	417.5	−93.7	49.3
$C_5H_{10}O$	2-Pentanone	7.7×10^{-2}	0.84 ± 0.08	55,400	49.26	−77.8	103.3
$C_6H_{14}O$	1-Hexanol	1.9×10^{-2}	2.03 ± 0.03	6000 ± 200	1.41	−51.6	157.0
C_6H_{12}	Cyclohexane	178 ± 20	3.44 ± 0.35	58 ± 4	125.7	6.5	80.7
C_6H_{12}	1-Hexene	413 ± 10	3.40 ± 0.10	53 ± 3	251.2	−98.5	66.0
C_6H_7N	Aniline	3.3×10^{-3}	0.90 ± 0.05	36,000	1.28	−6.2	184.4
$C_6H_6Cl_6$	Lindane	3.2×10^{-3}	3.72	7.48	8.3×10^{-5}	112.9	—
C_6H_6ClN	2-Chloroaniline	7.4×10^{-3}	1.90	3,800 (20°C)	0.22	—	208.8
C_6H_5ClO	2-Chlorophenol	5.6×10^{-4}	2.15	20,600	0.09	7.0	174.5
C_6H_5ClO	3-Chlorophenol	5.6×10^{-4}	2.50	21,800	0.10	32.5	214.0
C_6H_5ClO	4-Chlorophenol	5.6×10^{-4}	2.39	27,000	0.12	42.0	220.0
$C_6H_5NO_3$	*o*-Nitrophenol	14.2×10^{-3}	1.79	2,100 (20°C)	1.32	45	214.5
$C_6H_4Cl_2$	1,2-Dichlorobenzene	1.6	3.38 ± 0.01	145	1.55	−17.6	179.0
$C_6H_4Cl_2O$	2,4-Dichlorophenol	3.23×10^{-3}	3.06	4400	1.32	45.0	210.0
$C_6H_3Cl_3$	1,3,5-Trichlorobenzene	10.8	4.04 ± 0.03	5.37	0.32	52.5	218.5
$C_6H_3Cl_3O$	2,4,5-Trichlorophenol	11.0×10^{-3}	3.72	948	5.3×10^{-2}	62	251.8

$C_6H_2Cl_4$	1,2,3,4-Tetrachlorobenzene	258	4.6	4.3	5.14	46.5	254.0
$C_6H_2Cl_4O$	2,3,4,6-Tetrachlorophenol	—	4.10	183.3	—	69.5	275.0
C_6HCl_5O	Pentachlorophenol	8.7×10^{-3}	5.12	14.0	1.5×10^{-4}	191	309
$C_7H_7NO_2$	2-Nitrotoluene	5.5×10^{-2}	2.30 ± 0.25	650 (30°C)	0.26	−4.1	222.3
C_7H_8O	*o*-Cresol	1.6×10^{-3}	1.98 ± 0.05	31,000 (40°C)	0.20	30.8	190.8
C_7H_{14}	2-Heptene	418 ± 39	3.20	150	694	—	98.5
C_7H_{16}	*n*-Heptane	2270 ± 490	4.50 ± 0.25	2.4 ± 0.2	61.7	−90.6	98.4
C_8H_8	Styrene	2.6	3.05 ± 0.30	250 ± 60	6.25	−30.6	145.2
C_8H_{10}	*p*-Xylene	6.7	3.15	185	11.58	13.3	138.3
$C_6H_3Cl_2(C_2H_3O_3)$	2,4-D	1.4×10^{-7}	2.81	890	5.5×10^{-7}	138	215.0
$C_8H_{18}O$	2-Octanol	4.29×10^{-2} (32.8°C)	2.90 ± 0.25	4,000	1.32 (32.8°C)	−38.6	178.5
$C_8H_{14}ClN_5$	Atrazine	—	2.56	66.67	—	174.0	—
C_9H_{12}	1,2,3-Trimethylbenzene	3.18 ± 0.20	3.60 ± 0.20	69 ± 5	1.83	−25.5	176.1
C_9H_{20}	*n*-Nonane	414	5.65 ± 0.6	0.17 ± 0.08	5.75	−53.7	150.8
$C_9H_{20}O$	1-Nonanol	1.7×10^{-2}	4.02 ± 0.30	140	0.10	−5	213.5
$C_{10}H_{14}NO_5PS$	Parathion	0.4×10^{-3}	3.81	17.14	2.2×10^{-5}	6.1	—
$C_{10}H_8O$	1-Naphthol	—	2.84 ± 0.20	846	—	96	288.0
$C_{10}H_9N$	1-Naphthylamine	—	2.25 ± 0.25	1,700	—	50	306.1
$C_{11}H_{24}$	*n*-Hendecane	1830	5.58	4.0×10^{-3}	4.7×10^{-2}	−25.6	195.8
$C_{12}H_{26}O$	Dodecyl alcohol	9.0×10^{-2}	5.13 ± 0.30	4.0	1.9×10^{-3}	24	259.0
$C_{12}H_{10}$	Biphenyl	0.276 ± 0.02	3.98 ± 0.10	7.2 ± 0.2	0.010	69.5	254.9
$C_6H_5C_6H_4Cl$	2-Chlorobiphenyl	2.9	4.54	1.3	0.020	34	374.0
$C_{12}H_8Cl_6O$	Dieldrin	11.2×10^{-3}	5.48	0.22	6.6×10^{-6}	175.0	—
$2(C_6H_3Cl_2)$	2,2′,4,4′-Tetrachlorobiphenyl	0.9	5.90	6.8×10^{-2}	2.0×10^{-4}	83	—
$2(C_6H_2Cl_3)$	2,2′,4,4′,6,6′-Hexachlorobiphenyl	8.2×10^{-3}	7.00	7×10^{-4}	1.6×10^{-7}	114	—
$C_{12}H_4Cl_4O_2$	Dioxin	20.8×10^{-4}	7.02	3.17×10^{-4}	2×10^{-9}	295	—
$C_{13}H_{28}$	*n*-Tridecane	2313	6.65	1.04×10^{-3}	13.1×10^{-3}	−6.2	234.0
$C_{13}H_{10}$	Fluorene	8.4×10^{-2}	4.47	1.9 ± 0.1	7.1×10^{-4}	116	295.0
$C_{14}H_{30}$	*n*-Tetradecane	1,130	8.0 ± 1.0	2.2×10^{-3}	12.6×10^{-3}	5.5	252.5
$2(C_6H_4Cl)C_2HCl_3$	DDT	2.3×10^{-5}	6.19	3.0×10^{-3}	5.0×10^{-3}	108.5	—
$C_{15}H_{32}$	*n*-Pentadecane	479	7.72	7.6×10^{-5}	1.7×10^{-4}	10	270.5
$C_{16}H_{34}$	*n*-Hexadecane	228	8.25	9.0×10^{-4}	1.9×10^{-3}	18.5	287.5
$C_{16}H_{34}O$	Cetyl alcohol	1.7×10^{-2}	—	0.035	2.4×10^{-6}	49.3	344.0

APPENDIX A.6.1 (*Continued*)

Formula	Compound	$\mathcal{K}_H$ $\left(\frac{atm - m^3}{mol} \times 10^3\right)$	log $\mathcal{K}_{o,w}$	C_S (mg/L)	P^v (atm) × 10^3	T_m (°C)	T_b (°C)
$C_{16}H_{10}$	Fluoranthrene	10.5×10^{-3}	5.22	0.24	1.2×10^{-5}	110.8	202.3
$C_{17}H_{36}$	*n*-Heptadecane	53.4	8.79	5.5×10^{-6}	1.2×10^{-6}	22.5	303.0
$C_{18}H_{38}$	*n*-Octadecane	8.81	9.32	1.4×10^{-6}	2.1×10^{-4}	28.2	317.0
$C_{18}H_{38}O$	1-Octadecanol	1.1×10^{-2}	—	0.11 (34°C)	4.4×10^{-6}	58.5	349.5
$C_{18}H_{12}$	Benz[*a*]anthracene	5.8×10^{-3}	5.91	0.01	2.8×10^{-7}	159.8	435.0
$C_{19}H_{40}$	*n*-Nonadecane	2.89	9.86	4×10^{-7}	4.3×10^{-7}	32.1	330.0

[a]The numerical values given in this table are derived from a number of sources (see Section 6.9.1). In some cases they have been transformed in units, and in other cases calculated from tabulated data. Some values for Henry's constant, for example, have been calculated from vapor pressure and solubility data, while values for vapor pressure have been similarly calculated from Henry's constant. Values are given for 25°C unless otherwise stated. It should be noted that significant variations in the numerical values of such constants reported in different literature sources are common. When sufficient numbers of values were available, typical ranges of variation are reported in these tables.

7

RATE RELATIONSHIPS: CONCEPTS AND APPLICATIONS TO HOMOGENEOUS SYSTEMS

7.1 RATE CONCEPTS

The relationships presented in Chapters 5 and 6 enable us to determine the distribution of species involved in any reaction for a condition or state of *equi-*

librium. Complete characterization of a reaction requires that we know more than just the equilibrium distribution of species, however; we must know as well the *rate* at which this condition is approached. In practice, a reaction or process may be energetically favorable but not practicably feasible if it moves too slowly toward the equilibrium state. Analysis of *reaction kinetics* embodies the determination and interpretation of rates at which reactions occur, and facilitates prediction of species distributions as functions of time.

In our considerations of reaction kinetics we differentiate between homogeneous and heterogeneous reactions, as we did for reaction energetics in Chapters 5 and 6. We further differentiate between reversible and irreversible changes. In general terms, a *reversible* reaction is one that proceeds toward a final state in which there exists a well-defined distribution of reactant and product species, all being present in measurable quantities. Conversely, an *irreversible* reaction is one in which a stoichiometric combination of reactants leads to their essentially complete conversion to products. These general definitions are not rigorous, for according to the definitions of equilibrium, all reactions are, in principle, reversible. There are, however, numerous chemical reactions in which the extent of change in the reverse direction is so small that it cannot be detected experimentally over the period of observation or concern, and such reactions may for practical purposes be treated as irreversible. When applicable, the assumption of irreversibility greatly facilitates interpretation and representation of reaction rates in environmental systems.

As noted in Chapter 2, all mathematical models for describing process dynamics in environmental systems must account for rates of both transformation reactions and transport processes. Chapters 7 and 8 focus on methods for evaluation and representation of transformation rates.

7.1.1 Mass Law Relationships

Despite subsequent sophistication of kinetic theory, the *mass law* concepts set forth by Guldberg and Waage and by van't Hoff in the mid-nineteenth century still serve as a fundamental basis for analysis and interpretation of reaction rates. The law of mass action states that the rate of an elementary homogeneous chemical reaction is directly proportional to the product of the masses of the reacting species. The term *elementary reaction* refers to a reaction that proceeds directly as written, with no intermediate steps. For example, the stoichiometric equation for an *irreversible and homogeneous* elementary reaction between two chemical species, A and B, to form a new compound, AB, is written as

$$\mathrm{A} + \mathrm{B} \longrightarrow \mathrm{AB} \tag{7-1}$$

The law of mass action states that the rates of disappearance of A and B and production of AB for this condition are given by

$$r_{AB} = -r_A = -r_B = k[\mathrm{A}][\mathrm{B}] = kC_A C_B \tag{7-2}$$

where k is a constant of proportionality or rate constant, and $[A] = C_A$ and $[B] = C_B$ represent the *molar* (molecular) concentrations (more rigorously, activities) of A and B, respectively, at any time, t. *Mass* concentrations can be substituted for [A] and [B] with a resulting change in the units and magnitude of the rate coefficient. The basic relationships expressed in Equation 7-2 and ensuing mass law rate expressions are, however, predicated on *molecular*, and thus molar, proportionalities. The importance of this concept to our interpretations of reaction *orders* will be illustrated shortly.

Reaction rates and reaction rate constants are generally determined from measurements of C_A, C_B, and time, t. Completely mixed reaction vessels having no inflow or outflow are usually used to collect rate data. This type of system is referred to as a *completely mixed batch reactor* (CMBR). The explicit statement of conditions embodied by the term CMBR must be carefully considered; namely, it is a *closed system of homogeneous content*. Neither the simple term *batch reactor*, nor the commonly employed term, *continuous stirred tank reactor* (CSTR), adequately convey these explicit and important conditions. Indeed, the latter terminology (CSTR) implies an open (continuous flow) system rather than a closed (batch) system. If the adjective *continuous* is replaced by the adverb *continuously*, as is sometimes done, the reactor is known to be stirred continuously but not necessarily *completely mixed*. Moreover, a *continuously stirred reactor* conveys no information to determine if the system is *open* or *closed*. Different types of reactor systems are considered in detail in Chapters 9 through 11.

Because there is no inflow or outflow of reactant for a CMBR, and *because* the contents of the reactor can be considered homogeneous, the material balance expression developed in Chapter 2 reduces to

$$\begin{bmatrix}\text{net rate of accumulation}\\ \text{of AB within the CMBR}\end{bmatrix} = \begin{bmatrix}\text{net rate of production of}\\ \text{AB within the CMBR}\end{bmatrix} \tag{7-3}$$

or

$$V_{\text{CMBR}}\left(\frac{dC_{AB}}{dt}\right)_{\text{CMBR}} = r_{AB}V_{\text{CMBR}} \tag{7-4}$$

where V_{CMBR} is the volume of the contents of the reactor. Dividing through by V_{CMBR} and combining with Equation 7-2 then yields

$$\left(\frac{dC_{AB}}{dt}\right)_{\text{CMBR}} = r_{AB} = kC_AC_B \tag{7-5}$$

Mass conservation considerations allow the concentrations C_A and C_B in Equation 7-5 to be expressed in terms of their initial values, $C_{A,0}$ and $C_{B,0}$, less the concentration of product formed C_{AB}, or reactant *reacted*, C_R.

If $C_{AB,0} = C_{R,0} = 0$, introduction of these concentration parameters to Equation 7-5 yields:

$$\left(\frac{dC_{AB}}{dt}\right)_{\text{CMBR}} = \left(\frac{dC_R}{dt}\right)_{\text{CMBR}} = kC_A C_B = k(C_{A,0} - C_R)(C_{B,0} - C_R) \quad (7\text{-}6)$$

or, for the special case when $C_{A,0} = C_{B,0} = C_0$

$$\frac{dC_R}{dt} = k(C_0 - C_R)^2 \quad (7\text{-}7)$$

Integration of Equation 7-6 yields

$$kt = \frac{1}{C_{A,0} - C_{B,0}} \ln \frac{C_{B,0}(C_{A,0} - C_R)}{C_{A,0}(C_{B,0} - C_R)} \quad (7\text{-}8)$$

while the integrated form of Equation 7-7 is

$$kt = \frac{C_R}{C_0(C_0 - C_R)} \quad (7\text{-}9)$$

It is possible, in principle, to use mass law relationships to develop an appropriate rate equation for any chemical reaction in the manner in which Equations 7-8 and 7-9 were obtained.

For many simple irreversible reactions the rates of change in the reactant and product concentrations are directly proportional to the molecular concentration of only one reactant. For example, consider the decomposition of a single chemical substance, such as the common environmental oxidant hydrogen peroxide:

$$H_2O_2 \longrightarrow H_2O + \frac{1}{2} O_2 \quad (7\text{-}10)$$

This is termed a *monomolecular* reaction because it involves only one reactant molecule. The rate of disappearance of reactant according to the stoichiometry given in Equation 7-10 may be written

$$-r_A = kC_A = k(C_{A,0} - C_R) \quad (7\text{-}11)$$

where the subscript A refers to reactant, in the above case to the peroxide. The reactant concentration in a perfectly mixed closed system is found by integrating a material balance equation similar to Equation 7-4:

$$\frac{dC_A}{dt} = -kC_A \quad (7\text{-}12)$$

Upon integration, we observe that a monomolecular reaction is characterized by an exponential decrease in reactant concentration with time:

$$\frac{C_A}{C_{A,0}} = e^{-kt} \tag{7-13}$$

We can also replace C_A with $(C_{A,0} - C_R)$ in Equation 7-12 to express the rate of change in the concentration of reactant that has reacted:

$$\frac{dC_R}{dt} = k(C_{A,0} - C_R) \tag{7-14}$$

Integration of Equation 7-14 yields

$$kt = \ln \frac{C_{A,0}}{C_{A,0} - C_R} \tag{7-15}$$

The *fraction*, ϕ_r°, of peroxide reacted in time t is

$$\phi_r^\circ = \frac{C_R}{C_{A,0}} \tag{7-16}$$

Equation 7-15 can then be expressed as

$$k = \frac{-\ln (1 - \phi_r^\circ)}{t} \tag{7-17}$$

Thus, for a monomolecular reaction, a fixed fraction of the amount of material present at the start of the period is reacted for each fixed unit of time thereafter.

The decomposition of hydrogen peroxide also illustrates how rates of disappearance of reactants and appearance of products relate to stoichiometry. According to the stoichiometry shown in Equation 7-10, the rate of disappearance of hydrogen peroxide (A) is related to the rate of production of molecular oxygen (B) by

$$-r_A = 2r_B \tag{7-18}$$

and because $-r_A = kC_A$ and $r_B = k'C_A$, it follows that $k = 2k'$.

7.1.2 Reaction Orders

A monomolecular reaction is dependent on the concentration of only one reacting species and is thus said to be a *first-order* reaction. The general expression for the rate of an elementary reaction involving components A, B, C, and D may be written

$$r_{A,B,C,D} = k(C_A)^{\gamma_A}(C_B)^{\gamma_B}(C_C)^{\gamma_C}(C_D)^{\gamma_D} \tag{7-19}$$

This reaction is said to have an *overall order* of $n = (\gamma_A + \gamma_B + \gamma_C + \gamma_D)$; it is *nth-order* overall, γ_A *order* in C_A, γ_B *order* in C_B, and so on. The order of a reaction with respect to each reactant, i, corresponds to the *stoichiometric coefficient*, γ_i, of that reactant *only if the reaction is elementary*. Stated another way, for elementary reactions, and for elementary reactions *only*, the mechanism of reaction is expressed by the number of molecules involved. Corresponding to our prior use of the term *monomolecular* to describe H_2O_2 decomposition, the order of the reaction is also the *molecularity* of the reaction. For nonelementary reactions, however, the reaction order may be significantly different from that suggested by the stoichiometry.

The elementary reaction

$$2A \xrightarrow{k} B + C \tag{7-20}$$

has a molecularity of 2, and the rate of this reaction is thus second order:

$$-r_A = k(C_A)^2 \tag{7-21}$$

The stoichiometry of the reaction given in Equation 7-20 is preserved if it is written as

$$A \xrightarrow{k} \frac{1}{2}B + \frac{1}{2}C \tag{7-22}$$

Equation 7-22 does not express the proper molecularity, however, and thus does not represent the mechanism of the elementary reaction. In this case the reaction rate might mistakenly be interpreted to be first order when it is actually a bimolecular reaction.

Further examples of the relationship of reaction order to molecularity for elementary reactions are provided by Equations 7-6, 7-7, and 7-12. Equations 7-6 and 7-7 are rate equations for an elementary reaction which is second order overall. Equation 7-6 is applicable to an elementary reaction which is first order with respect to each of two reactants, whereas Equation 7-7 is additionally applicable to a reaction that is second order with respect to one reactant. Application of the latter equation to a reaction that is first order with respect to each of two reactants, however, is limited to the case where the initial concentrations of each are identical. Equation 7-12 is a rate expression for an elementary reaction which is first order.

The order of reaction must be an integer for elementary reactions because the mechanism of reaction necessarily involves entire molecules. On the other hand, the order of reaction can be fractional if the reaction is nonelementary. Rate equations for nonelementary reactions and complex processes may result either from the use of experimental rate data to devise an *empirical rate expression*, or from a characterization of reaction mechanism by analysis of the kinetics of each intermediate elementary reaction comprising the overall process. As will be illustrated later in this chapter, there are also many nonelementary reactions for which the reaction rate expression is not of the multiplicative form given by Equation 7-19. In these instances, an observed or *apparent order* of reaction developed from experimental rate data is very likely not to reflect the molecularity or stoichiometry of the reaction in a precise manner.

Rate equations expressing zero-order, first-order, and second-order dependence of rate on concentrations are of most practical interest for physicochemical and biological reactions in environmental systems. For nonelementary reactions, higher degrees of molecularity are probably involved. In many cases, however, the overall process occurs by way of a number of elementary reaction steps which are of first or second order in their dependence on reactant concentration. When these reaction steps are sequential, the appropriate rate expression is determined by the slowest single step: the *rate-limiting step*. Thus many reactions that are described by complicated stoichiometric equations may require only relatively simple expressions for a description of reaction rate.

The final test for reaction order must be experimental. A given set of experimental data that fit an assumed rate expression does not, however, prove that the actual molecularity of the reaction is given by the assumed rate expression. The only conclusion is that the data are consistent with that rate expression within the limits of observation.

The mass law hypothesis that stoichiometric coefficients characterize molecularity, and thus that they may be used as shown in Equation 7-19, is valid only if a reaction is elementary and if its stoichiometry represents its entire reaction scheme. *Regardless* of whether a reaction is elementary or nonelementary, however, *the law of mass action is always valid when relating the position of equilibrium for a reversible reaction to the point at which the forward and reverse reaction rates are equal.* For example, if the reaction given in Equation 7-1 is reversible such that

$$\mathrm{A} + \mathrm{B} \underset{k_2}{\overset{k_1}{\rightleftharpoons}} \mathrm{AB} \tag{7-23}$$

then, at equilibrium,

$$(r_{AB})_{\text{production}(\rightarrow)} = k_1 C_A C_B = (r_{AB})_{\text{depletion}(\leftarrow)} = k_2 C_{AB} \tag{7-24}$$

and

$$\frac{C_{AB}}{C_A C_B} = \frac{k_1}{k_2} = K \tag{7-25}$$

where K *is the mass law equilibrium constant*. We will make further use of the law of mass action after discussing the kinetics of reversible reactions.

7.1.3 Nonelementary Reactions

Nonelementary reactions are those in which the stoichiometry represents only the *net* result of a reaction, not its complete molecular scheme. Hence, nonelementary reactions involve one or more steps in which intermediate products that are not specifically identified in the stoichiometry of the reaction form and then disappear. Each intermediate step is an elementary reaction for which the kinetics are described rigorously by the law of mass action. The kinetics of the overall reaction then follow from analysis of the kinetics of each intermediate step.

Suppose that Equation 7-1 is not representative of an elementary reaction but represents instead the overall stoichiometry for a reaction that leads to formation of AB through an intermediate step involving the transformation of A to some short-lived species, A*. The overall reaction now has two rate steps:

$$\mathrm{A} \underset{k_2}{\overset{k_1}{\rightleftharpoons}} \mathrm{A}^* \qquad (7\text{-}26)$$

$$\mathrm{A}^* + \mathrm{B} \xrightarrow{k_3} \mathrm{AB} \qquad (7\text{-}27)$$

Two different approaches can be used to obtain rate expressions for the reaction scheme shown in Equations 7-26 and 7-27; these are the *stationary-state approximation* and the *rate-controlling-step* approach.

The stationary-state approximation assumes no net accumulation of the intermediate product, A*. The rate of formation and disappearance of A* for each elementary reaction must therefore sum to zero:

$$r_{A*} = k_1 C_A - k_2 C_{A*} - k_3 C_{A*} C_B = 0 \qquad (7\text{-}28)$$

The rate of formation of AB is

$$r_{AB} = k_3 C_{A*} C_B \qquad (7\text{-}29)$$

Equation 7-28 can be solved for C_{A*} and the result substituted into Equation 7-29 to obtain

$$r_{AB} = \frac{k_1 k_3 C_A C_B}{k_2 + k_3 C_B} \qquad (7\text{-}30)$$

This rate expression is of an entirely different form than Equation 7-2 which was for an *elementary* bimolecular reaction between A and B to form AB. Rate expressions can thus become increasingly complicated as more intermediate steps become involved. We may often be able to approximate the rate expression for nonelementary reactions with a simpler form, depending on the relative magnitude of the rate constants for the individual elementary reactions that are involved. For example, the rate expression given by Equation 7-30 may be approximated by

$$r_{AB} \approx k_1 C_A \qquad \text{for } k_3 C_B >> k_2 \tag{7-31}$$

$$r_{AB} \approx \frac{k_1 k_3}{k_2} C_A C_B \qquad \text{for } k_3 C_B << k_2 \tag{7-32}$$

The rate-controlling step approach involves direct assumptions about the relative rates of reaction given by Equations 7-26 and 7-27. If, for example, A* is produced by a nearly irreversible reaction but at a much slower rate than the formation of AB, that slower step is the rate-controlling step, and we conclude that

$$r_{AB} \approx k_1 C_A \tag{7-33}$$

Equation 7-33 is the special case of a rate expression given by the stationary-state approximation when $k_3 >> k_2$ (Equation 7-31). Alternatively, if the formation of AB is taken as rate limiting, then

$$r_{AB} \approx k_3 C_{A*} C_B \tag{7-34}$$

In this instance the first step is sufficiently fast to assume a near equilibrium condition, such that

$$\frac{C_{A*}}{C_A} \approx \frac{k_1}{k_2} = K \tag{7-35}$$

Substituting Equation 7-35 for C_{A*} in Equation 7-34 gives

$$r_{AB} \approx \frac{k_1 k_3}{k_2} C_A C_B \tag{7-36}$$

which is identical to the stationary-state approximation given in Equation 7-32.

Although stoichiometry does not provide a way to predict the rates of nonelementary reactions directly, we can start with an assumption that the reaction is elementary and use its stoichiometry to postulate a rate model. This model can then be tested against the experimental rate data to determine whether or not the assumption is reasonable. If the data do not agree with the model, the reaction is probably more complicated than the overall stoichiometry suggests (i.e., it is nonelementary). There are two alternative pathways in which to proceed for further analysis: (1) postulate intermediate reactions that lead to alternative rate expressions, or (2) fit an empirical rate model to the data without trying to characterize the reaction mechanisms. In many instances, the reaction may be too complex to unravel the reaction pathway easily, and an empirical rate model may be the only practical alternative. The limitations of such an approach must be recognized. Most important, *an empirical rate model should not be applied outside the range of reaction conditions for which it was formulated and has been calibrated.*

The reactions of hypochlorous acid (HOCl) with natural organic matter (NOM) in water to form trihalomethanes (THMs) illustrate a situation in which, because so much uncertainty exists about the fundamental mechanisms of such reactions, empirical overall reaction rate modeling has been used in lieu of reaction pathway models. In this instance, a relatively simplistic equation is used to describe the reaction:

$$\gamma_i(\text{HOCl}) + \gamma_j(\text{NOM}) = \gamma_k(\text{THMs}) + \text{other products} \qquad (7\text{-}37)$$

A stoichiometrically balanced equation for the reaction cannot be written because NOM represents a collection of molecular structures and molecular weights which are for the most part unknown. Moreover, the reactivity of each NOM fraction to chlorine is not well understood. Additionally, not all of the *products* may be identified, nor are the influences of reaction conditions such as pH and the presence of other halides, such as bromide, completely understood. Chlorination in the presence of bromide can lead, for example, to formation of bromoform. Accurate characterization of intermediate reactions is virtually impossible in such cases, yet a reliable rate relationship is needed for engineering practice. The answer lies in the development of an empirical rate model, as illustrated in Example 7-1.

Example 7-1

- **Situation:** A large regional water supply network employs a number of different raw water sources. To facilitate development of a uniform policy with respect to control of chlorine doses to minimize THM formation, the regional engineer has asked you, his laboratory director, to conduct a series of studies designed to develop an appropriate model for THM formation rates.
- **Question(s):** Design an experimental approach to the development of an appropriate empirical rate model for this situation.
- **Logic and Answer(s):**

1. Rates of trihalomethane formation are dependent on a variety of natural water conditions as well as on the dosage of chlorine applied. The first step in this development will thus be to perform a series of rate experiments with samples of each of the different natural waters involved.

2. Ambient temperature, pH, nonpurgeable organic carbon (NPOC) concentration and bromide concentration are different for the waters to be tested, so there must be some common denominator for the experiments. For example, the chlorine to NPOC ratio can be fixed to a given value for a series of baseline experiments in which THMs are measured with time for the different ambient conditions.

3. Additional experiments in which one parameter is varied while holding the others constant at the ambient condition should be conducted for each natural water.

4. Simple and multiple nonlinear regressions can then be performed to analyze the resulting data. One approach might utilize transformation of the independent and dependent variables into logarithmic forms followed by a multiple linear regression. Another might involve development of a nonlinear model without log transformation. A model of the former type is presented for illustration purposes:

$$\text{TTHM} = \gamma_1(\text{UVA} \cdot \text{TOC})^{\gamma_2}(\text{Cl}_2)^{\gamma_3}(t)^{\gamma_4}(T)^{\gamma_5}(\text{pH} - 2.6)^{\gamma_6}(\text{Br}^- + 1)^{\gamma_7}$$

where TTHM is total trihalomethane [i.e., the sum of the concentrations (mg/L) of chloroform, bromodichloromethane, dibromochloromethane, and bromoform]; UVA · TOC is the product of the ultraviolet light absorbance (UVA, cm^{-1}) at 254 nm and total organic carbon (TOC) concentration (mg/L), these both serving as surrogate measures of NOM; Cl_2 is the applied chlorine dose (mg/L); t is the reaction time (hours); T is temperature (°C); and Br^- is the bromide concentration (mg/L). This model also implies that TTHMs are not formed if $pH < 2.6$, an experimental, not theoretical, conclusion.

5. The values of γ_1 through γ_7 can then be determined by best fit of the pooled data after excluding results where no chlorine residual existed at the end of some specified time period, a practical problem in such kinetic analysis where the stoichiometric amount of chlorine needed is unknown.

6. Such modeling approaches must be subjected to sensitivity analyses to determine the level of accuracy required regarding information on the magnitude of specific dependent variables, and then subjected to verification using external data sets. [For a related application of the particular correlation given above, see Amy et al. (1987).]

7.2 APPLICATIONS OF FIRST-ORDER EQUATIONS

First-order rate equations are widely applicable for description of a number of different types of environmental reactions. In addition to true monomolecular reactions, such as those of radioactive decay and other self-decomposition reactions (e.g., Equation 7-10), mathematical expressions of the first-order type are often found useful for phenomenological description of (1) multimolecular reactions in which the concentrations of all but one of the reactants remains essentially constant; (2) complex processes, such as unsaturated microbial growth and disinfection, for which overall reaction rates may be represented approximately by a first-order equation; and (3) certain purely physical processes, such as the mass transfer processes associated with the dissolution of solids and gases in water, which follow this general mathematical form.

A reaction in which the concentrations of all but one of the reactants are so large that they change very little over the course of the reaction is termed *pseudo-first order*. Consider, for example, the case of the base catalyzed hydrolysis of 1,1,2,2 tetrachloroethane (TeCE) to trichloroethane:

$$C_2H_2Cl_4 + 2H_2O \longrightarrow C_2HCl_3 + H_3O^+ + Cl^- \tag{7-38}$$

In dilute aqueous solution the concentration of H_2O remains substantially constant at ~55.6 mol/L; consequently, the rate of hydrolysis for this bimolecular reaction is sensitive to, and therefore essentially dependent on, only the concentration of the TeCE.

One of the most familiar empirical applications of the first-order rate equation in water quality analysis is for description of rates of biochemical oxygen demand (BOD) exertion. Although BOD data may generally be described closely by a first-order equation, microbial utilization of oxygen during substrate oxidation is a complex process, certainly not monomolecular. Recognizing this, and further being unable to characterize lumped parameters such as BOD in terms of molar concentration units, we use mass concentration units in development of associated rate relationships. The differential and integrated forms of the first-order BOD equation are generally written as

$$r_R = \frac{dC_{R,\infty}}{dt} = k(C_{R,\infty} - C_R) \tag{7-39}$$

and

$$C_R = C_{R,\infty}(1 - e^{-kt}) = C_{R,\infty}(1 - 10^{-0.4343kt}) \tag{7-40}$$

respectively, where C_R is the BOD which has been exerted at any time, and $C_{R,\infty}$ is the ultimate *carbonaceous* (*first-stage*) BOD. In chemical reactions, $C_{R,\infty}$ would correspond to the initial concentration of a measurable reactant and thus be a known quantity. In the BOD test, however, the ultimate BOD cannot be measured at the outset of the experiment but must instead be determined along with the rate constant, k. For such situations, analysis of rate data by multiple nonlinear regression techniques is required to search optimal values for two unknown parameters (i.e., two-parameter search routines).

Another situation in which a simple first-order expression may be used in an empirical way to describe a complex biochemical process is for the bactericidal action of a disinfectant. This is evidenced by *Chick's law* for disinfection in a CMBR:

$$r_N = \frac{dC_N}{dt} = -kC_N \tag{7-41}$$

or, in integrated form,

$$\ln \frac{C_{N,0}}{C_N} = kt \qquad \log \frac{C_{N,0}}{C_N} = 0.4343kt \tag{7-42}$$

where C_N is the number of organisms per unit volume remaining at any time and $C_{N,0}$ is the original number. This first-order expression, however, is valid only if the disinfectant concentration is in excess. In most situations the disinfectant is not in excess, so the rate expression must also include the disinfectant concentrations raised to some order, which adds more empiricism.

The first-order expression is useful also for empirically describing certain purely physical processes, particularly mass transfer processes, under conditions of perfect mixing. Consider, for example, the dissolution of a solid with surface area $A°$ in a volume V_R of water in a CMBR; if the equilibrium saturation concentration (solubility) is C_s and the concentration at any time is designated as C, the rate of accumulation of C in solution is given by the material balance equation:

$$\frac{dC}{dt} = k_f \frac{A°}{V_R}(C_S - C) = k_f(a_s^\circ)_R(C_S - C) \tag{7-43}$$

where k_f is the *mass transfer coefficient* and $(a_s^\circ)_R$ the specific surface area per volume of reactor were defined in Chapter 4.

7.3 DEVELOPMENT AND ANALYSIS OF RATE DATA

The suitability of any empirical rate expression for describing a particular reaction can be ascertained only by collecting and analyzing experimental rate data. The proper rate model may not always be simple to find because many reactions are nonelementary. As a first step, all available information about the reactants and products must be assessed carefully. Not all reactants or products may be known, and this presents a problem at the outset of rate data analysis and modeling. Although rate expressions for reactions of interest may already have been reported in the literature, these should be verified. Each environmental system contains a unique mix of dissolved and particulate chemical species which might affect the rate of a specific type of reaction, by altering the rate constant or even the reaction mechanism.

Rate data are usually collected in CMBR systems operated at bench scale. The study of certain reactions may require more sophisticated experimental designs. For example, a reaction may depend strongly on temperature and light, in which case these parameters must be carefully controlled. Heterogeneous reactions also require special consideration with respect to the intensity of mixing between reacting phases.

In the majority of cases, rate analyses can be conducted in CMBR experimental systems and thus data collection is reasonably straightforward. Sample requirements are minimal since the systems are closed. The reactor itself may be as simple as a flask that is shaken or stirred continuously in a manner that ensures complete mixing of its contents. Samples are then withdrawn at specific time intervals to allow measurement of residual concentrations of reactants or products.

Regardless of how simple the system appears, the interpretation of rate data involves several assumptions. Further, the conditions associated with these assumptions must be satisfied in the experiment to make the interpretation valid. These assumptions are explicitly clear if the system truly fits the description of a completely mixed batch reactor (i.e., closed, so the volume of its contents is assumed to be constant, and completely mixed, so the contents are homogeneous). If these conditions are not met, the rate observation is *reactor specific*. More precisely stated, the material balance relationships presented in holistic form in Chapter 2 and in reduced form in Equation 7-3 must have the same meaning in a CMBR (i.e., the rate of accumulation or disappearance of substance in the reactor which is dC/dt observed, must be equal to dC/dt, corresponding to the reaction).

The ensuing discussion focuses on *interpretation* and *modeling* of experimental rate data collected in CMBRs. A rate expression can be obtained using either *integral* or *differential analysis*. Several special methods for *development* of rate data will also be considered; these are referred to commonly as the *half-life method*, the *method of excess*, and the *initial-rate method*.

7.3.1 Rate Data Analysis

7.3.1.1 Integral Method This method of analysis uses the integrated form of an appropriate reaction rate expression to characterize observed data regarding the concentration of reactant remaining, or product generated, with time. The general material balance relationship for a CMBR of volume V_R is written from Equation 7-4 as

$$V_R \frac{dC}{dt} = r V_R \tag{7-44}$$

where r is the reaction rate and C is the concentration of the species of interest. Except for a zero-order reaction, r depends on the concentration of the reacting species, and possibly on the concentration of other species as well.

The most common approach to analysis of rate data is to assume a particular rate expression and determine whether the data are fit adequately by that expression. To illustrate application of the integral analysis approach, assume that a particular reaction depends on the concentration of only the ith species, ignoring the possibility of a more complex reaction in which other components may be involved. We can further assume that k is constant if experimental conditions (e.g., temperature and pH) are held constant. For these assumed conditions, $r = k\phi(C)$, and Equation 7-44 can be integrated by separation of variables to yield

$$\int_{C_0}^{C} \frac{dC}{\phi(C)} = k \int_0^t dt = kt \tag{7-45}$$

where C_0 is the initial concentration. Experimental values of C at each value of t are used to calculate the value of the integral on the left-hand side of Equation 7-45. If the proper rate expression has been selected, a plot of this integral against time should yield a straight line, the slope of which yields the rate constant, k. A plot of this type is illustrated in Figure 7-1.

To demonstrate use of the integral method for testing the fits of several common rate models to experimental rate data, assume first that the rate of

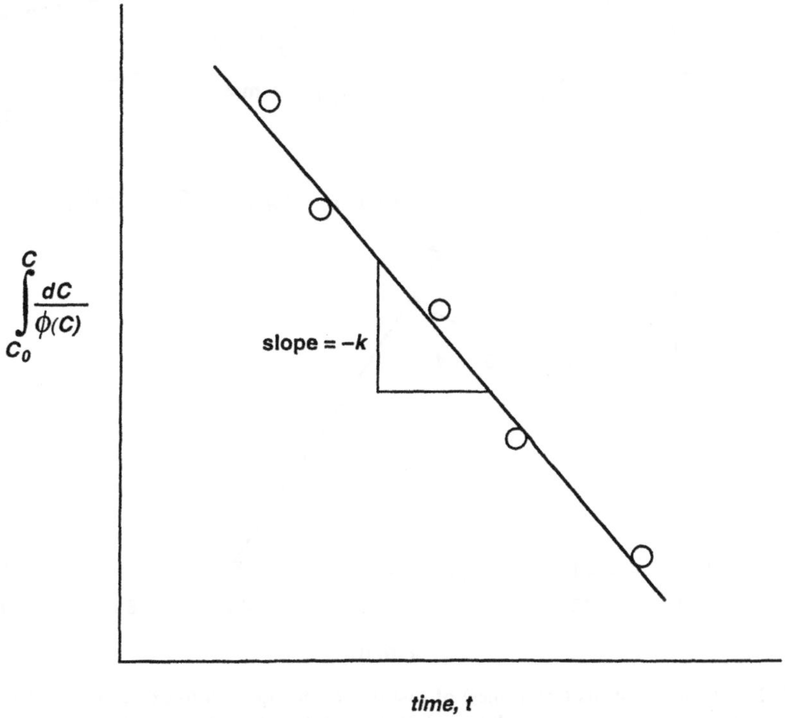

Figure 7-1 Integral method for rate data analysis.

disappearance of a reactant is first order. In this case Equation 7-45 becomes

$$kt = -\int_{C_0}^{C} \frac{dC}{C} = \ln \frac{C_0}{C} \tag{7-46}$$

or

$$\ln C = \ln C_0 - kt \tag{7-47}$$

If a plot of ln C (or log C) versus t for a given set of experimental data is found within experimental error to be linear, the objective order of the reaction may be taken as unity, and the rate constant may be calculated directly from the slope, $-k$, of the resulting trace. A plot of this type is illustrated in Figure 7-2 for the decomposition of ozone in buffered aqueous solutions of different carbonate concentration. In this example, the rate constant decreases with in-

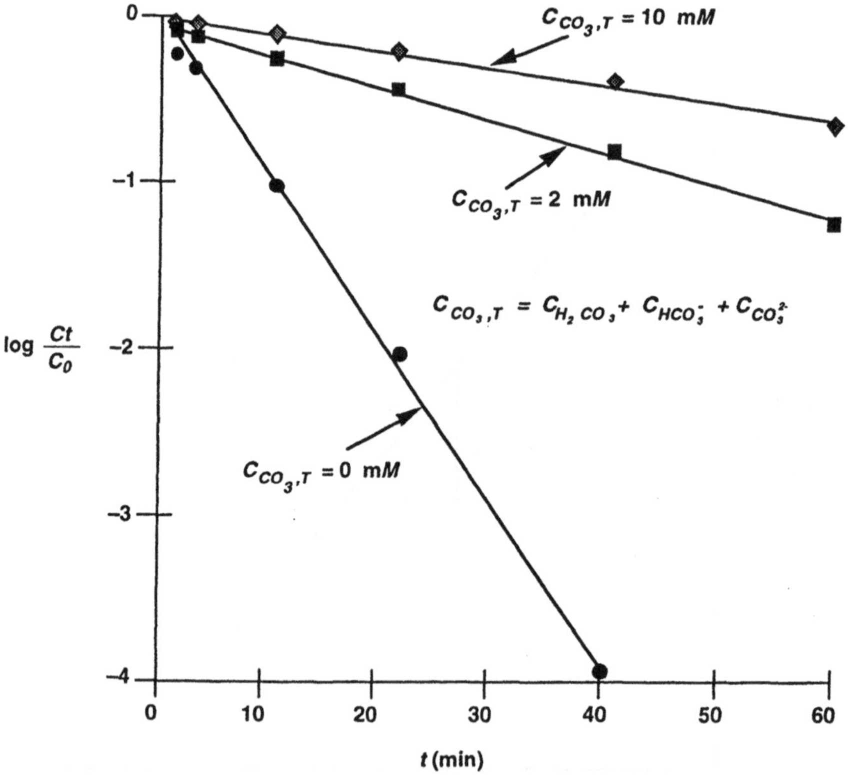

Figure 7-2 Statistical best fits (solid lines) of first-order rate equation to experimental data for decomposition of ozone in buffered solutions (pH 7) of different total carbonate concentration. [Adapted from Hull et al. (1992).]

creasing carbonate concentration because carbonate serves as a scavenger for hydroxyl radicals, which would otherwise promote ozone decomposition.

The first-order rate model may also be tested in terms of the amount reacted, $C_R = (C_0 - C)$, rather than C. Substitution of C_R for C in Equation 7-47 gives an expression that is identical in form to Equation 7-15:

$$kt = \ln \frac{C_0}{C_0 - C_R} \tag{7-48}$$

and a plot of $\ln(C_0 - C_R)$ or $\log(C_0 - C_R)$ versus t should therefore be linear.

Applications of first-order expressions for empirical description of certain rate processes (biochemical, chemical or physical) may require more than one parameter to be determined experimentally. The dissolution rate of a solid is one such situation, as illustrated in Example 7-2.

Example 7-2

- **Situation:** Polynuclear aromatic hydrocarbons (PAHs) are hydrophobic compounds with low water solubilities. Like PCBs, these compounds are commonly found associated with solids (e.g., soils, suspended solids, sludges). Their biodegradation is thus often limited by their desorption and solubilization. An experiment is devised to measure the rate of solubilization of phenanthrene, a representative PAH. Solid crystals of phenanthrene of a known surface area are placed in a flask and shaken. The data for dissolved concentration of phenanthrene with time are provided below [data from Grimberg et al. (1994)].

Time (min)	Concentration (μg/L)
0	0
21.5	235
56.5	508
169.5	950
299.5	1069
4298.5	1152

- **Question(s):** Postulate a rate expression to describe phenanthrene solubilization and to obtain the solubility limit.

- **Logic and Answer(s):**

1. The flask in which phenanthrene solubilization is measured can be considered a CMBR. One possible model for the rate of solubilization is that given in Equation 7-43, for which mass transfer is postulated as a pseudo-first-order reaction:

$$\frac{dC}{dt} = k_f(a_s^\circ)_R(C_S - C)$$

in which C is the concentration of phenanthrene in solution, C_S is its solubility limit, k_f is a mass transfer coefficient, and $(a_s^\circ)_R$ is the specific surface area of phenanthrene per unit volume of reactor.

2. The specific surface area of the phenanthrene is known to be 0.09 cm^{-1} initially. Although $(a_s^\circ)_R$ decreases with time, the solubility of phenanthrene is so small that this change can be considered insignificant over the period of observation. The integral method of rate data analysis then gives

$$\int_0^C \frac{dC}{C_S - C} = k_f(a_s^\circ)_R \int_0^t dt$$

or

$$-\ln \frac{C_S - C}{C_S} = k_f(a_s^\circ)_R t$$

Because both k_f and C_S are unknown, nonlinear regression analysis is most appropriate using the following form of the rate expression:

$$C = C_S\,(1 - e^{k_f(a_s^\circ)_R t})$$

3. The figure below shows the results of a nonlinear regression fit to the test data.

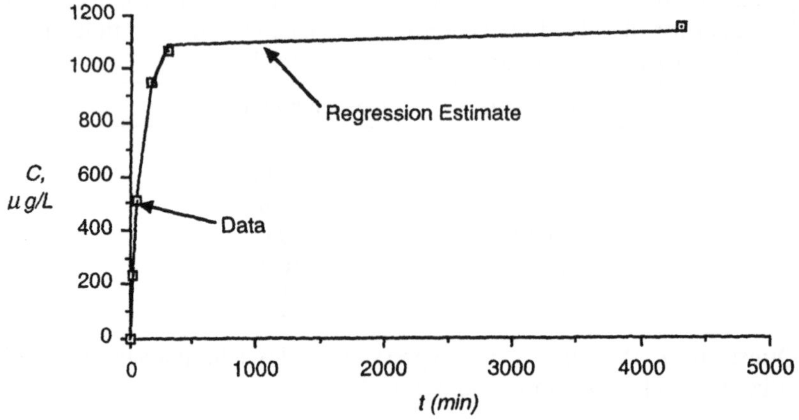

This nonlinear regression of the data yields values of C_S = 1.14 mg/L and k_f = 0.12 cm/min.

Observed concentration-time data for the product of a reaction may be used instead of that for the reactant. For example, it would be easier to measure a chlorinated organic product of a chlorine-substitution reaction than its unchlorinated precursor. Suppose that A reacts to form B. The rate of formation of B is then given by

$$r_B = kC_A \tag{7-49}$$

and application of the material balance relationship yields

$$\frac{dC_B}{dt} = kC_A \tag{7-50}$$

If the stoichiometry of the reaction specifies that 1 mole of B is formed for every mole of A reacted, then C_A can be replaced by $C_{A,0} - C_B$. Equation 7-50 can then be integrated to give

$$kt = -\ln \frac{C_{A,0} - C_B}{C_{A,0}} \tag{7-51}$$

or

$$\ln (C_{A,0} - C_B) = \ln C_{A,0} - kt \tag{7-52}$$

Because $C_A = C_{A,0} - C_B$, this equation is identical to Equation 7-47. We have shown therefore that the rate of formation of B is the same as the rate of disappearance of A. This must, in fact, be true to satisfy the stoichiometry of the reaction. Other stoichiometries will yield different relationships between the two rates. We can test the suitability of the first-order rate expression to describe the data by plotting $\ln(C_{A,0} - C_B)$ with time; the result should be a straight line with slope equal to the rate constant, k.

The rate expression for an elementary second-order reaction rate involving A and B was given by Equation 7-2. This can likewise be substituted into Equation 7-45 and integrated. Although we have introduced the formalism of rate data analysis by way of the *integral method*, the mathematical development is the same as that leading from Equation 7-6 to Equations 7-8 and 7-9. Example 7-3 illustrates an application of the second-order model when the rate of a complex reaction can be treated as first order in each of two principal components.

Example 7-3

- **Situation:** Ozone enjoys widespread use as a disinfectant and oxidant in drinking water treatment. Because ozone decomposes rapidly, however, it is generally necessary to add a secondary disinfectant, such as chlorine, to ensure residual disinfection after water leaves a treatment plant. Health concerns have been raised about some of the by-products formed when ozone reacts with natural organic matter: more specifically, about the potential further reactions of these by-products with chlorine added for residual protection. As director of laboratories for the regional water supply system discussed in Example 7-1, you conduct a series of tests which show that chlorine can react with acetaldehyde (AcA) produced by ozonation to form other disinfection by-products, pursuant to the general reaction scheme

$$\gamma Cl_2 + AcA \longrightarrow \text{products}$$

You suspect that the reaction pathway may involve an oxidation step followed by a series of chlorine substitution reactions leading to trichloroacetaldehyde (TCAcA), which would consume 3 mole of free chlorine per mole of acetaldehyde (i.e., $\gamma = 3$). A batch rate experiment with acetaldehyde in excess shows, however, that the reaction is first order in chlorine, which suggests that it is probably nonelementary (if it were elementary, the reaction should be third order with respect to chlorine, given the presumed stoichiometry for trichloroacetaldehyde formation). Another experiment is designed so that the initial concentrations of chlorine and acetaldehyde are in the stoichiometric ratio of 3 to 1. Measurements of free chlorine residual FCR, (i.e., not combined with products of acetaldehyde chlorination) with time are given below [data from McKnight and Reckow (1992)].

Time (h)	FCR Concentration (mg/L)
0.1	3.6
13	3.3
29	2.7
59	2.4
72	2.1

- **Question(s):** From the CMBR rate data provided, investigate the possibility of the reaction between chlorine and acetaldehyde being first order in each reactant and therefore second order overall.
- **Logic and Answer(s):** Assume from the hypothesis of trichloroacetaldehyde formation that

$$3Cl_2 + AcA \longrightarrow TCAcA$$

1. Letting C_R be the moles of chlorine reacted per liter, we postulate that

$$\frac{dC_R}{dt} = k(C_{Cl_2,0} - C_R)(C_{AcA,0} - C_R/3)$$

The CMBR experiment was initiated with

$$C_{Cl_2,0} = 3C_{AcA,0}$$

so that

$$\frac{dC_R}{dt} = \frac{k}{3}(C_{Cl_2,0} - C_R)(C_{Cl_2,0} - C_R)$$

2. We note that $C_R = C_{Cl_2,0} - C_{Cl_2}$, where C_{Cl_2} is the free chlorine residual remaining at time t. Thus the rate expression above can be written in the simpler form

$$-\frac{dC_{FCR}}{dt} = k'C_{FCR}^2$$

where $k' = k/3$. The rate data are now in a suitable form for testing by the integral method:

$$\int_{C_{FCR,0}}^{C_{FCR}} \frac{dC_{FCR}}{C_{FCR}^2} = -k' \int_0^t dt$$

$$\frac{1}{C_{FCR}} - \frac{1}{C_{FCR,0}} = k't$$

A plot of the reciprocal of C_{FCR} against time should yield a straight line from which the slope determines the rate constant, $k' = k/3$. As shown in the figure below, there is reasonably good agreement with the proposed second-order model.

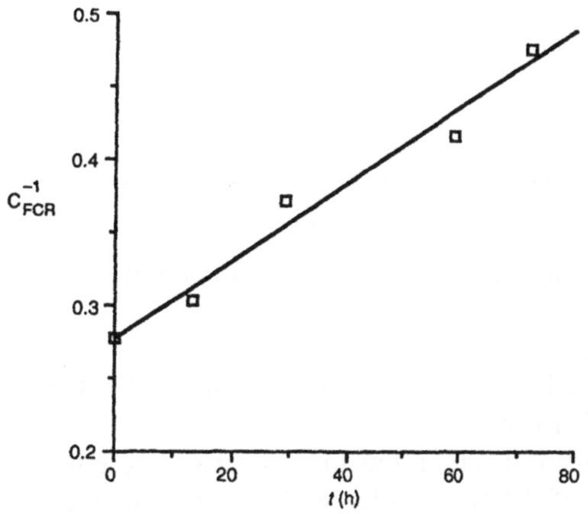

Equation 7-9 is equally applicable to a reaction that is second order with respect to the concentration of one reactant rather than first order with respect to each of two reactants. For testing of data in which we postulate the rate of disappearance of one reactant as second order, it is convenient to reexpress Equation 7-9:

$$kt = \frac{C_R}{C_0(C_0 - C_R)} = \frac{C_0 - C}{C_0 C} = \frac{1}{C} - \frac{1}{C_0} \tag{7-53}$$

Accordingly, the experimental data should be linearized by plotting C^{-1} versus t, and the slope of this plot yields the value of the rate constant, k. Figure 7-3 is such a plot.

The integral method applies equally well even when the assumed rate expression cannot be integrated analytically. In this instance, the right-hand

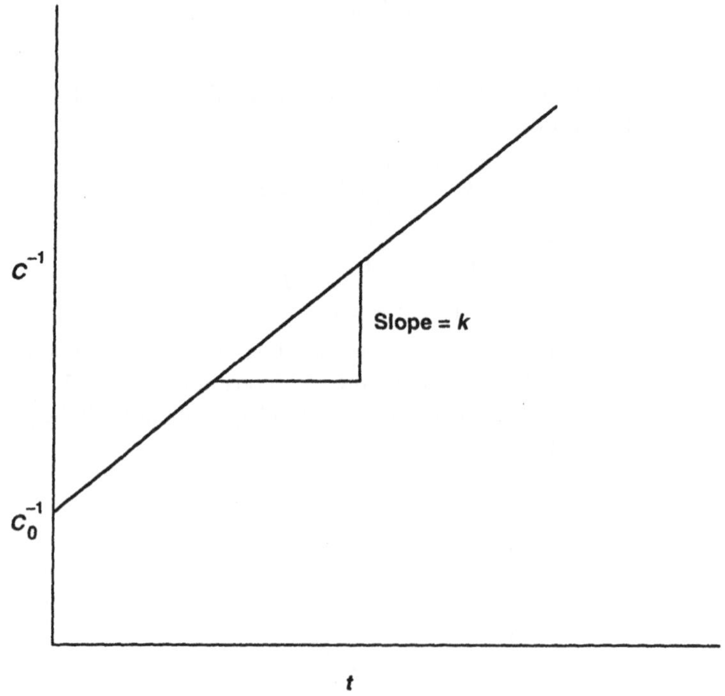

Figure 7-3 Linearization of a rate equation for a reaction that is second order with respect to one reactant.

side of Equation 7-45 can be integrated numerically from C_0 to any experimental value of C at time, t.

Rate expressions and integrated material balance equations in CMBRs for forward reactions of different order with respect to one reactant are summarized in Table 7-1. The first entry in this table is a zero-order reaction (i.e., a reaction in which the rate of change is independent of the concentration of the reactant). A zero-order reaction is unique in that the reaction rate is given simply by the rate constant, which in turn is given simply by the concentration reacted, C_R, over a given time ($ML^{-3}t^{-1}$). For other reaction orders, the dimensions of the rate constant vary and it is easy to show that they generalize to (concentration)$^{1-n}$ per unit time [$(ML^{-3})^{1-n}t^{-1}$]. The rate constant is dependent on the particular combination of reactants and products (reacted material C_R) and of course on the chemical nature of the reactants themselves, but it is independent of the initial concentration of reactant (to prove this, substitute $C_0 = C + C_R$ in the rate constant relationships given in Table 7-1). The rate constant also depends on temperature as is discussed in Section 7.4.

The integral method is also suitable for analysis of more complex reactions for which empirical rate expressions may be sought. Example 7-4 provides a practical illustration of both the method and its limitations for a situation involving an investigation of the reaction of chlorine with natural organic matter (NOM).

Table 7-1 Rate Relationships for Reactions of Different Order with Respect to One Reactant

Forward Reaction Rate, r	Rate Constant, k	Reaction Order, n
k	$\frac{C_R}{t}$	0
$kC^{0.5} = k(C_0 - C_R)^{0.5}$	$\frac{2}{t}[C_0^{0.5} - (C_0 - C_R)^{0.5}]$	0.5
$kC = k(C_0 - C_R)$	$\frac{1}{t}\ln\frac{C_0}{C_0 - C_R}$	1
$kC^2 = k(C_0 - C_R)^2$	$\frac{C_R}{C_0 t(C_0 - C_R)}$	2
$kC^3 = k(C_0 - C_R)^3$	$\frac{2C_0C_R - C_R^2}{2C_0^2 t(C_0 - C_R)^2}$	3

Example 7-4

- **Situation:** It was evident in Example 7-1 that TTHM production is a complex phenomenon that depends on many system variables. In the empirical rate model approach given in that example, no attempt was made to describe the kinetics in a fundamental way; that is, by postulating stoichiometry and investigating the dependence of the rate of reaction on reactant concentrations. Instead, reaction time was included in the regression analysis as an independent variable.

- **Question(s):** Can reactions between chlorine and natural organic matter be described by empirical rate models deriving from investigations of reaction rate as a function of reaction time?

- **Logic and Answer(s):** To address this question, you conduct a test on one of the natural raw water sources for your system.

  1. From this test you find that the overall stoichiometry (all species in molar units) can be described by the relationship

$$3\text{HOCl} + \text{TOC} \xrightarrow{k_n} \text{TTHM}$$

  where HOCl is hypochlorous acid; TOC is total organic carbon, a surrogate parameter for NOM; and TTHM is total trihalomethanes.

  2. The following empirical rate model can be postulated assuming a fixed temperature and pH:

$$\frac{dC_{\text{TTHM}}}{dt} = k_n C_{\text{TOC}} C^m_{\text{HOCl}}$$

3. Two further assumptions are reasonable: (1) that the TOC concentration remains constant even though a reaction occurs (i.e., NOM is present in excess of the stoichiometric amount needed to react with chlorine); and (2) as noted in Equation 7-37, other chlorinated products form in reactions of chlorine with NOM, and therefore the number of moles of TTHM produced per mole of HOCl consumed must be known. This is referred to as the *fractional yield*, ϕ_y° (typical values for TTHM yield range from 0.05 to 0.1). With these two assumptions, the proposed rate expression becomes

$$\frac{dC_{TTHM}}{dt} = k'\left[C_{HOCl,0} - \frac{3C_{TTHM}}{\phi_y^\circ}\right]^m$$

in which $C_{HOCl,0}$ is the initial concentration of hypochlorous acid and $k' = k_n(\text{TOC})$.

4. The rate expression is now in a form that can be integrated easily. For example, if $m = 3$, the result is

$$\frac{1}{\left(C_{HOCl,0} - \dfrac{3C_{TTHM}}{\phi_y^\circ}\right)^2} = \frac{6k't}{\phi_y^\circ} + \frac{1}{C_{HOCl,0}^2}$$

5. A test of this rate model is presented in the figure below, wherein the left-hand side of the rate equation is plotted against time, t [data from Kavanaugh et al. (1980)].

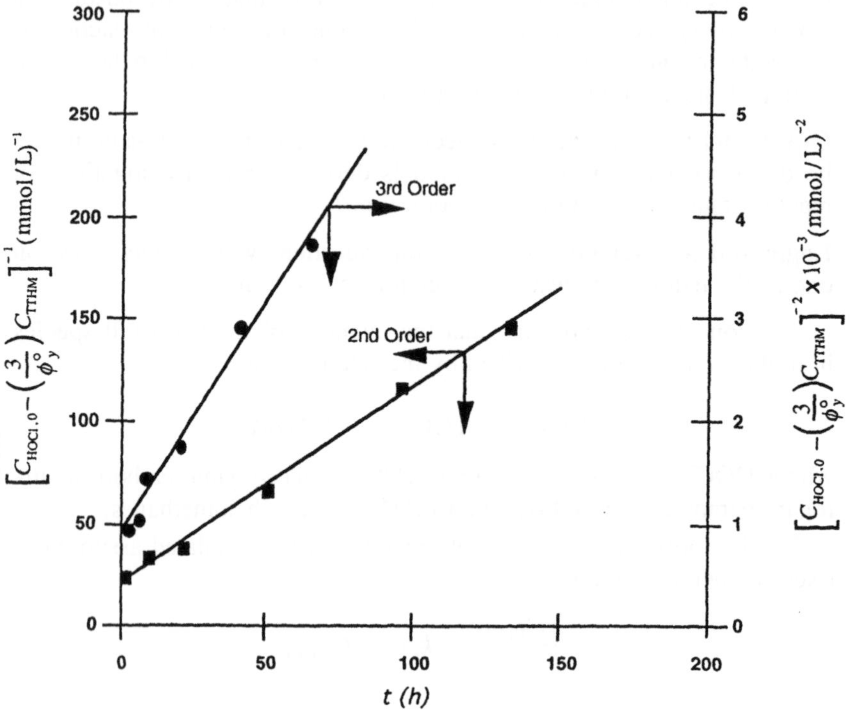

The resulting straight line suggests a fit to the proposed kinetic expression and the rate constant, k', can be determined from the slope. Although encouraging, this figure also shows that an equally good fit was obtained using a second-order dependence on HOCl. Such a dilemma is not unexpected when empirical rate expressions are postulated.

7.3.1.2 Differential Method The *differential method* of analysis involves approximating reaction rates from experimental measurements of changes in concentration over small increments in time. This method is especially convenient for complex rate expressions for which the integral method is cumbersome because numerical integration must be used, or where a rapid screening tool for a rate expression is sought. Values for the reaction rate, $\mathcal{r}$, which is given by dC/dt (or dC_R/dt), can be determined for different stages of a reaction as the slopes or tangents of a plot of C (or C_R) versus t. This procedure is depicted in Figure 7-4a. The amount and quality of the rate data are critical to a close estimate of dC/dt by $\Delta C/\Delta t$.

If the reaction rate expression is of the general form $\mathcal{r} = \phi(k, C)$ and k is a constant for the experimental conditions, the differential method gives

$$\frac{\Delta C}{\Delta t} \cong \frac{dC}{dt} = k\phi(C) \tag{7-54}$$

The values for $\Delta C/\Delta t$ (or $\Delta C_R/\Delta t$) are then plotted versus $\phi(C)$ to determine the validity of the rate model proposed. A first-order expression is confirmed if, as shown in Figure 7-4b, $\Delta C/\Delta t$ (or $\Delta C_R/\Delta t$) plots linearly against C. The slope of this plot is equal to the first-order rate constant, k. Several useful applications of this method will be illustrated in subsequent sections of this chapter.

7.3.2 Alternative Experimental Methods

7.3.2.1 Initial Rate Method The initial rate method is an experimental variation on the differential rate method of analysis. The objective is to estimate the reaction rate that obtains at the initial concentration in the reaction vessel. Data to test the proposed rate expression are obtained by measuring the initial reaction rate in a series of experiments at different initial concentrations. This is equivalent to measuring the rate at various stages of reaction in a single experiment. The data are described by

$$\left(\frac{\Delta C}{\Delta t}\right)_0 \cong \left(\frac{dC}{dt}\right)_0 = k\phi(C_0) \tag{7-55}$$

where $(\Delta C/\Delta t)_0$ is the experimental approximation of the initial rate; the expression is otherwise identical to Equation 7-54.

An inherent assumption associated with using the initial rate method is that

(a) Analysis of CMBR Rate Data

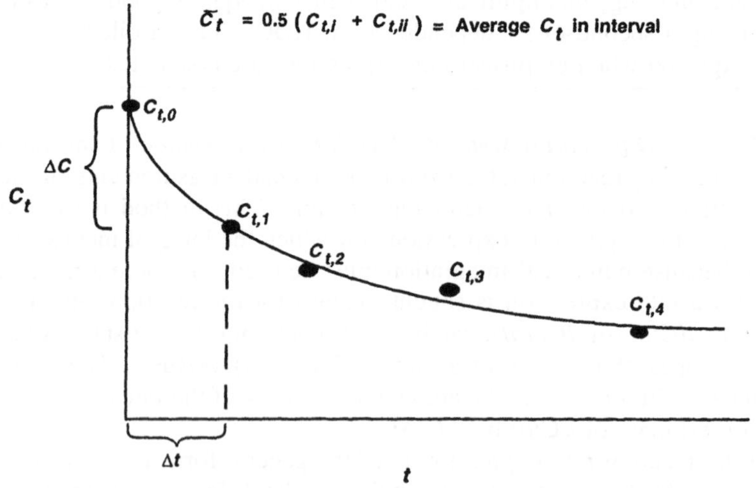

(b) Confirmation of First-Order Reaction Rate and Evaluation of Rate Coefficient

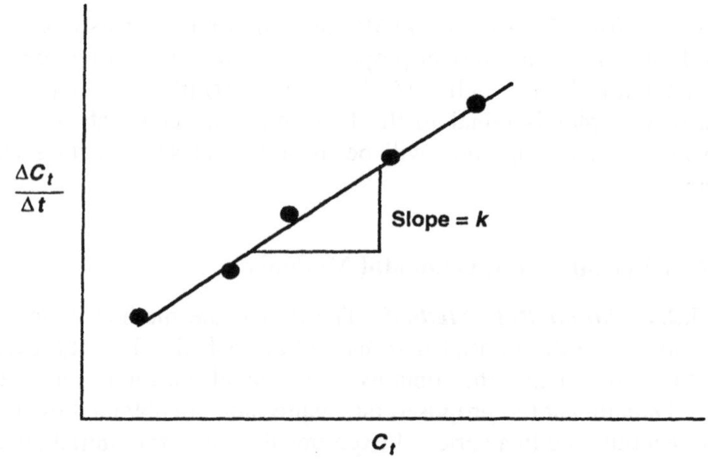

Figure 7-4 Differential method for rate data analysis.

ΔC is small enough to let $C = C_0$ over the entire period, Δt. An alternative form of this method involves measurement of the initial rate of product formation instead of reactant consumption. As long as the reaction stoichiometry is known, product analysis may be preferred if the product is initially present in low concentration relative to the reactant. In such cases, a significant change in concentration is measured, in contrast to measurement of a small change in the much larger concentration of reactant.

7.3.2.2 Half-Life Method As the name implies, the *half-life method* involves experimental determination of a rate expression by analysis of the time required to decrease the reactant concentration by 50%. This method is applicable to any reaction of the general form

$$-r_A = kC^n \tag{7-56}$$

The time to obtain a 50% reduction in C_0 is given by the integral method as

$$k\int_0^{t_{0.5}} dt = -\int_{C_0}^{0.5C_0} \frac{dC}{C^n} \tag{7-57}$$

Upon integration we obtain

$$\text{For } n = 1\text{: } t_{0.5} = \frac{\ln 2}{k} \tag{7-58}$$

$$\text{For } n \neq 1\text{: } \ln t_{0.5} = \ln \frac{2^{n-1} - 1}{(n-1)k} + (1 - n) \ln C_0 \tag{7-59}$$

We conclude from Equation 7-58 that for a first-order reaction rate ($n = 1$), the time required for the initial concentration to decrease by 50% is (1) independent of the initial concentration, and (2) a constant multiple of the first-order rate constant, k. The rate constant for such a reaction can thus be expressed in terms of its corresponding half-life. This is an important characteristic of first-order reactions. In other words, the time for the concentration to fall from C_0 to $0.5C_0$ is the same as the time for the concentration to fall from $0.5C_0$ to $0.25C_0$, and so on. The decay of a radioisotope is an example of this type of *exponential* process rate.

As shown by Equation 7-59, the half-life for an nth-order reaction other than $n = 1$ depends on the initial concentration. To test the suitability of this method for description of the rate data for any particular system, we need to conduct a series of half-life experiments in which the initial concentration is varied. If the resulting data are suitably described by this form of rate expression, a plot of $\ln t_{0.5}$ against $\ln C_0$ will be linear with a slope of $(1 - n)$, as illustrated in Figure 7-5. Although we cannot use Equation 7-59 to test a first-order reaction, the same series of experiments will produce a slope of zero because the half-life is independent of initial concentration.

7.3.2.3 Method of Excess This is an experimental technique designed and used to simplify the analysis of rates for reactions that involve two or more reactants. In this method, the concentrations of all but one reactant in the test vessel are maintained far in excess of stoichiometric requirements, and thus

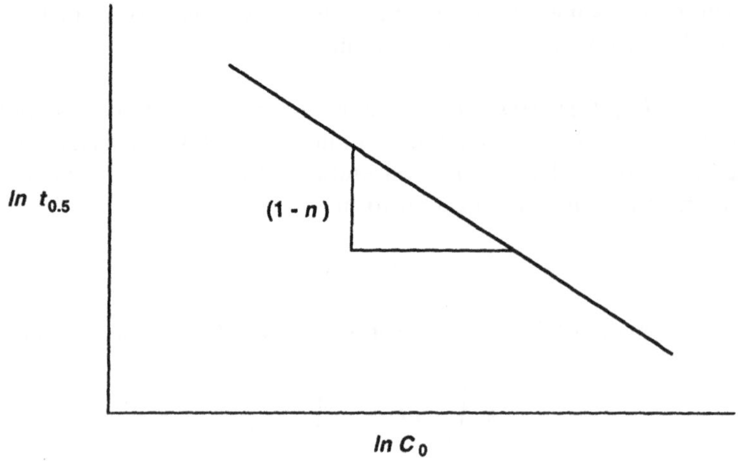

Figure 7-5 Evaluation of reaction rate order using the half-life method.

remain approximately constant over the course of the reaction. If the reaction rate expression is of the form

$$r_B = -kC_A^n C_B^m \tag{7-60}$$

and A is present in excess, then

$$r_B = -k_{\text{obs}} C_B^m \tag{7-61}$$

$$k_{\text{obs}} = kC_A^n \tag{7-62}$$

In other words, the effect of A on the reaction rate is accounted for in the *observed* reaction rate constant, and the reaction rate is *pseudo* mth order with respect to B. The order of reaction with respect to B can be determined using the integral or differential method. The dependency of rate on C_A in Equation 7-62 can be determined by conducting a series of experiments in which C_A is varied, but always kept far in excess of the stoichiometric amount required. The *intrinsic* rate constant, k, and the order of the reaction with respect to A can be determined from the logarithmic transform of Equation 7-62:

$$\ln k_{\text{obs}} = \ln k + n \ln C_A \tag{7-63}$$

A plot of $\ln k_{\text{obs}}$ against $\ln C_A$ should produce a straight line from which both n and k can then be estimated.

The method of excess is most appropriate for development of initial rate data because the reactant(s) which are added in excess remain so over the duration of the experiment. However, if more than initial rate data are used, either the integral or differential methods of analysis can be applied as long as the reactant remains in excess throughout the time of data collection.

Example 7-5 provides a practical illustration of how the method of excess can be incorporated in the overall rate assessment and model development for a complex reaction scheme.

Example 7-5

- **Situation:** In order to minimize waste generation, a new synthesizing reagent, A, is being considered for production of an organic chemical. Because it is used more efficiently in the synthesis process than the current reagent, less carries over to the wastewater treatment facility. However, pilot tests of the synthesis have shown that A also reacts slowly with an impurity, I, in the chemical stock feed to form a side product, P. This side product, although present in small concentrations, is of concern because it may require an additional waste treatment step. No information is available on the stoichiometry of the reaction. Two series of laboratory flasks with stir bars were used to measure concentrations of P with time. In series 1, the initial concentration of I was kept the same and the initial concentration of A was varied. In series 2, the reverse was true. The concentrations of both I and A did not change significantly during the first 12 h of either test series, but trace amounts of P were detected and measured accurately. The data are given below.

Rate Data for Series 1 $C_{I,0} = 2.0$ mmol/L		Rate Data for Series 2 $C_{A,0} = 240$ mmol/L	
C_P Formed in First 12 h (mmol/L × 10^3)	C_A Added Initially (mmol/L)	C_P Formed in First 12 h (mmol/L × 10^3)	C_I Added Initially (mmol/L × 10^3)
11.0	64.2	330	2500
40.9	139	250	1920
95.0	210	164	1360
106.0	204	109	1000
138.0	263	54.6	450

- **Question(s):** Use the rate data to determine reaction orders with respect to the reactant and the impurity as well as the intrinsic rate constant.

- **Logic and Answer(s):**

1. With no information available on stoichiometry, a general rate model of the following form must be tested:

$$\frac{dC_P}{dt} = kC_I^n C_A^m$$

2. The stirred laboratory flasks were sealed and well mixed and thus may be considered to have behaved as CMBRs. Because the concentra-

tions of I and A did not decrease significantly during the 12-h test, the initial rate method of analysis is appropriate for this very slow reaction. Thus, the rate of product formation is

$$\frac{\Delta C_P}{12} \cong \frac{dC_P}{dt}$$

3. If the proposed rate model is valid, the data from series 1 should plot linearly according to

$$\ln \frac{\Delta C_P}{\Delta t} = \ln(k_{obs})_1 + m \ln C_{A,0}$$

where

$$(k_{obs})_1 = kC_{I,0}^n$$

4. The graph below shows a linear plot with a slope of $m = 1.85$. Given the small amount of data, an integral order of 2 for the reaction with respect to A is assumed. The intercept value gives

$$(k_{obs})_1 = e^{-14.7} = 4.13 \times 10^{-7} \text{ L/mmol-h}$$

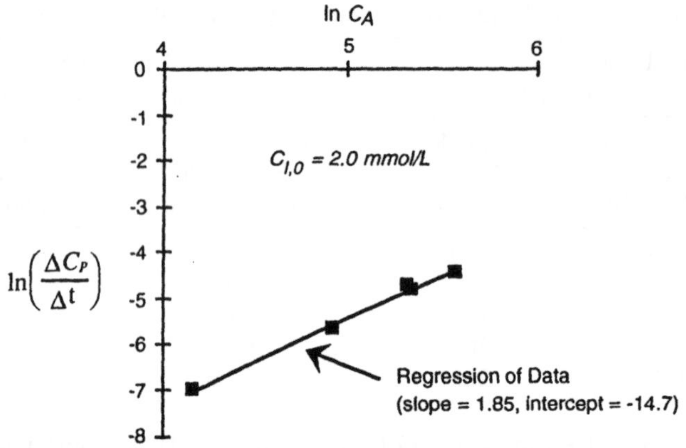

5. The proposed rate model can also be tested to determine the order of the reaction with respect to I from the data in series 2:

$$\ln\left(\frac{\Delta C_P}{\Delta t}\right) = \ln(k_{obs})_2 + n \ln(C_{I,0})$$

where

$$(k_{obs})_2 = kC_{A,0}^m$$

6. The resulting fit to the model is shown below

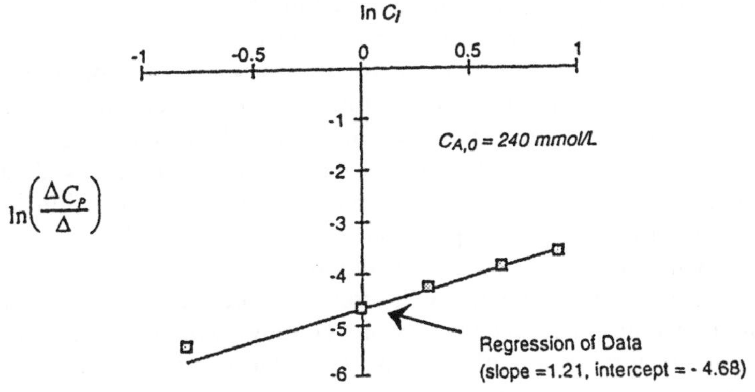

The slope of the plot is $n = 1.2$, which suggests an integer first-order dependency on I. The intercept value then gives

$$(k_{obs})_2 = e^{-4.68} = 9.28 \times 10^{-3}\ \text{h}^{-1}$$

7. The intrinsic rate constant, k, should be the same for the data obtained in both series.

$$k|_{\text{series 1}} = \frac{(k_{obs})_1}{C_{I,0}} = \frac{4.13 \times 10^{-7}}{2.0} = 2.07 \times 10^{-7}\ \text{L}^2/\text{mmol}^2\text{-h}$$

$$k|_{\text{series 2}} = \frac{(k_{obs})_2}{C_{A,0}^2} = \frac{9.28 \times 10^{-3}}{(240)^2} = 1.61 \times 10^{-7}\ \text{L}^2/\text{mmol}^2\text{-h}$$

Thus, values of the intrinsic rate constant are similar for both series. In summary, then, the following rate relationship is suggested:

$$\frac{dC_P}{dt} = kC_I C_A^2$$

- **Observation(s):** Given that the data were limited and that the agreement with the assumed elementary reaction was not exact, more rate data should be obtained before accepting the rate relationship above. To add further validity, series 1 could be repeated with a different $C_{I,0}$ and series 2 with a different $C_{A,0}$. These data would yield different values of k_{obs} in proportion to the values of $C_{I,0}$ and $C_{A,0}$ that were used.

7.3.3 Precautions in Choosing Rate Expressions

The suitability of any rate expression for description of a particular set of data is determined by calculating the value of k for a set of experimental values of

C (or C_R) and t. For example, a first-order rate expression is tested in the integral method by plotting $\ln C$ versus t and a second-order expression by plotting C^{-1} versus t. If the data are successfully linearized so that, within the limits of experimental error, each combination of C and t within the set is accounted for by the calculated value for k, a new set of experimental data should be obtained for the same temperature but for different initial concentration(s), C_0. A test of the rate expression for the new data should then be carried out and a second value for k obtained. If the second value of k is identical to that calculated from the first set of experimental data (again within limits of experimental error), the rate expression is applicable for description of the particular chemical reaction under consideration.

Experimental errors in rate data will often introduce some uncertainty in the selection of an appropriate rate expression. To judge whether the data conform to the linearized form of a particular rate expression, it is important to distinguish between *experimental error* and *systematic error*. When data are tightly but randomly scattered about the linear trace given by a normalized rate expression, experimental error is likely, but if the data consistently diverge from a linear trace as time increases, systematic error due to an inappropriate choice of rate model is suggested.

When a given set of data are only approximately normalized by the linear form of a rate model, alternative expressions should be tested to determine whether better agreement can be obtained. Rate models corresponding to simple elementary reactions (typically, first or second order) are the logical first choices for testing. However, we must also recognize that nonelementary reactions are possible, in which case trial-and-error evaluations of alternative rate expressions may be necessary.

7.4 TEMPERATURE AND ACTIVATION ENERGY

Analysis of the temperature dependence of reaction rate data can provide information necessary for process modeling and design. It may also be useful for interpretation of reaction mechanisms and identification of rate-limiting conditions in complex processes.

Chemical reaction rates generally increase monotonically with increasing temperature, in a manner similar to that illustrated by curve a in Figure 7-6. Rates that follow curve a can be described by the Arrhenius equation. Several important types of reactions in natural and engineered environmental systems are known to exhibit temperature dependence different from the Arrhenius type, however. Among those are certain oxidations which behave in the manner illustrated by curve b in Figure 7-6, and enzyme reactions, which have a temperature dependence similar to that illustrated by curve c. The Arrhenius equation in differential form is

$$\frac{d \ln k}{dT} = \frac{E_a}{\mathcal{R}T^2} \tag{7-64}$$

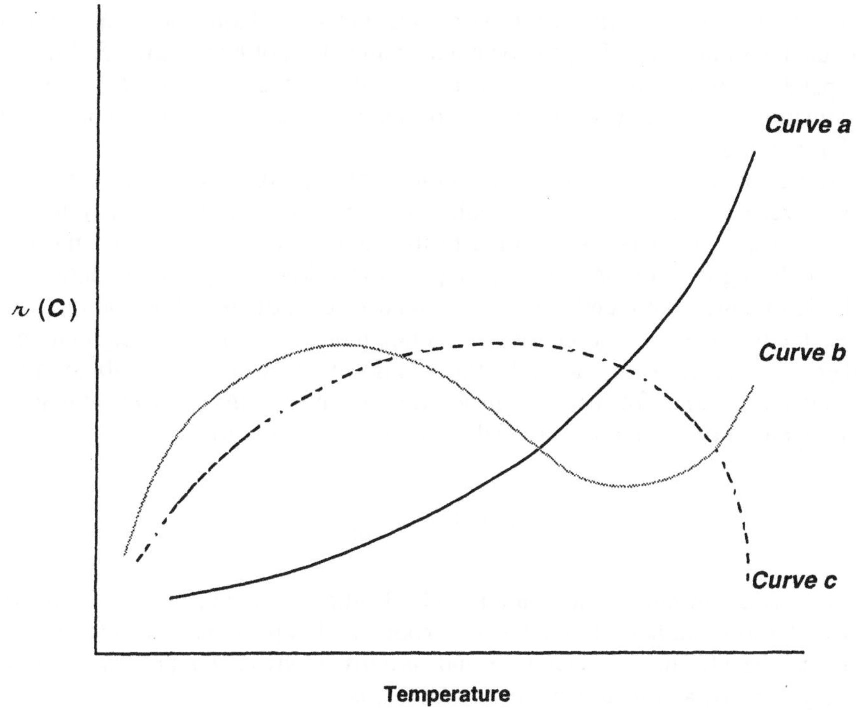

Figure 7-6 Different types of reaction rate dependence on temperature.

where k is the reaction rate coefficient, E_a the energy of activation, $\mathcal{R}$ the universal gas constant, and T the absolute temperature. Equation 7-64 has the same form as the van't Hoff equation (Equation 5-103) for expression of the variation of the equilibrium constant, K, with temperature. In the Arrhenius equation, the activation energy, E_a, replaces the standard-state reaction enthalpy term, dH_r°, of the van't Hoff equation. The Arrhenius equation can be justified by a probability analysis of molecular collisions that result in reaction (see Figure 7-8 and related discussion).

The activation energy, E_a, generally expressed in units of calories or joules per mole, is a characteristic constant for a particular reaction and is thus useful for identification of individual reaction steps that control overall rates in complex processes. The magnitude of the activation energy for a complex reaction involving several stages or steps is commonly determined by the slowest step. Thus an experimental evaluation of E_a for a complicated process can provide indication of the nature of the rate-limiting reaction. For example, four values of E_a occur commonly in complex biological processes: namely, 70, 67, 50, and 33 kJ/mol. These activation energies correspond respectively to rate-limiting reactions involving (1) oxidative dehydrogenation, (2) oxidative reactions catalyzed by iron, (3) OH^--catalyzed oxidative reactions, and (4) oxidative hydrolysis.

Given the similar form of the Arrhenius and van't Hoff equations, experimental determinations of E_a can be made from values of k measured at different temperatures in exactly the same manner as illustrated in Figure 5-13 for determination of enthalpy from measured values of the equilibrium constant at different temperatures.

Activation energy is a direct determinant of reaction rate; the larger is E_a, the slower is the reaction. For simple homogeneous gas-phase reactions, the energy of activation is proportional to the strength of the bond that must be broken during the rate-determining step of a reaction. Air-phase oxidations of volatile organic compounds by oxygen tend to be sluggish, for example, because the breaking of oxygen–oxygen bonds requires a relatively high energy. Other energetic factors (e.g., solvation energy) may strongly contribute to the energy of activation for reactions in solution, even for homogeneous reactions.

Integration of Equation 7-64 yields the indefinite integral

$$\ln k = \ln \kappa - \frac{E_a}{\mathcal{R}T} \tag{7-65}$$

where κ is a constant of integration and all other terms are as defined previously. A form that both illustrates the exponential dependence of the rate constant on the absolute temperature and assigns a physical significance to κ, namely $\kappa = A_f$, a *collision frequency* factor, is

$$k = A_f e^{-E_a/\mathcal{R}T} \tag{7-66}$$

The relationship above provides further insight to reaction rate mechanisms and a basis for a conceptual definition of activation energy. Reactions between molecules can occur only when they are in contact with one another. Reaction rates should thus be expected to relate to the frequency of contacts between molecules of reacting species, which in turn relates to the *kinetic energies* of these molecules. The number of such molecular contacts is extremely large in most systems, values of A_f typically being on the order of 10^{15} L/mol-s for a gas at 1 atm pressure, for example. If collisions were all that were required for reactions to occur, nearly all reactions would take place at practically instantaneous rates. The fact that most chemical reactions proceed at finite and measurable rates suggests that only a fraction of the collisions which occur lead directly to reaction; these may be termed *reactive collisions*.

A reactive collision is one that involves molecules possessing an energy of sufficient magnitude to permit reaction to occur. Only a fraction of the molecules in any particular system possess sufficient energy to accomplish a reactive collision. Such molecules are said to be in a reactive or *activated* state, and the energy required to achieve this state is termed the *activation energy*. Activation energy is not to be confused with the equilibrium energy states of reactants and products. The differences are illustrated graphically in Figure 7-7 for both exothermic (ΔH_r *negative*) and endothermic (ΔH_r *positive*) reactions

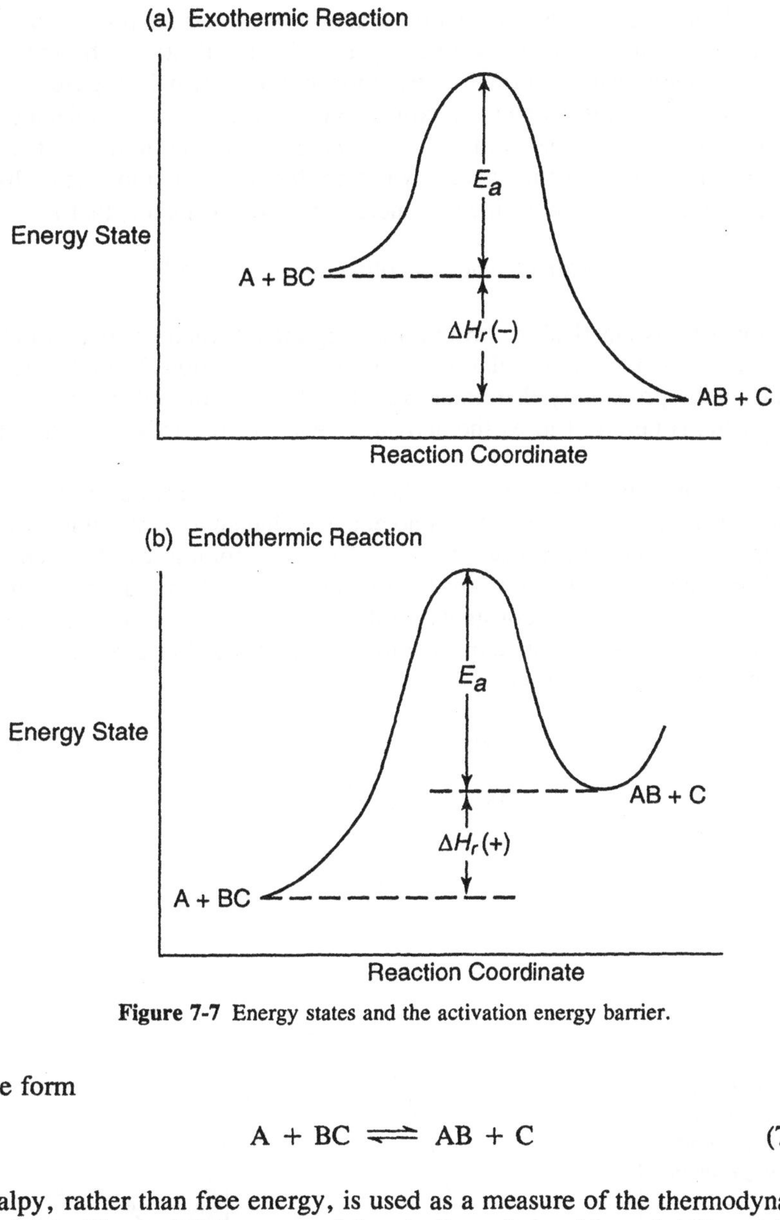

Figure 7-7 Energy states and the activation energy barrier.

of the form

$$A + BC \rightleftharpoons AB + C \tag{7-67}$$

Enthalpy, rather than free energy, is used as a measure of the thermodynamic potential in Figure 7-7 because of the similar relationships of ΔH_r and E_a to temperature. As shown in this figure, E_a is a potential energy barrier which must be overcome for a particular reaction to occur, whether the reaction is thermodynamically favorable or unfavorable. The higher the magnitude of this barrier, the slower is the reaction rate, because the lower is the fraction of molecules in a system possessing an energy E_a *in excess of the average energy* of the system.

Probability analysis reveals that the fraction of molecules possessing an energy in excess of a value E_a is given by $e^{-E_a/\Re T}$, as illustrated by the shaded area in the Gaussian or normal distribution curve shown in Figure 7-8. The quantity $e^{-E_a/\Re T}$ is the *Boltzmann* expression of the fraction of molecules having an energy E_a in excess of the average energy. The total number of reactive collisions in a system is then given by the product of the number of collisions, A_f, and the number of molecules of energy state E_a or higher; that is,

$$\text{number of reactive collisions} = A_f e^{-E_a/\Re T} \tag{7-68}$$

Equation 7-68 is identical in form to the integrated Arrhenius equation (Equation 7-66). For monomolecular reactions (e.g., Equation 7-10) the term A_f represents the probability that an activated molecule will undergo decomposition (reaction) before it loses the activation energy, E_a, to some other molecule.

One fundamental shortcoming of the collision theory for description of reaction mechanisms is that it assumes no entropy change for reversible reactions at temperatures other than absolute zero. This shortcoming can be seen in interpreting the equilibrium constant (and its associated thermodynamic properties) as the ratio of the rate constants for the forward and reverse reaction rates. We first write the Arrhenius expressions for the forward and reverse components of a reversible reaction

$$(k)_f = A_{f,f} e^{-E_{a,f}/\Re T} \tag{7-69}$$

$$(k)_r = A_{f,r} e^{-E_{a,r}/\Re T} \tag{7-70}$$

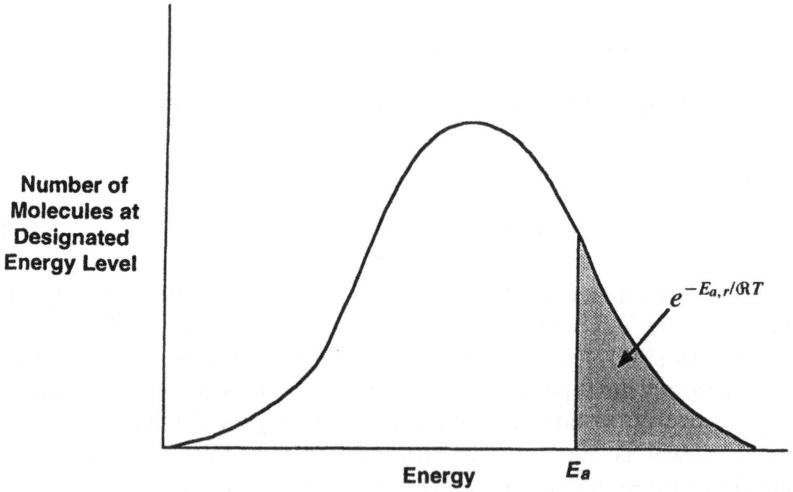

Figure 7-8 Energy distribution curve.

and then, the equilibrium constant, K, is

$$K = \frac{(k)_f}{(k)_r} = \frac{A_{f,f}}{A_{f,r}} e^{-(E_{a,f} - E_{a,r})/\mathcal{R}T} \tag{7-71}$$

It is reasonable to assume that the frequency factors for the forward and reverse reactions must be nearly the same, and that the ratio $A_{f,f}/A_{f,r}$ must therefore be approximately equal to unity. We observe that Equation 7-71 has the form of the van't Hoff equation in which case the difference between the forward and reverse activation energies, $(E_{a,f} - E_{a,r})$, becomes equivalent to the enthalpy difference between the products and the reactants. This can be true, however, only for *absolute zero* temperature or for a reaction involving *no entropy change*. Such reasoning leads us to conclude that the experimental activation energy calculated from Equation 7-66 is to some extent variable with temperature, and that the frequency factor, A_f, involves an entropy term.

Without knowing the value of A_f, the activation energy for a particular reaction can be calculated directly from measured values of the rate constant at two or more different temperatures assuming that A_f remains constant over the range of temperatures in question. From Equation 7-66 we can write

$$\ln \frac{(k)_1}{(k)_2} = \frac{E_a(T_1 - T_2)}{\mathcal{R}T_1 T_2} \tag{7-72}$$

A plot of $\ln k$ versus $1/T$ will thus give a straight line with a slope of $E_a/\mathcal{R}$. As noted earlier, this procedure is analogous to that presented in Chapter 5 for evaluation of ΔH°. Equation 7-72, written in terms of base 10 logarithms, is

$$\log \frac{(k)_1}{(k)_2} = \frac{0.4343\, E_a(T_1 - T_2)}{\mathcal{R}T_1 T_2} \tag{7-73}$$

For situations in which temperature varies over a relatively narrow range (10 degrees or so), as is often the case in natural waters and in treatment processes, the term $0.4343E_a/\mathcal{R}T_1T_2$ in Equation 7-73 remains reasonably constant, and can be approximated by the constant β_A, so that

$$\log \frac{(k)_1}{(k)_2} \cong \beta_A(T_1 - T_2) \tag{7-74}$$

where T_1 and T_2 may now be expressed in degrees Celsius rather than Kelvin because the difference (ΔT) is the same in both cases. The antilog of this equation is

$$\frac{(k)_1}{(k)_2} \cong 10^{\beta_A(T_1 - T_2)} \cong \beta_T^{T_1 - T_2} \tag{7-75}$$

where $\beta_T = 10^{\beta_A}$. Equation 7-75 is widely employed in the water quality field for characterizing the temperature dependence of the rate constant. For example, commonly quoted values of β_T are 1.047 for biochemical reactions and 1.024 for gas transfer. In essence, β_T is used instead of E_a to designate the temperature coefficient. However, β_T is not truly constant but is dependent on the two temperatures over which it is evaluated. The activation energy is thus a more rigorous parameter with which to characterize the temperature dependence of a chemical or biochemical reaction.

A common empirical rule is that reaction rates increase by a factor of 2 to 3 for each 10-degree rise. This rule applies, however, only if E_a is in the range of those measured for typical rate-limiting biological reactions (33 to 70 kJ/mol) discussed earlier and the temperature is in the common range of environmental systems (0 to 30°C, or 273 to 303 K).

7.5 COMPLEX REACTION KINETICS

7.5.1 Consecutive Reactions

Complex physicochemical and biochemical processes often proceed through a number of consecutive reaction stages or steps. The formation of a product, C, from a reactant, A, may, for example, involve the formation of an intermediate species, B. A reaction sequence of this type can be represented by the scheme

$$A \xrightarrow{r_1} B \xrightarrow{r_2} C \qquad (7\text{-}76)$$

where r_1 represents the rate of the reaction leading to the formation of the intermediate B from the reactant A, and r_2 the rate of the formation of the final product C from the intermediate B. The temporal distributions of species for a reaction scheme of the type given in Equation 7-76 are depicted in Figure 7-9. The successive chlorination of ammonia to form monochloramine and dichloramine is an example of such a reaction, as is the oxidation of cyanide to cyanate by chlorine followed by acid hydrolysis of the cyanate to ammonia.

For a series of consecutive first-order reactions in a closed system, the time rate of decrease of the reactant A may be expressed as

$$r_1 = \frac{dC_A}{dt} = -k_1 C_A \qquad (7\text{-}77)$$

and the rate of increase of the product C as

$$r_2 = \frac{dC_C}{dt} = k_2 C_B \qquad (7\text{-}78)$$

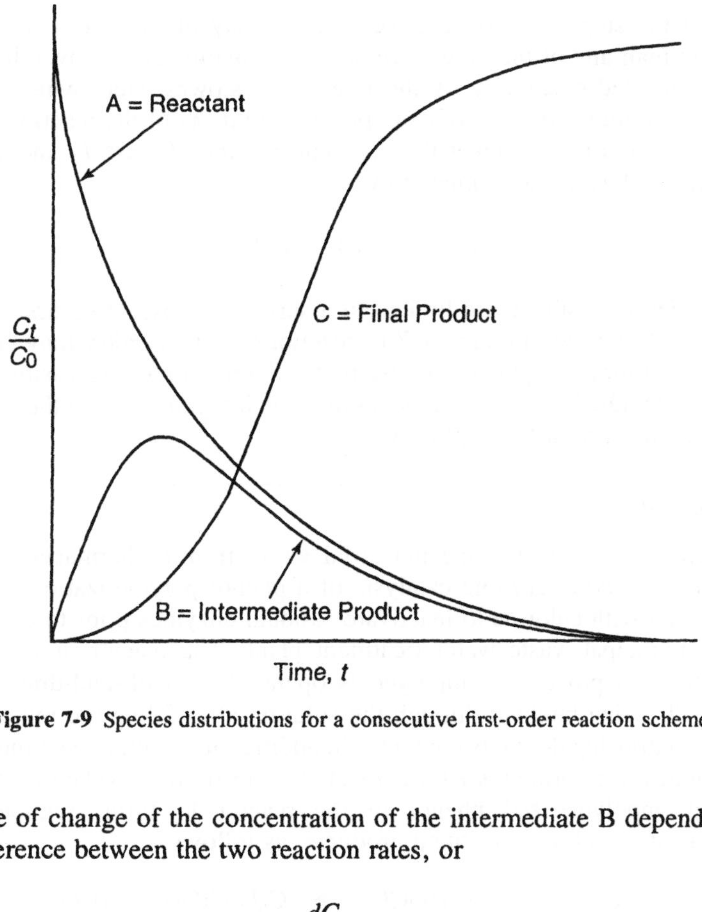

Figure 7-9 Species distributions for a consecutive first-order reaction scheme.

The rate of change of the concentration of the intermediate B depends on the net difference between the two reaction rates, or

$$r_1 - r_2 = \frac{dC_B}{dt} = k_1 C_A - k_2 C_B \tag{7-79}$$

The concentration of the reactant decreases steadily with time, whereas the concentration of final product increases steadily. The concentration of the intermediate, conversely, increases to some maximum value, beyond which it decreases steadily.

Upon integration, Equations 7-77, 7-78, and 7-79 can be written in terms of the initial concentration of the reactant A, as follows:

$$C_A = C_{A,0}\, e^{-k_1 t} \tag{7-80}$$

$$C_B = \frac{k_1 C_{A,0}}{k_2 - k_1}\left(e^{-k_1 t} - e^{-k_2 t}\right) \tag{7-81}$$

$$C_C = C_{A,0}\left(1 + \frac{k_1 e^{-k_2 t}}{k_2 - k_1} - \frac{k_2 e^{-k_1 t}}{k_2 - k_1}\right) \tag{7-82}$$

One of the steps in a consecutive reaction may often have a significantly lower rate than any of the others. Under such conditions, the overall reaction rate is controlled essentially by the rate of the slowest step, which then becomes *rate limiting*. By way of example, if the rate r_1 of the reaction given in Equation 7-76 is much slower than the rate r_2, then $k_1 << k_2$ and Equation 7-82 may be closely approximated by

$$C_C = C_{A,0}(1 - e^{-k_1 t}) \tag{7-83}$$

The considerations above are limited to a series of consecutive first-order reactions. As illustrated in Example 7-6, reactions of other orders may be treated in similar fashion, except that the resulting differential rate expressions are not as easily integrated. In such cases, numerical integrations or other approximations methods must be employed.

Example 7-6

- **Situation:** A small-volume industrial waste from a pharmaceutical operation involving enzyme catalysis of a phenol polymerization reaction is treated with chlorine to inactivate residual enzymes prior to discharge to a municipal waste water treatment plant. The reactor in which the deactivation process is employed is operated on a fill-and-draw basis, with Cl_2 being bubbled through the solution after filling the reactor and prior to drawing down its contents. In addition to effecting enzyme deactivation, the chlorine (as HOCl) reacts by substitutive oxidation with residual unpolymerized phenol in the reactor to form, sequentially, mono-, di-, and tri-substituted products, as follows:

$$\begin{array}{lcl}
\underset{\text{Phenol(Ph)}}{C_6H_5OH} + \underset{\text{Cl(I)}}{HOCl} & \xrightarrow{k_1} & \underset{\substack{\text{Monochlorophenol}\\ \text{(MCPh)}}}{C_6H_4ClOH} + H_2O \\
\underset{\text{MCPh}}{C_6H_4ClOH} + \underset{\text{Cl(I)}}{HOCl} & \xrightarrow{k_2} & \underset{\substack{\text{Dichlorophenol}\\ \text{(DCPh)}}}{C_6H_3Cl_2OH} + H_2O \\
\underset{\text{DCPh}}{C_6H_3Cl_2OH} + \underset{\text{Cl(I)}}{HOCl} & \xrightarrow{k_3} & \underset{\substack{\text{Trichlorophenol}\\ \text{(TCPh)}}}{C_6H_2Cl_3OH} + H_2O
\end{array}$$

- **Question(s):** Determine the amounts of the various chlorophenols that will be discharged as a function of the molar ratio of chlorine added and residual unpolymerized phenol present in the wastewater.

- **Logic and Answer(s):**

1. The reactor operation is not adequately described by either a completely mixed batch or a completely mixed flow reactor model.
2. Nonetheless, with reasonable assumptions about the volatilization

of unreacted gases, the system can be described as one of constant volume.

3. The appropriate expression for the rate of unpolymerized phenol transformation based on the first of the three reactions given above is, assuming stoichiometrically determined reaction orders,

$$-\frac{dC_{Ph}}{dt} = k_1 C_{Ph}\, C_{HOCl}$$

4. The net rates of formation of the three substituted species are

$$\frac{dC_{MCPh}}{dt} = k_1 C_{Ph}\, C_{HOCl} - k_2 C_{MCPh}\, C_{HOCl}$$

$$\frac{dC_{DCPh}}{dt} = k_2 C_{MCPh}\, C_{HOCl} - k_3 C_{DCPh}\, C_{HOCl}$$

$$\frac{dC_{TCPh}}{dt} = k_3 C_{DCPh}\, C_{HOCl}$$

5. The four rate equations given in steps 3 and 4 can be coupled with an appropriate mass balance relationship and solved for the respective yields of the three chlorinated phenols. Although the individual rate relationships are each second order, one of the reactants in the series is the same for each step: namely, HOCl. Thus, if any two of the rate equations are combined, C_{HOCl} is eliminated from the resulting equation.

6. For example, combining dC_{Ph}/dt and dC_{MCPh}/dt yields

$$\frac{dC_{MCPh}}{dC_{Ph}} = \frac{k_2}{k_1}\left(\frac{C_{MCPh}}{C_{Ph}} - 1\right)$$

This expression is a linear first-order differential equation for which an analytical solution can be developed.

7. If it is assumed that no chlorophenols are present at the outset of the reaction ($t = 0$), the analytical solution yields

$$\frac{C_{MCPh}}{(C_{Ph})_0} = \frac{k_2}{k_1 - k_2}\left[\left(\frac{C_{Ph}}{(C_{Ph})_0}\right)^{k_2/k_1} - \frac{C_{Ph}}{(C_{Ph})_0}\right]$$

which is, in fact, an expression for the yield of monochlorophenol.

8. Similarly, the expression for dC_{DCPh} can be combined with that for dC_{Ph} to yield

$$\frac{dC_{DCPh}}{dC_{Ph}} = \left(\frac{k_3}{k_1}\frac{C_{DCPh}}{C_{Ph}} - \frac{k_2}{k_1}\frac{C_{MCPh}}{C_{Ph}}\right)$$

9. If the problem is normalized per mole of phenol initially present [i.e., for $(m_{Ph})_0 = 1.0$], the relationships given in steps 7 and 8 can be

rewritten as

$$\frac{C_{MCPh}}{(C_{Ph})_0} = \frac{m_{MCPh}}{1.0} = m_{Ph}\frac{k_2}{k_1 - k_2}[(m_{Ph})^{k_2/k_1} - 1]$$

$$\frac{dm_{DCPh}}{dm_{Ph}} = \left(\frac{k_3}{k_1}\frac{m_{DCPh}}{m_{Ph}} - \frac{k_2}{k_1}\frac{m_{MCPh}}{m_{Ph}}\right)$$

10. The expressions given in step 9 can then be combined to replace m_{MCPh} with a function of m_{Ph}. This yields another linear first-order differential equation for m_{DCPh}, which can be integrated to obtain the yield of DCPh per mole of phenol.

11. The concentration of trichlorophenol per mole of phenol is then determined by mass balance; that is,

$$m_{Ph} + m_{MCPh} + m_{DCPh} + m_{TCPh} = 1$$

$$m_{TCPh} = 1 - (m_{Ph} + m_{MCPh} + m_{DCPh})$$

the three terms in parentheses having been determined as described above.

12. For solution of the equations above the individual rate coefficients—k_1, k_2, and k_3—need not be known. However, their relative magnitudes must be known [i.e., k_1/k_2 and k_2/k_3 (and, therefore k_1/k_3)]. A series of carefully designed experiments with different initial concentrations of the reacting species and their products should be performed to develop this information (Smith, 1981).

7.5.2 Parallel Reactions

A particular reactant may be involved in two or more reaction schemes concurrently in an environmental process. Chlorine and ozone, for example, can simultaneously oxidize both inorganic and organic compounds while also causing disinfection of microorganisms, and chlorine can react simultaneously with ammonia and phenol to form chloramines and chlorophenols. In other words, a reactant A might undergo two or more parallel and unrelated reactions to form multiple products, such as B and C in the essentially irreversible reaction scheme depicted below:

$$A \xrightarrow{r_1} B \quad (7\text{-}84)$$

$$A \xrightarrow{r_2} C \quad (7\text{-}85)$$

The overall rate of disappearance of reactant A is then given by the sum of the rates r_1 and r_2, or

$$\text{overall reaction rate} = r_1 + r_2 \quad (7\text{-}86)$$

If the reactions given in Equations 7-84 and 7-85 are both first order with respect to C_A, then Equation 7-86 may be written

$$-\frac{dC_A}{dt} = k_1C_A + k_2C_A = (k_1 + k_2)C_A \tag{7-87}$$

The overall reaction is thus first order with respect to C_A, with the effective rate constant given by the sum of the rate constants for the individual first-order reactions. This is true regardless of the number of parallel first-order reactions involved. Indeed, as long as the parallel reactions are of the same order, the overall rate constant for disappearance of the reactant(s) is always determined by adding together the individual rate constants. Accordingly, for the elementary parallel and essentially irreversible reactions,

$$a\text{A} + b\text{B} \xrightarrow{r_1} \text{C} \tag{7-88}$$

$$a\text{A} + b\text{B} \xrightarrow{r_2} \text{D} \tag{7-89}$$

$$a\text{A} + b\text{B} \xrightarrow{r_3} \text{E} \tag{7-90}$$

the overall reaction rate is

$$r = r_1 + r_2 + r_3 \tag{7-91}$$

or

$$-\frac{dC_A}{dt} = k_1C_A^aC_B^b + k_2C_A^aC_B^b + k_3C_A^aC_B^b = (k_1 + k_2 + k_3)C_A^aC_B^b \tag{7-92}$$

If the parallel reactions are not of the same order, the overall rate of change of the reactant of interest can still be written in terms of the sum of the rates of the individual reactions. The sequential reactions involved in the chlorination of phenol to form *ortho-*, *meta-*, and *para-*substituted chlorophenols discussed in Example 7-6 is also an example of an environmentally significant process involving parallel reactions of the type depicted schematically in Equations 7-88 through 7-92. In this process the parallelism results from chlorine substitution occurring either by an *ortho–meta–para* or *meta–para–ortho* substitution sequence.

7.5.3 Reversible Reactions

We have thus far assumed essentially irreversible reaction schemes, but they do not necessarily imply an assumption of absolute thermodynamic irreversibility. Rather, the particular state of the reaction under consideration is simply treated as being far removed from the thermodynamic equilibrium state. This assumption applies for the initial stages of most reactions and is especially valid for reactions in which the thermodynamic equilibrium state favors a high product/reactant ratio, that is, when the equilibrium constant for the reaction has a large value. In some instances, however, the reaction is not far enough removed from thermodynamic equilibrium or the equilibrium constant is not large enough to neglect rates of reverse reactions. This is particularly true when high relative concentrations of reactants remain at the final position of equilibrium.

As any reversible reaction proceeds toward an equilibrium state, the rate at which product species revert to reactant species becomes more and more significant. As the rate of the reverse reaction becomes appreciable, it has the net effect of decreasing the observed or *net rate* of reaction in the forward direction. For such cases it is necessary to employ an expression that takes account of reaction rates in both forward and reverse directions. Consider, for example, the reversible reaction

$$\mathrm{A} + \mathrm{B} \rightleftharpoons \mathrm{C} + \mathrm{D} \tag{7-93}$$

for which the net reaction rate is the difference between the forward rate, r_f, and the reverse rate, r_r, or

$$\text{net reaction rate} = r_f - r_r \tag{7-94}$$

Suppose that the forward and reverse reactions given in Equation 7-93 are pseudo first order with respect to A and D, respectively. The overall reaction rate for A is then

$$-\frac{dC_A}{dt} = (k)_f C_A - (k)_r C_D \tag{7-95}$$

The rate constants are subscripted outside the parentheses in this development to avoid confusion of $(k)_f$ with k_f, the latter being the notation used throughout the text for mass transfer coefficients. If $C_{R,A}$ represents the amount of reactant A converted to product D, and $C_{A,0}$ and $C_{D,0}$ represent the initial concentrations of A and D, respectively, then

$$\frac{dC_{R,A}}{dt} = (k)_f(C_{A,0} - C_{R,A}) - (k)_r(C_{D,0} + C_{R,A}) \tag{7-96}$$

or

$$\frac{dC_{R,A}}{dt} = [(k)_f + (k)_r]\left\{\left[\frac{(k)_f C_{A,0} - (k)_r C_{D,0}}{(k)_f + (k)_r}\right] - C_{R,A}\right\} \tag{7-97}$$

The constant quotient in brackets in Equation 7-97 expressing the initial condition for the reversible reaction can be replaced by a constant, β_R to give

$$\frac{dC_{R,A}}{dt} = [(k)_f + (k)_r]\,(\beta_R - C_{R,A}) \tag{7-98}$$

which upon integration yields

$$[(k)_f + (k)_r]t = \ln \frac{\beta_R}{\beta_R - C_{R,A}} \tag{7-99}$$

The form of the rate expression for this reversible reaction is the same as those presented earlier for irreversible reactions; similarly, the test of rate data is the same as well. Although both forward and reverse rate constants must be evaluated, they can be calculated from the equilibrium constant, K, which we know from previous discussions to be the ratio of the forward and reverse rate constants, $(k)_f/(k)_r$. The value of K can be obtained from separate equilibrium measurements, from prior knowledge of the equilibrium constant, or by calculation from the free-energy relationships described in Chapter 5. The rate constants in the term β_R in Equation 7-99 can be rewritten in terms of K. Further, we can use the definition of K to express the rate constants on the left-hand side of that equation in terms of K and either $(k)_f$ or $(k)_r$.

A convenient rate expression results when the initial concentration of the product, D, is zero and the equilibrium concentration of the reactant, A, is known. If we denote the equilibrium concentration by $C_{A,e}$ and recognize that $C_D = C_{A,0} - C_{A,e}$, then, at equilibrium, Equation 7-95 would yield

$$C_{A,0} = \frac{(k)_f + (k)_r}{(k)_r} C_{A,e} \tag{7-100}$$

Substituting for $C_{A,0}$ from above and setting $C_{D,0}$ equal to zero in Equation 7-97 gives

$$[(k)_f + (k)_r]t = \ln \frac{C_{A,0} - C_{A,e}}{C_A - C_{A,e}} \tag{7-101}$$

Equation 7-101 has the form of a first-order rate expression with an effective rate constant given by the sum of the rate constants for the forward and reverse reactions. By incorporating the equilibrium constant, the two rate constants may be separated and Equation 7-101 written in terms of the first-order rate constant for the forward reaction:

$$(k)_f t = \frac{K}{K+1} \ln \frac{C_{A,0} - C_{A,e}}{C_A - C_{A,e}} \tag{7-102}$$

The mathematical rate expression for the reaction given in Equation 7-93 becomes somewhat more complicated if the order of the reverse reaction is first order with respect to each of the products C and D. The differential rate expression for this case is

$$-\frac{dC_A}{dt} = (k)_f C_A - (k)_r C_C C_D \tag{7-103}$$

or if $C_{C,0} = C_{D,0}$ and C_R once again represents the amount reacted,

$$\frac{dC_R}{dt} = (k)_f(C_{A,0} - C_R) - (k)_r(C_{D,0} + C_R)^2 \tag{7-104}$$

When the initial concentrations of the two products are each zero, Equation 7-104 reduces to

$$\frac{dC_R}{dt} = (k)_f(C_{A,0} - C_R) - (k)_r C_R^2 \tag{7-105}$$

At equilibrium the forward and reverse reactions are equal, which introduces the equilibrium concentration, $C_{A,e}$, and the equilibrium value for the amount reacted, $C_{R,e}$:

$$(k)_f C_{A,e} = (k)_r C_{R,e}^2 \tag{7-106}$$

Substitution of Equation 7-106 into Equation 7-105 yields

$$\frac{dC_R}{dt} = (k)_f \left[C_{A,0} - C_R - C_{A,e} \left(\frac{C_R}{C_{R,e}} \right)^2 \right] \tag{7-107}$$

Equation 7-107 can then be integrated to solve for $(k)_f t$ in terms of the constants $C_{A,0}$ and $C_{R,e}$ and the single variable C_R.

If the reaction given in Equation 7-93 is second order with respect to each of the reactants and products in the forward and reverse directions, respectively, the problem becomes more complicated, but still tractable. For $C_{A,0} = C_{B,0}$ and $C_{C,0} = C_{D,0} = 0$ at $t = 0$, for example, we can describe $dC_R = dC_{R,B}$ in terms of $(k)_f$ and K as

$$\frac{dC_R}{dt} = (k)_f \left[(C_{A,0} - C_R)^2 - \frac{C_R^2}{K} \right] \tag{7-108}$$

Other expressions for reaction rates of various order in the forward and reverse directions can similarly be developed for reversible reactions, but such highly sophisticated rate characterizations have limited utility for environmental applications.

7.5.4 Kinetics and Thermodynamics

Some of the relationships between reaction rates and reaction energetics developed in Chapter 5 were reiterated in Section 7.5.3. Although the equilibrium constant is based on thermodynamic principles ($dG^\circ = -\mathcal{R}T \ln K$), it can also be calculated without the aid of thermodynamics by determining the ratio of forward and reverse rate constants. The kinetic interpretation of the equilibrium constant is intuitive for elementary reactions but not necessarily for nonelementary reactions. Relationships between reaction energetics and reaction rates enable use of tabulated thermodynamic information to provide valuable insight into certain kinetic aspects of a system that otherwise may not be feasible or easy to measure experimentally.

The nonelementary reaction scheme presented in Section 7.1.3 will be used here to demonstrate the relationship between kinetics and thermodynamics. In that instance, one reversible reaction step (Equation 7-26) and one irreversible reaction step (Equation 7-27) were given. The more general situation in which both steps are reversible will be analyzed here. We begin with Equation 7-26:

$$\mathrm{A} \underset{k_2}{\overset{k_1}{\rightleftharpoons}} \mathrm{A}^* \tag{7-109}$$

and from a modified version of Equation 7-27,

$$\mathrm{A}^* + \mathrm{B} \underset{k_4}{\overset{k_3}{\rightleftharpoons}} \mathrm{AB} \tag{7-110}$$

The condition of equilibrium for each elementary reaction can be expressed as

$$\frac{C_{A^*}}{C_A} = \frac{k_1}{k_2} \quad \text{and} \quad \frac{C_{AB}}{C_{A^*}C_B} = \frac{k_3}{k_4} \tag{7-111}$$

from which we obtain the equilibrium constant for the overall reaction:

$$K = \frac{C_{AB}}{C_A C_B} = \frac{k_3}{k_4}\frac{k_1}{k_2} \tag{7-112}$$

The equilibrium constant defined by Equation 7-112 can now be compared to that derived from kinetic considerations. Irrespective of the nature of the forward and reverse reaction rate expressions, these may be equated for the condition of thermodynamic equilibrium because equilibrium is defined (Chapter 5) as the state for which the net reaction velocity, $\mathcal{r}_{net}$, is zero ($\mathcal{r}_{net} = \mathcal{r}_f - \mathcal{r}_r = 0$). At equilibrium the net rate of change of each component is zero because the concentration of each component remains constant. The law of mass action also applies at equilibrium, so that

$$r_{A*} = 0 = k_1 C_A - k_2 C_{A*} - k_3 C_{A*} C_B + k_4 C_{AB} \tag{7-113}$$

$$r_A = 0 = -k_1 C_A + k_2 C_{A*} \tag{7-114}$$

Equation 7-113 can be rearranged to solve for C_{A*}, yielding

$$C_{A*} = \frac{k_1 C_A + k_4 C_{AB}}{k_2 + k_3 C_B} \tag{7-115}$$

and then substituted into Equation 7-114 to obtain

$$\frac{C_{AB}}{C_A C_B} = \frac{k_3}{k_4} \frac{k_1}{k_2} \tag{7-116}$$

Equation 7-116 is identical to Equation 7-112, proving that the equilibrium constant for a nonelementary reaction is defined by the ratio of the products of the rate coefficients associated with each step in the reaction scheme. Thus, for the condition or state of equilibrium, the equilibrium constant, K, is given by the ratio of the forward and reverse rate coefficients, regardless of the forms of the rate equations.

Examination of Equations 7-112 and 7-116 further reveals that the ratio of the forward and reverse reaction rates for a complex reaction can be directly related to the overall free energy for that reaction. This relationship becomes evident on substitution of the following terms for the reaction rate constants in Equation 7-112:

$$k_1 = \frac{r_1}{C_A} \qquad k_2 = \frac{r_2}{C_{A*}} \qquad k_3 = \frac{r_3}{C_{A*} C_B} \qquad k_4 = \frac{r_4}{C_{AB}} \tag{7-117}$$

Thus we have

$$K = \frac{r_3}{C_{A*} C_B} \frac{r_1}{C_A} \frac{C_{AB}}{r_4} \frac{C_{A*}}{r_2} \tag{7-118}$$

$$K = \frac{r_3 r_1}{r_4 r_2} \frac{C_{AB}}{C_A C_B} = \frac{r_3 r_1}{r_4 r_2} Q_r \tag{7-119}$$

Substitution of Equation 7-119 into Equation 5-79 then yields

$$dG_r = \mathcal{R} T \ln \frac{Q_r}{K} = \mathcal{R} T \ln \frac{r_4 r_2}{r_3 r_1} \tag{7-120}$$

or, more generally,

$$dG_r = \mathcal{R}T \ln \left(\frac{r_r}{r_f} \right)_{\text{overall}} \qquad (7\text{-}121)$$

We may thus conclude that the concepts developed in Chapter 5 and in Section 7.5.3 for interrelating reaction kinetics and thermodynamics are general and valid for application to interpretation of nonelementary reactions. We also observe that although the *net* overall reaction rate is zero at equilibrium, the individual rates of reaction in the forward and reverse directions may be large. Indeed, the magnitude of the equilibrium reaction rate is a measure of the degree of reversibility of a reaction; the higher that rate, the higher the degree of reversibility.

7.5.5 Modified First-Order Expressions

Nonelementary reaction schemes similar to those discussed above are frequently encountered in water quality transformations. A rate expression based on careful analysis of a reaction pathway may be obtainable from controlled laboratory experiments. There are many instances, however, where the information required for such rigorous treatment is not available, especially if field data must be interpreted. In such cases, the data constrain the analysis, and sophisticated interpretation is neither possible nor warranted.

We have already seen that the first-order rate equation is frequently employed empirically to represent observed rates of water quality transformations. This is often the only approach possible when data are limited and difficult to obtain. The approach is appropriate and valid as long as interpretations and generalizations are constrained accordingly. A common observation is that a first-order expression, or any other rate expression applied in an empirical way, provides a good fit of data over only a limited range, and that adjustments in rate parameter values are required to fit other ranges of the data. Gradual deviations of data trends from apparent first-order behavior are suggestive of nonelementary reactions having rates of progression that either gradually decrease or increase with time. Biochemical transformations of organic substrates often manifest this type of rate behavior.

A reaction for which rate decreases with time is termed *retardant*, while one that exhibits an increasing rate is termed an *accelerant* reaction. The effect of such behavior on the fitting of a first-order rate expression is either to decrease or increase the apparent (measured) value of the rate "constant," k, with increased period of experimental observation. It is possible to accommodate gradual deviations of retardant and accelerant reactions in an empirical way by incorporating a "sliding" factor in the elementary first-order rate equation.

For retardant reactions, the first-order equation can be modified by dividing the rate constant by a factor that increases with increasing time. Introducing a factor of this type to the first-order rate expression given in Equation 7-14 and writing a material balance equation in terms of the concentration of reactant that has reacted in a closed system gives

$$\frac{dC_R}{dt} = \frac{k}{1 + \alpha_r t}(C_0 - C_R) \tag{7-122}$$

where α_r is a reaction-characteristic rate coefficient. Upon integration we obtain

$$\frac{k}{\alpha_r} \ln \frac{1}{1 + \alpha_r t} = \ln \frac{C_0 - C_R}{C_0} \tag{7-123}$$

or

$$C_R = C_0[1 - (1 + \alpha_r t)^{-k/\alpha_r}] \tag{7-124}$$

Introduction of α_r requires that an additional unknown be determined. This can be done by employing nonlinear regression analysis and two-parameter search to find the best fit of the experimental data.

An accelerant reaction, as the name implies, is one for which rate appears to increase spontaneously as time progresses. An acceleration in rate is usually effected by the action of one of the products upon the reaction from which it was formed. If that product is in turn not changed or transformed by its reaction-accelerating role, the reaction can properly be termed *autocatalytic*. When reaction rates become dependent on the concentration(s) of product(s) formed as well as on the concentration(s) of reactant(s), the first-order expression may be modified to yield a differential rate equation of the form

$$\frac{dC_R}{dt} = k_1(C_0 - C_R) + k_2(C_0 - C_R)C_R \tag{7-125}$$

The first term on the right-hand side of Equation 7-125 is an elementary first-order equation. The second term expresses the rate acceleration factor as the product of a second rate constant, the remaining concentration of reactant, and the amount reacted or concentration of product formed. This term increases geometrically with time to a maximum value of 0.25 $C_0^2 k_2$, and decreases thereafter. The acceleration causes an increase in the forward rate of reaction which is especially noticeable if k_2 is much greater than k_1. However, the subsequent pattern of decrease of reaction rate from its maximum value will not be precisely the reverse of its pattern of approach to this maximum.

The integrated form of Equation 7-125 is

$$(k_1 + k_2C_0)t = \ln \frac{C_0(k_2C_R + k_1)}{k_1(C_0 - C_R)} = \ln \frac{C_0[k_2(C_0 - C) + k_1]}{k_1 C} \tag{7-126}$$

where C is the concentration of reactant remaining at time t. This equation is used for describing rate data which follow the general S-shaped pattern illustrated by the solid line in Figure 7-10. It is sometimes convenient and permissible to approximate accelerating rate behavior by neglecting the initial portion of the S-curve and considering the remainder as an elementary first-order reaction having a delayed starting point, as indicated by the dashed line in Figure 7-10. This approximation amounts to introducing a *lag period* in the first-order rate expression. If t_L is designated as the value of t at the end of the lag period, as noted in Figure 7-10, the first-order rate expression may be written as

$$k(t - t_L) = \ln \frac{C_0}{C_0 - C_R} \tag{7-127}$$

or

$$C = C_0 e^{-k(t - t_L)} \tag{7-128}$$

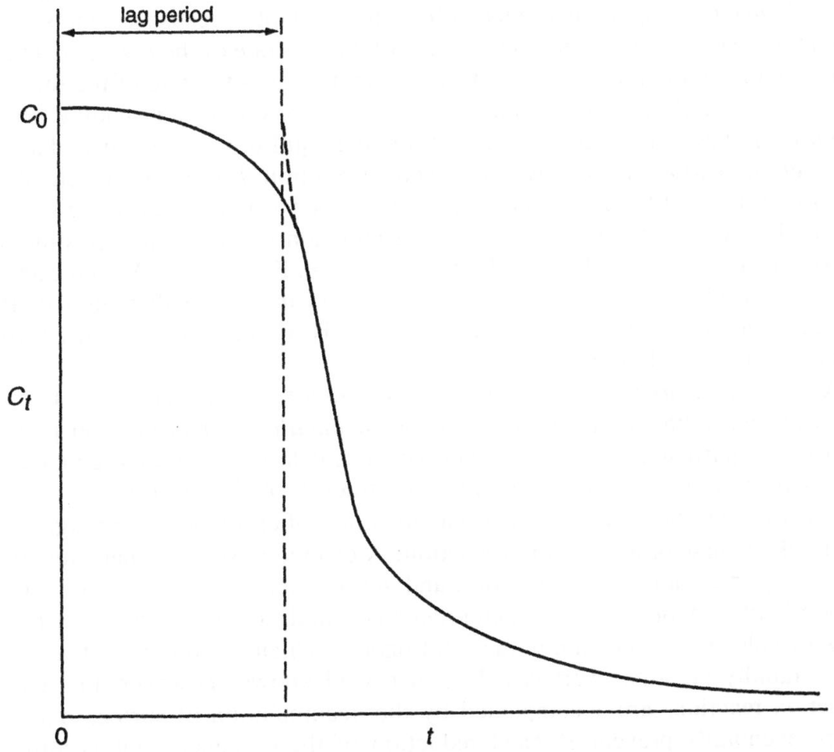

Figure 7-10 Approximation of an accelerant reaction by a first-order rate expression with a lag period.

where $t \geq t_L$. If the lag period, t_L, is constant for a given reaction, e^{kt_L} is also constant. Equation 7-128 may thus be written in terms of a reaction specific rate coefficient, β_r, such that

$$C = C_0 \beta_r e^{-kt} \tag{7-129}$$

As for the case of the retardant reaction, k, β_r, and if necessary, C_0 of Equation 7-129 may be solved for using a nonlinear regression technique for data analysis.

7.6 CATALYSIS

A catalyst may be defined in most simple terms as a substance that accelerates a chemical reaction but which itself remains the same in form and concentration after completion of the reaction as it was before the reaction occurred. The accelerating action of a catalyst is termed *catalysis*.

A catalyst alters the rate of approach of a reaction to its normal position of thermodynamic equilibrium. It does not have the capacity to initiate a reaction or to change the magnitude of an equilibrium constant. Because a catalyst does not alter the equilibrium constant, it follows that the rate of the reverse reaction must be affected to the same extent by the catalyst as is the rate of the forward reaction. A catalyst alters the rate of a chemical reaction by providing an *alternative pathway* for that reaction to approach equilibrium, a pathway having a *lower activation energy*. As discussed in Section 7.4, a lower activation energy means that the fraction of the total number of molecules in the system possessing sufficient energy for a reactive collision is greater and thus the reaction rate is greater, as illustrated schematically in Figure 7-11. We note again in Figure 7-11 that the catalyst has no effect on the thermodynamics of the system. The thermodynamics are determined only by the relative energy levels of the reactants and products.

A *heterogeneous catalyst* is one that exists in a separate phase from that of the reactants. Although a catalyst remains *chemically* unchanged after participation in a particular reaction, a heterogeneous catalyst must undergo a *physical* change. For example, the surfaces of smooth platinum, a catalyst commonly used for chemical oxidation reactions, are often found to be roughened during the course of an oxidation reaction. A change in surface characteristics suggests participation in the reaction and subsequent regeneration. Activated carbon is often thought of as a heterogeneous catalyst for the oxidation reactions of chlorine, monochloramine, and ozone, which are observed to be reduced rapidly at carbon surfaces. It is also well known, however, that these reduction reactions are accompanied by the accumulation of surface oxides, which eventually prevent efficient reduction of the oxidants. This is termed *catalyst deactivation* or *poisoning*.

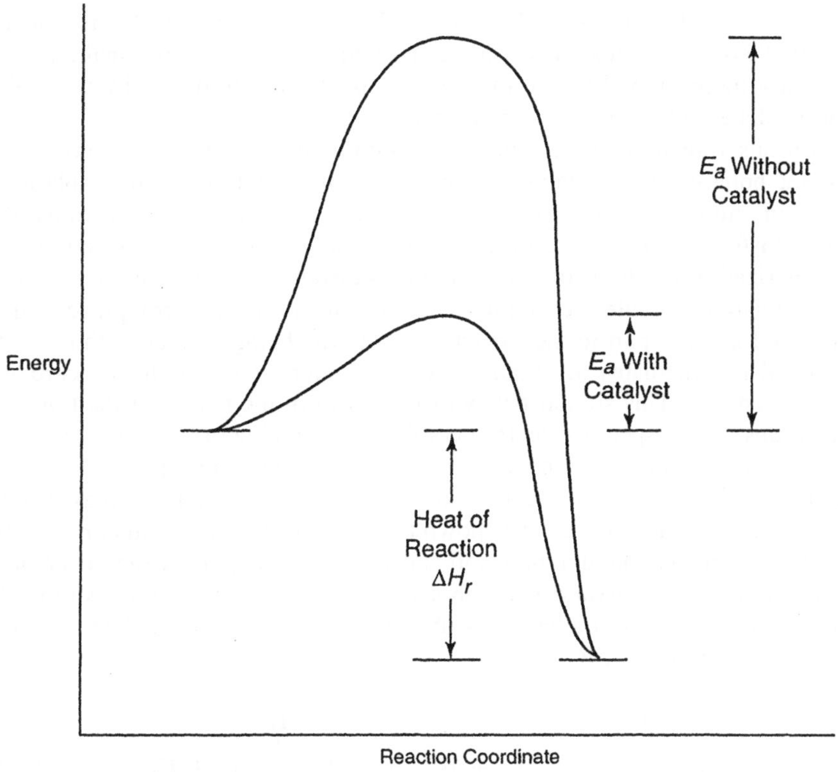

Figure 7-11 Catalysis and the activation-energy barrier for an exothermic reaction.

7.6.1 Homogeneous Catalysis

In homogeneous catalysis, a reaction proceeds through several stages or steps, thus providing a pathway of decreased resistance. This may be illustrated schematically for the following general reaction scheme:

$$A + B \rightleftharpoons AB \tag{7-130}$$

Although the thermodynamics for formation of AB from the two reactants may be favorable, the direct one-step reaction of A with B may involve such a high energy of activation that the reaction will proceed only at an imperceptible rate. If a catalyst, C*, is added to the system, however, the reaction may proceed by an alternative two-step pathway, as follows:

$$\text{Step 1: } A + C^* \rightleftharpoons AC^* \tag{7-131}$$

$$\text{Step 2: } AC^* + B \rightleftharpoons AB + C^* \tag{7-132}$$

Summation of Equations 7-131 and 7-132 gives the same results as Equation 7-130. Thus the thermodynamic position of equilibrium remains unaltered even though a more favorable rate mechanism has been introduced by way of the intermediate AC* compound or complex.

Various transition metal ions and coordination compounds serve as important homogeneous catalysts. Complexes of cobalt, palladium, platinum, rhodium, ruthenium, and other metals play important roles in various electron-transfer, hydration, hydrogenation, hydroformylation, oxidation, and polymerization reactions of many organic substances. The catalytic versatility of the relatively stable, yet highly reactive transition metal complexes can be attributed in large part to their reactivity patterns being similar to those of the principal intermediates of organic reactions, such as free radicals, carbenes, and carbanions. This similar behavior results from the nature of the transition metal ions with respect to electron configuration and coordination state.

Reductive dehalogenation reactions of organic compounds provide excellent examples of the roles that metals and organometallic complexes play as catalysts. These reactions have been shown to occur in certain natural communities of microorganisms found in both soil and water, although the exact mechanism is not known. The forward reaction scheme may involve either a simple *hydrogenolysis* reaction of a halogen-substituted carbon atom attached to a complex organic molecule, R:

$$\mathrm{R{-}\overset{\displaystyle H}{\underset{\displaystyle H}{\overset{|}{\underset{|}{C}}}}{-}Cl} + 2e^- + H^+ \longrightarrow \mathrm{R{-}\overset{\displaystyle H}{\underset{\displaystyle H}{\overset{|}{\underset{|}{C}}}}{-}H} + Cl^- \qquad (7\text{-}133)$$

or a *two-step reductive elimination*, in which two substituted halogen atoms are replaced by a carbon–carbon double bond:

$$\mathrm{R{-}\overset{\displaystyle H}{\underset{\displaystyle Cl}{\overset{|}{\underset{|}{C}}}}{-}\overset{\displaystyle Cl}{\underset{\displaystyle H}{\overset{|}{\underset{|}{C}}}}{-}R} + 2e^- \longrightarrow \mathrm{R{-}\overset{\displaystyle H}{\overset{|}{C}}{=}\underset{\displaystyle H}{\underset{|}{C}}{-}R} + 2Cl^- \qquad (7\text{-}134)$$

Although complete oxidation is not achieved, even partial dehalogenation can result in residual organic structures which are less toxic and more amenable to subsequent biological degradation. Reductive dehalogenation thus offers a potential way to treat hazardous chemicals.

Naturally occurring organometallic constituents of microbial cells, such as vitamin B_{12} (which contains cobalt) and hematin (which contains iron), serve as active catalysts in reactions of the type shown above. As depicted in Figure 7-12, these organometallic compounds serve as electron-transfer mediators (i.e., catalysts) for the dehalogenation reactions given by Equations 7-133 and 7-134. Support for the catalytic action of such organometallic compounds is

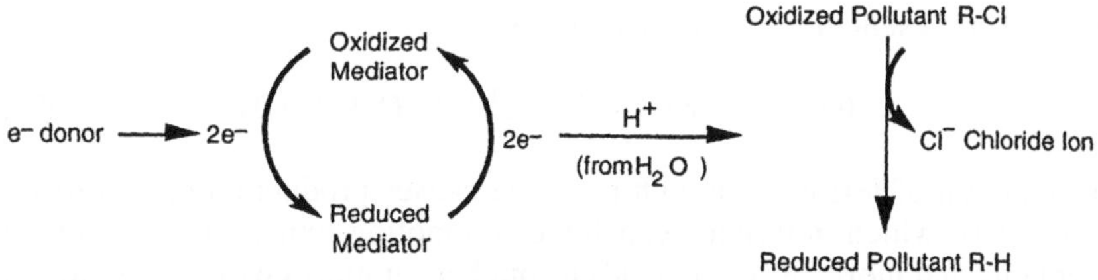

Figure 7-12 General scheme for a catalyzed reductive dechlorination reaction.

offered by observations of their ability to cause abiotic dehalogenation reactions by reducing agents which are unable to effect the reductions alone. In effect, these organometallic compounds are able to simulate the reducing capability of microorganisms. As noted in Figure 7-13, dehalogenation reactions of such poly-substituted compounds as hexachlorobenzene often follow consecutive reaction sequences not unlike those depicted in Figure 7-9.

Another catalytic reaction with potential for producing active reagents for oxidative treatment of resistant organic compounds is the *Fenton reaction*. The

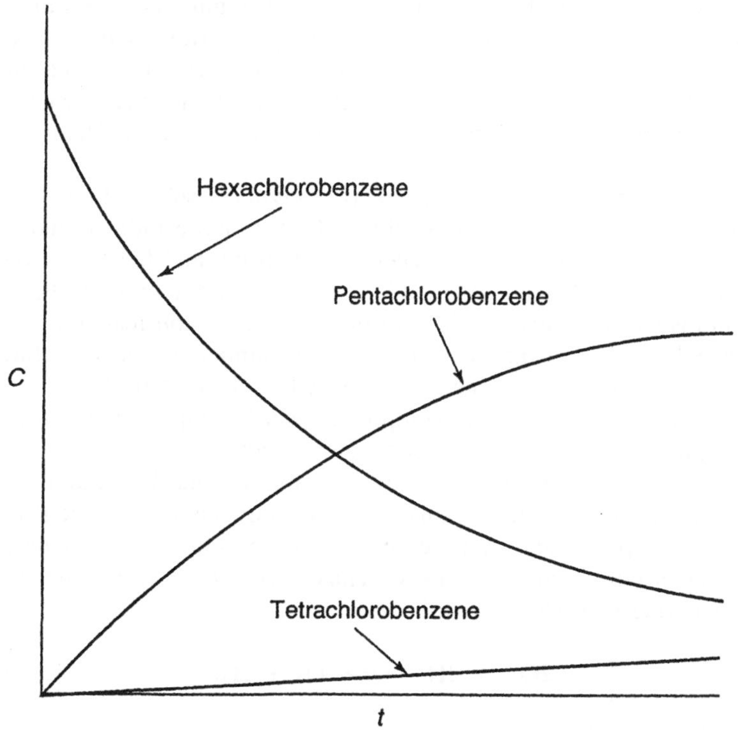

Figure 7-13 General reaction progress curves for vitamin B_{12}–catalyzed abiotic reductive dechlorination of hexachlorobenzene. [After Assaf-Anid et al. (1994).]

classical representation of this reaction is

$$Fe^{2+} + H_2O_2 \rightleftharpoons Fe^{3+} + {}^{\bullet}OH + OH^- \qquad (7\text{-}135)$$

The reaction of ferrous iron with peroxide causes production of hydroxyl radicals ($^{\bullet}OH$), which in turn are capable of promoting oxidation of a wide range of organic and inorganic compounds through a complex chain of electron transfer steps in which organic radicals ($^{\bullet}R$) and other oxygen radicals, such as superoxide ($O_2^{\bar{\bullet}}$), are produced. A simple example is:

$$^{\bullet}OH + RH \longrightarrow H_2O + R^{\bullet} \qquad (7\text{-}136)$$

$$R^{\bullet} \xrightarrow{O_2} R^+ + O_2^{\bar{\bullet}} \qquad (7\text{-}137)$$

$$O_2^{\bar{\bullet}} + Fe^{3+} \longrightarrow Fe^{2+} + O_2 \qquad (7\text{-}138)$$

The combination of Equations 7-138 and 7-135 represents the *Haber-Weiss* reaction. The reaction scheme depicted in Equations 7-135 through 7-138 is considered catalytic because the Fe^{2+} is ultimately regenerated chemically. The sequence also provides another example of a nonelementary process wherein the overall mechanism for oxidation of R must be determined by kinetic analysis of each intermediate elementary reaction. Various oxidations have been shown to be facilitated by addition of Fenton reagent, including those of chlorobenzenes, pentachlorophenol, chlorophenoxy herbicides, and dyes. Alternative pathways that do not include generation of $^{\bullet}OH$ may also be involved.

One modification of the Fenton reaction involves substitution of Fe^{3+} for Fe^{2+}. In this scheme, Fe^{3+} reacts with H_2O_2 through a radical chain mechanism to form O_2 and H_2O while generating hydroxy and hydroperoxyl ($HO_2^{\bullet}$) radicals, which can then attack organic chemicals. In the process, Fe^{2+} is produced and is subsequently oxidized to Fe^{3+} in the Fenton reaction.

Light enhances such radical reactions in a number of ways, including the photoreduction of Fe^{3+} to generate more hydroxyl radicals. Light- and iron-catalyzed oxidations of organic compounds can be important in atmospheric water droplets and in the near surface water of lakes and rivers.

The catalytic reaction schemes above are abiotic, but they can also be biologically mediated. Aerobic organisms almost universally produce superoxide ($O_2^{\bar{\bullet}}$) *in vivo*. Superoxide dismutase, an enzyme common to aerobic organisms, destroys superoxide, which otherwise tends to be lethal. In this reaction, hydrogen peroxide is produced as follows:

$$2O_2^{\bar{\bullet}} + 2H^+ \longrightarrow O_2 + H_2O_2 \qquad (7\text{-}139)$$

The addition of hydrogen peroxide to systems containing superoxide leads to radical formation in a manner similar to that given by addition of Equations 7-135 and 7-138 (i.e., the *Haber–Weiss reaction*):

$$O_2^{\cdot -} + H_2O_2 \longrightarrow {}^{\cdot}OH + OH^- + O_2 \tag{7-140}$$

The direct biological reaction given in Equation 7-140 is too slow to be of significance, while the nonelementary reaction pathway catalyzed by iron depicted in Equations 7-135 and 7-138 is very fast.

7.6.2 Heterogeneous Catalysis

Heterogeneous catalysis commonly involves presorption of reactant species at the surfaces of a solid catalyst, followed by formation of surface intermediates and subsequent completion of the overall reaction scheme. Except for a few unusual reactions, such as the interconversion of *ortho-* and *para-*hydrogen, surface accumulation of reactant species by strictly van der Waals or physical sorption does not suffice for heterogeneous catalysis. Instead, it is generally necessary for at least one reactant to be chemically adsorbed on the surface of the catalyst. The differences between physical and chemical adsorption are discussed in Section 6.4.3 of Chapter 6.

Complex reactions generally proceed by various series and parallel combinations of more simple reactions. Heterogeneous catalysts provide reactive surface sites that enhance the speed of a reaction by introducing new elementary intermediate steps, which in turn allow more rapid combination of reactant species. The precise nature of the reactive surface sites on a particular catalyst is rarely well known. Indeed, even for smooth metal catalysts there is some controversy as to whether all surface atoms possess the same catalytic reactivity, or whether reactivity is associated only with certain surface ridge atoms, dislocations, or point defects (i.e., *distributed reactivity*), as discussed with respect to adsorption in Chapter 6. Fortunately, the development of descriptive mechanisms for heterogeneous catalysis has been able to proceed without complete knowledge of the precise character and distribution of catalytic surface sites.

The usual sequence of steps associated with heterogeneous catalysis involves:

- Transport of reactant(s) to a reactive surface site
- Sorption of the reactant(s) on the surface site
- Surface-catalyzed reaction(s)
- Desorption of product(s) from the surface site
- Transport of product(s) away from the surface

The *Langmuir–Hinshelwood* method of analysis for such sequential rate steps in catalysis is discussed in detail in Chapter 8. In essence, the method includes (1) formulation of rate expressions for each step; (2) postulation of a rate-controlling step, from which a rate model for the overall process then emerges; and (3) test of the rate model against experimental data to confirm or modify the rate-limiting step postulation.

Microporous catalysts are often employed to provide greater surface areas for reaction. For such catalysts the first and fifth steps in the sequence above necessarily involve pore diffusion or intraparticle transport, which is generally quite slow. The retardation of these steps to the point where they become *rate limiting* often leads to development of severe concentration gradients in capillary pores, thus altering observed rate relationships, activation energies, and in extreme cases, even the apparent stoichiometry of reactions.

Heterogeneous photocatalysis semiconductors such as titanium dioxide (TiO_2) offer considerable promise for treatment of aqueous systems containing resistant dissolved organic chemicals. A schematic of a photocatalysis reaction on a semiconductor is presented in Figure 7-14. When illuminated, absorption of a photon of sufficiently high band gap energy causes excitation of electrons (e^-) to the *conduction band* (CB) and leaves an electron vacancy or *hole* (h^+) in the *valence band* (VB). Some of these electrons and holes recombine, but a significant number migrate to the surface of the catalyst and react with H_2O, O_2, and OH^-. The holes oxidize H_2O and OH^- to form hydroxyl radicals and the electrons reduce O_2 to form superoxide, which in turn forms more hydroxyl radicals or participates in other reactions. As discussed above, the hydroxyl radical goes on to oxidize a wide range of organic chemicals.

Heterogeneous semiconductor catalysts such as TiO_2 involve very complex pathways, making it virtually impossible to write rigorous mechanistically

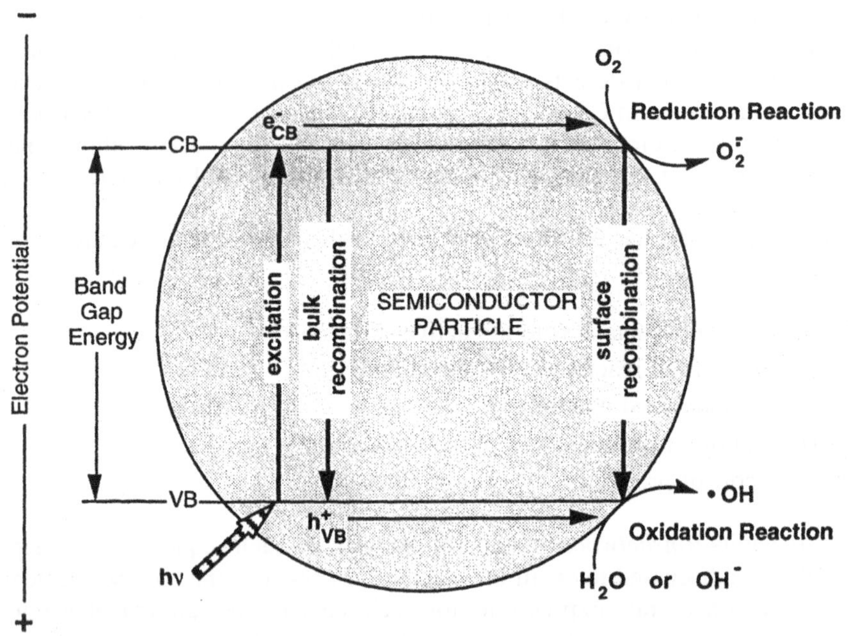

Figure 7-14 Functions of a photo-active semiconductor catalyst. [Adapted from Hand et al. (1991).]

based rate relationships. The mechanistic pathways of the biological catalysis of reactions by enzymes in microbial transformation processes are probably even more complex. Nonetheless, as we will now demonstrate, hypothetical reaction schemes have been prepared on which to base very workable rate expressions.

7.7 BIOLOGICAL CATALYSIS

As illustrated in Chapter 5 (Table 5-7 and related discussion), microbes are capable of carrying out a wide range of organic oxidation reactions using various ultimate electron acceptors (e.g., oxygen, nitrate, carbon dioxide, sulfate). Free-energy calculations show these reactions to be thermodynamically favorable. However, without *enzymatic catalysts* to provide alternative pathways of lower activation energy, such reactions are normally very slow.

Our previous discussions of the role of organometallic compounds such as vitamin B_{12} in reductive dehalogenation reactions and of biologically produced superoxide in radical generation and organic oxidations stresses the close link between abiotic and biotic transformation reactions. In those instances, metals and organometals were shown not only to be important catalysts but also to be present within the enzyme structures of common microbes.

Enzymes are chemical compounds produced by microbial synthesis. They are most commonly conjugated proteins, or combinations of large protein molecules and some other small organic molecules. The small prosthetic group is referred to as the *coenzyme*, while the protein molecule itself is termed the *apoenzyme*. Enzymes far exceed any known synthetic catalysts in their ability to accelerate chemical reactions. They are also highly specific in their catalytic activity. The enzyme urease, for example, catalyzes the hydrolysis of urea, $(NH_2)_2CO$, in concentrations as low as 1 mg of enzyme per liter but does not detectably alter the rate of hydrolysis of a substituted molecule such as methyl urea, $(NH_2)(CH_3NH)CO$. These specific biocatalysts, although produced by living cells, are often capable of functioning effectively outside the cell; that is, cell-free filtrates containing enzymes can act as catalytic solutions in many instances.

The importance of enzyme-catalyzed reactions in both natural and engineered systems cannot be overstated. They underlie all anaerobic and aerobic processes used to biostabilize organic matter in wastewater treatment and all biostabilization that occurs spontaneously or is induced (e.g., in situ bioremediation) in the natural environment. As more is learned about enzymatic reactions, new ways are being devised to take advantage of their specificity and speed for treatment of hazardous organic chemicals. For example, many fungi and plants possess peroxidase enzymes, which have been shown to be effective in mediating the oxidation of resistant aromatic structures by hydrogen peroxide.

The molecular diameters of enzymes range from approximately 100 Å to about 1000 Å, thus falling somewhere between the boundaries used to differentiate between homogeneous and heterogeneous catalysts; they are thus often referred to as *microheterogeneous* catalysts. Theoretical interpretation of the mechanism of enzyme catalysis can be based with about equal justification on either the formation of an intermediate compound between an enzyme and a substrate (i.e., a homogeneous reaction) or the adsorption of the substrate at the surface of the enzyme (i.e., a heterogeneous reaction). Most developments pursue the formation of an intermediate compound, so this has been chosen for our further examination of enzyme kinetics and rate expressions.

The importance of enzyme reactions in biological treatment of water and wastewater is more than ample reason by itself to explore the conceptual basis of their associated rate expressions. Another reason, however, is that rate expressions of mathematical forms similar to those developed for enzyme reactions can be used as well to describe many complex physicochemical reactions and processes that occur in environmental systems.

7.7.1 Rate Expressions

The general scheme for depiction of an enzyme reaction is

$$E + S \rightleftharpoons ES \rightleftharpoons (I)_n \rightleftharpoons EP \rightleftharpoons P + E \qquad (7\text{-}141)$$

where the enzyme, E, and substrate, S, interact to form an enzyme–substrate complex, ES. Chemical transformation of the enzyme–substrate complex, possibly over a number of intermediate steps, $(I)_n$, ultimately leads to an enzyme-bound product, EP, which in turn dissociates to yield free product, P, and enzyme.

The most common simplification of Equation 7-141 is referred to as the *Michaelis–Menten model*:

$$E + S \underset{k_2}{\overset{k_1}{\rightleftharpoons}} ES \overset{k_3}{\longrightarrow} P + E \qquad (7\text{-}142)$$

The catalytic nature of this two-step nonelementary reaction is indicated by the fact the enzyme is conserved. The assumptions associated with developments which follow from Equation 7-142 are that (1) the substrate is reversibly associated with the enzyme in the ES complex, (2) the transformation of the ES complex into the free product and enzyme is rate limiting, and (3) the reverse reaction between free product and enzyme is negligible during the initial course of the reaction. With these assumptions, two basic relationships may be set forth. First, the rate of increase of product concentration (C_{PR}) and decrease of substrate concentration (C_{SU}) are both functions of the concentration of the enzyme–substrate complex, C_{ES}; that is,

$$\frac{dC_{PR}}{dt} = \frac{-dC_{SU}}{dt} = k_3 C_{ES} \qquad (7\text{-}143)$$

Second, the conservation of mass relationship can be applied to the enzyme, such that

$$C_{EN,0} = C_{EN} + C_{ES} \tag{7-144}$$

where $C_{EN,0}$ is the concentration of enzyme present at the start of the reaction.

If the further assumption is made that after a brief transient initial stage, the ES complex attains a steady-state concentration, we may write

$$\frac{dC_{ES}}{dt} = k_1 C_{EN} C_{SU} - k_2 C_{ES} - k_3 C_{ES} = 0 \tag{7-145}$$

The forward step involving formation of the enzyme–substrate complex from the free enzyme and substrate has the general form of a second-order reaction, whereas both the reverse reaction and that for the formation of free product and enzyme from the enzyme–substrate complex have the general forms of first-order reactions. Solving Equation 7-145 for the steady-state concentration of enzyme–substrate complex gives

$$C_{ES} = \frac{k_1 C_{SU} C_{EN}}{k_2 + k_3} = \frac{k_1 C_{SU} C_{EN,0}}{k_1 C_{SU} + k_2 + k_3} \tag{7-146}$$

Combining this relationship with Equation 7-143 then leads to an expression for the time rate of decrease of substrate:

$$-\frac{dC_{SU}}{dt} = \frac{dC_{PR}}{dt} = k_3 C_{ES} = \frac{k_3 k_1 C_{SU} C_{EN,0}}{k_1 C_{SU} + k_2 + k_3} \tag{7-147}$$

The overall combination of rate constants given by $(k_2 + k_3)/k_1$ is termed the Michaelis constant, $\mathcal{K}_M$. Equation 7-147 may then be written

$$-\frac{dC_{SU}}{dt} = \frac{k_3 C_{SU} C_{EN,0}}{\mathcal{K}_M + C_{SU}} \tag{7-148}$$

The assumption of a steady-state concentration of enzyme–substrate complex is akin to an assumption that the first-order dissociation of the complex to free product and enzyme is rate controlling. This assumption is appropriate only for an extent of reaction $(C_{SU,0} - C_{SU}) > C_{ES}$ when $C_{SU} >> C_{EN,0}$ or $C_{ES} << C_{EN,0}$.

The Michaelis constant is a measure of enzyme–substrate affinity. That is, $\mathcal{K}_M$ is nearly equal to the dissociation constant (k_1/k_2) for an enzyme–substrate complex when the concentration of ES is in equilibrium with the concentrations of E and S [i.e., $k_2 >> k_3$ (see Equation 7-142)]. This is another way of stating that the rate-controlling step is conversion of the enzyme–substrate complex to enzyme and product as given by Equation 7-143.

The concentration of the enzyme in Equation 7-148 generally may be considered to remain constant. Thus, the rate of the reaction will depend only upon the concentration of the substrate. The effect of substrate concentration on the rate of the reaction for constant enzyme concentration is illustrated in Figure 7-15. At low substrate concentrations, the rate of substrate depletion is approximately first order with respect to substrate, whereas at high levels, the rate approaches zero order. A zero-order rate is explained by saturation of the enzyme with substrate such that $C_{EN,0} = C_{ES}$. It follows from Equation 7-142 that the maximum rate, $\mathscr{r}_{max}$, must equal $k_3 C_{EN,0}$. Therefore, we can replace $C_{EN,0}$ in Equation 7-148 to arrive at the final form of the Michaelis–Menten equation:

$$\frac{dC_{PR}}{dt} = -\frac{dC_{SU}}{dt} = \frac{\mathscr{r}_{max} C_{SU}}{\mathcal{K}_M + C_{SU}} \qquad (7\text{-}149)$$

The *Monod model* for biological process engineering uses an equation of the same form as Equation 7-149 for description of microbial growth rate and rate of substrate utilization as a function of substrate concentration. The Monod model is a useful empirical approach which, although it does not describe each of the enzyme–substrate steps involved in biodegradation, does have a fundamental origin.

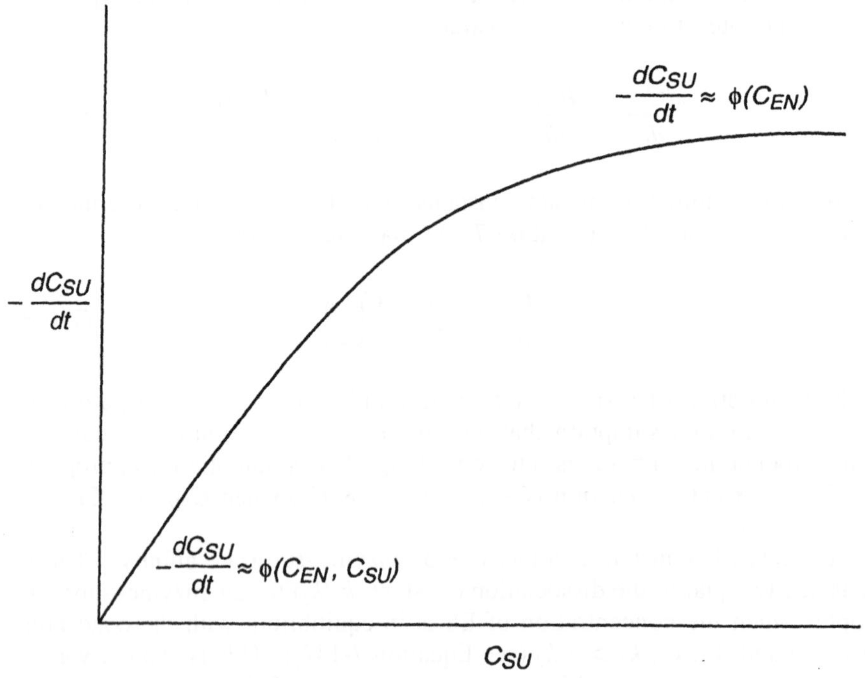

Figure 7-15 Substrate concentration effects on reaction rates in the Michaelis–Menten model.

The constants $\mathcal{K}_M$ and r_{max} of the Michaelis–Menten equation provide a way to characterize each enzyme system. Their values can be obtained from experiments in which the initial rate of product formation, dC_{PR}/dt, is determined for a number of different initial substrate concentrations (see Section 7.3.2.1). The constants in Equation 7-149 can be obtained either by using commonly available computer software to perform a multiple, nonlinear regression or by taking the reciprocal of both sides of Equation 7-149 to give the following linear expression:

$$\frac{1}{dC_{PR}/dt} = \frac{1}{r_{max}} + \frac{\mathcal{K}_M}{r_{max}} \frac{1}{C_{SU}} \tag{7-150}$$

A plot of $(dC_{PR}/dt)^{-1}$ versus C_{SU}^{-1} will give a linear trace with an intercept on the $(dC_{PR}/dt)^{-1}$ axis of $1/r_{max}$ and a slope of $\mathcal{K}_M/r_{max}$. This linearized form of the Michaelis–Menten equation is often referred to as a *Lineweaver–Burke plot*. An example of this type of plot is given in Figure 7-16. Although a multiple, nonlinear regression will provide a more accurate determination of the appropriate constants, the linearization procedure is convenient and gives an intuitive sense of whether the data conform to the model.

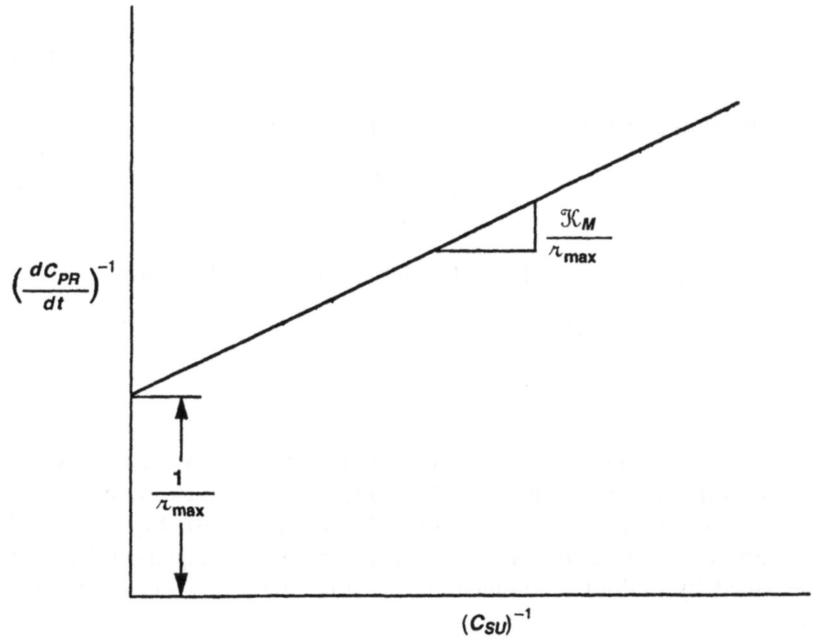

Figure 7-16 Linearized Michaelis–Menten equation.

7.7.2 Inhibited Reactions

Several different possibilities exist for the inhibition of enzymatic reactions. For systems in which inhibition occurs, appropriate modification of the corresponding rate equations is required for proper description of observed data. Three fundamentally different types of inhibition are common: substrate inhibition, competitive inhibition, and noncompetitive inhibition.

Substrate inhibition can be described as a reversible combination of the enzyme–substrate complex, ES, with the substrate, S, to form a substrate–enzyme–substrate complex SES, which is only capable of redissociation to S and ES. For this case, Equation 7-142 is rewritten as

$$
\begin{array}{ccccc}
\mathrm{E} + \mathrm{S} & \underset{k_2}{\overset{k_1}{\rightleftharpoons}} & \mathrm{ES} & \xrightarrow{k_3} & \mathrm{P} + \mathrm{E} \\
 & & + & & \\
 & & \mathrm{S} & & \\
 & & \upharpoonleft\downharpoonright & & \\
 & & \mathrm{SES} & &
\end{array}
\qquad (7\text{-}151)
$$

The conservation of mass expressions for this reaction scheme are

$$C_{EN,0} = C_{EN} + C_{ES} + C_{SES} \qquad (7\text{-}152)$$

$$C_{SU,0} = C_{SU} + C_{ES} + C_{SES} \qquad (7\text{-}153)$$

If the equilibrium constant for the SES complex is given as

$$\mathcal{K}_{SES} = \frac{C_{ES} C_{SU}}{C_{SES}} \qquad (7\text{-}154)$$

we can follow the same steps that led to Equation 7-149, to obtain

$$\frac{dC_{PR}}{dt} = \frac{r_{\max}}{1 + \dfrac{\mathcal{K}_M}{C_{SU}} + \dfrac{C_{SU}}{\mathcal{K}_{SES}}} \qquad (7\text{-}155)$$

This expression indicates that when the substrate concentration is very low, the third term in the denominator may be neglected and the expression reverts to the regular Michaelis–Menten equation (Equation 7-149). Equation 7-155 is also referred to as the *Haldane model*. The constants in Equation 7-155 can be determined by multiple, nonlinear regression or by the following linearization:

$$\frac{1}{dC_{PR}/dt} = \frac{1}{r_{\max}} + \left(\frac{\mathcal{K}_M}{r_{\max}} + \frac{C_{SU}^2}{\mathcal{K}_{SES}\, r_{\max}}\right)\frac{1}{C_{SU}} \qquad (7\text{-}156)$$

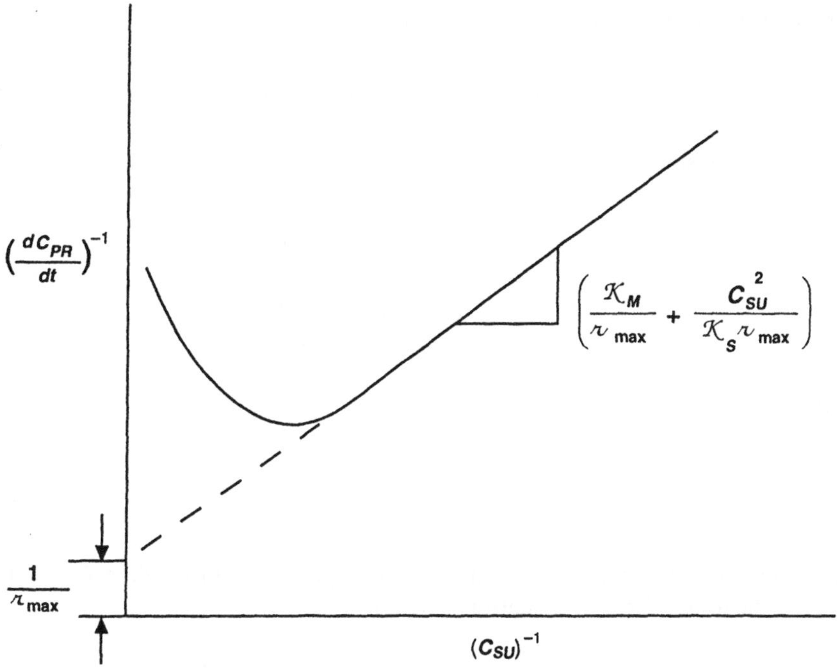

Figure 7-17 Substrate inhibition.

If substrate interference is large, the maximum velocity, r_{max}, may never be closely approached experimentally; this behavior is illustrated in Figure 7-17. However, the term $C_{SU}^2/\mathcal{K}_{SES}\, r_{max}$ in Equation 7-156 approaches zero at low values of C_{SU}, which means that the slope at high values of C_{SU}^{-1} approaches $\mathcal{K}_M/r_{max}$; as expected, this is the same as the linearized normal form given in Equation 7-150.

A second case of reaction inhibition, termed *competitive inhibition*, may occur as a result of a substance, IN, competing with the substrate for the same site on the enzyme. For this type of inhibition Equation 7-142 is modified to give

$$\begin{array}{l} E + S \underset{k_2}{\overset{k_1}{\rightleftharpoons}} ES \xrightarrow{k_3} P + E \\ + \\ IN \\ \updownarrow \\ EIN \end{array} \tag{7-157}$$

The conservation of mass expression for the enzyme is

$$C_{EN,0} = C_{EN} + C_{ES} + C_{EIN} \tag{7-158}$$

and with the same steps as those used to develop Equation 7-147, we can obtain

$$\frac{dC_{PR}}{dt} = \frac{r_{\max}}{1 + \dfrac{\mathcal{K}_M}{C_{SU}}\left(1 + \dfrac{C_{IN}}{\mathcal{K}_{EIN}}\right)} \tag{7-159}$$

where

$$\mathcal{K}_{EIN} = \frac{C_{EN}C_{IN}}{C_{EIN}} \tag{7-160}$$

Equation 7-159 reverts to the normal form (Equation 7-149) when $C_{IN} \approx 0$.

The linearized form of Equation 7-159 is

$$\frac{1}{dC_{PR}/dt} = \frac{1}{r_{\max}} + \frac{\mathcal{K}_M}{r_{\max}}\left(1 + \frac{C_{IN}}{\mathcal{K}_{EIN}}\right)\frac{1}{C_{SU}} \tag{7-161}$$

A plot of this equation is given in Figure 7-18, along with a comparative plot of the case for no inhibition. We observe that at infinitely high substrate concentration, or $C_{SU}^{-1} = 0$, the value for $r_{\max}$ is the same as that in the absence of competitive inhibition. Absence of competition should be expected at high substrate concentration; this provides the basis for definition of a competitive inhibitor.

The third type of reaction inhibition is caused by formation of reversible complexes between the inhibitor and enzyme and between the inhibitor and the enzyme–substrate complex. In contrast to the second type of inhibition, this is referred to as *noncompetitive inhibition* because the inhibitor–enzyme complex is independent of substrate concentration. The phenomenological depiction of this reaction scheme is

$$\begin{array}{ccccc} E + S & \underset{k_2}{\overset{k_1}{\rightleftharpoons}} & ES & \xrightarrow{k_3} & P + E \\ + & & + & & \\ IN & & IN & & \\ \updownarrows & & \updownarrows & & \\ EIN & & ESIN & & \end{array} \tag{7-162}$$

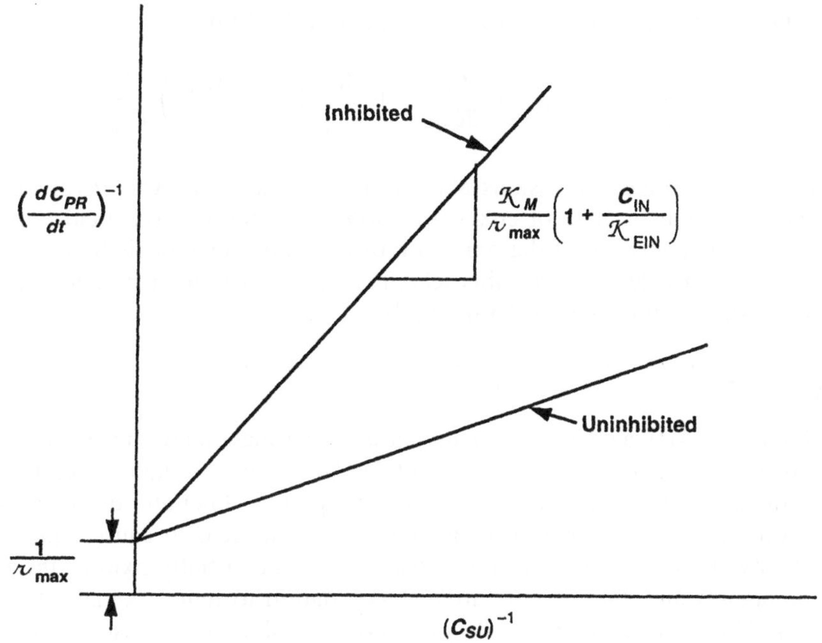

Figure 7-18 Competitive inhibition.

The initial amount of enzyme is now distributed among four different species at any time after initiation of the reaction or

$$C_{EN,0} = C_{EN} + C_{ES} + C_{EIN} + C_{ESIN} \tag{7-163}$$

If

$$\mathcal{K}_{EIN} = \frac{C_{EN}C_{IN}}{C_{EIN}} = \frac{C_{ES}C_{IN}}{C_{ESIN}} \tag{7-164}$$

then

$$\frac{dC_{PR}}{dt} = \frac{r_{max}}{\left(1 + \frac{\mathcal{K}_M}{C_{SU}}\right)\left(1 + \frac{C_{IN}}{\mathcal{K}_{EIN}}\right)} \tag{7-165}$$

When $C_{IN} = 0$, this equation reverts to Equation 7-149.

The rate expression given in Equation 7-165 can be linearized for graphic evaluation of the rate constants by rearranging in the form

$$\frac{1}{dC_{PR}/dt} = \frac{1}{r_{\max}}\left(1 + \frac{C_{IN}}{\mathcal{K}_{EIN}}\right) + \frac{\mathcal{K}_M}{r_{\max}}\left(1 + \frac{C_{IN}}{\mathcal{K}_{EIN}}\right)\frac{1}{C_{SU}} \tag{7-166}$$

In each of the rate expressions above, only one substrate was involved in the reaction. Another class of enzyme catalyzed reactions of practical importance involves mediation of the oxidation of hazardous chemicals by addition of hydrogen peroxide. In effect, this scheme includes enzymatic reactions with two substrates as illustrated in Example 7-7.

Example 7-7

- **Situation:** Activated sludge treatment at a dye manufacturing plant has proven fairly effective in reducing BOD. However, the aquatic toxicity of the effluent exceeds the state-mandated permit. Phenolic derivatives that are resistant to biological treatment are suspected to cause aquatic toxicity. Experiments have shown that ozone can partially oxidize these compounds and thus permit further biooxidation to reduce aquatic toxicity. However, ozonation is deemed too expensive in this case. Hydrogen peroxide was tried, but the oxidation reaction was far too slow to be practical. An alternative pathway for peroxide is provided by addition of peroxidase, an enzyme isolated from white rot fungi. The overall forward reaction scheme is

$$A + 2B \rightarrow 2P$$

where A is hydrogen peroxide, B is the phenolic derivative and P is partially oxidized product. (*Note:* The overall reaction is the same without the peroxidase enzyme, but the rate mechanism is different.) It has been established that the nonelementary pathway involves formation of two intermediate compounds, F and G, each of which reacts with B.

- **Question(s):** Develop a rate model for use in subsequent reactor design.
- **Logic and Answer(s):**

1. The series of elementary reactions that accounts for formation of two intermediates and the subsequent reaction of each with the phenolic derivative eventually to yield the oxidized product is

$$E + A \underset{k_2}{\overset{k_1}{\rightleftharpoons}} EA$$

$$EA \xrightarrow{k_3} F$$

$$F + B \underset{k_5}{\overset{k_4}{\rightleftharpoons}} FB$$

$$FB \xrightarrow{k_6} G + P$$

$$G + B \underset{k_8}{\overset{k_7}{\rightleftharpoons}} GB$$

$$GB \xrightarrow{k_9} E + P$$

where EA, FB, and GB are enzyme–substrate complexes. It should be noted that P is a radical that may undergo further reaction with either the parent compound, with itself, or with other intermediate products. Summing these reactions gives the overall stoichiometry shown in the forward reaction scheme above.

2. In order to quantify the rate of peroxidase-catalyzed oxidation, we seek an expression for the rate of production of oxidized organic compound (P) in terms of the reactants (A, B, and E_0) where

$$C_{E,0} = C_E + C_{EA} + C_F + C_{FB} + C_G + C_{GB}$$

and the initial rate of product formation is

$$\frac{dC_P}{dt} = k_6 C_{FB} + k_9 C_{GB}$$

3. Rate expressions for steady-state formation of each enzyme substrate complex and each of the intermediate products (EA, F, FB, G, and GB) can then be written for the reactions above. The algebraic manipulation leading to expressions for FB and GB is left to the reader in Problem 7-14. The final result is

$$\frac{dC_P}{dt} = \frac{r_{max} C_A C_B}{\alpha_{k,2} C_A + \alpha_{k,3} C_B + C_A C_B}$$

where

$$r_{max} = \alpha_{k,1} C_{E,0}$$

$$\alpha_{k,1} = \frac{2k_4 k_6 k_7 k_9}{k_6 k_7 k_9 + k_5 k_7 k_9 + k_4 k_6 k_9 + k_4 k_6 k_8}$$

$$\alpha_{k,2} = \frac{k_3[k_6 k_7 k_9 + k_5 k_7 k_9 + k_4 k_6 k_8 + k_4 k_6 k_9]}{k_4 k_7 [k_6 k_9 + k_3 k_9 + k_3 k_6]}$$

$$\alpha_{k,3} = \frac{k_6 k_9 [k_3 + k_2]}{k_1 [k_6 k_9 + k_3 k_9 + k_3 k_6]}$$

4. This expression is similar in form to that for the Michaelis–Menten model (Equation 7-149). However, it shows the effect of rate limitation caused by two substrates rather one substrate. If we divide through by C_A and then let C_A become large, we obtain the exact form of the Michaelis–Menten equation. In other words, if hydrogen peroxide is in ex-

cess, the rate is limited by one substrate (the organic compound) and not two. As for previous enzyme rate expressions, the initial rate data can be used to obtain the kinetic constants by taking the reciprocal of both sides of the rate expression:

$$\frac{1}{dC_P/dt} = \frac{1}{r_{max}}\left(\frac{\alpha_{k,2}}{C_B} + \frac{\alpha_{k,3}}{C_A} + 1\right)$$

7.7.3 Other Applications of Enzyme Rate Equations

Our discussion of rate expressions for biological catalysis began by emphasizing that the form of each of the enzymatic kinetic equations can also be applied—as appropriate—for description and modeling of a variety of complex physicochemical processes. The appropriateness of such rate expressions depends on the reaction schemes being of the same form as one of those given in Equations 7-142, 7-151, 7-157, 7-162, or as in Example 7-7, A + 2B → 2P.

7.8 SUMMARY

Rate relationships may take many different mathematical forms, which in turn can be used to describe transformations that are very different in their fundamental nature (i.e., chemical, biochemical, and even physical). Significant challenges are presented in the application of these rate relationships to transformations in environmental systems because (1) not all reactants may be known, so that chemicals must be lumped into surrogate or group parameters; (2) reaction pathways may be complex and poorly understood, so that empirical rate expressions are needed; (3) many reactions (some unknown) may occur simultaneously, so that only an overall rate can be expressed with confidence; and (4) reactions of interest may occur either too fast or too slow to allow convenient measurement, so that rate experiments must be designed carefully.

Contemporary approaches to environmental process engineering demand use of appropriate rate relationships in rigorous material-balance-based models. Thus the difficulties noted above must be either circumvented or overcome. Consider also the following examples of important and new areas of application for rate analysis. Waste minimization in the chemical synthesis industry requires knowledge of rates at which environmentally regulated chemicals may be produced in side reactions with trace contaminants of the feedstock or with industrial catalysis. Such reactions may have been heretofore ignored or considered unimportant because the concentrations of unwanted by-products were thought to be insignificant. Thermal processing to destroy or immobilize hazardous wastes requires that the role of temperature and pressure on reaction kinetics and production rates of unwanted by-products be quantified. Under these conditions, the general rule that reaction rate doubles for a 10°C rise in temperature is not applicable, due both to the nature of the oxidation reactions and the temperature ranges involved. More refined knowledge of catalyzed reactions (both biotic and abiotic) which offer potential for transformation of hazardous chemicals is essential to developing improved means for destruction of such materials as waste products, and for remediation of sites contaminated by them. The associated rate relationships are certain to be more complex than those given by simple models. The Michaelis–Menten saturation model is inappropriate, for example, if high concentrations of substrate cause inhibition or if two substrates are rate limiting rather than one.

The rate expressions developed in this chapter and the experimental methods described for their determination were illustrated using reactions of interest in dilute aqueous solutions. These principles can be extended to other important environmental conditions. Reactions in concentrated aqueous solutions (where concentration is adjusted for activity) and nonaqueous solutions (including air and organic solvents) can be analyzed similarly. Experimental determinations of rate relationships have been emphasized because this is state of the art for complex environmental systems. Rate expressions and rate constants reported in the literature for reactions which are similar, but not necessarily the same, are often useful starting points for such determinations. Theoretical analyses using chemical structure of the reactants and theories to describe transition states and reactive collisions can in some instances inform the experimental process and aid the development of more refined empirical rate models.

7.9 REFERENCES AND SUGGESTED READINGS

7.9.1 Citations and Sources

Amy, G. L., P. A. Chadik, and Z. K. Chowdury, 1987, "Developing Models for Predicting Trihalomethane Formation Potential and Kinetics," *Journal of the American Water Works Association*, *79*(7), 89–97. (Example 7-1)

Assaf-Anid, N., K. F. Hayes, and T. M. Vogel, 1994, "Reductive Dechlorination of Carbon Tetrachloride by Cobalamin(II) in the Presence of Dithiothreitol: Mechanistic Study, Effect of Redox Potential and pH," *Environmental Science and Technology*, *28*(2), 246–252 (Figure 7-13)

Grimberg, S. J., M. D. Aitken, and W. T. Stringfellow, 1994, "The Influence of a Surfactant on the Rate of Phenanthrene Mass Transfer into Water," *Water Science and Technology*. *30*(7), 23–30. (Example 7-2)

Hand, D. W., D. L. Perram, J. C. Crittenden, and M. E. Mullins, 1991, "Destruction of Disinfection By-Product Precursors Using Photoassisted Heterogeneous Catalytic Oxidation," *Proceedings of the 1991 American Water Works Association Annual Conference*, Philadelphia, PA, pp. 289–300. (Figure 7-14)

Hull, C. S., P. C. Singer, K. Saravan, and C. T. Miller, 1992, "Ozone Mass Transfer and Reaction: Completely Mixed Systems," *Proceedings of the 1992 American*

Water Works Association Annual Conference, Vancouver, British Columbia, pp. 457–465. (Problem 7-1 and Figure 7-2)

Kavanaugh, M. C., A. R. Trussel, J. Cromer, and R. R. Trussell, 1980, "An Empirical Kinetic Model of Trihalomethane Formation: Applications to Meet the Proposed THM Standard," *Journal of the American Water Works Association*, *72*(10), 578–582. (Example 7-4)

McKnight, A., and D. A. Reckhow, 1992, "Reactions of Ozonation Byproducts with Chlorine and Chloramines," *Proceedings of the 1992 American Water Works Association Annual Conference*, Vancouver, British Columbia, pp. 399–409. (Example 7-3)

Segel, I. H., 1975, *Enzyme Kinetics*, John Wiley & Sons, Inc., New York, pp. 606–608. (Example 7-7 and Problem 7-14)

Smith, J. M., 1981, *Chemical Engineering Kinetics*, 3rd Edition, McGraw-Hill Book Company, New York. (Example 7-6)

Snider, E. H., and F. C. Alley, 1979, "Kinetics of the Chlorination of Biphenyl Under Conditions of Waste Treatment Processes," *Environmental Science and Technology, 13*, 1244–1248. (Problem 7-2)

7.9.2 Background Reading

Bailey, J. E., and D. F. Ollis, 1986, *Biochemical Engineering Fundamentals*, 2nd Edition, McGraw-Hill Book Company, New York. Chapter 3 deals with the kinetics of enzyme-catalyzed reactions, various inhibition patterns. Chapter 8 covers micro- and macroscale transport phenomena that limit biochemical reactions and scaleup of mass transfer equipment.

Benson, S. W., 1960, *The Foundation of Chemical Kinetics*, McGraw-Hill Book Company, New York. Chapter 12 provides a very detailed comparison of collision and transition-state theory.

Benson, S. W., 1968, *Thermochemical Kinetics*, John Wiley & Sons, Inc., New York. The fundamental concepts for predicting rate expressions without experimental data are carried forward to provide procedures for estimating rate constants.

Frost, A. A., and R. G. Pearson, 1961, *Kinetics and Mechanism*, John Wiley & Sons, Inc., New York. An excellent, easy-to-read general reference.

Grady, C. P. L., Jr., and H. C. Lim, 1980, *Biological Wastewater Treatment*, Marcel Decker, Inc., New York. Chapter 10 provides coverage of various enzyme kinetic models, classical methods of data analysis to evaluate these models, and applications of enzyme kinetics to describe cell growth and substrate utilization in wastewater treatment.

Segel, I. H., 1975, *Enzyme Kinetics*, John Wiley & Sons, New York. Simple inhibition systems are discussed in Chapter 3, with many examples of various graphical methods to test kinetic data against models. Chapter 9 provides extensive coverage of multisubstrate systems for which steady-state kinetic models are developed.

Sigel, H., and A. Sigel (eds.), 1992, *Metal Ions in Biological Systems*, Vol. 28, Marcel Dekker, Inc., New York. Oxidative and reductive enzymatic pathways and, implicitly, the role of ligand-transition metal binding within these pathways, are discussed in a series of contributions for application to aerobic and anaerobic degradation of aromatic hydrocarbons and halogenated aliphatics.

Smith, J. M., 1981, *Chemical Engineering Kinetics*, 3rd Edition, McGraw-Hill Book Company, New York. In Chapter 3 a wide range of kinetic expressions is discussed, with many examples. The analysis of nonelementary reactions by the rate-controlling step and the stationary steady-state approximation for intermediates are compared. Collision theory and transition-state theory are explored as theoretical tools for prediction kinetics without experimentation.

7.10 PROBLEMS

7-1 The decomposition rate of ozone is assumed to adhere to first-order kinetics. The rate is also retarded by carbonate ions, which act as hydroxyl radical scavengers. Three sets of batch data collected at pH 7.0 with the same carbonate concentration (0.01 *M* total carbonate species) and with three different initial ozone concentrations are given below (Hull et al., 1992). Also shown is one set of data without addition of carbonate. Verify the first-order decomposition of ozone and the effect of the carbonate on the rate constant.

	C_{O_3} (mg/L)			
Time (min)	With Carbonate (0.01 *M*)			Without Carbonate
0	0.3	1.0	3.0	3.0
1	0.28	0.98	2.71	2.24
3	0.27	0.89	2.63	1.99
10	0.23	0.80	2.11	1.12
20	0.20	0.70	1.98	0.46
40	0.16	0.53	1.40	0.06
60		0.36	0.90	

7-2 Biphenyl (BP) is used as a dye carrier in the textile industry and is thus potentially present in the wastewaters from such facilities. The discharge of such an effluent to a publicly owned treatment facility where chlorine is used for odor control could allow for reaction with BP to form *o*-chlorobiphenyl (OCBP). Snider and Alley (1979) studied the kinetics of chlorination of BP and proposed that the rate of formation of OCBP is given by

$$\frac{d\, C_{\mathrm{OCBP}}}{dt} = k\, C_{\mathrm{BP}} C_{\mathrm{HOCL}}^2$$

where HOCl is undissociated hypochlorous acid and all concentrations are on a molar basis. The protolysis constant for HOCl is given by

$$K_A = \frac{C_{H^+} C_{OCl^-}}{C_{HOCl}} = 3.66 \times 10^{-8}$$

and

$$C_{HOCl_T} = C_{HOCl} + C_{OCl^-}$$

The rate of formation of OCBP is so slow that the initial rate method was used for verification of the reaction order. The initial rate is calculated by measuring C_{OCBP} after 12 h. Verify the rate relationship and show that the rate constant, k, is about the same in each data set.

Molecular weights (g/mol):
BP = 154; OCBP = 188.5; HOCl = 52.5

C_{BP} (mg/L)	C_{HOCl_T} (mg/L)	pH	C_{OCBP} (μg/L after 12 h)
3770	17.9	7.0	1.2
3770	38.5	7.0	5.7
3770	54.7	7.0	11.0
3770	74.7	7.0	23.0
3770	90.5	7.0	34.3
3400	247	6.78	222
2720	247	6.78	163
2040	247	6.78	122
1360	247	6.78	83.4
680	247	6.78	42.5
3590	304	7.06	187
3590	304	7.47	92.0
3590	304	8.04	18.4
3590	304	8.31	7.4
3590	304	9.17	3.4

7-3 Peroxidase, which has been isolated from the white rot fungus, is proposed to enhance the peroxide oxidation of phenol. A Michaelis–Menten rate relationship is postulated for the case of peroxide being present in large excess:

$$-r = \frac{k_3 C_{Ph} C_{EN,0}}{\mathcal{K}_M + C_{Ph}}$$

The initial enzyme concentration, $C_{EN,0}$, in a CMBR experiment designed to test the rate relationship is 0.01 mmol/L, and the initial phenol concentration, $C_{Ph,0}$, is 1 mmol/L. The rate data are:

Time (h)	C_{Ph} (mmol/L)
1	0.84
2	0.68
3	0.53
4	0.38
5	0.27
6	0.16
7	0.09
8	0.04
9	0.018
10	0.006

Determine whether these data can be reasonably fit by the Michaelis–Menten model, and if so, evaluate the rate constants.

Answers: $k_3 = 20.6\ h^{-1}$ *and* $\mathcal{K}_M = 0.3\ mmol/L.$

7-4 It has been reported that the chirping rate of crickets at night increases with temperature and can be described by the relationship

$$(\text{number of chirps in 15 s}) + 40 = (\text{temperature, °F})$$

Assuming that chirping rate is a direct measure of the metabolic rate, find the cricket activation energy corresponding to a temperature range 60 to 80°F. Discuss whether this activation energy is consistent with those found for typical rate-limiting steps in biochemical reactions.

7-5 The mechanism of manganese oxidation by oxygen is given by the following elementary reactions:

$$Mn^{2+} + O_2 \xrightarrow{k_1} MnO_2(s)$$

$$Mn^{2+} + MnO_2(s) \xrightarrow{k_2} \{Mn^{2+} \cdot MnO_2(s)\}$$

$$\{Mn^{2+} \cdot MnO_2(s)\} + O_2 \xrightarrow{k_3} 2\ MnO_2(s)$$

where $\{Mn^{2+} \cdot MnO_2(s)\}$ represents a short-lived intermediate product. Derive expressions for the rate of disappearance of Mn^{2+} and the rate of appearance of MnO_2 when O_2 is in excess.

Answers:
$$-\frac{dC_{Mn^{2+}}}{dt} = k_{ps} C_{Mn^{2+}} + k_2 C_{Mn^{2+}} C_{MnO_2(s)}$$

$$\frac{dC_{MnO_2(s)}}{dt} = k_{ps} C_{Mn^{2+}} + k_2 C_{Mn^{2+}} C_{MnO_2(s)}$$

where a pseudo-first order rate constant, k_{ps}, is given by

$$k_{ps} = k_1 C_{O_2}$$

7-6 The oxidation of Fe(II) by oxygen is carried out in a solution buffered at pH 7.0 and in contact with an airstream that contains an essentially constant partial pressure of oxygen (and, therefore, an essentially constant dissolved oxygen concentration) at 20°C. The following data are collected:

Time (min)	Fe(II) $\times 10^5$ M
0	3.3
5	2.1
10	1.3
15	0.82
20	0.59
25	0.32

The reaction is postulated to be first order each with respect to Fe(II) and P_{O_2} and second order with respect to OH^-. Verify the first-order dependency on Fe(II) and determine the rate constant.

7-7 Ozone decomposes in water according to the stoichiometric relationship

$$2O_3 \longrightarrow 3O_2$$

Measurements of ozone decomposition at pH 9 in the absence of any constituent which may be oxidized yield:

Time (s)	C_{O_3} (mg/L)
0	2.00
30	1.50
60	1.05
90	0.82
120	0.57

It is thought that the rate of ozone decomposition is nonelementary and given by

$$-\frac{dC_{O_3}}{dt} = k_1 C_{O_3}$$

Another set of experiments is conducted to measure the rate of ozone consumption by methanol oxidation. The reaction of ozone with methanol is thought to be elementary. In analyzing the results of this set of experiments it is necessary to consider the parallel reactions:

$$2O_3 \xrightarrow{k_1} 3O_2$$

$$O_3 + CH_3OH \xrightarrow{k_2} \text{products}$$

The experiments were conducted with excess amounts of methanol so that the methanol concentration, C_{Me}, could be considered constant during collection of the rate data. Several runs were made with different methanol concentrations to verify that the reaction was, in fact, first order with respect to methanol and second order overall. The experimental data are:

Run 1: $C_{Me} = 1.25 \times 10^{-4}$ mol/L

Time (s)	C_{O_3} (mg/L)
0	3.00
6	2.64
18	1.80
30	1.38

Run 2: $C_{Me} = 3.125 \times 10^{-4}$ mol/L

Time (s)	C_{O_3} (mg/L)
0	3.00
6	2.16
18	1.32
30	0.60

Run 3: $C_{Me} = 1.25 \times 10^{-3}$ mol/L

Time (s)	C_{O_3} (mg/L)
0	3.00
6	1.05
18	0.18
30	0.03

Verify the orders of the ozone oxidation of methanol and the ozone decomposition reactions, and calculate the rate constants for each.

7-8 Enzymes are known to deactivate naturally with time; alternatively, they may be deactivated by other components in the solution. An enzymatic process is being considered for treatment of a high-strength phenolic waste from an industry. This waste is generated in small volumes and the concentrations of phenols are so high that conventional biological treatment is impossible due to substrate inhibition. The enzymatic system requires, however, that hydrogen peroxide be added to promote hydroxyl free radicals but hydrogen peroxide also causes deactivation of the enzyme. A set of four rate tests were conducted to measure the rate of enzyme deactivation by hydrogen peroxide (in the absence of phenols). In each test, the starting concentration of enzyme was 1 μM and the hydrogen peroxide (H_2O_2) was added in excess at four different concentrations. The fractional remaining activity of the enzyme was measured with time to give:

	H_2O_2 in Excess in Each Test			
Time (min)	100 μM	500 μM	1000 μM	2000 μM
0	1.00	1.00	1.00	1.00
10	0.90	0.84	0.79	0.70
20	0.82	0.70	0.63	0.50
30	0.74	0.58	0.50	0.35
40	0.67	0.49	0.40	0.25
50	0.61	0.41	0.32	0.17
60	0.55	0.34	0.25	0.12
80	0.45	0.24	0.16	0.06

Find the rate constant and the order of the deactivation reaction with respect to enzyme and hydrogen peroxide concentrations.

7-9 Disinfection of anthrax spores in a CMBR yields the following data:

Time (min)	Number of Survivors
0	400
100	100
200	18
300	3
410	1

Assuming a first-order dependence on numbers of survivors, find the corresponding rate constant.

7-10 Show for an autocatalytic reaction the amount of reactant remaining as a function of time for assumed values of k_1, k_2, and C_0. By varying k_1 and k_2, illustrate the effect on the lag period.

7-11 The rate constant, k, for a second-order reaction between substances A and B is 0.135 L/mol-s at 69.4°C and 3.70 L/mol-s at 81.2°C. Determine the rate constant for the reaction at 55.0°C.

7-12 A radioactive waste contains 2 curies (Ci) of ^{60}Co per liter and 2 mCi of ^{45}Ca per liter. The waste can be discharged only if its total activity does not exceed 20 μCi/L. If the half-life of ^{60}Co is 10.7 min and that of ^{45}Ca is 152 days, determine how long the waste must be stored before it can be discharged.

Answer: 1010 d.

7-13 Nitrification is a fairly slow biochemical reaction in which ammonia is converted to nitrite by one genus of microorganism (e.g., *Nitrosomonas*) and to nitrate by another genus (e.g., *Nitrobacter*):

$$NH_4^+ \rightarrow NO_2^- \rightarrow NO_3^-$$

If ammonia is present in a surface water, there is the potential for nitrification in the water distribution system; nitrate formation is of concern because of health risks to infants. A batch kinetic study is undertaken to determine how fast nitrification occurs under worst-case conditions. A large number of cells of *Nitrosomonas* is added to a CMBR containing a 0.5 m*M* solution of NH_4^+ (9 mg/L); solution chemistry and temperature are similar to those of the drinking water. A first-order formation of nitrite is observed with a rate constant of 10^{-4} min^{-1}. A second experiment is conducted in which large cell numbers of both *Nitrosomonas* and *Nitrobacter* are added to a CMBR containing a 0.5 m*M* solution of NH_4^+ and nitrate production (m*M*) is measured with the following results:

Time (min × 10^{-3})	NO_3^- (m*M*)	Time (min × 10^{-3})	NO_3^- (m*M*)
0	0	7	0.16
1	0.01	8	0.19
2	0.02	9	0.21
3	0.05	10	0.24
4	0.07	12	0.28
5	0.10	14	0.32
6	0.13	18	0.38

Use this information to (a) determine the rate constant for first-order conversion of nitrite to nitrate, (b) investigate the sensitivity of results to the choice of rate constants for both nitrite and nitrate formation, and (c) draw a conclusion about which step is rate controlling.

Answer: (a) $k = 3 \times 10^{-4}$ min^{-1}.

7-14 The peroxidase-catalyzed reaction A + 2B → 2P was explored in Example 7-7. Derive the rate expression used in that example; that is,

$$\frac{dC_P}{dt} = \frac{r_{max} C_A C_B}{\alpha_{k,2} C_A + \alpha_{k,3} C_B + C_A C_B}$$

[*Hint:* Find EA, FB, and GB in terms of E, A, F, B, and G by writing the expression for each elementary reaction given in Example 7-7 and then use these results to obtain expressions for F and G in terms of E, A, and B. (see Segel (1975) for "Ping Pong Bi Bi System".]

7-15 An example of the mechanism given in Problem 7-14 is peroxidase-catalyzed oxidation of *p*-cresol (PC). Two substrates are involved: *p*-cresol and hydrogen peroxide. Initial rate data for disappearance of hydrogen peroxide (μmol/L-min) collected for various combinations of initial concentrations of hydrogen peroxide and *p*-cresol are as follows:

C_{0,H_2O_2} (μmol/L)	Initial Rate (μmol/L-min) for $C_{PC,0}$ of: 300 μmol/L	1400 μmol/L	4100 μmol/L
100	85	110	116
300	144	232	261
500	166	298	347
800	183	355	427
1800	201	430	541

Find r_{max} and the constants $\alpha_{k,2}$ and $\alpha_{k,3}$.

Answers: $r_{max} = 830\ \mu M/min$, $\alpha_{k,2} = 840\ \mu M$, $\alpha_{k,3} = 590\ \mu M$.

8

RATE RELATIONSHIPS: APPLICATIONS TO HETEROGENEOUS SYSTEMS

8.1 INTRODUCTION

Many reactions of importance in natural and engineered environmental systems involve more than one phase. We refer to these as heterogeneous reactions. The equilibrium principles discussed in Chapter 6 determine the ultimate distributions of constituents between two or more phases (e.g., between water and air, between water, air and sediment, or between water and activated carbon). It is essential in the analysis of system dynamics that we also characterize the rates at which equilibrium conditions are approached. In many instances, reaction rates themselves (i.e., *intrinsic* kinetics) may be very fast, but associated microtransport processes—that is, *interphase* mass transfer from solution to reactive phases or *intraphase* mass transfer within reactive phases—may be slow and therefore *rate determining*. The *adsorption* of organic solutes onto activated carbon surfaces, for example, is a very *fast reaction*, but *rates of uptake* of organic compounds from solution by microporous carbon are much slower because solutes must diffuse to interior adsorption sites before reacting.

Some reactions commonly thought of as homogeneous may in fact be heterogeneous, particularly when boundaries between phases are poorly defined. For example, chemical oxidations at colloid surfaces may be limited by transport of oxidants to those surfaces. Ozone must be transported to the surfaces of inorganic colloids before oxidation reactions occur with sorbed layers of natural organic material. Similarly, inactivation of bacteria in chemical disinfection processes may be limited by slow transport of disinfectant through cell walls before fast oxidation reactions with intracellular enzymes occur. The surface areas associated with enzymes are in the neighborhood of 1000 $Å^2$; thus, the formation of an enzyme–substrate complex may also involve surface reactions.

Heterogeneous reactions may involve mass transfer from gases or liquids to the external surfaces of solids, and if the solids are porous, internal mass transport as well. We refer to these as *microtransport processes* because they occur over spatial scales that are small relative to the spatial scales of mass transport through the reactor systems in which they take place. Three microtransport conditions are shown in Figure 8-1. In Figure 8-1a, reactants are first transported across a laminar fluid layer or film before reactions occur at the external surfaces of the solid. The laminar fluid layer is an *external impedance* or *resistance* to mass transfer. If as depicted in Figure 8-1b, the solid is porous and reaction sites are located internally, potential reactions encounter *internal resistance* to mass transfer as well, a transport process generally referred to as *intraparticle diffusion.* When reaction sites are located internally, *both external and internal impedances* in series often occur, as illustrated in Figure 8-1c. If the external and internal impedances are similar in magnitude, both are included in the formulation of heterogeneous rate relationships. If one impedance is much less than the other, it generally can be ignored. For example, if the external resistance to diffusive mass transfer is much less than the internal

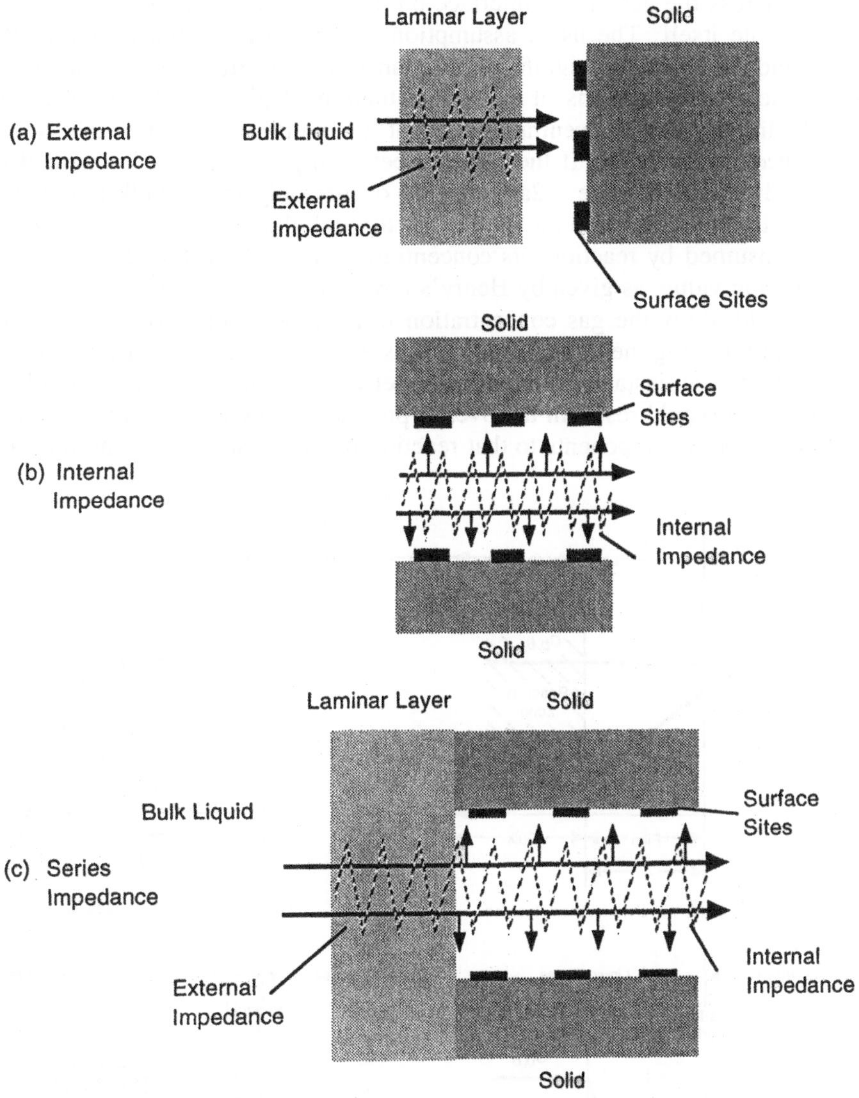

Figure 8-1 Microtransport involving external and internal mass transfer impedances.

resistance, the latter is said to be *controlling* and the rate analysis is simplified accordingly.

One important class of heterogeneous processes involves gas–liquid mass transfer accompanied by reaction. Process examples in this class include aqueous phase oxidations with gaseous oxidants such as chlorine, chlorine dioxide, ozone, and sulfur dioxide, all commonly used in water and wastewater treatment. As illustrated in Figure 8-2, the relative influence of mass transfer across the gas–liquid interface on overall process rate depends on the reaction rate itself. The usual assumption is that only the *liquid side* of the gas–liquid film provides significant resistance to mass transfer. Although this is not true for all situations, it is for the situations depicted in Figure 8-2, and thus the liquid-phase concentration of A at the interface ($x = 0$) is the saturation concentration, $C_{S,A}$. If the reaction between gas A and component B of the liquid is slow (Figure 8-2a), reaction occurs only in the bulk liquid, but the overall rate depends on the flux of gas through the liquid film. Because the gas is consumed by reaction, its concentration in the bulk liquid is less than its saturation value, as given by Henry's law. An even slower reaction (Figure 8-2b) would allow the gas concentration to approach saturation in the bulk, thereby minimizing the effect of mass transfer on overall reaction rate. At the other extreme, an instantaneous reaction between A and B at a *reaction plane* (Figure 8-2c) would result in the overall process rate being determined by the diffusion of both components to that reaction plane. Reactions of intermediate

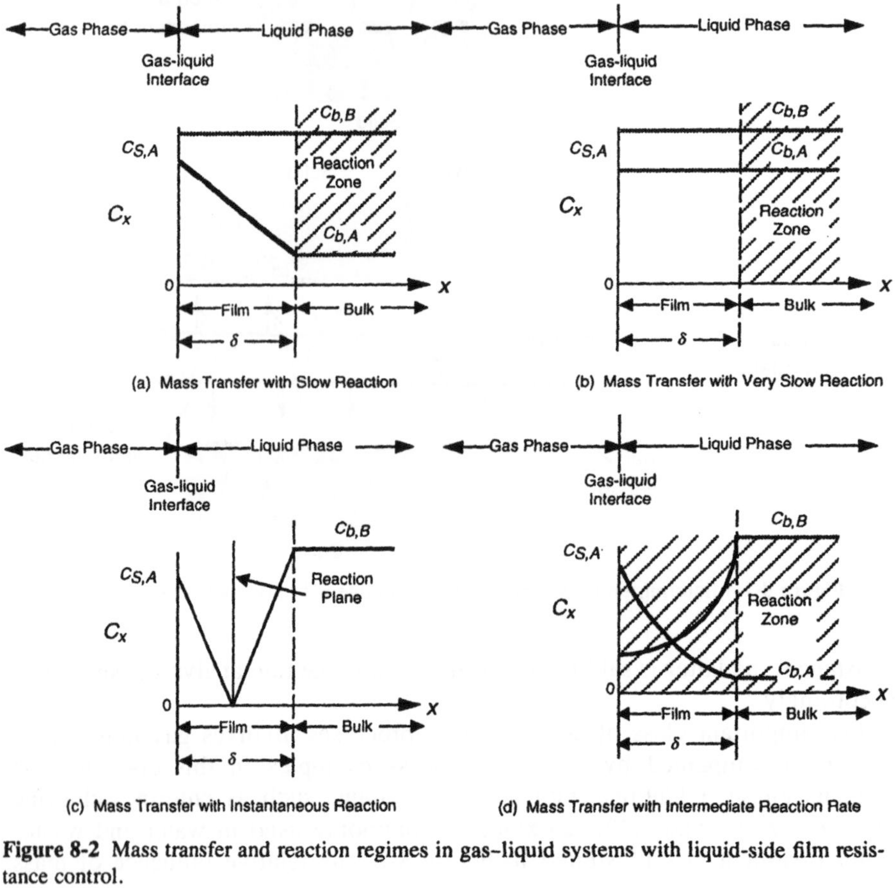

Figure 8-2 Mass transfer and reaction regimes in gas–liquid systems with liquid-side film resistance control.

rate can occur either in both the liquid film and bulk liquid, as depicted in Figure 8-2d, or entirely within the liquid film. The latter situation would lead to complete depletion of A within the film and thus no entry of that solute to the bulk solution.

Both macro- and microtransport processes occur in reactors designed to accommodate heterogeneous reactions. This is illustrated for solid–liquid and gas–liquid reactions in parts a and b, respectively, of Figure 8-3. *Macrotrans-*

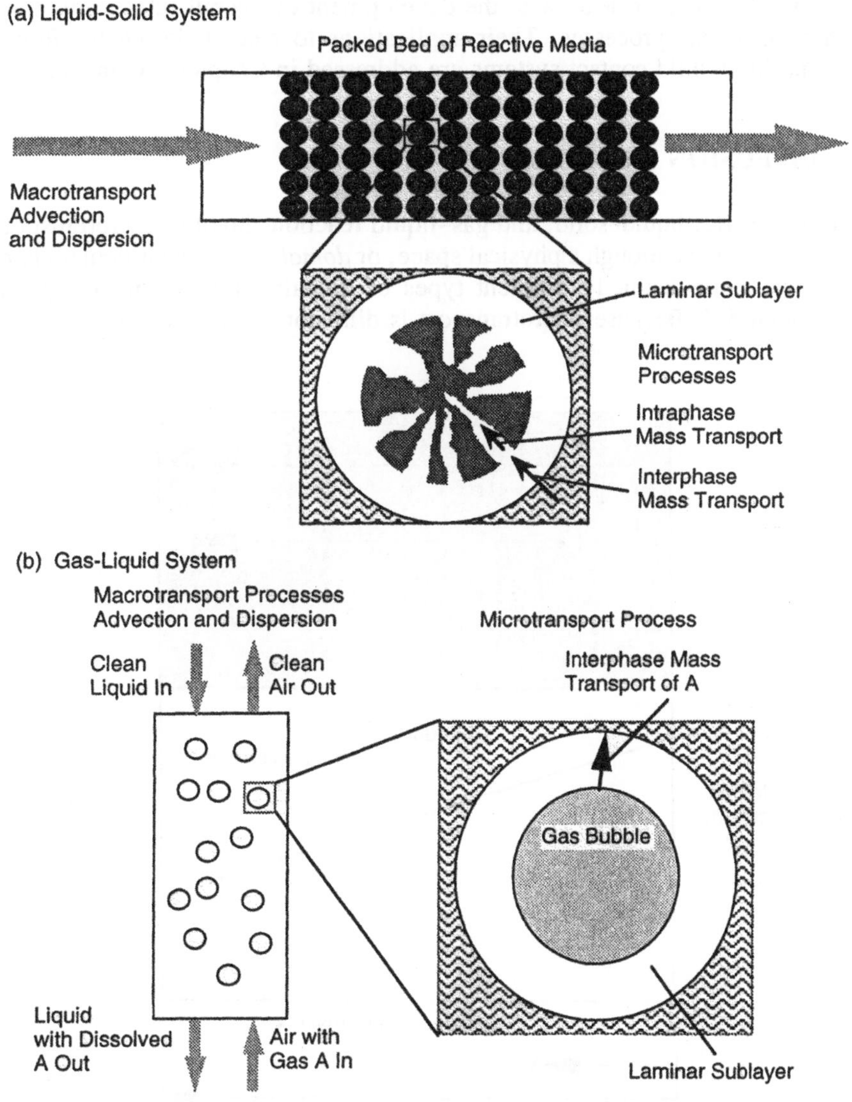

Figure 8-3 Macrotransport and microtransport processes in heterogeneous systems.

port is responsible for the movement of reactants in the axial direction through the reactor, either by advection alone or by a combination of advection and dispersion. Within any elemental volume, *microtransport* occurs in a radial direction and, in these examples, is responsible for mass transfer from the liquid to the external solid surface and within the solid in Figure 8-3a, and mass transfer from the gas bubbles to the liquid in Figure 8-3b. The micro-transport processes are incorporated into an "overall rate expression" for use in the material balance equation for an elemental volume of the reactor, in much the same way as is the appropriate rate relationship for a homogeneous reaction. This chapter deals with the development of overall rate expressions for heterogeneous processes. Their applications to reactor design for fluid–solid and fluid–fluid contact systems are addressed in Chapters 10 and 11.

8.2 DIFFUSION DOMAINS

All gas–solid, liquid–solid, and gas–liquid reactions involve microtransport of reactive species through a physical space, or *domain*, having particular characteristics. Four distinctly different types of domains are shown in Figures 8-4 through 8-7. Because microtransport is diffusion motivated, these are re-

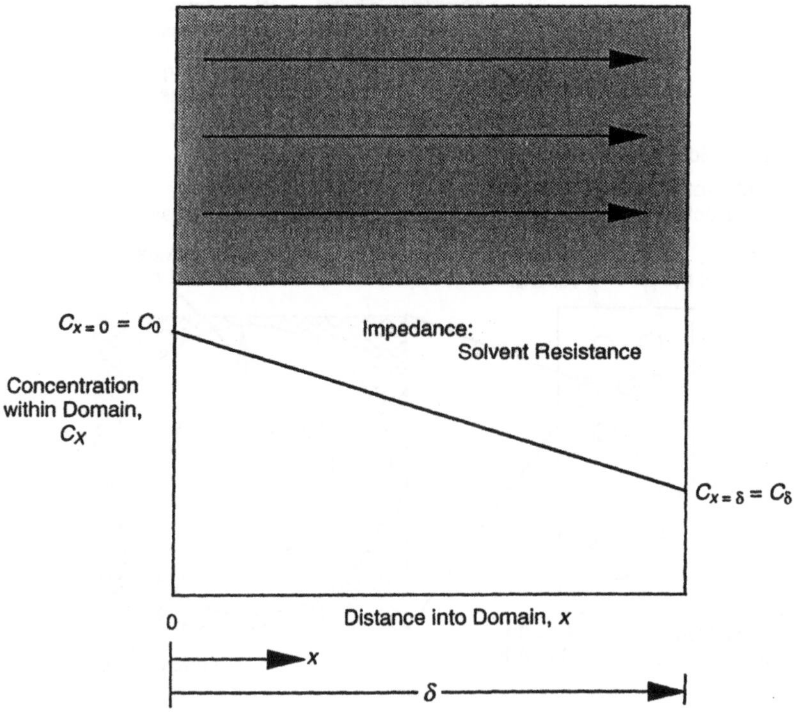

Figure 8-4 Solute migration and steady-state concentration profile for type I diffusion domains.

ferred to as *diffusion domains*. These domains are helpful in classifying the nature of microtransport process for various engineered and natural systems.

A *type I* domain (Figure 8-4) is one in which transport by molecular diffusion occurs without any concurrent reaction of the diffusing solute; the appropriate mathematical model for microtransport is thus simply Fick's law. The laminar boundary layer surrounding a reactive particle is an example of a type I domain. This layer acts as an *external impedance* to the diffusion of solutes attempting to reach the surfaces of the particle. Applications of type I domain models are not necessarily restricted to description of boundary layer microtransport processes, however. This is an appropriate model for use to describe diffusion of any conservative (nonreactive) substance through any homogeneous or single-phase domain (e.g., molecular diffusion through a quiescent water column).

If the domain is homogeneous but also involves a reaction of the species being transported, it is termed a *type II* domain (Figure 8-5). Here, both microtransport and reaction processes must be considered simultaneously in the mathematical formulation of a rate expression. A common example is a gas-liquid interface through which a reactive gas is transferred to a bulk liquid.

Transport in heterogeneous domains (multiphase) is described by either the

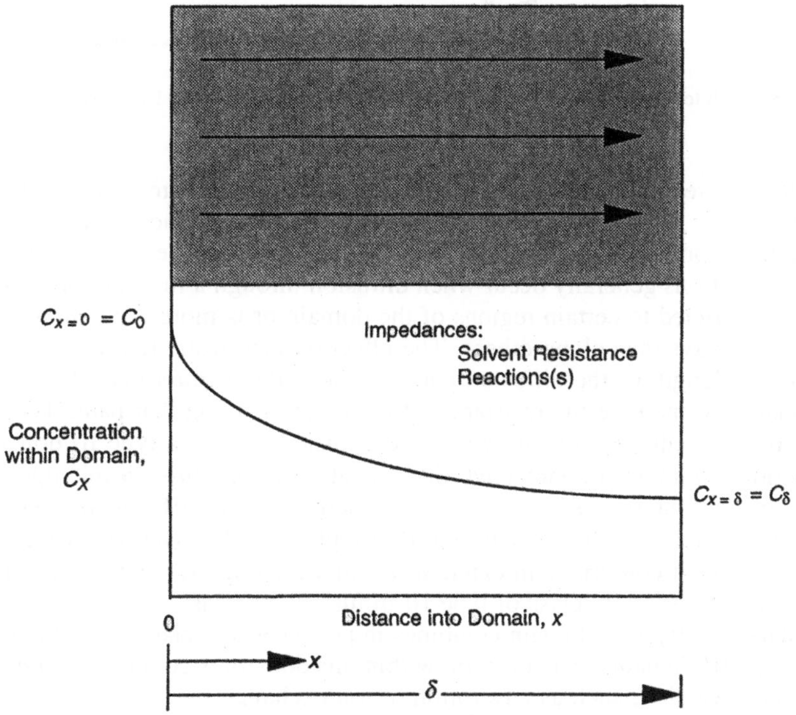

Figure 8-5 Solute migration and steady-state concentration profile for type II diffusion domains.

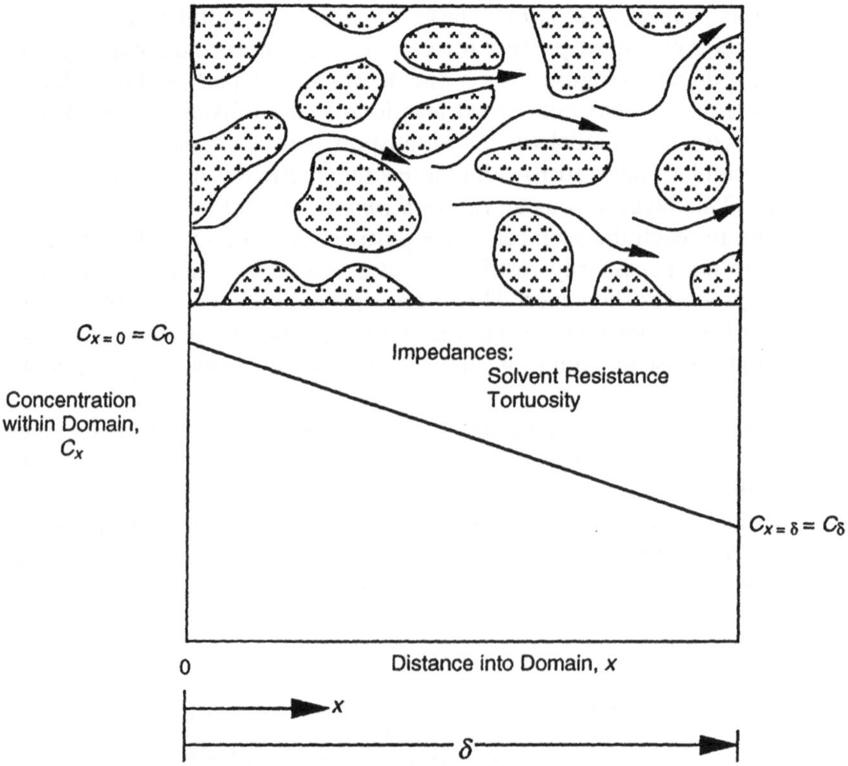

Figure 8-6 Solute migration and steady-state concentration profile for type III diffusion domains.

type III representation shown in Figure 8-6 or the *type IV* representation shown in Figure 8-7. The type III domain accounts for an impedance to nonreactive solute transport due to constraint of solute migration to specific flow paths. These conditions generally occur when diffusion through a multiple-phase medium is restricted to certain regions of the domain or is more rapid along particular pathways than along others. The effect of such restricted movement is to alter the length of the diffusion path, the so-called *tortuousity* effect, and, potentially, to increase the resistance of the medium along that path. The porosity of the medium may vary along the diffusion path of a type III domain, a situation common to nonreactive diffusional transport through microporous solids and the interstices of soils and sediments. A type III domain may lie between two bulk phases or between a bulk phase and a reactive surface; in each instance, it comprises an external impedance. For example, nonreactive membranes constitute a class of type III diffusion domains.

Finally, the type IV domain combines impedances associated with the type II and type III domains. The reaction within this domain is commonly a *phase-separation* reaction, such as adsorption or ion exchange.

The diffusion domain concept is useful in development of rate models for

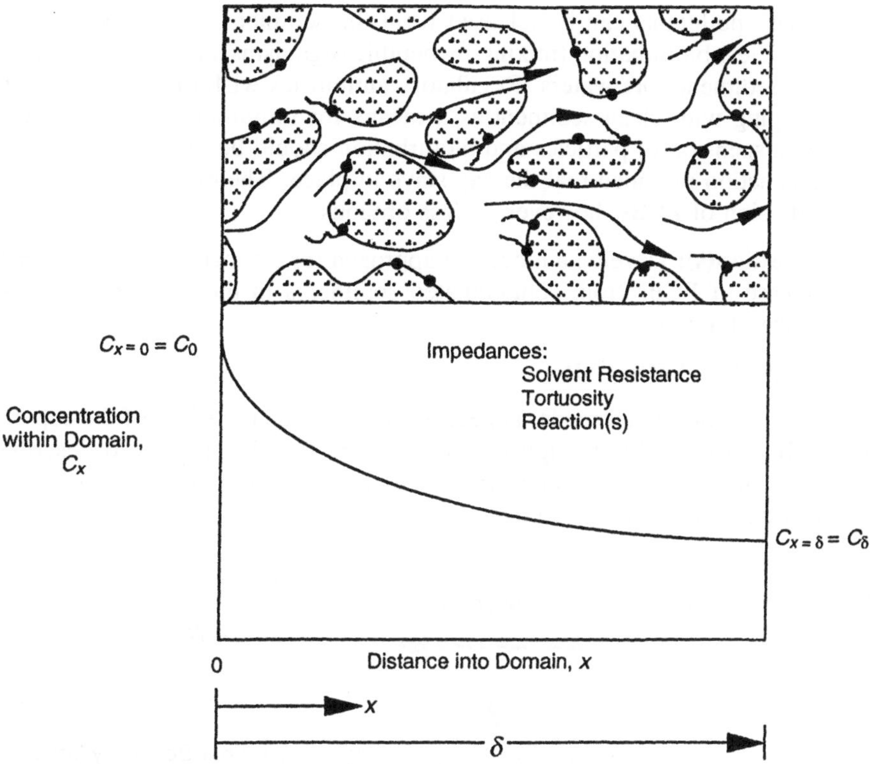

Figure 8-7 Schematic representation of solute migration and steady-state concentration profile for type IV diffusion domains.

heterogeneous systems, as demonstrated in this chapter for several specific types of environmental systems. With the exception of those involving sorption reactions, the models to be presented all involve steady-state microtransport. Moreover, all of the domains are *bounded*; that is, they have a *finite length. Unbounded-domain* descriptions may be required in physical situations for which a solute does not reach the downgradient boundary of the domain in the time period of concern. The assumption of an unbounded domain is frequently used to simplify the mathematical solution of microtransport and reaction models applied to process descriptions during unsteady-state conditions. These circumstances are discussed in Chapter 11.

Example 8-1 illustrates how several different types of diffusion domains may be involved in the characterization of transport phenomena in environmental systems.

Example 8-1

- **Situation:** Polychlorinated biphenyls (PCBs) have accumulated in the sediment of a harbor due to oil spills and leakages from barge traffic.

These pollutants can be released from the sediment by resuspension of particles, bioturbation (mixing by benthic organisms), and diffusion from the sediment pore waters. Regulatory authorities wish to evaluate *in situ* capping with clean sediment as a means for minimizing such releases. Resuspension of particles and bioturbation should be greatly reduced by this action provided that the cap stays in place. The effect of the cap on diffusion of PCBs, however, is less obvious.

- **Question(s):** Develop a modeling approach for estimating the diffusional release of PCBs from contaminated sediments in the presence and absence of a cap.
- **Logic and Answer(s):**

1. A conceptual model is needed to determine the effect of the cap on diffusion of PCBs through the pore water. The first step is to describe the macro- and microtransport processes that occur without a cap, as pictured schematically below.

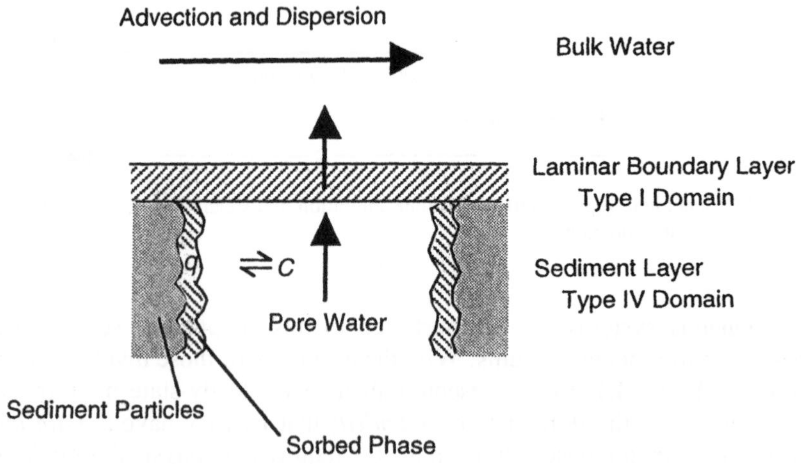

2. The sediment layer is comprised of solid particles and their interstitial pores. The sediment particles are coated with natural organic matter into which PCBs have sorbed (q = sorbed-phase concentration). On a simplified basis, the surfaces of these sediment particles can be thought of as forming the walls of the interstitial pores. PCBs can diffuse out of the sediment whenever the pore water concentration near the top of the sediment layer is lower than below.

3. Diffusion occurs simultaneously with sorption–desorption from the sediment particles. The simplest assumption to make with respect to the sorption process is that when PCB pore water concentration changes at

some position, reequilibration is instantaneous between the pore water and sorbed phase (i.e., $q = q_e$ and $C = C_e$).

4. Because pore diffusion and sorption (treated like a reaction) occur simultaneously, the sediment layer behaves as a type IV diffusion domain. [*Note:* A more complex microscale model might assume that sorption equilibrium is controlled by diffusion of PCBs through the matrix of natural organic matter in the layer surrounding the sediment particles and is thus not instantaneous; (i.e., a type III diffusion domain connected to a type IV diffusion domain).] Unsteady-state diffusion in this type IV domain can be described by Fick's second law modified to include accumulation terms in both the pore water and sorbed layer (i.e., $\partial C/\partial t$ and $\partial q/\partial t$); this modification is discussed in more detail later in the chapter.

5. PCB molecules which diffuse through the type IV domain reach the laminar boundary layer at the sediment–bulk water interface. This boundary layer is a type I diffusion domain, the thickness of which is determined by the hydrodynamic energy and mixing characteristics of the harbor. (*Note:* If PCBs also undergo biochemical or chemical reaction due to a shift in oxidation-reduction potential in the laminar boundary layer, that layer becomes a type II diffusion domain). Diffusion through a type I domain can be described by a simple film model (see Chapter 4).

6. For mass transfer of PCBs to occur from the sediment to the overlying water, the bulk water concentration must be less than the pore water concentration at the sediment layer–laminar boundary layer interface. In turn, the PCB concentration here is less than at all other locations within the sediment layer. The necessary mathematical boundary condition to solve Fick's second law for diffusion within the type IV domain is continuity of flux (i.e., the mass transfer flux of PCBs from the type I domain to the bulk water is equal to the diffusive flux from the type IV domain to the type I domain).

7. PCBs that diffuse through the type I domain reach the bulk water, where they are transported away from the sediment–water interface by advection and dispersion. If these macrotransport processes remain constant in time, the bulk concentration is also assumed to remain constant in time.

8. At the beginning of the diffusion process, the pore water concentration gradient will be very steep near the sediment–water interface, due to the loss of PCB molecules from only the topmost section of the sediment layer. As PCB molecules are lost from lower positions, the concentration gradient, and thus the diffusion flux, will gradually decline. A mathematical model based on the conceptual framework provided in steps 1 to 8 produces the general pattern of unsteady-state diffusion flux.

9. The presence of a cap adds an additional diffusion domain, as depicted below.

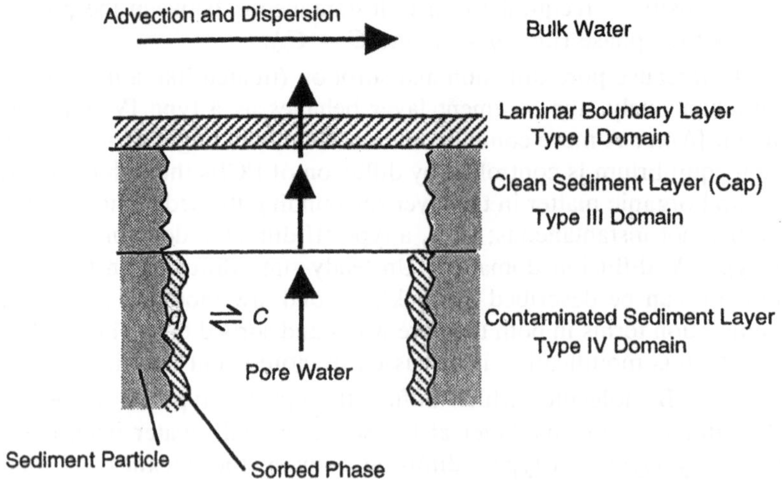

10. To simplify the analysis, sorption of PCBs is assumed to be negligible if the cap is comprised of clean inert sediment particles. The cap layer is thus described by a type III rather than a type IV diffusion domain.

11. The cap layer is assumed to be initially free of PCB molecules. In response to this initial condition, unsteady-state diffusion of PCBs will occur in both the cap and the contaminated sediment layer. (*Note:* The unsteady-state condition in the contaminated sediment layer may be treated as steady state early in the diffusion process because the loss of PCBs from the contaminated sediment layer is small; thus the concentration remains nearly constant at the contaminated sediment–cap interface.) Continuity relationships require that the flux leaving the contaminated sediment layer is equal to that entering the cap layer.

12. It may be reasonable to assume that the impedance to mass transport in the laminar boundary layer is relatively small compared to that in the cap. This assumption leads to an analytical solution to unsteady-state diffusion in both the contaminated sediment and cap layers (Thoma et al., 1993).

13. An intuitive analysis of PCB flux leaving the cap into the bulk water would suggest that the flux must increase initially as the concentration gradient develops in the cap layer. However, the concentration of PCBs at the contaminated sediment–cap layer interface will decrease, albeit relatively slowly due to replenishment by simultaneous desorption from the sediment particles. The long-term effect is a decline in the concentration gradient and flux at the contaminated sediment–cap layer interface, which eventually will produce a decrease in flux through the cap layer.

14. The general pattern of unsteady-state diffusive flux into the bulk water, which can be supported by a mathematical model based on steps 10 to 13, is shown below.

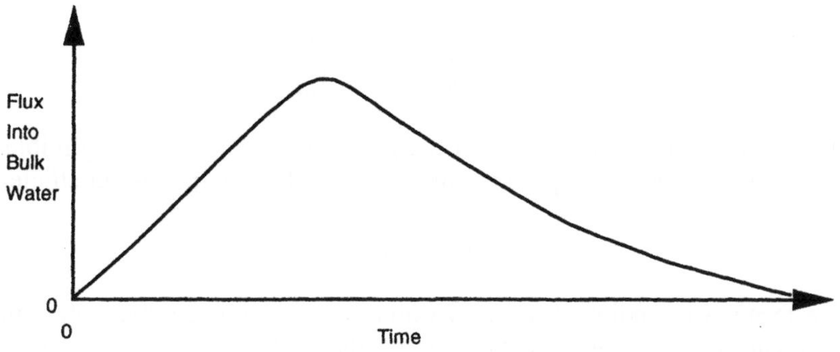

8.3 TYPE I DOMAINS: DIFFUSION WITH VARIOUS BOUNDARY CONDITIONS

As illustrated in Figure 8-4, a type I domain involves diffusional flux through a hydrodynamic boundary layer or film of some thickness, $x = \delta$, between two phases. Systems of greatest interest for the present discussion involve interfaces between aqueous phases and either solid surfaces or gas phases. As discussed in Chapter 4, diffusional flux through a boundary layer is generally defined by a mass transfer relationship lumping the diffusion coefficient and the boundary layer thickness into a mass transfer coefficient. The flux for systems involving mass transfer between water and a solid surface is given by

$$N = -\mathfrak{D}_l \frac{C_\delta - C_0}{\delta} = \frac{\mathfrak{D}_l}{\delta}(C_0 - C_\delta) \tag{8-1}$$

or

$$N = k_f(C_0 - C_\delta) \tag{8-2}$$

where k_f is the mass transfer coefficient, and C_0 and C_δ are the concentrations at the beginning and end of the diffusion domain, respectively. For systems involving mass transfer between water and a gas, the two-film theory defines the flux relationship as

$$N = \hat{k}_{f,l}(C_S - C_b) \tag{8-3}$$

where C_S is the saturation concentration of the gas in the liquid as calculated by Henry's law for a specified set of temperature and partial pressure conditions, C_b is the concentration in the bulk liquid, and $\hat{k}_{f,l}$ is the overall mass transfer coefficient relative to the liquid form of the diffusing solute, defined by

$$\hat{k}_{f,l} = \frac{1}{\dfrac{\mathcal{R}T}{k_{f,g}\mathcal{K}_H} + \dfrac{1}{k_{f,l}}} \tag{4-68}$$

If the resistance to mass transfer on the gas side of the film is negligible, $k_{f,g}$ is very large, and we can approximate the overall mass transfer coefficient as

$$\hat{k}_{f,l} \approx k_{f,l} \tag{8-4}$$

The termination point of a type I domain may be a reactive solid surface (e.g., a catalyst) or a reactive fluid (e.g., the bulk aqueous solution). These physical boundary conditions determine the nature of the rate expression for reactions that are preceded by the microtransport step (i.e., for processes involving *mass transfer and reaction in series*).

8.3.1 Fluid–Solid Interfaces with Surface Catalysis

Our analysis of various boundary conditions at the end of a type I domain begins with that of transformation reactions that occur on contact with surfaces. The rates of such reactions must be expressed in terms of *moles reacted per unit area of surface per unit time* because they occur on surfaces instead of in bulk liquids. To illustrate this heterogeneous process, consider a first-order surface reaction:

$$-r^\circ = k^\circ C^\circ \tag{8-5}$$

where r° represents the rate of surface reaction in the dimensions specified above (mol $L^{-2}t^{-1}$), and C° is the concentration of solute immediately adjacent to the surface (moles L^{-3}). To be consistent with this definition of reaction rate, the first-order rate constant, k°, has the dimensions of length per unit time (Lt^{-1}) instead of time^{-1} (t^{-1}) for a first-order homogeneous reaction. Continuity of flux requires that

$$k_f(C_b - C^\circ) = k^\circ C^\circ \tag{8-6}$$

The surface concentration usually cannot be determined, but this parameter can be eliminated from the rate relationship by expressing it in terms of C_b and the mass transfer and reaction coefficients; that is,

$$C^\circ = \frac{k_f C_b}{k_f + k^\circ} \tag{8-7}$$

Substitution of Equation 8-7 into Equation 8-5 gives

$$-r^{\circ} = \frac{k_f k^{\circ} C_b}{k_f + k^{\circ}} \tag{8-8}$$

Equation 8-8 can be used to examine the relative importance of the mass transfer coefficient and the reaction rate constant for such in-series processes. For very fast reactions, $k^{\circ} >> k_f$ and we obtain

$$-r^{\circ} \approx k_f C_b \tag{8-9}$$

The overall process rate can be increased in such instances only by increasing the rate of mass transfer (i.e., increasing either the mass transfer coefficient or the fluid-phase concentration). This is termed a *mass-transfer-limited* process. As discussed in Chapter 4, mass transfer correlations (see, for example, Table 4-2 and Appendix A.4.1) are available for various conditions, and these provide insight into how increases in k_f can be effected. Correlations for spheres, for example, show that k_f is increased by an increase in fluid velocity (i.e., an increase in $\mathfrak{N}_{Re}$) and/or by a decrease in particle diameter, d_p. The latter result is made apparent by inspection of the general form of the Sherwood, Reynolds, and Schmidt number interrelationships for mass transfer:

$$\mathfrak{N}_{Sh} = \psi_1(\mathfrak{N}_{Re})^{\psi_2}(\mathfrak{N}_{Sc})^{\psi_3} \tag{4-108}$$

Because the Sherwood and Reynolds numbers are both directly proportional to d_p, it follows that the mass transfer coefficient is proportional to $d_p^{\psi_2 - 1}$ and because $\psi_2 < 1$, the mass transfer coefficient (from the Sherwood number) must increase with decreasing d_p.

For very slow reactions, $k^{\circ} << k_f$, and Equation 8-8 reduces to

$$-r^{\circ} \approx k^{\circ} C_b \tag{8-10}$$

In this instance, the overall process is said to be *reaction-rate limited.* Processes that are mass-transfer limited can become reaction-rate limited if fluid velocity is increased or particle diameter is decreased sufficiently to make k_f large relative to k°.

Effectiveness factors have been developed as practical engineering tools to quantify the effects of microtransport on observed rates of reaction. The external reaction effectiveness factor, $\eta_{R,E}$, defines the ratio of the *observed rate* of surface reaction to the *intrinsic rate* in the absence of mass transfer limitations:

$$\eta_{R,E} = \frac{\text{observed rate of surface reaction}}{\text{intrinsic rate of surface reaction}} \tag{8-11}$$

The external reaction effectiveness thus lies between zero (severe mass transfer limitations) and 1 (no mass transfer limitations). For the first-order surface reaction described above, Equation 8-8 represents the observed rate and Equation 8-10 the intrinsic rate. The dimensionless external effectiveness factor is thus expressed as

$$\eta_{R,E} = \frac{k_f}{k^\circ + k_f} \tag{8-12}$$

Another important pair of dimensionless parameters is comprised by the group I and group II *Damkohler numbers*, $\mathfrak{N}_{\mathrm{Da(I)}}$ and $\mathfrak{N}_{\mathrm{Da(II)}}$ (see Appendix III). For surface reactions limited by external mass transfer or internal mass transfer, the Damkohler number is classified by standard nomenclature as being in the group II category. More specifically, for this application the Damkohler number is defined as

$$\mathfrak{N}_{\mathrm{Da(II)}} = \frac{\text{rate of reaction}}{\text{rate of mass transfer by molecular diffusion}} \tag{8-13}$$

and for a *first-order surface reaction and film diffusion model* is given by

$$\mathfrak{N}_{\mathrm{Da(II)}} = \frac{k^\circ C_b}{k_f C_b} = \frac{k^\circ}{k_f} \tag{8-14}$$

Combining Equations 8-12 and 8-14 yields the following relationship between the external reaction effectiveness factor and the group II Damkohler number:

$$\eta_{R,E} = \frac{1}{1 + \mathfrak{N}_{\mathrm{Da(II)}}} \tag{8-15}$$

We observe that when the process is mass-transfer limited (i.e., $k^\circ >> k_f$), $\mathfrak{N}_{\mathrm{Da(II)}}$ is large and the effectiveness factor approaches *zero*. At the opposite extreme, when the process is reaction-rate limited (i.e., $k^\circ << k_f$), $\mathfrak{N}_{\mathrm{Da(II)}}$ is small and the effectiveness factor approaches 1.

Other forms of the reaction effectiveness factor and $\mathfrak{N}_{\mathrm{Da(II)}}$ can be derived for various types of rate relationships (second order, saturation kinetics, etc.). The practical use of the $\eta_{R,E}$ and $\mathfrak{N}_{\mathrm{Da(II)}}$ dimensionless parameters is explored more fully in subsequent sections dealing with different boundary conditions and diffusion domains.

An approach to the analysis of potential rate-controlling steps in processes involving in-series transport and reaction phenomena is illustrated in Example 8-2.

Example 8-2

- **Situation:** A catalytic chemical oxidation/reduction treatment process is being tested for an industrial waste containing chromium(VI) and several recalcitrant synthetic organic chemicals (Prairie et al., 1993). The process, which is based on UV-light-catalyzed reactions at the surface of titanium dioxide (TiO_2), is expected in this application to reduce Cr(VI) to a less toxic and more readily removed form, Cr(III), and to partially oxidize the organic compounds to increase the effectiveness of downstream biological treatment. Although finely divided TiO_2 particles have been shown to be effective, gravity separation of these particles after treatment has proven difficult. A new technique has been developed to attach finely divided TiO_2 particles to larger inert particles which settle more readily, yielding a catalyst surface area of 10,000 cm^2/g. A series of rate tests is conducted using one concentration of catalyst particles (1 g/L) and different impeller speeds, ranging from 400 to 2000 rpm, in a CMBR. Measurements of the fractional remaining concentration of Cr(VI) with time $C(t)/C_0$ are shown below.

Time (min)	400 rpm	1000 rpm	1500 rpm	2000 rpm
0	1	1	1	1
2	0.55	0.38	0.34	0.32
4	0.30	0.15	0.11	0.10
6	0.17	0.06	0.04	0.03
8	0.09	0.02	0.01	0.01

- **Question(s):** A first-order surface reaction is proposed to account for the reduction of Cr(VI) on TiO_2. Determine whether the data above support this hypothesis.

- **Logic and Answer(s):**

1. The bulk-phase concentration, C_b, of a substance undergoing reaction in a CMBR decreases with time such that:

$$V_R \frac{dC_b}{dt} = -k_f(C_b - C^\circ)A_P^\circ = -k^\circ C^\circ A_P^\circ$$

where V_R is the volume of the solution in the reactor, A_P° is the total external surface area of the particles, and C° is the concentration of solute at the surface of the catalyst.

2. The right-hand side of the equality above can be solved for C° (see Equation 8-7) to give

$$\frac{dC_b}{dt} = -\frac{k^\circ k_f(a_s^\circ)_R}{k_f + k^\circ} C_b$$

where $(a_s^\circ)_R$ is the specific external surface area of particles per unit volume of solution in the CMBR. Letting $\hat{k}$ be the overall or lumped reaction rate coefficient,

$$\hat{k} = \frac{k^\circ k_f(a_s^\circ)_R}{k_f + k^\circ}$$

and integrating the CMBR rate expression gives

$$-\ln \frac{C_b}{C_{b,0}} = \hat{k}t$$

where $C_{b,0}$ is the initial ($t = 0$) concentration of Cr(VI).

3. If this model is applicable to this process, each set of rate data should give a straight line when plotted according to the expression above. As shown below, such plots indicate good agreement with the proposed model.

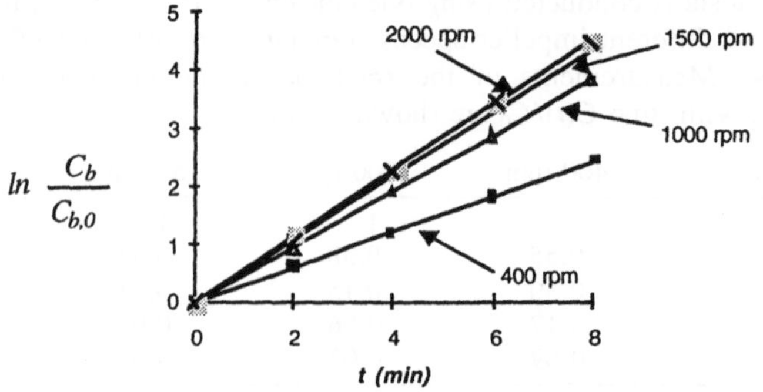

4. The CMBR data show that the rate of Cr(VI) reduction is nearly the same at stirring speeds of 1500 and 2000 rpm. This suggests that the surface reaction becomes almost entirely rate controlling ($k^\circ << k_f$) at 2000 rpm and thus that $\hat{k} \approx k^\circ(a_s^\circ)_R$. The slope of the line for the data collected at 2000 rpm determines the value of $\hat{k}$, and it follows that

$$k^\circ = \frac{\hat{k}}{(a_s^\circ)_R} = \frac{0.6(\text{min}^{-1}) \cdot (\text{min}/60\text{ s})}{10{,}000(\text{cm}^2/\text{g}) \cdot 0.001(\text{g/cm}^3)} = 0.001 \text{ cm/s}$$

5. Knowing $k^\circ(a_s^\circ)_R$, the rate data at the other stirring speeds can be used to determine k_f from the relationship

$$k_f = \frac{\hat{k}k^\circ}{k^\circ(a_s^\circ)_R - \hat{k}}$$

6. The Damkohler and external effectiveness factors calculated from Equations 8-14 and 8-15, respectively, and summarized below indicate that mass transfer controls the reaction rate to a significant extent unless the stirring speed is kept above 1500 rpm.

Stirring Speed (rpm)	k_f (cm/s)	k° (cm/s)	$\mathfrak{N}_{\text{Da(II)}}$	$\eta_{R,E}$
400	0.001	—	1	0.5
1000	0.004	—	0.25	0.8
1500	0.01	—	0.1	0.9
2000	—	0.001	—	—

8.3.2 Fluid–Solid Interfaces with Adsorption and Surface Catalysis

A type I domain may terminate at a catalytic surface where reactions take place only if *adsorption* occurs first. A simple illustration is provided in Figure 8-8. Here the reactant, A, is transported across a type I domain, adsorbs at a site, and reacts with the surface to form a product, Z. This product then desorbs and is transported back to the bulk solution through the type I domain. Light-catalyzed oxidation of organic molecules on semiconductor surfaces such as titanium dioxide (see Chapter 7), for example, may involve adsorption of organic molecules followed by reaction, or adsorption of hydroxyl radicals, which

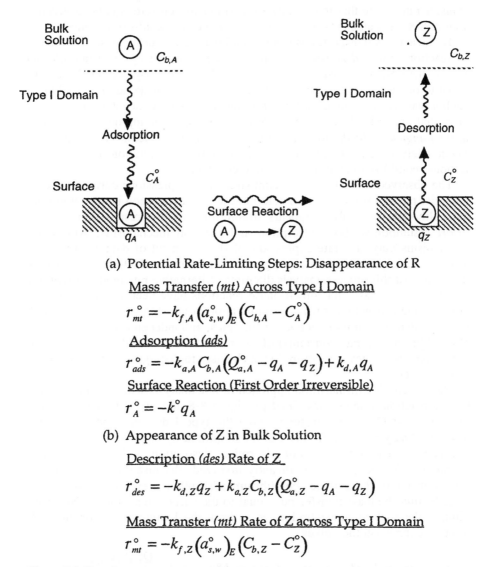

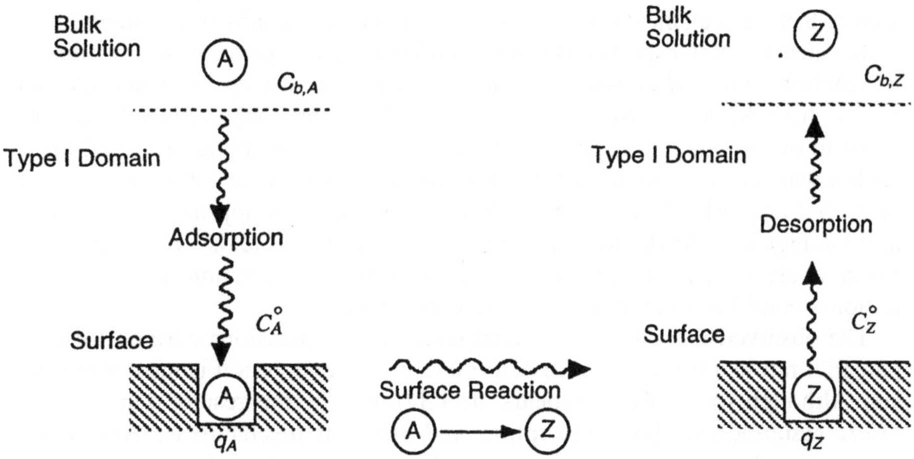

(a) Potential Rate-Limiting Steps: Disappearance of R

Mass Transfer *(mt)* Across Type I Domain

$$r_{mt}^\circ = -k_{f,A}\left(a_{s,w}^\circ\right)_E\left(C_{b,A} - C_A^\circ\right)$$

Adsorption *(ads)*

$$r_{ads}^\circ = -k_{a,A}C_{b,A}\left(Q_{a,A}^\circ - q_A - q_Z\right) + k_{d,A}q_A$$

Surface Reaction (First Order Irreversible)

$$r_A^\circ = -k^\circ q_A$$

(b) Appearance of Z in Bulk Solution

Description *(des)* Rate of Z

$$r_{des}^\circ = -k_{d,Z}q_Z + k_{a,Z}C_{b,Z}\left(Q_{a,Z}^\circ - q_A - q_Z\right)$$

Mass Transter *(mt)* Rate of Z across Type I Domain

$$r_{mt}^\circ = -k_{f,Z}\left(a_{s,w}^\circ\right)_E\left(C_{b,Z} - C_Z^\circ\right)$$

Figure 8-8 Type I domain transport process terminating with adsorption and surface reaction.

then react with the organic molecules. The product, Z, may be an intermediate of the photocatalytic reaction that occupies surface sites.

Transport of reactant and product through the type I domain is described by the mass transfer model given in Equation 8-2, rates of adsorption and desorption by the Langmuir model (see Chapter 6), and the rate of reaction by an appropriate rate expression (see Chapter 7), in this case a first-order expression. The surface reaction rate ($\mathcal{R}^\circ$) equations corresponding to these several steps, all expressed as *moles per mass of catalyst per unit time*, are included in Figure 8-8. The mass transfer flux must therefore be multiplied by the specific external surface area per unit weight of catalyst, $(a^\circ_{s,w})_E$, to obtain a mass transfer rate. Note that the adsorption and desorption rate equations given in Figure 8-8 are statements of the Langmuir *competitive* adsorption model.

Because mass transport, adsorption, and reaction all occur in series, the rate of reaction *observed* is essentially that of the slowest step. The rate of each step is then equal to that of the rate-controlling step. By setting these rates equal to one another, it is possible to combine the corresponding equations in such a way as to eliminate solid-phase and interfacial concentrations, parameters that are difficult to evaluate experimentally. This approach, referred to as the *Langmuir–Hinshelwood method of rate analysis*, yields a rate expression given solely in terms of measurable parameters (i.e., solution-phase concentrations, equilibrium constants, and rate constants).

The observed rate data can be compared with expressions derived for each potential rate-controlling step to determine which expression is most appropriate. Although the analysis appears straightforward, the results depend on *a priori* assumptions about adsorption and reaction mechanisms. Alternative mechanisms may give rate expressions that "fit" a set of data equally well. The photocatalytic oxidation of organic contaminants on TiO_2 provides a good illustration of this process analysis dilemma. Alternative rate models involving direct aqueous-phase and sorbed-phase reactions have been proposed for this process, and it is not clear which is more appropriate. More fundamental studies (e.g., determinations of activation energies, and independent measurements of equilibrium and rate constants) of the reaction than afforded by interpretation of observed rates are required in such cases to distinguish one mechanism from another.

The Langmuir–Hinshelwood approach to rate analysis will be demonstrated for the reaction schemes described in Figure 8-8 and two different rate-controlling steps: (1) mass transfer through the type I domain, and (2) surface reaction. Analysis of the former can be simplified by considering only initial rate data, so that the processes of adsorption, desorption, and mass transfer of the product, Z, are not yet significantly involved. Because the slowest step is mass transfer to the surface, equilibrium adsorption prevails and so $C^\circ_A = C^\circ_{e,A}$. Setting the mass transfer rate equal to the surface reaction rate and substituting the surface concentration, $q_{e,A}$, from the Langmuir equilibrium adsorption expression then gives

$$\mathcal{R}^\circ = k_f(a^\circ_{s,w})_E(C_{b,A} - C^\circ_{e,A}) = k^\circ q_{e,A} = k^\circ \frac{Q^\circ_{a,A}b_A C^\circ_{e,A}}{1 + b_A C^\circ_{e,A}} \tag{8-16}$$

where k° is the first-order rate constant for the surface reaction (time^{-1}) and $\mathcal{b}_A$ is the adsorption energy, or the ratio of adsorption-to-desorption rate constants $(k_{a,A}/k_{d,A})$ for A.

A Damkohler number can be defined to relate the *maximum* surface reaction rate to the *maximum* mass transfer rate, which occurs when $C^\circ_{e,A} = 0$:

$$\mathfrak{N}_{\mathrm{Da(II)}} = \frac{k^\circ Q^\circ_{a,A}}{k_f(a^\circ_{s,w})_E C_{b,A}} \tag{8-17}$$

It is possible to gain insight into the role of mass transfer control from the Damkohler number. Equations 8-16 and 8-17 are first combined, and then expressed conveniently in dimensionless form:

$$\frac{1 - c^\circ_A}{\mathfrak{N}_{\mathrm{Da(II)}}} = \frac{c^\circ_A}{\mathcal{b}^\circ_A + c^\circ_A} \tag{8-18}$$

where $c^\circ_A = C^\circ_{e,A}/C_{b,A}$ is the de-dimensionalized interfacial equilibrium concentration, and $\mathcal{b}^\circ_A = (\mathcal{b}_A C_{b,A})^{-1}$ is the de-dimensionalized Langmuir energy constant for A. The solution to the resulting quadratic equation for c°_A is

$$c^\circ_A = \frac{(-\mathcal{b}^\circ_A - \mathfrak{N}_{\mathrm{Da(II)}} + 1) \pm [(\mathcal{b}^\circ_A + \mathfrak{N}_{\mathrm{Da(II)}} - 1)^2 + 4\mathcal{b}^\circ_A]^{0.5}}{2} \tag{8-19}$$

We can examine the two boundaries of this solution. Strong mass transfer control is indicated when $\mathfrak{N}_{\mathrm{Da(II)}} >> 1$, and the quadratic equation above yields $c^\circ_A \approx 0$. The observed rate is thus given by Equation 8-16 with $C^\circ_{e,A}$ set equal to zero:

$$r^\circ \approx k_f(a^\circ_{s,w})_E C_{b,A} \tag{8-20}$$

The absence of mass transfer control corresponds to $\mathfrak{N}_{\mathrm{Da(II)}} << 1$, for which $c^\circ_A \approx 1$ (i.e., the bulk and interface concentrations are equal) and the observed rate from Equation 8-16 is

$$r^\circ = k^\circ \frac{Q^\circ_{a,A}\mathcal{b}_A C_{b,A}}{1 + \mathcal{b}_A C_{b,A}} \tag{8-21}$$

The analysis above shows that mass transfer control leads to an observed rate that is first order with respect to the bulk phase reactant concentration, whereas the observed rate in the absence of mass transfer control exhibits a variable-order dependence on bulk phase reactant concentration. This apparent shift in reaction order caused by mass transfer control in the diffusion domain is often referred to as *disguised kinetics*. Other examples of disguised kinetics are presented later in this chapter.

For situations where the mass transfer rate is high (for example, combinations of small-diameter particles and vigorous mixing), we can assume that the bulk and interfacial fluid phase concentrations are equal and the solid phase concentration that determines the surface reaction rate is that in equilibrium with the bulk fluid concentration. For the reaction mechanism in Figure 8-8, and its accompanying equations that describe adsorption-desorption rates pursuant to the Langmuir model, competitive adsorption between reactant and product is accounted for as follows:

$$C_{b,A}(Q^{\circ}_{a,T} - q_A - q_Z) = \frac{q_A}{b_A} \tag{8-22}$$

$$C_{b,Z}(Q^{\circ}_{a,T} - q_A - q_Z) = \frac{q_Z}{b_Z} \tag{8-23}$$

In the above equations, $Q^{\circ}_{a,T}$ is the total number of sorption sites available to either A or Z and therefore each sorbate occupies the *same* surface area per molecule. Different Q°_a values may be specified for each sorbate, however, as was done in section 6.5.5.1 where the competitive Langmuir adsorption model was first introduced. Specification of different Q°_a values implies (1) the surface area covered per molecule is implicitly different for each sorbate or (2) competition for sites is not completely mutual and the model is applied empirically. The other feature that distinguishes the above equation from the development in section 6.5.5.1 is elimination of the subscript "e" from the variables q and C_b. This is done to acknowledge that the bulk phase concentration, which determines adsorbed phase equilibrium concentration and thus the surface reaction rate, would decrease with time as adsorption proceeds. Equilibrium between q and C_b would transition through a continuous series of states along the corresponding Languir isotherm. The parenthetical term, a measure of available adsorption sites, is the same in each equation, and combining the two equations yields

$$q_Z = \frac{b_Z C_{b,Z}}{b_A C_{b,A}} q_A \tag{8-24}$$

Substitution of Equation 8-24 into Equation 8-22 and rearrangement yields the same expression for the solid phase concentration of A as the competitive Langmuir adsorption isotherm developed in Chapter 6 (Equation 6-95) subject to the assumptions noted above:

$$q_A = \frac{Q^{\circ}_{a,T} b_A C_{b,A}}{1 + b_A C_{b,A} + b_Z C_{b,Z}} \tag{8-25}$$

The solid-phase concentration, q_A, from above determines the surface rate of reaction:

$$r^{\circ} = k^{\circ} q_A = \frac{k^{\circ} Q^{\circ}_{a,T} b_A C_{b,A}}{1 + b_A C_{b,A} + b_Z C_{b,Z}} \tag{8-26}$$

A further simplification results if we assume that the reactant and product have similar adsorption energies and that the sum of their bulk-phase concentrations, $C_{b,A} + C_{b,Z} = C_{b,T}$, is nearly constant for the early and intermediate stages of reaction:

$$r^{\circ} \approx \frac{k^{\circ} Q^{\circ}_{a,T} C_{b,A}}{(1/b_A) + C_{b,T}} \approx k^{\circ}_{\text{obs}} C_{b,A} \tag{8-27}$$

where k°_{obs} is an observed first-order surface reaction rate constant given by

$$k^{\circ}_{\text{obs}} = \frac{k^{\circ} Q^{\circ}_{a,T}}{(1/b_A) + C_{b,T}} \tag{8-28}$$

According to Equation 8-28, this rate constant decreases as the initial concentration increases. It can be used to find k° (assuming that b_A and Q°_a are known) by varying the initial concentration in a series of CMBR tests and plotting $1/k^{\circ}_{\text{obs}}$ against $C_{b,T}$, as shown in Figure 8-9; a straight line should result, described by the inverse of Equation 8-28:

$$\frac{1}{k^{\circ}_{\text{obs}}} = \frac{1}{Q^{\circ}_{a,T} b_A k^{\circ}} + \frac{C_{b,T}}{k^{\circ} Q^{\circ}_{a,T}} \tag{8-29}$$

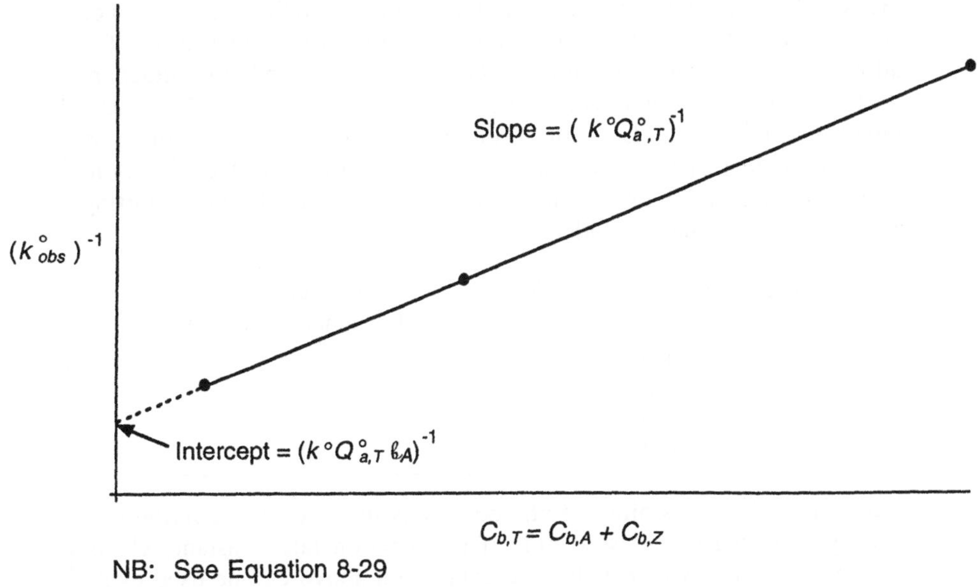

Figure 8-9 Evaluation of a model for competitive reactant and product adsorption with surface reaction rate control.

We can also evaluate a simpler mechanism of reaction wherein adsorption is followed by rate-controlled reaction without formation of a product that competes for adsorption sites. Equation 8-26 then reduces to

$$r^{\circ} = \frac{k^{\circ} Q^{\circ}_{a,T} b_A C_{b,A}}{1 + b_A C_{b,A}} \tag{8-30}$$

It can be observed here that the reaction rate approaches zero order at high concentrations of reactant and first order at low reactant concentrations. It is more than coincidental that the same rate expression obtains for enzyme–substrate (Michaelis–Menten) kinetics. In fact, the basis for the enzyme–substrate rate model is formation of an enzyme–substrate complex in a fast, reversible adsorption process, followed by a rate-controlling, irreversible step in which enzyme is released and product forms.

Example 8-3 provides an illustration of how an analysis of a set of rate data can be used to identify or isolate the rate-controlling step in a catalysis process involving coupled mass transfer and surface reaction.

Example 8-3

- **Situation:** CMBR tests are conducted to investigate rates of photocatalyzed oxidation of a pesticide on TiO_2 particles having a diameter (d_p) of 0.1 mm and a density (ρ_p) of 1.5 g/cm^3. The particles are agitated in an impeller-stirred baffled reactor for which the mass transfer characteristics are well known. The corresponding mass transfer coefficient determined from appropriate literature correlations is found to be 0.001 cm/s. The adsorption of the pesticide on the TiO_2 particles is found from independent experiments to conform to the Langmuir model, with $Q^{\circ}_a = 0.1$ mmol/g and $b = 0.1$ L/mmol. Data for organic chemicals similar in structure to that of the pesticide suggest a first-order surface reaction. The following initial rate measurements are made at different starting concentrations of the pesticide.

Initial Pesticide Concentration, $C_{b,0}$ (mmol/L)	Initial Surface Rate, r° (mmol/g−s) × 10^4
1.0	1.7
3.0	5.0
5.0	7.2
10.0	8.7

- **Question(s):** The feasibility of photocatalysis in large-scale continuous-flow systems can be assessed only if the reaction rate constant, k°, is known. Determine whether the rate data given above can be used evaluating k°.

- **Logic and Answer(s):**

1. The reaction rate constant cannot be determined if mass transfer is entirely responsible for rate control in these CMBR experiments. From Equation 8-20, complete mass transfer control corresponds to

$$\mathscr{r}^\circ \approx k_f(a^\circ_{s,w})_E C_{b,0}$$

2. A plot of the observed rate against the initial bulk-phase pesticide concentration is shown below. The fact that the data do not yield a linear trace suggests that mass transfer is not completely rate controlling.

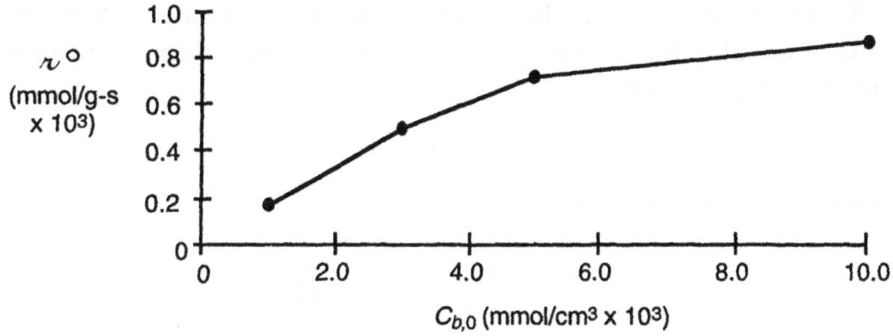

3. The complete absence of mass transfer control is another potential condition that can easily be checked. If mass transfer is not rate controlling, the rate constant can be determined directly from Equation 8-21:

$$\mathscr{r}^\circ = k^\circ \frac{Q^\circ_{a,T} \mathscr{b} C_{b,0}}{1 + \mathscr{b} C_{b,0}}$$

and a plot of reciprocal rate against concentration should yield a straight line:

$$\frac{1}{\mathscr{r}^\circ} = \frac{1}{k^\circ Q^\circ_{a,T} \mathscr{b}} \frac{1}{C_{b,0}} + \frac{1}{k^\circ Q^\circ_a}$$

The plot shown below is not linear, and we therefore conclude that rate control by mass transfer must be included in the rate analysis.

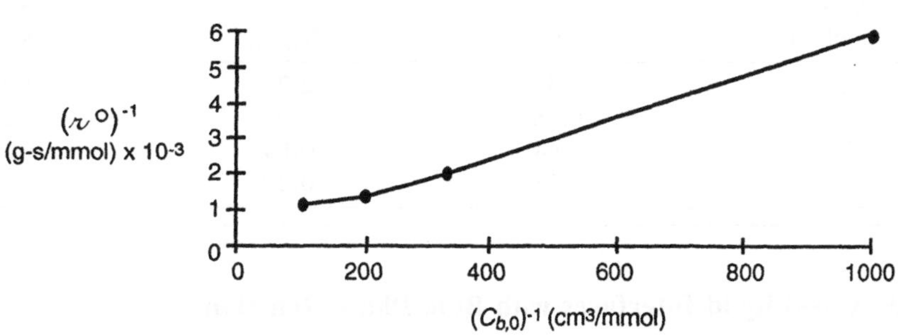

4. To find k° it is first necessary to calculate the dimensionless interfacial concentration, c°, for each rate test and then use an implicit solu-

tion technique to solve Equation 8-19 for the Damkohler number:

$$c^\circ = \frac{(-\mathscr{b}^\circ - \mathfrak{N}_{\mathrm{Da(II)}} + 1) \pm [(\mathscr{b}^\circ + \mathfrak{N}_{\mathrm{Da(II)}} - 1)^2 + 4\mathscr{b}^\circ]^{0.5}}{2}$$

Evaluation of k° follows directly from the definition of the Damkohler number (Equation 8-17):

$$k^\circ = \frac{k_f(a^\circ_{s,w})_E C_{b,0} \mathfrak{N}_{\mathrm{Da(II)}}}{Q^\circ_a}$$

5. The value of c° can be calculated because the rate of mass transfer must be equal to that of the reaction rate when these steps occur in series. The mass transfer rate is

$$r^\circ = k_f(a^\circ_{s,w})_E C_{b,0}(1 - c^\circ)$$

which upon solving for c° gives

$$c^\circ = 1 - \frac{r^\circ}{k_f(a^\circ_{s,w})_E C_{b,0}}$$

The external surface area per gram is calculated from the particle diameter, d_p, and density, ρ_p:

$$(a^\circ_{s,w})_E = \frac{\pi d_p^2}{(\pi d_p^3/6)\rho_p} = \frac{6}{(0.01)(1.5)} = 400 \text{ cm}^2/\text{g}$$

The mass transfer coefficient is also known, so that

$$k_f(a^\circ_{s,w})_E = (0.001)(400) = 0.4 \text{ cm}^3/\text{g-s}$$

6. The values of c°, $\mathfrak{N}_{\mathrm{Da(II)}}$, and k° are listed below. The values of c° are significantly greater than zero, as expected when surface reaction contributes to rate control. The values of k° obtained from the Damkohler numbers are in close agreement, which suggests that the proposed model is reasonable.

$C_{b,0}$ (mmol/L)	c°	$\mathfrak{N}_{\mathrm{Da(II)}}$	k° (s^{-1})
1	0.57	2.27	0.91
2	0.58	1.01	1.21
3	0.64	0.64	1.28
5	0.78	0.29	1.16

8.3.3 Gas–Liquid Interfaces with Bulk Phase Reaction

Gas–liquid mass transfer occurs in a type I domain that, according to the two-film theory, corresponds to a combination of gas and liquid sides of an interface. If the reaction of the gas is limited to the bulk liquid phase, such reaction produces another form of boundary condition for type I domains. The concentration profiles produced in the liquid side of the gas–liquid interface and in the bulk liquid were shown in a general way in Figure 8-2a. Although the reaction takes place entirely in the bulk liquid phase, the rate of gas transfer into the liquid is important because it limits the rate of reaction. Diffusion occurs in series with the reaction, and the rate of gas transfer into the solution must therefore equal the rate of reaction in solution:

$$NA_F^\circ = r V_R \tag{8-31}$$

where A_F° is the interfacial surface area of the liquid film and V_R is the volume of the bulk solution contained in the reactor. If we assume that gas flux is controlled by the *liquid side* of the film, and that the reaction rate is first order, we have

$$\hat{k}_{f,l}(a_s^\circ)_R(C_S - C_b) = kC_b \tag{8-32}$$

where, according to Equation 8-4, $\hat{k}_{f,l} \approx k_{f,l}$ and $(a_s^\circ)_R$ is the interfacial area per bulk volume of liquid in the reactor. This equation can be rearranged to give the bulk phase concentration, C_b:

$$C_b = \frac{\hat{k}_{f,l}(a_s^\circ)_R C_S}{k + \hat{k}_{f,l}(a_s^\circ)_R} \tag{8-33}$$

The smaller the reaction rate constant, k, the higher the bulk concentration for any specific value of the overall volumetric mass transfer coefficient, $\hat{k}_{v,l} = \hat{k}_{f,l}(a_s^\circ)_R$. Thus, as can be reasoned intuitively, for very slow reactions the bulk concentration approaches the saturation concentration, and mass transfer resistance is no longer important.

The general rate expression for mass-transfer limited reaction is given by

$$-r = kC_b \tag{8-34}$$

and, after substituting for C_b from above, we obtain

$$-r = \frac{\hat{k}_{f,l}(a_s^\circ)_R k}{k + \hat{k}_{f,l}(a_s^\circ)_R} C_S = \frac{C_S}{\dfrac{1}{\hat{k}_{f,l}(a_s^\circ)_R} + \dfrac{1}{k}} \tag{8-35}$$

We can see from this expression that the effect of mass transfer resistance is to lower observed reaction rate. If resistance to mass transfer is negligible $\hat{k}_{f,l}(a_s^\circ)_R$ is large and Equation 8-35 reduces to

$$r = -kC_S \tag{8-36}$$

In other words, the rate of reaction is determined by the intrinsic kinetics, for which the bulk concentration, C_b, is equal to the saturation concentration, C_S, in the bulk phase.

Example 8-4 applies the rate analysis procedures above to the decomposition of ozone in mixed gas–liquid phases in a completely mixed flow reactor (CMFR).

Example 8-4

- **Situation:** When dissolved in water, ozone decomposes to oxygen through a complex chain reaction in which hydroxyl radicals are produced. This chain reaction is initiated (catalyzed) by hydroxide ions; the rate of decomposition thus increases with increasing pH. Rates of ozone decomposition are important because the oxidizing powers of molecular ozone and the hydroxyl radical are different. Molecular ozone reacts more slowly and is more specific than highly reactive than the hydroxyl radical.

  The CMFR shown below is set up in the laboratory and operated at 20°C to investigate the rate of molecular ozone decomposition.

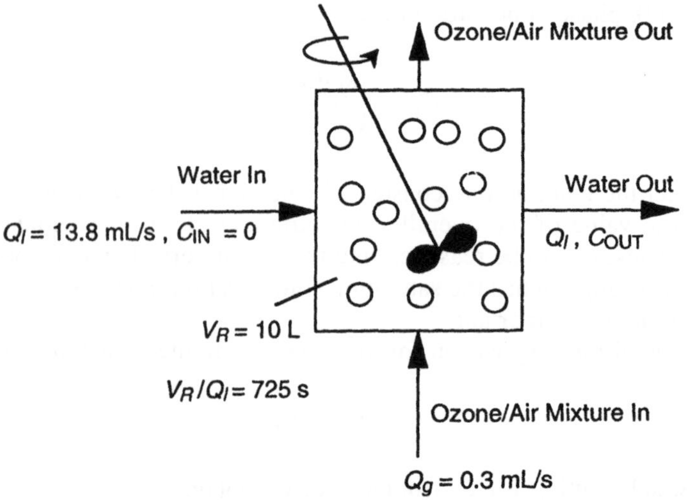

Three experiments are conducted in which the percent of O_3 in the feed air is varied from 1 to 3% while the pH is buffered at 8.0. The ozone concentration in solution and the percent ozone in the air leaving the reactor in each experiment at steady state are shown below.

Ozone in Air Mixture Entering Reactor (%)	Ozone in Air Mixture Leaving Reactor (%)	Aqueous-Phase Ozone Leaving Reactor, C_{OUT} (mmol/L)
1	0.8	0.08
2	1.6	0.16
3	2.4	0.24

The overall volumetric mass transfer coefficient, $\hat{k}_{v,l} = \hat{k}_{f,l}(a_s^\circ)_R$, was determined in separate experiments at pH 2, where ozone decomposition can be ignored because of the low OH^- concentration. The material balance at steady state (see Example 4-8) is

$$-QC_{\text{OUT}} + \hat{k}_{f,l}(a_s^\circ)_R(C_S - C_{\text{OUT}})V_R = 0$$

where C_S is the saturation concentration, C_{OUT} is the exit concentration, Q_l is the flow rate, and V_R is the volume of the reactor. Knowing C_S from Henry's law and Q_l, V_R, and C_{OUT} from each experiment, the value of $\hat{k}_{f,l}(a_s^\circ)_R$ is found to be 40 h^{-1} (0.011 s^{-1}).

- **Question(s):** Use the data collected at pH 8 to characterize ozone decomposition rates in these systems.
- **Logic and Answer(s):**

1. A search of the literature reveals that ozone decomposition is frequently described empirically as a first-order reaction. Other orders (e.g., 1.5 and 2.0) have also used empirically. We use the data from pH 8 to determine whether a first-order reaction rate formulation is reasonable. Moreover, we assume that the reaction is slow enough to be limited to the bulk solution.

2. For sparingly soluble gases such as ozone it is reasonable to assume that the liquid side of the gas–liquid film controls the rate of mass transfer.

3. If ozone decomposes slowly, the appropriate rate expression for the observed rate of ozone decomposition is given by Equation 8-35:

$$-r = \frac{1}{\dfrac{1}{\hat{k}_{f,l}(a_s^\circ)_R} + \dfrac{1}{k}} C_S$$

According to this model the rate of ozone decomposition is proportional to the saturation concentration of ozone.

4. The observed steady-state rate of ozone decomposition must equal the rate of ozone transfer, or

$$-r = \frac{Q_g}{V_R}\left(\frac{P_{\text{IN}}}{\mathcal{R}T} - \frac{P_{\text{OUT}}}{\mathcal{R}T}\right) - \frac{Q_l C_{\text{OUT}}}{V_R}$$

where $\mathcal{R} = 8.21 \times 10^{-5}$ atm-m^3/mol-K and $T = 293$ K.

5. The saturation concentration is determined from Henry's law:

$$C_S = \frac{P_{\text{OUT}}\ (\text{atm})}{\mathcal{K}_H\ (\text{atm-m}^3/\text{mol})}$$

The $\mathcal{K}_H$ value for ozone from Table 1-3 is 67.7×10^{-3} atm-m^3/mol. The partial pressure of ozone is that in the exit gas stream, assuming that the gas bubbles are completely mixed within the reactor. If the air is at

1 atm, the pressure of ozone in atm is given by the percent O_3 in the exit gas stream divided by 100.

6. Values for the rate of decomposition (calculated from step 4) and the ozone saturation (calculated from step 5) are tabulated below:

$-r$ (mmol/L-s $\times 10^4$)	C_S (mmol/L)
4.0	0.12
8.0	0.24
12.0	0.36

7. The figure below shows a linear relationship between decomposition rate and the saturation concentration, as predicted by Equation 8-35, with a slope of

$$\text{slope} = \frac{-r}{C_S} = \frac{1}{\dfrac{1}{\hat{k}_{f,l}(a_s^\circ)_R} + \dfrac{1}{k}} = 3.45 \times 10^{-3}\ \text{s}^{-1}$$

The value of k computed from the slope is $5 \times 10^{-3}\ \text{s}^{-1}$.

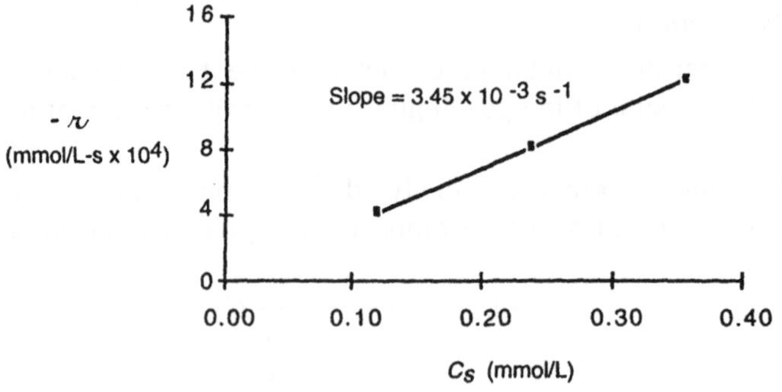

8. The effect of mass transfer on the observed rate of decomposition is given by Equation 8-34:

$$-r = kC_b = \frac{1}{\dfrac{1}{k_{f,l}(a_s^\circ)_R} + \dfrac{1}{k}} C_S = 3.45 \times 10^{-3} C_S$$

In the absence of mass transfer limitations, the rate would be higher; that is,

$$-r = kC_S = 5 \times 10^{-3} C_S$$

The more general analysis for the slowest rate regime for a second-order reaction between A and B in which the effects of both gas- and liquid-side resistances to gas transfer are included and the assumption of pseudo-first-order reaction no longer holds is somewhat more complex, but still manageable.

Here the rate of reaction of A in the bulk liquid is given by

$$-r_A = kC_{b,A}C_{b,B} \tag{8-37}$$

Once again, the rate of absorption of A from the gas phase into the bulk solution must equal the rate of reaction of A in bulk solution:

$$N_A(a_s^\circ)_R = -r_A \tag{8-38}$$

Substituting for N_A from Equation 8-3 and r_A from Equation 8-37 allows $C_{b,A}$ to be expressed as

$$C_{b,A} = \frac{C_{S,A}}{\dfrac{kC_{b,B}}{\hat{k}_{f,l}(a_s^\circ)_R} + 1} \tag{8-39}$$

Finally, combining Equation 8-39 with Equation 8-37 and simplifying the result yields the following expression for the observed rate of disappearance of A:

$$-r_A = \frac{C_{S,A}}{\dfrac{1}{\hat{k}_{f,l}(a_s^\circ)_R} + \dfrac{1}{kC_{b,B}}} \tag{8-40}$$

If mass transfer is not limiting, the overall volumetric mass transfer coefficient, $\hat{k}_{f,l}(a_s^\circ)_R$, is large and Equation 8-40 reduces to the intrinsic rate of reaction:

$$-r_A = kC_{S,A}C_{b,B} \tag{8-41}$$

Here the bulk concentration of A is equal to the saturation concentration, $C_{S,A}$, because mass transfer across the gas–liquid interface is rapid.

8.4 TYPE II DOMAINS: DIFFUSION WITH REACTION

Type II domains are essentially type I domains that include reaction of the diffusing solute within the domain. For instance, relatively fast reactions of dissolved gases occur within liquid boundary layers (i.e., liquid films) rather than in bulk solutions. Reaction and diffusion rates thus act as impedances to microtransport. As for type I domain processes, dimensionless process engineering parameters that aid understanding of a wide range of fluid–solid and gas–liquid mass transfer situations are available.

8.4.1 Thiele Modulus and the Damkohler Number

The Thiele modulus and the Damkohler number (already introduced in Section 8.3.1) are two important dimensionless process engineering parameters used to quantify the *relative contributions* of reaction and diffusion to impedance. They originate from development of rate expressions used to describe diffusion and simultaneous reaction. The appropriate material balance in one-dimensional rectangular coordinates is

$$-\frac{dN}{dx} + r = 0 \tag{8-42}$$

For diffusive flux and a first-order solute depletion reaction in the solution phase, Equation 8-42 becomes

$$\mathfrak{D}_l \frac{d^2C}{dx^2} - kC = 0 \tag{8-43}$$

The general solution to this linear homogeneous differential equation is

$$C = \kappa_1 e^{mx} + \kappa_2 e^{-mx} \tag{8-44}$$

where

$$m = \left(\frac{k}{\mathfrak{D}_l}\right)^{0.5} \tag{8-45}$$

The domain described schematically in Figure 8-5 shows that concentration remains constant at $x = 0$. If the domain ends at a surface, the flux must be zero at $x = L$. The mathematical statements of these boundary conditions are:

First boundary condition:

$$C = C_0 \quad \text{at } x = 0 \tag{8-46}$$

Second boundary condition:

$$\frac{dC}{dx} = 0 \quad \text{at } x = L \tag{8-47}$$

Accordingly, values of κ_1 and κ_2 can be found, and the final form of the solution to Equation 8-43 written in dimensionless form as

$$\frac{C}{C_0} = \frac{\cosh \mathfrak{N}_{\text{Tm}}(1 - \zeta^\circ)}{\cosh \mathfrak{N}_{\text{Tm}}} \tag{8-48}$$

where ζ° is dimensionless distance (x/L) along the diffusion path and $\mathfrak{N}_{\mathrm{Tm}}$ is the *Thiele modulus*, defined for a first order reaction by

$$\mathfrak{N}_{\mathrm{Tm}} = L\left(\frac{k}{\mathfrak{D}_l}\right)^{0.5} \tag{8-49}$$

As noted in Table III-A (Appendix III), the general form of the Thiele modulus for any reaction order is given by substituting $\mathcal{r}/C$ for k in Equation 8-49. Dimensionless concentration profiles in this diffusion domain for different values of the Thiele modulus are shown in Figure 8-10. When the rate of diffusion is slow relative to the intrinsic rate of reaction, the Thiele modulus is large. Thus, concentration decreases sharply with distance because diffusion limits the transport of solute (i.e., the overall process rate is *diffusion limited*). In contrast, a reaction that is intrinsically slow is characterized by a small Thiele modulus. Thus, unreacted solute penetrates much deeper into the diffusion domain and the overall process is *reaction-rate limited.* The importance of diffusion in limiting a reaction will be developed in more detail when we discuss type IV diffusion domains.

The group II Damkohler number also provides a convenient way to quantify diffusion limitations:

$$\mathfrak{N}_{\mathrm{Da(II)}} = \frac{\mathcal{r}}{\mathfrak{D}_l\left(\dfrac{C_0}{L^2}\right)} = \frac{L^2 k C_0^n}{\mathfrak{D}_l C_0} \tag{8-50}$$

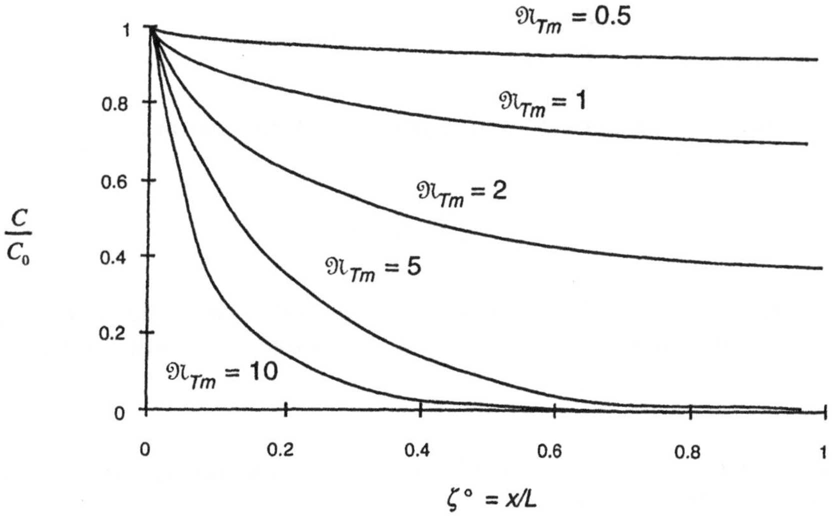

Figure 8-10 Dependence of the dimensionless concentration profile on the Thiele modulus for a type II diffusion domain of rectangular geometry and first order reaction.

A value of $\mathfrak{N}_{\text{Da(II)}}$ greater than 1 means that the reaction rate is faster than the diffusion rate. Comparing the group II Damkohler number given in Equation 8-50 with the general form of the Thiele modulus (Appendix III) reveals that, regardless of reaction order

$$\mathfrak{N}_{\text{Tm}} = (\mathfrak{N}_{\text{Da(II)}})^{0.5} \tag{8-51}$$

The physical meaning of the group II Damkohler number and the Thiele modulus are thus similar. Both provide a means by which to quantify the relative significance of chemical reaction and microtransport parameters.

As in a type I domain, it is possible to define an effectiveness factor to relate the observed reaction rate to the intrinsic rate. The observed rate is defined as

$$r_{\text{obs}} = \frac{\text{flux (surface area of reactive volume normal to the flux)}}{\text{reactive volume}} \tag{8-52}$$

which for rectangular geometry is

$$r_{\text{obs}} = \frac{NA_N}{A_N L} = -\frac{\mathfrak{D}_l}{L}\frac{dC}{dx}\bigg|_{x=0} \tag{8-53}$$

where A_N is the area normal to the flux. The concentration gradient, dC/dx, at the beginning of the domain ($x = 0$) is specific for the order of the reaction and the boundary conditions. For a first order reaction, the derivative of Equation 8-48 and its evaluation at $x = 0$ gives

$$\frac{dC}{dx}\bigg|_{x=0} = -\frac{C_0}{L}\mathfrak{N}_{\text{Tm}} \tanh \mathfrak{N}_{\text{Tm}} \tag{8-54}$$

Substitution of Equation 8-54 into Equation 8-53 provides an expression for the observed rate:

$$r_{\text{obs}} = -\frac{\mathfrak{D}_l C_0}{L^2}\mathfrak{N}_{\text{Tm}} \tanh \mathfrak{N}_{\text{Tm}} \tag{8-55}$$

The intrinsic rate, r_{int}, is simply the first-order rate that would occur if the concentration were C_0 everywhere within the microporous solid; that is,

$$r_{\text{int}} = -kC_0 \tag{8-56}$$

The ratio of Equations 8-55 and 8-56 defines the *reaction effectiveness factor* for a first order reaction as

$$\eta_R = \frac{\tanh \mathfrak{N}_{\text{Tm}}}{\mathfrak{N}_{\text{Tm}}} \tag{8-57}$$

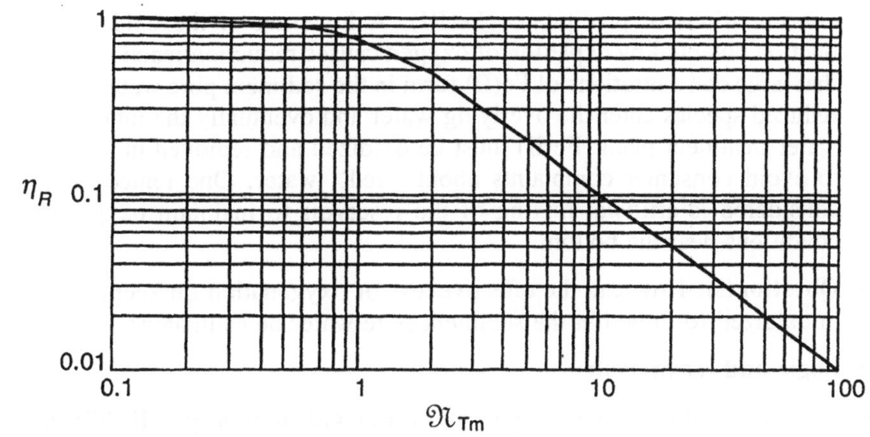

Figure 8-11 Reaction effectiveness factor as a function of the Thiele modulus for a type II diffusion domain of rectangular geometry for a first order reaction.

When the Thiele modulus is large, the reaction effectiveness factor may be approximated by

$$\eta_R \approx \frac{1}{\mathfrak{N}_{\mathrm{Tm}}} \tag{8-58}$$

The reaction effectiveness factor is plotted as a function of the Thiele modulus in Figure 8-11. We see that it approaches a value of 1 as the Thiele modulus approaches zero. A small Thiele modulus corresponds to a very slow reaction for which the rate constant is small relative to the diffusion coefficient (see Equation 8-49). Rates of reaction limit overall transformation rates for such conditions; that is, the intrinsic rate prevails and the reaction reaches its maximum "effectiveness." On the other hand, the effectiveness factor decreases to zero when the Thiele modulus is very large because the overall transformation process becomes diffusion limited; that is, microtransport prevents the process from reacting at its intrinsic rate and the "effectiveness" of the reaction decreases.

A typical environmental process involving solute transport through a type II domain is discussed in Example 8-5. Illustrations of the use of the Thiele modulus and the group II Damkohler number in the analysis of transport through a type II domain are provided in this example.

Example 8-5

- **Situation:** Thermal stratification occurs in a water supply reservoir during summer months. Vertical mixing ceases and oxygen concentration falls to zero near the bottom due to biodegradation of natural organic material in the sediment. The sediment is also known to contain large concentrations of precipitated Fe(III). A lowering of the oxidation-reduction potential by depletion of oxygen causes reduction reactions, such that large concentrations of Fe(II) form in the sediment pore water. These reduced species enter the overlying water and eventually the intake for a water treatment plant. Fe(II) must be oxidized and removed in the plant to avoid consumer complaints about "red" water. One remedy being considered is artificial mixing of the reservoir to maintain oxygenated conditions near the bottom.

- **Question(s):** How can the effectiveness of oxygenation on keeping iron in oxidized form within the sediment pore water be estimated?
- **Logic and Answer(s):**

1. The sediment pore water can be thought of as a type II diffusion domain into which oxygen diffuses and reacts with reduced iron. The corresponding conceptual model is shown below.

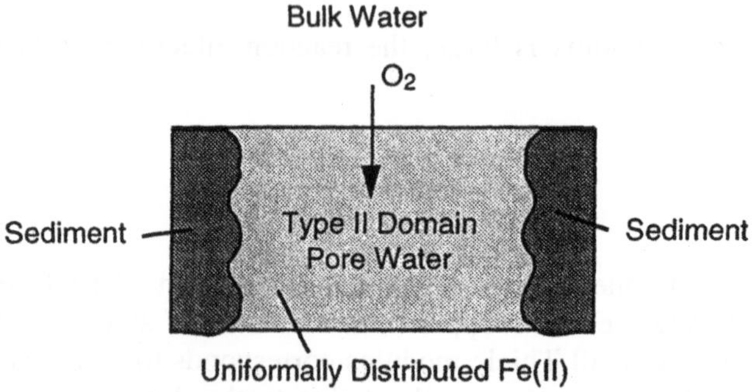

The impedance caused by the laminar boundary layer above the sediment is not included in this model (although it may well be important). We assume that the oxygen concentration remains constant at the surface of the sediment, Fe(II) is uniformly distributed throughout the depth of the pore water. The penetration of oxygen into the sediment is determined by the sum total of all oxidation reactions that might occur within the pore water.

2. Assume a temperature of 15°C and pH of 6.0 at the sediment–water interface.

3. The stoichiometry and kinetics of Fe(II) oxidation are well known (see Chapter 5). An estimate of the minimum rate of oxygen utilization follows from assuming Fe(II) as the sole reductant:

$$2Fe^{2+} + \frac{1}{2}O_2 + 5H_2O \rightleftharpoons 2Fe(OH)_3(s) + 4H^+$$

$$-\frac{d[\mathrm{Fe(II)}]}{dt} = k[\mathrm{Fe(II)}][\mathrm{OH^-}]^2 P_{O_2}$$

The k value is $8(\pm 2.5) \times 10^{13}$ L^2/(atm-min-mol^2) at 20°C (Stumm and Morgan, 1981). The pseudo-first-order rate constant k_{ps} at a temperature of 20°C, pH 6, and pressure of oxygen (P_{O_2}) of 0.2 atm is 1.6×10^{-1} min^{-1} (2.6×10^{-3} s^{-1}). The rate constant will be lower at lower temperature and P_{O_2}. We will use an order-of-magnitude estimate: $k_{ps} = 1.0 \times 10^{-3}$ s^{-1}. The stoichiometry of the reaction [4 mol Fe(II)/mol O_2] allows similar estimation of the corresponding rate constant, k_{ps,O_2}, for O_2 consumption, that is,

$$k_{ps,O_2} = 2.5 \times 10^{-4}\ \text{s}^{-1}$$

4. A search of the literature reveals rate data for sediment oxygen demand in the presence of Fe(II) (Howeler and Bouldin, 1971). An order-of-magnitude estimate for a first-order rate constant, $(k_{ps,O_2})_{sed}$, is

$$(k_{ps,O_2})_{sed} = 3 \times 10^{-3}\ \text{s}^{-1}$$

5. The one-order-of-magnitude difference in the foregoing estimates of the oxygen utilization rate constant provide a range for calculation of the oxic depth.

6. The concentration profile for diffusion with first-order reaction is quantified with the Thiele modulus ($\mathfrak{N}_{TM}$) (see Figure 8-10). Fractional oxygen concentration reaches zero approximately at the end of the diffusion domain for $\mathfrak{N}_{TM} = 5.0$. A more conservative estimate of an oxic layer corresponds to $\mathfrak{N}_{Tm} = 2$ for which the reaction effectiveness factor is calculated from Equation 8-57:

$$\eta_R = \frac{\tanh \mathfrak{N}_{Tm}}{\mathfrak{N}_{Tm}} = \frac{\tanh(2)}{2} = 0.48$$

The Damkohler number is calculated from Equation 8-51:

$$\mathfrak{N}_{Da(II)} = (\mathfrak{N}_{Tm})^2 = 4$$

Both of these process engineering parameters indicate strong diffusion control.

7. Estimating the free-liquid diffusivity of oxygen ($\mathfrak{D}_l$) to be 1×10^{-5} cm^2/s, the oxic layer thickness, L, is given from the definition of the Thiele modulus and the rate constants estimated from the literature in steps 3 and 4:

$$\mathfrak{N}_{Tm} = L\left(\frac{k}{\mathfrak{D}_l}\right)^{0.5}$$

$$L = \mathfrak{N}_{Tm}\left(\frac{\mathfrak{D}_l}{k}\right)^{0.5}$$

$$L(\text{based on } k_{ps,O_2}) = 2\left(\frac{1 \times 10^{-5}}{2.5 \times 10^{-4}}\right)^{0.5} = 0.40\ \text{cm}$$

$$L[\text{based on } (k_{ps,O_2})_{sed}] = 2\left(\frac{1 \times 10^{-5}}{3 \times 10^{-3}}\right)^{0.5} = 0.12 \text{ cm}$$

8. Step 7 shows that the faster the assumed oxidation rate, the smaller the oxic layer. The lower estimate of the oxic layer (1.2 mm) is probably more realistic.

8.4.2 Gas–Liquid Mass Transfer and the Hatta Number

The Hatta number is used to quantify flux enhancement at the gas–liquid interface when reaction accompanies mass transfer. It is analogous in function to the Thiele modulus in fluid–solid systems and incorporates the mass transfer coefficient as well as the diffusion coefficient and the reaction rate constant. As will be demonstrated, the Hatta number is useful as a process design parameter in wet scrubbing of contaminated gases or the transfer of gases into water.

The macro- and microtransport processes associated with gas–liquid mass transfer are illustrated in Figure 8-12. Here an interfacial liquid film surrounds

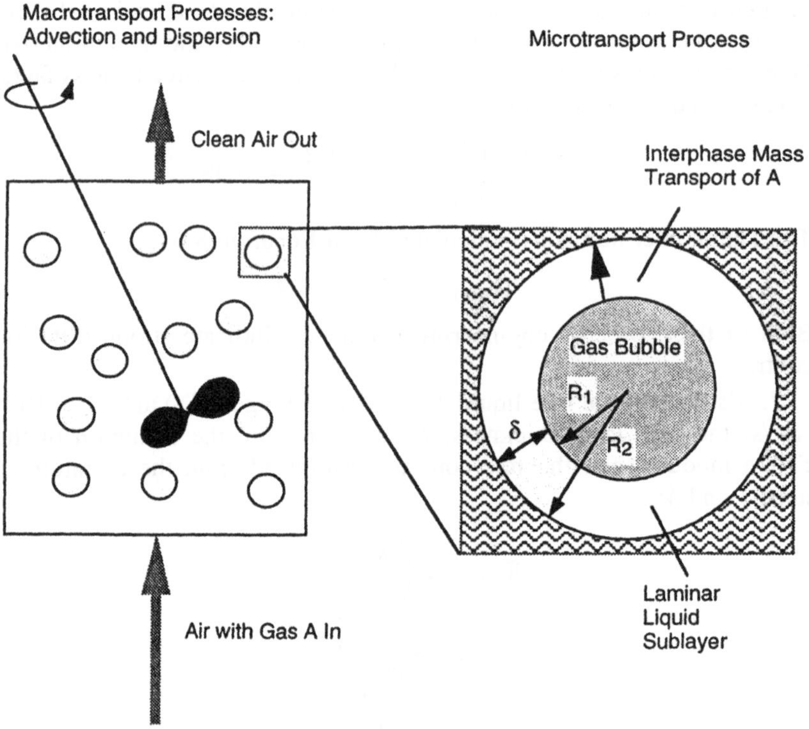

Figure 8-12 Transfer of component A from gas phase to liquid phase in a semibatch reactor.

gas bubbles being sparged into a liquid in a *semibatch* reactor. The term *semibatch reactor* is used because the solution volume is fixed but the gas bubbles are passed through it continuously.

A material balance can be written across the stagnant liquid film that surrounds the bubble using Equation 2-20 for spherical geometry:

$$-\frac{1}{r^2}\frac{\partial}{\partial r}(r^2 N) + r = \frac{\partial C}{\partial t} \tag{2-20}$$

The analysis can be simplified by ignoring the effects of bubble curvature. The validity of this common simplification can be appreciated by examining the steady state flux relationship for a nonreactive gas:

$$-\frac{1}{r^2}\frac{\partial}{\partial r}(r^2 N) = 0 \tag{8-59}$$

Equation 8-59 shows clearly that r^2N is constant. For thin films like those surrounding bubbles, the incremental radial distance, $\delta = R_1 \approx R_2$, across the film is very small, so that $R_1 \approx R_2$. From this we may conclude that $N_1 \approx N_2$ and that flux is therefore little changed by the curvature of the film. Accordingly, it is reasonable to use the material balance for one-dimensional transport and reaction in rectangular coordinates to describe radial transport in such cases.

To simplify the analysis further, we assume that (1) the concentration profile in the liquid film reaches a steady state soon after the bubble is formed, and (2) that the change in concentration of the dissolved gas in the bulk solution with time is slow enough to be neglected. For a first-order reaction, and with the dimension x defined by the interface of the bubble with the surrounding water, the material balance expression is that given by Equation 8-43, for which the general solution is defined in Equations 8-44 and 8-45. If the gas-side resistance is negligible, the concentration of gas at the liquid–gas interface is the saturation concentration, C_S, and the appropriate boundary conditions for the system depicted in Figure 8-12 are:

First boundary condition; $r = R_1$:

$$C = C_S \quad \text{at } x = 0 \tag{8-60}$$

Second boundary condition; $r = R_2$:

$$C = C_b \quad \text{at } x = \delta \tag{8-61}$$

The boundary condition at $x = \delta$ is different from the comparable position at $x = L$ in the earlier situation (see Equation 8-47). There we assumed that the concentration gradient was zero because flux must be zero at the end of the domain. However, in the present instance flux occurs at $x = \delta$ into the bulk solution; thus we need to define the bulk concentration as the boundary condition. The steady-state assumption further requires that the bulk concentration is constant. This is a reasonable assumption if the concentration is increasing slowly in the semibatch system shown in Figure 8-12.

The concentration profile within the liquid film is determine by evaluation of κ_1 and κ_2 in Equation 8-44 with the given boundary condition equations. The final form of the solution is

$$\frac{C}{C_S} = \frac{1}{\sinh m\delta}\left\{\frac{C_b}{C_S}\sinh m\delta\frac{x}{\delta} + \sinh m\delta\left(1 - \frac{x}{\delta}\right)\right\} \tag{8-62}$$

where the constant, m, is defined by Equation 8-45. Equation 8-62 describes the concentration profile within the liquid film. The general curvilinear nature of this profile was depicted earlier in Figure 8-2d without mathematical support. The corresponding equation to describe the concentration profile in the absence of reaction follows from integration of Equation 8-43 (with $k = 0$) twice between the limits of $x = 0$ and $x = \delta$, to yield

$$C = C_S - \frac{C_S - C_b}{\delta}x \tag{8-63}$$

The concentration trace within the liquid film is therefore defined by a line of constant slope. Concentration profiles for the cases of no reaction (type I domain) and reactions of two different velocities (type II domain) are illustrated in Figure 8-13. Comparison of these profiles shows that the effect of reaction is to steepen the concentration gradient and thus increase the flux of gas into

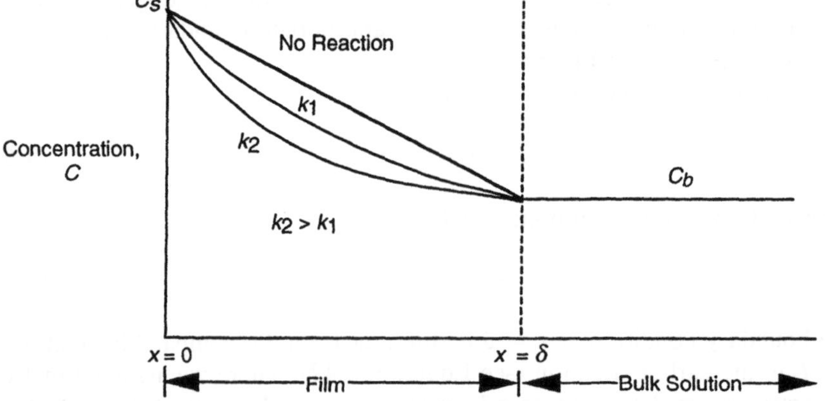

Figure 8-13 Concentration profiles in a film surrounding a bubble for diffusion with and without first-order reactions (rate constant, k).

the liquid–gas interface. The flux of gas to the liquid is

$$N = -\mathfrak{D}_l \frac{\partial C}{\partial x}\bigg|_{x=0} \tag{8-64}$$

The value of N can be evaluated by taking the derivative of Equation 8-62 at $x = 0$. In final form, this expression is

$$N = \frac{\mathfrak{D}_l C_S}{\delta} \frac{m\delta \cosh m\delta - m\delta(C_b/C_S)}{\sinh m\delta} \tag{8-65}$$

or, more simply:

$$N = \mathfrak{D}_l m \frac{C_S \cosh m\delta - C_b}{\sinh m\delta} \tag{8-66}$$

The dimensionless process engineering parameter known as the *Hatta number* can be obtained by starting with the definition of m (Equation 8-45) and the definition of the mass transfer coefficient $k_f = (\mathfrak{D}_l/\delta)$ given for the one-film model in Equation 4-53 or that corresponding to liquid-side control in the two-film model ($\hat{k}_{f,l} = \mathfrak{D}_l/\delta$) to obtain

$$\mathfrak{N}_{\text{Ha}} = m\delta = \frac{(k\mathfrak{D}_l)^{0.5}}{k_f} \approx \frac{(k\mathfrak{D}_l)^{0.5}}{\hat{k}_{f,l}} \tag{8-67}$$

The Hatta number is the equivalent of the Thiele modulus (Equation 8-49), in which the diffusion distance, L, is now the thickness of the liquid film, δ. Further, by explicitly incorporating the mass transfer coefficient, the Hatta number includes the effects of all three interacting processes: diffusion, reaction, and mass transfer between phases. The larger the Hatta number, the faster is reaction compared to mass transfer. Extremely fast reactions are characterized by Hatta numbers in excess of 3, while extremely slow reactions are characterized by Hatta numbers of less than 0.3. If the reaction is slow, a type I domain is justified for description of diffusion, wherein reaction occurs at the boundary of the domain (i.e., the bulk solution) instead of within the domain (i.e., within the liquid film layer); this situation was illustrated by Example 8-4.

Substituting the Hatta number, the definition of the mass transfer coefficient, and appropriate trigonometric identities into Equation 8-66 yields the following flux relationship for instances where diffusion and mass transfer occur with an accompanying reaction:

$$N_{\text{with reaction}} = \frac{\hat{k}_{f,l},\mathfrak{N}_{\text{Ha}}}{\tanh \mathfrak{N}_{\text{Ha}}} \left(C_S - \frac{C_b}{\cosh \mathfrak{N}_{\text{Ha}}} \right) \tag{8-68}$$

We use the two-film notation here to recognize the *potential* in some systems for impedance control by either, or both, the liquid and gas films.

This flux relationship can be compared to that which obtains when no chemical reaction occurs:

$$N_{\text{without reaction}} = \hat{k}_{f,l}(C_S - C_b) \tag{8-69}$$

The increase in flux caused by simultaneous chemical reaction can readily be appreciated by an expression of the ratio of these two fluxes for the steady-state condition and a bulk concentration of zero, a ratio known as the *enhancement* factor, $\mathfrak{F}_E$:

$$\mathfrak{F}_E = \frac{N_{\text{with reaction}}}{N_{\text{without reaction}}} = \frac{\mathfrak{N}_{\text{Ha}}}{\tanh \mathfrak{N}_{\text{Ha}}} \tag{8-70}$$

The flux-related enhancement factor is analogous in form to the reaction effectiveness factor (Equation 8-57), the latter being used to relate observed and intrinsic rates of reaction. For gas–liquid mass transfer, the enhancement factor relates the gas flux that occurs in the *presence of reaction* to that which occurs in the *absence of reaction*.

As the Hatta number increases, $\tanh \mathfrak{N}_{\text{Ha}} \rightarrow 1$ and the enhancement factor approaches the value of the Hatta number itself. An illustration of the practical significance and utility of the enhancement factor is given in Example 8-6.

Example 8-6

- **Situation:** A municipality plans to purchase gaseous chlorine in steel cylinders for use in the prechlorination of a water from a newly developed well supply. The water has a very low level of background organic matter. The chlorine gas will be bubbled into the water in an absorption tank located at the well field then. The water will be pumped to the water treatment plant some 2 miles away from the well field. It is important, for obvious reasons, that all of the chlorine gas be dissolved before the water is pumped into the transfer line.
- **Question(s):** As chief engineer on the project, you note in checking the design for the absorption tank that the staff engineer has sized the tank on the basis of a mass transfer coefficient of 1×10^{-2} cm/s but has not taken into account the reaction of dissolved chlorine gas with water to form hypochlorous acid (HOCl) and chloride ion. Is the absorber underdesigned or overdesigned? By approximately what percentage must the volume of the absorber be increased if it is underdesigned, or if it is overdesigned, what factor of safety is incorporated in the volume determined by the staff engineer?
- **Logic and Answer(s):**

1. We noted in Figure 8-13 that the effect of a reaction on a mass transfer process is to steepen the concentration gradient near the interface

of the gas phase and liquid film domain, thus increasing the flux of gas into this type II domain. Because the film was treated as a type I domain, the gas absorption tank is overdesigned; stated another way, its design incorporates an implicit safety factor.

2. The enhancement factor given in Equation 8-70 can be used to determine the magnitude of the implicit safety factor involved in the absorber design volume. A check of references of reaction rate data for chlorine indicates that the hydrolysis/redox reaction involved in converting dissolved $Cl_2(aq)$ to HOCl and Cl^- is a first-order reaction with a rate constant of $k = 11.0\ s^{-1}$. The free-liquid diffusivity of chlorine is determined to be $10^{-5}\ cm^2/s$. From Equation 8-67:

$$\mathfrak{N}_{\mathrm{Ha}} = \frac{(k\mathfrak{D}_l)^{0.5}}{\hat{k}_{f,l}} = \frac{[(11.0\ \mathrm{s}^{-1})(10^{-5}\ \mathrm{cm}^2/\mathrm{s})]^{0.5}}{10^{-2}\ \mathrm{cm/s}} = 1.05$$

and from Equation 8-63, the enhancement factor is

$$\mathfrak{F}_E = \frac{N_{\text{with reaction}}}{N_{\text{without reaction}}} = \frac{\mathfrak{N}_{\mathrm{Ha}}}{\tanh \mathfrak{N}_{\mathrm{Ha}}} = 1.34$$

3. The reaction thus enhances the mass transfer of chlorine gas by 34%. The enhancement factor of 1.34 represents the safety factor inherent in the staff engineer's design of the absorption tank.

The analysis above can be extended to reactions between dissolving gases and specific components already dissolved in solution phase. In the semibatch reactor depicted in Figure 8-12, we can picture a situation in which the initial concentration, $C_{b,B0}$, of component B, which reacts with the gas, A, in bulk solution phase is large enough that it does not change significantly over the period of observation. The reaction rate can then be treated as pseudo first order (see Chapter 7); that is, for $C_{b,B0} \gg C_{b,A}$:

$$-r_A = kC_{b,A}C_{b,B0} = k_{ps}C_{b,A} \tag{8-71}$$

and the rate constant, k_{ps}, is given by $kC_{b,B0}$. The rate of removal of component B from solution is related to the rate of removal of gas A according to the particular stoichiometry of the reaction. If, for example, the stoichiometry is

$$\gamma_A \mathrm{A} + \mathrm{B} \rightarrow \text{products} \tag{8-72}$$

then

$$-r_A = \gamma_A(-r_B) \tag{8-73}$$

The Hatta number is given by

$$\mathfrak{N}_{\mathrm{Ha}} = \frac{(kC_{b,B0}\mathfrak{D}_{l,A})^{0.5}}{\hat{k}_{f,l,A}} \tag{8-74}$$

The flux of gas into solution can again be determined from Equation 8-68, which can be greatly simplified for fast reactions (i.e., reactions described by Hatta numbers large enough that tanh $\mathfrak{N}_{\mathrm{Ha}} \rightarrow 1$). Further simplification results if we assume that the concentration of A in bulk solution is zero; this is reasonable if the reaction is fast and the initial concentration of B is far in excess of the stoichiometric amount needed to react with A. Under these conditions, Equation 8-68 reduces to

$$N_A = C_{S,A}(kC_{b,B0}\mathfrak{D}_{l,A})^{0.5} \tag{8-75}$$

The rate of absorption of A is related to the flux of A by

$$-r_A = (a_s^\circ)_R N_A \tag{8-76}$$

where $(a_s^\circ)_R$ is the surface area of interfacial film available per unit volume of solution in the reactor. Equation 8-73 relates the rate of absorption of A to the rate of removal of B, so that

$$-r_A = (a_s^\circ)_R C_{S,A}(kC_{b,B0}\mathfrak{D}_{l,A})^{0.5} = -\gamma_A r_B \tag{8-77}$$

Utilization of the relationships given in Equations 8-71 through 8-77 for analysis of rate control for reactions between dissolving gas-phase substances and substances already dissolved in the aqueous phase is illustrated in Example 8-7, with specific reference to the treatment of a water supply by ozone to remove organic contaminants.

Example 8-7

- **Situation:** A groundwater supply is found to have been contaminated by a *p*-nitrophenol (PNP) spill. Current treatment of this water involves fine bubble aeration for iron removal at pH 6.5 followed by filtration and chlorination. The pH is considerably lower than the pK_a for PNP (7.12) such that the undissociated form predominates. Oxidation with ozone is one logical treatment option.
- **Question(s):** Assess the feasibility of, and determine critical system design factors for this potential process application of ozone.
- **Logic and Answer(s):**

1. The use of ozone to oxidize the PNP appears attractive because an aeration system already exists. Ozone could thus be added in a mixture of oxygen to the fine bubble aerator. Moreover, the process could be used on an as-needed basis in the event the contamination problem is transient. A quick check of the literature reveals that ozone will in fact oxidize PNP effectively. The reaction is known to be nonelementary and to be first order in both ozone and PNP concentrations.

2. Based on the foregoing considerations, an experimental feasibility study of the ozonation process is warranted. The most critical information to be determined in this study is the rate of reaction, which is in turn determined by the conditions required for generation of ozone and its addition to the oxygen stream in the bubble aeration system.

3. The study is to be conducted as follows:

(a) A semibatch reactor such as that pictured in Figure 8-12 is to be used.

(b) A gas stream containing a mixture of ozone and air is to be delivered to the reactor at a constant rate.

(c) A series of experiments are to be made in which the initial concentration of PNP in the reactor is well in excess of the stoichiometric amount needed to react with ozone.

(d) The generation of ozone is to be varied to give different ozone concentrations in the gas bubbles.

Assumptions: The following assumptions are appropriate:

(a) The parallel decomposition reaction of ozone is assumed not to be important at the pH (6.5) of the groundwater, although this reaction increases in importance with increased pH.

(b) The reaction stoichiometry is

$$3O_3 + \text{PNP} \rightarrow \text{products}$$

(c) The reaction is nonelementary and first order with respect to ozone.

Analysis:

1. The initial rate of PNP oxidation $(-r_{\text{PNP}})$ in each experiment can be related by stoichiometry to the initial rate of ozone utilization. (*Note:* Although PNP is in excess, there is still a measurable decrease in its concentration.)

2. The available rate data on ozonation of PNP indicate that the reaction is fast; Equation 8-77 thus applies.

3. A plot of the initial rate of ozone disappearance, $(-r_{O_3})_0 = (-3r_{\text{PNP}})_0$, versus $C_{S,O_3}(C_{\text{PNP}_0})^{0.5}$ should thus yield a straight line with a slope equal to $(a_s^\circ)_R(k\mathfrak{D}_{l,O_3})^{0.5}$.

4. A 1-L semibatch reactor is used to collect the following initial rate data, with the conditions for ozone delivery as noted:

Q_{g,O_3} (L/min)	Q_{M,O_3} (mmol/min)	P_{O_3} (atm)	C_{PNP} (mmol/L)	$(-r_{\text{PNP}})_0$ (mmol/min)	$(-r_{O_3})_0$ (mmol/min)	C_{S,O_3} (mmol/L)
0.4	0.5	0.030	0.20	0.100	0.300	0.18
0.7	0.5	0.017	0.10	0.065	0.195	0.15
1.2	0.5	0.010	0.04	0.034	0.102	0.12

where Q_{g,O_3} is the ozone–air gas flow rate, Q_{M,O_3} is the molar rate of ozone delivery, and P_{O_3} is the pressure of ozone in the feed gas, $(Q_{M,O_3}/Q_{g,O_3})\mathcal{R}T$. The value of C_{S,O_3} (last column) is the ozone concentration that is in equilibrium with the ozone in the gas stream leaving the reactor, C_{g,O_3}, and this is determined from

$$C_{g,O_3} = \frac{Q_{M,O_3} - V_R\gamma(-r_{PNP})_0}{Q_g}$$

where the γ (3 in this case) relates to the stoichiometry of the reaction. Applying Henry's law (see Table 1-3 for K_H of ozone), we obtain

$$C_{S,O_3}\ (\text{mmol/L}) = \frac{C_{g,O_3}\mathcal{R}T}{\mathcal{K}_H} = \frac{C_{g,O_3}(0.0821 \times 10^{-3})(293)}{0.0677}$$

5. The rate data are plotted below to test the type II domain model.

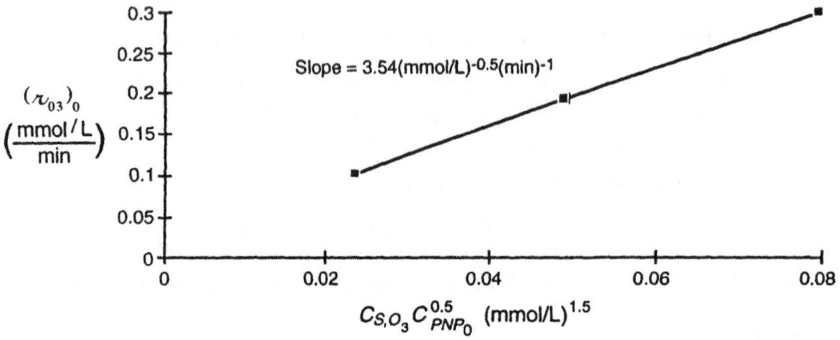

Evaluation of the slope of this plot, which is equal to $(a_s^\circ)_R(k\mathfrak{D}_{l,O_3})^{0.5}$, allows calculation of k after determining $(a_s^\circ)_R$ and $\mathfrak{D}_{l,O_3}$ (Beltran et al., 1992). Although $(a_s^\circ)_R$ is not easily measured independently, it can be estimated from experimental measurements of the overall volumetric mass transfer coefficient, $\hat{k}_{v,l} = \hat{k}_{f,l}(a_s^\circ)_R$, of ozone and a literature value for the mass transfer coefficient, $\hat{k}_{f,l}$. It is found that $\hat{k}_{v,l}$ is $2.6 \times 10^{-3}\ \text{s}^{-1}$ and $\hat{k}_{f,l}$ is 7×10^{-3} cm/s, from which $(a_s^\circ)_R$ is 0.37 cm^2/cm^3. The value of $\mathfrak{D}_{l,O_3}$ is also determined from the literature (about 1.1×10^{-5} cm^2/s). The resulting k is

$$k = \frac{(3.54 \times 10^{3\times 0.5}/60)^2}{(0.37)^2(1.1 \times 10^{-5})} = 2.3 \times 10^6\ M^{-1}\text{s}^{-1}$$

Conclusion(s):

1. Although the experimental data are limited, they do conform to the rate model for a fast pseudo-first-order reaction in a type II domain. Other initial concentrations and mass transfer conditions should be investigated to determine whether further information constrains this conclusion.

2. The Hatta number (Equation 8-67) can be evaluated to make sure that the reaction is indeed fast. The Hatta number corresponding to the lowest concentration of PNP (4×10^{-5} M) provides the most stringent check; that is,

$$\mathfrak{N}_{Ha} = \frac{(kC_{PNP,0}\mathfrak{D}_{l,O_3})^{0.5}}{k_{f,l,O_3}}$$

$$= \frac{(2.3 \times 10^6)(4 \times 10^{-5})(1.1 \times 10^{-5})^{0.5}}{7 \times 10^{-3}} = 4.6$$

This Hatta number is high enough to consider the reaction to be very fast, supporting use of the domain II model for diffusion with simultaneous reaction.

3. It is instructive to compare the Hatta number determined above to that obtained for the situation considered in Example 8-4, in which mass transfer occurs in series with a slow first-order reaction. We assume that the same value $\hat{k}_{f,l}$ exists such that the Hatta number for Example 8-4 is

$$\mathfrak{N}_{Ha} = \frac{(k\mathfrak{D}_{l,O_3})^{0.5}}{\hat{k}_{f,l}} = \frac{[(5 \times 10^{-3})(1.1 \times 10^{-5})]^{0.5}}{0.007} = 0.033$$

Logically, the Hatta number for a slow reaction is much smaller than that for a fast reaction.

8.4.3 Gas–Liquid Mass Transfer with Instantaneous Reaction

The last condition of interest in a type II domain is the instantaneous reaction of a component A from the gas phase with component B in the solution phase. Component A diffuses from the bulk gas phase through the liquid film and component B diffuses from the bulk solution phase through the liquid film. Because the reaction is instantaneous, A and B cannot coexist within the liquid film (i.e., A reacts with B on a *reaction plane* located at some position, x, within the liquid film of thickness, δ, as illustrated in Figure 8-2c). Because the reaction is assumed to be infinitely fast, its particular stoichiometry determines the relative rates of diffusion of the reactants through the liquid film. The order of the reaction, however, is unimportant. If the stoichiometry is given by

$$\gamma_A \text{A} + \text{B} \rightarrow \text{product} \tag{8-78}$$

then the rate of mass transfer (r_{mt}) of components A and B are related by

$$-r_{mt,A} = \hat{k}_{f,l,A}(a_s^\circ)_R(C_{S,A} - 0)\frac{\delta}{x} = \gamma_A(C_{b,B} - 0)\hat{k}_{f,l,B}(a_s^\circ)_R\frac{\delta}{\delta - x} \tag{8-79}$$

where it is once more assumed that mass transfer is controlled by the liquid-film side of the gas–liquid interface.

The fundamental definition of the liquid film mass transfer coefficient provides a way to express the movement of A and B within that film in terms of their respective diffusion coefficients:

$$\frac{\hat{k}_{f,l,A}}{\hat{k}_{f,l,B}} = \frac{\mathfrak{D}_{l,A}/\delta}{\mathfrak{D}_{l,B}/\delta} = \frac{\mathfrak{D}_{l,A}}{\mathfrak{D}_{l,B}} \tag{8-80}$$

The mass transfer coefficients in Equation 8-79 can thus be replaced by diffusion coefficients. Solving between the second and third equalities in Equation 8-79 for x gives

$$x = \frac{C_{S,A}\delta}{C_{S,A} + \dfrac{\gamma_A \mathfrak{D}_{l,B} C_{b,B}}{\mathfrak{D}_{l,A}}} \tag{8-81}$$

Isolating the first equality, we have

$$-\mathcal{r}_{mt,A} = \hat{k}_{f,l,A}(a_s^\circ)_R C_{S,A}\,\frac{\delta}{x} \tag{8-82}$$

and substituting for x from Equation 8-81 yields

$$-\mathcal{r}_{mt,A} = \hat{k}_{f,l,A}(a_s^\circ)_R C_{S,A}\left(1 + \frac{\gamma_A \mathfrak{D}_{l,B}}{\mathfrak{D}_{l,A}}\frac{C_{b,B}}{C_{S,B}}\right) \tag{8-83}$$

In this situation, the enhancement factor can be defined as

$$\mathfrak{F}_E = \frac{N_{\text{with reaction}}}{N_{\text{without reaction}}} = \frac{\hat{k}_{f,l,A}(a_s^\circ)_R C_{S,A}\left(1 + \dfrac{\gamma_A \mathfrak{D}_{l,A}}{\mathfrak{D}_{l,B}}\dfrac{C_{b,B}}{C_{S,A}}\right)}{\hat{k}_{f,l,A}(a_s^\circ)_R C_{S,A}} \tag{8-84}$$

or

$$\mathfrak{F}_E = 1 + \frac{\gamma_A \mathfrak{D}_{l,A}}{\mathfrak{D}_{l,B}}\frac{C_{b,B}}{C_{S,A}} \tag{8-85}$$

The enhancement factor shows that the flux of A into solution increases as (1) the concentration of B increases relative to the saturation concentration of A and (2) the diffusivity of A increases relative to the diffusivity of B.

Example 8-8 reconsiders the ozone oxidation problem examined in Example 8-7 for a condition in which the oxidation reaction is instantaneous.

Example 8-8

- **Situation:** Suppose that the groundwater treatment system discussed in Example 8-7 is to operate at pH 8 or higher instead of pH 6.5. The rate of oxidation is likely to be greater at higher pH due to production of hydroxyl radicals. In fact, it may be possible to consider the reaction as instantaneous, in which case the rate at which ozone reacts with *p*-nitrophenol (PNP) would not depend on the reaction kinetics but only on the mass transfer properties of the system.

- **Question(s):** Devise an experiment to verify the instantaneous reaction rate assumption.

- **Logic and Answer(s):**

1. Unlike Example 8-7, it is not necessary to design an experiment in which PNP is in excess because the rate relationship is no longer needed.

2. Once again, the most convenient experimental setup is the semibatch reactor (see Figure 8-12). Ozone gas is delivered at a constant partial pressure to establish a corresponding value of C_{S,O_3} in solution. To ensure that ozone is not reacting with secondary products, the rate of disappearance of PNP should be measured only during the initial stages of reaction.

3. The rate expression for instantaneous reaction (Equation 8-83) can be written as

$$r_{mt,O_3} = \gamma_{O_3}\left(-\frac{dC_{PNP}}{dt}\right) = k_{f,l,O_3}(a_s^\circ)_R C_{S,O_3}\left(1 + \frac{\gamma_{O_3}\mathcal{D}_{l,PNP}}{\mathcal{D}_{l,O_3}}\frac{C_{PNP}}{C_{S,O_3}}\right)$$

Integration of this expression by separation of variables (noting that C_{S,O_3} is contant in the semibatch reactor configuration gives

$$\ln \alpha = -\frac{\mathcal{D}_{l,PNP} k_{f,l,O_3}(a_s^\circ)_R}{\mathcal{D}_{l,O_3}} t$$

where

$$\alpha = \frac{C_{PNP} + \dfrac{\mathcal{D}_{l,O_3} C_{S,O_3}}{\gamma_{O_3}\mathcal{D}_{l,PNP}}}{C_{PNP,0} + \dfrac{\mathcal{D}_{l,O_3} C_{S,O_3}}{\gamma_{O_3}\mathcal{D}_{l,PNP}}}$$

If the rate data conform to the instantaneous rate regime, a plot of $-\ln \alpha$ versus time should be linear and $k_{f,l,O_3}(a_s^\circ)_R$ can be determined from the slope.

4. A semibatch reactor test is conducted at 20°C with an ozone pressure of 0.04 atm in the gas stream to the reactor. The resulting PNP–time data are:

Time (min)	PNP (mM)
0	10.50
10	3.86
20	1.42
30	0.52

5. Other information required to test the instantaneous reaction rate model are:

- $C_{S,O_3} = \dfrac{P_{O_3}}{\mathcal{K}_H} = \dfrac{0.04}{0.0677} = 0.59$ mM (28.4 mg/L)
- Diffusivities from the Wilke–Chang equation (see Chapter 4)
- $\mathfrak{D}_{l,O_3} = 1.1 \times 10^{-5}$ cm^2/s
- $\mathfrak{D}_{l,\text{PNP}} = 0.9 \times 10^{-5}$ cm^2/s
- $\gamma_{O_3} = 3$

6. To compute α in step 3, the constant value of the second term in both the numerator and denominator is

$$\frac{\mathfrak{D}_{l,O_3} C_{S,O_3}}{\gamma_{O_3} \mathfrak{D}_{l,\text{PNP}}} = \frac{(1.1 \times 10^{-5})(0.59)}{(3)\ (0.9 \times 10^{-5})} = 0.22$$

7. A fit of the data to the instantaneous reaction model is given in the graph below.

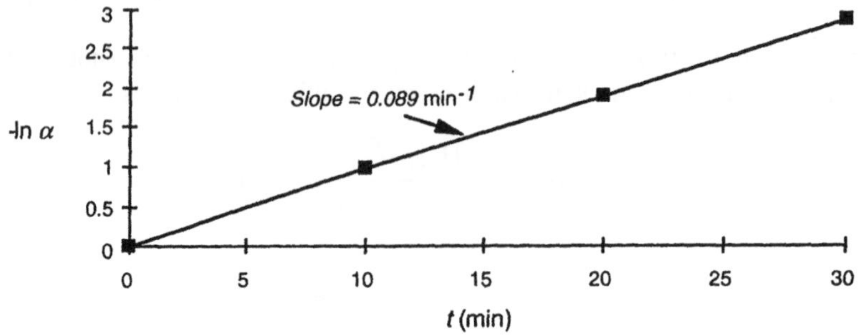

The predicted volumetric mass transfer coefficient for ozone is obtained from the slope of the graph as follows:

$$\hat{k}_{v,l} = \hat{k}_{f,l}(a_s^\circ)_R = \frac{\mathfrak{D}_{l,O_3}(0.089)}{\mathfrak{D}_{l,\text{PNP}}} = 0.11 \text{ min}^{-1}(6.6 \text{ h}^{-1})$$

8. Although the data appear to fit the model given in step 3, the experiment should be repeated at several initial concentrations of PNP to

be assured that the volumetric mass transfer coefficient remains the same. It is also necessary to confirm that the predicted overall mass transfer coefficient agrees with that measured in the absence of reaction. The latter is determined by examining ozone transfer at low pH to limit the decomposition reaction (see Example 8-4). The rate of mass transfer is given by

$$\frac{dC_{O_3}}{dt} = \hat{k}_{f,l}(a_s^\circ)_R(C_{S,O_3} - C_{O_3})$$

which, after integration gives

$$-\ln \frac{C_{S,O_3} - C_{O_3}}{C_{S,O_3}} = \hat{k}_{f,l}(a_s^\circ)_R t$$

A semilogarithmic plot of ozone concentration with time thus gives the volumetric mass transfer coefficient.

9. Care must be taken not to misinterpret an instantaneous reaction for a first-order reaction that is not limited by mass transfer. This interpretation derives from closer examination of the values presented below, with and without inclusion of the constant (0.22) from step 6.

	Values of α	
t (min)	With Constant	Without Constant
0	1.00	1.00
10	0.38	0.37
20	0.16	0.14
30	0.07	0.05

10. The constant is seen to have little effect on α for the range of PNP concentrations included in the data set. When the constant is ignored, the instantaneous reaction model given in step 3 thus predicts a first-order disappearance of PNP:

$$\ln \frac{C_{PNP}}{C_{PNP,0}} = -\frac{\mathfrak{D}_{l,PNP}\hat{k}_{f,l,O_3}(a_s^\circ)_R}{\mathfrak{D}_{l,O_3}} t$$

11. First-order models for disappearance of organic compounds in semibatch ozonation process are often reported in the literature without regard for the nature of the rate regime. As demonstrated here, the reaction can be of any order and the process still manifest first-order behavior if the reaction itself is rapid enough that the process is limited entirely by mass transfer to the reaction plane. It is therefore prudent to test proposed models for other rate regimes (see Examples 8-6 and 8-7) before drawing definitive conclusions.

8.5 TYPE III DOMAINS: DIFFUSION THROUGH MULTIPHASE MEDIA

Diffusion through multiphase media such as the internal structure of a porous solid is more complex than *free* diffusion through fluids. Pores may follow tortuous paths such that actual diffusion distances are greater than linear travel distances, x. The tortuosity factor, τ, is a parameter commonly used to adjust length dimensions to account for such effects. Substitution of τx for x in Fick's first law yields

$$N_p = -\frac{\mathfrak{D}_l}{\tau}\frac{\partial C}{\partial x} = -\mathfrak{D}_{p,e}\frac{\partial C}{\partial x} \tag{8-86}$$

where N_p is the flux within pores internal to the diffusion domain and $\mathfrak{D}_{p,e} = \mathfrak{D}_l/\tau$ is the *effective pore diffusivity.* Continuity relationships require that this internal flux be related to the flux at the external surfaces of a diffusion domain; that is,

$$\left(\mathfrak{D}_l \frac{\partial C}{\partial x}\right)\bigg|_{\text{surface}} = \epsilon_d\left(\mathfrak{D}_{p,e}\frac{\partial C}{\partial x}\right)\bigg|_{\text{within pore}} \tag{8-87}$$

where ϵ_d is the porosity of the diffusion domain (a dimensionless quantity). Flux is usually measured external to the diffusion domain, so that from Equation 8-87,

$$N = -\mathfrak{D}_e \frac{\partial C}{\partial x} \tag{8-88}$$

in which $\mathfrak{D}_e$ is the *effective* diffusivity, which is given by

$$\mathfrak{D}_e = \frac{\epsilon_d \mathfrak{D}_l}{\tau} \tag{8-89}$$

The effective diffusivity provides a convenient way to account for the effects of both porosity and tortuosity when developing a material balance relationship in terms of flux measured external to a particle.

8.6 TYPE IV DOMAINS: DIFFUSION THROUGH MULTIPHASE MEDIA WITH SIMULTANEOUS REACTION

A *type IV domain* is one in which diffusion is impeded by tortuosity, reaction, and sorption. Domains of this type are exemplified by porous adsorbent or catalyst particles and by soil particles and aggregates. The point form of the continuity equation for one-dimensional transport into such porous particle domains is given by

$$-\frac{\partial N}{\partial x} + \mathcal{r}^{\circ}(a_s^{\circ})_{I,P} = \epsilon_p \frac{\partial C}{\partial t} + (1 - \epsilon_p)\rho_s \frac{\partial q}{\partial t} \tag{8-90}$$

where N is the flux external to the particle, $(a_s^{\circ})_{I,P}$ is the internal surface area per unit volume of a porous particle (necessary to account for a surface reaction rate), ρ_s is the density of the porous particle, ϵ_p is its porosity, and q is the concentration of solute associated with the solid phase, typically given in units of mass of solute per mass of sorbent. The diffusive flux is determined by the effective diffusion coefficient, such that Equation 8-90 is

$$\mathfrak{D}_e \frac{\partial^2 C}{\partial x^2} + \mathcal{r}^{\circ}(a_s^{\circ})_{I,P} = \epsilon_p \frac{\partial C}{\partial t} + (1 - \epsilon_p)\rho_s \frac{\partial q}{\partial t} \tag{8-91}$$

8.6.1 Catalytic Surface Reactions

Consider first a simplification of Equation 8-91 in which a *catalytic reaction* occurs internal to the diffusion domain. Because the reaction is catalytic, reaction sites within the porous reactive material are not depleted. Reaction rate within a selected control volume thus remains the same for all time. Unsteady-state diffusion with reaction will be introduced in Section 8.6.3 and dealt with in greater detail in Chapter 11. Under steady-state conditions, Equation 8-91 reduces to

$$\mathfrak{D}_e \frac{\partial^2 C}{\partial x^2} + \mathcal{r}^{\circ}(a_s^{\circ})_{I,P} = 0 \tag{8-92}$$

An expression for diffusion accompanied by reaction in spherical particles can be derived in analogous fashion. The point form of the material balance relationship for radial transport through a porous sphere (see Chapter 2) is

$$-\left(\frac{\partial N}{\partial r} + \frac{2}{r} N\right) + \mathcal{r}^{\circ}(a_s^{\circ})_{I,P} = 0 \tag{8-93}$$

Radial diffusion is expressed by Fick's first law, so that Equation 8-93 becomes

$$\mathfrak{D}_e\left(\frac{\partial^2 C}{\partial r^2} + \frac{2}{r}\frac{\partial C}{\partial r}\right) + \mathcal{r}^{\circ}(a_s^{\circ})_{I,P} = 0 \tag{8-94}$$

The solutions to Equation 8-92 for diffusion and reaction in rectangular geometry or Equation 8-94 in spherical geometry depend on the nature of the reaction rate expression and on the initial and boundary conditions applied. Various examples will be presented in subsequent sections of this chapter.

8.6.2 Internal Reaction Effectiveness Factors

The internal reaction effectiveness factor, $\eta_{R,I}$, is used to describe the ratio of the *observed* and *intrinsic* rates of reactions for *type IV* domains; that is,

$$\eta_{R,I} = \frac{\text{observed rate}}{\text{intrinsic rate}} \tag{8-95}$$

As used here, $\eta_{R,I}$ reflects the extent to which diffusion through a microporous domain limits the surface reaction rate within it so that observed rates are lower than if all reactant surfaces were immediately accessible to the solute(s). Useful process design tools incorporating the internal effectiveness factor are developed in the following sections for rectangular and spherical geometries and different reaction orders.

8.6.2.1 First-Order Reactions and Rectangular Geometries Beginning with the general form of the diffusion-reaction equation (Equation 8-92), a first-order reaction at the surface with no local mass transfer limitation (i.e., $C^\circ = C$) yields the following specific relationship:

$$\mathfrak{D}_e \frac{\partial^2 C}{\partial x^2} - k^\circ (a_s^\circ)_{I,P} C = 0 \tag{8-96}$$

The solution is identical in form to that presented earlier for diffusion and first-order reaction in type II domains (see Equation 8-48). However, the Thiele modulus is now given in terms of a *surface reaction* within the domain rather than a fluid-phase reaction, and the *effective diffusivity* rather than the free-liquid diffusivity; that is,

$$\mathfrak{N}_{\text{Tm}} = L\left(\frac{k^\circ (a_s^\circ)_{I,P}}{\mathfrak{D}_e}\right)^{0.5} \tag{8-97}$$

These modified definitions of the parameters in the Thiele modulus do not change the reaction effectiveness factor that was provided in Equation 8-57.

The design of the reactor must be based on the observed rate. We can rearrange Equation 8-95 to express the observed rate as

$$r_{\text{obs}} = \eta_{R,I} r_{\text{int}} \tag{8-98}$$

Replacing r_{int} with $k^\circ (a_s^\circ)_{I,P} C$ leads to

$$r_{\text{obs}} = \eta_{R,I} k^\circ (a_s^\circ)_{I,P} C \tag{8-99}$$

The effectiveness factor provides a mathematical tool to account for reaction rate that varies with length in type IV domains due to declining concentration. In other words, the effectiveness factor is used in conjunction with measureable concentrations at the *entrance* to such domains to account for the reaction rate declining *internal* to that domain. Note that the oberved rate of the surface reaction is still first order. If the form of the type II domain definition of the reaction effectiveness factor given in Equation 8-57 is adopted for $\eta_{R,I}$, then

$$\mathcal{r}_{obs} = \frac{\tanh \mathfrak{N}_{Tm}}{\mathfrak{N}_{Tm}} k^\circ (a_s^\circ)_{I,P} C \tag{8-100}$$

For small values of the Thiele modulus $\tanh \mathfrak{N}_{Tm} = \mathfrak{N}_{Tm}$, and the rate constant is unaltered by diffusion (i.e., the internal reaction effectiveness factor is 1). This is consistent with the correspondence of small Thiele moduli to reaction-controlled conditions. For large values of the Thiele modulus (i.e., strong diffusive transport limitations), $\tanh \mathfrak{N}_{Tm}$ approaches 1. The result is an effectiveness factor that can be approximated by $1/\mathfrak{N}_{Tm}$ (and thus $\eta_{R,I} << 1$). Accordingly, the effective rate constant is greatly reduced and the observed rate can then be written as

$$\mathcal{r}_{obs} = \frac{k^\circ (a_s^\circ)_{I,P}}{\mathfrak{N}_{Tm}} C = \frac{[k^\circ (a_s^\circ)_{I,P} \mathfrak{D}_e]^{0.5}}{L} C \tag{8-101}$$

Note that the observed rate is still first order when diffusion controls, but the *intrinsic rate constant* is replaced by a constant to account for diffusion control through incorporation of $\mathfrak{D}_e$ and L.

8.6.2.2 First-Order Reactions and Spherical Geometries Diffusion processes accompanied by reaction in spherically shaped particles are of frequent interest in engineered and natural systems. Concentration profiles within spherical type IV domains are obtained by solving Equation 8-94 for the following boundary conditions and again for no local mass transfer limitation ($C^\circ = C$):

First boundary condition:

$$C = C_0 \qquad \text{at } r = R \tag{8-102}$$

Second boundary condition:

$$C \text{ is finite} \qquad \text{at } r = 0 \tag{8-103}$$

The solution for these boundary conditions is

$$\frac{C}{C_0} = \frac{R}{r} \frac{\sinh r[k^\circ (a_s^\circ)_{I,P}/\mathfrak{D}_e]^{0.5}}{\sinh R[k^\circ (a_s^\circ)_{I,P}/\mathfrak{D}_e]^{0.5}} \tag{8-104}$$

The *Thiele modulus* is defined in a way similar to that given for rectangular coordinates (Equation 8-97):

$$\mathfrak{N}_{Tm} = R\left(\frac{k^\circ (a_s^\circ)_{I,P}}{\mathfrak{D}_e}\right)^{0.5} \tag{8-105}$$

where $(a_s^\circ)_{I,P}$ is the internal specific surface area per unit volume of spherical particle. After making radial distance dimensionless ($\zeta^\circ = r/R$), Equation 8-104 can be written

$$\frac{C}{C_0} = \frac{1}{\zeta^\circ} \frac{\sinh \mathfrak{N}_{Tm}\zeta^\circ}{\sinh \mathfrak{N}_{Tm}} \tag{8-106}$$

The influence of the Thiele modulus on concentration profiles within a pore of a spherical particle is shown in Figure 8-14. These profiles are similar to those given in Figure 8-10 for rectangular geometry.

An effectiveness factor can also be easily developed for radial diffusion and reaction. The observed rate is

$$r_{obs} = \frac{N(4\pi R^2)}{4\pi R^3/3} = -\frac{3}{R}\mathfrak{D}_e \left.\frac{dC}{dr}\right|_{r=R} \tag{8-107}$$

After taking the derivative of Equation 8-106 at $\zeta^\circ = 1$ and substituting back into Equation 8-107, we obtain

$$r_{obs} = \frac{3\mathfrak{D}_e}{R^2}(-1 + \mathfrak{N}_{Tm} \coth \mathfrak{N}_{Tm})\, C_0 \tag{8-108}$$

The observed rate is thus still first order, with the several terms preceding C_0 expressing the effective rate constant. Recalling that the effectiveness factor is

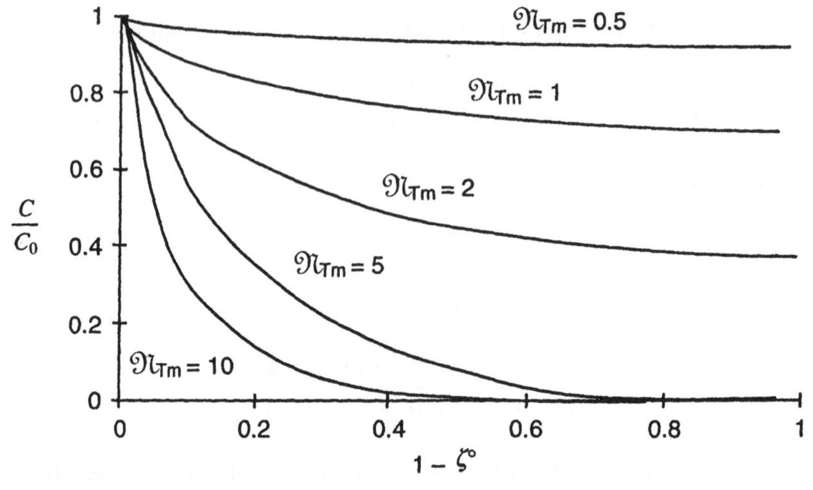

Figure 8-14 Dependence of the dimensionless surface concentration profile on the Thiele modulus for type IV diffusion domains of spherical geometry with first order reaction and no local mass transfer limitation.

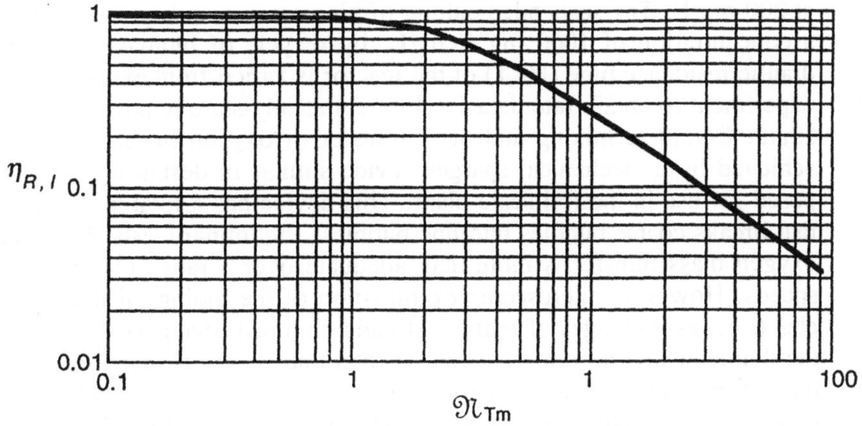

Figure 8-15 Internal surface reaction effectiveness factor as a function of the Thiele modulus for type IV diffusion domains of spherical geometry for same conditions as in Figure 8-14.

the ratio of the observed to intrinsic surface rate, $k^\circ(a_s^\circ)_{I,P}C_0$, we obtain

$$\eta_{R,I} = \frac{3}{\mathfrak{N}_{\text{Tm}}^2}(-1 + \mathfrak{N}_{\text{Tm}} \coth \mathfrak{N}_{\text{Tm}}) \tag{8-109}$$

The plot presented in Figure 8-15 shows that the effectiveness factor approaches 1 as the Thiele modulus becomes smaller. Conversely, as the Thiele modulus becomes larger, coth $\mathfrak{N}_{\text{Tm}}$ approaches 1, and the effectiveness factor is approximated by

$$\eta_{R,I} \approx \frac{3}{\mathfrak{N}_{\text{Tm}}} \tag{8-110}$$

It is clear from the similarity between Equations 8-110 and 8-58 that the geometry of the diffusion path does not change the general form of the relationship between the effectiveness factor and the Thiele modulus.

An illustration of how experimental rate data and related information can be analyzed to determine the intrinsic kinetics associated with a complex process involving concurrent diffusion and reaction processes in spherical type IV domains is given in Example 8-9 in the context of a biological waste treatment operation.

Example 8-9

- **Situation:** The biological treatability of a wastewater expected from a new food-processing plant is investigated using a completely mixed flow reactor (CMFR). The feed concentration of substrate (measured by the surrogate parameter of chemical oxygen demand, COD) is kept constant at 200 mg/L. The suspended culture of biomass is also kept constant, at a concentration of 1000 mg COD/L, by recycle of biomass. The hydraulic residence time (HRT) of the reactor is varied from experiment to experiment to obtain rate data for different steady-state removals of substrate. Complete mixing and mass transfer of oxygen in the reactor is achieved by a mechanical sparger device similar in design to that proposed for use in the full-scale plant. This device has a rotating impeller that breaks coarse bubbles into fine bubbles. Preliminary tests show that oxygen mass transfer is more than adequate over a range of rotational speeds. However, microscopic examination of the biological floc particles (i.e., assemblages of individual cells clumped together) shows that the diameter (d_p) becomes smaller as rotational speed is increased. The data below indicate that floc size affects rate of substrate removal because different exit substrate concentrations are obtained, even though the feed substrate concentration, biomass concentration, and HRT are the same.

		$C_{SU,\mathrm{OUT}}$ (mg COD/L)	
$C_{SU,\mathrm{IN}}$ (mg COD/L)	HRT, $\bar{t}$ (min)	Low Mixing Intensity, $d_p = 0.04$ cm	High Mixing Intensity, $d_p = 0.02$ cm
200	20	17.1	10.5
200	40	8.9	5.4
200	60	6.0	3.6

- **Question(s):** How can these rate data be used to obtain information about the intrinsic kinetics of the biological oxidation process?
- **Logic and Answer(s):**

1. Biodegradation requires diffusion of substrate and oxygen through the floc structure to reach individual cells (i.e., individual reaction sites). The observed biodegradation rate in each experiment is

$$(r_{\mathrm{obs}})_{SU} = \frac{C_{SU,\mathrm{IN}} - C_{SU,\mathrm{OUT}}}{\bar{t}}$$

This rate may be different than the intrinsic rate if diffusion is rate controlling.

2. The intrinsic rate of biodegradation $(r_{\mathrm{int}})_{SU}$ is proportional to the intrinsic rate of biomass growth $(r_{\mathrm{int}})_{BM}$:

$$(r_{\mathrm{int}})_{SU} = \frac{-(r_{\mathrm{int}})_{BM}}{\phi_y^\circ}$$

where ϕ_y° is the yield coefficient (COD of biomass produced per unit COD of substrate removed). The intrinsic rate of biomass growth is first order in biomass concentration, C_{BM}, but it also depends on substrate concentration (see Chapter 7 for a discussion of Michaelis–Menten enzyme kinetics and their extension to microbial cell level in Monod kinetics). At low substrate concentration, it may be possible to approximate the dependence as first order:

$$(r_{\text{int}})_{BM} = k^{\circ}(a^{\circ}_{s,w})_{I,P} C_{SU} C_{BM}$$

(*Note:* The reaction rate is second order overall, and the internal specific surface area, $(a^{\circ}_{s,w})_{I,P}$, is defined per unit weight of biomass, so the units of the surface-reaction-rate constant, k°, must therefore be cm/mg COD-s. C_{BM} is constant in these experiments, which makes the reaction rate a pseudo-first-order expression.) Substituting the intrinsic rate of biomass growth into the expression for substrate biodegradation gives

$$(r_{\text{int}})_{SU} = -k^{\circ}_{ps} C_{SU}$$

where k°_{ps} is a pseudo-first-order surface reaction rate constant:

$$k^{\circ}_{ps} = \frac{k^{\circ}(a^{\circ}_{s,w})_{I,P} C_{BM}}{\phi^{\circ}_{y}}$$

3. The observed rate for diffusion with first-order reaction in spherical particles is given by Equation 8-108:

$$(r_{\text{obs}})_{SU} = \frac{3\mathfrak{D}_e}{R^2}(-1 + \mathfrak{N}_{\text{Tm}} \coth \mathfrak{N}_{\text{Tm}}) C_{SU_0}$$

$$\mathfrak{N}_{\text{Tm}} = R\left(\frac{k^{\circ}_{ps}(a^{\circ}_{s})_{I,P}}{\mathfrak{D}_e}\right)^{0.5}$$

where $(a^{\circ}_{s})_{I,P} = \rho_{bm}(a^{\circ}_{s,w})_{I,P}$ and ρ_{bm} is the biomass density. A reasonable value for ρ_{bm} is ~ 1.05 g/cm^3. The rate model above can be tested by plotting $(r_{\text{obs}})_{SU}$ against C_{SU} in the reactor. [*Note:* C_{SU} is assumed equal to $(C_{SU})_{\text{OUT}}$ for a CMFR.] If a straight line results, the slope of the plot can be used to calculate the Thiele modulus once the effective diffusivity, $\mathfrak{D}_e$, and the floc diameter, d_p, are specified. The figure shown below indicates that the rate data conform to a first-order expression.

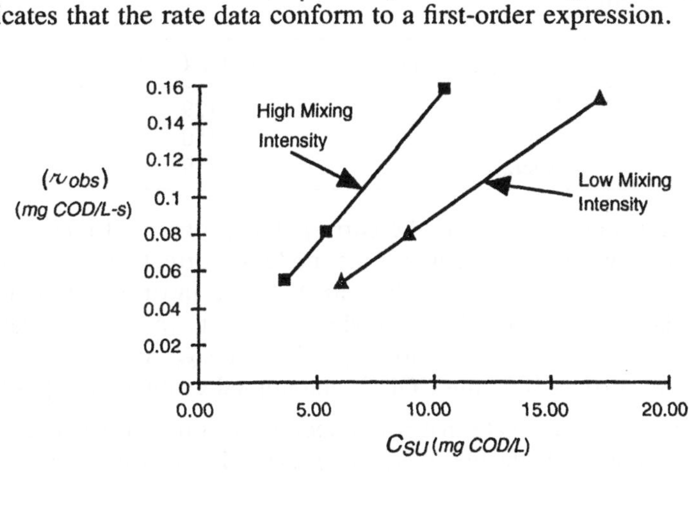

4. The effective diffusivity in a porous particle is given by Equation 8-89:

$$\mathfrak{D}_e = \frac{\epsilon_p \mathfrak{D}_l}{\tau}$$

A search of available literature on substrate diffusion through biomass reveals that a typical value for effective diffusivity is 1×10^{-6} cm^2 (i.e., about one order of magnitude smaller than free-liquid diffusivity). Computing the slope of the line in the figure above and substituting $\mathfrak{D}_e$ and $R = 0.5d_p$ into the definition of slope given by Equation 8-108 (step 3) allows calculation of $\mathfrak{M}_{\text{Tm}}$. The internal effectiveness factor, $\eta_{R,I}$, can then be calculated from Equation 8-109:

$$\eta_{R,I} = \frac{3}{\mathfrak{M}_{\text{Tm}}^2}(-1 + \mathfrak{M}_{\text{Tm}} \coth \mathfrak{M}_{\text{Tm}})$$

The results are tabulated below for the high- and low-mixing-speed experiments. The internal effectiveness factor is considerably smaller at the low mixing speed, indicating that diffusion is rate controlling.

Mixing Intensity	Slope (s^{-1})	d_p (cm)	$\mathfrak{M}_{\text{Tm}}$	$\eta_{R,I}$
High	0.015	0.020	1.29	0.9
Low	0.009	0.040	3.67	0.6

5. Having determined the Thiele modulus, the intrinsic rate constant from each set of experiments can be calculated:

$$k_{ps}^{\circ}(a_s^{\circ})_{I,P} = \frac{\mathfrak{D}_e}{(0.5d_p)^2}(\mathfrak{M}_{\text{Tm}})^2$$

The results are:

Mixing Intensity	$k_{ps}^{\circ}(a_s^{\circ})_{I,P}$ (s^{-1})
High	0.0129
Low	0.0150
Average	0.0139

These values are higher than the corresponding observed rate constants (line slope), which must be true if the model is correct. Further, the two experimental series give very close to the same intrinsic rate constant, which also must be true if the model is correct. [*Note:* The value of $k^{\circ}(a_s^{\circ})_{I,P}$ can be determined from $k_{ps}^{\circ}(a_s^{\circ})_{I,P}$ if C_{BM} and ϕ_y° are known (C_{BM} was stated as 1000 mg COD/L and a good estimate of ϕ_y° is 0.5).]

6. The results indicate that it is important to minimize floc diameter in order to maximize the observed rate in the full-scale treatment process.

The floc diameter for any internal effectiveness factor can be calculated once the corresponding Thiele modulus is known from Equation 8-109 (see step 4). If, for example, $\eta_{R,I} = 0.95$, $\mathfrak{N}_{\text{Tm}} = 0.9$, the floc diameter needed is

$$d_p = 2\mathfrak{N}_{\text{Tm}}\left[\frac{\mathfrak{D}_e}{k^\circ(a_s^\circ)_{I,P}}\right]^{0.5} = 2(0.9)\left(\frac{1 \times 10^{-6}}{0.0139}\right)^{0.5} = 0.015 \text{ cm}$$

8.6.2.3 Generalization to nth-Order Reactions The analysis for diffusion with a simultaneous reaction of nth order with no local mass transfer limitation (i.e., $C^\circ = C$) begins by generalizing the material balance relationship given by Equation 8-92:

$$\mathfrak{D}_e \frac{d^2C}{dx^2} - k^\circ(a_s^\circ)_{I,P}C^n = 0 \tag{8-111}$$

The solution to this differential equation leads to a generalized definition of the Thiele modulus:

$$\mathfrak{N}_{\text{Tm}} = L\left[\frac{k^\circ(a_s^\circ)_{I,P}C_0^n}{\mathfrak{D}_e C_0}\right]^{0.5} \tag{8-112}$$

where C_0 is again the concentration at the beginning of the domain (i.e., at $x = 0$ or $r = R$).

It has been shown that for different reaction order, plots of $\log \eta_{R,I}$ versus $\log \mathfrak{N}_{\text{Tm}}$ are approximately parallel and linear with slopes of -1 when the Thiele modulus is large, as evident in Figure 8-11 for a first-order reaction. This parallel behavior is approximately described in terms of reaction order, n, as $\eta_{R,I}\mathfrak{N}_{\text{TM}} = [2/(n+1)]^{0.5}$ which for first-order reaction is simply $\eta_{R,I}\mathfrak{N}_{\text{TM}} = 1$. The observed rates can thus be obtained for various orders of reaction and rectangular geometry. These are listed in Table 8-1 for zero-, first-, and second-order reactions.

Similarly, the material balance relationship for diffusion and nth-order reaction in spherical geometry is

$$\mathfrak{D}_e \frac{d^2C}{dr^2} + \mathfrak{D}_e \frac{2}{r}\frac{dC}{dr} - k^\circ(a_s^\circ)_{I,P}C^n = 0 \tag{8-113}$$

from which the general form of the Thiele modulus is determined to be

$$\mathfrak{N}_{\text{Tm}} = R\left[\frac{k^\circ(a_s^\circ)_{I,P}C_0^n}{\mathfrak{D}_e C_0}\right]^{0.5} \tag{8-114}$$

As for rectangular geometry, it has been shown that plots of $\log \eta_{R,I}$ versus $\log \mathfrak{N}_{\text{TM}}$ for different reaction order are approximately parallel and linear when

Table 8-1 Intrinsic and Observed Rate Expressions for Different Reaction Orders and Different Coordinate Systems Involved in Diffusive Transport with Simultaneous Surface Reaction with No Local Mass Transfer Limitations in Type IV Particle Domain

	Order of Intrinsic Reaction		
	Zero	First	Second
Intrinsic rate, r_{int}	$k^\circ (a_s^\circ)_{I,P}$	$k^\circ (a_s^\circ)_{I,P} C_0$	$k^\circ (a_s^\circ)_{I,P} C_0^2$
Effectiveness factor, $\eta_{R,I}$, for rectangular geometry[a]	$\frac{1}{L}\left[\frac{2\mathfrak{D}_e C_0}{k^\circ (a_s^\circ)_{I,P}}\right]^{0.5}$	$\frac{1}{L}\left[\frac{\mathfrak{D}_e}{k^\circ (a_s^\circ)_{I,P}}\right]^{0.5}$	$\frac{1}{L}\left[\frac{2\mathfrak{D}_e}{3k^\circ (a_s^\circ)_{I,P} C_0}\right]^{0.5}$
Observed Rate, r_{obs}, for rectangular geometry[b]	$\frac{[2k^\circ \mathfrak{D}_e (a_s^\circ)_{I,P}]^{0.5}}{L} (C_0)^{0.5}$	$\frac{[k^\circ \mathfrak{D}_e (a_s^\circ)_{I,P}]^{0.5}}{L} (C_0)$	$\frac{1}{L}\left[\frac{2\mathfrak{D}_e k^\circ (a_s^\circ)_{I,P}}{3}\right]^{0.5} (C_0)^{1.5}$
Effectiveness factor, $\eta_{R,I}$, for spherical geometry[a]	$\frac{4.24}{R}\left[\frac{\mathfrak{D}_e}{k^\circ (a_s^\circ)_{I,P}}\right]^{0.5} (C_0)^{0.5}$	$\frac{3}{R}\left[\frac{\mathfrak{D}_e}{k^\circ (a_s^\circ)_{I,P}}\right]^{0.5}$	$\frac{2.45}{R}\left[\frac{\mathfrak{D}_e}{k^\circ (a_s^\circ)_{I,P}}\right]^{0.5} (C_0)^{-0.5}$
Observed rate, r_{obs}, for spherical geometry[b]	$\frac{4.24}{R} [\mathfrak{D}_e k^\circ (a_s^\circ)_{I,P}]^{0.5} (C_0)^{0.5}$	$\frac{3}{R} [\mathfrak{D}_e k^\circ (a_s^\circ)_{I,P}]^{0.5} C_0$	$\frac{2.45}{R} [\mathfrak{D}_e k^\circ (a_s^\circ)_{I,P}]^{0.5} (C_0)^{1.5}$

[a]Effectiveness factors are restricted to large values of the Thiele modulus where $\eta_{R,I}\mathfrak{N}_{TM} = [2/(n+1)]^{0.5}$ for rectangular geometry and $\eta_{R,I}\mathfrak{N}_{TM} = 3[2/(n+1)]^{0.5}$ for spherical geometry.

[b]$r_{obs} = \eta_{R,I} r_{int}$ (mol L^{-3} t^{-1}).

the Thiele modulus is large. For spherical geometry, these parallel traces have slopes of −3. The resulting relationship, $\eta_{R,I}\mathfrak{N}_{TM} = 3[2/(n+1)]^{0.5}$, reduces to $3/\mathfrak{N}_{TM}$ for a first-order reaction (Equation 8-110). The resulting effectiveness factor and observed rate expressions for different orders of reactions are shown in Table 8-1.

The observed rate expressions in Table 8-1 illustrate that diffusion alters intrinsic reaction rates in two important ways: (1) the rate constant is always modified to include the effect of diffusion length (L or R) and effective diffusivity, $\mathfrak{D}_e$; and (2) the observed rate is of order $(n + 1)/2$. On this basis, a first-order reaction still appears to be first order. The observed rate constant, however, is not equal to the intrinsic rate constant. Moreover, both the rate constant and the observed order of reaction are altered for reactions that are intrinsically other than first order. The implications for reactor design are important. If diffusion limitations are ignored, the performance of a reactor cannot be predicted accurately. Microtransport limitations can also affect interpretation of rate data done prior to reactor design because such data are generally intended, and presumed, to measure *intrinsic* rates. If diffusional resistances are not eliminated or minimized in experiments used to collect the rate data, the observed rate will not equal the intrinsic rate; under such circumstances apparent rates thus incorporate an effect known as *disguised kinetics.*

Rates of microbial degradation in suspended growth systems provide a practical examples of differences between observed and intrinsic rates. The intrinsic kinetics are more complex than described in Table 8-1 because they derive from the enzyme–substrate saturation kinetic model (see Chapter 7), which is a "variable-order" rate model. Nonetheless, tractable observed rate expressions are possible to account for diffusion limitations. Experiments have shown that observed rates of microbial degradation of substrate in suspended growth systems decrease with increasing floc (biomass) size (see Example 8-9). If we take the viewpoint that the process involves biological catalysis, the framework for rate analysis is the same as presented above, although the intrinsic rate expressions may be more complex. Diffusion of substrate into floc particles containing individual bacterial cells (i.e., reaction sites) may very well limit observed rates of reaction.

It may be possible to infer whether microtransport is a rate-limiting process by measuring observed rates of reaction at different temperatures. Chemical processes usually have higher activation energies than physical processes. Reaction rates that do not show marked increases with temperature may therefore be more indicative of diffusion limitations than reaction control.

8.6.2.4 Combined External and Internal Impedances As indicated schematically in Figure 8-1c, external and internal resistances to transport often occur in series. In the most simple treatment of such circumstances, an overall reaction effectiveness factor, $\eta_{R,O}$, can be applied to describe the combined effects of external and internal impedances to observed rates of reaction, such that

$$\eta_{R,O} = \frac{\text{observed rate}}{\text{intrinsic rate based on bulk concentration, } C_b} \tag{8-115}$$

$$r^{\circ}_{\text{obs}} = \eta_{R,O} r^{\circ}_{\text{int}} \tag{8-116}$$

Equation 8-116 is analogous to that derived earlier for cases in which only internal resistance was included in the effectiveness factor. Here $\eta_{R,O}$ includes impedances to reaction caused by both external and internal resistances. The same precautions regarding design of reactors and interpretation of laboratory rate data hold as for internal impedance alone.

Based on our previous analysis of external impedance control (see Equation 8-1), the rate (mol/time) of mass transport to the reactive external surface of a particle is

$$N(a_s^{\circ})_{E,P} = k_f(C_b - C^{\circ})(a_s^{\circ})_{E,P} \tag{8-117}$$

where C° is the surface concentration of solute immediately adjacent to the external surface and $(a_s^{\circ})_{E,P}$ is the specific *external area per unit volume* of a porous particle. The observed rate, r°_{obs}, for a first-order surface reaction accompanied by internal resistance to mass transport is

$$r^{\circ}_{\text{obs}} = \eta_{R,I} k^{\circ} C^{\circ} \tag{8-118}$$

As noted earlier, $\eta_{R,I}$ accounts for the decline in rate that occurs as a result of declining concentration along the length of a type IV domain; it does so by simply applying a retardation effect to the corresponding fixed concentration at the entrance to the domain. The units for the observed rate are mass (or moles) per unit area per unit time. Flux continuity (and rate) conditions under steady state require that

$$k_f(C_b - C^{\circ})(a_s^{\circ})_{E,P} = \eta_{R,I} k^{\circ}(a_s^{\circ})_{I,P} C^{\circ} \tag{8-119}$$

The surface concentration, C° is obtained by rearrangement of Equation 8-119:

$$C^{\circ} = \frac{k_f(a_s^{\circ})_{E,P} C_b}{\eta_{R,I} k^{\circ}(a_s^{\circ})_{I,P} + k_f(a_s^{\circ})_{E,P}} \tag{8-120}$$

It is now possible to express the observed rate of reaction in terms of the measurable bulk concentration, C_b, by substituting Equation 8-120 into Equation 8-118 to obtain

$$r^{\circ}_{\text{obs}} = \eta_{R,I} k^{\circ} \left[\frac{k_f(a_s^{\circ})_{E,P}}{\eta_{R,I} k^{\circ}(a_s^{\circ})_{I,P} + k_f(a_s^{\circ})_{E,P}} \right] C_b \tag{8-121}$$

Comparison of the internal reaction effectiveness factor based on discrete external and internal diffusion effects with the overall factor based on a "lumped-parameter'' treatment of two types of diffusion gives

$$\eta_{R,O} = \frac{\eta_{R,I}}{1 + \eta_{R,I}[k^\circ(a_s^\circ)_{I,P}/k_f(a_s^\circ)_{E,P}]} \tag{8-122}$$

The relative importance of external and internal impedances can be understood from Equation 8-122. If external mass transport is fast, then $k_f(a_s^\circ)_{E,P} >> \eta_{R,I}k^\circ(a_s^\circ)_{I,P}$, and Equation 8-121 reduces to

$$r^\circ_{\mathrm{obs}} = \eta_{R,I}k^\circ C_b \tag{8-123}$$

which indicates that the process is controlled by internal resistance. Evaluation of the overall effectiveness factor (Equation 8-122) also permits examination of the effects of specific surface area (external and internal) and intrinsic reaction rates (through the effectiveness factor) on observed rates. As cited earlier and elaborated upon in Chapter 10, expressions of this type are significant for reactor design.

Example 8-10 expands upon the biological oxidation process examined in Example 8-9 to include the effects of combined external and internal mass transfer resistances in observed rates of biodegradation.

Example 8-10

- **Situation:** Assume that external mass transfer is also suspected as a potential rate-controlling step in the biological process described in Example 8-9. This situation corresponds to a type I domain in series with a type IV domain. Assume further that mass transfer coefficients determined from available correlations for mechanical spargers are 0.001 and 0.005 cm/s for the experiments conducted with low and high mixing speeds, respectively.
- **Question(s):** How does external mass transfer influence calculation of the intrinsic rate constants from the data provided in Example 8-9?
- **Logic and Answer(s):**

1. The observed rate of biodegradation is

$$(r_{\mathrm{obs}})_{SU} = \eta_{R,O}(r_{\mathrm{int}})_{SU} = \eta_{R,O}k^\circ(a_s^\circ)_{I,P}C_{SU}$$

where C_{SU} is the bulk concentration of substrate and $\eta_{R,O}$ is the overall effectiveness factor defined by Equation 8-122 to account for both external and internal impedances in series; that is,

$$\eta_{R,O} = \frac{\eta_{R,I}}{1 + \eta_{R,I}[k^\circ(a_s^\circ)_{I,P}/k_f(a_s^\circ)_{E,P}]}$$

2. Other parameters required for this analysis are

$$\mathfrak{N}_{\mathrm{Tm}} = R\left[\frac{k^\circ(a_s^\circ)_{I,P}}{\mathfrak{D}_e}\right]^{0.5}$$

$$(a_s^\circ)_{E,P} = \frac{\text{external surface area}}{\text{volume of floc}} = \frac{3}{R}$$

3. Combining the expressions above, we can write the observed rate as

$$(r_{\text{obs}})_{SU} = \frac{\eta_{R,I}}{[1/k^\circ(a_s^\circ)_{I,P}] + [\eta_{R,I}/k_f(a_s^\circ)_{E,P}]} C_{SU}$$

$$= \frac{\eta_{R,I}(\mathfrak{N}_{\text{Tm}})^2}{(R^2/\mathfrak{D}_e) + [R(\mathfrak{N}_{\text{Tm}})^2/3k_f]} C_{SU}$$

where the internal effectiveness factor and the Thiele modulus are related by Equation 8-109:

$$\eta_{R,I} = \frac{3}{\mathfrak{N}_{\text{Tm}}^2}(-1 + \mathfrak{N}_{\text{Tm}} \coth \mathfrak{N}_{\text{Tm}})$$

4. According to the external-internal impedance model given in step 1, a straight line should result from a plot of the observed rate against the substrate concentration in the reactor for each experimental series. The slope of the line (see data and figure in Example 8-9) can then be used to calculate the Thiele modulus because all the other parameters [$\mathfrak{D}_e$, R, k_f, and $(a_s^\circ)_{E,P}$] are known. An implicit solution technique can be used to find the value of $\mathfrak{N}_{\text{Tm}}$. The results are tabulated below. They show that external mass transfer accounts for a small but nevertheless measurable effect on the overall effectiveness factor. Moreover, the values of the Thiele modulus are significantly larger than calculated in Example 8-9.

Mixing Intensity	Slope	k_f (cm/s)	d_p (cm)	$\mathfrak{N}_{\text{Tm}}$	$\eta_{R,O}$	$\eta_{R,I}$
High	0.015	0.005	0.020	1.3	0.89	0.90
Low	0.010	0.001	0.040	4.0	0.54	0.56

5. The intrinsic rate constant can be calculated from the Thiele modulus for each experimental series and compared to those calculated in Example 8.9:

	$k^\circ(a_s^\circ)_{I,P}$ (s^{-1})	
Mixing Intensity	This Example	Example 8-9
High	0.0169	0.0129
Low	0.0178	0.0150
Average	0.0173	0.0139

6. The data above show that inclusion of the effect of external mass transfer control in the rate model increases the calculated intrinsic rate constant by about 24%. External mass transfer may therefore be an important factor to consider when attempting to determine intrinsic rate constants in such heterogeneous systems.

8.6.3 Sorption Reactions

There are many important examples in which rates of transformation or separation in natural and engineered environmental processes involve simultaneous diffusion and sorption. In subsurface systems, for example, hydrophobic organic molecules diffusing within aggregates and microporous soil particles associate simultaneously with soil organic matter. Water quality specialists concerned with transport and fate predictions for contaminants in such systems must develop quantitative descriptions of these combined rate phenomena. In engineered systems, microporous granules of activated carbon or polymeric resins are used to sorb contaminants from water flowing past them in a variety of different reactor configurations, again requiring an understanding of heterogeneous rate phenomena to optimize reactor configuration.

Heterogeneous media of interest are typically solid structures containing labrynths of pores. From a transport point of view, such structures are characterized as type IV domains. Movement of solute can occur by diffusion of solute in fluids contained in the pores, *pore diffusion*, and by migration of sorbate along the wall surfaces of the internal pores, *surface diffusion.* When sorbed-phase concentrations are sufficiently high, transport of solute by the latter mechanism can be a significant, even dominant, part of the overall *intraparticle* or *intra-aggregate* solute flux. In contrast to the catalytic reactions discussed to this point, rates of sorption reactions are inherently unsteady because surface sites are depleted continually. This complicates mathematical analysis of diffusion with reaction because two dependent variables, diffusion distance and time, must be considered.

In this chapter we provide an overview of sorption rate phenomena and relationships, but we limit mathematical developments to simple, one-dimensional diffusion in rectangular coordinates, as illustrated schematically in Figure 8-16. Other important physical descriptions of diffusion domains are dealt with in greater detail in Chapter 11, including simultaneous pore and surface

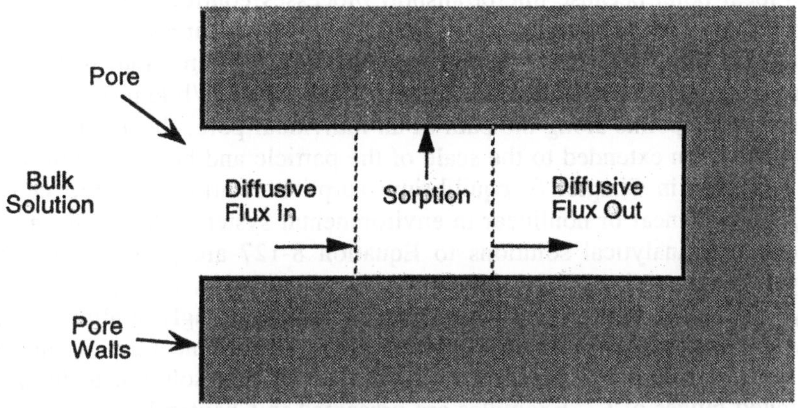

Figure 8-16 One-dimensional diffusion with sorption in an elemental pore volume.

diffusion, external and internal impedances in series, and spherical coordinate systems.

Microtransport by internal diffusion accounts for flux through the pore spaces of microporous particles and other porous domains. After incorporating the definition of effective diffusivity within the pore of a particle, we obtain

$$\mathfrak{D}_e \frac{\partial^2 C}{\partial x^2} = \epsilon_p \frac{\partial C}{\partial t} + (1 - \epsilon_p)\rho_s \frac{\partial q}{\partial t} \tag{8-124}$$

An alternative definition of effective diffusivity is the effective pore diffusivity $\mathfrak{D}_{p,e}$ which is obtained by dividing both sides by the porosity of the particle or domain:

$$\mathfrak{D}_{p,e} \frac{\partial^2 C}{\partial x^2} = \frac{\partial C}{\partial t} + \frac{(1 - \epsilon_p)\rho_s}{\epsilon_p} \frac{\partial q}{\partial t} \tag{8-125}$$

where

$$\mathfrak{D}_{p,e} = \frac{\mathfrak{D}_l}{\tau} \tag{8-126}$$

The two dependent variables, C and q, must be related in applying Equation 8-125. The chain rule for derivatives can be used for this purpose, yielding

$$\mathfrak{D}_{p,e} \frac{\partial^2 C}{\partial x^2} = \frac{\partial C}{\partial t} + \frac{(1 - \epsilon_p)\rho_s}{\epsilon_p} \frac{\partial q}{\partial C} \frac{\partial C}{\partial t} \tag{8-127}$$

Solution of this equation requires that the relationship between the liquid and solid phase concentrations, $\partial q/\partial C$, be known at all locations along the pore length. A common assumption is that equilibrium exists between the two phases at each local point because microtransport processes control arrival of solute at sorption sites but the sorption reaction itself is very fast at each specific site. This assumption is referred to as the *local equilibrium* assumption. Its validity decreases sharply as the scale of the term *local* increases. While probably valid for microscopic points along the path of an individual pore, the assumption is rarely valid when extended to the scale of the particle and bulk solution.

As discussed in Chapter 6, equilibrium sorption relationships between q_e and C_e may be linear or nonlinear in environmental systems. Most are in fact nonlinear, and analytical solutions to Equation 8-127 are possible only for cases of linear sorption. It is possible in some situations to approximate a weakly nonlinear sorption by a linear relationship, or a strongly nonlinear sorption by an irreversible isotherm expression, over narrow concentration ranges. When valid, these assumptions permit use of analytical solution techniques over limited ranges of C_e. Examples are presented in Chapter 11.

For purposes of our present discussion of sorption rates, we will simplify mathematical treatments of illustrations by assuming that *linear local equilibrium* (LLE) *conditions* exist as soon as a solute reaches a specific reactive site; that is, a linear equilibrium is attained instantaneously. The linear equilibrium sorption model, presented in Chapter 6, has the form

$$q_e = \mathcal{K}_D C_e \tag{6-51}$$

According to Equation 8-127, the derivative, $\partial q_e/\partial C_e$, is needed. This is obtained by differentiating Equation 6-51, to obtain

$$\frac{\partial q_e}{\partial C_e} = \mathcal{K}_D \tag{8-128}$$

Substitution into Equation 8-127 and simplification then yields

$$\mathfrak{D}_a \frac{\partial^2 C}{\partial x^2} = \frac{\partial C}{\partial t} \tag{8-129}$$

in which the *apparent diffusivity*, $\mathfrak{D}_a$, is given by

$$\mathfrak{D}_a = \frac{\mathfrak{D}_{p,e}}{1 + \dfrac{(1-\epsilon_p)\rho_s}{\epsilon_p}\mathcal{K}_D} = \frac{\mathfrak{D}_e/\epsilon_p}{1 + \dfrac{(1-\epsilon_p)\rho_s}{\epsilon_p}\mathcal{K}_D}$$

$$= \frac{\mathfrak{D}_l/\tau}{1 + \dfrac{(1-\epsilon_p)\rho_s}{\epsilon_p}\mathcal{K}_D} \tag{8-130}$$

Equation 8-129 has the same mathematical form as Fick's second law for unsteady-state free diffusion in a liquid. The difference between the two lies in the definition of the diffusion coefficient; that is, the *free-liquid diffusivity* in Fick's law is replaced by an *apparent diffusivity* in Equation 8-129. Apparent diffusivity accounts for impedances to microtransport caused by three factors: (1) free-liquid diffusion, (2) sorption to solid phases, and (3) restrictions in diffusion pathways through a solid.

Many different solutions to Equation 8-129 are possible, depending on the initial and boundary conditions. For illustrative purposes, consider a *type IV diffusion domain* that begins at $x = 0$ and is of infinite length in the positive x-direction. The concentration of solute is initially zero in both liquid and solid phases throughout the domain. A constant concentration, C_0, is then applied at $x = 0$ for all time. The initial and boundary conditions required to solve Fick's second law are:

Initial condition:

$$C(x > 0, t = 0) = 0 \tag{8-131}$$

First boundary condition:

$$C(x = 0, t > 0) = C_0 \tag{8-132}$$

Second boundary condition:

$$C(x \to \infty, t > 0) = 0 \tag{8-133}$$

The following solution is obtained by Laplace transforms:

$$\frac{C(x, t)}{C_0} = \operatorname{erfc}\left[\frac{x}{2(\mathfrak{D}_a t)^{0.5}}\right] \tag{8-134}$$

in which erfc is the *complementary error function.*

The shape of the concentration profile produced by simultaneous diffusion and sorption is important to understand because it determines the flux of solute at the entrance to the diffusion domain (i.e., at $x = 0$). To appreciate the solution given by Equation 8-134, we must consider the nature of the complementary error function of an arbitrary variable z, which is related to the *error function* (erf) of that same variable by

$$\operatorname{erfc}(z) = 1 - \operatorname{erf}(z) \tag{8-135}$$

The erf(z), a standard mathematical function having broad applications in mathematics and physics, is defined by

$$\operatorname{erf}(z) = \frac{2}{\pi^{0.5}} \int_0^z \exp(-z^2)\, dz \tag{8-136}$$

Both the error function and complementary error function are depicted graphically in Figure 8-17 and discussed further and tabulated in Appendix A.8.1. The error function is encountered frequently in statistical analysis, where it constitutes the standardized *normal cumulative distribution* (i.e., the integral of the standardized *normal distribution*). This reinforces the notion of diffusion as the resultant of a *random* movement of solute molecules through a fluid, and thus as a probability predictable by a *probability function.* As mentioned in Chapter 4, Einstein first developed a probability theory to predict the position of particles migrating by Brownian motion. The error function and its use in describing diffusion is discussed further in Chapter 11.

Concentration profiles within the diffusion domain described above, and calculated with Equation 8-134, are shown in Figure 8-18 for two different

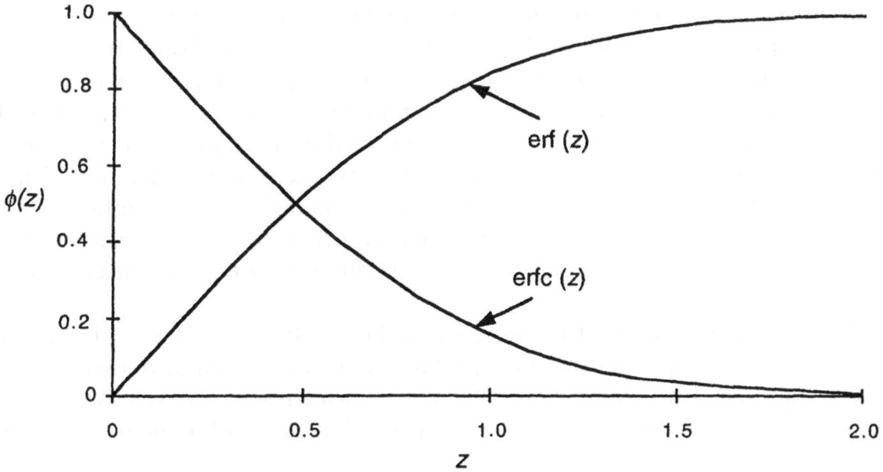

Figure 8-17 Error function and the complementary error function for a variable z.

values of apparent diffusivity and a fixed time (three years) after diffusion begins. We observe that the larger value of apparent diffusivity produces greater penetration into the porous sorbent. A larger apparent diffusivity corresponds to a smaller $\mathcal{K}_D$ and thus less sorption (see Equation 8-130). In other words, less strongly sorbed solutes migrate more rapidly. Diffusion is nevertheless a

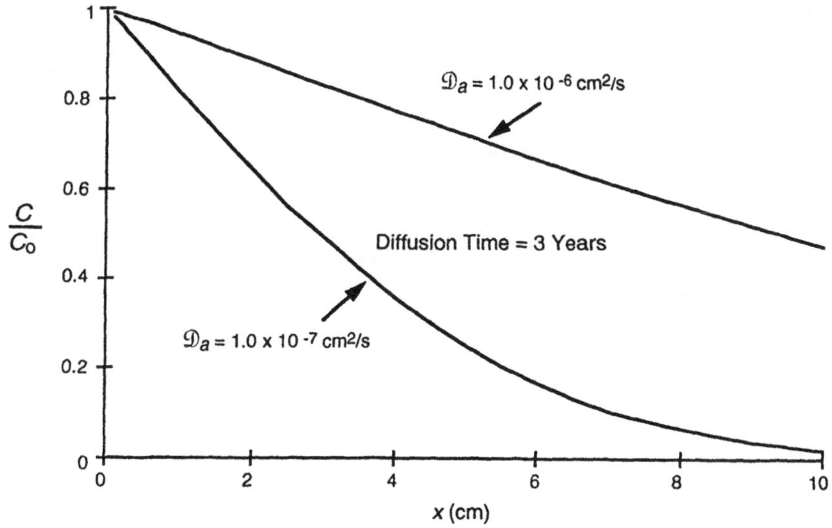

Figure 8-18 Dimensionless concentration profiles for type IV domains having different apparent diffusivities.

slow process, as indicated by the fact that it takes three years for concentrations to rise to the values shown in Figure 8-18 over a distance of only 10 cm.

The restrictions under which Equation 8-134 applies must be noted carefully. In reality, no sorption domain has infinite length. However, analytical solutions based on this assumption are reasonably accurate up to the times when the concentration rises significantly above zero at the end of a finite diffusion length, L. More complex analytical and numerical solutions are available to handle the general case of finite-length media, for which concentration at L cannot be assumed to be zero. Some of these solutions are discussed in Chapter 11.

From a process engineering standpoint, the sorption rate at $x = 0$ is of as much interest as the concentration profile within the diffusion domain. This might represent, for example, the net rate at which solute is removed from bulk solution as it flows past a sorbent surface. Diffusional flux is given in this case by

$$N = -\mathfrak{D}_a \left. \frac{\partial C}{\partial x} \right|_{x=0} \tag{8-137}$$

The derivative, $\partial C/\partial x$, is determined by first writing Equation 8-134 as

$$\frac{C(x, t)}{C_0} = c^\circ = 1 - \frac{2}{\pi^{0.5}} \int_0^z \exp(-\zeta^\circ)^2 \, d\zeta^\circ \tag{8-138}$$

in which it is assumed that the concentration at the beginning of the diffusion domain is the same as the bulk-phase concentration (i.e., $C_0 = C_b$) and where ζ°, a diffusional length ratio is

$$\zeta^\circ = \frac{x}{2(\mathfrak{D}_a t)^{0.5}} \tag{8-139}$$

The derivative of interest is obtained by the chain rule:

$$\frac{\partial c^\circ}{\partial x} = \frac{\partial c^\circ}{\partial \zeta^\circ} \frac{\partial \zeta^\circ}{\partial x} \tag{8-140}$$

It can be shown that

$$\left. \frac{\partial c^\circ}{\partial \zeta^\circ} \right|_{\zeta^\circ = 0} = -\frac{2}{\pi^{0.5}} \tag{8-141}$$

and thus we have

$$\left. \frac{\partial c^\circ}{\partial x} \right|_{x=0} = -\frac{1}{(\pi \mathfrak{D}_a t)^{0.5}} \tag{8-142}$$

Finally, substitution of C/C_0 for $c°$ gives the desired derivative in Equation 8-137, from which the flux is calculated as

$$N|_{x=0} = \mathfrak{D}_a C_0 \frac{1}{(\pi \mathfrak{D}_a t)^{0.5}} = C_0 \left(\frac{\mathfrak{D}_a}{\pi t} \right)^{0.5} \tag{8-143}$$

The mass of solute, $M(t)$, sorbed as a function of time is calculated by integrating the foregoing flux expression over time:

$$\frac{M(t)}{A_N^\circ} = \int_0^t N|_{x=0}\, dt = C_0 \left(\frac{\mathfrak{D}_a}{\pi} \right)^{0.5} \int_0^t t^{-0.5}\, dt \tag{8-144}$$

The result after integration is

$$\frac{M(t)}{A_N^\circ} = 2C_0 \left(\frac{\mathfrak{D}_a t}{\pi} \right)^{0.5} \tag{8-145}$$

where A_N° is the sorbent surface area normal to the direction of diffusion. Unsteady-state diffusion accompanied by sorption thus causes the uptake of mass to be proportional to the *square root of time*. Adherence of experimental data to this relationship is frequently cited as evidence of a diffusion controlled sorption process. The relationship is valid, however, only if (1) the concentration of solute at the beginning of the diffusion domain remains constant, and (2) the diffusion domain can be approximated as having an infinite length. These conditions may hold reasonably well during the initial stages of solute uptake by microporous sorbent particles.

A practical illustration of an application of the foregoing concepts regarding sorption processes in microporous media with accompanying transport and reaction phenomena is given in Example 8-11. The illustration is put in the context of an operating and maintenance error in a water treatment plant for the relatively simple circumstance in which sorption can be described by a local linear equilibrium condition (e.g., Equation 6-51).

Example 8-11

- **Situation:** After performing routine maintenance on two off-line aeration tanks used for taste and odor control at a water treatment plant, an inexperienced assistant operator introduces a solvent to a flow of water used to flush the tanks for removal of accumulated residues. He realizes too late that the solvent is a chlorinated compound, dichloroethylene (DCE), which may be sorbed from the flush water and retained by the porous-stone aerators in the tank. The sorbed solvent might then be released to the drinking water when the aeration units are put back on line. After realizing the error, the assistant stops the flow, drains the solvent containing flush water from the tanks, and refills them with clean rinse water. The two tanks were not flushed simultaneously and so the exposure time of their aerators to DCE were different (10 min and 1 h). Having read Chapter 6, the chief operator instructs the assistant to repeat the draining and filling process several times until the DCE is desorbed from the diffusers completely. The assistant is further instructed to measure DCE concentrations in the rinse water each time and to do a mass balance to ensure complete desorption.

- **Question(s):** How can the assistant operator determine that all of the DCE has been removed from the diffusers?

- **Logic and Answer(s):**

1. The amount to be removed must equal the amount sorbed, which can be estimated if the following information is known.

2. The diffusers are configured as flat circular plates having a wall thickness of 10 cm. They are known also to have a porosity of 0.4, a pore tortuosity factor of 2.0, and a domain density (exclusive of pores) of $\rho_d = 1.30\ \mathrm{g/cm^3}$. The free-liquid diffusivity of the DCE is estimated as $10^{-5}\ \mathrm{cm^2/s}$, and the sorption of this material by the diffuser stone material is taken as a linear process having an estimated $\mathcal{K}_D$ value of 10 L/g. The DCE concentration to which the diffusers were exposed is estimated as 100 mg/L.

3. Each diffuser had a surface area of 100 cm^2 and the total number in each tank was 1000.

4. Equation 8-145 can be used to determine uptake, provided that Equation 8-134 is the appropriate solution to the transient diffusion problem. Equation 8-145 is appropriate if the concentration at a distance of 10 cm remains close to zero after 1 h. $\mathfrak{D}_a$ is

$$\mathfrak{D}_a = \frac{\mathfrak{D}_{p,e}}{1 + \dfrac{(1 - \epsilon_d)\rho_s}{\epsilon_d}\mathcal{K}_D}$$

$$= \frac{(1 \times 10^{-5}\ \mathrm{cm^2/s})/2}{1 + \left[\left(\dfrac{0.6}{0.4}\right)\left(1.3\ \dfrac{\mathrm{g}}{\mathrm{cm^3}}\right)\left(10{,}000\ \dfrac{\mathrm{cm^3}}{\mathrm{g}}\right)\right]} = 0.26 \times 10^{-10}\ \mathrm{cm^2/s}$$

Substitution into Equation 8-134 gives

$$\frac{C(\text{at 10 cm, 1 h})}{C_0} = \operatorname{erfc}\frac{x}{2(\mathfrak{D}_a t)^{0.5}}$$

$$= \operatorname{erf}\frac{10\ \mathrm{cm}}{2\left[\left(0.26 \times 10^{-10}\ \dfrac{\mathrm{cm^2}}{\mathrm{s}}\right)(3600\ \mathrm{s})\right]^{0.5}}$$

$$\approx \operatorname{erfc}(\infty) = 0$$

The concentration at 10 cm is thus still approximately zero, so Equation 8-145 can be accepted as a valid predictor of uptake. For the tank that was flushed for 10 min, the prediction is

$$\frac{M(t)}{A_N^\circ} = 2C_0\left(\frac{\mathfrak{D}_a t}{\pi}\right)^{0.5} = 2\left(0.1\ \frac{\text{mg}}{\text{cm}^3}\right)$$

$$\cdot\left[\frac{(0.26 \times 10^{-10}\ \text{cm}^2/\text{s})(600\ \text{s})}{\pi}\right]^{0.5}$$

$$M(t) = \left(1.4 \times 10^{-5}\ \frac{\text{mg}}{\text{cm}^2}\right)(10^5\ \text{cm}^2) = 1.4\ \text{mg}$$

and for 1 h,

$$M(t) = \left(3.43 \times 10^{-5}\ \frac{\text{mg}}{\text{cm}^2}\right)(10^5\ \text{cm}^2) = 3.43\ \text{mg}$$

The mathematical relationships that develop from Equation 8-127 become more complex when sorption equilibria cannot be described by linear isotherms. As noted in Chapter 6, the Freundlich equation is one of the most frequently used nonlinear equilibrium adsorption models. Applying Equation 6-67 and assuming local equilibrium conditions ($q = q_e$), we can write

$$q = q_e = \mathcal{K}_F C_e^n \tag{6-67}$$

The derivative needed for substitution into Equation 8-127 is thus

$$\frac{\partial q}{\partial C} = n\mathcal{K}_F C_e^{n-1} \tag{8-146}$$

which leads for microporous particle domains to

$$\mathfrak{D}_{p,e}\frac{\partial^2 C}{\partial x^2} = \left[1 + \frac{(1-\epsilon_p)\rho_s}{\epsilon_p}(n\mathcal{K}_F C_e^{n-1})\right]\frac{\partial C}{\partial t} \tag{8-147}$$

Unlike Equation 8-129, which resulted from the assumption of linear equilibrium, the expression above cannot be solved analytically, and numerical techniques (e.g., finite difference, orthogonal collocation, or finite element methods) are required.

8.7 RATES OF PARTICLE COLLISION AND AGGREGATION

Collision and subsequent coalescence of particles of colloidal size to form larger particles is a common phenomenon in environmental systems. The process is broadly referred to as *coagulation* as discussed in section 6.6.4, an understanding of which entails among other things, an understanding of *particle growth or particle aggregation rates.* Increases in the average sizes of environmentally significant particles accompanied by shifts in their size distributions have profound effects on the transport and fate of such particles in aquatic environments. Moreover, the engineering of particle removal processes in water and waste treatment, most commonly by sedimentation and filtration, depends strongly on knowing the extent to which particle size distribution is changed by pretreatment processes to promote particle collisions and to enhance aggregation. The rates at which particles collide with one another can be described in a manner analogous to that used to describe rates at which molecular-level species collide in homogeneous systems. We assume for purposes of this introduction to particle dynamics that all particle *collisions* lead to particle aggregations. This is clearly not the case for most practical environmental applications any more than it is the case that all molecular-level collisions result in reaction (see Chapter 7).

The rate of collisions occurring between particles having diameters i and j can generally be considered analogous to that of a bimolecular reaction; that is,

$$\frac{dn_{ij}}{dt} = \beta_c n_i n_j \tag{8-148}$$

where n_i and n_j are the respective concentrations (numbers of particles per unit volume) of each diameter of particle, β_c, is the particle collision frequency function (analogous to the rate constant of a second-order reaction) and dn_{ij}/dt is the rate of particle collisions (number $L^{-3}t^{-1}$). The magnitude and form of the collision frequency function, β_c, is determined by one or more of three particle collision mechanisms: (1) random motion caused by diffusion of particles, or *Brownian motion*; (2) velocity gradients in laminar flow fields, which allow particles in adjacent flow lines to contact one another, or *laminar shear*; and (3) *differential settling*, by which settling particles are overtaken by more rapidly settling particles. The net rate of change in the number concentration of any particle size, k, is the difference between the rate of collisions that result in formation of k-sized particles from smaller particles and the rate of collisions of particle k size with other particles to form particles larger than k:

$$\frac{dn_k}{dt} = 0.5 \sum_{i+j=k} \frac{dn_{ij}}{dt} - \sum_{i=1}^{\infty} \frac{dn_{ik}}{dt} \tag{8-149}$$

or, substituting the rate expression for particle collisions,

$$\frac{dn_k}{dt} = 0.5 \sum_{i+j=k} \beta_c(d_i, d_j) n_i n_j - \sum_{i=1}^{\infty} \beta_c(d_i, d_k) n_i n_k \tag{8-150}$$

The factor 0.5 in the first term of Equations 8-149 and 8-150 adjusts for the fact that the summation would otherwise represent each collision twice. We also note that regardless of the mechanism responsible for collisions, β_c is always some function of particle diameter, and therefore must be included within the summation terms.

Equation 8-149 can be simplified for the special case in which (1) coagulation occurs by Brownian motion only, and (2) the particles are initially *monodispersed* (i.e., all particles are initially of the same size). If we restrict the rate analysis to a short time period over which the change in particle diameters is not large, it is possible to approximate β_c as a constant, β_C. Equation 8-149 then becomes

$$\frac{dn_k}{dt} = 0.5\beta_C \sum_{i+j=k} n_i n_j - \beta_C n_k \sum_{i=1}^{\infty} n_i \tag{8-151}$$

To find the rate of change in the number concentration of all particles, we can then write

$$\sum_{k=1}^{\infty} \frac{dn_k}{dt} = 0.5\beta_C \sum_{k=1}^{\infty} \sum_{i+j=k}^{\infty} n_i n_j - \beta_C \sum_{k=1}^{\infty} n_k \sum_{i=1}^{\infty} n_i \tag{8-152}$$

Expansion of the first summation reveals that Equation 8-152 is simply

$$\frac{dn_T}{dt} = -0.5\beta_C n_T^2 \tag{8-153}$$

where n_T is the total number concentration of all particle sizes. This expression can be integrated to yield

$$n_T = \frac{[n_T]_0}{1 + 0.5[n_T]_0 \beta_C t} \tag{8-154}$$

Changes in the number concentration of any particle size can be described similarly. The size of the particles comprising the originally monodisperse suspension is given by the counter, $i = 1$, and the counter k is equal to 1. From Equation 8-151 we see that the number concentrations of these particles can only decrease with time due to collisions resulting in the formation of larger particles:

$$\frac{dn_1}{dt} = -\beta_C n_1 n_T \tag{8-155}$$

After substituting for n_T from Equation 8-154 and integrating, we obtain

$$\frac{n_1}{[n_T]_0} = \frac{1}{[1 + 0.5[n_T]_0\beta_C t]^2} \tag{8-156}$$

The general result is

$$\frac{n_k}{[n_T]_0} = \frac{(t/\kappa)^{k-1}}{[1 + (t/\kappa)]^2} \tag{8-157}$$

where $\kappa = 2/\beta_C[n_T]_0$.

A plot of the total number concentration, n_T, and the number concentration of individual size particles, n_k, is given in Figure 8-19. The total number concentration decreases with time due to particle collisions to form a fewer number of larger particles, and n_1 decreases in time due to formation of n_2. Although n_2 increases initially, it later decreases as a result of collisions to form larger particles. The number concentration pattern of all larger particles is then similar to that of the n_2 particles.

Example 8-12 is intended as a simplified analysis of particle growth kinetics in which we begin with a system that is monodisperse and assume further that

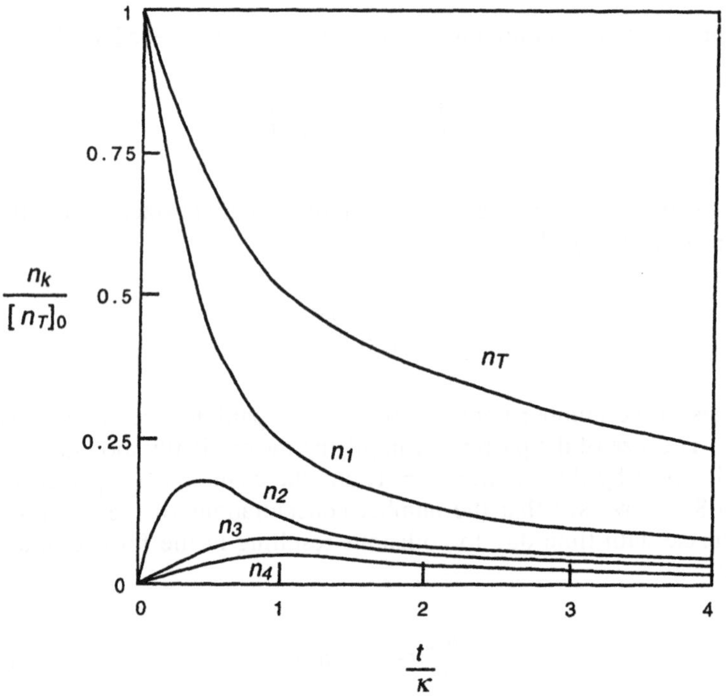

Figure 8-19 Variations in the dimensionless concentrations of variously sized particles with dimensionless time for an initially monodisperse suspension. [After Friedlander (1977).]

it remains so. In fact, β_c is a function of particle diameter and thus its magnitude would depend on the combination of particle sizes that interact. Therefore, Equation 8-150 would need to be integrated numerically because β_c cannot be factored out of the summation terms as a constant. However, once having lifted this restriction on analysis of particle growth, the particle growth model for monodispersed systems can be applied to heterodispersed systems.

Example 8-12

- **Situation:** A consulting firm plans to conduct a series of bench-scale experiments to evaluate the efficacy of various coagulants for use in treating the water supplied from a new reservoir. Rather than using turbidity, the traditional surrogate parameter for suspended particle concentration, the client wishes to have information on actual particle numbers. The new generation of particle-size counters permits rapid measurements. A key factor in the efficacy of a coagulant is the initial rate at which particles aggregate. The bench experiment is designed to use the conventional "jar test" apparatus, in which aggregation of particles is brought about by laminar shear forces created by a rotating paddle. Samples are withdrawn with time in order to measure the concentration of particles remaining.
- **Question(s):** It is desired to use the data collected in these bench-scale experiments to estimate the time required to reach 50% reduction in the concentration of particles with each coagulant.
- **Logic and Answer(s):**

1. An expression for β_C, the rate constant for particle collisions, is needed. Smoluchowski (1917) derived the following expression:

$$\beta_C = \frac{4}{3}\frac{dv}{dz}(r_i + r_j)^3$$

where dv/dz is the velocity gradient in the fluid induced by paddle stirring, r_i is the radius of particle i, and r_j is the radius of particle j.

2. All collisions are assumed to be 100% efficient (i.e., they all result in the colliding particles sticking together) but, the collision efficiency factor, η_c, is included to account for circumstances in which only a fraction of the collisions resulting in particle growth. If the simplifying assumption is made that particles at the beginning of flocculation are monodisperse with a uniform radius R, the final form for β_C is

$$\beta_C = \eta_c \frac{32}{3} \overline{G} R^3$$

where $\overline{G}$ is the mean value of the velocity gradient, dv/dz. The rate expression that follows from Equation 8-153 is

$$\frac{dn_T}{dt} = -\frac{16}{3}\eta_c \overline{G} R^3 n_T^2$$

3. The particle radius can be replaced by a parameter that is more convenient to measure, the volume fraction of particles (volume of particles per unit volume of suspension), ϕ_v°:

$$\phi_v^\circ = \frac{4}{3}\pi R_0^3 [n_T]_0$$

where, as usual, the subscript 0 signifies initial values. We assume that the volume of particles is conserved during particle growth such that

$$R^3 = \frac{3\phi_V^\circ}{4\pi n_T}$$

and replacement of R in the rate expression gives

$$\frac{dn_T}{dt} = -\frac{4}{\pi}\eta_c \overline{G} \phi_V^\circ n_T$$

This is essentially a first-order rate equation, which can be integrated to yield

$$\ln \frac{[n_T]}{[n_T]_0} = -\frac{4}{\pi}\eta_c \overline{G} \phi_V^\circ t$$

4. A plot of the data in the foregoing form should yield a straight line from which η_c can be obtained, given that $\overline{G}$ and ϕ_V° can be determined independently. The assumption that the system is monodisperse is highly restrictive, and thus only initial rate data should be used. Correlations between rotational speed of the paddles and $\overline{G}$ are available for standard geometries. Particle counters can be used to measure n_T.

5. Given a colloid having a radius of 10 μm, an initial number concentration of 10^8 particles per liter, and a $\overline{G}$ value of 10 s^{-1}, all of which are reasonably representative of water and waste treatment conditions, it can be shown that the time to reach a 50% reduction in particles is about 16 min if $\eta_c = 1$ and proportionally longer if η_c is smaller. The objective of chemical coagulation is to cause destabilization of particles (i.e., to make η_c as close to 1 as possible). By varying coagulation conditions in bench-scale experiments, it should be possible to optimize the process based on particle aggregation rate.

8.8 SUMMARY

We have explored the rate characteristics of many types of heterogeneous reactions that are important in natural and engineered environmental systems. In each instance, we have been able to arrive at a useful rate formulation to account for the effects of microtransport processes on overall reaction rate. These formulations are essential for the application of reactor design principles to heterogeneous systems, as discussed in Chapters 10 and 11. Our purpose here has been to demonstrate that a fundamental understanding of mass transport and chemical rate principles, coupled with appropriate material balance relationships, provide the basic tools to tackle any number of environmental process problems, regardless of their complexity. Greater detail in this context can be found in traditional chemical engineering texts.

8.9 REFERENCES AND SUGGESTED READINGS

8.9.1 Citations and Sources

Beltran, F. J., V. Gomez-Serranno, and A. Duran, 1992, "Degradation Kinetics of *p*-Nitrophenol Ozonation in Water," *Water Research*, *26*(1), 9–17. (Example 8-7 and Problem 8-10)

Friedlander, S., 1977, *Smoke, Dust and Haze*, John Wiley & Sons, Inc., New York. (Figure 8-19)

Gould, J. P., and G. V. Ulirsch, 1992, "Kinetics of the Heterogeneous Ozonation of Nitrated Phenols," *Water Science and Technology*, *26*(1–2), 169–180. (Problems 8-11 and 8-12)

Howeler, R. H., and D. R. Bouldin, 1971, "The Diffusion and Consumption of Oxygen in Submerged Soils," *Soil Science Society of America Proceedings*, *35*, 202–208. (Example 8-5)

Porcella, D. B., W. B. Mills, and G. L. Bowie, 1986, "A Review of Modeling Formulations for Sediment Oxygen Demand," pp. 120–138 in *Sediment Oxygen Demand, Processes, Modeling, and Measurement* (K. J. Hatcher, Ed.), Institute of Natural Resources, University of Georgia, Athens, GA. (Problem 8-5)

Pramauro, E., and M. Vincenti, 1993, "Photocatalytic Degradation of Monuron in Aqueous TiO_2 Dispersions," *Environmental Science and Technology*, *27*, 1790–1795. (Problem 8-3)

Prairie, M. R., L. R. Evans, B. M. Stange, and S. L. Martinez, 1993, "An Investigation of TiO_2 Photocatalysis for the Treatment of Water Contaminated with Metals and Organic Chemicals," *Environmental Science and Technology*, *27*, 1776–1782. (Example 8-2)

Smoluchowski, M., 1917, "Versuch einer mathematischen Theorie der Koagulations: Kinetik kolloider Lösunger," *Zeitschrift fuer Physikalische Chemie*, *92*. (Example 8-12)

Stumm, W., and J. J. Morgan, 1981, *Aquatic Chemistry*, 2nd Edition, Wiley-Interscience, New York. (Example 8-5)

Thoma, G. J., D. D. Reible, T. V. Kalliat, and L. J. Thibodeaux, 1993, "Efficiency of Capping Contaminated Sediments in Situ. 2. Mathematics of Diffusion-Adsorption in the Capping Layer," *Environmental Science and Technology*, *27*, 2412–2419. (Example 8-1)

8.9.2 Background Reading

Danckwerts, P. V., 1960, *Gas–Liquid Reactions*, McGraw-Hill Book Company, New York. This classic text develops the mathematical models required to describe diffusion and reaction (both steady and unsteady state) for a wide variety of boundary conditions.

Fogler, H. S., 1992, *Elements of Chemical Reaction Engineering*, 2nd Edition, Prentice Hall, Englewood Cliffs, NJ. Detailed analysis of diffusion and reaction in porous catalysts (emphasis on gas-phase reactions) is given in Chapter 11, with many examples of the effectiveness factor and Thiele modulus for spherical geometry.

Froment, G., and K. B. Bischoff, 1979, *Chemical Reactor Analysis and Design*, John Wiley & Sons, Inc., New York. Chapter 3 provides a good discussion of the Thiele modulus and effectiveness factors for solid catalysts and gas-phase reactions. A derivation is given for generalization of effectiveness factors for *n*th-order reactions.

Levenspiel, O., 1972, *Chemical Reaction Engineering*, 2nd Edition, John Wiley & Sons, Inc., New York. The various kinetic regimes encountered in fluid–fluid reactions and rate equations for various cases (slow, intermediate and fast reactions with and without gas-phase resistance) are discussed in Chapter 13. Chapter 14 covers solid-catalyzed reactions (emphasis on gas-phase reactions) and includes equations for various rate-limiting steps. Methods of collecting and interpreting experimental rate data are included.

Ollis, D. F., and H. Al-Ekabi, Eds., 1993, *Photocatalytic Purification and Treatment of Water and Air*, Elsevier, Amsterdam, The Netherlands. *Proceedings of the First International Conference on TiO_2 Photocatalytic Purification and Treatment of Water and Air.* A summary of the state of the art in application of heterogeneous reactions involving photocatalysis. Various papers include treatment of kinetic data.

Serpone, N., and E. Pelizetti, Eds., 1989, *Photocatalysis Fundamentals and Applications*, John Wiley & Sons, Inc., New York. Presentation of various mechanistic models for heterogeneous reactions involving photocatalysts. Numerous applications to environmental engineering are presented with a discussion of Langmuir–Hinshelwood kinetics.

Smith, J. M., 1981, *Chemical Engineering Kinetics*, 3rd Edition, McGraw-Hill Book Company, New York. Chapter 9 covers kinetic analysis of reactions of gases on catalytic surfaces. Langmuir–Hinshelwood kinetic models are developed for different rate-limiting steps.

Weber, W. J., Jr., P. M. McGinley, and L. E. Katz, 1991, "Sorption Phenomena in Subsurface Systems: Concepts, Models and Effects on Contaminant Transport," *Water Research*, *25*(5), 499–528. The concept of type I, II, III, and IV diffusion domains is presented. Various examples are given for steady- and unsteady-state analysis with emphasis on contaminant transport in subsurface systems.

Examples of Numerical Solutions to Nonlinear Diffusion–Adsorption Models

Ames, W. F., 1992, *Numerical Methods for Partial Differential Equations*, 3rd Edition, Academic Press, Inc., San Diego, CA. Major classes of equations (parabollic, elliptic, and hyperbolic) and various numerical methods are presented for solution. Methods are illustrated with many important physical examples, including numerical solution methods for the Blasius boundary layer equation.

DiGiano, F. A., and W. J. Weber, Jr., 1973, "Sorption Kinetics in Infinite-Bath Experiments," *Journal of the Water Pollution Control Federation, 45*, (4), 713–725.

Finlayson, B. A., 1980, *Nonlinear Analysis in Chemical Engineering*, McGraw-Hill Book Company, New York. Numerical solutions for both ordinary and partial differential equations are given. Partial differential equations are solved by finite difference, orthogonal collocation, and Galerkin finite element method.

Gear, G. W., 1971, *Numerical Initial Value Problems in Ordinary Differential Equations*, Prentice Hall, Englewood Cliffs, NJ.

Huyarkorn, P. S., and G. F. Pinder, 1983, *Computational Methods in Subsurface Flow*, Academic Press, Inc., New York. Finite element and finite difference methods are applied to solute transport in the subsurface.

Miller, C. T., and W. J. Weber, Jr., 1986, "Sorption of Hydrophobic Organic Pollutants in Saturated Soil Systems," *Journal of Contaminant Hydrology, 1*, 243–261.

Pedit, J. A., and C. T. Miller, 1988, "The Advantage of High-Order Basis Functions for Modeling Multicomponent Sorption Kinetics," in *Proceedings of the 7th International Conference, Computational Methods in Water Resources*, Vol. 2, pp. 293–298, *Numerical Methods for Transport and Hydrologic Processes* (M. A. Celia, L. A. Ferrand, C. A. Brebbia, W. G. Gray, and G. F. Pinder, Eds.), Computational Mechanics Publications, Southhampton, Hampshire, England.

Weber, W. J., Jr., and J. C. Crittenden, 1975, "MADAM I: A Numeric Method for Design of Adsorption Systems," *Journal of the Water Pollution Control Federation, 47* (5), 924–940.

8.10 PROBLEMS

8-1 Powdered titanium dioxide is added as a catalyst to solutions of different concentrations of dichloroethane (DCA) in a series of CMBR experiments in which UV irradiation is used to oxidize the DCA. Prior isotherm measurements have indicated that the Langmuir monolayer capacity of the TiO_2 for adsorption of DCA is 0.5 mmol/g. Data from oxidation experiments yields the following information regarding initial rates of DCA removal:

Initial DCA Concentration (mmol/L)	Initial Rate (mmol/g TiO_2-min)
500	0.66
100	0.35
56	0.25
47	0.20

Determine the surface rate constant, k° (s^{-1}) and the Langmuir adsorption energy, $\mathscr{b}$ (L/mol), from these rate data, assuming that the surface reaction is rate controlling and the oxidation products do not compete with DCA for adsorption on the TiO_2.

8-2 The photocatalytic reduction of Cr^{6+} to Cr^{3+} on a catalyst is thought to involve a direct surface reaction (i.e., adsorption does not have to occur first). The external surface area of the catalyst is 2000 cm^2/g. The following CMBR rate data are collected using a catalyst dosage of 1 g/L and a stirring speed sufficient to eliminate any potential mass transfer limitations:

Time (s)	Cr^{6+} Concentration (mg/L)
0	60.0
60	41.9
120	29.2
180	20.4
240	14.2

Another CMBR test is then conducted at a much lower stirring speed, with the following results:

Time (s)	Cr^{6+} Concentration (mg/L)
0	40
60	34.6
120	30.0
180	26.0
240	22.4

Determine the surface reaction rate constant $k°$ (cm/s); the mass transfer coefficient k_f (cm/s); the Damkohler number, $\mathfrak{N}_{Da(II)}$; and the effectiveness factor, $\eta_{R,E}$.

Answers: $k° = 0.003$ *cm/s;* $k_f = 0.002$ *cm/s;* $\mathfrak{N}_{Da(II)} = 1.5$; $\eta_{R,E} = 0.4$.

8-3 Monuron is a widely used surface-applied herbicide having potential to leach into groundwater. The rate of its photocatalytic degradation in the presence of a catalyst has been reported to be controlled by adsorption and to be slowed by competition for sites with intermediate products formed during the oxidation reaction (Pramauro et al., 1993). Data from three CMBR rate tests, each with 100 mg/L of catalyst present, are given below.

Time (min)	Monuron Concentration (mg/L)		
	Test 1	Test 2	Test 3
0	8	20	35
0.5	6	17	31
1	5.1	16	30
1.5	3.6	13	26

Assume that (1) the monolayer sorption coverage of Monuron (MW = 184 g/mol) is 1×10^{-4} mol/g; (2) the adsorption energies of Monuron and its oxidation intermediates are similar; and (3) the sum of the molar concentrations of the Monuron and its intermediates remains constant over the course of the test. Determine (a) the first-order rate constant for the surface reaction k° (s^{-1}), (b) the maximum rate of surface reaction (mol/g-s), and (c) the Langmuir adsorption energy, b (L/mol). [*Hint:* The units for the observed rate, r°, in Equation 8-27 are mol/weight catalyst-time, or

$$r^\circ = \frac{dC}{D_o\, dt} = -k_{obs}\, C$$

where D_o is the catalyst dosage (e.g., g/L). Different units are reported by Pramauro et al., (1993).]

Answers: *(a)* $k^\circ = 0.065\ s^{-1}$; *(b)* maximum rate = 6.5×10^{-6} *mol/g-s; (c)* $b = 3.2 \times 10^4$ *L/mol.*

8-4 Chlorine is reduced to chloride by reaction with the surface of activated carbon. A model in which chlorine first adsorbs and then reacts with the surface, and in which no sorption competition with intermediates occurs, is proposed for description of this process. Equilibrium adsorption experiments give the following Langmuir model constants: $b = 5$ L/mmol and $Q_a^\circ = 0.1$ mmol/g. A series of CMBR rate tests are conducted to determine the value of the surface reaction rate constant, k°. The same dose of activated carbon (0.5 g/L) is added in each batch. The diameter and density of these particles is 0.01 cm and 1.5 g/cm^3, respectively. Each solution is agitated with an impeller at a rate that yields a mass transfer coefficient for the uptake of chlorine from solution of 0.005 cm/s. The initial rate of surface reaction can be determined from the data given below.

Initial Cl_2 (mmol/L)	Cl_2 Remaining After 5 min (mmol/L)
0.1	0.08
0.2	0.17
0.3	0.26
0.5	0.43

(a) Determine $k°$ (s^{-1}), assuming that both mass transfer and surface reaction contribute to rate control. Tabulate the Damkohler numbers and solute concentrations at the surface of the particle, ignoring the effects of internal diffusion.

(b) Use $k°$ to calculate the surface reaction rate, $r°$ (mmol/g-s), corresponding to reaction rate control. Discuss the difference between this rate and the rate actually measured.

(c) Test the model for rate control by surface reaction using the experimental data. Compare the $k°$ value obtained to that determined in part (a).

Answers: (a) Average $k° = 0.023\ s^{-1}$; *(b) e.g, at* $C_0 = 0.5$ *mmol/L,* $r° = 1.64 \times 10^{-3}$ *mmol/g-s compared to* 7.8×10^{-4} *mmol/g-s, which was measured; (c) the* $k° = 0.0035\ s^{-1}$ *obtained from the best fit to the "linearized" plot is much lower than that in part (a) because the concentration at the surface is less than in the bulk solution due to mass transfer limitations, and the bulk concentration was used in calculating the surface rate constant in part (c); this is an example of disguised kinetics.*

8-5 Sediment oxygen demand (SOD; oxygen flux from the water column to the sediment surfaces) values for lakes and reservoirs in the Great Lakes area have been reported by Porcella et al. (1986) to range from about 0.01 (Lake Superior) to 5.0 g O_2/m^2-day (Horseshoe Lake, Illinois). An interpretation of oxygen demand by diffusion and simultaneous first-order reaction was presented in Example 8-5, with rate constants and corresponding oxic layer depths given as

k (s^{-1})	L (cm)
2.50×10^{-4}	0.4
3.00×10^{-3}	0.12

(a) Use the data above and an oxygen concentration of 2 mg/L (0.002 mg/cm^3) at the sediment–water interface to calculate a steady-state oxygen flux (SOD). Compare these values to the range reported by

Porcella et al. and discuss. [*Hint:* Flux is determined by the concentration gradient at $x = 0$.]

(b) Change the Thiele modulus to 5.0, determine values for the rate constant and oxic depth which yield a flux of 2 g O_2/m^2-day, and calculate the associated internal reaction effectiveness factor.

Answers: (a) *The low value of the rate constant gives 0.08 g* O_2/m^2*-day and the high value gives 0.29 g* O_2/m^2*-day; (b)* $k = 0.135\ s^{-1}$, $L = 0.43$ *mm,* $\eta_{R,I} = 0.2$.

8-6 Monochloramine is an attractive alternative to chlorine for disinfection of public water supplies because it does not as readily form THM by-products. However, excessive amounts of monochloramine in the dialysis baths of artificial kidney machines causes oxidation of hemoglobin to methemoglobin, resulting in damage to the red blood cells of patients. Catalytic reduction of monochloramine on the surfaces of activated carbon is a potential treatment for water to be used in dialysis baths. In contrast to the model used in Problem 8-4 for chlorine reduction, assume that this reaction proceeds *without* adsorption and that a model of internal diffusion with simultaneous reaction is more appropriate. A series of CMBR rate tests are conducted in which the initial removal rate of monochloramine per unit weight of activated carbon is found to increase linearly with initial monochloramine concentration. At 10 mg/L (0.01 mg/cm^3) of monochloramine, the observed initial rate is 1.11×10^{-5} mg/g-min.

(a) Find the intrinsic surface reaction rate constant, $k°$ (cm/s), and the internal effectiveness factor if (1) the effective diffusivity of monochloramine, $\mathfrak{D}_e$, is 0.2×10^{-5} cm^2/s; (2) the diameter of the activated carbon particles, d_p, is 0.1 cm; (3) the internal surface area per unit weight of particle, $(a^\circ_{s,w})_{I,P}$, is 1000 m^2/g; and (4) the dosage of activated carbon, D_o, is 0.5 g/L. [*Hint:* To convert the observed rate, r_{obs}, to mass reacted per volume of catalyst per unit time, multiply by the density of the catalyst, 1.5 g/cm^3.]

(b) Determine the concentration (C_{OUT}) of monochloramine leaving a column of activated carbon if the feed concentration is 10 mg/L; the packed density of the carbon particles is 0.45 g/cm^3 of bed; and the empty-bed hydraulic residence time (volume of empty bed/flow rate) is 2 min. [*Hint:* The point form of the material balance equation is appropriate for this reactor configuration (see Chapter 2). The reaction rate term for a differential volume of bed consists of $k°$ multiplied by the internal particle surface area per volume of reactor $(a^\circ_{s,w})_{I,R}$, which is obtained by multiplying $(a^\circ_{s,w})_{I,P}$ by the packed-bed density.]

Answers: (a) $k° = 6.7 \times 10^{-9}$ *cm/s,* $\eta_{R,I} = 0.24$*; (b)* $C_{OUT} = 1.7$ *mg/L.*

8-7 The observed rate of monochloramine reduction in Problem 8-6 can be affected significantly by resistance to mass transfer at the exterior surfaces of the adsorbent particles. If a mass transfer correlation for a packed bed of spheres gives $k_f = 0.002$ cm/s, and the external surface area per unit weight of particles is $(a^\circ_{s,w})_{E,P} = 80$ cm^2/g, determine the concentration leaving the packed bed and the overall effectiveness factor, $\eta_{R,O}$. Discuss the results relative to those obtained in Problem 8-6.
Answers: $C_{OUT} = 2.8$ *mg/L and* $\eta_{R,O} = 0.17$.

8-8 The situation given in Example 8-4 can be approached by writing a material balance for ozone gas transfer and reaction in a CMBR:

$$-QC_{\mathrm{OUT}} + k_{f,l}(a_s^\circ)_R(C_S - C_{\mathrm{OUT}})V_R - kC_{\mathrm{OUT}}V_R = 0$$

where C_{OUT} is the exit concentration of ozone in solution. Use this material balance equation and the data provided in the example problem to obtain the value of k. Use Equation 8-68 to explain why this material balance should give the same result as that obtained in the example problem. [*Hint:* The discussion of mass transfer and reaction processes in completely mixed flow reactors (Chapter 10) may be helpful.]

8-9 Flue gas (sulfur dioxide) is to be used in an industrial wastewater treatment operation for pH adjustment. SO_2 undergoes hydrolysis at pH 8 to produce sulfuric acid (H_2SO_3), which in turn dissociates. The hydrolysis reaction is pseudo first order with respect to SO_2, with a rate constant of 3.4×10^6 s^{-1}. To what extent can this reaction be expected to enhance the absorption of SO_2 into the waste stream?

8-10 A semibatch reactor is set up to measure initial rates of oxidation of *p*-nitrophenol (PNP) by ozone at a number of different partial pressures and a temperature of 20°C. The Henry's constant for ozone is 6.78×10^{-2} atm-m^3/mol. The results of the measurements are summarized below.

Initial Rate of PNP Oxidation (mmol/L-min × 10^2)	Pressure of O_3 in Feed Gas (atm × 10^3)
0.38	5.0
0.54	7.0
0.77	10.0
1.15	15.0

Assume that (1) the oxidation reaction is very slow (which is true at very low pH), (2) the stoichiometric ratio of ozone to PNP is 3 : 1, and (3) the amount of ozone reacted is small compared to the amount fed. PNP is added in excess so that the initial rate of reaction is pseudo first order with respect to ozone concentration. Because PNP is in excess, it is also reasonable to assume that the bulk ozone concentration is zero. Develop and test a model to determine the volumetric mass transfer coefficient for ozone. [*Hint:* See Beltran et al. (1992).]

Answer: $0.0156\ min^{-1}$

8-11 Studies on the ozonation of 2,6-dimethyl-4-nitrophenol (2,6-Me-4-NP) have been reported by Gould and Ulirsch (1992). The pressure of ozone in the feed gas to a semibatch reactor was 0.033 atm. Two different initial concentrations of 2,6-Me-4-NP were used to obtain the following results:

	2,6-Me-4-NP Concentration (mmol/L)	
t (min)	$C_{0,1}$	$C_{0,2}$
0	0.164	0.116
0.5	0.146	0.098
1.0	0.128	0.080
1.5	0.110	0.062
2.0	0.092	0.044
2.5	0.074	0.026
3.0	0.056	0.008

The reaction was thought to be fast enough to assume instantaneous reaction. Independent measurements of the volumetric mass transfer coefficient produced a value of about 20 h^{-1}. The stoichiometric ratio of ozone to 2,6-Me-4-NP is 4 : 1. Determine whether the rate data support the use of an instantaneous reaction model.

Answer: There is a slight, but nevertheless systematic trend of the rate data away from linearity when plotted according to the instantaneous reaction model. In addition, the predicted overall mass transfer coefficient is about 50% less than observed in independent experiments. These results suggest that the model is not appropriate.

8-12 A zero-order reaction model was proposed by Gould and Ulirsch (1992) to provide a better explanation of the data than given in Problem 8-11. The reaction is restricted to the bulk solution phase and fast enough that the ozone concentration is zero at the film–bulk solution interface throughout the rate experiment. The oxidation rate (r_{ox}) for 2,6-Me-4-NP is

$$r_{ox} = -\frac{\hat{k}_{v,l}}{\gamma} C_{S,O3}$$

Justify the form of this model and use it to reexamine the rate data given in Problem 8-11. Compare the predicted and measured values of the volumetric mass transfer coefficient, $\hat{k}_{v,l}$. [*Hint:* The integral form of the rate expression should be used.]

8-13 A deep lake becomes stratified during the summer months. Oxygen concentrations near the sediment surface drop to zero due to lack of mixing with oxygen-rich water above. The onset of cooler weather in the fall causes the lake to destratify. The dissolved oxygen concentration in the near sediment water suddenly increases from zero to 5 mg/L and remains at this concentration for a protracted period. To what depth of sediment will oxygen reach a level of 1 mg/L at the end of 30 days if the free-liquid diffusion coefficient is 1×10^{-5} cm^2/s, sediment pore tortuosity is 2.0, porosity is 0.5, and there is no reaction within the sediment?

Answer: 6.48 cm.

8-14 An artificially created lagoon is underlain by a thick layer of sorptive soil. The lagoon is used for storage of an industrial waste prior to treatment. A question has been raised concerning the potential penetration of trichlorophenol (TCP) and trichloroethylene (TCE) through this soil layer into the underlying groundwater over the 10-year period since the lagoon's construction. Batch experiments with soil particles and water containing these two contaminants show that partitioning between the water and soil is linear, with $\mathcal{K}_D = 30$ cm^3/g for TCP and 0.3 cm^3/g for TCE.

(a) Determine the concentration profiles for each contaminant in the soil pore water after 10 years if the concentration of each is constant at 10 mg/L at the surface of the soil, the porosity and tortuosity of the soil are 0.4 and 2, respectively, the soil particle density is 2 g/cm^3, and the free-liquid diffusivities of TCP and TCE are 8.4×10^{-6} and 7.2×10^{-6} cm^2/s, respectively.

(b) Determine the total mass of each contaminant that has penetrated the soil at the bottom of the lagoon if the surface area of the bottom is 5000 m^2.

Answer: (b) 12 kg of TCP and 13 kg of TCE.

APPENDIX A.8.1 ERROR FUNCTIONS AND ERROR FUNCTION COMPLEMENTS

A.8.1.1 Definition

The error function, erf(z), of an arbitrary value, z, and the error function complement of z, erfc(z), relate to the frequency and probability of errors associated with observations and measurements of values of z. In an *infinitely large set* of measurements, *random errors* are negative as often as positive, and thus have little effect on the *arithmetic mean value*, μ_m, of the set, referred to as a *true* or *population mean* because the set is infinitely large. All other errors are *systematic* and, if having the same cause, act to bias the mean value. If all systematic errors are eliminated, repeated measurements or observations can be used to obtain a *best estimate*, $\hat{\mu}_m$, of its true arithmetic mean of a quantity, as well as an assessment of the degree of reproducibility of the measurements made. The result of such a series of repeated measurements or observations is the quantity $\hat{\mu}_m \pm L_m$, where L_m is the *characteristic limit* of variation associated with a certain risk of error.

A.8.1.2 Relative Frequency of Errors

Figure A.8.1-1 illustrates the *Gauss–Laplace* or *normal* error frequency distribution. In this figure, the ordinate variable y represents the proportionate numbers of errors having value x. The variable y is then given by

$$y = \frac{1}{\sigma(2\pi)^{0.5}} e^{-x/2\sigma^2} \qquad \text{(A.8.1-1)}$$

where σ is the *standard deviation* of the set of observations from its true or population arithmetic mean value, μ_m, and σ^2 is the *variance* of that set from μ_m. The true or population arithmetic *mean* value of a distribution is given by the statistical *first moment* of the distribution about its *origin*:

$$\mu_m = \frac{1}{n} \sum_{i=1}^{n} m_i \qquad \text{(A.8.1-2)}$$

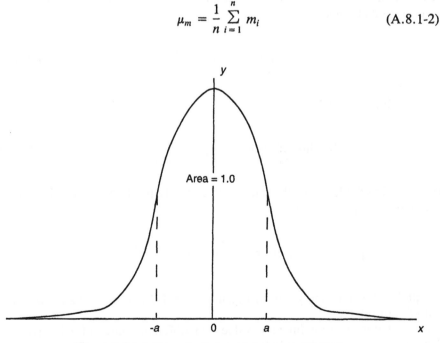

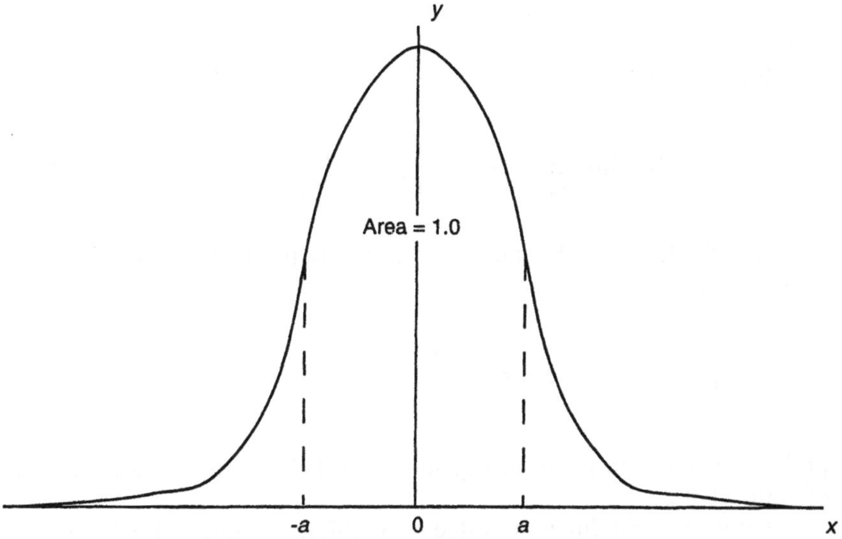

Figure A.8.1-1 Gauss–Laplace error frequency distribution.

the *variance* of the distribution is the statistical *second moment* about its *mean*:

$$\sigma^2 = \frac{1}{n} \sum_{i=1}^{n} (m_i - \mu_m)^2 \tag{A.8.1-3}$$

where the residuals or deviations of m_i from μ_m are taken as the *errors of observations*. The *standard deviation* of the distribution about its mean value is given by the square root of its variance:

$$\sigma = \left[\frac{1}{n} \sum_{i=1}^{n} (m_i - \mu_m)^2\right]^{0.5} \tag{A.8.1-4}$$

If, as is most commonly the case, the mean value employed in evaluating a finite set or *sample* of the population is not the *population* (*true*) mean but a *sample mean*, $\mu_{m,s}$, n should be replaced in the definition of the standard deviation by $n - 1$ to reflect the degree of freedom lost by using $\mu_{m,s}$ as an estimate of μ_m.

A.8.1.3 Probability of Error and the Error Function

Examination of Figure A.8.1-1 reveals that the fraction of the total number of errors having values between $x = -a$ and $x = a$, and thus the probability of an error in that range, is

$$\underset{-a<x<a}{\text{probability of error}} = \int_{-a}^{+a} y\, dx \tag{A.8.1-5}$$

or, from Equation A.8.1-1,

$$\underset{-a<x<a}{\text{probability of error}} = \frac{1}{\sigma(2\pi)^{0.5}} \int_{-a}^{+a} e^{-x^2/2\sigma^2}\, dx \tag{A.8.1-6}$$

Transforming the limits of the integral to a strictly positive domain, $0 < x < z$, where $z = a[\sigma(2)^{0.5}]$, gives

$$\underset{0<x<z}{\text{probability of error}} = \frac{2}{\pi^{0.5}} \int_{0}^{z} e^{-x^2/2\sigma^2}\, d\left(\frac{x}{\sigma(2)^{0.5}}\right) \tag{A.8.1-7}$$

The probability function given in Equation A.8.1-7 is termed the *error function of z*, or erf(z).

A tabulation of error function values for different values of z is given in Table A.8.1-1. Values for this function can be approximated using the expan-

Table A.8.1-1 Values of the Error Function of an Arbitrary Variable z

z	0.0	.01	.02	.03	.04	.05	.06	.07	.08	.09
0.0		.01128	.02256	.03384	.04511	.05637	.06762	.07886	.09008	.10128
.1	.11246	.12362	.13476	.14587	.15695	.16800	.17901	.18999	.20094	.21184
.2	.22270	.23352	.24430	.25502	.26570	.27633	.28690	.29742	.30788	.31828
.3	.32863	.33891	.34913	.35928	.36936	.37938	.38933	.39921	.40901	.41874
.4	.42839	.43797	.44747	.45689	.46623	.47548	.48466	.49375	.50275	.51167
.5	.52050	.52924	.53790	.54646	.55494	.56332	.57162	.57982	.58792	.59594
.6	.60386	.61168	.61941	.62705	.63459	.64203	.64938	.65663	.66378	.67084
.7	.67780	.68467	.69143	.69810	.70468	.71116	.71754	.72382	.73001	.73610
.8	.74210	.74800	.75381	.75952	.76514	.77067	.77610	.78144	.78669	.79184
.9	.79691	.80188	.80677	.81156	.81627	.82089	.82542	.82987	.83423	.83851
1.0	.84270	.84681	.85084	.85478	.85865	.86244	.86614	.86977	.87333	.87680
1.1	.88021	.88353	.88679	.88997	.89308	.89612	.89910	.90200	.90484	.90761
1.2	.91031	.91296	.91553	.91805	.92051	.92290	.92524	.92751	.92973	.93190
1.3	.93401	.93606	.93807	.94002	.94191	.94376	.94556	.94731	.94902	.95067
1.4	.95229	.95385	.95538	.95686	.95830	.95970	.96105	.96237	.96365	.96490
1.5	.96611	.96728	.96841	.96952	.97059	.97162	.97263	.97360	.97455	.97546
1.6	.97635	.97721	.97804	.97884	.97962	.98038	.98110	.98181	.98249	.98315
1.7	.98379	.98441	.98500	.98558	.98613	.98667	.98719	.98769	.98817	.98864
1.8	.98909	.98952	.98994	.99035	.99074	.99111	.99147	.99182	.99216	.99248
1.9	.99279	.99309	.99338	.99366	.99392	.99418	.99443	.99466	.99489	.99511
2.0	.99532	.99552	.99572	.99591	.99609	.99626	.99642	.99658	.99673	.99688
2.1	.99702	.99715	.99728	.99741	.99753	.99764	.99775	.99785	.99795	.99805
2.2	.99814	.99822	.99831	.99839	.99846	.99854	.99861	.99867	.99874	.99880
2.3	.99886	.99891	.99897	.99902	.99906	.99911	.99915	.99920	.99924	.99928
2.4	.99931	.99935	.99938	.99941	.99944	.99947	.99950	.99952	.99955	.99957
2.5	.99959	.99961	.99963	.99965	.99967	.99969	.99971	.99972	.99974	.99975
2.6	.99976	.99978	.99979	.99980	.99981	.99982	.99983	.99984	.99985	.99986
2.7	.99987	.99987	.99988	.99989	.99989	.99990	.99991	.99991	.99992	.99992
2.8	.99992	.99993	.99993	.99994	.99994	.99994	.99995	.99995	.99995	.99996
2.9	.99996	.99996	.99996	.99997	.99997	.99997	.99997	.99997	.99997	.99998
3.0	.99998	.99999	.99999	1.0000						

sion

$$\mathrm{erf}(z) \approx \frac{2}{\pi}\left[z - \frac{1}{1!}\left(\frac{z^3}{3}\right) + \frac{1}{2!}\left(\frac{z^5}{5}\right) - \frac{1}{3!}\left(\frac{z^7}{7}\right) + \cdots\right] \quad \text{(A.8.1-8)}$$

As discussed earlier in this chapter, a useful adjunct to erf(z) is the *complementary error function of* z, erfc(z), values for which can be calculated from Table A.8.1-1 using the relationship

$$\mathrm{erfc}(z) = 1 - \mathrm{erf}(z) \quad \text{(A.8.1-9)}$$

The results of measurements can also be expressed in terms of *probable error*, e_r [i.e., that number (z) which the actual error may be greater than or less than with equal probability]. The value of *erfz* for that case is 0.50. From the table given below we see that erf(z) = 0.49375 for $z = 0.47$ and 0.50275 for $z = 0.48$. Interpolation indicates that the value of z for erf(z) = 0.50 should be 0.47694. Thus the probable error e_r for a distribution having a standard deviation of σ is

$$e_r = 0.47694[\sigma(2)^{0.5}] = 0.6745\sigma \quad \text{(A.8.1-10)}$$

9

REACTOR ENGINEERING: STEADY-STATE HOMOGENEOUS SYSTEMS

9.1 INTRODUCTION

Reactor principles are used routinely in the field of chemical engineering to describe a broad range of industrial transformation processes involving gases, liquids, and solids. These same principles can be applied for description of the behavior of natural systems and to the design of engineered systems. It must be recognized at the outset, however, that the reactions involved in environmental systems are typically complex, demanding that the assumptions and limitations of reactor principles be well understood. The intent of this chapter is to develop performance equations and models for reactor design that can be applied to a wide variety of processes of interest, with a strong emphasis on developing an appreciation of the utility and limitations of such models.

9.2 DEFINITION OF A REACTOR

There are several important distinctions between the conditions under which reactors operate in the chemical industry and in environmental systems. Foremost, the precise composition of environmental "feedstocks," such as raw water supplies and wastewater streams, is rarely known. For example, such surrogate *lumped* measures of constituent concentrations as BOD, TOC, and conductivity must often be used in place of specific chemical entities in reaction rate expressions, thereby immediately introducing empiricism. In contrast, reactor design in the chemical industry generally deals with the conversion of well-characterized chemical feed stocks to well-known products. Not only are the specific compositions of environmental systems largely unknown, they are also commonly variable, which is true as well for rates of flow. Temporal variabilities complicate process descriptions by necessitating consideration of reactor performance under non-steady-state conditions. Such circumstances, which are not generally dealt with in elementary texts on chemical reactor theory, must be fully understood in order to recognize the proper application of process performance models in environmental systems.

A reactor is defined here as any device in which an incoming constituent undergoes chemical (or biochemical) transformation, phase transformation, or phase separation. A constituent is defined again in its broadest sense to include soluble, colloidal, and particulate substances. The material balance relationship described in Equation 2-5 for a *general* control volume is the starting point for analysis of reactor performance:

net rate of transport of i through the control volume	$\pm$	net rate of transformation of i within the control volume	=	net rate of change of i within the control volume

(2-5)

molar rate of input of i across the control volume boundaries	$-$	molar rate of output of i across the control volume boundaries	$\pm$	molar rate of chemical reaction of i within the control volume	=	net molar rate of accumulation (or depletion) of i within the control volume

Mathematical expressions appropriate for each *specific* type of reactor system must then be developed for the input, output, and reaction terms in this equation. The input and output (i.e., mass transport) terms must accommodate mass continuity through the reactor and thus account for advection, dispersion, and diffusion processes within the bulk fluid(s) in the reactor. These processes were discussed in Chapters 3 and 4. The reaction rate term for homogeneous reactions may be any of the rate expressions discussed in Chapter 7. Heterogeneous reactions, on the other hand, such as those described in Chapter 8, require consideration not only of macroscopic transport within the bulk fluid but also microscopic transport between and within other phases contained in the reactor. Descriptions of the latter are generally incorporated in the reaction rate expression term of the material balance equation, as discussed in detail in Chapter 10.

Given the broad definition of a constituent with which we have begun, it is necessary to analyze each term of the material balance relationship carefully to ensure its appropriate use. Strictly speaking, for example, the principles of chemical kinetics described in Chapter 7 are intended to apply to atoms and molecules in homogeneous phases. As discussed in Chapter 8, however, rates of interaction of particulate constituents and rates of particle capture by collectors are often described by rate expressions developed in a fashion *analogous* to that employed for molecular reactions. Many reactors employed in environmental systems are in fact designed to provide for *phase* transformation or *phase* separation processes rather than for transformation of *chemical species*. The reaction rate term in the material balance equation is thus appropriately replaced by an interphase mass transfer term. The objective of broadening the definition of a reactor is to provide a unified approach to process design and system description.

The flow pattern through a reactor determines the residence time of fluid elements within that reactor, and thus the time available for transformations and separations to occur. At one particular *extreme* or *limiting* condition of reactor behavior, all fluid elements spend the same time in the system, and mass transport occurs by advection only. This condition is termed *ideal plug flow*. At the other extreme, a reactor is uniformly mixed, producing an infinite extent of dispersion. This condition is referred to as *completely mixed*, for which the residence time of fluid elements within the reactor varies conceptually from zero to infinity. In practice, ill-described intermediate fluid mixing patterns are frequently encountered, with mass transport occurring by a combination of dispersion and advection somewhere between the two extreme conditions. These flow patterns are referred to as *nonideal*. If the precise nature of the nonideal flow pattern is not known, the residence time distribution must be estimated from experimental measurements. These measurements are then interpreted in the context of an *ideal model* for nonideal flow (i.e., a model having specific hypothesized advection and dispersion characteristics). The fact that experimental data on flow patterns must be obtained before the extent of transformation or separation can be calculated complicates prediction of reactor performance under nonideal flow conditions.

Ideal reactor models represent approximations to reality. In particular, assumptions about concentration gradients and mixing patterns allow simplification of the dependence of reaction rates on spatial position within a reactor volume. The assumptions are critical in model *formulations*, and the restrictions they impose must be recognized in model *applications*. Three of the four most common classifications of ideal reactors are those of the extreme condition types identified above: the *completely mixed batch reactor* (CMBR), the *completely mixed flow reactor* (CMFR), and the *plug flow reactor* (PFR). The fourth type of "ideal" reactor model is the hypothesized advection–dispersion model used to mimic the observed nonideal behaviors of arbitrary flow systems. This is referred to as the *plug flow with dispersion reactor* (PFDR) model. The PFDR model provides a quantitative approach to describing nonideal flow patterns in order to facilitate prediction of reactor performance. The immediate discussion in Section 9.3 focuses on the first three types of ideal reactor models. The PFDR model is discussed in Section 9.5.3.

Continuous flow reactors are usually designed to operate under steady-state conditions and the accumulation term in the material balance relationship can be taken as zero in such instances. However, there are many environmental applications of reactor principles in which unsteady-state conditions exist. Indeed, it is common practice to employ equalization basins designed specifically to smooth unsteady flow and concentration conditions existing at the headworks of treatment facilities. The treatment of unsteady-state conditions in reactor design and analysis is largely left to Chapter 11.

9.3 IDEAL REACTORS

9.3.1 Completely Mixed Batch Reactors

A CMBR has no inflow or outflow. Thus, as a reaction proceeds, the composition and the relative distribution of reactants and products change with time. The term *completely mixed* stipulates that no concentration or thermal gradients exist. A CMBR is well suited for bench-scale experimental studies of reaction rates (see Chapter 7). Moreover, it can be applied effectively for treatment of small quantities of materials, particularly these which are relatively expensive, toxic, or otherwise hazardous to deal with in flow-through reactors.

Vigorous mixing is employed in CMBRs to ensure that the concentrations of reactants and products are *uniform* and that the *reaction rate* is therefore *the same everywhere within the reactor volume*. The material balance relationship for an ideal CMBR thus reduces to

$$r V_R = V_R \frac{dC}{dt} \tag{9-1}$$

As written, it is assumed implicitly that the reactor volume (volume of fluid in the reactor), V_R, remains constant for all time. In a later section we deal with time-variable volumes, a condition that applies, for example, when a reaction occurs while a reactor is being filled. This mode of operation is sometimes referred to as *semibatch*.

Recognizing that $\mathscr{r}$ is generally a function of concentration, Equation 9-1 can be written as

$$\int_0^t dt = \int_{C_0}^{C_t} \frac{dC}{\mathscr{r}} \tag{9-2}$$

The CMBR design equation resulting from integration of Equation 9-2 can be used to determine the time required to change the concentration from C_0 to C_t if $\mathscr{r}$ is known; that is,

$$t = \int_{C_0}^{C_t} \frac{dC}{\mathscr{r}} \tag{9-3}$$

For a reaction in which the concentration decreases with time, $\mathscr{r}$ is negative. When there is production or generation of the reactant of interest, $\mathscr{r}$ is positive. Example 9-1 illustrates an application of the foregoing relationship to the selection of an appropriate CMBR detention time for a specified level of treatment of a waste of known composition and known reaction characteristics.

Example 9-1

- **Situation:** An enzyme treatment process is to operate on batches of wastewater that contain 50 μM of an aromatic substrate (SU). The enzyme (EN) involved accomplishes partial oxidation (ring opening), which makes the substrate amenable to further biological treatment. Bench-scale studies have shown that the initial enzyme concentration decreases with time due to thermal deactivation in a process described by the rate relationship

$$C_{EN} = C_{EN,0}e^{-kt}$$

Michaelis–Menten kinetics apply in the following modified form to account for the loss of enzyme with time:

$$-\mathscr{r} = \frac{\mathscr{r}_{max}e^{-kt}C_{SU}}{\mathscr{K}_M + C_{SU}}$$

where $\mathscr{r}_{max} = 10\ \mu$M/min, $\mathscr{K}_M = 20\ \mu$M, and $k = 0.1$ min^{-1}.

- **Question(s):** Determine the CMBR holding or residence time required for 80% removal of the substrate.

- **Logic and Answer(s):**

1. The material balance for a CMBR is given by Equation 9-1, and the reaction rate term derives directly from the reactor design equation:

$$-\mathscr{r} = \frac{dC_{SU}}{dt}$$

2. Substitution of the modified Michaelis–Menten expression for r, followed by separation of variables, gives

$$\frac{\mathcal{K}_M + C_{SU}}{C_{SU}} dC_{SU} = -r_{\max} e^{-kt} dt$$

This equation can be integrated analytically to yield

$$(\mathcal{K}_M \ln C_{SU} + C_{SU}) \Big|_{C_{SU,0}}^{C_{SU,t}} = \frac{r_{\max}}{k} e^{-kt} \Big|_0^t$$

3. An expression for the time needed for treatment is obtained by rearrangement:

$$t = \frac{-\ln \left[\frac{k\mathcal{K}_M}{r_{\max}} [\ln(C_{SU,t}/C_{SU,0}] + \frac{k}{r_{\max}} (C_{SU,t} - C_{SU,0}) + 1 \right]}{k}$$

4. Substitution of the numerical values given for the respective rate constants and the initial and final substrate concentrations yields

$$t = \frac{-\ln \left[\frac{2}{10} [\ln(10/50)] + \frac{0.1}{10} (10 - 50) + 1 \right]}{0.1} = 12.8 \text{ min}$$

5. The reader is left to explore the sensitivity of treatment time to the input data.

The analysis above applies to reactors operated at constant volume. This is the typical condition under which bench-scale experiments are conducted to determine the form of the rate expression for r, as discussed in Chapter 7. Constant-volume CMBR design equations may also be appropriate for certain treatment applications wherein sporadically generated waste streams are collected over time, and the reactants necessary to convert the contaminants added only when a specific fixed volume of waste is accumulated.

9.3.2 Sequencing Batch Reactors

A *sequencing batch reactor* (SBR) is an operational variation on the CMBR which allows for continuous treatment of a process stream while providing the flexibility of being able to adjust treatment time when feed conditions or reaction environments change. A schematic representation of the stages of operation of an SBR is presented in Figure 9-1. Figure 9-1a depicts the SBR in the *fill* stage, with reaction occurring simultaneously. Figure 9-1b shows the SBR in the *reaction* stage at constant volume, and Figure 9-1c shows the *draw-down* stage, in preparation for either an *idle* or another *fill* stage. A series of three SBRs operated in the alternative stages depicted in Figure 9-1 can thus

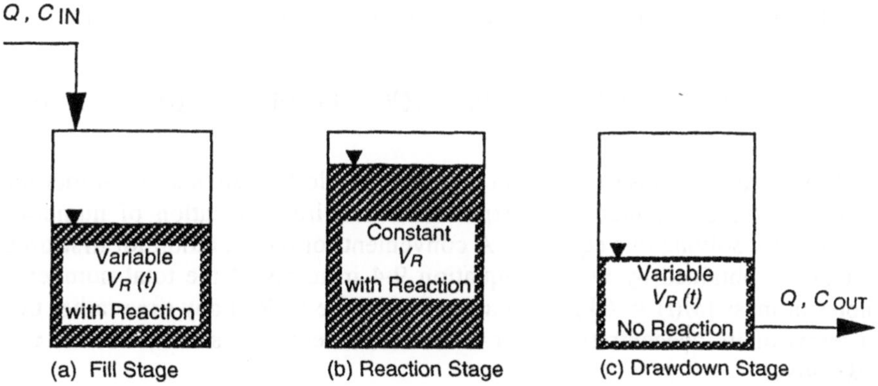

Figure 9-1 Operating stages of a sequencing batch reactor.

provide for essentially continuous treatment of a water or waste stream. Adjustments to the time spent in each stage can be made by using more than three reactors and by providing for flexibility in the fill and drawdown volumes.

In practice, the greatest utility of the SBR is for systems that are *heterogeneous* in character and involve combinations of treatment by species transformation *and* phase separation. For example, the settleability of biomass in waste treatment may vary due to a number of factors that are difficult to control. In conventional continuous-flow operation, for example, periods of slow settling in a secondary clarifier may lead to carryover of biomass into the effluent from the treatment process and thus potentially compromise treatment objectives. In the SBR configuration this problem can be overcome by delaying the onset of the drawdown period (now a decanting period), or by extending this period sufficiently to ensure the desired degree of solids settling and separation.

Process designs of SBRs for homogeneous systems begin with the writing of material balances for both the fill and reaction stages. The material balance during the fill stage must account for the molar or mass rate of entry of a reactant and the resultant increase in volume of the reactor with time:

$$QC_{\text{IN}} + \kappa V_R(t) = \frac{dCV_R(t)}{dt} \tag{9-4}$$

where Q is the flow rate during filling, C_{IN} is the concentration of the influent to the reactor, $V_R(t)$ is the volume of liquid at any time of the fill stage, and C is the corresponding concentration within that volume. The volume of liquid is given by

$$V_R(t) = Qt + V_R(0) \tag{9-5}$$

where t is time from beginning of the fill operation, and $V_R(0)$ is the volume of liquid initially in the reactor (i.e., at $t = 0$). Substitution of Equation 9-5 into Equation 9-4 followed by differentiation of the right-hand side yields

$$QC_{IN} + r[Qt + V_R(0)] = [Qt + V_R(0)]\frac{dC}{dt} + QC \tag{9-6}$$

Analytical solutions to Equation 9-6 are possible for zero and first-order rate relationships, but higher-order expressions require application of numerical methods for solving this equation. A convenient form of solution for first-order kinetics is obtained by writing Equation 9-4 in terms of the total number of moles or mass $[M(t) = CV_R(t)]$ of a reactant in the SBR at any time rather than in terms of its concentration. For a first-order reaction, substitution of $r = -kC$ gives

$$Q_{M,IN} - kM(t) = \frac{dM(t)}{dt} \tag{9-7}$$

where $Q_{M,IN} = QC_{IN}$ is the mass rate of reactant addition to the reactor during the fill stage. Separation of variables and integration of Equation 9-7 yields

$$\int_{M(0)}^{M(t)} \frac{dM(t)}{Q_{M,IN} - kM(t)} = \int_0^t dt \tag{9-8}$$

where $M(0)$ is the mass of reactant present in the reactor initially at the start of the fill stage (e.g., the mass in the residual volume left from the previous cycle at $t = 0$. Integration of Equation 9-8 gives

$$\ln \frac{Q_{M,IN} - kM(t)}{Q_{M,IN} - kM(0)} = -kt \tag{9-9}$$

or

$$M(t) = \frac{Q_{M,IN}}{k}(1 - e^{-kt}) + M(0)e^{-kt} \tag{9-10}$$

The final form of the solution in terms of reactant concentration is

$$C = \frac{C_{IN}}{(t + \bar{t}_0)k} - \left[\frac{C_{IN}}{\bar{t}_0 k} - C(0)\right]\frac{\bar{t}_0}{\bar{t}_0 + t}e^{-kt} \tag{9-11}$$

where $\bar{t}_0 = V_R(0)/Q$ and $C(0)$ is the initial concentration of reactant in the volume present in the reactor at the beginning of the fill stage. Figure 9-2 illustrates a sample solution for Equation 9-11. At the beginning of the fill period, the time spent in the reactor is short and the concentration increases because the rate at which mass enters is greater than that at which it can be

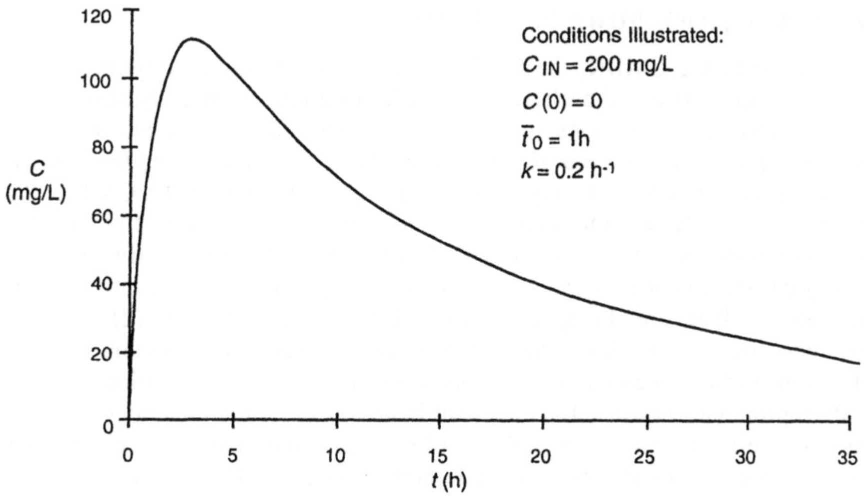

Figure 9-2 Performance characteristics of a sequencing batch reactor during the fill stage.

eliminated by the first-order reaction. The concentration reaches a maximum and then declines as residence time and volume increase.

A major objective of any process design is to determine the total residence time required to achieve a desired final concentration. Equation 9-11 provides the relationship between concentration in the reactor and fill time, t, for a given ratio of the initial volume, $V_R(0)$, to fill rate, Q. The fill rate is determined by the waste generation rate, and the fill time by the maximum volume selected for the reactor, the latter usually being constrained by certain economic and/or operational considerations. In most instances, reactions that occur during the fill time will not be sufficient by themselves to achieve the desired reduction in feed concentration, thus requiring that additional time be provided in the reaction stage when the reactor is operating at constant volume. The time required in this stage is calculated using the CMBR design equation given in Equation 9-2, the limits of integration being from C_{FILLED}, the concentration at the *end of the fill stage*, to C_{REACTED}, the desired final concentration at the *end of the reaction stage*, and the corresponding time requirement for the reaction stage being t_R. For a first-order reaction, this time is given by

$$t_R = -\frac{1}{k} \ln \frac{C_{\text{REACTED}}}{C_{\text{FILLED}}} \tag{9-12}$$

Various combinations of fill and reaction times can be employed to achieve a given effluent concentration. To approach an essentially continuous overall operation with a system of parallel reactors, short fill times and long reaction times require a larger number of reactors than long fill times and short reaction times.

9.3.3 Completely Mixed Flow Reactors

In contrast to the CMBR, the CMFR involves a continuous flow of feed and product water. As noted earlier, the CMFR designation correctly describes the explicit physical conditions required by the associated ideal reactor model: namely, a *flow-through reactor*, the contents of which are *completely mixed.* Many process reactors closely approximate the behavior of the CMFR. Two common examples are the rapid-mix tanks used to introduce chemicals in coagulation processes and the aeration tanks of completely mixed activated-sludge biological treatment processes. Two noteworthy features of reactors that behave like CMFRs are (1) rapid dilution of influent reactants, and (2) smoothing of time-variable input flow and concentrations. These are important considerations in reactors such as those involved in the rapid-mix operations of coagulation processes because the chemical behavior of coagulants depends markedly on their concentrations and on solution pH; moreover, chemical dosing must be paced with incoming particle concentrations. These same considerations are important for reactor systems used in activated sludge processes because sudden changes in substrate concentrations and chemical compositions, or the sudden appearance of toxicants, can adversely affect biochemical reactions and therefore process performance. Although the primary application of CMFRs is for full-scale treatment processes, the configuration is useful also for bench-scale experiments designed to evaluate process rates, as discussed in Section 9.3.5.

The CMFR configuration is illustrated schematically in Figure 9-3. Under normal operating conditions the flow rates entering and leaving the reactor are equal, and the volume in the reactor is thus constant. As for the CMBR, the CMFR is perfectly mixed, and the concentration, C_R, thus uniform throughout the volume; moreover, for the condition of complete mixing, C_R must be equal

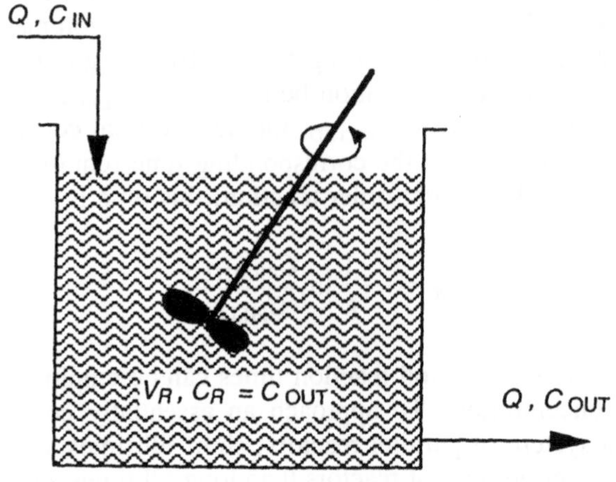

Figure 9-3 Ideal completely mixed flow reactor.

to the exit concentration, C_{OUT}. It follows that the reaction rate, r, is also uniform throughout the reactor and, as important, controlled by the effluent concentration [$r = \phi(C_R) = \phi(C_{OUT})$]. The material balance equation is written

$$QC_{IN} - QC_{OUT} + rV_R = V_R \frac{dC_{OUT}}{dt} \tag{9-13}$$

The first two terms on the left-hand side of Equation 9-13 are the molar or mass flow rates of component entering (IN) and exiting (OUT) the reactor, respectively, and the difference betwen them is the *net mass or molar flow rate* into the reactor, $Q_{M,IN}$.

Equation 9-13 is an ordinary differential equation describing the unsteady-state behavior of a CMFR. It can be used to predict the dynamic response of such reactors to transients in Q and C_{IN}, provided that the initial value of C_{OUT} and the reaction rate are known. Selected unsteady-state applications are discussed in Chapter 11.

The design of CMFRs for use in engineered systems is most commonly based on a given influent condition, a fixed effluent objective, and steady performance (i.e., steady-state operation). For this condition the rate of accumulation is zero, and

$$QC_{IN} - QC_{OUT} + rV_R = 0 \tag{9-14}$$

The steady-state value of the exit concentration, C_{OUT}, is conveniently expressed in terms of its ratio with respect to the influent concentration, C_{IN}, or

$$\frac{C_{OUT}}{C_{IN}} = 1 + r\frac{V_R}{QC_{IN}} \tag{9-15}$$

The reactor volume required to achieve a given steady-state value of the exit concentration is then given by

$$V_R = \frac{Q(C_{IN} - C_{OUT})}{-r} \tag{9-16}$$

An alternative form of this equation incorporates the definition of the *mean hydraulic residence time* (HRT) *or space time*, $\bar{t}$, of the reactor:

$$\bar{t} = \frac{V_R}{Q} \tag{9-17}$$

Equations 9-16 and 9-17 can be combined to yield the general CMFR design equations:

$$\frac{C_{OUT}}{C_{IN}} = 1 + \frac{r\bar{t}}{C_{IN}} \tag{9-18}$$

or,

$$\bar{t} = \frac{C_{IN} - C_{OUT}}{-r} \tag{9-19}$$

It will be shown later that $\bar{t}$ in a CMFR is also equal to the *mean value of the residence times* of fluid elements in the reactor.

Once the rate relationship for a particular reaction is known, a specific design equation can be written. For example, if a given component is transformed in a first-order reaction, $r = -kC_{OUT}$, Equations 9-18 and 9-19 become

$$\frac{C_{OUT}}{C_{IN}} = \frac{1}{1 + k\bar{t}} \tag{9-20}$$

$$\bar{t} = \frac{\dfrac{C_{IN}}{C_{OUT}} - 1}{k} \tag{9-21}$$

We note from Equation 9-20 that the influent-normalized exit concentration (C_{OUT}/C_{IN}) of the reactor becomes smaller as either the rate constant or the HRT increases. Similarly, we note from Equation 9-21 that the HRT needed to achieve a given normalized exit concentration increases as the rate constant decreases, and for a given rate constant, increases as the normalized exit concentration decreases. These effects are illustrated in Example 9-2 for an empirically derived half-order reaction involving the use of ozone for oxidation of color bodies in a raw water supply source.

Example 9-2

- **Situation:** Field-scale tests have been conducted to determine the effectiveness of ozone in oxidizing color-producing compounds in a water treatment operation. Ozone is sparged into a tank having an HRT of 10 min. The following empirical rate expression is thought to best describe the process:

$$r = -kC^{0.5}$$

where C is a lumped-parameter measure of the concentration of color-causing compounds expressed in terms of intensity as color units. The color units entering and leaving the ozone contactor have been measured in three separate tests with the results shown below.

Test Number	Color Units In	Color Units Out
1	8	1.9
2	15	4.8
3	22	8.0

These levels of removal are deemed unsatisfactory. The color entering the treatment plant typically averages about 10 color units, and the design goal is to achieve a product water of 1 color unit.

- **Question(s):** It is proposed to add another ozone contactor in parallel to the original tank and split the flow evenly between them to lengthen hydraulic residence time and obtain greater color removal. Will this scheme meet the treatment objective?
- **Logic and Answer(s):**

1. The ozone contactor is assumed to behave as a CMFR.

2. Before removal can be predicted at HRTs longer than those provided by the original tank, it is necessary to verify the rate expression from the data provided and to obtain an appropriate rate constant.

3. The material balance for a CMFR operated at steady state (Equation 9-14) is

$$QC_{\mathrm{IN}} - QC_{\mathrm{OUT}} + r V_R = 0$$

4. After substitution of the proposed empirical rate expression, the material balance can be rearranged to determine the rate constant for each of the three tests:

$$k = \frac{C_{\mathrm{IN}} - C_{\mathrm{OUT}}}{\bar{t} C_{\mathrm{OUT}}^{0.5}}$$

As shown below, the three values of k are reasonably close, suggesting that the proposed empirical rate expression is valid.

Test Number	k (color units$^{0.5}$/min)
1	0.44
2	0.47
3	0.50

5. The new HRT will be 20 min if a parallel ozone contactor of the same size as the original tank is employed. The CMFR design equation can be used to solve for C_{OUT} with the new $\bar{t}$ or, alternatively, the $\bar{t}$ needed to reach the new treatment objective of 1 color unit can be determined to evaluate whether this is greater or less than 20 min. Using an average k value of 0.47 color units$^{0.5}$/min the latter approach gives

$$\bar{t} = \frac{C_{IN} - C_{OUT}}{kC_{OUT}^{0.5}} = \frac{10 - 1}{0.47(1)^{0.5}} = 19.1 \text{ min}$$

6. As long as $C_{IN} \leq 10$ color units the proposed parallel reactor scheme therefore provides a hydraulic residence time which is just sufficient to achieve the treatment objective. However, the evident variability around the average value for C_{IN} suggests that the treatment objective frequently will not be met unless a longer HRT is used.

9.3.4 CMFRs in Series

CMFRs may be connected in series in various treatment operations so that the reaction rate in only the last reactor in the series is controlled by the final effluent concentration. CMFRs in series increase overall process efficiency and thus reduce the total reactor volume and detention time needed to meet a required effluent concentration. As detailed in Section 9.4, the analysis of *ideal* CMFRs in series also provides a useful way to model the performance of a single, *nonideal* reactor [i.e., a reactor in which the degree of mixing is intermediate between complete mixing (CMFR) and no mixing (PFR)]. A typical CMFR series arrangement, in which the exit concentration of one reactor is the feed concentration of the next reactor in the series, is illustrated schematically in Figure 9-4. At steady state, the normalized effluent concentration for the nth reactor in the series is

$$\frac{C_{n,OUT}}{C_{IN}} = \frac{C_1}{C_{IN}} \frac{C_2}{C_1} \cdots\cdots \frac{C_{n,OUT}}{C_{n-1}} \tag{9-22}$$

which can be determined by multiplicative solution of Equation 9-20, beginning with the first reactor in series.

The design equation for CMFRs of equal volume in series for a first-order

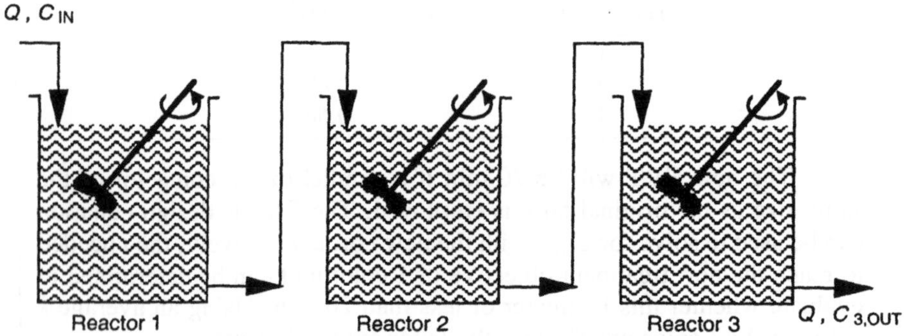

Figure 9-4 Completely mixed flow reactors in series.

reaction is obtained by substituting Equation 9-20 into 9-22 to yield

$$\frac{C_{n,\text{OUT}}}{C_{\text{IN}}} = \left(\frac{1}{1 + k\bar{t}}\right)^{n} \tag{9-23}$$

Note that the term $\bar{t}$ in this design refers to the HRT *of each specific reactor*. In this development, all $\bar{t}$ values must be equal. However, a more general approach to CMFRs in series is illustrated in Example 9-3. Taking the nth root of both sides and rearranging provides the complementary design equation for $\bar{t}$:

$$\bar{t} = \frac{\left(\dfrac{C_{\text{IN}}}{C_{n,\text{OUT}}}\right)^{1/n} - 1}{k} \tag{9-24}$$

The *total* HRT, $n\bar{t}$, required for a first-order reaction to achieve a given fractional concentration in the exit stream is

$$n\bar{t} = n\,\frac{\left(\dfrac{C_{\text{IN}}}{C_{n,\text{OUT}}}\right)^{1/n} - 1}{k} \tag{9-25}$$

Example 9-3

- **Situation:** Two completely mixed lagoons are connected in series to provide wastewater treatment for a small community. The HRTs are 10 days in the first lagoon and 5 days in the second. Along with BOD removal, these lagoons achieve significant reduction of fecal coliform organisms through natural die-off. The die-off rate is considered to be first order in character. Rate constants reported from a number of different lagoon systems, however, suggest that die-off rate may depend also on hydraulic residence time, $\bar{t}$, as shown by the empirical formulation below:

$$k = 0.2\bar{t} - 0.3$$

- **Question(s):** An estimate of the fecal coliform die-off is needed to determine if a new regulatory discharge standard of 99.9% reduction of coliforms can be met without adding a specific disinfection process to the treatment system. In the event the standard cannot be met in the existing system, it is of interest to explore the possibility of increasing the percent reduction by adding a third lagoon in series.

- **Logic and Answer(s):**

1. The general process design model for two CMFRs in series is given by Equation 9-22 as

$$\frac{C_{2,\mathrm{OUT}}}{C_{\mathrm{IN}}} = \frac{C_1}{C_{\mathrm{IN}}}\frac{C_{2,\mathrm{OUT}}}{C_1}$$

2. Equation 9-23 cannot be used to solve this problem because neither the HRT nor the first-order rate constant are the same for the two CMFRs. Instead, the solution proceeds by determining the product of the fractional reduction in fecal coliform in each CMFR:

$$\frac{C_{2,\mathrm{OUT}}}{C_{\mathrm{IN}}} = \left[\frac{1}{1 + k_1\bar{t}_1}\right]\left[\frac{1}{1 + k_2\bar{t}_2}\right]$$

3. From the relationship provided between the rate constant and the hydraulic detention time, the values of k_1 and k_2 are 1.7 and 0.7 day^{-1}, respectively. The fraction of fecal coliforms remaining is

$$\frac{C_{2,\mathrm{OUT}}}{C_{\mathrm{IN}}} = \left[\frac{1}{1 + (1.7)(10)}\right]\left[\frac{1}{1 + (0.7)(5)}\right] = 0.0123$$

and the percent reduction is $(1 - 0.0123) \times 100 = 98.77\%$, which is less than the requirement of 99.9% ($C_{\mathrm{OUT}}/C_{\mathrm{IN}} = 0.001$).

4. The effect of adding a third lagoon to bring the total reduction to 99.9% can be expressed by

$$\frac{C_{3,\mathrm{OUT}}}{C_{\mathrm{IN}}} = \left[\frac{C_2}{C_{\mathrm{IN}}}\right]\left[\frac{C_{3,\mathrm{OUT}}}{C_2}\right] = 0.001$$

The fraction of fecal coliforms remaining through treatment in the third lagoon would then

$$\frac{C_{3,\mathrm{OUT}}}{C_2} = 0.001\left(\frac{1}{0.0123}\right) = 0.081 \quad \text{or} \quad 8.1\%$$

5. The $\bar{t}$ required in the third lagoon to achieve this level of treatment can be found from Equation 9-21, which describes the performance of a CMFR with a first-order reaction rate; that is,

$$k\bar{t} = \frac{C_2}{C_{3,\mathrm{OUT}}} - 1 = 11.35$$

In this instance k depends on $\bar{t}$, as noted above.

6. The quadratic equation resulting from substitution of $k = 0.2\bar{t} - 0.3$ into Equation 9-21 can be solved for $\bar{t}_3$ to give

$$\bar{t}_3 = \frac{0.3 + [(0.3)^2 + 4(0.2)(11.35)]^{0.5}}{2(0.2)} = 8.3 \text{ days}$$

7. To evaluate fully the two alternative methods for meeting the new fecal coliform standard, the cost of constructing the third lagoon would need to be compared to the cost of adding appropriate disinfection facilities.

9.3.5 Determination of Reaction Rate Expressions in CMFRs

The methods for determining reaction rate expressions given in Chapter 7 were all based on measurement of data in CMBRs. An alternative approach is to use CMFRs operated at steady state. The usual process control variables are C_{IN} and $\bar{t}$, while C_{OUT} is the measured response variable. The CMFR is experimentally more complex than a CMBR. It requires a feed tank and a system to regulate flow. The reactor must be monitored to verify that steady state is maintained, and several experiments must be conducted for testing a rate expression to obtain a range of C_{OUT} values spanning the desired concentration range. Nevertheless, the CMFR may provide more reliable rate information because it better simulates the full-scale reactor environments in which treatment processes will eventually occur.

The relationship between reaction rate, process control variables, and the measured response variable is given by rearrangement of Equation 9-19:

$$-r = \frac{C_{IN} - C_{OUT}}{\bar{t}} \tag{9-26}$$

Several experiments using different values of C_{IN} and $\bar{t}$ will result in a range of observations of C_{OUT}, from which r may be calculated using Equation 9-26. The objective is to test the rate data to determine a reaction rate expression of the general form

$$r = \phi(C_{OUT}) \tag{9-27}$$

Zero- and first-order rate relationships can be determined from an arithmetic-scale plot of r against C_{OUT}, as shown in Figure 9-5. The reaction rate is zero order if r is independent of C_{OUT} and the rate constant, k, is simply the value of r. A first-order reaction is indicated if r increases linearly with C_{OUT} (i.e., $r = -kC_{OUT}$), and k is obtained from the slope of the graph. A reaction rate higher than first order is indicated if r increases more than proportionally with C_{OUT}.

Reactions having rate relationships other than zero and first order can also be verified for alternative formulations. The second-order rate expression is given by

$$r = -kC^2 \tag{9-28}$$

for which a plot of r versus $(C_{OUT})^2$ data yields a straight line. If no initial insights into reaction order are available, the more general rate expression

$$r = -kC^n \tag{9-29}$$

can be postulated, and the values of k and n determined by a logarithmetic

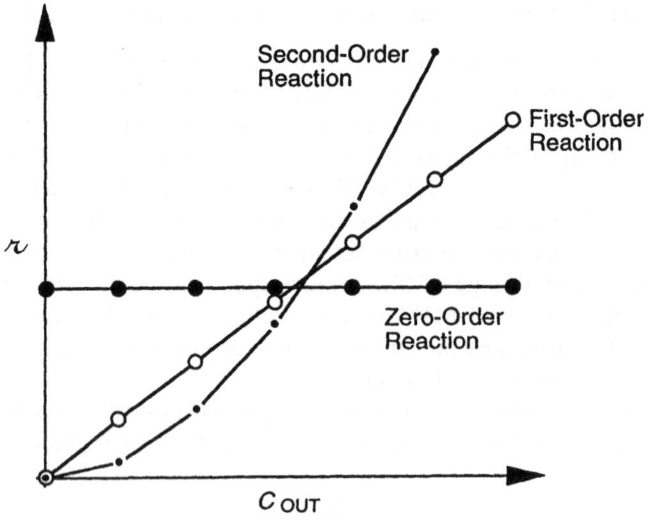

Figure 9-5 Determination of reaction order from data measurements in CMFRs.

transformation of the data to give

$$-\ln r = \ln k + n \ln C_{OUT} \tag{9-30}$$

The *method of excess* described in Chapter 7 for reaction rates which are dependent on the activities or concentrations of two chemical species can also be applied to CMFR data. For example, suppose that

$$r_i = -kC_i^n C_j^m \tag{9-31}$$

and the feed concentration, $C_{j,IN}$, is high enough that $C_j \approx C_{j,IN}$. The resulting log-arithmetic transformation of the rate data for different $C_{i,IN}$ for different $\bar{t}$ values for the CMFR is

$$-\ln r = \ln k_{obs} + n \ln C_{i,OUT} \tag{9-32}$$

where $k_{obs} = kC_{j,OUT}^m$. Thus the order of the reaction, n, with respect to component i is the slope of the ln r_i versus ln $C_{i,OUT}$ plot. The experiment can be repeated with component i in excess to determine the order of the reaction, m, with respect to component j, and k follows from determination of k_{obs}.

Rate relationships for enzyme and other catalysis reactions can also be examined conveniently in CMFRs. The rate expression given in Chapter 7 for an enzyme reaction is

$$-\frac{dC_{SU}}{dt} = \frac{C_{SU} r_{max}}{\mathcal{K}_M + C_{SU}} \tag{7-149}$$

which for the case at hand can be written from Equation 7-150 as

$$\frac{1}{\mathcal{r}} = \frac{\mathcal{K}_M}{\mathcal{r}_{max}}\left(\frac{1}{C_{SU}}\right)_{OUT} + \frac{1}{\mathcal{r}_{max}} \tag{9-33}$$

Rate data adhering to this relationship will yield a linear trace when plotted as $1/\mathcal{r}_{max}$ versus $1/(C_{SU})_{OUT}$ for a series of CMFR experiments in which $(C_{SU})_{IN}$ or $\bar{t}$ is varied.

As is always true, verification of rate expressions by plotting rate data in a linearized form of the expression are not necessary because nonlinear regression methods could be used. However, there are practical advantages in linearization techniques because by quick inspection they can provide an assessment of whether the proposed rate expression is reasonable.

The advantages of CMFRs over CMBRs for rate analysis are perhaps most evident in their applications to biological oxidation reactions, (e.g., rate relationships like those given in Equations 7-149 and 9-33). A CMFR better approximates the metabolic process environment of biological treatment operations than does a CMBR because it operates at steady state; in such applications, the CMFR is commonly referred to as a *chemostat*. For the unsteady-state conditions existing in a CMBR, metabolic processes tend to shift as substrate concentrations change with time, and microbial growth rates change accordingly. Such shifts in biomass activity in turn influence the results of rate experiments and measurements.

9.3.6. Plug Flow Reactors

A PFR, often referred to as a *tubular reactor*, is pictured schematically in Figure 9-6. Flow is assumed in this reactor model to be sufficiently turbulent that *fluid velocity is uniform* throughout any cross section. Elements of fluid are assumed to proceed through the reactor in an orderly and uniform manner, such that each cross-sectional element or *plug* of fluid is transported along the reactor length without intermixing with any other plug. Thus the time spent by each fluid element in the reactor is the same. It is important to recognize that *laminar flow in a tubular reactor does not produce a PFR condition* because *velocity is parabolically distributed* rather than uniform across the cross section (see Example 3-2); hence the residence times of individual fluid elements in the reactor are not equal. Residence time distributions for this and other conditions in nonideal reactors are discussed later in this chapter.

As shown in Figure 9-6, concentrations of nonconservative reactants and products vary continually along the length of a PFR because reaction time is proportional to reactor length. Reaction rates depend on concentration for other than zero-order reactions, and thus $\mathcal{r}$ varies along the length of the reactor. Because $\mathcal{r}$ is not constant, the integral form of the material balance that is used for the CMFR does not apply to a PFR. Instead, the material balance relationship is written for a differential volume element (dV_R) in which the reaction rate is uniform. This was referred to in Equation 2-16 as the *point form* of the

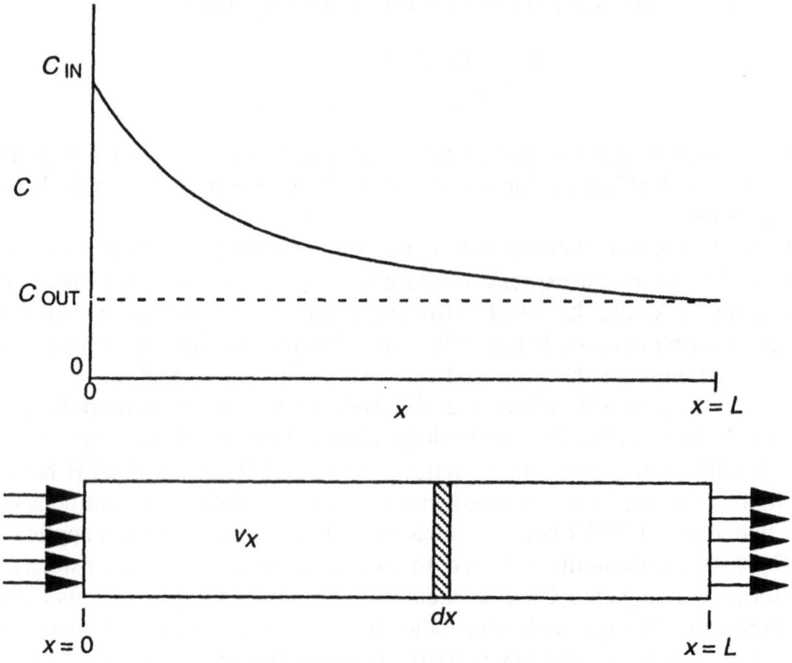

Figure 9-6 Ideal plug-flow reactor.

material balance equation. If advective transport is one-directional steady-state, Equation 2-16 can be simplified to:

$$-v_x \frac{dC}{dx} + r = 0 \tag{9-34}$$

The variables may be separated and the differential equation integrated over the length, L, of the reactor:

$$\int_0^L \frac{1}{v_x}\, dx = \int_{C_{IN}}^{C_{OUT}} \frac{dC}{r} \tag{9-35}$$

or

$$\frac{L}{v_x} = \bar{t} = \int_{C_{IN}}^{C_{OUT}} \frac{dC}{r} \tag{9-36}$$

Note that the term $\bar{t}$ in Equation 9-36 is the travel time or residence time of each fluid element in the reactor. Example 9-4 provides an illustration of the analysis of process behavior in reactors having the flow properties of ideal PFRs.

Example 9-4

- **Situation:** Cyanide is used in the production of intermediates in the pharmaceutical industry. A small resulting waste stream containing an organic cyanide, R—CN, is to be treated by a high-temperature hydrolysis process involving the reaction

$$R{-}CN + 2H_2O \rightarrow NH_3 + R{-}\underset{\displaystyle OH}{\underset{|}{C}}{=}O$$

The reaction has been studied at temperatures between 170 and 250°C and found to be first order with respect to R—CN concentration. The activation energy is 11 kcal/mol and the rate constant at 170°C is 8 h^{-1} (Midler et al., 1992). The goal of treatment is to reduce R—CN from 2000 mg/L to less than 5 mg/L.

- **Question(s):** A convenient means for carrying out an hydrolysis reaction of this type is in heated pipes. Compare the lengths of heated pipe required at temperatures of 170° and 250°C if the flow rate is 30 L/min and the inside pipe diameter is 15.2 cm (6 in.).

- **Logic and Answer(s):**

1. The pipeline reactor can be assumed to behave as a PFR and the general process design equation written from Equation 9-36 as

$$\bar{t} = \int_{C_{IN}}^{C_{OUT}} \frac{dC}{r}$$

2. The specific process design equation for a first-order rate of cyanide destruction is

$$\bar{t} = \int_{C_{IN}}^{C_{OUT}} \frac{dC}{-kC} = \frac{\ln (C_{IN}/C_{OUT})}{k}$$

$$\bar{t}\ (\mathrm{h}) = \frac{\ln (2000/5)}{k} = 6\, k^{-1}$$

3. The rate constant (k_2) at 250°C is obtained from the activation energy and the rate constant (k_1) at 170°C as follows:

$$\ln \frac{k_2}{k_1} = \frac{E_a(T_2 - T_1)}{RT_1T_2}$$

$$k_2 = 8 \exp\left[\frac{11000\ (523 - 443)}{1.98\ (523)(443)}\right] = 54.5\ \mathrm{h}^{-1}$$

4. Once the HRT is calculated, the length of the pipeline, L, is determined from

$$L = \frac{Q\bar{t}}{\pi d^2/4} = \frac{30(10^3)(60)(4)\bar{t}}{\pi(15.2)^2(100)} \quad \text{meters}$$

5. A summary of the two alternative designs is presented below. As expected, increasing the temperature increases reaction rate and greatly reduces the length of pipe required to meet the treatment objective.

Temperature (°C)	k (h^{-1})	$\bar{t}$ (h)	L (m)
170	8.0	0.75	74.3
250	54.5	0.11	10.9

9.3.7 Comparison of Ideal Reactor Performances

A comparison of performance predictions for CMBRs, CMFRs, CMFRs in series, and PFRs offers valuable insights into important reactor engineering concepts. Inspection and comparison of the design equations for a CMBR (Equation 9-3) and a PFR (Equation 9-36) reveals that the same residence time is needed in both to achieve a given extent of reaction. The relationship between these two reactors can be appreciated if we picture each cross sectional element of fluid entering a PFR as comprising an independent and infinitesimally thin CMBR which travels down the length of the reactor with the bulk flow velocity, so that its residence time in the PFR is $\bar{t}$.

Design equations for the hydraulic residence time requirements for CMFRs and PFRs are given in Equations 9-19 and 9-36, respectively. Substitution of appropriate rate expressions for reactions of zero-, first-, second-, and nth-order for κ in these equations leads to the design relationships given in Table 9-1. The hydraulic residence time requirements for CMFRs and PFRs can also be compared by graphical representation of these design relationships, as in-

Table 9-1 Hydraulic Residence Time Requirements for Reactions of Different Order in CMFRs and PFRs

	HRT Required to Obtain a Specified Conversion (C_{OUT}/C_{IN})	
Reaction Order	$\bar{t}_{CMFR}$	$\bar{t}_{PFR}$
0	$\frac{1}{k}(C_{IN} - C_{OUT})$	$\frac{1}{k}(C_{IN} - C_{OUT})$
1	$\frac{1}{k}\left(\frac{C_{IN}}{C_{OUT}} - 1\right)$	$\frac{1}{k}\left(\ln \frac{C_{IN}}{C_{OUT}}\right)$
2	$\frac{1}{kC_{OUT}}\left(\frac{C_{IN}}{C_{OUT}} - 1\right)$	$\frac{1}{kC_{IN}}\left(\frac{C_{IN}}{C_{OUT}} - 1\right)$
$n(n \neq 1)$	$\frac{1}{kC_{OUT}^{n-1}}\left(\frac{C_{IN}}{C_{OUT}} - 1\right)$	$\frac{1}{k(n-1)(C_{IN})^{n-1}}\left[\left(\frac{C_{IN}}{C_{OUT}}\right)^{n-1} - 1\right]$

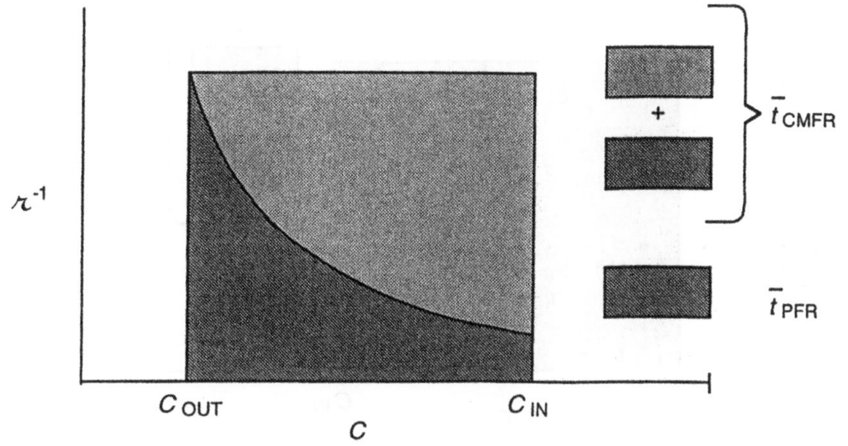

Figure 9-7 Performance comparison for a CMFR and a PFR.

dicated by the plot of r^{-1} versus C_{OUT} in Figure 9-7 for reaction orders higher than zero (i.e., reactions for which rate increases with increasing concentration). The hydraulic residence time for a PFR is the area under the curve between C_{OUT} and C_{IN}, while that for a CMFR is the area enclosed in the rectangle having a height of r^{-1} and a width of $(C_{IN} - C_{OUT})$. The area under the curve will always be smaller than the area enclosed by the rectangle. This illustrates that the residence time required to achieve a given extent of reaction will always be smaller for a PFR than for a CMFR for any reaction order other than zero. The importance of reaction order can also be observed. A higher reaction order means that r depends more strongly on C. Consequently, r^{-1} decreases more steeply with increasing concentration, and the area under the curve (the detention time of the PFR) becomes an increasingly smaller fraction of the area enclosed by the rectangle (the detention time of the CMFR). From this graphical analysis we conclude that the advantage of a PFR over a CMFR to achieve a given amount of conversion becomes greater as the reaction rate order increases. In other words, the higher the order of reaction, the faster the rate of reaction near the *influent end* of a PFR compared to that at its *effluent end*, which in turn is the rate at which the reaction occurs throughout the *entire volume* of a CMFR.

The advantage of CMFRs in series over a single CMFR of equal total volume is illustrated in Figure 9-8. CMFRs in series are represented graphically by a series of rectangles, each having a height determined by the reaction rate at the exit concentration of a given reactor and a width determined by the extent of reaction in that reactor. The total hydraulic residence time, the area enclosed by the series of rectangles, will always be less for reaction orders other than zero than that for one rectangle representing a single CMFR. Phenomenologically, the total residence time of CMFRs in series is less than one CMFR because the concentration, and thus the reaction rate, is higher in all but the

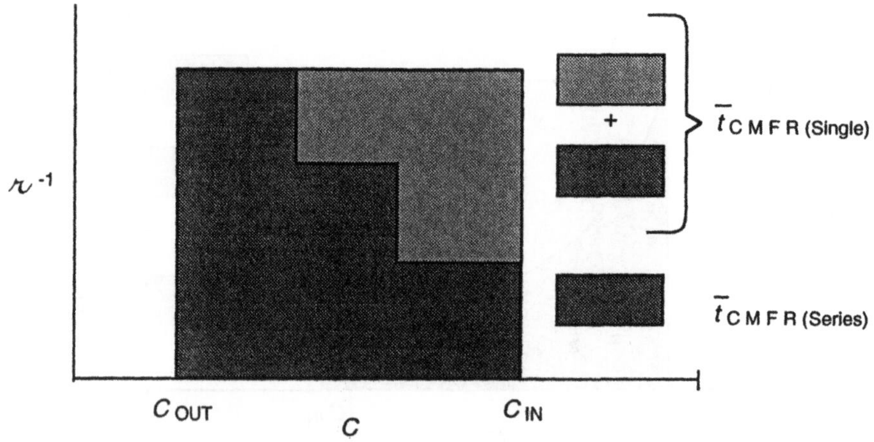

Figure 9-8 Performance comparison for a CMFR and CMFRs in series.

last reactor in the series than it is in a single CMFR of equal total volume. The consequence of increasing the number of CMFRs in series also can be appreciated readily from Figure 9-8; the greater the number of CMFRs in series, the more closely the total area of the rectangles approximates the area below the r^{-1} versus C curve for the PFR shown in Figure 9-7.

The hydraulic residence time and total volume required for a given degree of conversion for a specific reaction in an infinite number of CMFRs in series can be proven to be mathematically the same as that for a PFR. For example, the limit of Equation 9-25 as n approaches infinity for a first-order reaction is given by

$$\lim_{n \to \infty} n\bar{t} = \frac{\left(\dfrac{C_{IN}}{C_{n,OUT}}\right)^{1/n} - 1}{k} = -\frac{\ln\left(\dfrac{C_{n,OUT}}{C_{IN}}\right)}{k} \tag{9-37}$$

which is identical to the design equation for a PFR (see Table 9-1):

$$\int_{C_{IN}}^{C_{OUT}} \frac{dC}{r} = \int_{C_{IN}}^{C_{OUT}} \frac{dC}{-kC} = -\frac{\ln\left(\dfrac{C_{OUT}}{C_{IN}}\right)}{k} \tag{9-38}$$

The comparative performances of CMFRs and PFRs is dependent on the extent of reaction as well as the order of reaction, the latter shown in Figure 9-7. *Fractional removal* is a convenient way to express reactor performance. For example, the fractional removal for a first-order reaction that takes place in a CMFR can be compared to that in a PFR by rearrangement of Equations

9-21 and 9-38. The fractional removal for a first-order reaction in a CMFR is

$$\frac{C_{IN} - C_{OUT}}{C_{IN}} = \frac{k\bar{t}}{k\bar{t} + 1} \tag{9-39}$$

and that for a first-order reaction in a PFR is

$$\frac{C_{IN} - C_{OUT}}{C_{IN}} = 1 - e^{-k\bar{t}} \tag{9-40}$$

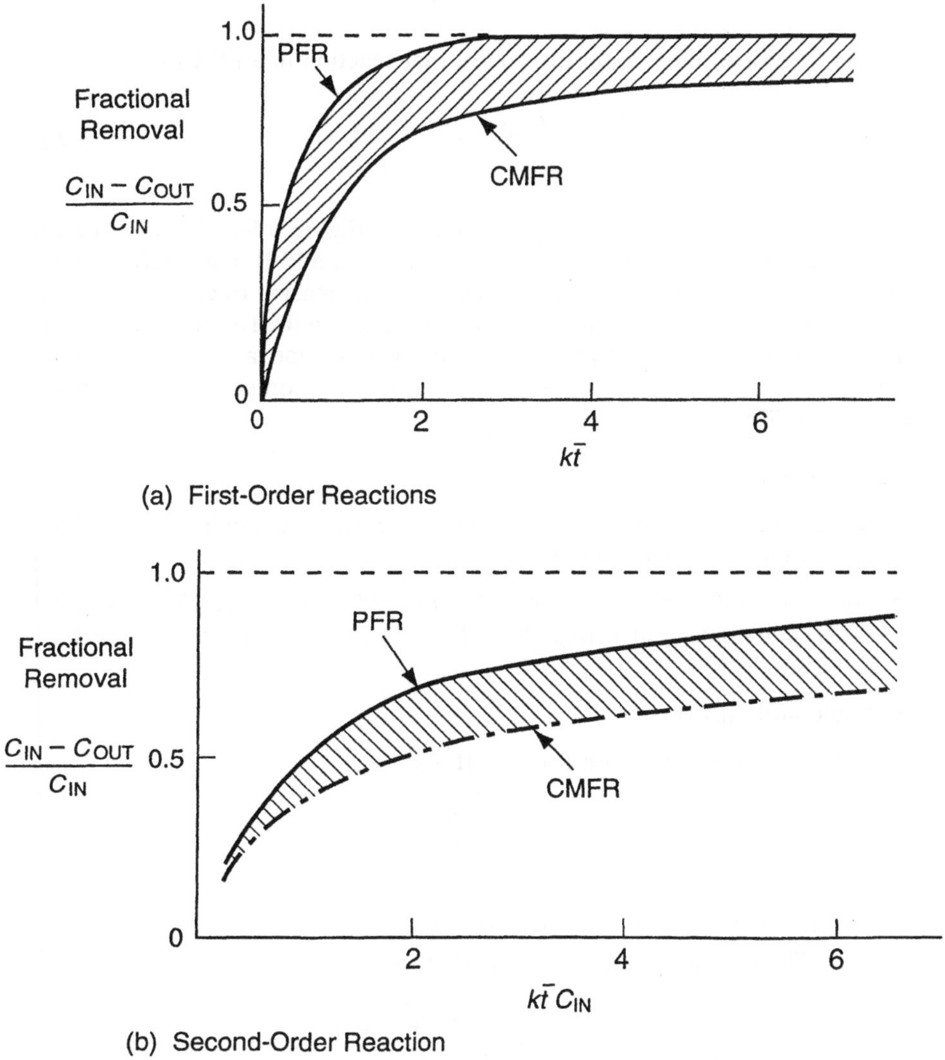

Figure 9-9 Fractional removal as a function of dimensionless reaction parameters for first- and second-order reactions.

Fractional removal for these two reactor configurations are plotted against the dimensionless reaction parameter, $k\bar{t}$, for first-order reactions in Figure 9-9a. When reactor performance is compared in this way, the advantage of a PFR over a CMFR as measured by the detention time required for a given conversion or removal is seen to become more significant as the fractional removal increases. A similar comparison is provided in Figure 9-9b for a second-order reaction, for which the fractional removal in a CMFR is given by

$$\frac{C_{IN} - C_{OUT}}{C_{IN}} = \left(\frac{C_{OUT}}{C_{IN}}\right)^2 kC_{IN}\bar{t} \tag{9-41}$$

and the fractional removal for a second-order reaction in a PFR by

$$\frac{C_{IN} - C_{OUT}}{C_{IN}} = \frac{k\bar{t}C_{IN}}{1 + k\bar{t}C_{IN}} \tag{9-42}$$

Fractional removals are plotted against $k\bar{t}C_{IN}$ in Figure 9-9b. As was true for a first-order reaction (Figure 9-9a), the advantage of a PFR over a CMFR with the same feed concentration, C_{IN}, increases with increasing fractional removal.

A comparison of the hydraulic residence times required in reactor systems of different configuration to obtain a given level of removal of a contaminant from a waste stream by a process controlled by a first-order reaction is given in Example 9-5.

Example 9-5

- **Situation:** A contaminant is removed from a waste stream by a first-order reaction having a rate constant, k.
- **Question(s):** Compare the relative HRTs needed to achieve 95% removal of this contaminant in a single CMFR, an in-line series of three CMFRs, and a single PFR.
- **Logic and Answer(s):**

1. The design equation for a CMFR is

$$\bar{t} = \frac{\dfrac{C_{IN}}{C_{OUT}} - 1}{k}$$

which, when $C_{OUT}/C_{IN} = 0.05$, gives $\bar{t} = 19k^{-1}$.

2. The design equation for CMFRs in series is

$$n\bar{t} = n\frac{\left(\dfrac{C_{IN}}{C_{n,OUT}}\right)^{1/n} - 1}{k}$$

and for the same extent of reaction with $n = 3$, $n\bar{t} = 5.13k^{-1}$.

3. Finally, the design equation for a PFR is

$$\bar{t} = \frac{\ln\left(\dfrac{C_{OUT}}{C_{IN}}\right)}{k}$$

and thus $\bar{t} = 3k^{-1}$.

4. The results, summarized below, show the advantage of CMFRs in series over a single CMFR, and that of a PFR over both CMFRs in series and a single CMFR.

Reactor Configuration	HRT
CMFR	$\bar{t} = 19k^{-1}$
Three CMFRs in series	$\bar{t} = 5.14k^{-1}$
PFR	$\bar{t} = 3k^{-1}$

9.3.8 Recycle Reactors

Recycle reactors offer advantages over other reactor configurations in certain circumstances, particularly those involving more than one phase. Batch recycle reactors are commonly used for bench-scale studies of fluid–solid mass transfer rates or fluid–solid reaction rates, whereas recycle of either fluids or solids is commonly employed in continuous-flow reactors to improve treatment performance.

9.3.8.1 Batch-Recycle Reactors So-called *batch-recycle reactors* are actually combinations of small CMFR or PFR reaction systems with large completely mixed tanks in which concentration equalization rather than reaction occurs. In this regard, the term *batch* is really a relative term. The *large* size of the completely mixed equalization tank in a recycle loop relative to the *small* size of the recycle flow to and from a *small* associated CMFR or PFR makes it possible to treat the entire system as essentially a CMBR.

A typical *batch-recycle reactor* scheme is illustrated in Figure 9-10. The reaction section of this system is a small PFR in which a reactive phase, typically a solid, is retained [i.e., a *fixed-bed reactor* (FBR)]. While the batch-recycle concept can accommodate virtually any reactor configuration, its greatest advantages apply to its use in conjunction with FBRs, for reasons discussed below.

The return flow from the FBR shown in Figure 9-10 passes into a large *completely mixed equalization tank* (CMET), where its concentration is smoothed by dilution; this flow is then recycled to the FBR. The most common objective of this type of experimental design is to provide *differential reactor operation* in the FBR section of the system. That is, the recycle rate and the relative sizes of the FBR and CMET components of the system are selected to make the residence time per pass through the FBR section short enough that only a small (*differential*) concentration change occurs due to either mass

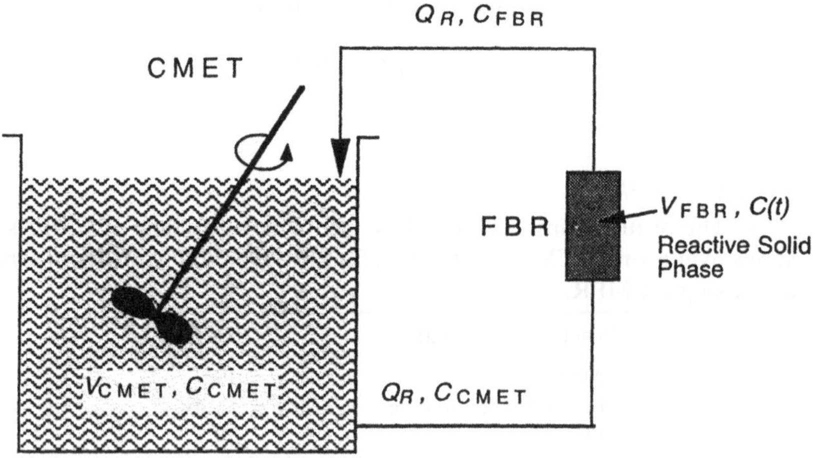

Figure 9-10 Coupling of a PFR with recycle to a completely mixed equalization tank.

transfer or reaction. The concentration in the CMET thus changes only very slowly for a given reaction volume and rate. Alternatively, the concentration in the reservoir can regularly be adjusted to maintain an essentially constant value, either continuously or in frequent stepwise fashion. By such adjustment, the rate process in the FBR may be considered approximately constant, just as for a CMFR operated at steady state. In either mode of rate data collection (slowly decreasing concentration or stepwise addition to maintain a constant concentration), reaction rates can be measured much more easily than by small differences in concentration across the reactive PFR section. This type of *chemostat* provides an especially attractive way to study rates of fast, fluid–solid phase processes.

The material balance around the system pictured in Figure 9-10 is

$$\int_0^{V_{\mathrm{FBR}}} r\, dV_{\mathrm{FBR}} = V_{\mathrm{FBR}} \frac{dC_{\mathrm{FBR}}}{dt} + V_{\mathrm{CMET}} \frac{dC_{\mathrm{CMET}}}{dt} \tag{9-43}$$

If a true *differential* reactor condition has been achieved, $C_{\mathrm{FBR}} \approx C_{\mathrm{CMET}}$ and the rate, r, is essentially constant throughout the volume, V_{FBR}. With these assumptions, Equation 9-43 simplifies to

$$r = \frac{V_{\mathrm{FBR}} + V_{\mathrm{CMET}}}{V_{\mathrm{FBR}}} \left[\frac{dC_{\mathrm{CMET}}}{dt} \right] \tag{9-44}$$

The reaction rate is estimated by determining the instantaneous slope of a plot of C_{CMET} versus time data. Analysis of the dependency of reaction rate on concentration follows the same procedure as for other reactor configurations.

Inspection of Equation 9-44 shows that the effect of V_{CMET} is to "magnify" a small rate of change of concentration to obtain the reaction rate.

The establishment of *differential-reactor* operation must be determined by a material balance around the FBR component rather than around the entire system:

$$Q_R C_{CMET} - Q_R C_{FBR} + \int_0^{V_{FBR}} r \, dV_{FBR} = V_{FBR} \frac{dC_{FBR}}{dt} \tag{9-45}$$

where Q_R is the recycle flow rate and C_{FBR} is the exit concentration from the differential reactor. Substitution of the integral from Equation 9-43 gives

$$C_{CMET} - C_{FBR} = -\frac{V_{CMET}}{Q_R} \frac{dC_{CMET}}{dt} \tag{9-46}$$

A differential-reactor condition is obtained when $C_{CMET} \rightarrow C_{FBR}$. This condition was assumed in deriving Equation 9-44 and must apply as well to Equation 9-46. Thus dC_{CMET}/dt may be replaced by Equation 9-44 to yield

$$C_{CMET} - C_{FBR} = -\frac{V_{CMET}}{Q_R} \left[\frac{V_{FBR}}{V_{FBR} + V_{CMET}} \right] r \tag{9-47}$$

Equation 9-47 can be used to design a differential FBR-batch-recycle system as well as to evaluate whether an existing device conforms to the assumptions of such a reactor. To design a differential FBR-batch-recycle system we need an *a priori* estimate of r and the volumes of the reactor and reservoir. The recycle flow rate can then be selected to make $C_{CMET} \rightarrow C_{FBR}$. Inspection of Equation 9-47 shows that $C_{CMET} \approx C_{FBR}$ as the recycle flow rate increases and as the FBR volume decreases. To check the validity of an assumption of differential reactor operation, rate data must be collected to determine r and to use Equation 9-47 to calculate $C_{CMET} - C_{FBR}$.

9.3.8.2 Continuous-Recycle Reactors Continuous-flow-recycle reactors are common in many treatment processes. Recycling of the fluid phase provides performance characteristics between that of a PFR and a CMFR (i.e., similar to that of CMFRs in series) and provides a means for improving process performance for fixed-bed reactor systems. For example, effluent recycling to attached-growth FBR bioreactors helps to dilute feed concentrations to prevent overloading. Similarly, recycling of solids is used to increase reactive mass concentrations and rates of reaction in CMFR and PFR processes such as chemical precipitation, adsorption by powdered forms of activated carbon, and suspended-growth biotreatment processes.

The development of recycle reactor design equations begins with steady-state analysis of the performance of a PFR with fluid-phase recycle, as illus-

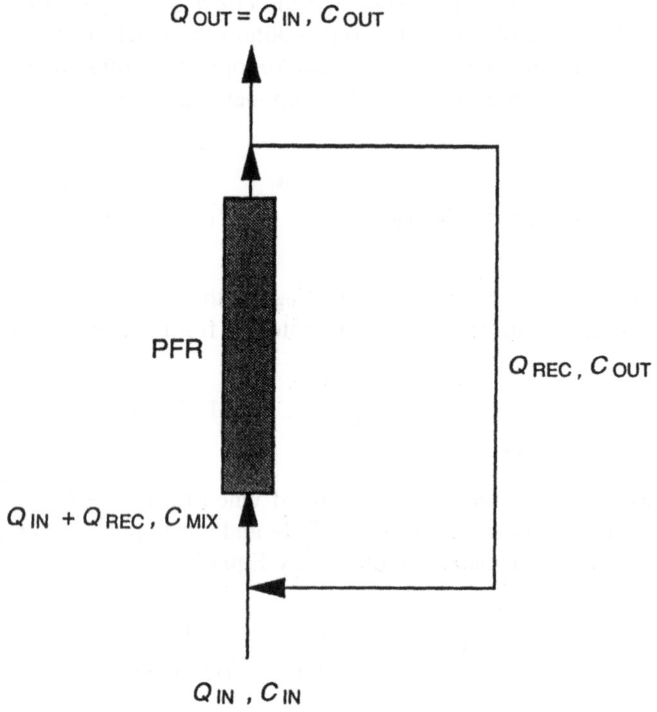

Figure 9-11 Ideal PFR with recycle.

trated in Figure 9-11. The general point form of the material balance (Equation 2-16) was given earlier:

$$-v_x \frac{dC}{dx} + r = 0 \tag{9-34}$$

In this instance, the flux of a specific constituent is given by

$$v_x = \frac{Q_{IN} + Q_{REC}}{A_N} \tag{9-48}$$

where Q_{IN} is the flow rate of the feed stream to the reactor, Q_{REC} the effluent recycle flow rate, and A_N the cross-sectional area of the reactor normal to the incoming flow. The steady-state mass balance equation for this condition is

$$\int_{C_{MIX}}^{C_{OUT}} \frac{dC}{r} = \frac{1}{Q_{IN} + Q_{REC}} \int_0^L A_N \, dx = \frac{V_R}{Q_{IN} + Q_{REC}} \tag{9-49}$$

where C_{MIX} is the concentration after blending the feed and recycle streams and V_R is the volume of the reactor. The recycle ratio, the ratio of recycle flow to influent flow, is usually defined as

$$R_Q = \frac{Q_{REC}}{Q_{IN}} \tag{9-50}$$

and the HRT as

$$\bar{t} = \frac{V_R}{Q_{IN}} \tag{9-51}$$

Substituting into Equation 9-49, we obtain the following design equation for a *PFR with recycle*:

$$\bar{t} = (1 + R_Q) \int_{C_{MIX}}^{C_{OUT}} \frac{dC}{r} \tag{9-52}$$

When operated without recycle ($R_Q = 0$), $C_{MIX} = C_{IN}$ and the design equation is equivalent to that for a PFR (Equation 9-36). The effect of recycle is to increase the required HRT, although the extent cannot be determined directly from Equation 9-52 until the value of C_{MIX} is calculated. The concentration of the blended feed and recycle flow, C_{MIX}, is necessarily less than the feed concentration, C_{IN}, and the value of the integral is therefore smaller than that for a similar PFR operated without recycle.

The blend concentration C_{MIX} is given by the following mass balance:

$$(Q_{IN} + Q_{REC})C_{MIX} = Q_{IN}C_{IN} + Q_{REC}C_{OUT} \tag{9-53}$$

$$C_{MIX} = \frac{C_{IN} + R_Q C_{OUT}}{1 + R_Q} \tag{9-54}$$

As recycle increases, C_{MIX} decreases, and the concentration change across the reactor becomes smaller. At very high recycle rates, the reaction rate becomes nearly constant throughout the reactor. Equation 9-52 may then be written as

$$\bar{t} = (1 + R_Q) \frac{C_{OUT} - C_{MIX}}{r} \tag{9-55}$$

After substitution for C_{MIX} from Equation 9-54, this expression reduces to the design equation for a CMFR:

$$\bar{t} = -\frac{C_{IN} - C_{OUT}}{r} \tag{9-56}$$

The analysis above supports a conclusion that can be arrived at by intuitive reasoning; that is, a continuous-flow recycle reactor is identical to a PFR when $R_Q = 0$ and to a CMFR when R_Q approaches infinity. Between these extremes, the detention time required for recycle reactors is intermediate to those of PFRs and CMFRs, just as was shown for CMFRs in series.

An application of Equation 9-52 for the design of a PFR with recycle is given in Example 9-6, and the results compared to those obtained in Example 9-5 for single CMFRs and PFRs and for CMFRs in series. The comparison illustrates the design flexibility afforded by different reactor configurations and arrangements.

Example 9-6

- **Situation:** As in Example 9-5, a contaminant is removed from a waste stream by a first-order reaction.
- **Question(s):** Compare the HRTs for recycle reactors with $R_Q = 1$ and $R_Q = 3$ to those calculated in Example 9-5 for a single CMFR, three CMFRs in series, and a single PFR.
- **Logic and Answer(s):**

1. The design equation for a first-order reaction follows from Equation 9-52:

$$\bar{t} = (1 + R_Q)\int_{C_{MIX}}^{C_{OUT}} \frac{dC}{r} = (1 + R_Q)\frac{-\ln \dfrac{C_{OUT}}{C_{MIX}}}{k}$$

2. C_{MIX}/C_{IN} is given by dividing both sides of Equation 9-54 by C_{IN}:

$$\frac{C_{MIX}}{C_{IN}} = \frac{1 + R_Q \dfrac{C_{OUT}}{C_{IN}}}{1 + R_Q}$$

3. For $C_{OUT}/C_{IN} = 0.05$, the values of C_{MIX}/C_{IN} (from Equation 9-54), C_{OUT}/C_{MIX} [which is given by $(C_{OUT}/C_{IN}) \times (C_{IN}/C_{MIX})$], and $\bar{t}$ (from Equation 9-52) are presented below. Note by comparison with the findings of Example 9-5 that the $\bar{t}$ values are intermediate between a single PFR ($3k^{-1}$) and a single CMFR ($19k^{-1}$); a recycle ratio of 1 corresponds roughly to three CMFRs in series.

R_Q	C_{MIX}/C_{IN}	C_{OUT}/C_{MIX}	$\bar{t}$
1	0.525	0.095	$4.7k^{-1}$
3	0.287	0.174	$6.98k^{-1}$

A CMFR with recycle is represented schematically in Figure 9-12. The material balance on a component in the fluid phase during steady-state opera-

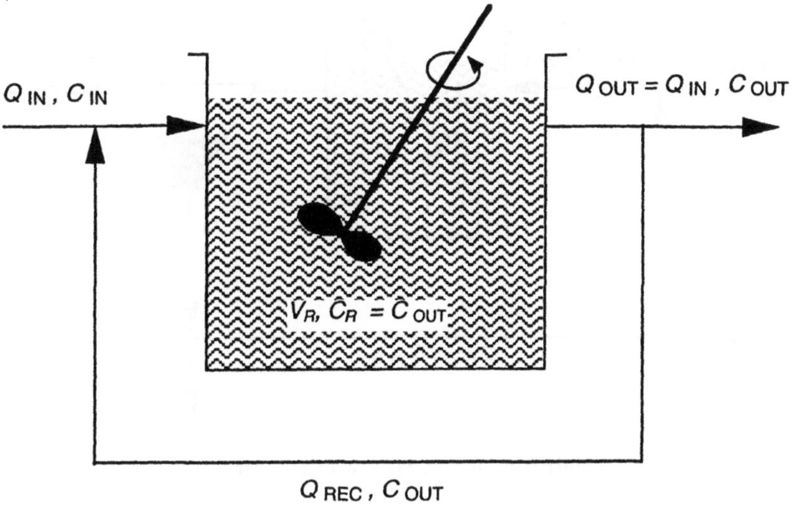

Figure 9-12 Ideal CMFR with recycle.

tion is

$$Q_{IN} C_{IN} + Q_{REC} C_{OUT} - (Q_{IN} + Q_{REC}) C_{OUT} + r V = 0 \qquad (9\text{-}57)$$

After simplification, the design equation can be written

$$\bar{t} = -\frac{C_{IN} - C_{OUT}}{r} \qquad (9\text{-}58)$$

which we recognize again as the design equation for a CMFR. Recycle does not change the required volume of the reactor, which, from the definition of $\bar{t}$, is calculated from the feed flow rate, Q_{IN}. Instead, the effect of recycle is to decrease the *detention time per pass* (t_p) through the CMFR:

$$t_p = \frac{V_R}{Q_{IN} + Q_{REC}} \qquad (9\text{-}59)$$

Although t_p becomes smaller with increasing Q_{REC}, the number of passes increases, such that each fluid element still spends time $\bar{t}$ in the reactor and the extent of reaction is the same as that without recycle.

Solids recycle in CMFRs is commonly practiced to enhance rates of solids-catalyzed reactions. For example, rates of precipitation may be described by $r = -k(C - C_S) C_{ps}$, where $(C - C_S)$ is the dissolved-phase concentration of precipitant in excess of the saturation concentration C_S, and C_{ps} is the dispersed-phase concentration of preformed (existing) solids. As illustrated in

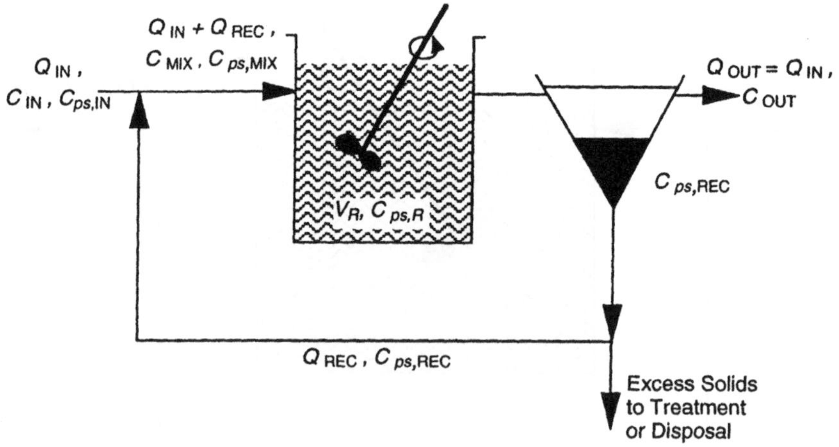

Figure 9-13 Ideal CMFR with solids recycle.

Figure 9-13, a settling tank following the CMFR is generally incorporated in the design of solids recycle systems to concentrate the solids for recycle. Solids are frequently the by-products of designed treatment processes. For example, chemical precipitates are produced in phosphorus removal and lime softening. Similarly, biomass is produced in most biological treatment processes. These solids must be *wasted* (treated, reused, disposed of) at a rate equal to their production rate in order to maintain constant steady-state concentrations in designed treatment processes. With reference to Figure 9-13, the solids concentration entering the CMFR is obtained from a solids material balance written after blending the feed and recycle rates:

$$Q_{\text{IN}} C_{ps,\text{IN}} + Q_{\text{REC}} C_{ps,\text{REC}} - (Q_{\text{IN}} + Q_{\text{REC}}) C_{ps,\text{MIX}} = 0 \tag{9-60}$$

$$C_{ps,\text{MIX}} = \frac{Q_{\text{IN}} C_{ps,\text{IN}} + Q_{\text{REC}} C_{ps,\text{REC}}}{Q_{\text{IN}} + Q_{\text{REC}}} = \frac{C_{ps,\text{IN}} + R_Q C_{ps,\text{REC}}}{1 + R_Q} \tag{9-61}$$

Usually, $C_{ps,\text{IN}}$ is so much smaller than $R_Q C_{ps,\text{REC}}$ that it can be neglected.

The mean solids residence time (SRT) or "age," $\bar{t}_S$, in such systems is given by

$$\bar{t}_S = \frac{C_{ps,R} V_R}{M_W} \tag{9-62}$$

where $C_{ps,R} V_R$ is the average mass of solids in the CMFR and M_W is the mass wastage rate. If the rate of solids build-up is small, $C_{ps,R} \sim C_{ps,\text{MIX}}$. The age of the solids can be controlled by the recycle rate, as seen from Equation 9-61.

The control of *solids age* can have important implications for reactor performance.

Example 9-7 provides an illustration of a solids recycle system in which an ancillary solids treatment process is incorporated in the recycle loop. Schemes of this type are sometimes used to "select" microorganisms with good settling characteristics (i.e., to eliminate poor settling characteristics) for return to the aeration tank.

Example 9-7

- **Situation:** An activated sludge treatment system is used to treat 7500 m^3/day of an industrial wastewater. The success of the process depends in large measure on maintaining a steady-state biomass concentration of 1500 mg/L in the aeration tank, which has an HRT of 10 h. Biomass is settled, the excess growth is wasted, and the remainder is recycled to the head of the aeration tank. The recycle ratio, R_Q, is 0.4. It has been discovered that a steady-state concentration cannot be maintained because a significant fraction of the bacteria present in the recycle flow are species of the genus *Nocardia*. These filamentous bacteria form a low-density, poorly-settling floc which allows biomass to escape the final clarifier in the effluent; moreover, the suspended solids concentration is in excess of the discharge permit. The net growth rate of *Nocardia* is first order with respect to the microorganism concentration, C_N, and the growth rate constant, k_G, is 0.006 h^{-1}. Hydrogen peroxide is known to inactivate *Nocardia*. The rate of inactivation is also first order with respect to *Nocardia* and the reaction rate constant, k_R, is 4 h^{-1}. A well-mixed tank (i.e., a CMFR) having an HRT of 0.5 h is constructed on the sludge recycle line to accommodate the hydrogen peroxide process, as indicated below.

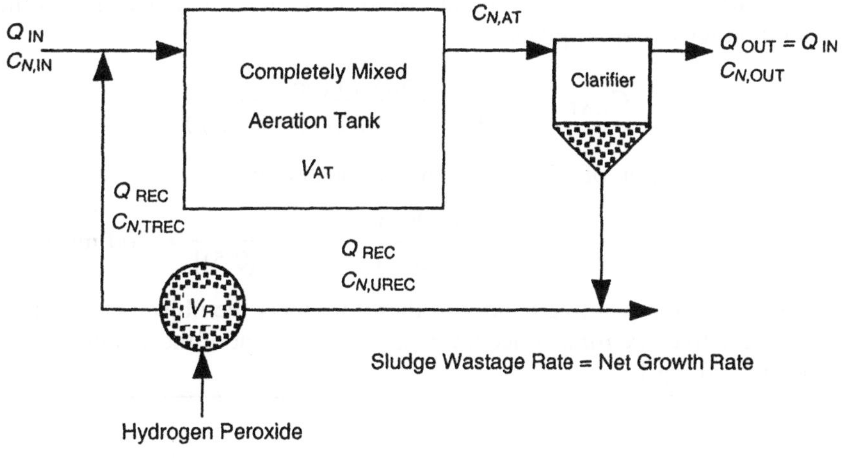

- **Question(s):** Data from other treatment plants indicate that good settling will result if the *Nocardia* can be kept below 400 mg/L in the effluent from the aeration tank. The active *Nocardia* concentration in the untreated recycled sludge entering the hydrogen peroxide reactor, $C_{N,\text{UREC}}$, is 1200 mg/L. Will the *Norcadia* concentration in the treated recycled sludge, $C_{N,\text{TREC}}$, meet the stated objective?
- **Logic and Answer(s):**

1. The steady-state material balance for active *Nocardia* species written around the aeration tank is

$$Q_{\text{IN}} C_{N,\text{IN}} + Q_{\text{REC}} C_{N,\text{TREC}} - (Q_{\text{IN}} + Q_{\text{REC}}) C_{N,\text{AT}} + k_G C_{N,\text{AT}} V_{\text{AT}} = 0$$

where $C_{N,\text{IN}} = 0$. (*Note*: To use a steady-state analysis, we assume that the growth rate of *Nocardia* is similar to that of all other microorganisms, for which case the wastage rate is exactly equal to the growth rate.)

2. Solving for the concentration of *Nocardia* leaving the aeration tank gives

$$C_{N,\text{AT}} = \frac{R_Q C_{N,\text{TREC}}}{1 + R_Q - k_G \bar{t}_{\text{AT}}}$$

where $R_Q = Q_{\text{REC}}/Q_{\text{IN}}$ and $\bar{t}_{\text{AT}}$ is the HRT in the aeration tank.

3. The material balance on *Nocardia* around the hydrogen peroxide reactor is

$$Q_{\text{REC}} C_{N,\text{UREC}} - Q_{\text{REC}} C_{N,\text{TREC}} - k_R C_{N,\text{TREC}} V_R = 0$$

from which

$$C_{N,\text{TREC}} = \frac{C_{N,\text{UREC}}}{1 + k_R \bar{t}_R}$$

where $\bar{t}_R$ is the HRT of the recycle reactor.

4. Returning the expression above to the material balance around the aeration tank (step 2) gives

$$C_{N,\text{AT}} = \frac{R_Q C_{N,\text{UREC}}}{(1 + R_Q - k_G \bar{t}_{\text{AT}})(1 + k_R \bar{t}_R)}$$

5. Substitution of all the known parameters gives

$$C_{N,\text{AT}} = \frac{0.4(1200)}{[1 + 0.4 - (0.006)(10)][1 + (0.4)(0.5)]} = 300 \text{ mg/L}$$

6. The proposed treatment will therefore meet the objective of maintaining active *Nocardia* concentrations in the effluent of the aeration basin below 400 mg/L.

9.4 NONIDEAL REACTORS

Ideal CMFR and PFR conditions and characteristics, while usually the intent of design practice, are often *closely approximated* but rarely *achieved* in most full-scale process applications. Deviations in flow patterns, and thus in the residence times and extent of reaction of constituents of concern, can be caused by short-circuiting, by recycle, or by the presence of stagnant zones within reactors, as illustrated in Figure 9-14. To determine how reactor performance will deviate from ideal conditions in such cases, it is usually necessary to determine reactor flow and mixing characteristics experimentally.

The basic approach to treatment of nonideal reactor behavior is to obtain and analyze information on how long individual fluid elements reside in a par-

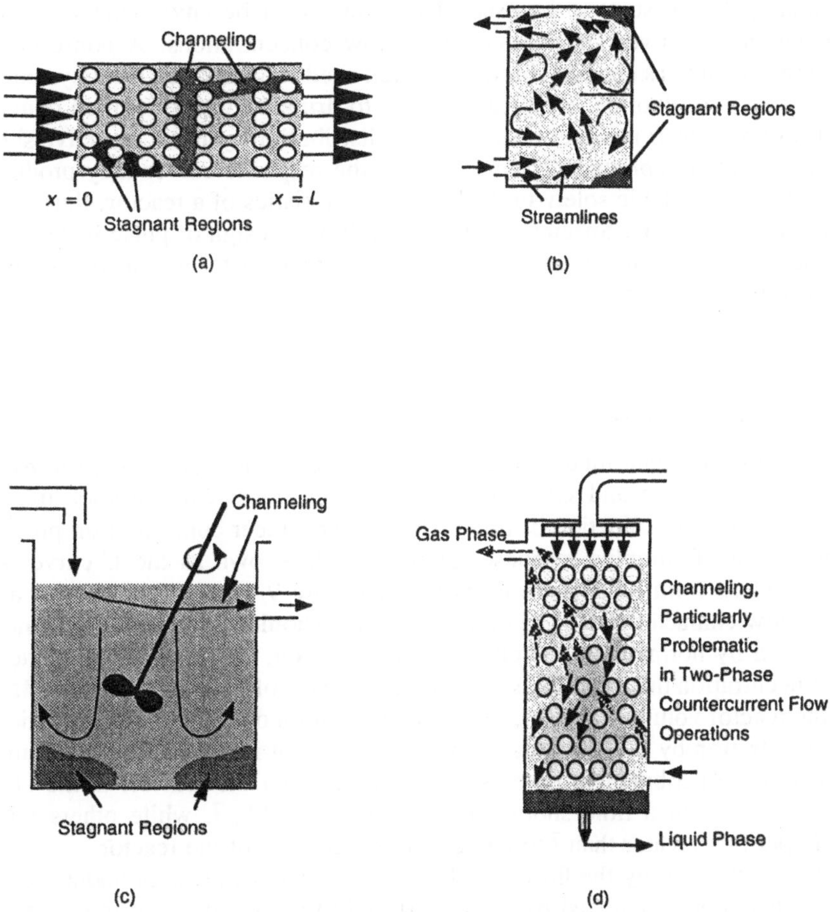

Figure 9-14 Examples of nonideal reactor behavior.

ticular reactor. This information can be expressed in terms of the distribution of the "ages" of fluid elements within the reactor at any time, or of the ages of fluid elements contained in specific volumes exiting the reactor per unit of time. The ages of these fluid elements are referred to terms of either *internal* or *exit age distribution functions*, respectively. These functions are determined by a procedure termed *residence time distribution* (RTD) analysis. An RTD is usually obtained by application of stimulus–response analyses involving introduction to reactors of readily detectable tracers, as either pulse or step inputs, and observation of reactor responses to these stimuli. A *pulse or delta input* results from virtually instantaneous injection of a *fixed mass* of tracer to the influent of a reactor. A *step input* is achieved by suddenly introducing and subsequently maintaining a *constant concentration* of tracer in the reactor influent.

The design of a tracer test begins with selection of the tracer and the type of input (pulse or step) to be used. The tracer must be environmentally acceptable, nonreactive, and measurable at low concentrations. A nonreactive dye or a salt such as sodium chloride is often used in waste treatment systems, whereas sodium fluoride, a typical additive to drinking water, is often convenient for tracer tests in water treatment and distribution systems. Reactive substances are inappropriate as tracers because the response curves they produce will not be attributable solely to the flow characteristics of a reactor, thus confounding the analysis. Special care is needed in the design of pulse input tests to ensure that a sufficient amount of tracer is injected to give a measurable response in the exit stream.

9.4.1 *C* and *E* Curves

When a pulse or delta input is used, the response of the reactor is given by a pattern of increasing and subsequently decreasing tracer concentration in the effluent stream as a function of time. The effluent tracer concentration profile for this type of stimulus–response relationship is known as the *C* curve. A dimensionless form of *C* curve is shown in Figure 9-15a for a reactor of arbitrary flow characteristics. The ordinate is dimensionless concentration which is obtained by normalizing the effluent concentration, C_{OUT}, by the hypothetical concentration that represents instantaneous entry of the mass of tracer, M_T, into the reactor volume, V_R (i.e., $C_\Delta = M_T/V_R$). With or without normalization of concentration by C_Δ, the *C* curve provides a visual impression of how fluid elements are distributed in time as they pass through the reactor. Some elements will exit in a time shorter than the mean HRT, $\bar{t}$, while others will require periods greater than $\bar{t}$ to move through and out of the reactor.

When multiplied by the flow rate through the reactor, the area under the *C* curve that is not normalized by C_Δ should yield the total mass of tracer, M_T, introduced in the pulse input:

$$M_T = Q \int_0^\infty C_{OUT}(t)\, dt \tag{9-63}$$

This simple mass balance provides a good check on the efficiency of tracer recovery and the assumption of nonreactivity.

The ***E** curve*, the *exit age or residence time distribution curve*, is repre-

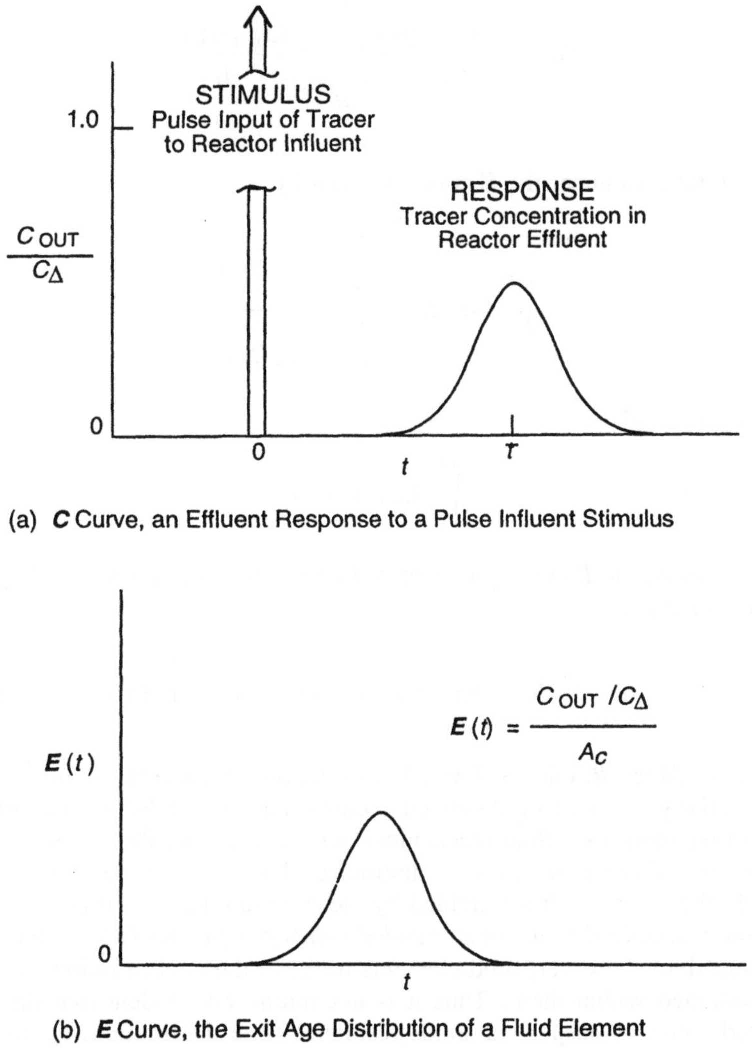

Figure 9-15 Stimulus–response relationships for pulse inputs of nonreactive tracers to reactors of arbitrary flow characteristics.

sented in Figure 9-15b. This curve is a further normalization of the **C** curve obtained by dividing each value of C_{OUT}/C_{Δ}, by the total area, A_C, under the **C** curve. From the normalized ***C curve*** shown in Figure 9-15a, the area under the **C** curve is

$$A_C = \frac{1}{C_{\Delta}} \int_0^{\infty} C_{OUT}(t)\, dt \tag{9-64}$$

The **E** curve is thus defined as

$$\boldsymbol{E}(t) = \frac{C_{OUT}(t)/C_{\Delta}}{\displaystyle\int_0^{\infty} \frac{C_{OUT}(t)}{C_{\Delta}}\, dt} = \frac{C_{OUT}(t)/C_{\Delta}}{A_C} \tag{9-65}$$

and the total area under the **E** curve is given by

$$\int_0^{\infty} \boldsymbol{E}(t)\, dt = \frac{\displaystyle\int_0^{\infty} C_{OUT}(t)\, dt}{\displaystyle\int_0^{\infty} C_{OUT}(t)\, dt} \tag{9-66}$$

which means that

$$\int_0^{\infty} \boldsymbol{E}(t)\, dt = 1 \tag{9-67}$$

In other words, the **E** curve produces a time-normalized or fractional age distribution such that

$$\text{fraction of fluid elements with age} < t = \int_0^{t} \boldsymbol{E}(t)\, dt \tag{9-68}$$

9.4.1.1 Mean Residence Time, Variance, and Skewness Fluid elements are essentially *carriers* of dissolved solutes and suspended solids. From a transporting fluid (i.e., fluid mechanics) perspective, one fluid element is the same as any other and the mean hydraulic residence time of the fluid (carrier) is simply the reactor volume divided by the flow rate through it (i.e., $\bar{t} = V_R/Q$). From a *dissolved solute* or *suspended solids perspective* (i.e., a *fluid quality* perspective), however, fluid elements differ in the concentrations of materials contained within them. Thus it is not intuitively evident that the mean dissolved solute or suspended solids residence time distribution, t_m (i.e., the mean constituent residence time), should be the same as the mean hydraulic or fluid residence time, $\bar{t}$.

The *mean value* of the residence time of a material dissolved or suspended in a fluid is given by the *first moment* of the centroid of the RTD curve about the origin (zero-time axis) of the distribution:

$$t_m = \frac{\int_0^\infty t\, E(t)\, dt}{\int_0^\infty E(t)\, dt} \tag{9-69}$$

From Equation 9-67, the denominator of Equation 9-69 is equal to unity, so that

$$t_m = \int_0^\infty t\, E(t)\, dt \tag{9-70}$$

This integral is approximated from experimental measurements of tracer concentration at the reactor exit as a function of time.

The *standard deviation*, σ_t, of a distribution is the square root of the *variance* of the distribution about its mean value. *Variance* quantifies the extent of spreading of the tracer "cloud" produced by a pulse input and is thus a measure of the *variation* in residence times of fluid elements about t_m. Variance is given by the *second moment* of the centroid of the RTD about the *mean* of the distribution:

$$\sigma_t^2 = \int_0^\infty (t - t_m)^2 E(t)\, dt \tag{9-71}$$

A more convenient form of the variance relationship is obtained by expansion of the integrand in Equation 9-71 and application of the definitions provided by Equations 9-67 and 9-70:

$$\sigma_t^2 = \int_0^\infty t^2 E(t)\, dt - t_m^2 \tag{9-72}$$

The integral above is calculated from tracer data in a manner analogous to that for mean residence time. Standard deviation is often a more convenient measure of variation in the sense it has the same units of measurement as the data.

A parameter involving the *third moment* of the centroid of the RTD is often employed as a measure of the *skewness* of the RTD. This parameter, s_3, is related to the third moment taken about the mean as follows:

$$(s_3)^3 = \frac{1}{\sigma_t^{1.5}} \int_0^\infty (t - t_m)^3 E(t)\, dt \tag{9-73}$$

Skewness, which unlike the third moment itself is independent of the units of measurement, gives further insight into deviations of reactor performance from ideal behavior. For example, channeling through a packed-bed reactor may cause the RTD to be skewed toward times shorter than the mean residence time.

9.4.1.2 Dimensionless E Curve The numerical value of E depends on both the flow characteristics and the mean HRT of a reactor. Because the time required to observe a given tracer concentration at the exit of any reactor will depend on the value of t_m for that reactor, it follows from inspection of Equation 9-72 that reactors of identical flow characteristics but different sizes will have numerically different values for their respective variances. To make it possible to compare flow characteristics directly, the RTD is normalized for the effect of hydraulic residence time. That is, the E curve in Equation 9-65, which has units of time^{-1}, can be made dimensionless for scaling purposes by expressing time as the ratio of real fluid element time, t, to the mean HRT, $\bar{t}$, to yield the variable $\theta° = t/\bar{t}$. Then, if $\bar{t} = t_m$,

$$E(\theta°) = \frac{C_{OUT}(t)}{\displaystyle\int_0^\infty C_{OUT}(t)\, d\left(\frac{t}{t_m}\right)} = t_m E(t) \tag{9-74}$$

and the area under the dimensionless E curve is

$$\int_0^\infty E(\theta°)\, d(\theta°) = \int_0^\infty t_m E(t)\, d\left(\frac{t}{t_m}\right) = 1 \tag{9-75}$$

We conclude that two reactors of different size that produce identical dimensionless plots of $E(\theta°)$ against $\theta°$ have identical flow characteristics. Once again, as emphasized in Chapter 4, the importance of *dimensionless variables for process scaleup* is evident.

A dimensionless form of the variance is obtained by dividing Equation 9-72 through by the square of the mean constituent residence time, t_m:

$$\sigma_{\theta°}^2 = \int_0^\infty (\theta°)^2\, E(\theta°)\, d\theta° - 1 \tag{9-76}$$

where

$$\sigma_{\theta°}^2 = \frac{\sigma_t^2}{t_m^2} \tag{9-77}$$

An illustration of the application of RTD analysis for determination of the performance characteristics of a system of sand filters in a water treatment plant is given in Example 9-8.

Example 9-8

- **Situation:** A parallel arrangement of six sand filters at a water treatment plant operate at a total flow rate of 52,920 m^3/day (14 mgd). The HRT in the filter box (volume above the filters), the filter media, and the underdrain is 50 min. Chlorine is added just ahead of these filters to prevent microbial growth on the media, which would otherwise cause premature clogging and reduce the time between required backwashings. The municipality is interested in determining the residence time distribution characteristics of water in the filters because this information can be used to estimate the disinfection of *Giardia* cysts occurring as a result of chlorine dosing. The most convenient point of addition of a tracer is in the pipeline leading to a common header for all six filters. A pulse input (2.9 kg) of sodium chloride is introduced. (*Note:* Sodium fluoride was ruled out because of adsorption onto aluminum hydroxide precipitate trapped within the filter). Following the pulse input, the sodium concentration is measured in the manifold, which collects the water from all six filters. The resulting data for time and sodium chloride concentration, C_t (after subtracting the background concentration), relating to a measurement time, t, are given in the first two columns of the spreadsheet developed below.
- **Question(s):** Characterize the residence time distribution (RTD) for the filters.
- **Logic and Answer(s):**

1. The response data for this pulse input can be used to analyze only the average RTD for the six filters operating in parallel. If significant differences are expected in the RTDs of individual filters, a tracer test should be conducted for each filter.

2. A spreadsheet is set up as illustrated below to summarize the time [column (1)] and concentration [column (2)] data and to determine the average concentration $\overline{C}_t$ in a time interval [column (3)], the mass of Na recovered from the tracer test [column (4)], the E curve [column (5)], the mean residence time [column (6)], the variance [column (7)], and the skewness [column (8)].

Data		E_t Calculations					
(1) t (min)	(2) C_t (mg/L)	(3) $\overline{C}_t$ (mg/L)	(4) $Q\,\overline{C}_t\Delta t$ (kg)	(5) E_t (min^{-1} $\times 10^3$)	(6) $tE_t\,\Delta t$ (min)	(7) $t^2E_t\,\Delta t$ (min^2)	(8) $(t - t_m)^3$ $E_t\,\Delta t$ (min^3)
0	0	0	0.00	0.0	0.00	0.00	0.00
10	0.1	0.05	0.02	0.67	0.07	0.67	−331.15
20	1.4	0.75	0.28	10.1	2.01	40.27	−1911.43
30	1.7	1.55	0.57	20.8	6.24	187.25	−965.15

Data		E_t Calculations					
(1) t (min)	(2) C_t (mg/L)	(3) $\overline{C}_t$ (mg/L)	(4) $Q\,\overline{C}_t\Delta t$ (kg)	(5) E_t (min^{-1} $\times 10^3$)	(6) $tE_t\,\Delta t$ (min)	(7) $t^2E_t\,\Delta t$ (min^2)	(8) $(t - t_m)^3$ $E_t\,\Delta t$ (min^3)
40	1.5	1.6	0.59	21.5	8.59	343.62	−63.95
50	1.2	1.35	0.50	18.1	9.06	453.02	6.64
60	0.7	0.95	0.35	12.8	7.65	459.06	301.50
70	0.4	0.55	0.20	7.4	5.17	361.74	936.51
80	0.2	0.3	0.11	4.0	3.22	257.72	1489.92
90	0.1	0.15	0.06	2.0	1.81	163.09	1637.06
100	0.1	0.1	0.04	1.3	1.34	134.23	2035.01
110	0.05	0.075	0.03	1.0	1.11	121.81	2556.07
120	0	0.025	0.01	0.33	0.40	48.32	1322.79
130	0	0	0.0	0.0	0.00	0.00	0.00
Sums	7.45	7.45	2.74	100	46.68	2570.81	7013.82

The calculations given in this table and their use in RTD analysis are outlined in steps 3 to 8 as given below.

3. Column (3): $\overline{C}_t = \dfrac{C_{t-1} + C_t}{2}$

The plot of the ***C*** curve given below is typical of nonideal flow reactors. Note that concentration, C_t, is used here because the ***C*** curve is not dimensionless.

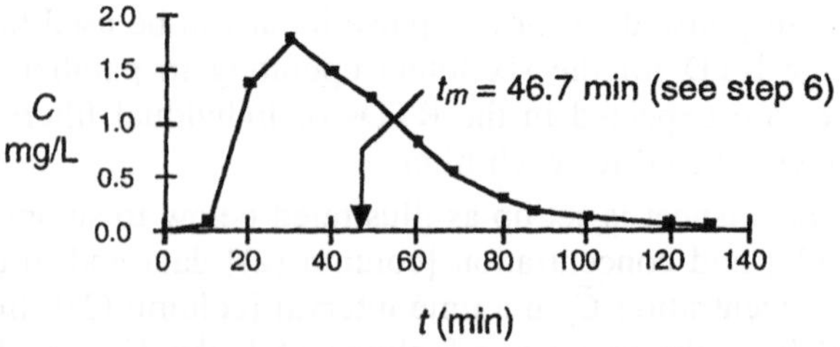

4. Column (4): $\sum_{i=1}^{n} Q\overline{C}_t\,\Delta t$ = total tracer mass recovered

$$Q = \frac{52{,}920\ \text{m}^3/\text{day}}{1440\ \text{min/day}} \times 10^3\ \text{L/m}^3$$

2.74 kg was recovered of 2.9 kg added, or 95% recovery, which indicates that the tracer behaved conservatively.

5. Column (5) from Equation 9-66:

$$E_t = \frac{C_{\text{OUT}}(t)}{\sum_{0}^{\infty} C_{\text{OUT}}(t)\,\Delta t}$$

We can also note from Equation 9-67 that

$$\sum_{i=1}^{n} E_t \, \Delta t = 1$$

which from the data provided can be checked as follows:

$$\sum_{i=1}^{n} E_t \, \Delta t = (\Delta t) \text{ [column (5) summation]}$$

$$= 10(100 \times 10^{-3}) \text{ min}^{-1} = 1$$

The $E(t)$ curve shown below indicates that the behavior deviates considerably from plug flow. This behavior can be explained by mixing that occurs in the filter box above the filter, and channeling and dispersion that occur in flow through the porous filter media.

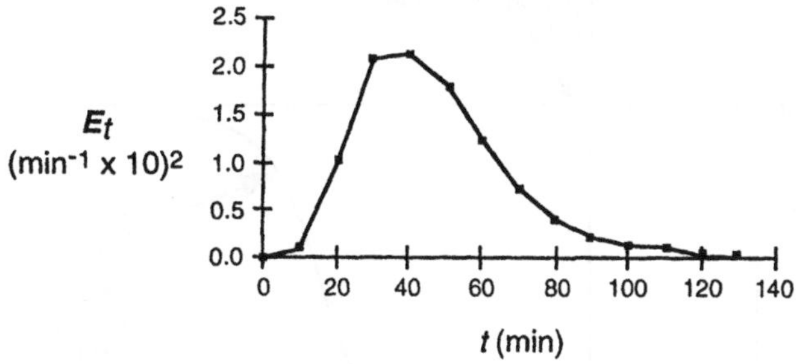

6. Column (6) (from Equation 9-70): $t_m = \Sigma_{t=t_1}^{t_n} tE_t \, \Delta t = 46.7$ min The mean residence time is slightly shorter than the hydraulic residence time (50 min), which indicates the possibility of dead space (i.e., volume in the reactor through which flow does not pass).

7. Column (7), from Equation 9-72:

$$\sigma_t^2 = \sum_{t=t_1}^{t_n} t^2 E_t \, \Delta t - t_m^2 = 2571 - (46.7)^2 = 392 \text{ min}^2$$

$$\sigma_t = 19.7 \text{ min}$$

or 68% of fluid elements spend between 27 and 66.4 min in the reactor (i.e., $t_m \pm \sigma_t$).

8. Column (8), from Equation 9-73:

$$(s_3)^3 = \frac{1}{\sigma_t^{1.5}} \sum_{t=t_1}^{t_n} (t - t_m)^3 E_t \, \Delta t = \frac{7014}{(19.8)^{1.5}} = 79.6 \text{ min}^3$$

$$\text{skewness, } s_3 = 4.3 \text{ min}$$

A positive value of s_3 indicates that the residence time distribution is skewed toward residences greater than the mean value.

9.4.2 *F* Curve

The ***F*** curve, like the ***C*** curve, is obtained from the response of a reactor to an influent tracer. In the case of the ***F*** curve, the tracer is introduced as a step-function stimulus. Because the tracer is fed continuously to the reactor after the step input is initiated, the exit concentration, C_{OUT}, must approach the feed concentration, C_{IN}, as time increases, as illustrated in Figure 9-16a. Figure 9-16b illustrates the ***F*** curve, which normalizes the effluent tracer concentration, C_{OUT}, by the feed concentration, C_{IN}. The value of ***F*** at any time, *t*, thus represents the fraction of tracer molecules having an exit age younger than *t*. This provides an important conceptual linkage between the ***F*** and ***E*** curves. Recalling that exit ages younger than *t* are also defined by the ***E*** curve (Equa-

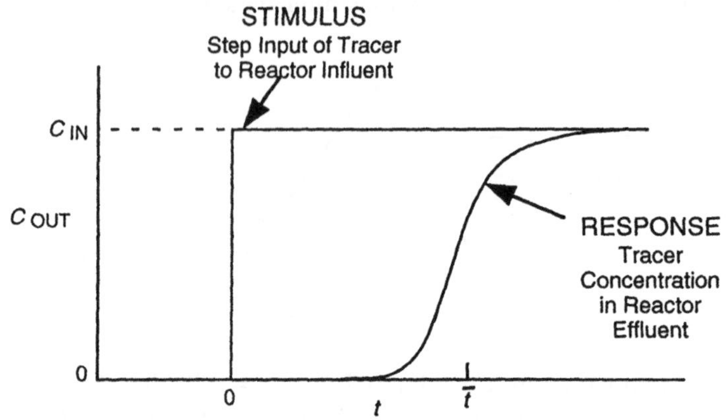

(a) Effluent Concentration Response to a Step Stimulus

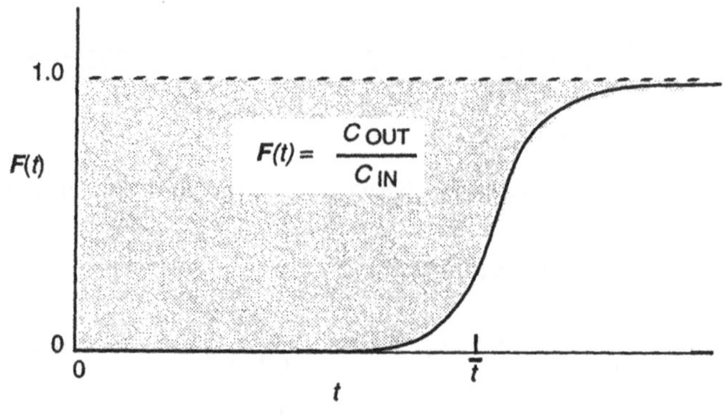

(b) *F* Curve, a Feed-Normalized Effluent Concentration

Figure 9-16 Stimulus–response relationships for step inputs of nonreactive tracers to reactors of arbitrary flow characteristics.

tion 9-68), if follows that the F and E curves are related by

$$F(t) = \int_0^t E(t)\,dt \tag{9-78}$$

or

$$\frac{dF(t)}{dt} = E(t) \tag{9-79}$$

Tracer curve relationships thus reveal that the derivative of the experimentally determined F curve should be the expected E curve and, conversely, that the integral of the experimentally determined E curve should be the expected F curve.

Because the F and E curves are related mathematically, the F curve can be made dimensionless in the same manner discussed in Section 9.4.1.2 for the E curve. The statistical properties and analyses described in Section 9.4.1.1 for the E curve apply as well for the F curve. The mean residence time and variance of a reactor, for example, can be determined from its F curve as well as from its E curve. The mathematical relationship between the F and E curves, given by Equation 9-78, provides a convenient substitution into the expressions for t_m and σ_t^2 derived from the experimental E curve (Equations 9-70 and 9-71, respectively); that is,

$$t_m = \int_0^{\infty} t\,E(t)\,dt = \int_0^1 t\,dF(t) \tag{9-80a}$$

and

$$\sigma_t^2 = \int_0^1 t^2 dF(t) - t_m^2 \tag{9-80b}$$

The values of t_m and σ_t^2 are determined readily from calculations related to the area above the F curve.

9.4.3 Residence Time Distribution Analysis for Ideal Reactors

RTD data can be used to compare the performances of real reactors to those of ideal reactors. We need first to develop the RTD patterns theoretically expected for ideal reactors (i.e., for CMFRs, PFRs, and CMFRs in series), and then to analyze mathematically the responses of these reactors to either step or pulse inputs of nonreactive tracers. The response to either type of stimulus is inherently transient (i.e., unsteady state) because exit concentrations vary with time. This can be seen from inspection of the C, E, and F curves presented in Figures 9-15 and 9-16. In contrast to earlier applications of the material balance principle for steady-state conditions, we must here analyze unsteady conditions.

9.4.3.1 RTD Analysis for CMFRs The response of a CMFR to a pulse input of a nonreactive tracer follows from its material balance relationship:

$$-QC_{\mathrm{OUT}} = V_R \frac{dC_{\mathrm{OUT}}}{dt} \tag{9-81}$$

In contrast to a general material balance for a CMFR (Equation 9-13), Equation 9-81 includes a mass rate term only for the tracer *leaving* the reactor because the tracer is injected into the reactor instantaneously. Further, because the tracer is nonreactive, the reaction rate term is zero. Integration of Equation 9-81 yields the C curve.

The initial condition must be specified to find the appropriate solution to Equation 9-81. For a pulse input, instantaneous injection of tracer mass, M_T, into a completely mixed reactor at $t = 0$ produces an initial concentration of tracer in the reactor, C_Δ, which is also the exit concentration at $t = 0$:

$$\frac{M_T}{V_R} = C_\Delta = C_{\mathrm{OUT}}\big|_{t=0} \tag{9-82}$$

Equation 9-81 can be integrated by separation of variables

$$-\int_{C_\Delta}^{C_{\mathrm{OUT}}} \frac{dC_{\mathrm{OUT}}}{C_{\mathrm{OUT}}} = \int_0^t \frac{dt}{\bar{t}} \tag{9-83}$$

which yields

$$-\ln \frac{C_{\mathrm{OUT}}}{C_\Delta} = \frac{t}{\bar{t}} \tag{9-84}$$

or

$$\frac{C_{\mathrm{OUT}}}{C_\Delta} = e^{-t/\bar{t}} \tag{9-85}$$

The C curve described by Equation 9-85 and illustrated in Figure 9-17 is characterized by an exponential decay of tracer concentration (i.e., a *washout* of the tracer initially injected into the reactor).

A convenient way to test whether a real reactor behaves as an ideal CMFR is to compare the experimentally determined C curve with that predicted for an ideal CMFR. According to Equation 9-84, a plot of experimental data in terms of $\ln(C_{\mathrm{OUT}}/C_\Delta)$ versus t for a reactor that behaves as an ideal CMFR should yield a straight line, the slope of which is the inverse of the HRT ($\bar{t} = V_R/Q$). If the predicted HRT is smaller than the experimentally determined value, some fraction of the reactor volume is most likely not fully active; that is, there is unused volume, or *dead space*, in the reactor.

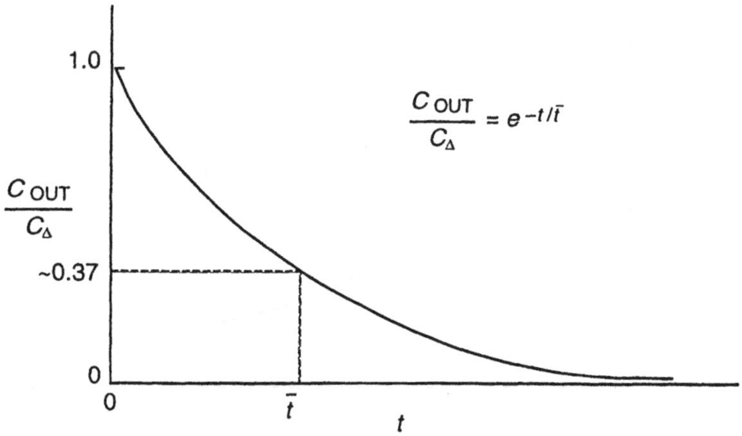

Figure 9-17 *C* curve for an ideal CMFR.

The ***E*** curve is then generated from the ***C*** curve by

$$\boldsymbol{E}(t) = \frac{\boldsymbol{C}\text{ curve}}{\text{area under the }\boldsymbol{C}\text{ curve}} = \frac{C_{\text{OUT}}/C_{\Delta}}{\int_0^{\infty} e^{-t/\bar{t}}\, dt} \tag{9-86}$$

or

$$\boldsymbol{E}(t) = \frac{C_{\text{OUT}}}{C_{\Delta}\bar{t}} \tag{9-87}$$

Substitution of the value for C_{OUT} given by Equation 9-85 yields

$$\boldsymbol{E}(t) = C_{\Delta}e^{-t/\bar{t}}\frac{1}{C_{\Delta}\bar{t}} = \frac{e^{-t/\bar{t}}}{\bar{t}} \tag{9-88}$$

The resulting ***E*** curve shown in Figure 9-18 describes the RTD for a CMFR. The area under the ***E*** curve is

$$\int_0^{\infty} \boldsymbol{E}(t)\, dt = \int_0^{\infty} \frac{e^{-t/\bar{t}}}{\bar{t}}\, dt = 1 \tag{9-89}$$

which agrees with the fundamental description of the RTD as the fractional-exit-age distribution (see Equation 9-67).

Equation 9-88 can be expressed in dimensionless form by introducing Equation 9-74 and assuming that $\bar{t} = t_m$ to give

$$\boldsymbol{E}(\theta^{\circ}) = e^{-\theta^{\circ}} \tag{9-90}$$

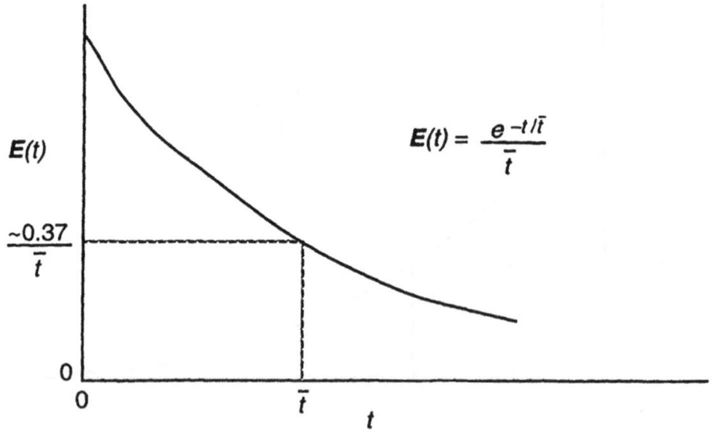

Figure 9-18 *E* curve for an ideal CMFR.

Here again we see the value of dimensionless expressions of reactor properties and characteristics; namely, while different $E(t)$ versus t curves will result for CMFRs of different size (Equation 9-88), all will produce the same dimensionless E curve (Equation 9-90).

Another concept made clearer by RTD analysis is that the *mean constituent residence time*, t_m, for a CMFR is exactly equal to its *mean hydraulic detention time*, $\bar{t}$. The proof follows from the definition of mean residence time given in Equation 9-70 and substitution of $E(t)$ for a CMFR from Equation 9-89,

$$t_m = \int_0^{\infty} t \frac{e^{-t/\bar{t}}}{\bar{t}} \, dt = \bar{t} \tag{9-91}$$

Thus, the mean HRT for a CMFR, which we calculate from two known physical quantities, V_R and Q, is also the mean constituent residence time.

The F curve for a CMFR follows from the material balance relationship:

$$QC_{\text{IN}} - QC_{\text{OUT}} = V_R \frac{dC_{\text{OUT}}}{dt} \tag{9-92}$$

The stimulus is now an instantaneous step input up to a new constant feed concentration, C_{IN}, of tracer; this is the origin of the first term on the left-hand side of Equation 9-92. The initial condition corresponds to the absence of tracer in the reactor. Integration by separation of variables yields

$$\int_0^{C_{\text{OUT}}} \frac{dC_{\text{OUT}}}{C_{\text{IN}} - C_{\text{OUT}}} = \int_0^t \frac{dt}{\bar{t}} \tag{9-93}$$

and

$$-\ln \frac{C_{IN} - C_{OUT}}{C_{IN}} = \frac{t}{\bar{t}} \tag{9-94}$$

or

$$F(t) = \frac{C_{OUT}}{C_{IN}} = 1 - e^{-t/\bar{t}} \tag{9-95}$$

The resulting ***F*** curve for an ideal CMFR is depicted in Figure 9-19. Note from Equation 9-80a that the shaded area above the ***F*** curve in Figure 9-19 is the mean constituent residence time. This expression can also be used to prove that the constituent residence time and HRT are equal for an ideal CMFR.

An experimentally measured ***F*** curve can be tested to determine whether the behavior of a reactor conforms to that of an ideal CMFR by plotting the data according to the logarithmic form of Equation 9-95. Behavior similar to that of a CMFR is indicated if the data fit a straight line; the HRT is then given by the inverse slope of that line.

9.4.3.2 RTD Analysis for PFRs The responses of an ideal PFR to pulse and step inputs of tracer can be reasoned intuitively. Fluid elements do not intermix as they travel down the length of an ideal PFR. The expected response of such a reactor to a pulse input of tracer is thus a spike in concentration at a response time equal to the hydraulic residence time, $\bar{t}$, of the reactor. This ***C*** curve is depicted in Figure 9-20a as a spike of infinite height and zero width, with a unit enclosed area. The expected response of an ideal PFR to a step input is an instantaneous step increase in the exit concentration from zero to a value equal to the feed concentration at time $\bar{t}$. For the ***F*** curve depicted in Figure 9-20b this step increase is from 0 to 1 at $\bar{t}$. A more rigorous mathematical approach utilizing the *Dirac delta function* to describe a pulse input of tracer leads to the same result.

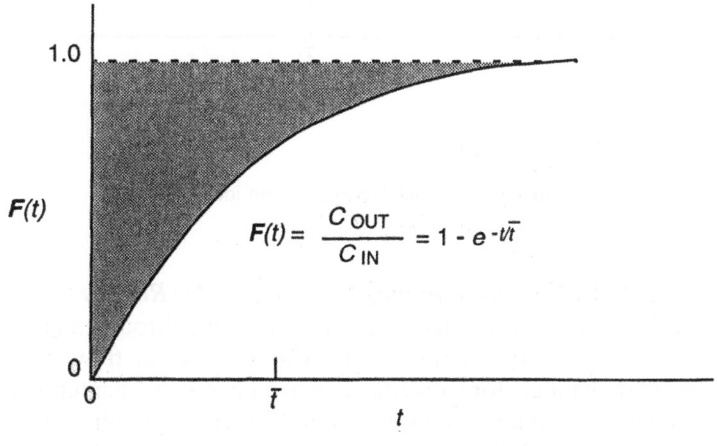

Figure 9-19 ***F*** curve for an ideal CMFR.

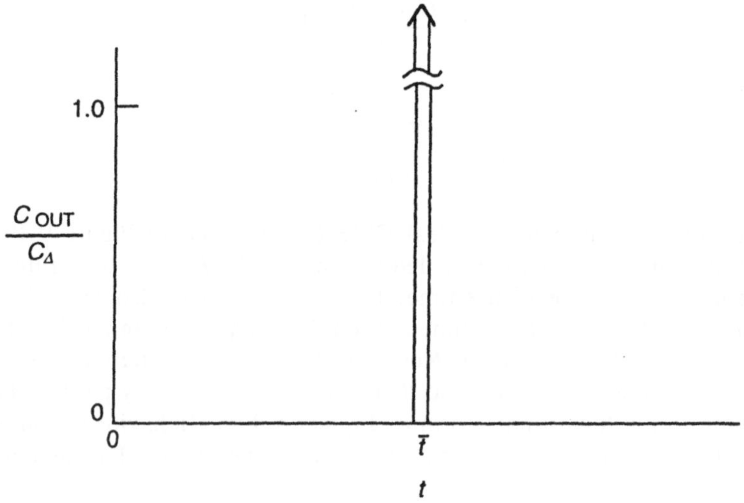

(a) Dimensionless **C** Curve for a PFR

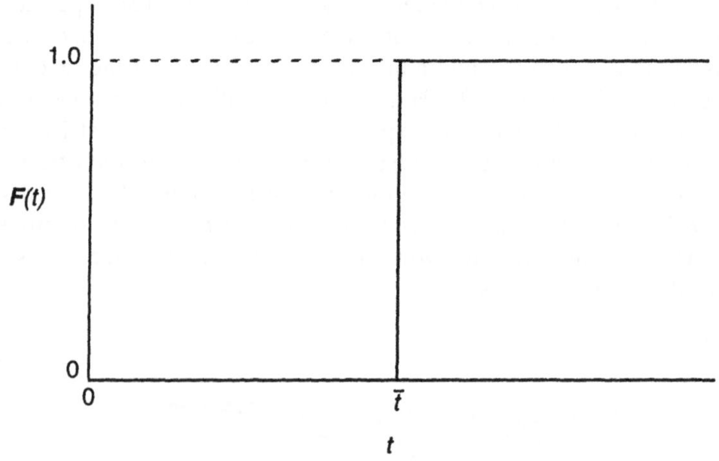

(b) **F** Curve for a PFR

Figure 9-20 *C* and *F* curves for an ideal PFR.

9.4.3.3 RTD Analysis for Laminar Flow in Tubular Reactors When flow through any device is truly laminar, the velocity profile through a cross section is parabolic. This velocity profile was derived for laminar flow through parallel, rectangular channels for subsurface environments in Chapter 3 (Example 3-2) by applying the Navier–Stokes equation. The corresponding velocity profile for laminar flow in tubular reactors can be shown (see Problem 3-7) to give

$$v_r = 2\overline{v}\left(1 - \frac{r^2}{R^2}\right) \tag{9-96}$$

where v_r is the velocity at radial distance r, R is the radius of the tube, and $\overline{v}$ is the average velocity. Because velocity varies with radial position, the residence time of fluid elements is also a function of radial position. The resulting RTD is therefore not the same as that for a PFR, in which all fluid elements spend the same time in the reactor.

The RTD can be derived by analyzing the behavior of a step input concentration, C_{IN}. To simplify the analysis, we will ignore radial diffusion. A material balance on the tracer to account for the fact that its advective rate of transport is a function of radial position gives

$$QC_{IN} = \overline{v}\pi R^2\overline{C} = C_{IN}\int_0^R 2\overline{v}\left(1 - \frac{r^2}{R^2}\right) 2\pi r\, dr \tag{9-97}$$

where $\overline{C}$ is the average concentration in a cross section measured at any length along the tubular reactor, and r is any radial position in that cross section. Integration of Equation 9-97 yields

$$\overline{v}\pi R^2\overline{C} = 4\pi C_{IN}\overline{v}(r)^2\,\frac{2R^2 - (r)^2}{4R^2} \tag{9-98}$$

The velocity, v_r, given in Equation 9-96 can be expressed in terms of distance traveled, x, and time of travel, t, such that radial position, r, can be written as

$$r = R\left(1 - \frac{x}{2t\overline{v}}\right)^{0.5} \tag{9-99}$$

Substitution of Equation 9-99 into Equation 9-98 leads to an expression for the F curve:

$$\frac{\overline{C}}{C_{IN}} = 1 - \left(\frac{x}{2t\overline{v}}\right)^2 \tag{9-100}$$

or, writing the average velocity, $\overline{v}$, as $x/\bar{t}$ gives

$$\frac{\overline{C}}{C_{IN}} = 1 - \left(\frac{\bar{t}}{2t}\right)^2 \tag{9-101}$$

The F curve for laminar flow (Equation 9-101) is plotted in Figure 9-21. Concentration in the exit stream begins to rise at $t = 0.5\bar{t}$ and approaches the feed concentration asymptotically. The definition of the mean constituent res-

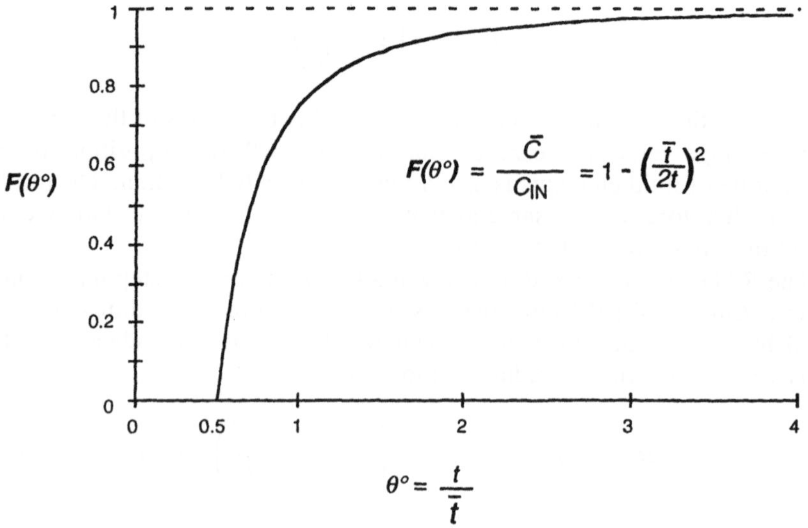

Figure 9-21 *F* curve for laminar flow in a tubular reactor.

idence time is

$$t_m = \int_{0.5\bar{t}}^{\infty} tE(t)\, dt \tag{9-102}$$

The *E* curve is obtained from Equation 9-101 as

$$E(t) = \frac{dF}{dt} = \frac{\bar{t}^2}{2t^3} \tag{9-103}$$

It can be shown by substitution of Equation 9-103 in Equation 9-102 and integration that t_m, the constituent residence time for a laminar flow tubular reactor, equals the HRT, $\bar{t}$. The same conclusion was reached from analysis of the RTD for a CMFR; this further reinforces the meaning of hydraulic residence time.

9.4.3.4 RTD Analysis for CMFRs in Series Mathematical descriptions of response curves for CMFRs in series is somewhat more complex than that for a single CMFR. The general form can be illustrated by evaluating the response of the first of two equal-volume (V_R) CMFRs in series to a pulse input. The material balance around the second reactor is in this case

$$QC_{1,\mathrm{OUT}} - QC_{2,\mathrm{OUT}} = V_R \frac{dC_{2,\mathrm{OUT}}}{dt} \tag{9-104}$$

We know that the exit concentration of the first reactor is given by

$$\frac{C_{1,\mathrm{OUT}}}{C_\Delta} = e^{-t/\bar{t}} \tag{9-105}$$

where $\bar{t}$ is the hydraulic residence time of *each* of the two equal-volume reactors. Substitution of this expression into Equation 9-102 produces a nonhomogeneous ordinary differential equation:

$$\frac{dC_{2,\mathrm{OUT}}}{dt} + \frac{1}{\bar{t}}\, C_{2,\mathrm{OUT}} = \frac{C_\Delta}{\bar{t}} e^{-t/\bar{t}} \tag{9-106}$$

the solution to which is the dimensionless **C** curve:

$$\frac{C_{2,\mathrm{OUT}}}{C_\Delta} = \frac{te^{-t/\bar{t}}}{\bar{t}} \tag{9-107}$$

The general solution for the **C** curve for n equal-volume CMFRs in series is

$$\frac{C_{n,\mathrm{OUT}}}{C_\Delta} = \frac{e^{-t/\bar{t}}}{(n-1)!}\left(\frac{t}{\bar{t}}\right)^{n-1} \tag{9-108}$$

Note again that $\bar{t}$ in this equation is the hydraulic residence time of each CMFR. In a development analogous to that of Equation 9-86, it can be shown that the corresponding **E** curve is

$$E(t) = \frac{e^{-t/\bar{t}}}{\bar{t}(n-1)!}\left(\frac{t}{\bar{t}}\right)^{n-1} \tag{9-109}$$

The **F** curve can also be generated because it is the time-derivative of the **E** curve:

$$F(t) = 1 - e^{-nt/\bar{t}_n}\left\{\sum_{i=1}^{n} \frac{1}{(i-1)!}\left(\frac{nt}{\bar{t}_n}\right)^{i-1}\right\} \tag{9-110}$$

where the total mean hydraulic residence time is $\bar{t}_n = n\bar{t}$.

The dimensionless **E** curve based on the total mean hydraulic residence time is $\bar{t}_n E(t)$. The equation for this **E** curve is obtained from Equation 9-109.

$$E(\theta_n^\circ) = \frac{n(n\theta_n^\circ)^{n-1}}{(n-1)!}\, e^{-n\theta_n^\circ} \tag{9-111}$$

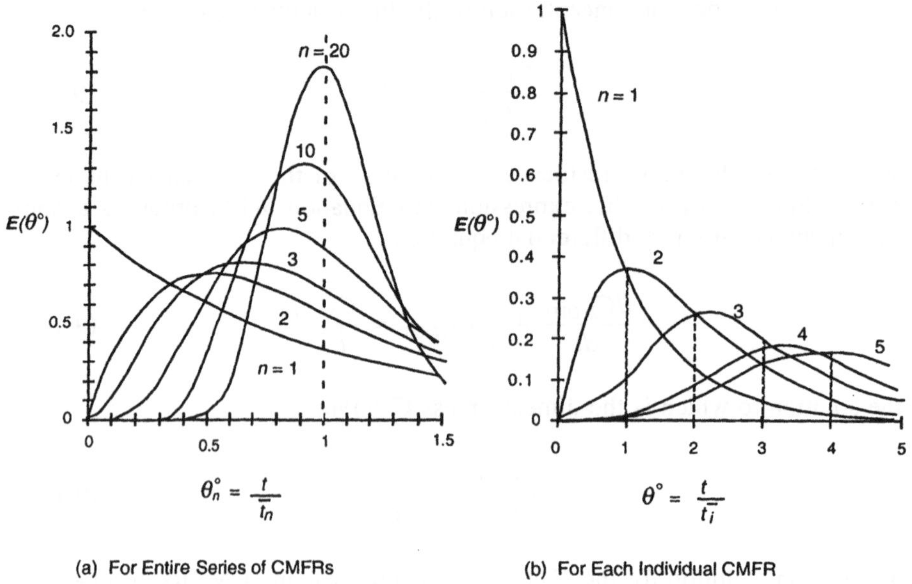
2.0
1.5
1
0.5
0
$E(\theta°)$
$n = 20$
10
5
3
2
$n = 1$
0
0.5
1
1.5
$\theta_n° = \frac{t}{\bar{t}_n}$
(a) For Entire Series of CMFRs
1
0.9
0.8
0.7
0.6
0.5
0.4
0.3
0.2
0.1
0
$E(\theta°)$
$n = 1$
2
3
4
5
0
1
2
3
4
5
$\theta° = \frac{t}{\bar{t}_i}$
(b) For Each Individual CMFR

Figure 9-22 Dimensionless exit age distributions obtained from RTD analysis of CMFRs in series.

where dimensionless time, θ_n°, refers to the series of n reactors, and is given by time, t, divided by $\bar{t}_n$. These $\boldsymbol{E}$ curves are shown in Figure 9-22a for the respective entire series ($\theta_n^\circ = t/\bar{t}_n$) of reactor systems having different numbers (n) of CMFRs in series. A similar set of $\boldsymbol{E}$ curves is shown in Figure 9-22b with θ° based on the mean hydraulic residence time of one reactor ($\theta_i^\circ = t/\bar{t}$). Each $\boldsymbol{E}$ curve is characterized by a rising and a falling limb. This shape arises because of the pulsed nature of the tracer input, which produces washout from each sequential reactor. As n gets larger, the shape of the $\boldsymbol{E}$ curve sharpens and becomes symmetrical about the hydraulic residence time of the reactor, approaching that for an ideal PFR as n approaches infinity.

9.5 PERFORMANCE OF NONIDEAL REACTORS

Most real reactors have RTDs that differ somewhat from those of ideal CMFR or PFR systems of seemingly similar volume, configuration, and flow. The performance of such real reactors thus cannot be determined directly from the design equations presented earlier for ideal CMFR, CMFRs in series, or PFR systems. Three different methods are commonly used to arrive at alternative design equations by utilizing the experimental information gained from RTD analyses. These are (1) the segregated flow model, (2) approximation of observed behavior with a CMFRs in-series model, and (3) application of a modified PFR model that can account quantitatively for observed deviations from true plug flow behavior. The third of these alternatives is similar to the second but uses the opposite extreme of ideal reactor behavior as a starting or reference point from which to make ideal model modifications to account for nonideal behavior.

9.5.1 Segregated Flow Model

In the *segregated flow* or *late mixing* model, we assume that fluid elements of different age remain segregated until they *exit* a reactor. A fluid comprised of such elements is often referred to as a *macrofluid* (i.e., a fluid matrix in which individual flowing elements remain associated with each other in an amorphous but fixed "mesh"). The flow is envisioned as being split into parallel PFRs, which, to give a distribution of ages of the fluid elements, have different HRTs. These different flows are brought together at the end of the reactor in a (hypothetical) process referred to as *late mixing*. It is, admittedly, somewhat difficult to visualize this segregated flow model for fluids having viscosities as low as that of water.

A conceptually similar but physically juxtapositioned alternative to the complete segregation or late mixing model is the *maximum mixedness* or *early mixing model*, wherein molecular scale mixing (*complete micromixing*) is envisioned as occurring at the *inlet* of a reactor. A fluid exhibiting this behavior is thought of as a *microfluid* because individual fluid elements are free to move about relative to one another. This flow concept is in fact identical to that envisioned as providing instantaneous mixing in a CMFR. Segregated flow following initial mixing can be modeled by again splitting the flow into parallel PFRs having different HRTs. Alternatively, partially segregated flow can be modeled by envisioning a nonideal reactor as a series of CMFRs, as discussed in Section 9.5.2.

Unfortunately, RTD analysis is of no assistance in distinguishing the extent of either early or late mixing, as quickly revealed by comparing an RTD for a series combination of a PFR followed by a CMFR (late mixing) to that of a CMFR followed by a PFR (early mixing). These RTDs are shown to be identical (the reader is left to perform this proof in Problem 9-10). Without knowing the extent of microscale mixing from an RTD, it is impossible to know the concentration of reactants within adjacent fluid elements. Without a model that either *predicts* or *simulates the effects of* micromixing, therefore, any reaction in which conversion depends on concentration cannot be predicted accurately.

Fortunately, fractional conversion for first-order reactions is independent of concentration (see Equation 7-17), and one important class of nonideal reactor problems can thus be addressed using the segregated flow model. The reader is asked to explore this point further in Problems 9-11 and 9-12. In this case, tracer data, the $\boldsymbol{E}$ curve, and a first-order rate expression are used to calculate conversion in nonideal reactors. Because the $\boldsymbol{E}$ curve represents the age distribution of fluid elements leaving the reactor, it follows that the average exit concentration is the weighted average of the residual concentrations in fluid elements which have had different amounts of time to react:

$$\overline{C}_{OUT} = \int_0^\infty C_{OUT}\boldsymbol{E}(t)\, dt \tag{9-112}$$

where C_{OUT} is the concentration of that element of fluid of exit age equal to t. The residual concentration within a fluid element spending time, t, and undergoing a first-order reaction in the reactor is then

$$C_{OUT} = C_{IN}e^{-kt} \tag{9-113}$$

Substitution into Equation 9-112 gives

$$\overline{C}_{OUT} = \int_0^\infty C_{IN}e^{-kt}\boldsymbol{E}(t)\, dt \tag{9-114}$$

To reemphasize the limitation of the segregated flow model, prediction of conversion is accurate only for first-order reactions, but this limitation does not preclude its use to approximate performance when other reaction orders are of interest. The maximum error associated with use of this model in such cases is not necessarily large. For example, the difference in estimated conversion for the segregated flow and maximum mixedness model may be only 10% for a second-order reaction. This error can be calculated by comparing the results of the model application in any specific instance with those obtained by applying the maximum mixedness model [see, for example, Chapter 10 in Levenspiel (1972)].

9.5.2 CMFRs-in-Series Model

The performance of CMFRs in series is known to fall between that of single-CMFR and single-PFR systems. Deviations from ideal flow behavior in an arbitrary reactor system can be simulated and phenomenologically quantified by selecting that number of CMFRs in series which best describes the *observed* response of a system to a pulse input of tracer. The CMFRs-in-series approach allows for perfect micromixing within each CMFR but no mixing between individual reactors in the series. In this way the model falls somewhere between the segregated flow and maximum mixedness models. Once again, because fractional conversion for first order reactions is independent of concentration within adjacent fluid elements, the performance can be predicted using the RTD of a CMFR in the segregated flow model given by Equation 9-112 (see Problem 9-12 for this proof).

The CMFRs-in-series model, with its inclusion of microscale mixing, provides a more realistic way to predict conversion for reaction rates that are concentration dependent. It is not restricted to use with first-order rate relationships as is the segregated flow model. Having determined n, the performance of the nonideal reactor can be estimated using the appropriate design equations for CMFRs in series (e.g., Equation 9-23 for steady-state analysis of conversion by a first-order reaction).

There are two ways to determine the best n value for description of an experimental C curve with a CMFRs-in-series model. One is by trial-and-error matching of the theoretical C curves for various numbers of CMFRs in series against the experimentally measured C curve for a nonideal reactor, thus to find the "best" n value. The second approach is by an analysis of tracer data for a pulse input. Beginning with the dimensionless form of the variance given by Equation 9-76 and substituting Equation 9-111 for $E(\theta_n^\circ)$ corresponding to n CMFRs in series, we get

$$\sigma_{\theta_n^\circ}^2 = \frac{n^n}{(n-1)!} \int_0^\infty (\theta_n^\circ)^{n+1} e^{-n\theta^\circ} d\theta_n^\circ - 1 = \frac{1}{n} \tag{9-115}$$

The integral shown in Equation 9-115 is available from a table of standard integrals (i.e., $\int_0^\infty x^n e^{-ax}\, dx = n!/a^{n+1}$). A residence time distribution analysis when represented by $E(\theta_n^\circ)$ therefore provides the value of n directly. An equivalent equation for determining n follows from the definition of dimensionless variance for n reactors in series:

$$\sigma_{\theta_n^\circ}^2 = \frac{\sigma_t^2}{(\bar{t}_n)^2} \tag{9-116}$$

Combining this result with Equation 9-115 gives

$$n = \frac{(\bar{t}_n)^2}{\sigma_t^2} \tag{9-117}$$

Both terms on the right-hand side are known from the C curve.

Once the behavior of a nonideal reactor has been characterized by residence time distribution analysis to be essentially the same as that for n CMFRs in series, its performance is predictable. For example, if a component disappears by a first-order reaction, the exit concentration can be obtained by

$$\frac{C_{n,\mathrm{OUT}}}{C_{\mathrm{IN}}} = \left(\frac{1}{1 + k\bar{t}}\right)^n \tag{9-23}$$

9.5.3 PFR-with-Dispersion Model

A plug flow reactor model that incorporates a hypothetical but empirically (experimentally) quantifiable dispersion term to account for deviation from ideal PFR behavior provides a conceptually satisfying and general way to quantify the fluid mixing characteristics of real reactors. As discussed in Chapter 3, dispersive mass transport is due to intermixing of fluid elements by large-scale eddies. The total flux in one-dimensional transport is the sum of advective and dispersive transport processes:

$$N_x = v_x C - \mathfrak{D}_d \frac{\partial C}{\partial x} \tag{3-74}$$

We can apply this model to characterize the hydraulic response of a nonideal reactor to a pulse input of tracer and then use the characterization to quantify the performance of that reactor with respect to anticipated process efficiency. In the first instance the component of interest is the nonreactive tracer, and its flux is affected only by advective and dispersive mass transport. We can thus write

$$-v_x \frac{\partial C}{\partial x} + \mathfrak{D}_d \frac{\partial^2 C}{\partial x^2} = \frac{\partial C}{\partial t} \tag{9-118}$$

We refer to this as an *ideal plug flow with dispersion reactor* (PFDR) model because the dispersion component, $\mathfrak{D}_d(\partial^2 C/\partial x^2)$, is simply a hypothetical term having the form of Fick's second law added to the ideal PFR model as a means for *phenomenologically* incorporating observed deviations from ideal PFR behavior. It has no particular mechanistic significance because use of the model requires empirical evaluation of the dispersion term. In other words, the dispersion characteristics of a reactor may thus be measured and *correlated* with other system properties, *but* they cannot be related in a rigorous mechanistic way to those other properties.

Both the distance and time variables in Equation 9-118 can be made dimensionless as follows

$$\zeta^\circ = \frac{x}{L} \quad \text{and} \quad \theta^\circ = \frac{v_x t}{L} = \frac{t}{\bar{t}}$$

These dimensionless variables can then be introduced to Equation 9-118, which, after rearrangement, becomes

$$-\frac{\partial C}{\partial \zeta^\circ} + \mathfrak{N}_d \frac{\partial^2 C}{\partial (\zeta^\circ)^2} = \frac{\partial C}{\partial \theta^\circ} \tag{9-119}$$

where $\mathfrak{N}_d$, the *dispersion number*, was defined in Chapter 3 as

$$\mathfrak{N}_d = \frac{\text{dispersive mass transport}}{\text{advective mass transport}} = \frac{\mathfrak{D}_d}{v_x L} \tag{3-83}$$

As the dispersion number increases, transport by dispersion increases relative to transport by advection. It is possible to show that the solution to Equation 9-119 for *infinite dispersion* results in a C curve identical to that for a CMFR. As the dispersion number decreases and approaches zero, the flow characteristics and resulting C curve approach those of a PFR.

The solution of Equation 9-119 yields a description of the concentration of a tracer as a function of time and distance in a *hypothetical reactor* having specific advective and dispersive transport characteristics. Several solutions are possible, depending on the extent of dispersion and the assumed boundary conditions at the inlet and outlet of the reactor. The unsteady-state condition of most interest is the response of a reactor to a pulse input of tracer.

For small amounts of dispersion (i.e., dispersion numbers smaller than ~0.01) a response such as that shown in Figure 9-23 can be envisioned for a pulse input. The mass of tracer begins to spread gradually due to dispersion as it is transported down the reactor by advection. As the mass spreads and travel time, t_i, increases, concentrations produced by the pulse become lower as indicated by the lighter shading in Figure 9-23. If the dispersion number is sufficiently low, the spreading of the tracer curve is small enough that its shape can be considered essentially constant over the measurement interval at the exit end of the reactor, and the tracer curve thus considered essentially symmetrical. Equation 9-119 can be solved assuming the so-called *open condition at the boundaries of the reactor*; that is, there is no transition from plug flow to plug flow superimposed by dispersion at the inlet nor a return to plug flow at the exit. Instead, the following boundary conditions are imposed over an

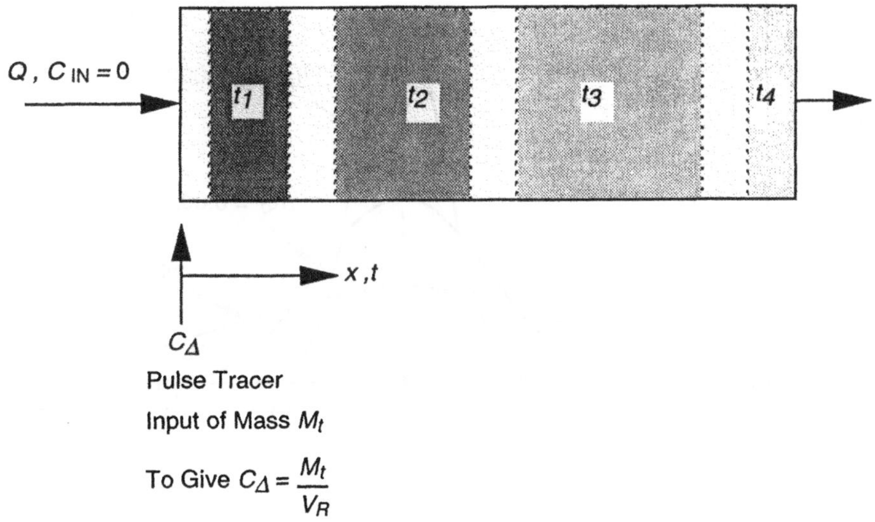

Figure 9-23 Time-series representation of a tracer pulse moving through a nonideal reactor by advection and dispersion.

infinite length of reactor:

$$C = 0 \quad \text{at } x = -\infty \quad \text{for } t \geq 0$$

$$C = 0 \quad \text{at } x = \infty \quad \text{for } t \geq 0$$

For small amounts of dispersion, the analytical solution to Equation 9-119 for dimensionless tracer concentration at position L, corresponding to the end of the reactor, is expressed as a function of dimensionless time by

$$\frac{C_{\text{OUT}}}{C_{\Delta}} = \frac{1}{(4\pi \mathfrak{N}_d)^{0.5}} \exp\left[-\frac{(1-\theta^\circ)^2}{4\mathfrak{N}_d}\right] \tag{9-120}$$

where C_Δ is given by the ratio of M_T, the mass of tracer injected as a pulse, to V_R, the volume of the reactor. Equation 9-120 is thus a dimensionless form of the C curve, solutions for which are presented in Figure 9-24 for different values of the dispersion number. Note that all of these C curves are Gaussian in shape and symmetrical about $\theta^\circ = 1$. This is true, however, only for small amounts of dispersion. The extent of spreading of the C curves is seen to increase with increasing values of $\mathfrak{D}_d$, which is intuitively reasonable.

The spreading of the tracer cloud can be related to the variance of a standard Gaussian function with reference to Figure A.8.1-1. Equation 9-120 has the same form as the Gauss–Laplace distribution functions in Appendix A.8.1,

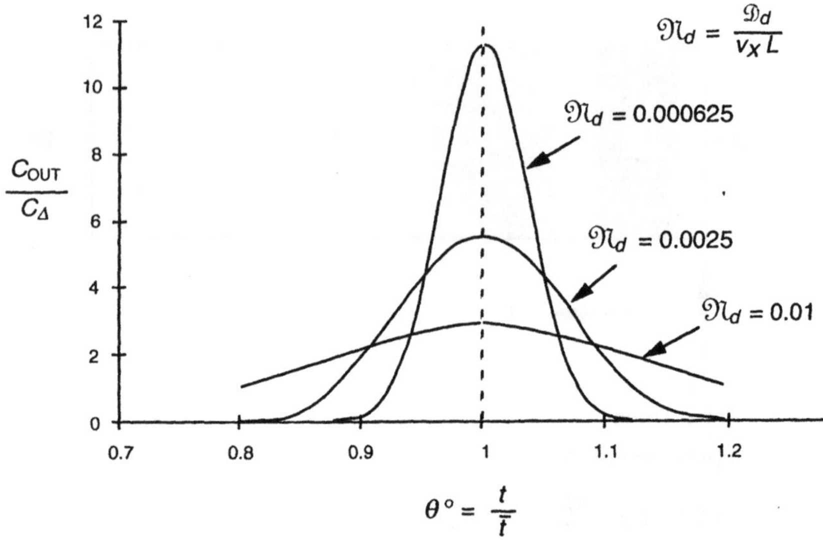

Figure 9-24 Symmetrical form of dimensionless C curves for low dispersion numbers in ideal PFDRs.

and thus we can write:

$$\phi(x) = \frac{1}{(2\pi)^{0.5}\sigma} \exp\left[-\frac{(x - \mu_m)^2}{2\sigma^2}\right]$$

$$f(\theta^\circ) = \frac{C_{\text{OUT}}}{C_\Delta} = \frac{1}{(2\pi)^{0.5}\sigma_{\theta^\circ}} \exp\left[-\frac{(\theta^\circ - 1)^2}{2\sigma_{\theta^\circ}^2}\right] \qquad (9\text{-}121)$$

where the mean value of the distribution is $\theta^\circ = 1$. The Gaussian equation can then be exactly matched to Equation 9-120 by substituting:

$$\sigma_{\theta^\circ}^2 = 2\mathfrak{N}_d \qquad (9\text{-}122)$$

Dimensionless variance is thus a measure of dispersion. In other words, having once determined the first and second moments of the RTD from tracer response data, values of σ_t^2 and t_m^2 can be used to calculate the dispersion number. However, as noted earlier, this procedure is valid only if the amount of dispersion is small and the **C** curve is reasonably close to symmetrical.

For large amounts of dispersion, the RTD is no longer Gaussian and alternative approaches to Equation 9-120 are needed for determining the dispersion number from the variance of the tracer response data. A wider range of dispersion numbers can be described by using different boundary conditions to solve the PFDR model. One such solution is obtained for a finite-length reactor, which is considered to conform to the *open condition* at the inlet but to be described at the outlet by closed condition; e.g., a type 3 boundary condition such as that given in Equation 2-35

$$\left[v_x C - \mathfrak{D}_d\left(\frac{\partial C}{\partial x}\right)\right]_{x=L} = v_x C_{\text{OUT}} \qquad (2\text{-}35)$$

where C represents the concentration of tracer just inside the reactor at the exit. By further assuming that $C = C_{\text{OUT}}$, the boundary condition becomes simply

$$\frac{\partial C}{\partial x} = 0 \quad \text{at } x = L \quad \text{for } t > 0$$

The solution provided over 50 years ago to describe mixing in aeration tanks is

$$\frac{C_{\text{OUT}}}{C_\Delta} = 2\sum_{n=1}^{\infty} u_n \frac{\alpha_{\mathfrak{D}} \sin u_n + u_n \cos u_n}{\alpha_D^2 + 2\alpha_D + u_n^2} \exp\left[\alpha_D - \frac{(\alpha_D^2 + u_n^2)\theta^\circ}{2\alpha_D}\right] \qquad (9\text{-}123)$$

where

$$\alpha_D = 0.5\mathfrak{N}_d^{-1} \qquad \text{and} \qquad u_n = \cot^{-1}\left[0.5\left(\frac{u_n}{\alpha_D} - \frac{\alpha_D}{u_n}\right)\right]$$

Such series solutions are typical when solving the transport equation that includes the derivative of the concentration gradient (i.e., the dispersive transport term). In this instance, the series is formed by the trigonometric roots, u_n, in the expression above.

The ***C*** curves predicted by Equation 9-123 are given in Figure 9-25. When the dispersion number begins to exceed ~0.25, the ***C*** curve begins to skew significantly from the typically symmetrical shape of a Gaussian distribution. The maximum skewness is reached when the dispersion number approaches infinity. This condition, in fact, yields the ***C*** curve for a CMFR, reemphasizing

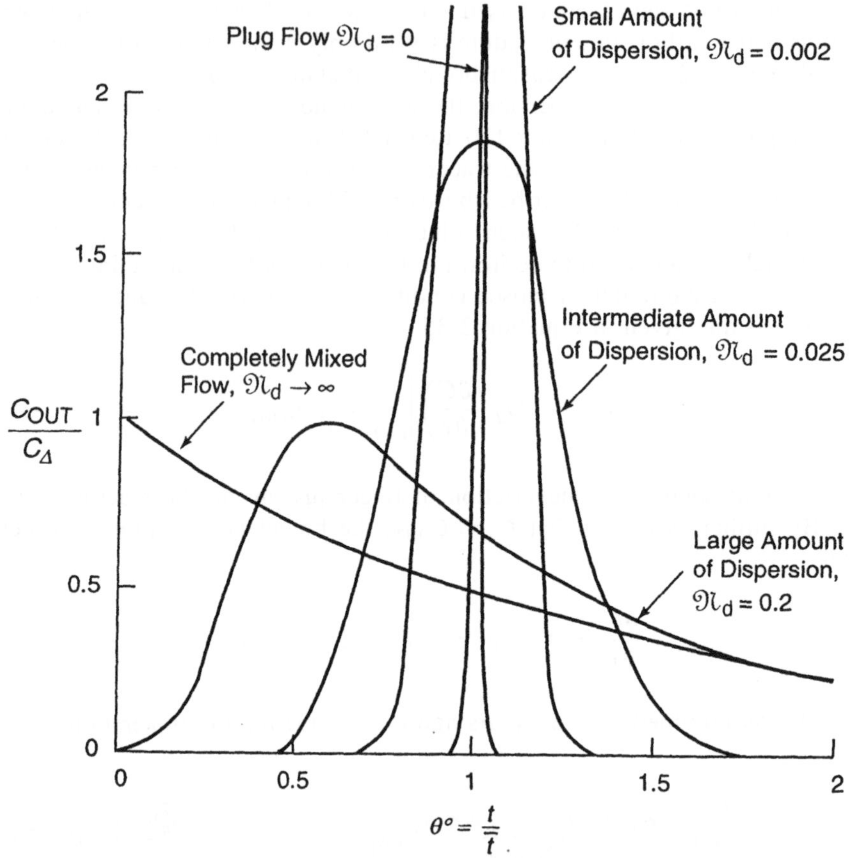

Figure 9-25 Deviations of dimensionless ***C*** curves from symmetry as dispersion numbers for ideal PFDRs increase.

use of the term *complete mixing* to imply *infinite dispersion*. A trial-and-error procedure can be used to fit Equation 9-123 to an experimentally determined **C** curve to determine the dispersion number.

A second approach to analyzing a wider range of dispersion conditions is to solve the PFDR relationships given in Equations 9-118 and 9-119 for a *step input* of tracer. A *closed-form mathematical solution* can be obtained by treating the stimulus as an injection of tracer into an open tube which extends from the point of injection to infinity in both directions, so that the *initial* and *boundary* conditions are:

$$C = 0 \quad \text{for } x > 0 \quad \text{at } t = 0 \qquad C = C_{IN} \quad \text{for } x < 0 \quad \text{at } t = 0$$

$$C = 0 \quad \text{at } x = \infty \quad \text{for } t \geq 0 \qquad C = C_{IN} \quad \text{at } x = -\infty \quad \text{for } t \geq 0$$

where C_{IN} is the step concentration of the tracer. Many other sets of boundary conditions can be assumed, and these lead to varying degrees of mathematical complexity in the solution to the **F** curve (see Example 2-5 for illustrations of two other possibilities). In this particular instance, the length of the dispersion domain is unbounded, and the mathematical solution to Equation 9-118 for determination of concentration as a function of position and time is given by the *error function* (see Chapters 8 and 11 for more on this function). When the position is fixed as the length of the reactor, the specific result is

$$\frac{C_{OUT}}{C_{IN}} = F(t) = 0.5\left\{1 - \text{erf}\left[0.5\, \mathfrak{N}_d^{-0.5}\, \frac{1 - (t/\bar{t})}{(t/\bar{t})^{0.5}}\right]\right\} \qquad (9\text{-}124)$$

The **F** curves predicted for different dispersion numbers are given in Figure 9-26. It can be observed again that the **F** curve broadens as dispersion increases, and in the limit of infinite dispersion is identical to that of a CMFR. A fit of tracer data to this equation provides another method for estimating the dispersion number.

A third alternative utilizes available relationships between $\sigma_{\theta^\circ}^2$ and the dispersion number derived for different reactor entry and exit conditions. We recall from Equation 9-77 that $\sigma_{\theta^\circ}^2$ is calculated from the variance and mean residence time obtained from tracer data. As discussed earlier in this section with regard to analysis and interpretation of **C** curves, two types of reactor entry and exit conditions are commonly used as *boundary conditions* to solve the PFDR model given in Equation 9-119. These lead to description of the reactor as either *closed* or *open* with respect to the conditions existing at its entry and exit.

A *closed* vessel is one in which *plug flow conditions are assumed to exist immediately before and after the vessel entrance and exit sections*, respectively. This could be described by a type 3 boundary condition (Robin condition) given by Equation 2-35 being applied at both the inlet and outlet rather than just at the outlet as used to derive Equation 9-123. This situation might

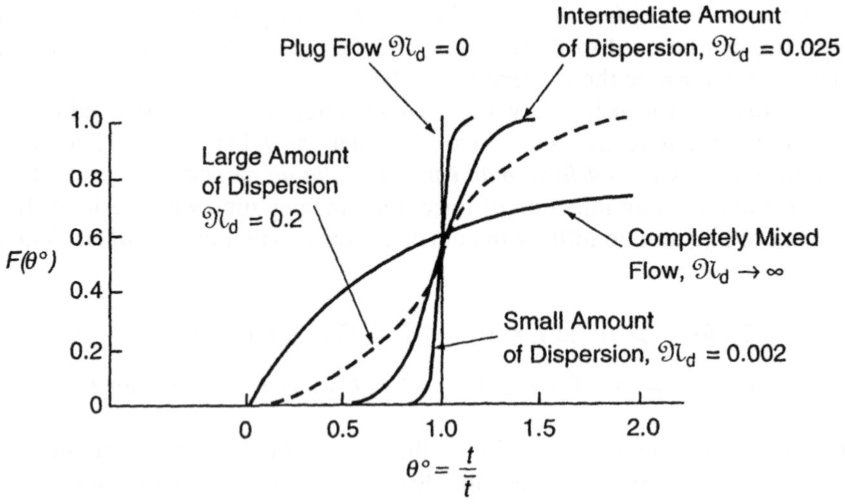

Figure 9-26 Dimensionless ***F*** curves for ideal PFDRs having different dispersion numbers.

occur if flow enters and leaves by a small pipe. An exact analytical solution to Equation 9-119 for these boundary conditions is not possible. A numerical solution of the equation gives the following approximate expression for relating experimentally measured values for $\sigma^2_{\theta°}$ to the dispersion number, $\mathfrak{D}_d/v_x L$:

$$\sigma^2_{\theta°} = 2\mathfrak{N}_d - 2\mathfrak{N}_d^2 [1 - \exp(-\mathfrak{N}_d^{-1})] \qquad \text{(closed vessel)} \qquad (9\text{-}125)$$

In contrast to a closed vessel, an *open* vessel is not characterized by plug flow at either end; it represents a section of a larger vessel in which dispersion is measured. An analytical solution to Equation 9-119 is possible for these boundary conditions, as given by

$$\frac{C_{\text{OUT}}}{C_\Delta} = \frac{1}{(4\pi\theta°\mathfrak{N}_d)^{0.5}} \exp\left[-\frac{(1-\theta°)^2}{4\theta°\mathfrak{N}_d}\right] \qquad (9\text{-}126)$$

The corresponding relationship between $\sigma^2_{\theta°}$ and the dispersion number is

$$\sigma^2_{\theta°} = 2\mathfrak{N}_d + 8\mathfrak{N}_d^2 \qquad \text{(open vessel)} \qquad (9\text{-}127)$$

In solutions to the advection–dispersion equation the dispersion number shown in Equations 9-125 and 9-127 is an implicit function of $\sigma^2_{\theta°}$. Thus, a trial-and-error solution is needed once $\sigma^2_{\theta°}$ has been calculated from an experimental tracer response curve.

Example 9-9 provides an illustration of the application of the three different approaches to nonideal reactor analysis (i.e., segregrated flow model, CMFRs in series model and the PFDR model) by expanding on the situation described earlier in Example 9-8.

Example 9-9

- **Situation:** An RTD analysis was performed in Example 9-8 on a set of filters in a water treatment plant to assess the potential for disinfection of *Giardia* cysts by addition of chlorine ahead of the filters. It is of interest now to evaluate the disinfection process itself in more detail. Disinfection rates are often described by the Chick–Watson equation:

$$-\frac{dC_N}{dt} = kC_{Cl_2}$$

where in this case C_N is the remaining cyst concentration (number/L) and C_{Cl_2} the chlorine concentration (mg/L). This equation, together with the results of a step input tracer test, can be used in an empirical way to estimate the disinfection efficiency of the filters, which behave as non-ideal reactors. Suppose, for example, that t_{10} is the time at which the concentration in the tracer response reaches 0.1 of the step concentration. We can then write the integrated form of the Chick–Watson equation as

$$\frac{C_{N,\text{OUT}}}{C_{N,\text{IN}}} = e^{-(kC_{Cl_2}t_{10})}$$

The equation above would "adjust" the HRT for nonideal behavior by effectively reducing its value to the t_{10} value while still using the PFR design equation. As the RTD of the actual reactor approaches that of an ideal PFR, t_{10} approaches the HRT and the maximum possible disinfection efficiency is achieved. Results of batch disinfection experiments found in the technical literature provide the value of k, which for the pH and temperature of the filter operation is 0.08 L/mg-min. The chlorine concentration applied to the filters is 2 mg/L.

- **Question(s):** Use the pulse input tracer data developed in Example 9-8 to compare the disinfection efficiency predicted from the empirical relationship above and a step tracer test to that predicted by the more rigorous engineering approaches to nonideal reactor behavior we have discussed above.

- **Logic and Answer(s):**

1. The empirical method for calculating *Giardia* inactivation requires measuring the response to a step input [i.e., generation of the $F(t)$ curve]. However, pulse input data and the resulting $E(t)$ curve (Example 9-8) can be used to generate this $F(t)$ curve. From Equation 9-78,

$$F(t) = \int_0^{\infty} E(t)\, dt = \sum_{t=t_1}^{t_n} E_t\, \Delta t$$

2. The spreadsheet from Example 9-8 is extended below [column (3)] to obtain the $F(t)$ curve.

(1) t (min)	(2) E_t ($\text{min}^{-1} \times 10^3$)	(3) $F(t) = \sum_{t=t_1}^{t_n} E_t \, \Delta t$	(4) $\frac{\overline{C}_{N,\text{OUT}}}{C_{N,\text{IN}}} \times 10^3$
0	0.0	0.00	0.00
10	0.67	0.01	1.36
20	10.1	0.11	4.10
30	20.8	0.32	1.71
40	21.5	0.53	0.36
50	18.1	0.71	0.06
60	12.8	0.84	0.01
70	7.4	0.91	0.0
80	4.0	0.95	0.0
90	2.0	0.97	0.0
100	1.3	0.99	0.0
110	1.0	1.00	0.0
120	0.33	1.00	0.0
130	0.0	1.00	0.0
Sums	100.0		7.60

A plot of the $F(t)$ curve shows the expected behavior of the nonideal reactor described in Example 9-8.

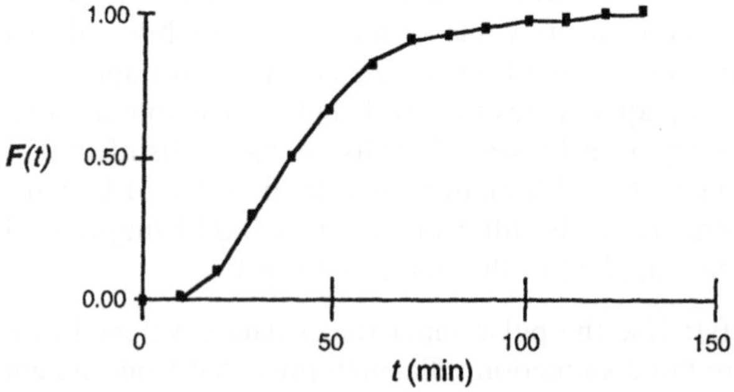

3. The value of kC_{Cl_2} is assumed to remain constant at 0.08×2 or 0.16 min^{-1} even though parallel oxidation reactions probably cause depletion of the chlorine concentration and a lowering of the observed pseudo-first-order rate constant, kC_{Cl_2} as the fluid packets travel through the filter. With this simplification, it is possible to proceed with three alternative calculations of disinfection efficiency.

4. Segregated flow model. Because the disinfection reaction is assumed as pseudo first order, the segregated flow model (Equation 9-114) is appropriate:

$$\overline{C}_{N,\text{OUT}} = \int_0^\infty C_{N,\text{IN}} e^{-kt} E(t) \, dt$$

Fractional removal of *Giardia* cysts is thus given by column (4) of the spreadsheet:

$$\frac{\overline{C}_{N,\mathrm{OUT}}}{C_{N,\mathrm{IN}}} = \sum_{t=t1}^{tn} e^{-kt} E_t \, \Delta t$$

The standard expression for removal is in log inactivation units, that is,

$$\text{log inactivation} = \log \frac{C_{N,\mathrm{IN}}}{\overline{C}_{N,\mathrm{OUT}}} = 2.12$$

5. CMFRs-in-series model. The equivalent number of CMFRs in series is obtained from the tracer data using Equation 9-117:

$$n = \frac{(\bar{t}_n)^2}{\sigma_t^2} = \frac{(46.7)^2}{(19.7)^2} = 5.6$$

To make a conservative estimate of disinfection, the n value is rounded downward to $n = 5$:

$$\frac{(C_{N,\mathrm{OUT}})_n}{C_{N,\mathrm{IN}}} = \frac{1}{(1 + k\bar{t})^n}$$

$$\frac{(C_{N,\mathrm{OUT}})_n}{C_{N,\mathrm{IN}}} = \frac{1}{[1 + 0.16(50/5)]^5} = 0.00841$$

$$\text{log inactivation} = \log \frac{C_{N,\mathrm{IN}}}{(C_{N,\mathrm{OUT}})_n} = 2.07$$

Note that the CMFRs-in-series model agrees very well with the segregated flow model in this case.

6. PFDR model. Assume that the closed vessel condition describes the inlet and outlet to the filters (i.e., there is a transition from advective pipe flow to advection with dispersion within the filter box and again a transition back to advective pipe flow upon leaving the bottom of the filter). The dispersion number is calculated from Equation 9-125

$$\sigma_{\theta^\circ}^2 = 2\mathfrak{N}_d - 2\mathfrak{N}_d^2 \, [1 - \exp(-N_d^{-1})]$$

where

$$\sigma_{\theta^\circ}^2 = \frac{\sigma_t^2}{(\bar{t}_n)^2} = \frac{(19.8)^2}{(46.7)^2} = 0.18$$

A trial-and-error procedure gives $\mathfrak{N}_d = 1.17$. A PFDR in which a first order reaction occurs is described by Equation 9-129

$$\frac{C_{N,\mathrm{OUT}}}{C_{N,\mathrm{IN}}} = \frac{4\beta_D \exp(0.5\, \mathfrak{N}_d^{-1})}{(1 + \beta_D)^2 \exp(0.5\, \beta_D\, \mathfrak{N}_d^{-1}) - (1 - \beta_D)^2 \exp(-0.5\, \beta_D\, \mathfrak{N}_d^{-1})}$$

$$\beta_D = [1 + 4k\bar{t}\mathfrak{N}_d]^{0.5}$$

After substitution for $\mathfrak{N}_d(1.17)$, k (0.16 min^{-1}), and $\bar{t}$ (50 min), we find

$$\frac{C_{N,\text{OUT}}}{C_{N,\text{IN}}} = 0.023$$

and

$$\text{log inactivation} = \log \frac{C_{N,\text{IN}}}{C_{N,\text{OUT}}} = 1.64$$

The PFDR model provides a more conservative estimate of *Giardia* inactivation than either the segregated flow or CMFRs-in-series model.

7. Empirical Model. From the ***F***-curve plot, the t_{10} value is 20 min; thus

$$\frac{C_{N,\text{OUT}}}{C_{N,\text{IN}}} = e^{-k(C_{\text{Cl}_2})t_{10}} = e^{-0.16(20)} = 0.0041$$

$$\text{log inactivation} = \log \frac{C_{N,\text{IN}}}{C_{N,\text{OUT}}} = 1.39$$

8. We conclude that the empirical approach provides a more conservative estimate of *Giardia* inactivation than any of the three nonideal reactor models. These results, however, are dependent upon the nonideality of the reactor (e.g., the n value), the observed disinfection rate constant, and the hydraulic detention time. At some higher n and $k\bar{t}$ value, the empirical model gives more inactivation than predicted by the reactor-based models (Lawler and Singer, 1993).

Our discussion of macrotransport processes in Chapter 3 ended with the intuitive observation that dispersion and fluid turbulence should be related. A general correlation of the dispersion number with Reynolds number for flow of gases and liquids in pipes is presented in Figure 9-27, where it can be noted that the characteristic length in both dimensionless numbers is the pipe diameter. Dispersion increases with the Reynolds number in regions of streamlined flow; that is, microscale mixing causes dispersion to increase. It is also apparent that the Schmidt number is important within this range. Recall from Chapter 4 that the Schmidt number is the ratio of momentum diffusivity to molecular diffusivity. At a given extent of microscale turbulence, therefore, dispersion increases as shear-gradient effects overwhelm molecular concentration-gradient-induced movement. For turbulent flow, Figure 9-27 shows that the dispersion number decreases with increasing Reynolds number due to the now overwhelming effect of macroscale mixing, which tends to level out concentrations along the axial dimensions of the pipe.

Practical interrelationships between the dispersion number and other dimensionless groups having significance for mass transfer and reaction processes are illustrated in Example 9-10 for an FBR activated carbon system.

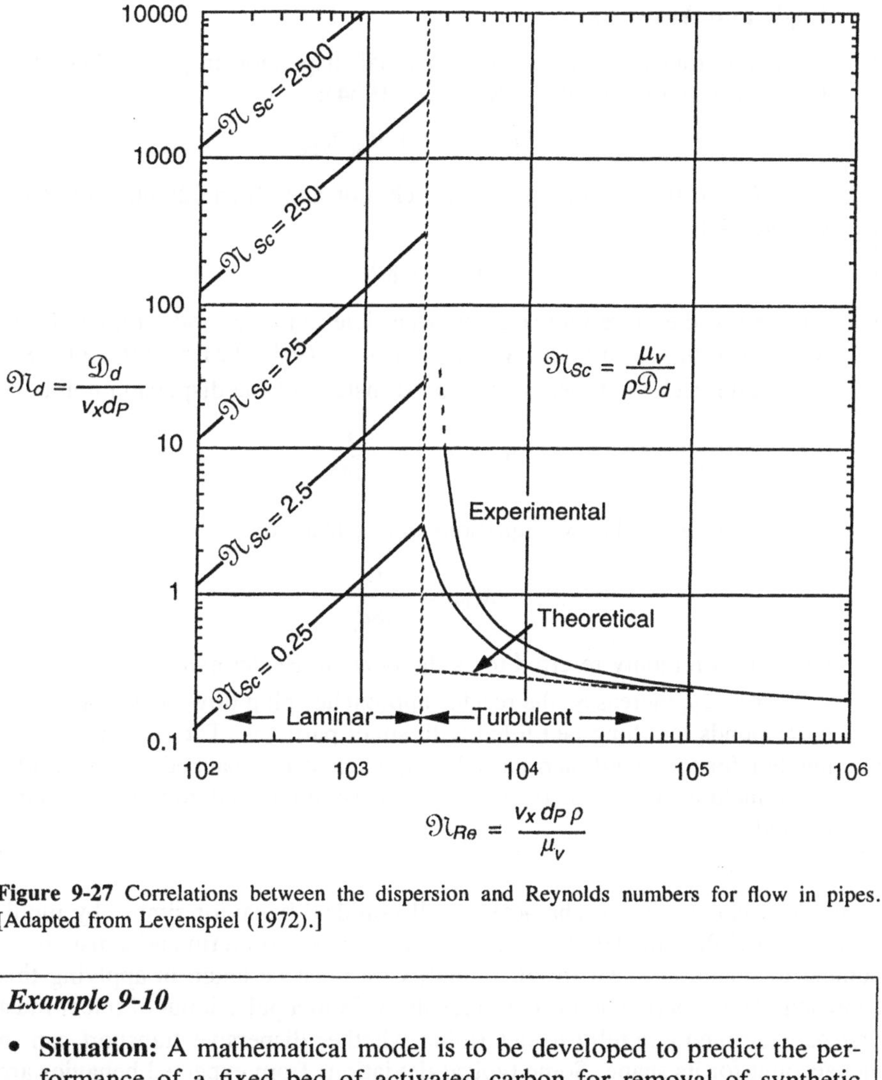

Figure 9-27 Correlations between the dispersion and Reynolds numbers for flow in pipes. [Adapted from Levenspiel (1972).]

Example 9-10

- **Situation:** A mathematical model is to be developed to predict the performance of a fixed bed of activated carbon for removal of synthetic organic chemicals. The complexity of the model depends on how various mass transport steps (advection and dispersion in the bulk water, interfacial mass transfer, and internal diffusion) are considered and on the choice of the adsorption equilibrium model. One important mathematical simplification results if dispersion can be ignored; this also eliminates one of the parameters that must be evaluated empirically during model calibration.

- **Question(s):** Determine whether dispersion is a significant transport mechanism for the following conditions: particle diameter, $d_p = 0.1$ cm; axial velocity, $v_x = 0.3$ cm/s (4.5 gal/min-ft^2); and length, $L = 2.7$ m.

- **Logic and Answer(s):**

1. The following correlation for axial dispersion in packed beds has been given in the literature (Ruthven, 1984):

$$\mathfrak{D}_d = 1.8\mathfrak{D}_l \mathfrak{N}_{Sc} \mathfrak{N}_{Re}$$

2. Given the definitions of the Schmidt and Reynolds numbers (see Chapter 4):

$$\mathfrak{D}_d = 1.8 d_p v_x$$

Dispersion therefore increases with particle diameter and with axial velocity. For this adsorption system, $\mathfrak{D}_d = 1.8(0.1)(0.3) = 0.054\ \text{cm}^2/\text{s}$.

3. The Péclet number is often used instead of the dispersion number:

$$\mathfrak{N}_{Pe} = \frac{1}{\mathfrak{N}_d} = \frac{v_x L}{\mathfrak{D}_d}$$

4. Combining the two equations above leads to

$$\mathfrak{N}_{Pe} = \frac{L}{1.8 d_p}$$

which is commonly referred to as the *particle Péclet number.*

5. As $\mathfrak{N}_{Pe}$ increases, the reactor approaches plug flow. This condition corresponds to a long bed filled with small particles. The particle Péclet number for the length and particle diameter of this packed bed is 1500. We conclude that axial dispersion is very small and thus can be neglected.

While it is tempting to characterize all nonideal behavior simply by modifying an ideal PFR model with a hypothetical dispersion term and a dispersion number to obtain an ideal PFDR model, care must be taken in applying this approach. The experimental C curve resulting from a pulse input of tracer must first be examined carefully to determine whether dispersive transport can in fact account for its shape. Some typical deviations from expected behavior are shown in Figure 9-28. If these deviations cannot be explained by experimental error, dispersion analysis is not appropriate. A C curve like that in Figure 9-28a indicates little deviation from the PFDR model. We note that the mean constituent residence time, $\bar{t}$, is closely approximated by the mean constituent residence time, t_m, which is obtained from analysis of the RTD. However, the response shown in Figure 9-28b shows that $t_m < \bar{t}$ and is characterized by a long tail. Such data are typically obtained for reactors or reaction systems having stagnant backwater regions. Another type of commonly observed deviation from the PFDR model occurs due to internal recirculation. This is suggested by the sequence of declining peaks shown in the response curve in Figure

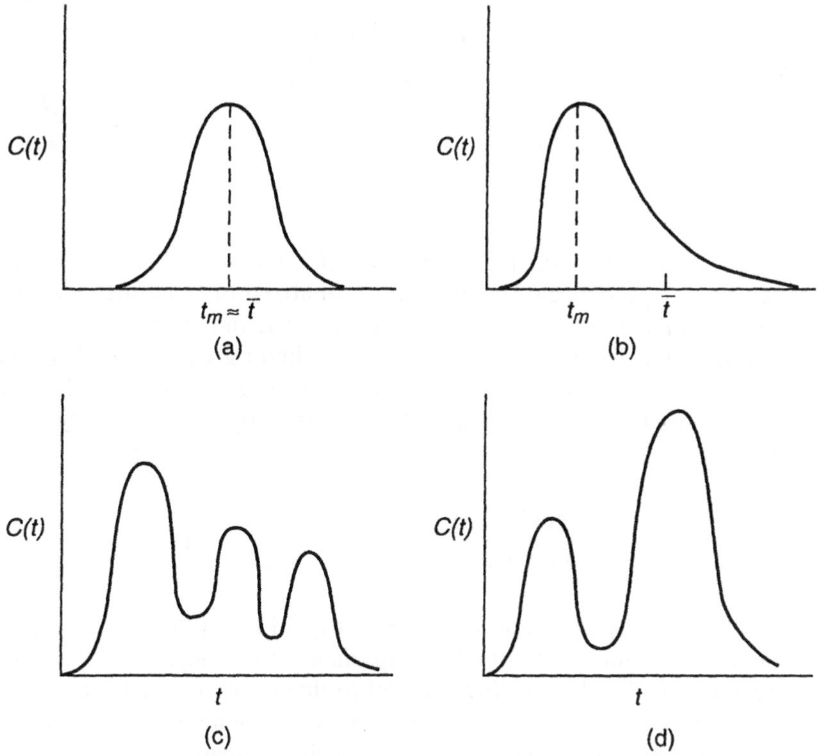

Figure 9-28 Deviations of *C*-curve behavior from ideal PFDR model predictions.

9-28c. Finally, the effect of parallel channeling is depicted in Figure 9-28d; this causes two peaks in the response curve, one due to rapidly moving fluid elements and the other due to slowly moving elements.

A design equation to predict the performance of a PFDR under steady-state operating conditions follows after determination of the dispersion coefficient as described above. Addition of a general reaction rate term in the PFDR model equation gives

$$-v_x \frac{dC}{dx} + \mathfrak{D}_d \frac{d^2C}{dx^2} - kC^n = 0 \tag{9-128}$$

Various solutions to Equation 9-128 can be found for different assumptions concerning the inlet and outlet conditions of the reactor, of infinite or finite length. It has been demonstrated, however, that the following analytical solution is a very reasonable approximation regardless of inlet and outlet conditions:

$$\frac{C_{\text{OUT}}}{C_{\text{IN}}} = \frac{4\beta_D \exp\left(0.5 \dfrac{v_x L}{\mathfrak{D}_d}\right)}{(1 + \beta_D)^2 \exp\left(0.5\beta_D \dfrac{v_x L}{\mathfrak{D}_d}\right) - (1 - \beta_D)^2 \exp\left(-0.5\beta_D \dfrac{v_x L}{\mathfrak{D}_d}\right)} \tag{9-129}$$

where $\beta_D = [1 + 4k\bar{t}(\mathfrak{D}_d/v_x L]^{0.5}$ for a *first-order reaction*. This design equation reveals the dependence of the influent-normalized or fractional concentration remaining on two dimensionless groups, $\mathfrak{D}_d/v_x L$ and $k\bar{t}$. We recognize the former group as the dispersion number, $\mathfrak{N}_d$. The latter group is the *Damkohler group I number*, $\mathfrak{N}_{\text{Da(I)}}$ for a *first-order reaction* (see Chapter 8 and Section III.A in Appendix III). More generally, for bulk flow by advection only, $\mathfrak{N}_{\text{Da(I)}}$ is given by

$$\mathfrak{N}_{\text{Da(I)}} = \frac{\text{chemical reaction rate}}{\text{advective flow rate}} = \frac{r L}{v_x C} = k\bar{t}C^{n-1}$$

The convenient graphical representation of Equation 9-129 shown for a first-order reaction in Figure 9-29 allows comparison of the required volume of a "real" reactor simulated by a PFDR model to that of an ideal PFR to achieve the same extent of reaction (i.e., the same fractional solute concentration leaving the reactor). An equivalent way to express the ordinate scale is

$$\frac{k\bar{t}_{\text{PFDR}}}{k\bar{t}_{\text{PFR}}} = \frac{V_{\text{PFDR}}}{V_{\text{PFR}}} \tag{9-130}$$

The value of $C_{\text{OUT}}/C_{\text{IN}}$ for each value of the Damkohler group I number ($k\bar{t}_{\text{PFDR}}$) is determined from Equation 9-129 for a constant dispersion number. The $k\bar{t}$ values for these same $C_{\text{OUT}}/C_{\text{IN}}$ values in a PFR are then calculated from Equation 9-38. Taking the ratio of the Damkohler numbers leads to the ordinate-scale values in Figure 9-29 and to the resulting plots (solid lines) for each dispersion number. The dashed lines derive from holding $k\bar{t}_{\text{PFDR}}$ constant while determining the relationship between $C_{\text{OUT}}/C_{\text{IN}}$ and the dispersion number from Equation 9-129. A similar plot generated for second-order reactions is shown in Figure 9-30.

The solid lines in Figures 9-29 and 9-30 illustrate that for any specific fractional concentration remaining after reaction, the volume of a real reactor having a specified set of ideal PFDR-simulated properties increases relative to that of an ideal PFR as the dispersion number increases. The performance of a reactor characterized by a specific dispersion number can also be determined. Each intersection of a dashed line on the plot with a solid line gives the con-

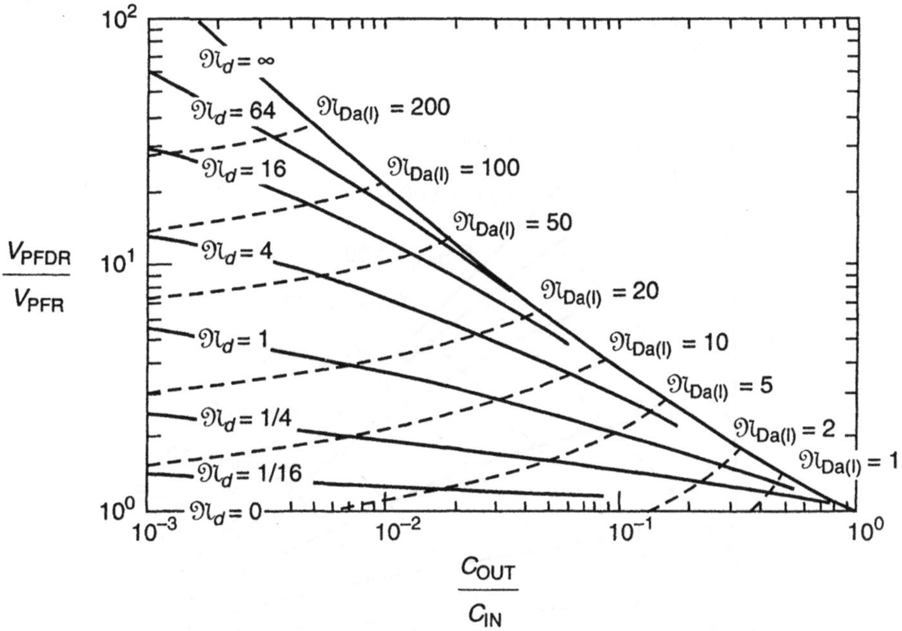

Figure 9-29 Comparison of PFR and PFDR performance characteristics for first-order reactions. [After Levenspiel and Bischoff (1959, 1961).]

version obtained in a PFDR for a unique combination of dispersion number (solid line) and Damkohler number (dashed line). For any dispersion number corresponding to a particular solid line, therefore, the extent of conversion increases as the Damkohler number increases (i.e., as the reaction rate increases relative to advective velocity).

Figures 9-29 and 9-30 emphasize the relative importance of dispersion and reaction rate in determining the effectiveness of engineered reactors for water and waste treatment. In some processes, dispersion is neither necessary nor desirable. For instance, reactor systems for inactivation of microorganisms (disinfection processes) should be designed to minimize dispersion so that inactivation is maximized. In other instances complex mixing may be a necessity, even though the reactor will perform less efficiently; this is true, for example, in air sparging to maintain aerobic biological treatment. A stimulus-response tracer test and a related residence time distribution analysis is essential for determining the phenomenological extent of dispersion in any reactor and for making modifications to reduce the observed dispersion (e.g., by addition of baffles) in cases where it is not desired. An illustration of the sensitivity of process performance to reactor dispersion characteristics is given in Example 9-11 for a set of treatment conditions typical of drinking waters and wastewaters.

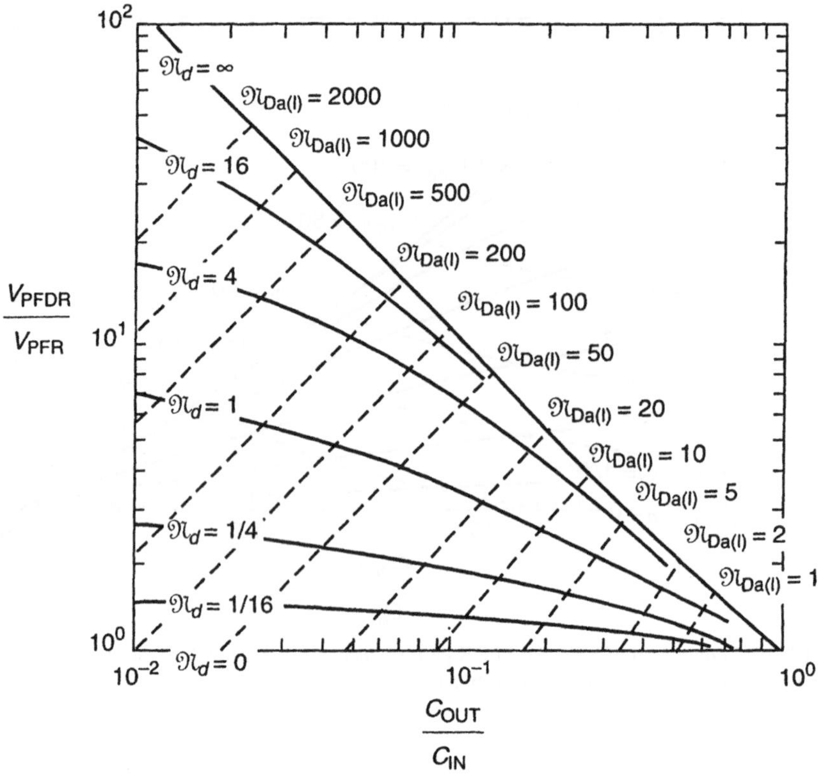

Figure 9-30 Comparison of PFR and PFDR performance characteristics for second-order reactions [After Levenspiel and Bischoff (1959, 1961).]

Example 9-11

- **Situation:** An enzymatically induced partial oxidation of the pesticide diazinon has been shown in laboratory batch studies to reduce its aquatic toxicity. The reaction can be approximated as first order with respect to diazinon concentration over the range of interest. This process is to be applied for treatment of wastewater from a pesticide production operation. A full-scale demonstration study (flow rate of 10,000 m^3/day and hydraulic detention time of 25 min) is then conducted in which the enzyme is added to an existing unbaffled rectangular reactor and 90% removal of diazinon is observed. A tracer study is performed and the dispersion number for the reactor is found to be 1.

- **Question(s):** Regulatory requirements demand 95% removal of diazinon. How can this goal be met with the existing reactor?

- **Logic and Answer(s):**

1. The rate constant for the reaction is not specified, but it can be found from the ideal PFDR model (Equation 9-129) or, more conve-

niently, from the graphical representation of this model given in Figure 9-29.

2. The intersection of $C_{OUT}/C_{IN} = 0.1$ and $\mathfrak{N}_d = 1$ in Figure 9-29 corresponds to $\mathfrak{N}_{Da(I)} = 5$, from which

$$k = \frac{5}{\bar{t}} = \frac{5}{25} = 0.2 \text{ min}^{-1}$$

Note that the ordinate value corresponding to the intersection between C_{OUT}/C_{IN} and $\mathfrak{N}_d$ is 2.2. Thus the real reactor is 2.2 times as large as an ideal PFR capable of achieving the same degree of treatment. The reader is asked to verify the graphical solution by solving Equation 9-129.

3. To increase diazinon removal to 95% without increasing the volume of the existing reactor will require reducing the dispersion. It is possible to calculate the required dispersion number, $\mathfrak{N}_d$, for 95% removal by entering the graph on the abscissa at $C_{OUT}/C_{IN} = 0.05$ and finding the intersection with $\mathfrak{N}_{Da(I)} = 5$. This intersection, which corresponds to a graphical solution to Equation 9-129, yields a dispersion number of 0.25.

4. A logical approach to decreasing the dispersion number from 1 to 0.25 is to install baffles. The success of different baffling arrangements in reducing the dispersion number may be explored on a pilot-plant scale. The dimensions of the existing reactor and flow rate must be properly scaled down in the pilot plant (i.e., dimensional scaling of flow conditions). Tracer tests will then determine the extent to which dispersion is reduced by different baffling schemes.

Although the PFDR model is most often employed for analysis of engineered systems, it may be used also to describe fate and transport of contaminants in natural systems. Caution must be exercised, however, when using a simple one-dimensional steady-state model such as that given by Equation 9-129. Dispersive transport in natural environment systems is often multidimensional; this applies to both surface and subsurface systems. In addition, steady state is established only if the contaminant source term remains constant over time. If these cautions are properly observed, the dispersion and Damkohler numbers can be important and useful dimensionless groups in the modeling and assessment of contaminant fate and transport in natural environmental systems.

9.6 SUMMARY

Material balance, macrotransport, and chemical reaction rate principles developed in previous chapters have been brought together here to describe reactor performance. We have seen that the extent of reaction is determined by the manner in which flow passes through a specific volume defining a "reactor" in any particular case, and by the physical configuration of the reactor volume.

The reactor may be a device in an engineered treatment system or simply a volume segment in a natural system. Complete (perfect) mixing and no intermixing (plug-like flow) of fluid elements provide the boundaries of "ideal" reactor behaviors to which those of real reactors can be compared. They define the minimum and maximum extent of reaction possible for any entrance condition, volume configuration, and exit condition. To quantify deviations in the behavior of real reactors from that of ideal reactors requires that we be able to characterize the residence time distributions (RTDs) of the real reactors, which can be done experimentally by means of tracer tests. RTD analysis is a powerful tool for performance prediction. It leads directly to structured descriptions of nonideal reactor behavior using ideal reactor models, such as those for CMFRs in series or PFDRs.

In a way the reliance on the experimental analysis of macrotransport behavior in a reactor (i.e., the RTD) is analogous to reliance on experimental analysis of transformation behavior (i.e., the development of reaction rate expressions). In both instances, theory provides the framework required to develop a working understanding of an observed process. The experimental observations that can be made, however, usually do not allow a direct theoretical interpretation of the processes. RTD analysis is not sufficient, for example, to allow specification of the extent to which fluid elements of different age mix together to alter local concentrations. This effect is referred to as *micromixing*, which may occur early or late in a reactor. Although not important for first-order reactions, micromixing phenomena and models are of interest in other situations. The reader is referred to the chemical engineering literature for further details.

9.7 REFERENCES AND SUGGESTED READINGS

9.7.1 Citations and Sources

Lawler, D. F., and P. C. Singer, 1993, "Analyzing Disinfection Kinetics and Reactor Design: A Conceptual Approach Versus the SWTR," *Journal of the American Water Works Association*, *85*(11), 67–76. (Example 9-9)

Levenspiel, O., 1972, *Chemical Reaction Engineering*, 2nd Edition, John Wiley & Sons, Inc., New York (Section 9.5.1 and Figure 9-27)

Levenspiel, O., and K. B. Bischoff, 1959, "Backmixing in the Design of Chemical Reactors" *Industrial Engineering Chemistry*, *51*, 1431. (Figures 9-29 and 9-30)

Levenspiel, O., and K. B. Bischoff, 1961, "Reaction Rate Constants May Modify the Effects of Backmixing" *Industrial Engineering Chemistry*, *53*, 313. (Figures 9-29 and 9-30)

Levenspiel, O., and K. B. Bischoff, 1963, *Advances in Chemical Engineering*, *4*, 95. (Equations 9-120, 9-125 to 9-127)

Midler, M., Jr., C. M. Bagner, A. S. Wildman, and E. S. Venkataramani, 1992, "Destruction of Cyanides by Alkaline Hydrolysis in a Pipeline Reactor," *Environmental Progress*, *11*(4), 251–255. (Example 9-4)

Ruthven, D., 1984, *Principles of Adsorption and Adsorption Processes*, John Wiley & Sons, Inc., New York. (Example 9-10)

Thomas, H. A., Jr., and J. E. McKee, 1944, "Longitudinal Mixing in Aeration Tanks," *Water and Sewage Works*, *16*(1), 42. (Equation 9-123)

Wehner, J. F., and R. H. Wilhelm, 1956, "Boundary Conditions for Flow Reactors," *Chemical Engineering Science*, *7*, 187. (Equation 9-120)

9.7.2 Background Reading

Bear, J., 1979, *Hydraulics of Groundwater Flow*, McGraw-Hill Book Company, New York, Chapter 7. Provides convenient solutions for Equation 9-118 using various boundary conditions.

Fogler, H. S., 1992, *Elements of Chemical Reaction Engineering*, 2nd Edition, Prentice Hall, Englewood Cliffs, NJ. A general approach to reactor design, including extensive discussion of micromixing. Classification of ideal reactors is presented in Chapters 1 and 2. Design equations for unsteady-state operation of semibatch (or sequencing-batch) reactors with first- and second-order reactions are presented in Chapter 4. Laboratory reactors for investigating reaction rates are covered in Chapter 5. Residence time distribution analysis and its applications to ideal and nonideal reactors are presented in Chapters 13 and 14. Derivation and examples of the segregation model and maximum mixedness models, CMFRs in series, and dispersion models (with discussion of the Peclét and Damkohler numbers) are provided.

Hill, C. G., 1977, *An Introduction to Chemical Engineering Kinetics and Reactor Design*, John Wiley & Sons, Inc., New York. Design equations for various ideal reactors (including those with recycle streams) are developed in Chapter 8. Mathematical details are given in Chapter 11 for different boundary conditions used with the PFDR model to describe nonideal flow reactors; similar detail is given for the *C* and *F* curves of CMFRs in series.

Levenspiel, O., 1972, *Chemical Reaction Engineering*, 2nd Edition, John Wiley & Sons, Inc., New York. The fundamentals of ideal and nonideal reactors are covered in Chapters 4 and 9. The concepts of fluids mixing (microfluids and macrofluids), degrees of segregation, and early and late mixing are developed in Chapter 10.

Smith, J. M., 1981, *Chemical Engineering Kinetics*, 3rd Edition, McGraw-Hill Book Company, New York. Ideal reactors are presented in Chapter 4 and include detail on recycle reactors (batch-recycle and continuous-flow recycle). Nonideal reactors (residence time distributions, advection–dispersion models, and CMFRs-in-series models) are covered in Chapter 6.

9.8 PROBLEMS

9-1 A CMBR is to provide for disinfection of a well water at a small, remote military installation. Time spent in the CMBR must be sufficient to reduce *Giardia* cysts to 0.001 of their original concentration. The rate of disinfection is given by

$$\frac{dC_N}{dt} = -k_d C_{\mathrm{HOCl}} C_N$$

where C_N is the number concentration (number/L) of *Giardia* and the concentration of the disinfectant, C_{HOCl}, is in mg/L. In a laboratory-scale CMBR, 2 mg/L of HOCl was added to distilled water containing *Giardia* cysts (no inorganic species present) and the fractional *Giardia* concentration remaining ($C_N/C_{N,0}$) after 30 min was 0.027. Although HOCl remained constant in this test, it will not be constant in the field-scale CMBR because simultaneous oxidation reactions occur that involve inorganic constituents (primarily H_2S, Fe, and Mn) in the well water. The rate expression for these reactions is given by

$$\frac{dC_{HOCl}}{dt} = -k_{ox}C_{HOCl}\left(\sum_i^n C_i\right)$$

where the summation term refers to all reduced inorganic species that can be oxidized. In another laboratory-scale CMBR, 2 mg/L of HOCl was added to the well water to be disinfected and 50% of the HOCl disappeared after 20 min; the concentrations of the reduced inorganic species did not decrease substantially due to their presence in excess of the chlorine added. Assume that all HOCl depletion is due to inorganic oxidation reactions and that inorganic species remain in excess.

(a) How much time is required to achieve the treatment goal in the proposed CMBR if the starting HOCl concentration is 5 mg/l?

(b) What is the chlorine residual concentration after this disinfection time?

(c) How much time would be required to achieve the same cyst removal if the inorganic species could be eliminated before adding the disinfectant?

Answers: (a) 46 min; (b) 1 mg/L; (c) 23 min.

9-2 An enzyme treatment process is proposed for partial oxidation of *o*-cresol using a sequencing batch reactor (SBR) configuration. The Michaelis–Menten equation describes the rate:

$$-\frac{dC_{SU}}{dt} = \frac{k_3 C_{EN,0} C_{SU}}{\mathcal{K}_M + C_{SU}}$$

The enzyme is assumed to remain in the reactor after decanting of the treated wastewater. The enzyme concentration at the beginning of each fill cycle is 2 μM, the *o*-cresol concentration is 2 μM, and the volume is 1000 L. The SBR is filled at a rate of 500 L/min for 18 min and the feed *o*-cresol concentration is 10 μM. The half-saturation constant, K_M, is 50 μM, and thus it is assumed that the reaction rate can be approximated by a first-order expression. If the value of k_3 in the Michaelis–Menten kinetic expression is 10 min^{-1}, find the time required in the react cycle to reach 2 μM. [*Hint:* The rate expression in the material balance must include dilution of the initial enzyme concentration as the tank fills. This leads to a different solution than given by Equation 9-11 and can be solved easily in terms of concentration rather than mass of reactant.]

Answers:

$$C_F = \frac{\kappa_2}{\kappa_1}\left\{1 - \frac{[V_R(0)]^{\kappa_1/Q}}{[V_R(0) + Qt]^{\kappa_1/Q}}\right\} + \frac{C(0)[V_R(0)]^{\kappa_1 Q}}{[V_R(0) + Qt]^{\kappa_1/Q}}$$

$$\kappa_1 = Q + \frac{k_3 C_{EN,0}(0)}{\mathcal{K}_M} V_R(0)$$

$$\kappa_2 = QC_{IN}$$

$$t_{reacted} = 25.1 \text{ min}$$

9-3 A water storage tank in a distribution system is filled at a rate of 300 m^3/h. The chlorine concentration of the water entering the tank from the treatment plant is 6 mg/L. Chlorine reactions with organic and inorganic constituents are first order with respect to chlorine and the average rate constant is 0.004 h^{-1}. Find the chlorine concentrations at: (a) the end of filling the storage tank if the volume at the beginning of the fill period is 2000 m^3, the chlorine concentration in that volume is 3 mg/L (due to chlorine demand exerted during previous storage), and the total storage volume is 4000 m^3; (b) the end of the next storage period of 10 h (assume no withdrawal); and (c) as a function of time during withdrawal from the storage tank if the rate of withdrawal is 500 m^3/h.

Answers: (a) 4.46 mg/L; (b) 4.28 mg/L.

9-4 A process used to treat a specific chemical waste is packaged in prefabricated reactor modules of fixed volume. The performance of each reactor module is close to that of an ideal CMFR. The reaction is first order and the existing treatment facility, which consists of one such module, achieves 90% removal. By how much can the volumetric treatment capacity be increased and still achieve 90% removal if a second unit is installed (a) in parallel to the first unit; (b) in series with the first unit? Do a similar comparison for modules having behavior close to that of ideal PFRs.

9-5 A CMFR has an HRT of 10 min. The steady-state feed concentration increases by tenfold from 10 to 100 mg/L.

(a) What is the anticipated increase in effluent concentration if the reaction in the CMFR is first order with a rate constant of 0.1 min^{-1}?

(b) What is the anticipated increase in effluent concentration if the reaction is second order with a rate constant of 0.1 L/mg-min? Discuss the difference between these results.

9-6 The following rate expression describes the oxidation of iron Fe(II) to Fe(III) by oxygen:

$$-\frac{d[\text{Fe(II)}]}{dt} = k[\text{Fe(II)}]P_{O_2}[\text{OH}^-]^2$$

in which $k = 8.0 \times 10^{13}$ L^2/mol^2-min-atm and P_{O_2} is the oxygen pressure. If a groundwater at pH = 6.5 contains 1.0 mg/L of dissolved Fe(II), how long a detention time following aeration must be provided for sufficient Fe(II) oxidation to insoluble Fe(III) to meet the EPA National Secondary Drinking Water Regulation of 0.3 mg/L? Assume that pH remains constant. Justify your choice of reactor type to describe this situation.

9-7 A fluid containing a radioactive element having a very short half-life of 20 h is passed through two CMFRs in series. If the flow rate is 100 L/h and the volume of each reactor 40,000 L, what is the decay in activity?

9-8 An enzyme (with a molecular weight of 40,000 g/mol) is proposed for oxidation of cresote in a hazardous waste treatment facility. A series of CMFR experiments are conducted at a feed enzyme concentation of 4 mg/L. The following Michaelis–Menten rate expression is to be tested:

$$-\frac{dC_{SU}}{dt} = \frac{k_3 C_{EN,0} C_{SU}}{\mathcal{K}_M + C_{SU}}$$

The steady-state exit concentration of substrate (cresote) is measured for four different combinations of HRT and substrate feed concentration, with the following results:

$\bar{t}$ (min)	C_{IN} (mmol/L)	C_{OUT} (mmol/L $\times 10^2$)
2	1.0	8.44
8	1.0	1.41
2	2.0	40.0
6	2.0	4.82

Find the values of k_3 and $\mathcal{K}_M$.

Answers: $k_3 = 10^4$ mmol substrate/mmol enzyme-min; $\mathcal{K}_M = 0.1$ mmol/L.

9-9 Hexachlorocylcohexanes (HCHs) are organochlorine pesticides frequently found in lakes and oceans. HCH compounds have been shown in laboratory tests to undergo a first-order base hydrolysis reaction. Measurements of HCHs in a small impoundment with a hydraulic retention time of 20 days indicate 38% removal. This impoundment behaves like two CMFRs in series. What would be the expected removal in another impoundment of the same pH which has a hydraulic retention time of 90 days and behaves like three CMFRs in series?

Answer: 83%.

9-10 Determine the residence time distribution (RTD) that results when a CMFR with an HRT of $\bar{t}_1$ is followed by a PFR with an HRT of $\bar{t}_2$. Reverse the order of the two reactors and again determine the RTD. Based on these results, can RTD analysis be used to distinguish between early and late mixing in a nonideal reactor?

9-11 Find the effluent concentration for each of the two reactor combinations in Problem 9-10 if the feed concentration is 10 mg/L, $\bar{t}_1$ is 2 min, $\bar{t}_2$ is 4 min, and the reaction is first order with a rate constant of 0.2 min^{-1}. Repeat the determination of effluent concentrations for a second-order reaction with a rate constant of 0.2 L/mg-min. What do you conclude about the importance of early versus late mixing and reaction order?

9-12 An RTD analysis reveals that a reactor behaves as a CMFR. Use this information and the completely segregated flow model (macrofluid behavior) to derive an equation for predicting the fractional removal expected for a first-order reaction. How does this equation compare to that derived from a material balance written for a CMFR in the usual way [i.e., when the exit concentration is taken as the average of the concentration within fluid elements (microfluid behavior)]?

9-13 A step input of fluoride to the clear well of a water treatment plant produces the following results:

Time After Step Input (h)	C_{OUT}/C_{IN}	Time After Step Input (h)	C_{OUT}/C_{IN}
0	0.000	3.00	0.551
0.5	0.000	3.33	0.700
1.00	0.000	3.67	0.800
1.33	0.017	4.00	0.876
1.67	0.106	4.33	0.915
2.00	0.220	4.67	0.957
2.33	0.313	5.00	0.987
2.67	0.431	5.50	1.00

The volume of the clear well is 37,500 m^3 (9.9 mg) and the flow rate during the tracer test was 285,000 m^3/d (75 mgd).

(a) Compare constituent residence time (t_m) values obtained from the F and E curves.
(b) Compare constituent residence time with the HRT.
(c) Calculate the dispersion number assuming a closed vessel configuration.
(d) Calculate the equivalent number of CMFRs in series.

9-14 Chlorine is usually added ahead of the clear well described in Problem 9-13 to provide for disinfection. Chlorine is also consumed in various oxidation reactions such that the chlorine concentration is reduced from 4.6 mg/L entering the clear well to 2.6 mg/L in the exit stream.
(a) Assume that the rate of chlorine demand is first order and estimate the rate constant, k, using the dispersion model.
(b) Repeat the calculation of k for the CMFRs-in-series model.
(c) The so-called C-t_{10} rule for adequate disinfection in U.S. drinking water regulations requires that the product of the exit disinfectant concentration (C) and the time (t_{10}) for 10% of the tracer to exit after a step input exceed 60 mg/L-min (for the pH of this water). Will this regulation be met given the current feed chlorine concentration and the chlorine demand kinetics at flow rates of 285,000 m^3/d and 570,000 m^3/d MGD? State any assumptions you need to make to adjust the estimate of t_{10} for the higher flow rate.

9-15 A regulated organic compound is treated by chemical oxidation in an existing industrial waste treatment facility. The oxidation process is first order with respect to the organic compound, with $k = 0.5\ \text{min}^{-1}$. The reactor in which the process is carried out is characterized by a dispersion number of 0.25 and an HRT of 10 min. Expansion of plant operations are planned which will generate an additional waste stream that must be sent to the same reactor. Through waste minimization, the flow rate of the additional waste stream is 25% of the existing waste stream and the compound concentration is 50% of that in the existing waste stream. In anticipation of this expansion, a tracer test is conducted at the increased flow rate ($1.25Q$), with the following results:

$$\sigma_t = 5.3\ \text{min}$$

$$t_m = 7\ \text{min}$$

Find the fractional change in the effluent concentration [i.e., $(C_{\text{OUT}})_{\text{new}}/(C_{\text{OUT}})_{\text{exist}}$], when the new waste stream is added to the existing waste stream.

9-16 A synthetic organic chemical, A, is found in a river section having a length of 3000 m, a velocity of 0.05 m/s, and a dispersion coefficient of 9.4 m^2/s. This chemical is susceptible to rapid photochemical degradation. The reaction rate at a given light intensity is thought to be first order with respect to the concentration of A, but the rate constant has never been measured. If the concentration is reduced by 98% in this stretch of river, how much decomposition can be expected in a more turbulent reach (with similar photochemical reaction characteristics) which has a length of 1500 m, a flow velocity of 0.08 m/s, and a dispersion coefficient of 30 m^2/s?

9-17 Effluent disinfection by chlorine is to be abandoned in favor of UV irradiation at a wastewater treatment plant to eliminate aquatic toxicity problems related to chlorinated organics. Banks of ultraviolet lamps are installed directly in the existing chlorine contact chamber, which has an HRT of 30 min. After startup, it is found that only a two-log reduction (i.e., $\log (C_{IN}/C_{OUT}) = 2$) in fecal coliform could be obtained, whereas a three-log reduction is needed. A tracer test reveals that $\sigma_t = 19$ min and the constituent residence time (t_m) is 26 min. What is the maximum improvement in disinfection efficiency that can be expected if the UV disinfection rate is first order with respect to fecal coliform concentration and the reactor can be modified with baffles to more closely approach plug flow behavior?

9-18 A state regulatory agency requires that all water treatment plants conduct a step input tracer test on clear wells to determine the t_{10} (i.e., the time for the tracer concentration to reach 10% of the input concentration). The value of t_{10} is multiplied by the exit disinfectant concentration to determine if adequate disinfection is being achieved. Personnel at a particular water treatment plant decided that it is more convenient to use a "negative step tracer input," whereby they interrupted the feed of fluoride to the clear well instead of increasing the feed rate to produce a positive step tracer input. The state regulatory agency is concerned that the test procedure will not yield the same t_{10} as by the conventional procedure. Are the results equivalent?

10

REACTOR ENGINEERING: STEADY-STATE HETEROGENEOUS SYSTEMS

10.1 INTRODUCTION

This chapter deals with the development of process models to describe and design reactors in which components are removed or produced by heterogeneous reactions. These models are developed in much the same way as are those for homogeneous reactions in the sense that they always include characterization of transport processes that affect the movement and distribution of reactive components. Transport is usually described at the *macroscopic* or *reactor* scale when a system is homogeneous. When reactions in heterogeneous systems are involved, however, it is necessary also to consider transport between and within reacting phases (i.e., transport processes at the *microscopic* scale).

Process models for heterogeneous reactors lend themselves to description of both natural and engineered systems. The behavior of many contaminants in surface and subsurface systems is affected by both *interphase* and *intraphase* mass transfer, and both processes must be modeled accurately to provide reliable predictions of species fate and transport. Heterogeneous-phase process modeling has a somewhat different goal in its applications to engineered reactor systems. Here we wish to model interphase and intraphase transport processes to determine contactor designs that maximize rates of mass transfer and thus maximize process rates. Various reactor configurations are possible, each having specific mass transfer properties. The amounts of surface area available for interphase mass transfer and the path lengths for intraphase transport within reacting media are features that contribute to determination of rates of mass transfer. Fluid velocity is also important because it influences the thickness of boundary layers between phases and thus associated impedances to mass transfer.

Examples of certain types of contactor configurations were given in Figure 8-3. Many other configurations are possible for fluid–solid and gas–liquid reactors. Simple illustrations are provided in this chapter to demonstrate applications of reactor principles to heterogeneous systems.

10.2 REACTORS FOR TWO-PHASE PROCESSES INVOLVING FLUIDS AND SOLIDS

Fluid–solid reactions require contact of *liquids* or *gases* with reactive surfaces. The rate relationships presented in Chapter 8 emphasize that observed rates of reaction incorporate the effects of microtransport processes. *External resistances* to mass transfer encountered in boundary layers between fluids and solid surfaces may limit rates at which components arrive at reactive surfaces. Similarly, for microporous and aggregated solids, *internal resistances* to mass transfer may limit the arrival of components at reactive surface sites incorporated within the solids.

10.2.1 Mass Transfer at the External Surfaces of Solid Phases in Fixed-Bed Reactors

A typical fixed-bed reactor is illustrated schematically in Figure 10-1. The "packing" material pictured is comprised by spherical beads, although other shapes are common. A fixed-bed reactor (FBR) is one in which the solid phase is retained within the reactor. The material used to pack the bed may (1) remain packed during operation (a *packed-bed* FBR) if the approach velocity (v_x) of

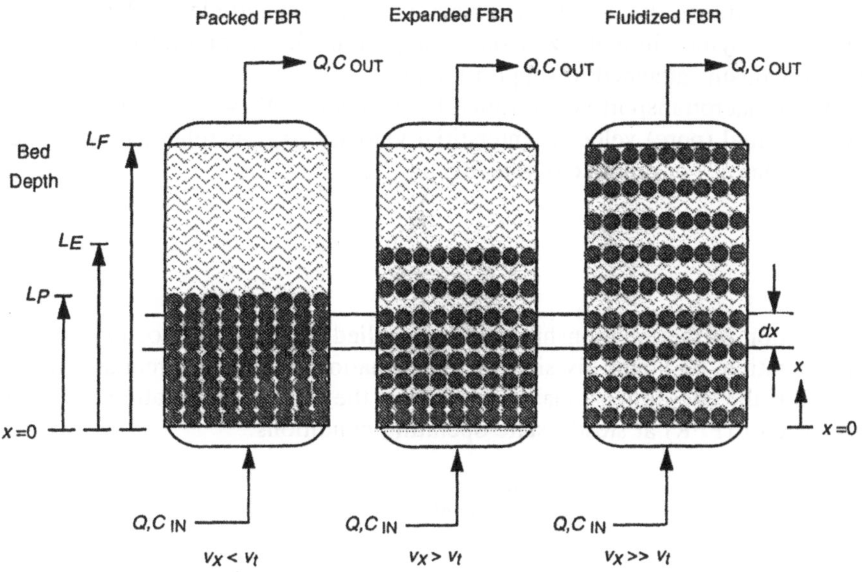

Figure 10-1 Fixed-bed reactor configurations.

the fluid is less than the terminal settling velocity (v_t) of the solid, (2) be expanded (an *expanded-bed* FBR) if $v_x > v_t$, or (3) be fluidized by the flow through the bed (a *fluidized-bed* FBR) if $v_x >> v_t$. The terms L_P, L_E, and L_F in Figure 10-1 represent the respective heights (depths) of the packed bed of solids, the expanded bed of solids, and the fluidized bed of solids. Regardless of which of these packing conditions prevails, the one-dimensional (x) form of the steady-state material balance for a differential control volume defined by $A_R\, dx$ is

$$dN_x \epsilon_B A_R + N^\circ (a_s^\circ)_{E,R} A_R\, dx = 0 \tag{10-1}$$

where N_x is the macrotransport flux along the length of the reactor, ϵ_B is the bed porosity, N° is the microtransport flux at and normal to the surface of the spheres, $(a_s^\circ)_{E,R}$ is the specific external surface area available for reaction *per volume of reactor*, and A_R is the cross-sectional area of the reactor. For spheres of diameter, d_p,

$$(a_s^\circ)_{E,R} = \frac{6(1 - \epsilon_B)}{d_p} \tag{10-2}$$

We note that Equation 10-1 is predicated on an assumption of steady-state conditions, which exist in an FBR *only* if the surface reaction is truly catalytic in nature (i.e., the net rate of reaction at any location does not change with time). This is the case, for example, for gas-phase or dissolved contaminants undergoing transformation upon contact with catalyst particles such as titanium dioxide, enzymes immobilized on a support medium, or biocatalysts (i.e., microorganisms attached to support media).

When macrotransport is controlled by advection, $N_x = v_{p,x}C$, where $v_{p,x}$ is the interstitial (pore) velocity. Interstitial velocity, $v_{p,x}$, is related to the superficial ("Darcy" or approach) velocity, $v_{s,x}$, by

$$v_{p,x} = \frac{v_{s,x}}{\epsilon_B} \tag{10-3}$$

Substitution of this relationship and the implied superficial velocity $v_x = v_{s,x}$ into Equation 10-1 and division of that equation by the differential volume, dx, yields the one-dimensional point form of the continuity equation for a plug flow reactor (PFR) at steady-state operating conditions:

$$-v_x \frac{dC}{dx} + N^\circ(a_s^\circ)_{E,R} = 0 \tag{10-4}$$

The flux *normal* and *immediately adjacent* to the surface of the sphere, N°, can be described by the film model for *interphase* mass transfer:

$$N^\circ = k_f(C - C^\circ) \tag{10-5}$$

where C and C° are the concentrations in the bulk solution and immediately adjacent to the surface, respectively. Substituting the expression above into the material balance relationship given in Equation 10-4 yields

$$-v_x \frac{dC}{dx} + k_f(a_s^\circ)_{E,R}(C - C^\circ) = 0 \tag{10-6}$$

To solve Equation 10-6 it is necessary to relate C° to C. At steady state, the flux at (and normal to) the surface must equal the rate of surface reaction. It was shown in Chapter 8 for a first-order surface reaction ($r^\circ = -k^\circ C^\circ$), for example, that

$$C^\circ = \frac{k_f C}{k_f + k^\circ} \tag{8-7}$$

When external resistance controls (i.e., $k_f << k^\circ$ and $C^\circ << C$), we can approximate $(C - C^\circ)$ in Equation 10-6 by C. This allows separation of variables and integration to yield

$$\frac{C}{C_{IN}} = \exp\left[-\frac{k_f(a_s^\circ)_{E,R}}{v_x}\right]x \tag{10-7}$$

where C_{IN} is the bulk phase concentration entering the reactor and C is the bulk phase concentration at any distance x along the reactor length. For a reactor of length L, we note that

$$\frac{L}{v_x} = \bar{t} \tag{10-8}$$

where $\bar{t}$ is again the mean hydraulic residence time (HRT). In the case of reactors containing solids, such as fixed-bed reactors, the mean HRT for the reactor if it were empty of solids is referred to as the *empty bed contact time* (EBCT). The extent of reaction given by Equation 10-7 can then be expressed as

$$\frac{C_{OUT}}{C_{IN}} = \exp[-k_f(a_s^\circ)_{E,R}\bar{t}] \tag{10-9}$$

A similar analysis can be made even when external resistance to flux at the surface is not entirely rate controlling, as long as k_f and k° are of similar magnitude. Equation 8-7 shows that C° would then be a significant fraction of the bulk concentration, C. Substitution for C° from Equation 8-7 into Equation 10-6 leads to

$$\frac{C_{OUT}}{C_{IN}} = \exp\left[-\frac{k_f k^\circ (a_s^\circ)_{E,R}\bar{t}}{k_f + k^\circ}\right] \tag{10-10}$$

The foregoing analysis can be extended to nonideal reactors, wherein macrotransport along the length of the reactor occurs by both advection and dispersion. The material balance relationship given by Equation 10-1 still holds, but the flux, N_x, along the length of the reactor is now the sum of advection and dispersion, yielding the steady-state plug flow with dispersion reactor (PFDR) model described in Chapter 9:

$$-v_x \frac{dC}{dx} + \epsilon_B \mathfrak{D}_d \frac{d^2C}{dx^2} + k_f(a_s^\circ)_{E,R}(C - C^\circ) = 0 \tag{10-11}$$

Substituting for C° for a first-order surface reaction and defining a dimensionless length ($\zeta^\circ = x/L$) yields

$$-\frac{dC}{d\zeta^\circ} + \frac{\epsilon_B \mathfrak{D}_d}{v_x L}\frac{d^2C}{d(\zeta^\circ)^2} + \frac{k_f(a_s^\circ)_{E,R}k^\circ}{k_f + k^\circ}\frac{L}{v_x}C = 0 \tag{10-12}$$

Two dimensionless groups emerge: the *dispersion number*, $\mathfrak{N}_d = \epsilon_B \mathfrak{D}_d / v_x L$, and the *product of the observed rate constant* and the *empty-bed hydraulic residence time*, L/v_x. The solution to this ordinary differential equation has a form similar to that given in Equation 9-128 for a first-order homogeneous reaction:

$$\frac{C}{C_{\text{IN}}} = \frac{4\beta_D \exp\left(0.5 \dfrac{v_x L}{\epsilon_B \mathfrak{D}_d}\right)}{(1 + \beta_D)^2 \exp\left(0.5\beta_D \dfrac{v_x L}{\epsilon_B \mathfrak{D}_d}\right) - (1 + \beta_D)^2 \exp\left(-0.5\beta_D \dfrac{v_x L}{\epsilon_B \mathfrak{D}_d}\right)} \tag{10-13}$$

The differences between the two lie in the definitions of the rate constants and the HRTs in the dimensionless groups that appear in β_D. For a reaction that is completely controlled by external mass transfer,

$$\beta_D = \left[1 + 4k_f(a_s^\circ)_{E,R}\bar{t}\left(\frac{\epsilon_B \mathfrak{D}_d}{v_x L}\right)\right]^{0.5} \tag{10-14}$$

and for a reaction in which external mass transfer is significant but not entirely rate controlling,

$$\beta_D = \left[1 + 4\left(\frac{k_f(a_s^\circ)_{E,R}k^\circ}{k_f + k^\circ}\right)\bar{t}\left(\frac{\epsilon_B \mathfrak{D}_d}{v_x L}\right)\right]^{0.5} \tag{10-15}$$

The models above are based on steady-state conditions, which, as noted earlier, can exist for catalytic surface reactions as long as the surface concentration remains constant over time at every location in the reactor. This circumstance is illustrated in the model applications described in Example 10-1 for a catalytic oxidation process. Conversely, for *sorption processes* in fixed-bed reactors, steady state cannot obtain because sorption leads to *accumulation* of sorbate on the surface, and the *surface concentration* thus *changes in time*. For FBR adsorbers, the exit concentration of the reactor increases in time, eventually reaching the feed concentration as the surface becomes saturated. The analysis of this type of unsteady condition is discussed in Chapter 11.

Example 10-1

- **Situation:** A UV-light-activated catalyst material is being developed for oxidation of atrazine, a pre-emergent herbicide occasionally found in raw water supply sources. The surface reaction rate (mol/cm^2-s) has been found in CMBR experiments to follow a first-order model with a rate constant, k°, of 0.01 cm/s. The presence of certain other constituents in the water, however, can cause catalyst poisoning, which reduces k° to 0.001 cm/s. Although the catalyst material is available only as very fine powder, it can be coated onto an inert support medium comprised by spherical particles having diameters in the range 0.8 to 1.5 mm. The proposed design is a fixed-bed reactor. A first-cut analysis of reactor design is needed to determine the effect of reactor hydrodynamics, support-medium size, and catalyst poisoning on the empty-bed contact time (EBCT) required to achieve 95% removal of atrazine.

- **Question(s):** Develop a general approach for analysis of the proposed reactor system.
- **Logic and Answer(s):**

1. An ideal plug flow reactor assumption would yield the lowest estimate of required EBCT, but attainment of ideal plug flow under these circumstances is not realistic. Reactor hydrodynamics will be influenced by the size of the support media and by the interstitial velocity. A PFDR model would thus appear more reasonable for the analysis. To assess the differences between PFR model predictions and those given by incorporation of dispersion effects (i.e., a PFDR model), we will perform both analyses.

2. The PFR model is considered first. To begin the analysis, we need to determine the external mass transfer coefficient, k_f, by way of a mass transfer correlation. A number of such correlations exist for a variety of different conditions (see, for example, Table 4-2 and Appendix A.4.1), one of which is the Gnielinski correlation (Appendix A.4.1):

$$\mathfrak{N}_{\text{Sh}} = (2 + 0.644\, \mathfrak{N}_{\text{Re}}^{0.5} \mathfrak{N}_{\text{Sc}}^{0.333})[1 + 1.5(1 - \epsilon_B)]$$

3. We assume or determine the following values for this application of the Gnielinski correlation:

$\epsilon_B = 0.4$ (typical value)

$d_p = 0.8$ mm (measured)

$v_x = 4$ m/h (typical value)

$\mathfrak{D}_l = 6.3 \times 10^{-6}$ cm^2/s (obtained from Wilke–Chang correlation)

$\nu_v = 8.96 \times 10^{-3}$ cm^2/s (given)

4. For these conditions, the Reynolds number is

$$\mathfrak{N}_{\text{Re}} = \frac{d_p v_{p,x}}{\nu_v} = \frac{d_p(v_x/\epsilon_B)}{\nu_v} = 2.5$$

the Schmidt number is

$$\mathfrak{N}_{\text{Sc}} = \frac{\nu_v}{\mathfrak{D}_l} = 1422$$

and the Sherwood number is determined from the Gnielinski correlation to be

$$\mathfrak{N}_{\text{Sh}} = 25.4$$

5. The mass transfer coefficient is then calculated as

$$k_f = \frac{\mathfrak{N}_{\text{Sh}}\mathfrak{D}_l}{d_p} = 2 \times 10^{-3} \text{ cm/s}$$

6. The steady-state effluent concentration relationship for a PFR with first-order reaction at the surface of the solid is given by Equation 10-10:

$$\frac{C_{\text{OUT}}}{C_{\text{IN}}} = \exp\left[-\frac{k_f k^\circ (a_s^\circ)_{E,R}\bar{t}}{k_f + k^\circ}\right]$$

Rearranging to solve for the empty-bed contact time (EBCT) in an ideal fixed-bed PFR gives

$$\bar{t} = \frac{k_f + k^\circ}{k_f k^\circ (a_s^\circ)_{E,R}} \ln \frac{C_{\text{IN}}}{C_{\text{OUT}}}$$

where $(a_s^\circ)_{E,R} = 6(1 - \epsilon_B)/d_p = 45$. Thus, the EBCT for 95% removal is

$$\bar{t} = \frac{0.002 + 0.01}{0.002 \times 0.01 \times 45} \ln 20 = 40 \ s$$

Changing the approach velocity and particle diameter will change k_f and $(a_s^\circ)_{E,R}$.

7. Incorporation of the effect of dispersion in the packed bed requires that Equations 10-13 and 10-15 for steady-state conditions in a PFDR model be used:

$$\frac{C}{C_{\text{IN}}} = \frac{4\beta_D \exp\left(0.5 \dfrac{v_x L}{\epsilon_B \mathfrak{D}_d}\right)}{(1 + \beta_D)^2 \exp\left(0.5\beta_D \dfrac{v_x L}{\epsilon_B \mathfrak{D}_d}\right) - (1 + \beta_D)^2 \exp\left(-0.5\beta_D \dfrac{v_x L}{\epsilon_B \mathfrak{D}_d}\right)}$$

$$\beta_D = \left[1 + 4\left(\frac{k_f (a_s^\circ)_{E,R} k^\circ}{k_f + k^\circ}\right) \bar{t}\left(\frac{\epsilon_B \mathfrak{D}_d}{v_x L}\right)\right]^{0.5}$$

An implicit solution technique is needed to find the EBCT for a given percentage removal. Alternatively, a convenient graphical form of the solution is provided in Figure 9-29. The rate constant required to use the graph is given by

$$k = \frac{k_f k^\circ (a_s^\circ)_{E,R}}{k_f + k^\circ} = 0.075 \text{ s}^{-1}$$

8. It would be necessary in practice to determine the dispersion number experimentally for the packed reactor at different flow rates. To illustrate the effect of dispersion, we assume that the dispersion number is 1. With the residual fractional concentration being 0.05, Figure 9-29 shows that $k\bar{t} \approx 8$, and then

$$\bar{t} = \frac{8}{0.075} = 107 \text{ s}$$

The effect of dispersion is thus to increase the EBCT from 40 s to 107 s.

9. The effect of superficial velocity, particle diameter, and catalyst poisoning (i.e., a decrease in $k°$) on the EBCT can be explored for both the PFR and PFDR models using the approach shown above. For simplicity, we assume that bed porosity, ϵ_B, does not change with particle diameter. The results are shown in the table below, in which the entries are EBCTs in seconds.

	PFR Model		PFDR Model	
Velocity and Particle Diameter	Normal Catalyst, $k° =$ 0.01 cm/s	Poisoned Catalyst, $k° =$ 0.001 cm/s	Normal Catalyst, $k° =$ 0.01 cm/s	Poisoned Catalyst, $k° =$ 0.001 cm/s
$v_x = 4$ m/h				
$d_p = 0.8$ mm	40	100	107	258
$d_p = 1.5$ mm	55	115	141	297
$v_x = 8$ m/h				
$d_p = 0.8$ mm	32	92	82	236
$d_p = 1.5$ mm	42	102	109	264

10. The effect of superficial velocity is to increase the Reynolds number, and the mass transfer coefficient, k_f, is thus increased. An increase in k_f decreases the required EBCT, but the amount is determined by the extent of rate control offered by external resistance. When the catalyst is not poisoned, the external resistance is almost entirely rate controlling ($k_f << k°$) and a doubling of superficial velocity reduces the EBCT by 20 to 24% (depending on particle diameter and type of reactor). The lower rate constant associated with the poisoned catalyst causes rate control to shift to the surface reaction. Thus, doubling the superficial velocity reduces the EBCT by only about 8 to 11%. Particle size affects the magnitude of the Reynolds number and the specific external surface area available for reaction, $(a_s°)_{E,R}$. The net effect is that increasing the particle diameter increases the EBCT for all situations examined in the table. The effect of catalyst poisoning is to increase the EBCT. The increase in EBCT is greater at the higher superficial velocity, because k_f has increased and rate control has shifted more toward the surface reaction.

10.2.2 Mass Transfer Internal to Solid Phases in Fixed-Bed Reactors

The reactive surfaces of microporous sorbents, ion-exchange media, and heterogeneous catalysts are contained for the most part within individual particles. The internal surface areas of such materials are typically 300 to 1000 m^2/g, while their external surface areas are generally small by comparison. For a particle having a diameter of 0.2 cm and a density of 1.5 g/cm^3, for example, the external surface area is only ~ 30 cm^2/g. The extent of reaction per unit volume of particle, and thus per unit volume of a fixed-bed reactor, can therefore be much greater for porous materials than for nonporous materials.

Porous catalysts have been used for years in a variety of industrial processes, and a number of porous materials that catalyze specific oxidation or reduction reactions are commercially available for water and wastewater treatment applications. Many important fixed-bed reactor applications of microporous media that *inherently* involve catalysis-type reactions are found in water and wastewater treatment processes. Activated carbon adsorbents with microorganisms attached to their external surfaces, for example, comprise biological catalysts. In this case substrates adsorbing from solution undergo biodegradation while diffusing through the assemblage of attached microorganisms. The *adsorption and oxidation* of free and combined chlorine on the surfaces of activated carbon provides another example. Strictly speaking, chlorine reduction by carbon is not a true catalytic process because surface reaction sites are slowly depleted; nonetheless, the general notion of this being a *diffusion-with-reaction* process is valid. The diffusive intraparticle transport of solutes that simply adsorb on surfaces within the micropore structures of noncatalytic adsorbents and ion-exchange materials is similarly a diffusion-with-reaction process, although it is one that involves a much more complex, nonsteady, equilibrium-approaching reaction between solid and liquid phases. Unsteady-state FBRs of this type, including many natural systems involving microporous adsorption or ion-exchange media (e.g., zeolites, clays, shales), configured in a way that allows them to be modeled as FBRs, are considered in Chapter 11.

The steady-state material balance over an element of an FBR in which a surface reaction occurs internal to the particles of a packing material is given by

$$dN_x \epsilon_B A_R + r^{\circ}_{\text{obs}} (a^{\circ}_s)_{I,P} (1 - \epsilon_B) A_R \, dx = 0 \tag{10-16}$$

where r°_{obs} is the *observed surface reaction rate* in mol/internal area-time, $(a^{\circ}_s)_{I,P}$ is the specific *internal surface area per unit particle volume*, and

$$1 - \epsilon_B = \frac{\text{total particle volume}}{\text{volume of bed}} \tag{10-17}$$

If we assume that advection is the only operating mechanism for macrotransport in the FBR, the resulting point form of Equation 10-16 is

$$-v_x \frac{dC}{dx} + \mathscr{r}^{\circ}_{\text{obs}}(a_s^{\circ})_{I,P}(1 - \epsilon_B) = 0 \tag{10-18}$$

or

$$-v_x \frac{dC}{dx} + \mathscr{r}_{\text{obs}}(1 - \epsilon_B) = 0 \tag{10-19}$$

The *reaction effectiveness factor*, η_R, was developed in Chapter 8 to relate *observed* and *intrinsic* rates of reaction. The *internal reaction effectiveness factor*, $\eta_{R,I}$, accounts for the retardation effect of internal diffusion on the intrinsic rate (i.e., $\mathscr{r}_{\text{obs}} = \eta_{R,I}\mathscr{r}_{\text{int}}$. Thus, a general expression for an nth-order surface reaction in which the reactant disappears is

$$-v_x \frac{dC}{dx} - (1 - \epsilon_B)\eta_{R,I}k^{\circ}(a_s^{\circ})_{I,P}C^n = 0 \tag{10-20}$$

Approximate expression for $\eta_{R,I}$ for zero-, first-, and second-order reactions are available (see Table 8-1). These expressions apply only to a process in which overall rates are controlled *strongly* by intraparticle diffusion impedances. Corresponding exact mathematical relationships that cover the entire range of internal impedance control can be derived for zero-, first-, and second-order reactions using the material balance relationships given in Chapter 8 for diffusion with simultaneous reaction.

First-order reactions accompanied by diffusion into microporous particles are involved in a variety of chemical and biochemical processes. This circumstance serves well to illustrate the reactor engineering approach. For particles that are spherical, or can be closely approximated as being so, the appropriate expression for the effectiveness factor derived in Chapter 8 is

$$\eta_{R,I} = \frac{3}{\mathfrak{N}_{\text{Tm}}^2}(-1 + \mathfrak{N}_{\text{Tm}} \coth \mathfrak{N}_{\text{Tm}}) \tag{8-109}$$

where the Thiele modulus, $\mathfrak{N}_{\text{Tm}}$, for radial diffusion into a sphere of radius R is defined as

$$\mathfrak{N}_{\text{Tm}} = R\left(\frac{k^{\circ}(a_s^{\circ})_{I,P}}{\mathfrak{D}_e}\right)^{0.5} \tag{8-105}$$

If the Thiele modulus is large, $\eta_{R,I} \approx 3/\mathfrak{N}_{\text{Tm}}$, Equation 8-109 is approximated by

$$\eta_{R,I} = \frac{3}{R}\left(\frac{\mathfrak{D}_e}{k^{\circ}(a_s^{\circ})_{I,P}}\right)^{0.5} \tag{10-21}$$

Substituting Equation 10-21 into the PFR material balance relationship given by Equation 10-20 yields

$$-v_x \frac{dC}{dx} - (1 - \epsilon_B) \frac{3}{R} [\mathfrak{D}_e k^\circ (a_s^\circ)_{I,P}]^{0.5} C = 0 \qquad (10\text{-}22)$$

A more compact expression can be written by recognizing that the product of $(1 - \epsilon_B)$ and $3/R$ is $(a_s^\circ)_{E,R}$ the *external surface area of a sphere per volume of reactor* (see Equation 10-2).

Integration of Equation 10-22 by separation of variables gives

$$\frac{C}{C_{\text{IN}}} = \exp[-(a_s^\circ)_{E,R}(\mathfrak{D}_e k^\circ (a_s^\circ)_{I,P})^{0.5} \bar{t}] \qquad (10\text{-}23)$$

This solution is analogous to the PFR model for external impedance with simultaneous reaction in packed FBRs (see Equation 10-9). The effects of important process design parameters on reactor performance are clear. For a given reactant, increased removal may be accomplished by increasing (1) the surface area per unit volume of bed $(a_s^\circ)_{E,R}$, (2) the internal surface area per unit volume of particle $(a_s^\circ)_{I,P}$, or (3) the empty-bed mean residence time $(\bar{t})$. External surface area can be increased by employing particles of smaller diameter, but internal surface area can be increased only by changing the physical character of the catalyst material. Equation 10-23 also indicates that removal is *reactant dependent*; that is, for a given reaction energy, reactants that diffuse more rapidly are removed more effectively. The dependence on diffusion rate is expected because the process model assumes that internal mass transport impedence is dominated by intraparticle diffusion control.

A more general PFR design model for internal mass transport with simultaneous first-order reaction can be developed with no assumption as to the extent of diffusion control. In this case, Equation 8-109 is used instead of the approximate relationship for the effectiveness factor given by Equation 10-21, which holds only for strong diffusion control. The resulting form of the material balance for an FBR is

$$-v_x \frac{dC}{dx} - (1 - \epsilon_B)\eta_{R,I} k^\circ (a_s^\circ)_{I,P} C = 0 \qquad (10\text{-}24)$$

which, after integration, yields

$$\frac{C}{C_{\text{IN}}} = \exp[-(1 - \epsilon_B)\eta_{R,I} k^\circ (a_s^\circ)_{I,P} \bar{t}] \qquad (10\text{-}25)$$

Reactor analysis when macrotransport includes advection, dispersion, and internal impedence control parallels that presented earlier for external impedance control. The general form of the PFDR model is now given by

$$-v_x \frac{dC}{dx} + \epsilon_B \mathfrak{D}_d \frac{d^2C}{dx^2} + r^{\circ}_{\text{obs}}(1 - \epsilon_B) = 0 \tag{10-26}$$

A specific application for a reaction that is intrinsically first-order and for which the internal process is strongly diffusion controlled is characterized by the relationship

$$-v_x \frac{dC}{dx} + \epsilon_B \mathfrak{D}_d \frac{d^2C}{dx^2} - (a_s^{\circ})_{E,R}(\mathfrak{D}_e k^{\circ}(a_s^{\circ})_{I,P})^{0.5} C = 0 \tag{10-27}$$

This equation is of the same form as Equation 10-11, and its solution thus has the same general form as given in Equations 10-13 through 10-15, the only difference relating to definition of the rate constant, which in Equation 10-27 is given by the expression $(a_s^{\circ})_{E,R}(\mathfrak{D}_e k^{\circ}(a_s^{\circ})_{I,P})^{0.5}$.

A practical application of the foregoing relationships to the characterization and quantification of coupled diffusion and reaction in microporous media for a typical water and waste treatment process is given in Example 10-2.

Example 10-2

- **Situation:** One means for removing ammonia from waters and wastewaters involves reacting it with chlorine to form monochloramine in a homogeneous substitutive oxidation reaction, with subsequent complete oxidation of nitrogen to N_2 by heterogeneous catalysis on the surfaces of activated carbon. The second reaction can be envisioned as a two-step process in which monochloramine reacts with active surface sites, S_{AC}, on the carbon in one step to produce surface oxide sites, $S_{AC}O$:

$$NH_2Cl + H_2O + S_{AC} \rightarrow NH_3 + H^+ + Cl^- + S_{AC}O$$

The surface oxide sites in turn react with more monochloramine to produce N_2 and regenerated active sites:

$$2NH_2Cl + S_{AC}O \rightarrow N_2 + H_2O + 2Cl^- + S_{AC}$$

These reactions take place at sites internal to the activated carbon. Pore diffusion thus often dominates overall transformation rates. Rate experiments have shown that the surface reaction is first order with respect to monochloramine and that the surface reaction rate constant, k°, is 1.6×10^{-9} cm/s.

- **Question(s):** To evaluate the feasibility of this process for a particular application you decide to build a small pilot-scale reactor with materials that are readily available in the plant laboratory. These include a bag of fairly uniform activated carbon having a mean particle diameter of 0.15 cm and a specific internal surface area per unit weight, $(a^{\circ}_{s,w})_P$, of 1000 m^2/g, and a fixed-speed pump and a column of specific capacity and size, which together provide a superficial velocity of 4 m/h and an empty-bed contact time (EBCT) of 8 min. Estimate the removal of monochloramine that might be expected at steady state in the pilot plant. What would the anticipated monochloramine removal have been had you employed a similar grade of activated carbon having a particle diameter of 0.05 cm?

- **Logic and Answer(s):**

1. For the purpose of simplicity, assume that the pilot column behaves as an ideal PFR.

2. The following information is available:

- Bed porosity (ϵ_B) = 0.4 (measured)
- Solid phase density (ρ_s) = 2.1 g/cm^3 (measured)
- Particle porosity (ϵ_p) = 0.66 (measured)
- Effective diffusivity ($\mathfrak{D}_e$) of monochloramine = 5.5×10^{-6} cm^2/s (assumes a tortuousity factor of 2)

3. The effectiveness factor for a first-order reaction in a spherical particle of radius R is given in Equation 8-109 as

$$\eta_{R,I} = \frac{3}{\mathfrak{N}_{\text{Tm}}^2}(-1 + \mathfrak{N}_{\text{Tm}} \coth \mathfrak{N}_{\text{Tm}})$$

and the Thiele modulus, $\mathfrak{N}_{\text{Tm}}$, in Equation 8-105 as

$$\mathfrak{N}_{\text{Tm}} = R\left(\frac{k^\circ(a_s^\circ)_{I,P}}{\mathfrak{D}_e}\right)^{0.5} = R\left(\frac{k^\circ(a_{s,w}^\circ)_{I,P}(1 - \epsilon_p)\rho_s}{\mathfrak{D}_e}\right)^{0.5}$$

and

$$(a_s^\circ)_{I,P} = 10^7 \times (1 - 0.66) \times 2.1 = 7.14 \times 10^6 \text{ cm}^{-1}$$

For a particle diameter of 0.15 cm (R = 0.075 cm),

$$\mathfrak{N}_{\text{Tm}} = 0.075\left(\frac{1.6 \times 10^{-9} \times 7.14 \times 10^6}{5.5 \times 10^{-6}}\right)^{0.5} = 3.42$$

Substituting this value into Equation 8-109 (see above) gives an effectiveness factor of $\eta_{R,I} = 0.62$.

4. Note that the approximate solution to Equation 8-109,

$$\eta_{R,I} = \frac{3}{\mathfrak{N}_{\text{Tm}}}$$

is not valid because the Thiele modulus is not large enough to neglect the term, $-3/\mathfrak{N}_{\text{Tm}}^2$. Therefore, Equation 10-21 cannot be used. Instead, the general form of the ideal PFR model given in Equation 10-25 must be employed:

$$\frac{C_{\text{OUT}}}{C_{\text{IN}}} = \exp[-(1 - \epsilon_B)\eta_{R,I}k^\circ(a_s^\circ)_{I,P}\bar{t}]$$

and

$$\frac{C_{OUT}}{C_{IN}} = \exp[-(1 - 0.4)(0.64)(1.6 \times 10^{-9}) \cdot (7.14 \times 10^6)(8)(60)] = 0.13$$

or an expected removal of 87%.

5. Repeating the calculations above for a particle diameter of 0.05 cm instead of 0.15 cm yields a Thiele modulus of 1.14, an effectiveness factor of 0.92, and a fractional monochloramine residual (Equation 10-25) of 0.05. Thus, the effect of reducing the particle diameter would be nearly to eliminate the internal mass transfer limit on overall reaction rate and increase the expected removal to 95%.

10.2.3 Coupled External and Internal Mass Transfer in Fixed-Bed Reactors

The relationship for the observed surface reaction rate when external and internal impedances act in series was derived in Chapter 8:

$$\overset{\circ}{r}_{obs} = \eta_{R,O}\overset{\circ}{r}_{int} \tag{8-116}$$

where, for a first-order reaction, the overall effectiveness factor is

$$\eta_{R,O} = \frac{\eta_{R,I}}{1 + \eta_{R,I}[k^\circ(a_s^\circ)_{I,P}/k_f(a_s^\circ)_{E,P}]} \tag{8-122}$$

A useful PFR design equation for a packed FBR is obtained by substituting this expression into Equation 10-20:

$$-v_x\frac{dC}{dx} - \eta_{R,O}k^\circ(a_s^\circ)_{I,P}(1 - \epsilon_B)C = 0 \tag{10-28}$$

The solution is again obtained by separation of variables:

$$\frac{C}{C_{IN}} = \exp[-\eta_{R,O}k^\circ(a_s^\circ)_{I,P}(1 - \epsilon_B)\bar{t}] \tag{10-29}$$

This process model is of the same form as those for external resistance alone (Equations 10-9 and 10-10) and for internal resistance alone (Equation 10-23). We note that the model is valid for any extent of internal diffusion control in that the general form of the effectiveness factor relationship (Equation 8-109) is used to determine $\eta_{R,O}$.

The PFDR model appropriate for cases in which both external and internal impedances contribute significantly to rate control is of the same form pre-

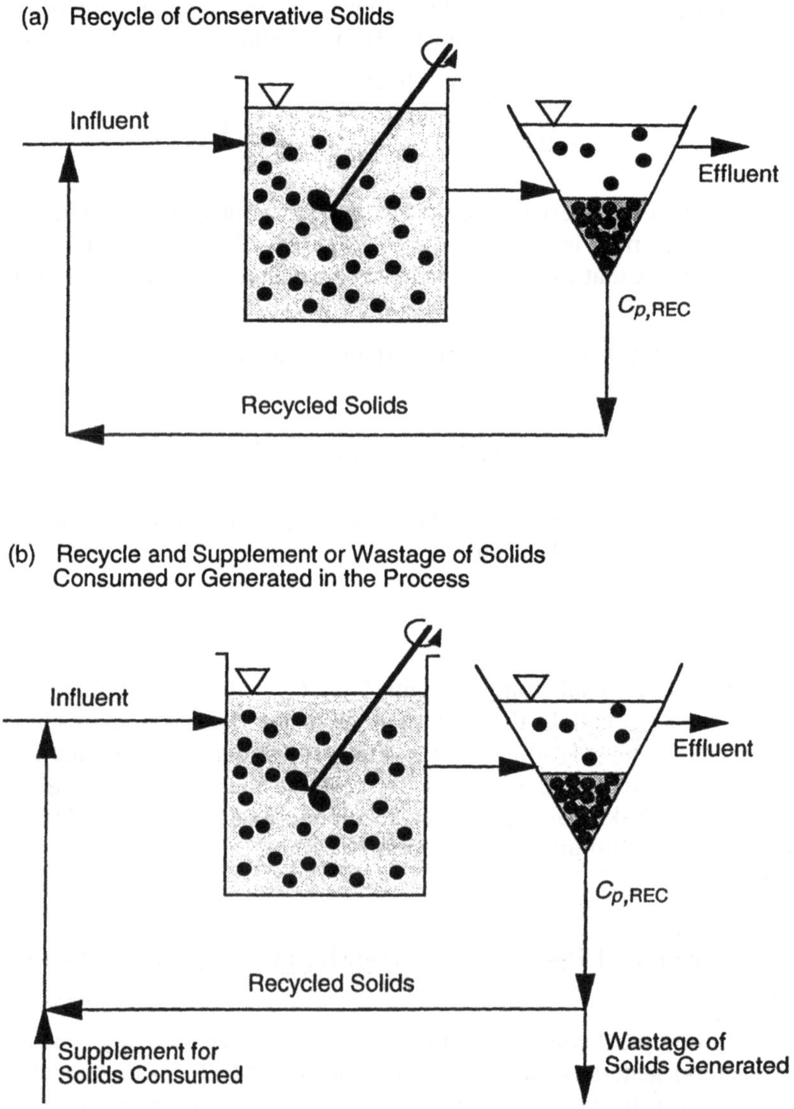

Figure 10-2 CMFR models for continuous-flow slurry reactors.

the cases cited above, solids are generated in the process. To maintain the same concentration of catalyst, and thus steady-state operation, an amount of the settled solids equivalent to the growth of microorganisms or production of precipitated solids is discharged or *wasted* from the system on each cycle, as illustrated in Figure 10-2b. The intrinsic microbial rate of reaction is often represented by Monod kinetics (see Section 7.7). However, external impedances to mass transport at the surfaces of the floc and precipitate structures and internal impedances within the porous matrices of those structures often cause observed rates of reaction to be lower than the associated intrinsic rates, and potentially of different order.

Another common slurry reactor application for water and waste treatment involves the addition of powdered activated carbon to the influent of a CMFR. With the exception of truly catalytic surface reactions, the analysis of this type of slurry reactor is more complex because the process requires consideration of internal diffusion accompanied by adsorption on surface sites. This circumstance is dealt with in Chapter 11.

The general CMFR steady-state material balance for a continuous-flow slurry reactor in which external resistances limit catalytic surface reactions is given by

$$QC_{\mathrm{IN}} - QC_{\mathrm{OUT}} + r^{\circ}_{\mathrm{obs}}(a^{\circ}_s)_{E,R}V_R = 0 \tag{10-31}$$

The external surface area per unit volume of reactor, $(a^{\circ}_s)_{E,R}$, is related to the *surface area per volume of particle*, $(a^{\circ}_s)_{E,P}$, the particle density, ρ_p, and the concentration of catalyst particles, C_p (mass/volume), in the reactor by

$$(a^{\circ}_s)_{E,R} = (a^{\circ}_s)_{E,P}\frac{C_p}{\rho_p} \tag{10-32}$$

Alternatively, if the process involves only *internal* diffusion and reaction, $(a^{\circ}_s)_{I,R}$, the internal particle surface area per volume of reactor is substituted directly for $(a^{\circ}_s)_{E,R}$ to give

$$QC_{\mathrm{IN}} - QC_{\mathrm{OUT}} + r^{\circ}_{\mathrm{obs}}(a^{\circ}_s)_{I,R}V_R = 0 \tag{10-33}$$

Like $(a^{\circ}_s)_{E,R}$, $(a^{\circ}_s)_{I,R}$ is related to the catalyst particle concentration; that is,

$$(a^{\circ}_s)_{I,R} = (a^{\circ}_s)_{I,P}\frac{C_p}{\rho_p} \tag{10-34}$$

Finally, the appropriate steady-state CMFR material balance equation for *external* resistance *in series* with *internal* resistance is

$$QC_{\mathrm{IN}} - QC_{\mathrm{OUT}} + \eta_{R,O}r^{\circ}_{\mathrm{int}}(a^{\circ}_s)_{I,R}V_R = 0 \tag{10-35}$$

where $\eta_{R,O}$ is again the overall effectiveness factor (e.g., see Equation 8-122 for a first-order reaction.)

The CMFR material balance relationships presented above, together with the definitions of $(a^{\circ}_s)_{E,R}$ and $(a^{\circ}_s)_{I,R}$ given in Equations 10-32 and 10-34, respectively, provide the fundamental equations required for design of continu-

Table 10-1 Steady-State CMFR Design Equations for Intrinsic First-Order Reactions in Spherically Shaped Microporous Catalyst Particles

Dominant Mass Transfer Impedance	Design Equation
Extraparticle	$\dfrac{C_{\mathrm{OUT}}}{C_{\mathrm{IN}}} = \left[1 + \dfrac{k_f(a_s^\circ)_{E,P}C_p}{\rho_p}\bar{t}\right]^{-1}$
Intraparticle (general)	$\dfrac{C_{\mathrm{OUT}}}{C_{\mathrm{IN}}} = \left[1 + \dfrac{\eta_{R,I}k^\circ(a_s^\circ)_{I,R}C_p}{\rho_p}\bar{t}\right]^{-1}$
Intraparticle (strong control)	$\dfrac{C_{\mathrm{OUT}}}{C_{\mathrm{IN}}} = \left[1 + \dfrac{3C_p(\mathfrak{D}_e k^\circ(a_s^\circ)_{I,P})^{0.5}}{R_p\rho_p}\bar{t}\right]^{-1}$
Extraparticle and intraparticle in series	$\dfrac{C_{\mathrm{OUT}}}{C_{\mathrm{IN}}} = \left[1 + \dfrac{\eta_{R,O}k^\circ(a_s^\circ)_{I,R}C_p}{\rho_p}\bar{t}\right]^{-1}$

ous-flow slurry reactors operated at steady-state conditions. Appropriate expressions for the intrinsic reaction rate and effectiveness factor must be supplied. Design equations for processes involving intrinsic first-order reactions coupled with various types of external and internal impedances are presented in Table 10-1. An application of these relationships in the interpretation of rate limitations and formulation of design equations for a biological waste treatment process is illustrated in Example 10-4.

Example 10-4

- **Situation:** A suspended-growth biomass system is used to treat an industrial wastewater. The intrinsic surface rate of substrate utilization, r_{int}° (mass of substrate degraded per surface area of biomass per unit time, $ML^{-2}t^{-1}$), is given by Monod (saturation) kinetics as

$$-r_{\mathrm{int}}^\circ \frac{r_{\max}^\circ C}{\mathcal{K}_s + C}$$

where $r_{\max}^\circ$ is the maximum surface substrate utilization rate, C is the substrate concentration in solution (ML^{-3}), and $\mathcal{K}_s$ is the half-saturation constant (ML^{-3}). At substrate concentrations substantially higher than $\mathcal{K}_s$, the reaction rate becomes substrate independent. For substrate concentrations substantially smaller than $\mathcal{K}_s$, it is common to approximate the reaction rate by a first-order expression:

$$-r_{\mathrm{int}}^\circ = \frac{r_{\max}^\circ}{\mathcal{K}_s} C$$

- **Question(s):** Develop a design equation for this system relating substrate removal to detention time and biomass concentration.

- **Logic and Answer(s):**

1. The appropriate material balance relationship is given by Equation 10-33:

$$QC_{\text{IN}} - QC_{\text{OUT}} + r^{\circ}_{\text{obs}}(a_s^{\circ})_{I,R}V_R = 0$$

The internal biomass surface area available per unit volume of reactor (Equation 10-34) is equivalent to the surface area of cells contained within the floc present in a unit volume of reactor, that is,

$$(a_s^{\circ})_{I,R} = (a_s^{\circ})_{I,P}\frac{C_p}{\rho_p}$$

2. Typical values of $r^{\circ}_{\max}$ and $\mathcal{K}_s$ are 1.7×10^{-6} mg COD/cm^2-s and 0.01 mg COD/cm^3, respectively (Schroeder, 1977). The floc particle diameter, d_p, measured microscopically in bench-scale treatability tests is 0.1 cm. Estimates of other design parameters are (a) diameter of an individual cell within the floc, $d_c = 2 \times 10^{-4}$ cm; (b) floc density, ρ_p = 1.05 g/cm^3; (c) effective diffusivity of substrate within the floc volume, $\mathfrak{D}_e = 1 \times 10^{-6}$ cm^2/s.

3. The internal area per volume of floc, $(a_s^{\circ})_{I,P}$, is then

$$(a_s^{\circ})_{I,P} = \left(\frac{\text{floc volume}}{\text{cell volume}} \times \frac{\text{surface area}}{\text{cell}}\right) \div \text{floc volume}$$

$$= \left(\frac{d_p^3}{d_c^3} \times \pi d_c^2\right)\left(\frac{6}{\pi d_p^3}\right)$$

or

$$(a_s^{\circ})_{I,P} = \frac{6}{d_c} = \frac{6}{2 \times 10^{-4}} = 30{,}000 \text{ cm}^{-1}$$

4. The internal reaction effectiveness factor for a first-order rate expression is given in Equation 8-109 as

$$\eta_{R,I} = \frac{3}{\mathfrak{N}_{\text{Tm}}^2}(-1 + \mathfrak{N}_{\text{Tm}} \coth \mathfrak{N}_{\text{Tm}})$$

and the Thiele modulus, $\mathfrak{N}_{\text{Tm}}$, is

$$\mathfrak{N}_{\text{Tm}} = R\left[\frac{k^{\circ}(a_s^{\circ})_{I,P}}{\mathfrak{D}_e}\right]^{0.5} = R\left[\frac{(r^{\circ}_{\max}/\mathcal{K}_s)(a_s^{\circ})_{I,P}}{\mathfrak{D}_e}\right]^{0.5}$$

$$= 0.05\left(\frac{(1.7 \times 10^{-4} \times 30{,}000)}{1 \times 10^{-6}}\right)^{0.5} = 113$$

thus, from Eq. 8-109

$$\eta_{R,I} = 0.0263$$

5. Accordingly, the design equation listed in Table 10-1 for first-order reactions with strong internal mass transfer control is appropriate:

$$\frac{C_{\text{OUT}}}{C_{\text{IN}}} = \left[1 + \frac{3C_p[\mathfrak{D}_e(\mathring{r}_{\max}/\mathcal{K}_s)(a_s^\circ)_{I,P}]^{0.5}}{R\rho_p}\bar{t}\right]^{-1} = [1 + 0.129C_p\bar{t}]^{-1}$$

where C_p is the biomass concentration in g/cm^3 and $\bar{t}$ is the HRT in seconds.

6. The same result should be obtained using the general form of the design equation given in Table 10-1 for first-order reactions with internal impedances:

$$\frac{C_{\text{OUT}}}{C_{\text{IN}}} = \left[1 + \frac{\eta_{R,I}(\mathring{r}_{\max}/\mathcal{K}_s)(a_s^\circ)_{I,P}C_p}{\rho_p}\bar{t}\right]^{-1}$$

$$= \left[1 + [0.0263(1.7 \times 10^{-4}) \times (30{,}000)]C_p\bar{t}/(1.05)]\right]^{-1}$$

$$\frac{C_{\text{OUT}}}{C_{\text{IN}}} = [1 + 0.128C_p\bar{t}]^{-1}$$

which is in fact nearly equal to the value computed in step 5.

7. For a typical C_p value of 2000 mg/L (2×10^{-3} g/cm^3) and a detention time of 3 h, for example, the fractional substrate removal is

$$\text{fraction removal} = 1 - \frac{C_{\text{OUT}}}{C_{\text{IN}}} = 1 - 0.26 = 0.74$$

8. The removal efficiency would be improved greatly by an increase in the effectiveness factor. Suppose, for example, that the floc diameter can be decreased to 0.02 cm. The new Thiele modulus then has a value of

$$\mathfrak{N}_{\text{Tm}} = 0.01\left[\frac{(1.7 \times 10^{-4})(30{,}000)}{1 \times 10^{-6}}\right]^{0.5} = 23$$

and the internal effectiveness factor is, from the approximate form of Equation 8-109,

$$\eta_{R,I} \approx \frac{3}{\mathfrak{N}_{\text{Tm}}} = 0.13$$

The expression for fractional concentration is then

$$\frac{C_{\text{OUT}}}{C_{\text{IN}}} = \frac{1}{1 + 0.646\ C_p\bar{t}}$$

and the same biomass concentration and detention time used above now gives a fractional removal of

$$\text{fractional removed} = 1 - \frac{C_{OUT}}{C_{IN}} = 1 - 0.07 = 0.93$$

9. Mathematically more complex design equations result when Monod expressions are used in lieu of linear approximations to describe microbial rates. However, the analysis procedure is the same, beginning with the material balance for a CMFR (Equation 10-33). Effectiveness factor expressions for Monod kinetics are available in the literature [see, for example, Grady and Lim (1980)].

10.3 REACTORS FOR TWO-PHASE PROCESSES INVOLVING GASES AND LIQUIDS

Reactors involving gas–liquid exchanges and reactions are frequently employed in water and wastewater treatment, and commonly found in natural systems. In engineered systems, for example, gaseous reagents such as ozone, chlorine dioxide, and chlorine are dissolved into water to serve as oxidants and disinfectants. Stripping operations using air for removal of unwanted dissolved gases such as carbon dioxide, ammonia, hydrogen sulfide, and volatile organic chemicals, either by sparging air through water or by dispersing water droplets into air, are standard treatment procedures.

We confine our discussion in this chapter to the general principles of reactor analysis and draw upon a few salient illustrations, focusing on *dissolution of gases accompanied by reaction*. Ozone is an especially interesting example because it undergoes auto-decomposition while simultaneously oxidizing other species. The intrinsic rates of such reactions may range from very slow to very fast, with interphase mass transfer being important in both cases. Reactions that are slow occur in bulk solution but are controlled by interphase mass transfer. Fast reactions occur simultaneously with diffusion through interfacial regions. The rate relationships developed in Chapter 8 are applied here for different reactor configurations.

Various devices and reactor configurations are used for absorption and stripping of gases and volatile organic chemicals. Some of the more common reactor schemes are illustrated in Figure 10-3. The choice of reactor configuration is usually driven by economic considerations. A CMFR (Figure 10-3a) is commonly used to sparge gas into water. Referred to as a bubble sparger, this type of reactor is used to transfer gaseous forms of oxidants (e.g., ozone) into solution for oxidation and disinfection reactions, or to transfer volatile organic chemicals from solution into air bubbles. *Mechanical surface aeration* (Figure 10-3b) may be used to transfer oxygen from the atmosphere into solution to accommodate chemical oxidations of such inorganic species as iron(II), biochemical oxidations of various organic chemicals, or to strip volatile organic

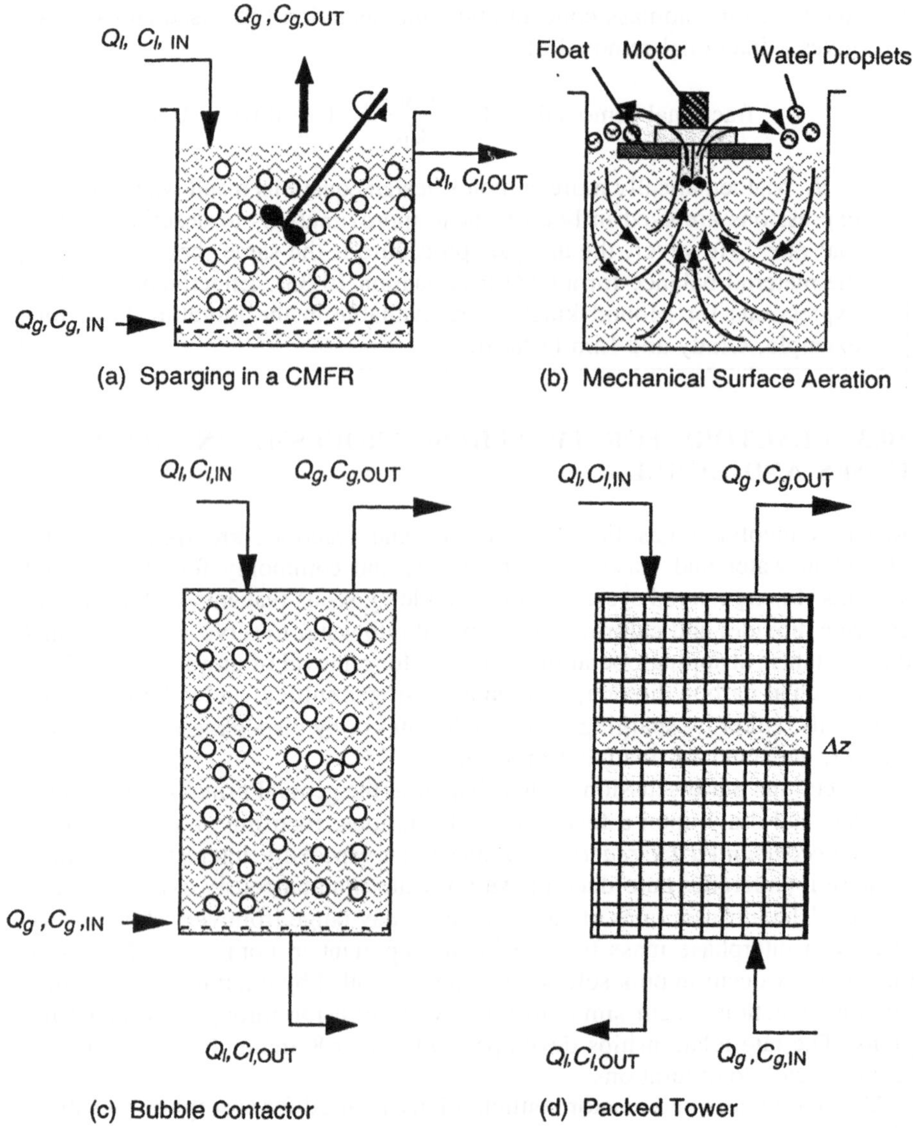

Figure 10-3 Typical gas-liquid contact systems.

chemicals from solution. A *bubble contactor* (Figure 10-3c) involving either co-current or countercurrent flow of water and gas bubbles in a tower configuration can be used for the same purposes. Advection and dispersion may be important in both water and gas phases in this type of reactor. The packed-tower contactor shown in Figure 10-3d is often used for gas absorption or stripping of volatile contaminants. Again, flow of water and air can be either co-current or countercurrent. A variety of commercial packing materials is available to create surface area and promote mass transfer. While often modeled as PFRs, dispersion may be important in both the gas and liquid phases in packed FBR systems of this type.

10.3.1 Mass Transfer and Reaction Processes in Completely Mixed Flow Reactors

The general liquid-phase material balance for a CMFR such as that depicted in Figure 10-3a is

$$Q_l C_{l,\mathrm{IN}} - Q_l C_{l,\mathrm{OUT}} + N_l (a_s^\circ)_R V_R + r V_R = 0 \tag{10-36}$$

in which N_l is the flux of component i into or out of the liquid phase, V_R the volume of the reactor, and $(a_s^\circ)_R$ the specific mass transfer surface area per unit volume of reactor. If N_l is positive, gas is transferred into solution (i.e., absorbed); conversely, if it is negative, gas (or a volatile organic chemical) is transferred out of solution (i.e., stripped).

Gas absorption phenomena accompanied by relatively fast reactions require consideration of simultaneous diffusion and reaction in liquid films. This approach to reactor modeling will be illustrated for a first-order reaction. The expression to account for enhancement of gas flux by simultaneous reaction within liquid films was given by Equation 8-68, from which we may write

$$(N_l)_{\text{with reaction}} = \frac{\hat{k}_{f,l}\, \mathfrak{N}_{\mathrm{Ha}}}{\tanh \mathfrak{N}_{\mathrm{Ha}}} \left(C_{l,e} - \frac{C_{l,\mathrm{OUT}}}{\cosh \mathfrak{N}_{\mathrm{Ha}}} \right) \tag{10-37}$$

The concentration in the bulk of the liquid phase is written as $C_{l,\mathrm{OUT}}$ to be consistent with the CMFR assumption of complete mixing, and the saturation concentration as $C_{l,e}$, the solution phase concentration corresponding to an equilibrium with the gas phase concentration. This implies complete mixing of both the liquid *and* gas phases. However, depending on mixing intensity, there may be situations in which an assumption of complete mixing in the gas phase is not valid. Other modeling approaches are then needed to quantify the extent of gas mixing through residence time distribution analysis (see Chapter 9).

An additional model assumption for purposes of the following analysis is that the overall mass transfer rate is controlled by the liquid side of the interface (i.e., $\hat{k}_{f,l} = k_{f,l}$). The *Hatta number*, $\mathfrak{N}_{\mathrm{Ha}}$, is for this situation given by

$$\mathfrak{N}_{\mathrm{Ha}} = \frac{(k \mathfrak{D}_l)^{0.5}}{\hat{k}_{f,l}} \tag{8-67}$$

The resulting design equation for relatively fast reactions is obtained by substituting Equation 8-67 and Equation 10-37 into Equation 10-36:

$$Q_l C_{l,\mathrm{IN}} - Q_l C_{l,\mathrm{OUT}} + \hat{k}_{f,l} (a_s^\circ)_R \frac{\mathfrak{N}_{\mathrm{Ha}}}{\tanh \mathfrak{N}_{\mathrm{Ha}}} \left(C_{l,e} - \frac{C_{l,\mathrm{OUT}}}{\cosh \mathfrak{N}_{\mathrm{Ha}}} \right) V_R - k C_{l,\mathrm{OUT}} V_R = 0 \tag{10-38}$$

The influent flow generally does not contain the target dissolved gas; thus $C_{l,\mathrm{IN}} = 0$. Rearranging Equation 10-38 then gives

$$C_{l,\mathrm{OUT}} = \frac{C_{l,e}}{\dfrac{k}{\hat{k}_{f,l}(a_s^\circ)_R}\dfrac{\tanh \mathfrak{N}_{\mathrm{Ha}}}{\mathfrak{N}_{\mathrm{Ha}}} + \dfrac{1}{\hat{k}_{f,l}(a_s^\circ)_R\bar{t}}\dfrac{\tanh \mathfrak{N}_{\mathrm{Ha}}}{\mathfrak{N}_{\mathrm{Ha}}} + \dfrac{1}{\cosh \mathfrak{N}_{\mathrm{Ha}}}} \tag{10-39}$$

As the Hatta number increases to sufficiently large values, $\mathfrak{N}_{\mathrm{Ha}}/\tanh \mathfrak{N}_{\mathrm{Ha}}$ can be approximated by $\mathfrak{N}_{\mathrm{Ha}}$, and $\cosh \mathfrak{N}_{\mathrm{Ha}} \to \infty$, yielding

$$C_{l,\mathrm{OUT}} \approx \frac{C_{l,e}}{\dfrac{k}{\hat{k}_{f,l}(a_s^\circ)_R\mathfrak{N}_{\mathrm{Ha}}} + \dfrac{1}{\hat{k}_{f,l}(a_s^\circ)_R\bar{t}\,\mathfrak{N}_{\mathrm{Ha}}}} \approx \frac{\hat{k}_{f,l}(a_s^\circ)_R\,\mathfrak{N}_{\mathrm{Ha}}C_{l,e}}{k + \dfrac{1}{\bar{t}}} \tag{10-40}$$

or, substituting for the Hatta number (Equation 8-67),

$$C_{l,\mathrm{OUT}} \approx \frac{(a_s^\circ)_R(k\mathfrak{D}_l)^{0.5}C_{l,e}}{k + \dfrac{1}{\bar{t}}} \tag{10-41}$$

The physical sense of the expression above is that the reactor exit concentration for a given intrinsic rate of reaction increases with larger values of (1) the interfacial surface area per unit volume of solution in the reactor $(a_s^\circ)_R$, (2) free-liquid diffusivity $(\mathfrak{D}_l)$, and (3) the mean HRT $(\bar{t})$.

If the Hatta number is small, the reaction rate constant, k, is small relative to the mass transfer coefficient, $\hat{k}_{f,l}$, and Equation 10-37 reduces to

$$N_{l,\,\text{with reaction}} \approx \hat{k}_{f,l}(C_{l,e} - C_{l,\mathrm{OUT}}) \tag{10-42}$$

This is another way of stating that reaction within the interfacial film is negligible. Diffusion occurs in series with reaction. Accordingly, for slow reactions Equation 10-36 becomes

$$Q_lC_{l,\mathrm{IN}} - Q_lC_{l,\mathrm{OUT}} + \hat{k}_{f,l}(a_s^\circ)_R(C_{l,e} - C_{l,\mathrm{OUT}})V_R - kC_{l,\mathrm{OUT}}V_R = 0 \tag{10-43}$$

If we assume further that the feed stream does not contain the dissolved gas, Equation 10-43 can be written as

$$C_{l,\mathrm{OUT}} = \frac{\hat{k}_{f,l}(a_s^\circ)_RC_{l,e}\bar{t}}{1 + k\bar{t} + \hat{k}_{f,l}(a_s^\circ)_R\bar{t}} \tag{10-44}$$

or, more conveniently,

$$C_{l,\mathrm{OUT}} = \frac{C_{l,e}}{\dfrac{1}{\hat{k}_{f,l}(a_s^\circ)_R \bar{t}} + \dfrac{k}{\hat{k}_{f,l}(a_s^\circ)_R} + 1} \tag{10-45}$$

Inspection of Equation 10-45 reveals that as $\hat{k}_{f,l}(a_s^\circ)_R$ increases, the exit concentration of the reactor approaches the saturation or solubility limit for the partial pressure of the gas in the system as given by Henry's law. Substituting $C_{l,e}$ for $C_{l,\mathrm{OUT}}$ in Equation 10-43 shows clearly that the rate of reaction is no longer limited by mass transfer and the reaction rate within the CMFR is simply $kC_{l,e}$. The same conclusion was reached in Chapter 8 when examining mass transfer in series with a slow reaction in a semibatch reactor (see Section 8.4.2). We note also that the exit concentration is increased by increasing the hydraulic detention time for a given mass transfer system [i.e., for a given $\hat{k}_{f,l}(a_s^\circ)_R$]. Because reaction occurs in the bulk solution, however, $C_{l,e}$ can never be achieved. Increasing the reaction rate constant for a given $\hat{k}_{f,l}(a_s^\circ)_R$ and $\bar{t}$ causes the exit concentration to decrease.

Our analyses of mass transfer and reaction in CMFRs thus produces two different design equations: Equation 10-45, for slow reactions, and Equation 10-40 for fast reactions. It is also possible to modify these relationships for reactions of dissolved gas A with component B. For example, if component B is in excess, the new definition of the rate constant is $kC_{B,0}$ and, as shown in Chapter 8, the Hatta number is

$$\mathfrak{N}_{\mathrm{Ha}} = \frac{(kC_{B,0}\mathfrak{D}_{l,A})^{0.5}}{\hat{k}_{f,l,A}} \tag{8-74}$$

Design considerations associated with a sparged CMFR like that shown in Figure 10-3a in which simultaneous gas transfer and reaction occur are illustrated in Example 10-5.

Example 10-5

- **Situation:** You are asked to critique the design of a CMFR in which ozone is to be bubbled through a water to accommodate fast oxidation reactions with inorganic species. CMBR rates tests indicate that the reactions can be described by pseudo-first-order rate expressions. A critical feature of the design is that a residual ozone concentration of 0.5 mg/L is required in the reactor effluent to be available for slower oxidation reactions which occur in a downstream storage tank.
- **Question(s):** Determine whether the ozone contactor design will provide a 0.5-mg/L residual, and if not, what can be done to modify the design.

- **Logic and Answer(s):**

1. You first determine that the following design features and intended operating conditions exist: water flow rate, $Q_l = 0.22\ \text{m}^3/\text{s}$ (5.07 mgd); volume of reactor, $V_R = 44.0\ \text{m}^3$; hydraulic residence time, $\bar{t} = 200$ s; ozone gas flow rate, $Q_g = 0.15\ \text{m}^3/\text{s}$; bubble diameter, $d_b = 0.1$ cm; average bubble residence time, $\bar{t}_b = 30$ s; average equilibrium saturation concentration of ozone, $C_{l,e} = 9.0$ mg/L (this corresponds to 1.27% ozone in air); molecular diffusivity of ozone, $\mathfrak{D}_l = 1.7 \times 10^{-5}\ \text{cm}^2/\text{s}$; pseudo-first-order rate constant, $k = 5\ \text{s}^{-1}$.

2. The appropriate material balance relationship is given by

$$Q_l C_{l,\text{IN}} - Q_l C_{l,\text{OUT}} + N_l (a_s^\circ)_R V_R - k C_{l,\text{OUT}} V_R = 0$$

3. Assume that the flux, N_l, must account for first-order reactions in the aqueous film surrounding the bubbles. For this case $C_{l,\text{OUT}}$ can be determined from Equation 10-39:

$$C_{l,\text{OUT}} = \frac{C_{l,e}}{\dfrac{k}{\hat{k}_{f,l}(a_s^\circ)_R}\dfrac{\tanh \mathfrak{N}_{\text{Ha}}}{\mathfrak{N}_{\text{Ha}}} + \dfrac{1}{\hat{k}_{f,l}(a_s^\circ)_R \bar{t}}\dfrac{\tanh \mathfrak{N}_{\text{Ha}}}{\mathfrak{N}_{\text{Ha}}} + \dfrac{1}{\cosh \mathfrak{N}_{\text{Ha}}}}$$

4. The controlling liquid-side mass transfer coefficient, $\hat{k}_{f,l}$, can be estimated from an appropriate correlation provided in Appendix A.4.1. For example, the Calderbrook and Moo-Young (1961) correlation gives

$$\frac{\hat{k}_{f,l}}{\mathfrak{D}_l} = \frac{2.0}{d_b} + 0.31\left[\frac{(\rho_l - \rho_g)g}{\mu_{v,l}\mathfrak{D}_l}\right]^{0.333}$$

5. Using the $\mathfrak{D}_l$ (m^2/s) and d_b (m) values provided, and $\rho_g = 1.205\ \text{kg/m}^3$, $\rho_l = 1000\ \text{kg/m}^3$, $\mu_{v,l} = 10^{-3}$ kg/m/s, and $g = 9.8\ \text{m/s}^2$, we find that

$$\hat{k}_{f,l} = 9.7 \times 10^{-5}\ \text{m/s} = 9.7 \times 10^{-3}\ \text{cm/s}$$

6. The Hatta number (Equation 8-67) is

$$\mathfrak{N}_{\text{Ha}} = \frac{[5(1.7 \times 10^{-5})]^{0.5}}{9.7 \times 10^{-3}} = 0.95$$

which is large enough that reaction within the film surrounding the bubble cannot be ignored (i.e., $\tanh \mathfrak{N}_{\text{Ha}} < \mathfrak{N}_{\text{Ha}}$ and $\cosh \mathfrak{N}_{\text{Ha}} > 1$) and Equation 10-39 cannot be simplified further.

7. The interfacial bubble area per unit volume of reactor, $(a_s^\circ)_R$, is also needed:

$$(a_s^\circ)_R = \frac{\text{surface area of bubbles in reactor}}{\text{volume of reactor}} = \frac{Q_g \bar{t}_b \pi d_b^2}{(\pi d_b^3/6)V_R}$$

$$(a_s^\circ)_R = \frac{6\bar{t}_b Q_g}{d_b V_R} = \frac{6 \times 30 \times 0.15}{(0.1 \times 10^{-2})44} = 614\ \text{m}^{-1} = 6.14\ \text{cm}^{-1}$$

8. Substitution of the parameters determined above into Equation 10-39 gives

$$C_{l,\mathrm{OUT}} = \frac{9.0}{\dfrac{5}{(9.7 \times 10^{-3})(6.14)}\left(\dfrac{0.74}{0.95}\right) + \dfrac{1}{(9.7 \times 10^{-3})(6.14)(200)}\left(\dfrac{0.74}{0.95}\right) + \dfrac{1}{2.04}}$$

$$= 0.14 \text{ mg/L}$$

9. The reactor design thus fails to provide for an ozone residual of 0.5 mg/L. Inspection of the result above shows that effluent concentration can be increased by increasing the interfacial area, $(a_s^\circ)_R$. Given the magnitude of each term in the denominator, a reasonable approximation is

$$C_{l,\mathrm{OUT}} = \frac{C_{l,e}}{\dfrac{k}{\hat{k}_{f,l}(a_s^\circ)_R}\dfrac{\tanh \mathfrak{N}_{\mathrm{Ha}}}{\mathfrak{N}_{\mathrm{Ha}}}}$$

from which

$$(a_s^\circ)_R \approx \frac{kC_{l,\mathrm{OUT}}\dfrac{\tanh \mathfrak{N}_{\mathrm{Ha}}}{\mathfrak{N}_{\mathrm{Ha}}}}{\hat{k}_{f,l}C_{l,e}} = \frac{(5)(0.5)\dfrac{0.74}{0.95}}{(9.7 \times 10^{-3})(9)} \approx 21.1 \text{ cm}^{-1}$$

10. From step 7, this interfacial area can be obtained by increasing the ozone flow rate from 0.15 m^3/s to 0.52 m^3/s.

The *stripping* or release of gases or volatile organic chemicals from solution phase to gas phase is analyzed in a manner similar to the *absorption* of gases from vapor phase into solution, except that the concentration in solution phase now *decreases* as a result of the mass transfer process. The appropriate material balance relationship for mass transfer in the absence of reaction is then

$$Q_l C_{l,\mathrm{IN}} - Q_l C_{l,\mathrm{OUT}} + \hat{k}_{f,l}(a_s^\circ)_R(C_{l,e} - C_{l,\mathrm{OUT}})V_R = 0 \qquad (10\text{-}46)$$

where the condition $C_{l,e} < C_{l,\mathrm{OUT}}$ is necessary for the direction of flux to be into the gas phase.

If a surface aerator is used to bring droplets of water into contact with the open atmosphere, as illustrated in Figure 10-3b, the gas-phase concentration of the volatile chemical is assumed to remain at zero due to its large dilution in the atmosphere; thus, $C_{l,e}$ in Equation 10-46 must also be zero. The resulting design equation is

$$\frac{C_{l,\mathrm{OUT}}}{C_{l,\mathrm{IN}}} = \frac{1}{1 + \hat{k}_{f,l}(a_s^\circ)_R\bar{t}} \qquad (10\text{-}47)$$

An application of this design equation was given in Example 4-9. Conversely, if a bubble contactor (Figure 10-3c) is used, the gas-phase concentration of the volatile compound is *zero only at the bubble inlet* position. As the air bubbles rise, their gas-phase concentrations increase and $C_{l,e}$ increases correspondingly. Bubble contactors are analyzed in the next section.

10.3.2 Absorption and Stripping in Bubble Contactors

The mass transfer rate of a dissolved gas or volatile chemical from solution into a single rising bubble in a bubble contactor such as that shown in Figure 10-3c is

$$V_B \frac{dC_g}{dt_b} = \hat{K}_{f,g} A_B^{\circ}(C_g - C_{g,e}) \tag{10-48}$$

where t_b is the travel time of the rising bubble, C_g is the concentration of the volatile compound in the gas phase (mol/L^3), $C_{g,e}$ its concentration in the gas phase were it in equilibrium with the liquid-phase concentration at that instant in time, V_B the volume of the bubble, A_B° the surface area of the bubble and $\hat{K}_{f,g}$ the overall mass transfer coefficient relative to the gas film (see Equation 4-72). Note that by following the change in gas-phase concentration as the bubble rises, we adopt a Lagrangian analytical perspective.

The projected surface area of all the bubbles in the contactor is

$$(A_B^{\circ})_T = \frac{Q_g \bar{t}_b}{V_B} A_B^{\circ} \tag{10-49}$$

in which Q_g is the gas flow rate (L^{3}t^{-1}) and $\bar{t}_b$ is the average residence time of the rising bubbles, assuming that they rise in plug flow fashion. Substitution of Equation 10-49 into Equation 10-48 gives

$$\frac{dC_g}{dt_b} = \hat{K}_{f,g} \frac{(A_B^{\circ})_T}{Q_g \bar{t}_b}(C_g - C_{g,e}) \tag{10-50}$$

The change in C_g as a function of vertical position in the contactor rather than as a function of time is obtained from

$$t_b = \frac{z}{v_t} \tag{10-51}$$

where v_t is the terminal rise velocity of the bubble and z is the vertical distance traveled. In turn, the bubble velocity can be expressed as

$$v_t = \frac{-Z_B Q_g}{(V_B)_T} = -\frac{Z_B}{\bar{t}_b} \tag{10-52}$$

where $(V_B)_T$ is the total volume of bubbles in a contactor having a bubble travel length Z_B. Substitution of v_t from Equation 10-52 into Equation 10-51, taking the derivative with respect to time, and then substituting this result into Equation 10-50 gives

$$\frac{d(C_g)}{dz} = -\hat{k}_{f,g} \frac{(A_B^\circ)_T}{Z_B Q_g} (C_g - C_{g,e}) \tag{10-53}$$

A more general definition of the interfacial area available for mass transfer is given by $(a_s^\circ)_R$, the specific area per unit volume of the contactor or reactor. If V_R is the volume of the contactor, then $(a_s^\circ)_R = (A_B^\circ)_T/V_R$, and

$$\frac{d(C_g)}{dz} = -\hat{k}_{f,g}(a_s^\circ)_R \frac{V_R}{Z_B Q_g} (C_g - C_e) \tag{10-54}$$

The concentration in the gas phase depends on the local pressure on the gas bubble, in this case the hydrostatic pressure. The ideal gas law dictates that

$$C_g = \frac{P_H}{\mathcal{R}T} X_g \tag{10-55}$$

where P_H is the hydrostatic pressure and X_g is the mole fraction of the volatile substance in the gas phase. Substituting the foregoing definition for C_g (and likewise for $C_{g,e}$) into Equation 10-54, we have

$$\frac{X_g}{P_H}\frac{dP_H}{dz} + \frac{dX_g}{dz} = -\hat{k}_{f,g}(a_s^\circ)_R \frac{V_R}{Z_B Q_g} (X_g - X_{g,e}) \tag{10-56}$$

To simplify the mathematical solution to this ordinary differential equation, we will ignore the change in hydrostatic pressure with depth (i.e., $dP_H/dz = 0$). If $X_{g,e}$ is taken as a constant mole fraction of the volatile compound in the gas phase in *equilibrium* with the *average* liquid-phase concentration across the depth of the contactor, Equation 10-56 then simplifies to

$$\frac{dX_g}{dz} = -\beta_Z (X_g - X_{g,e}) \tag{10-57}$$

where

$$\beta_Z = \frac{\hat{k}_{f,g}(a_s^\circ)_R V_R}{Z_B Q_g} \tag{10-58}$$

is an expression of the spatial rate of change of the mole fraction of the volatile compound in a rising bubble.

The general solution to Equation 10-58 can be obtained by separation of variables, yielding

$$X_g = X_{g,e} + \kappa \exp(-\beta_Z z) \tag{10-59}$$

where κ is a constant of integration. The particular solution for either gas absorption or stripping is then obtained by an appropriate mathematical statement of the gas-phase concentration in the bubble at the inlet to the reactor. The solution for stripping of a volatile compound into a gas bubble is determined by the following condition:

$$X_g = 0 \qquad \text{at } z = Z_B \tag{10-60}$$

For this condition, Equation 10-59 becomes

$$X_g = X_{g,e}[1 - \exp(-\beta_Z(Z_B - z))] \tag{10-61}$$

The mole fraction of the volatile compound in the gas bubble when it reaches the top of the contactor ($z = 0$) is more conveniently expressed through the ideal gas law as

$$X_g = C_{g,\text{OUT}} \frac{\mathcal{R}T}{P_H} \tag{10-62}$$

where $C_{g,\text{OUT}}$ is the gas-phase concentration of the compound in the bubbles leaving the contactor. Substitution of this relationship for X_g in Equation 10-61 gives

$$C_{g,\text{OUT}} = C_{g,e}[1 - \exp(-\beta_Z Z_B)] = C_{g,e}\left[1 - \exp\left(-\frac{\hat{k}_{f,g}(a_s^\circ)_R V_R}{Q_g}\right)\right] \tag{10-63}$$

A design equation to determine the loss of the compound from the liquid phase follows directly from the development above if we assume that the bubble contactor (liquid phase) behaves as a CMFR. The material balance in the liquid phase is

$$Q_l C_{l,\text{IN}} - Q_l C_{l,\text{OUT}} - Q_g C_{g,\text{OUT}} = 0 \tag{10-64}$$

The concentration, $C_{g,\text{OUT}}$, is given by Equation 10-63. We note further from Equation 10-63 that $C_{g,e}$ is the gas-phase concentration that exists in equilibrium with the liquid-phase concentration within the contactor, which, for a CMFR, is equal to $C_{l,\text{OUT}}$. The appropriate Henry's law relationship is

$$C_{g,e} = \frac{\mathcal{K}_H}{\mathcal{R}T} C_{l,\text{OUT}} \tag{10-65}$$

Accordingly, Equation 10-64 can be written

$$\frac{C_{l,\text{OUT}}}{C_{l,\text{IN}}} = \left[1 + \frac{Q_g \mathcal{K}_H^\circ}{Q_l}\left(1 - \exp\left(-\frac{\hat{k}_{f,g}(a_s^\circ)_R V_R}{Q_g}\right)\right)\right]^{-1} \tag{10-66}$$

in which, as noted in Table 6-2, $\mathcal{K}_H^\circ$ is the dimensionless Henry's constant, $\mathcal{K}_H/\mathcal{R}T$, obtained when concentrations are expressed in terms of molar ratios. The gas–liquid mass transfer coefficient for such systems is usually determined from measurements of liquid-phase concentrations rather than gas-phase concentrations and so, the overall mass transfer coefficient is $\hat{k}_{f,l}$. Recalling from Chapter 4 (Equation 4-73) that $\hat{k}_{f,g} = \hat{k}_{f,l}\mathcal{R}T/\mathcal{K}_H$, Equation 10-66 can be expressed in terms of the overall mass transfer coefficient relative to the liquid phase as follows:

$$\frac{C_{l,\mathrm{OUT}}}{C_{l,\mathrm{IN}}} = \left[1 + \frac{Q_g\mathcal{K}_H^\circ}{Q_l}\left(1 - \exp\left(-\frac{\hat{k}_{f,l}(a_s^\circ)_R V_R}{Q_g\mathcal{K}_H^\circ}\right)\right)\right]^{-1} \tag{10-67}$$

The two dimensionless parameter groups that appear in Equation 10-67 are

$$R_S = \frac{Q_g\mathcal{K}_H^\circ}{Q_l} \tag{10-68}$$

the so-called *stripping factor*, and

$$\beta_S = \frac{\hat{k}_{f,l}(a_s^\circ)_R V_R}{Q_g\mathcal{K}_H^\circ} \tag{10-69}$$

a *stripping parameter* that incorporates the mass transfer rate.

Stripping efficiency $(C_{l,\mathrm{IN}} - C_{l,\mathrm{OUT}})/C_{l,\mathrm{IN}}$ as a function of β_S is shown for various values of R_S in Figure 10-4. The stripping factor term characterizes the

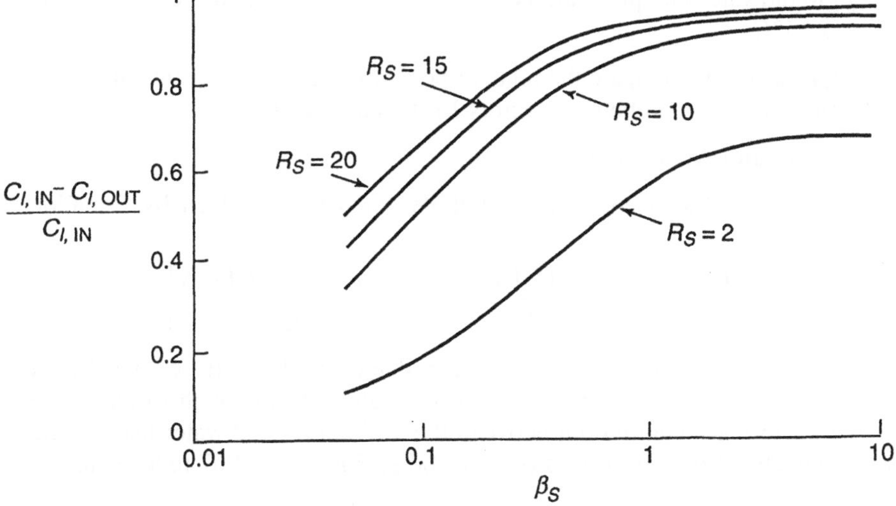

Figure 10-4 Stripping of volatile compounds in a bubble contactor.

equilibrium partitioning properties of the compound of interest, $\mathcal{K}_H^\circ$, and the ratio of gas to liquid flow rates. The parameter β_S accounts for the rate of mass transfer in the contactor, incorporating as it does the *overall volumetric mass transfer coefficient* $\hat{k}_{v,l} = \hat{k}_{f,l}(a_s^\circ)_R$. Increasing β_S while holding R_S constant causes stripping efficiency to increase, asymptotically, approaching a limiting value determined by R_S. Increasing R_S values, accomplished by increasing either the gas-liquid flow ratio or the Henry's constant, yield higher limiting stripping efficiencies. The effects of each of these major design parameters on stripping are logical and intuitive. The overall mass transfer coefficient or the volume of the contactor (and thus the liquid *holdup time*) must be increased to increase β_S without changing R_S. An increase in these parameters might be expected to improve stripping. It is an oversimplification, however, to discuss the impacts of the overall mass transfer and the gas and liquid flow rates as if they are independent parameters. In fact, as evidenced in Appendix A.4.1, experimental correlations show that the overall mass transfer coefficient depends on the gas and liquid flow rates. Example 10-6 provides a detailed illustration of how the principles and derivative relationships above can be applied, in this particular case to analysis of the anticipated performance characteristics of two different air stripper design configurations being considered for a groundwater decontamination application.

Example 10-6

- **Situation:** A pump-and-treat operation involving air stripping is planned for removal of trichloroethylene (TCE) from a contaminated groundwater. A bubble contactor in which the sparging device generates bubbles having a diameter of 0.1 cm is being considered for this application. The TCE concentration must be reduced from 50 μg/L to 5 μg/L. The groundwater pumping rate is 0.0438 m^3/s (1 mgd) and the temperature is 15°C.

- **Question(s):** Compare the relative performance characteristics of single- and three-stage bubble contactors for this application.

- **Logic and Answer(s):**

1. The design equation for bubble contactors is given by Equation 10-67:

$$\frac{C_{l,\mathrm{OUT}}}{C_{l,\mathrm{IN}}} = \left[1 + \frac{Q_g\mathcal{K}_H^\circ}{Q_l}\left(1 - \exp\left(-\frac{\hat{k}_{f,l}(a_s^\circ)_R V_R}{Q_g\mathcal{K}_H^\circ}\right)\right)\right]^{-1}$$

2. As a first-cut analysis, we can determine the ratio of gas–liquid flow rates required if the mass transfer rate is fast enough to allow the TCE to reach equilibrium between the air bubbles and water leaving the contactor. For large values of $\hat{k}_{f,l}(a_s^\circ)_R V_R$, Equation 10-67 reduces to

$$\frac{C_{l,\mathrm{OUT}}}{C_{l,\mathrm{IN}}} = \left(1 + \frac{Q_g\mathcal{K}_H^\circ}{Q_l}\right)^{-1}$$

Note that this relationship can also be obtained by writing the following material balance for the equilibrium situation:

$$Q_l C_{l,\mathrm{IN}} - Q_l C_{l,\mathrm{OUT}} = Q_g \mathcal{K}_H^\circ C_{l,\mathrm{OUT}}$$

3. A value of $\mathcal{K}_H = 11.7 \times 10^{-3}$ atm-m^3/mol is known for TCE for a temperature of 25°C (see Table 6-1), while the groundwater of interest is at 15°C. The temperature dependence of equilibrium constants was discussed in Chapter 5, and the associated van't Hoff equation relating the equilibrium constant for a reaction to the enthalpy of that reaction was given in Equation 5-103. Montgomery (1985) has reported enthalpy values (ΔH_r°) (expressed in kcal/mol) and constants of integration (κ) for Henry's constant (expressed in units of atmospheres) for several organic compounds of common environmental interest, as listed below:

Compound	ΔH_r° (kcal/mol)	κ
$CHCl_3$	4.00	9.10
CCl_4	4.05	10.06
$CCHCl_3$	3.41	8.59
C_2Cl_4	4.29	10.38
C_6H_6	3.68	8.68

Thus, we can estimate that $\mathcal{K}_H$ for TCE at 15°C is

$$\log \mathcal{K}_H = -\frac{\Delta H_r^\circ}{\mathcal{R}T} + \kappa = \frac{-3.41}{1.987 \times 10^{-3}\,(288)} + 8.59 = 2.63$$

$$\mathcal{K}_H = 426.6 \text{ atm} \quad \text{or} \quad 426.6 \times 18 \times 10^{-6} = 7.7 \times 10^{-3} \text{ atm-m}^3/\text{mol}$$

4. The dimensionless form of Henry's constant needed above is

$$\mathcal{K}_H^\circ \text{ (dimensionless)} = \frac{\mathcal{K}_H \text{ (atm-m}^3/\text{mol)}}{\mathcal{R}T}$$

$$= \frac{7.7 \times 10^{-3}}{0.0821 \times 10^{-3} \times 288} = 0.326$$

5. Solving for the gas–liquid flow ratio for a single-stage contactor gives

$$\frac{Q_g}{Q_l} = \frac{1 - \dfrac{C_{l,\mathrm{OUT}}}{C_{l,\mathrm{IN}}}}{\mathcal{K}_H^\circ \dfrac{C_{l,\mathrm{OUT}}}{C_{l,\mathrm{IN}}}} = 28$$

6. From Equation 10-68, the stripping factor for the one-stage system is

$$R_S = \frac{Q_g \mathcal{K}_H^\circ}{Q_l} = 9$$

7. The gas–liquid flow ratio and stripping factor for a three-stage system can be found in a similar way. If the same Q_g/Q_l is used in each stage, and equilibrium between gas and liquid is reached in each, the fractional removal of TCE must be the same in each stage. The overall fractional removal is then given by

$$\frac{C_{l,\mathrm{OUT}}}{C_{l,\mathrm{IN}}} = \left(\frac{C_{l,i+1}}{C_{l,i}}\right)^3 = 0.1$$

Thus, the fractional removal per stage is 0.46.

8. The relationship given in step 5 can be used to find the gas–liquid flow ratio required for each stage:

$$\frac{Q_g}{Q_l} = \frac{1 - \dfrac{C_{l,\mathrm{OUT}}}{C_{l,\mathrm{IN}}}}{\mathcal{K}_H^{\circ} \dfrac{C_{l,\mathrm{OUT}}}{C_{l,\mathrm{IN}}}} = 3.5$$

9. The total gas–liquid flow ratio is thus 3 × 3.5 = 10.6, which is considerably less than the value of 28.0 for a single-stage bubble contactor. Similarly, the stripping factor is reduced from 9.0 to 3.5. The reason for the improvement in efficiency is analogous to that found when comparing a single CMFR to several CMFRs in series. Here, the fractional concentration in the exit liquid stream of each contactor is made larger by staging, and less is stripped into the gas flow in each stage, so that a lower Q_g/Q_l is required. Note that as more stages are added, a minimum Q_g/Q_l is obtained.

10. We can also determine the depth of contactor needed to approach equilibrium. First, we write Equation 10-67 in a form that includes the depth of the contactor by making the following substitutions:

$$(a_s^{\circ})_R = \frac{(A_B^{\circ})_T}{V_R} = \frac{[Q_g \bar{t}_b/(\pi d_b^3/6)]\pi d_b^2}{V_R}$$

and

$$\bar{t}_b = \frac{Z_B}{v_t}$$

leading to

$$\frac{C_{l,\mathrm{OUT}}}{C_{l,\mathrm{IN}}} = \left[1 + \frac{Q_g \mathcal{K}_H^{\circ}}{Q_l}\left(1 - \exp\left(-\frac{6 k_{f,l} Z_B}{v_t d_b \mathcal{K}_H^{\circ}}\right)\right)\right]^{-1}$$

11. As in Example 10-5, the mass transfer coefficient is obtained from a correlation provided in Appendix A.4.1:

$$\frac{k_{f,l}}{\mathfrak{D}_l} = \frac{2.0}{d_b} + 0.31\left[\frac{(\rho_l - \rho_g)g}{\mu_{v,l}\mathfrak{D}_l}\right]^{0.333}$$

Given that $\mathfrak{D}_l = 8.37 \times 10^{-10}$ m²/s for TCE, $\rho_g = 1.205$ kg/m³, $\rho_l = 1000$ kg/m³, $\mu_{v,l} = 10^{-3}$ kg/m-s, and $g = 9.8$ m/s², we can calculate

$$\hat{k}_{f,l} = 6 \times 10^{-5} \text{ m/s} = 6 \times 10^{-3} \text{ cm/s}$$

12. The bubble rise velocity, which can be calculated from Stokes law, depends upon the flow condition (laminar, transitional, or turbulent), which is in turn determined by the size of the rising bubbles. For a bubble diameter of 0.1 cm, transition region flow occurs and the appropriate expression is (Weber, 1972)

$$v_t = [0.072g(\rho_l - \rho_g)d_b^{1.6}\rho_l^{-0.4}\mu_{v,l}^{-0.6}]^{0.714} = 10.8 \text{ cm/s}$$

13. The depth required to obtain equilibrium between the exit gas and liquid to obtain is approached when the exponential term in the expression for $C_{l,\text{OUT}}/C_{l,\text{IN}}$ (step 10) becomes very small. All of the parameters within the exponential term are known except Z_B. If we let

$$\exp\left(-\frac{6\hat{k}_{f,l}Z_B}{v_t d_b \mathcal{K}_H^\circ}\right) = 0.01$$

then

$$\frac{6\hat{k}_{f,l}Z_B}{v_t d_b \mathcal{K}_H^\circ} = 4.6$$

and

$$Z_B = \frac{4.6 v_t d_b \mathcal{K}_H^\circ}{6\hat{k}_{f,l}} = 44.0 \text{ cm}$$

14. The calculation above shows that equilibrium is reached in a very short bubble travel distance. Design of a bubble contactor of any depth greater than that required is pointless. However, the depth to equilibrium increases with bubble diameter as noted in the equation for Z_B given above. According to Stokes law, rise velocity is proportional to d_b^2 for laminar flow, $d_b^{1.6}$ for transition flow, and $d_b^{0.5}$ for turbulent flow (Weber, 1972). The mass transfer coefficient may also increase somewhat, but the net effect predicted is that coarser bubbles require greater depths for equilibrium to obtain.

10.3.3 Absorption and Stripping in Fixed-Bed Reactors

The essential features for analysis of countercurrent operation of packed FBRs are shown in Figure 10-3d. For simplicity, we assume that advection is the only form of macroscopic transport through the tower, and that the component being absorbed from the gas stream or stripped from the liquid stream does not undergo reaction. The material balance for a component in the liquid phase taken over a differential control volume accounts for mass transfer into (absorption) or out of (stripping) the liquid:

$$dN_{l,z}A_R + N^\circ(a_s^\circ)_R A_R\, dz = 0 \tag{10-70}$$

where $N_{l,z}$ is the advective flux of the component in the liquid flow stream, A_R is the cross-sectional area of the reactor, N° is the flux of the component due to mass transfer, and $(a_s^\circ)_R$ is the interfacial mass transfer area per unit volume of reactor. Substituting the appropriate definitions for these fluxes, the resulting point form of the material balance equation is

$$-v_z \frac{dC_l}{dz} + \hat{k}_{f,l}(a_s^\circ)_R(C_{l,e} - C_l) = 0 \tag{10-71}$$

Separation of variables leads to

$$\int_0^Z dz = \frac{v_z}{\hat{k}_{f,l}(a_s^\circ)_R} \int_{C_{l,\mathrm{IN}}}^{C_{l,\mathrm{OUT}}} \frac{dC_l}{C_{l,e} - C_l} \tag{10-72}$$

The pre-integral ratio on the right side of Equation 10-72 is defined as the *height of a transfer unit* (HTU):

$$\mathrm{HTU} = \frac{v_z}{\hat{k}_{f,l}(a_s^\circ)_R} \tag{10-73}$$

The HTU term, which has the dimensions of *length*, incorporates $\hat{k}_{f,l}(a_s^\circ)_R$ and is thus a measure of the mass transfer capacity of the tower packing material. Two of the most commonly used mass transfer correlations for packed towers, the *Onda* and the *Sherwood–Holloway* correlations, are given in Appendix A.4.1. Key parameters in these correlations are the gas and liquid velocities in the tower, often referred to as the *gas* and *liquid loading rates*, respectively, because they derive from ratios of flow rates to reactor unit cross-sectional area. Other parameters to be specified in the correlation relate to the characteristics of the packing material.

The quantity within the right-hand integral of Equation 10-72 is termed the *number of transfer units* (NTU); that is,

$$\mathrm{NTU} = \int_{C_{l,\mathrm{IN}}}^{C_{l,\mathrm{OUT}}} \frac{dC_l}{C_{l,e} - C_l} \tag{10-74}$$

The NTU is a dimensionless quantity, the magnitude of which depends on the difference between the equilibrium ($C_{l,e}$) and actual (C_l) liquid-phase concentrations at each position in the tower. Adopting the definitions above for HTU and NTU, the height, Z_T, of a tower required for a given stripping or absorption process is given by Equation 10-72 as

$$Z_T = \text{HTU} \cdot \text{NTU} \tag{10-75}$$

To find the value of NTU, it is necessary to determine $C_{l,e}$ as a function of tower height. This is analogous to the mass transfer situation in bubble contactors, where $C_{l,e}$ depends on the local gas-phase concentration, C_g, at any height. The gas phase concentration of a component is in turn given as a function of its pressure contribution, P, by the ideal gas law:

$$C_g = \frac{P}{\mathcal{R}T} \tag{10-76}$$

or, as a function of the liquid-phase concentration, by Henry's law:

$$C_g = \frac{\mathcal{K}_H}{\mathcal{R}T} C_{l,e} = \mathcal{K}_H^\circ C_{l,e} \tag{10-77}$$

The relationship between gas- and liquid-phase concentrations within the packed tower is determined by an *overall or macroscopic* material balance which accounts for the gain or loss of component in each phase:

$$Q_g C_{g,\text{IN}} - Q_g C_{g,\text{OUT}} = Q_l C_{l,\text{OUT}} - Q_l C_{l,\text{IN}} \tag{10-78}$$

In countercurrent operation, $C_{g,\text{OUT}}$ refers to the gas-phase concentration *at the point of entrance of the liquid phase* to the tower and $C_{g,\text{IN}}$ to the gas-phase concentration *at the point of exit of the liquid phase* from the tower (see Figure 10-3d).

Both gas absorption and stripping operations in FBRs can be analyzed in much the same way using the material balance relationship presented above. The stripper design equation is developed here because it has such wide utility for removal of volatile organic chemicals from contaminated waters, wastewaters, and liquid hazardous waste streams. The reader is encouraged as an exercise to develop an analogous equation for the design of a gas absorption system.

Stripping in an FBR is generally accomplished by pumping air upward through a tower of packing material while water flows downward (i.e., a countercurrent operation), although other configurations are possible. A common and reasonable assumption by which the analysis can be simplified is that the concentration of the compound of interest entering with the *airstream* is *zero*. This is in fact an operating condition that would be normally selected for the design of a single-stage stripper. For $C_{g,\text{IN}} = 0$, the "end-to-end" material balance can be written as

$$\frac{Q_g}{Q_l} = \frac{C_{l,\text{IN}} - C_{l,\text{OUT}}}{C_{g,\text{OUT}}} \tag{10-79}$$

The gas-phase exit concentration is then given by rearranging Equation 10-79 to obtain

$$C_{g,\mathrm{OUT}} = \frac{Q_l}{Q_g}(C_{l,\mathrm{IN}} - C_{l,\mathrm{OUT}}) \tag{10-80}$$

When plotted as in Figure 10-5, the relationship given in Equation 10-80 defines the *operating line*. Any point (C_g, C_l) on the operating line is defined by

$$C_g = \frac{Q_l}{Q_g}(C_l - C_{l,\mathrm{OUT}}) + C_{g,\mathrm{IN}} \tag{10-81}$$

where for stripping it is common to assume that $C_{g,\mathrm{IN}} = 0$ (see Problem 10-17 for exceptions). The *maximum* liquid-phase concentration that could exist in equilibrium with the gas-phase concentration at any point on the operating line is determined from Henry's law:

$$C_{l,e} = \frac{\mathcal{R}T}{\mathcal{K}_H} C_g = \frac{C_g}{\mathcal{K}_H^\circ} \tag{10-82}$$

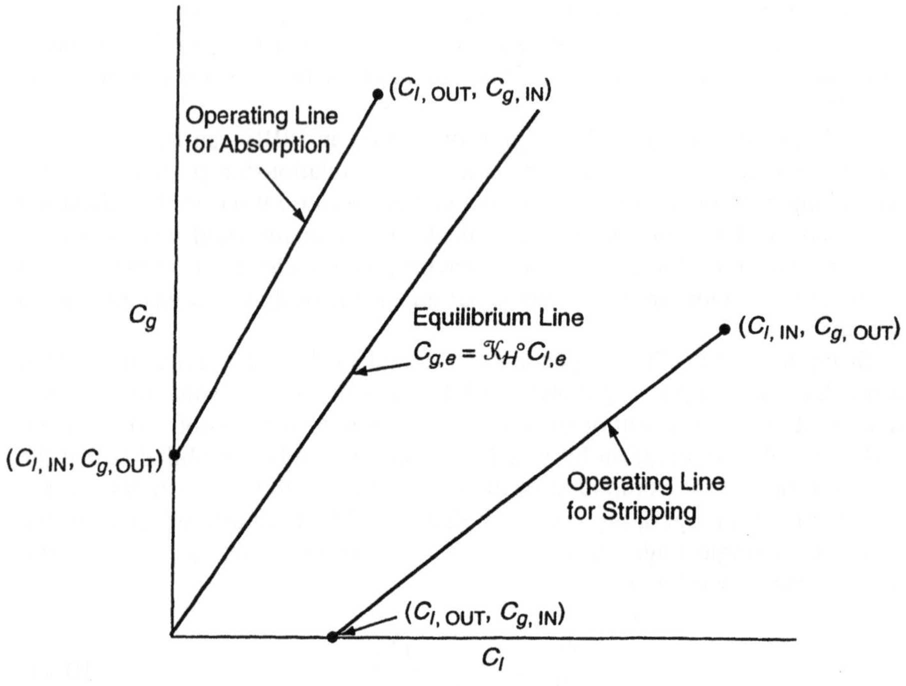

Figure 10-5 Equilibrium and operating lines for FBR gas strippers and absorbers.

The relationship given in Equation 10-82 determines the *equilibrium line* shown in Figure 10-5. The *horizontal distance* between the equilibrium and operating lines at any gas-phase concentration is $(C_{l,e} - C_l)$ the *driving force* for stripping. The inherently *negative value* of $(C_{l,e} - C_l)$ indicates *loss* from the liquid phase. This driving force expression is needed to solve Equation 10-72. A similar definition of the driving force exists for *absorption*, in which case, as shown in Figure 10-5, the operating line obtained from Equation 10-80 is above and to the left of the equilibrium line; accordingly, $(C_{l,e} - C_l)$ is *positive*.

Combining Equations 10-81 and 10-82 gives

$$C_{l,e} = \frac{Q_l}{\mathcal{K}_H^\circ Q_g}(C_l - C_{l,\text{OUT}}) \tag{10-83}$$

Substituting Equation 10-83 into Equation 10-72 yields

$$\int_0^Z dz = \frac{v_z}{\hat{k}_{f,l}(a_s^\circ)_R} \int_{C_{l,\text{IN}}}^{C_{l,\text{OUT}}} \frac{dC_l}{\dfrac{Q_l}{\mathcal{K}_H^\circ Q_G}(C_l - C_{l,\text{OUT}}) - C_l} \tag{10-84}$$

We observe that the value of the integral on the right-hand side of this equation (i.e., the NTU value) depends on the value of Henry's constant and on the ratio of gas to liquid flow rates selected. This integral can be simplified by combining the two terms in the denominator containing the variable, C_l, after which the integral is of the general form and solution

$$\int \frac{dC_l}{\kappa_1 + \kappa_2 C_l} = \frac{1}{\kappa_2}\ln(\kappa_1 + \kappa_2 C_l) \tag{10-85}$$

where the constants κ_1 and κ_2 are given by

$$\kappa_1 = -\frac{Q_l C_{l,\text{OUT}}}{\mathcal{K}_H^\circ Q_g} \quad \text{and} \quad \kappa_2 = \frac{Q_l}{\mathcal{K}_H^\circ Q_g} - 1 \tag{10-86}$$

The integral on the right side of Equation 10-84 is then

$$\text{NTU} = \frac{R_S}{R_S - 1}\ln\left(\frac{\dfrac{C_{l,\text{IN}}}{C_{l,\text{OUT}}}(R_S - 1) + 1)}{R_S}\right) \tag{10-87}$$

where R_S is the stripping factor defined in Equation 10-68.

Strictly speaking, the relationship given in Equation 10-87 holds only for situations in which the influent gas to the FBR is free of the contaminant to be stripped from the liquid. The reader is asked in Problem 10-17 to develop an equation of similar form for the more general case (i.e., for a finite contaminant level in the influent gas). The latter situation is normally of concern only for multistage stripping operations. The relationships given by Equation 10-87 between the fraction of a contaminant remaining in the liquid effluent of an FBR ($C_{l,\mathrm{OUT}}/C_{l,\mathrm{IN}}$) for a specific stripping factor (R_S) and the number of transfer units (NTU) provide important insights to the design of stripping towers. These relationships can be represented conveniently as shown in Figure 10-6.

Figure 10-6 can be used in many ways with Equation 10-75 to design and analyze FBR stripping and absorption tower performance. For example, the relationship of tower height, Z_T, to removal of a particular compound at a fixed gas–liquid flow rate, and thus a fixed R_S, is determined by first finding the NTU value corresponding to the fractional removal targeted in a particular application. Given Equation 10-75, and assuming that the HTU value is fixed

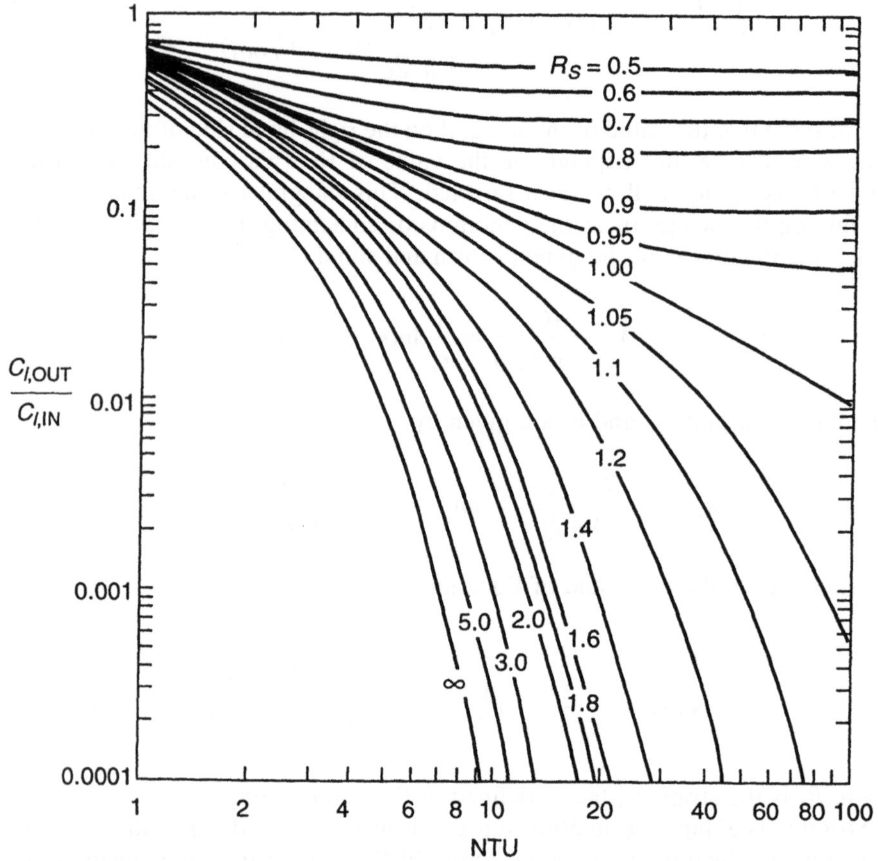

Figure 10-6 Relationships between removal efficiency and number of transfer unis for FBR gas strippers. [Adapted from Treybal (1980).]

by the packing material and the gas–liquid flow ratio, the value of Z_T can be calculated. According to Figure 10-6, NTU must increase as the desired fractional removal increases; this means that a tower of greater depth or height is needed. Tower height is also influenced by compound volatility. A more volatile compound gives a larger R_S for a given gas-liquid flow ratio. Thus, the tower height needed to achieve the same fractional removal as a less volatile compound is smaller because NTU is smaller. The gas–liquid flow ratio affects both the NTU and HTU values. A higher ratio will give a higher R_S value, and therefore the same fractional removal can be achieved at a lower NTU value, with a resulting savings in tower height. Increasing the gas–liquid flow ratio also affects the boundary layer between the two phases, usually leading to an increase in the mass transfer coefficient, as illustrated by the correlations presented in Table 4-2 and Appendix A.4.1. The HTU value is thus decreased, allowing a further reduction in tower height. There are economic and practical limits to changing gas flow rates, however. For instance, when the gas pressure drop becomes so large at high gas flow rates that it impedes the downward flow of water, tower *flooding* occurs, a failure condition in which the tower is literally flooded with water.

An application of the principles and relationships above to the design of an FBR stripping column is given in Example 10-7. The conditions specified in this example are the same as those given in Example 10-6, thus affording a direct comparison between the performance of a bubble contactor and a fixed-bed reactor.

Example 10-7

- **Situation:** The influent conditions and treatment objectives for removal of trichloroethylene (TCE) discussed in Example 10-6 are to be met using a countercurrent FBR employing 25-mm Intalox saddles, a widely used commercial packing material. The volumetric liquid loading rate, $Q_l/A_R = v_z$, to the FBR is specified as 0.035 m^3/m^2-s (51.5 gal/min-ft^2), and P_H is 1 atm.
- **Question(s):** Develop a design procedure for the stripping tower.
- **Logic and Answer(s):**

1. Alternative designs can be obtained for different stripping factors, R_S, and for different volumetric liquid loading rates. For this analysis, Q_l/A_R is specified as 0.035 m^3/m^2-s. Assume that $R_S = 3$. The dimensionless form of the Henry's constant for TCE at 15°C was determined in Example 10-6 to have a value of 0.326.

2. Using Equation 10-68, the gas–liquid flow ratio is determined to be

$$\frac{Q_g}{Q_l} = \frac{R_S}{\mathcal{K}_H^\circ} = \frac{3}{0.326} = 9.2$$

and the required volumetric flow of air is

$$Q_g = 9.2 \times 0.0438 \text{ m}^3/\text{s} = 0.40 \text{ m}^3/\text{s}$$

3. The number of transfer units is obtained from Equation 10-87:

$$\text{NTU} = \frac{R_S}{R_S - 1} \ln\left(\frac{\dfrac{C_{l,\text{IN}}}{C_{l,\text{OUT}}}(R_S - 1) + 1}{R_S}\right) = 2.92$$

4. The height of a mass transfer unit (HTU) requires selection of a mass transfer correlation to determine the overall volumetric mass transfer coefficient, $\hat{k}_{f,l}(a_s^\circ)_R$. The Sherwood–Holloway correlation (Appendix A.4.1) is used:

$$\frac{\hat{k}_{f,l}(a_s^\circ)_R}{\mathfrak{D}_l} = 10.76\psi_1\left(\frac{0.3048N_l}{\mu_{v,l}}\right)^{1-\psi_2}\left(\frac{\mu_{v,l}}{\rho_l\mathfrak{D}_l}\right)^{0.5}$$

in which the mass liquid flux or loading rate, N_l, is $(0.035 \text{ m}^3/\text{m}^2\text{-s}) \times 1000 \text{ kg/m}^3$. The other parameters needed are $\mu_{v,l} = 10^{-3}$ kg/m-s, $\rho_l = 10^3$ kg/m^3, and $\mathfrak{D}_l = 8.37 \times 10^{-10}$ m^2/s. The typical values of ψ_1 and ψ_2 are 63 and 0.28, respectively. The value of $\hat{k}_{f,l}(a_s^\circ)_R$ from this correlation is 0.016 s^{-1}.

5. Then, from Equation 10-73,

$$\text{HTU} = \frac{v_z}{\hat{k}_{f,l}(a_s^\circ)_R} = \frac{0.035 \text{ m}^3/\text{m}^2\text{-s}}{0.016 \text{ s}^{-1}} = 2.2 \text{ m}$$

6. The height of the tower can now be determined from Equation 10-75 as

$$Z_T = \text{HTU} \cdot \text{NTU} = 2.2 \times 2.92 = 6.5 \text{ m}$$

and the reactor cross-sectional area as

$$A_R = \frac{Q_l}{v_z} = \frac{0.0438 \text{ m}^3/\text{s}}{0.035 \text{ m}^3/\text{m}^2\text{-s}} = 1.25 \text{ m}^2$$

7. For different assumed values of the stripping factor, the liquid loading rate will yield different values for HTU and NTU, different tower heights, and different cross-sectional areas.

8. Other important aspects of tower design are the power requirements to lift the water to the top of the tower and to pull air through the tower. The gas pressure drop for a given packing material and temperature (referred to as the packing factor) has been shown by experiment to increase with the liquid-to-gas flow ratio (i.e., the inverse of the R_S value) and with the absolute value of the liquid loading rate (Treybal, 1980). With a packing factor of 33, the pressure drop for the system designed above is about 5 kg/m^2 per meter of height (see Appendix II-H for various expressions for pressure) or, for this design, the total pressure drop would be 6.4 m $\times$ 5 kg/m^3 = 32 kg/m^2 (1.25 in. H_2O).

10.4 MASS TRANSFER COEFFICIENTS

Mass transfer coefficients must be specified for every heterogeneous reactor discussed in this chapter. These coefficients depend on reactor configuration, specific operating conditions, physical characteristics of interfaces between the phases involved, and the physical–chemical properties of the component(s) being transferred. Mass transfer correlations such as those discussed in Chapter 4 are used extensively. Table 4-2 and Appendix A.4.1 are useful compendia of correlations applicable to a variety of natural and engineered environmental systems.

10.5 SUMMARY

Chapters 9 and 10 cover a wide range of reactor engineering applications under steady-state conditions. A general categorization of such applications is presented in Table 10-2. The material in these two chapters enables a performance or design equation to be obtained for each application listed in this table. The same general framework for process analysis can be used for more complex, steady-state situations [e.g., for the analysis of combined advective and dispersive transport with reaction in fluid–fluid systems (PFDR systems)]. The resulting design equations may not be as straightforward as those discussed

Table 10-2 Summary of Reactor Engineering Applications Addressed for Homogeneous and Heterogeneous Systems Operated at Steady State

Type of Reactor	Homogeneous Systems: Reaction Rate Relationships	Heterogeneous Systems: Reaction Rate Relationships Coupled with Mass Transfer Processes: Fluid–Solid Systems	Fluid–Fluid Systems
CMFR	All rate expressions	Zero-, first-, and second-order rate expressions; external, internal, and combined impedances	Gas-sparged systems; slow-to-fast first-order rate expressions
PFR	All rate expressions	Zero-, first-, and second-order rate expressions; external, internal, and combined impedances	Bubble contactors and countercurrent flow packed FBRs; mass transfer only
PFDR	Zero-, first-, and second-order rate expressions	Zero-, first-, and second-order rate expressions; external, internal, and combined impedances	

above, and numerical methods may be needed to solve the corresponding material balance equations. Such model extensions are discussed in Chapter 12. The analysis of *unsteady-state* situations, which is inherently more involved because concentrations of reactants are functions of both space and time, is addressed in Chapter 11.

10.6 REFERENCES AND SUGGESTED READINGS

10.6.1 Citations and Sources

Boyden, B. H., D. T. Banh, H. K. Huckabay, and J. B. Fernandes, 1992, "Using Cascade Aeration to Strip Chlorinated VOCs from Drinking Water," *Journal of the American Water Works Association*, *84*(5), 62–69. (Problem 10-8)

Grady, C. P., and H. C. Lim, 1980, *Biological Wastewater Treatment: Theory and Applications*, Marcel Dekker, Inc., New York. (Example 10-4)

Gurol, M. D., and P. C. Singer, 1982, "Kinetics of Ozone Decomposition: A Dynamic Approach," *Environmental Science and Technology 16*(7), 377–383. (Problem 10-11)

Letterman, R. D., M. Haddad, and C. T. Driscoll, 1991, "Limestone Contactors: Steady-State Design Relationships," *Journal of Environmental Engineering*, *117*, 339–358. (Problem 10-6)

Montgomery, J. M., 1985, *Water Treatment Principles and Design*, John Wiley & Sons, Inc., New York. (Example 10-6)

Schroeder, E. D., 1977, *Water and Wastewater Treatment*, McGraw-Hill Book Company, New York, pp. 240–245. (Example 10-4)

Treybal, R. E., 1980, *Mass Transfer Operations*, 3rd Edition, pp. 139–145. McGraw-Hill Book Company, New York, (Example 10-7, Figure 10-6)

Weber, W. J., Jr., 1972, *Physicochemical Processes for Water Quality Control*, Wiley-Interscience, New York, pp. 111–115. (Example 10-6)

10.6.2 Background Reading

Bennett, C. O., and J. E. Myers, 1982, *Momentum, Heat, and Mass Transfer*, McGraw-Hill Book Company, New York. In Chapter 31, individual and overall mass transfer coefficients and their use in the design of packed gas-absorption towers are introduced. Chapter 35 covers countercurrent contacting of immiscible phases for concentrated and dilute solutions. Equilibrium-stage contactor design for immiscible and partially miscible phases is discussed in Chapter 37 and 38.

Bird, R. B., W. E. Stewart, and E. N. Lightfoot, 1960, *Transport Phenomena*, John Wiley & Sons, Inc., New York. In this classic text in transport phenomena, the shell balance (differential) approach to developing the material balance equations and flux expressions for several different diffusion scenarios, including diffusion with reaction, diffusion and forced convection, and diffusion in porous catalysts are discussed in Chapter 17. In Chapter 22 a macroscopic mass balance is applied to steady-state problems, including a packed tower absorber. A short discussion of boundary conditions is also provided.

Cussler, E. L., 1984, *Diffusion: Mass Transfer in Fluid Systems*, Cambridge University Press, New York. Chapter 9 introduces mass transfer coefficients and their correlations and illustrates different contacting systems with examples, including gas scrubbing and aeration. Chapter 10 treats gas absorption of dilute and concentrated solutions in more detail. In Chapter 13, rate relationships for diffusion with heterogeneous chemical reaction are introduced and developed.

Fogler, H. S., 1992, *Elements of Chemical Reaction Engineering*, 2nd Edition, Prentice Hall, Englewood Cliffs, NJ. Reaction mechanisms and rate equations for catalysis are developed in Chapter 6, and the design of catalytic reactors is illustrated. Chapter 10 covers mass transfer coefficients, diffusion, and diffusion with reaction in heterogeneous systems; application to the design of steady-state packed beds is illustrated. Diffusion and reaction in porous catalysts are considered in Chapter 11, effectiveness factors are derived, and application to the design of steady-state packed beds is illustrated.

Grady, C. P., and H. C. Lim, 1980, *Biological Wastewater Treatment: Theory and Applications*, Marcel Dekker, Inc., New York. The influences of external and internal resistance on reactions within biofilms are discussed in detail in Chapter 14. The effectiveness factor concept is developed and applied to Monod kinetics.

King, C. J., 1980, *Separation Processes*, McGraw-Hill Book Company, New York. In Chapter 10 mass transfer is introduced, theoretical forms of the mass transfer coefficient based on laminar and turbulent flow theories are developed, interphase mass transfer is discussed, and the design of continuous countercurrent and cocurrent contactors is illustrated.

Levenspiel, O., 1972, *Chemical Reaction Engineering*, 2nd Edition, John Wiley & Sons, Inc., New York. The design of heterogeneous reacting systems is introduced in Chapter 11 and various reactor configurations are discussed. Chapter 13 covers design procedures for plug and mixed-flow countercurrent fluid-fluid contacting towers, including mass transfer and mass transfer with reaction. Rate expressions for solid catalyzed reactions are developed in Chapter 14, with examples of the design of plug flow reactors.

James J. Montgomery Consulting Engineers, 1980, *Water Treatment Principles and Design*, John Wiley & Sons, Inc., New York. Chapter 5 covers the principles of separation process engineering from a water treatment perspective; material balance equations for several different geometries are developed using the differential method, and applications to ideal reactor configurations, including air stripping reactors, are provided. The material balance approach is used to develop mathematical process models for adsorption processes in Chapter 8 and gas transfer processes in Chapter 11.

Thibodeaux, L. J., 1996, *Chemodynamics: Environmental Movement of Chemicals in Air, Water, and Soil*, 2nd Edition, John Wiley & Sons, Inc., New York. Steady-state chemical exchange rates between air and water are considered in Chapter 4, as are mass transfer between water and sediments in lakes and rivers for several steady-state scenarios in Chapter 5.

Treybal, R. E., 1980, *Mass Transfer Operations*, 3rd Edition, McGraw-Hill Book Company, New York. Chapter 5 covers intraphase mass transfer and develops material balance equations for cocurrent and countercurrent processes. Equipment for gas–liquid mass transfer operations is discussed in Chapter 6, together with correlations for bubble size, power requirements and mass-transfer coefficients for sparged and mechanically aerated systems, and correlations for pressure drop and mass transfer coefficients for tray and packed-tower systems (and data for different packing materials). Chapters 7, 9, and 10 cover applications to humidification, distillation, and liquid extraction processes, respectively. A detailed analysis of gas absorption processes for dilute and concentrated solutions, including multicomponent and nonisothermal systems, is provided in Chapter 8. Steady-state moving-bed adsorption processes and Michaels' model, which considers a steady-state analysis of the adsorption zone, are treated in Chapter 11. Applications to drying and leaching are discussed in Chapters 12 and 13, respectively.

10.7 PROBLEMS

10-1 A first-order reaction having the following intrinsic rate relationship takes place at internal sites of spherically shaped microporous catalyst particles of radius R_p:

$$r_{\text{int}} = -k^\circ (a_s^\circ)_{I,P} C^\circ \text{ (mol/cm}^3 \text{ catalyst-s)}$$

Show that the criterion corresponding to essentially negligible pore resistance (i.e., an effectiveness factor >0.8) is given by

$$-r_{\text{int}} \frac{R_p^2}{C^\circ \mathfrak{D}_e} < 11.1$$

10-2 Oxidation of zero-valent iron by trichloroethene (TCE), essentially a corrosion reaction, has been shown to cause reductive dechlorination of this organic compound. The intrinsic rate of the surface reaction is first order, with a rate constant of 1.8×10^{-7} cm/s. As an application of this technology, it is proposed to excavate a trench in the path of a contaminated groundwater plume at the site of a TCE spill, and backfill it with iron filings as a permeable barrier to provide *in-situ* dechlorination of the TCE. The filings have a mean diameter of 5 mm and their porosity in the trench is 0.4. The longitudinal dispersion coefficient has been estimated as 1 m^2/d and the pore velocity as 0.3 m/s. A published mass transfer correlation yields an estimate of 1×10^{-3} cm/s for the mass transfer coefficient between the groundwater and the surfaces of the iron filings. Determine the width of barrier (in the flow direction) required to achieve 95 percent dechlorination of the TCE. *Answer:* ~1.5 m.

10-3 The uppermost 1 cm of sediment in a small impoundment contains organic matter that is biodegrading aerobically. Oxygen diffusing through this sediment layer is consumed in the biodegradation reaction. If the impoundment behaves as a CMFR, find the steady-state, outlet oxygen concentration given the following data:

flow rate = 40 m^3/day
volume of impoundment = 1000 m^3
inlet oxygen concentration = 6 mg/L
first-order biodegradation reaction rate constant,

$$k = 0.64 \times 10^{-4}\ s^{-1}$$

diffusion coefficient for oxygen in sediment = 1×10^{-5} cm^2/s
sediment surface area to impoundment volume ratio = 0.5 m^{-1}

10-4 Bacteria are known to regrow in water distribution systems. They form biofilms on pipeline walls even in the presence of disinfectant residuals. Once established, such biofilms may be responsible for chemical reduction of disinfectants in a process that may be conceptualized as diffusion into the biofilm accompanied by reaction within that film. Find the fraction of disinfectant remaining at the first tap in a distribution system for the following circumstances: (1) flow in the pipeline is essentially plug flow; (2) external resistance to mass transport in the laminar boundary layer adjacent to the biofilm is negligible; (3) the biofilm thickness is steady at 0.01 cm; (4) reduction of the chemical disinfectant within the biofilm is a first-order surface reaction ($k° = 2.25 \times 10^{-3}$ cm/s); (5) the internal area of reactive sites per volume of biofilm is $(a_s^\circ)_{I,B} = 10\ cm^2/cm^3$; (6) the effective diffusivity within the biofilm is $\mathfrak{D}_e = 1 \times 10^{-6}\ cm^2/s$; (7) the pipe diameter is 50 cm; and (8) the HRT in the pipeline between the treatment plant and the first tap is 720 min. [*Hint:* Consider that the biofilm is very thin in comparison to the pipe radius such that the change in diffusional flux in the radial direction can be ignored. The diffusional flux (and the effectiveness factor) is written in one-dimensional rectangular coordinates rather than in radial coordinates. See Chapter 2 for a discussion of the material balance when accounting for wall diffusion normal to the advective direction in conduits].
Answer: Fraction remaining = 0.625.

10-5 In the absence of prechlorination, a thin biological film has grown around anthracite particles used as filter media in a water treatment plant. Chlorine is applied to oxidize and inactivate the biofilm. During the initial period following application of chlorine, the inactivation rate is pseudo first order with respect to chlorine because the number of cells in the biomass is large and thus changes slowly. Determine the concentration of chlorine remaining in the exit stream of the filter if (1) the chlorine concentration in the feed is 2 mg/L; (2) k_f is 0.03 cm/min; (3) the empty-bed contact time is 5 min; (4) the diameter of the filter media particles is 0.15 cm; (5) the bed porosity is 0.4; and (6) the inactivation reaction is limited to the surface of the biofilm and is sufficiently fast that external mass transfer is entirely rate controlling.

10-6 FBR limestone contactors may be used to treat acidic waters for pH adjustment and corrosion control. The rate of $CaCO_3$ dissolution in slightly acidic-to-alkaline solutions is controlled by interfacial calcium mass transfer and first-order surface reaction in series. The rate of mass transfer is modeled as

$$r = \hat{k}(a_s^\circ)_L(C_e - C)$$

where mass transfer and reaction are included in the overall rate coefficient $\hat{k} = \left(\frac{1}{k_f} + \frac{1}{k^\circ}\right)^{-1}$, $(a_s^\circ)_L$ is the specific surface area per unit volume of interstitial fluid, and C_e is the equilibrium calcium concentration. The surface reaction rate has been correlated as

$$k^\circ(\text{cm/s}) = 1.6 \times 10^{14}[\text{H}^+]_e^{1.7}$$

where $[\text{H}^+]_e$ is the equilibrium hydronium ion concentration (see Problem 10-7).

(a) Size a contactor of length L to treat a mildly acidic water having a pH of 4.4 to a desired effluent pH of 8.3. The design flow and loading rates are 0.001 m^3/s and 1.2 cm/s, respectively. The equilibrium concentration of calcium is 4.13 mg/L, and the equilibrium pH is 9.65. The effluent calcium concentration corresponds to an effluent pH of 8.3 and is equal to 1.68 mg/L. For this design, assume a calcium particle diameter, d_p, of 1 cm, a bed porosity, ϵ_B of 0.4, a calcium free-liquid diffusivity of 1.2×10^{-5} cm^2/s, and neglect axial dispersion. Assume an influent calcium concentration, $C_{Ca,0}$, of zero. Estimate the mass transfer coefficient from an appropriate correlation for liquids in packed beds from Chapter 4.

(b) Compare your design in part (a) with the following design equation given by Letterman et al. (1991), which includes the effects of dispersion:

$$(C_{Ca})_{z=L} = C_{Ca,e} + \left\{\exp\left[\frac{\hat{k}(a_s^\circ)_{I,L}L\epsilon_B}{v_p}\right. \right. \\ \left.\left. + \left(\frac{\hat{k}(a_s^\circ)_{I,L}L\epsilon_B}{v_p}\right)^2 \mathfrak{N}_d\right](C_{Ca,0} - C_{Ca,e})\right\}$$

where v_p is the interstitial velocity and $\mathfrak{N}_d$ is the dispersion number, approximated by $2d_p/L$.

10-7 Consider the FBR limestone contactor of Problem 10-6 further.

(a) Data were collected to investigate the effect of influent calcium ion concentration on the effective overall rate constant and on the relative importance of mass transfer and surface reaction. $[H^+]_e$ was measured for several influent calcium concentrations, $C_{Ca,0}$, ranging from zero to 5×10^{-3} M; these data are tabulated below. To illustrate the magnitude of reaction rate control in this system, plot the fraction of the total impedance due to reaction, $k°/(k° + k_f)$, as a function of pH_e and influent calcium concentration. Assume a value of k_f equal to 0.002 cm/s.

$C_{Ca,0}$ (mg/L)	pH_e
0.00	9.65
0.80	9.61
4.01	9.48
20.0	9.12
40.1	8.89
120	8.48
200	8.27

(b) Discuss the plot resulting from part (a) in the context of the proposed mechanism of $CaCO_3$ dissolution, and discuss how the presence of C_{T,CO_3} and metal ions other than Ca^{2+} in the influent would affect the reaction rate and the relative importance of mass transfer resistance.

10-8 A process in which water flows over a corrugated surface forming an inclined channel is proposed for removal of volatile organic chemicals from a water supply. The turbulence generated by sheet flow over this surface enhances mass transfer at the air–water interface. The overall volumetric mass transfer coefficient for this type of system has been correlated with Henry's constant, $\mathcal{K}_H$ (atm) and the angle of incline, ϕ to give

$$\hat{k}_{v,l} = \hat{k}_{f,l}a_s^\circ = 0.0191 + 0.0913\,(\mathcal{K}_H)^{0.25}(\sin\phi)^{1.33}$$

for liquid loading rates (volume per time per unit width of channel) ranging from 0.001 to 0.003 m^3/s-m (Boyden et al., 1992).

(a) Assuming that dispersion is negligible, develop a design equation to predict the concentration of VOCs remaining in the water as a function of water velocity and distance along the incline. Discuss other assumptions that may be involved.

(b) Determine the width and length of a cascade aerator required to reduce the VOC concentration in a raw water supply for a small community (1000 people) from 300 μg/L to 5 μg/L if (1) the flow rate is 0.00525 m^3/s (0.12 mgd), (2) the loading rate is 0.001 m^3/

s-m; (3) the angle of inclination is 45°, and (4) the velocity is 0.75 m/s.

(c) The correlation above is found to explain observations from pilot plant studies that the overall volumetric mass transfer coefficient increases very little for angles of inclination beyond 60° if Henry's constant is greater than 300 atm, whereas this coefficient increases significantly for VOCs having lower Henry's constants. At steep incline angles, an increase in turbulence in the airflow around the water is noted, due to the tendency of the water to separate from the channel surface. How do these observations relate to the liquid- or gas-side control of mass transfer in the two-film theory?

10-9 An ozone sparger system in a water treatment plant provides complete mixing of both gaseous and aqueous phases. The bubble diameter is 0.15 cm, the specific surface area for mass transfer is 10 cm²/cm³, and the mass transfer coefficient is 0.01 cm/s. The HRT for the sparger system is 600 s. Ozone is consumed by various reduced species in reactions that can be modeled as pseudo first order with respect to ozone concentration, although the rate constant varies considerably from one compound to another. If an oxidation reaction is slow enough, the exit concentration of ozone can be predicted by ignoring reactions within the liquid film; that is,

$$-Q_l C_{\mathrm{OUT}} + \hat{k}_{f,l}(a_s^\circ)_R(C_{l,e} - C_{l,\mathrm{OUT}})V_R - kC_{l,\mathrm{OUT}}V_R = 0$$

However, if a reaction is fast, it may be necessary to account for reaction within the liquid film when predicting the exit concentration.

(a) Determine the value of the rate constant for which the error introduced by using the design equation above to estimate exit concentration is less than 10%. Assume that the diffusion coefficient for ozone is 1.7×10^{-5} cm²/s and the equilibrium concentration of ozone is $C_{l,e} = 9$ mg/L.

(b) How is the value of the rate constant that meets the criterion above affected by the magnitude of the HRT?

10-10 A groundwater containing dissolved Fe(II) is to be aerated to form Fe(III), which can then be precipitated and removed as $Fe(OH)_3(s)$. Studies of Fe(II) oxidation rates at different pH and oxygen gas-phase pressures indicate the following rate relationship:

$$-\frac{d[\mathrm{Fe(II)}]}{dt} = k[\mathrm{Fe(II)}]P_{O_2}[\mathrm{OH}^-]^2$$

in which $k = 1.5 \times 10^{13}$ L^2/mol^2-min-atm. The aeration tank is described as a CMFR, pH is 7.5, and oxygen is to be maintained at 80% of saturation.

(a) Determine the HRT required to achieve a reduction in Fe(II) concentration from 6.25×10^{-5} M (3.5 mg/L) to 3.6×10^{-6} M (0.2 mg/L).

(b) Determine the overall volumetric mass transfer coefficient, $\hat{k}_{v,l} = \hat{k}_{f,l}(a_s^\circ)_R$, needed to maintain the stated oxygen concentration if the Henry's constant for oxygen at the water temperature of interest is 0.73 atm-m^3/mol and the incoming water is devoid of oxygen.

10-11 A CMFR is used to transfer ozone to water to accommodate oxidations that will occur in a downstream reactor. Assume that the only reaction of importance in the upstream CMFR is ozone decomposition, rate expressions for which have been given in a model proposed by Gurol and Singer (1982):

$$-r = k[\mathrm{OH}]^{0.55}[\mathrm{O_3}]^2$$

where $k = 1.4 \times 10^4$ s^{-1} $M^{-1.55}$. Determine the ozone concentration in the exit stream from the CMFR, with and without consideration of ozone decomposition, if the pH is 8; the overall volumetric mass transfer coefficient, $\hat{k}_{v,l} = \hat{k}_{f,l}(a_s^\circ)_R$, is 0.5 min^{-1}; and the HRT is 4 min. Use a Henry's constant for ozone of 0.0677 atm-m^3/mol and assume that the average ozone composition of the gas bubbles in the tank is 3%.

10-12 The general form of the design equation for a bubble contactor was given in Equation 10-59 as

$$X_g = X_{g,e} + \kappa \exp(-\beta_z z)$$

where X_g and $X_{g,e}$ are, respectively, the nonequilibrium and equilibrium mole fractions of a compound in the airstream, κ is a constant of integration, and β_Z is a mass transfer parameter for the contactor. Derive design equations for stripping and for absorption operations. Discuss the practical significance of the equations that result in each case when the overall mass transfer coefficient becomes very large.
[*Hint:* For stripping assume that $X_g = 0$ when air enters the contactor at Z_B. For absorption, begin with general solution for the mole fraction of ozone in a bubble X_g as a function of its height in the contactor (z), find the constant of integration and then $C_{g,\mathrm{OUT}}$, i.e., the concentration of ozone in the gas phase at $z = 0$. The conversion from mole fraction (X_g and $X_{g,e}$) to gas-phase concentration (C_g and $C_{g,e}$) is

$$X_g = \frac{P}{P_H} = C_g \frac{\mathcal{R}T}{P_H}$$

Assume for both X_g and $X_{g,e}$ that total pressure, P_H, is equal to the hydrostatic pressure at middepth. Note that although P_H cancels out of the final expression, we still need to know it for determination of $C_{g,\mathrm{IN}}$, the concentration of ozone in the inlet gas stream.]

Answer: The stripping design equation was developed earlier in this chapter; the design equation for absorption is

$$C_{g,\mathrm{OUT}} = \mathcal{K}_H^{\circ} C_{l,\mathrm{OUT}} + (C_{g,\mathrm{IN}} - \mathcal{K}_H^{\circ} C_{l,\mathrm{OUT}}) \exp(-\beta_S)$$

10-13 A field test of an ozone bubble contactor is made to determine the overall volumetric mass transfer coefficient, $\hat{k}_{v,l} = \hat{k}_{f,l}(a_s^{\circ})_R$. The incoming water has been stripped of all ozone-demanding substances, and the pH is sufficiently low that ozone decomposition is not significant. The resulting ozone concentration in the effluent water is 0.18 mol/m^3 (8.64 mg/L) at steady state. Determine $\hat{k}_{f,l}(a_s^{\circ})_R$ and the efficiency of ozone transfer, $(C_{g,\mathrm{IN}} - C_{g,\mathrm{OUT}})/C_{g,\mathrm{IN}}$, given the following information:

bubble contactor volume = 100 m^3
contactor depth = 3 m
water flow rate = 1200 m^3/h
percentage ozone in the gas feed = 4%
gas flow rate = 178 m^3/h
temperature = 20°C
universal gas constant, $\mathcal{R} = 8.314 \times 10^{-5}$ atm-m^3/mol-K
Henry's constant for ozone is $\mathcal{K}_H = 0.0677$ atm-m^3/mol

[*Hint:* Do all calculations of concentrations in both gas and liquid phases in mol/m^3 to avoid conversion factors. Calculate the unitless Henry's constant from the $\mathcal{K}_H$ value that is given. Calculate the inlet concentration of ozone in the gas stream using the mid-depth for determination of P_H i.e., P_H will be greater than 1 atm due to hydrostatic pressure on the gas bubble.]

Answers: $\hat{k}_{f,l}(a_s^{\circ})_R = 10\ h^{-1}$ and efficiency = 0.64.

10-14 Determine whether the ozone bubble contactor in Problem 10-13 can produce an effluent residual ozone concentration of at least 0.0417 mol/m^3 (2 mg/L) if it receives a water with an ozone demand of 0.208 mol/m^3 (10 mg/L).

[*Hint:* The additional equation required is

$$Q_g(C_{g,\mathrm{IN}} - C_{g,\mathrm{OUT}}) = Q_l C_{D,\mathrm{O}_3}$$

where C_{D,O_3} is the ozone demand. Assume that the demand can be satisfied in the residence time of the contactor.]

10-15 A countercurrent flow air stripping tower is to be designed to reduce trichloroethylene in a groundwater supply from 100 μg/L to 5 μg/L. The water flow rate is 0.044 m^3/s (1 mgd) and the temperature is 15°C. The air-to-water flow ratio (m^3 air/m^3 water) has been set at 11.4/1 and the water loading rate (velocity) at 0.05 m^3/m^2-s in order to have a pressure drop of 10 kg/m^2 per meter of tower height.

(a) Determine the tower height and diameter and justify your selection of an overall volumetric mass transfer coefficient.

(b) What would the TCE exit concentration be if the feed concentration increases to 200 μg/L?

(c) Will the existing design meet its treatment objective if benzene also appears in the groundwater at a concentration of 150 μg/L and it is necessary to reduce its concentration to 5 μg/L? If not, what new air-to-water flow ratio must be selected, and what are the implications for operating costs? [*Hint:* See Example 10-6 for the temperature dependence of $\mathcal{K}_H$ values for TCE and benzene.]

10-16 A wastewater is reaerated before discharge to a river by air spargers located along the bottom of a channel. The dissolved oxygen (DO) at the end of this channel is found to be too low. A decision is made to increase the aeration rate (i.e., increase the overall volumetric mass transfer coefficient) in the reaeration channel. We wish to determine the unsteady state response of DO within the channel as well as the new steady state. Assume that velocity and dispersion are constant and one-directional.

(a) Develop the dimensionless form of the partial differential equation that describes macrotransport of oxygen in the channel by advection and dispersion and absorption of oxygen from air sparging. Use the following dimensionless parameters:

$$c^\circ = \frac{C}{C_{\text{IN}}} \qquad c_S^\circ = \frac{C_S}{C_{\text{IN}}} \qquad \theta^\circ = \frac{t}{\bar{t}} \qquad \text{and} \qquad \zeta^\circ = \frac{z}{L}$$

where C_{IN} is the oxygen concentration at $z = 0$, C_S is the saturation concentration of oxygen in equilibrium with air, $\bar{t}$ is the HRT of the channel, and L is the length of the channel.

(b) Identify two dimensionless groups that result from the de-dimensionalization, describe their physical significance, and discuss how each group characterizes process performance.

10-17 The design relationship given in Equation 10-87 for the number of transfer units required to obtain a given stripping efficiency was predicated on an influent gas stream free of the contaminant to be stripped from a given water. This will not always be the case, and certainly not for gas streams entering downstream strippers in multistage operations. Develop an appropriate expression for NTU for cases in which $C_{g,\text{IN}} \neq 0$.

Answer:
$$NTU = \frac{R_S}{1 - R_S} \ln \left[\frac{C_{l,\text{IN}} - (C_{g,\text{IN}}/\mathcal{K}_H^\circ)}{C_{l,\text{OUT}} - (C_{g,\text{IN}}/\mathcal{K}_H^\circ)} \frac{R_S - 1}{R_S} + \frac{1}{R_S} \right].$$

11

REACTOR ENGINEERING: UNSTEADY-STATE SYSTEMS

11.1 INTRODUCTION

11.1.1 Unsteady Conditions: Causes and Effects

Unsteady conditions occur when temporal transients in the concentrations of substances within control volumes take place or when the size of control volumes change with time. The spatial scales of temporal transients encompass those of mega-, macro-, and microsystems. Despite their importance in the analysis of both natural and engineered systems, unsteady-state models are surprisingly underdeveloped in most textbooks. In this chapter we address conceptual approaches that find applications in a variety of problems at various spatial scales. The differences between unsteady-state conditions prevalent at the mega- and macroscale and those prevalent at the microscale are emphasized in several sections.

Transient conditions are reflected in the *accumulation term*, $d(VC)/dt$, of the material balance equation. For a transient to be produced there must be a *forcing function.* At the mega- and macroscales, the most common type of forcing function is a change in the rate of molar or mass input caused by variations in feed concentration and/or flow rate. Although less common, a reaction rate coefficient that does not remain constant is another form of forcing function at these scales. Rate coefficients can vary due to changing environmental conditions (e.g., temperature, sunlight, and surface turbulence if interfacial transport is important) or changing concentrations of other constituents that are directly involved in, catalyze, or inhibit the reaction(s) of interest. We are concerned with unsteady-state conditions at the microscale, largely in terms of diffusive mass transfer processes. The forcing function in such instances is commonly a change in concentration, either at the boundary of, or within, the diffusion domain.

11.1.2 Issues of Scale

Unsteady-state forcing functions at the mega- and macroscale vary greatly in terms of their dependence on time. Flow rates fluctuate daily, or more frequently in engineered systems because of operating conditions and use-generation cycles. In natural systems they may vary only seasonally, and even over longer-term cycles, due to changes in hydrologic and climatic conditions. Various stochastic models are used to determine most likely flow-rate patterns, and these require examination of historical records using such techniques as *time-series analysis.* Concentrations of components of interest may also vary. For example, irregular, accidental or otherwise uncontrolled releases of chemicals to environmental systems typically produce unsteady conditions. These conditions may differ considerably from one system to another. Rupture of an underground storage tank may produce a sudden release of contaminants to a groundwater at a virtual instant in time, while the release of contaminants from NAPL residuals in the vadose zones of subsurface systems are dependent on infiltration events and groundwater levels and are thus sporadic in time. In both instances, there are temporal variations in the concentrations of contaminants entering the groundwater. The subsequent transport and reactions (e.g., sorption onto soil grains, biodegradation, etc.) of such contaminants are thus also time dependent, or unsteady.

The responsible forcing function is generally more readily identified and more easily described in engineered systems than it is in natural systems. For example, variations in flow rate experienced daily in small wastewater treatment plants are cyclical due to diurnal patterns of water use. Repeatable patterns in the concentrations of industrial chemicals may also be experienced based on routine production schedules, although in addition there are random events due to accidental spills and batch discharges. The occurrence of *unsteady-state conditions* is of concern in engineered systems because treatment facilities and operations are generally designed on the basis of *steady-state or average conditions*. Overloading and underloading of systems designed for steady-state operations may each detract from process performance. In these situations, unsteady-state analysis can be used to design *flow* and *concentration equalization* facilities ahead of treatment units to stabilize the operation of those units at or near steady-state conditions, and thus improve their performance.

The design basis for certain types of treatment units may be intrinsically the unsteady state. The unsteady operating character of sequencing-batch reactors (SBRs), for example, was described in Chapter 9. The design basis for such systems is unsteady state because both bulk volumes and concentrations of reactants within the reactor(s) inherently vary in time. Unsteady conditions at the megascopic and macroscopic, or *reactor* scales, are discussed in Section 11.2.

Fixed-bed reactors (FBRs) containing activated carbon or ion-exchange materials are also designed for unsteady-state operation, although in this instance the unsteady condition occurs at the microscale of sorbent or ion-exchange particles and their surrounding boundary layers. In these unsteady microscopic processes, solute is transported by film diffusion through the hydrodynamic boundary layer and then by diffusion internal to the porous particles. Adsorption or ion-exchange reactions occur only when the diffusing components reach available sites within the particles. Diffusion rates are unsteady in such instances because the driving forces for solute separation decrease with time as available sites become exhausted. Unsteady conditions at the microscopic, or *reaction scale*, are discussed in Section 11.4.

Unsteady responses to certain forcing functions can sometimes be estimated using an approach referred to as *quasi-steady-state analysis*. For example, a series of steady states can be used to determine responses to long-term variations in flow if reaction rates are fast. In such situations a new steady state will be reached after the change in input concentration or flow rate to the control volume because the response time is short. A good approximation to the unsteady concentration leaving a control volume is thus obtained from a steady-state analysis at each flow rate. Quasi-steady state is also a useful assumption at the microscale. For instance, if the time rate of change of concentration at the entrance to a diffusion domain is much slower than the time rate of change of the concentration gradient within the domain, steady-state (Fick's first-law) rather than unsteady-state (Fick's second law) diffusion can be assumed. Quasi-steady state thus provides a convenient way of simplifying the analysis of an unsteady-state problem.

11.2 SIMPLE TRANSIENTS AT MACROSCALES AND MEGASCALES

Periodic changes in flow rate and concentration are smoothed or dampened in large well-mixed volumes. When this principle is applied to facilitate the design of treatment facilities for near-steady-state operation, the resulting process is referred to as *equalization.* The design objective of equalization is to provide enough volume and mixing in one or more upstream tanks or reservoirs to smooth variations in flow and concentration to obtain near-steady-state feeds to downstream reactors. The same principles involved in analyzing equalization processes for treatment operations apply to the analysis of connected bodies of natural waters, such as a string of lakes; each of these mixed volumes provides for dampening of changes in pollutant discharges and runoffs.

Flow equalization is a simple concept having many practical applications. It can be used to dampen variations in dry and wet weather flows at municipal wastewater treatment plants and to smooth flows entering stormwater treatment facilities. In addition, flow equalization is important in industrial wastewater treatment because flows and concentrations may vary widely due to batch operations in manufacturing. Some treatment units, while not being designed specifically as equalization facilities, also provide this important function. For example, pre-sedimentation tanks are used as a first step at water treatment plants on highly turbid river water supplies. These tanks serve to equalize fluctuations in the turbidities of raw waters. In a similar manner, primary sedimentation tanks in municipal wastewater treatment operations often function as equalization tanks thereby smoothing, for example, any intermittent loadings of industrial wastes.

11.2.1 Flow Transients and Equalization

Figure 11-1 shows two types of flow-equalization schemes. The on-line tank provides equalization continuously, whereas the off-line tank functions only when the flow rate exceeds a preset value, emptying later when the flow rate recedes. Whether the equalization volume has the mixing characteristics of a CMFR, PFR, or nonideal reactor is not important for flow equalization, but it is very important for concentration equalization.

The water balance for an on-line flow-equalization tank or "reactor" is

$$Q(t)_{\mathrm{IN}} - Q_{\mathrm{OUT}} = \frac{dV_R(t)}{dt} \tag{11-1}$$

where $Q(t)_{\mathrm{IN}}$ and Q_{OUT} are the respective flow rates in and out of the tank, respectively, and $V_R(t)$ is the volume of liquid (water) contained in the tank at any time, t. Once the time-dependent nature of $Q(t)_{\mathrm{IN}}$ and the desired Q_{OUT} are known, variations in tank volume can be determined. Cyclical variations

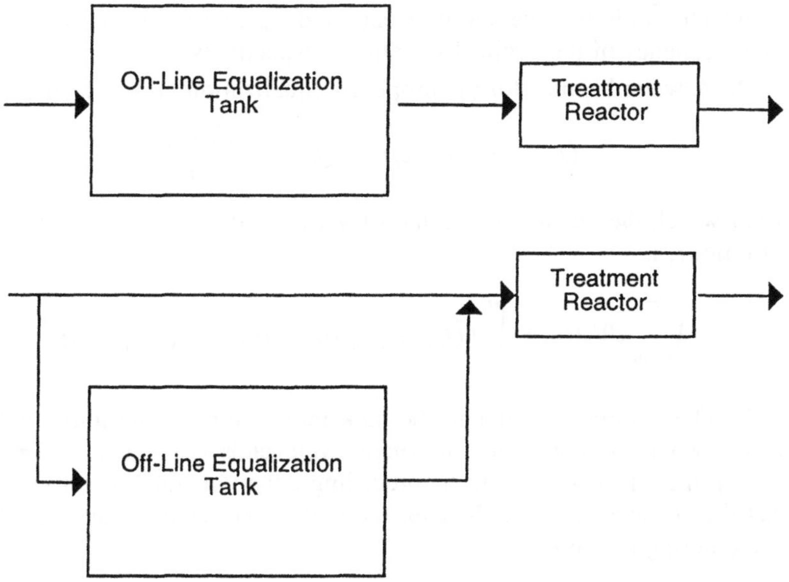

Figure 11-1 Alternative equalization tank arrangements.

in flow rate at treatment facilities can very often be approximated as sinusoidal, with the amplitude and frequency of the sinusoidal cycle depending on local conditions.

An illustration of a sinusoidal approach to the design of an on-line equalization tank is given in Example 11-1.

Example 11-1

- **Situation:** The design engineer for a municipal waste treatment facility wishes to explore the cost-effectiveness of placing an equalization tank ahead of a particular treatment unit. Determination of whether a net savings can be achieved by such a design depends on estimating the cost of equalization, and this, in turn, depends on determining the volume of the equalization tank.
- **Question(s):** A general model is sought whereby the sensitivity of the equalization volume to the amplitude and frequency of flow variations that approximate a sinusoidal cycle can be assessed.
- **Logic and Answer (s):**

1. A general sinusoidal model has the form

$$Q_{IN}(t) = \overline{Q}_{IN} + \alpha_m \sin \omega_f t$$

in which $\overline{Q}_{IN}$ is the average flow rate and α_m and ω_f are the amplitude and frequency of the cyclical variation, respectively.

2. A water balance for the proposed equalization tank is then

$$\overline{Q}_{IN} + \alpha_m \sin \omega_f t - Q_{OUT} = \frac{dV_R(t)}{dt}$$

from which the change in volume of water in the tank over an interval of time, t, is

$$\int_{V_R(0)}^{V_R(t)} dV_R(t) = \int_0^t (\overline{Q}_{IN} + \alpha_m \sin \omega_f t)\, dt - Q_{OUT} \int_0^t dt$$

3. The volume of water in the tank must return to the initial value, $V_R(0)$, when the complete sinusoidal cycle ends (i.e., at $t = 2\pi/\omega_f$). Setting the left-hand side of the preceding equation equal to zero proves that the constant flow rate leaving the tank must equal the average flow rate entering the tank:

$$Q_{OUT} = \frac{\overline{Q}_{IN} t \Big|_0^{2\pi/\omega_f} - \dfrac{\alpha_m}{\omega_f} \cos \omega_f t \Big|_0^{2\pi/\omega_f}}{2\pi/\omega_f} = \overline{Q}_{IN}$$

4. From the last two equations, the volume of water to be stored at any time, t, is

$$V_R(t) = \frac{\alpha_m}{\omega_f}(1 - \cos \omega_f t) + V_R(0)$$

and the maxima and minima in storage volume, V_S, are found by setting $dV_R/dt = 0$:

$$\alpha_m \sin \omega_f t = 0$$

5. The maximum storage requirement occurs at $t = \pi/\omega_f$ and the minimum at $t = 2\pi/\omega_f$. It is reasonable that a maximum occurs at $t = \pi/\omega_f$ because this time corresponds to the midpoint of the cycle when the flow rate begins to drop below the average value. The maximum water storage volume at $t = \pi/\omega_f$ is then

$$V_S = \frac{2\alpha_m}{\omega_f} + V_R(0)$$

The maximum volume required is thus determined by knowing the frequency of the cycle and its amplitude. The storage volume–time relationship for one cycle is shown below.

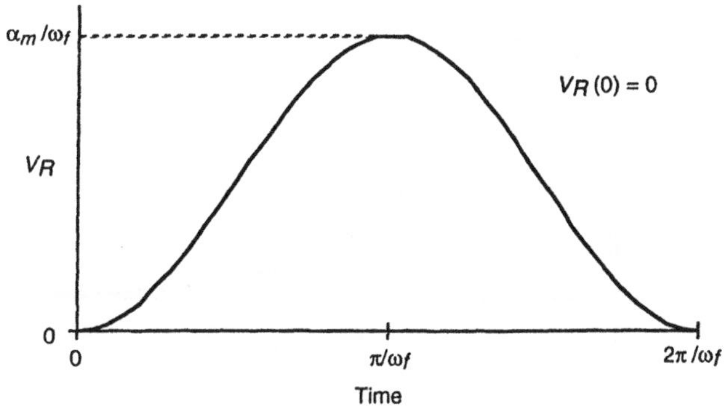

6. As an example, suppose that a daily cycle occurs such that $\omega_f = 2\pi\ \text{day}^{-1}$ and the amplitude of the variation in flow rate is two times the average flow rate, $\alpha_m = 2\overline{Q}_{\text{IN}}$. The storage volume needed for equalization is then

$$V_S = \frac{2\overline{Q}_{\text{IN}}}{2\pi} + V_R(0) = 0.64\overline{Q}_{\text{IN}} + V_R(0)$$

or 64% of the daily volumetric flow assuming that $V_R(0) = 0$. This equalization tank or reactor will be very large compared to those of other unit operations in the plant and thus its need must be considered carefully.

For flow equalization by off-line storage, which for example is often used to capture and treat combined sewer overflows, the water balance relationship is straightforward:

$$Q(t) = \frac{dV_R(t)}{dt} \tag{11-2}$$

where $Q(t)$ is the flow rate in excess of the average flow rate. If the nature of this forcing function is known, the storage volume can be determined by integration of Equation 11-2. An illustration is provided in Example 11-2.

Example 11-2

- **Situation:** A storage facility is to be designed to capture the entire volume of a maximum, 48-h combined sewer overflow event that occurs once a year. The combined sewer overflow rate–time relationship generated from historical records is shown below.

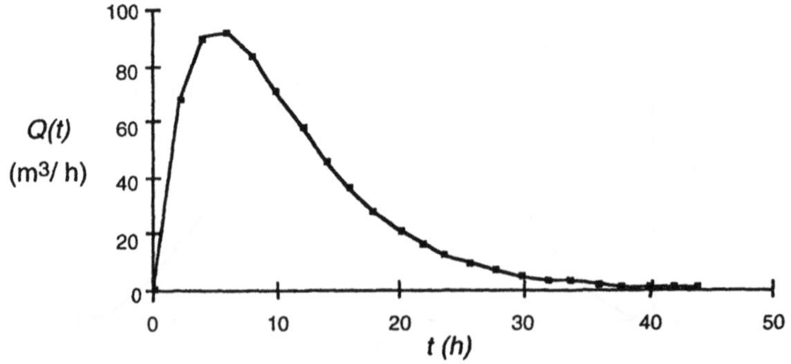

The following empirical relationship describes the event:

$$Q(t)(\mathrm{m^3/h}) = 500(e^{-0.15t} - e^{-0.25t})$$

in which t is in hours.

- **Question(s):** What is the storage volume required, and what flow rate is needed to empty the basin in 2 weeks' time?
- **Logic and Answer (s):**

1. The water balance is

$$\frac{dV_R(t)}{dt} = 500(e^{-0.15t} - e^{-0.25t})$$

2. When integrated, this material balance relationship gives the required storage volume, V_S:

$$V_S = 2000e^{-0.25\times 48} - 3333e^{-0.15\times 48} = 1330\ \mathrm{m^3}(352{,}078\ \mathrm{gal})$$

3. The flow rate required to empty the volume in 2 weeks is

$$Q_{\mathrm{OUT}} = \frac{1330\ \mathrm{m^3}}{14 \times 24\ \mathrm{h}} = 4\ \mathrm{m^3/h}\ \text{or}\ 1048\ \mathrm{gal/h}$$

11.2.2 Concentration Transients and Equalization

The conceptual framework for understanding constituent concentration equalization is closely related to that of residence time distribution analysis. As demonstrated in Chapter 9, RTD analysis provides a measure of the extent of mixing between fluid elements in a tank or reactor. Mixing is maximized in a CMFR, whereas no mixing occurs in an ideal PFR. Consequently, changes in exit concentrations corresponding to specific changes in feed concentration are minimized in systems having the mixing characteristics of a CMFR. It is for this reason that a CMFR may be selected over a PFR in certain treatment processes even though the PFR might require less tank volume. Real reactors have mixing characteristics between those of CMFRs and PFRs. The equalization capacities of these reactors will therefore be less than those of CMFRs but more than those of PFRs.

The analysis of concentration equalization will be presented in a general way that is applicable to a wide range of situations in both natural and engineered systems. In some situations it is of interest to determine the equalization capability of a tank or reservoir of fixed volume and geometry. This is true, for example, for a description of lakes and ponds which serve to equalize stormwater runoff. In design situations, conversely, the objective is to select a volume and geometry that can provide a specified extent of equalization; this is typically encountered in the dampening of industrial waste concentrations from different manufacturing processes.

We focus our attention first on the CMFR because it has the greatest equalization capability. A general form of the material balance relationship which accounts for all possible forcing functions can be written as

$$Q(t)_{IN}C(t)_{IN} - Q(t)_{OUT}C(t)_{OUT} + r(t)V_R(t) = \frac{d(V_R(t)C_{OUT})}{dt} \tag{11-3}$$

Simultaneous temporal variations in feed flow rate, $Q(t)_{IN}$, and feed concentration, $C(t)_{IN}$, can introduce complex transients. Temporal variations in feed flow rate may lead to transients in effective reactor (reaction) volume, $V_R(t)$ and exit flow rate, $Q(t)_{OUT}$; these depend on the type of outlet flow control device (e.g., overflow weir, submerged pipe, etc.). Transients are usually less complex to analyze when only one parameter (i.e., feed flow rate or feed concentration) changes in time. For all but zero-order reactions, any change in feed flow rate and/or concentration that results in a change in concentration within the CMFR produces a transient in reaction rate, $r(t)$; the rate expression is then of the form $r(t) = \phi(k, C(t)_{OUT})$. Moreover, as noted earlier, certain reactions may require inclusion of a variable rate coefficient, $k(t)$; in these instances, $k(t)$ acts as a forcing function even when feed flow rate and concentration are constant.

The ease with which Equation 11-3 is integrated to give C_{OUT} as a function of time depends on the mathematical complexity of the forcing functions, $Q_{IN}(t)$, $C_{IN}(t)$, and $k(t)$. These functions may vary in a well-known and physically explainable manner with time. Conversely, their time-dependent nature in real-world situations may be more stochastic than deterministic, thus creating a virtually limitless number of possible mathematical descriptions. Various classical transformations (Laplace, Fourier, Bessel) can be used effectively in a number of situations.

Our analysis of mass or concentration transients begins with a simple, yet practical illustration of an unsteady-state condition for which the mathematical solution is straightforward. Consider the case in which the forcing function is a sudden change in feed concentration while flow rate remains constant. This

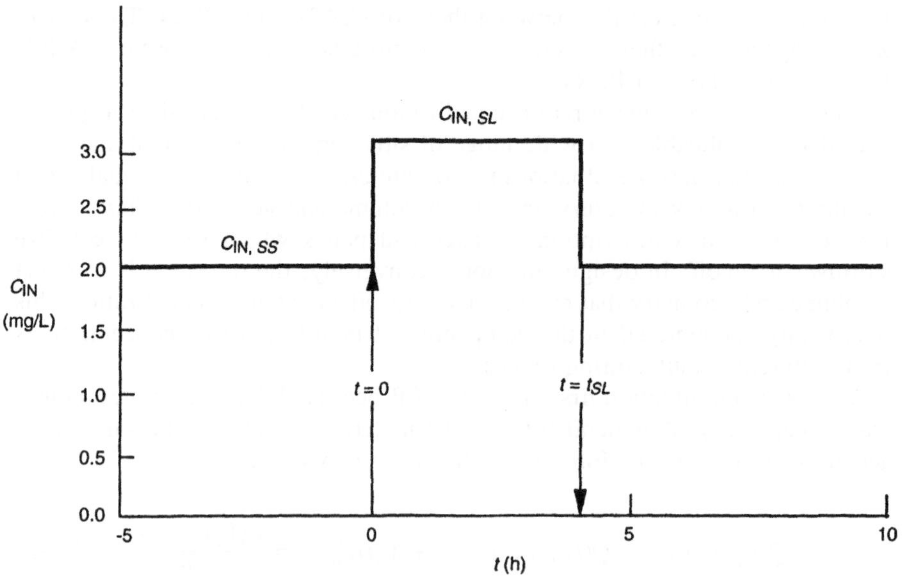

Figure 11-2 Shock load in feed concentration.

function is shown graphically in Figure 11-2. Prior to the change in feed concentration, the system is assumed to be at steady state. A forcing function described by such a step change in feed concentration is often referred to as *shock loading* (SL). The feed concentration quickly rises to a value of $C_{IN,SL}$ and then stays constant for a period of time, t_{SL}, before returning to the original steady-state value, $C_{IN,SS}$. We further assume for illustrative purposes that the reaction rate within the CMFR is first order.

The material balance written specifically for the time period of the shock load is

$$QC_{IN,SL} - QC_{OUT} - kC_{OUT}V_R = V_R \frac{dC_{OUT}}{dt} \tag{11-4}$$

This differential equation can be solved by Laplace transforms or by separation of variables. The solution is

$$-\frac{1}{1 + k\bar{t}} \ln\left[C_{IN,SL} - (1 + k\bar{t})C_{OUT}\right] = \frac{t}{\bar{t}} + \kappa \tag{11-5}$$

in which $\bar{t}$ is the HRT. The constant of integration is determined from the steady-state exit concentration that existed before the shock load event; that is,

$$C_{OUT} = C_{OUT,SS} = \frac{C_{IN,SS}}{1 + k\bar{t}} \quad \text{at } t = 0 \tag{11-6}$$

The specific solution needed for prediction of the unsteady-state exit concentration is thus

$$C_{\text{OUT}} = \frac{C_{\text{IN},SL}}{1 + k\bar{t}}\left\{1 - \exp\left[-\left(\frac{1 + k\bar{t}}{\bar{t}}\right)t\right]\right\} + C_{\text{OUT},ss}\exp\left[-\left(\frac{1 + k\bar{t}}{\bar{t}}\right)t\right] \quad \text{for } 0 < t < t_{SL} \tag{11-7}$$

A generalized plot of this response is given in Figure 11-3. The exit concentration is seen to increase during the shock load. If the shock loading continues, the exit concentration will asymptotically approach a new steady-state level defined by the new feed concentration:

$$C_{\text{OUT}} = \frac{C_{\text{IN},SL}}{1 + k\bar{t}} \tag{11-8}$$

The instantaneous return of the feed concentration to its original steady-state value, $C_{\text{IN},ss}$, at some time t_{SL} signals the end of the shock load. In response, the exit concentration begins a decline toward the original steady-state exit concentration (see Equation 11-6). To predict the decrease in exit concentration with time, we again write the material balance:

$$QC_{\text{IN},ss} - QC_{\text{OUT}} - kC_{\text{OUT}}V_R = V_R\frac{dC_{\text{OUT}}}{dt} \tag{11-9}$$

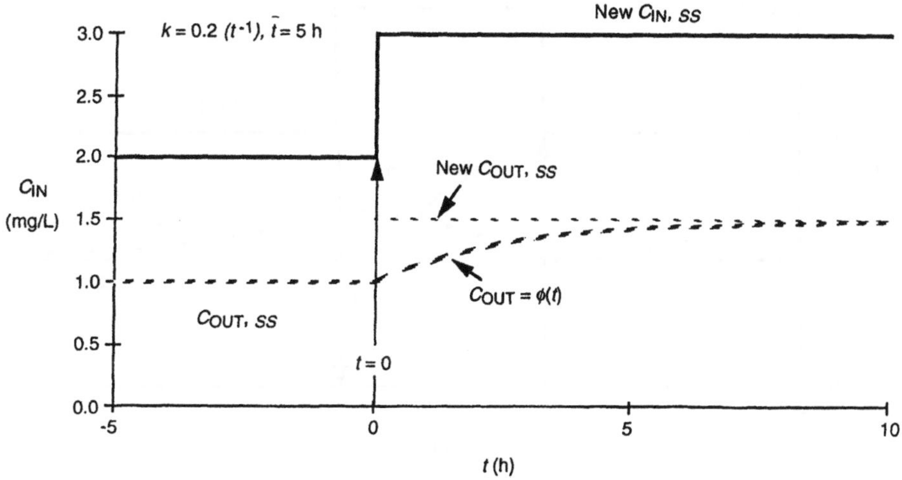

Figure 11-3 Response of a CMFR to a step shock load in feed concentration.

After separation of variables and integration, we have

$$-\frac{1}{1 + k\bar{t}} \ln \left[C_{IN,SS} - (1 + k\bar{t})C_{OUT}\right] = \frac{t}{\bar{t}} + \kappa \qquad (11\text{-}10)$$

The constant of integration κ is evaluated from the condition at the *end of the shock loading period:*

$$C_{OUT} = C_{OUT,t_{SL}} \qquad \text{at } t = t_{SL} \qquad (11\text{-}11)$$

Thus the concentration history *following the shock load* is

$$C_{OUT} = \frac{C_{IN,SS}\left[1 - \exp\left\{-(1 + k\bar{t})\dfrac{t_{SL} - t}{\bar{t}}\right\}\right]}{1 + k\bar{t}} + C_{OUT,t_{SL}} \exp\left\{-(1 + k\bar{t})\frac{t_{SL} - t}{\bar{t}}\right\} \qquad \text{for } t > t_{SL} \qquad (11\text{-}12)$$

The entire history of the exit concentration during (Equation 11-7) and after the shock load (Equation 11-12) is shown in Figure 11-4. From inspection of Equation 11-12, we can see that as $t \to \infty$, the original steady state (Equation 11-6) will be approached. This observation also leads to the conclusion that the original steady state will never be reached in a finite time period. Thus the exit concentration at the beginning of the next shock load will not be equal to that before the first shock load. The value of C_{OUT} needed to solve Equation

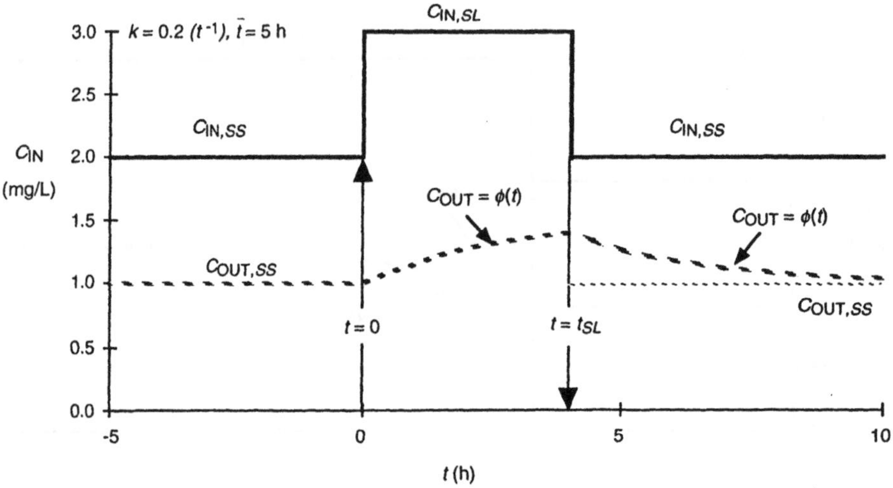

Figure 11-4 Response of a CMFR to a pulse shock load in feed concentration.

11-5 for the next shock load must be determined from Equation 11-12. Intuitively, the shorter the interval between repeated shock loads, the higher the peak concentration at the end of each shock load. It follows that equalization processes function best if the time interval between shock loads is long enough to allow exit concentrations to return fairly close to the original steady-state values.

The analysis of unsteady state response to a shock load is much simpler if there is no reaction in the control volume. This situation is illustrated in Example 11-3.

Example 11-3

- **Situation:** A water supply is located on a river that is subject to large and sudden variations in turbidity during heavy rainfalls. A pre-sedimentation basin having an HRT of 15 h typically provides for a reduction of 30% in turbidity. Alum is currently dosed into a rapid-mix tank following pre-sedimentation, but a polymer is being considered as a replacement for alum to improve coagulation and lessen sludge production. The appropriate polymer dosage is approximately related stoichiometrically to the concentration of particles, the latter being measured operationally by nephelometric turbidity units (NTUs).
- **Question(s):** Past experience suggests that the worst-case condition is a sudden increase in raw water turbidity from 10 NTU to 40 NTU, which then remains at that level for 12 h. The maximum turbidity that can be treated economically by the polymer is 20 NTU. Is sufficient equalization provided by the existing pre-sedimentation tank to prevent the turbidity from exceeding 20 NTU?
- **Logic and Answer(s):**

1. A material balance on turbidity must be written for the unsteady-state condition specified in the pre-sedimentation basin. No information is available on the mixing characteristics, but in order to make an estimate of equalization, we assume that it behaves as a CMFR. Although some turbidity removal is expected (30%) based on past performance data, the rate and mechanism of removal are unknown.

2. As a conservative calculation of equalization, we assume that no turbidity removal occurs. Accordingly, the reaction rate is zero and

$$Q_0 C_{\text{IN},SL} - QC_{\text{OUT}} = V_R \frac{dC_{\text{OUT}}}{dt}$$

3. A worst-case step increase or shock load in which turbidity increases from 10 NTU to 40 NTU and remains at this level for 12 h is then assumed. The general form of solution for this type of forcing function for a first-order reaction was given by Equation 11-7:

$C_{OUT} =$

$$\frac{C_{IN,SL}}{1+k\bar{t}}\left\{1 - \exp\left[-\left(\frac{1+k\bar{t}}{\bar{t}}\right)t\right]\right\} + C_{OUT,SS}\exp\left[-\left(\frac{1+k\bar{t}}{\bar{t}}\right)t\right]$$

4. The solution above can be used for a reaction rate of zero by setting $k = 0$ and, correspondingly, $C_{OUT,SS} = C_{IN,SS}$. Equation 11-7 then simplifies to

$$C_{OUT} = C_{IN,SL} + (C_{IN,SS} - C_{IN,SL})e^{-t/\bar{t}}$$

The same result would also be obtained by direct integration of the material balance expression given in step 2.

5. The objective is to determine whether the existing hydraulic residence time, $\bar{t}$, is sufficient to prevent C_{OUT} from exceeding 20 NTU at the end of the shock load period of 12 h. Solving the equation above for $\bar{t}$ gives

$$\bar{t} = \frac{t}{\ln\dfrac{C_{IN,SL} - C_{IN,SS}}{C_{IN,SL} - C_{OUT}}} = \frac{12}{\ln\dfrac{40 - 10}{40 - 20}} = 29.6 \text{ h}$$

6. This calculation shows that the existing pre-sedimentation basin with an HRT of 15 h will not accomplish the necessary equalization of turbidity. The turbidity at the polymer feed point will instead be

$$C_{OUT} = C_{IN,SL} + (C_{IN,SS} - C_{IN,SL})e^{-t/\bar{t}} = 40 - 30e^{-12/15} = 26.5 \text{ NTU}$$

7. Removal of turbidity by sedimentation was ignored in the calculations above. An estimate of this effect is difficult without more information on the rate process. However, a first-order rate expression with a rate coefficient, k_s, obtained from performance data for steady state can be used to make a first-cut estimate of removal rate. Using the CMFR design equation (Equation 11-6), an HRT of 15 h, and 30% removal of turbidity, we find that

$$k_s = \frac{(C_{IN,SS}/C_{OUT,SS}) - 1}{\bar{t}} = \frac{1/0.7 - 1}{15} = 0.028 \text{ h}^{-1}$$

Substituting $t = 12$ h, $\bar{t} = 15$ h, $k_s = 0.028$ h^{-1}, $C_{IN,SL} = 40$ NTU, and $C_{OUT,SS} = 0.7 \times 10 = 7$ NTU in Equation 11-7 gives

$$C_{OUT} = 21.3 \text{ NTU} \qquad \text{when } t = 12 \text{ h}$$

8. The results above are intuitively reasonable. The effect of turbidity removal within the presedimentation basin is to decrease the impact of the shock load. In this instance, accounting for removal reduces the maximum turbidity from 26.5 to 21.3 NTU. Hence, if turbidity removal were to be a first-order process, the turbidity following the pre-sedimentation basin is only slightly in excess of the maximum allowable for economical polymer dosing. More data would be required to determine if this assumption is reasonable.

11.3 COMPLEX TRANSIENTS AT MACROSCALES AND MEGASCALES

11.3.1 Harmonics in Flow Rates and Mass Concentrations

Equalization problems become more complex when considering random fluctuations in feed concentration or flow rate, and equalization basins that cannot be described as CMFRs. One approach to describing random fluctuations mathematically is to decompose them into *harmonic* or *cyclical* components of different frequencies and amplitude using a Fourier series. Such an analysis assumes that the cyclical patterns are independent and thus that their sum describes the total fluctuation. For example, temporal variations in concentration at an industrial waste treatment facility may result from a series of cyclical patterns of production and equipment washing that range from daily to monthly to seasonally cycles. Some fluctuations, however, are truly random and must be analyzed with more sophisticated mathematical tools. Once random fluctuations can be described mathematically, their attenuation in PFDRs and CMFRs can be predicted.

Fourier series analysis depends on finding the Fourier transform that describes the variation(s) in influent concentration, $C(t)_{\mathrm{IN}}$, or flow $Q(t)_{\mathrm{IN}}$, to an equalization device. The Fourier transform, for example, for $C(t)_{\mathrm{IN}}$ is

$$\begin{aligned} Z_F(\omega_f) &= \int_{-\infty}^{\infty} C(t)_{\mathrm{IN}} \cos(\omega_f t)\, dt + i \int_{-\infty}^{\infty} C(t)_{\mathrm{IN}} \sin(\omega_f t)\, dt \\ &= \alpha(\omega_f) + i\beta(\omega_f) \\ &= \alpha_m \cos \phi_s(\omega_f) + \alpha_m \sin \phi_s(\omega_f) \end{aligned} \tag{11-13}$$

where ω_f is the frequency (2π times the number of cycles per unit time), ϕ_s is the phase shift, and α_m is the amplitude, given by

$$\alpha_m = ([\alpha(\omega_f)]^2 + [\beta(\omega_f)]^2)^{0.5} \tag{11-14}$$

The inverse Fourier transform is thus

$$C(t)_{\mathrm{IN}} = \frac{1}{2\pi} \int_{-\infty}^{\infty} \alpha_m \cos [\omega_f - \phi_s(\omega_f)]\, d\omega_f \tag{11-15}$$

for which the integrand is a sinusoid having a specific frequency, phase shift, and amplitude.

In analysis of historical records, estimates of $Z_F(\omega_f)$ are obtained at particular values of ω_f based on values of $C(t)_{\mathrm{IN}}$; this is referred to as a *discrete Fourier transform.* The resulting Fourier inversion formula is represented by

$$C(t)_{\mathrm{IN}} = \overline{C}_{\mathrm{IN}} + \sum_{j=1}^{n} (\alpha_j \cos \omega_{f,j}t + \beta_j \sin \omega_{f,j}t) + e_r(t) \tag{11-16}$$

in which $\overline{C}_{IN}$ is the average feed concentration, $\omega_{f,j} = \pi_j/n$, and $e_r(t)$ is an error term that decreases as n increases. The objective is to determine the coefficients, α_j and β_j, of the Fourier series that best describes the feed concentration–time data set. Not all the coefficients need to be used in application of the Fourier series. An alternative and often preferable procedure is to evaluate the *spectral density* of $C(t)_{IN}$, identify a few values of $\omega_{f,j}t$ for which there are significant peaks, and regress $C(t)_{IN}$ on the corresponding cos $\omega_{f,j}\,t$ and sin $\omega_{f,j}t$ values. A practical problem that is often encountered in real data analysis is whether the length of record and quality of data are sufficient to assure a good predictive tool.

The usefulness of Fourier transforms can be illustrated by a simple hypothetical example. The continuous temporal variation in concentration given by the uppermost line in Figure 11-5a has been synthesized by summing together three sinusoidal functions, each having the same average concentration ($\overline{C}_{IN}$) but a different frequency (ω_f), amplitude ($C_{IN,MAX}$), and phase shift (ϕ_s):

$$C_{IN}(\omega_f, t) = \overline{C}_{IN} + C_{IN,MAX}[\sin \omega_f(t - \phi_s)] \tag{11-17}$$

Of the three frequencies in Figure 11-5a that having a value of $\omega_f = 0.1\ \text{h}^{-1}$, or one cycle approximately every 63 h, is the most important source of variance (σ^2) in concentration (Figure 11-5b) because its amplitude is twice that of the other two frequencies. However, more detail on overall variance is given by including the other two sinusoidal functions.

Analytical solutions for transients in exit concentration, $C(t)_{OUT}$, from a CMFR or PDFR are available when the forcing function is sinusoidal. These solutions show that the corresponding variations in exit concentration with time are also sinusoidal, but dampened due to concentration equalization. The extent of dampening of each sinusoidal function of frequency ω_f is quantified by the *amplitude gain factor*, $\alpha_g(\omega_f)$:

$$C_{OUT}(\omega_f, t) = \alpha_g(\omega_f)\, C(t)_{IN}(\omega_f, t) \tag{11-18}$$

The lower the amplitude gain factor, the more effective is dampening. The amplitude gain factor can be expressed as

$$\alpha_g(\omega_f) = \frac{\text{exit concentration variance}}{\text{feed concentration variance}} = \frac{\sigma^2_{C,OUT}}{\sigma^2_{C,IN}} \tag{11-19}$$

The variance in feed concentration over any selected time period, t, is simply

$$\sigma^2_{CIN} = \frac{1}{t}\int_0^t (C(t)_{IN} - \overline{C}_{IN})^2\, dt = \frac{(C_{IN,MAX})^2}{2} \tag{11-20}$$

The variance in exit concentration can be obtained from the analytical solutions for CMFRs and PFDRs with sinusoidal input functions. These definitions lead to a convenient mathematical expression of equalization capability in terms of

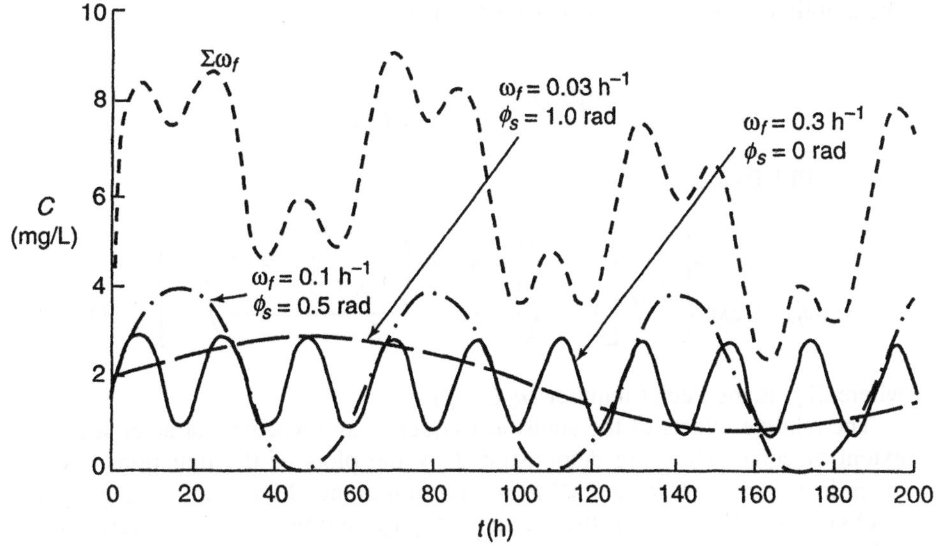

(a) Example of Fourier Series Analysis

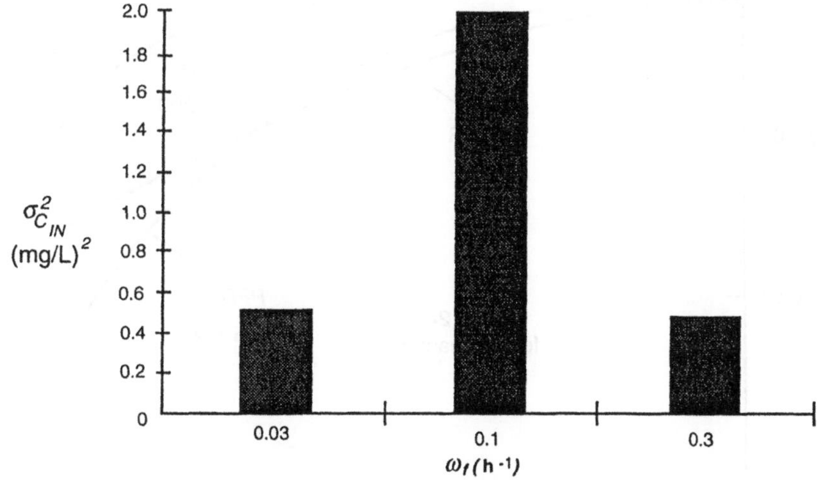

(b) Variance Associated with Each Frequency

Figure 11-5 Fourier series analysis.

the amplitude gain factor, which for CMFRs is,

$$\alpha_g(\omega_f) = \frac{1}{[1 + (\omega_f \bar{t})^2]^{0.5}} \tag{11-21}$$

and for PFDRs is,

$$\alpha_g(\omega_f) = \exp\left\{\frac{\mathfrak{N}_{Pe}}{2}\left[1 - \left(\frac{\left\{1 + 16\left(\frac{\omega_f \bar{t}}{\mathfrak{N}_{Pe}}\right)^2\right\}^{0.5} + 1}{2}\right)^{0.5}\right]\right\} \tag{11-22}$$

where $\mathfrak{N}_{Pe}$ is the Péclet number ($v_x L/\mathfrak{D}_d$).

The relative effects of the sinusoid frequency and reactor characteristics on extent of equalization are demonstrated by the plots of the equations above given in Figure 11-6 for a CMFR and two different PFDRs, all having equal HRTs of 24 h. Fluctuations that occur at long cycle times (i.e., one cycle every 628 h or at frequencies below 0.01 h^{-1}) compared to the HRT of 1 day are not significantly dampened by either a CMFR or a PFDR [i.e., $\alpha_g(\omega_f)$ is close to 1]. However, the extent of dampening increases sharply as the cycle time ap-

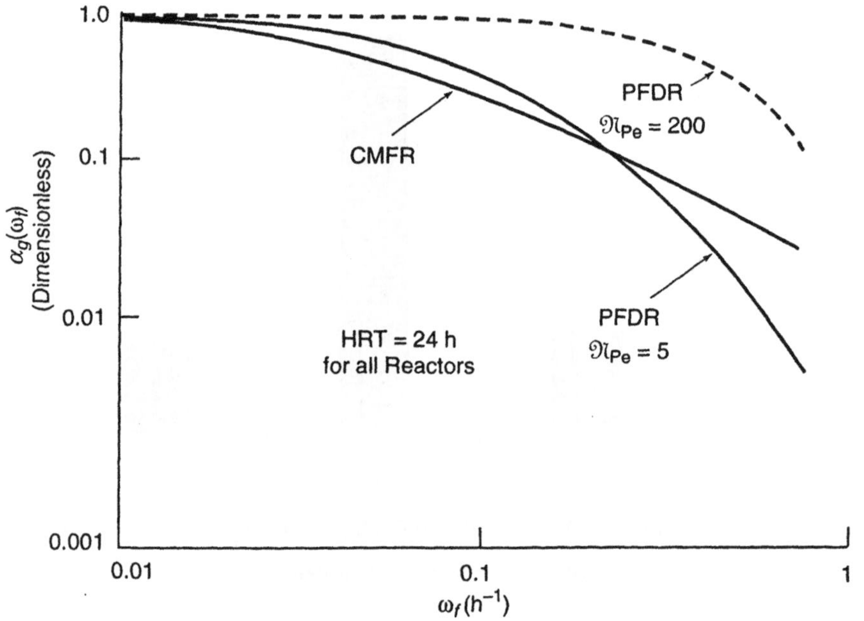

Figure 11-6 Dampening of sinusoidal variations in CMFRs and PFDRs for different degrees of mixing.

proaches the value of the HRT. For instance, the amplitude gain factor decreases to 0.1, corresponding to a 90% decrease in variance, for a frequency in the range of 0.4 h^{-1} (about 1.5 cycles/day). In other words, an equalization basin with an HRT that is 50% longer than the time for one cycle in feed concentration is sufficient to provide good equalization.

The effectiveness of equalization in a CMFR relative to that in a PFDR is also evident in Figure 11-6. Intuitively, equalization effectiveness is expected to decrease in the order CMFR, PDFR with low Péclet number (high dispersion), and PFDR with high Péclet number (approaching plug flow). In fact, this is the order predicted at low frequencies. However, the PFDR with a low Péclet number begins to outperform the CMFR at frequencies greater than about 0.2 h^{-1} (or about 0.8 cycles/day). This can be explained by the dispersive flux that occurs in PFDRs. When sinusoidal fluctuations in feed concentration are rapid enough so that the entire cycle of variation in feed concentration is contained within the HRT of the PFDR, large concentration gradients are produced. Dispersive flux then becomes large enough to lead to smoothing of concentration within the reactor.

11.3.2 Time-Variable Reaction Rates

Temporal variations in rates of transformation of constituent species are prevalent throughout natural environmental systems and industrial and municipal treatment processes. They may be random, as in wind-induced volatilization of gases from streams and lakes to the atmosphere, or deterministic, as in periodically operated chemical reactors with temperature-dependent reaction rates. The most general tool for modeling a system with time-variable transformation rates is dynamic, numerical simulation employing reaction rates that are determined from system conditions for which time histories are either known or calculated by the model. Even in such complex cases, however, simple models that lend themselves to analytical solutions can provide valuable insight. One method is to examine the time-varying reaction rate based on a probabilistic analysis of the residence time distributions for a reactor and the time series of the relevant reaction rate coefficient.

As an example, consider a plug flow reactor in which a first-order irreversible reaction with a constant rate coefficient, k, occurs. The fractional concentration remaining in the exit stream, as developed in Chapter 9, is given by

$$\phi_R^\circ = \frac{C_{\mathrm{OUT}}}{C_{\mathrm{IN}}} = E_{\tilde{t}}^\circ\,(e^{-k\tilde{t}}) \tag{11-23}$$

in which $E_{\tilde{t}}^\circ\,(e^{-k\tilde{t}})$ is the expected value of $e^{-k\tilde{t}}$ over the residence times, $\tilde{t}$. The probability density function for the HRT of a plug flow reactor is

$$\phi_D(\tilde{t}) = \{1 \text{ for } \tilde{t} = \bar{t};\ 0 \text{ for } \tilde{t} \neq \bar{t}\} \tag{11-24}$$

such that the fraction remaining is given by

$$\phi_R^\circ = e^{-k\bar{t}} \tag{11-25}$$

More generally stated, the fraction remaining for any residence time distribution is

$$\phi_R^\circ = \int_0^\infty e^{-k\tilde{t}}\,\phi_D(\tilde{t})\,d\tilde{t} \tag{11-26}$$

which reduces to Equation 11-25 for a plug flow reactor.

We now examine a first-order reaction for a time-dependent rate coefficient, k, which takes place in a PFR. The variation in k with time is shown in Figure 11-7. A parcel of fluid that enters the reactor at time t_1 and exits at time $t_1 + \tilde{t}$ experiences the rate k_1 for the first short interval of time (Δt) that it is in the reactor, k_2 for the second short interval, and so on, so that the conversion is

$$\phi_R^\circ = e^{-k_1\Delta t}e^{-k_2\Delta t}\cdots e^{-k_n\Delta t} \tag{11-27}$$

Noting that $\Delta t = \tilde{t}/n$ and taking the limit as n becomes large leads to

$$\phi_R^\circ = e^{-\bar{k}\tilde{t}} \tag{11-28}$$

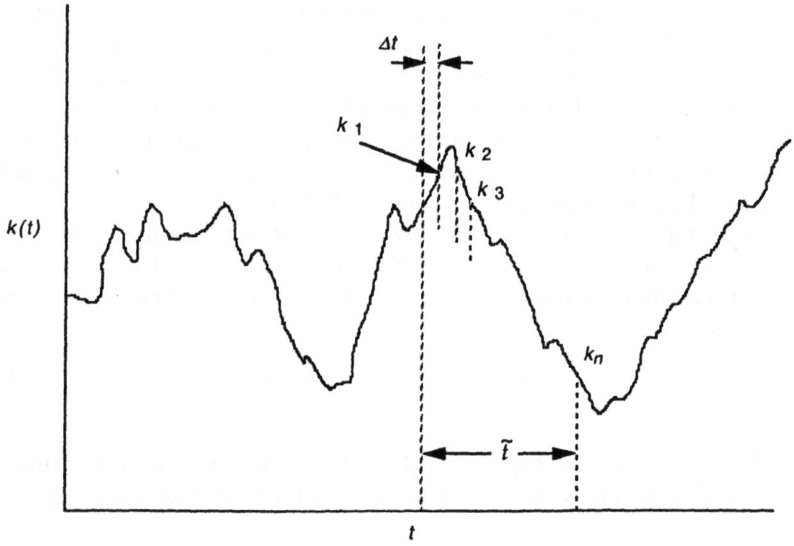

Figure 11-7 Randomly varying reaction rate.

where $\hat{k}$ is the moving average of $k(t)$ over the interval $t_1 < t < t_1 + \tilde{t}$. The mean conversion is determined by averaging the fraction remaining over all fluid elements, which is now an expected value over both $\hat{k}$ and $\tilde{t}$:

$$\phi^{\circ}_{R,k(t)} = E^{\circ}_{\hat{k},\tilde{t}}(e^{-\hat{k}\tilde{t}}) \tag{11-29}$$

Note that $\hat{k}$ is dependent on $\tilde{t}$. That is, the $\hat{k}$ associated with those fluid elements having long residence times relative to the predominant periods of variation in $k(t)$ will have moving averages tightly distributed around $\hat{k} = \bar{k}$, while elements with relatively short residence times are more likely to experience either higher or lower values of $\hat{k}$.

Consider next the sinusoidal variation in the rate for a first-order reaction shown in Figure 11-8 and represented below:

$$k(t) = \bar{k} + \alpha_m \sin \omega_f t \tag{11-30}$$

The average rate coefficient for a particular fluid element depends on the time, t, at which that element enters the reactor. The simplest mathematical way to determine the moving average value, $\hat{k}$, of the rate coefficient is to consider the midpoint of the residence of a fluid element in the reactor. If we define the midpoint of the residence time of the fluid element in a PFR as $\tilde{t}_m$, the moving

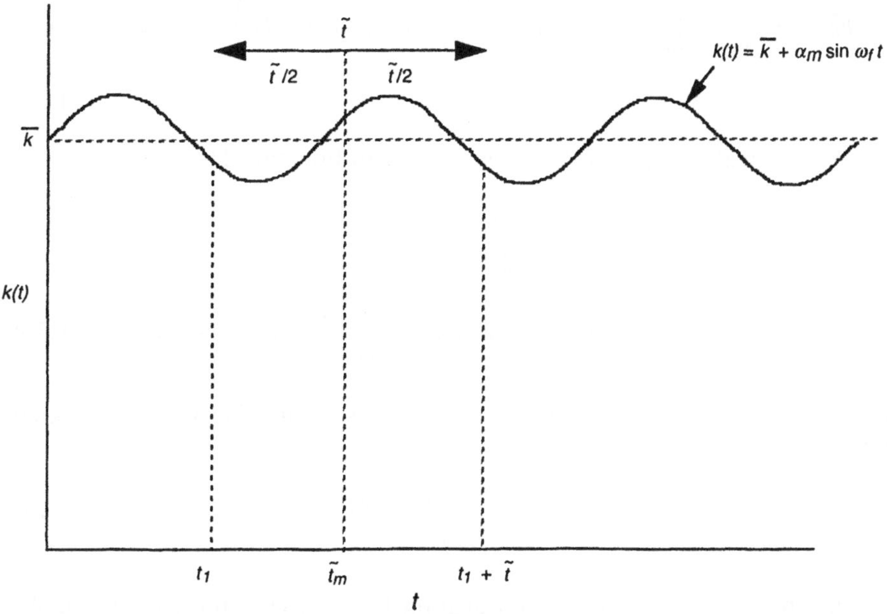

Figure 11-8 Sinusoidally varying first-order reaction rate in a PFR.

average $\hat{k}(\bar{t}, \bar{t}_m)$ can be defined as

$$\hat{k}(\bar{t}, \bar{t}_m) = \bar{k} + \int_{\bar{t}_m - \bar{t}/2}^{\bar{t}_m + \bar{t}/2} \alpha_m \sin(\omega_f t)\, dt \tag{11-31}$$

$$= \bar{k} + \frac{2\alpha_m}{\bar{t}\omega_f} \sin\frac{\omega_f \bar{t}}{2} \sin \omega_f \bar{t}_m \tag{11-32}$$

The expectation over $\hat{k}$ in Equation 11-29 is then equivalent to an expectation over $\bar{t}_m$ and $\bar{t}$. Substitution of Equation 11-32 into Equation 11-29 gives

$$\phi^{\circ}_{R,k} = E^{\circ}_{k(t),\bar{t}}\left\{\exp\left[-\left(\bar{k} + \frac{2\alpha_m}{\bar{t}\omega_f} \sin\frac{\omega_f\bar{t}}{2} \sin \omega_f \bar{t}_m\right)\bar{t}\right]\right\} \tag{11-33}$$

$$= e^{-\bar{k}\bar{t}}\, E\left\{\exp\left[-\left(\frac{2\alpha_m}{\omega_f} \sin\frac{\omega_f\bar{t}}{2} \sin \omega_f \bar{t}_m\right)\right]\right\} \tag{11-34}$$

The expected value, $E^{\circ}(\omega_f, \bar{t}_m)$, for a reactor having any residence time distribution, ϕ_D, is obtained by recognizing that for all fluid elements, $\bar{t}_m$ is uniformly distributed over 0 to $2\pi/\omega_f$: Thus the fractional concentration remaining in *any reactor* for a sinusoidal variation in first-order rate constant would be given by replacing Equation 11-26 with

$$\phi^{\circ}_{R,k(t)} = \int_{\bar{t}=0}^{\infty} \int_{\bar{t}_m=0}^{2\pi/\omega_f} e^{-\bar{k}\bar{t}} \exp\left[-\left(\frac{2\alpha_m}{\omega_f} \sin\frac{\omega_f\bar{t}}{2} \sin \omega_f\bar{t}_m\right)\right] \frac{\omega_f}{2\pi}\, d\bar{t}_m \phi_D(\bar{t})\, d\bar{t} \tag{11-35}$$

An analytical solution of the inner integral in Equation 11-35 yields

$$\int_0^{2\pi/\omega_f} e^{-\bar{k}\bar{t}} \exp(\beta \sin \omega_f\bar{t}_m) \frac{\omega_f}{2\pi}\, d\bar{t}_m$$

$$= \int_0^{2\pi/\omega_f} e^{-\bar{k}\bar{t}} \exp(\beta \sin \omega_f \bar{t}_m) \frac{d(\omega_f \bar{t}_m)}{2\pi} = e^{-\bar{k}\bar{t}}\, \phi_{B,0}(|\beta|) \tag{11-36}$$

in which $\phi_{B,0}(|\beta|)$ is a modified zero-order Bessel function of the first kind with

$$\beta = -2\left(\frac{\alpha_m}{\omega_f}\right) \sin\frac{\omega_f\bar{t}}{2} \tag{11-37}$$

The dimensionless exponent, $\bar{k}\bar{t}$, in Equation 11-36, is essentially the Damkohler I number, $\mathfrak{N}_{\mathrm{Da,I}}$, for a first-order reaction. Returning the analytical result for the inner integral to Equation 11-35 gives

$$\phi^{\circ}_{R,k(t)} = \int_{\tilde{t}=0}^{\infty} e^{-\bar{k}\tilde{t}}\, \phi_{B,0}(|\beta|)\phi_D(\tilde{t})\, d\tilde{t} \tag{11-38}$$

Equation 11-38 can now be used to determine the performance of a PFR. The probability density function of residence time, $\phi_D(\tilde{t})$, for an ideal PFR is equal to 1 because the residence time, $\tilde{t}$, of all fluid elements is simply the HRT, $\bar{t}$. Hence, the fraction remaining

$$\phi^{\circ}_{R,k(t)} = \phi_{B,0}(|\beta|)e^{-\bar{k}\bar{t}} = \phi_{B,0}\left[\left|-2\left(\frac{\alpha_m}{\omega_f}\right)\sin\frac{\omega_f\bar{t}}{2}\right|\right]e^{-\bar{k}\bar{t}} \tag{11-39}$$

In other words, for a sinusoidally varying reaction rate in a PFR, the mean fraction remaining is determined by the fraction remaining at the mean rate multiplied by a correction factor, $\mathfrak{F}_C = \phi_{B,0}(|\beta|)$; that is,

$$\phi^{\circ}_{R,k(t)} = \mathfrak{F}_C\phi^{\circ}_{R,\bar{k}} \tag{11-40}$$

The correction factor is obtained by evaluating the Bessel function, $\phi_{B,0}(|\beta|)$, from readily available tables and polynomial approximations. Inspection of β in Equation 11-37 shows that it depends on the dimensionless parameters, α_m/ω_f and $\sin(\omega_f\bar{t}/2)$. Because the $\phi_{B,0}(|\beta|)$ function equals 1.0 for $\beta = 0$, no correction is needed to the fractional conversion expected from the average rate coefficient, $\bar{k}$, when either α_m/ω_f or $\sin(\omega_f\bar{t}/2)$ is equal to zero. If $\alpha_m/\omega_f = 0$, the sine wave has zero amplitude, and, by definition, the rate coefficient does not vary. If $\sin(\omega_f\bar{t}/2) = 0$, the detention time of the reactor is an integer

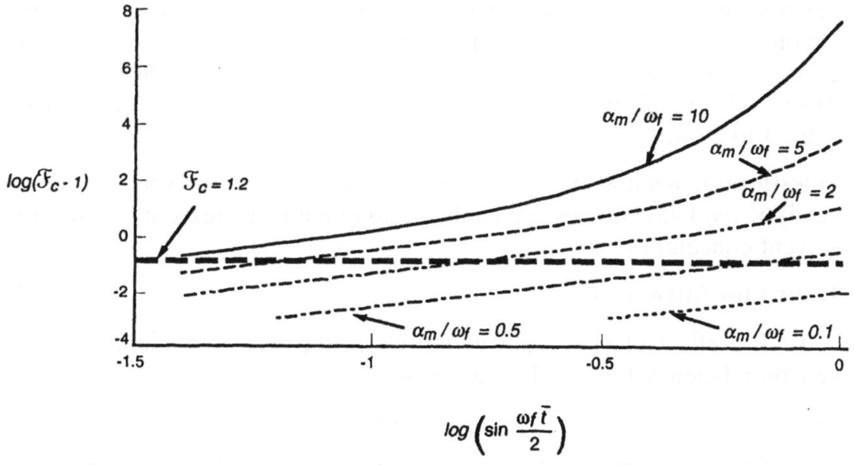

Figure 11-9 Correction factor for conversion by a first-order reaction with a sinusoidally varying rate coefficient in a PFR.

multiple of the period of the sinusoidal variation, so the rate coefficient for each fluid element is given by $\bar{k}$, regardless of when the element enters the reactor.

The dependence of the correction factor, $\mathfrak{F}_C$, on α_m/ω_f and $\omega_f\bar{t}$ is shown in Figure 11-9. This figure can be used in several ways. For instance, suppose that we wish to determine the conditions under which we could expect a decrease in conversion efficiency of 20% [i.e., an $\mathfrak{F}_C$ value of 1.2 (ordinate value of -0.7)]. This situation would occur if the range of amplitudes (α_m/ω_f) is from 5.0 to 0.5, which corresponds to a dimensionless frequency ($\omega_f\bar{t}$) in the range 0.2 to 3.14 (abscissa values of $-1 < |\log(\sin \omega_f\bar{t}/2)| < 0$). This range of α_m/ω_f values becomes larger at larger $\mathfrak{F}_C$ values. In other words, the greater the amplitude of the rate coefficient variation, the worse the performance of the reactor for a given frequency of sinusoidal variation in k and hydraulic detention time. Figure 11-9 provides a convenient way to calculate $\mathfrak{F}_C$, as illustrated in Example 11-4.

Example 11-4

- **Situation:** A process utilizing a fixed bed of titanium dioxide catalyst is being considered as a pretreatment step to partially oxidize aromatic chemicals prior to biological treatment at an industrial facility. Although the concentrations of these chemicals are fairly constant, the discharge of a solvent from another operation is cyclical, and solvent concentration thus varies in a more-or-less sinusoidal fashion. Bench-scale studies of catalyst have shown that the rate of ring cleavage of the aromatic organics is first order. The rate coefficient, however, was shown to decrease proportionally with the concentration of solvent added. Based on the average and range of solvent concentration to be expected in treatment, the average rate coefficient, $\bar{k}$, is 1.07 h^{-1} and its amplitude is 0.9 h^{-1}. A PFR has been designed to achieve 80% reduction in the parent aromatic ring chemicals using the average rate coefficient. The hydraulic residence time, $\bar{t}$, was obtained from $\bar{t} = -\ln \phi_R^\circ/\bar{k}$, which gives 1.5 h. One complete cycle in solvent concentration can be expected to take anywhere from 3 to 24 h.

- **Question(s):** What is the mean treatment efficiency to be expected in this catalytic oxidation process given the inhibition effect caused by a variable solvent concentration?

- **Logic and Answer(s):**

1. The effect of sinusoidal variations in the rate coefficient on conversion efficiency for a PFR is given by

$$\phi^\circ_{R,k(t)} = \mathfrak{F}_C \phi^\circ_{R,\bar{k}}$$

2. The minimum and maximum times for one cycle of solvent concentration are 3 h and 24 h, respectively, so the frequency, ω_f, values

are given by 2π/cycle time, or 2.09 and 0.26 h^{-1}, respectively. The two corresponding sets of values for $\omega_f \bar{t}$ and α_m/ω_f are then

$$(\omega_f \bar{t})_{\min} = 0.26 \times 1.5 = 0.39 \quad \text{and} \quad \left(\frac{\alpha_m}{\omega_f}\right)_{\min} = \frac{0.9}{0.26} = 3.46$$

$$(\omega_f \bar{t})_{\max} = 2.09 \times 1.5 = 3.14 \quad \text{and} \quad \left(\frac{\alpha_m}{\omega_f}\right)_{\max} = \frac{0.9}{2.1} = 0.43$$

The abscissa values of Figure 11-9 are

$$\log\left(\sin\left(\frac{\omega_f \bar{t}}{2}\right)_{\min}\right) = -0.7$$

$$\log\left(\sin\left(\frac{\omega_f \bar{t}}{2}\right)_{\max}\right) = 0$$

For $(\omega_f \bar{t})_{\max}$, we enter the graph with an abscissa value of 0 and $(\alpha_m/\omega_f)_{\max} = 0.4$ and find that

$$\log(\mathfrak{F}_C - 1) \approx -0.70 \quad \text{and} \quad \mathfrak{F}_C = 1.2$$

For $(\omega_f \bar{t})_{\min}$, the abscissa value is -0.7 and $(\alpha_m/\omega_f)_{\min} = 3.4$, such that

$$\log(\mathfrak{F}_C - 1) \approx -0.30 \quad \text{and} \quad \mathfrak{F}_C = 1.5$$

3. Exact and direct solutions are also possible by evaluating the Bessel function given in Equation 11-37:

$$\mathfrak{F}_C = \phi_{B,0}(|\beta|) = \left|-2\left(\frac{\alpha_m}{\omega_f}\right) \sin \frac{\omega_f \bar{t}}{2}\right|$$

The results obtained show that solvent concentration cycling decreases the mean conversion rate significantly (as measured by $\mathfrak{F}_C > 1$). $\mathfrak{F}_C$ increased from 1.2 to 1.5 when a tenfold lower cycle frequency was used in the calculations. It can be shown that any further decrease in cycle frequency has little effect on $\mathfrak{F}_C$. There thus exists a maximum $\mathfrak{F}_C$ value that is determined by the amplitude (a higher amplitude gives a higher maximum $\mathfrak{F}_C$); as amplitude increases, the maximum is not approached until the frequency is lowered.

The approach above can be generalized for application to random fluctuations in first-order reaction rates by use of discrete Fourier transforms (see Equation 11-16):

$$k(t) = \bar{k} + \sum_{j=1}^{n} (\alpha_j \cos \omega_{f,j} t + \beta_j \sin \omega_{f,j} t) + e_r(t) \tag{11-41}$$

By following the steps presented in Equations 11-33 to 11-39, it is possible to obtain an analytical expression for the mean conversion obtained in a PFR:

$$\phi^{\circ}_{R,k(t)} = e^{-\bar{k}\bar{t}} \prod_{j=1}^{n} \left\{ \phi_{B,0}\left[\left(2\,\frac{\alpha}{\omega_{f,j}} \right) \sin \frac{\omega_{f,j}\bar{t}}{2} \right] \right\} \left\{ \phi_{B,0}\left[\left(2\,\frac{\beta}{\omega_{f,j}} \right) \sin \frac{\omega_{f,j}\bar{t}}{2} \right] \right\} \tag{11-42}$$

The correction factor to account for fluctuations in k is simply the product of the values of $\mathfrak{F}_C$ associated with the individual sine and cosine components of the Fourier series. The mean conversion can therefore be estimated from the Fourier coefficients of $k(t)$ and from Equation 11-39 or Figure 11-9.

11.4 SIMPLE TRANSIENTS AT THE MICROSCALE

11.4.1 Domain Concept

Diffusive microscale transport is very often associated with unsteady-state operations. Sorption and ion-exchange processes are, for example, implicitly unsteady because the sorbed or exchanged component *accumulates* in the solid phase with time. These processes stand in contrast to those for solid-phase catalysis, which, unless the catalyst is poisoned, are steady state because catalyst sites are never depleted. Concentrations that change at the boundaries of diffusion domains also produce unsteady-state conditions. So-called *film diffusion* across a hydrodynamic boundary layer separating the bulk solution from a solid sorbent illustrates this point. If film diffusion is rate controlling, the overall process is not at steady state because the concentration on the inner side of the boundary layer changes with time due to accumulation of sorbate in the solid phase.

The nature of a domain in which unsteady-state diffusion occurs can be categorized using the approach introduced in Chapter 8 for steady-state systems. Different diffusion domains were described in Chapter 8 as:

- *Type I domains:* homogeneous without accompanying reaction
- *Type II domains:* homogeneous with accompanying reaction
- *Type III domains:* heterogeneous without accompanying reaction/sorption
- *Type IV domains:* heterogenous with accompanying reaction/sorption

The general nature of the unsteady state concentration profiles which obtain for diffusion in any of these domains is compared to that for steady-state diffusion in Figure 11-10. In the next four sections of this chapter we investigate various mathematical models used to describe the various unsteady-state conditions that occur in the different diffusion domains at the microscale level, as well as the resulting macroscale process models that incorporate these microscale models.

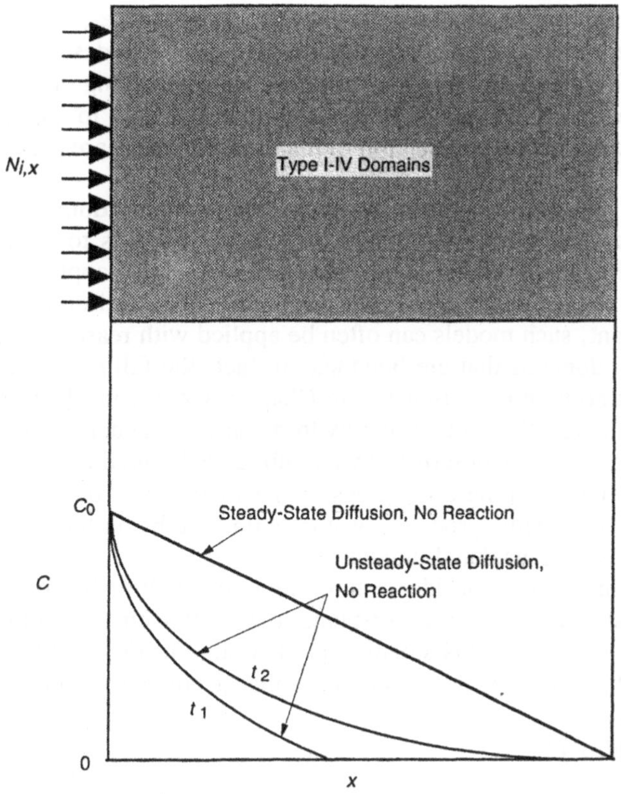

Figure 11-10 Concentration profiles for steady- and unsteady-state diffusion.

11.4.2 Type I Domains

The general principles for formulation of the material balance relationship given in Chapter 2 provide for description of unsteady-state diffusional flux in a homogeneous domain of rectangular, spherical, or cylindrical configuration. For simple film diffusion, the length of the domain is no longer than the thickness of a hydrodynamic boundary layer separating a bulk solution from a reactive medium. On the other hand, the length of the diffusion domain could be as large as the width of a cutoff barrier separating a contaminated subsurface region from a noncontaminated region. Diffusion domains thus range in size from micro- to macroscale.

The unsteady-state diffusion equation, derived and referred to in Chapter 4 as *Fick's second law*, is given in one-dimensional form and rectangular coordinates by

$$\mathfrak{D}_l \frac{\partial^2 C}{\partial x^2} = \frac{\partial C}{\partial t} \tag{4-6}$$

The particular mathematical solution to this equation [i.e., $C = \phi(x, t)$], depends on the initial and boundary conditions required to describe the physical system. Its form may be analytical or numerical, depending on the complexity of these conditions. Several sets of conditions that lead to simple analytical solutions yet have wide application to practical situations are presented below.

11.4.2.1 Unbounded Type I Domains Unbounded domain models provide a mathematical means to achieve tractable solutions to a wide range of complex diffusion problems. They can, of course, be applied to truly unbounded domains, which are rarely encountered in environmental systems. Most important, such models can often be applied with reasonable accuracy to real diffusion domains that are bounded. In fact, the falling liquid film in the Higbie penetration model discussed in Chapter 4 was considered unbounded, even though its length was that of a hydrodynamic boundary layer. Whether a bounded domain can be described by an unbounded domain model depends on the relative length and the time scales over which flux is of interest. Because diffusion is often a slow process, even domain lengths on the order of meters can sometimes be considered unbounded.

As a starting point, consider a finite amount of diffusing material that is instantaneously deposited in a y-z plane at $x = 0$ such that diffusion occurs only in the x-direction. This situation is illustrated in Figure 11-11a. The appropriate initial and boundary conditions that accompany Fick's second law are in this case:

Initial condition:

$$C\,(-\infty < x < \infty,\, 0) = 0 \tag{11-43}$$

First and second boundary conditions:

$$\frac{dC}{dx} = 0 \qquad \text{as } x \rightarrow \pm\infty \tag{11-44}$$

By use of a Laplace transformation, it can be shown that the general form of solution is

$$C = \frac{\kappa}{(t)^{0.5}} \exp\left(-\frac{x^2}{4\mathfrak{D}_l t}\right) \tag{11-45}$$

where κ is a constant of integration.

The value of κ is found by accounting for all of the initially deposited *specific mass*, or *mass per unit area*, M_s:

$$M_s = \int_{-\infty}^{\infty} C\,dx = \int_{-\infty}^{\infty} \frac{\kappa}{(t)^{0.5}} \exp\left(\frac{-x^2}{4\mathfrak{D}_l t}\right) dx \tag{11-46}$$

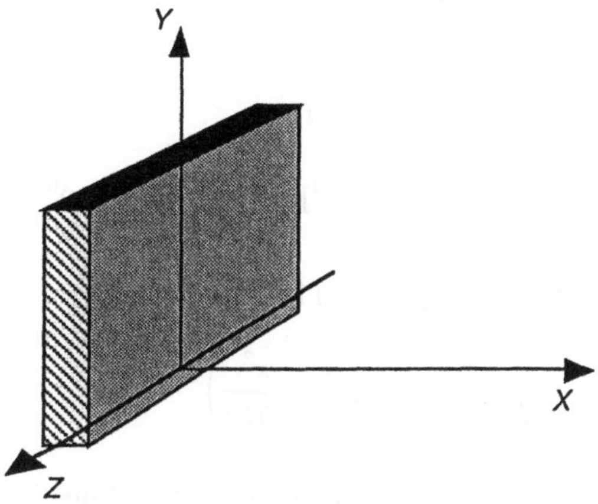

(a) Instantaneous y-z Planar Diffusion Source Deposited at $x = 0$

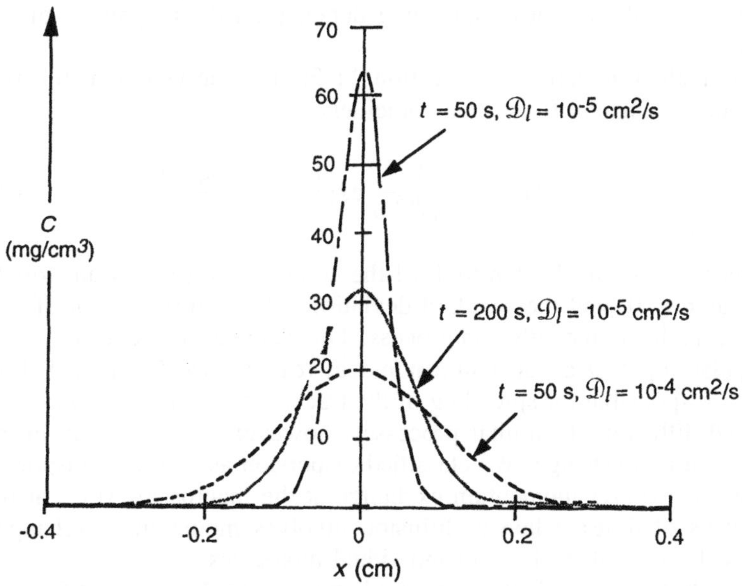

(b) Concentration Profiles at Different Times

Figure 11-11 Concentration profiles produced by one-dimensional diffusion from an instantaneous planar source in an unbounded domain.

The solution is made more general by creating a new dimensionless variable:

$$(\zeta^\circ)^2 = \frac{x^2}{4\mathfrak{D}_l t} \tag{11-47}$$

so that

$$dx = 2(\mathfrak{D}_l t)^{0.5}\, d\zeta^\circ \tag{11-48}$$

$$M_s = 2\kappa\,(\mathfrak{D}_l t)^{0.5} \int_{-\infty}^{-\infty} \exp[-(\zeta^\circ)^2]\, d\zeta^\circ = 2\kappa\,(\pi\mathfrak{D}_l t)^{0.5} \tag{11-49}$$

Solving for κ and substituting into Equation 11-45, we obtain

$$C = \frac{M_s}{2(\pi\mathfrak{D}_l t)^{0.5}} \exp\left(-\frac{x^2}{4\mathfrak{D}_l t}\right) \tag{11-50}$$

The general shape of the concentration profile described by Equation 11-50 is shown in Figure 11-11b. Diffusional transport produces a spreading of concentration that is symmetrical about the initial deposit position of the diffusing material, and the extent of spreading depends on the magnitude of the diffusivity.

The analogous nature of Equation 11-50 and the Gaussian function, first presented in Chapter 9, is not a coincidence:

$$\phi(x) = \frac{1}{(2\pi)^{0.5}\,\sigma} \exp\left[-\frac{(x - \mu_m)^2}{2\sigma^2}\right] \tag{9-121}$$

The mean value, or the centroid, of the concentration C in Equation 11-50 is found at $x = 0$, and the standard deviation relates to the extent of spreading that occurs due to the diffusion process. This behavior is identical to the spread of a pulse input of a tracer in a nonideal reactor (see Equation 9-120). The process responsible for spreading of the tracer was termed *dispersion*. Dispersion and diffusion are similar processes, however, differing only in the scale of momentum exchange. We classified dispersion as a macrotransport process because it involves momentum exchange at the macroscale (i.e., at the level of packets of fluid), whereas diffusion involves momentum exchange at the microscale (i.e., at the level of individual molecules).

The solution to Fick's law presented above is easily extended for an instantaneous source that is deposited initially in the y–z plane but which can diffuse only in the positive x direction. Because of the symmetry of diffusion from the source, the diffusing material moving into the negative x domain is simply added back into the positive x domain (the *reflection principle*), so that

$$C = \frac{M_s}{2(\pi\mathfrak{D}_l t)^{0.5}} \exp\left(-\frac{x^2}{4\mathfrak{D}_l t}\right) + \frac{M_s}{2(\pi\mathfrak{D}_l t)^{0.5}} \exp\left(-\frac{x^2}{4\mathfrak{D}_l t}\right) \tag{11-51}$$

or

$$C = \frac{M_s}{(\pi \mathfrak{D}_l t)^{0.5}} \exp\left(-\frac{x^2}{4\mathfrak{D}_l t}\right) \tag{11-52}$$

The resulting concentration profiles at different diffusion times for an initial deposition of mass at $x = 0$ are shown graphically in Figure 11-12. One pertinent example to which this relationship would apply is an *instantaneous* discharge of a chemical into the end of a stagnant channel of water. For diffusion to be one dimensional, the discharge must be uniform across the channel width and height.

The diffusion of an *infinite* (vis-à-vis *finite*) source of material deposited at $x = 0$ is also of considerable interest. Modification of the example above to illustrate this situation would involve a *continuous* discharge of a chemical at the end of the stagnant channel of water. The initial and boundary conditions needed to solve Fick's second law are:

Initial condition:

$$C(x > 0, t = 0) = 0 \tag{11-53}$$

First boundary condition:

$$C(x = 0, t > 0) = C_0 \tag{11-54}$$

Second boundary condition:

$$C(x \to \infty, t > 0) = 0 \tag{11-55}$$

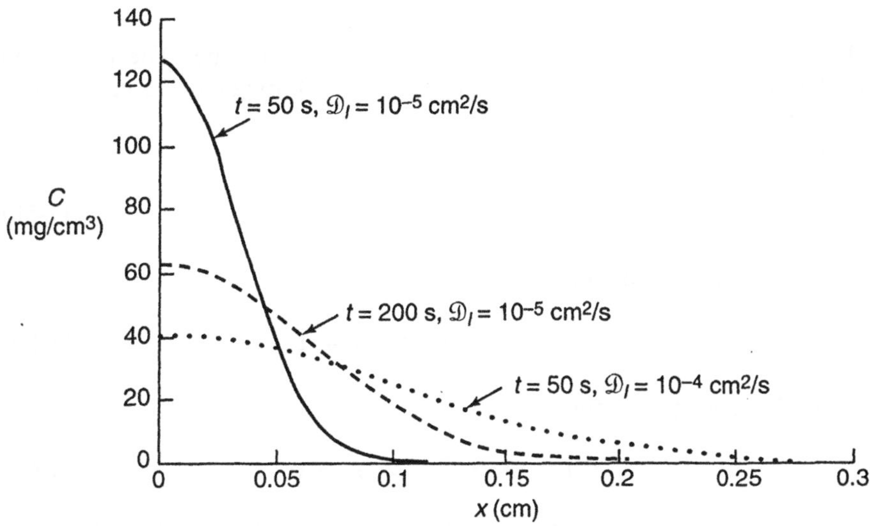

Figure 11-12 Concentration profiles produced by one-directional diffusion from an instantaneous planar diffusion source in an unbounded domain.

The following solution is obtained by a Laplace transformation:

$$\frac{C(x, t)}{C_0} = c^\circ(x, t) = \text{erfc}\left[\frac{x}{2(\mathfrak{D}_l t)^{0.5}}\right] \tag{11-56}$$

where erfc is the *complementary error function*. It may be recalled at this point that the same form of solution was presented in Chapter 8 to describe sorption rates in a type IV domain:

$$\frac{C(x, t)}{C_0} = \text{erfc}\left[\frac{x}{2(\mathfrak{D}_a t)^{0.5}}\right] \tag{8-134}$$

The only difference is in the definition of diffusivity, where $\mathfrak{D}_a$ is the apparent diffusivity, a modification of free-liquid diffusivity that accounts for impedances to transport caused by restrictions in diffusion pathways and sorption within the domain.

Fick's second law can also be written in dimensionless form as a way to generalize the solution:

$$\frac{\partial^2 c^\circ}{\partial(\zeta_x^\circ)^2} = \frac{\partial c^\circ}{\partial \tau^\circ} \tag{11-57}$$

where

$$c^\circ = \frac{C}{C_0} \qquad \zeta^\circ = \frac{x}{L} \qquad \text{and} \qquad \tau^\circ = \frac{\mathfrak{D}_l t}{L^2}$$

and τ° can be thought of as a dimensionless diffusion time. The initial and boundary conditions described in Equations 11-53 through 11-55 can then be rewritten as:

Initial condition:

$$c^\circ\,(\zeta^\circ > 0,\ \tau^\circ = 0) = 0 \tag{11-58}$$

First boundary condition:

$$c^\circ\,(\zeta^\circ = 0,\ \tau^\circ > 0) = 1 \tag{11-59}$$

Second boundary condition:

$$c^\circ\,(\zeta^\circ \to \infty,\ \tau^\circ > 0) = 0 \tag{11-60}$$

The dimensionless form of Equation 11-56 corresponding to the solution to Fick's second law is then

$$c^\circ(\zeta^\circ, \tau^\circ) = \operatorname{erfc}\left[\frac{\zeta^\circ}{2(\tau^\circ)^{0.5}}\right] \tag{11-61}$$

The value of the dimensionless representation of Fick's second law and particular set of initial and boundary conditions given by Equations 11-58 to 11-60 is that they accommodate a wide variety of diffusion situations. If a diffusion problem can be represented in this way mathematically, the solution is always given by Equation 11-61. Even though the length, L, of a real diffusion domain is not infinite, this solution may still be a good approximation for the early stages of a diffusion process. If the diffusion domain is bounded, the concentration at the end of the domain will not remain zero for all time, but will eventually increase as material is reflected back into the domain. As a result, the concentration gradient and the flux at any point in the domain will be different than those for an infinite domain, and the solution above is invalid. It may, however, take a very long time for enough material to reach the farthest distance of the real domain and thus alter the concentration gradient. Until this condition occurs, the infinite domain solution given in Equation 11-61 is approximately correct. Also, if the diffusion source at $x = 0$ is finite rather than infinite, concentration will decrease with time and another mathematical boundary condition is needed. However, the infinite source boundary condition is approximately correct for *small decreases* in the source concentration.

The flux of material at the boundary of the diffusion domain is as important as the concentration within the domain because it determines the loss of material from one region and the gain in another. For diffusion into a type I domain, the flux is given by

$$N|_{x=0} = -\mathfrak{D}_l \frac{\partial C}{\partial x}\bigg|_{x=0} \tag{11-62}$$

The derivative, $\partial C/\partial x$, is obtained from Equation 11-56. The steps involved in finding this derivative are identical to those presented in Chapter 8 for flux into a type IV domain (see Equations 8-138 to 8-142), and the resulting expression for flux is

$$N|_{x=0} = C_0 \left(\frac{\mathfrak{D}_l}{\pi t}\right)^{0.5} \tag{11-63}$$

The specific mass of contaminant, $M_s(t)$, released with time can also be calculated by integrating the flux over time:

$$M_s(t) = \frac{M(t)}{A_N} = \int_0^t N|_{x=0}\, dt = C_0 \left(\frac{\mathfrak{D}_l}{\pi}\right)^{0.5} \int_0^t t^{-0.5}\, dt \tag{11-64}$$

$$M_s(t) = \frac{M(t)}{A_N} = 2C_0 \left(\frac{\mathfrak{D}_l}{\pi}\right)^{0.5} t^{0.5} \tag{11-65}$$

in which A_N is the surface area normal to the flux into the domain across which material is transported. The relationship above is important because it shows that the loss of diffusing material into the diffusion domain is proportional to the *square root* of the *diffusion coefficient* and the *square root of time*. This relationship is commonly used to determine either the amount lost from the diffusion source or the amount gained by the diffusion domain.

Example 11-5 provides an illustration of how the concepts discussed above can be applied to microtransport processes in type I domains can be approximated as unbounded and how these descriptions are interfaced with macroscale transport models.

Example 11-5

- **Situation:** Submerged drums containing waste industrial chemicals have been found sitting upright in the bottom muds of a barge canal. The best estimate of the time that these drums have spent in the canal is 3 years. The covers of the drums have been punctured, allowing for diffusion of contaminants into the water flowing through the canal. Each drum is 1 m in height and on average the covers have 100 cm^2 of holes. Among the contaminants of concern are polychlorinated biphenyls (PCBs). The concentration of PCBs within the waste mixtures in the drums is originally believed to have been 1.0 mg/L. Due to dilution and partitioning, the concentration of PCBs in the canal water (C_{cw}) is uniformly 0.005 mg/L. This concentration is assumed to remain constant in time due to the continual passage of canal water around the drums.
- **Question(s):** Estimate the total mass of PCBs released to the river in the past 3 years.
- **Logic and Answer(s):**

1. The fluid within the drums is quiescent. However, diffusion of PCBs through the puncture holes occurs as a result of the lower concentration in the surrounding canal water. The diffusion domain is the length of the drum (1 m) as represented schematically below.

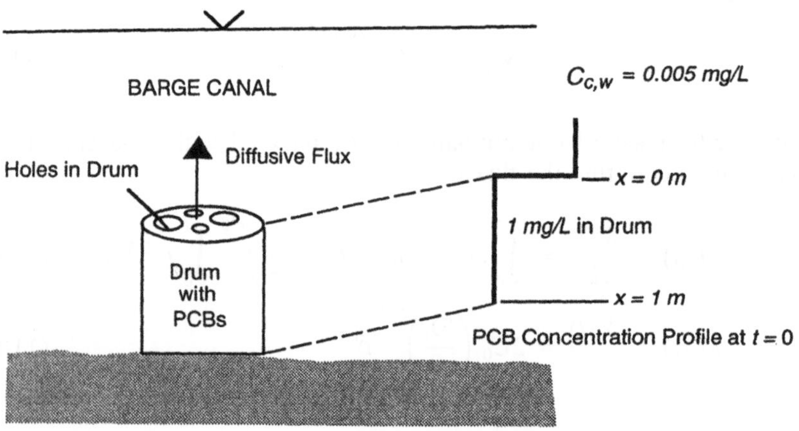

The initial condition within the domain ($0 \leq x \leq 1$ m) is the original concentration of PCBs, which was estimated as 1.0 mg/L. The position $x = 0$ corresponds to the end of the domain in contact with the canal water where the concentration of PCBs is 0.005 mg/L.

2. At first glance, the diffusion domain in the schematic above does not appear to be one for which Equation 11-56 can be used. First, the initial condition corresponds to 1.0 mg/L throughout the diffusion domain ($0 < x < 1$ m) rather than zero as needed to use Equation 11-56. Second, the diffusion length is not very long (1 m), and the assumption of an infinite diffusion domain with concentration remaining unchanged as $x \to \infty$ would thus not appear reasonable. However, both of these seemingly limiting circumstances need to be considered further.

3. If it is assumed that the concentration at depth $x = 1$ m within the domain has not decreased substantially, the diffusion domain is for all intents and purposes infinite (this assumption will be checked later). The initial and boundary conditions are:

Initial condition:

$$C(x > 0, t = 0) = C_0 = 1.0 \text{ mg/L}$$

Boundary conditions:

$$C(x = 0, t > 0) = C_{cw} = 0.005 \text{ mg/L}$$

$$C(x = 1 \text{ m}, t > 0) = C_0 = 1 \text{ mg/L}$$

4. The conditions above still do not match those required by Equations 11-58 to 11-60. However, we can define a dimensionless concentration variable:

$$c° = \frac{C_0 - C}{C_0 - C_{cw}}$$

from which Equation 11-57 is obtained after substitution into Fick's second law:

$$\frac{\partial^2 c°}{\partial (\zeta°)^2} = \frac{\partial c°}{\partial \tau°}$$

$$\zeta° = \frac{x}{L} \quad \text{and} \quad \tau° = \frac{\mathfrak{D}_l t}{L^2}$$

The new initial and boundary conditions associated with the dimensionless equation are given by:

Initial condition:

$$c°(\zeta° > 0, \tau° = 0) = 0$$

Boundary conditions:

$$c°(\zeta° = 0, \tau° > 0) = 1 \quad \text{and} \quad c° (\zeta° \to L \to \infty, \tau° > 0) = 0$$

These are exactly the same conditions described by Equations 11-58 to 11-60. The appropriate solution to Fick's second law is then Equation 11-61:

$$c^\circ(\zeta^\circ, \tau^\circ) = \mathrm{erfc}\left[\frac{\zeta^\circ}{2(\tau^\circ)^{0.5}}\right]$$

or

$$c^\circ(x, t) = \mathrm{erfc}\left[\frac{x}{2(\mathfrak{D}_l t)^{0.5}}\right]$$

5. An average free-liquid diffusion coefficient for a mix of various cogeners of PCBs can be estimated using the Wilke–Chang equation, as illustrated in Example 4-4. We assume an average value of 0.5×10^{-6} cm^2/s as a conservative (high) first cut. The resulting plot of C as a function of distance is given below for times of 1 and 3 years:

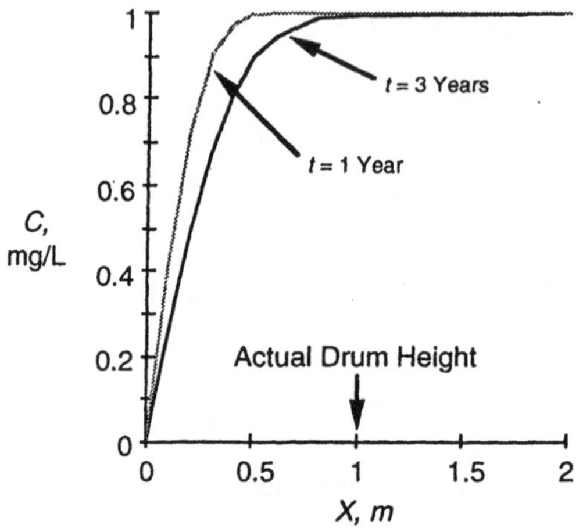

Although some depletion of concentration occurs at a distance of 1 m from the boundary of the drum with the canal water, the assumption of an infinite diffusion domain is reasonable; thus the proposed solution to Fick's second law is approximately correct. We can also observe from the small change in concentration gradient over 3 years that diffusion is a very slow process.

6. The flux of PCBs from the diffusion domain into the river at $x = 0$ is now given by

$$N|_{x=0} = \mathfrak{D}_l(C_0 - C_{cw})\left.\frac{dc^\circ}{dx}\right|_{x=0} = -\mathfrak{D}_l(C_0 - C_{cw})\left(\frac{\mathfrak{D}_l}{\pi t}\right)^{0.5}$$

7. The mass released per unit area after any time, t, is calculated from integrating the flux over time:

$$M_s(t) = \frac{M(t)}{A_N} = -2(C_0 - C_{cw}) \left(\frac{\mathfrak{D}_l}{\pi}\right)^{0.5} t^{0.5}$$

or

$$M_s(t) = -2(1 \times 10^{-3} - 5 \times 10^{-6}) \left(\frac{0.5 \times 10^{-5}}{3.14}\right)^{0.5} \cdot (3 \times 365 \times 24 \times 3600)^{0.5} = 0.024 \text{ mg/cm}^2$$

The total mass released per drum is then 0.024 × 100 cm^2 of holes, or 2.4 mg per drum cover. Note that the negative sign is a consequence of the direction of diffusion—in this instance being opposite to the convention on which Fick's law is written (i.e., diffusion in the positive-x direction).

11.4.2.2 Complex Bounded Type I Domains Numerous environmental situations exist in which diffusional transport occurs through complex type I domains. For instance, organic solvents may diffuse through cutoff barriers made of soil–bentonite mixtures or through plastic liners intended to contain them. Similarly, organic solvents in contaminated groundwaters may diffuse through the walls of submerged plastic pipes. Some membrane separation processes in which inorganic or organic constituents chemically associate with the membrane phase (i.e., "dissolve in") and diffuse through this barrier material into product water also involve nonaqueous phase diffusion. In each of these examples, concentrations in the diffusion domain must be expressed differently than in the aqueous phase. Moreover, assumptions of infinite domains are often inappropriate, even over short time periods.

For relatively low concentrations (chemical activities close to 1), solute transport through nonaqueous domains can be described by Fick's second law:

$$\mathfrak{D}_m \frac{\partial^2 C_m}{\partial x^2} = \frac{\partial C_m}{\partial t} \tag{11-66}$$

where $\mathfrak{D}_m$ is the diffusivity in the complex matrix or domain and C_m is the concentration of solute within that matrix. Expressing distance and time in their dimensionless forms gives

$$\frac{\partial^2 C_m}{\partial (\zeta^\circ)^2} = \frac{\partial C_m}{\partial \tau^\circ} \tag{11-67}$$

Concentrations of materials absorbed within the diffusion matrix can be related to those adjacent vapor phases (v) or aqueous phases (a) by the equilibrium partitioning principle:

$$C_m = \mathcal{K}_{m,v} C_v \tag{11-68}$$

$$C_m = \mathcal{K}_{m,a} C_a \tag{11-69}$$

In some situations, the diffusing component may be in the vapor phase on one side of the nonaqueous diffusion domain and in the aqueous phase on the other.

Various boundary conditions are involved in describing diffusion through finite domains. Consider first the diffusion of a vapor through a wall into a flowing water that transports the solute away from the wall so rapidly that concentration at the wall does not increase; this situation is represented schematically in Figure 11-13a. Note that for mathematical efficacy, we define the diffusion domain for such situations to begin at $x = L$ and end at $x = 0$. The initial and boundary conditions needed to solve Fick's second law are:

Initial condition:

$$C(\zeta^\circ, \tau^\circ = 0) = 0 \tag{11-70}$$

First boundary condition:

$$C(\zeta^\circ = 1, \tau^\circ > 0) = C_{v,L} \tag{11-71}$$

Second boundary condition:

$$C(\zeta^\circ = 0, \tau^\circ > 0) = 0 \tag{11-72}$$

The finite nature of the diffusion domain is stated explicitly in Equation 11-71; we note the difference from Equation 11-60. The partitioning relationships given by Equations 11-68 and 11-69 provide the necessary linkages between concentrations within the wall and the aqueous and vapor phases.

For these boundary conditions, the solution to Fick's second law is

$$\frac{C_m}{C_{m,L}} = \frac{C_m}{\mathcal{K}_{m,v} C_{v,L}} = c^\circ = \zeta^\circ + \sum_{n=1}^{\infty} \frac{2}{n\pi} (-1)^n \sin n\pi\zeta^\circ \exp(-n^2\pi^2\tau^\circ) \tag{11-73}$$

The concentration gradient within the diffusion domain generally is computed accurately using fewer than eight terms in the summation term in Equation 11-73. The dimensionless form of the concentration profile defined by evaluating Equation 11-73 at different ζ° values is shown in Figure 11-13b for several different values of τ°. As τ° becomes larger, steady-state diffusion is approached to accommodate the unchanging boundary values of concentration on either side of the domain, and the concentration gradient approaches linearity.

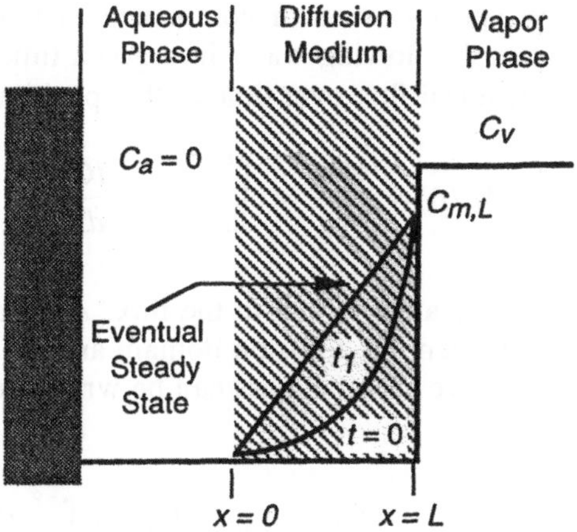

(a) Domains and concentration profiles

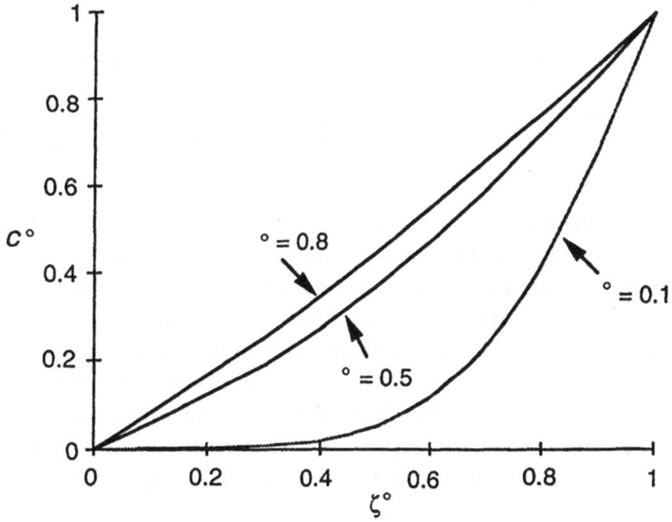

(b) Dimensionless concentration profiles at three different dimensionless
times resulting from unsteady-state diffusion in the bounded domain
described by Figure 11-13a; note approach to steady-state, linear concentration profile

Figure 11-13 Diffusion through a complex bounded domain into a finite aqueous volume with no concentration buildup.

The cumulative specific mass of component that penetrates the diffusion boundary and enters the flowing water in a given time period is obtained by integrating the diffusional flux over time at that position:

$$M_s(t) = \frac{M(t)}{A_N} = \int_0^t - \mathfrak{D}_m \left.\frac{dC_a}{dx}\right|_{x=0} dt \tag{11-74}$$

where A_N is the surface area normal to the flux. After accounting for equilibrium partitioning between the diffusion domain and the aqueous phase ($C_m = \mathcal{K}_{m,a} C_a$), the cumulative specific mass can be written as a function of dimensionless time (τ°)

$$M_s(\tau^\circ) = \int_0^t \frac{C_{m,L}}{\mathcal{K}_{m,a} L} \mathfrak{D}_m \left.\frac{dc^\circ}{d\zeta^\circ}\right|_{\zeta^\circ=0} dt = \frac{C_{m,L} L}{\mathcal{K}_{m,a}} \int_0^{\tau^\circ} \left.\frac{dc^\circ}{d\zeta^\circ}\right|_{\zeta^\circ=0} d\tau^\circ \tag{11-75}$$

Substitution of the derivative, $\left.\frac{dc^\circ}{d\zeta^\circ}\right|_{\zeta^\circ=0}$, from Equation 11-73 and rearrangement of Equation 11-75 to form a dimensionless uptake term gives

$$M_s^\circ(\tau^\circ) = \frac{\mathcal{K}_{m,a} M_s(\tau^\circ)}{C_{m,L} L} = \tau^\circ - \frac{1}{6} - \frac{2}{\pi^2} \sum_{n=1}^{\infty} \frac{(-1)^n}{n^2} \exp(-n^2 \pi^2 \tau^\circ) \tag{11-76}$$

where $M_s^\circ(\tau^\circ)$ is a dimensionless expression of the cumulative amount penetrating the bounded domain in the dimensionless diffusion time, τ°.

The plot of $M_s^\circ(\tau^\circ)$ against dimensionless diffusion time given in Figure 11-14 emphasizes two important characteristics of the diffusion of material through the nonaqueous domain. First, there is a *lag period* before penetration begins. Second, the amount that penetrates is eventually linearly dependent on time and thus at steady state (i.e., the slope of the plot becomes constant). The steady-state specific mass release is approximated from Equation 11-76 by noting that the summation terms approach 0 for τ° approaching infinity; thus

$$M_s(\tau^\circ) = \frac{C_{m,L} L}{\mathcal{K}_{m,a}} \left(\tau^\circ - \frac{1}{6}\right) \approx \frac{C_{m,L} L \tau^\circ}{\mathcal{K}_{m,a}} \tag{11-77}$$

The above equation shows that the lag period before mass penetrates the domain by diffusion is

$$\tau_{\text{lag}}^\circ = \frac{1}{6} \tag{11-78}$$

Use of the boundary conditions above for steady-state diffusional transport in a complex bounded domain is illustrated in Example 11-6.

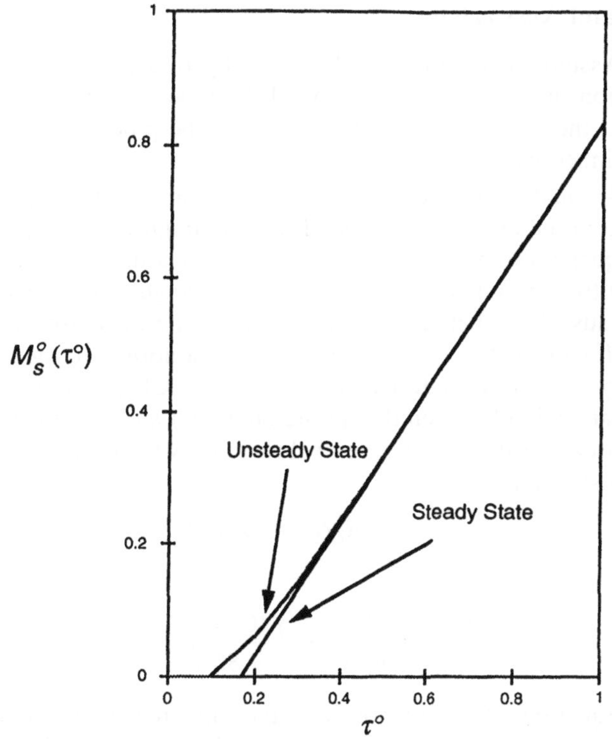

Figure 11-14 Dimensionless specific mass penetration through a complex bounded diffusion domain as a function of dimensionless diffusion time.

Example 11-6

- **Situation:** Polybutylene pipe has been installed to distribute reclaimed wastewater for lawn irrigation in a small community. The pipe diameter is 10.16 cm (4 in.) and the wall thickness is 0.3 cm (118 mils). An underground gasoline tank is discovered to be leaking in the vicinity of one of these pipes. The vapor-phase concentration of toluene in the surrounding soil is 4×10^{-8} mol/cm^3 of air. Toluene is known to penetrate this type of plastic pipe (Park et al., 1991). The diffusivity of toluene, $\mathfrak{D}_m$, in polybutylene and the partition coefficient, $\mathcal{K}_{m,v}$, for toluene between polybutylene and air have been measured experimentally as 4.5×10^{-9} m^2/day and 1200 (dimensionless), respectively. There is concern over the volatilization of toluene during lawn irrigation if its penetration of the pipe is significant.
- **Question(s):** Estimate the time before the toluene begins to appear in the water inside the pipe, and then estimate its steady-state release to the water.

- **Logic and Answer(s):**

1. Assume that the pipe wall is initially free of toluene, that the concentration in the vapor phase around the pipe remains constant in time, and that the aqueous concentration within the pipe is kept very low due to continued use of the water.

2. An analysis of radial diffusional flux through the pipe wall subject to these initial and boundary conditions is required for a precise answer to the questions above. However, for purposes of an estimate, a simplification to one-dimensional diffusion in rectangular coordinates is possible because flux changes very little due to the decrease in surface area with diffusion distance (i.e., the surface area normal to the flux direction is nearly constant with diffusion distance); see Chapter 2 for more details. The relationship between flux at the outer (O) and inner (I) edges of the pipe wall at steady state is given below to prove that the flux change with radial distance is minor:

$$4\pi R_O^2 N_O = 4\pi R_I^2 N_I$$

or

$$\frac{N_O}{N_I} = \left(\frac{R_I}{R_O}\right)^2 = \left(\frac{10.16}{9.86}\right)^2 = 1.06$$

From Equation 11-78 we note that the lag time predicted from Fick's second law in rectangular coordinates and the initial and boundary conditions is

$$\tau^\circ_{\text{lag}} = \frac{1}{6}$$

where

$$\tau^\circ = \frac{\mathfrak{D}_m t}{L^2}$$

$$t_{\text{lag}} = \frac{L^2 \tau^\circ}{\mathfrak{D}_m} = 333 \text{ days}$$

3. The steady-state mass of toluene released into the water per unit area of pipe surface per unit time is given by Equation 11-77:

$$M_s(\tau^\circ) = \frac{C_{m,L} L}{\mathcal{K}_{m,a}} \left(\tau^\circ - \frac{1}{6}\right) \approx \frac{C_{m,L} L \tau^\circ}{\mathcal{K}_{m,a}}$$

where $C_{m,L}$ is the concentration of toluene in the pipe at its outside edge.

4. If both the aqueous and vapor phases behave ideally,

$$\mathcal{K}_{m,v} C_v = C_m = \mathcal{K}_{m,a} C_a$$

and Henry's law relates the vapor (v) and aqueous (a) concentrations:

$$C_v = \frac{\mathcal{K}_H}{\mathcal{R}T} C_a$$

in which $\mathcal{K}_H$ for toluene is 6.6×10^{-3} atm-m^3/mol (see Table 6-1) and $\mathcal{R}$ is 8.2×10^{-5} (atm-m^3)/(mol-K). Combining the two relationships above, we find that

$$\mathcal{K}_{m,a} = \frac{\mathcal{K}_H}{\mathcal{R}T} \mathcal{K}_{m,v}$$

and that the concentration just inside the outer edge of the pipe wall can be expressed as

$$C_{m,L} = \frac{\mathcal{K}_{m,a}\mathcal{R}T}{\mathcal{K}_H} C_{v,L}$$

5. The mass released per unit area of pipe surface per unit of time is then calculated from Equation 11-77 after substituting $\mathfrak{D}_m\, t/L^2$ for $\tau°$:

$$M_s(t) = \frac{\mathfrak{D}_m \mathcal{R}TC_{v,L}}{\mathcal{K}_H L}$$

$$= \frac{(4.5 \times 10^{-9})(8.2 \times 10^{-5})(298)(4 \times 10^{-2})}{(6.66 \times 10^{-3})(0.3 \times 10^{-2})}$$

$$= 2.2 \times 10^{-7} \text{ mol/m}^2\text{-day}$$

6. The diffusivity of toluene in the pipe material is critical to the estimates given above. Polyvinyl chloride (PVC) pipe is characterized by a diffusivity that is several orders of magnitude lower than that used here for polybutylene pipe, and this will increase the lag time by the same factor. Although PVC has good structural integrity, it may be softened at extremely high concentrations of some solvents, and solute diffusivity may increase as a result.

Another environmentally important circumstance of transport in complex bounded domains is depicted in Figure 11-15a. This circumstance, which involves the unsteady-state diffusion of a substance through a finite domain followed by its accumulation in a finite volume, requires specification of a different set of boundary conditions. Because of accumulation, the concentration increases at $x = 0$ for this condition rather than remaining at 0 as it did in the circumstance discussed previously (see Figure 11-13a). The permeation of chemicals through a plastic liner into a very slowly moving lense of groundwater may fit this mathematical description of diffusion. The initial and boundary conditions needed to solve Fick's second law (Equation 11-67) are:

Initial condition:

$$C_m(\zeta°, \tau° = 0) = 0 \tag{11-79}$$

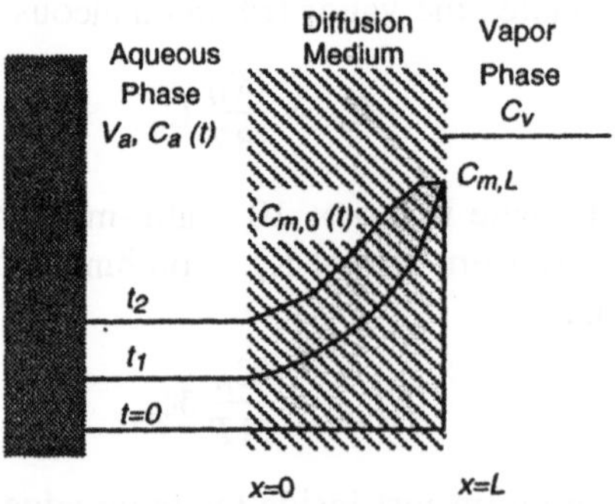

(a) Domains and Concentration profiles

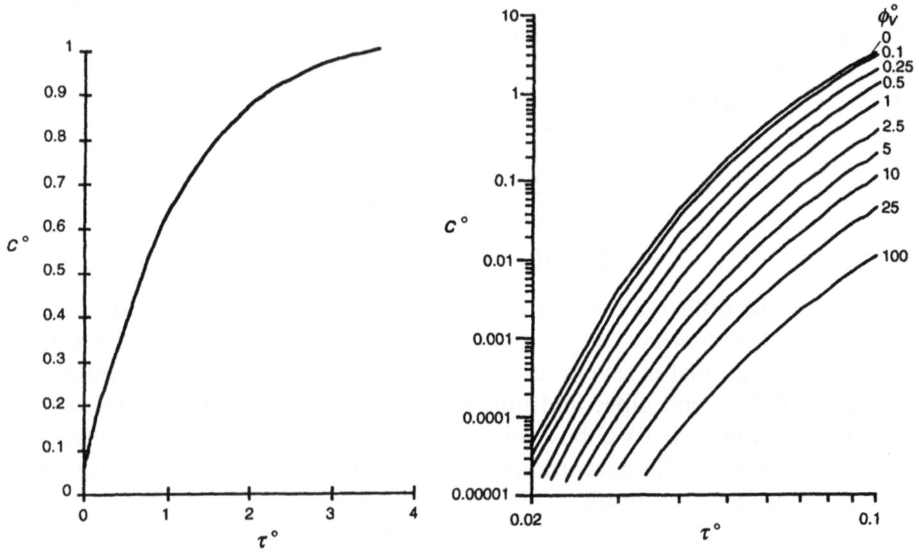

(b) Plot (left panel) of equation 11-85 for intermediate value of ϕ°_V(0.5) and plot right panel) of asymptotic expansion of

$c^\circ = C_{m,0}/C_{m,L} = 8(\tau^\circ/\pi)^{0.5}[\tau^\circ/(\phi^{\circ V} + 2\tau^\circ)]\exp(-0.25\tau^\circ)$, valid for low values of

τ° [after Selleck and Marinas (1991)]

Figure 11-15 Diffusion from vapor phase through a bounded nonaqueous type I domain with accumulation in an adjacent aqueous phase.

First boundary condition:

$$C_m(\zeta^\circ = 1, \tau^\circ > 0) = C_{m,L} \tag{11-80}$$

Second boundary condition:

$$C_m(\zeta^\circ = 0, \tau^\circ > 0) = C_{m,0}(\tau^\circ) \tag{11-81}$$

The second boundary condition requires a mathematical statement of the rate of accumulation of diffusing material in the finite volume, V_a, of the aqueous phase adjacent to the complex diffusion domain:

$$V_a \frac{dC_a}{dt} = A_N \mathfrak{D}_m \left.\frac{dC_m}{dx}\right|_{x=0} \tag{11-82}$$

The concentrations in the aqueous and complex medium, C_a and C_m, are related by the partitioning principle used in the previous diffusion model. Making use of this concept and dimensionless parameters, we have

$$\phi_V^\circ \frac{dC_m}{d\tau^\circ} = \left.\frac{\partial C_m}{\partial \zeta^\circ}\right|_{\zeta^\circ=0} \tag{11-83}$$

where

$$\phi_V^\circ = \frac{V_a}{A_N L \mathcal{K}_{m,a}} \tag{11-84}$$

The constant, ϕ_V°, is essentially a ratio of the volume in the aqueous phase to that in the diffusion medium (per unit length of pipe), divided by the partitioning coefficient between the nonaqueous and aqueous phases. For diffusion through a pipe wall, the ratio $V_a/A_N L$ is $R_I/2L$, where R_I is the inside radius of the pipe.

Solution to Fick's second law for the initial and boundary conditions above yields

$$c^\circ = \frac{C_{m,0}}{C_{m,L}} = 1 - 4 \sum_{n=1}^{\infty} \frac{\sin u_n \exp(-u_n^2 \tau^\circ)}{2u_n + \sin 2u_n} \tag{11-85}$$

where u_n is the nth root of

$$\phi_V^\circ u_n \sin u_n - \cos u_n = 0 \qquad n = 1, 2, 3, \ldots \tag{11-86}$$

This open, series form of mathematical solution is commonly encountered in diffusion problems, where material is being either gained or lost from a finite volume as a result of diffusion through a finite domain. A graphical representation is presented in Figure 11-15b, in which the ratio of the internal to external concentrations, $C_{m,0}/C_{m,L}$, is plotted as a function of dimensionless time, τ°, for an intermediate value (0.5) of ϕ_V°. Because $\tau^\circ = \mathfrak{D}_m t/L^2$, the higher the diffusivity and the thinner the diffusion domain, the shorter the time needed for a given extent of permeation. Additionally, the Figure 11-15c shows that the smaller the value of ϕ_V°, the faster the permeation. This makes intuitive sense from the definition of ϕ_V°, because the smaller the finite volume adjacent to the diffusion domain compared to the diffusion domain, the greater the concentration produced from a given diffusing mass.

11.4.3 Type II Domains

Type II domains include *impedance by simultaneous reaction*. For illustrative purposes, we analyze a situation in which a component disappears by first-order reaction while being transported by unsteady-state diffusion in one dimension. The corresponding steady-state solution was presented in Chapter 8 and used extensively for analysis of flux into spherical catalyst particles. Unlike steady state, the diffusion domain is initially free of the component such that flux changes at each position with time. Eventually, a steady-state condition will be reached.

The general point form of the material balance was given in Chapter 2:

$$-\nabla \cdot N_i + r_i = \frac{\partial C_i}{\partial t} \tag{2-16}$$

The one-dimensional point form for the problem at hand (dropping the subscript *i* for convenience of notation) is

$$\mathfrak{D}_l \frac{\partial^2 C}{\partial^2 x} - kC = \frac{\partial C}{\partial t} \tag{11-87}$$

We assume that the concentration being introduced into the diffusion domain remains constant through time, that the domain is free of the diffusing component initially, and that the diffusion domain is of infinite length. Once again, the length of the domain may be as short as the thickness of a hydrodynamic boundary layer. The mathematical solution is still valid for such a boundary layer as long as the concentration does not increase substantially above zero at the end of this diffusion domain. The following initial and boundary conditions apply:

Initial condition:

$$C(x = 0, t = 0) = 0 \tag{11-88}$$

First boundary condition:

$$C(x = 0, t > 0) = C_0 \tag{11-89}$$

Second boundary condition:

$$C(x \to \infty, t > 0) = 0 \tag{11-90}$$

The analytical solution is then

$$\frac{C}{C_0} = c^{\circ} = \frac{1}{2}\exp\left[-x\left(\frac{k}{\mathfrak{D}_l}\right)^{0.5}\right]\operatorname{erfc}\left[\frac{x}{2(\mathfrak{D}_l t)^{0.5}} - (kt)^{0.5}\right]$$

$$+ \frac{1}{2}\exp\left[x\left(\frac{k}{\mathfrak{D}_l}\right)^{0.5}\right]\operatorname{erfc}\left[\frac{x}{2(\mathfrak{D}_l t)^{0.5}} + (kt)^{0.5}\right] \tag{11-91}$$

We note that the complementary error function appears in the solution, as was also true for diffusion without reaction (i.e., a type I domain) of infinite length.

The cumulative specific mass of a component entering the diffusion domain after any time, t, is

$$M_s = \frac{M(t)}{A_N} = \int_0^t N|_{x=0}\, dt = C_0\mathfrak{D}_l \int_0^t \left(\frac{dc^{\circ}}{dx}\bigg|_{x=0}\right) dt \tag{11-92}$$

After evaluating the flux at $x = 0$ from the derivative of Equation 11-91, the solution can be written as

$$M_s^{\circ}(\tau^{\circ}) = \frac{M(t)}{A_N C_0}\left(\frac{k}{\mathfrak{D}_l}\right)^{0.5} = (kt + 0.5)\operatorname{erf}(kt)^{0.5} + \left(\frac{kt}{\pi}\right)^{0.5} e^{-kt} \tag{11-93}$$

where $M_s^{\circ}(\tau^{\circ})$ is the cumulative specific mass uptake in the diffusion domain in dimensionless form, and kt is dimensionless time, also referred to as the Damkohler (I) number (see Chapter 9 and Appendix III). Equation 11-93 is plotted in Figure 11-16. If the rate coefficient, k, is fixed, the instantaneous slope ($dM_s^{\circ}/k\,dt$) is proportional to the mass per unit area per unit time entering the diffusion domain or the unsteady-state flux.

The time to reach steady state can be estimated graphically from Figure 11-16. As t becomes large, the slope reaches a constant value (i.e., the flux reaches steady state). To find the steady-state flux, we note that the error function approaches 1 for large values of its argument (kt) and the exponential term is negligible; hence Equation 11-93 becomes

$$M_s^{\circ}(\tau^{\circ}) = C_0\left(\frac{\mathfrak{D}_l}{k}\right)^{0.5}(kt + 0.5) \tag{11-94}$$

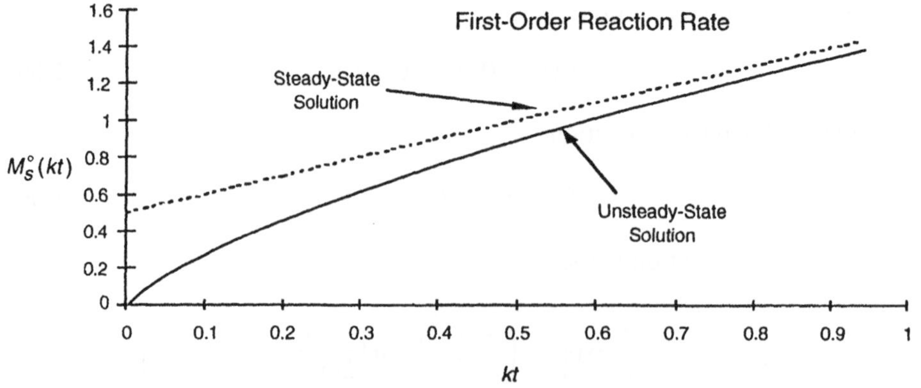

Figure 11-16 Cumulative dimensionless specific mass entering an unbounded type II domain as a function of dimensionless time [After Danckwerts (1960).]

The steady-state flux, N, is determined from the derivative with respect to time:

$$N|_{x=0} = C_0(\mathfrak{D}_l k)^{0.5} \tag{11-95}$$

As should be expected, relationships for steady-state diffusion and first-order reaction can also be derived beginning with the steady-state analysis presented in Chapter 8 (i.e., multiply Equation 8-54 by $\mathfrak{D}_l$ to obtain flux and note that $\tanh \mathfrak{N}_{TM} \rightarrow 1$ for large values of $\mathfrak{N}_{TM}$). The only difference is in the starting point of analysis. Here the component is initially absent from the diffusion domain, whereas in Chapter 8, a steady-state concentration profile already existed in the domain. For either unsteady- or steady-state conditions, the disappearance of a component by first-order reaction has the effect of increasing the rate at which that component enters the diffusion domain. This is intuitively reasonable because a reaction lowers the concentration, thus producing a steeper concentration gradient and a greater driving force for mass transport.

11.4.4 Type III Domains

Type III domain diffusion relationships are used to describe the *transport of a component through porous media*. The effect of domain heterogeneity is to alter the path length of diffusion from a straight line. Once again using a one-dimensional transport case for illustrative purposes, the general material balance for unsteady state is

$$\frac{\partial \epsilon_d N_p}{\partial x} = \epsilon_d \frac{\partial C}{\partial t} \tag{11-96}$$

in which N_p is the diffusional flux within the pores and ϵ_d is the domain porosity which, for the general case shown here, may not be constant with distance, x. Using Fick's law to describe diffusion, the resulting point form of the material balance is

$$\frac{1}{\epsilon_d}\frac{\partial}{\partial x}\left(\epsilon_d \mathfrak{D}_l \frac{\partial C}{\partial x}\right) = \frac{\partial C}{\partial t} \tag{11-97}$$

As explained in Chapter 8, alterations in path length domain III are accounted for by tortousity, τ. An *effective pore diffusivity*, $\mathfrak{D}_{p,e}$, relative to the pore flux is defined by

$$N_p = -\mathfrak{D}_l \frac{dC}{\tau\, dx} = -\mathfrak{D}_{p,e} \frac{dC}{dx} \tag{8-86}$$

in which

$$\mathfrak{D}_{p,e} = \frac{\mathfrak{D}_l}{\tau} \tag{11-98}$$

If porosity does not vary with length in the domain, ϵ_d can be taken out of the derivative in Equation 11-97 and canceled to yield

$$\mathfrak{D}_{p,e} \frac{\partial^2 C}{\partial x^2} = \frac{\partial C}{\partial t} \tag{11-99}$$

External flux, N, is of more interest than internal flux, N_p, for example, when dealing with loss of a component from solution to the diffusion domain of a porous particle. Continuity of flux requires that $NA_N = N_p(\epsilon_d A_N)$. Accordingly, if external flux is used, the definition of effective diffusivity, $\mathfrak{D}_e$, becomes that presented in Chapter 8:

$$\mathfrak{D}_e = \frac{\epsilon_d \mathfrak{D}_l}{\tau} \tag{8-89}$$

We can see that with the exception of the definition of diffusivity, the Fick's second law expression given in Equation 11-100 is identical to those derived for type I and type II domains. Thus, the same mathematical solutions that were found for each set of initial and boundary conditions can still be used. The only difference is that the diffusion coefficient is altered by the heterogeneity of the domain.

The steady-state condition in domain III is given by Fick's first law, just as it is in domain I. Flux can be expressed analogously to Equation 8-2 by

$$N = k_{f,e}(C_0 - C_\delta) \tag{11-100}$$

in which $k_{f,e}$ is the *effective* mass transfer coefficient, combining the effective diffusivity of the solute and the characteristic length of domain III (i.e., $k_{f,e} = \mathfrak{D}_e/L$). A mass transfer model for a heterogeneous domain may be appropriate, for example, in describing diffusion of nonbiodegradable solute through a biofilm, which is composed of water and cells, and then on to an adjacent solid phase.

11.4.5 Type IV Domains

Type IV domain processes include *two impedances to transport* of a component: (1) heterogeneity of the domain, and (2) simultaneous reaction or phase separation. The emphasis in this section will be on diffusion accompanied by sorption reactions because these are common in both natural and engineered environmental systems. In natural systems, for example, hydrophobic organic molecules associate with soil organic matter in the subsurface and thus diffuse within soil grains; an understanding of this process is required to predict contaminant fate and transport. In engineered systems, activated carbon is used to adsorb contaminants from solution; toward this end, process engineering involves design of optimal contact systems.

The topic of sorption rates was introduced in Chapter 8 with assumptions that made it possible to obtain an unsteady-state model rather easily. First, the diffusion domain was described in rectangular coordinates and was unbounded in length. Type IV domains in environmental systems are more often approximately spherical, however. This requires analysis of radial rather than planar diffusion, as well as consideration of a bounded rather than an unbounded domain. Second, sorption was introduced in the earlier development as a linear process when, in fact, nonlinear processes are more commonly encountered in environmental systems. Third, the only transport mechanism included was pore diffusion, but surface diffusion along pore walls, either alone or in combination with diffusion within pore fluids, may also occur. Lifting the simplifying assumptions used in Chapter 8 leads to more complex mathematical descriptions. Nevertheless, analytical solutions are still possible for a variety of important practical problems. Systems of equations that cannot be addressed by analytical solutions require numerical methods.

11.4.5.1 Pore Diffusion with Sorption A pore diffusion–sorption model was derived in Chapter 8 for a one-dimensional rectangular coordinate system. The same approach can be used to derive a model in which transport by pore diffusion is in the radial direction into a spherical particle (domain). A general material balance for radial transport was given in Chapter 2. The addition of an unsteady-state term to account for sorption by the solid phase within the particle gives

$$-\left(\frac{\partial N}{\partial r} + \frac{2}{r}N\right) = \epsilon_p \frac{\partial C}{\partial t} + (1 - \epsilon_p)\rho_s \frac{\partial q}{\partial t} \tag{11-101}$$

in which N is the flux measured external to the particle (note that $N = \epsilon_p N_p$), ρ_s is the solid phase density, ϵ_p is the particle porosity, and q is the concentration in the solid phase. If we include tortuousity, τ, to account for the alignments of pores that are not radial, the flux is

$$N = -\frac{\epsilon_p \mathfrak{D}_l}{\tau}\frac{\partial C}{\partial r} = -\mathfrak{D}_e \frac{\partial C}{\partial r} \tag{11-102}$$

Sorption processes may be linear or nonlinear. For linear *equilibrium* sorption, q_e and C_e are related by $q_e = \mathcal{K}_D C_e$ at any radial position. If a condition of *instantaneous local equilibrium* is assumed (i.e., $q = q_e$ and $C = C_e$), the chain rule for derivatives yields

$$\frac{\partial q}{\partial t} = \mathcal{K}_D \frac{\partial C}{\partial t} \tag{11-103}$$

Equation 11-101 can then be simplified to

$$\mathfrak{D}_a\left(\frac{\partial^2 C}{\partial r^2} + \frac{2}{r}\frac{\partial C}{\partial r}\right) = \frac{\partial C}{\partial t} \tag{11-104}$$

where $\mathfrak{D}_a$ is the apparent diffusivity, defined in Chapter 8 as

$$\mathfrak{D}_a = \frac{\mathfrak{D}_l/\tau}{1 + \dfrac{(1 - \epsilon_p)\rho_s}{\epsilon_p}\mathcal{K}_D} \tag{8-130}$$

The description of diffusion into particles accompanied by linear sorption thus reduces to an expression of Fick's second law wherein the diffusion constant is modified to account for the effects of porosity, tortuousity, and sorption. Analytical solutions to Fick's second law for radial diffusion are available for a number of initial and boundary conditions. Example 11-7 uses a well-known solution for analysis of adsorption rates in a CMBR.

Example 11-7

- **Situation:** Natural organic matter (NOM) remaining after chemical coagulation at a municipal water treatment plant reacts with chlorine in disinfection to yield chlorinated by-products that exceed federal limits. One strategy being explored for removal of the organic precursors is adsorption on activated carbon. Equilibrium experiments have shown that a linear isotherm provides a reasonable approximation of observed sorption data. Rate data are now needed to design an FBR. Although estimates of tortousity and free-liquid diffusivity are available from the literature, it also known that the chemical structure of the NOM can vary widely from site to site and that structure affects adsorption rates. Pore diffusion is thought to be a reasonable assumption for the microtransport mechanism. A simple CMBR rate test is planned to measure the effective diffusivity. A given dose of activated carbon will be added and measurements made of the residual NOM concentration in solution as a function of time.

- **Question(s):** Devise a model to test the validity of the pore diffusion model and to obtain the estimates of diffusivity for use in FBR models. What are the limitations to these data?

- **Logic and Answer(s):**

1. It is convenient to analyze sorption into a single sphere as shown below.

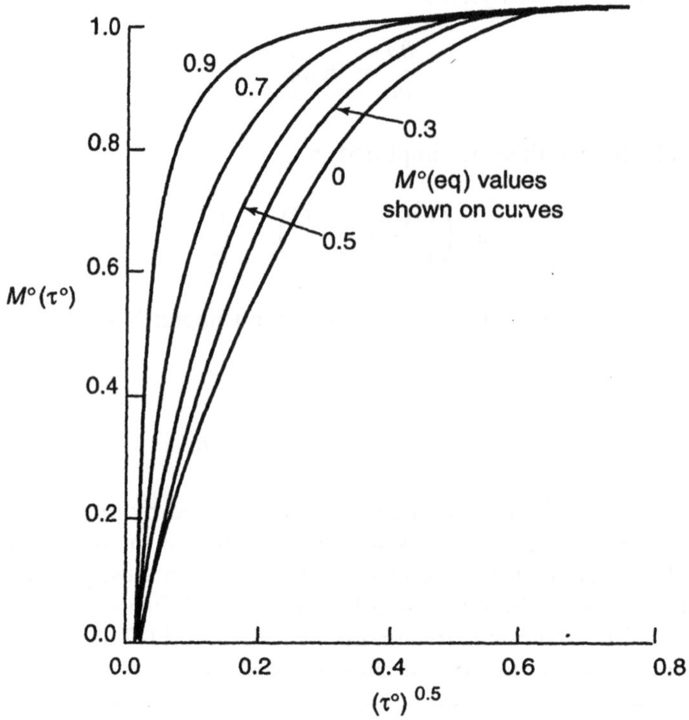

2. The volume, $V_{R,P}$, of solution in the reactor served by each spherical carbon particle can be calculated readily from the sorbent dosage. Fick's second law for radial diffusion must be solved:

$$\mathfrak{D}_a \left(\frac{\partial^2 C}{\partial r^2} + \frac{2}{r} \frac{\partial C}{\partial r} \right) = \frac{\partial C}{\partial t}$$

3. The following initial and boundary conditions are:

Initial condition:

$$C(r, t = 0) = 0$$

Boundary conditions:

$$V_{R,P} \frac{\partial C}{\partial t} = -4\pi R^2 \mathfrak{D}_a \left. \frac{\partial C}{\partial r} \right|_{r=R}$$

$$\mathfrak{D}_a \left. \frac{\partial C}{\partial r} \right|_{r=0} = 0$$

where R is the particle radius.

4. Making distance and time dimensionless, Fick's second law is written as

$$\frac{\partial^2 C}{\partial(\zeta^\circ)^2} + \frac{2}{\zeta^\circ}\frac{\partial C}{\partial \zeta^\circ} = -\frac{\partial C}{\partial \tau^\circ}$$

in which $\zeta^\circ = r/R$ and $\tau^\circ = \mathfrak{D}_a t/R^2$. Similarly, the dimensionless surface condition is

$$\frac{V_{R,P}}{4\pi R^3}\frac{\partial C}{\partial \tau^\circ} = -\left.\frac{\partial C}{\partial \zeta^\circ}\right|_{\zeta^\circ = 1}$$

(Note the similarity in form of this surface condition to that given by Equation 11-83 for diffusion through a pipe wall and accumulation inside the pipe; the physical situations are analogous.)

5. The material balance on solution phase solute concentration at equilibrium (C_e) is

$$C_0 V_{R,P} = C_e V_{R,P} + \rho_s \mathcal{K}_D \frac{4\pi R^3}{3} C_e$$

Equilibrium uptake, $M^\circ(\text{eq})$, is then defined as

$$M^\circ(\text{eq}) = \frac{C_0 - C_e}{C_0} = \frac{\rho_s\, \mathcal{K}_D \dfrac{4\pi R^3}{3} C_e}{C_0 V_{R,P}} = \frac{1}{1 + \phi_V^\circ}$$

where

$$\phi_V^\circ = \frac{3V_{R,P}}{4\pi R^3 \rho_s \mathcal{K}_D}$$

6. The analytical solution given by Crank (1975) for diffusion into spheres from a limited volume of solution is

$$M^\circ(\tau^\circ) = 1 - \sum_{n=1}^{\infty} \frac{6\phi_V^\circ(\phi_V^\circ + 1)\exp(-u_n^2 \tau^\circ)}{9 + 9\phi_V^\circ + u_n^2(\phi_V^\circ)^2}$$

where

$$M^\circ(\tau^\circ) = \text{fractional uptake} = \frac{C_0 - C_t}{C_0 - C_e}$$

and u_n is the nth root of $\dfrac{3u_n}{3 + \phi_V^\circ u_n^2} - \tan u_n = 0$. The first six roots of this function are given below for various values of ϕ_V° which in turn relate to specific values of equilibrium uptake, $M^\circ(\text{eq})$ as shown in step 5.

M° (eq)	ϕ_V°	u_1	u_2	u_3	u_4	u_5	u_6
0	∞	3.1416	6.2832	9.4248	12.5664	15.7080	18.9496
0.1	9.0000	3.2410	6.3353	9.4599	12.5928	15.7292	18.8671
0.2	4.0000	3.3485	6.3979	9.5029	12.6254	15.7554	18.8891
0.3	2.3333	3.4650	6.4736	9.5567	12.6668	15.7888	18.9172
0.4	1.5000	3.5909	6.5665	9.6255	12.7205	15.8326	18.9541
0.5	1.0000	3.7264	6.6814	9.7156	12.7928	15.8924	19.0048
0.6	0.6667	3.8711	6.8246	9.8369	12.8940	15.9779	19.0784
0.7	0.4286	4.0236	7.0019	10.0039	13.0424	16.1082	19.1932
0.8	0.2500	4.1811	7.2169	10.2355	13.2689	16.3211	19.3898
0.9	0.1111	4.3395	7.4645	10.5347	13.6133	16.6831	19.7564
1.0	0	4.4934	7.7253	10.9041	14.0662	17.2208	20.3713

7. The value of ϕ_V° is known in each experiment from step 5:

$$\phi_V^\circ = \frac{1 - M^\circ(\text{eq})}{M^\circ(\text{eq})} = \frac{C_e}{C_0 - C_e}$$

The value of C_e is obtainable directly from the material balance at equilibrium between the two phases since

$$q = \frac{C_0 - C_e}{D_o} = \mathcal{K}_D C_e$$

$$C_e = \frac{C_0}{1 + \mathcal{K}_D D_o}$$

where D_o is the activated carbon dosage (mass/volume).

8. Plots of fractional uptake, $M^\circ(\tau^\circ)$, against the square root of dimensionless time as calculated from the analytical solution in step 6 and the relationship between ϕ_V° and M° (eq) in step 7 are given below:

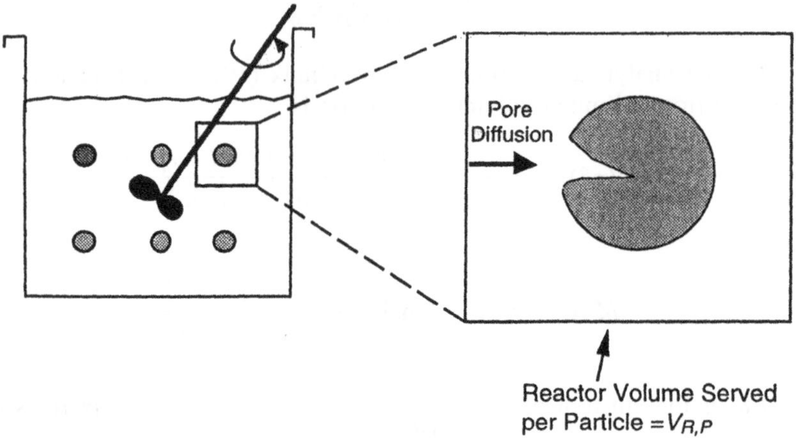

9. Note that initial uptake is nearly linear with the square root of τ°. This should be expected from our earlier discussion of unbounded diffusion (see Equation 11-65) because in the early stage of uptake the con-

centration outside the sphere is nearly constant and the physical boundary of the center of the sphere has not yet prevented further diffusion. Problem 11-12 provides an exercise in fitting data to this diffusion model. Selection of the proper scaling factor for the abscissa, i.e., the diffusivity value, will allow all experimental values of uptake $[M^\circ(\tau^\circ)]$ to lie on the appropriate curve corresponding to the equilibrium uptake $[M^\circ(\text{eq})]$ for the particular experimental system.

- **Limitations:** Whether the diffusion with linear sorption model is appropriate needs to be assessed by comparing the model to the experimental data over the entire spectrum of sorption time. Having determined the proper value of ϕ_V° from independent measurements, a single value of $\mathfrak{D}_a$ should provide a good match between the predictions and the experimental uptake data. However, a single value of $\mathfrak{D}_a$ may represent only an average value because NOM is actually a mixture of many different molecular species that differ in their diffusivity and sorption characteristics. These factors are not accounted for in the model being tested. Another concern is whether pore diffusion is the appropriate description of the true diffusion process. If it is, experiments conducted with different starting concentrations should all yield the same values for $\mathfrak{D}_a$. If $\mathfrak{D}_a$ is not constant, surface diffusion control is possible, as will be discussed in the next section.

If the sorption process is nonlinear and described by a Freundlich equation ($q_e = \mathcal{K}_F C^n$), the chain rule for relating derivatives of the solid- and solution-phase concentrations with respect to time in the material balance equation yields

$$\frac{\partial q}{\partial t} = \frac{\partial q}{\partial C}\frac{\partial C}{\partial t} = n\mathcal{K}_F C^{n-1}\frac{\partial C}{\partial t} \qquad (11\text{-}105)$$

Unlike the situation for linear sorption, this derivative, when returned to the material balance relationship (Equation 11-101), makes it impossible to define an apparent diffusivity and a simple Fick's second law expression. Instead, a nonlinear partial differential equation results:

$$\frac{\mathfrak{D}_l}{\tau}\left(\frac{\partial^2 C}{\partial r^2} + \frac{2}{r}\frac{\partial C}{\partial r}\right) = \left[1 + \frac{(1-\epsilon_p)\rho_s n\mathcal{K}_F C^{n-1}}{\epsilon_p}\right]\frac{\partial C}{\partial t} \qquad (11\text{-}106)$$

An analytical solution of Equation 11-106 is not possible, but various numerical methods are available. These methods include finite difference schemes and orthogonal collocation techniques (the latter generally being more efficient and stable). Many other initial and boundary conditions can be applied to describe diffusion in type IV domains for various settings. Analytical solutions that are available for type I, II, or III domains can often be adapted for type IV domains. Some further useful examples of analytical solutions are provided in Table 11-1.

Table 11-1 Useful Analytical Solutions to Unsteady-State Diffusion Relationships[a]

Situation	Schematic	Equations	Reference
A. Diffusion from "limited" volume (length, δ) into a finite domain (length, L) $C(L < x < L + \delta, 0) = C_0$ $C(0 < x < L, 0) = 0$ $\delta \frac{\partial C}{\partial t} = \mathfrak{D} \frac{\partial C}{\partial x}\Big\vert_{x=L}$ or for type III domain, $\delta \frac{\partial C}{\partial t} = \epsilon_d \mathfrak{D} \frac{\partial C}{\partial x}\Big\vert_{x=L}$	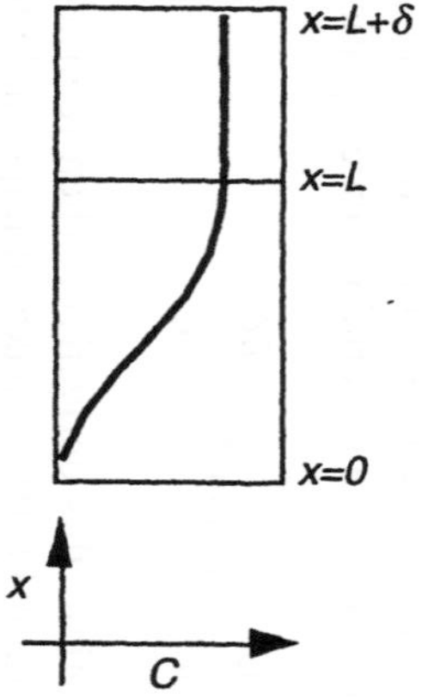	$M^\circ(t) = \frac{M(t)}{M(\mathrm{eq})} = 1 - \sum_{n=1}^{\infty} \frac{2\phi_V^\circ(\phi_V^\circ + 1)}{1 + \phi_V^\circ + (\phi_V^\circ)^2 u_n^2} \exp\left(\frac{-u_n^2 \mathfrak{D} t}{L^2}\right)$ where $M(\mathrm{eq})$ = equilibrium uptake of solute mass in length, L $\phi_V^\circ = \frac{\delta}{\rho \mathcal{K}_D L}$, or for porous diffusion domain, $\phi_V^\circ = \frac{\delta}{\mathcal{K}_D \epsilon_d L}$ $\mathcal{K}_D$ = distribution coefficient for type IV domain $\tan \mu_n = -\phi_V^\circ u_n$ [roots given by Crank (1975)]	Crank (1975)
B. Diffusion from an infinite domain into a "limited" volume (length δ) $C(x, 0) = C_\infty$ $C_\delta(0) = C_b(0) = 0$ $\delta \frac{\partial C}{\partial t} = \mathfrak{D} \frac{\partial C}{\partial x}\Big\vert_{x=L}$ or, for a type III domain,	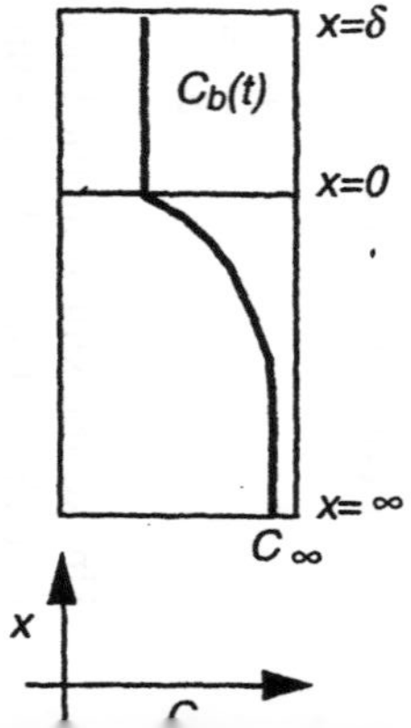	$\frac{C_b}{C_\infty} = 1 - \exp\left(\frac{\mathfrak{D} t}{\delta^2}\right) \mathrm{erfc}\left(\frac{(\mathfrak{D} t)^{0.5}}{\delta}\right)$ or, for a type III domain, $\frac{C_b}{C_\infty} = 1 - \exp\left(\frac{\epsilon_d^2 \mathfrak{D} t}{\delta^2}\right) \mathrm{erfc}\left(\frac{\epsilon_d (\mathfrak{D} t)^{0.5}}{\delta}\right)$	Van Rees et al. 1991

$$\delta \frac{\partial C}{\partial t} = \epsilon_d \mathfrak{D} \left. \frac{\partial C}{\partial x} \right|_{x=L}$$

$$\left. \frac{\partial C}{\partial x} \right|_{x=\infty} = 0$$

C. Diffusion from an infinite domain ($x = 0 \rightarrow \infty$) with mass transfer at a boundary layer ($x = 0 \rightarrow \delta$) between two phases $C(0 < x < \infty, 0) = C_\infty$ $C_{b,e}$ is the bulk concentration in phase II, defined by an equilibrium condition between the two phases (e.g., Henry's law) $\mathfrak{D} \left. \frac{dC}{dx} \right\|_{x=0} = k_f(C_\delta - C_{b,e})$	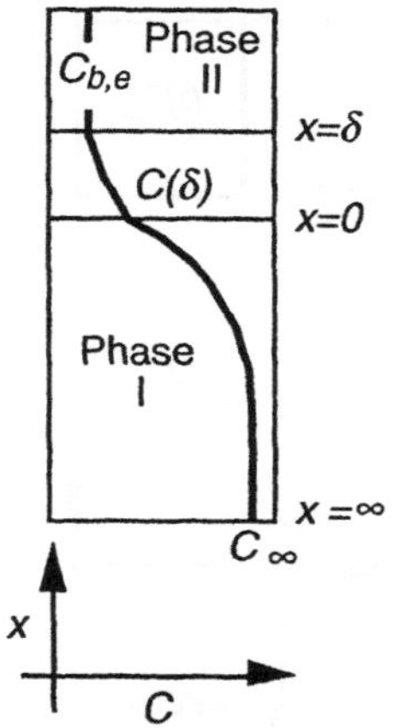	$M_s(t) = \frac{C_{b,e} - C_\infty}{\Psi_m} \cdot \left\{ [\exp(\Psi_m^2 \mathfrak{D} t)] \operatorname{erfc} [\Psi_m (\mathfrak{D} t)^{0.5}] - 1 + \frac{2\Psi_m (\mathfrak{D} t)^{0.5}}{\pi^{0.5}} \right\}$ where $M_s(t)$ = mass lost from diffusion source per unit area (i.e., specific mass) $\Psi_m = \frac{k_f}{\mathfrak{D}}$ $C_{b,e}$ = equilibrium bulk solution concentration (e.g., zero if diffusion of a volatile organic chemical from water to air is involved	Crank, 1975

Table 11-1 (*Continued*)

Situation	Schematic	Equations	Reference
D. Diffusion from a finite domain with mass transfer at the boundary between the two phases $C\,(0< x < L, 0) = C_L$ $C_{b,e}$ is the bulk concentration in phase II defined by equilibrium condition between the two phases (e.g., Henry's law) $\mathfrak{D}\left.\frac{dC}{dx}\right\| = k_f(C_\delta - C_{b,e})$	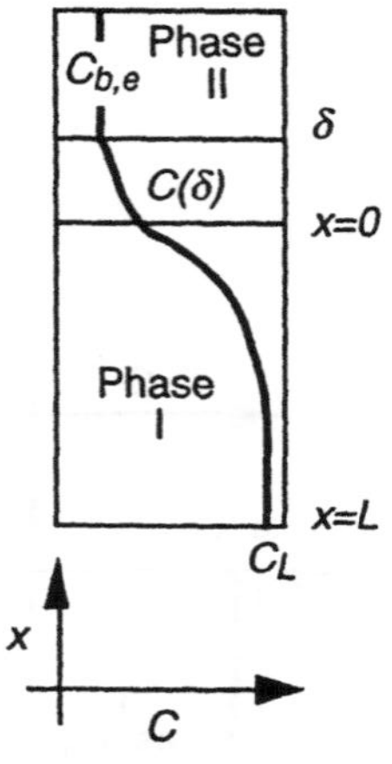	$M_s^\circ(t) = \frac{M_s(t)}{M_s(\text{eq})} = 1 - \sum_{n=1}^{\infty} \frac{2(\Psi_m^\circ)^2 \exp(-u_n^2\mathfrak{D}t/L^2)}{u_n^2(u_n^2 + (\Psi_m^\circ)^2 + \Psi_m^\circ)}$ where $M_s^\circ(t)$ = fraction lost by diffusion as a function of time $M_s(\text{eq})$ = mass lost per unit area at equilibrium Flux: $N\|_{x=0} = 2C_L k_f \Psi_m^\circ \sum_{n=1}^{\infty} \frac{\exp(-u_n^2\mathfrak{D}t/L^2)}{\Psi_m^\circ(\Psi_m^\circ + 1) + u_n^2}$ $u_n \tan u_n = \Psi_m^\circ$ [roots given by Crank (1975)] $\Psi_m^\circ = \frac{k_f L}{\mathfrak{D}}$	Crank, 1975 Carslaw and Jaeger, 1959
E. Diffusion from a "finite" volume (length, δ) into infinite type III domain $C_b(t = 0) = C_0$ $C(0 < x < \infty, 0) = 0$ $\delta \frac{dC}{dt} = \epsilon_d \mathfrak{D} \left.\frac{dC}{dx}\right\|_{x=0}$	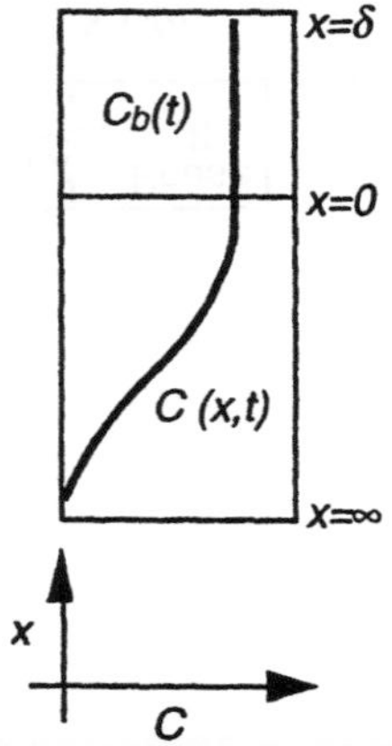	$\frac{C_b}{C_0} = \exp\left(\frac{\epsilon_d^2\mathfrak{D}t}{\delta^2}\right) \text{erfc}\left[\frac{\epsilon_d(\mathfrak{D}t)^{0.5}}{\delta}\right]$ $\frac{C(x, t)}{C_0} = \exp\left(\frac{x}{\delta} + \frac{\epsilon_d^2\mathfrak{D}t}{\delta^2}\right) \text{erfc}\left[\frac{xl}{2(\mathfrak{D}t)^{0.5}} + \frac{\epsilon_d(\mathfrak{D}t)^{0.5}}{\delta}\right]$	Van Rees, et al. 1991

F. *Case 1*: Diffusion from an "unlimited" volume (i.e., no depletion of concentration over time) into a domain of spherical geometry $C(R, t) = C_0$ $C(r, 0) = 0$ $\mathfrak{D} \left.\frac{\partial C}{\partial r}\right\|_{r=0} = 0$ *Case 2*: diffusion from a domain of spherical geometry into an "infinite" volume (i.e., no increase in concentration over time) $C(R, t) = 0$ $C(r, 0) = C_0$ $\mathfrak{D} \left.\frac{\partial C}{\partial r}\right\|_{r=0} = 0$	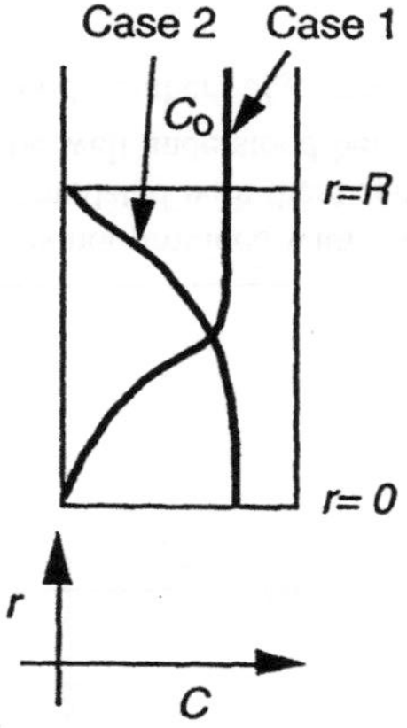	*Case 1*: $M^\circ(t) = \frac{M(t)}{M(\text{eq})} = 1 - \frac{6}{\pi^2} \sum_{n=1}^{\infty} \frac{1}{n^2} \exp\left(\frac{-n^2\pi^2\mathfrak{D}t}{R^2}\right)$ where $M(\text{eq})$ = equilibrium mass uptake by sphere (see Example 11-8) *Case 2*: $M^\circ(t) = \frac{M(t)}{M_0(\text{eq})} = \frac{6}{\pi^2} \sum_{n=1}^{\infty} \frac{1}{n^2} \exp\left(\frac{-n^2\pi^2\mathfrak{D}t}{R^2}\right)$ where $M_0(\text{eq})$ = initial equilibrium mass uptake by sphere	Crank, 1975

Table 11-1 (*Continued*)

Situation	Schematic	Equations	Reference
G. Diffusion from a "limited" reactor volume ($V_{R,P}$) served by a single spherical sorbent particle into a finite domain of spherical geometry (the particle) $C_b(t = 0) = C_0$ $C(0 < r < R, 0) = 0$ $\mathfrak{D}\left.\frac{\partial C}{\partial r}\right\|_{r=0} = 0$ $V_{R,P}\frac{dC}{dt} = -4\pi R^2\mathfrak{D}\left.\frac{\partial C}{\partial r}\right\|_{r=R}$	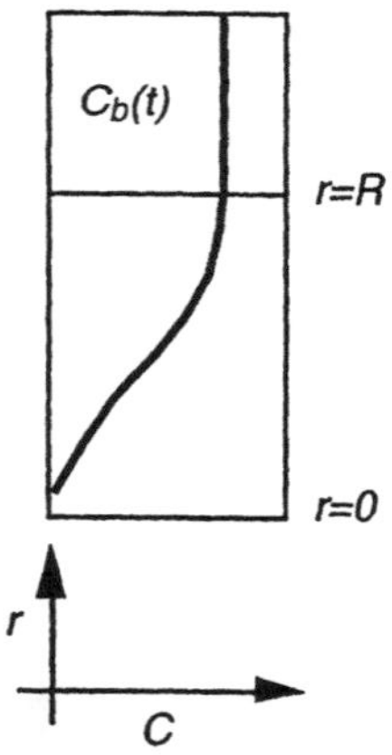	$M^\circ(\tau^\circ) = 1 - \sum_{n=1}^{\infty} \frac{6\phi_V^\circ(\phi_V^\circ + 1)\exp(-u_n^2\tau^\circ)}{9 + 9\phi_V^\circ + u_n^2(\phi_V^\circ)^2}$ $\tau^\circ = \frac{\mathfrak{D}_a t}{R^2}$ (for type IV domain) $M^\circ(\tau^\circ) = \text{fractional uptake} = \frac{C_0 - C_t}{C_0 - C_e}$ $\tan u_n = \frac{3u_n}{3 + \phi_V^\circ u_n^2}$ $\phi_V^\circ = \frac{C_e}{C_0 - C_e}$ (see Example 11-7)	Crank, 1975

[a]In this table, the diffusion coefficient, $\mathfrak{D}$, is not provided with a subscript in order to emphasize that tortuosity in type III diffusion domains and linear sorption in type IV diffusion domains can be accommodated with these same solutions if the appropriate definition of the diffusivity is used. It is advised, however, that the derivation of each analytical solution be well understood before altering the diffusivity term.

Note: Ψ_m° has the same form as the Sherwood number, $\mathfrak{N}_{Sh}$, but not the same physical meaning.

11.4.5.2 Surface Diffusion with Sorption Surface diffusion accounts for the transport of sorbed solute along the interfaces between solid and solution phases. This transport process is described by Fick's first law:

$$N_s = -\rho_s \mathfrak{D}_s \frac{\partial q}{\partial r} \tag{11-107}$$

where N_s is the surface flux and $\mathfrak{D}_s$ is the surface diffusivity. Pore and surface diffusion occur in parallel such that the total flux, N_T, parallel to the surface is

$$N_T = N_s + N_p = -\left(\rho_s \mathfrak{D}_s \frac{\partial q}{\partial r} + \mathfrak{D}_{p,e} \frac{dC}{dr}\right) \tag{11-108}$$

where $\mathfrak{D}_{p,e} = \mathfrak{D}_e/\tau$ as discussed in Chapter 8 (Equation 8-86). For transport processes that operate in parallel, the *faster* of the two is *rate determining*. However, the distinction between surface and pore diffusion may not easily be detected in sorption rate experiments. This can be appreciated by restating Equation 11-108 as

$$N_T = N_s + N_p = -\left(\rho_s \mathfrak{D}_s \frac{\partial q}{\partial C} + \mathfrak{D}_{p,e}\right) \frac{dC}{dr} \tag{11-109}$$

In this form it is apparent that total flux could be interpreted as being due solely to pore diffusion:

$$(N_s)_p = -(\mathfrak{D}_s)_p \frac{dC}{dr} \tag{11-110}$$

where $(N_s)_p$ is the "apparent" pore flux and $(\mathfrak{D}_s)_p$ is the "apparent pore diffusivity," in which the effects of surface diffusion are implicitly considered. If sorption is linear, Equation 11-109 shows that the apparent pore diffusivity is

$$(\mathfrak{D}_s)_p = \rho_s \mathfrak{D}_s \mathcal{K}_D + \mathfrak{D}_{p,e} \tag{11-111}$$

In other words, the effect of surface diffusion can be detected only if $(\mathfrak{D}_s)_p$ is larger than expected from the effective pore diffusion coefficient, $\mathfrak{D}_{p,e}$. The more that $(\mathfrak{D}_s)_p$ exceeds $\mathfrak{D}_{p,e}$, the more important is surface diffusion.

A different result is obtained if sorption is nonlinear. For instance, if the Freundlich equation is applied in Equation 11-109, then

$$(\mathfrak{D}_s)_p = \rho_s \mathfrak{D}_s \, n \mathcal{K}_F \, C^{n-1} + \mathfrak{D}_{p,e} \tag{11-112}$$

When $n < 1$, $(\mathfrak{D}_s)_p$ decreases with increasing solution-phase concentration. Thus, if a pore diffusion model is used to interpret a series of CMBR sorption

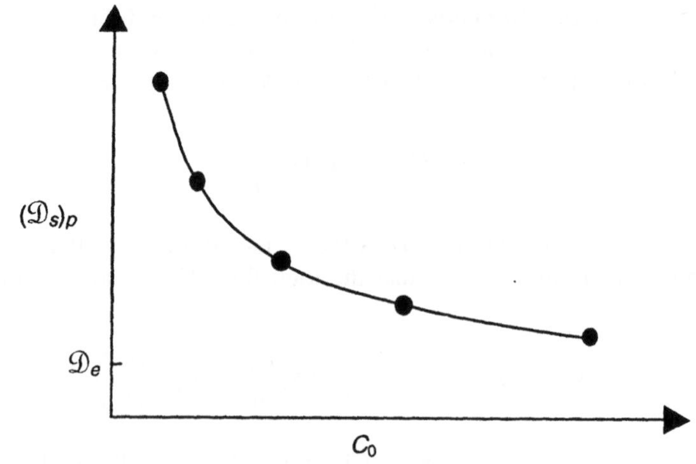

Figure 11-17 Variations in apparent pore diffusivity values with different initial solute concentrations in a CMBR.

rate experiments conducted at different initial solution-phase concentrations, C_0, it is likely that no single value of a pore diffusion coefficient will fit all of the data if significant surface diffusion is involved. An illustration of this effect is given in Figure 11-17. Caution is advised for any attempt to distinguish between different mechanisms of intraparticle transport simply on the basis of model fits to experimental data.

Another way to conceptualize intraparticle transport is in terms of *homogeneous surface diffusion*, as illustrated in Figure 11-18. Instead of considering parallel transport by surface and pore diffusion, the particle is viewed as a homogeneous solid phase, so that only surface diffusion is possible. The solid-phase concentration at the outer edge of the particle is thus established by equilibrium with the solution-phase concentration external to the particle. Because the particle is modeled as a homogeneous solid phase, a material balance is only needed for the solid phase:

$$-\left(\frac{\partial N_s}{\partial r} + \frac{2}{r} N_s\right) = \rho_s \frac{\partial q}{\partial t} \tag{11-113}$$

Substituting Fick's first law for flux, we arrive at

$$\mathfrak{D}_s \left(\frac{\partial^2 q}{\partial r^2} + \frac{2}{r}\frac{\partial q}{\partial r}\right) = \frac{\partial q}{\partial t} \tag{11-114}$$

This configuration of Fick's second law is referred to as the *homogeneous surface diffusion model* (HSDM). It can be solved with any appropriate initial and boundary conditions just as we have done with the pore diffusion model.

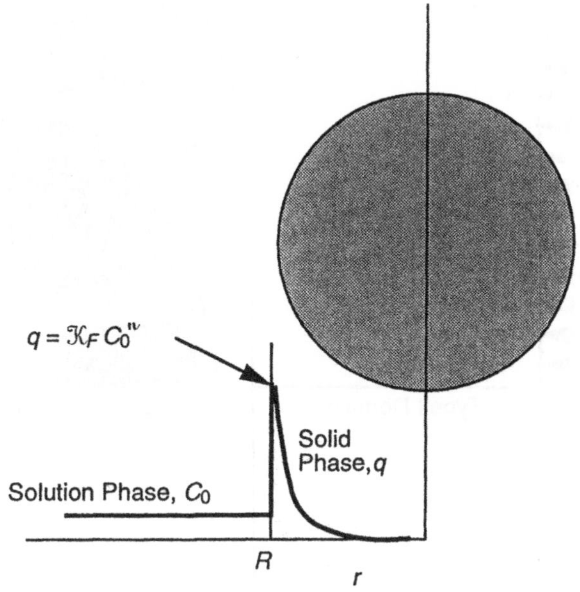

Figure 11-18 Schematic representation and concentration profile for the homogeneous surface diffusion model.

The HSDM has been used extensively, for example, to describe sorption on activated carbon in batch and continuous-flow systems. In fact, the HSDM provides an important computational advantage when sorption is nonlinear because, unlike the pore diffusion model (Equation 11-106), equilibrium between the two phases has been effectively decoupled from the transport equation; the equilibrium condition is needed only at the surface of the particle.

Computational advantages aside, however, the HSDM is not appropriate unless the microtransport mechanism is surface diffusion. For instance, it has been found that the sorption of high-molecular-weight natural organic matter on activated carbon causes blockage of surface diffusion by low-molecular-weight sorbates, shifting the principal mode of intraparticle transport of the latter to pore diffusion.

11.5 DOMAINS IN SERIES

The rate of diffusion through the hydrodynamic boundary layer (domain I) surrounding a porous particle (often referred to as *film diffusion*) may be of the same order of magnitude as that intraparticle diffusion (type IV domain) by either pore or surface diffusion. This situation is depicted schematically in Figure 11-19 and is referred to as the *external–internal series diffusion model* or the *dual-impedance model*. The same situation was analyzed in Chapter 8

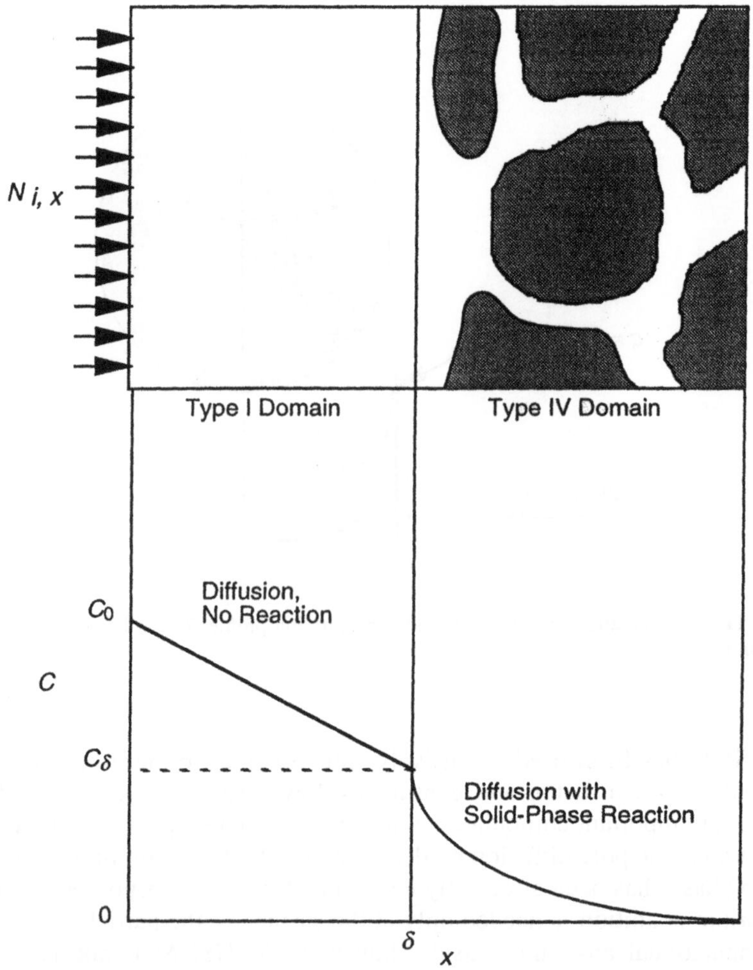

Figure 11-19 Schematic representation and concentration profile for a type I–type IV dual-impedance model.

but for steady-state conditions in both domains. The boundary condition at the surface of the particle provides the relationship between the solid- and solution-phase material balance equations:

Film-pore diffusion boundary condition:

$$k_f(C_0 - C_\delta) = \mathfrak{D}_e \left.\frac{\partial C}{\partial x}\right|_{x=\delta} \tag{11-115}$$

Film-surface diffusion boundary condition:

$$k_f(C_0 - C_\delta) = \mathfrak{D}_s \rho_s \left.\frac{\partial q}{\partial x}\right|_{x=\delta} \tag{11-116}$$

Unsteady state is produced because both the fluid- and solid-phase concentration gradients at $x = \delta$ are changing with time due to sorption in the type IV domain. Through either of these surface conditions, it is possible to eliminate mathematically the unknown and unmeasurable concentration, C_δ, in much the same way as shown in Chapter 8. Inclusion of a surface condition adds mathematical complication, but analytical solutions are still possible for unsteady state in certain situations, as will be illustrated when we discuss quasi-steady-state models.

Another interesting combination of domains is illustrated in Figure 11-20. In this instance, slow mass transfer by diffusion in domain III is followed by relatively rapid absorption and dispersal throughout an essentially homogeneous (completely mixed) domain. Such a combination is used, for example, to

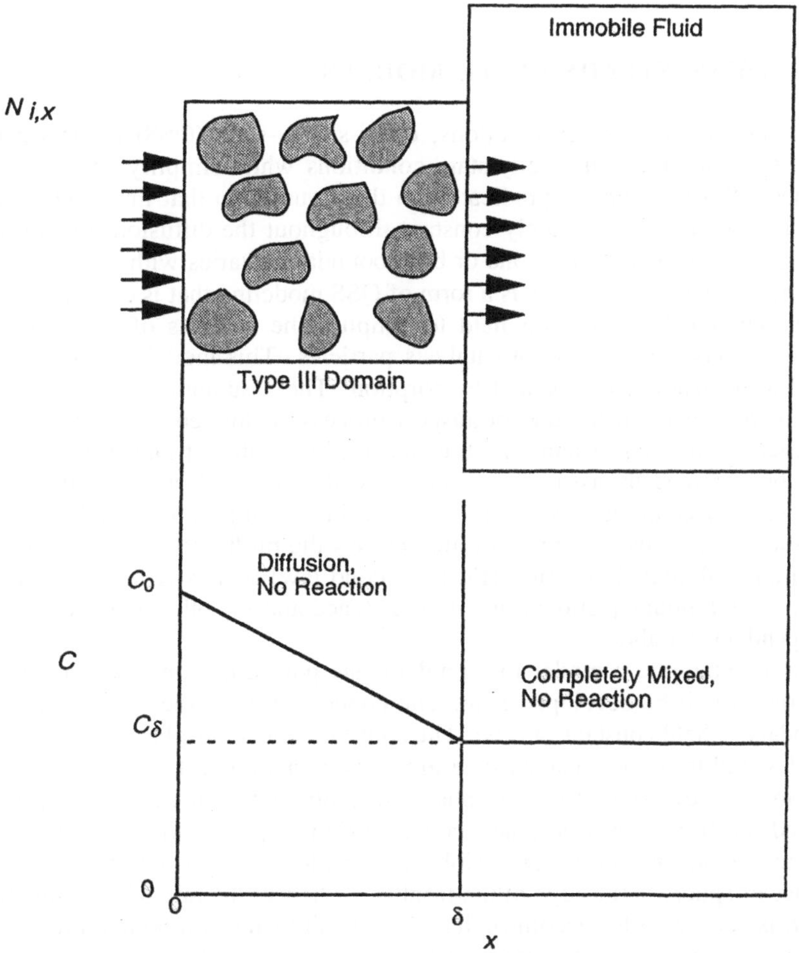

Figure 11-20 Schematic representation and concentration profile for the bicontinuum or mobile-immobile dual-domain model.

conceptualize transport of solute through nonsorptive external regions of soil aggregates and then through an *immobile* internal fluid region; this conceptualization is sometimes referred to as a *bicontinuum model*. The flux into the immobile region is

$$\frac{k_f A_e}{V_{IM}}(C_0 - C_\delta) = \hat{k}_{v,l}(C_0 - C_\delta) = \epsilon_d \frac{\partial C_\delta}{\partial t} + \rho_s(1 - \epsilon_d)\frac{\partial q_\delta}{\partial t} \tag{11-117}$$

in which A_e is the effective cross-sectional area (i.e., that portion of the total cross section of the domain that is available for fluid-phase diffusion into the completely mixed domain, V_{IM} is the volume of the immobile region, and $\hat{k}_{v,l}$ is the overall volumetric mass transfer coefficient at the interface between the domains.

11.6 QUASI-STEADY-STATE MODELS

In certain microtransport situations, quasi-steady-state (QSS) models can be used to approximate unsteady-state conditions while simplifying the mathematics. QSS modeling is predicated on the assumption that the concentration profile remains approximately constant throughout the diffusion domain even though the concentration at one or both boundaries varies with time.

The *shrinking core model* is a form of QSS modeling that is commonly used in the chemical engineering field to simplify the analysis of diffusion with instantaneous reaction within catalysis particles. This model can also be applied to diffusion accompanied by sorption. The assumption of quasi–steady state is reasonable if the microtransport process is slow compared to sorption or reaction. In these instances the concentration profile at any instant in time may be obtained by assuming steady-state diffusion. Unsteady state is then described by coupling this steady-state concentration profile at each instant in time with the boundary condition that causes the profile to change with time. A partial differential equation (Fick's second law) is thus reduced to two ordinary differential equations, one with distance and the other with time as the independent variable.

As an example, a model for simultaneous pore diffusion with irreversible adsorption will be developed here. The concentration at the solution–particle boundary is held constant in time. The solution- and solid-phase concentration profiles within a spherical particle are shown in Figure 11-21. Because adsorption is irreversible, the solid-phase concentration profile has a sharp front. Behind the front, the solid-phase concentration is q_e; ahead of the front $q = 0$. Because adsorption is irreversible, the solid-phase concentration from the surface of sphere to the edge of the front is q_e; hence, the rate of accumulation, $\partial q/\partial t$, is zero. As sites become exhausted, the diffusing component must move farther inward, causing the sorption front to also move slowly inward.

The material balance for radial diffusion–sorption within a spherical particle

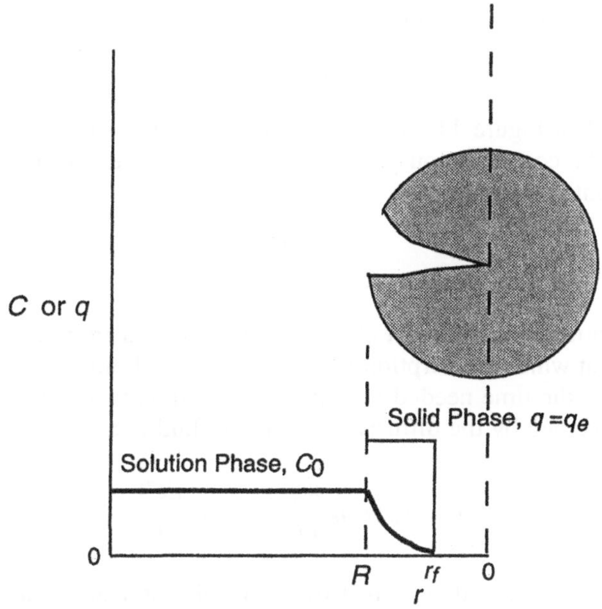

Figure 11-21 Schematic representation and concentration profile for a quasi-steady-state pore diffusion model with irreversible adsorption, no film diffusion and constant solute concentration in solution.

domain, presented earlier as Equation 11-101, can also be written as

$$\frac{1}{r^2}\frac{\partial}{\partial r}\left(r^2\mathfrak{D}_e\frac{\partial C}{\partial r}\right) = \epsilon_p\frac{\partial C}{\partial t} + (1 - \epsilon_p)\rho_s\frac{\partial q}{\partial t} \tag{11-118}$$

We have already noted that in advance of the sorption front, $\partial q/\partial t = 0$. The QSS assumption is then made, which sets $\partial C/\partial t = 0$ and yields a Fick's first law expression:

$$\frac{d}{dr}\left(r^2\mathfrak{D}_e\frac{dC}{dr}\right) = 0 \tag{11-119}$$

After twice integrating Equation 11-119, we obtain

$$\mathfrak{D}_e C = \frac{-\kappa_1}{r} + \kappa_2 \tag{11-120}$$

The constants of integration, κ_1 and κ_2, are determined from the following boundary conditions:

First boundary condition:

$$C = C_0 = C_R \qquad \text{at } r = R \text{ for } t \geq 0 \tag{11-121}$$

Second boundary condition:

$$C = 0 \qquad \text{at } r = r_f \tag{11-122}$$

As indicated in Figure 11-21, r_f is the radial position of the sorption front at any time. The concentration profile between the surface of the sphere and the sorption front is therefore

$$\frac{C}{C_R} = \frac{R(r - r_f)}{r(R - r_f)} \tag{11-123}$$

This concentration profile will change as the sorption front advances.

The rate at which the sorption front advances is of practical interest because it determines the time needed to exhaust the sorptive capacity. Its rate of advance is governed by the diffusive flux in the fluid phase at $r = r_f$:

$$(1 - \epsilon_p)\rho_s q_e \frac{dr_f}{dt} = -\mathfrak{D}_e \left.\frac{dC}{dr}\right|_{r=r_f} \tag{11-124}$$

The right-hand side is determined by taking the derivative, dC/dr, from the QSS analysis of the concentration profile (Equation 11-123) and evaluating it at $r = r_f$:

$$(1 - \epsilon_p)\rho_s q_e \frac{dr_f}{dt} = -\frac{R}{(R - r_f)r_f}\mathfrak{D}_e C_R \tag{11-125}$$

Separation of variables and integration leads to

$$3\left(\frac{r_f}{R}\right)^2 - 2\left(\frac{r_f}{R}\right)^3 = -\frac{6\mathfrak{D}_e C_R}{R^2(1 - \epsilon_p)\rho_s q_e} t + \kappa \tag{11-126}$$

The constant of integration κ is evaluated from the initial position of the sorption front ($r_f = R$). After evaluation of κ and rearrangement of Equation 11-126, the time for penetration of the sorption front any fractional distance (r_f/R) into the spherical particle is

$$t = \frac{R^2(1 - \epsilon_p)\rho_s q_e}{6\mathfrak{D}_e C_R}\left[1 - 3\left(\frac{r_f}{R}\right)^2 + 2\left(\frac{r_f}{R}\right)^3\right] \tag{11-127}$$

The time to reach saturation of the sorptive capacity, t_s, corresponds to the sorption front reaching the center of the sphere ($r_f = 0$), for which the expression above simplifies to

$$t_s = \frac{R^2(1 - \epsilon_p)\rho_s q_e}{6\mathfrak{D}_e C_R} \tag{11-128}$$

The QSS model above, derived for diffusion with simultaneous irreversible adsorption, shows that the time to saturation of sorptive capacity is proportional to the square of the radius. The same dependence on radius was seen in other rate analyses of diffusion-controlled processes, in both steady- and unsteady-state situations.

There are many ways to use QSS analysis. For example, even though the model above holds strictly for irreversible adsorption and constant solution concentration external to the sphere, it may be possible to approximate highly nonlinear adsorption (i.e., very small Freundlich n value) as irreversible. Moreover, a falling concentration in solution may be included as would be true for sorption in a CMBR. The effect of film diffusion can also be easily included by addition of the continuity of flux relationship:

$$k_f(C_0 - C_\delta) = \mathfrak{D}_e \left.\frac{\partial C}{\partial r}\right|_{r=R} = \frac{r_f}{(R - r_f)R} \mathfrak{D}_e C_\delta \tag{11-129}$$

where C_0 is the bulk fluid concentration (a constant). This relationship allows the unknown concentration, C_δ, to be expressed in terms of C_0, which then replaces C_R in Equation 11-125. Following the same steps thereafter that leads to Equation 11-128 in the analysis of pore diffusion, the time to saturation is

$$t_s = \frac{R^2(1 - \epsilon_p)\rho_s q_e}{6\mathfrak{D}_e C_0}\left(1 + \frac{2\mathfrak{D}_e}{k_f R}\right) \tag{11-130}$$

When film diffusion is fast relative to pore diffusion (i.e., $k_f R >> \mathfrak{D}_e$), Equation 11-130 becomes identical to that derived for pore diffusion. The effect of significant film diffusion is to lengthen the time to reach exhaustion of sorptive capacity.

Other QSS models are discussed in the following section. They are especially useful approximations when mass transfer limits the uptake of components in packed sorbent beds.

11.7 MACROTRANSPORT AND MICROTRANSPORT IN SORPTION PROCESSES

We thus far have considered movement of solute through each of four diffusion domains. These domains provide a convenient way to describe microtransport by molecular diffusion and the effects of additional impedances to solute movement caused by reaction, tortuousity, and sorption. In natural and engineered systems, diffusion domains are actually encountered as microsystems *within* a macrosystem. That is, the typical unsteady-state problem confronting the environmental engineer is to predict the concentration of solute in a control volume in which macrotransport is coupled with microtransport. Macrotransport accounts for the large-scale movement of solute through the control volume by advection and dispersion. On the other hand, microtransport accounts for rate control of transformation or separation processes occurring at the microscale within that control volume.

In engineered systems, we wish to analyze the unsteady-state performance of both slurry and fixed-bed reactors. Slurry reactors are often described as CMFRs into which sorbent particles are continuously dosed. At the macrosystem scale, the slurry reactor operates at steady state. However, at the microsystem scale, the reactor is not at steady state because the driving force for intraparticle diffusion (microtransport) within any sorbent particles decreases as it is transported through the reactor. Fixed-bed reactors similarly do not operate at steady state. These reactors can be described as either PFRs or PFDRs, just as we did for steady-state analysis of heterogeneous reactions in Chapter 10. The difference here, however, is that the microtransport process is once again unsteady by virtue of the finite sorptive capacity of particles at the microsystem scale; thus at any bed location, the driving force for sorption decreases in time. In fact, we can broaden our description of fixed-bed reactors to encompass description of control volumes in the subsurface environment and thus gain understanding of the fate and transport of contaminants in this very important natural system.

11.7.1 Sorption in Slurry Reactors

A CMFR being fed with fresh sorbent particles is shown schematically in Figure 11-22. If we assume that the sorbent particles are suspended in the fluid and transported with it, their exit age distribution, $E(t)$, is the same as that of the fluid in a CMFR:

$$E(t) = \frac{e^{-t/\bar{t}}}{\bar{t}} \qquad (9\text{-}88)$$

The time for diffusion and sorption is thus distributed according to this exit age function.

Two different reactor models can be derived fairly easily: (1) pore diffusion accompanied by linear sorption; and (2) homogeneous surface diffusion accompanied by nonlinear sorption. The appropriate initial and boundary conditions needed to solve Fick's second law for radial pore diffusion and linear sorption (Equation 11-104) are:

Initial condition:

$$C(r, t = 0) = 0 \qquad (11\text{-}131)$$

First boundary condition:

$$C(r = R, t > 0) = C_{\mathrm{OUT}} \qquad (11\text{-}132)$$

Second boundary condition

$$\mathfrak{D}_e \left.\frac{\partial C}{\partial r}\right|_{r=0} = 0 \qquad (11\text{-}133)$$

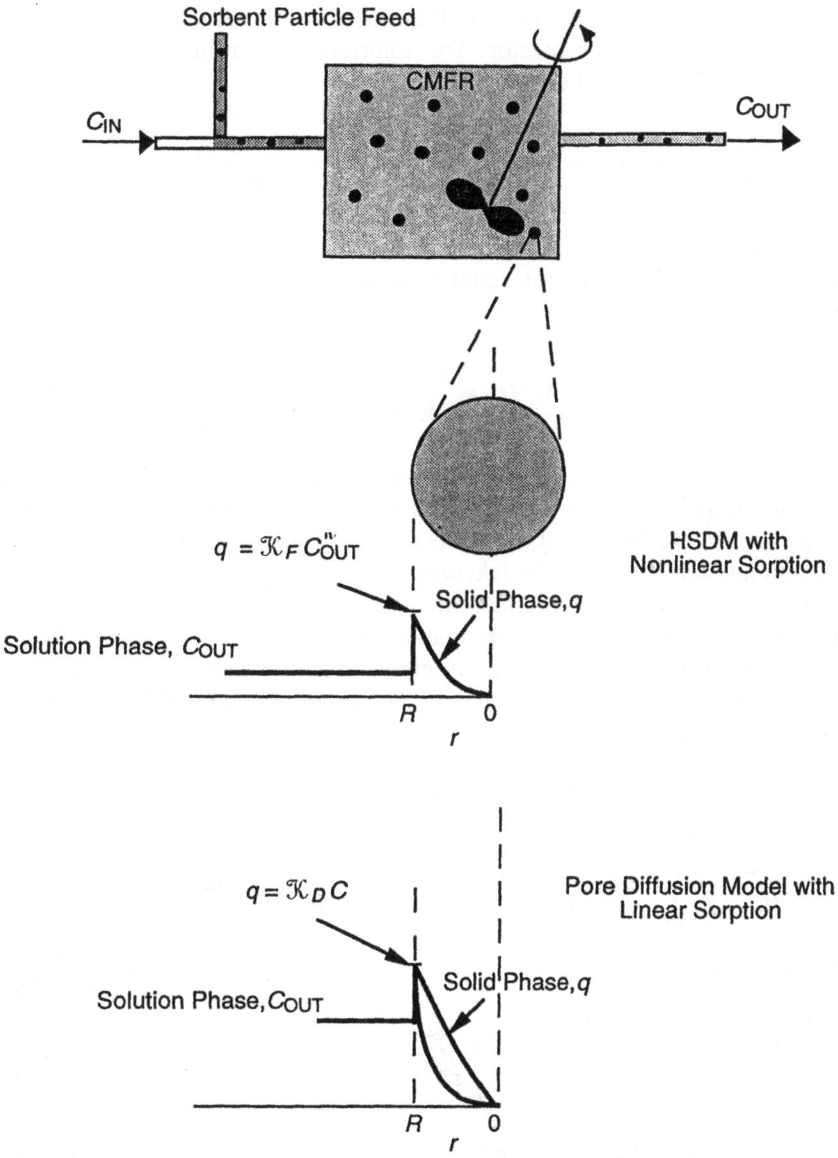

Figure 11-22 Application of unsteady-state sorption modeling to slurry reactors.

The first boundary condition reflects the definition of a CMFR, namely that the concentration everywhere in the reactor is the same as the exit concentration. If the dosage of sorbent and feed concentration are constant, the reactor is operating at steady state and C_{OUT} is constant even though the microtransport process within the particle is unsteady state. The analytical solution for sorption in slurry reactors when pore diffusion controls can be expressed as the fractional uptake, $M°(t)$, obtained in the time that an individual spherical sorbent particle spends in the reactor. The solution is given in Table 11-1 (entry F, case 1) and illustrated below:

$$M°(t) = 1 - \frac{6}{\pi^2}\sum_{n=1}^{\infty}\frac{1}{n^2}\exp\left(-\frac{\mathfrak{D}_a n^2\pi^2 t}{R^2}\right) \quad (11\text{-}134)$$

The average fractional uptake achieved by all particles in a CMFR follows from the material balance that equates solute loss from solution with solute gain by the sorbent particles:

$$\overline{M}°(t) = \frac{C_{IN} - C_{OUT}}{\mathcal{K}_D C_{OUT} D_o} \quad (11\text{-}135)$$

in which $\overline{M}°(t)$ is the average fractional uptake based on the age distribution of the sorbent particles and D_o is the sorbent dosage in the reactor (solid-phase concentration). The fractional uptake is a ratio of uptake by the sorbent to the maximum amount that could be taken up if equilibrium were reached with the solution-phase concentration in the exit stream.

The final reactor model requires that we account for the residence time of each particle in the reactor, which is given as t in Equation 11-134, and is determined from the exit age distribution. The average fractional uptake, $\overline{M}°(t)$, is thus defined as

$$\overline{M}°(t) = \int_0^{\infty} M°(t)\frac{e^{-t/\bar{t}}}{\bar{t}}\,dt \quad (11\text{-}136)$$

$$\overline{M}°(t) = \frac{1}{\bar{t}}\int_0^{\infty} e^{-t/\bar{t}}\,dt - \frac{6}{\bar{t}\pi^2}\int_0^{\infty}\sum_{n=1}^{\infty}\frac{1}{n^2}\exp\left[-\left(\frac{\mathfrak{D}_a n^2\pi^2}{R^2} + \frac{1}{\bar{t}}\right)t\right]dt \quad (11\text{-}137)$$

$$\overline{M}°(t) = 1 - \frac{6}{\pi^2}\sum_{n=1}^{\infty}\frac{1}{n^2\left(\dfrac{n^2\pi^2\bar{t}\,\mathfrak{D}_a}{R^2} + 1\right)} \quad (11\text{-}138)$$

Finally, a practical design equation is obtained by substitution of the definition of fractional uptake given by Equation 11-135 into Equation 11-138. After rearrangement to solve for the exit concentration, we have

$$C_{OUT} = \frac{C_{IN}}{1 + \mathcal{K}_D D_o\left[1 - \dfrac{6}{\pi^2}\displaystyle\sum_{n=1}^{\infty}\frac{1}{n^2\left(\dfrac{n^2\pi^2\bar{t}\mathfrak{D}_a}{R^2} + 1\right)}\right]} \quad (11\text{-}139)$$

Sorption in a slurry reactor can also be analyzed with the homogeneous surface diffusion model rather than the pore diffusion model. This modeling approach can provide an analytical solution for the more common situation where equilibrium sorption is a nonlinear process described by a Freundlich equation. The initial and boundary conditions that correspond to the appropriate Fick's second law expression (Equation 11-114) are:

Initial condition:

$$q(r, t = 0) = 0 \qquad (11\text{-}140)$$

First boundary condition:

$$q(r = R, t > 0) = q_e \qquad (11\text{-}141)$$

Second boundary condition:

$$\mathfrak{D}_s \left.\frac{\partial q}{\partial r}\right|_{r=0} = 0 \qquad (11\text{-}142)$$

These mathematical statements are analogous to those for pore diffusion. However, the surface diffusion coefficient does not account for simultaneous sorption; rather, sorption takes place only at the outermost edge of the particle followed by diffusion takes in the sorbed phase. The analytical solution is, therefore, identical in form to Equation 11-138:

$$\overline{M}°(t) = 1 - \frac{6}{\pi^2} \sum_{n=1}^{\infty} \frac{1}{n^2 \left(\dfrac{n^2 \pi^2 t \mathfrak{D}_s}{R^2} + 1 \right)} \qquad (11\text{-}143)$$

The average fractional uptake is now defined by

$$\overline{M}°(t) = \frac{C_{\text{IN}} - C_{\text{OUT}}}{\mathcal{K}_F C_{\text{OUT}}^n D_o} \qquad (11\text{-}144)$$

After combining Equations 11-143 and 11-144, we obtain the following reactor design equation:

$$C_{\text{OUT}} = C_{\text{IN}} - \mathcal{K}_F C_{\text{OUT}}^n D_o \left[1 - \frac{6}{\pi^2} \sum_{n=1}^{\infty} \frac{1}{n^2 \left(\dfrac{n^2 \pi^2 t \mathfrak{D}_s}{R^2} + 1 \right)} \right] \qquad (11\text{-}145)$$

In contrast to the design equation for pore diffusion with linear sorption (Equation 11-139), we observe that nonlinear sorption requires an implicit solution for C_{OUT}, but this is, nevertheless, a tractable solution procedure. The design of a CMFR for adsorption of a synthetic organic chemical using powdered activated carbon is illustrated in Example 11-8.

The same model can apply for measurement of sorption rate in a differential-batch-recycle reactor (see Figure 9-10). In this reactor, the concentration external to the particles can be held constant by addition of sorbate. The fractional uptake, $M°(t)$, is simply the cumulative amount of substrate added divided by the amount to be added to reach equilibrium. The time, t, in Equation 11-134 is the elapsed time in the batch experiment and $\mathfrak{D}_s$ replaces $\mathfrak{D}_a$ for the HSDM. The objective of the experiment is to test the HSDM through calculation of a surface diffusivity by measuring the mass of solute sorbed with time. If the sorbate concentration is high enough, it may even be possible to modify the reactor to include a sensitive quartz spring balance that will measure uptake directly and continuously by weighing the sorbent particles.

Example 11-8

- **Situation:** A groundwater has been found to be contaminated with 400 μg/L of trichlorophenol (TCP). The proposed treatment system is powdered activated carbon (PAC) addition to a CMFR, followed by separation of the carbon particles by ultrafiltration (UF). UF membranes have pore sizes that are considerably smaller than carbon particles, but they are much too large to reject TCP. Economic considerations suggest a maximum carbon dosage of 25 mg/L and an HRT of 1 h in the CMFR to achieve the treatment goal of ≤ 10 μg/L of TCP. An adsorption isotherm test has been conducted and the data are found to fit a Freundlich isotherm with the following parameter values:

$$\mathcal{K}_F = 20{,}000\ (\mu\text{g/g})\ (\mu\text{g/L})^{-0.3}$$

$$n = 0.3$$

- **Question(s):** Explore whether the adsorption process can meet the design objectives.
- **Logic and Answer(s):**

1. We assume that the homogeneous surface diffusion model (HSDM) is appropriate for describing the adsorption of TCP and thus Equation 11-145 is the appropriate design equation for PAC addition to an ideal CMFR (Najm, et al., 1990).

2. To use Equation 11-145, we first need to determine the ratio $R^2/\pi^2\bar{t}\mathfrak{D}_s$. Commercial PAC is comprised of a mix of particles that range in diameter from about 20 to 100 μm. For a first-cut feasibility study, we are interested in bracketing the removal effectiveness of PAC by calculating the removal if the entire PAC mixture were comprised of one or the other of these two diameters (the actual removal effectiveness will lie in between).

3. A search of the literature on adsorption rates reveals that the surface diffusivity of TCP is about 1×10^{-11} cm²/s.

4. A spreadsheet is set up to evaluate the series ($n = 12$ is more than sufficient) within Equation 11-145 and then to solve implicitly for C_{OUT} in μg/L. The results for a dosage of 25 mg/L, a range of HRTs (10, 20, 40, and 60 min) and two different particle radii ($R = 0.001$ and 0.005 cm) are shown below:

	C_{OUT} (μg/L)	
$\bar{t}$ (min)	$R = 0.001$ (cm)	$R = 0.005$ (cm)
10	51.0	237
20	23.4	211
40	9.8	172
60	6.0	145

5. The HSDM indicates that adsorption of TCP is much more rapid if the particle radius is 0.001 cm instead of 0.005 cm; this is expected from diffusion principles. Although the treatment objective can be achieved with the smaller size of PAC, it is not clear whether a distribution of sizes between the two used in these calculations will also be successful. To determine this, the appropriate weighted average of the C_{OUT} values for each particle size would need to be calculated.

6. It is also instructive to determine the equilibrium uptake of TCP for the PAC dosage that is applied. A material balance across the CMFR gives

$$C_{IN} - C_{OUT} = \mathcal{K}_F C_{OUT}^n D_o$$

An implicit solution is required to find C_{OUT} given the Freundlich parameters, the carbon dosage, and C_{IN}. We find that $C_{OUT} = 0.5$ μg/L and conclude that an HRT of 60 min is insufficient to approximate equilibrium with either PAC particle size, albeit the smaller size approaches it much more closely.

11.7.2 Sorption in Fixed-Bed Reactors

A control volume of a packed FBR is shown schematically in Figure 11-23. Macrotransport is responsible for movement of solute through the bed, while microtransport (both interphase and intraparticle) controls the sorption (phase transfer/exchange) rate. We can envision various engineered system applications that involve a variety of sorptive processes, activated carbon beds being but one. Moreover, we can easily view this control volume to be that taken from a natural system (e.g., a region of subsurface or lake sediment).

Desorption is as important as sorption in packed beds. Desorption is caused

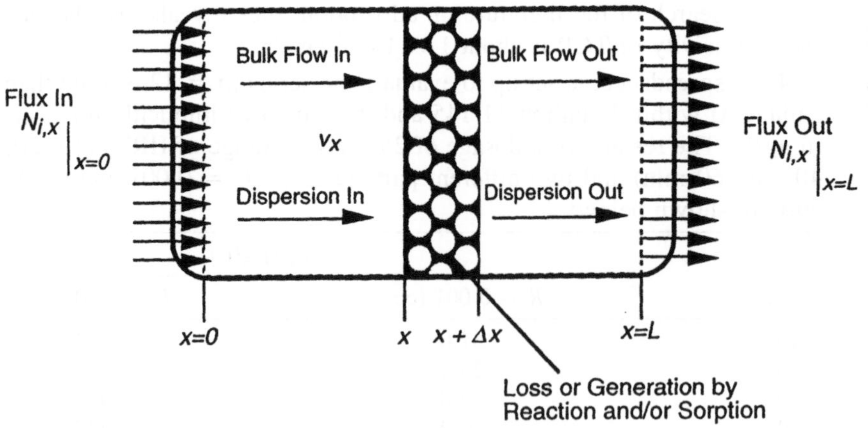

Figure 11-23 Control volume for one-dimensional transport through porous media.

in an engineered system by a decrease in the feed concentration to the packed bed after it has been in operation for some time and sorption sites have been occupied. In response to a lower feed concentration, sorbed solute desorbs in an attempt to achieve equilibrium with this lower fluid-phase concentration. Desorption is similarly important in the subsurface environment. Contaminants can desorb from soil particles if the concentration of that contaminant in the groundwater flow is reduced. A practical illustration of the desorption concept is provided by careful analysis of the pump and treat strategy for remediation of the subsurface. In this instance, the natural system is "engineered" by the pumping of clean water through the subsurface in an attempt to "flush" contaminants. Because much of the contaminant burden may be sorbed onto soil particles, this flushing action also causes desorption; the rate at which desorption occurs, however, may be very slow.

The fundamental steps of macrotransport, microtransport, and reaction-sorption are the same for sorption and desorption processes in a packed bed; it is only the initial and boundary conditions for a control volume that must change. As will be shown in ensuing sections, the level of complexity of the resulting mathematical model depends on what assumptions are made concerning these macrotransport, microtransport, and reaction steps.

The appropriate point form of the material balance for one-dimensional transport of solute and simultaneous phase transfer/exchange is:

$$-\epsilon_B \frac{\partial N}{\partial x} = \epsilon_B \frac{\partial C}{\partial t} + (1 - \epsilon_B)\rho_s \frac{\partial q}{\partial t} \qquad (11\text{-}146)$$

in which ϵ_B is the porosity of the bed and ρ_s is the density of the solid phase. The flux term accounts for macrotransport through the reactor, which includes advection and perhaps dispersion. Assuming that both mechanisms are important, Equation 11-146 becomes

$$-v_x \frac{\partial C}{\partial x} + \mathfrak{D}_d \frac{\partial^2 C}{\partial x^2} = \left(\frac{\partial C}{\partial t}\right)_x + \frac{(1 - \epsilon_B)\rho_s}{\epsilon_B} \left(\frac{\partial q}{\partial t}\right)_x \tag{11-147}$$

The accumulation terms carry the subscript x to emphasize their dependence on position along the length of the bed. This is often referred to as the *advection–dispersion–sorption* (ADS) *model*; it is evident that the ADS model has exactly the same form as an ideal PFDR model.

The one-dimensional PFDR model may be quite adequate for representation of many engineered systems. However, it is only an approximation for description of the subsurface, where transport can occur in all directions. For this situation, macroscopic models for transport are generally structured on principles of mass conservation applied on a volume-averaged or otherwise statistically averaged basis. When the control volume is generated on a differential scale and adsorption is treated like a reaction, the continuity relationship yields the following advection–dispersion–reaction equation alluded to in Chapter 3:

$$-v \text{ grad } C + \text{div } (\mathfrak{D}_d \cdot \text{grad } C) + S(C) + \mathscr{r} = \frac{\partial C}{\partial t} \tag{11-148}$$

In Equation 11-148 v is the pore-velocity vector, $\mathfrak{D}_d$ is the second-rank hydrodynamic dispersion tensor, $S(C)$ is a solution-phase solute source term, and $\mathscr{r}$ is the reaction rate for any reaction occurring at the microscale. In fact, the microscale reaction process includes adsorption, which may in turn be represented by reaction rate models or by models describing microscale mass transfer processes. Further development of the approach is beyond the scope of this book.

Recognizing the limitations of a one-dimensional transport model, it nevertheless serves to illustrate the impact of different assumptions regarding the kinetics of microtransport into the sorbent media. The sorption term, $\partial q/\partial t$, in Equation 11-147 can assume a variety of forms comprising different rate and equilibrium components. Several examples of one-dimensional macroscopic transport models incorporating various equilibrium and microscopic rate models are summarized in Table 11-2. These have applications to both natural and engineered systems.

11.7.2 Linear Local Equilibrium Model The most simplistic approach to representation of sorption phenomena in contaminant transport models or reactor engineering models is to assume equilibrium at any point x and adsorption as a linear process. This is referred to as the *local linear equilibrium model* (LLEM). Local equilibrium implies that the time scales associated with the microscopic process of diffusion and sorption are very much smaller than those associated with the macroscopic processes of fluid transport. Moreover,

Table 11-2 One-Dimensional Forms of the Advection–Dispersion–Reaction Equation Incorporating Different Sorption Equilibrium and Rate Expressions (after Weber et al. 1991).

A. *General Equation:*

$$\frac{\partial C}{\partial t} = \mathfrak{D}_d \frac{\partial^2 C}{\partial x^2} - v_x \frac{\partial C}{\partial x} - \rho_s \frac{1-\epsilon_B}{\epsilon_B}\frac{\partial q}{\partial t} \tag{i}$$

B. *Local Equilibrium* (see solutions in item H):

$$\frac{\partial C}{\partial t} = \mathfrak{D}_d \frac{\partial^2 C}{\partial x^2} - v_x \frac{\partial C}{\partial x} - \rho_s \frac{1-\epsilon_B}{\epsilon_B}\frac{\partial q}{\partial C}\frac{\partial C}{\partial t} \tag{ii}$$

With a linear isotherm:

$$\frac{\partial C}{\partial t} = \mathfrak{D}_d \frac{\partial^2 C}{\partial x^2} - v_x \frac{\partial C}{\partial x} - \rho_s \frac{1-\epsilon_B}{\epsilon_B}\mathcal{K}_D \frac{\partial C}{\partial t} \tag{iii}$$

With a Freundlich isotherm:

$$\frac{\partial C}{\partial t} = \mathfrak{D}_d \frac{\partial^2 C}{\partial x^2} - v_x \frac{\partial C}{\partial x} - \rho_s \frac{1-\epsilon_B}{\epsilon_B}\mathcal{K}_F n C^{n-1} \frac{\partial C}{\partial t} \tag{iv}$$

C. *First-order sorption $(k)_f$ and desorption $(k)_r$:*

$$\frac{\partial C}{\partial t} = \mathfrak{D}_d \frac{\partial^2 C}{\partial x^2} - v_x \frac{\partial C}{\partial x} - (k)_f C + \rho_s \frac{1-\epsilon_B}{\epsilon_B}(k)_r q \tag{v}$$

D. *Equilibrium plus first-order sorption $(k)_f$ and desorption $(k)_r$:*

$$\frac{\partial C}{\partial t} = \mathfrak{D}_d \frac{\partial^2 C}{\partial x^2} - v_x \frac{\partial C}{\partial x} - \rho_s \frac{1-\epsilon_B}{\epsilon_B}\frac{\partial q}{\partial C}\frac{\partial C}{\partial t} - (k)_f C + \rho_s \frac{1-\epsilon_B}{\epsilon_B}(k)_r q \tag{vi}$$

E. *Mobile* (M)*/immobile* (IM) *dual-domain model with quasi-steady-state linear driving force and sorption*:

$$\phi^\circ_{V,M}\frac{\partial C_M}{\partial t} = \phi^\circ_{V,M}\mathfrak{D}_d \frac{\partial^2 C_M}{\partial x^2} - \phi^\circ_{V,M} v_x \frac{\partial C_M}{\partial x} - \rho_s \frac{1-\epsilon_B}{\epsilon_B}\phi^\circ_{V,M}\frac{\partial q_M}{\partial t}$$
$$- \frac{\phi^\circ_{V,IM}}{\epsilon_B}\hat{k}_{v,l}(C_M - C_\delta) \tag{vii}$$

where

$$\hat{k}_{v,l}(C_M - C_\delta) = \rho_s(1-\epsilon)\frac{\partial q_\delta}{\partial t} + \epsilon\frac{\partial C_\delta}{\partial t}$$

$(\epsilon)_M = (\epsilon)_{IM} = \epsilon$
$(\rho_s)_M = (\rho_s)_{IM} = \rho_s$
$\phi^\circ_{V,M}$ = volume of mobile phase per total volume
$\phi^\circ_{V,IM}$ = volume of immobile phase per total volume

F. *Internal diffusion*:

$$\frac{\partial C}{\partial t} = \mathfrak{D}_d \frac{\partial^2 C}{\partial z^2} - v_x \frac{\partial C}{\partial z} - \rho_s \frac{1-\epsilon_B}{\epsilon_B}\frac{\partial q_{\text{avg}}}{\partial t} \tag{viii}$$

where

$$q_{\text{avg}} = \frac{3}{R^3}\int_0^R qr^2\,dr$$

$$\frac{\partial q}{\partial t} = \frac{1}{r^2}\frac{\partial}{\partial r}\left(r^2\mathfrak{D}_s\frac{\partial q}{\partial r}\right)$$

G. *Dual impedance*:

$$\frac{\partial C}{\partial t} = \mathfrak{D}_d\frac{\partial^2 C}{\partial z^2} - v_x\frac{\partial C}{\partial z} - \frac{3}{R}\frac{1-\epsilon_B}{\epsilon_B}k_f(C - C_\delta) \qquad \text{(ix)}$$

where

$$\frac{k_f}{\rho_s}(C - C_\delta) = \mathfrak{D}_s\frac{\partial q}{\partial r} \qquad \text{at } r = R$$

$$\frac{\partial q}{\partial t} = \frac{1}{r^2}\frac{\partial}{\partial r}\left(r^2\mathfrak{D}_e\frac{\partial q}{\partial r}\right)$$

H. *Addendum*: Solutions for local equilibrium conditions and dispersed flow [after Crittenden et al. (1986)]

For the following initial and boundary conditions:

$$C(x > 0, t = 0) = 0$$

$$v_x[C_{\text{IN}} - C(x = 0+, t)] = \mathfrak{D}_d\frac{\partial C(x = 0+, t)}{\partial x}$$

$$\frac{\partial C(x = L, t)}{\partial x} = 0$$

For $\mathfrak{N}_{\text{Pe}} > 40$

$$\frac{C(\zeta^\circ = 1, \tau^\circ)}{C_{\text{IN}}} = 0.5\,\text{erfc}\left[\frac{(\mathfrak{N}_{\text{Pe}})^{0.5}}{2}\frac{1-\tau^\circ}{(\tau^\circ)^{0.5}}\right] + \frac{1}{\pi^{0.5}}\left(\frac{\tau^\circ}{\mathfrak{N}_{\text{Pe}}}\right)^{0.5}\frac{\tau^{\circ 2} + 4\tau^\circ - 1}{(\tau^\circ + 1)^3}\exp\left[\frac{-\mathfrak{N}_{\text{Pe}}(1-\tau^\circ)^2}{4\tau^\circ}\right]$$

$$\zeta^\circ = \frac{x}{L}$$

$$\mathfrak{N}_{\text{Pe}} = \frac{v_x L}{\mathfrak{D}_d}$$

$$\tau^\circ = \frac{v_x t}{L[1 + (1-\epsilon_B)\rho_s\mathcal{K}_D/\epsilon_B]}$$

For the following alternative boundary conditions:

$$C(x > 0, t = 0) = 0$$

$$C(x = 0, t \geq 0) = C_{\text{IN}}$$

$$C(x = \infty, t > 0) = 0$$

$$\frac{C(\zeta^\circ, \tau^\circ)}{C_{\text{IN}}} = 0.5\left\{\text{erfc}\left[\frac{(\mathfrak{N}_{\text{Pe}})^{0.5}}{2}\frac{\zeta^\circ - \tau^\circ}{(\tau^\circ)^{0.5}}\right] + \exp(\mathfrak{N}_{\text{Pe}}\zeta^\circ)\,\text{erfc}\left[\frac{(\mathfrak{N}_{\text{Pe}})^{0.5}}{2}\frac{\zeta^\circ + \tau^\circ}{(\tau^\circ)^{0.5}}\right]\right\}$$

it ignores sorption rate control by the microtransport steps. Thus local equilibrium is achieved so that the time rate of change of the sorbed phase concentration, q, at any point x is reflected instantaneously in the time rate of change of the solution-phase concentration, C, at that point. Given the very slow movement of groundwater (on the order of meters per day), the local equilibrium assumption may have appeal at least as a first-cut approximation of the impact of sorption processes. However, it is a poor assumption in engineered process applications where advective transport is much faster than in the subsurface.

The linear sorption assumption is used in Equation 11-147 in just the same way as presented earlier in this chapter for analysis of diffusion–sorption problems:

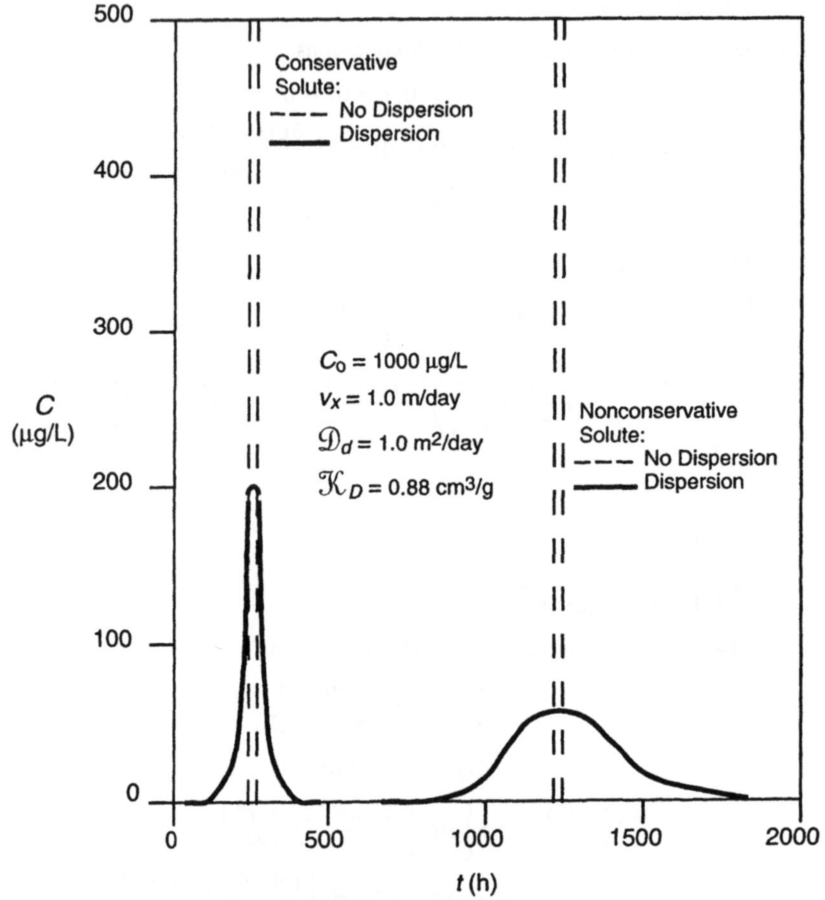

Figure 11-24 Ten-meter field scale simulations for transport of a moderately hydrophobic contaminant in a subsurface system. [After Weber et al. (1991).]

$$-v_x \frac{\partial C}{\partial x} + \mathfrak{D}_d \frac{\partial^2 C}{\partial x^2} = \left(1 + \frac{(1 - \epsilon_B)\rho_s}{\epsilon_B} \mathcal{K}_D\right) \left(\frac{\partial C}{\partial t}\right)_x = \mathfrak{F}_R \left(\frac{\partial C}{\partial t}\right)_x \quad (11\text{-}149)$$

in which $\mathfrak{F}_R$ is the *retardation factor*. Analytical solutions are available for a variety of boundary conditions (e.g., entry H in Table 11-2). The results of using the LLEM to simulate the concentration–time profiles resulting from subsurface transport of a moderately sorbing solute under representative conditions of fluid flow and hydrodynamic dispersion (both assumed constant) are presented in Figure 11-24. These profiles represent concentration patterns at a point 10 m downgradient of a pulse input of contaminant as a function of time after its addition. Concentration–time profiles are also presented neglecting sorption (i.e., a conservative solute) and/or dispersion. The effect of sorption is to retard the appearance of the solute at this downgradient location; increases in the sorption coefficient, $\mathcal{K}_D$, will cause further retardation. Without dispersion, the solute appears as an instantaneous spike at the downgradient point, directly reflecting the manner in which it was introduced. The effect of dispersion is to introduce spreading in the concentration–time profile for both the conservative and sorbing solutes. While this example serves to demonstrate the important idea that solute transport is retarded by sorption, it has become increasingly apparent that this model frequently fails to provide adequate representation of the effects of sorption on contaminant transport. Inclusion of more sophisticated, nonlinear equilibrium models, as well as rate models that incorporate nonlinear equilibrium, often provide better representation of sorption phenomena. Example 11-9 illustrates various levels of mathematical sophistication in predictive models that can result from fundamental analysis of the desorption process in a pump-and-treat operation.

Example 11-9

- **Situation:** Pump and treat technology has been employed widely for groundwater remediation. The concept is simple; pump clean water through an aquifer to flush contaminants out of this subsurface reservoir. Flushing, however, must also account for desorption of contaminants from soil particles as concentrations in the surrounding fluid phase are lowered. Conceptual models at the micro- and macroscale are needed to predict the effectiveness of pump and treat strategies (i.e., to predict the time required to decontaminate the water and soil to a given level).

- **Question(s):** As a consultant, you are asked to advise your client about the various levels of sophistication that are possible in a model to predict the effectiveness of pump and treat remediation technologies.

- **Logic and Answer(s):**

1. The simplest (and most unrealistic) model would ignore spatial variations in concentrations of contaminants within the groundwater,

whereupon the integral form of the material balance relationship is used instead of the point form. Further simplification results from assuming that the subsurface behaves as a CMFR of volume V_R:

$$QC_{\mathrm{IN}} - QC_{\mathrm{OUT}} = \epsilon_B V_R \frac{dC_{\mathrm{OUT}}}{dt} + (1 - \epsilon_B)\rho_p V_R \frac{dq}{dt}$$

2. If mass transfer is not rate limiting and sorption is linear (the linear, local equilibrium model), we have

$$QC_{\mathrm{IN}} - QC_{\mathrm{OUT}} = \epsilon_B V_R \frac{dC_{\mathrm{OUT}}}{dt} + (1 - \epsilon_B)\rho_p V_R \frac{dq}{dC_{\mathrm{OUT}}} \frac{dC_{\mathrm{OUT}}}{dt}$$

$$QC_{\mathrm{IN}} - QC_{\mathrm{OUT}} = [\epsilon_B + (1 - \epsilon_B)\rho_s \mathcal{K}_D] V_R \frac{dC_{\mathrm{OUT}}}{dt}$$

3. The following simple model is obtained if the water used to flush the subsurface does not intially contain any of the contaminant of concern:

$$-\frac{dC_{\mathrm{OUT}}}{C_{\mathrm{OUT}}} = \frac{Q}{\epsilon_B V_R \mathfrak{F}_R} dt$$

where $\mathfrak{F}_R$ is the retardation factor (see Equation 11-149):

$$\mathfrak{F}_R = 1 + \frac{(1 - \epsilon_B)\rho_s}{\epsilon_B} \mathcal{K}_D$$

4. Upon integration, this flushing model can be written as

$$-\ln \frac{C_{\mathrm{OUT},t}}{C_{\mathrm{OUT},0}} = \frac{V_V}{\mathfrak{F}_R}$$

or

$$\frac{C_{\mathrm{OUT},t}}{C_{\mathrm{OUT},0}} = e^{-V_V/\mathfrak{F}_R}$$

where V_V is the total number of pore or void volumes flushed in time, t:

$$V_V = \frac{Qt}{\epsilon_B V_R}$$

According to this simple model, the flushing process produces an exponential decay in the concentration of contaminant. The rate of remediation is directly proportional to the pumping rate, Q, and inversely related to the retardation factor, $\mathfrak{F}_R$ (i.e., contaminants that are more strongly sorbed will take longer to remove).

5. The model is simplistic in many ways. At the very least, the one-directional spatial gradient of concentration (in the principal direction of

flow caused by pumping) should be considered. If both advection and dispersion are included along with local, linear equilibrium, Equation 11-149 must be solved for some reasonable set of initial and boundary conditions. For example, analytical solutions are available for a semi-infinite subsurface domain:

$$C(x > 0, t = 0) = C_0$$

$$C(x < 0, t = 0) = 0$$

$$C(x = 0, t > 0) = 0$$

b. The next level of sophistication would lift the restriction of linear sorption. For example, ignoring dispersion in Equation 11-149 and applying the same initial and boundary conditions as given above would give (Crittenden et al., 1986)

$$C(x, t) = C_0$$

$$\text{for } t \leq \frac{x}{v_x}\left[1 + \frac{1 - \epsilon_B}{\epsilon_B}(\epsilon_B + \rho_s \mathcal{K}_F n C_0^{n-1}\right]$$

$$C(x, t) = \left\{\left[\left(\frac{v_x t}{x} - 1\right)\frac{\epsilon_B}{1 - \epsilon_B} - \epsilon_B\right][\rho_s \mathcal{K}_F n]^{-1}\right\}^{1/(n-1)}$$

$$\text{for } t \geq \frac{x}{v_x}\left[1 + \frac{1 - \epsilon_B}{\epsilon_B}(\epsilon_B + \rho_s \mathcal{K}_F n C_0^{n-1})\right]$$

Further mathematical complexity would be added if boundary conditions to describe a finite rather than infinite domain were used and if dispersion were included.

7. Mathematical modeling complexity will generally increase as the following additional effects are included: (a) mass transfer limitations to desorption (internal diffusion and mass transfer at the surface of the particles; (b) two- and three-dimensional flow; and (c) random spatial distribution of porosity and soil sorptive characteristics. Numerical methods become necessary. Discretization of the space and time domain to accommodate multidimensions and random variations in soil characteristics can challenge the most powerful computational systems available today.

8. From a pragmatic viewpoint, a judgment must be made as to whether predictions for decision making are improved enough to warrant the cost of additional model sophistication. For example, the simple flushing model in step 1) may give enough guidance for initial planning. From a conceptual viewpoint, mathematical sophistication should not necessarily be equated with a "truer" representation of the phenomenon, and thus of reality; this will depend on the validity of all process descriptions at the microscale.

Equilibrium sorption concepts can also be used for analysis of FBR performance in treatment plants. The objective is to obtain a first-cut estimate of the time before the sorptive capacity of a sorbent bed is exhausted (i.e., the *bed life*). In contrast to the subsurface environment, where macrotransport processes in nature are inherently slow and the notion of local equilibrium has some validity, the intent of engineered reactors is to compress the time scale of treatment by using much greater flow rates. Moreover, relatively large-diameter granular particles are used to minimize head-loss development in fixed-bed sorption systems, and this leads to slower diffusion. Thus short detention times and large particles shift rate control to the microtransport step and make it unrealistic to assume that local equilibrium exists. The result is a gradual instead of step increase in exit concentration, eventually reaching the feed concentration, as shown in Figure 11-25. The prediction of this *breakthrough curve* requires dynamic modeling of the microtransport steps.

Essential elements of dynamic modeling are presented in the following sections. The resulting models typically require numerical methods for their solution. The Background Reading list for Chapter 8 provides useful sources for details of various numerical methods commonly used.

11.7.2.2 External Film Diffusion Model As indicated in Figure 11-19, the first step in microtransport involves solute mass transfer across an immobile boundary layer of fluid (a type I diffusion domain) immediately adjacent to the external surfaces of a sorbent. If this step is slower than internal microtransport, we ignore internal diffusion in modeling; this is referred to as the *external film model*. The flux through the film is expressed for a type I diffusion domain

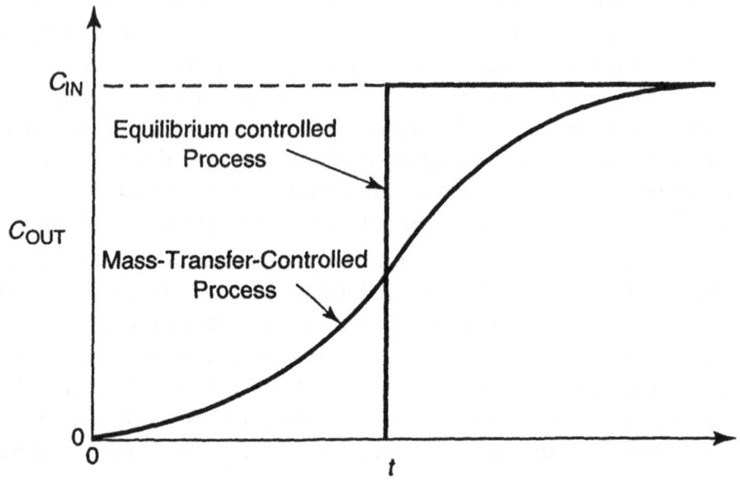

Figure 11-25 Comparison of mass transfer and equilibrium control of sorption in a fixed-bed reactor.

of thickness δ and having a cross-sectional area controlled by the dimensions of the particle surface. For a spherical particle, we have

$$\frac{4\pi R^3}{3}\rho_p\frac{dq}{dt} = 4\pi R^2 k_f(C_0 - C_\delta) \qquad (11\text{-}150)$$

which simplifies to

$$\frac{dq}{dt} = \frac{3}{R\rho_p}k_f(C_0 - C_\delta) \qquad (11\text{-}151)$$

The concentration, C_δ, increases with time due to exhaustion of sorptive capacity, and thus the driving force for mass transfer varies as sorption proceeds. However, if the gradient in concentration across the film is steep and the sorbent strongly sorbing, the use of a quasi-steady-state model for describing flux across the film region is not unreasonable. To obtain a solution with the film model for concentration as a function of time and distance in a packed bed, Equation 11-151 replaces $\partial q/\partial t$ in Equation 11-147. Values of k_f can in some instances be measured experimentally, or estimated by means of semiempirical correlations with physical system properties, as discussed in Chapter 4.

11.7.2.3 Intraparticle Diffusion Models If *surface diffusion* is rate controlling and external diffusion can be ignored, a convenient quasi-steady-state approach employs the mean solid-phase concentration, $\bar{q}$, in calculation of surface flux:

$$N_s = -\rho_p\mathfrak{D}_s\frac{\partial q}{\partial r} \approx k_{f,s}\rho_p(q|_{r=R} - \bar{q}) \qquad (11\text{-}152)$$

where $k_{f,s}$ is treated as an internal mass transfer coefficient. Written in terms of the homogeneous surface diffusion model (HSDM), the rate of adsorption is then

$$\frac{d\bar{q}}{dt} = \frac{3}{R}k_{f,s}\,(q_e|_{r=R} - \bar{q}) \qquad (11\text{-}153)$$

The value of q at the external surface of the particle is determined by the equilibrium relationship with the fluid phase (e.g., $q_e = \mathcal{K}_F C^n$. If both external and internal diffusion are both of the same order of magnitude, a combined *film-surface diffusion model* is appropriate where

$$N_s = k_f(C_0 - C_\delta) = k_{f,s}\rho_p(q|_{r=R} - \bar{q}) \qquad (11\text{-}154)$$

$$q|_{r=R} = \mathcal{K}_F C_\delta^n \qquad (11\text{-}155)$$

The surface and film-surface diffusion models provide two alternative means for characterization of the microtransport at the particle scale within fixed-bed adsorbers. These models are used to replace $\partial q/\partial t$ in Equation 11-147. A numerical methods procedure is required to solve the resulting equation and thus concentration as a function of time and distance along the packed bed.

Entry G in Table 11-2 describes both external and internal impedances in a more rigorous fashion than is provided in Equation 11-154, where averaging of the solid-phase concentration was used. The material balance for the macroscale of the fixed bed and the microscale of the sorbent particle shown in Table 11-2 lead to formulation of a widely used dimensionless parameter known as the mass transfer *Biot number*, $\mathfrak{N}_{Bi,M}$:

$$\mathfrak{N}_{Bi,M} = \frac{k_f R(1 - \epsilon_B)}{\mathfrak{D}_s \phi_D^\circ \epsilon_B} \tag{11-156}$$

where ϕ_D° is the solute distribution parameter $= \rho_B q_{IN}/\epsilon_B C_{IN}$; ρ_B, the bulk density of the bed $= \rho_p(1 - \epsilon_B)$; ρ_p, the apparent density of the sorbent particles; and the sorbent particle are assumed to be spherical in shape. Inspection of the Biot number reveals that it represents the ratio of external to internal impedance. As a general rule, $\mathfrak{N}_{Bi,M} > 30$ implies that surface diffusion controls, whereas $\mathfrak{N}_{Bi,M} < 0.5$ implies that film transfer controls.

11.8 SUMMARY

More often than not, unsteady state is the situation encountered in natural and engineered systems. The material balance principle is essential, just as it is in the analysis of steady-state problems. Obviously, the mathematical complexity of unsteady-state analysis can be much greater than discussed here. Nevertheless, the starting point for solution of the problem is the same: preparation of a material balance; selection of the appropriate equilibrium and rate models for mathematical description of the variation in feed concentration and/or flow; and description of reactor configuration(s).

Appropriate conceptualization of the problem is the foremost concern. Once done, mathematical solution of model equations is relatively straightforward. A number of useful analytical solutions have been presented for unsteady-state diffusion. These find widespread use in natural and engineered systems, as illustrated in a number of places in this chapter. In many situations we will recognize immediately that rigorous analytical solutions are not possible and that numerical methods (see Background Reading list in Chapter 8) must be used. However, in certain situations, reasonable first-cut analytical solutions may be obtained by approximating the initial and boundary conditions. Such approaches are encouraged to gain an understanding of the importance of various physical constants before spending time developing more rigorous numerical solutions.

11.9 REFERENCES AND SUGGESTED READINGS

11.9.1 Citations and Sources

Carslaw, H. S., and J. C. Jaeger, 1959, *Conduction of Heat in Solids*, Clarendon Press, Oxford. (Table 11-1)

Crank, J., 1975, *Mathematics of Diffusion*, Pergamon Press Ltd., Oxford. (Example 11-7, Equations 11-73 and 11-138, and Table 11-1)

Crittenden, J. C., N. J. Hutzler, D. G. Geyer, J. L. Oravitz, and G. Friedman, 1986, "Transport of Organic Compounds with Saturated Groundwater Flow: Model Development and Parameter Sensitivity," *Water Resources Research*, *22*(3), 271–284. (Table 11-2 and Example 11-9)

Danckwerts, P. V., 1960, *Gas–Liquid Reactions*, McGraw-Hill Book Company, New York. (Equations 11-88 through 11-93 and Figure 11-16)

Khudenko, B. M., 1985, "Development and Analysis of Equalization Basins," *Journal of the Environmental Division, 111*(6), 907–922. (Problems 11-5 and 11-6)

Mott, H., and W. J. Weber, Jr., 1991, "Diffusion of Organic Contaminants Through Soil–Bentonite Cut-Off Barriers," *Journal of the Water Pollution Control Federation*, *63*(2), 166–176. (Problem 11-8)

Najm, I. N., V. L. Snoeyink, M. T. Suidan, C. H. Lee, and Y. Richard, 1990, "Effect of Particle Size and Background Natural Organics on the Adsorption Efficiency of PAC," *Journal of the American Water Works Association*, *72*(1), 65–72. (Equations 11-138 to 11-145, Example 11-8)

Novotony, V., and R. M. Stein, 1976, "Equalization of Time Variable Waste Loads," *Journal of the Environmental Engineering Division*, *102*(EE3), 613–625. (analytical solutions leading to Equations 11-21 and 11-22)

Park, J. K., L. Bontoux, T. M. Holsen, D. Jenkins, and R. E. Selleck, 1991, "Permeation of Polybutylene Pipe and Gasket Materials by Organic Chemicals," *Journal of the American Water Works Association*, *83*(10), 71–78. (Example 11-6)

Selleck, R. E., and B. J. Marinas, 1991, "Analyzing the Permeation of Organic Chemicals Through Plastic Pipes," *Journal of the American Water Works Association*, *83*(7), 92–97. (Figure 11-15 and Equation 11-85).

Van Rees, K. C. J., E. A. Sudicky, P. Suresh, C. Rao, and K. R. Reddy, 1991, "Evaluation of Laboratory Techniques for Measuring Diffusion Coefficients in Sediments," *Environmental Science and Technology*, *25*(9), 1605–1611. (Table 11-1)

Weber, W. J., Jr, P. M. McGinley, and L. E. Katz, 1991, "Sorption Phenomena in Subsurface Systems: Concepts, Models, and Effects on Contaminant Transport," *Water Research*, *25*, 5, 499–528. (Figure 11-24 and Table 11-2)

11.9.2 Background Reading

Bear, J., 1979, *Hydraulics of Groundwater Flow*, McGraw-Hill Book Company, New York. Analytical solutions for advective–dispersive transport through porous media are presented in Chapter 7. Solutions are provided for continuous-flow and slug injection of solute into one-dimensional, infinite, and semi-infinite-length media. Solutions are also provided for simultaneous sorption, for dispersion in two dimensions, and for radial dispersion.

Carslaw, H. S., and J. C. Jaeger, 1959, *Conduction of Heat in Solids*, Clarendon Press, Oxford. The diffusion of heat and mass are analogous processes and the mathematics to describe them are identical. This classic text covers a wide range of unsteady-state situations (different initial and boundary conditions) and provides a very useful compendium of analytical solutions which can be applied directly to diffusion of mass. Transport through different geometries and diffusion in series with mass transfer at a boundary are included.

Crank, J., 1975, *The Mathematics of Diffusion*, Clarendon Press, Oxford. This often-cited classic text gives analytical solutions that are suitable for many environmental systems. Solutions for diffusion in infinite and semi-infinite media and in bounded systems (e.g., plane sheets, cylinders, and spheres suspended in a finite volume) are provided. Various important boundary conditions (e.g., to describe mass transfer at an interface) are included. More complex diffusion problems that involve simultaneous reaction or adsorption and diffusion coefficients that vary spatially are also covered. An introduction is given to numerical methods approaches for those diffusion situations that do not result in an analytical solution.

Danckwerts, P. V., 1960, *Gas–Liquid Reactions*, McGraw-Hill Book Company, New York. A classic text that develops the mathematical models to describe diffusion and reaction (both steady and unsteady state) for a wide variety of boundary conditions.

Weber, W. J., Jr., P. M. McGinley, and L. E. Katz, 1991, "Sorption Phenomena in Subsurface Systems: Concepts, Models and Effects on Contaminant Transport," *Water Research*, *25*(5), 499–528. The concept of type I, II, III, and IV diffusion domains is presented. Various examples are given for steady- and unsteady-state analysis, with emphasis on contaminant transport in groundwater.

11.10 PROBLEMS

11-1 A 4000-L CMBR is to be used for oxidation of a small, sporadically generated industrial waste stream. If the reaction rate coefficient is $k = 3.5\ \text{day}^{-1}$, determine whether a 90% reduction in waste concentration can be achieved in the 4-hour time estimated before receiving the next batch.

11-2 Surface aerators have been installed in a lagoon used to treat wastewater from a pulp mill. The variation in wastewater flow rate is described by the relationship $Q(\text{m}^3/\text{h}) = 250\ [\sin(\pi/12)t + 1]$. The surface area of the lagoon remains constant at 2000 m^2, and its mean hydraulic residence time is 12 h. Determine the range of variations in water depth that the surface aerators must be designed to accommodate if the wastewater is withdrawn at a constant rate that will result in no net change in the lagoon volume from cycle to cycle.

11-3 A large metropolitan area draws its water from an upland reservoir. Sodium fluoride is added to the inlet of the pipeline that carries the water to a smaller storage reservoir closer the city. Before reaching the storage reservoir, which has a volume of 7.56×10^5 m^3, the water travels a distance of 10 km in the pipeline, which has a carrying capacity of 2.2 m^3/s (50 mgd). Officials are concerned about the worst-case condition that would exist if the fluoride pump were to malfunction and produce a steep increase in concentration from the nominal value of 1 mg/L up to 5 mg/L. How would the fluoride concentration vary with time at the exit point of the storage reservoir following a pump malfunction that lasts for two 8-h work shifts? Assume that the pipeline behaves as a PFR and the reservoir as a CMFR.

11-4 An on-line equalization tank has been designed to smooth concentration peaks caused by periodic discharges of metal-laden particulates. Some settling occurs in the equalization basin such that the removal of these particulates is approximated by a zero-order reaction, $k = 5$ mg/L-h.

(a) Find the concentration leaving the equalization basin as a function of time after a step increase from 100 to 250 mg/L if the HRT is 6 h and the shock load period is 4 h

(b) What would be the response if particulate agglomeration is promoted in the equalization tank by addition of a chemicals such that the reaction rate is better described as first order with $k = 0.1\ h^{-1}$?

11-5 An industrial process produces a constant flow of wastewater containing complex aromatic structures that interfere with the performance of the treatment system for the plant. Suppose that the concentration of these aromatics varies sinusoidally and that the frequency (1 h^{-1}) is constant in time. *Two-channel, spectral-selective equalization* has been proposed (Khudenko, 1985) whereby 50% of the flow passes through a plug flow reactor while the other 50% is bypassed. What HRT is needed to achieve complete smoothing of the sinusoidal variation in concentration?

11-6 Suppose that the pattern of variation in concentration of aromatics in the situation described in Problem 11-5 is more complex. A Fourier series analysis reveals three important frequencies of sinusoidal variations (1 h^{-1}, 0.5 h^{-1}, and 0.25 h^{-1}). *Multichannel spectral selective equalization* (Khudenko, 1985) involves dividing the flow equally into parallel plug flow reactors of different HRT to achieve equalization. Explain the concept involved and design considerations required for smoothing the input concentration pattern described.

11-7 An abandoned industrial site contains a five-by-five stack of 55-gallon drums of cresol that are located several meters beneath the surface of the ground and 100 m from a river bank. These drums constitute a contained planar source of cresol oriented parallel to the river bank and having an effective concentration of 10 g/m^2. Suppose that the groundwater table rises seasonally and covers the drums. The drums begin to corrode and cresol diffuses out. Assume that transport by advection is very slow and that dispersion is approximated by molecular diffusion ($\mathfrak{D}_l = 1 \times 10^{-5}$ cm^2/s) with a tortuousity factor of 2 and a porosity of 0.4.

(a) Find the concentration of cresol as a function of distance after 50 and 100 days.

(b) Determine the flux of cresol (g/m^2-s) at a distance of 1 m after 50 and 100 days.

(c) Discuss the validity of all assumptions.

11-8 Retardation of the horizontal transport of carbon tetrachloride (CTET) in contaminated subsurface zones by soil–bentonite barriers amended with fly ash has been studied (Mott and Weber, 1991). Lengthening of the diffusive pathway by irregular void spaces and the impedances to transport by small constrictions within the pore matrices were accounted for with the relationship $\mathfrak{D}_{p,e} = \epsilon_d^{1.33}\mathfrak{D}_l$, where $\mathfrak{D}_{p,e}$ is the effective pore diffusivity and $\mathfrak{D}_l$ is the free-liquid diffusivity (1.07×10^{-5} cm^2/s). The Freundlich equation was used to fit the adsorption isotherm data, $q_e = 1.53\ C_e^{0.581}$, where q_e is in mg/g and C is in mg/L.

(a) Estimate the cumulative penetration of CTET through the barrier per unit area normal to the direction of diffusive transport, $M(t)/A_N$, for the following assumptions: thickness of the barrier = 1 m; density of barrier material = 2.24 g/cm^3; porosity = 0.48; $C = 1$ mg/L on the entrance side of the barrier; $C \approx 0$ on the exit side of the barrier; and a linear approximation to the Freundlich isotherm for $0 < C < 1$ mg/L.

(b) Compare the results in part (a) with those for a barrier that does not contain fly ash and has no sorptive capacity for CTET.

[*Hint:* The mathematical solution for these same boundary condition was presented in this chapter (see Section 11.4.2.2) and can be adapted for this problem (type IV domain) by assuming that $\mathcal{K}_{P,a} = 1$ and incorporating retardation by linear sorption into the apparent diffusivity, as in other type IV problems. The reader should verify that the solution given by Mott and Weber is equivalent to that given in this chapter.

11-9 The interstitial pore water of a lake sediment contains 100 μg/L of methyl mercury, to a depth of 10 cm. Under reducing conditions, this methyl mercury is released into the lake water.

(a) Determine the total mass of methyl mercury released in 60 days using the following assumptions: relationship between effective pore diffusivity and free-liquid diffusivity as given in Problem 11-8; porosity = 0.4; $\mathfrak{D}_l = 1 \times 10^{-5}$ cm^2/s; negligible sorption onto sediment particles; concentration in the lake water remains at zero; surface area of the sediment = 1×10^7 m^2; and the diffusion domain is semi-infinite (surface of sediment is $x = 0$).

(b) Examine the validity of the semi-infinite domain assumption.

[*Hint:* Define a dimensionless concentration that makes the boundary conditions conform to those given in Equations 11-58 to 11-60. Note also that mass is released only from that fraction of the surface area of the sediment composed by pores.]

11-10 Reconsider Problem 11-9 as diffusion of methyl mercury from a finite type III diffusion domain (length = 10 cm) into a finite, well-mixed volume (lake depth = 3 m) (i.e., a pond with no inflow or outflow). Find the concentration in the pond as a function of time. [*Hint:* Table 11-1 provides an appropriate analytical solution.]

11-11 Chloroform is produced in water treatment as a result of chlorination. Two clear wells in parallel provide storage prior to distribution. One of these clear wells is filled and then taken off-line for 4 days (no inflow or outflow). Estimate the chloroform concentration remaining given that mass transfer occurs at the air–water interface ($\hat{k}_{f,l} = 0.005$ cm/s); the starting chloroform concentration is 50 μg/L; the depth and surface area of the clear well are 5 m and 100 m^2, respectively; and the diffusivity of chloroform is 1×10^{-5} cm^2/s. [*Hint:* Assume the clear well is quiescent and that the concentration of chloroform in the air above does not increase significantly. Table 11-1 provides an appropriate analytical solution.]

11-12 The sorption of trichlorobenzene (TCB) on a particular soil is approximated by a linear isotherm: $q_e = 0.078C_e$, where q_e is in mg/g and C_e is in mg/L. The diffusivity of TCB into the soil particles (diameter = 0.1 cm, density = 2.67 g/cm^3, and internal porosity = 0.02) is studied in a CMBR. The initial solution-phase concentration of TCB is 1000 μg/L, and enough soil particles are added to produce an equilibrium solution-phase TCB concentration of 90 μg/L. The fractional concentration remaining with time is given below.

t (h)	C/C_0
0	1.0
50	0.37
100	0.30
150	0.26
250	0.20
500	0.15

(a) Determine the apparent diffusivity ($\mathfrak{D}_a$) and the effective diffusivity ($\mathfrak{D}_e$) assuming a pore diffusion model and a free-liquid diffusivity of 6.4×10^{-6} cm^2/s.

(b) Assuming that the more appropriate model is surface diffusion in parallel with pore diffusion, determine the corresponding surface diffusivity (assume a tortuosity factor of 2 to estimate the effective pore diffusion coefficient).

[*Hint:* The dosage of particles and thus the volume of solution surrounding each particle, $V_{R,P}$, is known implicitly from the information provided.]

Answers: (a) Apparent diffusivity $= 2.9 \times 10^{-11}$ *cm^2/s, effective diffusivity* $= 3 \times 10^{-7}$ *cm^2/s; (b) surface diffusivity* $= 1.13 \times 10^{-9}$ *cm^2/s.*

11-13 The equilibrium adsorption of methylisoborneol (MIB), a taste- and odor-causing compound found in drinking water, is approximated by a linear isotherm in the very low concentration range; $q_e = 100C_e$, where q_e is in μg/g and C_e is in μg/L. The rate of adsorption onto powdered activated carbon (PAC) particles (diameter = 0.001 cm; density = 1.5 g/cm^3; internal porosity = 0.5) is assumed to be controlled by internal diffusion. Assume that the free-liquid diffusivity of MIB is 0.8×10^{-5} cm^2/s and that the tortuosity factor is 2.

(a) Find the concentration remaining after 40 min of contact in a series of baffled mixing tanks that approximate a PFR if the initial concentration of MIB is 10 μg/L and the PAC dosage is 25 mg/L. (*Hint:* Take the Lagrangian viewpoint in which the time for adsorption is equal to the travel time of a PAC particle through the tank.)

(b) Compare this value to the equilibrium value.

Answer: 3.7 μg/L; equilibrium concentration = 2.9 μg/L.

11-14 Assume that the reactor in Problem 11-13 is described more appropriately as a CMFR. Furthermore, suppose that the homogeneous surface diffusion model is more appropriate than the pore diffusion model and that the isotherm is better described by a Freundlich isotherm, $q_e =$

$380C_e^{0.45}$, where q_e is in μg/g and C_e is in μg/L. The surface diffusion coefficient is 1.3×10^{-12} cm^2/s.

(a) Find the concentration of MIB in the exit stream of the reactor and compare it to the equilibrium value.

(b) Discuss the reasons for the difference in answers in Problems 11-13 and 11-14.

Answer: *6.4 μg/L; equilibrium concentration = 0.9 μg/L.*

12

RETROSPECTIONS AND PERSPECTIVES

12.1 PROCESS DYNAMICS

The book has focused on development of a comprehensive conceptual framework within which to understand environmental change. More specifically, it has examined in depth the two principal classes of processes that govern the dynamics of change in environmental systems, reaction and transport processes. This is the central theme for each chapter of the book and for the book in general. The roots of process dynamics are embedded in well-established physical and chemical laws. Mathematical formulations or *models* can be developed by coupling explicit statements of these laws with appropriate mass, energy, and momentum balances. These models may then be calibrated and used to elucidate, interpret, and design natural and engineered systems in a myriad of different environmental settings.

The overarching goal of our treatment of process dynamics has been to demonstrate that its principles, once mastered, can be generalized to handle specificity in both the settings and constituents of different types of environmental systems and the changes that occur within them. This goal is essential to achieve because contemporary environmental scientists and engineers must be able to cope with and solve a wide range of problems in a broad array of systems. In this final chapter we review the salient principles of process dynamics that have been covered. We then present a synopsis of processes and models that not only builds upon these principles but extends them to an even broader array of environmental systems than presented thus far. Several detailed examples are given to demonstrate the facility with which familiar process models can be modified as necessary to fit the specifics of different settings and situations.

12.2 PROCESS THERMODYNAMICS

Gathering information about changes in one or more of the multitude of chemical, physical, and biological constituents comprising environmental systems begins with an analysis of the thermodynamic feasibilities of reactions in which those constituent(s) are likely to participate. The results of such an analysis generally reveal *energetically feasible* distributions of reactants and products. (see Chapters 5 and 6). Some caution must be practiced in this exercise. Unlike chemical synthesis in industrial facilities, the settings surrounding changes in environmental systems are often complex and ill defined. A given constituent may react by several different and parallel pathways, some involving homogeneous reactions and others involving heterogeneous reactions and associated microscale mass transfer processes. It is difficult in many of these settings to apply thermodynamics principles in a totally comprehensive and rigorous way. Nonetheless, the goal of establishing at least the *energy domains* in which a given process may occur, or can be induced to occur, is meaningful and worthwhile. A first-cut thermodynamic analysis yields initial estimates of equilibrium distributions of constituents among various macroscopic environmental compartments. It provides insight, guidance, and a logical basis upon which to structure more detailed analyses. The approaches presented in Chapter 6 are particularly useful in this context.

12.3 PROCESS RATES

A complete understanding of environmental change under thermodynamically favorable conditions also requires knowledge of the rate(s) at which that change may occur. Our ultimate *scientific* goal is to quantify rates of change based on a molecular-scale understanding of *reaction kinetics*; that is, on the basis of each elementary step associated with a reaction. As we have seen in Chapter 7, however, reaction pathways are in many situations too complex or ill defined to permit us to attain this goal. Nonetheless, useful *empirical rate relationships* that allow interpretation of data within structured conceptual frameworks can be established from experimental data. The robustness of data bases on which empirical rate relationships are developed (e.g., ranges of concentrations, temperatures, and other important variables included in experimental analyses) determines the breadth of their applicability to different environmental settings.

For a variety of practical reasons, rate data for process models may have to be obtained in experimental systems having physical characteristics different from those of the systems to which the information is to be eventually applied. Differences in the physical characteristics of particular systems and settings can markedly affect both the *intrinsic* and *observed* rates at which processes occur. The most obvious examples are temperature and conditions of mixing, but others, such as exposure to light and the presence of species that may act as catalysts or deterrents must also be recognized. Material balance models which incorporate appropriate mass transport relationships, internal source (input) and sink (output) terms, and boundary conditions relating to each of the different types of systems in question provide the means by which rates of change in one type of system (e.g., bench scale) can be applied to another (e.g., field scale).

12.3.1 Rate Relationships and Time Scales

Applicable time scales are very important in developing rate relationships for use in process analysis. Approximations of reaction "half-lives" (see Chapter 7) provide a commonsense and mathematically tractable means for estimating time scales associated with processes, especially for first- or second-order reactions. Half-lives for various reactions of environmental concern range from seconds to years. Estimates of these values, together with some rudimentary knowledge of the transport phenomena operative in a given setting, can often facilitate valuable first-cut assessments of the significance of changes in particular system constituents or characteristics.

12.3.2 Microscale Transport Processes

The behavior of a transporting fluid and its effects on homogeneous and heterogeneous reactions is another important consideration for process dynamics. Prediction of the extent of change of species involved in truly homogeneous reactions (e.g., the autodecomposition of dissolved ozone in a completely mixed reactor) requires a good understanding only of macroscale transport phenomena. This is because the rates of such reactions do not depend on interactions among fluid elements (i.e., only *intrinsic* rates are important). In contrast, rates of heterogeneous reactions are determined either in part or entirely by microscale transport processes and reaction boundaries (e.g., reactions at solid interfaces or within porous catalyst solids). In such instances, the rates *observed* are specific to the environmental setting and to the reacting medium. Transport phenomena in various types of microscale impedance domains are quantified in terms of mass transfer and diffusion coefficients. These concepts were developed in Chapters 4 and 8 and applied to reactor engineering in Chapters 10 and 11. Mass transfer coefficients depend on the hydrodynamics of transporting fluids, the diffusivities of reacting species, and the characteristics of phase interfaces. The diffusion coefficients or diffusivities of reacting species depend in turn on temperature, the molecular character of the species involved, and the physical structure of the reactive phase. Much of our knowledge base relative to parameter–property interrelationships exists in the form of empirical correlations. Such correlations are not only useful but often essential for predicting the behavior of microscale processes.

Microscale transport is particularly important to our understanding of constituent partitioning between the phases and compartments of systems. The natural environmental fate of anthropogenic compounds serves as a vivid example. While fugacity is a useful concept for predicting thermodynamic phase-partitioning equilibria, it is of limited value unless rate information is also available. Moreover, there is an important interplay between thermodynamic and rate-controlling principles in multiphase systems. Mass transfer relationships associated with the partitioning of components between gaseous and liquid phases provide a simple yet important illustration. Rates of gas–liquid mass transfer depend not only upon the properties of boundary layers but also upon the ultimate equilibrium distributions of components between the two phases (e.g., Henry's law for dilute systems). These equilibrium distributions are determined by the energy available to drive the reaction. Thus the interplay between energy (a thermodynamic function) and mass transfer (a rate function) becomes apparent.

Another form of interplay exists between mass transfer and thermodynamic factors in the important class of phase partitioning referred to broadly as sorption. As introduced in Chapter 8 and expanded upon in Chapter 11, sorption equilibrium relationships of the types described in Chapter 6 are often attained slowly, on the order of days, weeks, or longer—in some cases years. Because driving forces for sorption reactions are proportional to differences between equilibrium concentrations dictated by thermodynamics and actual or existing nonequilibrium concentrations dictated by mass transfer rates, knowledge of equilibrium relationships is critical. Unlike gas–liquid partitioning, sorption equilibrium and rate relationships are not always evident, nor are they straightforward to obtain. In many environmental situations, for example, there is competition among the components of mixtures for sorption sites, thus sharply increasing the complexity of associated process formulations.

12.4 REACTORS AND PROCESS MODELS

The thermodynamic and rate aspects of processes, including microtransport processes (Chapter 4), comprise the essential elements of process dynamics at very small scale; that is, at the *microscale*. We have seen that these processes are *mass dependent*. To translate knowledge at this level to engineering practice we must be able to quantify *mass distributions* of reacting substances at the larger *macroscale* (*system scale*), where they are determined by macrotransport processes (Chapter 3). The overall extent of change that occurs in a reaction system, or *reactor*, is significantly influenced by macroscale mass transport.

12.4.1 Conceptual Models

As demonstrated in Chapters 9 through 11, equilibrium and rate relationships, coupled with appropriate material balances to account for macroscale transport processes, form the foundations of conceptual models for environmental change and for the design of engineered systems.

We have developed simplified transport models in the context of *ideal reactors* that behave in a conceptually predictable and experimentally reproducible manner (e.g., CMFR, CMFR in series, PFR, PFDR). These models have great practical utility. Of course, "real reactors" do not necessarily conform exactly to ideal reactor models. We determine the degree of conformance with the aid of an experimental and statistical *residence time distribution* (RTD) *analysis* on the real reactor. Once determined to be suitable for describing the behavior of a real reactor, an ideal reactor model affords insight into the influence of transport processes on changes that occur in the real reactor. It provides a reliable way in a wide range of situations to interpret processes in natural systems and to scale-up and design engineered systems.

Despite the advantages of simplicity offered by ideal conceptual models, there are situations in which transport can be understood only by returning to fundamental *momentum balance* concepts. We should in fact recognize that all ideal models eventually derive from momentum balances on fluid elements. Sophisticated mathematical tools such as the Navier–Stokes equation may be especially useful for describing flow fields in two and three dimensions, as discussed in Chapter 3. Because of their inherent complexity, however, these approaches have limitations for direct application in reactor design. Although mathematical complexity can be overcome with sufficient computational power or capability, it is often difficult or impossible to quantify many of the terms required for solution of the resulting equations. Even if these equations are solved successfully, the resulting model may not yield predictions that are any more accurate than those of much simpler models.

Despite the precautions mentioned above, the momentum balance may in many instances provide the only approach to modeling natural systems, where two- and three-dimensional descriptions of transport are often essential. Large lake systems, for example, do not lend themselves to the relatively simple reactor characterizations given by ideal CMFR, CMFRs in series, PFR, or PFDR models. Similar problems are encountered in describing transport phenomena in subsurface environments. In fact, it is generally impossible in such situations even to conduct meaningful RTD analyses for purpose of selecting an ideal reactor model. Experiments with physical models can provide alternatives to fundamental mathematical descriptions of two- and three-dimensional transport in such situations. The objective of such experiments is to achieve hydrodynamic similitude between test-scale and full-scale systems.

The material balance principle has been employed throughout this book as the central algorithm for coalescing descriptions of relevant reaction and transport phenomena. We add here for emphasis that a thorough understanding of the material balance principle and facile use of the mathematical relationships that derive from it are essential skills for environmental scientists and engineers. The material balance principle allow analysis of every problem involving environmental change from a common starting point, regardless of system characteristics or applicable temporal and spatial scales. It is both the most significant unifying concept and most important working tool available for tackling the quantification of process dynamics.

12.4.2 Synopsis of Processes and Models

A synopsis of processes and models is presented in Tables 12-1 through 12-3 and Appendix A.12.1 according to various types of systems. Process models for *homogeneous and quasi-homogeneous reactions* in both *flow and nonflow reactor systems* are given in Table 12-1. Those for *heterogeneous reactions and mass transfer in nonflow reactor systems* are given in Table 12-2. The more common process models involving *heterogeneous reactions in flow reactor systems* are given in Table 12-3. The models presented in Table 12-3 are integrated with others in Appendix A.12.1 to cover an even wider array of environmental settings involving heterogeneous reactions in flow reactor systems. These tables include descriptions of the most common reactor configurations and key parameters that must be quantified to predict environmental change, which in most cases is a change in the concentration of a particular constituent between the inlet and outlet of a reactor. The form (i.e., *point* or *integral*) of the appropriate material balance relationship for each model is also given, along with the general type of equation its mathematical solution must address (i.e., *algebraic*, *ordinary differential*, or *partial differential*).

Relatively few of the 90 models identified in the tables described above have been developed fully in the preceding chapters. Those that have not, however, can be derived by using various combinations of phase conditions, rate relationships, and reactor configurations described throughout the book. In fact, an even larger number and greater diversity of specific models than those listed in the tables can be developed in this manner.

Nearly every model developed for one type of ideal reactor configuration has a counterpart for every other configuration. This is illustrated in Table 12-4, in which applications drawn from specific developments in previous chapters involving homogeneous or quasi-homogeneous reactions in three common reactor configurations—CMBR, CMFR and PFR—are listed. The list of reaction types developed for a CMBR can of course take place in a CMFR or in a PFR, and an appropriate model can thus be developed using the same principles as those employed for the CMBR models. Moreover, many of the application conditions described for these three reactor configurations can be extended directly to PFDRs and to various series and parallel arrangements of CMFR, PFRs, and PFDRs; this is illustrated in the first (Example 12-1) of a series of examples presented at the end of this chapter as Appendix A.12.2. Example 12-1 extends the development of unsteady application conditions for a CMFR presented in Section 11.2.2 (Model 4-A in Table 12-4) to a particular application condition for CMFRs in series (Model 6 in Table 12-1); producing a model not developed specifically in any preceding chapters.

Extrapolation of a model developed for one reactor configuration for application to another is similarly possible when there is an unsteady rather than steady state, heterogeneous rather than homogeneous reaction, porous rather than nonporous solid phase, gas–liquid rather than liquid–liquid mass transfer, cocurrent rather than countercurrent phase flow, and so on. A specific model can be rather easily modified to accommodate change(s) in the combinations of conditions, once the rationale underlying rate and phase relationships, reactor configurations, and initial and boundary conditions has been defined.

Four additional illustrations of model extensions are given in Examples 12-2 through 12-5 in Appendix A.12.2. Example 12-2 extrapolates the essential elements of SBRs presented in detail in Section 9.3.2 and summarized generically in Models 18 through 24 in Table 12-2 and Model 2 in Table 12-1. The example also illustrates an analysis of cycle time for the *heterogeneous-phase reactions* of particle aggregation and sedimentation. Example 12-3 extrapolates Model 27 of Appendix A.12.1 for heterogeneous catalysis at the external surfaces of solids in a CMFR when both phases are fed *continuously* (see Section 10.2.4 for details), to a circumstance in which the solids are *retained* and accumulated in the reactor; the result is Model 29 in Appendix A.12.1. Example 12-4 extrapolates Model 32 of Table 12-3, developed for sorption on the *internal surfaces of a porous solid* in a CMFR when both phases are fed continuously (see Section 11.7.1 for details), to a situation in which sorption occurs on the *external surfaces of a nonporous solid* and sorption rate is controlled by external mass transfer; the result is Model 31 of Table 12-3. Example 12-5 further extends the development of CMFR Model 31 to obtain Model 61 for a PFR and its equivalent PFDR model (see Models 70 through 90).

A number of reactor configuration and process pairs identified in Tables 12-1 through 12-4 and Appendix A.12.1 are identical in all pertinent features except for the state (i.e. *steady or unsteady*) at which the system(s) they describe are operated. Chapters 9 and 10 discussed the modeling of various reactor configurations and conditions for operations conducted at steady state, while modeling for systems operating under unsteady conditions were covered in Chapter 11. The terms *steady* and *unsteady* are nonetheless subject to some degree of interpretation and latitude in many practical process model applications. That is, the relative *steadiness* of an operation relates in part to the time and space *scales* over which the system or process of interest is considered or modeled. It is often feasible to employ the *steady-state* form of a model to approximate the behavior of an inherently *unsteady system* if the *temporal and spatial scales of unsteadyness* (e.g., changes in temperature, input flows, concentrations, etc.) are large with respect to the scales over which the model is applied.

Unsteady behavior over long periods of time can be handled by iterative application of the steady-state form of a model over sequential periods. These periods are chosen to coincide with *acceptably steady values* of parameters or characteristics of a system that change with time. Large lakes that undergo major changes in key input parameters, internal source and sink terms, or boundary conditions only on a time scale of *seasons* may often be modeled, for example, with a *series of time-sequenced steady-state models*. Each time-sequenced model is appropriately interfaced with others to describe long-term events and system behavior. This approach is termed *temporal model segmentation*.

A similar segmentation approach can be taken to describing complex systems having spatially variable characteristics (e.g., velocities, boundary configurations, mixing conditions, internal mass distributions, etc.). Various simplified models (e.g., CMFR, PFR, PFDR) can in such cases be applied in sequence or series to physical segments of smaller spatial scale over which they capture reasonably steady local physical conditions. This approach, commonly employed to model the behavior of large river reaches and estuaries, is termed *spatial model segmentation*.

Table 12-5 is a closure to this chapter and to the book. It contains the complete list of general process descriptions for natural and engineered environmental systems given at the outset of this book in Table 1-1. The critical information that has been added is a corresponding list of process models and application conditions taken from Tables 12-1 to 12-3 that are typically useful in each case. Variations of these general models, such as provided in Appendix A.12.1, for example, could also be included in many of the entries to Table 12-5. The reader is encouraged to examine alternative models from Appendix A.12.1 in this regard as a final exercise to enhance understanding of the concepts stressed in this book. In summary, Table 12-5 emphasizes that a variety of models can be employed to describe a particular type of system or process. Different models originate from different considerations of temporal and spatial scales. Their validity depends on the appropriateness of implicit and explicit assumptions underlying each model's application to a particular system.

We have stressed throughout this book the importance of a working knowledge of model assumptions in judging their validity. Toward this end, we hope to have provided a coherent language for translating principles from varied fields of science into sound engineering approaches to process dynamics in environmental systems.

Table 12-1 Process Dynamics Involving Homogeneous and Quasi-homogeneous[a] Reactions in Flow and Nonflow Reactors

Reactor Configuration	Model Number	Operating State	Form of Material Balance Equation	Mathematical Solution	Key Reactor Parameters	Relevant Chapters
CMBR	1	**Unsteady** $C(t)$	**Integral**	**ODE** $C(t)$	V_R and t	1, 2, 7, 8, 9
SBR	2	**Unsteady** $C(t)$ and $V(t)$	**Integral**	**ODE** $C(t)$ and $V(t)$	V_R and t	7, 9
CMFR	3	**Steady**	**Integral**	**Algebraic**	V_R and Q	2, 9
	4	**Unsteady** $C_{IN}(t)$, $Q_{IN}(t)$, and $k(t)$	**Integral**	**ODE** $C_{OUT}(t)$	V_R and Q	2, 9, 11
CMFRs in Series	5	**Steady**	**Integral**	**Algebraic** C_{OUT}	V_R and Q $\mathfrak{N}_d$ also needed to model behavior of a single CMFR as a series of CMFRs	9
	6	**Unsteady** $C_{IN}(t)$ and $Q_{IN}(t)$	**Integral**	**ODE** $C_{OUT}(t)$		9, 11
PFR	7	**Steady**	**Point**	**ODE** $C(x)$	V_R and Q	3, 9
	8	**Unsteady** $C_{IN}(t)$ and $Q_{IN}(t)$	**Point**	**PDE** $C(x, t)$	V_R and Q	9, 11
PFDR	9	**Steady**	**Point**	**ODE** $C(x)$	V_R, Q, and $\mathfrak{N}_d$	3, 9
	10	**Unsteady** $C_{IN}(t)$ and $Q_{IN}(t)$	**Point**	**PDE** $C(x, t)$	V_R, Q, and $\mathfrak{N}_d$	2, 9, 11

[a]Dispersions of micron- or submicron-size particles (e.g., bacteria) or macromolecules (e.g., humic substances) can under particular hydrodynamic conditions be considered to respond in a manner similar to dissolved substances with respect to the influence of mixing and other macroscopic transport phenomena.

Table 12-2 Process Dynamics Involving Heterogeneous Reactions and Mass Transfer in Nonflow Reactors

Reactor Configuration and Reaction Type	Model Number(s)[a]	Operating State	Form of Material Balance Equation	Mathematical Solution	Microtransport Step(s)[b]	Relevant Chapters
CMBR Gas–liquid reactions with gas fed only at start [e.g., by pressurizing and supersaturating all or part of liquid in reactor (sealed headspace, V_H)]	11 (SBR 18)	**Unsteady**	**Integral**	**Two simultaneous ODEs** $C_l(t)$ and $C_g(t)$ since $C_{S,l}(t)$	**Film diffusion**	2, 4, 7, 8, 10
CMBR (semibatch) Gas–liquid reactions with gas fed continuously	12 (SBR 19)	**Unsteady**	**Integral**	**ODE** $C_l(t)$	**Film diffusion**	2, 4, 7, 8, 10
CMBR Catalysis on the external surfaces of solid particles	13 (SBR 20)	**Unsteady**	**Integral**	**ODE** $C(t)$	**Film diffusion**	8
CMBR Catalysis on the internal surfaces of solid particles	14 (SBR 21)	**Unsteady**	**Integral; point form within solid phase**	**PDE** $C(t, r)$	**Film and intraparticle diffusion**	8, 11
CMBR Sorption/desorption on the external surfaces of solid particles	15 (SBR 22)	**Unsteady**	**Integral**	**PDE** $C(t, r, q)$; open-form solutions or numerical methods	**Film diffusion**	11

CMBR Sorption/desorption on the internal surfaces of porous particles	16 (SBR 23)	**Unsteady**	**Integral; point form within solid phase**	**PDE** $C(t, r, q)$; open-form solutions or numerical methods	**Film and intraparticle diffusion**	11
CMBR Particle growth and particle–particle interactions (e.g., precipitation and aggregation)	17 (SBR 24)	**Unsteady**	**Integral**	**Simultaneous ODEs** *$n_k(t)$ ODEs, where $k = 1, \ldots, n$;* numerical method required for simultaneous integration of particle collision rate expression	**Particle collisions and collision efficiency**	8
SBR All combinations of CMBR models 11 through 17 with variable-volume conditions (i.e., reactions during fill and/or draw stages)	18–24	**Unsteady** $C(t)$, $V(t)$, and $r(t)$	**Integral**	**ODE** $C(t)$, $V(t)$, and $r(t)$	**Various types and combinations of mass transfer processes**	9, 10

[a]Model number for a corresponding reaction type in an SBR are given in parentheses to facilitate cross reference to Table 12-5.
[b]Relative phase velocities, phase interface characteristics (e.g., roughness), and shear and mixing at interfaces between phases are implicitly key parameter-determining reactor factors for heterogeneous systems. These factors are usually captured in system specific mass transfer correlations.

Table 12-3 Selected Illustrations of Process Dynamics Involving Heterogeneous Reactions and Mass Transfer in Flow Reactors[a]

Reactor Configuration and Reaction Type	Model Number(s)[b]	Operating State	Form of Material Balance Equation	Mathematical Solution	Microtransport Step(s)[c]	Relevant Chapters
CMFR Gas–liquid reactions; both phases fed continuously	26 (CMFR in Series 38)	**Steady**	**Integral**	**Algebraic** $C_{g,\mathrm{OUT}}$ *and* $C_{l,\mathrm{OUT}}$; *observed rate determined by Hatta number*	**Film diffusion**	4, 8, 10
CMFR Catalysis on the external surfaces of nonporous particles; both phases fed and removed continuously.	27 (CMFR in Series 39)	**Steady**	**Integral**	**Algebraic** C_{OUT}; *observed rate determined by external effectiveness factor*	**Film diffusion**	8, 10
CMFR Catalysis on the internal surfaces of porous particles; both phases fed and removed continuously	28 (CMFR in Series 40)	**Steady**	**Integral; point form within solid phase**	**Algebraic** C_{OUT}; *observed rate determined by internal effectiveness factor*	**Film and intraparticle diffusion**	10, 11

CMFR Sorption/desorption on the external surfaces of nonporous particles; both phases fed and removed continuously	31 (CMFR in Series 43)	**Steady fluid phase; unsteady solid phase**	**Integral**	**Algebraic** C_{OUT} *(must find average solid-phase loading based on distribution of residence times for particles in CMFR)*	**Film diffusion**	6, 11
CMFR Sorption/desorption on the internal surfaces of porous particles; both phases fed and removed continuously	32 (CMFR in Series 44)	**Steady fluid phase; unsteady solid phase**	**Integral; point form within solid phase**	**Algebraic** C_{OUT}; *open form of solution to PDE for solid phase*	**Film and intraparticle diffusion**	6, 11
CMFR Particle growth and particle–particle interactions; particles fed (or generated) and removed continuously	35 (CMFR in Series 47)	**Steady**	**Integral**	**Simultaneous ODEs** $n_k(t_D)$ *ODEs where* t_D *is discrete value from distribution of residence times in CMFR and* $k = 1, \ldots, n$; *numerical method required for simultaneous integration of particle collision rate expression* $0 < t < t_D$	**Particle collisions and collision efficiency**	8, 9, 10

Table 12-3 (*Continued*)

Reactor Configuration and Reaction Type	Model Number(s)[b]	Operating State	Form of Material Balance Equation	Mathematical Solution	Microtransport Step(s)[c]	Relevant Chapters
CMFRs in series All combinations of CMFR models 25 through 35	37–47	**Steady or unsteady**	**Integral; point form within solid phase where appropriate**	**Algebraic, ODE, or PDE** $C_{g,\mathrm{OUT}}$ *or* $C_{g,\mathrm{OUT}}(t)$; $C_{l,\mathrm{OUT}}$ *or* $C_{l,\mathrm{OUT}}(t)$; *observed rate determined by Hatta number for gas–liquid reactions and effectiveness factor for catalytic solid–liquid reactions; PDE(r, t) for solid phase if sorption is involved; all unsteady-state situations yield ODE(t)*	**Film and/or intraparticle diffusion**	8, 9, 10, 11
PFR Gas–liquid reactions with continuous cocurrent gas feed and removal; packed bed reactor	51 (PFDR 72)	**Steady**	**Point**	**ODE** $C_l(x)$ *and* $C_g(x)$	**Film diffusion**	4, 10

PFR Gas–liquid reactions with countercurrent gas feed and removal; fixed-bed reactor	53 (PFDR 74)	**Steady**	**Point**	**ODE** $C_l(x)$ *and* $C_g(x)$	**Film diffusion**	4, 10
PFR Catalysis on the external surfaces of nonporous particles; continuous cocurrent catalyst feed and removal	54 (PFDR 75)	**Steady**	**Point**	**ODE** $C_l(x)$	**Film diffusion**	8, 10
PFR Catalysis on the external surfaces of nonporous particles; fixed-bed reactor	56 (PFDR 77)	**Steady**	**Point**	**ODE** $C(x)$; *observed rate determined by external effectiveness factor*	**Film diffusion**	8, 10
PFR Catalysis on the internal surfaces of porous particles; fixed-bed reactor	59 (PFDR 80)	**Steady**	**Point**	**ODE** $C(x)$; *observed rate determined by internal effectiveness factor*	**Film and intraparticle diffusion**	8, 10
PFR Sorption/desorption on the external surfaces of nonporous particles; continuous sorbent feed and removal	60 (PFDR 81)	**Steady**	**Point**	**ODE** $C(x)$	**Film diffusion**	6, 10

Table 12-3 (*Continued*)

Reactor Configuration and Reaction Type	Model Number(s)[a]	Operating State	Form of Material Balance Equation	Mathematical Solution	Microtransport Step(s)[b]	Relevant Chapters
PFR Sorption/desorption on the external surfaces of nonporous particles; fixed-bed reactor	62 (PFDR 83)	**Unsteady**	**Point**	**PDE** $C(x, t)$; *numerical solutions for nonlinear equations*	**Film diffusion**	6, 10, 11
PFR Sorption/desorption on the internal surfaces of porous particles; fixed-bed reactor	66 (PFDR 87)	**Unsteady**	**Point**	**PDE** $C(x, t, r)$; *numerical solutions for nonlinear equations*	**Film and intraparticle diffusion**	6, 10, 11

[a]See Appendix A.12.1 for complete listing.
[b]Model numbers for a corresponding reaction type in CMFR-in-Series or PFDR configurations are given in parentheses for CMFR and PFR configurations, respectively, to facilitate cross reference to Table 12-5.
[c]Relative phase velocities, phase interface characteristics (e.g., roughness), and shear and mixing at interfaces between phases are implicitly key parameter-determining reactor factors for heterogeneous systems. These factors are usually captured in system specific mass transfer correlations.

Table 12-4 Model Applications Involving Homogeneous and Quasi-homogeneous[a] Reactions in Flow and Nonflow Systems

Reactor Configuration	Model Number	Application Conditions	Relevant Equations	Relevant Figures	Relevant Examples
CMBR	1	A. General	7-3 through 7-9, 7-44		1-1, 1-2, 2-2
Unsteady state		B. First-order reactions	7-12	7-2	
		C. First-order reactions with lag periods	7-127 through 7-129	7-10	
		D. First-order reactions with particle aggregation			8-12
		E. Pseudo first-order reactions	7-38 through 7-43		7-2
		F. Second-order reactions	7-5	7-3	7-3
		G. Complex reaction orders	7-30		7-4, 7-5, 7-7
		H. Accelerant and retardant reactions	7-122 through 7-129	7-10	
		I. Catalytic reactions (see also K and L)	7-130 through 7-140	7-11 through 7-14	
		J. Consecutive reactions	7-76 through 7-82	7-9	7-6
		K. Michaelis–Menten kinetics	7-142 through 7-149	7-15, 7-16	7-7
		L. Inhibited enzyme reactions	7-151 through 7-165	7-17, 7-18	
		M. Parallel reactions	7-84 through 7-92		
		N. Rate data analysis	7-44 through 7-63	7-1 through 7-5	7-1 through 7-5
		O. Rate relationships for different reaction orders		Table 7-1	
		P. Reversible reactions	7-93 through 7-108		

Table 12-4 (*Continued*)

Reactor Configuration	Model Number	Application Conditions	Relevant Equations	Relevant Figures	Relevant Examples
CMFR	3	A. General	2-12, 9-13 through 9-19, 11-1	9-3	2-2
Steady state		B. Half-order reactions			9-2
		C. First-order reactions	9-20, 9-21, 9-39		9-5, 9-7
		D. Flow recycle	9-57 through 9-62	9-12, 9-13	9-7
		E. Hydraulic residence times (HRTs) for different reaction orders		Table 9-1	
		F. Performance characteristics	9-39 through 9-42	9-7 through 9-9	9-5, 9-6
		G. Rate analysis	9-26 through 9-33	9-5	9-2
CMFR Unsteady state	4	A. Equalization and load dampening	11-1 through 11-12, 11-21		11-1 through 11-3
		B. Residence time distribution (RTD) analysis	9-81 through 9-95, 9-104 through 9-111	9-17 through 9-19, 9-22	

PFR	7	A. General	2-16, 9-34 through 9-36	9-6	9-4, 9-9
Steady state		B. First-order reactions	9-40		9-4, 9-5
		C. Flow recycle	9-43 through 9-56	9-10, 9-11	9-6
		D. Hydraulic residence times (HRTs) for different reaction orders		Table 9-1	
		E. Performance characteristics	9-40	9-7, 9-9, 9-29, 9-30, 11-9	9-5
PFR Unsteady state	8	A. General	9-120, 9-123, 9-124, 9-126 (for small $\mathfrak{N}_d$)	9-20, 9-25, 9-26	
		B. Time-variable reaction rates	11-23 through 11-46	11-9	11-4
		C. Type I diffusion domains; unbounded	11-50 through 11-65	8-4, 11-11, 11-12	8-1, 11-5
		D. Type I diffusion domains; bounded	11-73 through 11-86	11-13, 11-14, 11-15	11-6
		E. Type II diffusion domains	11-87 through 11-95	11-16	8-1
		F. Type III diffusion domains	11-96 through 11-100		8-1
		G. Type IV diffusion domains	11-101 through 11-130	8-7, 11-7, 11-18, 11-19, 11-21, 11-22	8-9 through 8-11, 11-7, 11-8
		H. Residence time distribution (RTD) analysis	9-96 through 9-103	9-15, 9-16, 9-20, 9-21	9-8

[a]Dispersions of micron- or submicron-size particles (e.g., bacteria) or macromolecules (e.g., humic substances) can under particular hydrodynamic conditions be considered to respond in a manner similar to dissolved substances with respect to the influence of mixing and other macroscopic transport phenomena.

Table 12-5 Models Applicable to Common Processes in Natural and Engineered Environmental Systems

Process	Engineered Systems	Commonly Applicable Model(s)	Natural Systems	Commonly Applicable Model(s)	Nature of Reaction(s)	Common Interphase Mass Transfer Phenomena
Absorption	Aeration of biological treatment units (e.g., activated sludge systems)	12, 19, 50–53, 71–74	Uptake of atmospheric oxygen by lakes, streams, and estuaries	26, 38	Gas–liquid mass transfer and dissolution of molecular oxygen	Molecular diffusion of oxygen at air–water interfaces
Adsorption	Removal of organic contaminants from waters and wastes by activated carbon treatment	15, 16, 31–34, 60–66, 81–87	Uptake of organic contaminants from groundwaters by soil mineral surfaces (e.g., sand)	62, 66, 83, 87	Interactions and accumulation of solutes at solid surfaces	Interfacial and intraparticle diffusion of dissolved solutes
Biochemical oxidation	Degradation of organic contaminants in biological treatment units (e.g., activated sludge, trickling filter systems)	2, 3, 5, 9, 13, 14, 20, 21, 27, 28, 56, 59	Exertion of biochemical oxygen demand (BOD) in surface waters	3, 5, 9, 77, 80	Enzyme-mediated transformations of chemical species by electron transfer	Diffusion of substrates, nutrients, and metabolic products across microbial cell walls
Chemical oxidation	Transformation of cyanide to cyanate by chlorine in industrial waste treatment systems; oxidation of organic compounds by ozone in contaminated surface or subsurface water supplies	2, 3, 7, 9	Photooxidation of dissolved organic contaminants in surface waters	3–10	Homogeneous phase transformations of chemical species by electron transfer reactions	None
Chemical oxidation with catalysis	Transformation of cyanide to CO_2 by oxygen in the presence of copper and activated carbon	20, 21, 27, 28, 56, 59, 77, 80	Photooxidation of organic compounds sorbed at mineral surfaces	27, 28, 77, 80	Heterogeneous phase transformation of chemical species by electron transfer reactions (e.g., oxidations at reactive surfaces)	Molecular diffusion at the interfaces of and within solid catalysts

Chemical reduction	Transformation of hexavalent chromium to trivalent chromium for subsequent precipitation	2, 3, 7, 9	Decomposition of ozone by CFCs in the upper atmospheres	3–10	Homogeneous phase transformations of chemical species by electron transfer	None
Coagulation	Destabilization of turbidity and suspended solids by iron, aluminum, lime, or organic polyelectrolyte, coagulants in water and wastewater treatment	31, 33, 81	Destabilization of colloids by natural salts in marine estuaries, or by natural biopolymers in fresh waters	33, 45, 81	Modifications of particle surface and near-surface chemistry to reduce particle–particle repulsions	Molecular diffusion of dissolved species into particle double layers and to particle surfaces
Disinfection (pathogenic organisms) and sterilization (all organisms)	Destruction or inactivation of organisms using chemicals, heat, or shortwave irradiation	2, 3, 7, 9, 27, 50, 71	Destruction or inactivation of organisms by naturally occurring chemical conditions, heat, or irradiation (e.g., sunlight)	3, 7, 9, 27, 39	Enzyme inactivation, protein denaturing, or cell lysis	Mass or heat transfer across cell membranes
Electrochemical transformations	Galvanic and anodic protection of metal components employed in heat transfer and reactor systems	25, 49, 56, 62	Dissolution of metallic contaminants from ores, pipes, and other metal structures	13, 25, 77, 83	Heterogeneous phase transformations of chemical species by electron transfer at electrode surfaces	Diffusion of ions and corrosion inhibitors through inorganic and organic films at metal–water interfaces
Equalization	Moderation of flow or concentration transients to facilitate downstream process control	2, 4, 6, 10	Natural chemical and flow moderations effected by lakes, embayments, and estuaries receiving river or stream discharges	2, 4, 6, 10	Flow and mass buffering provided by expanded spatial and temporal scales	None
Extraction; liquid–liquid	Removal of organic contaminants from aqueous streams using immiscible solvents	26, 50	Dissolution of organic contaminants from nonaqueous phase liquids (NAPLs) phases in surface water or groundwater systems	49, 70	Phase transfer of species by dissolution	Mass transfer at liquid–liquid interfaces

Table 12-5 (*Continued*)

Process	Engineered Systems	Commonly Applicable Model(s)	Natural Systems	Commonly Applicable Model(s)	Nature of Reaction(s)	Common Interphase Mass Transfer Phenomena
Extraction; liquid–solid	Soil washing with surfactant solutions; extraction of contaminants from solids using supercritical fluids (e.g., CO_2)	31, 32, 60, 62, 65,66	Leaching of minerals and organic species from natural solids (e.g. soil, rocks) by water	83, 87	Phase transfer of species by dissolution	Mass transfer at liquid–solid interfaces
Filtration	Removal of suspended solids from waters and wastes by deep-bed or septum filtration	62, 69, 83, 90	Clogging of aquifers by individual and aggregated bacteria and other colloids in subsurface systems	69, 90	Particles or aggregate interception and accumulation at solid surfaces	Microscopic particle transport and interfacial deposition
Flocculation	Aggregation of chemically destabilized colloids induced by fluid mixing processes	24, 35, 47, 67	Aggregation of bacteria and other colloids in surface water and groundwater systems	35, 47, 88	Particle–particle collisions	Particle–particle transport and interactions in macroscopic flow fields
Flotation	Removal of greases, oils, and low-density particles in water and waste treatment	26, 35, 36, 50, 68	Flotation of particulate solids attached to gas bubbles in natural aquatic systems	50, 68	Physical separation of immiscible phases	Particle–particle interactions in near-quiescent flow fields
Hydrolysis	Decomposition of cyanate to ammonia and CO_2 under acidic conditions	2, 3, 7	Hydrolysis of metal ions discharged or dissolved into natural water bodies	3, 7, 9	Transformations of chemical species involving proton and/or hydroxide interactions	None
Incineration	High-temperature oxidation of organic wastes in gas phase	27, 28, 50, 54	Volcanic eruptions and forest fires	27, 70, 72, 75	Transformation of chemical species by electron transfer reactions at elevated temperatures in heterogeneous phase systems	Gas–solid mass and heat transfer

Ion exchange	Removal of metals from water and wastes by ion exchange (e.g., softening, demineralization, recovery of precious metals)	32, 66	Multivalent cation uptake and retardation by clayey soils (e.g., Ca^{2+}, Cd^{2+}, Pb^{2+})	62, 83, 66, 87	Exchange of ions of like but unequal charge at charged surfaces	Interfacial and intraparticle ion diffusion for solid ion-exchange materials; None for liquid ion exchange materials, which are subsequently removed by ultrafiltration, adsorption, or precipitation
Membrane separation	Desalination of marine and brackish waters by reverse osmosis and electrodialysis	62, 66	Separation of dissolved oxygen from water by the gill membranes of fish	62, 66	Selective separations of molecular species by microporous barriers	Molecular diffusion at solid-water interfaces and within microporous membranes
Neutralization	Adjustment of pH from excessively low or high values for downstream process control or product quality	3, 7, 9, 26, 38, 50, 56, 71	Acid-base reactions of carbonic species (e.g., CO_2, H_2CO_3, HCO_3^-, and CO_3^{2-}), which function to poise the natural waters at pH levels near ~8.3	26, 56, 71, 77	Proton and related coordination-partner transfer reactions	None for homogeneous phase (liquid) acid or base additions; interfacial mass transfer when gaseous or solid phase acids or bases are used
Polymerization	Generation of metal-ion and activated silica polyelectrolytes in water and waste coagulation	3, 31, 60	Formation of polynuclear humic and fulvic substances in concentrated sediment pore waters and at solid surfaces	9, 82, 83	Homogeneous interactions of dissolved monomeric species to form polymers or macromolecules	Diffusion of monomers across polymer–water interfaces and within amorphous polymer matrices
Precipitation	Removal of heavy metals and phosphates from wastewaters; removal of hardness (e.g., Ca^{2+} and Mg^{2+}) in water treatment operations	31, 33, 35, 36, 48, 60, 67	Precipitation of iron oxides at surfaces and wetted interfaces near groundwater outcrops; deposition of calcium carbonates and magnesium silicates on submerged surfaces in natural aquatic systems	33, 36, 45, 48, 61, 69, 90	Phase transformations of chemical species by coordination-partner exchange reactions	Interfacial and intraparticle diffusion associated with crystallization and particle growth phenomena

Table 12-5 (*Continued*)

Process	Engineered Systems	Commonly Applicable Model(s)	Natural Systems	Commonly Applicable Model(s)	Nature of Reaction(s)	Common Interphase Mass Transfer Phenomena
Supercritical water oxidation	Oxidation of organic wastes at high temperatures and pressures in supercritical water phase	1, 3, 7, 9	Geothermal activity in deep waters (e.g., beneath ocean floors)	1, 3, 7, 9	Chemical transformations of chemical species by electron transfer at elevated temperatures in homogeneous phase systems	None
Thermal desorption	Volatilization of organic matter adsorbed to soils and other solid phases	60, 62, 64, 65, 66, 81, 85, 87	Sun-baking of surface soils	15, 16	Phase transformation by volatilization at elevated temperatures	Mass and heat transfer at phase interfaces
Volatilization	Stripping of taste and odor compounds from drinking waters, ammonia from wastes, and volatile organic contaminants from groundwaters by vigorous mixing	26, 50–53, 71–74	Escape of hydrogen sulfide from benthic deposits and overlying waters into the atmosphere	11, 26, 38	Phase transformations by volatilization at moderate (ambient) temperature	Mass transfer at phase interfaces

APPENDIX A.12.1 ILLUSTRATIONS OF PROCESS DYNAMICS INVOLVING HETEROGENEOUS REACTIONS AND MASS TRANSFER IN FLOW REACTORS

Reactor Configuration and Reaction Type	Model Number(s)[a]	Operating State	Form of Material Balance Equation	Mathematical Solution	Microtransport Step(s)[b]	Relevant Chapters
CMFR Liquid–solid reactions between reactor walls and bulk fluid phase (e.g., corrosion of reactor surfaces)	25 (CMFR in Series 37)	**Steady** if wall reaction is catalytic; **unsteady** if wall reaction sites are depleted with time	**Integral; point form possible for wall film**	**Algebraic** if catalytic reaction; **ODE** if wall reaction sites are depleted C_{OUT}; *observed rate determined by external effectiveness factor if catalytic*	**Film diffusion**	8, 10
CMFR Gas–liquid reactions; both phases fed continuously	26 (CMFR in Series 38)	**Steady**	**Integral**	**Algebraic** $C_{g,OUT}$ *and* $C_{l,OUT}$ *observed rate determined by Hatta number*	**Film diffusion**	4, 8, 10
CMFR Catalysis on the external surfaces of solid particles; both phases fed and removed continuously	27 (CMFR in Series 39)	**Steady**	**Integral**	**Algebraic** C_{OUT}; *observed rate determined by external effectiveness factor*	**Film diffusion**	8, 10

APPENDIX A.12.1 (*Continued*)

Reactor Configuration and Reaction Type	Model Number(s)[a]	Operating State	Form of Material Balance Equation	Mathematical Solution	Microtransport Step(s)[b]	Relevant Chapters
CMFR Catalysis on the internal surfaces of porous particles; both phases fed and removed continuously	28 (CMFR in Series 40)	**Steady**	**Integral; point form within solid phase**	**Algebraic** C_{OUT}; *observed rate determined by internal effectiveness factor*	**Film and intraparticle diffusion**	10, 11
CMFR Catalysis on the external surfaces of nonporous particles; particles fed continuously and retained in reactor	29 (CMFR in Series 41)	**Unsteady** Surface area increases with time	**Integral**	**Algebraic** C_{OUT}; *observed rate determined by external effectiveness factor*	**Film diffusion**	8, 10
CMFR Catalysis on the internal surfaces of porous particles; particles fed continuously and retained in reactor	30 (CMFR in Series 42)	**Unsteady** Surface area increases with time	**Integral; point form within solid phase**	**Algebraic** C_{OUT}; *observed rate determined by internal effectiveness factor*	**Film and intraparticle diffusion**	8, 10
CMFR Sorption/desorption on the external surfaces of nonporous particles; both phases fed and removed continuously	31 (CMFR in Series 43)	**Steady fluid phase; unsteady solid phase**	**Integral**	**Algebraic** C_{OUT}; *average solid-phase loading based on particle residence time distribution*	**Film diffusion**	6, 11

CMFR Sorption/desorption on the internal surfaces of porous particles; both phases fed and removed continuously	32 (CMFR in Series 44)	**Steady fluid phase; unsteady solid phase**	**Integral; point form within solid phase**	**Algebraic** C_{OUT}; *open form of solution to PDE for solid phase*	**Film and intraparticle diffusion**	6, 11
CMFR Sorption/desorption on the external surfaces of nonporous particles; particles fed continuously and retained in reactor	33 (CMFR in Series 45)	**Unsteady** Surface area increases with time	**Integral**	**ODE** $C_{OUT}(t)$ *and* $q(t)$ *(both increasing)*	**Film diffusion** V, Q, $D_o(t)$, $(a_s^\circ)_E$, and sorption equilibrium parameters	6, 11
CMFR Sorption/desorption on the internal surfaces of porous particles; particles fed continuously and retained in reactor	34 (CMFR in Series 46)	**Unsteady** Surface area increases with time	**Integral; point form within solid phase**	**PDE** $C(t, q)$ *and* $q(r, t)$	**Film diffusion** V, Q, $D_o(t)$, $(a_s^\circ)_I$, and sorption equilibrium parameters	6, 11
CMFR Particle growth and particle–particle interactions; particles fed (or generated) and removed continuously	35 (CMFR in Series 47)	**Steady**	**Integral**	**Simultaneous ODEs** $n_k(t_D)$ *ODEs, where* t_D *is discrete value from distribution of residence times in CMFR and* $k = 1, \ldots, n$; *numerical method required for simultaneous*	**Particle collisions and collision efficiency**	8, 9, 10

APPENDIX A.12.1 ***(Continued)***

Reactor Configuration and Reaction Type	Model Number(s)[a]	Operating State	Form of Material Balance Equation	Mathematical Solution	Microtransport Step(s)[b]	Relevant Chapters
				integration of particle collision rate expression $0 < t < t_D$		
CMFR Particle growth and particle–particle interactions; particles fed (or generated) continuously and retained in reactor	36	**Unsteady**	**Integral**	**Simultaneous ODEs** $n_k(t)$; $k = 1, \ldots, n$, *where* n_k *must include term to account for unsteady conditions resulting from retention of particles as well as gain by large particle breakup or by aggregation of smaller particles*	**Particle collision and collision efficiency**	8, 9
CMFRs in Series All combinations of CMFR models 25 through 35	37–47	**Steady or unsteady**	**Integral; point form within solid phase where appropriate**	**Algebraic, ODE, or PDE** $C_{g,\text{OUT}}$, *or* $C_{g,\text{OUT}}(t)$; $C_{l,\text{OUT}}$ *or* $C_{l,\text{OUT}}(t)$; *observed rate determined by Hatta number for gas–liquid*	**Film and/or intraparticle diffusion**	8, 9, 10

				reactions and effectiveness factor for catalytic solid–liquid reactions; PDE (r, t) *for solid phase if sorption is involved; all unsteady-state situations yield ODE*(t)		
CMFRs in Series Particle growth and particle–particle interactions; particles fed (or generated) continuously and retained in last reactor (i.e., combinations of models 35 and 36)	48	**Unsteady**	**Integral**	**Simultaneous ODEs** *Same solution conditions and requirements as for model 35, but, as for model 36,* n_k *for the last reactor must include term to account for unsteady conditions resulting from retention of particles as well as gain by large particle breakup or by aggregation of smaller particles*	**Particle collisions and collision efficiency**	8, 9

APPENDIX A.12.1 ***(Continued)***

Reactor Configuration and Reaction Type	Model Number(s)[a]	Operating State	Form of Material Balance Equation	Mathematical Solution	Microtransport Step(s)[b]	Relevant Chapters
PFR Liquid–solid reactions between reactor walls and bulk fluid phase	49 (PFDR 70)	**Usually steady**	**Point**	**ODE** $C_l(x)$	**Film diffusion**	2, 4, 11
PFR Gas–liquid reactions with continuous cocurrent gas feed and removal	50 (PFDR 71)	**Steady**	**Point**	**ODE** $C_l(x)$ *and* $C_g(x)$	**Film diffusion**	10
PFR Gas–liquid reactions with continuous cocurrent gas feed and removal; fixed-bed reactor	51 (PFDR 72)	**Steady**	**Point**	**ODE** $C_l(x)$ *and* $C_g(x)$	**Film diffusion**	4, 10
PFR Gas–liquid reactions with continuous countercurrent gas feed and removal	52 (PFDR 73)	**Steady**	**Point**	**ODE** $C_l(x)$ *and* $C_g(x)$	**Film diffusion**	4, 10
PFR Gas–liquid reactions with continuous counter-current gas feed and removal; fixed-bed reactor	53 (PFDR 74)	**Steady**	**Point**	**ODE** $C_l(x)$ *and* $C_g(x)$	**Film diffusion**	4, 10

PFR Catalysis on the external surfaces of nonporous particles; continuous cocurrent catalyst feed and removal	54 (PFDR 75)	**Steady**	**Point**	**ODE** $C_L(x)$	**Film diffusion**	8, 10
PFR Catalysis on the external surfaces of nonporous particles; continuous catalyst feed and accumulation in reactor	55 (PFDR 76)	**Unsteady** Surface area of catalyst increases with time	**Point**	**ODE** $C(x, t)$; $V_R(t)$	**Film diffusion**	8, 10, 11
PFR Catalysis on the external surfaces of nonporous particles; fixed-bed reactor	56 (PFDR 77)	**Steady**	**Point**	**ODE** $C(x)$; *observed rate determined by external effectiveness factor*	**Film diffusion**	10
PFR Catalysis on the internal surfaces of porous particles; continuous cocurrent catalyst feed and removal	57 (PFDR 78)	**Steady fluid phase; unsteady within catalyst**	**Point**	**PDE** $C_{OUT}(x)$; *analogous to CMBR; solve PDE for* $C(r, t)$, *evaluate* $C(R, t)$ *and use Lagrangian view to translate diffusion time to distance traveled in reactor*	**Film and intraparticle diffusion**	8, 10

APPENDIX A.12.1 (*Continued*)

Reactor Configuration and Reaction Type	Model Number(s)[a]	Operating State	Form of Material Balance Equation	Mathematical Solution	Microtransport Step(s)[b]	Relevant Chapters
PFR Catalysis on the internal surfaces of porous particles; continuous cocurrent catalyst feed with accumulation in reactor	58 (PFDR 79)	**Unsteady** Surface area increases with time	**Point**	**ODE** $C_{OUT}(t)$; $V_R(t)$; *observed rate determined by internal effectiveness factor*	**Film and intraparticle diffusion**	8, 11
PFR Catalysis on the internal surfaces of porous particles; fixed-bed reactor	59 (PFDR 80)	**Steady**	**Point**	**ODE** $C(x)$; *observed rate determined by internal effectiveness factor*	**Film and intraparticle diffusion**	10
PFR Sorption/desorption on the external surfaces of nonporous particles; continuous sorbent feed and removal	60 (PFDR 81)	**Steady**	**Integral**	**ODE** $C(x)$	**Film diffusion**	6, 10
PFR Sorption/desorption on the external surfaces of nonporous particles; continuous sorbent feed and accumulation in reactor	61 (PFDR 82)	**Unsteady**	**Point**	**PDE** $C(x, t)$ *and* $q(x, t)$; $V_R(t)$	**Film diffusion**	4, 6, 10

PFR Sorption/desorption on the external surfaces of nonporous particles; fixed-bed reactor	62 (PFDR 83)	**Unsteady**	**Point**	**PDE** $C(x, t)$; *numerical solutions for nonlinear equations*	**Film diffusion**	10
PFR Sorption/desorption on the internal surfaces of porous particles; continuous sorbent feed and removal	63 (PFDR 84)	**Steady**	**Point**	**ODE** $C(x)$ *and* $q(x)$; *open form of solution to PDE or numerical methods for solid phase and use Lagrangian view view to translate diffusion time to distance traveled in bed*	**Film and intraparticle diffusion**	6, 8, 10, 11
PFR Sorption/desorption on the internal surfaces of porous particles; continuous cocurrent sorbent feed and accumulation in reactor	64 (PFDR 85)	**Unsteady**	**Point**	**PDE** $C(x, t)$ *and* $q(x, t)$; $V_R(t)$	**Film diffusion**	4, 6, 10
PFR Sorption/desorption on the internal surfaces of porous particles; continuous countercurrent sorbent feed and removal	65 (PFDR 86)	**Steady**	**Point**	**ODE** $C_{OUT}(x)$ *and* $q_{OUT}(x)$; *open form of solution to PDE or numerical methods for solid phase and Lagrangian view*	**Film and intraparticle diffusion**	6, 8, 10, 11

APPENDIX A.12.1 (*Continued*)

Reactor Configuration and Reaction Type	Model Number(s)[a]	Operating State	Form of Material Balance Equation	Mathematical Solution	Microtransport Step(s)[b]	Relevant Chapters
				to translate diffusion time to distance traveled in bed		
PFR Sorption/desorption on the internal surfaces of porous particles; fixed-bed reactor	66 (PFDR 87)	**Steady fluid phase; unsteady solid phase**	**Point**	**PDE** $C(x, t, r)$; *numerical solutions for nonlinear equations*	**Film and intraparticle diffusion**	6, 10, 11
PFR Particle growth and particle–particle interactions; continuous cocurrent particle feed (or generation) and removal	67 (PFDR 88)	**Steady**	**Point**	**Simultaneous ODEs** $n_k(t)$; $k = 1, \ldots, n$, *where* n_k *must include term to account for steady rate of introduction of particles as well as gain by large particle breakup or by aggregation of smaller particles*	**Particle collision and collision efficiency**	8, 9
PFR Particle growth and particle–particle interactions; continuous cocurrent particle feed	68 (PFDR 89)	**Unsteady**	**Point**	**Simultaneous ODEs** $n_k(t_D)$ *ODEs, where* t_D *is discrete value from distribution of residence times*	**Particle collisions and collision efficiency**	8, 9

(or generation) and retention in reactor				*in PFR and $k = 1, \ldots, n$; n_k must include term to account for unsteady conditions resulting from retention of particles as well as gain by large particle breakup or loss by aggregation of smaller particles*		
PFR Particle growth and particle–particle interactions; continuous particle feed (or generation) and retention in fixed-bed reactor	69 (PFDR 90)	**Unsteady**	**Point**	**PDE**	**Particle diffusion, collision, sedimentation, attachment, and detachment (related to fluid-shear forces)**	8, 9, 11
PFDR	70–90	**Conditions and considerations applicable to PFDR models 70 through 90 are identical to those of their respective counterparts among PFR models 49 through 69 with additional relationships to account for dispersive flow conditions**				

[a]Model numbers for a corresponding reaction type in CMFR-in-Series or PFDR configurations are given in parentheses for CMFR and PFR configurations, respectively, to facilitate cross reference to Table 12-5.

[b]Relative phase velocities, phase interface characteristics (e.g., roughness), and shear and mixing at interfaces between phases are implicitly key parameter-determining reactor factors for homogeneous systems. These factors are usually captured in system specific mass transfer correlations.

reactor:

$$C_{\text{OUT},2}(t) = C_{\text{OUT},SS,2}e^{-t\alpha_2} + \frac{\beta}{\alpha_2}(1 - e^{-t\alpha_2})$$

$$+ \frac{\beta}{\alpha_2 - \alpha_1}(e^{-t\alpha_2} - e^{-t\alpha_1}) + \frac{C_{\text{OUT},SS,1}}{\bar{t}_2(\alpha_2 - \alpha_1)}(e^{-t\alpha_1} - e^{-t\alpha_2})$$

where

$$\beta = \frac{C_{\text{IN},SS,1}}{\bar{t}_2(1 + k\bar{t}_1)}$$

7. For the condition $\alpha_1 = \alpha_2$, the following solution obtains for the concentration profile from the second reactor:

$$C_{\text{OUT},2}(t) = C_{\text{OUT},SS,2}e^{-t\alpha_2} + \frac{\beta}{\alpha_2}(1 - e^{-t\alpha_2})$$

$$- \beta te^{-t\alpha_2} + \frac{C_{\text{OUT},SS,1}}{\bar{t}_2} te^{-t\alpha_2}$$

8. Note that the equations in steps 6 and 7 have the same steady-state solution:

$$C_{\text{OUT},SS,2} = \frac{\beta}{\alpha_2} = C_{\text{IN},SS,1}\left(\frac{1}{1 + k\bar{t}_1}\right)\left(\frac{1}{1 + k\bar{t}_2}\right)$$

or if $\alpha_1 = \alpha_2$, the same result presented in Chapter 9, Equation 9-23, obtains:

$$C_{\text{OUT},SS,2} = \frac{\beta}{\alpha} = C_{\text{IN},SS,1}\left(\frac{1}{1 + k\bar{t}}\right)^n$$

- **Conclusions**

1. The model developed above is an *application condition* for *Model 6 in Table 12-1*, a model not fully developed earlier in the book.

2. The largest increase in influent BOD would be from 50 mg/L to 150 mg/L, which occurs during a 6-h period every other day. Let us consider that 50 mg/L represents the steady-state influent concentration, $C_{\text{IN},SS,1}$. We are concerned that the concentration of toxicant in the aeration tank does not exceed 25 mg/L during this same 6-h period. The effluent BOD concentrations for the two different reactor configurations during a 6-h shock load of 150 mg/L are plotted below.

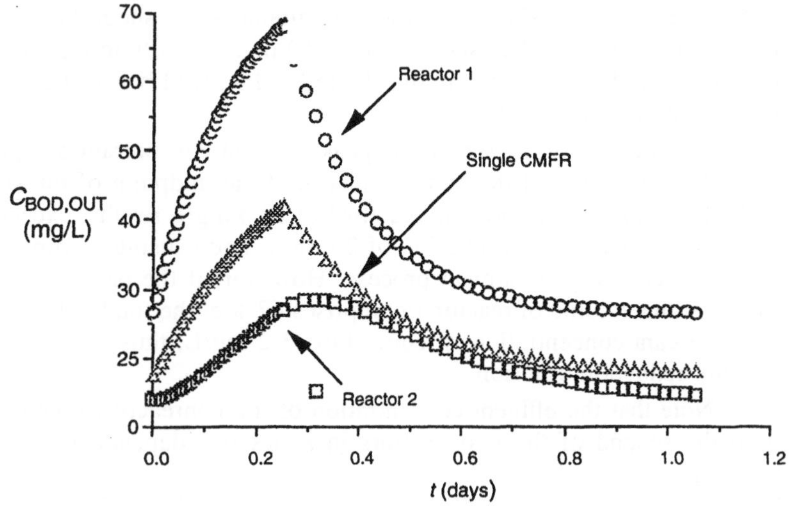

Comparison of single CMFR to two CMFRs in series.

3. Under steady-state conditions, corresponding to $t = 0$, the single CMFR can meet the 20-mg/L discharge requirement for BOD. However, under transient conditions, two CMFRs in series are required to meet the 30-mg/L peak discharge limit. This illustrates the enhanced treatment performance of reactors in series.

4. The effluent concentration of toxicant during the shock-load period is described by the same model as that for effluent BOD, except that the rate constant is now zero. The figure below provides the effluent toxicant concentration, $C_{TOX,OUT}$, as affected by different baffle placements.

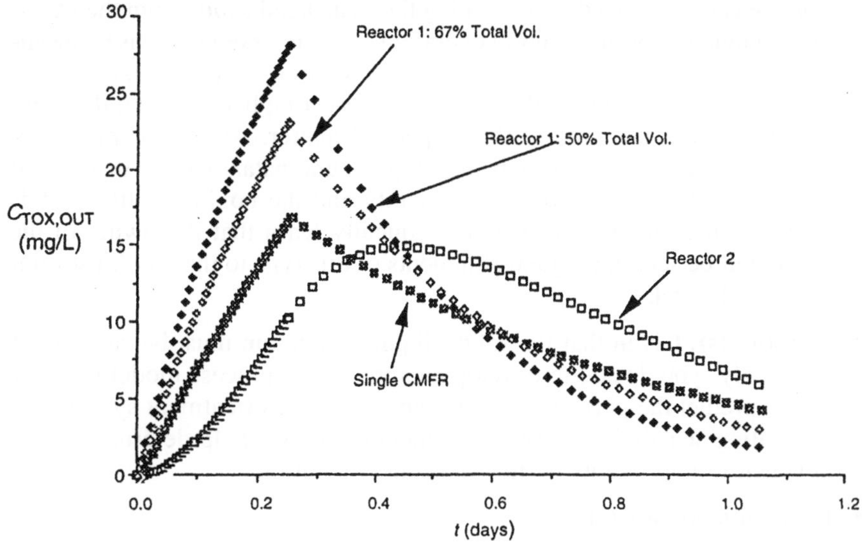

Responses to shock load and effect of baffle placement.

5. The single CMFR is better able to attenuate the shock load of the nonreactive toxicant. The shock load of 50 mg/L results in a peak concentration of about 17 mg/L in the single CMFR, well below the critical concentration of 25 mg/L.

6. For the reactors in series, the peak toxicant concentration depends on baffle placement. If the baffle is placed at the midpoint of the single CMFR, a peak concentration in reactor 1 of 28.3 mg/L results, exceeding the critical toxicant concentration of 25 mg/L and causing disruption to the activated sludge treatment process. However, if the baffle is placed such that the volume of reactor 1 comprises 67% of the total volume, the peak toxicant concentration is reduced to 23.2 mg/L, thus satisfying the toxicant reduction criterion.

7. Note that the effluent concentration of the nonreactive component from the second of the two reactors in series is independent of baffle position.

Example 12-2

- **Situation:** The operator of a tanker truck containing a hazardous waste consisting of an aqueous dispersion of suspended organic matter is intercepted as he is about to discharge the contents of the tanker to a secure cell of a land disposal site. The site supervisor and the owner of the waste have made a last-minute decision that the dispersion should be concentrated prior to discharge to the cell. This decision is predicated on (1) the high cost per unit volume (or weight) of waste discharged and (2) the suspended organic matter being solely responsible for the hazardous nature of the waste. Thus significant savings in both disposal costs and cell volume can be realized by separating the suspended solids from the water. As an engineer for the disposal operation, you are asked to devise a means of solid–liquid separation. You decide to employ a spare Baker tank (a portable waste storage tank) at the site to accomplish sequential coagulation and sedimentation of the suspended matter. At the completion of this treatment, the relatively benign liquid supernatant can be discharged to an on-site biological treatment facility and the noxious settled solids placed in the secure cell. You subsequently learn that the owner of the waste will be sending regular shipments of this type to the landfill several times each week.

- **Question(s):** Given that the solids–liquid separation must be carried out repeatedly, you decide to develop an appropriate process model to allow you to predict the performance of your improvised treatment system. By facilitating sensitivity analysis this model will also help determine ways in which to improve its performance and operation.

- **Logic and Answer(s):**

1. Model 2 in Table 12-1 was developed in Section 9.3.2 for sequencing batch reactor (SBR) applications to homogeneous systems. We must

now develop an SBR model for a heterogeneous system involving particle aggregation and settling (i.e., *Model 24 in Table 12-2*).

2. Rates of particle aggregation were discussed in Chapter 8, and a simplified example was provided (Example 8-12). It was assumed in Example 8-12 that the system initially contained monodisperse particles and that the total particle volume remained constant, two rather restrictive and unrealistic assumptions. In the present case, we must develop an approach to analyzing heterodisperse, heterogeneous systems.

3. The general mathematical expression for particle aggregation is given by Equations 8-149 and 8-150. From those equations the rate of change of the number of primary particles per unit volume (n_1) in the system can be written as

$$\frac{dn_1}{dt} = -\beta_{c,1,1}n_1n_1 - \beta_{c,1,2}n_1n_2 - \beta_{c,1,3}n_1n_3 - \cdots - \beta_{c,1,j}n_1n_j$$

where $n_2, n_3, \cdots, n_j$ are the number concentrations of the secondary, tertiary, . . . , jth-order particles, respectively, and $\beta_{c,i,j}$ is a particle collision frequency function which depends on particle diameter.

4. If β_c is taken as a constant, β_C, Equation 8-155 is obtained:

$$\frac{dn_1}{dt} = -\beta_C n_1 n_T$$

where n_T is the total number concentration of particles in the system. This equation reveals that the number of primary particles always decreases as particles of other size are formed. Secondary particles are formed by collisions of primary particles and are decreased in number by collisions with other particles, as shown in Figure 8-19; therefore, the rate of change of secondary particles is

$$\frac{dn_2}{dt} = \beta_C n_1 n_1 - \beta_C n_2 n_1 - \beta_C n_2 n_2 - \cdots - \beta_C n_2 n_j$$

5. Similarly for tertiary and higher-order particles:

$$\frac{dn_3}{dt} = \beta_C n_1 n_2 - \beta_C n_3 n_1 - \beta_C n_3 n_2 - \cdots - \beta_C n_3 n_j$$

$$\frac{dn_4}{dt} = \beta_C n_1 n_3 + \beta_C n_2 n_2 - \beta_C n_4 n_1 - \beta_C n_4 n_2 - \cdots - \beta_C n_4 n_j$$

6. The discrete settling of particles can be modeled as a first-order process; the rate coefficient is equal to a particle settling velocity (Lt^{-1}) divided by an average settling distance. A reasonable settling distance might be half the depth of the SBR, for example. The terminal settling velocity is given by Stokes' law. For particle Reynolds numbers below

0.3, the settling velocity is

$$v_t = \frac{g(\rho_p - \rho)d_p^2}{18\mu_v}$$

where g is the acceleration due to gravity, ρ_p is the particle density, ρ is the fluid density, d_p is the particle diameter, and μ_v is the fluid viscosity. The first-order settling rate coefficient is obtained by dividing the settling velocity by half the SBR tank depth, H_D:

$$k_s = \frac{2g(\rho_p - \rho)d_p^2}{18\mu_v H_D}$$

7. The material balance for the SBR during the filling period is given by Equation 9-6:

$$(V_R(0) + Qt)\frac{dn_k}{dt} + Qn_k = Qn_{k,\mathrm{IN}} + (V_R(0) + Qt)r$$

where the subscript k refers to the kth size particle and r is a reaction rate which here involves a particle collision rate and a particle settling rate. For the purpose of illustration, we make a simplifying assumption that only tertiary particles form in the system under consideration.

8. By substituting the appropriate rate expressions into the SBR material balance equation, we obtain a system of ODEs which can be solved for particle concentrations as a function of time:

$$(V_R(0) + Qt)\frac{dn_1}{dt} + Qn_1 - (V_R(0) + Qt)$$

$$\cdot\left(\left.\frac{dn_1}{dt}\right|_{\mathrm{reaction}} - k_{s,1}n_1\right) = Qn_{1,\mathrm{IN}}$$

$$(V_R(0) + Qt)\frac{dn_2}{dt} + Qn_2 - (V_R(0) + Qt)$$

$$\cdot\left(\left.\frac{dn_2}{dt}\right|_{\mathrm{reaction}} - k_{s,2}n_2\right) = Qn_{2,\mathrm{IN}}$$

$$(V_R(0) + Qt)\frac{dn_3}{dt} + Qn_3 - (V_R(0) + Qt)$$

$$\cdot\left(\left.\frac{dn_3}{dt}\right|_{\mathrm{reaction}} - k_{s,3}n_3\right). = Qn_{3,\mathrm{IN}}$$

9. The system of equations above can be solved numerically to yield particle concentrations, particle size distributions, and settling velocity

distributions as functions of time. The problem can be simplified if we make the assumption that the filling time is short and that no appreciable particle aggregation occurs during filling; in this case the SBR can be treated as a CMBR. We can also assume that the SBR is operated in such a way that a mixing period, during which particle aggregation occurs exclusively, is followed by a quiescent period, during which settling takes place.

10. The following system of equations is then solved to determine particle concentrations as functions of time during the mixing period:

$$-\frac{dn_1}{dt} = \beta_C n_1 n_1 - \beta_C n_1 n_2 - \beta_C n_1 n_3$$

$$\frac{dn_2}{dt} = \beta_C n_1 n_1 - \beta_C n_2 n_1 - \beta_C n_2 n_2 - \beta_C n_2 n_3$$

$$\frac{dn_3}{dt} = \beta_C n_1 n_2 - \beta_C n_3 n_1 - \beta_C n_3 n_2 - \beta_C n_3 n_3$$

11. The solution to this system of equations provides the initial condition for the settling period. Particle concentrations as functions of time during settling can be determined by solving the following first-order equations:

$$\frac{dn_1}{dt} = -k_{s,1} n_1$$

$$\frac{dn_2}{dt} = -k_{s,2} n_2$$

$$\frac{dn_3}{dt} = -k_{s,3} n_3$$

- **Conclusions:** This example demonstrates that as restrictions and model assumptions are relaxed, a model more closely approximates real conditions, but the complexity of the model and the associated mathematical techniques required to solve the model equations increases. Each modeling situation should be evaluated on a case-by-case basis to identify that model which is as simple as possible and as accurate as necessary.

Example 12-3

- **Situation:** Cyanide, a common component of wastes from the steel industry and many other industrial operations, can be oxidized by chlorine under alkaline conditions (see Example 5-8). Cyanide can also be oxidized by molecular oxygen (note: demonstrate as an exercise in the application of free-energy concepts that the cyanide reaction with O_2 is thermodynamically feasible) if an appropriate catalyst is present. Specifically, activated carbon, which adsorbs both cyanide and dissolved oxygen, has been shown to function as a catalyst in this oxidation reaction, the process being further facilitated by complexation of the cyanide with copper(II) to enhance its adsorption (Weber, W. J., Jr. and O. Corapcioglu, *Proc. 36th Purdue Industrial Waste Conference*, Sec. 11, pp. 500–508, 1982).

- **Question(s):** Develop an appropriate process model for a cyanide treatment operation in which the waste stream is dosed with oxygen, copper, and powdered activated carbon, then fed continuously to a CMFR in which the activated carbon is retained (i.e., does not exit with the effluent flow).

- **Logic and Answer(s):**

1. Activated carbon is a microporous adsorbent. The rates of processes in which it is used are generally controlled by either intraparticle pore or surface diffusion (e.g., see Section 8.6.3 and 11.4.5)

2. The general CMFR material balance equation is given by

$$V_R \frac{dC}{dt} = QC_{\mathrm{IN}} - QC_{\mathrm{OUT}} - \eta_{R,O} r^{\circ}_{\mathrm{int}} (a^{\circ}_s)_{I,R} V_R$$

where $\eta_{R,O}$ is the overall effectiveness factor, r°_{int} the intrinsic surface reaction rate (on a surface-area basis), and $(a^{\circ}_s)_{I,R}$ the surface area (on a reactor-volume basis).

3. Because the carbon is retained in the CMFR and the internal surface area available for reaction increases continuously, a *nonsteady* condition exists. The situation is thus appropriately described by *Model 29 in Appendix A.12a*, a model that was not specifically developed in the text. Model 27 in Table 12-3 describes a *steady-state* situation in which the solution and the catalyst (in this case the catalyst mediator) are fed and removed simultaneously; this model, therefore, can be used as a starting point.

4. For the steady-state condition of Model 27, in which the waste stream and the activated carbon are fed and removed continuously, $(a^{\circ}_s)_{I,R}$, is constant, and given by the product of the instantaneous carbon dose, D_o, and the internal surface area per unit weight of carbon, $(a^{\circ}_{s,w})_I$:

$$(a^{\circ}_s)_{I,R} = (a^{\circ}_{s,w})_I D_o$$

where

$$D_o = \frac{\text{cumulative weight of carbon added}}{\text{cumulative volume of waste stream}} = \frac{W_g(t)}{V_L(t)}$$

5. Again for the steady-state condition assumed in Model 27, we have for $r^{\circ}_{\text{int}} = k^{\circ} C_{\text{OUT}}$

$$0 = QC_{\text{IN}} - QC_{\text{OUT}} - \eta_{R,o} k^{\circ} C_{\text{OUT}} (a^{\circ}_{s,w})_I D_o V_R$$

or

$$C_{\text{OUT}} [Q + \eta_{R,o} k (a^{\circ}_{s,w})_I D_o V_R] = QC_{\text{IN}}$$

$$C_{\text{OUT}} = \frac{C_{\text{IN}}}{1 + \eta_{R,o} k^{\circ} (a^{\circ}_{s,w})_I D_o \bar{t}}$$

which, with minor rearrangement and replacement of terms, can be shown to be essentially equivalent to the steady-state design equation given in Table 10-1 for the case of general intraparticle mass transfer rate control.

6. For the unsteady condition that accompanies Model 29, the concentration of the activated carbon catalyst increases with time as a function of the cumulative carbon dose, $D_o(t)$, such that

$$D_o(t) = \frac{W_g(t)}{V_R}$$

where V_R is the reactor volume. The dosage of carbon is assumed small enough relative to the waste flow that the reactor volume remains essentially constant over the period of carbon accumulation.

7. The number of active sites or effective surface area is now a function of time:

$$(a^{\circ}_s)_{I,R}(t) = (a^{\circ}_{s,w})_I \, D_o(t) = (a^{\circ}_{s,w})_I \frac{W_g(t)}{V_R}$$

8. The material balance equation for the cyanide can then be written as

$$V_R \frac{dC_{\text{OUT}}}{dt} = QC_{\text{IN}} - QC_{\text{OUT}} - \eta_{R,o} k C_{\text{OUT}} (a^{\circ}_{s,w})_I D_o(t) V_R$$

$$= QC_{\text{IN}} - C_{\text{OUT}} \left[Q - \eta_{R,o} k^{\circ} (a^{\circ}_{s,w})_I \frac{W_g(t)}{V_R} V_R \right]$$

9. Because $W_g(t) = D_o V_L(t) = t D_o Q$, dividing through by V_R yields

$$\frac{dC_{\text{OUT}}}{dt} = \frac{C_{\text{IN}}}{\bar{t}} - \frac{C_{\text{OUT}}}{\bar{t}} [1 - \eta_{R,o} k^{\circ} (a^{\circ}_{s,w})_I (D_o)(t)]$$

$$= \frac{1}{\bar{t}} \{C_{\text{IN}} - C_{\text{OUT}} [1 - \eta_{R,o} k^{\circ} (a^{\circ}_{s,w})_I (t)]\}$$

- **Conclusion:** The model developed above for the nonsteady-state catalysis of cyanide oxidation is essentially *Model 29 in Appendix A.12.1.* The differential equation comprising this model can be solved most readily by numerical integration.

Example 12-4

- **Situation:** Powdered activated carbon is fed cocurrently with a warm dye bath effluent to a CMFR for removal of color prior to reuse of the water in other dye bath operations. This recycle step conserves energy, reducing the amount of heat required to bring the dye bath to its required process temperature. The carbon is also recycled, after biological regeneration in an off-line CMFR. Mass transfer rates for use in the CMFR design model were found by experiment to be controlled by intraparticle diffusion. *Model 32 in Table 12-3* was thus developed and applied to the CMFR process design. After a few weeks of operation, however, the biological regeneration process was discovered to leave a film on the external surfaces of the carbon particles, creating an external impedance greater than that offered by the internal diffusion process.
- **Question(s):** Develop a modified process model to accommodate the change in impedance caused by the residual film on the activated carbon particles. This process model is required to determine process changes (e.g., HRTs, carbon dose, etc.) required to meet the original performance criteria. The model may also help determine whether a second-stage treatment in the off-line carbon regeneration step is needed to eliminate the residual film and its associated mass transfer impedance.
- **Logic and Answer(s):**

1. Two material balance equations are required; one for the liquid phase, and one for the solid phase.

2. The material balance equations for adsorption on dispersed non-porous solids are:

Solution phase: $$V_R \frac{dC_{\mathrm{OUT}}}{dt} = QC_{\mathrm{IN}} - QC_{\mathrm{OUT}} - \frac{\text{mass adsorbed}}{\text{time}}$$

Solid phase: $$\frac{dq}{dt} = \frac{\text{mass adsorbed}}{(\text{mass of adsorbent})(\text{time})}$$

The mass of adsorbent in the reactor is given by $D_o V_R$, where D_o = dose M/L^3 and V_R = reactor volume, or

$$D_o = \frac{\text{rate of adsorbent addition}}{Q}$$

3. These two material balance equations can be coupled to yield

$$V_R \frac{dC_{\text{OUT}}}{dt} = QC_{\text{IN}} - QC_{\text{OUT}} - D_o V_R \frac{dq}{dt}$$

4. For the given situation mass transfer impedance across an external film is rate controlling, we can therefore rewrite the equation above as

$$V_R \frac{dC_{\text{OUT}}}{dt} = QC_{\text{IN}} - QC_{\text{OUT}} - k_f (a_s^\circ)_E V_R (C - C_e)$$

where $(a_s^\circ)_E V_R$ is the external surface area available in the reactor, or the external surface area available per PAC particle times the number of PAC particles in the reactor and C_e is the concentration that should (based on an adsorption isotherm) exist in equilibrium with the actual amount adsorbed, q_e. Because the process is operated at steady state, this surface area is constant.

5. The steady-state number of particles in the reactor, $(n_p)_R$, is given by

$$(n_p)_R = \frac{\text{total mass of particles}}{\text{mass of a single particle}} = \frac{D_o V_R}{4\pi R_p^3 \rho_p / 3} = \frac{3 D_o V_R}{4\pi R_p^3 \rho_p}$$

6. The quantity $(a_s^\circ)_E V_R$ is then given by

$$(a_s^\circ)_E V_R = \frac{(4\pi R_p^2)(3 D_o V_R)}{4\pi R_p^3 \rho_p} = \frac{3 D_o V_R}{R_p \rho_p}$$

where $4\pi R_p^2$ is the surface area of one particle.
or

$$(a_s^\circ)_E = \frac{3 D_o}{\rho_p R_p}$$

7. The material balance relationship given in step 4 can then be rewritten:

$$V_R \frac{dC_{\text{OUT}}}{dt} = QC_{\text{IN}} - QC_{\text{OUT}} - \frac{3 k_f D_o V_R}{R_p \rho_p} (C_{\text{OUT}} - C_e)$$

8. The rate of mass transfer through the film must be equal to the rate of adsorption, thus:

$$\frac{3 k_f D_o V_R}{R_p \rho_p} (C_{\text{OUT}} - C_e) = D_o V_R \frac{dq}{dt}$$

9. The solution and solid-phase material balance relationships (the latter given in word form) in step 2 can now be expressed in terms of

more quantifiable parameters:

$$V_R \frac{dC_{\text{OUT}}}{dt} = QC_{\text{IN}} - QC_{\text{OUT}} - \frac{3k_f D_o V_R}{R_p \rho_p}(C_{\text{OUT}} - C_e)$$

and

$$\frac{dq}{dt} = \frac{3k_f}{R_p \rho_p}(C_{\text{OUT}} - C_e)$$

10. For steady-state conditions in the liquid phase,

$$0 = QC_{\text{IN}} - QC_{\text{OUT}} - \frac{3k_f D_o V_R}{R_p \rho_p}(C_{\text{OUT}} - C_e)$$

where C_{OUT} is constant.

11. Consider for simplicity of illustration that the sorption process can be described by a linear relationship,

$$q_e = \mathcal{K}_D C_e$$

or,

$$C_e = \frac{q_e}{\mathcal{K}_D}$$

12. We can now solve for $q(t)$. First we substitute $q_e/\mathcal{K}_D$ for C_e in the solid-phase mass balance developed in step 9 and rearrange to obtain

$$\int \frac{dq}{\dfrac{3k_f C_{\text{OUT}}}{R_p \rho_p} - \dfrac{3k_f q}{R_p \rho_p \mathcal{K}_D}} = \int dt$$

This equation can then be integrated to yield

$$-\frac{\mathcal{K}_D R_p \rho_p}{3k_f} \ln \left(\frac{3k_f C_{\text{OUT}}}{R_p \rho_p} - \frac{3k_f q}{R_p \rho_p \mathcal{K}_D} \right) = t + \kappa$$

13. The constant of integration, κ, can be obtained by solving the equation above for the initial conditions $t = 0$ and $q = 0$:

$$-\frac{\mathcal{K}_D R_p \rho_p}{3k_f} \ln \frac{3k_f C_{\text{OUT}}}{R_p \rho_p} = \kappa$$

14. Substitution of κ from step 13 into the equation in step 12 gives

$$\frac{\mathcal{K}_D R_p \rho_p}{3k_f} \ln \frac{\dfrac{3k_f C_{\text{OUT}}}{R_p \rho_p}}{\dfrac{3k_f C_{\text{OUT}}}{R_p \rho_p} - \dfrac{3k_f q}{R_p \rho_p \mathcal{K}_D}} = t$$

Rearrangement of this relationship yields

$$\frac{C_{\mathrm{OUT}}}{C_{\mathrm{OUT}} - \dfrac{q}{\mathcal{K}_D}} = \exp\,(3k_f t/\mathcal{K}_D R_p \rho_p)$$

Thus

$$C_{\mathrm{OUT}} - \frac{q}{\mathcal{K}_D} = C_{\mathrm{OUT}} \exp\,[-(3k_f t/\mathcal{K}_D R_p \rho_p)]$$

or

$$\frac{q}{\mathcal{K}_D} = C_{\mathrm{OUT}}(1 - \exp\,[-(3k_f t/\mathcal{K}_D R_p \rho_p)])$$

We see that these expressions are correct for the limiting time conditions; that is, when $t = 0$, $q = 0$ and $C_e = 0$, and when $t = \infty$, $q = q_e$ and $C_e = C_{\mathrm{OUT}}$.

15. For a CMFR with an HRT of $\bar{t}$:

$$q(\bar{t}) = \frac{1}{\bar{t}} \int_0^\infty q(t) e^{-t/\bar{t}}\, dt$$

(*Note:* See Equation 11-136, which is analogous.) Substituting the expression developed in step 14 for $q(t)$, gives

$$q(\bar{t}) = \frac{\mathcal{K}_D C_{\mathrm{OUT}}}{\bar{t}} \int (1 - e^{-\alpha t}) e^{-t/\bar{t}}\, dt$$

where

$$\alpha = \frac{3k_f}{\mathcal{K}_D R_p \rho_p}$$

$$q(\bar{t}) = \frac{\mathcal{K}_D C_{\mathrm{OUT}}}{\bar{t}} \int_0^\infty e^{-t/\bar{t}} - \frac{\mathcal{K}_D C_{\mathrm{OUT}}}{\bar{t}} \int_0^\infty e^{-(\alpha + 1/\bar{t})t}$$

16. From a standard table of integrals we can determine that

$$\int_0^\infty e^{-\alpha x}\, dx = \frac{1}{\alpha}$$

Thus

$$q(\bar{t}) = \frac{\bar{t}\mathcal{K}_D C_{\mathrm{OUT}}}{\bar{t}} - \frac{\mathcal{K}_D C_{\mathrm{OUT}}}{\bar{t}\left(\alpha + \dfrac{1}{\bar{t}}\right)} = \mathcal{K}_D C_{\mathrm{OUT}}\left(1 - \frac{1}{\alpha\bar{t} + 1}\right)$$

17. Finally, since the mass adsorbed per unit time in the CMFR is given by

$$q(\bar{t})D_oQ\left(\frac{\text{mass adsorbed}}{\text{mass adsorbent}}\ \frac{\text{mass adsorbent}}{\text{volume}}\ \frac{\text{volume}}{\text{time}}\right),$$

we can write the steady-state material balance relationship as

$$0 = QC_{\text{IN}} - QC_{\text{OUT}} - q(\bar{t})D_oQ$$

Substituting for $q(\bar{t})$ from step 16 and dividing through by Q gives

$$C_{\text{IN}} = C_{\text{OUT}} + \mathcal{K}_D C_{\text{OUT}}\left(1 - \frac{1}{\alpha\bar{t} + 1}\right)D_o$$

$$= C_{\text{OUT}}\left[1 + \mathcal{K}_D D_o\left(1 - \frac{1}{\alpha\bar{t} + 1}\right)\right]$$

Rearrangement of this equation and substitution of the expression given for α in step 15 yields

$$C_{\text{OUT}} = C_{\text{IN}}\left[1 + \mathcal{K}_D D_o\left(1 - \frac{1}{\dfrac{3k_f\bar{t}}{\mathcal{K}_D R_p \rho_p} + 1}\right)\right]^{-1}$$

- **Conclusions:** Several intuitive facts regarding the operation of this process are confirmed and quantified by the design equation arrived at above; namely:

  1. C_{OUT} can be lowered by increasing $\mathcal{K}_D$, D_o, k_f, and/or $\bar{t}$.
  2. C_{OUT} can be lowered by decreasing ρ_p and/or R_p.
  3. As $\bar{t} \to \infty$, $C_{\text{OUT}} \to C_e \to C_{\text{IN}}/(1 + \mathcal{K}_D D_o)$. Thus,

$$C_e + C_e\mathcal{K}_D D_o = C_{\text{IN}}$$

$$C_e + q_e D_o = C_{\text{IN}}$$

$$q_e = \frac{C_{\text{IN}} - C_e}{D_o}$$

  4. The model developed above is essentially *Model 31 in Table 12-3*.

Example 12-5

- **Situation:** As discussed in Example 12-4, the rate of adsorption of color in the dye bath effluent is decreased due to the increased external mass transfer impedance caused by a residual biofilm on the external surfaces of the adsorbent. You discover, to your dismay, that biological regeneration is required by the process patent, and that you cannot change other parameters enough to meet process objectives. You are subsequently charged by your group chief with developing more effective reactor designs. The premise is that replacement of the existing CMFR is cost-prohibitive.

- **Question(s):** How can you still use the existing CMFR and maintain effective treatment?
- **Logic and Answer(s):**

1. One possible solution is to baffle the existing system to more closely approaching plug flow behavior. This would increase color removal for a fixed HRT.

2. A model is needed to prove that baffling will lead to improved performance.

3. You note that Appendix A.12.1 identifies two models that are potentially applicable: *Model 60* and its PFDR equivalent, *Model 81*, in Appendix A.12.1. These models, however, are not developed in the preceding chapters.

4. PFR Model 60 and its equivalent PFDR model, both in Appendix A.12.1, can be developed from Model 31, which itself was just developed in Example 12-4. To derive them, start with development of the PFR design model for a fixed-bed adsorber. The first step is to write the appropriate material balance equation(s). Two phases are involved (the liquid waste and the adsorbent), so there must be two material balance equations. Describing the sorption reaction in word form first, we can write the liquid-phase material balance relationship as

$$\frac{dC}{dt} = -v\frac{dC}{dx} - \left\{\begin{matrix}\text{mass of color bodies adsorbed per unit} \\ \text{volume of liquid treated per unit time}\end{matrix}\right\}$$

The word-form right-hand term of this material balance relationship will have to be developed in equation form. Two different but equal equation forms are possible: (a) as a function of C in the reaction term of the liquid-phase material balance relationship; and (b) as a function of q in the solid-phase material balance relationship.

5. To express the amount of color adsorbed, we first need a way to express the mass of adsorbent present in the reactor. We begin by incorporating the concept of porosity from fixed-bed reactor modeling, in which the volume of liquid per volume of reactor is given by the bed void ratio, ϵ_B

$$\frac{\text{volume of liquid}}{\text{total reactor volume}} = \epsilon_B$$

The volume of adsorbent particles per volume of reactor is similarly given by

$$\frac{\text{volume of carbon}}{\text{total reactor volume}} = 1 - \epsilon_B$$

Combining the two relationships above leads to

$$\frac{\text{volume of carbon}}{\text{volume of liquid}} = \frac{1 - \epsilon_B}{\epsilon_B}$$

Converting from volume of adsorbent particles to mass of adsorbent particles yields

$$\frac{\text{mass of carbon}}{\text{volume of liquid}} = \rho_p \frac{1 - \epsilon_B}{\epsilon_B}$$

6. From step 5 we can write

$$\frac{\text{amount of color sorbed}}{(\text{volume of liquid})(\text{time})} = \rho_p \frac{1 - \epsilon_B}{\epsilon_B} \frac{dq}{dt}$$

7. The "effective activated carbon dose" in a fixed-bed reactor is given as

$$D_o = \frac{\text{mass of carbon}}{\text{volume of liquid}}$$

Dosage can also be related to reactor porosity using the information in step 5, that is,

$$\frac{D_o}{\rho_p} = \left(\frac{\text{mass of carbon}}{\text{volume of liquid}}\right)\left(\frac{\text{volume of carbon}}{\text{mass of carbon}}\right)$$

and

$$\frac{D_o}{\rho_p} = \frac{\text{volume of carbon}}{\text{volume of liquid}} = \frac{1 - \epsilon_B}{\epsilon_B}$$

8. Returning the definition of dosage from step 7 to step 6 gives:

$$\rho \frac{1 - \epsilon_B}{\epsilon_B} \frac{dq}{dt} = D_o \frac{dq}{dt}$$

This equation provides a way to express the solid-phase material balance that will be useful in our final modeling and analysis of a baffled CMFR as a PFDR.

9. The reaction term in the liquid-phase material balance can be expressed as a function of the solution-phase concentration, C, which decreases with time and depth in the fixed-bed reactor due to transfer of solute mass to the adsorbent:

$$\frac{dC}{dt} = -v \frac{dC}{dx} - k_f(a_s^\circ)_{E,R}(C - C_e)$$

where $(a_s^\circ)_{E,R}$ is the specific external particle area per unit volume of liquid in the reactor.

10. We can now rearrange terms and develop a convenient form of the required equality between the right-hand mass transfer rate term in step 9 and the solid-phase material balance relationship developed in step 8:

$$\frac{\text{area of carbon}}{\text{volume of liquid}} = \left(\frac{\text{mass of carbon}}{\text{volume of liquid}}\right)\left(\frac{\text{volume of carbon}}{\text{mass of carbon}}\right)\left(\frac{\text{area of carbon}}{\text{volume of carbon}}\right)$$

$$= \frac{D_o}{\rho_p}\frac{4\pi R_p^2}{(4/3)\,\pi R_p^3} = \frac{3D_o}{R_p\rho_p}$$

The liquid-phase material balance is now

$$\frac{dC}{dt} = -v\frac{dC}{dx} - k_f\frac{3D_o}{R_p\rho_p}(C - C_e)$$

which can also be expressed as

$$\frac{dC}{dt} = -v\frac{dC}{dx} - D_o\frac{dq}{dt}$$

The following equality results:

$$D_o\frac{dq}{dt} = k_f\frac{3D_o}{R_p\rho_p}(C - C_e)$$

$$\frac{dq}{dt} = \frac{3k_f}{R_p\rho_p}(C - C_e)$$

11. Considering only steady-state conditions in the liquid phase, $C(x)$ is constant in time, which gives:

$$0 = -v\frac{dC}{dx} - \frac{3D_o k_f}{R_p\rho_p}(C - C_e)$$

$$\frac{dC}{dx} = -\frac{3D_o k_f}{vR_p\rho_p}(C - C_e)$$

$$\frac{dq}{dt} = \frac{dq}{dx}\frac{dx}{dt} = v\frac{dq}{dx} = \frac{3k_f}{R_p\rho_p}(C - C_e)$$

The above is a Lagrangian perspective with respect to carbon particle transport along the reactor length at a velocity v; i.e. the solid phase concentration increases along the parth of particle transport.

12. To complete development of *Model 61 in Appendix A.12.1* for the PFR system it remains only to solve two ordinary differential equations simultaneously:

$$\frac{dC}{dx} = -\frac{3D_o k_f}{vR_p\rho_p}(C - C_e)$$

$$\frac{dq}{dx} = \frac{3k_f}{vR_p\rho_p}(C - C_e)$$

The solution technique is straightforward, particularly for linear partitioning.

13. To extend the development to obtain a corresponding PFDR model, we now have to incorporate the dispersion term in the liquid-phase material balance relationship. Thus

$$\frac{dC}{dt} = -v\frac{\partial C}{\partial x} + \mathfrak{D}_d\frac{\partial^2 C}{\partial x^2} - D_o\frac{\partial q}{\partial t}$$

$$\frac{\partial C}{\partial t} = -v\frac{\partial C}{\partial x} + \mathfrak{D}_d\frac{\partial^2 C}{\partial x^2} - \frac{3D_o k_f}{R_p\rho_p}(C - C_e)$$

and

$$\frac{dq}{dx} = \frac{3k_f}{vR_p\rho_p}(C - C_e)$$

14. As determined for the PFR system, all that remains for completion of *Model 81 in Appendix A.12.1* for the PFDR system is the simultaneous solution of two ordinary differential equations; in this case,

$$\frac{d^2C}{dx^2} = \frac{v}{\mathfrak{D}_d}\frac{dC}{dx} + \frac{3D_o k_f(C - C_e)}{R_p\rho_p\mathfrak{D}_d}$$

$$\frac{dq}{dx} = \frac{3k_f}{vR_p\rho_p}(C - C_e)$$

GENERAL APPENDICES

APPENDIX I PARAMETER NOMENCLATURE AND DIMENSIONS

I-A English Alphabet

a = chemical activity
subscript:
i = component i

a_s° = specific surface area, per unit volume basis (L^{-1})
subscripts:
B = biofilm
E = external
I = internal
L = per liquid volume
N = NAPL (nonaqueous phase liquid)
P = per particle or bubble volume
R = per reactor volume
T = total
W = wetted

$a_{s,w}^\circ$ = specific surface area, per unit weight basis (L^2M^{-1})
subscripts: see specific area, per unit volume basis, a_s°

atm = atmospheres, a unit of pressure, stress, or force per unit area ($ML^{-1}t^{-2}$)

A = ampere, a unit of electric current (Qt^{-1})

A = area (L^2)
subscripts:
E = effective
C = under curve
N = normal to flux
P = particle
R = reactor

A_f = Arrhenius collision frequency factor ($L^3M^{-1}t^{-1}$)

A° = surface or interstitial area (L^2)
subscripts:
B = of bubble(s)
F = of film
N = normal to flux
P = of particle(s)

b = coefficient, constant, general variable (arbitrary dimensions)

$\mathcal{b}$ = Langmuir energy-related isotherm parameter (L^3M^{-1} or L^3 mol^{-1})

$\mathcal{b}_g$ = generalized Langmuir energy-related isotherm parameter (L^3M^{-1} or L^3 mol^{-1})

$\mathcal{b}^\circ$ = de-dimensionalized Langmuir energy constant $(\mathcal{b}/C_e)^{-1}$

Btu = British thermal unit, a unit of power (ML^2t^{-3})

$\mathcal{B}$ = BET energy-related isotherm parameter (dimensionless)

c = coefficient, constant, general variable (arbitrary dimensions)
subscripts:
a, h = adsorption enthalpy related

a, s = adsorption entropy related
d = diffusion reaction parameter
p = partitioning

c° = concentration ratio (dimensionless)
subscripts: see concentration, C

C = coulomb, a unit of charge, quantity of electricity, and dielectric flux (Q)

°C = temperature in degrees Celsius or centigrade

C = concentration, mass (ML^{-3}), molar (mol L^{-3}), or number (L^{-3})
subscripts:

a = aqueous phase
b = bulk phase
e = equilibrium
H = heat ($ML^{-1}t^{-2}$)
δ = at distance δ from a specified point of reference
g = gas phase
i = solute i
IN = influent
I = within or at an interface
l = liquid phase
L = at distance $x = L$ from $x = 0$
m = in nonaqueous matrix or domain
n = in nth reactor
N = number concentration
o = octanol phase
OUT = effluent
p = particle
ps = preformed or precipitated solids
r = at a radial distance r from $r = 0$
R = concentration reacted or in a reactor, as specified
s = sorbed or solid
sr = mass-ratio solubility (dimensionless, c° only)
ss = suspended solids
S = saturation (equilibrium) or solubility limited
SL = shock load
SS = steady-state
T = total
v = vapor phase
x = at a distance x from $x = 0$
y = at a distance y from $y = 0$

0 = initial (spatially or temporally)
δ = film or boundary layer
Δ = hypothetical initial value from pulse input M_T/V_R

superscripts:

M = mixed-solute system
S = single-solute system

$C(t)$ = concentration as a function of time (ML^{-3}, mol L^{-3} or, L^{-3})
subscripts: see concentration, C

C° = concentration adjacent to a surface (ML^{-3})
subscripts: see concentration, C

C_D = drag coefficient (dimensionless)

$C(t)$ = C curve = C_{OUT}/C_Δ as a function of time for a *pulse* or *delta* input of nonreactive tracer to a reactor of arbitrary flow characteristics

d = diameter (L)
subscripts:

b = bubble
c = cell within floc
d = droplet
f = floc
i = component i
imp = impeller
mo = molecular
p = particle
P = pipe
R = reactor
S = stirrer

dyn = dyne, a unit of force (MLt^{-2})

D_o = dosage (ML^{-3})

$D_o(t)$ = cumulative dosage to time t (ML^{-3})

$\mathfrak{D}$ = diffusion or dispersion coefficient (L^2t^{-1})
subscripts:

a = apparent diffusion
ax = axial dispersion
b = binary bulk diffusion
d = mechanical or hydrodynamic dispersion
e = effective diffusion
ϵ = eddy dispersion
g = gas-phase free molecular diffusion
k = Knudsen diffusion
l = liquid-phase free molecular diffusion

m = diffusion in a nonaqueous matrix or domain
p = pore diffusion
s = surface diffusion

$\mathfrak{D}^\circ$ = diffusivity to a surface or into an interface (L^2t^{-1})
e = base of Napierian logarithms = 2.71828 . . .
e° = elementary electric charge, 1.602×10^{-19}C
e_m = energy per unit mass (L^2t^{-2})
e_r = probable error (arbitrary dimensions)
erg = a unit of energy, work, or heat (ML^2t^{-2})
E = energy (ML^2t^{-2})

subscripts:

a = activation energy
A = attractive potential
c = compact-layer potential
DL = $E_{R,O}/e$ for low values of $E_{R,O}$
n = net sorption energy
R = repulsive potential
S = reference state for sorption energy (lowest physically realizable value)
Z = zeta potential
0 = characteristic sorbent site energy or surface potential

$\boldsymbol{E}(t)$ = $\boldsymbol{E}$ curve, an exit age or residence time (t) distribution function for a reactor of arbitrary flow characteristics = $(C_{OUT}/C_\Delta)A_c^{-1}$.
$\boldsymbol{E}(\theta)$ = dimensionless $\boldsymbol{E}$ curve, a dimensionless exit age or residence time (θ°) for a reactor of arbitrary flow characteristics = $(C_{OUT}/C_A)t_m A_c^{-1}$.
$E^\circ(\cdot)$ = expected value of $(\cdot)$
f = fugacity ($ML^{-1}t^{-2}$)

subscripts:

a = aqueous phase
b = biotic phase
c = colloidal phase
cs = crystalline solid state
l = liquid state
o = octanol phase
p = pure phase (liquid or solid)
s = sorbed phase
v = vapor state

superscripts:

M = mixed-solute system
S = single-solute system

f° = standard fugacity ($ML^{-1}t^{-2}$) subscripts: see fugacity, f

F = force (MLt^{-2})

subscripts:

C = coulombic (Q^2L^{-2})

mo, x, i = diffusive transport force per molecule of i in x direction

$F(t)$ = **F** curve = C_{OUT}/C_{IN} as a function of time (t) for a *step* input of nonreactive tracer to a reactor of arbitrary flow characteristics.

$F(\theta)$ = dimensionless **F** curve = C_{OUT}/C_{IN} as a function of dimensionless time (θ°) for a *step* input of nonreactive tracer to a reactor of arbitrary flow characteristics.

Fd = faraday, a unit of charge, quantity of electricity, and dielectric force (Q)

$\mathfrak{F}$ = Faraday constant: given by the product of Avogadro's number ($\mathfrak{N}_{Av} = 6.022 \times 10^{23}\ mol^{-1}$) and the elementary charge ($e^\circ = 1.602 \times 10^{-19}$ coulombs); i.e., 96,485 C/mol; 96,485 Joules (23,060 calories) per volt-equivalent.

$\mathfrak{F}_C$ = correction factor (dimensionless)

$\mathfrak{F}_E$ = enhancement factor (dimensionless)

$\mathfrak{F}_F$ = fanning friction factor (dimensionless)

$\mathfrak{F}_P$ = power factor (dimensionless)

$\mathfrak{F}_R$ = retardation factor (dimensionless)

g = gravitational acceleration constant (Lt^{-2}) 32.2 ft/s^2; 760 cm/s^2; 6.67 $\times 10^{-8}$ (dyn-cm)^2/g^2

g_s° = specific gravity (dimensionless)

subscript:

l = liquid

G = free energy (ML^2t^{-2})

subscripts:

a = adsorption

f = formation

l = liquid phase

r = reaction

v = vapor phase

G° = standard-state free energy (ML^2t^{-2})

subscripts: see free energy, G

$\overline{G}$ = mean value of "shear" velocity gradient, dv/dz (t^{-1})

h = hydraulic head (L)

subscripts:

l = head loss

se = specific energy head

hp = horsepower, a unit of power (ML^2t^{-3})
H = enthalpy (ML^2t^{-2})
subscripts:
- a = adsorption
- f = formation
- l = liquid phase
- r = reaction
- v = vapor phase

H_D = height or depth (L)
subscript:
- c = critical depth

H° = standard-state enthalpy (ML^2t^{-2})
subscripts: see enthalpy, H

HTU = height of a (mass) transfer unit (L)

I_u = uniformity index (dimensionless)

I° = impedance; resistance(s) of a system to the flow or movement of a particular component subjected to an applied force or potential (MLt^{-2})
subscripts:
- f = frictional
- g = gas phase
- l = liquid phase

J = joule, a unit of energy, heat, or work (ML^2t^{-2})

J_i = diffusive flux of component i; flux of component i relative to the flux of a mixture of which it is a part ($ML^{-2}t^{-1}$)

$\bar{J}_i$ = average diffusive flux of component i ($ML^{-2}t^{-1}$)

$\mathfrak{J}_D$ = Colburn mass transfer factor (dimensionless)

$\mathfrak{J}_L$ = Leverett function (dimensionless)

k = reaction or mass transfer rate coefficient; if not specifically subscripted, k denotes a reaction rate coefficient; its dimensions are dependent on the corresponding reaction order
subscripts:
- a = adsorption
- d = desorption
- e = effective
- f = mass transfer (Lt^{-1})
- f, s = internal mass transfer (Lt^{-1})
- $(\cdot)_f$ = forward reaction rate coefficient
- h = heat transfer ($MT^{-1}t^{-3}$)
- obs = observed

ps = pseudo-order rate coefficient
$(\cdot)_r$ = reverse reaction rate coefficient
s = settling rate coefficient (t^{-1})
v = volumetric mass transfer (t^{-1})

k_c = thermal conductivity ($MLT^{-1}t^{-3}$)

$\hat{k}$ = overall (lumped) reaction rate coefficient (arbitrary dimensions)

$\hat{k}_f$ = overall two-film mass transfer coefficient (Lt^{-1})
subscripts:
g = relative to the gas-side concentration of a component
l = relative to the liquid-side concentration of a component
m = for non-aqueous, non-gaseous medium (e.g., NAPL)

$\hat{k}_v$ = overall volumetric mass transfer coefficient (t^{-1})
subscripts: see overall (two-film) mass transfer coefficient, $\hat{k}_f$

k° = first-order surface reaction rate coefficient (Lt^{-1})
subscript:
obs = observed
ps = pseudo first order

K = degrees Kelvin (a unit of absolute temperature)

K = equilibrium constant (arbitrary dimensions)
subscripts:
a = acidity or protolysis
f = formation
superscripts:
p = parent
s = substituted

$\mathcal{K}$ = equilibrium-related constant (arbitrary dimensions)
subscripts:
B = biota or bioconcentration
D = distribution
F = Freundlich
H = Henry's (for dimensions, see Table 6-1)
i, j = generic i–j phase partition
M = Michaelis
o, w = octanol–water partition
P = partition
s = half-saturation
S = solubility

$\mathcal{K}^\circ_{o,w}$ = modified octanol–water partition coefficient (dimensionless)

l_m = Prandtl mixing length (L)
lbl = poundal, a unit of force (MLt^{-2})
L = a length dimension (L)
L = length (L)
subscripts:
c = characteristic
e = entry
E = expanded bed
F = fluidized bed
mfp = mean free path
P = packed bed
R = reactor
x = distance from $x = 0$ along x coordinate
m = general variable (arbitrary dimensions); integer number (dimensionless)
m_v = molecular mobility, molecular velocity of solute per unit field force (Mt^{-1}/molecule)
subscript:
i = component i
$\mathfrak{m}$ = generalized Langmuir isotherm heterogeneity parameter (dimensionless)
M = a mass dimension
M = mass or number of moles of a substance (M or mol)
subscripts:
i = in phase i
j = in phase j
ss = suspended solids
T = total
M_s = specific mass accumulated, deposited or released on a mass per unit area basis (ML^{-2})
M_W = mass wastage rate (MT^{-1})
$M(\mathrm{eq})$ = mass accumulated at equilibrium (M)
$M(t)$ = mass accumulated in time t (M)
$M(0)$ = initial mass at $t = 0(M)$
$M_s(t)$ = specific mass accumulated or released in time t (ML^{-2})
$M(\tau^\circ)$ = mass accumulated in dimensionless time τ°
$M^\circ(\mathrm{eq})$ = de-dimensionalized mass accumulated at equilibrium (dimensionless)
$M^\circ(t)$ = $M(t)/M(\mathrm{eq})$, ratio of mass accumulated in time t to ultimate accumulated mass at equilibrium (dimensionless)

$M^\circ(\tau^\circ)$ = $M(\tau^\circ)/M(\mathrm{eq})$ ratio of mass accumulated in dimensionless time τ° to accumulated mass at equilibrium (dimensionless)

$M_s^\circ(\tau^\circ)$ = de-dimensionalized specific mass accumulated or released in dimensionless time τ° (dimensionless)

$\overline{M}^\circ(t)$ = average fractional uptake based on age distribution of particles in slurry reactors (dimensionless)

n = general variable (arbitrary dimensions); integer number (dimensionless)
subscripts:
- e = number of electrochemical equivalents per mole (mol^{-1})
- i, j, k = number concentrations (L^{-3})
- p = number of particles (dimensionless)
- T = total number concentration (L^{-3})

n° = number of moles per unit surface area ($\mathrm{mol\ L}^{-2}$)
subscript:
- i = component i

superscripts:
- M = mixed-solute system
- S = single-solute system

$\boldsymbol{n}$ = normal vector (dimensions associated with vector quantity involved)

$\mathcal{n}$ = Freundlich isotherm parameter and generalized Langmuir isotherm heterogeneity parameter (dimensionless)

N = newton, a unit of force (MLt^{-2})

N = flux ($\mathrm{ML^2t^{-1}}$ for mass flux; Mt^{-3} for heat or energy flux)
subscripts:
- g = gas
- H = heat
- i = component i
- j = component j
- l = liquid
- mo = molar ($\mathrm{mol\ L^{-2}t^{-1}}$)
- p = pore (in a pore fluid)
- s = surface (along a surface or interface)
- T = total = $N_p + N_s$

$N(t)$ = flux as a function of time ($\mathrm{ML^{-2}t^{-1}}$ for mass flux)

$\overline{N}$ = average flux ($\mathrm{ML^{-2}t^{-1}}$ for mass flux)

N° = flux *to or into* a surface or interface ($\mathrm{ML^{-2}t^{-1}}$ for mass flux)

N = flux vector ($ML^{-2}t^{-1}$ for mass flux)
NTU = number of transfer units (dimensionless)
$\mathfrak{N}_{Ar}$ = Archimedes number (dimensionless)
$\mathfrak{N}_{Av}$ = Avogadro's number (6.022×10^{23} mol^{-1})
$\mathfrak{N}_{Bd}$ = Bond number (dimensionless)
$\mathfrak{N}_{Bi,H}$ = Biot heat transfer number (dimensionless)
$\mathfrak{N}_{Bi,M}$ = Biot mass transfer number (dimensionless)
$\mathfrak{N}_{Bo}$ = Bodenstein number (dimensionless)
$\mathfrak{N}_{Bq}$ = Boussinesq number (dimensionless)
$\mathfrak{N}_{Ca}$ = capillary number (dimensionless)
$\mathfrak{N}_{d}$ = dispersion number (dimensionless)
$\mathfrak{N}_{Da,I}$ = Damkohler number I (dimensionless)
$\mathfrak{N}_{Da,II}$ = Damkohler number II (dimensionless)
$\mathfrak{N}_{Eu}$ = Euler number (dimensionless)
$\mathfrak{N}_{Fl}$ = fluidization number (dimensionless)
$\mathfrak{N}_{Fr}$ = Froude number (dimensionless)
$\mathfrak{N}_{Ga}$ = Galileo number (dimensionless)
$\mathfrak{N}_{Gr}$ = Grashof number (modified) (dimensionless)
$\mathfrak{N}_{Ha}$ = Hatta number (dimensionless)
$\mathfrak{N}_{Kn,I}$ = Knudsen number I (dimensionless)
$\mathfrak{N}_{Kn,II}$ = Knudsen number II (dimensionless)
$\mathfrak{N}_{Ne}$ = Newton number (dimensionless)
$\mathfrak{N}_{Nh}$ = Nusselt heat number (dimensionless)
$\mathfrak{N}_{Nm}$ = Nusselt mass number (dimensionless)
$\mathfrak{N}_{Pe}$ = Péclet number (dimensionless)
$\mathfrak{N}_{Pr}$ = Prandtl number (dimensionless)
$\mathfrak{N}_{Re}$ = Reynolds number (dimensionless)
$\mathfrak{N}_{Re,R}$ = Reynolds rotational number (dimensionless)
$\mathfrak{N}_{Re,E}$ = Reynolds entry number (dimensionless)
$\mathfrak{N}_{Sc}$ = Schmidt number (dimensionless)
$\mathfrak{N}_{Sh}$ = Sherwood number (dimensionless)
$\mathfrak{N}_{St}$ = Stanton number (dimensionless)
$\mathfrak{N}_{Tm}$ = Thiele modulus (dimensionless)
$\mathfrak{N}_{We}$ = Weber number (dimensionless)
ohm = a unit of electrical resistance ($MQ^{-2}L^{2}t^{-1}$)
p = partial pressure (dimensionless; see definition of P_i)
subscript:
i = component i

p^v = partial vapor pressure (dimensionless; see definition of P_i^v)
subscripts:
- i = component i
- l = liquid
- s = solid

P = pressure ($ML^{-1}t^{-2}$)
subscripts:
- b = bulk phase
- C = capillary
- F = frictional (pressure drop)
- H = hydrostatic
- i = component i; commonly referred to also as the *partial* (contributing) pressure of component i in a gaseous mixture
- L = at $x = L$
- S = saturation (equilibrium)
- T = total
- 0 = at $x = 0$

P° = standard state pressure ($ML^{-1}t^{-2}$)
subscripts: see pressure, P

P^v = vapor pressure ($ML^{-1}t^{-2}$)
subscripts:
- i = component i; commonly referred to also as the *partial* (contributing) vapor pressure of component i in a solid or liquid mixture
- l = liquid
- s = solid

P_w = power input or consumption (ML^2t^{-3})

Pa = Pascal, a unit of pressure, stress, or force per unit area ($ML^{-1}t^{-2}$)

q = amount of solute sorbed per unit weight of sorbent (MM^{-1} or mol M^{-1})
subscripts:
- e = equilibrium
- i = component i

superscripts:
- M = mixed-solute system
- S = single-solute system

q° = dimensionless representation of the cumulative amount of material entering a diffusion domain (dimensionless)

Q = a measure of the quantity of electricity (Q)
Q = volumetric flow rate (L^3t^{-1})
F = in fractures or pores
g = gas
l = liquid
T = total
$Q(t)$ = volumetric flow rate as a function of time (L^3t^{-1})
subscripts: see volumetric flow rate, Q
Q_E = electric charge, quantity of electricity or dielectric flux (Q)
Q_H = heat (ML^2t^{-2} or mol L^2t^{-2}, per mole or unit mass)
subscripts:
m = melting (fusion)
v = vaporization
Q_M = mass (Mt^{-1}) or molar (mol t^{-1}) flow rate
subscripts: see volumetric flow rate, Q
Q_r = reaction quotient (arbitrary dimensions)
$Q°$ = capacity
subscripts:
a = Langmuir isotherm parameter (MM^{-1} or mol M^{-1})
f = fugacity ($L^{-2}t^2$)
g = generalized Langmuir isotherm parameter (MM^{-1} or mol M^{-1})
H = specific heat ($ML^2t^{-2}T^{-1}$ or mol $L^2t^{-2}T^{-1}$; per mole or unit mass)
H, V = volumetric heat ($ML^{-1}t^{-2}T^{-1}$)
r = characteristic radial distance (L) or direction (dimensionless)
subscripts:
c = cylindrical coordinates
f = radial position of sorption front
s = spherical coordinates
$\mathcal{r}$ = reaction rate (arbitrary dimensions; e.g., $ML^{-3}t^{-1}$, Mt^{-1}, mol $L^{-3}t^{-1}$, etc.)
subscripts:
a = adsorption
d = desorption
f = forward reaction
i = component i
int = intrinsic
mo = molar (mol $L^{-3}t^{-1}$)

mt = overall mass transfer rate
obs = observed
ox = oxidation
r = reverse reaction
v = volatilization

$\mathscr{r}(t)$ = reaction rate as a function of time (arbitrary dimensions; e.g., $ML^{-3}t^{-1}$, mol $L^{-3}t^{-1}$, etc.)
subscripts: see reaction rate, $\mathscr{r}$

$\mathscr{r}^\circ$ = surface reaction rate (arbitrary dimensions; e.g., $ML^{-2}t^{-1}$, mol $L^{-2}t^{-1}$, etc.)
subscripts: see reaction rate, $\mathscr{r}$

R = radius (L)
subscripts:
H = hydraulic
I = inner
O = outer
p = particle
P = pipe

R_L = length scaling factor (dimensionless)

R_Q = recycle ratio (dimensionless)

R_S = stripping factor (dimensionless)

R° = resistance (dimensions dependent upon those of force or potential applied; e.g., electrical resistance to applied voltage has the units of ohms and the dimensions of $MQ^{-2}L^2t^{-1}$)
subscript:
F = molecular friction factor

$\mathcal{R}$ = universal gas constant ($ML^2t^{-2}T^{-1}$) 8.314 J/mol K, 8.314 Pa-m^3/mol-K, 1.987 cal/mol K, 8.21×10^{-5} atm-m^3/mol-K

s_3 = the skewness of an empirical frequency distribution

S = entropy (ML^2t^{-2})
subscripts:
i = phase or component i
j = phase or component j
k = phase or component k
r = reaction

S° = standard-state entropy (ML^2t^{-2})
subscripts: see entropy, S

S_C = control surface (L^2)

t = a time dimension

t = time (t)

subscripts:

- b = bubble residence
- c = contact
- D = discrete value obtained from a distribution of residence times in a CMFR
- L = lag
- m = mean constituent residence
- p = residence per pass (recycle reactors)
- r = residence
- R = reaction stage (SBR)
- s = to saturation of a second phase
- S = solids residence
- SL = shock load period

$\bar{t}$ = average residence time, mean hydraulic residence time, HRT (t)
subscripts: see time, t

$\tilde{t}$ = residence times comprising an arbitrary residence time distribution (t)

$\tilde{t}_m$ = midpoint of the residence time of a fluid element in which an ongoing reaction has a time-variable rate coefficient (t)

torr = a unit of pressure, stress, or force per unit area ($ML^{-1}t^{-2}$)

T = temperature (T)
subscripts:

- F = fluid
- S = surface

$\mathfrak{J}_u$ = von Kármán's universal turbulence constant (dimensionless)

u = general variable (arbitrary dimensions)
subscripts:

- n = root of a trigonometric function
- 0 = initial value

U = internal energy (ML^2t^{-2})

v = velocity (Lt^{-1})
subscripts:

- a = aqueous phase
- b = bulk
- c = characteristic
- d = droplet
- f = fluidization
- i = component i
- j = component j
- k = component k

l = liquid
m = mixture
mo = molar
p = pore or interstitial
r = in radial direction r
s = superficial or "Darcy"
t = terminal settling or rise (e.g., particles or bubbles)
x = in direction x

$v(t)$ = velocity as a function of time (Lt^{-1})
subscripts: see velocity, v

$\tilde{v}$ = velocity fluctuations (Lt^{-1})
subscripts: see velocity, v

V = volt, a unit of electric potential ($MQ^{-1}L^2t^{-2}$)
subscripts: see velocity, v

V = volume (L^3)
subscripts:

a = aqueous phase
B = bubble
C = control
g = gas phase
H = head space
I = interstitial
IM = immobile phase
l = liquid phase
M = mobile phase
mo = molar (L^3mol^{-1})
p = pure phase
P = particle
R = reactor (generally, the volume of liquid in a reactor)
S = storage
T = total
v = vapor phase
w = water
V = pore or void volume

$V(t)$ = volume as a function of time (L^3)
subscripts: see volume, V

W = work (ML^2t^{-2})
subscripts:

F = flow
S = shaft

W = watt, a unit of power (ML^2t^{-3})
W_g = weight (MLt^{-2})
subscripts:
e = equivalent (M eq^{-1})
mo = molecular (M mol^{-1})
mo,i = component i (M mol^{-1})
mo,s = solute (M mol^{-1})
s = specific, (weight per unit volume) (ML^{-3})
x = general variable, characteristic distance (L) or directional coordinate (dimensionless)
X = mole fraction (dimensionless)
subscripts:
a = aqueous phase
g = gas phase
i = component i
l = liquid phase
L = at distance $x = L$ from $x = 0$
m = mass fraction
o = octanol phase
p = pure phase
r = reacted
s = sorbed or solid phase
S = at saturation or solubility limit
v = vapor phase
0 = at distance $x = 0$
superscripts:
M = mixed-solute system
S = single-solute system
$X(t)$ = mole fraction as a function of time (dimensionless)
subscripts: see mole fraction, X
y = general variable, characteristic distance (L) or directional coordinate (dimensionless)
Y = finite length in the y direction (L)
subscript:
W = width in y direction
z = general variable, characteristic distance (L) or directional coordinate (dimensionless)
$z°$ = electronic charge (Q)
Z = finite length in the z direction (L)
subscript:
B = bubble travel distance

T = tower height = HTU × NTU
W = width in z direction

I-B Greek Alphabet

α = coefficient, constant, general variable (arbitrary dimensions)
subscripts:
- a = aqueous-phase solute activity coefficient
- a, h = adsorption-enthalpy parameter
- a, s = adsorption-entropy parameter
- d = diffusion-reaction parameter
- D = $0.5\, \mathfrak{N}_d^{-1}$
- f = fugacity coefficient
- g = amplitude gain factor
- i = activity coefficient for component i
- j = Fourier series coefficient
- k = complex rate coefficient parameter
- m = amplitude of a cyclical variation
- o = octanol-phase solute activity coefficient
- p = pure solute-phase activity coefficient
- P = empirical parameter relating the partition constant of a solute to its octanol–water partition constant
- r = reaction-characteristic rate coefficient

β = coefficient, constant, general variable (arbitrary dimensions)
subscripts:
- A = $0.4343\, E_a/\mathcal{R}T_1T_2$
- a, h = adsorption-enthalpy parameter
- a, s = adsorption-entropy parameter
- c = variable particle collision frequency factor ($M^{-1}\, L^3\, t^{-1}$)
- C = constant particle collision frequency factor ($M^{-1}\, L^3\, t^{-1}$)
- d = diffusion reaction parameter
- D = $[1 + 4k\bar{t}\mathfrak{N}_d]^{0.5}$ for a first-order homogeneous reaction
- D = $[1 + 4k_f\, (a_s^{\circ})_{E,R}\, \bar{t}\epsilon_B\mathfrak{N}_d]^{0.5}$ for a heterogeneous reaction controlled by external impedance(s)
- $D = \left[1 + 4\left(\dfrac{k_f(a_s^{\circ})_{E,R}k^{\circ}}{k_f + k^{\circ}}\right)\bar{t}\epsilon_B\mathfrak{N}_d\right]^{0.5}$ for a heterogeneous reaction controlled by a combination of external impedance and surface reaction

e = characteristic Langmuir enthalpy constant (dimensionless)
j = Fourier series coefficient
P = empirical parameter relating the partition constant of a solute to its octanol–water partition constant
r = reaction-characteristic rate coefficient
R = initial condition parameter for a reversible reaction
S = stripping parameter relating mass transfer rate to gas flow rate (dimensionless)
$T = 10^{\beta_A}$
Z = change in the mole fraction of a gas in a bubble per unit travel distance z

γ = stoichiometric coefficient (dimensionless)
Γ = surface excess (mol L^{-2})
subscript:
i = component i

δ = depth, thickness, incremental distance (L)
subscripts:
L = laminar boundary layer
L, s = laminar boundary sublayer
m = medium or membrane
p = incremental particle movement
T = turbulent boundary layer
Δ = see Appendix I-C

ϵ = porosity or void ratio (dimensionless)
subscripts:
B = bed
d = domain
p = particle

ζ = arbitrary coordinate position
ζ° = length ratio or $x/2(\mathfrak{D}_a t)^{0.5}$ (dimensionless)
η = efficiency, effectiveness, efficiency factor (dimensionless)
subscripts:
c = particle collision efficiency factor
R, E = external reaction efficiency factor
R, O = overall (external-internal) reaction efficiency factor
R, I = internal reaction efficiency factor

θ° = ratio of specific time to average time $t/\bar{t}$ (dimensionless)
θ = angle (degrees or radians)
subscript:
c = contact

κ = arbitrary constant (arbitrary dimensions)

κ_B = Boltzmann constant ($ML^2t^{-2}K^{-1}$) (1.38×10^{-16} erg/molecule-K)

κ_D = dielectric constant (dimensionless)

κ_{DL} = reciprocal of diffuse double layer thickness, the distance from a surface at which the potential is $E_{R,DL} = E_{R,0}/e$

$\kappa°$ = correlation parameter

μ = chemical potential (ML^2t^{-2})

subscripts:

- i = component i
- l = liquid state
- s = sorbed or solid state
- v = vapor state

superscripts:

- M = mixed-solute system
- S = single-solute system

$\mu°$ = standard chemical potential (ML^2t^{-2})

subscripts and superscripts: see chemical potential, μ

μ_m = arithmetic mean value, first moment of the centroid of an empirical frequency distribution about its origin (arbitrary dimensions)

μ_v = dynamic viscosity ($ML^{-1}t^{-1}$)

subscripts:

- a = aqueous phase
- e = eddy viscosity
- l = liquid
- n = nonaqueous phase

ν_v = kinetmatic viscosity (L^2t^{-1})

subscripts: see dynamic viscosity, μ_v

ξ = permeability (L^2)

π = 3.14159 . . .

π = see Appendix I-C

Π_S = spreading pressure (ML^{-2})

$\Pi°$ = parameter group in the Buckingham Pi theory (dimensionless)

ρ = density (ML^{-3})

subscripts:

- a = aqueous phase
- b = biota (apparent)
- B = bulk (bed)
- bm = biomass (apparent)
- d = droplet or bubble
- g = gas

l = liquid
m = suspending medium
mo = molar
n = nonaqueous phase
p = particle or object (apparent)
s = solid phase (apparent)
w = water

σ = standard deviation, root-mean-square value of the deviations of a set of observations from the mean value of those observations, the square root of the second moment of the centroid of an empirical frequency distribution about its mean (arbitrary dimensions)

σ^2 = variance, the second moment of the centroid of an empirical frequency distribution about its mean (arbitrary dimensions)
subscripts:
C_{IN} = influent concentration
C_{OUT} = effluent concentration

σ_n = normal shear stress ($ML^{-1}t^{-2}$)

σ° = surface tension (Mt^{-2})
subscripts:
a = aqueous phase
n = nonaqueous phase
s = monolayer (saturated)
0 = initial (pure solvent)

Σ = see Appendix I-C

τ = tortuosity (dimensionless)

τ_s = shear stress ($ML^{-1}t^{-2}$)
subscript:
t = turbulent
superscript:
$^\circ$ = surface

τ° = dimensionless diffusion time $\mathfrak{D}_l t/L^2$ (dimensionless)

ϕ = angle (degrees or radians)

ϕ_s = phase shift (radians)

ϕ° = fraction, ratio, or coefficient (dimensionless)
subscripts:
d = diameter ratio
D = solute distribution parameter
f = fugacity coefficient
g = fractional gas hold-up
m = of monolayer adsorbed or surface covered
N = non-aqueous phase volumetric fraction

oc = organic carbon
r = reacted
R = remaining
S = solvent association parameter
V = volume ratio or fraction
vm = viscosity mobility
y = yield
z = gas-liquid mass transfer parameter

$\phi(\cdot)$ = function of $(\cdot)$
subscripts:
$B, 0$ = zero-order Bessel function
D = probability or frequency distribution function
G = Gaussian (normal) distribution function

$\Delta\Phi$ = electrical potential gradient ($MQ^{-1}L^2t^{-2}$)
ψ = system specific coefficient (arbitrary dimensions)
Ψ_m = $k_f/\mathfrak{D}$ (L^{-1})
Ψ_m° = $k_fL/\mathfrak{D}$ (dimensionless)
ω = frequency or angular velocity (t^{-1})
subscripts:
f = frequency of cyclical variation
r = rotation rate, angular velocity

Ω = ohm, a unit of electric resistance ($MQ^{-2}L^2t^{-1}$)

I-C Mathematical Notation

$\frac{dx}{dy}$ = ordinary derivative of x with respect to y

$\frac{\partial x}{\partial y}$ = partial derivative of x with respect to y

$\frac{D'x}{D'y}$ = substantial (total) derivative of x with respect to y

$\mathrm{erf}(\cdot)$ = error function of $(\cdot)$

$\mathrm{erfc}(\cdot)$ = error function complement of $(\cdot)$

Δ = finite increment of change in a variable (dimensions related to the variable)

$\bullet$ = dot product of vector quantities

$\Sigma_i^n\ (\cdot)$ = sum of quantities $i \ldots n$

$\Pi_i^n\ (\cdot)$ = product of quantities $i \ldots n$

∇ = del operator, three-dimensional gradient of a variable (dimensions associated with variable)

$Z_F(\cdot)$ = Fourier transform of $(\cdot)$

APPENDIX II CONVERSION FACTORS

Table II-A Length (L)

Multiply Number of → **by Factor** to Obtain ↓	**mm** (millimeters)	**cm** (centimeters)	**in** (inches)	**ft** (feet)	**yd** (yards)	**m** (meters)
mm	1	10^1	2.540×10^2	3.048×10^2	9.144×10^2	10^3
cm	10^{-1}	1	2.540	3.048×10^1	9.144×10^1	10^2
in	3.937×10^{-2}	3.937×10^{-1}	1	12	3.600×10^1	3.937×10^1
ft	3.281×10^{-3}	3.281×10^{-2}	8.333×10^{-2}	1	3	3.281
yd	1.094×10^{-3}	1.094×10^{-2}	2.778×10^{-2}	3.333×10^{-1}	1	1.094
m	10^{-3}	10^{-2}	2.540×10^{-2}	3.048×10^{-1}	9.144×10^{-1}	1

Note: The SI base unit of length is the *meter*, defined as 1,650,763.73 wavelengths of the orange-red line of the spectrum of krypton-86 in a vacuum.

Table II-B Area (L^2)

Multiply Number of → **by Factor** to Obtain ↓	**mm^2**	**cm^2**	**in^2**	**ft^2**	**yd^2**	**m^2**
mm^2	1	10^2	6.452×10^2	9.290×10^4	8.361×10^5	10^6
cm^2	10^{-2}	1	6.452	9.290×10^2	8.361×10^3	10^4
in^2	1.550×10^{-3}	1.550×10^{-1}	1	1.440×10^2	1.296×10^3	1.550×10^3
ft^2	1.076×10^{-5}	1.076×10^{-3}	6.944×10^{-3}	1	9	1.076×10^1
yd^2	1.196×10^{-6}	1.196×10^{-4}	7.716×10^{-4}	1.111×10^1	1	1.196
m^2	10^{-6}	10^{-4}	6.452×10^{-4}	9.290×10^{-2}	8.361×10^{-1}	1

Note: The SI derived unit of area is the *square meter*. *1 Acre* (U.S.) = $4047\ m^2 = 4840\ yd^2 = 43{,}560\ ft^2$

Table II-C Volume (L^3)

Multiply Number of → by Factor to Obtain ↓	cm^3	in^3	L (liters)	gal (gallons)	ft^3	yd^3	m^3
cm^3	1	1.639×10^1	10^3	3.785×10^3	2.832×10^4	7.646×10^5	10^6
in^3	6.102×10^{-2}	1	6.102×10^1	2.310×10^2	1.728×10^3	46.656×10^3	6.102×10^4
L	10^{-3}	1.639×10^{-2}	1	3.785	28.32	764.6	10^3
gal	2.642×10^{-4}	4.329×10^{-3}	2.642×10^{-1}	1	7.481	2.020×10^2	2.642×10^2
ft^3	3.531×10^{-5}	5.787×10^{-4}	3.531×10^{-2}	1.337×10^{-1}	1	2.700×10^1	3.531×10^1
yd^3	1.308×10^{-6}	2.143×10^{-5}	1.308×10^{-3}	4.951×10^{-3}	3.704×10^{-2}	1	1.308
m^3	10^{-6}	1.639×10^{-5}	10^{-3}	3.785×10^{-3}	2.832×10^{-2}	7.646×10^{-1}	1

Note: The SI derived unit of volume is the *cubic meter (m^3)*. *1 Acre* (U.S.)*-foot = 1234 m^3 = 43,560 ft^3 = 325,851 gal.*

Table II-D Linear Velocity (Lt^{-1})

Multiply Number of → by Factor to Obtain ↓	$cm \cdot s^{-1}$	$ft \cdot (min^{-1})$	$m \cdot (min^{-1})$	$km \cdot h^{-1}$	$ft \cdot s^{-1}$	$m \cdot s^{-1}$	$km \cdot (min^{-1})$
$cm \cdot s^{-1}$	1	5.080×10^{-1}	1.667	2.778×10^1	3.048×10^1	10^2	1.667×10^3
$ft \cdot (min^{-1})$	1.969	1	3.281	5.468×10^1	6.0×10^1	1.968×10^2	3.281×10^3
$m \cdot (min^{-1})$	0.6	3.048×10^{-1}	1	1.667×10^1	1.829×10^1	6×10^1	10^3
$km \cdot h^{-1}$	3.600×10^{-2}	1.829×10^{-2}	6×10^{-2}	1	1.097	3.6	6×10^1
$ft \cdot s^{-1}$	3.281×10^{-2}	1.667×10^{-2}	5.468×10^{-2}	9.113×10^{-1}	1	3.281	5.468×10^1
$m \cdot s^{-1}$	10^{-2}	5.080×10^{-3}	1.667×10^{-2}	2.778×10^{-1}	3.048×10^{-1}	1	1.667×10^{-1}
$km \cdot (min^{-1})$	6×10^{-4}	3.048×10^{-4}	10^{-3}	1.667×10^{-2}	1.829×10^{-2}	6×10^{-3}	1

Note: The SI derived unit for velocity is the *meter per second ($m \cdot s^{-1}$)*. The *second (s)* is the defined unit of time, the duration of 9,192,631,770 cycles of the radiation associated with a specified transition of the cesium ion. *1 mile/h = 1.609 km/h = 26.8 m/s.*

Table II-E Mass (M) and Weight

Multiply Number of → **by Factor** to **Obtain** ↓	**mg** (milligrams)	**gr** (grains)	**g** (grams)	**oz** (ounces)	**lb** (pounds)	**kg** (kilograms)	**tn** (short tons)
mg	1	6.481×10^1	10^3	2.835×10^4	4.536×10^5	10^6	9.072×10^8
gr	1.543×10^{-2}	1	1.543×10^1	437.5	7000	1.543×10^4	1.4×10^7
g	10^{-3}	6.481×10^{-2}	1	2.835×10^1	4.536×10^2	10^3	9.072×10^5
oz	3.527×10^{-5}	2.286×10^{-3}	3.527×10^{-2}	1	16	3.527×10^1	3.2×10^4
lb	2.205×10^{-6}	1.429×10^{-4}	2.205×10^{-3}	6.250×10^{-2}	1	2.205	2×10^3
kg	10^{-6}	6.481×10^{-5}	10^{-3}	2.835×10^{-2}	4.536×10^{-1}	1	9.072×10^{-2}
tn	1.102×10^{-9}	7.143×10^{-8}	1.102×10^{-6}	3.125×10^{-5}	0.500×10^{-3}	1.102×10^{-3}	1

Note: The SI derived unit for mass is the *kilogram*, defined as the mass of a specific cylinder of platinum-iridium alloy maintained in Paris. The *weight*, W_g, of an object is not the same as its *mass*. Weight is a measure of force given by the product of mass and the gravitational acceleration constant, g (i.e., $W_g = M \cdot g$). The conversion factors given above apply to the *gravitational* units of force having the corresponding names. The dimensions of these units when used as gravitational units of force are MLt^{-2}; see Table II-G. One metric *tn* = 2,204.6 *lb* = 10^3 *kg*.

Table II-F Density or Mass per Unit Volume (ML^{-3})

Multiply Number of → **by Factor** to **Obtain** ↓	**kg · m^{-3}**	**lb · ft^{-3}**	**g · cm^{-3}**	**lb · in^{-3}**
kg · m^{-3}	1	1.602×10^1	10^3	2.768×10^4
lb · ft^{-3}	6.243×10^{-2}	1	6.243×10^1	1.728×10^3
g · cm^{-3}	10^{-3}	1.602×10^{-2}	1	2.768×10^1
lb · in^{-3}	3.613×10^{-5}	5.787×10^{-4}	3.613×10^{-2}	1

Note: The SI derived unit for density is the *kilogram per cubic meter* ($kg \cdot m^{-3}$).

Table II-G Force (MLt^{-2} or F)

Multiply Number of → by Factor to Obtain ↓	dyn (dynes)	g	lbl (poundals)	N (newtons)	lb	kg	$J \cdot cm^{-1}$ (joules per centimeter)
dyn	1	9.807×10^{2}	1.383×10^{4}	10^{5}	4.448×10^{5}	9.807×10^{5}	10^{7}
g	1.020×10^{-3}	1	1.410×10^{1}	1.020×10^{2}	4.536×10^{2}	10^{3}	1.020×10^{4}
lbl	7.233×10^{-5}	7.093×10^{-2}	1	7.233	3.217×10^{1}	7.093×10^{1}	7.233×10^{2}
N	10^{-5}	9.807×10^{-3}	1.383×10^{-1}	1	4.448	9.807	10^{2}
lb	2.248×10^{-6}	2.205×10^{-3}	3.108×10^{-2}	2.248×10^{-1}	1	2.205	2.248×10^{1}
kg	1.020×10^{-6}	10^{-3}	1.410×10^{-2}	1.020×10^{-1}	4.536×10^{-1}	1	1.020×10^{1}
$J \cdot cm^{-1}$	10^{-7}	9.807×10^{-5}	1.383×10^{-3}	10^{-2}	4.448×10^{-2}	9.807×10^{-2}	1

Note: Conversion factors between absolute and gravitational force units apply only under standard acceleration due to Earth's gravitational field. The SI derived unit of force is the *newton* ($kg \cdot m \cdot s^{-2}$). Coulombic force can be expressed in terms of electrical quantities and units as $F_c = Q^2 m^{-2}$ and $N = 1.11 \times 10^{-10}$ $C^2 m^{-2}$.

Table II-H Pressure, Stress, or Force per Unit Area ($ML^{-1}t^{-2}$ or FL^{-2})

Multiply Number of → by Factor to Obtain ↓	$dyn \cdot cm^{-2}$	Pa (pascals)	$kg \cdot m^{-2}$	torr (mm Hg at 0°C)	in (H_2O) (inches of water at 4°C)	$lb \cdot in^{-2}$	atm (atmospheres)
$dyn \cdot cm^{-2}$	1	10^{1}	9.806×10^{1}	1.333×10^{3}	2.491×10^{3}	6.895×10^{4}	1.013×10^{6}
Pa	10^{-1}	1	9.806	1.333×10^{2}	2.491×10^{2}	6.895×10^{3}	1.013×10^{5}
$kg \cdot m^{-2}$	1.020×10^{-2}	1.020×10^{-1}	1	1.359×10^{1}	2.540×10^{1}	7.031×10^{2}	1.033×10^{4}
torr	7.501×10^{-4}	7.501×10^{-3}	7.357×10^{-2}	1	1.869	5.172×10^{1}	7.599×10^{2}
in (H_2O)	4.015×10^{-4}	4.015×10^{-3}	3.937×10^{-2}	5.351×10^{-1}	1	2.768×10^{1}	4.067×10^{2}
$lb \cdot in^{-2}$	1.450×10^{-5}	1.450×10^{-4}	1.422×10^{-3}	1.934×10^{-2}	3.612×10^{-2}	1	1.469×10^{1}
atm	9.869×10^{-7}	9.869×10^{-6}	9.681×10^{-5}	1.316×10^{-3}	2.459×10^{-3}	6.808×10^{-2}	1

Note: The SI derived unit of pressure, stress, or force per unit area is the *pascal* ($kg \cdot m^{-1} \cdot s^{-2}$). The atmosphere is used as the standard unit in this book because atmospheric pressure (and thus the pressures associated with most environmental systems) is defined in terms of the atmosphere (*1 atm* = 1.013×10^{5} *Pa*) and the term thus has a more familiar and convenient numerical value.

Table II-I Energy, Work, and Heat (ML^2t^{-2} or FL)

Multiply Number of → **by Factor** to **Obtain** ↓	**ergs** (dyne-centimeters)	**J** (joules or watt-seconds)	**ft-lb** (foot-pounds)	**Btu** (British thermal units)	**kcal** (kilocalories)	**hp-hr** (horsepower-hours)	**kWh** (kilowatt-hours)
ergs	1	10^7	1.356×10^7	1.055×10^{10}	4.184×10^{10}	2.684×10^{13}	3.6×10^{13}
J	10^{-7}	1	1.356	1.055×10^3	4.184×10^3	2.684×10^6	3.6×10^6
ft-lb	7.367×10^{-8}	7.376×10^{-1}	1	7.783×10^2	3.087×10^3	1.980×10^6	2.655×10^6
Btu	9.480×10^{-11}	9.480×10^{-4}	1.285×10^{-4}	1	3.969	2.545×10^3	3.413×10^3
kcal	2.389×10^{-11}	2.389×10^{-4}	3.239×10^{-4}	2.520×10^{-1}	1	6.413×10^2	8.600×10^2
hp-hr	3.722×10^{-14}	3.722×10^{-7}	5.050×10^{-7}	3.929×10^{-4}	1.559×10^{-3}	1	1.341
kWh	2.778×10^{-14}	2.778×10^{-7}	3.766×10^{-7}	2.930×10^{-4}	1.163×10^{-3}	7.457×10^{-1}	1

Note: The SI derived unit for energy, work, and heat is the *joule* ($kg \cdot m^2 \cdot s^{-2}$).

Table II-J Power or Rate of Doing Work (ML^2t^{-3} or FLt^{-1})

Multiply Number of → **by Factor** to **Obtain** ↓	**erg · s^{-1}**	**ft-lb · s^{-1}**	**Btu · (min^{-1})**	**kcal · (min^{-1})**	**hp** (horsepower)	**kW** (kilowatts)
erg · s^{-1}	1	1.356×10^7	1.758×10^8	6.977×10^8	7.457×10^9	10^{10}
ft-lb · s^{-1}	7.376×10^{-8}	1	1.297×10^1	5.144×10^1	5.500×10^2	7.376×10^2
Btu · (min^{-1})	5.689×10^{-9}	7.712×10^{-2}	1	3.969	4.241×10^1	5.689×10^1
kcal · (min^{-1})	1.433×10^{-9}	1.943×10^{-2}	2.520×10^{-1}	1	1.069×10^1	1.433×10^1
hp	1.341×10^{-10}	1.818×10^{-3}	2.357×10^{-2}	9.355×10^{-2}	1	1.341
kW	10^{-10}	1.356×10^{-3}	1.785×10^{-2}	6.977×10^{-2}	0.7457	1

Note: The SI derived unit for power or rate of doing work is the *watt* ($kg \cdot m^2 \cdot s^{-3} = J \cdot s^{-1}$).

Table II-K Charge, Quantity of Electricity, and Dielectric Flux (Q)

Multiply Number of → **by Factor** to **Obtain** ↓	**statcoulomb**	**C** (coulomb)	**abcoulomb**	**A-h** (ampere-hour)	**Fd** (faraday)
statcoulomb	1	2.998×10^{9}	2.998×10^{10}	1.080×10^{13}	2.893×10^{14}
C	3.335×10^{-10}	1	10^{1}	3.600×10^{3}	9.649×10^{4}
abcoulomb	3.335×10^{-11}	10^{-1}	1	3.600×10^{2}	9.649×10^{3}
A-h	9.259×10^{-14}	2.778×10^{-4}	2.778×10^{-3}	1	2.680×10^{1}
Fd	3.457×10^{-15}	1.036×10^{-5}	1.036×10^{-4}	3.731×10^{-2}	1

Note: The SI derived unit for quantity of electricity, electric charge, or dielectric flux is the *coulomb* ($A \cdot s$). Expressed in terms of coulombic force, $Q = LF_c^{0.5}$ and $C = 9.48 \times 10^{4}\ m \cdot N^{0.5}$.

Table II-L Electric Current (Qt^{-1})

Multiply Number of → by Factor to Obtain ↓	statampere	A (ampere)	abampere
statampere	1	2.998×10^9	2.998×10^{10}
A	3.335×10^{-10}	1	10^1
abampere	3.335×10^{-11}	10^{-1}	1

Note: The SI base unit for electric current is the *amphere*, defined as the magnitude of the current which results in the development of a force of 2×10^{-7} *newton per meter of wire length* between two long parallel wires separated by 1 meter in free space.

Table II-M Electric Potential and Electromotive Force ($MQ^{-1}L^2t^{-2}$ or $FQ^{-1}L$)

Multiply Number of → by Factor to Obtain ↓	abvolt	mV (millivolt)	V (volt)	statvolt
abvolt	1	10^5	10^8	2.998×10^{10}
mV	10^{-5}	1	10^3	2.998×10^5
V	10^{-8}	10^{-3}	1	2.998×10^2
statvolt	3.335×10^{-11}	3.335×10^{-6}	3.335×10^{-3}	1

Note: The SI derived unit for electrical potential and electromotive force is the *volt* (kg $m^2 \cdot s^{-3}$ $= J\ A^{-1}\ s^{-1}$).

Table II-N Electric Resistance ($MQ^{-2}L^2t^{-1}$ or $FQ^{-2}Lt$)

Multiply Number of → by Factor to Obtain ↓	abohm	μΩ (microohm)	Ω (ohm)	statohm
abohm	1	10^3	10^9	8.988×10^{20}
μΩ	10^{-3}	1	10^6	8.988×10^{17}
Ω	10^{-9}	10^{-6}	1	8.988×10^{11}
statohm	1.112×10^{-21}	1.112×10^{-18}	1.112×10^{-12}	1

Note: The SI derived unit for electric resistance, $R°$, is the *ohm* (Ω = kg $m^2\ s^{-3}\ A^{-2} = V\ A^{-1}$). The SI derived unit for electric conductance ($F^{-1}QL^{-1}t^{-1}$) is the *siemens*, where 1 *siemens* = Ω^{-1}. The older name for the siemens unit is the *mho*.

APPENDIX III DIMENSIONLESS NUMBERS, COEFFICIENTS, AND FACTORS

Table III-A Dimensionless Numbers

Dimensionless Number	Symbol	Definition	Ratio Represented	Applications
Archimedes	$\mathfrak{N}_{\text{Ar}}$	$\frac{g d_p^3 \rho_p}{\mu_{l'}^2}(\rho_p - \rho_l)$ *for a particle or droplet in a liquid*	Gravitational to viscous forces times $\mathfrak{N}_{\text{Re}}$	Particle or droplet fluidization, motion of liquids resulting from density differences
Biot (heat transfer)	$\mathfrak{N}_{\text{Bi, H}}$	$\frac{k_h L_c}{k_c} \equiv \mathfrak{N}_{\text{Nh}}$	Conductor to surface film heat transfer resistances	Unsteady-state heat transfer
Biot (mass transfer)	$\mathfrak{N}_{\text{Bi, M}}$	$\frac{k_f \delta}{\mathfrak{D}_m}$	Interfacial (thickness δ) to interior domain mass transfer rates	Mass transfer between fluids and solids
Bodenstein	$\mathfrak{N}_{\text{Bo}}$	$\frac{v_x L_x}{\mathfrak{D}_{ax}}$	Advective to axial dispersive transport in a tubular reactor	Dispersion in tubular reactors
Bond	$\mathfrak{N}_{\text{Bd}}$	$\frac{(\rho_p - \rho_m) L_c^2 g}{\sigma^\circ}$ for a particle or droplet in a fluid medium $\equiv \mathfrak{N}_{\text{We}}/\mathfrak{N}_{\text{Fr}}$ as $\rho_p - \rho_m \approx \rho_p$	Gravitational to surface tension forces	Atomization, motion of bubbles and droplets
Boussinesq	$\mathfrak{N}_{\text{Bq}}$	$\frac{v}{(2 g R_H)^{0.5}} \equiv \mathfrak{N}_{\text{Fr}}^{0.5}$	Square root of inertial to gravitational forces	Wave behavior in open channels

Capillary	$\mathcal{N}_{Ca}$	$\frac{v\mu_v}{\sigma^\circ} \equiv \mathcal{N}_{We}/\mathcal{N}_{Re}$	Viscous to surface-tension forces	Atomization, two-phase flow
Damkohler I	$\mathcal{N}_{Da,I}$	$\frac{\imath L_x}{v_x C}$ $\frac{kL_x}{v_x}$ *for a first-order reaction*	Chemical reaction to bulk mass flow rates	Chemical reaction, momentum, and heat transfer
Damkohler II	$\mathcal{N}_{Da,II}$	$\frac{\imath L_c^2}{\mathfrak{D}_l C} \equiv \mathcal{N}_{Tm}^2$ $\frac{kL_c^2}{\mathfrak{D}_l}$ for a first-order reaction	Chemical reaction to molecular diffusion rates	Chemical reaction, momentum, and heat transfer
Dispersion	$\mathcal{N}_d$	$\frac{\mathfrak{D}_d}{v_x L_x}$	Dispersive (longitudinal) to advective mass transport in a tubular reactor	Mass transport and reactor behavior
Euler	$\mathcal{N}_{Eu}$	$\frac{\Delta P_F}{\rho v^2}$	Friction to velocity heads	Fluid friction in conduits
Fluidization	$\mathcal{N}_{Fl}$	$\frac{v_b}{v_f}$	Bulk flow to initial fluidization velocities	Particle fluidization
Froude	$\mathcal{N}_{Fr}$	$\frac{v^2}{gL_c}$	Inertial to gravitational forces	Wave and surface behavior
Galileo	$\mathcal{N}_{Ga}$	$\frac{g\rho_p^2 d_p^3}{\mu_v^2}$ *for a particle or droplet in a fluid medium*	Inertial to viscous times gravitational to viscous forces	Circulation of dense fluids

Table III-A (*Continued*)

Dimensionless Number	Symbol	Definition	Ratio Represented	Applications
Grashof (modified)	$\mathfrak{N}_{Gr}$	$\dfrac{g\rho^2 L_c^3}{\mu_v^2}$	Inertial to viscous times buoyant to viscous forces	Circulation of buoyant fluids
Hatta	$\mathfrak{N}_{Ha}$	$\dfrac{(k\mathfrak{D}_l)^{0.5}}{\hat{k}_{f,l}}$ *for a first-order reaction*	Reaction to diffusion rates	Gas absorption with chemical reaction
Knudsen I	$\mathfrak{N}_{Kn,I}$	$\dfrac{L_{mfp}}{L_c}$	Mean free path to characteristic length	Low-pressure gas flow
Knudsen II	$\mathfrak{N}_{Kn,II}$	$\dfrac{\epsilon_B \mathfrak{D}_b}{\mathfrak{D}_k \tau}$	Binary bulk to Knudsen diffusion rates	Gaseous diffusion in packed beds
Newton	$\mathfrak{N}_{Ne}$	$\dfrac{I_f^\circ}{\rho v_x^2 L_x^2}$	Frictional impedance (resistance) to inertial force	Friction in fluid flow and agitation
Nusselt (heat)	$\mathfrak{N}_{N,H}$	$\dfrac{k_h L_c}{k_c} \equiv \mathfrak{N}_{Bi,H} \equiv (\mathfrak{N}_{Re}\mathfrak{N}_{St})$	Total to conductive heat transfer	Forced convection
Nusselt (mass)	$\mathfrak{N}_{N,M}$	$\dfrac{k_f \delta}{\mathfrak{D}_l} \equiv \mathfrak{N}_{Sh}$	Interfacial film (thickness, δ) mass flux to molecular diffusion	Mass transfer
Péclet	$\mathfrak{N}_{Pe}$	$\dfrac{v_x L_x}{\mathfrak{D}_d} \equiv (\mathfrak{N}_{Re}\mathfrak{N}_{Sc}) \equiv \mathfrak{N}_d^{-1}$	Advective to dispersive (longitudinal) mass transport in a tubular reactor	Mass transfer

Prandtl	$\mathfrak{N}_{\mathrm{Pr}}$	$\dfrac{Q_H^\circ \mu_v}{k_c}$	Momentum to thermal diffusivities	Forced and free convection
Reynolds	$\mathfrak{N}_{\mathrm{Re}}$	$\dfrac{v_x \rho L_x}{\mu_v}$	Inertial to viscous forces	Dynamic similitude in fluid flow
Reynolds (entry)	$\mathfrak{N}_{\mathrm{Re,E}}$	$\dfrac{L_e}{2R_H}\mathfrak{N}_{\mathrm{Re}} = \dfrac{v\rho L_e}{\mu_v}$	Inertial to viscous forces	Entry or inlet processes
Reynolds (rotating)	$\mathfrak{N}_{\mathrm{Re,R}}$	$\dfrac{\omega_r \rho L_c^2}{\mu_v}$	Inertial to viscous forces	Impellers and agitation processes
Schmidt	$\mathfrak{N}_{\mathrm{Sc}}$	$\dfrac{\mu_v}{\rho_l \mathfrak{D}_l}$	Dynamic viscosity to molecular diffusivity	Diffusion in flowing liquids
Sherwood	$\mathfrak{N}_{\mathrm{Sh}}$	$\dfrac{k_f L_c}{\mathfrak{D}_l} \equiv \mathfrak{N}_{\mathrm{N,M}}$	Overall mass to molecular diffusivities; known also as the Taylor number	Mass transfer
Stanton	$\mathfrak{N}_{\mathrm{St}}$	$\dfrac{k_h}{Q_H^\circ \rho V} \equiv \mathfrak{N}_{\mathrm{N,H}}/\mathfrak{N}_{\mathrm{Re}}\mathfrak{N}_{\mathrm{Pr}}$	Heat transferred to fluid heat capacity	Forced convection
Thiele modulus	$\mathfrak{N}_{\mathrm{Tm}}$	$\left(\dfrac{r L_c^2}{\mathfrak{D}_l C}\right)^{0.5} \equiv (\mathfrak{N}_{\mathrm{Da,II}})^{0.5}$ $\left(\dfrac{k L_c^2}{\mathfrak{D}_l}\right)^{0.5}$ *for a first-order reaction*	Surface reaction to intraparticle diffusion rates. Square root of the Damkohler II number	Diffusion in microporous solids
Weber	$\mathfrak{N}_{\mathrm{We}}$	$\dfrac{v^2 \rho L_c}{\sigma^\circ}$	Inertial to surface tension forces	Bubble formation, particle migration

Table III-B Dimensionless Coefficients and Factors

Dimensionless Parameter	Symbol	Definition	Basis	Applications
Drag coefficient	C_D	$\frac{(\rho_p - \rho_m) g L_c}{\rho_p v^2}$	Gravitational and inertial forces	Free settling velocities, impedance to objects moving through fluids
Fanning friction factor	$\mathfrak{F}_F$	$\frac{d_P \Delta P_F}{2\rho v_x^2 L_x}$	Shear stress at pipe wall expressed as number of velocity heads	Fluid friction in conduits
Power factor	$\mathfrak{F}_P$	$\frac{P_w}{\rho \omega_r^3 L_c^5}$	Drag or inertial force on agitator impeller	Power consumption by agitators, fans, and pumps
Colburn mass transfer factor	$\mathfrak{J}_D$	$\frac{k_f \delta \mathfrak{N}_{Sc}^{2/3}}{v_x L_x} \equiv \mathfrak{N}_{Sh}(\mathfrak{N}_{Sc})^{-1/3}(\mathfrak{N}_{Re})^{-1}$	Viscous, inertial, and diffusion impedances	Mass transfer
Leverett function	$\mathfrak{J}_L$	$\frac{P_C(\xi/\epsilon)^{0.5}}{\sigma^\circ \cos \Theta_c}$	Characteristic surface curvature to pore dimensions	Two-phase flow in porous media

Table III-C Selected Parameters and Related Dimensionless Groups

Parameter	Symbol	Dimensions	Power	Dimensionless Groups
		Dimensionless Parameters		
Tortuosity	τ	—	−1	$\mathfrak{N}_{\mathrm{Kn,II}}$
Porosity	ϵ	—	−0.5	$\mathfrak{N}_{\mathrm{Le}}$
			+1	$\mathfrak{N}_{\mathrm{Kn,II}}$
		Parameters Involving One Fundamental Unit		
Angular velocity	ω_r	t^{-1}	−3	$\mathfrak{N}_{\mathrm{Re,R}}$; $\mathfrak{F}_P$
Characteristic linear dimension	L_c	L	−2	$\mathfrak{N}_{\mathrm{Ne}}$
			−1	$\mathfrak{N}_{\mathrm{Fr}}$; $\mathfrak{N}_{\mathrm{Kn,I}}$; $\mathfrak{F}_F$
			−0.5	$\mathfrak{N}_{\mathrm{Bq}}$
			+1	$\mathfrak{N}_{\mathrm{Bi,H}}$; $\mathfrak{N}_{\mathrm{Bi,M}}$; $\mathfrak{N}_{\mathrm{Da,I}}$; $\mathfrak{N}_{\mathrm{N,H}}$; $\mathfrak{N}_{\mathrm{N,M}}$; $\mathfrak{N}_{\mathrm{Pe}}$; $\mathfrak{N}_{\mathrm{Re}}$; $\mathfrak{N}_{\mathrm{Re,E}}$; $\mathfrak{N}_{\mathrm{Sh}}$; $\mathfrak{N}_{\mathrm{Tm}}$; $\mathfrak{N}_{\mathrm{We}}$; C_D
			+2	$\mathfrak{N}_{\mathrm{Bd}}$; $\mathfrak{N}_{\mathrm{Da,II}}$
			+3	$\mathfrak{N}_{\mathrm{Gr}}$
Dimension of impeller	d_{imp}	L	−5	$\mathfrak{F}_P$
			+2	$\mathfrak{N}_{\mathrm{Re,R}}$
Pipe diameter	d_P	L	+1	$\mathfrak{F}_F$
Mean free path	L_{mfp}	L	+1	$\mathfrak{N}_{\mathrm{Kn,I}}$
Particle diameter	d_p	L	+3	$\mathfrak{N}_{\mathrm{Ar}}$; $\mathfrak{N}_{\mathrm{Ga}}$
Reactor length	L_R	L	+1	$\mathfrak{N}N_{\mathrm{Bo}}$
Permeability	ξ	L^2	+0.5	$\mathfrak{N}_{\mathrm{Le}}$
Volume	V	L^3	+1	$\mathfrak{N}_{\mathrm{Ar}}$
		Parameters Involving Mass and Length		
Density	ρ	ML^{-3}	−1	$\mathfrak{J}_D$; $\mathfrak{N}_{\mathrm{Eu}}$; $\mathfrak{N}_{\mathrm{Ne}}$; $\mathfrak{N}_{\mathrm{Sc}}$; $\mathfrak{N}_{\mathrm{St}}$; C_D, $\mathfrak{F}_F$; $\mathfrak{F}_P$
			−0.667	$\mathfrak{J}_D$

Table III-C ***(Continued)***

Parameter	Symbol	Dimensions	Power	Dimensionless Groups
			+0.333	$\mathcal{J}_D$
			+1	$\mathcal{N}_{Ar}$; $\mathcal{N}_{Bd}$; $\mathcal{N}_{Re}$; $\mathcal{N}_{Re,R}$; $\mathcal{N}_{Re,E}$; $\mathcal{N}_{We}$; C_D
			+2	$\mathcal{N}_{Ga}$; $\mathcal{N}_{Gr}$
Concentration	C	ML^{-3}	−1	$\mathcal{N}_{Da,I}$; $\mathcal{N}_{Da,II}$
			−0.5	$\mathcal{N}_{Tm}$
		Parameters Involving Mass and Time		
Surface tension	σ°	Mt^{-2}	−1	$\mathcal{N}_{Bd}$; $\mathcal{N}_{Ca}$; $\mathcal{N}_{Le}$; $\mathcal{N}_{We}$
		Parameters Involving Mass, Length, and Time		
Viscosity	μ_v	$ML^{-1}t^{-1}$	−2	$\mathcal{N}_{Ar}$; $\mathcal{N}_{Ga}$; $\mathcal{N}_{Gr}$
			−1	$\mathcal{N}_{Re}$; $\mathcal{N}_{Re,R}$; $\mathcal{N}_{Re,E}$
			+0.667	$\mathcal{J}_D$
			+1	$\mathcal{N}_{Ca}$; $\mathcal{J}_D$; $\mathcal{N}_{Pr}$; $\mathcal{N}_{Sc}$
Impedance (resistance)	I°	MLt^{-2}	+1	$\mathcal{N}_{Ne}$
Power	P_w	ML^2t^{-3}	+1	$\mathcal{F}_P$
Pressure	P	$ML^{-1}t^{-2}$	+1	$\mathcal{N}_{Le}$
Pressure drop	ΔP	$ML^{-1}t^{-2}$	+1	$\mathcal{N}_{Eu}$; $\mathcal{F}_F$
		Parameters Involving Mass, Temperature, and Time		
Overall heat transfer coefficient	k_h	$MT^{-1}t^{-2}$	+1	$\mathcal{N}_{Bi,H}$; $\mathcal{N}_{N,H}$; $\mathcal{N}_{St}$
		Parameters Involving Mass, Length, Temperature, and Time		
Thermal conductivity	k_c	$MLT^{-1}t^{-3}$	−1	$\mathcal{N}_{Bi,H}$; $\mathcal{N}_{N,H}$; $\mathcal{N}_{Pr}$

		Parameters Involving Length and Time		
Diffusivity (free liquid)	$\mathfrak{D}_l$	L^2t^{-1}	−1	$\mathfrak{N}_{Bi,M}$; $\mathfrak{N}_{Bo}$; $\mathfrak{J}_D$; $\mathfrak{N}_{Da,II}$; $\mathfrak{N}_{N,M}$; $\mathfrak{N}_{Pe}$; $\mathfrak{N}_{Sc}$; $\mathfrak{N}_{Sh}$
			−0.667	$\mathfrak{J}_D$
			−0.5	$\mathfrak{N}_{Ha}$; $\mathfrak{N}_{Tm}$
			+1	$\mathfrak{N}_{Kn,II}$
Diffusivity (Knudsen)	$\mathfrak{D}_k$	L^2t^{-1}	−1	$\mathfrak{N}_{Kn,II}$
Fluid velocity	v	Lt^{-1}	−2	$\mathfrak{N}_{Eu}$; $\mathfrak{N}_{Ne}$; C_D; $\mathfrak{F}_F$
			−1	$\mathfrak{N}_{Da,I}$; $\mathfrak{N}_{Fl}$
			+1	$\mathfrak{N}_{Bo}$; $\mathfrak{N}_{Bq}$; $\mathfrak{N}_{Ca}$; $\mathfrak{N}_{Fl}$; $\mathfrak{N}_{Pe}$; $\mathfrak{N}_{Re}$; $N\mathfrak{N}_{Re,R}$
			+2	$\mathfrak{N}_{Fr}$; $\mathfrak{N}_{We}$
Mass transfer coefficient	k_f	Lt^{-1}	+1	$\mathfrak{N}_{Bi,m}$; $\mathfrak{N}_{Nm}$; $\mathfrak{N}_{Sh}$; $\mathfrak{J}_D$
Gravitational acceleration	g	Lt^{-2}	−1	$\mathfrak{N}_{Fr}$
			−0.5	$\mathfrak{N}_{Bq}$
			+1	$\mathfrak{N}_{Ar}$; $\mathfrak{N}_{Bd}$; $\mathfrak{N}_{Ga}$; $\mathfrak{N}_{Gr}$; C_D
		Parameters Involving Length, Time, and Temperature		
Heat capacity	Q_H°	$L^2t^{-2}T^{-1}$	−1	$\mathfrak{N}_{St}$
			+1	$\mathfrak{N}_{Pr}$

INDEX

ENVIRONMENTAL SCIENCE AND TECHNOLOGY

A Wiley-Interscience Series of Texts and Monographs

Edited by JERALD L. SCHNOOR, *University of Iowa*
ALEXANDER ZEHNDER, *Swiss Federal Institute for Water Resources and Water Pollution Control*

PHYSICOCHEMICAL PROCESSES FOR WATER QUALITY CONTROL
Walter J. Weber, Jr.

pH and pION CONTROL IN PROCESS AND WASTE STREAMS
F. G. Shinskey

AQUATIC POLLUTION: An Introductory Text
Edward A. Laws

INDOOR AIR POLLUTION: Characterization, Prediction, and Control
Richard A. Wadden and Peter A. Scheff

PRINCIPLES OF ANIMAL EXTRAPOLATION
Edward J. Calabrese

SYSTEMS ECOLOGY: An Introduction
Howard T. Odum

INTEGRATED MANAGEMENT OF INSECT PESTS OF POME AND STONE FRUITS
B. A. Croft and S. C. Hoyt, Editors

WATER RESOURCES: Distribution, Use and Management
John R. Mather

ECOGENETICS: Genetic Variation in Susceptibility to Environmental Agents
Edward J. Calabrese

GROUNDWATER POLLUTION MICROBIOLOGY
Gabriel Bitton and Charles P. Gerba, Editors

CHEMISTRY AND ECOTOXICOLOGY OF POLLUTION
Des W. Connell and Gregory J. Miller

SALINITY TOLERANCE IN PLANTS: Strategies for Crop Improvement
Richard C. Staples and Gary H. Toenniessen, Editors

ECOLOGY, IMPACT ASSESSMENT, AND ENVIRONMENTAL PLANNING
Walter E. Westman

CHEMICAL PROCESSES IN LAKES
Werner Stumm, Editor

INTEGRATED PEST MANAGEMENT IN PINE-BARK BEETLE ECOSYSTEMS
William E. Waters, Ronald W. Stark, and David L. Wood, Editors

PALEOCLIMATE ANALYSIS AND MODELING
Alan D. Hecht, Editor

BLACK CARBON IN THE ENVIRONMENT: Properties of Distribution
E. D. Goldberg

GROUND WATER QUALITY
C. H. Ward, W. Giger, and P. L. McCarty, Editors

TOXIC SUSCEPTIBILITY: Male/Female Differences
Edward J. Calabrese

ENERGY AND RESOURCE QUALITY: The Ecology of the Economic Process
Charles A. S. Hall, Cutler J. Cleveland, and Robert Kaufmann

AGE AND SUSCEPTIBILITY TO TOXIC SUBSTANCES
Edward J. Calabrese

ECOLOGICAL THEORY AND INTEGRATED PEST MANAGEMENT PRACTICE
Marcos Kogan, Editor

AQUATIC SURFACE CHEMISTRY: Chemical Processes at the Particle Water Interface
Werner Stumm, Editor

RADON AND ITS DECAY PRODUCTS IN INDOOR AIR
William W. Nazaroff and Anthony V. Nero, Jr., Editors

PLANT STRESS-INSECT INTERACTIONS
E. A. Heinrichs, Editor

INTEGRATED PEST MANAGEMENT SYSTEMS AND COTTON PRODUCTION
Ray Frisbie, Kamal El-Zik, and L. Ted Wilson, Editors

ECOLOGICAL ENGINEERING: An Introduction to Ecotechnology
William J. Mitsch and Sven Erik Jorgensen, Editors

ANTHROPOD BIOLOGICAL CONTROL AGENTS AND PESTICIDES
Brian A. Croft

AQUATIC CHEMICAL KINETICS: Reaction Rates of Processes in Natural Waters
Werner Stumm, Editor

GENERAL ENERGETICS: Energy in the Biosphere and Civilization
Vaclav Smil

FATE OF PESTICIDES AND CHEMICALS IN THE ENVIRONMENT
J. L. Schnoor, Editor

ENVIRONMENTAL ENGINEERING AND SANITATION, Fourth Edition
Joseph A. Salvato

TOXIC SUBSTANCES IN THE ENVIRONMENT
B. Magnus Francis

CLIMATE-BIOSPHERE INTERACTIONS
Richard G. Zepp, Editor

AQUATIC CHEMISTRY: Chemical Equilibria and Rates in Natural Waters, Third Edition
Werner Stumm and James J. Morgan

PROCESS DYNAMICS IN ENVIRONMENTAL SYSTEMS
Walter J. Weber, Jr., and Francis A. DiGiano

ENVIRONMENTAL CHEMODYNAMICS: Movement of Chemicals in Air, Water, and Soil, Second Edition
Louis J. Thibodeaux

Printed in the USA/Agawam, MA
July 15, 2022

795761.003